COMPENDIUM OF SOIL FUNGI

COMPENDIUM OF SOIL FUNGI

Volume 1

K. H. DOMSCH
Institute of Soil Biology
Federal Agricultural Research Centre
Braunschweig
Federal Republic of Germany

W. GAMS
Centraalbureau voor Schimmelcultures,
Baarn
The Netherlands

TRAUTE-HEIDI ANDERSON
Institute of Soil Biology
Federal Agricultural Research Centre
Braunschweig
Federal Republic of Germany

1980

ACADEMIC PRESS

A Subsidiary of Harcourt Brace Jovanovich, Publishers

London   New York   Toronto   Sydney   San Francisco

ACADEMIC PRESS (LONDON) LTD
24/28 Oval Road
London NW1

United States Edition published by
ACADEMIC PRESS INC.
111 Fifth Avenue
New York, New York 10003

British Library Cataloguing in Publication Data

Domsch, Klaus Heinz
  Compendium of soil fungi.
  Vol. 1
  1. Soil fungi
  I. Title   II. Gams, Walter   III. Anderson, T-H
  589'.2'09148   QR111   80-41403

  ISBN 0-12-220401-8

Printed and bound in Great Britain
at The Pitman Press, Bath

Preface

This compendium presents condensed data on selected species of soil fungi. In order to facilitate access to relevant sources of information, we previously compiled ecological and physiological data for 209 species ("Pilze aus Agrarböden", 1970; "Fungi in Agricultural Soils" [English translation], 1972). This documentation has been continued and considerably expanded during the last ten years to enable a more comprehensive and more representative compilation to be published now.

In recent years information retrieval has been enhanced by computerized data banks. However, unlike computer-assisted searches, this project represents a systematic arrangement, specification and condensation of mycological data into a coherent body of information.

In contrast to available subject-orientated texts on soil mycology, the emphasis here is exclusively on *individual species*. The literature has been evaluated to a point where the information presented will be sufficient to facilitate the retrieval of additional data from the original texts. In addition to a great deal of specialized literature, "Abstracts of Mycology" were examined up to September 1976, when the literature search was formally ended; relevant taxonomic information has, however, been incorporated up to January 1979. Much valuable information contained in the Symposium "Soil-borne Plant Pathogens" (B. Schippers and W. Gams, eds., Academic Press, London, 1979) could thus not be incorporated. Species, for which little information had been available, were computer-searched using the BIOSIS and CAB data bases back to 1970. In spite of these efforts we have certainly not succeeded in covering all the literature for the species treated. Nevertheless, we trust that most salient features have been mentioned.

Throughout the compilation it was assumed that the identifications of publishing authors were correct, although in many cases doubts exist. Taxonomic publications generally proved to be the most reliable source of data on distribution. When identifications were obviously uncertain or incorrect, the information was excluded. Information on collective species, for example the mono-verticillate *Penicillium* species and *Fusarium roseum,* was omitted. After working through about 7500 references, those giving no additional information were eliminated.

January 1980

KHD
WG
T-HA

v

Acknowledgements

We are most indebted to Dr D. L. Hawksworth, Kew, who acted as an editor, and painstakingly corrected our texts linguistically, and also improved the content in many ways, in particular by providing additional information on *Chaetomium*. The final version was also read and corrected by Dr J. P. E. Anderson, Braunschweig. We are indebted to our mycological colleagues at CBS who contributed information from their special fields; amongst them, we particularly thank Dr R. A. Samson and Miss A. C. Stolk for contributions on *Penicillium* and *Aspergillus* (and their teleomorphs) which enabled us to produce unconventional keys which, we hope, will prove easier to use than ones hitherto available; Dr R. A. Samson further contributed some of his excellent scanning electron micrographs; and Mrs A. J. van der Plaats-Niterink corrected the Oomycetes texts and added to these. Prof. W. Gerlach, Berlin, worked through the *Fusarium* and *Cylindrocarpon* texts and provided illustrations of *Nectria ventricosa*. We also thank Mr G. H. Boerema, Wageningen, for checking the texts on *Phoma* and supplying originals of his illustrations; we are further grateful to the copyright holders of these illustrations and those of *Aphanomyces, Neotestudina, Phymatotrichopsis* and *Typhula* for permitting their reproduction. Gustav Fischer Verlag, Stuttgart, granted permission to reproduce some figures from our previous book on soil fungi. We are also indebted to the members of the chemistry section of the Centraalbureau voor Schimmelcultures for additional documentation, Dr G. W. van Eijk on metabolites and Dr A. C. M. Weijman on cell wall constituents; in addition, Dr van Eijk kindly checked the chemical nomenclature. We thank the Director of the Commonwealth Mycological Institute, Kew, for permitting us to extract information from their herbarium and records, and some mycologists at that institute for reading parts of the manuscript. We thank Mrs D. L. Brand-de Heer, University Library, Utrecht, for effecting the computer search for additional information.

The careful assistance of Miss Marianne Nieuwstad in the preparation of specimens for, and handling of, the scanning electron microscope is also acknowledged. The pencil drawings were inked by the skilful hands of Mrs G. Lafrenz-Dölker and Mrs I. Garbe (illustrations from "Pilze aus Agrarböden"), Mrs H. Kruggel, Miss S. Storm and Miss I. ten Hoedt.

KHD
WG
T-HA

Contents

Introduction

Each individual species monograph is divided into three main sections: Taxonomy, ecology and physiology. The general format adopted for the texts is tabulated below.

  T.-H.A. was responsible for sorting and abstracting the information, the taxonomic contributions were prepared by W.G., and those on ecology and physiology were written by K.H.D.

Arrangement of data in the monographs

TAXONOMIC SECTION
      Name and synonyms
      References to descriptions
      Brief description
      Characters differentiating the species from its relatives
      Cytology
      Genetics
      Ultrastructure
      Cell constituents
      DNA analysis

ECOLOGICAL SECTION
      General remarks on the principal substrate
      Special techniques for isolation
      Overall distribution
      Vegetation and soil types
      Unusual sites
      Soil conditions (including pH, depth, temperature, moisture etc.)
      Microhabitats (including rhizosphere, roots, seeds, dung etc.)
      Substrates other than soil
      Means of distribution
      Means of survival
      Susceptibility to antagonism
      Susceptibility to predation
      Biological control

PHYSIOLOGICAL SECTION
      Physico-chemical conditions for reproduction
      Physico-chemical conditions for propagule germination
      Physico-chemical conditions for vegetative growth
      Utilization of carbon and nitrogen sources and other nutrients

Requirements of trace elements, growth factors
Capacities for enzymatic degradation
Metabolites (including toxins, antibiotics and their actions)
Pathogenicity
Tolerance to chemical and physical stresses

Selection of the species

In our previous book the 209 species were selected on the basis of our own isolates from two wheat-field soils near Kiel. In the present "Compendium", however, we have endeavoured to incorporate the most common species occurring in all kinds of soils, particularly in the temperate zone, in addition to most of the species treated in the previous book. This selection was achieved by fixing an arbitrarily chosen minimal number of records in a preliminary selection of the literature, and in discussions with experts on various groups. Most species likely to be encountered in soil fungal analyses are consequently now included.

Keys and taxonomic texts

We considered it appropriate to concede to the users' needs and provide both more taxonomic information and keys to the genera and species treated.

A key to a limited selection of species may in many cases lead to misidentifications. A warning about the incompleteness of the keys is particularly appropriate in the larger genera; for these the monographs cited should be referred to. To permit a reliable comparison of unknown isolates with the species considered in this book, we have sometimes included more information in the keys (with occasional omission of alternatives) than might at first seem necessary. In some genera (e.g. *Chaetomium, Mortierella, Trichoderma*) more soil-borne species are keyed out than are treated in the text; otherwise it would be difficult to arrive at a species identification with the available literature. The identification reached from the keys should be verified by comparison with the morphological descriptions and notes on differential characters provided in the taxonomic sections of the text. Several *Fusarium* keys will be available in most laboratories, but species identification remains difficult here because of the variability of many characters; in order to overcome this difficulty, a synoptic key allowing a free choice of characters to be compared, is provided.

Generic introductions are given separately when more than one species is treated, and incorporated into the taxonomic sections where a single species is discussed. Except for the Deuteromycetes, we have indicated the appropriate family for each genus, followed by its most distinctive features, the overall distribution, and the most characteristic substrates. The principal taxonomic studies of each group are cited and the keys are usually derived from these publications; in some cases (indicated) they have been taken directly from them.

For each species we have tried to use the most appropriate name which has been applied in modern taxonomic studies. We have included the years of publication of all scientific names and were thus led to correct some currently used names and author citations according to the International Code of Botanical Nomenclature. We have listed only the more important synonyms to be found in current literature.

In the case of pleomorphic fungi, the terms "anamorph" and "teleomorph" are used consistently to indicate imperfect (conidial) and perfect (ascigerous or basidial) states (forms of propagation), respectively, as suggested by Hennebert and Weresub (2404). The name of the teleomorph is generally given preference over that of the anamorph. Nevertheless, we have also keyed out the anamorphs in their appropriate form-genera. We have listed a fungus under its anamorph name in cases where the connection with a particular teleomorph is equivocal from the conidia alone, e.g. in *Alternaria alternata, Botrytis cinerea, Gliocladium roseum*, and the *Trichoderma* species. In *Acremonium butyri, Cylindrocarpon magnusianum, Fusarium flocciferum* and *Verticillium tenerum*, the teleomorphs have been observed very rarely or only once; in view of the variability of the anamorphs it would be unwise to conclude that the connection was indisputable.

In the text we refer to the most adequate descriptions of each species and repeat only as much detail as is necessary for the confirmation of the identification. Much emphasis is placed on maintenance under optimum conditions for development of the typical colonies (Hochkultur). To give a rough impression, data on growth rates are generally given, from either the available literature or our own observations. The terminology used in describing conidiogenous structures partly follows that proposed in Kendrick (2921) and by Cole and Samson (1109).

The illustrations are an important part of the information and we do not usually repeat all the depicted features in words. Drawings are usually preferred to light micrographs, and these have been supplemented by scanning electron micrographs (SEM) and Nomarski interference contrast (NIC) micrographs. Light micrographs from unstained water mounts were taken with apochromate optics by K.H.D., while the other illustrations were prepared by W.G. (unless otherwise indicated). *The smallest interval shown on the scale bars represents 10 μm* (unless otherwise indicated). CBS numbers are given of isolates maintained in the CBS collection.

Where information on cytology, ultrastructure, cell wall constituents and DNA analysis is available, this is included in the text; ultrastructure has attained particular importance in relation to the mode of conidiogenesis. Available data on protein analysis and genetics are usually only briefly mentioned.

Ecological texts

We have attempted to include as much data on the geographical distribution as possible. For less common species the wealth of information contained in the herbarium records of the Commonwealth Mycological Institute (IMI) at Kew in the U.K. has also been evaluated and included in the text. In a few cases information available from cultures kept at the Centraalbureau voor Schimmelcultures (CBS) is also incorporated. Where distributional data for particular regions, soils or habitats were available in more than ten publications, no detailed citation of sources is given.

In collating data from different authors, different techniques of isolation, commonly uncritically used, provide an especial problem and often make comparisons between investigations impossible. Although different soil fungal patterns are mainly quantitative, we have had to omit quantitative data in most instances and confine ourselves to qualitative statements of presence. Where necessary the reader will be able to reassess the reports listed for a given species, select the comparable investigations, and integrate these data at a higher level.

As far as soil types are concerned, no attempt has been made to equate terms authors employed with modern soil nomenclatural practice; this is usually impossible to do because of incomplete information in the original papers. More important than geographical distribution, is the occurrence of fungi in particular soil micro-habitats. These are commonly poorly defined, but this disadvantage is frequently balanced by information which can be derived from physiological potentials and requirements. We have generally tried to place observations made in soil in the ecological sections and those obtained with pure cultures in the physiological sections.

Scientific names of soil animals have been cited as originally published, while some of the botanical names have been modernized, or replaced by common names.

Physiological texts

The relatively wide range of temperature requirements given does not exclude the possibility that individual isolates have a much narrower optimal range, especially when they originate from areas with rigid climatic conditions. Where not explicitly stated, sugars (and sugar alcohols) are usually the D-isomers and amino acids the L-isomers. Where no indication was given in the original publication, we have not added this information.

The ability of fungi to grow under conditions of reduced water potential, either on natural or artificial substrates, has been expressed in units of negative pressure (−bars); many of the original data have consequently been recalculated into these units. In most cases, however, the soil literature does not contain the information necessary to determine the water availability.

The numerous pure-culture studies on the effects of pesticides on the germination and growth of fungi have not been evaluated here, although population analyses in soil related to such studies have been included. The nomenclature of enzymes has been adapted to the *1972*-Recommendation of the Commission on Biochemical Nomenclature. We have also attempted to reduce confusions in the names of metabolic products by referring to recent publications, such as Laskin and LeChevalier (3231), but inevitably, some synonyms may still be present.

In general, data from phytopathological literature have not been considered in cases where the reference was exclusively to host–parasite complexes or means of chemical control.

References

As quotations from a scattered literature are increasingly condensed, the titles of the original publications rise in importance as they give some indication of the type of study carried out. Titles of articles are therefore printed in full in Volume 2. References are listed numerically after having been sorted and printed with the help of computer programs.

Abbreviations used in the text

CMA	cornmeal agar
CMC	carboxymethyl cellulose
CzA	Czapek's agar
GC	guanidine + cytosine
MEA	malt extract agar
OA	oatmeal agar
PCA	potato–carrot agar
PDA	potato–dextrose agar
PSA	potato–sucrose agar
V 8	commercial extract from 8 vegetables
YpSs	Emerson's yeast–starch medium

CBS   Centraalbureau voor Schimmelcultures, Baarn
IMI   Commonwealth Mycological Institute
NIC   Nomarski Interference contrast
SEM   Scanning electron microscopy (micrograph)
TEM   Transmission electron microscopy (micrograph)
q.v.   quod videas (= see under this name)

The symbols = and ≡ are used to distinguish facultative and obligate synonyms, respectively.

Figures printed in italics indicate years, those printed in Roman type point to the references.

Absidia van Tieghem *1876*

The genus *Absidia* of the Mucoraceae Dumort. is characterized by a differentiation of the hyphae into arched stolons bearing more or less verticillate sporangiophores in the raised part, and rhizoids formed after renewed contact with the substrate. The globose or pyriform sporangia are supported by a funnel-shaped apophysis (contrary to the genus *Gongronella* q.v.). Most species are heterothallic and the zygospores are usually surrounded by finger-shaped arched appendages arising from one, or both, suspensors. The genus is divided into the subgenera *Absidia* and *Mycocladus* Hesseltine & J. J. Ellis *1964*; the latter, here represented only by *A. corymbifera*, has zygospores without any such appendages. For routine identification purposes it is, however, more useful to distinguish three sections based on spore shape which have been named but not validly published by Zycha *et al.* (6588). Those authors have summarized the important work of Hesseltine and his coworkers on this genus (1592, 2429). *Absidia* currently contains 21, mostly soil-borne species. The five commoner species, discussed here, can be distinguished as follows.

Key to the species treated

1	Sporangiophores arising singly from stolons and bearing several branches; spores oval or subglobose (sect. *Repens*); growing well at 37–42°C, maximum 48–52°C	
		A. corymbifera (p. 8)
	Sporangiophores arising in whorls from stolons, mostly unbranched; spores either globose or cylindrical; mesophilic species with a temperature maximum at *c.* 30°C	**2**
2(1)	Spores regularly cylindrical (sect. *Cylindrospora*)	**3**
	Spores regularly globose (sect. *Glauca*)	**4**
3(2)	Heterothallic species; colonies light grey; spores 2·2–3·5 μm diam *A. cylindrospora* (p. 10)	
	Homothallic species; colonies brownish grey; spores 2·0–2·5 μm diam *A. spinosa* (p. 14)	
4(2)	Colonies greyish green; columella with or without a short conical or finger-like projection	
		A. glauca (p. 12)
	Colonies violet; columella with a finger-like projection 4–6 $\times$ 3 μm *A. coerulea* (p. 7)	

Absidia coerulea Bain. *1889*

   = *Absidia tieghemii* Deckenbach *1896*
   = *Absidia orchidis* (Vuill. *1903*) Hagem *1908*

For further synonyms see Ellis and Hesseltine (1592).

DESCRIPTION: Ellis and Hesseltine (1592). — Colonies blue to violet blue, later becoming deep greyish olive. Sporangiophores mostly in whorls of two to four, occasionally branched.

Apophyses pigmented, separated from the sporangiophores by a septum. Spores smooth-walled, globose, 3–5, mostly $3.5\,\mu$m diam. Heterothallic. Zygospores coarsely roughened, with 8–20 finger-like appendages arising from the larger suspensor. — *A. coerulea* is close to *A. glauca* (q.v.), but only a weak imperfect interspecific mating reaction occurs between isolates of opposite mating type (1592). — The mycelium contains glucose, fructose, sucrose, malic and tartaric acids and nine identified amino acids (5055).

The available data indicate that its main distributional area lies in the temperate zone. There are reports from the Ukrainian SSR (4474), Poland (269, 272, 1592, 1978), Norway (2206), the British Isles (307, 745, 2736, 4445, 4448, 5222, 6182), the Netherlands (1592), Germany (2467), Italy (3449, 3538), Spain (3446, 3447), Canada (540), the USA (1028, 1592, 2385) and India (4997, 5058). It has been frequently isolated from conifer forests (269, 1028, 2385, 2779, 3447, 4445) in which it occurred on both mineral and organic particles (307), beech forest soils (272) including ones in alpine sites (1876, 3538), peaty soil (4474), uncultivated soil (540), heathland (2736, 5220, 5221, 5222), acid sand dunes (745), acid sites in grasslands (6182), nursery soils (1978), at a reduced incidence in sewage-treated land (2467), and at a low frequency in arable soils (4997). This fungus has been found on a wide range of decaying substrates including *Betula ulmifolia* leaf litter (3449), rotting roots of *Pinus taeda* (3159), excrements of *Falco tinnunculus* (6223), and also isolated from seedling roots of *Pinus nigra* var. *laricio* (4448) and heath plants (5220). To judge from various reports it occurs preferentially in the A_0- and A_1-horizons (1028, 2736, 4445, 5220) but in one case there was no difference in frequency in the range from 4 cm to below 20 cm (269). — The mycelium can serve as food for *Acanthoscelides obtectus* and *Lepinotus reticulatus* (5381), the springtail *Hypogastrura tullbergi* (3830) and the mites *Acarus gracilis* (5380) and *Acotyledon rhizoglyphoides* (5379).

Good growth on mannose, glycogen, fructose, lactose and galactose (5053), no utilization of sorbose, melibiose, raffinose, melezitose, inulin, dulcitol, nitrate and no requirement for thiamine (5126) were observed. From mixtures of amino acids, asparagine, methionine and leucine were taken up preferentially (5054); good growth is also obtained on media containing glutamic acid, tyrosine (5056), K_2SO_4, $KHSO_4$, $K_2S_2O_5$, $MgSO_4$, and particularly good sporulation on ones including ammonium salts, $NaHSO_4$, $K_2S_2O_3$, $K_2S_2O_5$, glycine and glutamic acid (5056, 5057). Usnic acid can be degraded by *A. coerulea* (322). In liquid culture citric, tartaric, succinic and glutaric acids are produced in the presence of primary minerals and glucose (4009) with modest liberation of K^+ ions from these minerals (4008). Accumulation of an unknown morphogenic staling substance in the culture medium has been demonstrated (4425).

Absidia corymbifera (Cohn) Sacc. & Trotter *1912* — Fig. 1.

$\equiv$ *Mucor corymbifer* Cohn in Lichtheim *1884*
= *Absidia ramosa* (Lindt) Lendner *1908*
$\equiv$ *Mucor ramosus* Lindt *1886*
= *Absidia lichtheimii* (Lucet & Cost. *1901*) Lendner *1908*
= *Absidia ornata* Sarbhoy *1965*

DESCRIPTIONS: Nottebrock *et al.* (4221), and CMI Descriptions No. 521, *1977*. — Two or more *Absidia* species with fast-growing colonies (growing well at 37–42°C) and ovoid spores have usually been distinguished: *A. ramosa* with regularly elongate ovoid to ellipsoidal, but never globose, spores, and *A. corymbifera* with short ovoid and often globose spores. Scholer and Müller (5126), Vogt (6114) and Nottebrock *et al.* (4221), after studying numerous isolates, showed convincingly that it was impossible to draw a line amongst the numerous intermediate isolates with spores of an average length:width ratio between 1·05 and 1·67, the average measurements of which fall within the range $3·4–4·6 \times 2·8–3·8\,\mu m$; compatible crossings often yielded zygospores independently of spore shape. The zygospores are reddish brown, slightly roughened, with up to 3(–5) equatorial ridges, the suspensors are devoid of finger-like appendages. — The unusual occurrence of ambivalent strains compatible with both (+) and (−) strains was also recorded. Zygospores were formed on a yeast extract agar particularly at 31–34°C; strains with white mycelium formed more zygospores with compatible strains than those with lower and darker aerial mycelium, showing a reciprocal tendency to sexual and asexual sporulation (4221).

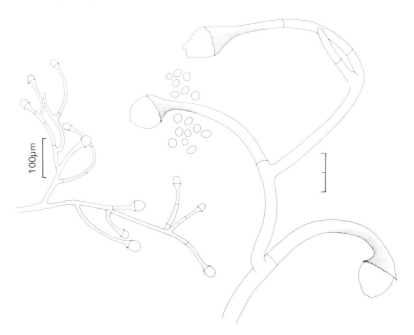

FIG. 1. *Absidia corymbifera*, sporangiophores with columellae and sporangiospores, CBS 102.48.

The worldwide distribution of this species is evidenced by reports from the British Isles (163, 166, 168, 4038), Poland (273), Germany (1424, 3051, 3127, 4055), Belgium (4816), France (3941), Italy (4538, 6085), Hungary (6551), Tunisia (4055), Egypt (8, 4962), Israel (2764, 2768, 2772), Somalia (5048), the USA (1028, 1166), Pakistan (57, 61, 2297), India (2854, 3518, 3732, 4698, 4700, 4998, 5058, 5349) and Indonesia (578). It has been isolated from a variety of substrates including waterlogged and dry grassland (164, 166, 4538), swamps (163), mangrove mud with salty water (4700), dunes (3941), forest soils (273, 2854, 5048, 5512), rarely in compost soil (1424), soil in a citrus plantation (2764), arable land under groundnut (2768, 2772) and sugar cane (3732), soils overlying limestone (3127), sewage

(1166), mines in Czechoslovakia (1703), a bat cave in Hungary (6551), the crater rim of the volcanic island of Surtsey (2406), coal spoil tips (1665), and air in the British Isles (1666, 2588). It was shown to be present in rhizosphere soil of an oak forest (4816), in the rhizosphere and the rhizoplane of wheat (4727), barley, oat (3004) and groundnut (2768), on dung of rabbits (6085), cattle (1399) and mules (1028), in earthworm casts (2857), on bat guano (6551), on plant remains in birds' nests (168), in nests and on feathers of birds (2575) and on decaying *Pteridium aquilinum* litter (1821). *A. corymbifera* is an early colonizer of wheat straw compost (953) and has been isolated from composted municipal waste (3051), hay (1399), stored grain of wheat and rice (2297, 4492), barley (1751, 4038), *Sorghum vulgare* (57), cotton (5009), pecans (2572), decaying seeds of *Phaseolus radiatus* (578), dry fodder (3258) and mouldy flour (3930). There seems to be a preference for neutral or weakly alkaline pH conditions (1665, 3732, 4055, 4538, 4700, 5048, 5349). The fungus is reported to be dominant at a soil depth of 30–45 cm in a black clayey Dwaba soil in India (5349), but it also occurs in higher horizons (163, 164, 273, 4998, 5048). It is affected by the application of soil fungicides (1426).

This fungus has been grouped as psychrotolerant thermophilic (160, 168, 517, 6114) with an optimum growth temperature of 35–37°C (3051, 5430, 6114). The pH optimum is in the range 3·8–6·8 (6114). — Assimilation of raffinose, melibiose, melezitose, lactose, dulcitol has been demonstrated but there is no utilization of sucrose, L-sorbose, cellulose, inulin, xylan, ethanol, or meso-erythritol (5126, 6114). It is able to degrade chitin (1430), arabinoxylan and (with moderate enzymatic activity) CMC (1757, 3051); β-glucosidase and β-xylosidase are produced (1758). Urea (3127), amides, peptone, nitrate and ammonium salts (6114) can be utilized as N sources, and tryptamine oxidized to indole-3-acetic acid (1522). The herbicide linuron is metabolized to 3,4-dichlorophenyl-1-methoxyurea (5824). The species can colonize hairs internally without decomposing the keratin (1633). A thiamine requirement was found (6114) in all of 46 isolates tested (5126), and a significant positive growth response to Zn ions occurs, with an even stronger response to Fe, Mn and Cu ions; the effects of trace elements are positively related to N concentration (6114). — Antibiotic activity against *Salmonella typhosa* is reported (3812). — This species is the most common causal agent of mucormycosis of the lungs, nasal sinuses, the cornea and other organs in both man and warm-blooded animals, and also of mycotic abortion in cattle (63, 912, 3932, 4221).

Absidia cylindrospora Hagem *1908* var. *cylindrospora* — Fig. 2.

DESCRIPTION: Hesseltine and Ellis (2429). — This species is the most frequent of the *Absidia* species with cylindrical spores. It shows a radial growth of 5 cm in 5–7 days at 25°C. Typical diagnostic characters are the absence of zygospores in single-sporangium transfers, light grey colonies growing well at 25°C and up to 31°C, scanty whorls (up to four branches) of sporangiophores, and spores 3·3–5·5 × 2·2–3·5 μm. Zygospores covered with blunt stellate projections to 4·5 μm long and with finger-like appendages arising from the larger, or from both suspensors. The varieties *nigra* Hesseltine & J. J. Ellis *1964* and *rhizomorpha* Hesseltine & J. J. Ellis *1961* are distinguished from var. *cylindrospora* by the production of dark grey colonies and the formation of rhizomorph-like hyphal strands, respectively. — DNA analysis gave a GC content of 40·5 to 41·5% (5600). — Analysis of the polysaccharides in the hyphal

Fɪɢ. 2. *Absidia cylindrospora*, sporangiophores with intact and deliquesced sporangia showing columellae and sporangiospores, CBS 101.37.

walls revealed ʟ-fucose, ᴅ-galactose and ᴅ-gluconic acid in the ratios 3·9 : 1·0 : 1·5 (3890) in addition to mannose, glucose, *N*-acetylglucosamine and *N*-acetylgalactosamine (3892).

A. cylindrospora has a worldwide distribution and has been reported, for example, from Norway (2206), Greenland (4174), Denmark (2745), Belgium (4816), Czechoslovakia, Yugoslavia (3798, 5534), Turkey (4245), Tunisia (4055), Somalia (5048), South Africa (1555, 1556, 1559, 4407), the Canary Islands (1827), the Bahamas (2006), Brazil (4547), Peru (2005), India (5058), Nepal (1826), Central America (2429, 3360), Argentina (1827), Hawaii (6138), Tasmania (5930) and the USA (1028, 1032, 1166, 2385, 2573). It has been found in various soil types and detailed reports are available for podzol soils (1876, 3416, 5534). It has been isolated very frequently from forest soils and also from grassland (159, 982, 2573, 6182) to 2000 m elevation in Ceylon (4021), from soils with steppe type vegetation (1559, 1700, 4407), arable soil (2719), fen (5559), mor type soil (2745), estuarine sediments and marsh soils (2741, 5316), a guano cave (4312), sewage (1155), and rarely in acid mine drainage streams (1166). In contrast to other Mucorales, *A. cylindrospora* occurs even at depths to 60–100 cm in the soil (159, 164, 1556, 3975, 6182). It tolerates a wide pH range from acid to slightly alkaline, although it possibly has a preference for acid conditions; high salinity is also tolerated (3446). The optimum humidity for vegetative growth lies at saturation, the minimum at −45 bars water potential (3109). The species appears, however, both in soils of arid steppes (1032, 1700) and in soils of high humidity (164, 4021). It has also been reported from the rhizosphere of various oak forest plants (4816). Roots have been found to be colonized in *Picea abies* (3573), *Pinus nigra* (3447), various other trees (3142, 4814) and Australian heathland plants (5819). It has been isolated from fragments of grass roots (982), corn (4553), leaf litter of *Eucalyptus maculata* (1558), spoiled Brazil nuts (2515), and feathers, nests and pellets of free-living birds (2575). The fungus is inhibited *in vitro* by tannin-containing extracts from freshly fallen oak leaves (2282).

A. *cylindrospora* forms abundant sporangiospores on all culture media tested. The optimum temperature for growth is 25°C and the maximum 30°C (1204, 1827). Amylolytic and proteolytic activities have been described (1827, 2719). Assimilation of sorbose, sucrose, raffinose and inulin has been observed and the production of phenol oxidase and the utilization of lignosulphonate, humic and fulvic acids reported (934). *A cylindrospora* can metabolize the herbicide linuron to 3,4-dichlorophenyl-1-methoxyurea *in vitro* (5824). It has no thiamine requirement (5126). — It is sensitive to soil steaming (6183).

Absidia glauca Hagem *1908* — Fig. 3.

DESCRIPTION: Ellis and Hesseltine (1592). — Colonies at first white, later becoming glaucous and finally slate olive. Sporangiophores solitary or more usually in whorls of 2–4. Apophyses pigmented, separated from the sporangiophore by a septum. Spores smooth-walled, globose, 2·5–5 (mostly 3·0) μm diam. Heterothallic. Zygospores coarsely roughened, with 12–20 finger-like projections arising from the larger suspensor. The size of the sporangia and the shape of the tips of substrate hyphae were found to be temperature dependent (5789). Sporangiospores are uninucleate; they increase to about double their size before germ tubes will protrude; at that stage spore cytoplasm and nuclei stream towards the hyphal tips. In paired cultures zygospore formation is abundant on oatmeal agar (3199). Zygophore initials can be induced by extracts from mixed heterothallic cultures but not by those from individual (+) or (−) strains (5995). In the course of three decades, a pair of compatible (+) and (−) strains was found to change under laboratory conditions from an isogamous condition to a heterogamous one (2330). — DNA analysis gave a GC content of 44% (3232).

A. *glauca* has a worldwide distribution, particularly in the temperate zone. There are reports from Norway (2206), Denmark (2745), the British Isles (163, 164, 165, 3567), Germany (4055), Poland (269, 273, 650, 1422), Austria (1592), the Ukrainian SSR (4474), Hungary (6551), Italy (271, 3452, 4537), Turkey (4245), Egypt (4962), Israel (2764, 4759), Canada (1592, 3959, 5363), the USA (1028, 1032, 1166, 1592, 2007, 4711), Australia (3720, 5930), New Zealand (1592, 5930), India (5058), Japan (3267) and Central America (1697). It has been frequently isolated from soils under broad-leaved forests such as beech (270, 271, 272, 273, 2745, 3138, 4537, 4989), or aspen (3959), also from willow and cottonwood lowland (2007), from coniferous forests (269, 1032, 3267, 3570, 5363), arable soil (2161, 2162, 4703), grassland (159, 164, 165, 540, 982, 1700, 4152, 4177, 4313, 5814, 6182), particularly in *Festuca* tussocks (5812, 5813), swamps and marshes (164), peat fields (4474), peat of raised bogs (861, 6520), fen peat (5559), heath soils (2736), sand dunes (745, 861, 1477), silt (5316), chalk soils (1259, 4152, 5559, 6182), bat guano (6551), and rarely from various acid mine drainage streams (1166). It was found to be stimulated by manuring (2764) and occurs equally in soils with high (164) and low humidity (1700). Seasonal variations with a maximum in November have been reported in Wisconsin (2004). There is no apparent restriction with respect to either soil depth (164, 269, 273) or soil pH (269, 271, 273, 1032, 4055, 4245). *A. glauca* has also been found associated with the ambrosia beetle, *Crossotarsus niponicus* (4102), and on disintegrated hair baits in soil (1422), in nests, on feathers and in pellets of free-living birds (2575). Additional recorded habitats include roots of *Festuca* (5812), *Lolium perenne* and *Trifolium repens* (5815), various cereal species (3004, 3567, 4443), ash (3142, 3156), heath (5819) and forest (4814) plants, the rhizosphere of wheat (3567), barley, oat (3004), poplar (3452), freshly fallen *Fagus crenata* leaves (4990) and litter of heathland plants (3720), coniferous (681) and broad-leaved trees (650), and wheat seeds (3491).

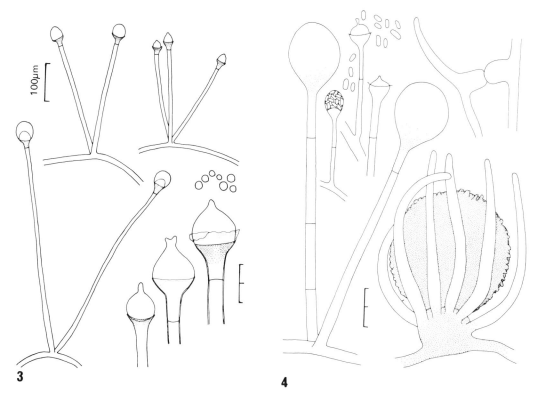

3   **4**

FIG. 3. *Absidia glauca*, sporangiophores with intact and deliquesced sporangia showing columellae, and sporangiospores.

FIG. 4. *Absidia spinosa*, sporangiophores with intact and deliquesced sporangia showing columellae, and sporangiospores; gametangia and zygospores, CBS 106.08.

Production of sporangiospores has been found to be stimulated with increasing salinity of the medium (up to 60% seawater) (833). The optimal temperature for growth is 24°C, the minimum 2°C, and the maximum 30°C (5789). Growth rates on solid media lie between 74 and 91 mm in ten days with a mean doubling time in liquid culture of 5·8 h (5890); light in the long wavelength range has no measurable effect on growth characteristics (1654). — This fungus is autotrophic for thiamine (4860, 5133) and in comparison with other fungi, it produces relatively large amounts of folic acid in continuous culture (3673). In contrast to other *Absidia* species (e.g. *A. corymbifera*), it does not assimilate sugars like melibiose or raffinose (5126). It is capable of liquefying gelatine (2719) and decomposing pectin (3127). It has been shown with *A. glauca* that ^{32}P is absorbed and transported only by growing hyphal tips (3424). *A. glauca* can metabolize the herbicide linuron to 3,4-dichlorophenyl-1-methoxyurea *in vitro* (5824). — The growth of *Chalara elegans* can be inhibited *in vitro* (2753). — *A. glauca* is sensitive to tannin contents in the culture medium (1206). It is tolerant of >15% NaCl in the culture medium (5881), which corresponds to approx. −140 bars water potential. In a chronically irradiated soil it survived to intensities of 230 R per day (2009).

Absidia spinosa Lendner *1907* — Fig. 4.

DESCRIPTION: Hesseltine and Ellis (2429). — This is one of the three homothallic species of *Absidia* with cylindrical spores (3·5–5 × 2·0–2·5 μm). In addition to its homothallic character it differs from *A. cylindrospora* in having brownish grey colonies and somewhat narrower sporangiospores, 3·5–5·0 × 2·0–2·5 μm. Zygospores are covered with blunt stellate projections to 4·5 μm long; finger-like appendages arise from the larger or both suspensors. — Boedijn (578) described a var. *azygospora* from soil in Java which produces azygospores. Among the other homothallic species, *A. parricida* Renner & Muskat ex Hesseltine & J. J. Ellis *1964* grows more slowly and is a parasite of *Mucor* species, whilst *A. anomala* Hesseltine & J. J. Ellis *1964* has characteristic violet to purple colonies. — DNA analysis gave a GC content of 41% (5600).

This species is typically soil-borne but, although generally less common than *A. cylindrospora*, it has a worldwide distribution and is particularly frequently recorded from soils in the temperate zone. There are reports from the British Isles (163, 745, 1598, 3567, 4429, 4646, 5222), Poland (273, 1978), Czechoslovakia (1700, 1702), Germany (3127, 4055), the Netherlands (CBS), France (2161), Austria (3413, 4179), Yugoslavia (5534), Italy (270, 271, 3538), the USA (660, 1040, 1164, 2482, 2573, 4313) and Canada (5363, 6352). It has been frequently isolated from soils in Israel (2764, 2768, 2777, 4749) and India (1315, 1525, 2854, 3868, 4030, 4995, 5000, 5349, 5512, 5513) but more rarely from ones in the Bahamas (2006), Japan (3267), Australia (977), Egypt (4962) and South Africa (1559). It has also been recorded from New Guinea (3020), Indonesia (2429), New Zealand (4947, 4948), Nepal (1826), Chile (1824) and Peru (2005). *A. spinosa* has a marked preference for soils with a high organic content, particularly for the humus layer in forest soils: under beech (270, 271, 273, 1702, 3538, 3570, 4947), *Eugenia* (5512), *Casuarina* (2006), teak (2854), and coniferous trees (1598, 3267, 4179, 5363, 6352). It has also been reported from moor (163), salt-marsh (4646), soils with steppe type vegetation (1559, 1700, 1702, 4313), heathland (4055, 5222), and as dominating in non-irrigated grassland (4948). This fungus has also been found in sand dunes (745), soil overlying sandstone (3127), lateritic soil (5000), pseudogley, semipdozol (3414), podzol (3413), soils from a granite outcrop (660), loamy sand from lava under chaparral vegetation (1164), uncultivated land (4030) and under wheat and beet (2161), in a citrus plantation (2764), groundnut field (2768) and forest nursery soil (1978). It occurs at soil depths of 15 cm (163, 1700, 4995, 5349), down to 100 cm (600). It colonizes substrates with an acid reaction (163, 4055, 5363) but it is more frequently recorded from soils with a pH of around 7 (270, 1315, 1559, 5000, 5349). This species is also known from the rhizospheres of wheat (3567), barley, oat (3004), coriander (1523) and groundnut (2768); it is also found on fresh groundnut kernels (2765), and earthworm casts (2857). It has also been isolated from litter of beech (273), *Eugenia heyneana* (5513) and *Pinus sylvestris* (2344), and organic detritus in unpolluted streams (4429). An increased occurrence after NP fertilization is recorded (2777). — The fungus can serve as food for the stored-product mites *Glycyphagus domesticus* and *Acarus siro* (5382).

No active growth occurs at humidities below saturation (977, 2095). The carbohydrates sorbose, sucrose, raffinose and inulin are well utilized. Phenylamide herbicides can be metabolized by oxidative transformations (5824). This species has no thiamine requirement (5126), can utilize pectin and urea (3414), produce β-glucosidase (6159) and shows antibiotic activity against *Fusarium oxysporum* (3570) and *Achromobacter* sp. (4853). — No growth occurs under anaerobic conditions (1244). Compared with non-mucoraceous fungi, *A. spinosa* is sensitive to γ-irradiation; 90% inactivation occurred at 11 krd h^{-1} in spore germination and at 18 krd h^{-1} in mycelial growth (5748).

Acremonium Link ex Fr. *1821*

= *Cephalosporium* Corda *1839* sensu auct.
? = *Hyalopus* Corda *1838*
= *Gliomastix* Guéguen *1905*

Teleomorphic genera: *Nectria* (Fr.) Fr. *1849, Emericellopsis* van Beyma *1940, Mycoarachis* Malloch & Cain *1970*

Gams (1882) in his monograph of the *Cephalosporium*-like Hyphomycetes, concluded that as the name *Cephalosporium acremonium* Corda (the type species of *Cephalosporium* Corda) was of uncertain application, *Acremonium* Link ex Fr. (typified by the identifiable *Acremonium alternatum* Link ex Gray) should be taken up for the genus containing most moulds hitherto placed in *Cephalosporium*. The formation of conidia in slimy heads or in dry chains, traditionally used to separate *Acremonium* and *Cephalosporium*, proved untenable as a generic criterion. Similarly, the pigmentation of the conidia lacks sufficient weight to be used for generic segregation.

Acremonium is therefore now considered as characterized by hyaline, generally slow-growing fine mycelia which produce mostly simple, awl-shaped, erect (orthotropic) phialides from the substratum (phalacrogenous) or from fasciculate aerial hyphae (plectonematogenous or synnematogenous); if compound conidiophores occur, branching is confined to the lower part (basitonous). Conidia are one-celled, exceptionally two-celled, hyaline or pigmented, in slimy heads or in dry chains.

The genus is distinguished from hyaline isolates of *Phialophora* by the absence or very limited development of a collarette on the phialides and the predominant formation of well differentiated, awl-shaped phialides with a basal septum. Reduced, short "adelophialides" occur only submerged in the agar, while they are dominant in the *Phialophora hoffmannii* (van Beyma) Schol-Schwarz group. Some *Cephalosporium*-like fungi were placed in *Verticillium* Nees ex Link sect. *Prostrata* W. Gams *1971* which has usually fluffy white to yellowish colonies, bearing very slender, awl-shaped phialides on single, prostrate aerial hyphae, singly or in whorls, as in *V. lecanii. Paecilomyces* Bain. (q.v.) is very similar to *Acremonium*, but here the conidiophores are generally more strongly branched and the phialides basally inflated. *Monocillium* S.B. Saksena (q.v.) always has simple phialides with a more or less pronounced wall thickening in the lower part; in the middle they may show an inflation where the wall starts to become thinner. The anamorph of *Plectosphaerella cucumerina* (q.v.) grows some-what faster than most *Acremonium* species and has wider hyphae, the conidia are slightly curved and, at least on OA, mostly two-celled. Microconidial *Fusarium* isolates may be confused with *Acremonium,* but they grow faster and have colonies with a characteristic fluffy appearance.

For identification, *Acremonium* isolates are generally grown in Petri dishes with 2% MEA, where conidia are streaked out in two lines; for the study of older cultures OA or CMA slants are most convenient.

Gams (1882) described 82 species of *Acremonium*, since then 18 further species have been added. The genus is subdivided into three sections: sect. *Acremonium* (≡ sect. *Simplex* W. Gams) with mostly unbranched conidiophores without any recognizable internal wall-thickening and collarette at the phialide apex; sect. *Nectrioidea* with generally wider hyphae and often basitonously branched conidiophores, the phialides often undulate in the upper part, with progressive wall-thickening, and sometimes with a narrow collarette; and sect. *Gliomastix* which includes not only species with pigmented conidia, but also some with hyaline conidia; the main distinctive feature is the presence of thick-walled, chondroid hyphae which render the colonies tough. *Sagenomella* W. Gams *1978* (q.v.) has been segregated from *Acremonium* because of its biapiculate conidia forming "connected" chains and often irregularly proliferating phialides.

In Gams's monograph (1882) a special key deals with the soil-borne species; the nine considered here can be distinguished as follows:

Key to species (and related genera) treated

1   Ascomata produced in culture   **2**
           Ascomata absent in culture   **4**

2(1)   Colonies forming hyaline cleistothecia; asci globose, containing pigmented, ellipsoidal asco-spores with longitudinal wing-like appendages   cf. *Emericellopsis* (p. 272)
       Colonies forming ostiolate perithecia   **3**

3(2)   Perithecia dark brown; asci clavate, usually with two rows of hyaline, two-celled ascospores
                                     cf. *Plectosphaerella* (p. 660)
       Perithecia light orange-brown; asci cylindrical, with a single row of subglobose, pale orange-ochre ascospores   cf. *Neocosmospora* (p. 509)

4(1)   Colonies light coloured; conidia hyaline, always in slimy heads   **5**
       Colonies more or less dark due to the presence of pigmented conidia; conidia in slimy heads or in dry chains   **10**

5(4)   Hyaline chlamydospores produced on OA after 12 days; conidia cylindrical   *A. kiliense* (p. 22)
       Chlamydospores absent   **6**

6(5)   Conidia with pointed ends, slightly asymmetrical along the longitudinal axis, often two-celled, at least on OA   cf. *Plectosphaerella cucumerina* (p. 660)
       Conidia with rounded ends or symmetrically fusiform, generally one-celled   **7**

7(6)   Conidia short-cylindrical, strongly staining in aniline blue (or cotton blue) and other dyes; phialides with a rather long, narrow collarette, on aging often proliferating subapically
                                        A. furcatum (p. 20)
       Conidia elongate-cylindrical, hardly staining in aniline blue etc.; phialides usually not proliferating   **8**

8(7)   Colonies turning yellow-green to olive-green; conidiophores often basitonously verticillate
                          A. butyri (teleomorph: *Nectria viridescens*) (p. 18)
       Colonies pale pink to salmon; conidiophores not verticillate   **9**

9(8)   Conidia mostly intensely salmon; phialide base with chromophilic granulations (staining strongly in aniline blue)   *A. rutilum* (p. 27)
       Colonies pale pink; phialide base hardly chromophilic, not granulose   *A. strictum* (p. 28)
       (If colonies almost white and fluffy and sclerotia present, at least on lupin stems
                                       A. slerotigenum)

10(4) Conidia partly ochraceous brown or light olivaceous-brown, partly hyaline   **11**

Conidia at maturity uniformly dark olivaceous **12**

11(10) Conidia partly pigmented and fusiform, smooth-walled, partly globose, hyaline and roughened *A. fusidioides* (p. 21)

Conidia partly pigmented, egg-shaped or onion-shaped, coarsely warted, partly more slender, almost hyaline and smooth-walled cf. *Sagenomella diversispora* (p. 710)

12(10) Conidia fusiform with truncate ends, in chains *A. luzulae* (p. 23)

Conidia not fusiform **13**

13(12) Conidia guttuliform, smooth-walled, always in slimy heads *A. cerealis* (p. 19)

Conidia globose to ellipsoidal, with warty incrustations, in irregular chains or heads *A. murorum* (p. 25)

Acremonium butyri (van Beyma) W. Gams *1971* — Fig. 5.

= *Tilachlidium butyri* van Beyma *1938*
= *Cephalosporium mycophilum* sensu Tubaki *1955* [non *Hyalopus mycophilus* (Corda) Corda *1838*]
= *Cephalosporium khandalense* Thirum. & Sukap. *1966*
= *Gliomastix lavitskiae* Zhdanova *1966*
= *Cephalosporium viride* Grebenyuk *1971*

Teleomorph: *Nectria viridescens* C. Booth *1959*

DESCRIPTIONS: Tubaki (5924), and Gams (1882). — A very variable soil fungus, the teleomorph of which has never been obtained in pure culture with soil isolates and is only known to be formed on twigs (636). Because of its heterogeneity it appears to be safer to use the name of the anamorph for soil isolates. Colonies reach 1–2 cm diam in ten days at 20°C on MEA. *A. butyri* is the only *Acremonium* so far known from soil with an intense olive-green coloration to the centrally ± floccose colonies. The basitonous, sometimes verticillate ramification of the conidiophores is typical of the section *Nectrioidea*; conidia slightly asymmetrical, elongate ellipsoidal to globose, 3·5–5·8 (–8·1) × 1·5–2·5 µm. — This fungus has been the subject of genetic investigations: auxotrophic and colour mutants have been isolated from a heterokaryotic isolate, but they gave no prototrophic progeny from parasexual recombination (5953).

Although there are only very few published reports on its occurrence, this species must rank as ubiquitous and very frequent. It has been found in soils of forest nurseries, particularly those that have become acid (3908), forest soils with white cedar in Canada (505), tundra soils in Alaska (1750, 3062), and soils in Tasmania (5930) and India (2885). In the American tropics it has been frequently observed on the stromata of ascomycetes (5032). The following reports of soil isolates have been confirmed by examination in culture: a *Thuja–Abies* swamp in Wisconsin, river banks in Utah and Idaho, arable soils in France (2161), soils under wheat in Germany; soils cultivated with potato, sugar beet or grass in the Netherlands (1614, 1616); it has also been reported from the British Isles, Pakistan, India, and Singapore (IMI). It has frequently been isolated from samples of various tree species, dead wood, leaf litter (IMI) and is also known from rabbit dung (2279). It was found to be associated with the ambrosia beetle, *Crossotarsus niponicus* (4102), and has frequently been found on agarics in Japan (3085, 5924). This also applies to Central Europe where it occurs regularly on polypores and lignicolous ascomycetes (1882). It has also been isolated from the rhizosphere of *Lamium galeobdolon* in Belgium (4814) and stained birch wood in Japan (5061). — Its growth *in vitro* can be inhibited by *Pholiota nameko* (3085).

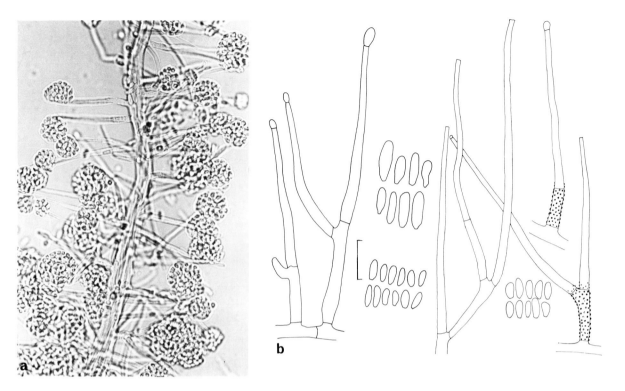

Fig. 5. *Acremonium butyri.* a. Hyphal bundle with conidiophores and conidial heads, × 500, CBS 148.62; b. conidiophores and conidia of various isolates, showing the variation in size (longer conidia in young, shorter conidia in older cultures), chromophilic incrustation drawn on the conidiophores to the right.

A. butyri is autotrophic for growth substances and weakly proteolytic (1425, 4570, 5926) with no amylase production (3566); no cellulose degradation (4191); weak chitin and pectin degradation (5926) and growth on gallic acid (1750). The olive-green pigment, which is both localized in vacuoles and superficially encrusts the hyphae, has been identified as the dinaphtho-dihydropyranon cephalochromin (2311).

Acremonium cerealis (Karst.) W. Gams *1971* — Fig. 6.

≡ *Coniosporium cerealis* Karst. *1887*
≡ *Gliomastix cerealis* (Karst.) Dickinson *1968*
= *Gliomastix guttuliformis* J. C. Brown & Kendrick *1958*

DESCRIPTIONS: Dickinson (1374), and Gams (1882). — Sect. *Gliomastix, A. murorum* series. — Colonies reaching 1·0–1·4 cm diam in ten days on MEA at 20°C, olivaceous-black, finely floccose, reverse often brown. Conidia smooth-walled, guttuliform, olivaceous-black, 4·0–5·3 × 2·5–3·7 μm, formed in slimy heads.

The wide distribution of *A. cerealis* in the British Isles, Finland, Germany, and Venezuela has been documented thoroughly (1374, 1882). Further records come from the Windward Islands, India, Egypt (IMI) and central Africa (3063). It has been found in sand dunes (745, 4655) and forest soils (1374, 3063), and down to the C-horizon of a podzol under *Pinus nigra* (2923, 4445). *A. cerealis* has not, however, been found to occur with great frequency anywhere. In soils under wheat in Germany it was obtained only five times in a series of over 20 000 isolates (CBS). Preferential substrates are rotting organic materials (1374, 1882) but it has also been isolated from *Musa* (6158), *Nephelium litchi*, and damaged potatoes (IMI).

On the whole, *A. cerealis* shows little difference between individual isolates in its capacity to decompose xylan (1432). It exerts a relatively high degree of antagonistic influence on *Rhizoctonia solani*, *Pythium ultimum*, *Fusarium culmorum*, and *Pseudocercosporella herpotrichoides* (1431). *A. cerealis* caused marked damage to seedling roots of wheat, rape and pea (1430).

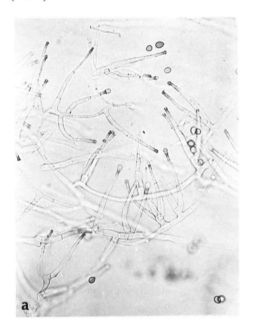 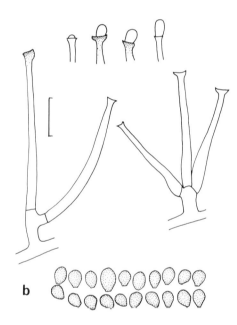

FIG. 6. *Acremonium cerealis*. a. Phialides with darkened apices, × 500, CBS 393.66; b. details of conidiophores and conidia, CBS 208.70.

Acremonium furcatum F. & V. Moreau ex W. Gams *1970* — Fig. 7.

≡ *Cephalosporium furcatum* F. & V. Moreau *1941* (nom. inval., Art. 36)

DESCRIPTION: Gams (1882). — Sect. *Nectrioidea*. — Validation of the name in Gams and Domsch (1887). Colonies reaching 1·8–2·4 cm diam in ten days at 20°C on MEA, pale ochre-coloured, usually with a pungent odour. Phialides often in whorls, often subapically proliferating without cross-walls (schizophialides); conidia short, cylindrical chromophilic, 3·0–3·7 × 1·6–2·1 μm.

A. furcatum is a very common and ubiquitous soil fungus, particularly in various sandy soils and ones with increased salt concentrations (1882, 3941); it is also very common in cultivated soil and on decaying plant remains. *A. furcatum* has been reported from the Netherlands (1614, 1616), Germany, Belgium, France, the British Isles, Portugal, Italy, Nigeria, Hong Kong (1882), Turkey (4245), India (5358) and Nepal (1826). This fungus was found to increase in frequency after repeated cropping of beets and potatoes (1615, 1616); it has been isolated from sugar cane fields and wilted sugar canes (5358), and has increased in occurrence after cultivation of rape (1433). On washed particles from wheat fields, it occurred in about 50% of the samples studied (1889). It is also commonly found on decaying plant substrates, including stems of *Angelica archangelica*, leaves of *Canna indica*, beechwood, cork, fruit-bodies of a *Gymnopilus* sp., and is even known from moist walls (1882).

Rapid sporulation occurs on all culture media. Of the six isolates tested, each strongly inhibited the growth of *Chlorella pyrenoidosa*. Of five pathogenic soil fungi investigated, *Rhizoctonia solani* proved to be the most sensitive fungus to the antagonistic influences of *A. furcatum* (1431). — Utilization of cellulose (4191), CMC and xylan is recorded with little variability between isolates (1432).

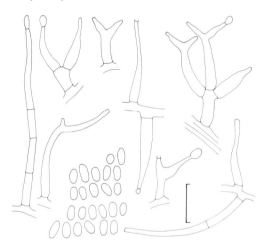

FIG. 7. *Acremonium furcatum*, conidiophores, some with sympodially proliferating phialides (schizophialides), and conidia.

Acremonium fusidioides (Nicot) W. Gams *1971* — Fig. 8.

≡ *Paecilomyces fusidioides* Nicot *1968*
= *Fusidium coccineum* sensu Tubaki *1954* [non Fuckel *1870*]

DESCRIPTIONS: Gams (1882), and Nicot (4165). — The true *Fusidium coccineum*, which is correctly designated as *Ramularia coccinea* (Fuckel) Westergr. *1899*, causes leaf spots on

Veronica officinalis and has blastoconidia. — Sect. *Acremonium, A. alternatum* series (≡ *A. terricola* series). — Colonies reaching 8–10 mm diam in ten days at 20°C on MEA, ochraceous brown, powdery; conidia catenulate, of two kinds: (a) predominantly slightly pigmented, fusiform with truncate ends, 4·9–6·4 × 1·2–2·1 μm, (b) globose, hyaline, slightly warty, 3·4–4·7 μm in diam. — DNA analysis gave a GC content of 37·7–38·4% (2908).

A. fusidioides is a widespread, but not very common soil fungus, recorded from Canada, the Netherlands (1616, 1882), northern Italy (6082), India (IMI) and, on excrements, from Japan and Africa (1882, 4165).

Good growth occurs on media based on rice, corn, oat and others (3186). This fungus produces the antibiotic substance fusidic acid and its sodium salt fucidin (= ramycin) both of which have a high specific activity against Staphylococci and other Gram positive bacteria (1882, 2015, 2981). The sodium salt forms micelles similar to the bile salt sodium taurocholate (894). The biosynthesis of the protosterol fusidic acid from mevalonic acid has been described (1533, 1534, 4037). Isolates differ in their fusidin production (3186); those with high production have a retarded sporulation and a reduction of various metabolic activities (376).

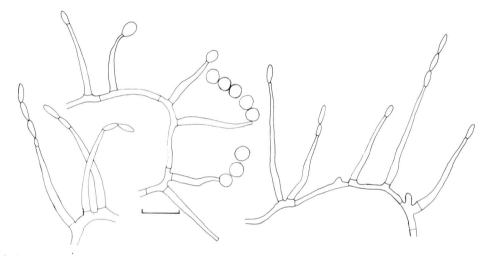

FIG. 8. *Acremonium fusidioides*, phialides producing both fusiform and globose, catenulate conidia, CBS 991.69.

Acremonium kiliense Grütz 1925 — Fig. 9.

= *Cephalosporium acremonium* auct. p.p. (particularly in medical mycology)

For numerous other synonyms see Gams (1882).

DESCRIPTION: Gams (1882). — Sect. *Simplex*. — Colonies reaching 1·8–2·3 cm diam in ten days at 20°C on MEA; pinkish, ± moist, sometimes centrally tufted. Sabouraud agar, and to a lesser extent MEA, is stained brown. Numerous hyaline chlamydospores with chromophilic walls, 4–8 μm diam, appear after 12 days (particularly on OA). Simple phialides mostly arising from submerged hyphae, very thin-walled. Conidia cylindrical, 3·1–5·8 × 1·0–1·6 μm. — The similar *A. tubakii* W. Gams *1971* produces many more chlamydospores, often in chains, and does not cause any brown discoloration of the media. — DNA analysis gave a GC content of 32·5% (2908).

A. kiliense is a common, ubiquitous soil fungus, reported from various European countries (1882), and also from India (521, 1882), Egypt and Nigeria (1882). It has been isolated from soil in grassland (6082), cereal fields, the rhizosphere of *Cassia occidentalis* (6067) and alder, as an aerial contaminant, from tomatoes in a greenhouse, endosperm of rye grains, hay (1882), apples (1198), *Cassia occidentalis* (521, 6068), *Canna indica* and pineapples, groundnuts and bagasse pulp (IMI). — As one of the causal agents of gummatous ulcers (known as maduromycosis and mycetoma), particularly in warmer climates, it has received much attention (1200, 2720, 3103).

Phenoloxidase tests with tyrosine and Dopa were positive, mainly inside the hyphae (1882). *A. kiliense* can oxidize manganese in the soil (1882); it produces an alkaline proteinase (2438, 2439, 2440) and amylase production varies between isolates (3566). Norethisterone is converted into its 1α-hydroxy-derivative (122). — *A. kiliense* has probably an antagonistic effect against *Fusarium oxysporum* f. sp. *lycopersici* (4543).

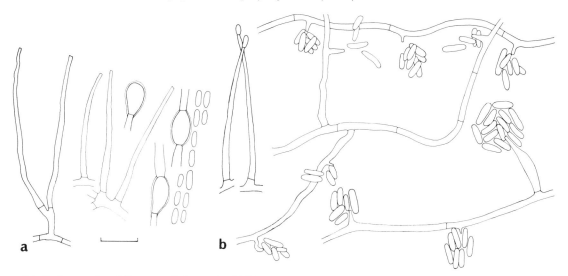

Fɪɢ. 9. *Acremonium kiliense*. a. Orthotropic phialides, conidia and chlamydospores of various isolates; b. reduced phialides produced in the agar.

Acremonium luzulae (Fuckel) W. Gams *1971* — Fig. 10.

≡ *Torula luzulae* Fuckel *1870*
≡ *Gliomastix luzulae* (Fuckel) Mason ex Hughes *1958*

= *Fusidium viride* Grove *1885*
= *Oospora virescens* sensu Sacc. *1881* [non (Link) Wallr. *1833*]

DESCRIPTIONS: Dickinson (1374), and Gams (1882). — Sect. *Gliomastix, A. luzulae* series. — Colonies reaching 8–10 mm diam in ten days on MEA, dark olivaceous, powdery; phialides coarsely warted over the whole surface; conidia fusiform with truncate ends, darkest in the middle part, 5·2–8·2 × 1·9–2·8 μm. — *A. luzulae* has a counterpart with hyaline conidia in *A. longisporum* (Preuss) W. Gams *1971*.

The distributional data include a wide range of geographical regions and habitats from salt-marshes in the British Isles (413) to paddy fields in India (1519). It has further been reported from Libya (4084), the USA (1166, 2822), New Zealand, Borneo, Germany, Belgium, Ghana, Sierra Leone, Togo (1374, 2603) and central Africa (3063). *A. luzulae* has been found in the rhizospheres of *Artemisia herba-alba* (4084) and some plants growing in sand dunes (4371). It also occurred commonly on dried decaying and green leaves of cotton (5231) and bananas (3063) and has been isolated from wheat leaves (4548), stems of *Asparagus, Petasites,* corn, potato, pineapple, banana, *Luzula sylvatica, Urtica dioica, Heracleum sphondylium, Gunnera scabra, Oenanthe crocata, Borassus aethiopum* and *Rhopalostylis sapida* (IMI), hair baits (1422) and various decaying woods (1374, 1882). Forest burning had no influence on the number of propagules of *A. luzulae* in soil (2822).

A. luzulae produces ergosterol in the presence of methionine (6045) as well as ergosterol peroxide and cerevisterol (930). The diterpenes virescenol A, B and C give rise to isovirescenol B (1181), the glucosidic virescenosides A, B and C (836) and the acidic virescenosides F and G (929); a further terpenoid metabolite is ascochlorin (835).

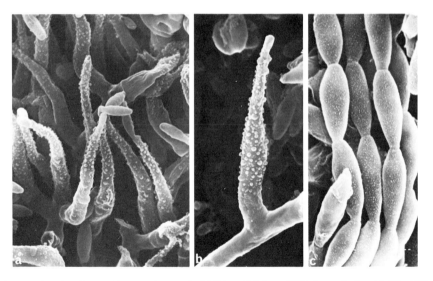

FIG. 10. *Acremonium luzulae*. a. Group of warted phialides, SEM, × 1850; b. a single phialide, SEM, × 2500; c. conidial chains, SEM, × 4000; CBS 494.67.

Acremonium murorum (Corda) W. Gams *1971* — Fig. 11.

≡ *Torula murorum* Corda *1839*
≡ *Gliomastix murorum* (Corda) Hughes *1958*
= *Gliomastix chartarum* (Corda *1840*) Guéguen *1905*
= *Gliomastix convoluta* (Harz *1871*) Mason *1941*
≡ *Torula convoluta* Harz *1871*

For many other synonyms see Gams (1882).

DESCRIPTIONS: Guéguen (2150), Mason (3661), Hughes (2602), Dickinson (1374), Hammill (2240), and Gams (1882). — Sect. *Gliomastix, A. murorum* series. — Colonies reaching 1·8–2·8 cm diam in ten days at 20°C on MEA, olivaceous-black, mostly strongly tufted and powdery, reverse often brown. Conidia ellipsoidal to subglobose, olivaceous-black, mostly coarsely warted, 3·4–5·7 × 2·0–3·7 μm.

Dickinson retained the subdivision of *Gliomastix murorum* into var. *murorum* (spherical conidia in chains) and var. *felina* (March.) Hughes (oval-cylindrical conidia in heads), although this division cannot be effected with complete accuracy. The var. *felina* appears to be the more

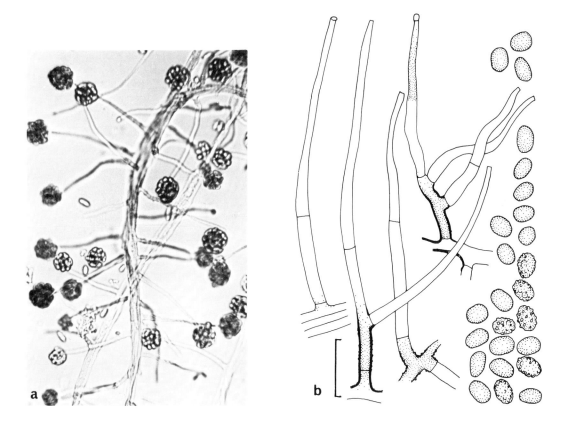

FIG. 11. *Acremonium murorum*. a. Hyphal bundle with numerous phialides and conidial heads, × 500; b. single conidiophores showing chromophilic base, and conidia.

usual in soil analyses, whereas var. *murorum* is very common on plant substrates. In view of the wide range of variation in this and other *Acremonium* species, however, this varietal separation was not retained by Gams (1882). — The closest relative to *A. murorum, A. polychromum* (van Beyma) W. Gams *1971*, has smooth-walled ovoid conidia, 3·8–4·9 × 1·8–2·9 μm, with an apiculate base, which cohere in regular chains. *A. roseogriseum* (S. B. Saksena) W. Gams *1971* has dacryoid conidia measuring 4·9–6·5 × 2·6–4·0 μm. *A. murorum* has a hyaline counterpart in *A. persicinum* (Nicot) W. Gams *1971*. — DNA analysis gave a GC content of 57% (2908).

A. murorum is a saprophyte with a worldwide distribution on an extremely wide range of substrates, especially in soils (1374, 1882). This fungus is easily isolated with the aid of dilution plates (1379, 2161), being found rather rarely on buried cellophane (5961) and chitin (2068); on washed soil samples (1889) it occurred predominantly in organic fraction particles (1614). — Besides its frequent occurrence in the temperate zone, it has also been isolated in tropical habitats. There are too many reports from Europe to New Zealand (2603) to cite here individually but among these are included deciduous and coniferous forest soils, grassland, arable and garden soils, further soils in citrus and banana plantations (2035, 2764, 3630, 3632), moorland soils (3918, 5559), alpine pastures (3445), heathland (3720), podzol of volcanic material (1021), sand dunes (745, 3941, 4162, 4655), salt-marsh (413, 1379), children's sandpits (1421), river (1162) and marine sediments (5316) and water of high salinity (5249). It has also been found on organic detritus (4429) and wood baits (5249) in running water (1154, 1157). In soil it has been found at varying depths ranging from 0 to 100 cm (159, 2161, 3630, 5048, 6184, 6351). It appears to be largely independent of soil pH (159, 6182) but is favoured by manuring in the order NPK > NP > NK (2162, 2163) with no further rise by application of farmyard manure (2161). In some other trials, however, it was found non-responsive to N fertilizers (3498). Its frequency is reduced by several soil fungicides (6136). There are also isolates from sewage (1166) and sewage-treated land (1163, 2467). Its presence has been demonstrated in the root regions of banana (2035), grasses (1614), wheat (1614, 2816, 3567), barley, bean (3006, 4451, 4452), groundnuts (2532), clover (5815), alfalfa (3982), *Ammophila arenaria* (1882), *Halimione portulacoides* (1379), and various forest plants (4448, 4814, 4816). *A. murorum* is, however, rarer in the wheat rhizosphere than in control soil (2816, 3567); in beans too, no rhizosphere effect was observed (4452). It was shown to be present on mycorrhiza of *Corylus avellana* (1794) and in truffle soils (3451); also on seeds of wheat (3491), oat (3556), *Avena fatua* (2961), common grasses (3512), on bananas (6158), feathers of free-living birds (2575) and cysts of the nematode *Globodera rostochiensis* (3090). After two weeks incubation, 98% and 81% of the conidia survived in mull and mor soils, respectively (633). An ecological monograph on *A. murorum* has been published (6376).

Optimum growth occurs at a water potential of −2·3 bars, but more than −9 bars are tolerated (5316). — In contrast to many other fungi, growth on D-arabinose is good; mycelium production was greater on L-arabinose than on glucose (3328). It also utilizes mannose and cellobiose (5387) as well as starch (1827) and various forms of cellulose (386, 2706, 2736, 4191, 6085, 6326), though not filter paper (386, 2706, 3618). The optimum pH for cellulase activity is said to be 9 and the optimum temperature 29°C (with a rapid decline towards 35°C). Organic N compounds (e.g. glutamic acid) proved particularly favourable for cellulase production (5388); both chitin (1425, 2068, 2706) and wool (6330) are attacked; the capacity for xylan decomposition is good, with little variability between different isolates (1432); no particular requirements for growth substances are known (1425). — Antagonistic activity towards bacteria or fungi ranges from low or entirely absent (2736) to highly active towards *Gaeumannomyces graminis* (3567) and *Poria weirii* (4123). — This fungus can grow at CO_2 concentrations of up to 10% (784) but there is complete inhibition at 100% CO_2; good growth occurs at 100% N_2 (5602). It is sensitive to CS_2 treatment of the soil (3987).

Acremonium rutilum W. Gams *1971* — Fig. 12.

≡ *Cephalosporium roseum* Oudem. *1884* [non *Acremonium roseum* Petch *1922*]

DESCRIPTION: Gams (1882). — Sect. *Nectrioidea*. — Colonies reaching 1·0–2·0 cm diam in ten days at 20°C on MEA, usually intensely pink, velvety, with more yellow reverse. Phialides usually simple, with highly chromophilic granular secretions at their base. Conidia ellipsoidal, $3·0–6·0 \times 2·0–3·3 \mu$m. — Similar soil-borne species are *A. biseptum* W. Gams *1971*, regularly with a second septum in the conidiophores and more ochraceous to pale pink colonies, and *A. verruculosum* W. Gams & Veenbaas–Rijks *1971* with verruculose conidia. — DNA analysis gave a GC content of 55·2% (2908).

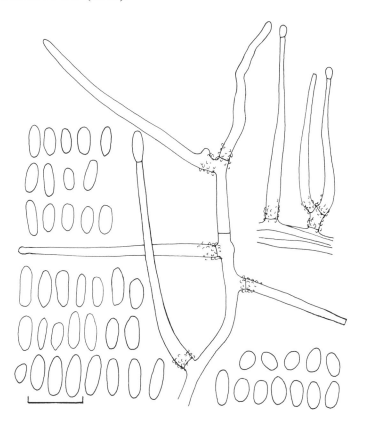

FIG. 12. *Acremonium rutilum*, phialides showing chromophilic warts at their base, and conidia, CBS 394.66.

A. rutilum is frequent in various types of soils. It also occurs as a secondary fungal colonizer on rotting plant substrates. From the available data, a worldwide distribution can be assumed (1882). There are reports of soil isolates from forests (1828, 3413), *Salix–Populus* flood planes (1882), fen peat (5559), cultivated soils (159, 3790, 5392), saline beach (4162), land flooded with brackish water (4477), rocks of carst caves (3364, 3365) and soil around human dwellings (1420). It has also been found on stored wheat seeds (2716) and fruits of *Punica granatum* (IMI).

Decomposition of pectin is demonstrable but in comparison to other fungi not strong (1432); cellulose decomposition has been shown on textiles (1828, 3618) and on CMC (1432). — *A. rutilum* possesses antagonistic activity against *Rhizoctonia solani, Gaeumannomyces graminis, Pythium ultimum, Fusarium culmorum, Pseudocercosporella herpotrichoides* (1431) and has also been shown to be antagonistic to *Botrytis cinerea* and *Aspergillus niger* (4139); isolates from diseased apples gave positive infections on reinoculation (735).

Acremonium strictum W. Gams *1971* — Fig. 13.

= *Cephalosporium acremonium* auct. mult. p.p.
= *Tilachlidium medietatis* Novobranova *1972*

DESCRIPTION: Gams (1882). — Sect. *Acremonium.* — Colonies reaching 1·6–2·5 cm diam in ten days at 20°C on MEA, pink, usually moist and smooth. Simple phialides arising from submerged or slightly fasciculated aerial hyphae. Conidia cylindrical, mostly 3·3–5·5 × 0·9–1·8 μm. *A. strictum* possesses the least number of differential characters, i.e. it does not form chlamydospores or sclerotia, or stain Sabouraud agar brown. — The colonies are moist to slimy in contrast to the similar *Acremonium sclerotigenum* (F. & V. Moreau ex Valenta) W. Gams *1971* which has whitish densely floccose colonies; particularly on lupin stems the latter produces hyaline, globose sclerotia, 15–50 (–90) μm diam. The phialides of *A. strictum* have a thin wall unlike the otherwise similar *Monocillium tenue* W. Gams *1971*. A yellow fungus often referred to as *Cephalosporium acremonium* which produces cephalosporin C, is now named *Acremonium chrysogenum* (Thirum. & Sukap.) W. Gams *1971*. *A. strictum* is not identical with the fungus known by medical mycologists as *Cephalosporium acremonium* (causal organism of gummatous ulcers), which is now called *Acremonium kiliense* Grütz (q.v.) (1882). — DNA analysis gave a GC content of 56·5–57·3% (2908).

Because species identifications in *Cephalosporium*-like fungi have generally been inadequate, it seemed unwarranted to reproduce the data presented in countless publications on *C. acremonium*. Nevertheless, our own observations and the specimens deposited in IMI have shown that the species has a worldwide distribution and, in fact, represents the most frequent *Acremonium* species inhabiting the soil. — Cultures and specimens have been seen from Germany, France, the British Isles, Egypt, Zaïre, the USA, India, Australia (1882), Nepal (1826), South Africa (4408), New Zealand, Papua, Malaysia, Pakistan, the Windward Islands, Israel, Iraq, and Turkey (IMI). It has frequently been isolated from prairie and arable soils (2161); in a soil under wheat in Germany, it occurred regularly though not frequently, and in

South Africa it was isolated from wheatfield debris (4408). It is often found in the rhizosphere of *Ammophila arenaria* (1882), *Cicer arietinum, Musa* (IMI) and in the poplar rhizosphere in truffle grounds (3452); also on leaf surfaces of vascular plants (296), *Argemone mexicana* (6068), and on cotton bolls (5009). It is frequently observed on other primary fungal colonizers like rusts or powdery mildews; also on agarics, *Scleroderma*, polypores and stromata of *Hypoxylon*, and as a hyperparasite on *Helminthosporium* spp. (2930). It also occurs on excrements, hay, stained wood and as an ambrosia fungus (1882). In addition, it is widely distributed in the atmosphere (1882), and as a contaminant of fuel and fuel filters (IMI). *A. strictum* can serve as food for the mite *Pygmephorus mesembrinae* (3104).

Decomposition of xylan is good with little variability between different isolates (1432). With only one exception (3957), all reports mentioning production of cephalosporin C in *Cephalosporium acremonium* refer to *Acremonium chrysogenum* (Thirum. & Sukap.) W. Gams. Norethisterone is converted to its 1*a*-hydroxy-derivative (122).

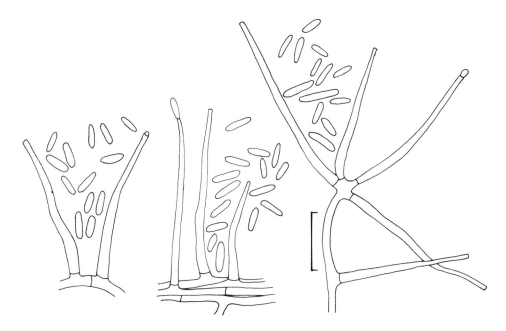

FIG. 13. *Acremonium strictum*, phialides and conidia from various isolates.

Acrophialophora Edward *1959*

Acrophialophora fusispora (Saksena) Samson *1970* — Fig. 14.

≡ *Paecilomyces fusisporus* Saksena *1953*
≡ *Acrophialophora fusispora* (Saksena) M. B. Ellis *1971*
= *Masoniella indica* Salam & Ramarao *1960*

DESCRIPTIONS: Brown and Smith (743), Samson and Tariq Mahmood (5030), and Ellis (1603). — The genus *Acrophialophora* with the type species *A. nainiana* Edward *1959*, is close to *Paecilomyces* but differs in the presence of pigmented, warted conidiophores, verticillate phialides in limited numbers with a narrowly tapering tip and frequent sympodial proliferation. — Amongst the three species recognized, *A. fusispora* is distinguished by pigmented fusiform conidia ornamented with spiral bands, measuring 5–12 × 3–6 μm. Colonies reaching 4–7 cm diam in 14 days at 25°C on MEA.

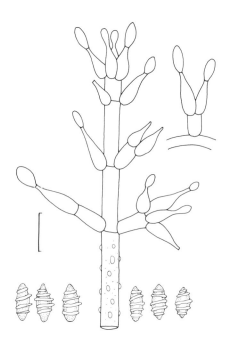

FIG. 14. *Acrophialophora fusispora*, conidiophore and conidia; orig. R. A. Samson.

With the exception of records from Czechoslovakia (1703), Singapore (3331), the Tuamotu Archipelago (6084), Egypt (8), and Pakistan (4855), this fungus has been almost exclusively reported from India. It is known from uncultivated soils (4030, 4698), field soils (3868) under leguminous plants (1317, 2164) or grasses (3863, 3865, 4933), forest soils (1525, 2854, 4995, 5512) and salt-marshes (8). *A. fusispora* is common on the root surface and in the rhizospheres of grasses (3271, 3331), cotton (5009), *Euphorbia* spp. (3866), *Carica papaya* (5345), and on rhizoids of *Funaria* sp. (2860). In the rhizosphere of *Trigonella foenum-graecum*, its frequency increased following foliar sprays with the antibiotic subamycin (2187). It has also been isolated from earthworm casts (2857) and wood exposed to seawater (4546). There seems to be a distinct preference for deeper (30–45 cm) soil horizons in India (1317, 1525, 3865, 4995, 5349). *In vitro* growth was found to be stimulated by root exudates and extracts of *Cassia tora* (5638).

The temperature minimum is 14°C, the optimum 40°C and the maximum 50°C (1665).

Actinomucor Schost. *1898*

Actinomucor elegans (Eidam) C. R. Benjamin & Hesseltine *1957* — Fig. 15.

≡ *Rhizopus elegans* Eidam *1884*
= *Mucor corymbosus* Harz *1871* [non Wallr. *1833*]
= *Actinomucor repens* Schost. 1898
= *Mucor botryoides* Lend. *1910*

DESCRIPTIONS: Benjamin and Hesseltine (447), and Zycha *et al.* (6588). — *Actinomucor* Schost. *1898* is a monotypic genus of the Mucoraceae Dumort., characterized by differentiation into stolons and acrotonously verticillate sporangiophores which arise mostly from rhizoids. Zygospores are unknown. *A. harzii* (Berl. & De Toni) Rosenberg *1959* was allegedly distinguished by its larger sporangia but falls within the range of variation of *A. elegans*. — Primary sporangia mostly 60–80 μm, secondary sporangia 20–50 μm diam, with a spiny wall which ruptures near the columella so that the sporangia fall off with most of the spores still included (2648). Spores globose, hyaline, smooth-walled to faintly roughened, 6–8 μm diam. Chlamydospores sparsely produced, singly or in short chains, elongate, hyaline, 8–16 μm diam. — Mycelial extracts have been shown to contain glucose, fructose, and sucrose, succinic (4009), malic and tartaric acids (5055), and the amino acids aspartic and glutamic acid, serine, glycine, a-alanine, lysine and tyrosine (4009).

Distribution is worldwide but principally in cultivated soils of the temperate and subtropical zones. It has been reported from the British Isles, Ireland (IMI), France (2161), Czecho-slovakia (1700), Poland (1421), Yugoslavia (4573), Hungary (6551), Israel (654, 2764, 4759), South Africa (1559), Kenya, Tanzania, Egypt (IMI), Pakistan (4855), Kuwait (4001), India (1315, 1317, 3732, 4995, 4997, 4998, 5622), Nepal (1826), Australia (3020, 6186), the Bahamas (4312), Brazil (6007), Chile (1824), Canada (5363), and the USA (1165, 1166, 3732). It is preferentially and very frequently found in arable and grassland soils (4573, 6182) but also known from soils with steppe type vegetation (1700), open savannah (1559), saline soils (4001, 4645, 4646), forest soils (273, 3138, 5363), caves (4312, 6551), children's sandpits (1421), and garden compost (1425, 3041). It occurs at soil depths from 0–30 cm (1317, 2161, 4995, 4998) and was found to be stimulated by NPK fertilization (2161). To judge from the available data, its habitats tend to be neutral to slightly alkaline (1315, 1559, 3732, 4055, 6184). With moisture requirements for maximum and minimum growth at −5 bars and −90 bars water potential, respectively, the species can be classed as relatively xerophilic (3109). It has been isolated from the rhizospheres of oat and barley (3004), corn (4553), wheat (2949) and alfalfa (3982, 3983), cotton seedling roots, groundnut pods, barley seed (346), hay and chaff (4548), sugar cane, mouse and rabbit dung (IMI), and is also reported from nests, feathers and droppings of free-living birds (2575), and from dead bees and combs (402). — The hyphae can be parasitized by *Trichoderma viride* and *Myrothecium verrucaria* (1511).

The mean doubling time in submerged culture is 4 h (5890). Good utilization of arabinose, galactose, maltose (4910), glucose, sucrose, L-xylose, starch (447, 654), glycogen, sorbitol, mannose and sorbose is reported (5053). Poor C sources are arbutin, dulcitol, malic and tartaric acids (5053); lactose, glycerol and cellulose are not utilized (447); good N sources are KNO_3, $NaNO_3$, $(NH_4)_2SO_4$ (5056), urea (447) and gelatine (654); nitrites can be utilized (5056). From a mixture of four amino acids (each 350 mg N·l^{-1}), *A. elegans* took up leucine completely after three, arginine after five and valine after eight days, whilst glutamic acid was utilized only very slowly (5054). Suitable S sources are $MgSO_4·7H_2O$, $(NH_4)_2SO_4$ and $K_2S_2O_5$ (5057). It has an inducible nitrate reductase and is able to reduce Fe^{3+} to Fe^{2+} (4330), produces alcohol dehydrogenase (1999) and relatively high amounts of proteinase and phosphatase (5309); optimal proteinase production occurs on solid substrates (wheat bran) with a 50–63% moisture content after three days incubation at 20°C (6173). Optimum proteolytic activity is reported at 32°C and pH 7 (3001). Formation of humic acid-like substances occurs during autolysis of the mycelium (4470). *A. elegans* appears to have growth-substance requirements (1425). — Weak antibacterial (6172) and antifungal activities against *Cochliobolus sativus* (868) have been reported. The development of *Chlorella pyrenoidosa* is inhibited *in vitro*, and the growth of corn seedlings stimulated (4553).

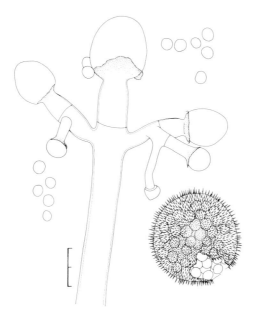

FIG. 15. *Actinomucor elegans*, verticillate sporangiophore with columellae of various sizes, sporangium and sporangiospores

Alternaria Nees ex Fr. *1821*

= *Macrosporium* Fr. *1832*

Type species: *Alternaria alternata* (Fr.) Keissl. (≡ *Alternaria tenuis* Nees)

Monographic treatments: Neergaard (4118), Joly (2792), and Ellis (1603, 1604); key to the species: Joly (2795); typification of the genus: Simmons (5332). Some 44 species are recognized in this genus (1603, 1604).

Hyphae, conidiophores and conidia light to dark brown. Conidiophores mostly simple and usually becoming geniculate by sympodial elongation. Poroconidia formed singly or in acropetal chains, consisting of an ovoid or ellipsoidal body with a broadly rounded base and an apical beak, muriform with several transverse and fewer longitudinal septa.

Alternaria contains numerous plant parasites which are mostly host-specific and often seed-borne; a few species are ubiquitous and also very frequently soil-borne; *A. alternata* is the commonest of these.

Because isolates soon tend to loose their capacity to sporulate, cultivation on natural media (e.g. OA, CMA, PCA) and incubation under near-UV light is recommended (e.g. 2758, 3241). Two of the most ubiquitous species with catenulate conidia are treated here.

Key to the species treated

1   Conidia formed in long chains of 10 or more conidia; conidia irregularly shaped, with a short beak not exceeding one third of the conidial length, medium to dark brown

A. alternata (p. 34)

    Conidia formed in chains of 3–5 (rarely to 10) conidia; conidia usually obclavate, tapering gradually into a beak of half the conidial length, ± pale golden brown   *A. tenuissima* (p. 38)

Alternaria alternata (Fr.) Keissler *1912* — Fig. 16.

≡ *Alternaria tenuis* Nees *1816* (devalidated name)
≡ *Torula alternata* Fr. *1832*
? = *Alternaria tomato* (Cooke) Brinkman *1931*
≡ Macrosporium tomato Cooke *1883*

Teleomorphs: *Clathrospora diplospora* (Ellis & Everh. *1894*) Wehm. *1954*, *Clathrospora elynae* Rabenh. *1854*, *Leptosphaeria heterospora* (de Not. *1863*) Niessl *1972*

DESCRIPTIONS: Neergaard (4118), Joly (2792), Malone and Muskett (3556), Simmons (5332) (with nomenclature), Verona and Firpi (6074), Matsushima (3680), and Ellis (1603). — Colonies reaching 6 cm diam in seven days at 22–30°C on MEA. Conidiophores and conidia usually medium golden brown. Conidiophores usually simple, straight or curved, 1–3-septate,

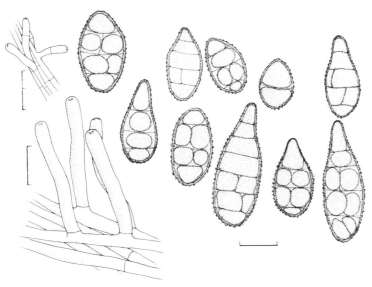

Fɪɢ. 16. *Alternaria alternata*, hyphal bundle with conidiophores, detail showing the apical pores, and conidia, CBS 326.65.

to 50 μm long, 3–6 μm wide, with one or several apical conidial pores. Conidia formed in long, often branching chains, ovoid, obclavate, obpyriform, or more rarely ellipsoidal, with a conspicuous basal pore, with or without a short conical or cylindrical apical beak not exceeding one third of the conidial length, medium brown, smooth-walled or warted, slightly constricted at the three to eight transverse septa, in the lower part each portion has one or two longitudinal septa, 18–63 × 7–18 μm; TEM and SEM studies of the conidia have been carried out (5412). — The similar anamorph of *Pleospora infectoria* Fuckel *1870* which is sometimes regarded as a synonym (6238), differs from *A. alternata* in having mostly obclavate conidia which taper gradually towards a beak which is up to half the total length, dense transverse septa in the lower part and coarsely verrucose walls, 20–70 × 9–13 μm. *Alternaria citri* Ellis & Pierce *1902*, which occurs only on citrus fruits and leaves, has broader conidia to 24 μm wide; *A. tenuissima* (Kunze ex Pers.) Wilts. *1933* (q.v.) has shorter conidial chains and paler conidia with longer beaks. In the equally similar and ubiquitous *Ulocladium chartarum* (Preuss *1848*) Simmons *1967* the conidia are commonly formed in chains of 2–10, predominantly ellipsoidal to obovoid in shape, not exceeding 38 μm in length, and often with false and distinctly paler beaks. — Vegetative cells in *A. alternata* are multinucleate; the conidial nuclei are derived from a single nucleus which migrates into the primordium (2290). Mycelia of some isolates have been found to contain virus-like particles (674, 2510, 2677). — Chemical analyses of mycelia revealed mannitol, a number of fatty acids and fatty acid esters including myristic, palmitic, stearic, oleic and linoleic acids (1173) and 14 protein-bound and 11 free amino acids (6091).

A. alternata is an extremely common and cosmopolitan species occurring on many kinds of plant and other substrates including soil, foodstuffs and textiles (1603). Virulence to different host plants does not appear to be correlated with any morphological traits (3389). A comprehensive survey of the distribution of this species was published in *1945* (4118); this can now be supplemented for soil by the following reports from widely separated areas: the White

Sea (3368), Canada and Ireland (1446, 5363), recently deglaciated soils in Alaska (1171), alpine habitats (3976), Israel with high frequencies (2764, 2768, 2772), Somalia (5048, 5049), Turkey (4245, 4246), Spain (3446), Iraq (92), Kuwait (4001), Syria (5392), Libya (4083, 4084, 6510), Egypt (3988, 3992), Pakistan (61, 4855), Nepal (1827), India (numerous records), Sudan (4222), Zaïre and the Ivory Coast (4159), central Africa (3063), South Africa (1559, 4407), Jamaica (4886), Honduras (2031), Chile (1824), Brazil (4547) and Japan (2532). *A. alternata* has often been found in forest and cultivated soils, particularly in the upper layers of very varied soil types (3414, 3416, 4742, 5349). It has also been isolated from soils under, for example, garlic (2159), paddy rice (774), sugar cane (2861), alfalfa (3982), flax (5826), *Spartina townsendii* (5389), pear (5499), citrus (3988), *Euphorbia* species (3866), oak (963, 5394), various forest plants (4814), various halophytes (3375, 4665), and *Artemisia herba-alba* (4084). Its frequency in soil was reduced by cropping with corn (1330). Further known habitats include sand dunes (3941, 4162, 4371, 4655), desert soils (654, 1512, 4083, 4733), open savannah (1559, 4407), Pamir soils at high elevations (3856), a guano cave (4312), clay and limestone in carst caves (3364, 3365), a uranium mine (1703), peat bogs (1376, 4474), moorland soils (3234), estuarine sediments (655), salt-marsh soils (8, 1378, 4646, 5316), other saline soils (92, 3446, 4000, 4001), saline lakes (1292), seawater (4918), fresh water (1166, 3809), sewage (1154, 1155, 1157, 1166, 1172), fields treated with sewage sludge (1163), compost (1528, 3041), coarse fodder (4548), corn silage (1092), rotten wood (1199, 3568, 5016), wood exposed to water (4955) or buried in soil (2833), wood pulp (2792, 3294, 4118, 6168), the slime of paper mills (5060), brown coal (4474), aircraft fuel (4935), bee combs (402), the ear of a white-tailed deer (5011), rabbit dung (6085), nests, feathers and pellets of free-living birds (2575, 2577, 4649, 6223), food materials and textiles (2792, 3707, 5205, 5980). It can penetrate to shallow or intermediate depths in various soils (1736, 1823, 2161, 2948, 6347) and occurs in soils of very different pH, including alkaline ones (1259); at pH 8·6, however, it was less frequently isolated than at 6·5 (1713). Dry habitats are frequently inhabited (1824, 3417, 4313). Due to its preference for rotting organic substrates, an increased frequency was observed after applications of farmyard manure as compared with NPK (2762). Reports from rhizospheres are very numerous and include, for example, those of wheat (2023, 2816, 3567, 5514), oats, barley (3006), corn (4553, 6563), coriander (1523), groundnuts (2532), poplar (3452), and *Carica papaya* (5345); it has also been isolated from the rhizoplane of *Adiantum* sp. (2856), *Aristida coerulescens* (4083), *Abutilon indicum*, *Trichodesma amplexicaule, Cynodon dactylon* (2855, 3864), and sugar cane (4886), and roots of sugar beet (4559), strawberry (4133), red clover (6493), potato (1615) and *Halimione portulacoides* (1378). *A. alternata* is also known from the geocarposphere of groundnuts (2768) and leaves of *Adiantum, Polypodium, Cyclosorus, Dryopteris* (2859), *Pseudoscleropodium purum* (2969), *Pteridium aquilinum* (2013), *Brachypodium pinnatum* (2970, 2971), *Hippophaë rhamnoides* (3352, 3353), *Nothofagus truncata* (4946), *Fagus sylvatica* (2500), several Hawaiian vascular plants (296), tobacco (4214), cotton (5009), citrus (3988), culms and leaves of *Dactylis glomerata* (6237, 6238, 6243), *Setaria glauca* (5233), and *Elytrigia repens* (2590), the culm bases of rye, wheat and barley (3219, 4136), *Carex paniculata* (4644), stems of *Heracleum sphondylium* (6462), decaying shoots of *Bothriochloa pertusa* (2955), *Urtica dioica* (6463) and *Typha latifolia* (4663). On *Phragmites australis*, it can persist for some time on collapsed floating culms (5713). Its colonization of seed material is of particular importance, and extensive lists of affected seeds have been compiled (2792, 4118). An immunofluorescence technique to detect its mycelium in barley seeds has been devised (6200). Reports of its occurrence on wheat and barley grains, groundnuts, pecans, clover and grasses are particularly numerous. — *A. alternata* has been isolated from storage places of gerbils (5393), and the intestine and integuments of *Dendrolimus sibiricus* caterpillars (2848);

it can serve as food for the springtail *Hypogastrura tullbergi* (3830), the mites *Acarus gracilis* (5380), *A. siro*, *Tarsonemus waitei*, and *Acotyledon redikorzevi* (5379), the bean weevil *Acanthoscelides obtectus* and the psocid *Lepinotus reticulatus* (5381). — *A. alternata* contributes to the stabilization of soil aggregates (225). It is inhibited *in vitro* by *Memnoniella echinata* (4), *Bacillus subtilis* (2393) and the phytoalexin *a*-tomatine (182). Under dry conditions, more than 40% of the conidia can survive for 300 days (2499) and some remain viable for more than ten years on wheat seed (3492) although the number of viable conidia drops considerably during the first years of storage (4951). — Violent conidium discharge has been observed when sporulating substrates were transferred from humid to dry conditions (3763). The conidia are released in a diurnal rhythm from 12.00–18.00 h (4483).

Optimal growth occurs at (22–)25–28(–30)°C, the maximum temperature for growth is ⩾31–32°C and the minimum in the range 2·5–6·5°C (4118, 4487, 4517), except for isolates adapted to cold with minima at 0°C, −2 or −5°C (191, 2761, 4487); no growth occurs at ⩾36°C (2310, 4958). The thermal death point in apple juice has been determined as at 63°C for 25 min (3620). The minimum humidity for vegetative growth is at −210 bars water potential, for sporulation −70 bars (4369), for conidial germination at −100 bars (6464) or dew deposition (4483). Sporulation was better in the dark than in full white light (1742), but was improved in the red spectrum of light (1654, 1742), and particularly in the near-UV range (3241), under the influence of other fungi (2210), and in the presence of Ca, Zn, Cu, Mo, B, Fe and Mn ions (2085, 5372); zonation in cultures is induced by alternating light/dark (1865). Optimal sporulation occurs at 25–27°C, with a minimum at 15°, and a maximum at 33°C (4483) on PDA or a synthetic medium of pH 6 (5364). The formation of sectors is influenced by both glucose and light (1442). Optimal growth occurs at pH 4 (4320) or 5·4, but growth is possible between pH 2·7 and 8 (2310). At 0·25% (v/v) O_2 in the atmosphere, growth is reduced to a quarter (1793); over the range 4–21% (v/v) O_2 concentration and growth response are proportional (6281); the maximum $CO_2:O_2$ ratio tolerated is 3:1 (2096). — Suitable C sources include maltose, sucrose, raffinose (283, 965, 4958, 6074) and D-galactose (1844). *A. alternata* can utilize starch (654, 1827, 6074), xylan and arabinoxylan with differences between various isolates (1432, 1753) and pectin (654, 6356), polygalacturonase and pectin-methylesterase (2239, 4370) being produced. β-Galactosidase can be produced in submerged culture (5821). The capacity for cellulose decomposition has been noted by numerous authors on a wide range of substrates and increased activity found to follow the addition of yeast or corn extracts (3383). Wool (6330) and wood (1199, 4955) are slightly utilized and growth on phenol-lignin (1744) and kerosene (4935) has been reported. The following N sources can be utilized: $Mg(NO_3)_2$, $Ca(NO_3)_2$, peptone, D-alanine, L-phenylalanine, ammonium acetate and oxalate, glycine, acetamide, L-asparagine, L-aspartic acid, L-glutamic acid, urea, ammonium nitrate and tartrate (283, 2084, 4958, 5509); leucine was not utilized and histidine gave poor growth (5737). No specific sulphur sources are required (5348). The insecticide bromophos is effectively metabolized in liquid culture (5558). The activity of glyoxalate transaminase (1819) and phenol-oxidase (4894) has been investigated and compared with that of other fungi; ferric ions are enzymatically reduced (4330). The mycelium grows chemotropically towards phosphate granules but does not visibly dissolve them (3074). Progesterone is 1,2-dehydrogenated in the A-ring (876). Of the metabolic products so far investigated (4912), the toxic tenuazonic acid (1924, 2831, 3766, 3815, 5281, 5578) with larvicidal activity on *Lucilia sericata* (1112), the structurally related altenuene (4513), the toxic tentoxin (3793, 4959) causing chlorosis in cucumber and many other dicotyledonous seedlings by inhibition of phosphorylation (2228), altenuisol (4511), altenusin and dehydroaltenusin (1173), and particularly the isocoumarin alternariol with antifungal activity and able to induce chlorosis (as its monomethyl ether) in tobacco (4512), have evoked further interest (796, 1923, 4706, 5187, 5395, 5802, 5803). Chemically unidentified substances stabilizing soil aggregates are also produced (224). — Severely infected cereals are toxic to warm-blooded animals (4514, 6412); mycotoxins causing toxic leucopenia in man have been reported (1815). Rabbits and guinea pigs responded to inoculations with a suspension of particles of this fungus by producing antibodies and superficial skin lesions (4763). Mycelial extracts proved toxic to *Paramecium caudatum* (1456), brine shrimps, chicken embryos and rats (1388), and caused lesions in the mucous membranes of horses (2050). — The growth of *Chlorella pyrenoidosa* is inhibited by *A. alternata in vitro*; antagonistic effects on numerous fungi and bacteria have been observed *in vitro* (1092, 2210, 2753, 3304) as well as on *Cochliobolus sativus* on wheat roots (1231). Culture filtrates stimulated the growth of corn plants (4553) but inhibited or killed seedlings of alfalfa and *Poa pratensis* (1225, 3982, 3983) and discoloured rape and pea roots (2210), while barley suffered no adverse effects (293). Fruit rots due to *A. alternata* have been reported in citrus, apple and pear (4118, 5958). Other hemiparasitic effects of *A. alternata* include a yellowing of jute cotyledons (4688), chlorosis in young apple leaves (5761), and albinism in citrus seedlings and numerous other plants (1845, 4956). The germination of spores of dwarf and covered smuts on rye is said to be favoured by *A. alternata* (1922). — It is able to tolerate 0·1% calcium cyanamide (6077) and has a high resistance to UV (1512) and γ-radiation (3854).

Alternaria tenuissima (Kunze ex Pers.) Wilts. *1933* — Fig. 17.

≡ *Helminthosporium tenuissimum* Kunze ex Pers. *1822*
≡ *Macrosporium tenuissimum* (Kunze ex Pers.) Fr. *1832*
? = *Alternaria humicola* Oudem. *1902*

DESCRIPTIONS: Neergaard (4118), Joly (2792), Verona and Firpi (6074), Matsushima (3680), and Ellis (1603). — Colonies reaching 6–7 cm diam in seven days at 25–30°C on MEA. Conidiophores pale or medium pale brown, with one or several conidial scars, to 115 μm long and 4–6 μm thick. Conidia solitary or in short chains of 3–5(–10) conidia, obclavate or with an ellipsoidal body tapering gradually into a paler beak 2–4 μm wide which constitutes not more than half of the total conidium length and is often slightly swollen to 4–5 μm at the tip, ± pale golden brown, mostly smooth-walled but sometimes minutely verruculose, muriform with four to seven transverse and a few portions with one or two longitudinal septa, 22–95 × 8–19 μm. — *A. tenuissima* is rather similar to *A. alternata* (q.v.) but has shorter chains of paler conidia; the similar *A. longipes* (Ellis & Everh.) Mason *1928*, restricted to tobacco, was regarded as a synonym of *A. alternata* by Lucas (3422). — An analysis of the mycelium yielded aspartic and glutamic acids, alanine, leucine and valine (2792).

A. tenuissima is very common and cosmopolitan and has been recorded on an extremely wide range of plants, where it usually occurs as a secondary invader (1603, 2792). In India it is the second most common species of *Alternaria* besides *A. alternata* (4742). Soil isolates have been obtained from arable soils in France (2161) and Central Europe (3414, 3416), wet prairies in Wisconsin (4313), grassland and *Sphagnum* bogs in the Italian Alps (6082), saline soil in Kuwait (4000, 4001), deserts in Egypt (3996) and the USA (4733), a uranium mine in Czechoslovakia (1703), soils in Pakistan (61, 2297), Nepal (1826), South Africa (1555, 1556,

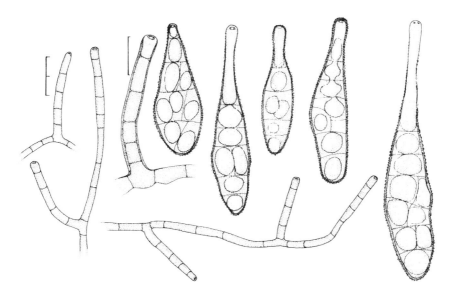

FIG. 17. *Alternaria tenuissima*, conidiophores at different magnifications and conidia, CBS 795.72.

1559), Venezuela, Cuba and the USSR (IMI). In addition to numerous occurrences on live and dead leaves, stems and branches it has been found on stored wheat grain (2297, 5293), in fresh water (4229), muddy brook sediments, air (IMI), on the tick *Ixodes ricinus* (5018), and on feathers of free-living birds (2575). It survived a storage in water for ≥7 years (605).

Light stimulates sporulation (2758), the optimum temperature for conidium germination is 28°C (3803), and a pH of about 5 (3996) and a saturated atmosphere (4493) allow optimal growth. — Utilization of starch (1827), hemicellulose, pectin, and urea (3414) and the production of proteinase (2758) with great variability between isolates (2760) have been reported. Progesterone is 1,2-dehydrogenated in the A-ring (876).

Amorphotheca Parbery 1969

Amorphotheca resinae Parbery *1969* — Fig. 18.

Anamorph: *Hormoconis resinae* (Lindau) v. Arx & de Vries *1973*

    ≡ *Hormodendron resinae* Lindau *1906*
    ≡ *Cladosporium resinae* (Lindau) de Vries *1955*
  = *Cladosporium avellaneum* de Vries *1952*

DESCRIPTIONS of the ANAMORPH: de Vries (6121), Ellis (1603), Nicot and Zakartchenko (4172), Parbery (4412), von Arx (202), and Sheridan *et al.* (5263, 5264); description and TEM studies: Sheridan and Troughton (5265); TELEOMORPH: Parbery (4412), and Sheridan and Steel (5262). — Colonies relatively fast-growing, reaching 2·0–3·0 cm diam in ten days at 20°C on MEA, pale greyish brown to olive-green; reverse dark olivaceous-brown. Conidiophores 30–1200(–2000) μm long, 3–4(–6) μm wide, straight or flexuose, smooth-walled or warted (5265), somewhat darker than the vegetative hyphae, with acropleurogenous ramoconidia and conidial chains arising at several levels; not sympodially elongating. Conidia hyaline to pale brown, smooth-walled, mostly ellipsoidal without prominent scars, mostly 3–7 × 2–4 μm. Chlamydospores present, globose to ellipsoidal, 9–13 μm diam. — Four different forms have been derived from one conidial isolate (6121): f. *avellaneum* (de Vries *1952*) de Vries *1955* with brown colonies, a conidiophore habit like *Cladosporium herbarum*, and relatively short conidia 4–6 × 3–4 μm (Fig. 18a), f. *resinae* (= f. *viride* de Vries *1952*) with olive-green cultures, conidiophores with single slender branches, and elongate conidia 3–12 × 2–4 μm (Fig. 18b) (occurring less frequently in nature); and f. *albidum* (de Vries) de Vries *1955* and f. *sterile* (de Vries) de Vries *1955*, with all possible combinations, which do not normally occur in nature. The anamorph has been transferred from *Cladosporium* to a separate genus (202) as the conidia lack prominent scars and the teleomorph is totally different from that known in other *Cladosporium* species.

The teleomorph of *Amorphotheca* (Eurotiales, fam. Amorphothecaceae Parb. *1969*) is produced by about 20% of the isolates of the first two forms mentioned above on cherry decoction agar or on OA-salts medium within 2–3 weeks at 25°C; on other media the process takes 5–9 weeks. The teleomorph is also produced in mineral oil overlaying MEA slants after several months (5262), and on creosoted matchsticks in contact with soil (4414). The species seems to be heterothallic (4417). Ascomata are small, black, sclerotioid, usually deeply immersed in the agar but sometimes superficial, usually ± spherical, but on soil, matchsticks or mineral oil often with funnel-shaped outgrowths in the apical region, 70–130 μm diam. Ascomata develop from spirally coiled ascogonia with or without a slender central antheridium and subsequently form clusters of radiating hyphae which become enclosed in a dark amorphous mass. Asci ovoid, almost sessile, thin-walled, evanescent, leaving the cavity filled with ascospores; ascospores ellipsoidal, thick-walled, at first hyaline but later becoming brown, 7–9·5 × 3·5–5·5 μm. — The hyphal cells are normally uninucleate and the conidia plurinucleate (6121). — DNA analysis gave a GC content of 54% (5600). — The forms *avellaneum* and *resinae* seem to be genetically stable; saltation occurs only from the former to the latter and not in the reverse direction; heterokaryosis commonly occurs and then f. *avellaneum* is apparently dominant (4417). — A TEM analysis showed indole melanins to be localized in the outermost layers of the hyphal walls (1587). — The major fatty acids contained in the phospholipids are palmitic and one or more C_{18} unsaturated ones (2873).

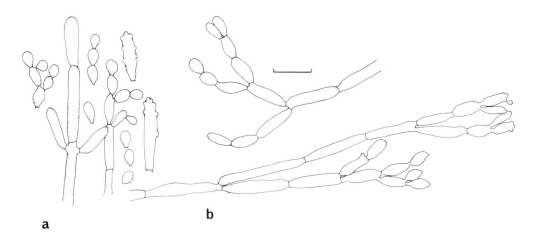

FIG. 18. *Amorphotheca resinae*. a. Conidiophores and conidia of *Hormoconis resinae* var. *avellanea*, CBS 406.68; b. conidiophores and conidia of var. *resinae*, CBS 624.70.

This, the "kerosene" (2379, 4411) or "creosote" fungus (3616), has attracted much attention in the last two decades because of its occurrence in petroleum products, aviation kerosene, creosoted wood (1037, 3616), wood treated with copper-, chrome- or arsenic preservatives (4193), coniferous resin, asphalt pavements (1037) and similar substrates. It is the commonest fungus in aviation fuel where it causes damage by clogging filters and corroding pumps and tanks (1037, 5261). However, it is also a ubiquitous soil fungus and the studies by Parbery in Australia (*1967* on) and Sheridan in New Zealand (*1970* on) have given a detailed picture of its ecology (5259). It was not isolated frequently prior to the development of a selective isolation technique involving baiting by creosote-soaked matchsticks or similar material for 7–10(–21) days at 25–30°C (4411, 4935, 5257, 5263). Its occurrence has now been established in Australia, the British Isles, France, Germany, Sweden, New Zealand (5256, 5260, 5263), the USA and the Chatham Islands/S. Pacific (5257). Over one third of the soil samples taken from near creosoted poles, road sides, petrol stations, or under conifers yielded the fungus (4413). It does not normally occur in the litter layer but rather in the soil below (4413). The initial lag phase of growth increased in isolates from increasing soil depths (to 45 cm) where the fungus was apparently less adapted to hydrocarbon substrates (4417). Other reports on its occurrence are available from Iraq (92), Syria (5392), desert soils in the Negev (654) and Pamir (3856), a sandy river bed in Australia (IMI), larch forest in Japan (3267), mangrove swamp soil in Hawaii (3264), a uranium mine in Czechoslovakia (1703), soil along a railway line (5257) and polluted streams (1166) in the USA. It has also been found on plant remains (3113), wood of *Picea abies*, pine stumps (IMI), and keratinous substrates such as feathers, nails and hairs (2575, 4413, 5256). The conidia can be disseminated by birds

(6198). This species is, like *Cladosporium* species, commonly air-borne (2293) and can be isolated from air on a selective medium (5256, 5258). The conidia germinate on cellophane over soil in the presence of creosote vapours (4415); it appears that the creosote vapours encourage the germination of *A. resinae* conidia and at the same time suppress the growth of most other fungi. Only a few isolates are able to grow directly on creosoted wood and the effect of creosote is consequently considered to be due to a suppression of normally competing micro-organisms (4417).

A. resinae grows well on conventional media (3616, 3842), but for the maintenance of its physiological properties a V8 juice agar with 0·1% creosote is recommended (5256); this medium also provides the thiamine, riboflavin and pyridoxin required (3838). — Conidium production is greater in wet than dry air (2294), but conidia can be trapped in both wet and dry weather (5256). The optimal temperature for growth is close to 30°C with considerable growth still occurring at over 35°C, retarded growth at 10° and 40°C, and none at 0 and 50°C (1037, 4416, 5263). *A. resinae* can grow over the wide range of pH 3–8 (1037). — Suitable C sources are maltose, D-xylose, D-glucose, D-mannose, ß-methyl-D-glucoside, D-galactose, L-arabinose and D-glucuronolactone (5770). Suitable N sources are peptone, ammonium sulphate and nitrate (3616). An optimal medium for growth with kerosene contained 20 mg $MgSO_4·7H_2O$, 2 mg $CaCl_2$, 100 mg KH_2PO_4, 950 mg NH_4NO_3, and traces of $FeCl_3$ per litre; this medium allows 98% of the isolates tested to grow on fuel (4418). Not all natural isolates are adapted to grow on creosote (4412) and the lag phase for growth on fuel lasts between 2 and 33 days with different isolates; this capacity can be induced in single isolates or transferred by heterokaryosis (4417). It can also be lost when the fungus is grown on various other media, including MEA (4412), but not on a 2% glucose-salts medium. Creosote is tolerated to over 3% in the presence of another C source (5263). The capacities to degrade creosote, hydrocarbon fuels and resins are all correlated (4415). *A. resinae* can grow on phenol (3842) and benzene but not on cyclohexane or cyclohexene (1093). The oxidation of hexadecane is stimulated and uptake increased in the presence of *p*-xylene and toluene (6153). The utilization of carbohydrates can be inhibited by short-chain fatty acids; alcohols (C_5–C_7) prove to be toxic (5768, 5770) and inhibit the constitutive glucose transport system (5769). Hydrocarbons with chains of C_9 to C_{19} are utilized by a constitutive enzyme system more rapidly than long chain alkanes (1093, 1175, 6151), but *n*-alkanes with short chains ($<C_{14}$) gave a suboptimal mycelium production (3308). Other substrates degraded include agar (4413), vegetable oils, *Eucalyptus* leaf waxes, β-hydroxybenzoate (4413) and lignin (3616). Keratin and chitin both support the growth of this fungus but are not readily degraded themselves (4413, 4415). A few isolates can decompose cellulose (4413). The detergent-homologue 1-phenylundecane-*p*-sulphonate and its analogue dodecane derivative are decomposed in all their moieties (6369). Cells grown on glucose contain C_7–C_{36} aliphatic hydrocarbons, mostly pristane and *n*-hexadecane; when grown on glutamic acid they contain C_7–C_{23} compounds, mostly *n*-tridecane, *n*-tetradecane, *n*-hexadecane and pristane; and when grown on C_{10}–C_{16} compounds the cells contain mainly the compound on which they were grown amongst a range of other C_{10}–C_{32} compounds (6150). The major pathway for initial oxidation of *n*-alkanes is via the primary alcohol, aldehyde, and monoic acid (6152). Enzymes oxidizing *n*-alkanes are regarded as constitutive in isolates able to grow on these substrates without lag phases; whole cells and cell-free preparations oxidize dodecane and hexadecane and their primary alcohols and aldehydes; cell-free preparations also oxidize hexane, hexanol and hexanal which are toxic to living cells (6151). Cells grown on *n*-dodecane contain the highest concentration of total lipids, those on *n*-hexadecane have the lowest amount, those on glucose the maximum of phospholipids (2873), while the extracellular lipids on all three of them

consist mainly of neutral triglycerides, containing dodecanoic and less tetra-, hexa- and octadecanoic acids as well as some phospholipids (5385); hyphae grown on n-dodecane or kerosene release acetic and other fatty acids as well as some ketoacids and (iso)citric acid (4141). *A. resinae* produces amylase, xylanase (4192), a-1,3-glucanase, mycodextranase and a-1,3-glucosidase (1364). In a model experiment aluminum foils were corroded (2379).

Aphanoascus Zukal *1890*

Aphanoascus fulvescens (Cooke) Apinis *1968* — Fig. 19.

≡ *Badhamia fulvescens* Cooke *1875*
≡ *Eurotium fulvescens* (Cooke) Cooke *1879*
≡ *Anixiopsis fulvescens* (Cooke) de Vries *1969*
= *Anixiopsis stercoraria* (Hansen) Hansen *1897*
≡ *Eurotium stercorarium* Hansen *1876*
≡ *Anixiopsis fulvescens* var. *stercoraria* (Hansen) de Vries *1969*
= *Aphanoascus cinnabarinus* Zukal *1890*
= *Anixiopsis reticulispora* Routien *1967*

Anamorph: *Chrysosporium keratinophilum* (D. Frey) Carm. *1962*

≡ *Aleurisma keratinophilum* D. Frey *1961*

DESCRIPTIONS of the TELEOMORPH: Stolk (5592), Routien (4922), de Vries (6123), Frey (1835), and Cain (840); ANAMORPH: Frey (1833), Carmichael (902), and Matsushima (3680). — Colonies growing moderately rapidly, reaching 2·5–3·5 cm diam on phytone-yeast extract agar in two weeks at 25°C, flat, dense, dry, with a somewhat granular surface, white at first but later becoming cream to sulphur yellow; reverse cream to tan. Conidia formed solitarily at the tips of the hyphae, on short or long pedicels or intercalarily, pyriform with a truncate base, hyaline, smooth-walled to sparsely roughened, mostly 10–12 × 7–8 μm. Distinct chlamydospores are present. Ascoma initials consisting of subglobose to slightly elongated and curved ascogonia; a slender hypha growing out from the basal cell develops closely attached to the ascogonial cell and branches to form the peridium. Mature ascomata ± globose, yellow-brown, later becoming dark brown (when aggregated on hairs forming black clusters), 100–over 400 μm diam; peridium consisting of a few layers of flattened cells, the outer-most thick-walled. Asci (sub)globose, 8-spored, thin-walled, 8·5–13 μm diam; ascospores ± ovoid, spinulose, with ± reticulate arrangement of the ornamentation, yellowish brown, 5·0–8·5 × 4·0–5·5 μm. Homothallic (*A. reticulispora* is reported as heterothallic). — Because the original description of *Aphanoascus cinnabarinus* Zukal *1890* is sufficiently clear to recognize this fungus, we adopt the generic name *Aphanoascus* as being the oldest. De Vries (6123) distinguished the varieties *fulvescens* and *stercoraria*, the latter having less regularly reticulate ornamentation but this difference is gradual as shown by SEM (Fig. 19b). — *A. peruviana* Cain *1957* (correctly designated *Xanthothecium peruvianum* (Cain) v. Arx & Samson *1973*), however, differs from *A. fulvescens* by darker ascomata and smaller (2·5–3 × 2·0–2·5 μm), echinulate, lighter, ascospores, and the absence of conidia and keratinolytic capacities. — Teleomorphic isolates of *A. fulvescens* differ from anamorphic isolates only in their sexual capacity and this criterion is not taken as sufficient for their taxonomic separation. *Anixiopsis* differs from all other genera of the Gymnoascaceae by its closed, reddish brown ascomatal peridium

 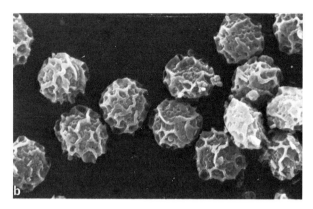

FIG. 19. *Aphanoascus fulvescens*. a. Attached and liberated conidia of *Chrysosporium keratinophilum*; b. ascospores, SEM, × 3000, orig. R. A. Samson.

consisting of angular, flattened cells, but its keratinolytic behaviour, the presence of racquet hyphae and a *Chrysosporium* anamorph place it in this family.

A. fulvescens (as *Ch. keratinophilum*) has usually been isolated from soil only by selective techniques for keratinophilic fungi (6038). It was originally reported from Australia, New Guinea, Canada, the USA and Germany from various soils, a horse pasture, pig and chicken yards, a garden, and under a tree frequented by sparrows, as well as from dog, fox, bear, racoon, and coyote dung (840, 902, 5592, 6123). Subsequent reports include ones from Spain (1216), Italy (80), France (1197, 6106, 6107), Hungary (6550), Czechoslovakia (1014, 4325), Poland (1421, 4230, IMI), Siberia (5228), Yugoslavia (5549), Romania (118), Turkey, Afghanistan, Iran, Pakistan (2366), Israel (1733), India (1903, 4932), New Zealand (4922), Chile (4569), Colombia (4896), Taiwan (6119) and Easter Island (75). It has been isolated from arable soil, peaty soil (3530), soils close to human dwellings (1421, 4932), garden and forest soil, beach sand (1733), grassland (1419, 2573), caves (1216), and has also been observed in soils of dog and mink farms (4230), birds' nests (2577, 4650), on birds' feathers (2575, 4649, 5229), skin scrapings of domestic animals (2155) and cysts of the nematode *Globodera rostochiensis* (3090). *A. fulvescens* can survive ten years storage in non-sterile soil (117).

Growth at 37°C is slow but microscopically similar to that at 25°C (902, 5592). Proteinase activity is low compared to that of other fungi, but phosphatase activity is strong; in one study, after 77 days, more than 50% of the keratin provided in a 0·5% suspension of hair keratin was decomposed (585). — In contrast to the majority of keratinophilic fungi, it does not appear to be pathogenic to guinea pigs (1197).

Aphanomyces de Bary *1850*

Aphanomyces euteiches Drechsler *1925* — Fig. 20.

DESCRIPTION: Jones and Drechsler (2810); monographic treatment of the genus: Scott (5169).

The genus *Aphanomyces* (family Saprolegniaceae Kützing *1843*, type species *A. stellatus* de Bary) is characterized by delicate, non-septate hyphae, which form (without further differentiation) filamentous zoosporangia of variable length in which a single row of biflagellate zoospores differentiates; after an apical rupture of the sporangium, the zoospores normally remain encysted at its orifice until they emerge for a second swarming phase. Oogonia are formed on short lateral stalks, are fertilized by one to several monoclinous or diclinous antheridia, and contain a single oospore.

Twenty-eight species are known; these have been divided between three subgenera according to their oogonium ornamentation. *A. euteiches* has been placed in subgenus *Aphanomyces* (5169) which is characterized by smooth outer oogonial walls; however, the type species of the genus, *A. stellatus,* has been placed in subgenus *Asperomyces* Scott (5169), so some nomenclatural changes at sectional rank appear necessary. — *Aphanomyces* species occur as saprophytes on plant or animal debris in soil or water, particularly on dead insects; some species can parasitize protozoa, rotifers, crustacea, other fungi or filamentous algae; five species are root parasites.

Colonies reaching $3\cdot0$–$3\cdot5$ cm diam in three days at 20°C on OA. *A. euteiches* has 4–10 μm wide, moderately branched hyphae and produces an unpleasant odour which is characteristic of *Aphanomyces* and some other Saprolegniales. Zoosporangia arise from the conversion of large hyphal segments and discharge through one to several lateral branches which taper distally to 4 μm. Hundreds of primary zoospore cysts 8–11 μm diam remain attached to each opening and release secondary zoospores through a protruding papilla. Oogonia 25–35 μm diam, thick-walled $(1\cdot0$–$2\cdot5(-5)\,\mu$m$)$, with a sinuous inner contour; oospores hyaline, $(18-)20$–$23(-25)\,\mu$m diam, thick-walled $(1\cdot2$–$1\cdot8\,\mu$m$)$; antheridia 1–5, stalks diclinous in origin (but apparently homothallic), entwining the oogonium before delimiting the clavate antheridial cells. The oospores may germinate immediately by the formation of one to several germ tubes or a germ sporangium. — The differentiation of vegetative hyphae into zoosporangia with zoospores has been studied using a micro-growth chamber perfused with a salt solution (2474). — The hyphal walls contain cellulose. — TEM studies of hyphae and encysted zoospores have been carried out (4393). — Growing hyphae have numerous irregularly shaped nuclei; EM studies revealed the existence of microtubules in addition to the usual organelles (5243).

The five root-parasitic *Aphanomyces* species all have an undulate inner oospore wall and several, mostly diclinous, antheridia. Further species differentiation is mainly based on oogonial size which is 19–26 μm diam in *A. camptostylus* Drechsler *1929* (on *Avena sativa*), 20–29 μm diam in *A. cochlioides* Drechsler *1928* (on sugar beet), 25–35 μm diam in *A. euteiches* Drechsler *1925* (on *Pisum* and other leguminous hosts), and 32–45 μm diam in *A. raphani* J. B. Kendrick *1927* (on *Raphanus sativus*). *A. cladogamus* Drechsler *1929*, parasitic on

Lycopersicon esculentum and several other hosts, has oogonia 19–33 μm diam and more frequently monoclinous antheridia. *A. euteiches* also resembles the aquatic *A. laevis* de Bary *1860* and *A. helicoides* Minden *1915* which both have thinner oogonium walls which are, however, smooth inside. An exhaustive review of the taxonomy, distribution, ecology, pathogenicity, physiology and economic importance of *A. euteiches* and *A. cochlioides* is given by Papavizas and Ayers (4393).

Aphanomyces species are not normally isolated by the standard techniques used for soil fungi, mainly because of their sensitivity to many antibacterial antibiotics and their dependence on a water phase for propagation. They can be baited from soil by placing pieces of snake skin, split boiled hemp seeds or grass blades on a covering water layer for 5–7 days and subsequent transfer of the blotted baits onto water agar. Washed infected root fragments have been plated directly onto various agar media (2810) but were more successfully baited by corn kernels for 4–12 h, subsequently washed and transferred through several water changes to soil extract agar (4393). Amongst several antibacterial antibiotics tested, only streptomycin and aureomycin are tolerated up to 50 ppm in CMA (4393). Oospores can be detected in soil by soil washing and pouring the supernatant through a 75 μm sieve on which they can be counted microscopically (2810).

A. *euteiches* is a most destructive pathogen of pea roots, causing the common root rot in them (accounting for e.g. 80% of losses in Minnesota) but it also occurs on *Lathyrus odoratus, Lupinus luteus, Vicia, Trifolium* and *Phaseolus* species, and possibly several other plants (4393), including conifer seedlings; it has been found in most parts of the USA but causes most problems in the eastern and central states; it is also known in Norway (5169), Sweden, France, Denmark, the British Isles, the USSR (4393), Canada, Jamaica, Tasmania and Australia (4393). The different leguminous hosts harbour different races of the pathogen (1577, 5468) and isolates from the same host also vary in their sporulation, formation of sexual organs and ability to infect different pea varieties (2993). — *A. euteiches* is dependent on high soil moisture levels for its development, and temperatures of 22–28°C are suitable (4393). It can persist in soil for many years without susceptible crops (5772, 5773). It is most abundant in the top 5 cm, and is found down to not more than 15 cm (4393) in clay and some other heavy soils of pH 5·4–7·5. One study showed that it does not occur in soils never cropped with peas (4393). *A. euteiches* is strongly associated with organic soil particles (3883). Hyphae do not grow in non-sterile soil unless it is amended with pea or wheat straw (5274). — Oospores were viable after two years of alternate freezing and thawing or continuous freezing in dry, moist or saturated both non-sterile and sterile soils (5274). After adding leaves and stems of cabbage, kale or mustard to infested soils, the infection of peas was reduced (4383, 4406); this effect is partially caused by volatile, sulphur-containing degradation products from crucifers which adversely affect the development of the oospores and mycelial growth (3313). Zoospores are attracted by roots of different leguminous hosts but the degree of chemotaxis is not correlated to host susceptibility (1237). Oospore germination is normally suppressed by soil mycostasis (4393).

Zoospore production can be induced in cultures growing on any agar medium after transfer through several changes of tap water, lake water, dilute NaCl or other salt solutions (1236, 3385, 3884). Aeration further increases the amount of zoospores produced (900, 1236, 3385). The optimal temperature for zoospore formation is 24°C. Ca^{2+} ions promote spore differentiation, Mg^{2+} zoospore motility, and K^+ and Zn^{2+} germ tube development (3884, 4393, 5134), while Na^+ at 5×10^{-3} mol l^{-1} or greater concentrations inhibits secondary spore emergence (3884). Zoospores can germinate on dilute nutrient media, root tissue or even water agar

48   COMPENDIUM OF SOIL FUNGI

(4393); their formation and germination can be inhibited by β-methylaspartic acid on media with NH_4Cl as an N source (3438). Oospore production occurs on complex agar media, such as CMA, MEA, PDA or a synthetic medium containing DL-methionine or L-cystine within a pH range of 4·9–5·4 (4399, 6480), and best C sources are D-glucose and D-fructose (4390); peptone inhibits oogonium formation (4393) while aeration with 5 and 20% O_2 in mixtures with 1·5 and 5% CO_2 stimulated oogonium production (3880). The type of oospore germination (i.e. germ sporangium vs. germ tubes) depends on the oxygen (5084) and nutrients available, washing promoting germ tube formation. Chymotrypsin and pronase (6478) or acidification to pH 3·3–5·1 (with HCl in tap water) enhance germination (4284). The optimal temperature for vegetative growth is in the range 25–30°C, while retarded development occurs below 15°C and above 30°C (6481). Good growth is obtained in the pH range 5·2–7·7, and optimal growth at pH 5·4–6·5; no growth occurs below pH 4·0 or above pH 8·0 (4398, 5444, 6481). — Vegetative growth is possible on most currently used media; maximal mycelium production occurs on PDA, PSA, potato agar, or CMA; PDA and maltose-peptone agar are mostly used for its propagation while CMA is best for its maintenance (4393). Peptone, peptides, casein hydrolysate or several amino acids (2207, 3884, 6480) can all serve as C sources, but carbohydrates are preferred; D-glucose and maltose were by far the

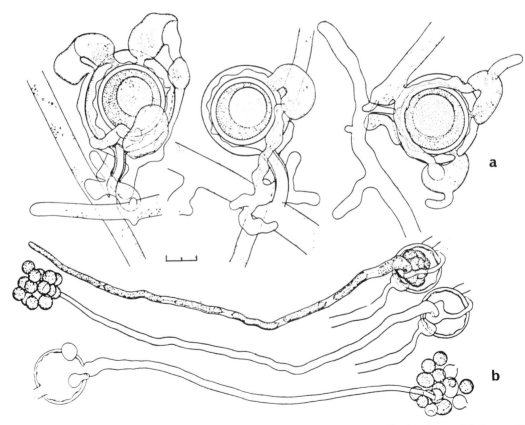

FIG. 20. *Aphanomyces euteiches*. a. Oogonia with antheridia; b. oospores germinating in van Tieghem cells, germination sporangia before and after spore discharge; from Drechsler (2810).

most suitable, followed by glycerol, D-galactose, D-fructose, dextrin, soluble starch and cellobiose (4390, 4393), whilst several other monosaccharides and sugar alcohols were not utilized (4390). In most experiments, only organic nitrogen compounds have been found to be utilized but NH_4^+ salts can be suitable if a decrease in pH is prevented (4398). — Most critical are the sulphur sources, amongst which sulphate is not utilized and methionine is the most suitable, although it can be replaced by L-cystine, L-homocystine or L-cysteine; thioglycolic acid is less suitable; in sodium thiosulphate the reduced sulphur atom can be utilized but not the oxidized one, and thiourea can also support growth. It has been concluded that only sulphur in the reduced or molecular states can be assimilated (1287, 2208, 4393). One of the best synthetic media developed for the vegetative growth of *A. euteiches* contains, per litre, 5·5 g D-glucose, 4 g DL-asparagine, 0·1 g glutathione, 0·7 g KH_2PO_4, 0·4 g K_2HPO_4, 0·4 g $MgCl_2 \cdot 6H_2O$ and 20 mg $CaCl_2 \cdot 2H_2O$ (6481); calcium ions are required, as is at least one of the elements Mg, Fe and Zn, but no vitamins are needed (4393, 4398). Reduced O_2 partial pressure (0·05 atm) is more favourable to vegetative growth than a normal atmosphere (5273), but little growth occurs with 0·3 or 0·01% O_2 (4393). — The enzymes C_x (optimally at pH 6) and endopolygalacturonase have been detected (259, 4393). Mycelial mats of *A. euteiches* have been found to accumulate considerable amounts of dieldrin, DDT and PCNB from treated soil suspensions within an hour (3057). — The fungus was found to inhibit *in vitro* both *Aureobasidium pullulans* and the anamorph of *Nectria radicicola* (6029).

Apiosordaria v. Arx & W. Gams *1966*

Apiosordaria verruculosa (Jensen) v. Arx & W. Gams *1966* var. *verruculosa* — Fig. 21.

≡ *Pleurage verruculosis* Jensen *1912*
≡ *Sordaria verruculosa* (Jensen) Trotter *1928*

Anamorph: *Cladorrhinum* sp.

DESCRIPTIONS: von Arx and Gams (209), and Udagawa and Furuya (5974, 5975). — *Apiosordaria* v. Arx & W. Gams is a monotypic genus of the Sordariaceae Wint., characterized by broadly clavate, unequally two-celled ascospores, 26–29 × 11–17 μm, the upper cell of which is opaque and often coarsely ornamented and the lower one almost hyaline and smooth-walled; an apical germ pore is seen in SEM studies (3746). In var. *verruculosa* the upper cell is covered with coarse, distinct warts whilst in var. *maritima* (Apinis & Chesters) v. Arx & W. Gams *1966* the warts become confluent so as to provide an almost smooth surface (5974). Both varieties are interfertile. — The anamorph belongs to *Cladorrhinum* Sacc. & Marchal *1885* (q.v.) and is distinct from other species in this genus in its oval conidia, 2·5–3·5 × 1·7–2·5 μm. Conidia are not able to germinate but act as spermatia in the heterothallic species. Colonies grow rather rapidly, reaching 5·5 cm diam in five days at 20°C on OA, which is the best medium for inducing the formation of both conidia and perithecia.

The species has rarely been mentioned, identification probably being hampered by its heterothallism. It is reported from agricultural soils in New York/USA (2740), soils in Canada (505), the British Isles (3567, IMI) and Japan (5975), wheat rhizosphere (3567), sandy grassland (6182), sandy soil of a forest nursery (6184), soil in a white cedar forest (505), beech wood soil (272), and *Spartina* humus in salt-marsh soils (167); further, from larvae of the lovebug, *Plecia nearctica* (3012), from fox, hen (4917), and rabbit dung, mangold seeds (IMI), and straw from oats infected with foot rot (4916).

In alternating light and dark regimes, ascospores are released only in the dark phase; this has been interpreted as a delayed light reaction (2652). — *A. verruculosa* utilizes cellulose and lignin when growing on straw and can cause considerable losses in both weight and tensile strength in maple-wood strips (2211).

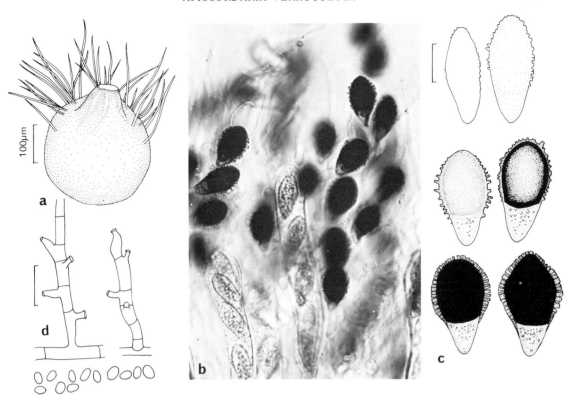

FIG. 21. *Apiosordaria verruculosa*. a. Perithecium; b. crushed perithecium showing 4-spored asci, × 500; c. maturation stages of ascospores; d. *Cladorrhinum*-type conidiophores and conidia; CBS 184.66 and others.

Apiospora Sacc. *1875*

Apiospora montagnei Sacc. *1875* — Fig. 22.

Anamorph: *Arthrinium* sp.

≡ *Papularia arundinis* (Corda) Fr. *1849*
≡ *Coniosporium arundinis* (Corda) Sacc. *1880*

DESCRIPTIONS: teleomorph: Hudson (2585); anamorph: Ellis (1601, 1603), and Ellis *et al.* (1605); conidiogenesis: Cole and Samson (1109). — While the teleomorph is known only from the natural substrate (*Saccharum, Arundo, Bambusa* etc.), where it occurs together with masses of conidia, the anamorph is easily obtained in culture. Colonies reaching 4·5–5·0 cm diam in ten days at 20°C on OA; appearance like in *Arthrinium phaeospermum* (q.v.). The elongating part of the conidiophores is 0·5–1·0 μm wide and provided with hyaline septa (in

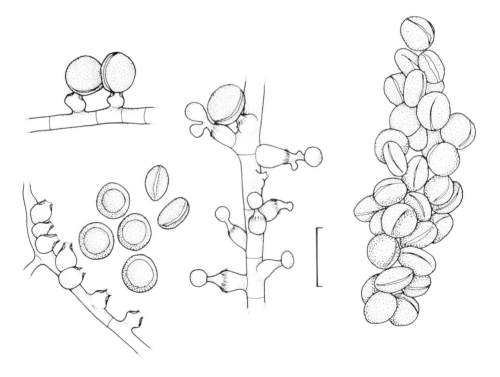

FIG. 22. *Apiospora montagnei*, conidiophores of the *Arthrinium* anamorph without and with basauxic elongation and conidial cluster, CBS 372.67.

the similar *A. sacchari* (Speg.) M. B. Ellis *1965* $1·0–1·5\,\mu$m wide, with brown septa); the basauxic growth of the conidiophores and the firstly holoblastic and later enteroblastic production of conidia has been studied by TEM (864). The conidia are lenticular, $5·5–8·0 \times 3·0–4·5\,\mu$m or sometimes irregularly elongate, light brown, provided with an equatorial germ slit. In a TEM study of conidiogenesis and conidium germination the conidia were shown to split along the periphery (864). — Ascospores hyaline, unequally two-celled, usually $19–33 \times 7–11\,\mu$m, not distinguishable from those of related species (2589).

The worldwide general distribution in natural habitats follows that of the genus *Arthrinium* (q.v.). Reports of soil isolates are less numerous (1601, 1605, 3063), including forest litter (2411), mor in beech woods (2745), alkaline peat (5559), cultivated soils (540, 1614, 1616, 2816, 4647), forest nursery soils (6184), and dunes (745, 4655, 6414); it has also been found in the root region of *Ammophila breviligulata, Andropogon scoparius* var. *septentrionalis* (6414) and *Populus* sp. (5898). Long host lists are available for its occurrence on plant remains (1601, 1605, IMI), and it is also reported from rotten wood (1199, 2338), feathers of live birds (4649) and seeds of oat, barley (3556), *Dactylis glomerata* (5940), rice, lettuce, pea, coriander and melon, rotting grass, hay, straw, dung (IMI), and germinating corn grains (3781).

The selective isolation on cellulose agar (4647, 4655) and degradation of cardboard samples (2665) suggest cellulolytic activity. Erosion of birch wood (4191) and weight loss in beech and pine wood have been demonstrated (1199).

Arachniotus Schröt. *1893*

Arachniotus ruber (van Tiegh.) Schröt. *1893* — Fig. 23.

≡ *Gymnoascus ruber* van Tiegh. *1877*.

DESCRIPTIONS: Kuehn and Orr (3155), Apinis (162) and, with a revision of the five other species of the genus, by von Arx (200). — Colonies reaching 1·7 cm diam in ten days at 20°C on OA. Ascoma initials consisting of two equal coiled gametangia. Peridial hyphae present or absent, asci scattered in orange to red patches throughout the aerial mycelium. Ascospores yellow to orange, oblate with two equatorial rims separated by a furrow, smooth-walled, 5·5–7·0 × 4·0–5·0 µm. Conidial state consisting of scanty arthro-aleurioconidia. SEM studies show the ascospores to have a fine wart-like ornamentation (6105). — The pigmentation and equatorial rims of the ascospores distinguish this species from others of its genus.

A. *ruber* has thus far been isolated only rarely and it is known from the British Isles, India (IMI) and Italy (1794). A summary of ecological observations on it has been published (3155). A fairly frequent occurrence has been observed only in alluvial grassland soils (159), but it is also known from the litter of salt-marsh plants (162), grassland and arable soils (162, IMI), wheat rhizosphere (3567), *Corylus* mycorrhiza (1794), flowers of *Tagetes patula* (IMI), freshwater streams (1166), and dog, goat, sheep, cat and rat dung (3155).

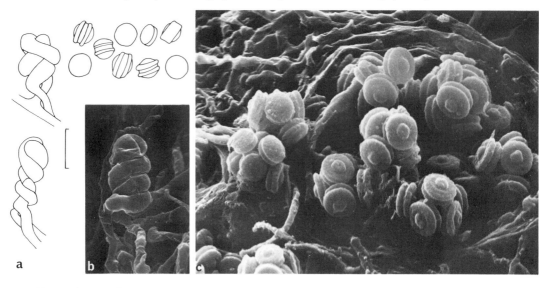

FIG. 23. *Arachniotus ruber*. a. Ascogonial coils and ascospores; b. ascogonial coil, SEM, × 2350; c. asci on the surface of ascoma, SEM, × 2000; CBS 112.69.

Arthrinium Kunze ex Fr. *1821*

= *Camptoum* Link *1824*
= *Papularia* Fr. *1825*

For further synonyms see Ellis (*1601*).

Type species: *Arthrinium caricicola* Kunze ex Ficinus & Schubert

MONOGRAPHIC TREATMENT: Ellis (1601, 1603). — Basauxic conidiophores arising from a ± distinct basal inflated conidiophore mother cell, bearing terminal and subsequently formed intercalary clusters of blastoconidia. Conidia one-celled, pigmented, provided with a germ slit; in the two species treated here the conidia are lenticular. — Sporulation is enhanced on media like OA, CMA and by illumination with near-UV. Species of the genus *Arthrinium* are widely distributed saprophytes on dead plant material (IMI), particularly in swampy habitats where whole culms of grasses are often entirely blackened by masses of conidia; their occurrence in soil is of secondary importance.

Key to the species treated

1	Conidia $5 \cdot 5–8 \cdot 0 \, \mu$m diam in face view, light brown	*Apiospora montagnei* (p. 52)
	Conidia $8–12 \, \mu$m diam in face view, dark brown	*A. phaeospermum* (p. 55)

Arthrinium phaeospermum (Corda) M. B. Ellis *1965*

≡ *Gymnosporium phaeospermum* Corda *1837*
= *Stilbospora sphaerosperma* Pers. ex Gray *1821*
≡ *Papularia sphaerosperma* (Pers. ex Gray) Höhn. *1916* [non *Arthrinium sphaerospermum* Fuckel *1873*]
= *Botryoconis sanguinea* Tubaki *1952*
= *Papularia rosea* Grebenjuk & Kuznetzowa *1971*

Teleomorph: *Apiospora* sp. (2589).

DESCRIPTIONS: Ellis *et al.* (1605), Ellis (1601, 1603), and Gjaerum (1998); conidiogenesis: Hashmi (2305). — Colonies fast-growing, reaching $5 \cdot 0$ cm diam in five days at 20°C on OA, with floccose whitish aerial mycelium, often producing a red pigment in the agar; sporulation generally localized in some dark spots in the aerial mycelium; elongating part of the conidiophores $1 \cdot 0–1 \cdot 5 \, \mu$m wide, with hyaline or pale brown septa (in the similar *A.*

saccharicola Stevenson *1917*, 2–4 μm wide and with brown septa); conidia lenticular, dark brown, 8–12 × 5–7 μm, provided with an equatorial germ slit. — The teleomorphs of *A. phaeospermum*, *Apiospora montagnei* (q.v.) and *A. camptospora* Penz. & Sacc. (*Pteroconium* state = *Papularia vinosa* (Berk. & Curt.) Mason) cannot be distinguished on ascospore size or shape but only on the anamorph (2589). — Hyphal cells and conidia are uninucleate, nuclei 1·3 μm diam.

A. phaeospermum has a worldwide distribution, and is particularly frequently found on plant material, mainly Gramineae (1601, 1959, 2338, IMI). In Europe it occurs primarily in swampy habitats, and less frequently in soil. It can be isolated by various techniques (5812), including sterile hair baits (2091). It has been found in soils in the British Isles (745, 1601, 2338), the Netherlands (1614, 1616), Germany (1424), Poland (3138), France (2159, 2161), Italy (3452, 3453, 3975, 3976, 4538), the USA (1163, 1959, 2004, 2007, 2008, 2573, 4540, 6414), Australia (3720), New Zealand (5812), Bangladesh (2712), and India (1524). It has been reported from forest soils (2004, 2007, 2008, 3138, 3975), heathland (3720), grassland (1524, 2573, 5812, 6182), alpine habitats (3976), agricultural soils (1614, 1616, 2159, 2161, 2712, 3446, 4538), garden compost (1424), water-logged soil (IMI), soils treated with sewage sludge (1163), fresh water (4229, 4429), estuarine sediments (655), sand, dunes (745, 3941, 4162, 4540, 6414), and caves (3453). It is mostly isolated from the upper soil layers (2161, 3975). Seasonal variations have been observed in Anatolia/Turkey (4246) and in Varanasi/India (3863, 5233), increased sporulation occurring in the winter months, whereas on reeds and other plants in Britain maximum sporulation is reported in summer (1605). Further habitats include the root regions of sugar beet, wheat (1614), rice (2712), onions (2159), broad beans (6134), *Ammophila breviligulata* (6414), flax (5826) and poplar (3452, 5898), rotten wood (1199, 3568) and bamboo stems (3063), wood pulp and paper (6168), the fruiting-body of *Psathyrella obtusata* (2688), coarse fodder (4548), nests, feathers and pellets of free-living birds (2575), the integuments of *Dendrolimus sibiricus* caterpillars (2848), pellets of *Glomeris marginata* (4154), rabbit dung (6085), seeds of *Avena fatua* (2961), wheat, oats, barley (1752), common grasses (3512), and pecans (2572).

The capacity to decompose cellulose can be inferred both from its habitat and observations on *Apiospora montagnei* (1199). Metabolites include the tetrahydroanthraquinone pigment bostrycin, succinic acid, ergosterol, a phenolic compound $C_{18}H_{20}O_5$ (1562) and the antibiotically active L-threo-β-hydroxyaspartic acid (2688). *In vitro* the growth of *Chalara elegans* was inhibited (2753).

Arthrobotrys Corda *1839*

= *Didymozoophaga* Soprunov & Galiulina *1951*

Type species: *Arthrobotrys superba* Corda

MONOGRAPHS AND KEYS: Cooke and Godfrey (1151), Jarowaja (2722), and Haard (2196). — Colonies spreading, thin, hyaline or pale pink. Conidiophores erect, arising from the substrate or from fasciculate aerial hyphae, simple or branched, producing apical clusters of two-celled (in some species one- or more-celled), hyaline conidia in sympodial succession on broad denticles; conidial heads often becoming intercalary by renewed growth of the conidiophore. Most species capture nematodes by means of an adhesive network, adhesive knobs or constricting hyphal rings. — Two very closely related genera have been distinguished by Rifai and Cooke (4843), viz. *Genicularia* (correctly *Geniculifera* Rifai *1975* [non *Genicularia* Rouss. ex Desv. *1808*]) with single or clusters of few conidia and long internodes, and *Candelabrella* Rifai & R. C. Cooke *1966* with dense clusters of long, branch-like denticles bearing the conidia. These two genera have been recombined with *Arthrobotrys* by Jarowaja (2722). Previously, species with one- or three- (four-) celled conidia were excluded from this genus, and it contained 24 species, while now, with the inclusion of species with more-celled conidia, 47 species are listed (5089). The four species most commonly found in soil are discussed here; these are closely related, form conidia mostly in several whorls at swollen nodes along the conidiophore, and can produce adhesive networks in the presence of nematodes.

Key to the species treated

1 Conidiophores sometimes branched; conidia elongate-ellipsoidal, with parallel sides for most part of their length and an equal or slightly longer apical cell, not constricted at the septum, 17–30 × 10–14 μm, length/width 1·8–2·1   *A. arthrobotryoides* (p. 58)
  Conidiophores unbranched; conidia of other sizes or shapes   **2**

2(1) Conidial cells of equal length, conidia elongate-ellipsoidal with parallel sides, not constricted at the septum, 12–22 × 6–11 μm, length/width 1·5–2·0; colonies rose   *A. superba* (p. 63)
  Distal cell of the conidia longer than the basal cell; conidia with convex sides and slightly constricted at the septum; colonies almost hyaline   **3**

3(2) Conidia obovoid to pyriform, 22–27 × 8–15 μm, length/width 1·5–2·0   *A. oligospora* (p. 60)
  Conidia elongate-obovoid with the distal cell swollen, 18–36 × 8–15 μm, length/width 2·0–3·0   *A. conoides* (p. 59)

Arthrobotrys arthrobotryoides (Berl.) Lindau *1907* — Fig. 24.

≡ *Cephalothecium roseum* Corda var. *arthrobotryoides* Berl. *1888*

DESCRIPTIONS: Drechsler (1460), Haard (2196), and Jarowaja (2722). — Colonies reaching 4·0 cm diam in five days at 20°C on OA, hyaline to pale pink; conidiophores arising from the substrate or fasciculate aerial hyphae, unbranched, or occasionally branched, 300–450 μm tall, bearing clusters of conidia on one to several nodes with numerous short blunt denticles. Conidia ellipsoidal with almost parallel sides, divided into two ± equal cells, 17–30 × 10–14(–16) μm. Chlamydospores, 15–17 μm diam, present in old cultures. — This species is very similar to *A. oligospora* (q.v.) and has probably often been confused with it.

A. *arthrobotryoides* has been observed frequently in investigations of nematode-trapping fungi, for example, in the USSR (3745, 5480, 5481), Poland (2722), but other data from soil are sparse. In Californian citrus soils it was reported to be the most frequent nematode-trapping fungus, particularly in the citrus rhizosphere (3575, 3578). Near Mariazell/Austria it was found in soils under *Geranium sanguineum* and *Vaccinium myrtillus* (4179), in the Ukrainian SSR in soils under oat and barley (3006), and in Germany it was isolated rarely from washed soil particles in wheat fields. Further reports include ones from leaf litter (1460), straw (4548) cow dung (5983) and (in the original description of the species) rotten mulberry wood. Conidium germination in soil is inhibited by mycostasis (3574).

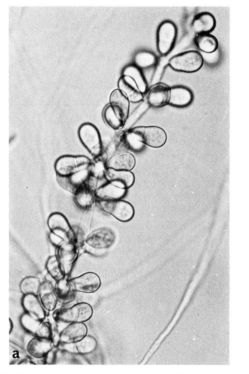 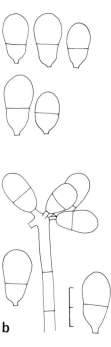

FIG. 24. *Arthrobotrys arthrobotryoides*. a. Old conidiophore with numerous proliferations, × 500; b. details of sympodial conidium formation.

The utilization of pectin, cellulose and chitin and proteolytic activity has been reported (5926).

Arthrobotrys conoides Drechsler *1937* — Fig. 25.

DESCRIPTIONS: Drechsler (1458), Duddington (1479), Shepherd (5253), Haard (2196), and Jarowaja (2722). — Colonies reaching 5·0 cm diam in ten days at 20°C on OA, white to yellow. Conidiophores arising from the substrate, 150–400 μm tall, bearing several conidiiferous nodes with densely clustered small denticles. Conidia pyriform to clavate, the upper cell longer and larger than the basal one, 18–36 × 8–15 μm. Yellow chlamydospores 7–17 μm diam present in old cultures.

In studies on nematode-trapping fungi, *A. conoides* has regularly been reported, although less commonly than the other species mentioned here; for example, in various parts of the USSR (3745, 5480), Poland (2722), the British Isles (1479), the USA (3575, 3578), New Zealand (1809) and India (4964). Immunofluorescent staining has been applied to the detection of this fungus in soil (1646). It is rather common on dung (1479, 1480) and in nematode-infested soils (3575) but is also reported from soil under sugar cane in Taiwan (1047). In a microbiologically active soil its hyphal growth and conidium production were not observed (1645) although its conidia germinated freely in contact with soil. — Conidia may stick to nematodes, form "adhesive swellings" and germinate directly into the prey (4515) so that the fungus is endoparasitic rather than predaceous. The formation of any adhesive trapping network occurs only in the presence of nematodes (4612).

A number of substances have been found to induce trap formation *in vitro* including an unidentified metabolite named nemin (4613), D- or L-valine, the latter as the inducing principle of yeast extract (6439). Valine is actively taken up with an optimum at pH 5–6 (2181) and incorporated into cellular protein but for the greater part transformed to various carboxylic and other amino acids (2182). Other trap-inducing substances include specific C sources like L-(+)rhamnose, L-(+)arabinose and D-(−)ribose; very few traps were produced in the presence of glucose, mannose, galactose, xylose or fructose (2715). The formation of traps *in vitro* occurs at pH 6·3–8·05 (3575). The fungus grows well on complex media (e.g. cornmeal agar) or in the presence of nematodes (1479, 5090); on glucose-mineral salt media, supplementation with yeast extract proves necessary, the essential components of yeast extract being zinc ions, biotin and thiamine in concentrations of 5, 100 and 400 μg l^{-1}, respectively (1194). Optimal growth was observed at 20–28°C, and no development found to occur at either 4°C or 37°C; trap formation is reduced at 32°C and none occurs at 37°C (1047). The optimal pH for mycelial growth lies between 4·0 and 6·0, and an initial pH of 5·0 is favoured (1194, 3575).

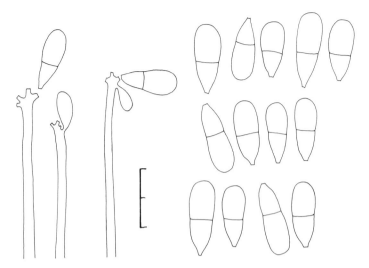

F<small>IG</small>. 25. *Arthrobotrys conoides*, conidiophores and sympodially formed conidia, CBS 109.52.

Arthrobotrys oligospora Fres. *1850* — Fig. 26.

D<small>ESCRIPTIONS</small>: Drechsler (1457, 1458), Duddington (1479), Haard (2196), Jarowaja (2722), and Juniper (2828). — Colonies reaching 3·8 cm diam in five or 6·5 cm diam in ten days (1143) at 20°C on OA, hyaline to pale pink or yellow. Conidiophores arising from the substrate or from fasciculate aerial hyphae, unbranched, 350–450 μm tall, bearing several conidiiferous nodes where the conidia are borne on short blunt denticles. Conidia obovoid to pyriform, broadest in the longer apical cell, 22–27 × 8–15 μm. Yellow chlamydospores, 9–20 μm diam, are found in old cultures. — The nematode-trapping adhesive network has been illustrated by SEM (4207).

A. oligospora is by far both the commonest and most cosmopolitan of nematode-capturing fungi and the first species in which this ability was recognized. It has been used in many ecological and physiological studies on predaceous fungi and has repeatedly been used in trials to assess its potential for the biological control of nematodes. In studies on predaceous fungi it has been frequently recorded, for example, in the British Isles (1481), various parts of the USSR (3745, 5480, 5481), Poland (2722), Canada (IMI), New Zealand (1809, 6432) and India (4964); it is known from forest-steppe soil (1700), soils of a mixed forest (IMI), and a mediterranean brown soil of pH 6·9–8·0 (2963). Most reports of A. oligospora, however, relate to cultivated soils (1047, 1481, 1514, 1614, 1616, 4451, 4964, 4997, 5253), but it has also been noted in permanent grassland, shrubland, plantations, and sheep and cattle yards (1809). A marked increase in frequency has been observed in the rhizosphere of soya bean (4525) but in that of wheat the frequency was reduced and no isolates were obtained from washed root segments (4451). With citrus, on the other hand, its occurrence was found to be restricted to the rhizosphere (3578). There are also reports for the rhizospheres of barley, bcans (4451) and sugar bcct (1614). It has also been isolated from beech and pine wood baits buried in the surface layer of soils (5237). According to various reports, a frequently colonized

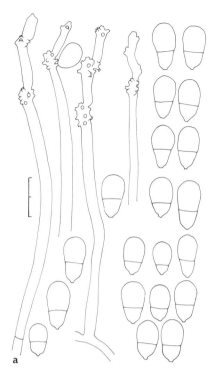

FIG. 26. *Arthrobotrys oligospora*. a. Conidiophores and sympodially formed conidia, CBS 879.68, b. conidial clusters, CBS 106.49, SEM, × 750.

habitat is partially decomposed plant material, leaves, roots, moss and dung (1458, 1480, 1809, 2828, 2966, 2967, 6223); it has also been isolated from larvae of the lovebug. *Plecia nearctica* (3012) and cysts of the nematode *Heterodera schachtii* (3090). — Growth in soil has been recorded between vapour saturation and −40 bars water potential (3109). *A. oligospora* is generally found at a soil pH of above 5·5, but in intensely cultivated soils can occur down to pH 4·5 (1809). — It has a relatively high competitive saprophytic ability (1143) compared to other nematode-trapping fungi with constricting rings but was found to be sensitive to antagonistic influences by various soil organisms (3744). Its saprophytic existence in soil is enhanced by a capacity to degrade durable substrates such as cellulose (5065). Conidium germination in contact with soil is 100%, but usually no trapping organs are formed (1145). The influence of the soil moisture level in remoistened air-dried soil on restoration of soil mycostasis was studied using *A. oligospora* as a test organism (3110). Its nematophagous activity in soil can be increased by adding green manure or carbohydrates (1144). The nematode content of a soil has occasionally been successfully reduced by the addition of *A. oligospora* (1514), although very variable results have so far prevented any practical application on a large scale; in one case the soil nematode population increased after its introduction as the hyphae serve as food for some nematodes (1142).

The optimal temperature for growth without nematodes is 20°C but with them 25°C or more (3911), and the optimal pH is in the range 5·0–7·0. The formation of nematode-trapping adhesive networks *in vitro* occurs only at temperatures below 37°C (1047) and between pH 4·9 and 8·1 (after the addition of nematodes). In permanent darkness, growth is somewhat less than in continuous light (3575, 4290); vegetative extension is stimulated in the presence of

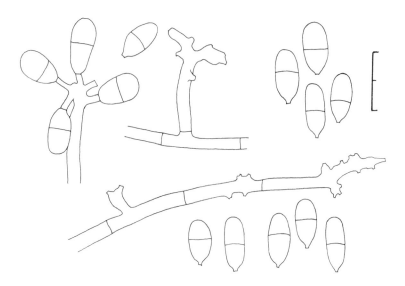

FIG. 27. *Arthrobotrys superba*, conidiophores and conidia.

nematodes (3911). A synthetic culture medium has been described (545); the nematode-trapping activity is optimal in the range 100–200 mg l^{-1} sucrose and 10–20 mg l^{-1} NH$_4$-nitrogen in a basal medium (4291). In addition to the presence of nematodes, trap formation can be induced by a nematode-extract "nemin" (4612) and peptides from hydrolysed casein (4208); peptides of 2–7 amino acid residues with a high proportion of non-polar and aromatic residues were the most active in this respect (4209). For the parasitic phase, further suitable C and/or N sources include cellobiose and L-asparagine, L-arginine, and DL-glutamic acid. L-Arabinose and L-sorbose, L-cystine, L-histidine, DL-tryptophan and urea are, however, utilized poorly (546, 2347, 5064). Nitrite, nitrate and ammonium salts are suitable inorganic N sources (5064). Pectin, chitin and cellulose are utilized (5926) and the cellulases C$_x$ and C$_1$ have been shown to be present (4779). Saprophytic growth is very good on D-xylose, D-mannose, and cellobiose, but is further stimulated by the addition of biotin, thiamine (5063), Ca-pantothenate, and *p*-amino benzoic acid, while D-ribose, L-rhamnose, lactose, D-mannitol, D-dulcitol, inulin, citrate, malate, glycine, and DL-lysine proved to be unsuitable substrates (546). Proteolytic activity has been demonstrated (5926). Parasitized nematodes contain a substance that paralyses healthy nematodes; the presence of an unstable fungal toxin is assumed (4289). The pink hyphal pigments have been analysed as β-carotene with minor portions of γ-carotene, neurosporoxanthin and torulene (6033).

Arthrobotrys superba Corda *1839* sensu Drechsler — Fig. 27.

≡ *Arthrobotrys drechsleri* Soprunov *1958*

DESCRIPTIONS: Drechsler (1458), Peach (4482), Haard (2196), and Jarowaja (2722). — *A. superba* as understood by Drechsler is a nematode-trapping species with pink to salmon colonies reaching 5·0 cm diam in five days at 20°C on OA, with erect, 100–300 μm tall conidiophores, bearing several closely spaced conidiiferous nodes, and conidia which are almost cylindrical centrally and have nearly equal cells, 12–22 × 6–11 μm. Chlamydospores unknown. Trapping nematodes by means of adhesive loops or network. Because of some discrepancies with Corda's original diagnosis, Drechsler's fungus was renamed *A. drechsleri* Soprunov (5480); Corda's fungus would be non-predaceous with thick-walled conidia of about the same size (2722). Unfortunately, no type material is existent to solve the problem and all reports below refer to the species as interpreted by Drechsler. — DNA analysis gave a GC content of 50·5% (5600).

A. superba appears fairly frequently in analyses of predaceous fungi, for example, in the USSR (5480, 5481), and Poland (2722), but is also reported in analyses of saprophytic soil fungi, for example, in the L- and F-horizons of beech and pine forests in Poland (269, 272, 273, 650), under beech in Italy (271), a dry soil in a mixed forest in India (4716), and cultivated soils in France (2161) and Israel (2772). It is also known from the British Isles (4482), Finland (4943) and Zaïre (2966, 2967). Plant remains (2966, 3113, 4548) and leaf litter (650, 3448, 4482) are the more frequent habitats but there are occasional isolations from wheat seeds, germinating seeds of clover (4120), rabbit dung (6085) and other nematode-infested materials (4943). Its occurrence in the upper 5–10 cm of soil (269, 273, 2967) must be interpreted in relation to the distribution of the nematodes attacked.

The optimal temperature for linear growth is in the range 20–22°C (4290). It sporulates freely on cornmeal and peameal agars (4482). Weight losses when grown on different woods have been reported (4191). Phenol oxidases are produced and humic acid, fulvic acid and lignosulphonate are utilized (934). The pink coloration is caused by the carotenoids neurosporoxanthin, γ-carotene, β-carotene and torulene (202).

Arthroderma Berk. *1860*

Type species: *Arthroderma curreyi* Berk.

Anamorphic genera: *Trichophyton* Malmst. (incl. *Keratinomyces* Vanbr.), *Chrysosporium* Corda

MONOGRAPHIC TREATMENT: Padhye and Carmichael (4343). — Fam. Gymnoascaceae Baran. — Ascoma initials consisting of a clavate antheridium surrounded by a spiralled ascogonium. Ascomata (gymnothecia) white to pale yellow, globose; peridium consisting of a network of hyaline, septate hyphae which branch dichotomously (never verticillately) and divergently (uncinately) curved over the ascoma, consisting of cells which are thick-walled, densely echinulate, strongly constricted symmetrically (dumb-bell shaped) or asymmetrically and often bearing distinct protuberances at the ends. The dumb-bell shaped hyphae usually bear terminal, slender, straight or spiralled appendages. Asci globose to subglobose, 8-spored, evanescent. Ascospores small (to $3 \cdot 5 \mu$m diam), oblate, smooth-walled, hyaline, yellow in mass. — *Arthroderma* is distinguished from *Nannizzia* Stockd. (q.v.) by both the strongly constricted peridial cells which often bear protuberances and the smooth-walled *Trichophyton* macro-conidia.

All 15 known *Arthroderma* species are keratinophilic and none is cellulolytic. The species are subdivided into geophilic ± saprophytic and more strictly parasitic (zoophilic and anthropophilic) (1521, 4326). Reviews of the dermatophytic species are given by Ajello (73) and Dey (1365). Keratinophilic fungi also include *Nannizzia* (with *Microsporum* anamorphs), *Ctenomyces* and some *Chrysosporium* species (q.v.). Species of these genera are common on hairs and feathers, on the soil surface, and especially near animal burrows and birds' nests. Two *Arthroderma* species (*A. benhamiae* Ajello & Cheng *1967*, and *A. uncinatum* Dawson & Gentles q.v.) as well as several *Trichophyton* species for which no teleomorph is known, are pathogenic to man and other mammals.

Arthroderma is perhaps the most representative genus of keratinophilic and dermatophilic fungi and the techniques mentioned below for its study apply equally to other representatives of this ecological group. For isolation from soil, a hair bait technique using horse or children's hair exposed above moist soil incubated at 24°C (6038) is in general use. For subsequent purification Sabouraud-glucose agar with chloramphenicol and cycloheximide is recommended (1521). For maintenance "Pablum" cereal agar, OA or PDA have been recommended; on very rich media the isolates tend to degenerate. Thirteen of the fifteen species known are heterothallic and have been discovered since *1960*; mating tests are required for a reliable identification of atypically sporulating conidial isolates and species which have indistinguishable anamorphs. For such tests, sterile horse or children's hair is placed on sterilized or non-sterile moist soil, soil extract agar, Weitzmann and Silva-Hutner's oatmeal-salt agar, or cornmeal-glucose agar with powdered hair, and inoculated with conidial suspensions of the two isolates to be tested and incubated at 24°C; isolates of different species are always incompatible (4347). Heterokaryotic macro-conidia containing both + and − nuclei have been found in some species after mating (3189).

Seven common saprophytic geophilic species are treated here.

Key to the species treated

1	Multiseptate macro-conidia always absent	**2**
	Multiseptate macro-conidia present, although sometimes scanty	**4**
2(1)	Conidia less than 4 µm diam, smooth-walled	**3**
	Conidia more than 6 µm diam, densely tuberculate; peridial cells symmetrically constricted	*A. tuberculatum* (p. 72)
3(2)	Conidia borne singly on vegetative hyphae; homothallic; peridial cells symmetrically constricted without protuberances	*A. curreyi* (p. 67)
	Conidia borne generally in clusters; heterothallic; peridial cells asymmetrically constricted with 3–4 protuberances on each end	*A. cuniculi* (p. 66)
4(1)	Macro-conidia predominating, fusiform with rounded tip, wall mostly over 2 µm thick; peridial cells symmetrically constricted	*A. uncinatum* (p. 73)
	Macro-conidia not usually predominating, cylindrical, wall less than 1·2 µm thick	**5**
5(4)	Peridial cells symmetrically constricted	*A. lenticulare* (p. 69)
	Peridial cells asymmetrically constricted	**6**
6(5)	Peridial cells bearing two outwardly facing protuberances at each end	*A. quadrifidum* (p. 70)
	Peridial cells without protuberances	*A. insingulare* (p. 69)

Arthroderma cuniculi Dawson *1963* — Fig. 28.

Anamorph: *Chrysosporium* sp.

DESCRIPTIONS: Dawson (1310), Apinis (162), and Padhye and Carmichael (4343). A heterothallic species. Colonies reaching 4·5–5·5 cm diam in two weeks at 25°C, flat, dense, floccose to downy, yellowish to buff in the centre; reverse yellow to brown. Ascomata on hair-soil media globose, pale yellow, 180–380 µm diam (excluding the appendages). Peridial hyphae pale yellow, septate, uncinately branched, usually outwardly; cells dumb-bell shaped at first, 6–11 × 3–6 µm, distinctly echinulate, at maturity with 3–4 protuberances around each end of the dumb-bell, to 3·5 µm long; spiral appendages terminal or lateral, coiled 4–16 times. Asci subglobose, 8-spored. Ascospores hyaline, yellow in mass, smooth-walled or finely roughened, lenticular, 2·2–2·9 × 1·4–2·0 µm. Conidia borne on short, undifferentiated pedicels or sessile, in dense groups, one-celled or occasionally 2- or 3-celled, hyaline, clavate, smooth-walled, 2·4–5·3 × 1·2–2·0 µm. — The similar *A. multifidum* Dawson *1963* has more frequently 1–2-septate sparsely roughened conidia, 9–20 × 6–12 µm and 4–8 protuberances on each of the peridial dumb-bells.

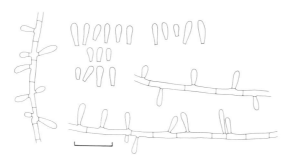

FIG. 28. *Arthroderma cuniculi*, micro-conidia, CBS 492.71.

A. cuniculi has been found in soil in Scotland and Australia (4343); further habitats include animal hair, reptile scales, birds' pellets, nests and feathers (162, 2575, 2577, 4650). It does not grow at 37°C (4343).

Arthroderma curreyi Berk. *1860* — Fig. 29.

≡ *Illosporium curreyi* (Berk.) Sacc. *1886*
= *Ctenomyces xylophilus* Höhn. *1905*

Anamorph: *Chrysosporium* sp.

DESCRIPTIONS: Benjamin (449), Dawson and Gentles (1312), Apinis (162), Orr and Kuehn (4314), Udagawa and Furuya (5976), and Padhye and Carmichael (4343). — A homothallic species. Colonies reaching 6·0–7·0 cm diam in two weeks at 25°C, white to cream or buff. Ascomata forming on hair-soil or ordinary agar media in fresh isolates, 100–450 μm diam (excluding appendages); peridial hyphae with a generally spiral configuration, bearing single branches directed towards the outside of the spiral; cells thick-walled, densely asperulate, ± symmetrically constricted, dumb-bell shaped; comb-like hyphae also sometimes occur. Asci subglobose, 8-spored; ascospores lenticular, 2·5–3·5 × 1·5–2·5 μm. Conidia arising singly and laterally on undifferentiated hyphae, sessile or on short pedicels, clavate-pyriform, usually smooth-walled, 2·5–6 × 1·5–2·0 μm. — The similar and also homothallic *A. ciferrii* Vars. & Ajello *1964* has vinaceous to brown colonies and sometimes 2–3 celled, longer (to 12 μm) conidia.

The distribution of *A. curreyi* is probably worldwide; it has been reported from the Netherlands (1616), France (6106), Czechoslovakia (2577, 3174), Hungary (6552), the British Isles (4343, 4662), Austria, Germany, the USA, and Panama (3149). It is mainly found on feathers and hairs: in Queensland (4772), Czechoslovakia (2575) and northern Europe (4649, 4650) it has been found very frequently on wild birds, but not on domestic hens (4773), it is particularly common on feathers of several *Turdus* species in the British Isles (4648, 4653), also in soil from rabbit burrows (1310), and on bat guano (6552). It has also been found frequently in coastal soils and leaf litter (162, 4662). *A. curreyi* grows well in sterilized soil; its spread in non-sterile soil depends on the availability of keratinous substrates but ascomata survive for a long time in soil (4648).

During growth in liquid mineral salt media, the pH rises from 5·2 to 8·1. Keratin is degraded to >40% within 77 days; a high proteinase activity and an alkaline phosphatase have been demonstrated (585).

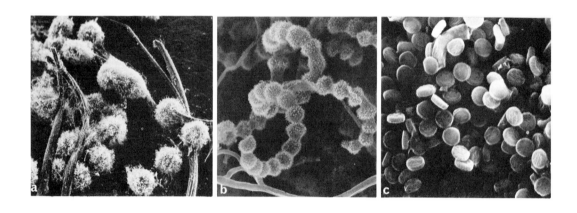

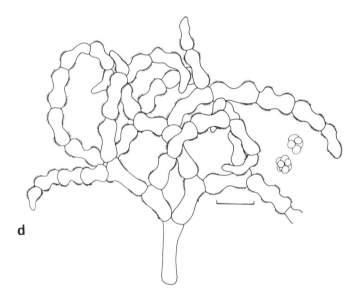

FIG. 29. *Arthroderma curreyi*. a. Ascomata formed on hairs spread on agar, SEM, × 20; b. peridial appendages, SEM, × 1000; c. ascospores, SEM, × 2500; d. peridial appendage and two asci; all CBS 130.70.

Arthroderma insingulare Padhye & Carm. *1972* — Fig. 30.

Anamorph: *Trichophyton terrestre* Durie & D. Frey *1957* complex

DESCRIPTION: Padhye and Carmichael (4344). — Colonies as in *A. quadrifidum* (q.v.) but often with a pink to vinaceous reverse, reaching 3·7 cm diam in nine days at 24°C on 2% MEA. Heterothallic. Ascomata 250–500 μm diam. Peridial cells asymmetrically constricted, without protuberances at the ends, 8–12 × 5–6 μm. Ascospores as in *A. quadrifidum*. Conidia predominantly 2-celled (to 6-celled), 6–30 × 2–5 μm. — The conidia resemble those of *A. ciferrii* Varsavsky & Ajello *1964* and *A. cuniculi* (q.v.).

A. insingulare is known from Hungary, Czechoslovakia, Canada and the USA, from soil, rodent hair and chicken feathers (4344).

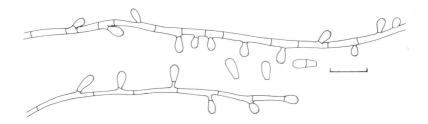

FIG. 30. *Arthroderma insingulare*, micro-conidia, CBS 521.71.

Arthroderma lenticulare Pore, Tsao & Plunkett *1965* — Fig. 31.

Anamorph: *Trichophyton terrestre* Durie & D. Frey *1957* complex

DESCRIPTIONS: Pore *et al.* (4601), and Padhye and Carmichael (4343). — Colonies as in *A. quadrifidum* (q.v.) but with a yellow reverse on Sabouraud agar, reaching 3·6 cm diam in nine days at 24°C on 2% MEA. Heterothallic. Ascomata 300–600 μm diam. Peridial cells dumb-bell shaped with symmetrical constrictions and no protuberances, 7–10 × 5·5–8·5 μm. Ascospores and conidia as in *A. quadrifidum*.

A. lenticulare is so far known only from California from soil, keratinous substrates and animal burrows (4343).

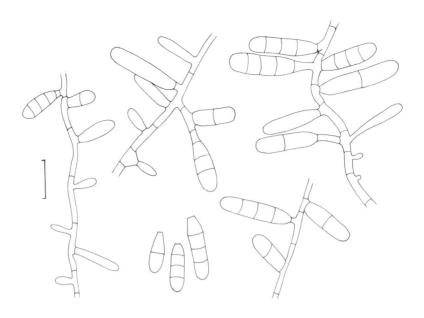

FIG. 31. *Arthroderma lenticulare*, micro- and macro-conidia, CBS 307.63.

Arthroderma quadrifidum Dawson & Gentles *1961* — Fig. 32.

Anamorph: *Trichophyton terrestre* Durie & D. Frey *1957* complex

= *Ctenomyces trichophyticus* Szathmary & Herpai *1960*
= *Trichophyton thuringiense* H. A. Koch *1969*

DESCRIPTIONS: Durie and Frey (1506); anamorph: Dawson and Gentles (1312), Dvořák and Otčenášek (1521), and Padhye and Carmichael (4343); conidiogenesis: Galgóczy (1862). — A heterothallic species. Colonies reaching 4·0–5·0 cm diam in two weeks at 25°C, white at first but later becoming pale yellowish; yellowish brown in reverse. Ascomata form on hair-soil medium but also on Czapek-Dox with peptone, not above 25°C. Ascomata pale buff, 400–700 μm diam (excluding the appendages). Peridial hyphae uncinately branched, pale yellow, septate; cells thick-walled, strongly echinulate, dumb-bell shaped when young but at maturity with two protuberances at each end, 8–13 × 5–9 μm, bearing spiralled appendages laterally and terminally. Asci subglobose, 8-spored; ascospores hyaline, yellow in mass, smooth-walled or finely roughened, lenticular, 2·0–3·0 × 1·8–2·5 μm; TEM of ascospores shows that the wall is thicker in the valves (1696). Macro-conidia cylindrical or rarely clavate, thin- and smooth-walled, sessile, 2–6 celled, 8–52 × 4–5 μm. Micro-conidia borne singly or in groups, elongate pyriform, 3–6·5 × 1·5–3·5 μm. — In the anamorph *A. lenticulare* (q.v.) and *A. insingulare* (q.v.) can hardly be distinguished from *A. quadrifidum*, and reports referring only to anamorphs are thus compiled here under this, the commonest of the three species. — Investigations of protein patterns revealed that *Trichophyton terrestre* belongs in a group together with *T. tonsurans, T. mentagrophytes* and *T. megninii* (3223). — The vegetative hyphae are multinucleate (2–8) (2560); the nuclear behaviour has been studied during mitosis when no typical spindles or spindle bodies were identified; the chromosome number in the haploid stage is 4 (2560). Intrahyphal growth and also fusion of those hyphae has been observed (1695).

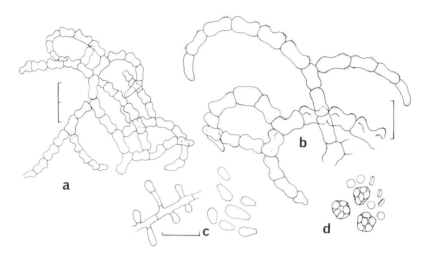

Fig. 32. *Arthroderma quadrifidum*. a, b. Peridial appendages at different magnification; c. micro-conidia; d. asci and ascospores; CBS 613.74 × 2523.

A. quadrifidum has a worldwide distribution (4343) in soil, on feathers, hair, etc., and especially on the coats of small mammals (2548, 4324). It is regarded as non-pathogenic (4326), although it has been isolated twice from papular lesions of a horse (1312) and infection experiments in mammals were positive (305). *A. quadrifidum* seems to occur preferentially in subtropical and temperate zones (4327); reports include the British Isles (1312, 4662, 5585), the Netherlands (CBS), Finland (5015), Belgium (79), Czechoslovakia (1013, 1014, 2577, 4324, 5338), the USSR (6533, 6534), Germany (580, 582, 584, 3743, 4536, 5521), France (2548, 6106), Italy (80), Yugoslavia (5549), Hungary (2541), Bulgaria (306), Romania (118, 119), Turkey, Iran, Afghanistan (2366), Israel (1733), Tunisia (2721), Iran (103), Colombia (4896), Chile (4569, 6545), Argentina (6047), Brazil (4897), the Bahamas (6117), the USA (88, 1256, 2573, 3053, 4315), Canada (305), Taiwan (6119), New Zealand and Polynesia (3613, 3614, 5470). In addition to numerous records from a variety of mostly cultivated soils, this fungus has been isolated from alpine regions up to 3400 m (3742, 4569), urban areas (2541, 3742, 4368, 6117), beach sand (1733), nests, pellets and feathers of free-living birds (2575, 4649, 4650), birds' feathers on tide-washed coastal areas (1256), the area around bird traps and birds' nests (2577), hair and skin scrapings of wild animals (119, 762, 3053, 4324), cysts of the nematode *Globodera rostochiensis*, live hedgehogs, pigs, opossums, and mice (IMI), and stables (306, 6533). An isolate taken from lesions of a human scalp caused superficial dermatitis on inoculation into guinea pigs (305). A case of a mycotic granuloma in man caused by *A. quadrifidum* has been reported (2120). An isolate from hedgehog was, in contrast to soil isolates, not able to grow in sterile soil (3615). The fungus can survive in non-sterile soil for as long as seven years (117).

The optimal temperature for growth is 25°C, and the maximum 31–35°C for different cream-coloured isolates and up to 36°C for red isolates (4967). High temperatures (1312, 1313) and light (1313) inhibit sexual reproduction. Ascoma production was better on non-sterile than on autoclaved soil (1313). In contrast to soil isolates, which seem to tolerate the wide pH range 5–8 (1313), one pigmented isolate was characterized by a narrow pH tolerance (3615). The total lipid content and composition of phosphoglycerides of a normal "wild" isolate and one exposed to space flight conditions showed remarkable differences (5074). Keratin is degraded to >50% within 77 days; proteinase, alkaline phosphatase (585) and elastase activity (4849) have been demonstrated and the process of hair degradation has been followed by EM and light microscopy (2562). *A. quadrifidum* can grow on media with 10% NaCl but no growth occurs at 15% NaCl (2875).

Arthroderma tuberculatum Kuehn *1960* — Fig. 33.

Anamorph: *Chrysosporium* sp.

≡ *Chrysosporium tuberculatum* (Kuehn) Dominik *1967*

DESCRIPTIONS: Kuehn (3151), and Padhye and Carmichael (4343); anamorph: Carmichael (902). — A heterothallic species (5190). Colonies growing moderately fast, reaching 4·0–5·0 cm diam in two weeks at 25°C, flat, dense and dry, centrally buff and powdery due to the conidia. Ascomata appearing on YpSs agar in three weeks at 28°C, globose, cream, 300–700 μm diam excluding the appendages; peridial hyphae uncinately branched, almost hyaline; cells thick-walled, densely asperulate, 10–27 × 6·5–7·0 μm, usually symmetrically constricted to 4 μm, infrequently ending with a short and curved or longer spiralled slender appendage. Asci ovoid, 8-spored; ascospores lenticular, bright yellow, smooth-walled, 2·5–3·0(–3·5) × 1·5–1·8 μm. Conidia abundantly produced, mostly lateral, ± pedicellate or sessile, subglobose to ovoid, thick-walled, warted, cream, 10–15 × 7–10 μm, the basal hilum 2·2 μm diam. — SEM of conidia and peridial hyphae (6104) shows vermicular projections. A disc-gel electrophoretic comparison of proteins and some enzymes in single and mated isolates gave 1·5–2·0 times more bands in mated cultures (5191). — The similar *A. multifidum* Dawson *1963* has ± smooth-walled conidia.

A worldwide distribution is evidenced by reports from the USA (4315), Canada, Panama, Argentina, the British Isles, Czechoslovakia (4343), Belgium (79), Hungary (6552), Italy (80), France (6106, 6107), Finland (5015), Mozambique (IMI), India (4343), Taiwan (6119), and Australia (1835, 4772). It has been mostly isolated from feathers or hairs (79, 162, 1835, 3151, 4772, 4773, 5229), birds' nests (2575, 2577), bat guano (6552), or soil (4343) but once was obtained from psoriatic lesions on a finger (2907).

A. tuberculatum fails to grow at 37°C. The optimum initial pH for growth is 7·0–8·0 and ascomata are produced on soil extract and hair agar, oatmeal-mineral salt agar, diluted Pablum agar, YpSs agar, and soil and hair medium (4347). The following C sources support good growth (in descending order): succinic acid, mannose, itaconic acid, glucose and sucrose; a recommended growth medium contains 6% monosodium glutamate (or glutamic acid), 4% glucose, 0·1% K_2HPO_4, 0·05% $MgSO_4 \cdot 7H_2O$ (3152).

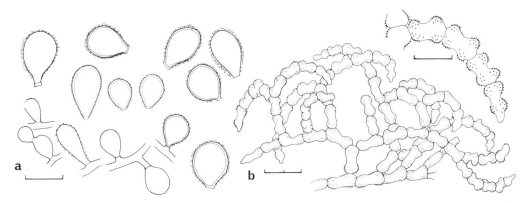

FIG. 33. *Arthroderma tuberculatum.* a. Attached and liberated conidia; b. peridial appendage with detail at higher magnification; CBS 314.65.

Arthroderma uncinatum Dawson & Gentles *1961* — Fig. 34.

Anamorph: *Trichophyton ajelloi* (Vanbr.) Ajello *1968*

≡ *Keratinomyces ajelloi* Vanbr. *1952*

DESCRIPTIONS: teleomorph: Vanbreuseghem (6038), Georg *et al.* (1938) and Dawson and Gentles (1311); anamorph: Dawson and Gentles (1312), Dvořák and Otčenášek (1521), and Padhye and Carmichael (4343); conidiogenesis: Galgóczy (1862). — A heterothallic species, with self-fertile isolates sometimes occurring. Colonies reaching 7·0–7·5 cm diam in two weeks at 25°C; two colony types occur, one cream to tan, and the other with diffusing dark vinaceous pigment; these are not correlated with the compatibility types (4342). Ascoma initials consisting of a club-shaped antheridium surrounded by a spiralled ascogonium. Ascomata pale buff, 300–900 μm diam excluding the appendages. Peridial hyphae pale yellow, septate, uncinately branched (usually to one side); cells thick-walled, strongly echinulate, symmetrically dumb-bell shaped, 7–11 × 4–7 μm; appendages consisting of septate spirals or macroconidia. Asci subglobose, 8-spored; ascospores hyaline, yellow in mass, smooth or finely roughened, lenticular, 2·3–2·7 × 1·4–1·8 μm. Macro-conidia abundantly produced, mostly arising laterally on the hyphae, sessile or pedicellate, cylindrical-fusiform with rounded tips, 8–12-septate, thick- (over 2 μm) and smooth-walled, 24–60 × 7–12 μm, containing on average about three nuclei per cell (2367). Micro-conidia present in some isolates (1938), pyriform to obovate, 6–13 × 3·5–7·0 μm. — The nuclei contain four chromosomes (6268). Contrary to other *Arthroderma* species, *A. uncinatum* gives a moderate sexual reaction (ascoma initials) with opposite mating type isolates of *A. simii* Stockd. *et al. 1965* (5588). Investigations of protein patterns indicated that *A. uncinatum* isolates belong to a monospecific group (3223).

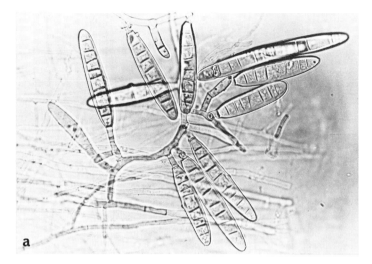
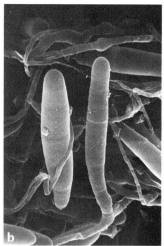

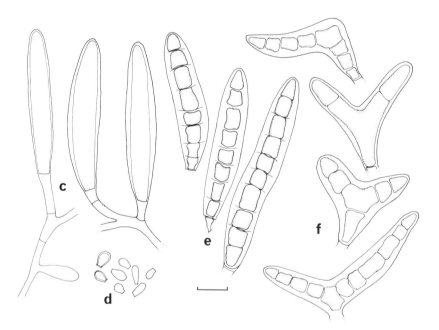

FIG. 34. *Arthroderma uncinatum*. a. Macro-conidia attached to a hypha, × 500; b. smooth-walled macro-conidia, SEM, × 1000; c. young attached macro-conidia; d. micro-conidia; e. mature macro-conidia; f. irregularly shaped macro-conidia; CBS 315.65.

A. uncinatum is the most widely distributed keratinophilic soil fungus and occurs mainly in the temperate region (4327, 4343, 4368). It has been known ever since the hair bait technique came into use (6038). Of 150 types of hair tested, that of male baboons was found superior for trapping keratinophiles (90). A fluorescent antibody technique has been developed for the direct observation of *A. uncinatum* in the soil and used to show that it can grow through non-sterile soil and sporulate there only when there is a keratinous substrate in the vicinity (2631). — Reports include ones from the British Isles (1264, 1311, 1312, 4659, 5585), Sweden (4368), Norway (3351), Finland (5015), the USSR (6533, 6534), Poland (1421, 1422, 1423, 3530), Germany (582, 584, 3743, 4536, 5521), France (6106), the Netherlands (CBS), Czechoslovakia (1013, 1014, 2577, 4324, 5338), Austria (3742) where it is very frequent in rural areas, France (1197), Belgium (79, 6038), Italy (893), Yugoslavia (5549), Hungary (2541, 6552), Bulgaria (306), Romania (118, 119), Turkey (2366), Egypt, Ethiopia (5766), Kenya, South Africa (90), Colombia (4896), Chile (6544, 6545), Argentina (6047), Brazil (4897), Easter Island (75), the USA (67, 3053, 4315), Canada (305, 4661), India, Australia (1835, 1836, 2091, 4348), Polynesia (3614) and New Zealand (3613, 5470). It has been isolated from cultivated soils (1423), forest soils (118, 582, 4368), grassland (893, 1013, 1014, 5521), garden soils (118, 2541), sandy soils (118, 3530, 5521), muddy clay in a cave (6552), soils from urban areas (1421, 1422, 2541, 3742, 5521) and soils from uninhabited mountain regions (3742). It is said to occur mainly in acid soils (1835). Hair baits are colonized in the moisture range between vapour saturation and −40 bars water potential (2095) and its frequency increased with increasing soil humidity (1013). *A. uncinatum* is particularly frequent on substrates with a high keratin content, for example birds' nests (2575, 2577, 4650) and under starling roosts (79), and has also been found in stables (306, 6533) and animal confinements (4536) and isolated from hair and skin scrapings of bison (3053) and small mammals (119, 762, 4324) and silage (6533). — The fungus has a high competitive saprophytic ability with respect to colonization of keratin substrates (145). It is able to survive in non-sterile soil for >8 years (117).

Ascoma formation is better on non-sterile than autoclaved soil and occurs at temperatures in the range 15–30°C. Light inhibits ascoma formation (1313); *in vitro*, good ascoma production occurs on oatmeal-mineral salt agar or soil and hair medium (4347). The germination of newly produced conidia occurs at vapour saturation (4660) and is stimulated by feather fat from pheasants but not from pigeons (4659, 4660). Light has an inhibitory effect on the germination of conidia and on growth of mycelia, particularly at short wavelengths (761); no growth occurs at 37°C (4343), but this species has a broad optimal temperature range of 15–28°C (1313). — The following C sources can be utilized (in descending order): glucose, fructose, galactose, arabinose, maltose, sucrose, lactose, raffinose, mannitol, sorbitol, inulin (2365) and also starch (6570). According to other reports, maltose has a variable effect and galactose reduced growth, while trehalose and D-xylose were not very suitable (4694). The following N sources can be utilized: L-phenylalanine, L-tyrosine, L-isoleucine, L-valine, L-threonine, L-alanine (2541), L-asparagine, casein hydrolysate, L-glutamic acid, urea, $(NH_4)_2SO_4$, NH_4NO_3 and KNO_3 (2364, 2365). Phenylalanine was found to be by far the most suitable amino acid for growth and sporulation, in contrast to findings for other dermatophytes (4695). Keratin (3666, 4692) can serve as the sole C and N source for *A. uncinatum* (145, 985). Phosphatases and proteinases are both produced during the process of keratin decomposition (585); the latter enzymes, when formed in the presence of casein or keratin, are not uniform (4693). During the decomposition of wool, the pH rises to a maximum of 8·7 rendering the keratin more accessible to enzymatic degradation; the liberation of *S*-sulphocysteine has been demostrated (4942) and that of tyrosine can be taken as a measure of the degree of keratin decomposition (985). — Skin infections have been observed rarely in animals and in isolated cases in man (1554). — *A. uncinatum* can grow on media with 10% NaCl but no growth occurs at 12% NaCl (2875).

Aspergillus Mich. ex Fr. *1821*

= *Sterigmatocystis* Cramer *1859*
= *Aspergillopsis* Speg. *1911*

Teleomorphic genera: *Eurotium* Link ex Fr. (=*Edyuillia* Subram. ≡*Gymnoeurotium* Malloch & Cain), *Emericella* Berk. & Br., *Dichlaena* Mont. & Durieu, *Hemicarpenteles* Sarbhoy & Elphick (=*Sclerocleista* Subram.), *Chaetosartorya* Subram. (≡*Harpezomyces* Malloch & Cain), *Warcupiella* Subram. (≡*Sporophormis* Malloch & Cain), *Neosartorya* Malloch & Cain, *Petromyces* Malloch & Cain (≡*Syncleistostroma* Subram.), *Fennellia* Wiley & Simmons.

MONOGRAPHIC TREATMENT: Raper and Fennell (4744); physiology and genetics: Smith and Pateman (5439). — The hyphomycete genus *Aspergillus* can be briefly characterized by the presence of erect conidiophores which terminate in a vesicle covered with either a palisade-like layer of phialides (often termed sterigmata) (uniseriate) or a layer of subtending cells (metulae) which bear small whorls of phialides (the so-called biseriate structure). The metulae and phialides are blown out synchronously and the latter produce basipetal chains of conidia which often form their surface ornamentation in a maturation phase after complete separation from the conidiiferous cytoplasm. At that stage the conidial ends are usually differentiated into connectives.

The subsequent account mainly follows Raper and Fennell (4744) except that some author citations have been corrected to accord with the current International Code of Botanical Nomenclature. Raper and Fennell recognize 132 species and 18 varieties and in the last decade, 90 additional taxa have been described, of which 34 are accepted at CBS (Samson, *1979**).

Aspergillus species occur commonly in soil in warmer climates, in compost, decaying plant matter, stored grain etc. A number of species, particularly of the genus *Eurotium* (*A. glaucus* group) and of the *A. restrictus* group, are osmophilic. *A. fumigatus* is thermotolerant, while several other species occur preferentially at somewhat raised temperatures. Various aspects of differentiation in *Aspergillus* were exhaustively reviewed by Smith and Anderson (5438).

The teleomorphs of *Aspergillus* (and *Penicillium*) belong to the family Trichocomaceae Fischer (=Eurotiaceae Clem. & Shear); the often cited name Aspergillaceae Link is incorrect as *Aspergillus* is strictly an anamorphic genus. Following studies by Benjamin (446), the teleomorphic genera have been reviewed and further differentiated recently by Subramanian (5627) and Malloch and Cain (3552, 3554). Amongst the eight recognized genera with an *Aspergillus* anamorph, species of four are included here and these can be distinguished as follows:

Emericella Berk. & Br. *1857* (*Aspergillus nidulans* group): Ascomata surrounded by thick-walled hülle cells; ascospores usually red to purple; conidiophores with phialides borne on metulae, conidia in columns.

*Stud. Mycol. Baarn **18**: 38 pp.

Fennellia Wiley & Simmons *1973* (*A. flavipes*): Ascomata developing within masses of yellow thick-walled elongate to helicoid cells; conidiophores with phialides borne on metulae, conidia in columns.

Eurotium Link ex Fr. *1829* (*A. glaucus* group): Ascomata mostly yellow with a one cell thick, cellular, smooth wall; phialides arising directly from the vesicle (not on metulae) in a radiating manner.

Neosartorya Malloch & Cain *1972* (*A. fumigatus* group): Ascoma wall pseudoparenchymatous or hyphal, several cells thick; phialides borne directly on the vesicle (not on metulae), forming a conidial column.

The standard techniques for species identification in *Aspergillus* are the same as used in *Penicillium* (q.v.); primary descriptions are based on cultures growing on Czapek's agar at 25°C, but higher temperatures (30°, 37°, 45°C) and more concentrated media (CzA or MEA with 20% sucrose, with 40% sucrose = Harrold's medium, or with 12 and 25% NaCl, respectively) are often used, particularly for inducing the teleomorph.

For the study of ascospore ornamentation, scanning electron microscopy is particularly valuable (3394, 3395). Spore measurements from SEM mounts may deviate from those obtained in light-microscopic work which are reported here. An attempt to use the long-chain fatty acids content of 14 species in a computerized cluster analysis showed that this type of numerical classification required rigidly controlled experimental conditions (3266).

Key to the groups and species treated:

1   Conidial heads some shade of green during development, metulae absent or present   **2**
    Conidial heads some other colour; metulae mostly present   **10**

2(1)   Vesicles clavate; philiades borne directly on the vesicles; conidial heads blue-green at first
        becoming light grey with age   *A. clavatus* group
        Conidiophores smooth-walled, uncoloured, 2–4 mm long   *A. clavatus* (p. 86)
        If vesicles subclavate and conidia darken with age   (*A. ornatus* group)
    Vesicles not clavate; metulae present or absent   **3**

3(2)   Conidial heads bright yellow-green when young, sometimes becoming brown with age, loosely
        radiating or columnar; metulae mostly present, at least in some conidiophores; conidiophores
        colourless and usually roughened   *A. flavus* group (p. 78)
    Conidial heads other shades of green; conidiophores colourless or pigmented but usually
        smooth-walled   **4**

4(3)   Colonies mostly showing naked yellow ascomata of *Eurotium* and yellow or red encrusted
        hyphae; conidiophores hyaline or shades of brown; metulae absent; typically osmophilic
                A. glaucus group; cf. *Eurotium*, p. 289
    Ascomata, if present, surrounded by hülle cells or several layers of pseudoparenchymatous cells
        and lacking yellow or red encrusted hyphae   **5**

5(4)   Conidial heads markedly columnar   **6**
    Conidial heads globose, radiating, or loosely columnar   **8**

6(5)   Metulae absent; phialides borne directly on the vesicles   **7**
        Metulae present; conidiophores more or less brown, usually sinuous; globose to subglobose hülle
            cells commonly present; ascomata of *Emericella* occurring in several species; ascospores
            orange-red to violet   *A. nidulans* group; cf. *Emericella* p. 264

7(6)   Conidial heads long, narrow (often twisted) or irregular; conidia usually formed as cylindrical
            segments from the phialides; conidiophores short, smooth-walled or nearly so, usually hyaline,
            sometimes greenish; typically osmophilic   *A. restrictus* group
        Heads strictly columnar; colonies growing very restrictedly on CzA, but spreading broadly
            on MEA with 40% sucrose, dark olive-green   *A. restrictus* (p. 109)
        Conidial heads compact and typically uniform in diameter; conidiophores commonly in shades of
            green, smooth, commonly of uniform diameter; conidia not formed as cylindrical segments;
            ascomata of *Neosartoria* present in some species; not typically osmophilic; spreading broadly
            on most media   *A. fumigatus* group (p. 79)

8(5)   Vesicles small, variable in shape; metulae regularly present   **9**
        Vesicles large, strictly globose; conidiophores constricted below the vesicles; metulae sometimes
            absent   (*A. sparsus* and *A. cremeus* groups)

9(8)   Conidial heads blue-green, dull yellow-green, or grey blue-green, radiating to loosely columnar;
            conidiophores mostly uncoloured or sometimes brown, usually smooth-walled; hülle cells, if
            present, globose to subglobose   *A. versicolor* group (p. 79)
        Conidial heads olive, olive-grey, drab, to light brown; radiating to broadly columnar; co-
            nidiophores smooth-walled, shades of brown; hülle cells elongate, often twisted *A. ustus* group
        Conidiophores with upright vesicles; conidial heads ± columnar at maturity; hülle cells
            scattered throughout the colony, not associated with yellow mycelium   *A. ustus* (p. 117)

10(1)   Growth very sparse and sporulation poor on CzA   (*A. cervinus* group)
        Growth and sporulation usually abundant on CzA   **11**

11(10) Heads loosely to compactly columnar; conidiophores smooth-walled; aleurioconidia present
                                                          A. flavipes group (p. 80)
        Heads globose to radiating; conidiophores smooth-walled or roughened; aleurioconidia absent
                                                                              12

12(11) Heads persistently white; larger heads definitely globose or radiating; conidiophores smooth-
            walled, colourless or apically slightly yellow; conidia smooth, globose to subglobose
                                              A. candidus group; *A. candidus* (p. 83)
        Heads not white   **13**

13(12) Heads black or dark brown; conidiophores hyaline to brown, mostly smooth-walled; conidia
            mostly pigmented and ornamented; metulae absent in some species   *A. niger* group (p. 80)
        Heads yellow, ochraceous or light brownish   **14**

14(13) Heads sulphur yellow to ochraceous; conidiophores faintly or conspicuously pigmented, yellow
            to brown, smooth-walled to coarsely roughened; conidia hyaline or light-coloured, usually
            smooth-walled; sclerotia present in several species   *A. alutaceus* group (p. 80)
        Heads yellow-brown to dull buff, conidiophores colourless; metulae sometimes absent; conidia
            yellow to yellow brown, rough-walled; sclerotia absent   *A. wentii* group
        Conidiophores smooth-walled or slightly granular, several mm long; metulae mostly present
                                                          A. wentii (p. 122)

Aspergillus flavus group:

The four species treated, out of a total of about ten can be distinguished by the following key
modified from Murakami (4046, 4047).

1   Conidia 5–6 μm or more diam; old colonies more or less brown; conidial mass on CzA + 0·5% anisic acid regularly pink (*A. oryzae* subgroup)   **2**

         Conidia 4–5 μm or less diam; old colonies remaining yellow-green or greenish brown; conidial mass sometimes turning pink on anisic acid (*A. flavus* subgroup)   **4**

2(1)   Metulae absent; conidiophores smooth or almost so, conidia prominently echinulate (*A. sojae*)

         Metulae present in many conidiophores; conidiophores generally roughened; conidia not prominently echinulate   **3**

3(2)   Young colonies yellow-green; conidia finely echinulate   *A. oryzae* (p. 105)

         Young colonies brown; conidia tuberculate   *A. erythrocephalus* (p. 88)

4(1)   Metulae absent; conidia prominently echinulate; sclerotia absent   *A. parasiticus* (p. 108)

         Metulae present in many conidiophores; conidia finely echinulate; brown, irregularly shaped sclerotia sometimes present   *A. flavus* (p. 90)

Aspergillus fumigatus group:

Two out of ten species are treated here:

1   Ascomata absent   *A. fumigatus* series

         Conidial heads erect, compact and strongly columnar; conidia globose, echinulate

                         A. fumigatus (p. 94)

         Ascomata of *Neosartorya* present   *A. fischeri* series

         Ascomata and enveloping hyphae white to cream, not yellow to orange; ascospores with two distinct equatorial crests   *Neosartorya fischeri* (p. 511)

Aspergillus versicolor group:

Of the 17 species of this group treated by Raper and Fennell, two are discussed both of which have conidial heads of one colour, and globose to somewhat elongate vesicles which are fertile over most of the vesicular surface; globose to subglobose hülle cells may be present, but sclerotia are absent; the conidia do not exceed 4·0 μm diam and are consistently globose to subglobose.

1   Conidial heads variable in colour: pale yellow-green, buff to orange-yellow, or occasionally flesh coloured   *A. versicolor* (p. 120)

         Conidial heads always blue-green when young   *A. sydowii* (p. 112)

Aspergillus flavipes group:

This group comprises four species.

1   Heads compactly columnar, avellaneous to cinnamon; conidiophore stipes colourless; colonies
     normally velvety, cinnamon to orange-brown, without sclerotia on MEA *A. terreus* (p. 114)
     Heads loosely columnar, white, flesh, or cream-buff; conidiophore stipes yellow to brown or
     uncoloured   **2**

2(1) Conidiophores definitely pigmented, yellow to light brown; conidial heads usually white to very
      pale buff, occasionally darker and almost avellaneous
                                              Fennellia flavipes (anamorph *A. flavipes*) (p. 303)
      Conidiophores unpigmented or very slightly yellowish   **3**

3(2) Conidial heads persistently white   *Fennellia (Emericella) nivea* (p. 269)
      Conidial heads at first white but later becoming vinaceous-fawn   *A. carneus* (p. 85)

Aspergillus niger group:

Thirteen species are recognized by Raper and Fennell, but these are often difficult to
distinguish; the following three are most commonly encountered.

1   Phialides borne on metulae; conidial heads black; conidiophores $1 \cdot 0 – 3 \cdot 0 (–4 \cdot 0)$ mm long with
     vesicles $46–65(–75)\,\mu$m diam   **2**
     Phialides borne directly on the vesicle; colonies purple-brown to purple-black; conidiophores
     mostly $500–1000\,\mu$m long with vesicles $20–35\,\mu$m diam; conidia globose to subglobose,
     conspicuously echinulate, mostly $3 \cdot 0 – 3 \cdot 5\,\mu$m diam (excluding the spines) *A. japonicus* (p. 98)

2(1) Conidia at maturity flattened, mostly $3 \cdot 0 – 3 \cdot 5\,\mu$m diam, with longitudinal bands or striations;
      colonies dark brown to black   (*A. phoenicis*)
      Conidia at maturity globose, mostly $4 \cdot 0 – 5 \cdot 0\,\mu$m diam, irregularly roughened, with conspicuous
      ridges and echinulations not arranged as longitudinal striations; colonies black
                                                                                      A. niger (p. 100)

Aspergillus alutaceus group:

(*Aspergillus ochraceus* group, 4744).

Of the nine species of this group three are discussed here.

1   Conidial heads pale pure yellow; sclerotia white to cream to pale pink, produced singly,
     $1 \cdot 0 – 1 \cdot 5$ mm diam   *A. sclerotiorum* (p. 110)
     Conidial heads dull yellowish cream, buff or ochraceous   **2**

2(1) Sclerotia abundantly produced, commonly $400–500\,\mu$m, pure yellow but later becoming brown;
      conidia globose, subglobose or ellipsoidal, $2 \cdot 7 – 3 \cdot 5\,\mu$m diam   *A. melleus* (p. 99)
      Sclerotia scattered, developing late, commonly $500–1000\,\mu$m, pink to vinaceous-purple when
      mature; conidia globose to subglobose, $2 \cdot 5 – 3 \cdot 0\,\mu$m diam   *A. alutaceus* (p. 81)

Aspergillus alutaceus Berk. & Curt. *1875* — Fig. 35.

= *Aspergillus ochraceus* Wilhelm *1877*

DESCRIPTIONS: Raper and Fennell (4744), Subramanian (5626), and Hirayama and Udagawa (2464); nomenclature following Subramanian (5626); conidiophores in SEM: Fennell *et al.* (1718). — The *Aspergillus alutaceus* group includes species with teleomorphs in *Petromyces* Malloch & Cain *1972* (=*Syncleistostroma* Subram. *1972*, nomen confusum). — Colonies reaching 3–4 cm diam in 10–14 days at 24–26°C on CzA, on MEA 4·5–5 cm, light ochraceous-buff to buff from the abundant conidial heads, which are at first globose but later split into two or three divergent columns; conidiophores 1·0–1·5 mm long, thick-walled, dull yellow to light brown, coarsely roughened; vesicles globose, with metulae all over the surface; conidia globose to subglobose, delicately roughened, mostly 2·5–3·0 μm diam. The rodlet pattern of the conidial surface as seen in freeze-etch replicas is slightly interlacing, consisting of very prominent broad bundles of rodlets (2422). — Sclerotia often present, more abundant on CzA than on MEA, borne singly, white to pale pink at first but later becoming vinaceous-purple, globose or irregularly shaped, to 1 mm diam. — DNA analysis gave a GC content of 52·5–54·0% (5600).

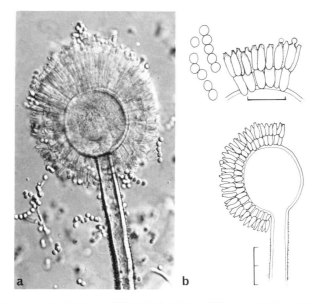

FIG. 35. *Aspergillus alutaceus.* a. Conidial head, CBS 626.78, NIC, × 400; b. conidial head with detail of phialides and conidia, CBS 547.65.

This fungus is particularly widely distributed in India and is also known in Pakistan (4855), Israel (654, 2764, 4759), Syria (277, 5392), Iraq (92), Kuwait (4000, 4001), Libya (6510), Egypt (8, 100, 3988, 3993, 4962), central Africa (3063), South Africa (1415, 1559), Morocco (3415), Jamaica (4886), Central America (1697), Peru (2005), Argentina (6402, 6403, 6404, 6405), the USA (976, 1163, 1166, 3817, 4733, 6294), Australia (978), Turkey (4183), Spain (3417), the British Isles (2587), the USSR (2871), Czechoslovakia (1700, 1703), and Poland (1421). Most of the reports are from cultivated soils but it also occurs in forest soil (3063), uncultivated land (3868, 4855), desert soils (100, 654, 3415, 4733), salt-marsh (8, 4000), under steppe type vegetation (1559, 1700, 3417, 4407), a bat cave in the Bahamas (4312), children's sandpits (1421), a uranium mine (1703), running water, an acid mine drainage stream, a waste stabilization pond, sewage (1166), activated sludge (1165, 1172), and fields treated with sewage sludge (1163). A relative increase in numbers occurred after soil fumigation (3631). Further known microhabitats include rhizospheres (and pods) of groundnuts (2768), clover (3429) and other legumes, wheat (45), *Pennisetum typhoideum* (3867), *Linosyris tatarica* (3376), *Euphorbia thymifolia, E. hirta* at all its growth stages (3866), ferns (2856, 2858), roots of sugar cane (4886), citrus where it causes a reduction in root elongation (4744), achenes of *Platanus occidentalis* (1683), green coffee beans (3302), seed of spoiled corn (4831), shelled corn with a moisture content of at least 15·6% (3069), seeds of sorghum (3989), wheat (2097, 3989, 4492) in which the embryo is decayed (4744), *Pennisetum typhoideum* (57), cotton seed stored at water potentials between −140 and −230 bars at 30°C (4744), pecans (2572, 5092), seeds of rice (3181), milled rice, wheat and fish products (2464, 5980), ground black and white pepper (3644), stored sweet potatoes (3066), dry fodder (3258), coarse fodder (4548), fruit juices (5205), moderately common also in some other foodstuffs (4986, 6146), damaged flue-cured tobacco (6282), wood pulp (3294), wood in seawater (4546), cotton fabrics and leather (6326), feathers and pellets of free-living birds (2575), dead imagoes of the Colorado beetle (291), and different genera of wild and domesticated bees (402). It was found with a low incidence, 0·7% and 1·4% respectively, in the air spora in France (3548) and the British Isles (2587, 3548). — Conidia survived after 36 months of storage on silica gel at −20°C (435). — *A. alutaceus* can serve as food for the woodlouse *Acanthoscelides obtectus* (5381).

Conidia germinate in the temperature range 12–37°C and at a water potential of −270 bars (3873); good growth is obtained at 27°C (291). The thermal death point was determined at 62°C after 20 min exposure (3620). For sporulation under submerged conditions, 2·5% NaCl and beet molasses or barley extract are required in the basal medium (6088). — The degradation of starch (2782), straw from *Pennisetum typhoideum* (4611), cotton duck and wool (3618, 4744, 4775, 6326), high proteolytic (276, 3429, 5843), lipolytic (1676) and pectolytic activities (4744), and utilization of *o*-, *m*-, *p*-xylene and hydrocarbons from fuel oil have all been observed (4239). *A. alutaceus* produces ketones when grown on *n*-alkanes (4497, 4498), hydroxylates progesterone and other steroids (1490), and produces the volatile flavour compounds 1-octen-3-ol and 2-octen-1-ol (2867), riboflavin (4762), indole acetic acid and tryptophol when grown on tryptamine (1522). Fat production by *A. alutaceus* has also been studied (4642) and urea found to be a suitable N source (5353). The fat contains large amounts of saturated acids, especially palmitic acid, and shows a high linoleic acid content as compared to other microbial fat products (5479). One of the more intensely studied metabolites of this species is the isocoumarin ochratoxin (5042, 5576) which is lethal to mice (4831), HeLa cells (4986), chicks (1444), and young ducklings (5158). The toxicity of ochratoxin A is comparable to that of aflatoxin B_1 (3777); three derivatives of ochratoxin (A, B, C) have now been identified (3776). The conditions for optimum production of ochratoxin include suitable organic substrates, e.g. rice, yeast extract, cocos flakes, wheat or rye bread (2432, 4789, 6470), trace element (Cu, B, Fe, Mn, Mo, Zn) requirements (3212, 5543), stimulatory effects of L- or DL-glutamic acid, proline, L-phenylalanine (1728, 6470), and the influences of age, pH (5042), and temperature (1055). Concurrently with ochratoxin, aflatoxin (6146) and penicillic acid have frequently been found (1055, 2888, 4112), with low temperatures favouring the production of the latter (1055). Further metabolites include flavacol, neoaspergillic acid, deoxyneo-β-hydroxy-aspergillic acid (6472), 6-methoxy-8-hydroxyisocoumarin-3-carboxylic acid (6473), emodine, secalonic acid, the latter with mammalian toxicity and antibiotic properties (6471), mellein, 4-hydroxymellein (1117, 3917), L-prolyl-L-leucine anhydride, L-prolyl-L-valine anhydride, the cyclodepsipeptides destruxin A and B toxic to silkworms (3068), and aspochracin (a cyclotripeptide) also with insecticidal properties (949, 4060). Antimycotic activity against several pathogenic fungi was observed (569). — *A. alutaceus* can attack some insects (e.g. silkworms, Colorado beetles) directly, entering them through the hind gut (4744). It has exceptionally also been isolated from sputum of patients with pulmonary disturbances (114). — The fungus has a high tolerance for allyl alcohol in soil (3991) and also for NaCl concentrations up to 30% (976, 1475, 3166, 5881).

Aspergillus candidus Link ex Link *1824* — Fig. 36.

DESCRIPTIONS: Raper and Fennell (4744), Subramanian (5626), and Hirayama and Udagawa (2464). — This is the only species of the *Aspergillus candidus* group. It differs little from the *A. niger* group, except for the absence of pigmentation and roughening of the conidia. *A. candidus* also superficially resembles *Emericella nivea* (q.v.) which has much smaller conidial heads. — Colonies relatively slow-growing, reaching 1·5–3·0 cm diam in 14 days at 24–26°C on CzA but somewhat more rapid with the addition of 20% sucrose, white, at most becoming cream to yellowish cream. Conidiophores typically with large globose vesicles, bearing inflated, club-shaped metulae and narrow phialides. Conidia globose to ellipsoidal, smooth-walled, 2·5–3·5(–4·0) μm diam, uninucleate (6514), the chains sometimes sliming down. Purple or black sclerotia are produced by some isolates.

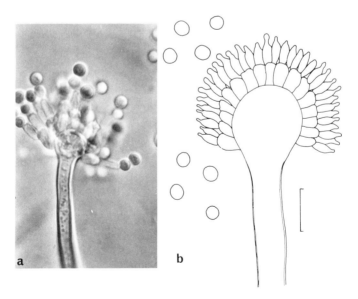

FIG. 36. *Aspergillus candidus*. a. Conidial head of small type, CBS 102.13, NIC, × 1000; b. typical conidial head and conidia, orig. A. C. Stolk.

A. candidus has its main distributional range in subtropical and tropical regions. There are numerous reports from India, where it has been found in cultivated fields (44, 1317, 1519, 4610), grassland (3865, 4933), forest soils (2179, 2186, 4716, 4995), uncultivated soil (4030), lateritic soil, chernozem (5000), black clay (4736), alkaline soil (968), sand dunes (4371), highly saline soils (4477), soils with different water regimes (1315), and mangrove mud (4700); a reduced occurrence in the hot summer months is reported (2179). Records are also available from Bangladesh (2712), Pakistan (4855), Kuwait (4000, 4001), Sri Lanka (4021), South Africa (1415), Somalia (5049), equatorial West Africa (1420), the Chad area and the Sahara (2974, 3415), Libya (4084), Egypt (8, 3993, 4962), Syria (5392), Israel (2764, 2772), Argentina (6402, 6403, 6404, 6405), the Bahamas (4312), New Guinea, the Solomon Islands (5983), China (3475), Central America (1697), Chile (4569), the USSR (1482, 2871, 3850, 4474, 4548), Nepal (1826), and the USA (2482, 3550). It is a rare component of the soil mycoflora in European countries but it is known from soils in Spain (3417, 3447), Italy (1794, 3445), Hungary (1712, 2165), Austria (3413, 3418), Czechoslovakia (1702, 1703), Germany (1424, 1712, 5316), France, Britain (3548), and Ireland (1376). In addition to the Indian records cited above, the fungus has been isolated from forest soils (1712, 2482, 2854, 3447), soils with steppe type vegetation (1702, 3415, 5049), desert soils (2973, 4084), garden compost (1424), peat (1376, 4474), rendzina (3413, 3414), caves and mines (1703, 4312), polluted water (1482), saline soil (8, 4000, 4001), marine environments (4918) and the air (3548). It is remarkable that *A. candidus* in soil is almost exclusively found either on seeds or in the rhizosphere and not in root-free soil. The fungus occurs on seeds of wheat, oats and barley (57, 1029, 1752, 1757, 2716, 3989, 4038, 4492) of which it inhibits the germination to some degree (4306), mouldy sorghum (795, 3989), corn (3989, 4831), rice (1337, 2297, 2464, 3181), green coffee beans (3302), in the rhizospheres of groundnut (2768, 4735), *Trifolium alexandrinum* (1963), wheat (45), corn (4555), rice (2712), *Abutilon indicum* (3864), *Cassia occidentalis* (6067), *Euphorbia* spp. at all growth stages (3866), various steppe plants (3376), in the ectomycorrhiza of *Corylus avellana* (1794), on litter of *Pinus ponderosa* (681), in a wheatstraw compost (953), rarely on coarse fodder (3258, 4548), most frequently in dough products (2067), also on frozen fruit cakes (3153), nuts and dried fruits (2258, 2572, 6252), various other foodstuffs (4986, 5980), damaged flue-cured tobacco (6282), wood exposed to seawater (4546), fresh water and sewage (1166), pig (5983), rabbit (6085) and vole (IMI, 2338), dung, nests, feathers and pellets of free-living birds (2575), and dead bees (402). Some reports suggest a preferential occurrence in neutral or slightly alkaline habitats (968, 3417, 3445, 4700, 5000, 5049); and it is also found in habitats with temperatures rising to 50–55°C (5000, 5049, 6273). *A. candidus* may cause the self-heating of stored wheat grain on which substrate it has a strong competitive ability at very low water potentials (1036, 4744); in vitro optimal growth occurs at −30 bars, minimum at about −300 bars water potential (261).

Optimal growth occurs at 25–28°C, the minimum at 11–13°C and the maximum at 41–42°C (517, 1827), but in some reports these temperatures have been given as 32°, 10° and 44°C, respectively (261). Optimal growth was obtained at 0·016 atm O_2 partial pressure (1428). — Growth occurs on media containing *p*-hydroxy-benzoic aldehyde (2165), lignin sulphonate (1425) or tannin (1260). *A. candidus* is reported to use straw lignin (2165), utilize arabinoxylan modestly (1757) and degrade cellulose (4186, 5674), with $(NH_4)_2SO_4$ as the most suitable N source at a C/N ratio of 10:1 (5675). In stored corn it causes an increase in free fatty acids due to its lipolytic activity (4744). Proteolytic, amylolytic and pectolytic processes have also been studied (1827, 4545). It can oxidize tryptamine to tryptophol (1522) and the conidia can produce D-mannitol from glucose and other sugars (4124). There is evidence showing the production of ethylene (2641), citrinin (6005), kojic acid (5278) and two antibiotics: chlorflavonin acting against fungi (4042, 4832) and candidulin against (myco)bacteria (512, 5532); other authors, however, have not observed any antibacterial activity (5974). Further metabolic products include dechlorochlorflavonin and *p*-terphenyl derivatives (3597, 3598) which were shown to act on HeLa cells (5689). Various disorders due to *A. candidus* have been observed in pigs (3932). In mixed culture, *A. candidus* suppressed growth and aflatoxin production of *A. parasiticus* (621) and showed antimycotic activity against *Aspergillus fumigatus* (569). Riboflavin is apparently not produced, contrary to many other *Aspergillus* species (1620). Significant stimulation of flax seedlings has been observed in greenhouse experiments (1427). — *A. candidus* shows optimal growth on media with 10% NaCl (1428), it tolerates 20% NaCl (976) or up to 40% sucrose (4001) in nutrient media and 10% CO_2 in the atmosphere (4029). Cellulose degradation by *A. candidus* proved to be highly sensitive to the herbicide paraquat (5676).

Aspergillus carneus Blochwitz *1933* — Fig. 37.

? = *Sterigmatocystis carnea* van Tiegh. *1877*

DESCRIPTIONS: Raper and Fennell (4744), and Subramanian (5626). — Blochwitz described *A. carneus* as a new species and excluded van Tieghem's fungus because he regarded its original description as inadequate; the epithet "*carneus*" in *Aspergillus* thus cannot be ascribed to van Tieghem as has been done by many authors (4744).

Aspergillus flavipes group. — Colonies are rather fast-growing, reaching 4–5 cm diam in 14 days at 24–26°C on CzA, on MEA growing more slowly, at first white but later becoming vinaceous-buff due to the conidia. Masses of thick-walled sterile hyphae may occur. Conidial heads loosely columnar; conidiophores hyaline or slightly yellow, smooth-walled; vesicles hemispherical, covered with metulae; conidia globose to subglobose, thin- and smooth-walled, mostly $2 \cdot 4 – 2 \cdot 8 \, \mu$m diam. TEM photographs of the conidia are available (4600). Botryose aleurioconidia similar to those of *A. terreus* (4598) are formed; these have a two-layered wall (4600) and are also produced in animal tissue (4599).

A. carneus is almost completely restricted to tropical and subtropical areas. It was originally isolated in Java (4744) and has subsequently been found in India in cultivated (44, 1519, 2861, 3737, 3868, 4997) and uncultivated soils (2179, 4030, 4698), in a teak forest (2854), a podzolic forest soil in Hong Kong (1021), a mangrove swamp in Hawaii (3264), open savannah in South Africa (1559, 4407), and savannah (5049) and forest soil (5048) in Somalia; it is also reported from Argentina (6402, 6403, 6404, 6405), Kuwait (4000, 4001), Egypt (8, 3993), Italy (4538), desert soil (4733), cultivated soil (3817) and forest nurseries (2482) in the Southern States of the USA, also from Canada (5363), the USSR (2871, 5297) and Poland (1978). In rhizosphere studies, it is rarely reported but occurs on wheat with and without urea foliar sprays (45); it has also been reported on *Trigonella foenum-graecum* and *Abelmoschus esculentus* (3733). A preferential occurrence in slightly alkaline soils can be deduced from several reports (1559, 4538, 4698, 5048, 5049). It has also been isolated from wheat seeds (3989), wild bees and their fermented provision (402).

Optimal growth occurs at 41–42°C, the minimum at ≤6–7°C and the maximum at 46–48°C (517, 4598). — Cellulose is degraded (3618) and considerable amounts of dextranases (I and II) are produced on media containing 1% dextran (2462, 5918). It has a relatively high salt tolerance and is able to grow well at low water potentials (976, 4001, 5781). — Cereal and legume products infested with *A. carneus* proved toxic to young ducklings (5158). This species can cause lung aspergilloma (3955, 4744) and after intravenous inoculation, produce cerebral aspergilloma in mice (4599).

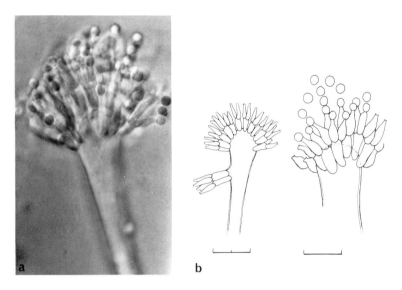

FIG. 37. *Aspergillus carneus.* a. Conidial head and conidia, NIC, × 1000; b. conidial head with detail; CBS 494.65.

Aspergillus clavatus Desm. *1834* — Fig. 38.

DESCRIPTIONS: Raper and Fennell (4744), Hirayama and Udagawa (2464), Sugiyama (5633), and Subramanian (5626). — *Aspergillus clavatus* group. — Amongst the three species of the *A. clavatus* group, *A. clavatus* is characterized by blue-green clavate conidial heads which split into several divergent columns, and conidiophores 1·5–3·0 mm tall, arising from specialized widened hyphal cells which become the branching foot cells (3026, 5365). Colonies reaching 3–3·5 cm diam in 10 days at 24–26°C on CzA, blue-green, felty due to the very long conidiophores. The conidia are smooth-walled, $3·0–4·5 \times 2·5–3·5\,\mu$m and uninucleate (6514); the phialide development and conidium formation has been studied by TEM (2257, 5633); the first-formed conidium has a wall continuous with that of the phialide as seen by SEM (2256). Heterokaryotic conidial heads with mixed conidial colours were produced after recombination with an albino mutant (4743). — DNA analysis gave a GC content of 52·5–55% (5600). Soluble wall carbohydrates consist, as in other fungi, of mannitol and arabitol, but not ribitol and sorbose as claimed by some authors (2508).

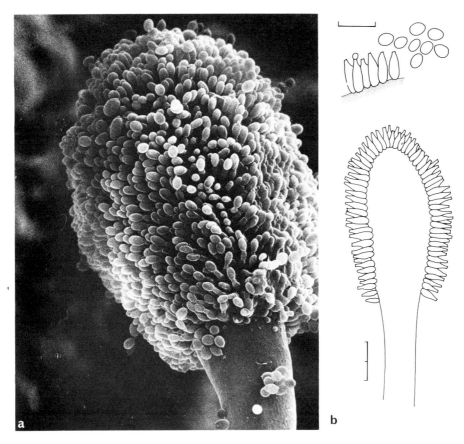

FIG. 38. *Aspergillus clavatus*. a. Conidial head, SEM, × 1000; b. conidial head and detail with phialides and conidia; CBS 114.48.

With only a few exceptions, *A. clavatus* appears to be tropical, subtropical and mediterranean in its distribution. It has been reported from India (40, 44, 1315, 1317, 1519, 2854), although at low frequencies in some soils there (2179), Bangladesh (2712), Sri Lanka (4021), Hong Kong (1021), Jamaica (4886), Brazil (399), Argentina (6402, 6403, 6404, 6405), South Africa (2816), the Ivory Coast (4719), Egypt (100), Libya (4084), Turkey (4245), Greece (4184), Italy (3913), the USA (3817), Japan (2532, 4987, 5633), the USSR (2871, 3364, 3850, 4474) and Czechoslovakia (1703). It was found on clay or rocks in a carst cave (3364) and in stratigraphic core samples down to 1200 m in central Japan but otherwise has been almost exclusively isolated from cultivated soils, including ones under cotton, potatoes (2871), sugar cane (4886), legumes (1317), paddy (2712, 3913), and *Artemisia herba-alba* (4084). It is also reported from soil under burnt steppe type vegetation (4719), desert soil (100), potato residues (3113), the rhizospheres of banana (2030), clover (3429), groundnuts (2532), *Ocimum basilicum* (35), rice (2712), and wheat where it was the dominant fungus after repeated wheat cropping (2816). It is stimulated by nitrogen (2816) and NPK (2712) fertilizers, and has been found in the ripe compost of municipal waste (3041). It is also known from various stored food products, including sorghum seeds (795), milled rice (2464), flour and dough products (2067, 5980), nuts and dried fruits (2258, 6252). *A. clavatus* has been isolated from insects (4744), particularly dead adult bees and honeycombs (402), and also from feathers and droppings of free-living birds (2575). — Germination *in vitro* can be inhibited by emanations from leaves of *Origanum majorana* (35). *A. clavatus* can serve as food for Collembola (4007). It has been found to be parasitized by *Fusarium solani* (5668).

On MEA colonies arising from one conidium were found to contain after four days 5×10^7 conidia, and after six days 25×10^7 conidia (3940). The elongation of the conidiophores is stimulated by light (1654). Starch, dextrin, glycogen and particularly fructose are more favourable C sources than glucose (43); cellulose (2718, 3618, 4775, 5059, 5076) and usnic acid (322) are degraded. Considerable amounts of lipids can be produced (4642). In liquid culture, production of riboflavin (1620), ribonuclease, acid phosphodiesterase and acid phosphatase (2693, 3954) is known. *A. clavatus* is able to oxidize tryptamine to indole acetic acid (1522), assimilate hydrocarbons from fuel oil (4239), utilize metaphosphate (4544) and produce ethylene (2639, 2641), clavatol (2314) and kojic acid (6388), the latter with variability between isolates (4458). Culture filtrates may have antifungal (716, 718) and/or antibacterial (2126, 6141, 6357) activity; mycotoxins produced are ascladiol (2425), patulin (=clavacin) (467, 3929, 6005, 6143), a reduced derivative of patulin (5660), cytochalasin E (769) and two tremorgenic compounds called tryptoquivaline and tryptoquivalone (1062). A mycotoxin for mammals (733), causing hyperkeratosis in calves (898), and a substance highly toxic to poultry when occurring in infested grain (1799, 5158) have been described. — *A. clavatus* may cause allergic alveolitis in man (957). The growth of barley seedlings has been stimulated by the presence of *A. clavatus* in experiments *in vitro* (293). — It can tolerate NaCl concentrations of >15% (5881), corresponding to about −140 bars water potential.

Aspergillus erythrocephalus Berk. & Curt. *1869* — Fig. 39.

= *Aspergillus tamarii* Kita *1913*

DESCRIPTIONS: Raper and Fennell (4744), Hirayama and Udagawa (2464), Subramanian (5626), and Murakami (4046); nomenclature following Subramanian (5626). — *Aspergillus flavus* group (*A. oryzae* subgroup). — Colonies reaching 5–6 cm diam in 10–14 days at 24–26°C on CzA, on MEA 6–7 cm, brown from the beginning and not green. Conidia 5–6·5 μm diam, coarsely tuberculate as shown in SEM (4907), on CzA with anisic acid turning pink; 3–4-nucleate. Heterokaryotic conidial heads were obtained after recombination with an albino strain (4743). — DNA contains 26·1% adenine, 24·4% guanine, 23·6% cytosine and 25·9% thymine (3232). — Polyacryl-gel electrophoresis gave 31 fractions of soluble proteins, 14 of which were shown to be homologous with *A. parasiticus*, *A. flavus* and *A. oryzae*, 13 with *A. leporis* and *A. fumigatus,* and 12 with *A. alutaceus* (3165). Pyrolysis gas-liquid chromatograms (6094) of the conidia showed a close similarity to those of *A. flavus* (6093).

A. erythrocephalus is known from India (40, 1315, 1317, 1519, 2854, 2861, 3545, 3732, 4716, 4995, 5349), Bangladesh (2712), Malaysia (6046), China (3475), Japan (2464, 4046), Pakistan (61, 969, 4855), Somalia (5048), the Chad area (3415), Gambia (1984), Iraq (92), Syria (5392), Israel (2764, 2772), Libya (6510), Spain (3446), the Bahamas (2006), Peru (2005), and the USA (655, 1166, 3817). It has most commonly been reported from cultivated

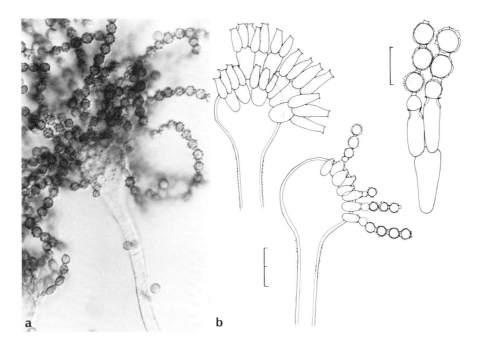

FIG. 39. *Aspergillus erythrocephalus*. a. Conidial head, NIC, × 400; b. uniseriate and biseriate conidial heads; CBS 109.63.

soils, including rice (2712), wheat (3446), a citrus plantation (2764), a coconut grove (2006), and soils under legume crops (1317, 2772), but occurs in tropical dry forest (4716), natural rain forests (2854, 6046), Himalayan forest soils (6046) and estuarine sediments (655). It is one of the later invaders of grass compost (*Diplachne fusca*) after the compost temperature has fallen to *c*. 35°C (3546). *A. erythrocephalus* has also been isolated from the rhizospheres of wheat (45), barley (3526), corn (4555), sugar cane (2861), rice (2712), groundnuts (2768), euphorbiaceous plants at all growth stages (3866), *Trigonella foenum-graecum , Abelmoschus esculentus, Raphanus sativus, Brassica oleracea* (3733), *Cucumis sativus* (4726), *Polypodium* sp., *Cyclosorus* sp. (2856), and *Ophioglossum reticulatum* (5510), seeds of wheat, sorghum, rice (2297, 2464, 3181, 3989, 4492), corn (57), castor beans (3172), soya (4975), fresh and stored groundnuts (1984, 2765, 3220), stored sweet potatoes (3066), copra and cacao (1676), pecans (2572), rotting Brazil nuts (2515), damaged flue-cured tobacco (6282), ground black and white pepper (3644), tamari sauce (the original source of *A. tamarii*), rarely from other food products (4986, 5980), and coarse fodder (4548). It also occurred in the pulp of a paper mill (3294), on cotton fabrics (6326), *Spodoptera littoralis* (Lepidoptera) (2931) and was found to be associated with bees of different genera (402) and other insects (4744).

For growth and germination, the optimal temperature and water potential were found to be 33°C and −30 bars (261). It can attack cellulose and wool (4775, 4779, 6326), and tests for β-glucosidase also proved positive (4779, 6159). Amylase and proteinase were studied and separated chromatographically (5334), and lipase production found to occur on coconut oil emulsion media (1676); *A. erythrocephalus* can produce indole acetic acid (294), oxidize tryptamine to indole-3-acetic acid (1522), has a strong phosphatase activity (2731), and utilizes *o-, p-* and *m*-xylene as well as hydrocarbons from fuel oil (4239). *A. erythrocephalus* hydroxylates cinerone (5681). Urease production prior to conidiation allows efficient utilization of urea; urease may serve as storage protein (6547). Little selectivity for either L- or D-malic acid is shown but there is a preference for L-tartaric over D-tartaric acid (2052). Production of kojic acid (2730, 4458) on C sources, including mono- and polysaccharides and compounds containing three and more carbon atoms (530), has frequently been demonstrated, while rice koji is not browned (4046). Ascorbic acid is also produced (3231). Some antibiotic activity against *Pythium irregulare* (2945), *Escherichia coli* and *Staphylococcus aureus* has been described (5974). — *A. erythrocephalus* can also cause seedling virescence in corn (3073), seedling losses in soya (4975), a rot in apples (2581) and injuries of pine seeds on wounding (1975), and one isolate was found to be toxic to mice (4986).

Aspergillus flavus Link ex Gray *1821* — Fig. 40.

DESCRIPTIONS: Raper and Fennell (4744), Murakami (4047), Subramanian (5626), Hirayama and Udagawa (2464), Fandialan and Ilag (1685), and CMI Descriptions No. 91, *1966*; conidiogenesis: Cole and Samson (1109); SEM: Fennell et al. (1718). — The species of the *Aspergillus flavus* group are often treated as *A. "flavus-oryzae"* because they are difficult to distinguish. Several "micro-species" can be recognized (4046, 4047) and this subdivision becomes important from a toxicologic standpoint. — Colonies reaching 3–7 cm diam in ten days at 24–26 °C on CzA, on MEA 6–7 cm, characteristically yellow-green. Conidiophores hyaline, 0·4–1·0 mm long and rough-walled. Conidial heads radiating; on larger conidiophores a layer of metulae supports the phialides, whilst on smaller ones metulae are absent, although intermediate conditions occur. The ultrastructure of the conidiophores has been studied by TEM (608). Conidia globose to subglobose, finely roughened to echinulate, mostly 3·5–4·5 μm diam and 1–3 nucleate (6514). SEM photographs of conidia (1591) show irregularly curved short ridges. Some isolates produce light to dark red or brown sclerotia, mostly 400–700 μm diam, and more are formed at 34°C than 24°C (4047), especially with 3% sucrose and 0·5% $NaNO_3$ in the medium (2431). — A parasexual cycle with heterokaryosis, diploidization, meiotic recombination and haploidization has been described (4375), and seven linkage groups were identified by means of the parasexual cycle; crosses between markers assigned to the same linkage group indicated a low frequency of mitotic crossing-over (4376). — The main components of the fungal fat fraction were oleic (25·3%), palmitic (24·0%), linoleic (23·7%), and stearic acids (21·5%) (5351). The sterols cholesterol, ergosterol and 5,7-ergostadienol were found in aflatoxin producing and non-producing isolates (4720). Mycelia were found to contain virus particles in some isolates (674). — DNA analysis gave a GC content of 50–51% (5600). — Analysis of soluble proteins (47 fractions) from *A. flavus* and *A. parasiticus* did not give support to the assumption of a close relationship (3165). Enzyme patterns (4117) and pyrolysis gas-liquid chromatograms of conidia (6094) have been compared with other species.

This fungus is worldwide in its distribution but found mainly in tropical an subtropical regions. Because of its toxin production, occurrences on food have received particular attention. For selective isolation from soils, groundnuts and bread, MEA with 10 ppm botran (2,6-dichloro-4-nitroaniline) and 50 ppm streptomycin (434) or a basal minimal medium with 10%

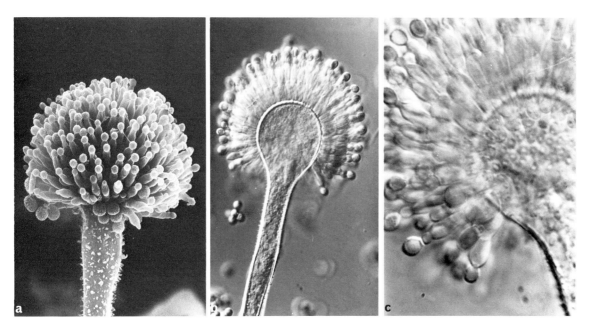

Fig. 40. *Aspergillus flavus*. a. Conidial head, SEM, × *c*. 500, orig. R. A. Samson; b, c. biseriate conidial heads, NIC, × 500 and × 1100.

NaCl and rose bengal (2103) have been recommended. A diagnostic medium consisting of tryptone, yeast extract and ferric citrate has been described (661) and an immunofluorescence test developed to detect it in seeds (6200). — Reports of its geographical occurrence are too numerous to be listed in detail here. About one third of all reports come from either India or the temperate zone and it has been found in such extreme areas as soils in southeast Alaska (1171), an ice-free antarctic lake (5637), mangrove mud (4700), Pamir soil at 4000 m elevation (3850), wet mountain grasslands in Sri Lanka (4021), and extremely dry areas in Chile (1824). The majority of records are by far from cultivated soils. In soils under groundnut in Virginia/USA the fungus did not exceed 10^2 propagules g^{-1} soil (2104). Some reports indicate a tolerance of relatively high salt concentrations, for example, isolations from salt-marshes (8, 413, 4000), saline soil in Kuwait (4001), the eulittoral range of the Florida coast (4918), and estuarine sediments (655, 4477, 5316). It has also been isolated from activated sludge (1387), fields treated with sewage sludge (1163) and from river and pond beds (3868). In soil, the distribution does not seem to be limited by pH or soil depth; *A. flavus* was frequently isolated from samples taken at 45 cm depth (1317, 3865, 5000, 5349) and colonized hair baits exposed to soil in the range between saturation and −140 bars (977). It has a relatively high competitive ability (2791). — It has been found in the rhizosphere and/or rhizoplane of wheat (4727), particularly after urea foliar sprays (45), and in that of *Trigonella foenum-graecum* with increased occurrence after foliar application of the antibiotic subamycin (2187); further in the rhizospheres of barley, oats (3003), corn (4555), rice (2712), *Ammophila breviligulata* (6414), cotton (6511), sugar beet (4559), *Aristida coerulescens* (4083), *Abutilon indicum* (3864), sugar cane (2861, 4886), *Tephrosia purpurea, Cassia tora, C. obtusifolia* (3869), *Euphorbia* spp. (3866), healthy and virus-infected *Carica papaya* (5345), *Coffea arabica* (5313), tomato, onion, *Abelmoschus esculentus,* radish, coriander

(1523, 3733), groundnuts (2532), peas (3870), *Pennisetum typhoideum* (3867), *Dichanthium annulatum, Bothriochloa pertusa, Setaria glauca* (3271, 3272), various steppe plants (3376), some bryophytes (2860) and in the phyllosphere (2859) and "rhizosphere" of some ferns (2856, 2858). Many isolations originate from seeds immediately after harvest or after storage; the fungus was found in thorough investigations on groundnuts in various groundnut-producing countries, on seeds of peas (6407), soybean (1370, 4975), *Cyamopsis tetragonoloba* (5350), castor bean (3172), green coffee beans (3302), corn (57, 3645, 3989, 4831), wheat (1752, 2297, 2716, 3989, 4492), barley (4038, 6199), rice (1337, 2464), sorghum (3989), *Pennisetum typhoideum, Sorghum vulgare* (57), common grasses (3512), cotton (2227, 5339, 6058) and tomato (5899), on olives (4545), sweet potatoes (3066), Brazil nuts (2515), pecans (2258, 2572, 3326), other nuts and dried fruits (6252), ground black and white pepper (3644), and various foodstuffs in Canada (6146) and Japan (4986, 5980). As to be expected from its frequent colonization of seeds, it has also been found in flour (3930, 5980) and dough products (2067, 3153). Only occasionally has it been found on litter, for example, leaves of *Eucalyptus maculata* (1558), green and decaying leaves of cotton (5009, 5231), decaying shoots of *Bothriochloa pertusa* (2955) and rotting tubers of manihot (1568). Other reports include coarse fodder (4548), damaged flue-cured tobacco (6282), wood exposed to seawater (4546), maple leaves placed into running water (274), wood pulp (3294), earthworm cultures (1961) and casts (2857), nests, feathers and pellets of free-living birds (2575, 5067), excrements of arctic animals (4174), leather, cotton fabrics (6326), and frescoes of a monastery (2666). It has been found in the alimentary canal of man and elk (5943), and also commonly on insects (corn earworm, corn borer etc.) (2263) and on sclerotia of *Sclerotinia sclerotiorum* (4699). — *A. flavus* is influenced by a mycostatic substance produced by *Fusarium oxysporum* (4882), the toxic action of rubratoxin B (from *Penicillium rubrum*) with morphogenic effects (4808) and also by *Penicillium nigricans in situ* (5002). Condia did not germinate and hyphae failed to grow in non-sterile soil (2582). Self-parasitism by the formation of intrahyphal hyphae can occur (617) and growth be inhibited by cuticular lipids from insects (3079). *A. flavus* can serve as food for a variety of soil-dwelling invertebrates (4007, 5380, 5381).

The germination of conidia was not (self-)inhibited by high densities (5861), but in the range $0.1–12.5 \times 10^6$ conidia ml^{-1} the decrease in germination was linear (3924). An exogenous C source is required for germination, a mixture of amino acids was more suitable than glucose and NH$_4$Cl (4460, 4461). Conidia germinate at temperatures between 12° and 37°C, with a minimum water potential at -270 bars (3873, 4461); they do not survive -350 bars at 29°C and -270 bars at 45°C (5771); the growth-limiting water potential is at -310 bars, the optimum at -30 bars (1192). Optimal growth occurs within the broad range 25–42°C (517, 978, 1686, 1827, 2700), with a minimum of 17–19°C and a maximum of 47–48°C. Best pH for growth is 7·5 and for conidium production 6·5; light has an inhibitory influence (4292). — Of eight C sources tested, sucrose was superior for mycelial growth and conidium production followed by glucose; galactose was utilized; amongst various N sources, L-glutamic acid gave the best growth while ammonium nitrate favoured conidium production (4292). The following alcohols can be used for growth and sporulation: sorbitol, D-mannitol, dulcitol and glycerol; poor growth occurs on methanol, ethanol, isopropanol and butanol; none occurs on monocarboxylic acids, but succinic and citric acids are relatively suitable substrates (331). A strong preference for L-tartaric over D-tartaric acid is reported (2052). Amylases (1527, 2431, 2582, 2782, 4545), endo-1,3-β-xylanase (1753, 1757), rutinase (2342), pectinases (2239, 4238, 4545, 6089), proteinases (1284, 1527, 3049, 4545), urate oxidase (3202), tryptic enzymes, urease, asparaginase, lipase (1676, 2582), phosphatase (2731) and a moderately active phytase (5280) have been demonstrated. Chitin synthetase was assessed in cell-free particulate preparations (2728). Lauric acid is oxidized to undecanone-2 (3139), tryptamine to indole acetic acid (1522), and indole acetic acid is also produced in liquid culture (294). Ester formation occurs on decanol and pentadecane (4497). Cellulose (2718, 4186, 4292, 4775, 4779, 5076, 6326), straw from rice and *Pennisetum typhoideum* (4611) are degraded; the complex of cellulases has been separated into eight components (4293). A variety of plasticizers (acids, polyhydric alcohols, esters) can be used as C sources (473) and the methylthio group of the herbicide prometryne can be used as a sulphur source (4053). Urea-formaldehyde fertilizer can be degraded and used as N source (4011). The insecticide carbaryl is hydroxylated *in vitro* to 1-naphthyl-*N*-hydroxy carbamate and 4- and 5-hydroxy-1-naphthylmethylcarbamate (610). *A. flavus* has a high tolerance for tannic acid (1260, 3052), can grow on gallotannin (3315), sinigrin (4774), *n*-alkanes (4498), and utilize hydrocarbons from fuel oil, and *o*- and *p*-xylene (4239). Known metabolic products include riboflavin (1620), the volatile compounds 3-methyl-butanol, 3-octanone, 1-octene-3-ol, 1-octanol, *cis*-2-octane-1-ol (2866) and ethylene (2641), free D-amino acids (506), kojic acid abundantly (512, 530, 2431, 4048, 4458, 6388), oxalic acid (2730), β-nitropropionic acid (803, 6387), deferriferrichrysine (4050), asperflavin and anhydroasperflavin (2133), the ergoline alkaloids ergocryptine, agroclavine and elymoclavine (1609), the isocoumarine derivative asperentin (from an entomogenous strain) (2132), 4'- and 5'-hydroxyasperentin and a methylether derivative, the anthraquinone pigments physcion and questin (2134), sterigmatocystin derivatives (1113, 3231), and humic compounds including fulvic acid (6102). — An interesting feature of *A. flavus* is its capability of denitrification (6502) and growth-rate-linked heterotrophic nitrification (2029, 2529, 3622, 3905) on peptone, casein (1675), L-asparagine, L-glutamine and nitropropionate (1453, 1454, 2319, 3904); there was no nitrification on urea, $(NH_4)_2SO_4$ (1675), NH_2OH, $NaNO_2$ (1453), sulphur-containing amino acids (2319), and in the absence of sucrose or $(NH_4)_2HPO_4$ in sterilized soil (1455). — The growth of grasses can be inhibited by *A. flavus* (3272) as well as seedlings, hypocotyls and roots of several other plants (1221, 2276). — More pronounced and more important are the toxic effects on animals: mycosis of bees (402, 5863), the cockchafer *Melolontha melolontha* (2620), *Tenebrio molitor* (4809), *Laphygma exempta* (3628), *Cimex lectularius, Schizoneura lanuginosa* (3257), the pine beetle *Dendroctonus frontalis* (3915, 3916), *Pyrausta nubilalis* (5864), as well as fatal effects on rabbits and guinea pigs (2405). Later it was recognized that cereal grain and groundnuts contaminated with *A. flavus* were highly toxic to pigs, mice (792), tumbler pigeons (63), rats, poultry (63, 1799, 3222, 5158), wild birds (64) and cattle (86). Spawning rainbow trouts fed with mouldy cotton seed acquire epidemic liver cancer (6455). An acute lethal poisoning with mouldy spaghetti (6146) and a first case of maduromycetoma (3521) have been reported. Besides the antibacterial metabolite aspergillic acid (512, 1498, 6321), the discovery of the aflatoxins gave rise to intensive studies on their production within different genera of soil fungi (3167, 3874, 4458), within *Aspergillus* (2431, 2767), and within *A. flavus* itself (1385), which proved to be the most common and strongest producer; significant differences in the production of aflatoxins have been found in UV-induced mutants (3250). Generally, some 50% of the naturally occurring isolates of *A. flavus* produce aflatoxin as assayed in a duckling test; no correlation between this property and any other morphological or physiological character has so far been found in the numerous isolates investigated (2431, 3250, 4050). Active isolates show their greatest aflatoxin production on peanuts, cotton seed cake, rice flour medium (2430, 2431), and a synthetic medium containing 5% glucose, 0·3% $(NH_4)_2SO_4$, 1% KH_2PO_4, 0·2% $MgSO_4·7$ H_2O and trace elements (33); the influence of trace elements on aflatoxin production has been investigated (3515). Several review articles have been published on aflatoxins (648, 733, 1638, 2430, 4490, 6413), of which the types B_1, B_2, G_1, G_2, M, the hydroxy-derivatives B_{2a}

and G_{2a} (1520) and the derivative aspertoxin (4889) have become known. Aflatoxin can be produced in a temperature range of 20–40°C (414, 1386), with specific optima for the different compounds and derivatives (5093, 5483); reports on its production at low temperatures are contradictory (1386, 6144) but aeration appears to favour aflatoxin production directly rather than growth (2346). Phenolic amino acids and tryptophan can act as precursors in aflatoxin formation (33, 4077). — A substance with hemagglutinating action on rabbit erythrocytes (5823) and an anticoagulant, aspergillin O (86), have been described. A new toxin, flavotoxin, was isolated from a non-aflatoxin-producing strain (3009); other toxic fluorescent metabolites have been described as unsaturated fatty acids (381). — Though less commonly than *A. fumigatus*, *A. flavus* can cause pulmonary infections in birds which can progress into generalized infections; in man lung infections by *A. flavus* are rare, and endocarditis and bladder infections even rarer. *A. flavus* has been observed several times in the human ear and eye sockets; the extent of any involvement in skin infections is uncertain (2154, 4744). Larvae of the corn borer, *Chilo partellus,* are reported as hosts of *A. flavus* from India (231). — *A. flavus* caused fruit rot of *Litchi chinensis* (4615) and seedling losses in guar (5350) and soya (4975) after artificial inoculation. — The fungus has a relatively high tolerance for allyl alcohol and formalin (3991). It has been found to grow on agar with 40% sucrose (4001) or even to 25% NaCl (976, 1991, 3166, 4697, 5881) with the optimal growth occurring at <5% NaCl (3998). Growth is retarded in gas mixtures with 70% and 87% CO_2 (2020) and, while none occurs in 100% nitrogen (187), low oxygen partial pressures are tolerated (3814). *A. flavus* is sensitive to γ-irradiation (4027, 4585), depending on the age of the conidia or mycelium (3903).

Aspergillus fumigatus Fres. *1863* — Fig. 41.

DESCRIPTIONS: Raper and Fennell (4744), CMI Descriptions No. 92, *1966,* Hirayama and Udagawa (2464), Sugiyama (5633), Udagawa and Takada (5984), and Subramanian (5626); conidiogenesis: Cole and Samson (1109). — *Aspergillus fumigatus* group. — Colonies reaching 7 cm diam in ten days at 24–26°C on CzA, spreading broadly, thin, bluish green, with strictly columnar conidial heads. Pigmented conidiophores with clavate vesicles arising from clearly differentiated thick-walled foot cells (5365). Conidia globose to subglobose, echinulate, mostly 2·5–3·0 μm diam. — An earlier TEM study of the conidia (5633) was amended by a subsequent more thorough TEM investigation (1968) which showed that the crustose outer and innermost electron-dense wall layers are laid down after conidium separation; after freeze-etching, an intricate rodlet pattern becomes visible which includes localized raised areas with no underlying conidium surface showing (2422). Formation of phialides and conidia, and discharge of conidia have been studied by TEM (4039). During germination the outermost wall layer disintegrates and the inner one gives rise to the germ tube (859). — Heterokaryotic mixed conidial colours were observed on single conidiophores after recombination with an albino mutant (4743). Conidia are normally uninucleate but sometimes may be 2- or 4-nucleate (6124, 6514). The ultrastructure of hyphae has been studied (858). — DNA analysis gave a GC content of 53% (5600); the nucleotide ratios in both mesophilic and thermophilic isolates have been studied (1674). — Electrophoresis of the soluble proteins exhibits 48 fractions, the highest number of all *Aspergillus* species tested (4117). Enzyme pattern have been studied in this and other *Aspergillus* species (3165). The mycelial walls contain 7·23% glucosamine (3232).

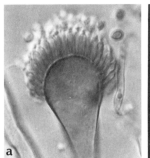

FIG. 41. *Aspergillus fumigatus*. a. Conidial head, CBS 192.65, NIC, × 1000; b. conidial head, SEM, × 1450, orig. R. A. Samson.

A. fumigatus is a thermotolerant fungus with a worldwide distribution. Due to a rather wide temperature range for good growth, it is not limited to habitats with permanently high temperatures, even though these are obviously very frequently reported, and incubation at 40°C facilitates isolation. *A. fumigatus* often contaminates cultures set up for the isolation of other thermophilic fungi (1174). — In investigations on the air spora it has been reported from India (5784); in that of the British Isles (1666, 2587, 2588, 3204, 3548) and France (3548), it was by far the commonest *Aspergillus* found, a fact which may be related to its cosmopolitan distribution. *A. fumigatus*, which can be easily identified, is listed regularly, but never as a dominating species in soils. Numerous reports, which cannot be cited here individually, are available from most parts of the world. The highest number of individual reports refer to plantations, garden soils and cultivated fields; *A. fumigatus* is apparently less frequently found in uncultivated soils (3868, 4030, 4698) and in grasslands (163, 166, 1551, 3865, 4538), but occurs from coastal (8, 161, 468, 4000, 5316) and marine habitats (413, 655, 1256, 4646, 4918) to mountainous areas (4021, 4569, 6082), and from a penguin rookery (6346) in Antarctica (possibly as a human contaminant) to non-acidic thermal habitats in the Yellowstone National Park/USA (5745). Remarkable is its repeatedly proven existence in peaty habitats: open bogs (1039, 6082), peat-podzol transitions (2502), peat soil (4180, 4474), and cut and uncut peat (1375, 1376, 1438, 3157, 4365); numerous finds in forest soils include nurseries (1978, 2482), stands with beech (1978), *Salix-Populus* (2007, 2008), floodplain hardwood (4226, 6192), wet-mesic forests (1040), teak (2854), pine (3447), and larch forests (3267), tropical dry mixed forests (4716), Himalayan forests (2858, 4995), and soils under *Acacia mollissima* (2186), and *A. karroo* (4407). It is also known from the litter layer (272, 3538), the F- or H-layer (5671, 6352), the mineral horizon (273), mull and mor type humus (2745, 4815), and extreme habitats such as stratigraphic drilling cores in central Japan at 1800–2100 m depth (middle Miocene!) (5633), sand blows (4540), sand beaches (3014), saline soils (4001), and mangrove swamps (3264, 4700). It has also been isolated from sewage (1163, 1165, 1166, 1170), activated sludge (1172, 1387), slime of a paper mill (5060), and fresh water (1166, 3809, 4429). Further habitats include an alpine soil with long snow coverage (3976), desert soils in Israel (405, 654), Egypt (100), Africa (2974) and the USA (4733), soils under steppe type vegetation (1559, 3415, 4719, 5049, 6347), with tallgrass (1632) and fescue (1702), heathland (3720, 3721, 5047), coal spoil tips (1664, 1665) and caves and mines (1703, 3367, 4137, 4312). Available data on the pH of the soils investigated reveal an occurrence in the pH range 5–8·5 with no distinct preference; soil depth between 0 and 45 cm also seems to be without influence. It seems to survive a cyanamide treatment of the soil

(3050). It has a high competitive saprophytic ability (2791). In rhizosphere studies, *A. fumigatus* has been isolated from pines (3572), *Coffea arabica* (5313), flax (3870), clover (3429), groundnuts (2532), onion (3733), *Pennisetum typhoideum* (3867), rice (2712), corn (4555), barley, oat (3003, 3006), wheat where its density increased after urea foliar sprays (45), several sand-dune plants (4371), steppe plants (3376), peat bog plants (1376), ferns (2856, 2858), around the rhizoids of moss (2860), roots of strawberry (4133) and broad beans (6134), leaves of peas (1373) and grasses during senescence (161), bolls (5009) and green and ageing leaves of cotton (5231). It is known from decaying plant materials, including *Pteridium aquilinum* (1821) and *Adiantum* (2859), barley, cabbage, potato (3113), *Ammophila* (163), pine (681, 2343, 2344, 2345), *Abies grandis* (681), *Picea sitchensis*, sclerotia of *Sclerotinia sclerotiorum* (4699), plant debris from birds' nests (168, 2575, 5067), birds' roosts and feathers (682, 2575), birds' droppings (1256, 1557, 2575), chicken pens (2156), alligator nesting material (5742), nests and storage places of gerbils (5393), garden compost (1424, 1557), mushroom compost at the end of its pasteurization phase (1723), straw and grass compost (953, 1528, 3546), composted municipal waste (2874, 3041, 3051, 5618), dung of cattle and horses (2338, 2409), wood chips (1756, 5445), self-heated hay (4821), self-heated corn (252), corn silage (1092, 5423), coarse fodder (4548), seeds of wheat, barley, oats, rice, sorghum and corn (57, 1751, 1752, 1755, 2297, 2464, 2716, 3181, 3989, 4492, 6199), grasses (3512), cotton (6058), groundnuts (5, 3220, 3702), castor beans (3172), pecans (2572), tomato (5899), cacao beans whose taste is spoiled (5138), stored sweet potatoes (3066), tobacco (1174, 4607), rarely in flour (4986), and common on cotton fabrics in the tropics (6326). — Soil aggregates are stabilized by the production of a hydrophobic metabolite in the presence of sucrose (2280). — The fungus can serve as food for collembola (4007) and mites (3104, 5380). The growth *in vitro* was found to be inhibited by *Memnoniella echinata* (4) and morphogenically influenced by *Pseudomonas aeruginosa* (570, 571). Hyphae, when incubated in soil, show a medium sensitivity to soil mycostasis (2559). The germination of conidia is reportedly inhibited on peat (1436).

Conidia require an external C source for germination (859). The propensity of hyphal tips to undergo apical bursting after osmotic or heat shock is explained by low resistance against the activity of their lytic enzymes (370). Good growth is obtained between pH $3 \cdot 7$ and $7 \cdot 8$ (2741). *A. fumigatus* is characterized by the ability to grow in the range between 12° and 57°C with an optimum between 37° and 43°C (1220, 1827); it tolerates temperatures during pasteurization to 63°C for 25 min (613, 3620) and causes heating to 50°C, for example in cracked corn (1174). Soil steaming seems to favour the fungus (6183). *A. fumigatus* grows and sporulates well in sun-heated temperate soil (2699); it is, however, strictly to be excluded from the thermophilic fungi because it also grows well at 20°C (1174). — Glucose, mannose, maltose, fructose, sucrose, xylose, arabinose, trehalose, lactose, cellobiose, inulin, starch, gluconate, mannitol, dulcitol and gum arabic, (less suitable dextrin, galactose, glycerol), peptone, casein, asparagine, alanine, glycine, phenylalanine, proline, cystine, methionine, isoleucine, serine, valine, glutamic acid, urea, cyanamide and nitrates were all found to support good growth (1375, 1758, 1827, 4212, 4329, 5107, 6124). No growth factors (6124), but about 1 ppm Fe and Zn ions are required in culture media (4862). Pectin is utilized (2239, 3051, 3538, 4779) as is cellulose in different forms, including straw from paddy, *Pennisetum typhoideum* (4611), oats, rye (4212) and wheat (4213), chitin (1425), tannin (1425, 3315) with the formation of gallic acid; fructosans are degraded, excluding those produced by *Acetobacter* sp. and *Bacillus* sp. (3404), and also 1,2-β-glucan (4780), mannan, mannose (4781), arabinoxylan (1757), catechin (3315), and sinigrin (4774). Sodium humate is reportedly utilized (499), but fulvic acid is not (3671). *A. fumigatus* causes losses in tensile strength of woollen fabrics (6330). Slow growth occurs on hair without the development of boring hyphae (1633); wood of aspen, birch (4191), spruce and pine is degraded (4247, 4248) in the latter with medium weight loss (4248); bark is degraded (4329). Also utilized are hydrocarbons from fuel oil, *o*-, *m*-, *p*-xylene (4239, 4935) and the herbicides atrazine (2892) and simazine (2894, 2906). DDT is degraded by cell-free filtrates of 7-day-old cultures (134), as are soft PVC materials of different molecular weight (5148). Proteolytic enzymes (654, 2719, 2741) with variability between isolates (2759), amylase, 1,3-β-glucanase, xylanase (1756), extracellular *a*- and β-glucosidases (1758, 4779, 4936, 4937, 5598, 6159), a specific guanyloribonuclease (2003), glyoxylate aminotransferase (1819), lipases (1676) and phosphatase with modest activity (2731) are produced. Metaphosphate may be utilized (4544), phenylacetic acid converted to homogentisic and *o*-hydroxyphenylacetic acids (5990), tryptamine oxidized to indole acetic acid and tryptophol (1522) and tyramine formed from *N*-acetyltyramine (2691). A strong preference for L-tartaric acid over D-tartaric acid is reported (2052). — The following organic acids are produced by *A. fumigatus*: citric (2730), orsellinic, epoxysuccinic (3231), kojic with variability between isolates (4458), melilotic and *o*-coumaric from 4-hydroxy coumarin and dicoumarol (829) and helvolic which has antibacterial properties (529, 940, 1230, 2002, 3748). The production of lipids on mineral media (4642) and the synthesis of fatty acids in relation to growth phases have been studied (4341, 6190). Riboflavin is also produced (1620, 4762). — Metabolites with antibiotic activity produced are gliotoxin (2002), and the sideramines fusigen (1380) and ferricrocin (6355, 6528). An antibiotic, fumigacin, active against both gram positive and negative bacteria has been isolated (6139, 6142), and this has also been found in or on human and animal pulmonary tissues. The benzoquinone fumigatin (151, 4339), biosynthesized via an orsellinic acid precursor (4529), and a number of related toluquinones, including phyllostine and spinulosin, have been thoroughly studied (3231, 6468). Other known metabolites are the anthraquinones emodin, 2-chloro-emodin and 2-chloro-citreorosein; the hydroxybenzenes *m*-cresol, orcinol, tri- and tetrahydroxytoluene; the hydroxybenzophenone methylsulochrin; the grisan trypacidin; the ergot alkaloids festuclavine, fumigaclavine A, B and C, elymoclavine, and chanoclavine I; and the sesquiterpene fumagillin (3231). Toxins, fatal to chicken embryos (1304) and causing tremor in various animals (577, 1561, 6474), as well as nephrotoxins (2405, 5817, 5823) causing fatal uremia, and ones with haemolytic properties (6460) are known to occur in *A. fumigatus*. — *A. fumigatus* is the most important causal agent of systemic mycosis other than *Histoplasma capsulatum, Coccidioides immitis* and *Ajellomyces (Blastomyces) dermatitidis* (4744). The most common sites of attack are the lungs and the respiratory tracts (246, 1399, 1571, 1624, 4309) where it can cause acute or chronic infections; it occurs in cattle, lambs, rodents, piglets, poultry where it once caused 90% mortality in chicken, wild geese and herring gulls (64, 425), and in lungs of pregnant cows it caused minute lesions (246). In man infection is rarely acute but after certain predispositions infection may become generalized and also attack the cardiovascular and urinary systems and the brain; it is also commonly found in the human ear but is then mostly regarded as a secondary colonizer although it can cause keratitis of the eye (2154); in skin infections it is generally regarded as secondary invader. In addition, it is a common agent of egg infection in poultry and mycotic placentitis and abortion in cattle (912) and horses (1182) and may occur as a facultative parasite in bees (2117). Like some other *Aspergillus* species it is a well-known allergen (4744). No correlation between the pathogenicity of numerous isolates and any morphological or physiological characters has been found (1507, 6124). — *A. fumigatus* was found to accelerate the growth of spruce seedlings (3588) and culture filtrates can inhibit the germination of cotton seeds (6058), the growth of barley seedlings (293), several bacterial cultures (1092, 5842), the germination of conidia of *Drechslera rostrata*, growth and sclerotium formation of *Verticillium dahliae* (920), and growth of *Venturia inaequalis* (6205). — Conidia have a medium, and mycelium a high, sensitivity to γ-irradiation (5748); *A. fumigatus* can grow in a 100% N_2 atmosphere in the presence of a reductant, biotin and thiamine (5680); 10% CO_2 is also tolerated *in vitro* (4029).

Aspergillus japonicus Saito *1906* — Fig. 42.

DESCRIPTIONS: Raper and Fennell (4744), Subramanian (5626), and Hirayama and Udagawa (2464). — *Aspergillus niger* group. — Colonies reaching 5–6 cm diam in ten days at 24–26°C on CzA, on MEA 7–8 cm, with a purple-black layer of conidia. Conidiophores 500–1000 μm tall, with globose or elongate, somewhat pigmented vesicles. Conidia mostly globose, 3·0–3·5 μm diam, strongly echinulate, as seen by TEM (2638), uninucleate (6514). White to cream sclerotia are common in some isolates. — DNA analysis gave a GC content of 54·5% (5600).

Distributional data indicate a preferential occurrence outside the temperate zone. It has been reported from soils in Japan (2464, 3267), India (2730, 2854, 3863, 4030, 4716, 4933), Pakistan (IMI), Brazil (399, IMI), the Bahamas (2006), Tahiti (6084), Sierra Leone, Ghana (IMI), South Africa (1555), and Turkey (4245). In India it was isolated moderately frequently from forest, cultivated grassland, and uncultivated soils (2179, 4933). It has also been found in the rhizospheres of wheat (3737) and ferns (2856), leaf litter of *Eucalyptus maculata* (1558), *Polypodium* sp. and *Cyclosorus* sp. (2859), on *Theobroma cacao* (IMI), fruits of *Anisophyllea laurina,* stored sweet potatoes (3066), the geocarposphere of groundnuts (IMI), and milled rice (2464).

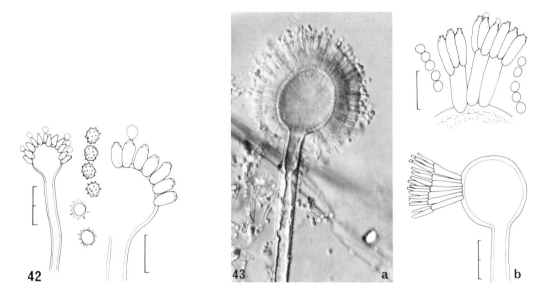

FIG. 42. *Aspergillus japonicus*, uniseriate conidial head and detail, CBS 568.65.
FIG. 43. *Aspergillus mellus*. a. Conidial head, NIC, × 400; b. conidial head and detail with conidia; CBS 546.65.

Conidia were produced under light over a temperature range of 22–30°C (2354). Production of β-glucosidase (2635, 2636), endo-pectin lyase and endo-polygalacturonase (2685, 2686) have been reported. One isolate was found to decompose cellulose in cotton duck (3618), while others did not (6331). Phosphatase activity (2731) and production of citric and oxalic acids has been demonstrated (2730).

Aspergillus melleus Yukawa *1911* — Fig. 43.

? = *A. quercinus* (Bain.) Thom & Church *1926*
  ≡ *Sterigmatocystis quercina* Bain. *1881*

Although the epithet *"quercinus"* predates that of *A. melleus*, the latter name is retained, as sclerotia were not mentioned in Bainier's original diagnosis and some doubt remains about its identity.

DESCRIPTIONS: Raper and Fennell (4744), Subramanian (5626), and Hirayama and Udagawa (2464); conidiophores in SEM: Fennell *et al.* (1718). — *Aspergillus alutaceus* group. — Colonies reaching 2·5–3 cm diam in ten days on CzA, on MEA 6–7 cm, creamy buff to pinkish cinnamon at first but later becoming ochraceous-buff from the presence of sclerotia; globose conidial heads somewhat paler. Conidiophores usually 0·5–2 mm but rarely to 3 mm tall, thick-walled, pigmented, conspicuously roughened. Conidia globose to subglobose, smooth-walled or irregularly roughened, 2·8–3·5 µm diam. The rodlet pattern of the conidial surface as seen in freeze-etch replicas is very prominent and often has raised bundles (2422). Sclerotia are sometimes absent but usually form in a dense layer; they are globose, somewhat elongate or irregular, *c*. 400 µm diam. — The lipid content of the mycelium has been studied (4642).

According to the available data, *A. melleus* appears to be widespread in soils of tropical and subtropical regions. There are reports from Israel (654, 2764, 2768, 2772, 4759); in India it was found to be rare in forest soils (2179). *A. melleus* is also known from Somalia (5049), South Africa (1415), Sierra Leone, Nigeria, Ghana, Egypt, Kenya, Sudan, India, Pakistan (IMI), Burma (2464), Peru (2005), and Argentina (6402, 6403, 6404, 6405). It is frequently found in the rhizosphere, geocarposphere (2768) and seeds of groundnut (2765, IMI), in the rhizosphere of *Cicer arietinum*, on seeds of soya beans (1599), rice and corn, in milled rice (2464), on dried salted food (4744), and isolated from *Dioscorea alata* and *Theobroma cacao* (IMI).

The temperature range for growth is 20–45°C, the optimum at 35°C (5028). Strong proteinase and lipase activities are recorded (4744). Metabolites include penicillic acid, ochratoxin A and B (1055, 2432, 5278) and the sideramine ferrichrysin (1532, 6528). Specific growth conditions are recorded for the production of both ochratoxin A (3212) and ferrichrysin (6355). At 10–20°C mainly penicillic acid, at 28°C more ochratoxin A was produced (1056). In addition, it produces mellein (ochracin), xanthomegnin, viopurpurin and viomellein (1508). *A. melleus* is used for several steroid conversions (4744). — It can attack and kill *Melolontha melolontha* (2620). Seeds of soya beans inoculated with *A. melleus* produced stunted and chlorotic seedlings (1599). *A. melleus* tolerates 40% sucrose (5028) or 20% NaCl in the medium (976), the latter corresponding to −210 bars osmotic potential.

Aspergillus niger van Tieghem *1867* — Fig. 44.

DESCRIPTIONS: Raper and Fennell (4744), CMI Descriptions No. 94, *1966,* Subramanian (5626), and Hirayama and Udagawa (2464). — *Aspergillus niger* group. — Colonies reaching 2·5–3 cm diam in ten days at 24–26°C on CzA, on MEA 5–6 cm, typically black powdery. Conidiophores arising from long, broad, thick-walled, mostly brownish, sometimes branched foot cells (5365), commonly 1·5–3·0 mm tall. Conidia in large, radiating heads, mostly globose, irregularly roughened, 4·0–5·0 μm diam, uninucleate (6514). — SEM and TEM studies of young conidiophores show the synchronous development of metulae and phialides (5732, 5848). The wall of the first-formed conidium is said to be continuous with that of the phialide while subsequent conidia are formed in a typical phialidic manner. Other TEM photographs of conidia have been published (5916). During germination, a new inner wall layer is formed which gives rise to the germ tube (2332). The rodlet pattern of the conidium surface as seen in freeze-etch replicas is prominent and interlacing with prominent localized raised areas (2422). Germinating conidia swell and shed their irregular outermost layer as seen in TEM (5915). The fine structure of hyphae has also been studied (5731). — Heterokaryotic conidial heads were observed after recombination with an albino strain (4743). — Some isolates produce sclerotia (4847), particularly at 30°C and to a lesser extent at 20°C; sclerotium production is inversely correlated to conidium production. — Some mycelia were found to contain virus particles (674, 2510). DNA analysis gave a GC content of 50·1–52% with nearly equal portions of adenine, guanine, cytosine and thymine (3232). — Cell wall hydrolysates contain glucose, galactose, mannose, arabinose, glucosamine and galactosamine (1223, 2790, 3232, 3640); glucuronic acid and 2-amino-2-deoxy-D-galactose also occur in young mycelia (2049). The cell walls and mycelial extracts contain the alkali-soluble polysaccharides nigeran and pseudonigeran (2535, 5941, 6515). A similar polymer is attached to chitin in the alkali-insoluble fraction (5517); conidiophore walls contain cellulose (2049). The cell walls contain at least 17 amino acids with serine, threonine, glycine, glutamic and aspartic acids dominating (3232, 3878). The lipid content of mycelia (4642) and conidia has been determined; conidia containing 4·7% (dry weight) with palmitic, stearic, oleic, linoleic and linolenic acids predominating (2169, 2312) and also choline sulphate (5696) as the major one of two sulphur-containing compounds.

The occurrence of this easily recognizable fungus is documented from all parts of the world, but, in contrast to many other *Aspergillus* species, not more than 50% of the finds are within the tropics. It nevertheless is more common in somewhat warmer regions and on south-exposed sites (4181, 4847). — Selective isolation is possible by adding 20% tannin to the nutrient medium (4520, 4847). — Considering the distribution with respect to climate, vegetation and soils, a representative list should include the following: recently glaciated soils in Alaska (1171), lowland soils (2008, 3417), pasture and grassland (3417, 3445, 3863, 3865, 4021, 4538, 4933), dry soils with steppe type vegetation (1559, 4313, 5049, 6403), coniferous forests (140, 269, 3447, 6351, 6352), deciduous forests (without individual citations), truffle soil (3451), cultivated soils (without individual citations), heathland (3720, 5047), desert soils (100, 405, 654, 2974, 3415, 3996, 4083, 4084, 4733), sand dunes (4371, 6414), salt-marshes (8, 1958, 4000, 4646, 4987), estuaries (655, 4477, 4850, 5316), mangrove mud (3264, 4700), other marine environments (468, 1256, 3014, 4918), fresh (3809, 4229, 4429, 4850) and polluted waters (1163, 1166), and river and pond beds (3868); isolations are also reported from vineyards with high copper content (4921), sandstone (3127, 4736), frescoes of a monastery (2666), fields with disintegrated limestone (6294), gravel slopes (399), loess (4759), chernozem, rendzina (3413, 3414, 3418), and podzol (1021). — *A. niger* was found in soil depths from 0–45 cm, but possibly preferentially in the upper 15 cm. According to numerous reports, there seems to be no distinct influence of soil pH on its distribution in the pH range 4–8; there was an increase in relative numbers after soil fumigation (3629, 3631), after manuring and NPK fertilization (2762). It can tolerate high concentrations to 35 mg g^{-1} of the herbicides dalapon and 2,4-D on soil plaques (6050). — *A. niger* has been found on fresh litter of *Panicum coloratum* (1560), *Carex paniculata* (4644), bolls (5009) and decaying leaves of cotton (5231), *Spartina alterniflora* (1958), *Eugenia heyneana* (5513), *Adiantum* sp., *Polypodium* sp. (2859), pine litter (269, 681), as a late colonizer of senescent culms and leaves of grasses (161), decaying yam tubers (30), *Manihot utilissima* (1568), decaying fruit-bodies of *Serpula lacrymans* (3130), sclerotia of *Sclerotinia sclerotiorum* (4699), bark and leaves of *Citrus* (3988), wood of broad-leaved trees (1295), damaged flue-cured tobacco (6282), roots of corn (4553), sugar beet (4559), rice (1282), cotton (1965) and strawberry plants (4133). *A. niger* has been found in the rhizospheres of banana (2030, 2035), wheat (45, 4727, 5825), barley, oats (3003), corn (4555), rice (2712), sugar cane (2861), *Cynodon dactylon*, *Trichodesma amplexicaule* (3864), *Ammophila breviligulata* (6414), *Atriplex cana*, *A. frigida* (3376), *Pennisetum typhoideum* (3867), *Dichanthium annulatum*, *Bothriochloa pertusa*, *Setaria glauca* (3271, 3272), cotton, *Vicia faba* (6511), clover (3429), peas (3870), groundnuts (2532, 2768, 4735), *Aristida coerulescens* (4083), *Ophioglossum reticulatum* (5510), *Abutilon indicum* (3864), young *Euphorbia* plants (3866), poplar (3452), *Carica papaya* (5345), *Coffea arabica* (5313), coriander (1523), wilted pineapple plants (6210), cucumber, *Luffa cylindrica* (4726), *Polypodium* spp. and other ferns (2856), and rhizoids of *Riccia* sp. (2860). In the rhizosphere of *Trigonella foenum-graecum* it occurred in higher frequency after foliar spray

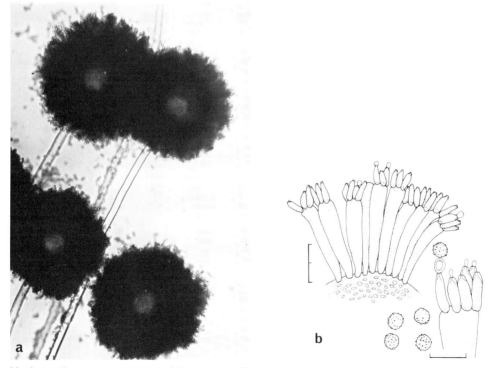

FIG. 44. *Aspergillus niger*. a. Intact conidial heads, × 50; b. part of conidial head with long metulae and short phialides, CBS 117.48.

with the antibiotic subamycin (2187). — Another plant substrate frequently colonized by *A. niger* are seeds, and it has been isolated from those of wheat (1752, 2297, 2716, 3989, 4013, 4492, 5293), *Avena sativa* (1752), *A. fatua* (2961), sorghum (57, 795, 3989), corn (57, 3989, 4831) when stored at or above 18·3% moisture (3069), rice (1337, 2297, 2464, 3181, 3584), barley (57, 1752), onions (5467), flax, peas, red clover (4013, 6407), green coffee beans (3302), groundnuts (5, 346, 1984, 2765, 2772, 3220, 3369, 3702, 4602), soya (4975), cucumber, *Luffa cylindrica* (4726), castor beans (3172), some common grasses (3512), and cotton (2227, 5339, 6058, 6407). It is frequently found on dried fruits and nuts (2258, 2572, 5092, 6252), in fruit juice (3428, 5205), on miso and starch products (4986), sweet potatoes (3066), onion scales (4447) and causes decay in litchi fruits (4615), apples (2581) and a well-known rot in stored onions where it is spread by mites; it is also rarely found on coarse fodder (3258, 4548). It can be a strong competitor on buried wheat straw (5961) and plant materials containing high residues of the herbicide diquat (6367). *A. niger* causes decay in cotton bolls after insect wounding, "smut" of white fig, and is a potential facultative parasite of potatoes (5227). It has also been isolated from queen larvae of bees (4631), the tick *Ixodes ricinus* (5018), earthworm casts (2857), rabbit dung (3267), nests, droppings and feathers of free-living birds (2575, 5067), buried sticks (1956, 5237), cotton cellulose (4186), grass compost (*Diplachne fusca*) at late stages of its decomposition (3546), raw and composted municipal waste (3041), sewage sludge (1165, 1167), commonly in the slime of paper mills (5060), on wood pulp (3294), detritus in fresh water (4429), maple leaves placed in running water (274), and boards in sauna baths (5016). The fungus is distributed via the air (969, 2587, 3204) and was shown to be a fast recolonizer of steam-sterilized soil (3330); it contributes to the water stability of soil aggregates (225). The spread on organic substrates is clearly related to both humidity (30, 977) and temperature (978, 2700). — *A. niger* is to be grouped as a soil invader rather than a soil inhabitant (4421). It is reported to be inhibited by *Chaetomium globosum* (956), *Penicillium nigricans* (5002), *Fusarium oxysporum* (527), *Memnoniella echinata* (4), *Trichoderma polysporum* (1017), culture filtrates of *Bacillus subtilis* (2393), coumarin at high (>50 ppm) concentrations (1391, 3055), and the antibiotic cochliodinol (704), while rubratoxin B (from *Penicillium rubrum*) has a strong morphogenic influence (4808). The germination of conidia *in vitro* can be inhibited by emanations from *Origanum majorana* (35). Dead conidia were relatively resistant to lysis in soil (1049). — The conidia are a suitable food for collembola and mites (4007, 5380, 6063).

The production of conidia depends on both the concentration and type of the C and N sources (4148); maximum production in liquid media with 1% NaCl (3773). It is induced by glutamate and stimulated by 3-phosphoglycerate, pyruvate (6036), Krebs cycle intermediates, nitrate and amino acids, but inhibited by high NH_4^+ concentrations (1859) and thiourea (1015). Enzyme activities of the TCA and glyoxylate cycles have been studied during different stages of conidiation (1860). Cu^{2+} and molybdate ions are also required for conidium formation (4035). Conidial pigmentation is inhibited by substances with S=O groups, particularly dimethyl sulphoxide (896). During early conidiophore development, glycogen and trehalose accumulate in the hyphae but at the time of phialide formation these levels drop and trehalase activity increases (4150). The optimal temperature for conidium swelling is 37°C (130, 3726). Respiratory enzymes are induced during germination (4296) and catalase activity in resting conidia is higher than in germinating ones (502), polyphosphates and phospholipids are mobilized (4195), the synthesis of phosphatase (503), esterase (3388), fructose-biphosphate aldolase, phosphoglucomutase and glucosephosphate isomerase in germinating conidia have been observed (504); germination was found to depend on CO_2 (4848) and to be optimal at 0·5% CO_2 (6032) and 1 molar sugar solutions (4988). The development of conidiophores directly from conidia (microcyclic conidiation) depends on the glutamate concentration and temperature (129); enzyme induction during microcyclic development has been tested (4149). — The minimum temperatures for mycelial growth are 11–13°C, the optimum in the range 17–42°C, and the maximum 47–48°C (261, 517, 1974); it survives 60°C in vitro (2974), 65°C during composting (3041) and has a thermal death point (in apple juice) at 63°C and 25 min. exposure (3620). The mean doubling time in submerged culture was found to be 5·9 h (5890) with filamentous growth occurring below pH 3 (1861). On agar, a water potential of −30 bars is optimal, and between −210 and −310 bars are tolerated (261, 976, 1192, 1475, 3166, 4697, 5881). Growth is optimal at 3% NaCl and still possible on 5% NaCl in liquid medium (3773). — Good C sources for mycelial growth are D-mannose, D-glucose, D-fructose, L-sorbose, cellobiose and methyl-β-D-glucoside; but D-arabinose (3328), D-mannitol, D-lactose and D-galactose (3603, 5387, 5552) are less suitable or ineffective; tannin is utilized as a C source (3127, 3315, 4184), to >20% in solution (1260, 3052, 4847) are tolerated, but the high tolerance of tannin can be lost in prolonged culture although it is restorable by passage over increasing concentrations of it (4181). In the carbohydrate metabolism of *A. niger*, a larger number of enzymes have been studied, including glucose-oxidase (1894) induced by glucose, fructose, mannose, D-xylose, L-arabinose and sucrose (1820), amylase induced by maltose, glucose, and starch (372, 1896, 3354, 3471), in the latter case without any accumulation of glucose (2782) and with variability between isolates in submerged culture (3284), glucosidases (2497) acting on myrosine (4259), salicin (3137, 6159), quercetin (2231), sinigrin (4260) and scilliroside (5598); exo- 1,4-α-glucosidase (4481), α-mannosidase (3681), β-mannosidase (5912); β-galactosidase (2497); α- and β-D-galactopyranosidase (3270), β-D-xylopyranosidases acting on aryl and alkyl β-D-xylopyranosides, oligomers of D-xylose, L-serine β-D-xyloside (4777), phenyl β-D-xylose and indolyl β-D-xylopyranoside (3231); α-L-arabinofuranosidase (2842), 1,3-β-glucanases (3231), 1,4-β-glucanase (1071), D(-)mandelate oxidase (4717), endo-1,3-β-xylanase (5701), exo-1,4-β-xylosidase (5702), pullulanase (4992, 4993, 4994); D-iditol dehydrogenase (1352), hydrolases acting on raffinose (2497) or phloretin (3835); polygalacturonases (2239, 3127, 3471, 4184, 4238, 5914), producing galacturonic acid as the main product (657), chitinase (2902) (chitin synthetase was assessed in cell-free particulate preparations (2728)), *m*-hydroxybenzoate-4-hydroxylase (4628), *o*-pyrocatechuic acid carboxylase (5620), a *trans*-glucosylase which transfers α-D-glucopyranosyl residues to phenolic or alcoholic hydroxy groups (2528); phenol oxidase acting on benzidine (934); lipase (1676, 2719), proteinases (658, 2719, 2948), ribonucleases, phosphomonoesterase and -diesterase (3231, 4067), phosphatases and polymetaphosphatase (915, 3343, 5280) with considerable variability between isolates (2731). — Good N sources for mycelial growth are urea (3127), NO_3^- (588), NH_4^+ and salts of nitrohydroxylaminic acid (5551), cyanamide (5107), alanine, arginine, aspartic acid, glutamic acid, glycine, hydroxyproline and proline (5553), but no growth occurs on nitrite, hyponitrite, hydrazine or azide. *A. niger* has been used extensively for trace element studies for about fifty years; it requires ions of iron, copper, zinc (653, 988, 4176, 5553), manganese (4176, 5553, 6312), molybdenum (5051, 5550, 5553), boron (1953), vanadium (490), gallium (5551, 5553), and scandium (5552). Detailed studies revealed the function of Mo ions in activating nitrate reductase (5550) and of Zn ions in activating glucose-6-phosphate dehydrogenase (1957). Cerium and six other rare earth cations were found to interfere with glucose uptake and respiration (5712). — Growth is also stimulated by phenolic substances (2320), and sterols related to cholesterol or ergosterol among others (2734). In the presence of montmorillonite, biomass and citric acid production are accelerated (2212), but respiration was inhibited by concentrations >4% of the clay mineral in nutrient solutions (5604). — The fungicides PCNB (935) and nitrofungin (880), and the insecticides endosulfan

(1613), carbaryl (610) and bromophos (5558) are metabolized; the herbicides diquat and paraquat can be degraded (5447) and paraquat accumulated in the mycelium (5446); phenoxyacetic and β-phenoxypropionic acids are hydroxylated (575, 830, 1077), as well as MCPA and 2,4-D (1705, 1706); other hydroxylations are performed with 17-norkauran-16-one, 16-norphyllocladan-16-one (127), (−) and (+)-cavone (4204) and cinerone (5681). Mercury is methylated at concentrations below 40 ppm $HgCl_2$ (6120). Utilization of Ca-lignosulphonate (934, 3260), and humic and fulvic acids (499, 934) is reported. During the degradation of taurine a new sulphonic acid, isethionic acid, was found but this is not used as a C source (690). The fungus is able to grow on numerous plasticizers and related compounds (473), oxidize n-alkanes (4498), degrade several hydrocarbons from fuel oil (4239) and urethane rubber (251). It releases metallic and silicate ions from various minerals and rocks but not from montmorillonite and quartz (2378), liberates potassium from orthoklase and oligoklase in small amounts (4008), corrodes copper, steel and aluminium wire under certain conditions (85), dissolves iron from iron-containing minerals with considerable differences between individual isolates (189), and degrades and utilizes urea-formaldehyde fertilizers (4011). — The production of organic acids is well documented (5278); in particular, the conditions for the optimal synthesis of citric acid by it have been thoroughly investigated: C/N ratio, KNO_3 concentration (4017), suitable N source (281, 5078), temperature (4018), improvement by UV-induced mutants (5179), pH and age of culture (279), mitochondrial activity (58) and trace element requirements (280). High citrate concentrations can lead to the accumulation of oxalate (4019). There are also reports on the production of oxalic acid (280, 4016, 4017, 4018, 4509) and also on a correlation between this metabolite and the pathogenicity of A. niger to groundnut seedlings (1974). Other products include gluconic (4017, 4018), 5-keto-D-gluconic (3231), glutaric, trans-glutaconic, cis-aconitic, glycolic, glyoxylic, pyruvic, hydroxypyruvic, α-ketoglutaric (2443, 4713, 4714, 6156), galactonic, D-mannonic, saccharic (3231), acetic (4509), L-ascorbic (1866, 1935), and L-xyloascorbic acids (3231), ethanol (4509), 1-octen-3-ol, 3-octanone (2867), riboflavin (4762, 5662) in nutrient solutions low in Mg (1620, 5050) and humic acid-like substances (2879, 5240) with a high proportion of aliphatic compounds (5124). α-, β- and γ-hydroxy-butyrates occur when A. niger is grown on acetate or sucrose (4236); a glycolipid, mono-glucosyloxyoctadecenoic acid has also been identified (3214), homogentisic (3048) and/or 2-hydroxyphenylacetic acids (1707) are produced from phenylacetic acid; indole acetic acid (294), and tryptophol or 5-hydroxyindolyl-3-acetic acid from tryptamine (1522); tryptophan is converted to 2,3-dihydroxybenzoic acid which acts as a growth factor (5619). The biosynthesis of organic sulphur compounds (3231, 6475), and the occurrence of nitrogen-containing metabolites (3862, 3968) have been studied. Other metabolic products reported include kojibiose, isomaltose, panose (3231), aspergillin and flavosperone (= asperxanthone) (3439), auras-perone A, asperyellone (= asperenone) (2733, 3231), asperrubrol (4685), the antibiotic jawaherene (725), and the sideramine ferrichrome (6528). — Antagonistic properties have occasionally been described (2766, 2872, 4859), but seem not to be very pronounced. Aflatoxin can be detoxified, and its producton by A. flavus inhibited (791). A sporostatic factor, identified as nonanoic acid, occurs (1905), as does a nematicidal substance possibly identical with oxalic acid (3577), and an insecticidal metabolite, nigragillin (2689). The pentapeptide malformin A (which causes deformations in mammals (6504) and plants) was detected in vitro (576) and found in infected onion scales (1251). Culture filtrates inhibited the growth of seedlings of barley (293), soya (4975), Dichanthium annulatum, Setaria glauca (3272) and Picea excelsa (3588), and inhibited the germination of cotton seed (6058). A stimulatory effect was observed on the growth of protonemata in Funaria hygrometrica (2209). An indirectly beneficial influence on plant growth is the stabilization of soil aggregates

(224, 6228) by this fungus. — *A. niger* was claimed to induce a mycosis in *Melolontha melolontha* (2620) and was found to be toxic to larvae of *Tenebrio molitor* (4809), brine shrimps and chicken embryos (1303). It has also been isolated from both dead adult bees (*Apis*) and their mummified larvae (402, 2117), and *Phorbia coarctata* (Diptera) (3257). It is also known from paranasal sinuses of man in northern Sudan (5037), and is rarely isolated from lung tissue; any possible involvement in aspergillosis is uncertain. *A. niger* is more commonly found in the outer and middle ear; toxicity to pigs noted may be caused by oxalate (4744); it is also reported to cause death in mice (4831, 4986), and muscular convulsion and paralysis in rabbits when fed with conidia or mycelium (5943). — Both conidia and mycelium are relatively resistant to γ-irradiation (5748). Good growth is said to occur in an atmosphere of 10% CO_2 (4029), but the degree of inhibition at higher CO_2 concentrations is counteracted by increasing temperatures (2020). Contradictory results are reported for growth under anaerobic conditions, for example, both no development (1244) and good growth (5680) in an atmosphere of pure nitrogen; formation of intrahyphal hyphae was observed under anaerobic conditions (3813). It may be concluded from other studies that oxygen is required at a rather low level: *c.* 0·56 mg per 1000 ml in liquid media (3814) depending on the oxygen solubility in the medium (679), and at least 4·2% oxygen in the atmosphere (2020). *A. niger* survived when exposed to 100% oxygen at a pressure of 10 atm. for ten days (848). — It has a relatively low sensitivity to the fungicide benomyl (1541) and the fumigants allyl alcohol and formaldehyde (3991).

Aspergillus oryzae (Ahlburg) Cohn *1884* — Fig. 45.

≡ *Eurotium oryzae* Ahlburg *1876*

DESCRIPTIONS: Raper and Fennell (4744), Murakami (4047), Subramanian (5626), Hirayama and Udagawa (2464), and Sugiyama (5633); conidiophores in SEM: Fennell *et al.* (1718). — *A. flavus* group. — Colonies reaching 5–6 cm diam in ten days at 24–26°C on CzA, on MEA faster, similar to those of *A. flavus* (at 34°C sometimes slower than at 24°C) and tending to be a light brownish green; conidiophores warty or rarely smooth, longer than in *A. flavus*; conidia slightly roughened or finely echinulate, 5–6 μm or more diam, 1–3-nucleate (6514). A TEM study of conidia has been carried out (5633); SEM studies showed either almost smooth (1591) or coarsely echinulate conidia (5848); the change in conidium ultrastructure during germination has also been studied (5730). — Sclerotia are only very rarely formed. — Heterokaryotic conidial heads were obtained after recombination with an albino strain (4743) and nutritional markers have also proved the occurrence of heterokaryons (2687), although these are unstable. — Members of the *A. flavus* group (the *A. oryzae* sub-group as understood by Murakami (4047)), the so-called "koji moulds", are important in food fermentation in Asia where they are used for making koji and its derivatives, sake, shoyu and miso, and also soy sauce. Numerous isolates were compared (4048) morphologically and physiologically and showed a great deal of intergradation. Isolates from fermenting food are mainly of the *A. oryzae*, *A. sojae* and *A. erythrocephalus* types (4051) and do not produce aflatoxin (4049); *A. flavus* itself occurs in these fermentation processes only as a contaminant. The isolates were subdivided by Murakami (4047) into eight groups according to their production of kojic acid, pink coloration of conidia on anisic acid-containing media, production of hydroquinone-monomethyl ether, and browning of rice koji. All these tests are mostly positive in *A. oryzae* (3183). The koji isolates produce fluorescent compounds other than aflatoxin (4052). — DNA analysis of the mycelium gave a GC content of 52·5% (5600); in dormant (and in parenthesis germinating) conidia, 28·2% (27·2%) adenine, 20·5% (22·0%) guanine, 24·7% (25·5%) cytosine, and 26·5% (25·4%) thymine were found (3232). — The following enzymes were detected in conidia: glucose and mannitol dehydrogenase (2534, 5384), β-fructofuranosidase, amylase, proteinase, urease, glucosidase, catalase, peroxidase, nuclease, lecithinase, and phytase (1527, 5384, 5639). Analyses of soluble proteins yielded 47 fractions, 25% of which were homologous with *A. flavus,* 19% with *A. parasiticus,* 17% with *A. fumigatus* and *A. alutaceus,* and 14% with *A. leporis* and *A. erythrocephalus* (3165). Based on the mobility of alkaline proteinases (4109), acid proteinase, pectin lyase and cellulase (4110) electrophoretic patterns, *A. oryzae* seems to be clearly distinct from *A. sojae.* Pyrolysis gas-liquid chromatograms (6094) of conidia are also clearly distinct from those of *A. flavus* (6093). — Mycelial walls contain 48% glucosamine and 54% neutral sugars, while in conidia about half those amounts were found (3232); dominant sugars are glucose, galactose and mannose (3640); hydrolysis of young mycelium gave 2-amino-2-desoxy-D-galactose; conidia contain glycogen (2049) and spore coats polysaccharides consisting of glucose, galactose, mannose, glucosamine, and a β-1,3-linked laminarin-like material in addition to phosphate, protein and nucleic acid (2533).

This fungus is reported from cultivated (3732, 4996, 4997), grassland (2179) and forest soils (2854, 4995) in India and also from the USSR (2871, 2948), Czechoslovakia (1700), Japan (3267), Tahiti (6084), Peru (2005), Syria (5392), Italy (4538), and air in the USA (1166) and the British Isles (2587). It has been isolated from stratigraphic cores in central Japan down to 2700 m (Middle Miocene) (5633), the rhizospheres of wheat (45), cotton (264)

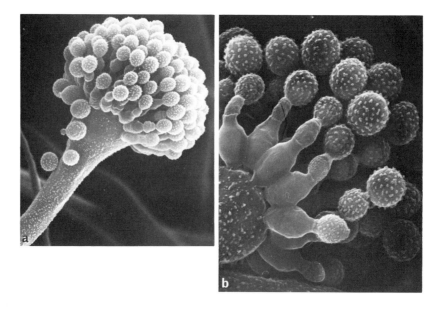

FIG. 45. *Aspergillus oryzae.* Uniseriate conidial heads, SEM, a. × 1000, b. × 2000, CBS 134.52.

and other cultivated plants (3737), milled rice in East Asia (2464), seeds of wheat (2716) and *Pennisetum typhoideum* (57), pods of groundnut (346), olives (4545), fermented food, dried cereals and legumes (5980), dried fruits and nuts (2572, 6252), nesting material of birds (5067), coarse fodder (4548), wood pulp (3294), wood exposed to seawater (4546) and cotton fabrics (6326). — Conidium germination is sensitive to soil mycostasis (6284) and the cell walls are hydrolysed by enzymes of a *Streptomyces* species (1049, 5408).

The optimum growth temperature is 32–36°C (4051) and the pH range is given with $\leqslant 2$–$\geqslant 8$ (2784). Conditions for the maximum production of amylases (3269) have been investigated, viz. variability of isolates (3284), media, pH, inoculum density, temperature (215, 1896, 3173, 3795, 5556), productivity of mutants (2978), influence of trace elements and suitable N sources (5556). A transglucosidase, effective in carbohydrate synthesis (4480), β-glucosidase, β-galactosidase, xylosidase, endo-1,3-β-xylanase and endo-1,3-glucanase (3231) have been described (2497) as has the degradation of β-glucosidic derivatives of N-acetyl-D-glucosamine (3163), sinigrin (4774) and scilliroside (5598). Glucose dehydrogenase production is induced by *p*-benzoquinone and hydroquinone; it catalyses the oxidation of D-glucose, D-xylose, D-fructose and D-mannose (292). Maltose can be hydrolysed to glucose and nigerose (3-O-a-D-glucopyranosyl-D-glucose) (4479). Pectin is degraded (5503) with galacturonic acid being the main breakdown product (657); cellulose is readily degraded (2948, 4186, 5076, 6326). The electrophoretic patterns of the cellulase and pectin lyase differ from those of these enzymes in *A. sojae* (4110). Tannin and gallic acid can both serve as C source (1260, 6083); a tannase (= tannin acyl hydrolase), producing glucose and gallic acid, was also described (2637, 6465). Proteinase (1284, 1715, 3022, 4091, 4092, 4093, 4211), glyoxalate aminotransferase (1819), aminopeptidase (cytosol) (4090), lysophospholipase, phospholipase D, phosphatases (1127, 2019, 5003), guanyloribonuclease (3231), two deoxyribonucleases (2890), a new nuclease (6006), aryl sulphatase (452, 4810) and general lipolytic activity are also recorded (4545). The degradation of sodium humate (499) and the utilization of tartaric acid (2052) have been described; oxalic acid can be utilized after pH adjustment to pH 6·4 (473), and of 25 polyhydric alcohols and esters (plasticizers) tested, 9 can be utilized by this species (473); the insecticide malathion is effectively degraded (3309). — Isoleucine, leucine and alanine were the most efficient N sources out of 14 amino acids tested (5717). *A. oryzae* produces 6-(a-D-glycosyl)-isomaltose and -maltose, the amino acids L-a-aminoadipic acid, betaine, and stachydrine, and furthermore urea on Czapek's medium (573), and nitrite and nitrate on peptone and/or ammonium salts (3622). L-Tyrosine is converted to L-3,4-dihydroxyphenylalanine (2251). Kojic acid is generally produced but to a lesser extent than in *A. flavus* (512, 530, 4048, 4458). Additional products include the hydroxybenzene maltoryzine, the piperazine derivative anhydroaspergillomarasmine, the porphyrin coproporphyrin I, uric acid (3231), pyruvic and ketoglutaric acids (4713), indole-3-acetic acid (294, 6115), the volatile flavour compounds 1-octen-3-ol, 3-octanone and 3-methylbutanol (2867), riboflavin (1620, 4762, 5662) and an uncharacterized staling substance with morphogenic properties (4425). Two fluorescent compounds are designated as asperopterin A and B (2876). *A. oryzae* can cause *Verticillium dahliae* to produce microsclerotia in the aerial mycelium (264). Isolates from tempeh or koji produced metabolites toxic to mice (2999); a strong producer of the toxin cyclopiazonic acid, originally published as *A. versicolor* (4255) has been reidentified as *A. oryzae* at CBS. Other isolates from foodstuffs in Japan, however, were not toxic to either HeLa cells or mice (4986). The metabolites hydroxyaspergillic (725), muta-aspergillic (4097), and β-nitropropionic acids (4096) exhibited antibacterial activity. Production of deferri-ferrichrysin (3231) also occurs but no aflatoxin is formed (2431). — *A. oryzae* is not pathogenic but it was isolated only once from brain tissue (4744). A stimulatory influence of culture filtrates on barley seedlings has been observed (293). Ions of the trace elements Fe and Zn are required for the optimal growth of *A. oryzae* (4862); it is still able to grow on a 4·1 M $MgSO_4$ salt concentration (2784). No growth was found in 100% nitrogen and sensitivity to carbon monoxide is assumed (5716).

Aspergillus parasiticus Speare *1912* — Fig. 46.

? = *Aspergillus sojae* Sakag. & Yamada *1944*, fide Raper and Fennell (4744) [but non sensu Murakami]

DESCRIPTIONS: Raper and Fennell (4744), Subramanian (5626), and Murakami (4047). — *Aspergillus flavus* group. A member of the *A. flavus* subgroup of Murakami (4047), who considers *A. sojae* to belong to the *A. oryzae* subgroup. — Colonies reaching 2·5–4 cm diam in 8–10 days at 24–26°C on CzA, on MEA 5–6 cm, with more pronounced green tints than in other species of the group. Conidiophores are coarsely roughened (smooth-walled in *A. sojae*), and metulae absent (the otherwise similar *A. toxicarius* Murakami *1971* is biseriate (4047)); conidia with prominently echinulate walls, 3–4-nucleate (6514), discoloured pinkish with anisic acid in the medium (remaining green in *A. sojae*). — SEM photographs of conidia (1591) show them to be coarsely spiny. Apical vesicles near the plasma membrane in germinating conidia have been demonstrated (2136). — Soluble proteins separated by polyacrylamide gel electrophoresis gave 43 distinct fractions, 22% of which were homologous with those of *A. flavus*, 19% with each of *A. oryzae* and *A. fumigatus*, 16% with *A. leporis*, 15% with *A. alutaceus* and 14% with *A. erythrocephalus* (3165). Conclusions from pyrolysis-gas-liquid chromatography were closely related to the conventional taxonomical placement of the species (6094).

A. parasiticus has occasionally been isolated from cultivated soils in India (1519, 2179, 3732), Japan (6210) and Italy (3913), and is also known from soils and air in Argentina (6403, 6404), activated sludge in the USA (1387), a steppe soil (1702) and a uranium mine (1703) in Czechoslovakia. It was found on rice (620), groundnuts (1984), pecans (2258, 2572, 3326, 5092), and in the rhizosphere of wilted pineapple plants (6210). Originally it was isolated from insects on sugar cane, subsequent records also indicate a frequent occurrence on insects including *Pseudococcus calceolariae* in Hawaii, *Diacrisia obliqua* in India, *Diatraea venosata* larvae in Sri Lanka (IMI), pupae of *Nomia* bees (402), a mealy bug in Trinidad (673), and the tick *Ixodes ricinus* (5018); it has also been found on feathers of free-living birds (2575).

The optimal growth temperature is 30°C, maximal aflatoxin production occurring at 25°C; optimal growth occurs with 10% glucose, but to 30% are required for high lipid and aflatoxin production (5286). Degradation of chitin (5019) and starch and proteolytic and lipolytic activities have been reported (1527, 2431, 5019). Culture filtrates may contain acetylated 2-amino-2-deoxy-D-galactan (2049). In an early report the production of an uncharacterized antibacterial principle "parasilicin" has been described (1130). Identified metabolites include the volatile flavour compounds 1-octen-3-ol, 2-octen-1-ol, 3-octanol, 3-methylbutanol (2867), kojic acid (512, 530, 3341, 4458, 5278), the pigment norsolorinic acid (3268), the aflatoxins B_1, B_2, G_1, G_2 (619, 3326, 4458, 6389), the B_1-related parasiticol (5617) and also parasiticolide A (2233). Conversely, aflatoxins are absent in *A. sojae* (4047). In *A. parasiticus* aflatoxin B_1 appeared as the precursor of all the other aflatoxins (3516), but it was also found to occur after B_2 (1356); highest production is obtained in still cultures at temperatures below

28°C with 20–30% glucose and 1% ammonium sulphate (5287). On stored rice at vapour saturation the toxin production (B_1, B_2, G_1, G_2) was highest at 30–35°C, at lower water potentials the temperature optimum was also decreased (622). Sterigmatocystin is produced and transformed to aflatoxin (2557) and averufin synthesized from acetate was identified as a precursor of aflatoxin B_1 (3339), B_2 and G_2 (3340); aflatoxins were found in inoculated rice when stored between vapour saturation and −140 bars water potential, and invasion of seeds and growth occurred even at water potentials of −350 bars (620). The absence of Mn^{2+} ions is

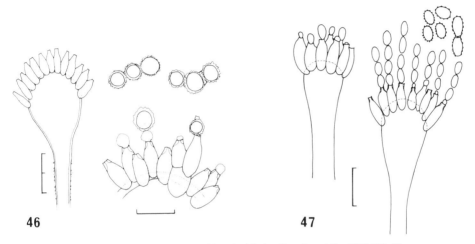

46   **47**

Fig. 46. *Aspergillus parasiticus*, uniseriate conidial head with detail and conidia, CBS 571.65.
Fig. 47. *Aspergillus restrictus*, conidial heads and conidial columns, CBS 116.50.

necessary for the induction of a yeast-like (arthrosporic) growth phase in shake cultures, both states producing aflatoxin (1357). Lack of O_2 inhibits aflatoxin production, and production is reduced at low oxygen levels (5285). There is a marked reduction in aflatoxin production when *A. parasiticus* is cultivated simultaneously with either *Eurotium chevalieri* (618) or *A. candidus* (621). Aflatoxin producing and non-producing isolates were found to equally contain cholesterol, ergosterol, and 5,7-ergostadienol (4720). — An acute lethal poisoning with mouldy Brazil nuts in a 45-year-old man was reported (604). — In contrast to other *Aspergillus* species, growth is restricted in the presence of 10% NaCl (3166).

Aspergillus restrictus G. Sm. *1931* — Fig. 47.

DESCRIPTIONS: Raper and Fennell (4744), and Hirayama and Udagawa (2464). — *Aspergillus restrictus* group. — Colonies reaching less than 1·5 cm diam after three weeks at 24–26°C on CzA, but growing more rapidly, 3–4 cm diam in 14 days on CzA with 20% sucrose, dark olive-green; conidiophores 75–200 μm long, hyaline and smooth-walled, with a flask-shaped or hemispherical vesicle, lacking metulae; conidial heads columnar; conidia narrowly ellipsoidal, echinulate, mostly 4·5–5·0 × 3·0–3·5 μm; uninucleate (6514). — DNA analysis gave a GC content of 52% (5600).

The scattered distributional data available do not suggest any particular adaptation to regions with higher temperatures: *A. restrictus* has been found in an ice-free lake in the Antarctic (5637), organic detritus in streams in Northern Ireland (4429) and acid mine drainage streams in the USA (1166). It has been reported from arable land in France (2161), soil in Australia (977), Honduras (2031), and India (4716) where it was found with low frequencies in forest, cultivated and grassland soils (2179), cultivated land in Pakistan (4855), salt-marsh in Kuwait (4000, 4001), and from Finland, the British Isles, Brazil, Argentina, Burma, Malaysia, Nigeria and Tanzania on various other substrates, including seeds of Gramineae (3512), such as corn, wheat and barley, *Colocasia esculenta, Theobroma cacao,* milled rice (2464), ground black and white pepper (3644), pecans (2572), cotton goods especially in the British Isles but not in the tropics (4744), moderately commonly on foodstuffs in Japan (4986, 5980), on dried sheep meat (IMI), fruit juices (5205), nests of free-living birds (2575), and in the air of Cambridge/England with a winter maximum (2587). In soil, *A. restrictus* requires water potentials between -140 and -350 bars (977, 4493) with a possible optimum at -270 bars (2095); very limited growth occurred on sunflower seeds below 6·5% moisture content (wet weight basis) (1033) and none on wheat seeds with a moisture content below 13·6% (1031) where a special medium with 18% NaCl, corresponding to -180 bars water potential, facilitates its detection (4744).

The optimal growth temperature is 25°C (978). Sucrose is utilized more efficiently than glucose (1018). Metabolic products include riboflavin (1620) and restrictocin which has some anti-tumour activity (4288). — *A. restrictus* reduces the germinability of wheat, peas, squash (2276) and corn seeds even at moisture contents below 15% and a temperature of 10°C; the infection of grain is apparently favoured by the action of weevils (4744).

Aspergillus sclerotiorum Huber *1933* — Fig. 48.

DESCRIPTIONS: Raper and Fennell (4744), and Subramanian (5626). — *Aspergillus alutaceus* group. — Colonies growing rather slowly on CzA, reaching 4–4·5 cm diam in 14 days on CzA or in ten days on MEA, with a creamy-yellow layer of conidia. Conidiophores usually 500–750(–1200) μm long, with light yellow, thick, echinulate walls; vesicles globose; conidial heads hemispherical or columnar, pale yellow; conidia globose, smooth-walled or rarely delicately roughened, 2–3 μm diam. Sclerotia borne singly, scattered over the colony, globose to subglobose, creamy white at first but later becoming flesh coloured, 1·0–1·5 mm diam. — In the related *A. alliaceus* Thom & Church *1926* the blackening sclerotia develop after several months into ascomata of the genus *Petromyces* Malloch & Cain *1972*.

The available distributional data show a preferential occurrence in subtropical and tropical zones. In India *A. sclerotiorum* has been reported from cultivated (1519, 2179, 4698, 4996, 5349), forest (1525, 4052, 4995, 5000) and uncultivated soils (2179), and in Israel from cultivated (2764, 2768, 2772, 4759) and desert (654) soils. It has been isolated from soil in Pakistan (2621), Brazil (399), Argentina (6403, 6404, 6405), Hawaii (mangrove swamps) (3264), Turkey (4245), Syria (5392), France (918), Germany (1827), and the Ukrainian SSR (5297). It is frequently observed to occur in soil depths down to 30–50 cm (1525, 4759, 5349), but it is also reported from surface soils. The fungus was isolated from the rhizospheres of groundnuts (2768), banana (2030) and corn (4555), and also from pods of groundnuts (346).

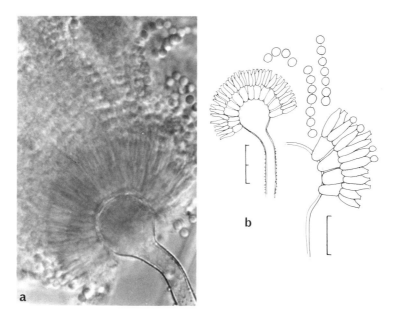

Fig. 48. *Aspergillus sclerotiorum*. a. Conidial head, NIC, × 1000; b. conidial head and detail with conidia; CBS 549.65.

Growth *in vitro* is inhibited by butanolic extracts from soils enriched with chitin (5453). *A. sclerotiorum* has also been found to be associated with fermented provisions of the alkali bee, *Nomia melanderi* (402).

The formation of sclerotia is inhibited by any oxygen deficiency (4701). Isolates from alkaline soils show a higher osmotolerance and produce more sclerotia at lower water potentials and lower H^+-ion (pH 7–10) concentrations compared to isolates from fertile soils (5781, 5782). Water potentials between −140 and −210 bars were tolerated in nutrient media (976, 4697). — Growth was better with starch, dextrin, glycogen or fructose as C source than with glucose (43). Metaphosphate can be utilized (4544) and *A. sclerotiorum* may attack wool (4775). It produces ochratoxin A and B, mainly at higher temperatures, with differences in efficiency occurring between isolates (2432); penicillic acid (1055), particularly at temperatures at and below 20°C (1056); and the antibiotics neohydroxyaspergillic (725, 6266), neoaspergillic (3796) and hydroxyaspergillic acids (4744, 5278). Riboflavin is also produced, the greatest yield being obtained in submerged culture on an oil-hydrocarbon and sucrose medium (4762). — *A. sclerotiorum* was found to cause a decay in apples and pears (4744).

Aspergillus sydowii (Bain. & Sart.) Thom & Church *1926* — Fig. 49.

≡ *Sterigmatocystis sydowii* Bain. & Sart. *1913*

The epithet is mostly cited as originally spelt *sydowi*, but should be corrected to *sydowii*.

DESCRIPTIONS: Raper and Fennell (4744), Subramanian (5626), and Hirayama and Udagawa (2464). — *Aspergillus versicolor* group. — This species is very close to *A. versicolor* but differs mainly in the intensely blue-green or greyish blue ("Delfts blue") pigmentation of the colonies which also grow slightly faster, reaching 3–4 cm diam in 14 days at 24–26°C on CzA and 4–5 cm on MEA; the conidia are also more strongly echinulate. — The lipid content of dried mycelium was determined as >5%, crude protein <27%, carbohydrates >70% and free fatty acids >8% (4642); to 1·36% sterol and 2% ergosterol contents of the mycelium are reported (4744). DNA analysis gave a GC content of 52·5% (5600).

A. sydowii, like *A. versicolor*, is a typical soil fungus with a worldwide distribution. It has been reported from such diverse habitats as recently deglaciated soils in Alaska (1171), alpine habitats (3445, 3976), palm stands in Malaysia (6046), mangrove mud in India (4700), and soil under chaparral type vegetation in California (1164). About half of all the available records come from India where it occurs in cultivated (1317, 1519, 1966, 2861, 3732, 3868, 4716, 4736, 4996, 4997), uncultivated (2179, 4030, 4698), and forest soils (2186, 2854, 4716, 4995, 5000), grassland (3863, 4933), and also humus from manure pits (968). In Australia it has been frequently found (977, 978, 3720), but it was comparatively rare in the alkaline soils of open-savannah in South Africa (1559); it has also been detected in estuarine sediments in North Carolina (655), in desert soil (4733), forest soil (2573, 3817) and forest nurseries (2482) in other states of the USA, an acid mine drainage stream, creek water, a sewage treatment plant, and a waste stabilization pond (1165, 1166). It is also known from various soils in Central America (1697), Brazil (379, 399), Peru (2005), Japan (2532, 5846), Singapore (3331), Argentina (6402, 6403, 6404, 6405), Pakistan (61, 4855), Kuwait (4000, 4001), Syria (5392), Egypt (8, 3993), equatorial West Africa (1420), Namibia (1827), Canada (540), the British Isles (1598, 5559), Austria (5678), Czechoslovakia (1700), the USSR (2871), Italy (4538) and Spain (3446, 3447); it has also been reported from a uranium mine in Czechoslovakia (1703), the air in India (969), and seawater and soil in the eulittoral zone (4918). There is no indication that soil depth influences its distribution, as *A. sydowii* is frequently reported to occur from the surface layers down to 45 cm (1317, 5349, 6046). Colonization of hair in soil has been observed in the range between vapour saturation and −280 bars water potential (977). — It recolonizes steam-sterilized soil relatively slowly (3330); a relative increase in numbers was observed in fumigated soils (3629, 3631). — *A. sydowii* is recorded in the rhizospheres of cotton (6511), groundnuts (2532), tomato (264), coriander (1523), *Cassia tora* (6067), *Brassica nigra* (3870), *Bothriochloa pertusa* (3271), *Ophioglossum reticulatum* (5510), wheat where it increases after urea foliar sprays (45), corn (4555), euphorbiaceous plants in all growth stages (3866), various steppe plants (3376), *Funaria* (2860), *Polypodium, Cyclosorus, Dryopteris,* and *Adiantum* (2856); it is also known on the rhizoplane of *Desmodium triflorum, Eragrostis elongata, Ischaemum indicum* (3331), on seeds of barley (57), wheat, corn, sorghum, rice (2297, 2464, 3989) and soya (4975), stored sweet potatoes (3066), copra and cacao (1676), litter and live cotton leaves (5231), decaying *Adiantum* and *Polypodium* fronds (2859), decaying shoots of *Bothriochloa pertusa* (2955) and other rotting plant remains (4548), in a termitarium (2857), on nests and feathers of free-living birds (2575), cotton fabrics, and leather in the tropics (6326), dead bees and combs of *Apis* (402), in apple and grape juice (5205), fermented foods in Japan (4986, 5980), and finally tobacco, drying food stuffs, and butter (2464).

FIG. 49. *Aspergillus sydowii*. a. Conidial heads and conidia; b. conidial head with intact conidial chains, SEM, × 2000; CBS 129.55.

Optimal growth *in vitro* is said to take place at −60 bars while minimum requirement *in vitro* is at −230 bars (3109) or lower (30% NaCl) water potentials (976). Certain clays added to the culture medium apparently enhance the total biomass, glucose utilization and CO_2 evolution (630). — *A. sydowii* can degrade starch (1425, 1827), pectin (2239) and cellulose substrates of various types (386, 1425, 4186, 4611, 4775, 4779, 6326); it produces a lipase when grown on coconut oil emulsion (1676), and a semi-alkaline proteinase (3231). It has an enzyme system converting sucrose into poly- and oligofructans (2049, 2900, 3404). Optimal glucose utilization or starch and protein degradation was found to take place at 25°C (1825, 1827). It can utilize both L- and D-malic acid but shows a strong preference for L-tartaric over D-tartaric acid (2052). *A. sydowii* can utilize or grow on sinigrin (4774), synthetic substrates such as tricresylphosphate (5859), polyethylenes and PVC plasticizer (5148). Up to 60% more lipid is formed in mass cultures than in flasks (4642). Hydrocarbons from fuel oil can be assimilated (4239) and tryptamine oxidized to indole-3-acetic acid and tryptophol (1522). *A. sydowii* produces riboflavin with the best yields when grown in submerged conditions on a hydrocarbon and sucrose medium (4762). Other metabolic products include choline sulphate (3231), sterigmatocystin (3929), the phenols orsellinic acid, *p*-hydroxycinnamic acid, 2,5-dimethylresorcinol, 6-methylsalicylic (2214) and sydowic acids (2237). Antibiotic activity against pathogenic fungi has been reported (569). It is the most efficient source of the flavour-improving thioglucosidase known (4744). — The fungus can inhibit mustard, pea and vetch seedlings (3852), induce the formation of chlamydospores in phialides of *Verticillium tricorpus* (264) and has rarely been found as a causal agent of onychomycosis (1195, 4744). — In a chronically irradiated pine-oak forest, it still occurred after exposure to ⩾1250 R per day (2009).

Aspergillus terreus Thom *1918* — Fig. 50.

DESCRIPTIONS: Raper and Fennell (4744), CMI Descriptions No. 95, *1966*, Subramanian (5626), and Hirayama and Udagawa (2464). — *Aspergillus flavipes* group with which the *A. terreus* group of Raper and Fennell may be best combined (4598). — Colonies growing rapidly, reaching 3–3·5 cm diam in ten days at 24–26°C on CzA, on MEA 6–7 cm, cinnamon to orange-brown or brown, velvety; conidiophores normally 100–250 μm long, smooth-walled, hyaline, with hemispherical vesicles; metulae present; conidial heads strictly columnar; conidia globose to slightly ellipsoidal, smooth-walled, mostly 1·8–2·4 μm diam, uninucleate (6514). — The sequential steps in the formation and wall structure of conidia have been studied by TEM (1775). Solitary or botryose globose lateral cells ("aleurioconidia") occur on the vegetative mycelium (4598, 4744) which are bigger (6–8 μm diam) and more regularly globose than in the other species of the *A. flavipes* group. Sclerotia are normally not produced but occur abundantly on MEA (light tan, 100–200 μm diam) in the var. *africanus* Fennell & Raper *1955*. — Heterokaryotic conidial heads were obtained after recombination with an albino mutant (4743). A high percentage of inter-strain heterokaryon incompatibility was found within *A. terreus* (921). — DNA analysis gave a GC content of 55–57% (5600). — Cell wall analysis yielded glucose, galactose, mannose and glucosamine as constituents (3640).

This is one of the *Aspergillus* species with a worldwide distribution in soils, but at the same time with a distinct prevalence in tropical and subtropical zones: it appears particularly commonly in India from which one third of the available recordings come. It is also known from Israel (654, 2764, 2772, 4759), Kuwait (4000, 4001), equatorial West Africa (1420), Syria (5392), Egypt (8, 100, 3993, 3996, 4962), Libya (4083, 4084), Spain (3417), France (242), Italy (5047), Turkey (4245), and Greece (4184); other areas include the Bahamas (2006), Honduras (2035), Peru (2005), Uruguay (3501), Brazil (398, 399), and Argentina (6402, 6403, 6404, 6405), and the USA are represented by isolates from Rhode Island (1958), Wisconsin (1205, 4226, 4313), Ohio (1166, 2573), Indiana (6414), Missouri (682), Florida (4918) and other southern states (2482, 3817, 4733). *A. terreus* is also known in soils in Australia (977, 978, 3720, 3721), Malaysia (6046), China (3475), Japan (2532), India (2186, 2854, 4995), Pakistan (61, 969, 2297, 4855), Bangladesh (2712), the Chad (3415), Somalia (5048, 5049), South Africa (1415), Zaïre (4159), Jamaica (4886), Tahiti (6084), the USSR (2871, 2872, 3003, 4474, 5297), Czechoslovakia (1703), France (4921) and the British Isles (4429, 4646, 4662, 5559). It is most frequently isolated from cultivated soils (40, 61, 968, 3868, 4716, 4962, 4996, 4997), ranging from paddy fields (1519, 2712) to citrus plantations (2764), coffee plantations (4159), legume fields (1317, 2164, 2772, 2871), cotton, vegetable (13, 2861, 2872), and sugar cane fields (2861, 3732, 4886) and vineyards (242) even with high copper content (4921) but also occurs in forest soils (very numerous reports), uncultivated soils (2179, 4030, 4698, 4855), under steppe type vegetation (1205, 3415, 4313, 5049), in desert soils (100, 654, 3996, 4083, 4733), sand dunes (4371, 6414), salt-marshes

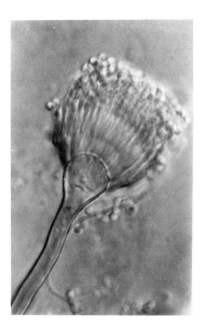

Fɪɢ. 50. *Aspergillus terreus*, conidial head, NIC, × 1000, CBS 383.75.

and estuaries (8, 1958, 4000, 4477, 4646, 4662), mangrove mud (4700), seawater from the eulittoral zone (4918), heathland (3720, 3721, 5047), peaty soil (5559), and peat bogs (4474). It seems to be relatively rare, however, in grassland soils (3363, 3417, 3863, 3865, 4933). More unusual habitats include coal spoil tips (1665), a uranium mine (1703), a river bed (3868), a waste pond (1166), and organic detritus from fresh (4429) and polluted waters (1166). Distribution within the soil profile ranges from the surface layers down to 45 cm (1317, 5349); soil pH does not seem to have any recognizable influence on its distribution. Steam-sterilized soil is recolonized at a medium rate (3330), and hair colonization in soil occurs between vapour saturation and −280 bars water potential (977, 978) with an optimum at 30°C (978). Only 5% of the conidia survive when exposed to 44–45°C and −270 bars (5771). — By far the most frequently listed micro-habitat of *A. terreus* is, however, the rhizosphere, and it has been found in that of wheat (45, 4727), barley (3003, 3526, 3870), corn (4555), rice (2712, 4726), oats (3003), potato (264), cotton (4744, 6511), *Brassica nigra* (3870), *Raphanus sativus* (3733), *Dichanthium annulatum*, *Bothriochloa pertusa*, *Setaria glauca* (3271), *Stipa lessingiana*, *Festuca sulcata* (3376), *Trichodesma amplexicaule* (3864), *Tephrosia purpurea*, *Cassia tora* and *C. obtusifolia*, particularly during flowering (3869), some euphorbiaceous plants (3866), *Carica papaya* (3869), pineapple (4744), banana (2030, 2035), groundnuts (2532, 2768), beans, *Origanum majorana* (35), *Aristida coerulescens* (4083), some sand-dune plants (4371), *Ammophila breviligulata*, *Andropogon scoparius* var. *septentrionalis* (6414), *Dactyloctenium egyptiacum*, *Digitaria bifasciculata* (2862), different ferns (2858), and *Riccia* sp. (2860). In contrast to some other rhizosphere fungi, its presence on *Pennisetum typhoideum* was not influenced by foliar sprays with leaf extracts of *Calotropis* and *Datura* (3867), but its occurrence was increased in *Trigonella foenum-graecum* after foliar spray with the antibiotic subamycin (2187). It has also been recorded from the leaves of some vascular plants (296), cotton bolls (5009), grass litter (166) and sclerotia of *Sclerotinia sclerotiorum*

(4699). It is common on stored seeds of crop plants, including sorghum (3989) when stored at high moisture levels (795), wheat, oats (1752, 3989, 4492, 4744, 4986), barley (57, 1752, 4038), corn (3989, 4831), rice (1337, 2297, 2464, 3181), cotton (6058), castor beans (3172), also stored sweet potatoes (3066), copra and cacao (1676), pecans (2258, 2572), groundnuts (5, 3220), nuts and dried fruits (6252), flour-type foodstuffs (5980), corn silages (1092), and rarely on coarse fodder (4548). The fungus has been found in nests of free-living birds (2575, 4650) and under (682) birds' roosts, on straw and grass compost (1528), *Spartina alterniflora* decomposing in a salt-marsh soil (1958), associated with wild bees at all developmental stages, hives and faeces (402), in lungs of small mammals (2153), on wood exposed to seawater (4546), pulp of a paper mill (3294), cotton fabrics and leather in the tropics (6326) and fresh faeces of man (5943).

The optimal temperature for linear growth is given as 35–40°C, the minimum 11–13°C, and the maximum 45–48°C (517, 1827, 4598), with an optimal growth at −40 bars water potential at 37°C (261), but −210 and −260 bars water potentials still allow growth (976, 5881); growth occurs in the pH range ≤ 2–≥ 8 (2784). In densely packed substrates (glass beads), CO_2 production is less than under conditions of good aeration (4457). Growth of the fungus *in vitro* was stimulated by the presence of corn and oat plants (1536), and root extracts of *Cassia tora* or *Crotalaria medicaginea* were more stimulatory than their root exudates (5638). — *A. terreus* shows little selectivity for either L- or D-malic acid, but a preference for L-tartaric acid (2052). Good growth is supported by the N sources peptone, casein (4212) and urea (3127); NH_4^+ ions being best assimilated at pH 8·4 and NO_3^- ions at pH 3·8 (2835); when grown on NO_3^--N or NO_2^--N, ninhydrine-positive compounds are released into the medium (4231). Nitrate has proved to be the most suitable N source for fat production by this species (5353); the fat contains high amounts of linoleic, linolenic and palmitic acids (5354). *A. terreus* is able to degrade starch (1827), pectin, tannin (3127), usnic acid (322), cellulose (2718, 3127, 3618, 4186, 4191, 4651, 4775, 6326), cellulosic plant materials including straw of oats or rye (4212), wheat (4213) and jute fibres (386). The fungus also has a strong ability to decompose arabinoxylan and CMC (1757); β-glucosidases are present (6159); lipase production occurs on coconut oil emulsion (1676); phosphatase activity was found to be strong (915, 2731, 6466) and specific phytase production comparatively weak (5280). Keratin (1633, 4662) and wool (4775) are also degraded. The carbohydrates galactose and mannose are excreted into liquid media (3640) and when grown in liquid culture with KH_2PO_4, $MgSO_4 \cdot 7H_2O$ and horn chips, NH_4MgPO_4 can be formed (2880). The influence of clay minerals on oxygen uptake has been studied (5604). Lauric acid is oxidized to undecanone-2 (3139), and the hydrocarbons decane, undecane, hexadecane, paraffin (3112) as well as hydrocarbons from fuel oil and *m*- and *p*-xylene (4239) are also efficiently oxidized. From 99 acids tested and their esters (plasticizers), 77 could be utilized, and also of 25 tested alcohols and esters 18 compounds could serve as C sources (473). The plasticizers dibutylphthalate (473) and tributylcitrate (3023) can also be utilized. The insecticide carbaryl is hydroxylated and its further metabolic degradation has been studied (610, 3379). — The metabolic products include unsubstituted dicarboxylic acids such as oxalic (845, 2730, 4705), succinic (845, 4705) and itaconic acids (845, 3397, 4002), the hydroxy acids itatartaric (5278) and citric acids (2730), phenol carboxylic acids such as gentisic, 3,5-dihydroxy-phthalic (3231), orsellinic (5278) and its dimethyl derivative (3231), 6-methylsalicylic (5278), and 4,6-dihydroxy-2,3-dimethylbenzoic acids (3231), the hydroxybenzaldehydes orsellin aldehyde (3231) and flavipin (4704), the hydroxydiphenyl ether asterric acid (5278), the hydroxybenzene orcinol (3231) and others (4089), hydroxybenzophenones such as dihydrogeodin, dechlorogeodin and sulochrin (5278), the geodin-related grisans erdin (844, 3601) and geodoxin (1247, 1248), the benzoquinone terreic acid (2881), the anthraquinones emodin and questin (3231), an isocoumarin derivative (2313), indole-3-acetic acid (294), patulin (2932), terrein (4705), terremutin (3819), terrecin (3231), citrinin (4705), several butenolides (4261, 4262), riboflavin (4762), L-proline (3862) and *m*-erythritol (5278). Antibiotic properties (716, 2126) have been ascribed to flavipin, erdin, geodin, patulin, terreic acid and the sideramine ferrichrysin (6528). An antiviral metabolite identified as a symmetrically substituted epidithiapiperazine-dione is also known (3820). *A. terreus* was found to inhibit the growth of several bacteria tested (1092), of *Trichophyton mentagrophytes, Sporothrix schenckii* (569) and the formation of microsclerotia in *Verticillium dahliae* (264). — The root and seedling growth of both *Setaria glauca* and *Dichanthium annulatum* can be inhibited by *A. terreus* (3272), but there is also a report of the stimulation of barley seedlings by culture filtrates (293). *A. terreus* causes a rot in apples (2581). Extracts of contaminated corn caused death of mice (4831) and HeLa cells (4986). It can also attack human skin and nails, and has been found as a parasite in the human ear (3501) and as a causal agent of bovine abortion (63, 4744), mastitis in cattle and aspergillosis in a parrot (63). Conidia and mycelium when fed to rabbits proved fatal (5943). *A. terreus* can attack larvae and prepupae of some bees (*Apis, Nomia*) (402). — The effects and mechanisms of ionizing radiation have also been investigated (2786, 5533). *A. terreus* was found to tolerate high concentrations (15 mg g^{-1}) NaCl on soil plaques (6050), growth on 4·1 M MgSO$_4$·7H$_2$O *in vitro* has also been reported (2784).

Aspergillus ustus (Bain.) Thom & Church *1926* — Fig. 51.

≡ *Sterigmatocystis usta* Bain. *1881*

DESCRIPTIONS: Raper and Fennell (4744), Sugiyama (5633), Subramanian (5626), and Hirayama and Udagawa (2464). — *Aspergillus ustus* group. — Colonies spreading broadly, reaching 4·5–6 cm in 10–14 days on CzA and 5·5–7 cm in 14 days on MEA at 24–26°C, olive-grey to drab. Conidiophores to 400 μm long, smooth-walled, brown; conidial heads loosely or broadly columnar at maturity; vesicles hemispherical or globose, upright, commonly pigmented, metulae covering most of the surface; conidia globose, roughened, 3·2–4·5 μm diam. Irregularly elongate, hyaline hülle cells typically present and scattered throughout the colony. — DNA analysis gave a GC content of 53% (5600).

A. ustus is one of the more ubiquitous soil-borne *Aspergillus* species (4744). A worldwide distribution is well documented, but to judge from 80% of all records its main occurrence is in the tropics and subtropics; with only a few exceptions it occurs locally at low frequencies. As with numerous other *Aspergillus* species, most isolations are reported from India; it has also been recorded in soil population studies from Pakistan (61, 969), Nepal (1826), Japan (2464, 2532), China (3475), Hawaii (3264), Polynesia (6084), the Bismarck Archipelago (3020), tropical Australia (977, 978), Central America (1697), Peru (2005), Chile (1824), Argentina (6402, 6403, 6404, 6405), Brazil (398, 399), Jamaica (4886), the Bahamas (4312), the USA (with numerous individual citations), Israel (654, 2764, 2768, 2772, 4759), Iraq (92), Kuwait (4000), Syria (277, 5392), Egypt (8, 3993, 4962), Libya (6510), Sahara, the Chad (3415), Somalia (5048, 5049), Zaïre (4159), South Africa (1559), Namibia (1827), Turkey (4245), Italy (270, 271, 3976, 4538), Spain (3417, 3446), France (242, 918, 2161, 3451), the British Isles (2923, 3918), Czechoslovakia (1700, 1702, 1703), Austria (1823), Poland (269), and the

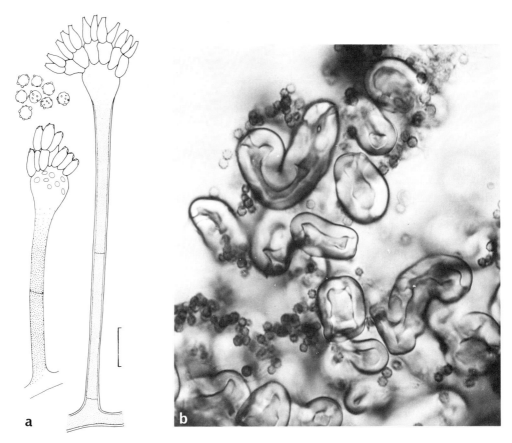

Fig. 51. *Aspergillus ustus*. a. Conidiophores and conidia, CBS 261.67; b. hülle cells, CBS 561.65, NIC, × 750.

USSR (2871, 2872, 3364, 3850, 4548, 5297). Records from cultivated soils dominate, including fields cropped with wheat and beet (918, 2161, 3446), paddy (1519), potatoes (2871), lupin, corn (4538), groundnut (2532, 2768, 2772), cotton, alfalfa (2871, 4744), sugar cane (4610, 4886) and under grape vines (242), citrus or coffee (4159); in the last case it was said to indicate a nutrient deficiency in the soil (2764). Finds in forest soils include ones in forest nurseries (2482), under pine (269), willow (2008), beech (270, 271, 2923), teak (2854), in Himalayan (4995, 4998), tropical (4716) and other forests (2851, 5000, 5048). It has been isolated from alkaline truffle soils (3451). There are also reports of occurrence in soils with steppe type vegetation (1559, 1700, 1702, 3415, 5049), desert soil (654, 3415), sand dunes (6414), uncultivated land (2179, 4698), estuaries (655), salt-marsh (8, 4000), and other maritime habitats (1256, 3264), alkaline soils (1559, 6510), Usar soils (4698, 2179), chernozem (3414), soils treated with sewage sludge (1163), a sewage treatment plant, a waste stabilization pond, an acid mine drainage stream (1166), and other unusual habitats, such as recently deglaciated soil in Alaska (1171), an alpine soil with long snow coverage (3976), carst (3364) and bat caves (4312), and a uranium mine (1703). — The soils investigated show an average pH range from 6–7·5, and a vertical distribution from 0–20 cm; the fungus apparently does not play any major role in the deeper soil layers but was reported from stratigraphic drilling cores in Japan down to 3200 m (5633). The colonization of hair baits in soil occurred in the range between vapour saturation and −140 bars water potential (977, 2095). In contrast to the relatively high number of soil isolations, there are few data on its distribution in plant microhabitats; examples are the rhizospheres of lupins (4400), groundnuts (2532, 2768, 4735), tomatoes (264, 3733), wheat (45), corn (4555), poplar (3452), and various steppe plants (3376), rhizoplane of sugar cane (4886), pine litter (681), cotton bolls (5009), leaves of several vascular plants (296), sclerotia of *Sclerotinia sclerotiorum* (4699); seeds of wheat (3989, 4492), groundnut (2765, 2772, 3220) and cotton (6058), pecans (2572), milled rice (2464) and coarse fodder (4548). It is also known on feathers of free-living birds (2575), rabbit dung (6085) and on cotton fabrics (6326), in the pulp of a paper mill (3294), leather and associated with nest contents of the alkali bee, *Nomia melanderi* (402). — Out of 52 fungi tested, *A. ustus* was the only one resistant to an antibiotic produced by *Stachybotrys chartarum* (823); it displays a relatively low sensitivity to soil mycostasis (2559). It was found to be an early colonizer of formaldehyde-treated soil (3908).

A. ustus sporulates freely in the light, but the production of hülle cells is entirely light sensitive (4045). The optimal temperature for linear growth is 25–28°C, the minimum 6–7°C, and the maximum 41–42°C (517, 1827); the thermal death point has been determined at 62°C and 25 min exposure (3620). It can grow on media with water potentials between −140 and −210 bars (976, 1475, 4697). — The utilization of sugar and the degradation of starch or protein are optimal at 25°C, weak at 15°C, but still good at 35°C (1825, 1827). Dextranase is produced but the production of the enzyme can be inhibited by the presence of other C sources (e.g. fructose, lactose, or some sugar alcohols) (2249). Cellulose in various forms is readily degraded (386, 3618, 4186, 4775, 6085, 6326). $(NH_4)_2SO_4$ is the most suitable N source (5674, 5675). *A. ustus* grows well on coconut oil (4744). When incubated with poplar wood blocks it causes 5·8% weight loss after 12 weeks (3767). It can utilize both L- and D-malic acid (2052). Three isolates used 68–74 compounds out of 99 tested acids and their esters commonly used as plasticizers (473); polyvinylacetate (5859), paraffin (4957), and simazine (2894) can also be utilized as C sources. The herbicide atrazine is dealkylated in the alkylamino group (2892) and a number of aniline-based herbicides are metabolized with aniline and/or chlorine ions as degradation products (2893). Scilliroside is cleaved to scillirosidin and glucose (5598). Tryptamine is oxidized to indole acetic acid (1522) and amino acid oxidases have been investigated (3900). *A. ustus* produces riboflavin (1620), pyridoxin (5677), and kojic (4458) and ustic acids (5278). Comparatively little ethylene is produced in a corn steep liquor and glucose medium (2641) but good fat production occurs on whey (4744). Culture filtrates contain a substance toxic to barley seedlings (293) and this, as well as conidium suspensions, reduce the germination and seedling vigour of cotton (6058); antibiotic activity against *Staphylococcus aureus* (3599) and also antimycotic activity against *Sporothrix schenckii, Scopulariopsis brevicaulis* (569), *Candida albicans* (3599), and *Sclerotinia sclerotiorum* (4699) have been reported; it was found to induce formation of chlamydospores in phialides of *Verticillium tricorpus* (264). When grown on cornmeal, several toxic diketopiperazines are produced, including prolyl-2-(1′,1′-dimethylallyl) tryptophyl diketopiperazine, 12,13-dehydroprolyl-2-(1′,1′-dimethylallyltryptophyl) diketopiperazine, 10,20-dehydro [12,13-dehydroprolyl]-2-(1′,1′-dimethylallyltryptophyl) diketopiperazine, austamide and 12,13-dihydroaustamide (5575). The main metabolite, austdiol, has been characterized as a benzopyran carbaldehyde (6110). In addition, versicolorin C, averufin and six xanthenones, called austocystins, were isolated (5577). — The fungus has a low tolerance for CO_2 (2096).

Aspergillus versicolor (Vuill.) Tiraboschi *1926* — Fig. 52.

≡ *Sterigmatocystis versicolor* Vuill. *1903*

DESCRIPTIONS: Raper and Fennell (4744), Subramanian (5626) and Hirayama and Udagawa (2464). — *Aspergillus versicolor* group. — Colonies reaching 2–3 cm diam on CzA and 4–5 cm diam on MEA in two weeks at 25°C; variable in colour, light yellowish, pink to flesh-coloured, ochre or orange yellow to yellowish green, with exudate and reverse of equally variable colour. Conidiophores colourless or yellowish, smooth-walled, to 500–700 μm long; vesicles elongate with metulae and phialides covering most of the surface; conidial heads radiating. Conidia globose, echinulate, mostly 2–3 μm diam, uninucleate (6514). Globose hülle cells sometimes present. A high frequency of inter-strain heterokaryon incompatibility was observed (921). — DNA analysis gave a GC content of 51·5–53% (5600).

An ecological and (mainly) physiological monograph of this species was published by Orth in *1973* (4318). — This is a particularly widely distributed species which occurs in almost all soils investigated, although often in small numbers (4744). While *Aspergillus* species are generally found predominantly in warm regions where they assume the role of *Penicillia* in temperate zones, *A. versicolor* penetrates farthest into the colder regions. Numerous observations relate to the most varied types of cultivated soils; there are fewer reports from forest soils. The following have been mentioned as extreme habitats: peat bogs (1376, 3918, 4474, 6520), an alpine pasture with *Nardus stricta* (3445), recently deglaciated soil in Alaska (1171), an ice-free lake in Antarctica (5637), salt-marsh (8, 413), estuarine sediments (655, 4477), mangrove mud (4700), desert soils (654, 1512, 4733), caves in the Bahamas (4312) and Italy (3453), carst caves in the USSR (3365) and loamy deposits of an ice cave (3367), and a uranium mine (1703). — *A. versicolor* has been reported from different soil depths: down to 20 cm (269, 2161, 2967, 5048), or 50 cm (273, 3447, 5349), and appears to be particularly frequent in the deeper layers (6182). Similarly, a very wide pH range is tolerated (5895) with a good tolerance of alkaline conditions (1559, 2179, 6510). The minimum water potential for vegetative growth in soil is at *c.* −350 bars and for sporulation −280 bars (2093); it was isolated from hair baits in the range between vapour saturation and −280 bars and found in cereals at a water potential of −230 bars (4318). The species was isolated from plots treated with farmyard manure more frequently than from unmanured ones (2762), and often occurs in garden compost (1425). Its frequency is said to be reduced in soils deficient in mineral nutrients (2764). After the fumigation of citrus soils, recolonization by *A. versicolor* took place rapidly (3629, 3631), and the fungus proves not to be very sensitive to soil-steaming (6183). It survived a cyanamide treatment (0·2%) of the soil (3050). Its hyphae are decomposed relatively quickly in the soil but the species can nevertheless be involved to a certain extent in the formation of stable soil aggregates (3633); a chemotropic reaction towards phosphate granules has been encountered, although they were not dissolved (3074). *A. versicolor* occurred with considerable frequency in the air in Cambridge/England (2587) and Paris, Marseille, Lyon/France (3548), and on frescoes of a monastery (2666). It is also known in more or less polluted river water (1154, 1155, 1157, 1482), a sewage treatment plant (1165, 1166), activated sludge (1387), seawater (4918), and wood pulp (6166, 6168). The species is also known on leaf litter of *Pinus radiata* (6080), decaying fronds of *Polypodium* (2859), rotting straw (953), sticks buried in soil (1956), the phylloplane of cotton (5231), aecidiospores of *Cronartium comandrae* on *Pinus contorta* var. *latifolia* (4608), the rhizoplane of banana (2035) and *Desmodium triflorum* (3331), the ectomycorrhiza of *Suillus luteus* with *Pinus strobus* (1794), the rhizospheres of pea (5825), alfalfa (3982), groundnuts (2532, 2768),

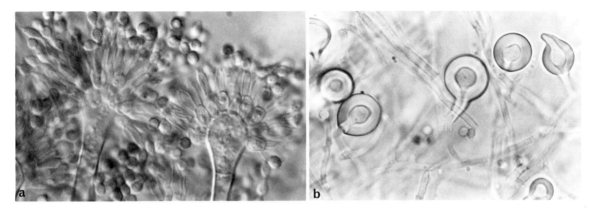

Fɪɢ. 52. *Aspergillus versicolor*. a. Conidial heads, NIC, × 1000; b. hülle cells, NIC, × 1000.

garlic (2159), barley (3003), oats (4443), corn (4553, 4555), rice (2712), flax (5826), oak (5394), tea (41), and some Australian heathland plants (5819). It is also known from cotton bolls (5009), frequently on stored grain (1752, 2097, 2297, 2716, 3181, 3645, 3989, 4831, 6273) when the moisture content exceeds 14% (4492), seeds of cultivated grasses (3512), tomatoes (5899) and castor beans (3172), fresh or stored groundnuts (1984, 2765), pecans (2572, 5092), mouldy hay (2081), coarse fodder (4548), dry fodder (3258), flue-cured tobacco (6282), fruit juices (5205), jam and various berries (4318), meat products and spices (2201, 3644), milled rice (2464) and other food products (3153, 4986, 5980). It is partly responsible for the rotting of military equipment and optical instruments in the tropics (2941, 4744). Other habitats include rabbit dung (6085), excrements of cattle and pigs (4318, 5983), nests, feathers and droppings of free-living birds (2575), and burrow tunnels (5393). The fungus has been isolated from the integuments of *Dendrolimus sibiricus* (2848) and was found to parasitize caterpillars of *Lymantria monacha* (2717). — It can serve as food for the Coleoptera species *Enicmus minutus* and *Microgramme arga* (5381).

Conidia germinate at temperatures between 12° and 37°C, and a minimum water potential of –290 bars (3873), but no germination occurs at 10°C and –100 bars (6344). Growth is possible with NaCl concentrations of up to 30% (976, 5881), and sucrose to 40% (4001, 4318, 5028) and *A. versicolor* can therefore be classed as very xerophilic. The minimum temperature for growth is 4°C, but little occurs below 10°C and the optimum is 21–22°C or 25–30°C (4687, 5028, 5895) depending on the geographical origin of the isolate (1827) and the maximum temperature is 40°C (4318). The formation of hülle cells is reported to be slightly light-sensitive (4045). Sporulation is favoured by glucose and sucrose but inhibited by urea; sporulation in liquid medium only occurs on a shallow layer in a stationary state (5895). — Starch (1827) is superior to cellulose or gum arabic as C source; galactose, glucose, fructose, maltose, sucrose and raffinose all support growth equally well (5894). ᴅ- or ʟ-malic acid can be utilized (2052). Decomposition of pectin is pronounced (3539) and of xylan very good (1432), but that of cellulose slight compared with many other fungi (386, 1425, 4213, 4775, 6085), particularly in tests using crystalline forms of cellulose (3618, 6326). Plasticizers are also attacked (473), including alcohols from C_{12} onwards (5518) and polyethylenes of decreasing molecular weight with increasing efficiency (5148); *A. versicolor* grows on *n*-paraffins up to C_{34} as sole C source (1841), utilizes hydrocarbons from fuel oil and *p*-xylene

(4239), but hardly grows with oil as a carbon source although lipase is present (4318); *A. versicolor* can grow on 0·2% sinigrin (4774) and slightly utilize atropine sulphate as the sole source of C and N (5108). Humic acids are also decomposed (3672); tryptamine is oxidized to indole-3-acetic acid (1522), five different aniline-based herbicides are degraded to aniline and/or chloride ions (2893); the action of β-thioglucosidase (4774), amylase, cellulase, lipase and proteinase was demonstrated (1827, 3944, 5896). *A. versicolor* is autotrophic for growth substances (1425) and can produce riboflavin (1620). Progesterone added to the culture medium was transformed mainly to testosterone (*c.* 24%), testolactone (16%), 4-androstene-3,17-dione (*c.* 13%) and hydroxy-6-androstene-3,17-dione (*c.* 6%) (11). Small amounts of the toxic cyclopiazonic acid (4255, 4256), cyclopiazonic acid imine and bisseco dehydrocyclopiazonic acid (4256) are produced, as are mannitol (4744), fructosan (3404), L-proline (3862), and ergosterol, the latter also when grown on *n*-alkanes (3337). The mycelium of *A. versicolor* contains numerous pigments (4744), especially anthraquinones (1611, 3231, 4744, 4818, 4867) and there are detailed reports for averantin (538), averufin (4676), hydroxyaverufin, averythrin, averufanin (238), with its methyl ethers, aversin (the methyl ether of versicolorin B) (778, 2512), norsolorinic acid (2236), versiconol (2323), versimide (742) with strong insecticidal activity (1112), the toxic and carcinogenic sterigmatocystin (779, 4318, 4687) also found in infested cornmeal, wheat, oat, apple juice, coffee beans, and sausages (4318), methoxysterigmatocystin, dihydromethylsterigmatocystin (778, 2322, 2503), demethylsterigmatocystin (1611), and sterigmatin (2234). Metabolic products found to have a high antibacterial activity include versicolorin A, B, C (2232, 2324), 6,8-*o*-dimethylversicolorin A (2321), and the siderochromes ferricrocin (6528) and ferrirhodin (3231). Versicolorin (1366, 1367, 1368) has specific antifungal properties. Simple culture filtrates, however, proved inactive against a series of test bacteria (716, 6365). — *A. versicolor* can cause a soft rot in peaches (4744). It was isolated relatively frequently from mycoses in domesticated animals (63) although it is clearly not an agent of keratomycoses (1992) and any certain pathogenicity has not been established (4744), but mycelial extracts proved toxic to chicken embryos (1388). — It can grow in a N_2 atmosphere and at very high CO_2 concentrations but with reduced vigour (4318).

Aspergillus wentii Wehmer *1896* — Fig. 53.

DESCRIPTIONS: Raper and Fennell (4744), Subramanian (5626), and Sugiyama (5633). — *Aspergillus wentii* group. — Colonies reaching 2–3·5 cm diam in ten days on CzA at 24–26°C, on MEA slightly less. This is one of the tallest *Aspergillus* species with conidiophores several mm long and radiating conidial heads to 500 μm diam, which are at first yellowish but later become brownish. Foot cells branched and thick-walled; vesicles globose, with metulae covering most of the surface. Conidia ellipsoidal at first but later becoming subglobose, smooth to broadly verrucose, mostly 4·5–5·0 μm diam, uninucleate (6514). TEM studies of conidia have been made (5633) and the rodlet pattern of their surface as seen in freeze-etch replicas is prominent and interlacing with slightly raised localized areas (2422). On CzA with 20% sucrose *A. wentii* grows faster and often sporulates more abundantly than on ordinary CzA. — Mycelium contains 8·9% glucosamine (3232).

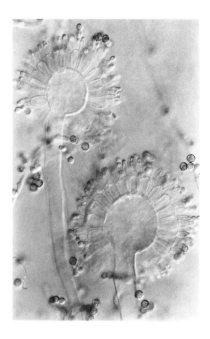

Fig. 53. *Aspergillus wentii*, conidial heads, NIC, × 400, CBS 226.67.

From the available distributional data it can be concluded that *A. wentii* is a rather common species with its main distribution in warmer climates. It has been found in cultivated soils in the USA (13, 4733), France (242, 918, 2161), Spain (3446), Italy (4538), Turkey (4245), Israel (654, 2762, 2768, 2772, 4759), Iraq (92), Egypt (3993, 4962), Libya (4085), Somalia (5048), South Africa (1559), Pakistan (4855), India (2179, 2854, 4030), Bangladesh (2712), Chile (1824), Peru (2005), Argentina (6403, 6404, 6405), the Bahamas (2006), China (3475), and Japan where it was isolated from stratigraphic drilling cores down to 1200 m (late Miocene) (5633). It has been isolated from uncultivated soil (4030), forest (2854), cultivated and grassland soils (2179); preferentially in dry soils (1824) including deserts (654, 4733); it was a comparatively rare species in an alkaline soil of an open savannah (1559); it is also known from peat bogs (4474), clay in a carst cave (3364), running fresh water as well as a waste stabilization pond, a sewage treatment plant (1166), estuarine sediments (655), and the eulittoral zone of seawater (4918). — Micro-habitats include the rhizospheres of barley, oat (3003), corn (4555), rice (2712), pineapple (4744), groundnuts (2768), and various steppe plants (3376), dried stems of *Coptis japonicus* (4744), leaf litter of pine and fir (681), seeds of groundnuts (2772), mouldy corn (3069, 3645, 3989, 4831), sorghum and wheat (3989, 4492), pecans (2572), decaying Brazil nuts (2515), tobacco (6326), birds (2575) and nests of gerbils (5393). It is a comparatively rare component of the air spora in Europe (3548).

The mean doubling time of the mycelium in liquid culture was determined to be 4·8 h (5890). It shows optimal growth on media with 6% NaCl or 40% sucrose (4697), corresponding to c. -50 bars water potential. — L- or D-malic acid can be utilized (2052). Starch is hydrolysed by conidia followed by an excess of glucose which appears in the medium (2782); pectolytic activity is particularly high on wheat bran (5503); amylase production varies between isolates (3284) and a dextranase produced is commercially exploited (4744). The majority of reports on cellulose degradation are positive (2006, 5076, 6326) and straw of both rice and *Pennisetum typhoideum* is degraded. When it is cultured in malt broth with salicin, very good production of β-glucosidase occurs (6159). Tryptamine is oxidized to indole-3-acetic acid (1522) and nitrate and nitrite are produced when the fungus is cultured with NH_4-N as N source (3622). Gluconic acid is produced together with smaller amounts of mannitol and glycerol on Czapek-Dox glucose solution (534, 3231). Preformed starved mycelium in contact with 0·01 M sodium arsenite in 0·1 M phosphate buffer produces ketoacids (4713). Further metabolites reported are citric (4744), kojic, 2-hydroxy-methylfuran-5-carboxylic (3929, 5278), and 1-amino-2-nitrocyclopentane-1-carboxylic acids (797, 3231), the latter with morphogenic effects and inhibiting chlorophyll synthesis in pea seedlings (714) and *Chrysanthemum* (6426); crude culture filtrates have been shown to be toxic to barley (293), wheat, pea and mustard (3852) seedlings. *A. wentii* can inhibit the growth of *Escherichia coli* and *Staphylococcus aureus* (5974). One out of five isolates tested produced aflatoxin B_1 (3167) and the production of traces of aflatoxin B_2 is also reported (5141). An intracellular toxin different from aflatoxin and with a strong action on chicken and mice has been isolated (6279, 6452). — The thermal death point was determined as 63°C for 25 min. (3620). The fungus was unable to resume growth after a ten days exposure to 10 atm of 100% oxygen (848).

Athelia Pers. *1822* emend. Donk *1957*

Athelia rolfsii (Curzi) Tu & Kimbrough *1978* — Fig. 54.

≡ *Corticium rolfsii* Curzi *1932*
≡ *Pellicularia rolfsii* (Curzi) E. West *1947*
≡ *Botryobasidium rolfsii* (Curzi) Venkatarayan *1950*

Anamorph: *Sclerotium rolfsii* Sacc. *1911*

DESCRIPTIONS: Tu and Kimbrough (5921), Stevens (5568), Goujon (2058, 2063), and CMI Descriptions No. 410, *1974*; teleomorph: Curzi (1252), and Kulkarni and Ahmed (3169); reviews: Jackson and Bell (2701), West (6299), and Aycock (256, 257); physiology: Watkins (6221), and Epps *et al.* (1643). — Colonies very fast growing, reaching about 9 cm diam in three days at 23°C, white, usually with many hyphal strands; primary conductive hyphae superficial, 4·5–9 μm wide, generally bearing clamp connections at widely spaced (about 240 μm distant) septa; secondary and tertiary hyphae narrower, 1·5–2·0 μm wide, usually lacking clamp connections, and often anastomosing. Sclerotia superficial, abundantly produced near the colony margin, ± globose, smooth-walled, cream to ochraceous or brown, 1–2 mm (average 1·2 mm) diam, consisting of a pigmented rind, faintly pigmented cortex and hyaline medulla of cells with unevenly thickened walls (993, 6371). Crustose basidial hymenia are rarely formed on agar media, and sometimes observed on natural substrates; basidia clavate to obovoid, 7–9(–15) × 4–5(–6) μm, sterigmata 2·5–4(–6) μm long, basidiospores obovoid to clavate, apiculate, curved 4·5–6·8 × 3·5–4·5 μm, but extremely variable in size and depending on the media used. Many colonies derived from single basidiospores will produce hymenia but some do so only after mating (CMI Descriptions No. 410). Because of the superficial hymenia with clavate basidia and the sclerotia, *Athelia* is the most suitable genus to accommodate the teleomorph.

FIG. 54. *Athelia rolfsii*, 2-week-old culture on MEA, CBS 214.52.

Two colony types have been distinguished: (1) type R forming relatively few sclerotia (80–100 per plate), and (2) type A with abundant sclerotia (500–600 per plate) after 6–7 days at 30°C on a synthetic medium (992). — The rather similar and equally polyphagous *Sclerotium delphinii* Welch *1924* has a less fluffy aerial mycelium, usually larger but more irregular sclerotia (average 2·2 mm diam) which often coalesce, are usually concave from below, and are supported by an erect hyphal strand; the medulla consists of longer cells. *S. coffeicolum* Stahel *1921* has a slightly slower growth rate and forms little aerial mycelium and much less radiating hyphae; the sclerotia are 2·3 mm diam on average and not differentiated into a cortex and medulla; setiform mycelial strands occur on host leaves (2062).

In *A. rolfsii* germ tubes arise from the sclerotial medulla (995) and penetrate the sclerotial wall at the darker spots (2058); these tubes consist of narrow (average 2·2 μm wide), clampless, rather short-celled hyphae, but after 15 h convert to wider (average 4·9 μm wide), rarely septate and clamp-bearing conductive hyphae from which narrower, clampless lateral hyphae subsequently arise (2058). Hyphal strands develop when the radiating hyphae encounter an obstacle or adverse substrates; efficient translocation is possible in both directions through these strands (2059). Sclerotia can be initiated not earlier than after 3–5 days, when growth is mechanically limited; in large dishes they appear after 8–11 days at the latest and over 15 cm from the inoculum (2060); mycelial growth can continue indefinitely through agar tubes (3031). The sclerotia themselves originate from coalescing branches of hyphal strands (5868).

The initial narrow hyphal cells contain about 10 nuclei, while those of radiating hyphae on average include 41. Vegetative nuclei divide perpendicularly to the longitudinal hyphal axis and several nuclei in a cell divide synchronously (758). — Hyphal and sclerotial walls were studied by X-ray analysis (991). Hyphal wall hydrolysates were found to contain 3·5% glucosamine and 67·5% neutral sugars (3232), including glucose and mannose (2802), whereas those of sclerotia had 1·6% glucosamine and 39·5% neutral sugars (3232). Hyphal walls can be digested by streptomycete culture filtrates or a mixture of chitinase and 1,3-β-glucanase (562). Electrophoretic patterns of total proteins showed differences between the hyphae and sclerotial initials on the one side, and young and mature sclerotia on the other (904). Palmitic and octadecadienoic acids were the major fatty acids (18 and 60%, respectively) in sclerotia (2555, 4563). The content of free and bound sterols (mainly ergosterol) in mature sclerotia is half that of the initials (4563).

A. rolfsii is a facultative parasite of stem bases, shoots and leaves of well over 200 vascular plants (256, 257, 6235). A vast phytcpathological literature has consequently been accumulated on it. It is found mainly in the tropical to warm temperate countries of all continents and is known from most states of the USA (2445, 6235), Canada, South Africa, the West Indies and Italy. It is also known from Germany and the USSR, various parts of Africa, Iran, India, Japan, Malaysia, Sri Lanka, Taiwan, Indonesia (6299), the Lebanon (1285), Nigeria (3702) and Bangladesh (2679). Soil isolates have been reported from open fallow soils (1315), mangrove mud (4700), and poorly drained soils (1078). *A. rolfsii* is of economic importance on various leguminous crops and numerous other cultivated plants including *Beta, Brassica, Citrus,* cucumber, tomato, banana, potato, rice, sugar cane and ornamentals (CMI Descriptions No. 410, 1285, 4963) but also on wheat and barley (2679), groundnuts (3702), and senescent leaves of cotton (5231). — A general decrease in activity with very high soil moisture has been observed (2024, 3669); poor development is recorded in fine-textured soils (1749, 2024), whilst acidic soils favour its growth (2024, 3669). It thrives particularly at 25–35°C (2024, 3669, 4059) with high moisture and preferentially attacks crowded plants on shady habitats, while in dry soils infection tends to occur further below the soil surface (CMI

Descriptions No. 410). — The principal propagules are the sclerotia which are disseminated by cultural practices and with seed; these can survive passage through the intestines of sheep and cattle (CMI Descriptions No. 410). A low sclerotial inoculum density can be discovered and assessed quantitatively by flotation in a molasse solution with specific gravity 1·073 and screening over a 0·25 mm sieve (3249, 4891); sclerotial viability is subsequently tested on a selective medium containing 160 mg gallic acid and 10 g potassium oxalate per litre (266). The mycelial state has been isolated down to 10 cm (5691) and sclerotia are mainly found in the upper 15 cm of soil (3249); sclerotia buried below 15 cm usually do not germinate (CMI Descriptions No. 410). The addition of inorganic phosphate to soil improved mycelial spread (2024), whilst ammonium nitrate (250) or 12 ppm of the herbicide atrazine (1239) acted mycostatically. *A. rolfsii* grows rapidly through soil with a comparatively high competitive saprophytic ability (4961) and colonizes buried baits within one week at 28°C (249). A few days of dry storage increases the sclerotial germinability and periods of high moisture following drought are particularly conducive to extensive plant damage (5419, 6221). Sclerotia survive in soil in a temperate climate less than one month (6375), while under dry laboratory conditions a third were still viable after five years storage (5787). *A. rolfsii* is very sensitive to low temperatures and only a slight frost will kill the mycelium; dormant sclerotia can, however, withstand −10°C for a short period (2445). At low temperatures, survival can be limited by the action of antagonists (1749, 6221). The melanized sclerotial rind determines the resistance to microbial attack (562, 993). Drying promotes the leakage of nutrients from sclerotia and their subsequent colonization by micro-organisms; these can induce sclerotial decay within two weeks (5419, 5420). A "mycosphere effect" occurs around the sclerotia, and soils previously amended with sclerotial leachates or sclerotia showed an accumulation of bacteria and were inhibitory to mycelial growth (3347). Nitrogenous amendments of soil decreased the germinability of sclerotia indirectly by stimulating antagonistic micro-organisms at the soil/sclerotium interface (2391). The addition of oat residues to the soil enhanced the degradation of the mycelium but promoted the production and germination of sclerotia (3887). Cropping with susceptible plants (e.g. beans) maintains a much higher sclerotial population than wheat or barley (3249). An antifungal substance active against *A. rolfsii* has been extracted from chitin-amended soil (5453). Culture filtrates of *Bacillus subtilis* inhibited growth in vitro (2393). A *Lactarius* species (4438), *Trichoderma, Alternaria, Aspergillus*, and *Geotrichum* species, some actinomycetes and bacteria have been reported as antagonists (2260, 6275, CMI Descriptions No. 410), whilst some actinomycetes, bacteria and *Penicillium* species can stimulate both growth *in vitro*, probably by supplying thiamine (3973), and sclerotium formation (484), but this is not so in soil. The proportion of potential antagonists in one particular soil was found to be rather low with 0·2% of bacterial, 1·7% of actinomycete and 3·5% of fungal isolates, but all these were capable of suppressing *A. rolfsii* in greenhouse tests (3973). In mixed cultures on sterilized sandy loam with the most commonly observed antagonist, *T. viride*, the pH was raised and the production of alkaline phosphatase, a-glucosidase and β-fructofuranosidase by *A. rolfsii* reduced (4890).

For the production of the teleomorph, media poor in nutrients and temperatures fluctuating between 16 and 22°C are most suitable (3877). — Sclerotium formation can be induced by mechanical barriers limiting linear mycelial extension (6315) or by mechanical damage (2392) and the concomitant production of an unidentified morphogenic factor (2061, 2063). Sclerotium formation is stimulated by light (12, 3697, 5883) when five times as many are formed as in darkness; their average size, however, is half as large, and their lipid content is also reduced in light by 25% (4563). The optimal temperature for sclerotium formation is 30°C, and the optimal pH about 3 (12). Sclerotia can form at O_2 contents above 15% and CO_2

of less than 4% (2099). Acid and heat-stable staling compounds can induce sclerotium formation (2609). L-threonine at 10 mM is immediately transformed to glycine and serine and stimulates sclerotium formation, whilst 15 mM sodium oxalate suppresses it (3132), as do glycine and NH_4Cl, acetyl-CoA, and L-cysteine; it is concluded that sclerotium formation requires an increased supply of carbohydrate intermediates which are mainly supplied by the glyoxylate pathway (3132). 2–4 mM mercaptoethanol and some other mercaptans completely inhibit both sclerotium formation and at >4 mM mycelial growth (1042). D-Glucose and thiamine are required for sclerotium production, whereas D-glucose at concentrations above 4–5% delays their development (6221, 6314). The addition of 0·5% lactose to a D-glucose-mineral medium induces sclerotium formation and β-D-galactosidase synthesis, but this effect is inversely related to the glucose concentration (4267); 0·5–3% (v/v) ethanol delays and reduces sclerotium formation, and lower concentrations stimulate mycelial growth, but ethanol is unsuitable as a sole C source (6313). Similarly, lactose enhances and ethanol suppresses the translocation of substrates through the radiating hyphae (4266). The colony type R is particularly sensitive to suppression of sclerotium formation by 0·2% 2-deoxy-D-glucose and other substances, whilst type A is not affected by them at all (992). In general, substances which inhibit sclerotium formation (e.g. sodium acetate, cysteine) favour the production of aerial hyphae at the colony margins, whereas those inducing sclerotium production (e.g. lactose, ethanol, L-threonine) lead to an extensive development of mycelium within the substrate (2394). In 6–7 day-old cultures translocation of labelled phosphorus takes place from old to young, but after 10–22 days the direction can be reversed; light enhances the translocation (6359). Sclerotium germination is possible only near vapour saturation (CMI Descriptions No. 410); abundant germination (91–98%) occurred after storage at 25 to 35% r.h. (6220). Punctured sclerotia germinate more rapidly because the intact rind apparently prevents nutrient uptake (990). Germinating sclerotia exude sugars and amino acids into the medium (995). A complete inhibition of germination was noted at pH values above 10 (2390). Sclerotia germinate only at O_2 concentrations above 6% and CO_2 values under 10% (2099). Germination is stimulated by soil aeration (12), volatiles of alfalfa (3346, 3347, 3348), and pretreatment in 0·5% NaOCl solution (3348) but only slightly inhibited (5%) by immersion for 4 h in a 150 ppm solution of $Ca(CN)_2$, although the percentage germination decreases with increasing immersion time or increasing concentration of $Ca(CN)_2$ (6079). Vegetative growth from sclerotia shows a sigmoid pattern with a two day lag phase on all media tested (394). — The optimal temperature for growth is 25–35°C (12, 2445, CMI Descriptions No. 410), the minimum 8°C and the maximum 40°C, although 37°C appeared to be the maximum tolerated for continuous normal growth (2445). A. rolfsii shows an extremely wide pH range (1·4–8·3) when grown on a beef extract-peptone medium, but the pH is invariably adjusted by the fungus itself to about 4 (2445); optimal growth occurs at pH 3·0–6·5 and neutral or alkaline media retard growth; light (except blue-green and ultra-violet) stimulates vegetative growth (CMI Descriptions No. 410). Iodoacetic acid induces hyphal strand formation, and L-threonine the formation of lateral hyphae (2394). — Of the numerous C sources tested the following were found to be the most suitable: D-glucose, D-fructose, D-mannose, maltose, sucrose, starch and pectin; poor C sources include galactose, D-arabinose, D-mannitol and citric acid (12, 394, 3328), while cellulose and rhamnose were not utilized (6221). In another report, citric acid was described as an excellent C source as was pure lemon juice (2445). A. rolfsii can grow well on lactose but the production of an extracellular soluble glucan was reduced with increasing lactose concentrations (4268). Most carbohydrates stimulate mycelial respiration, with the exception of D-rhamnose and D-xylose; L-lysine is stimulatory and L-cysteine and glycine inhibitory; indolyl-3-butyric and indolyl-3-acetic acids strongly enhance respiration; the antibiotics streptomycin, terramycin, neomycin, gramicidin, amphotericin and penicillin all are stimulatory but nystatin is not (6128); on the other hand, chloramphenicol, cycloheximide and actinomycin-D all delay or prevent sclerotium formation (2063). Organic substrates such as peptone or L-asparagine are the most suitable N sources, followed by ammonium salts and sodium nitrate (12). A medium supporting optimal mycelial production contains 1·5 mM $MgSO_4$, 4 mM K_2HPO_4, 2 mM KCl, 12·5 mM NH_4NO_3, 0·1 ppm thiamine hydrochloride, 4% glucose and 2 ppm of each iron, zinc and manganese salts (6221). Thiamine is required and 0·1 ppm already supports optimal growth; *A. rolfsii* can synthesize the thiazole moiety and combine it with externally supplied pyrimidine (2898, 4860). A C_x cellulase complex has been separated into three components all with an optimum pH of 4·0 (392). Pectin and CMC can induce endopolygalacturonase, polygalacturonase and pectinesterase formation (393, 394, 2841). After growth on a bean hypocotyl medium, the following enzymes were isolated: exogalactanase, endomannanase, a- and β-D-galactosidases and endoxylanase (2814, 2839, 6041); in addition, an arabanase (1096), a-L-rhamnosidase (2840), phosphatidase B (5910), acid phosphatase, and peroxidase (994) have been demonstrated; up to six polyphenoloxidase isozymes were found to be present in young and mature sclerotia but (except one) absent in hyphae (994). Tyrosinase (but not laccase) is active during sclerotium development and enhanced by both blue and white light (3821). Nine enzymes of the EMP pathway, five of the phosphogluconate and four of the tricarboxylic acid cycles have been demonstrated, whereby oxalate regulates the carbon flux through a glyoxylate bypass (3687). A high oxalic acid production occurs on CMC (394) and is mediated by glyoxylate dehydrogenase (3689); oxalate may also partially account for its phytotoxic actions (256). A combined glucose/organic acid or glucose/amino acid medium promoted high oxalate accumulation during the early growth phase (3688). *A. rolfsii* is a good producer of L-proline compared with other fungi (3862) and also of indole-3-acetic acid, even without tryptophan as precursor (2172); it is antibiotically active against several bacteria (1291, 4859). *A. rolfsii* does not seem to have developed any host specialization, although amongst five isolates with different origins the pathogenicity was found to be strongest towards the original host, except for an isolate from potato which attacked all hosts equally well (5203); isolates from different host plants showed only little variability in their growth rate (5203). — At over 0·03% CO_2 growth decreases steadily and at 20% CO_2 linear growth was found to be reduced by 55% (1504). While low concentrations of O_2 hardly affect *A. rolfsii* directly, it is likely that other competing micro-organisms may be at an advantage under these circumstances (6221). The mycelium is highly sensitive to ammonia which is lethal at 50 ppm (3248); nitrite inhibits growth at 20 ppm (256).

Aureobasidium Viala & Boyer *1891*

= *Kabatiella* Bubák *1907* emend. Karakulin *1923*
= *Polyspora* Lafferty *1921*
= *Pullularia* Berkh. *1923* emend. Wynne & Gott *1956*

Type species: *Aureobasidium pullulans* (de Bary) Arnaud (= *Aureobasidium vitis* Viala & Boyer)

MONOGRAPHIC TREATMENTS: Cooke (1159), and Hermanides-Nijhof (2412).

Colonies spreading, smooth, soon covered with slimy masses of conidia, cream or pink to light brown or black. Hyphal cells hyaline, frequently soon becoming brown and thick-walled, broad (often exceeding 10 μm in width). Conidiogenous cells intercalary, lateral or terminal in ± hyaline hyphae. Blastoconidia produced synchronously in dense groups from minute denticles at the apex of distinct hyphal projections or directly from undifferentiated hyphae; less commonly, they are discharged into emptied hyphal portions ("endoconidia"). Conidia hyaline, smooth-walled, one-celled, of variable shape; often producing secondary smaller conidia in a similar way. Dark hyphal portions may act as chlamydospores or fragment like arthroconidia.

Species of the melanconiaceous genus *Kabatiella* Bubák *1907*, known as causal agents of leaf spots on various phanerogamous hosts, behave in culture like *A. pullulans*, and this genus has therefore been subsumed with *Aureobasidium* (2412) which then comprises 14 accepted species of which *A. pullulans* is the only ubiquitous one sporulating abundantly in culture by straight conidia mostly 9–11 × 4–5·5 μm. Another similar ubiquitous soil fungus which has lunate conidia has been transferred to *Microdochium* as *M. bolleyi* (Sprague) de Hoog & Hermanides-Nijhof (q.v.) (2412). — In the similar genus *Hormonema* Lagerb. & Melin *1927* (type species *H. dematioides* Lagerb. & Melin), phialoconidia are produced in basipetal succession from both hyaline and pigmented, mostly intercalary hyphal cells through 1–2 usually scarcely prominent collarettes.

Aureobasidium pullulans (de Bary) Arnaud *1910* — Fig. 55.

≡ *Dematium pullulans* de Bary *1866*
≡ *Pullularia pullulans* (de Bary) Berkh. *1923*
≡ *Hormonema pullulans* (de Bary) Robak *1932*
= *Aureobasidium vitis* Viala & Boyer *1891*
= *Pullularia fermentans* Wynne & Gott *1956*

For numerous other synonyms see Hermanides-Nijhof (2412) and Citerri *et al.* (1061); some other uncertain or deviating taxa were listed as synonyms by Cooke (1156).

DESCRIPTIONS: Hermanides-Nijhof (2412), Dennis and Buhagiar (1342), Cooke (1159), Joly (2794), Butin (811), Durrell (1510), Subramanian (5626), and Ramos and García-Acha (4721); SEM study: Pechak and Crang (4486). — The teleomorph is unknown; all teleomorphic species so far reported can be related to *Hormonema* Lagerb. & Melin *1927* (2412). — Colonies reaching 4·0 cm diam in 7 days at 24°C on MEA; vegetative hyphae to 12 μm wide, pigmented hyphae markedly constricted at the septa. Conidia hyaline, ellipsoidal but of very variable shape and size, straight, (7·5–)9–11(–16) × (3·5–)4·0–5·5(–7) μm, often with an indistinct basal apiculation; secondary conidia and endoconidia similar but smaller. — Two varieties can be distinguished (2412): var. *pullulans* with colonies remaining pink or light brown for at least three weeks, and var. *melanogenum* Hermanides-Nijhof *1977* with colonies rapidly becoming dark olivaceous-green or black and with pigmented hyphae often seceding like arthroconidia.

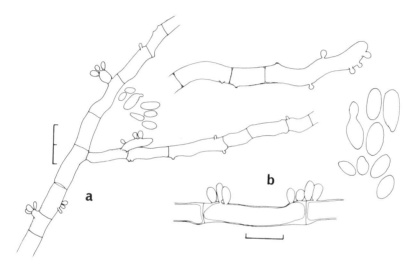

FIG. 55. *Aureobasidium pullulans*. a. Fertile hypha with discrete conidiogenous cells and sporulation from lateral denticles; b. detail of conidiogenesis and conidia.

A very similar species is *Aureobasidium lini* (Lafferty) Hermanides-Nijhof *1977* (≡ *Kabatiella lini* Lafferty *1922*) which grows specifically on *Linum* sp.; this was synonymized with *A. pullulans* by Levitin and Zhuravlev (3306) but has slightly larger conidia, 11–16 μm long, and pigmented hyphae hardly constricted at the septa. *Hormonema* species similar to *A. pullulans* include *H. prunorum* (C. Dennis & Buhagiar) Hermanides-Nijhof *1977* which reaches 4·0 cm diam in seven days only at 21°C (and less at 24°C) and has conidia 5–16·5 × 2·5–6·5 μm. Smaller conidia (4·5–)7–8(–12) × (2·5–)3–4(–4·5) μm and faster spreading colonies reaching 5·0–6·0(–7·0) cm diam in seven days at the optimal temperature of 24°C are found in *H. dematioides* Lagerb. & Melin *1927* (other anamorph (pycnidial) *Dothichiza pityophila* (Corda) Petr. *1923*; teleomorph *Sydowia polyspora* (Bref. & Tavel) E. Müll. *1953*), in which pycnidia can be obtained after 6–8 weeks growth on varnished pine wood blocks (811, 6461). *Trichosporon pullulans* (Linder *1895*) Didd. & Lodd. *1942* has permanently hyaline colonies and arthroconidia and is not related to *A. pullulans* (1342). — Wild isolates can have yellow, red or purple pigments which act as pH indicators (decolourizing in alkaline media) and change colour in combination with each other or with other micro-organisms (6345). — TEM studies of conidiogenesis show budding taking place after a localized wall hydrolysis so that the conidial wall is continuous only with an inner wall layer of the conidiogenous cell (4722). Hyphal cells contain between two and five nuclei, and heterokaryosis is assumed to occur (1156); the conidia are mostly uninucleate, but larger conidia may contain two or rarely three nuclei; the septa with a simple pore are of the ascomycete type (1510). — DNA analysis gave a GC content of 51·5% (3232). — Conidial cell wall hydrolysates were found to contain 1·9% glucosamine, 54% glucose, 9·3% mannose, 4·2% galactose, 1·8% glucuronic acid and 0·9% rhamnose, while those of the hyphae contain more glucosamine (2·7%), glucose (70%) and rhamnose (1·9%) (751, 3232). According to another report, however, the conidial walls contain 13% glucose and more mannose, galactose (and bound lipids) than given above, whilst the chlamydospore walls mainly consist of glucan (750). Xylose is absent in the walls (2524).

Contrary to other species of *Aureobasidium* (previously placed in *Kabatiella* Bubák), *A. pullulans* is a ubiquitous and cosmopolitan saprophyte (1158, 2412); it was originally described from golden spots on grapes in France and is most commonly encountered on leaf surfaces of plants (see below). It is probably due to an extreme sensitivity to heat (5406) that it does not usually appear in warm-poured soil dilution plates, but available data indicate that its main habitat are aerial plant parts. — The data compiled below are to be accepted with considerable reservation since species of *Hormonema* have not usually been distinguished from *A. pullulans*. — Isolation from soil is facilitated after enrichment with vanillin or *p*-hydroxybenzaldehyde (2375). Judging from the available records it is apparently most common in temperate zones with very numerous records both from the British Isles and the USA but also found in Canada (234, 1446, 3959, 3962, 5363, 6352), Alaska (1171), Antarctica (856), Denmark (2745), Germany (1424, 1736), the Netherlands (1614, 1616), Poland (3497), Austria (1823), Czechoslovakia (1703), and the USSR (2871, 3850, 3856, 4548). There are fewer reports of it from the mediterranean and arid zones which include ones from Italy (3913, 4537), France (242), Egypt (4962), Iraq (92), Pakistan (3546), South Africa (1559, 3630, 4407), and the southern USA (56, 660, 1164, 2482, 4918), and also scanty reports from tropical regions such as those from Brazil (6008), India (4371, 4698), Malaysia (6046), the Bismark Archipelago (3020), Hawaii (296), the Bahamas (2006, 4312) and Jamaica (4886). — *A. pullulans* has been frequently isolated from moorland, peat bogs or peat podzol (1039, 1376, 2502, 3234, 3850, 6082), and also from forest soils (1736, 2482, 2573, 2923, 3817, 5363), with stands of cottonwood, willow (2007, 2008), beech (2745, 4537), aspen (3959), larch (3267), pine (6352) and mixed conifer-hardwood (3962, 6351). Other noteworthy habitats include fresh water (1166, 3794, 4429), estuarine (56) and marine sediments and seawater (56, 4918), sewage, and other liquid waste materials (1163, 1166, 1172, 1672, 3294, 5060). On the other hand, its occurrences in acid sand dunes (745, 4371), heathland (5220, 5221, 5222), a recently deglaciated soil (1171), an open savannah (1559, 4407), soils under tundra (1446) as well as chaparral vegetation (1164), and in desert soil (4733) indicate that it does not necessarily depend on rich organic substrates. As far as data are available, the fungus has been found exclusively in the surface layers of soils (660, 1164, 1823, 2923, 3445, 3720, 4537, 6186, 6352) and in the pH range 3·6–6·5 (1032, 1164, 3331, 3445, 5363, 6351). An increased frequency was observed after N fertilization (3498); it shows a relatively high tolerance to methylbromide (1268) and several soil fungicides (1426). — *A. pullulans* has even been found in a tropical air mass over the mid-Atlantic (4351). It occurs very frequently as the dominant fungus on leaves, not merely as a surface (phylloplane) colonizer, but also as an invader of healthy leaf tissue as demonstrated in peas (1373), oak, hazel (2411), beech (2500, 5417), aspen, poplar (6101), maple (4656), *Hippophaë rham-*

noides (3352, 3353), *Typha latifolia* (4663), *Carex paniculata* (4644), *Abies firma* (4984), *Pseudotsuga menziesii* (5267), apple, peach (5417), *Spartina townsendii* (5389), and *Brachypodium pinnatum* (2971); it is also known from honeydew on fir (651), decaying leaf litter of pine (2344, 2345, 2375, 2925, 3801), birch (3449), oak, beech, hornbeam, hazel (4817, 4989), several heathland plants (3720), and various other dicotyledonous plants (2586), on fronds (2013) and decaying stalks of *Pteridium aquilinum* (1821, 1822), decaying stems of cocksfoot (6237), sorghum, wheat (3219) and *Heracleum sphondylium* (6462), compost of *Diplachne fusca* at a late stage of degradation (3546), and rotting straw (953). It is not regarded as a typical member of the root mycoflora, where it has been only rarely found, for example in the rhizosphere of grasses (1614), wheat (1614, 3567), corn (6563), and various sand-dune plants (4371), and the rhizoplane of barley, cabbage (4451), *Desmodium triflorum* (3331), sugar cane (4886), Corsican pine (4448), and both diseased and healthy strawberry plants (4133). It has been recorded on seeds of wheat, barley, oats (1751, 1752, 1754, 2630, 5293), *Dactylis glomerata* (5940), tomato (5899), freshly harvested pecans (2572), *Acer pseudoplatanus* achenes (1683), but does not belong to the characteristic flora of either fresh or stored seeds. However, it is in accordance with its enzymatic abilities (see below) that *A. pullulans* has been frequently isolated as an endophyte from bark, wood, twigs of *Acer pseudoplatanus* (4656), willow (2338), stained wood chips of *Betula maximowicziana* (5061) and other trees (812), stumps of alder, beech and willow (3568), pine poles buried in various forest soils (2833), and wood pulp (4857). Other substrates recorded are aecidiospores of *Cronartium comandrae* (4608), honeycombs (402), nests and feathers of free-living birds (2575), frozen fruit cake (3153), leather, cotton fabrics (3707, 6326), surfaces of concrete, paint, plastics (4824) and optical lenses (4833), and a human lymph node (2412). — In soil the mycelium is degraded at a slower rate than that of *Mucor hiemalis* (3787). An antifungal substance (2,3′,4,5′-tetrahydroxystilbene) from *Maclura pomifera* completely suppresses its growth *in vitro* (347). Dense populations of *A. pullulans* on leaves prevent or reduce an infection of beans by *Alternaria zinniae* (2435), of onion leaves by *Alternaria porri* (1790), and of rye leaves by *Cochliobolus sativus* (1789).

The temperature range for growth is 2–35°C, the optimum 25°C, and the maximum 35°C (4487, 5406); an adaptation to low temperatures has repeatedly been reported (3153) and human-pathogenic isolates were found to have slightly higher temperature characteristics (5406). — Light stimulates growth in the presence of polysaccharides, but is not effective in combination with monosaccharides; the best C sources are polysaccharides (dextrine, starch), followed by most oligo-, di- (except lactose) and monosaccharides (1342, 3357); only colourless mycelium is formed when grown on lactose, inulin, galactose or trehalose (3357). On ammonium salts as N source a mixture of filamentous and yeast-like growth is produced, while on nitrate the yeast-like growth form and chlamydospores predominate (750). These findings are in contrast to observations where different inorganic N sources did not alter yeast-like growth, though morphogenic changes to filamentous growth could be induced with phenylalanine and methionine medium (5219). High inoculum density in liquid media results in yeast-like growth and a low density in mycelial growth which can revert to the yeast-type on ageing (4721). *A. pullulans* has a pronounced ability to degrade pectin (3414, 5417), and the formation of polygalacturonase, polymethylgalacturonase, pectintranselimin-ase (pH optimum 6·5), and pectinesterase has been demonstrated (480, 2239, 4895). There are also reports of its production of L-rhamnose dehydrogenase (3382), laccase, mannanase (pH optimum 4), cellulase (on CMC) and xylanase (480, 1753, 4894, 4895). It grows well on hydrolysed rye grass with an optimum at pH 5 and 32°C (2248). The metabolism of lignin related compounds (*p*-hydroxybenzaldehyde, ferulic acid, syringaldehyde, and vanillin) by this fungus has repeatedly been studied (544, 2375, 2376, 2377); *A. pullulans* can penetrate into beech wood (5238) and cause a soft rot of timber (5072). *n*-Alkanes can be utilized as the sole C source, *n*-hexadecane being the most suitable (3762). In liquid cultures growth was found to be stimulated in the presence of DDT (2488). Amongst 99 tested acids and their esters (plasticizers and related compounds), 65 supported growth (473); dinonyladipate, epoxy-plasticizers and tricresylphosphate were all degraded (5859). Flavonoid compounds such as rutin, naringenin, taxifolin, quercetrin, and catechin are utilized (6311), and the glycoside scilliroside hydrolysed (5598). *A. pullulans* produces the extracellular 1,4-, 1,6-, and 1,3-a-glucan pullulan in low-pH media and in the presence of NH_4^+ ions (923, 924, 1578, 5683, 5684), the synthesis being induced by the addition of carbon sources to starving cells (925) or at low nitrogen availability in the filamentous stage of growth (926). Pullulan gives mainly maltotriose and 6·6% of a maltotetraose subunit after isoamylase action (928). A peak in trehalose production is associated with the formation of secondary conidia (in continuous culture) (927). Polysaccharides exuded into the culture medium are very similar to those contained in the cells and cell-walls (2968); another polysaccharide product with flocculating properties has been described (6530). A hyaluronic acid-like substance has also been reported (5849). In the outermost wall layer dark granules resembling melanin have been extracted and observed in TEM (750, 1510, 3357). On sucrose monopalmitate the production of a β-fructofuranosidase has been demonstrated (4776). — *A. pullulans* has been isolated from cheloid blastomycosis in man and found to cause comparable lesions on rabbit (6069). *A. pullulans* can stimulate the rhizomorph development of *Armillaria mellea* (4501) and the metabolite involved appears to be ethanol (4502). — It can tolerate anaerobic conditions in the presence of biotin and thiamine (5680), is relatively heat-sensitive, dark isolates being more resistant (5406), and has an unusually high resistance to both UV- and X-rays (3114). In a chronically irradiated soil it increased in numbers of viable cells at up to 820 R per day, and survived at ≥1250 R per day (2009).

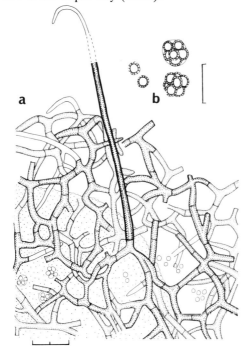

FIG. 56. *Auxarthron umbrinum.* a. Sector of ascoma with loosely woven peridium and uncinate appendage; b. asci and ascospores.

Auxarthron Orr & Plunkett *1963*

Auxarthron umbrinum (Boud.) Orr & Plunkett *1963* — Fig. 56.

≡ *Gymnoascus umbrinus* Boud. *1892*
= *Gymnoascus subumbrinus* A. L. Sm. *1917*
= *Myxotrichum emmonsii* Kuehn *1955*

DESCRIPTIONS: Orr *et al.* (4316) and Apinis (162). — Fam. Gymnoascaceae Baran. — Colonies reaching 1·5–2·2 cm diam in ten days at 20°C on OA. Peridial network consisting of thick-walled, red-orange to orange-brown, smooth to asperulate, anastomosing hyphae, 2·6–3·0 μm wide, usually slightly wider at the septa, with some short spines or some darker, to 1000 μm long, distally bent or coiled appendages. Ascospores ± globose, hyaline to orange-brown, echinulate, 2·1–3·6 μm diam. The anamorph consisting of scanty arthro-aleurioconidia. — *A. umbrinum* can be distinguished from the other five recognized species of the genus by the orange vegetative mycelium and sometimes sparse presence of long, distally curved peridial appendages provided with several basal septa. — The distinction of *Auxarthron* from *Gymnoascus* is justified by the ornamented, ± globose ascospores and the presence of long and dark peridial appendages; *Myxotrichum* has a still darker peridium with long appendages and ellipsoidal to fusiform or lenticular, smooth-walled ascospores. — SEM studies show a dichotomous branching of the peridial hyphae forming nets, and either smooth or granular, closely reticulate surfaces of the ascospores (6105).

This striking species has been isolated by various authors but not with any great frequency. With the exception of reports from Iraq (92), Chile (4569) and South Africa (1559, 3590, 4407), all observations relate to the northern temperate zone of Europe and America. The comprehensive documentation of *A. umbrinum* in the USA and the British Isles (162, 4315, 4316) includes reports from arable, grassland, swampy meadow, salt marsh and acid forest soils (989). Further substrates include straw and humus under *Agropyron pungens* (167), leaf litter of *Acacia karroo* (3590), wheat rhizosphere (3567), the mycorrhiza of *Pinus strobus* and *Larix leptolepis* (1794) and truffle soil (3451). *A. umbrinum* has also been isolated from excrements of bats and lizards and rat lungs (4316).

Beauveria Vuill. *1912*

Type species: *Beauveria bassiana* (Bals.) Vuill.

MONOGRAPHIC TREATMENTS: McLeod (3505), and de Hoog (2522). — Colonies growing slowly, mostly not exceeding 2 cm diam in ten days at 20°C, woolly, at first white but later often becoming yellow to pinkish. Conidiogenous cells arising either singly, in whorls from vegetative hyphae, or more commonly in clusters from swollen stalk cells; conidiogenous cells differentiated into a subglobose to ellipsoidal or cylindrical venter and a filiform, zig-zag shaped, denticulate rhachis arising by sympodial elongation. Conidia hyaline, globose or ellipsoidal, with a rounded or slightly pointed base. — A *Cordyceps*-like teleomorph has been reported (5081).

Three species are recognized, two of which are well known parasites of insects. Numerous taxa have been described, mostly from insects, but it has been recognized by McLeod (3505), Schaerffenberg (5081) and de Hoog (2522) that the considerable morphological variability which occurs is largely influenced by the cultural media and the age of the isolate; only two common entomogenous species are certainly distinct, one having globose and the other ellipsoidal conidia. The latter has usually been named *B. densa* or *B. tenella*, but examination of type material has shown that these epithets also referred to the globose-spored species; the oldest available epithet for the species with ellipsoidal conidia is in fact "*brongniartii* Sacc." A third, non-entomogenous species, *B. alba* (Limber) Saccas *1948* has usually been retained in *Tritirachium* Limber; because of its denticulate conidiiferous rhachis, however, it has been removed from *Tritirachium* (2522) which has a similar zig-zag shaped geniculate rhachis lacking prominent denticles, but is now referred to as *Engyodontium album* (Limber) de Hoog *1978*. A further species, *B. vermiconia* de Hoog & Rao *1975*, isolated from volcanic ash soil in Chile, resembles *B. bassiana* but has sickle-shaped conidia, about $2 \cdot 5 \times 1 \cdot 5 \, \mu$m. *Spicellum roseum* Nicot & Roqueb. *1974* is close to *Beauveria* but has more spreading pink colonies and conidiogenous cells with shorter geniculations.

Key to the species:

1   Conidiogenous cells slender, cylindrical to awl-shaped, to $2 \cdot 5 \, \mu$m wide, arising in whorls on erect or prostrate aerial hyphae; conidia globose   (*B. alba*)
         Conidiogenous cells usually broader, arising either in clusters from swollen stalk cells, or irregularly scattered on aerial hyphae   **2**

2(1)   Conidiogenous cells globose to flask-shaped, usually forming dense clusters; conidia globose or sickle-shaped   **3**
         Conidiogenous cells more slender, ellipsoidal to cylindrical, scarcely clustered; conidia ± ellipsoidal (rarely subglobose)   *B. brongniartii* (p. 139)

3(2)   Conidia globose or subglobose (rarely broadly ellipsoidal)   *B. bassiana* (p. 137)
       Conidia sickle-shaped   (*B. vermiconia*)

Beauveria bassiana (Bals.) Vuill. *1912* — Fig. 57.

  ≡ *Botrytis bassiana* Bals. *1835*
  = *Sporotrichum densum* Link ex Pers. *1822* [non (Ditm. ex Pers.) Fr. *1832*]
    ≡ *Beauveria densa* (Link ex Pers.) Picard *1914* (nomen confusum)
  = *Sporotrichum globuliferum* Speg. *1880*
    ≡ *Beauveria globulifera* (Speg.) Picard *1914*
  = *Beauveria bassiana* subsp. *tenella* Sacc. *1882*

For numerous other synonyms see de Hoog (2522).

DESCRIPTIONS: McLeod (3505), and de Hoog (2522). — Colonies attaining 0·6–2·3 cm diam in eight days at *c.* 20°C, woolly, floccose, often appearing powdery due to the abundant conidia, rarely forming synnemata to 5 mm tall; at first white but later often becoming yellow or pinkish. Conidiogenous apparatus forming dense clusters of swollen stalk cells which consist of a subglobose to flask-shaped venter, 3–6 × 2·5–3·5 μm, and a zig-zag shaped, denticulate rhachis to 25 μm long and 1 μm wide. After prolonged periods in culture, the conidiogenous cells tend to be more slender and less densely clustered. Conidia hyaline, smooth-walled, ± globose, mostly 2–3 μm diam. — A *Cordyceps*-like teleomorph has been reported (5081). — *B. bassiana* can easily be confused with the soil-borne *Tolypocladium inflatum* W. Gams *1971* which has similar colonies and conidiogenous cells, but produces phialoconidia. — A time-lapse study of conidiogenesis has been published (2926) and SEM and TEM investigations have also been carried out; the conidiogenous cells and conidia are surrounded by a double-layered wall, of which the outer is electron-opaque (4802) and contains polysaccharides and proteins. This species has been compared serologically with *B. brongniartii* (1692). — Auxotrophic mutants have been used in experiments on heterokaryosis (6517). Mutagenic effects of UV- and X-rays have also been assessed (3305); its virulence to insects is subjected to mutation (3010) and may be lost after prolonged periods in culture (2522). — The total, neutral and polar lipids contain mainly palmitic, oleic, linoleic and *a*-linolenic acids (5965).

B. bassiana is best known as the causal agent of the disastrous muscardine in silkworms (416). It is the most widely distributed species of the genus and is generally found forming white dusty raised tufts on Coleoptera, Lepidoptera, Diptera and other insects in both temperate and tropical areas (2522, 3505). Long host lists have been compiled for North America (960) and the British Isles (3257). Records from soil and other habitats are usually connected with parasitized insects. — In addition to Europe and North America, there are reports of its occurrence in Turkey (4245), the Ivory Coast (4718), equatorial West Africa (1420), central Africa (3063), South Africa (1415, 1559, 4407), the Bahamas (2006), Nepal (1826), East Siberia (4249), New Zealand (2705) and Japan (155, 157, 3267, 5846). Habitats range from an alpine soil with long snow coverage (3976) to heathland (5221, 5222), peat bogs (1376), soils with savannah type vegetation (4407), forest and cultivated soils (reports too numerous to cite individually), sand blows (4540) and dunes (745), desert soil (4733) and running water (1166). Incidental isolations from the rhizoplane of peat bog plants (1376), the rhizosphere of clover (3429) and dead bark (3063) cannot be regarded as typical. It has also been isolated from rodents (3505), and nests, feathers and droppings of free-living birds (2575). — *B. bassiana* is sensitive to soil-mycostatic factors (1075, 2582). Survival in soil is favoured by darkness and reductions in both temperature and soil moisture; at 8°C and under dry conditions, 90% of the conidia survive for over 635 days (1076, 6164); on silica gel conidia survived up to 36 months of storage at –20°C (435). In a chronically irradiated soil this fungus was still present after exposures to 2780 R per day for seven years (2009). The mite *Sancassania phyllognathi* is considered a vector of this fungus in soil, being resistant to fungal attack itself (5021).

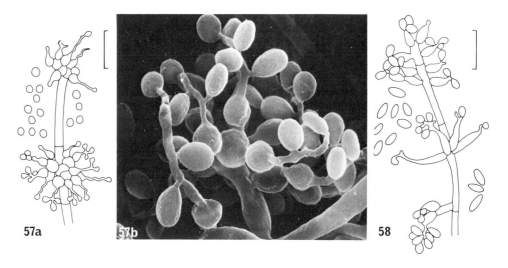

57a   57b   58

FIG. 57. *Beauveria bassiana*. a. Conidiophore with strongly sympodially elongating conidiogenous cells; b. details of conidiogenesis, SEM, × c. 4000, CBS 337.32; orig. R. A. Samson.
FIG. 58. *Beauveria brongniartii*, conidiophores and conidia; orig. R. A. Samson.

Because of its suitability for large-scale culture (4026) and broad spectrum of virulence to insects (310, 3257), it is potentially suitable for the biological control of insects (1499). Biological control experiments with it are numerous and have been directed particularly to the Colorado beetle and other Coleoptera and Lepidoptera. Various species of mosquito larvae can be killed by dusting conidia on the water surface (1070); conversely, bee larvae were not susceptible while adult bees succumbed (5863). It has also been isolated from the lungs of giant tortoises and box turtles affected with a pulmonary disease (1939); an increase in the air spora has also led to allergic responses in man (4866).

On agar media conidiogenesis starts after six days, while in liquid culture this takes only 3–4 days (5017). In stirred liquid cultures employed in mass production of this fungus, so-called "blastospores" developed which are thin-walled, larger (3–5 × 2–3 μm) and less resistant than conidia (4025); blastospores germinate in 6–10 h at 18–24°C, while conidia require 15–20 h (4025). The germination of conidia requires a saturated atmosphere and the optimal temperature for growth is in the range 25–30°C, minimum 10°C, and maximum 32°C, apparently depending on the geographic origin of the isolate (291, 2849, 6490); no germination occurs either below 10° or above 35°C; the thermal death point of conidia has been determined as 50°C for 10 min in water (6164). The optimal pH for growth is 5·7–5·9, and for conidium formation 7–8 (2036). — Good growth occurs on maltose and sucrose (2582) and, among others, on the N sources glutamic and aspartic acids, and ammonium oxalate, citrate or tartrate (156). A medium recommended for optimal growth contains 2% corn steep liquor, 2·5% glucose, 2·5% starch, 0·5% NaCl, and 0·2% $CaCO_3$ (5017). The fungus produces lipase, protease, urease, amylase, chitinase, cellulase (1074, 2582, 3293, 5019, 5020, 5324) and 1,2-β-glucanase (4780). Chitinase (1074) can be released into the medium during autolysis, but was also found to act jointly with other enzymes in decomposing insect integuments (5020). The production of a toxic substance (5863) said to act as a contact insecticide (5082) has been reported (3292) and is now known to be a proteinase complex consisting of two fractions with different molecular weights (3148). This toxic metabolite is produced most abundantly on complex media (e.g. cornmeal, yeast or beef extract); inorganic N sources are ineffective (3147). A red bibenzoquinone pigment, oosporein, with antifungal properties (6095) and the yellow pigments tenellin and bassianin (387, 3711) have been found. *B. bassiana* is able to hydroxylate progesterone and to split minor amounts of it to yield testosterone (875).

Beauveria brongniartii (Sacc.) Petch *1924* — Fig. 58.

≡ *Botrytis brongniartii* Sacc. *1892*
? = *Sporotrichum epigaeum* Brun. *1888* sensu Aschieri *1929*
  ≡ *Beauveria epigaea* (Brun.) Langeron *1936*
 = *Botrytis melolonthae* Sacc. *1912*
  ≡ *Beauveria melolonthae* (Sacc.) Cif. *1929*
 = *Beauveria tenella* (Sacc.) McLeod sensu McLeod *1954*
 = *Beauveria densa* (Link ex Pers.) Picard *1914* sensu auct.

DESCRIPTIONS: McLeod (3505), and de Hoog (2522). — Colonies as in *B. bassiana*, reaching 1·0–1·6 cm diam in eight days at 20°C on MEA. On the aerial hyphae small groups of lateral cells to 9 × 5 μm develop which give rise to either similar secondary cells or 1–3 conidiogenous cells. Conidiogenous cells with a flask-shaped to subcylindrical venter, 4–15 × 2·0–3·5 μm, and a zig-zag shaped denticulate rhachis to 25 μm long and 1·0–1·5 μm wide. Conidia hyaline, smooth-walled, ellipsoidal, sometimes with a pointed base, mostly 2·5–4·5 × 2·0–2·5 μm. — DNA analysis gave a GC content of 53% (5600).

B. brongniartii is somewhat less common than *B. bassiana* but also has a worldwide distribution and occurs as a pathogen on numerous insects as evidenced by host-lists for North America (960) and the British Isles (3257). A medium described for its selective isolation from insects includes glucose, peptone, oxgall, chloramphenicol and actidione (6053), and for selective isolation from soil, glucose, yeast extract, oxgall and antibiotics (2825). An immunofluorescence technique for its study in the soil has been developed (2824). —*B. brongniartii* has been reported from open bogs (1039) and alpine habitats (4711), forest soil in Hong Kong (1021), terra rossa in Greece (4647), a *Calluna* heath (5047), and an alpine grassland (3445) in Italy and sand dunes in the British Isles (4655). — Experiments on its possible application for the biological control of cockchafers, mosquitoes and other insects have been numerous. High concentrations (10^{10} conidia m^{-2}) of a virulent isolate gave some success against cockchafers in the second year (1731); in a pine forest soil, the moisture range 3·5–18·3%, pH 4·4–5·0, humus contents of 0·74–2·3%, and weak to medium concentrations of mobile P and K encouraged the infection of larvae (1650). In vitro the growth can be inhibited by *Penicillium frequentans, P. granulatum, P. verrucosum* var. *cyclopium* and *P. lanosum* (2848).

For the production of conidia, the best C source proved to be sucrose combined with asparagine or glycine; the growth of the mycelium, however, is favoured most by media with glucose or sucrose supplemented with extracts of yeast, corn or soya (3040); a C/N ratio of 10:1 gave maximum conidium production (3039). Proteolytic and lipolytic activities have been demonstrated (5019); the lipolytic activity of different isolates is evidently correlated to their virulence (4420). As in *B. bassiana*, a red pigment, oosporein, and the yellow pigments tenellin and bassianin were found (387, 3711); further metabolites described are the cyclodepsipeptides isarolide (2522) and beauvellide (1829). A specific protein has been described which is characteristic for pathogenic isolates (4419). *B. brongniartii* is able to hydroxylate progesterone and to split minor amounts of it to yield testosterone (875).

Bipolaris Shoem. *1959*

= *Drechslera* Ito *1930 pro parte*

Type species: *Bipolaris maydis* (Nisikado) Shoem.

Contributions to a MONOGRAPHIC TREATMENT: Shoemaker (5303), Ellis (1603, 1604), and Chidambaram *et al.* (1002), sub *Drechslera*.

Conidiophores erect or ascendent, pigmented, geniculate due to sympodial elongation, producing single conidia through conspicuous pores as shown in TEM (1099). Poroconidia ± fusiform to ellipsoidal, brown, smooth-walled, distoseptate (pseudoseptate), germinating only from the ends (bipolar). Two species are treated here under their teleomorph names in *Cochliobolus*.

The subdivision of the *Helminthosporium* segregates into *Drechslera* and *Bipolaris* by Shoemaker (5303) was not accepted by many mycologists as it was difficult to apply and *Bipolaris* was heterogeneous (1603, 5628). The recognition of a third genus, *Exserohilum* Leonard & Suggs *1973* (3290), however, removes much of the inconsistency and permits the more natural arrangement shown in Table I. *Bipolaris* now contains about 45 species.

TABLE I. Subdivision of *Helminthosporium* segregates

Anamorphs	Main characters	Teleomorphs	Main characters
Drechslera	Conidia cylindrical, germinating from any cell	*Pyrenophora*	Ascomata immersed and erumpent, not setose; ascospores very large, dictyoseptate, with a gelatinous sheath
Bipolaris	Conidia fusiform-ellipsoidal, central cells not much darker and broader than the distal ones, hilum not protuberant; germination bipolar	*Cochliobolus*	Ascomata superficial or erumpent, setose; ascospores vermiform, many-celled, coiled in the ascus
Curvularia	Conidia with 2–3 broader and darker central cells, often curved, with or without a prominent hilum; germination bipolar		
Exserohilum	Conidia fusiform-cylindrical to obclavate, with a protuberant hilum; germination bipolar	*Setosphaeria*	Ascomata superficial, setose; ascospores fusiform, phragmosporous, with a gelatinous sheath and similar distal appendages

Key to the species treated (under *Cochliobolus*):

1   Conidiophores usually flexuose and geniculate, bearing up to 6 conidia; conidia dark brown, ellipsoidal but gradually tapering towards the ends and often slightly bent, usually 60–100 × 18–23 μm, 6–9-septate   *C. sativus* (p. 216)
     Conidiophores with a long, densely geniculate rhachis bearing numerous conidia; conidia oblong-ellipsoidal, straight, usually 20–40 × 9–14 μm, 3-septate   *C. spicifer* (p. 220)

Botryosphaeria Ces. & de Not. *1863*

Botryosphaeria rhodina (Berk. & Curt.) v. Arx *1970* — Fig. 59.

≡ *Physalospora rhodina* Berk. & Curt. *1889*

Anamorph: *Lasiodiplodia theobromae* (Pat.) Griffon & Maubl. *1909*

≡ *Botryodiplodia theobromae* Pat. *1892*
= *Lasiodiplodia tubericola* Ellis & Everh. *1896*

For numerous other synonyms of both states see Petch (4518), Zambettakis (6535), and Goos *et al.* (2033).

DESCRIPTIONS: Petch (4518), Zambettakis (6535), Yokoyama *et al.* (6500), Alasoadura (84) (in culture), and CMI Descriptions No. 519, *1976*. — *Botryosphaeria* Ces. & de Not. (type species *B. quercuum* (Schw.) Sacc.) is characterized by rather large stromatic fruit-bodies containing one to several cavities and hardly prominent ostioles, and asci which are bitunicate, clavate and contain 8 one-celled, ± hyaline ascospores. Anamorphs belonging to various genera of the Sphaeropsidales. Ascomata of *B. rhodina* are known only from the natural substrate and contain (sub)hyaline, ellipsoidal, one-celled ascospores 24–42 × 10–18 μm. — The sphaeropsidaceous genus *Lasiodiplodia* Ellis & Everh. *1896* is characterized by stromatic, often clustered dark, hairy pycnidia, and brown, two-celled, ellipsoidal, striate conidia formed on elongate phialides. The striate conidia and the presence of paraphyses are the main features to distinguish *Lasiodiplodia* from *Botryodiplodia* Sacc. *1884*.

Colonies of *L. theobromae* are fast-growing, reaching over 9 cm diam in five days at 25°C on OA, and have abundant dark-brown aerial mycelium; the pycnidia arise in clusters on stromata, mostly embedded in a hairy felt and are dark olivaceous, spherical, with an elongated neck. Phialides cylindrical, 5–12 × 2–4 μm, intermixed with long paraphyses. Conidia two-celled, ellipsoidal, dark brown at maturity, with 7–9 lighter longitudinal striations, mostly 22–28 × 12–15 μm. Conidia maturing very slowly, remaining hyaline and one-celled for a long time. — The formation of the septum in the conidia and of germ tubes has been investigated by TEM (6293); germination of the conidia is preceded by the deposition of an internal wall layer. Conditions favouring the pigmentation and septation of the conidia have been studied (5988), and unpigmented, one-celled conidia are found to germinate faster than more mature ones (5987).

B. rhodina is a widespread soil-borne saprophyte or wound parasite in the tropics of all continents but has not been reported from the temperate zones of Europe and the USA. It is common on fruits, stems and roots of numerous tropical and subtropical plants, particularly *Theobroma*, *Castilla*, *Hevea*, *Citrus*, *Mangifera*, *Ficus* and *Musa* (2033, 3063, 4518, 6158); indeed, it is said to be seldom absent from dead parts of *Theobroma* or *Hevea* (4518). It has also been found in the rhizosphere of sand dune plants (4371), decaying stems of wheat

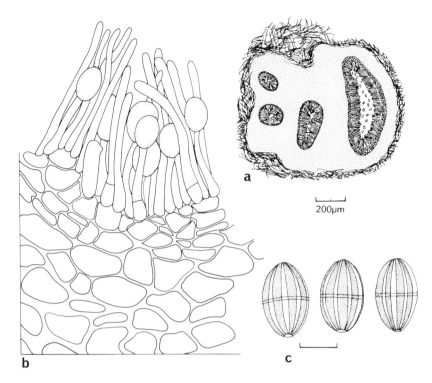

FIG. 59. *Botryosphaeria rhodina*. a. Section through pycnostroma with several conidiogenous cavities; b. detail of conidiogenous cavity with conidiogenous cells and paraphyses; c. mature conidia; CBS 456.78; orig. J. Veerkamp.

(3219), and groundnuts (2768, 3702), banana roots (5608), corms of *Colocasia esculenta* (2022), *Cassia occidentalis* (6067), as a secondary colonizer of buried cotton roots (1965), on wheat seeds (3989), pecans (2572), *Hevea* wood chips on soil, hardwood stumps, paper (IMI) and cotton fabrics (6326). *B. rhodina* has also been isolated in a case of mycotic keratitis in a human eye (3239).

Growth on natural substrates requires high relative humidities (1568). The formation of pycnidia varies according to the conditions of growth (particularly the nutrient concentration and humidity). The optimal temperature for pycnidium formation is in the range 23–30°C (5986). Cultures are positively phototropic, light in the near-UV range having a stimulatory influence on the formation of stromata and pycnidia (84, 1568, 1569, 6194). Three different developmental stages have been postulated leading to a light-induced and metabolically synchronized production of pycnidia (4740). Full-size pycnidia were obtained on media containing K_2HPO_4, $NaNO_3$ and sucrose; potassium ions were found to be essential for pycnidium formation (1566). — Good mycelial growth was obtained on sucrose and D-glucose; suitable N sources for this were found to be L-cystine, D,L-alanine, glycine, L-arginine, L-tyrosine, urea, NH_4^+-N and NO_3^--N (5308). The formation of pycnidia and mycelial growth were both strongly stimulated by the presence of B-vitamins and ascorbic acid (5311). Conidia germinate well at 20–37°C and in a water-saturated atmosphere (84, 5987). The optimal temperature for linear growth lies in the range 20–25°C (or higher), the minimum is 15°C, and no (or poor) growth is obtained at 35–40°C (30, 84, 5986). Polygalacturonase and cellulase activity have been described (178). Protein synthesis in different stages of conidium germination has been studied, the synthesized soluble polypeptides being quite different at the onset of germination and germ tube formation (6044). Known metabolites include ethyl hydrogen fumarate, jasmonic acid, indole-3-aldehyde, indole-3-carboxylic acid, isocoumarin, 4-hydroxymellein, lasiodiplodin, and desmethyllasiodiplodin (3231). An antibiotic, botryodiplodin, active against both gram positive and negative bacteria, has also been described (5204). Mycelial extracts were found to be toxic to chicken embryos (1388).

Botrytis Mich. ex Fr. *1821*

= *Polyactis* Link ex Steud. *1824*
= *Haplaria* Link ex Leman *1821*
= *Phymatotrichum* Bonord. *1851*

Type species: *Botrytis cinera* Pers. ex Nocca & Balb. (lectotype)

Teleomorph: *Botryotinia Whetzel* (=*Sclerotinia* Fuckel pro parte)

TAXONOMIC SURVEYS: Ellis (1603), Morgan (3948, 3949) (with numerical taxonomy), Hennebert (2397) (species on *Allium*), and Menzinger (3760); a review of its taxonomy, physiology and pathogenicity: Jarvis (2725).

Colonies spreading broadly, hyaline at first but soon becoming light grey to dark brown; hyphae hyaline to brown. Conidiophores erect, brown, mononematous, solitary or in groups, apically repeatedly alternately branching, with terminal swollen conidiogenous cells producing numerous blastoconidia simultaneously on short denticles, soon collapsing but sometimes proliferating; conidia pale brown, globose to ovate or ellipsoidal, smooth-walled or almost so, hydrophobic. Micro-conidia of *Myrioconium* H. Sydow *1912* produced on sporodochial (spermodochial) phialides; phialides short, flask-shaped, with or without a pronounced collarette, conidia hyaline, ± globose, uniguttulate, smooth-walled, small. Sclerotia commonly produced, consisting of a black cortical layer and a white medulla, plano-convex, flattened or pulvinate, rounded or elongate, smooth-walled or nodulose.

Of the about 380 taxa referred to *Botrytis* at various times, only about 25 species can be recognized in it since it has been redefined by Hennebert (2400). *B. cinerea* is by far the commonest species of the genus growing on a wide range of host plants as a parasite or saprophyte; most other species of the genus have a more restricted host range.

The teleomorphic genus *Botryotinia* Whetzel *1945* (Sclerotiniaceae, Helotiales) has been segregated from *Sclerotinia* Fuckel *1870* on account of the *Botrytis* anamorph and the compact, gelatinized sclerotial medulla; it now contains 13 species. — For adequate development of both conidiophores and sclerotia, colonies are grown on PDA under intermittent near-UV ("black") light for about one week at about 20°C; sparsely sporulating isolates produce more conidia on poorer media such as PCA or hay decoction agar. Sclerotium formation can be favoured by incubation on richer media (containing glucose) in darkness. Apothecia have been obtained (2141, 2402) in culture from sclerotia developed at 5°C on wheat grains by incubating them on sterile sand at first in darkness at 0°C (one month) and then at 5°C (2–6 weeks), spermatized with the opposite mating type, a further incubation for three weeks at 10–14°C, and finally retention in the same temperature range in a greenhouse.

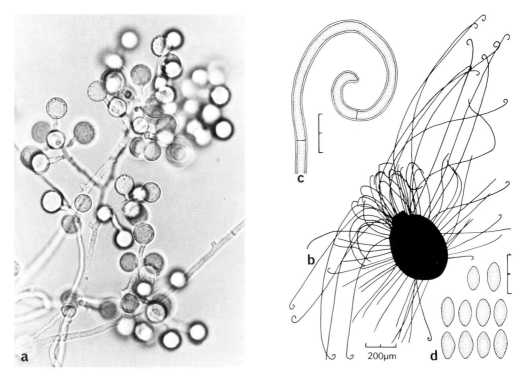

Fig. 60. *Botryotrichum piluliferum*. a. Conidiophores with aleurioconidia, × 500; b. perithecium of *Chaetomium piluliferum*; c. detail of terminally uncinate hair; d. ascospores.

appropriate. — It has a worldwide distribution. There are numerous reports, too many to cite here individually, from cultivated soils and it has also been found in fields treated with sewage sludge (1163), dunes (4655, 4759), dry soils (654), salt marshes (8), slate slopes (566), carst caves (3364), mountainous regions (4569), soils under steppe (1700), beech (3138) and white cedar forests (505), cedar and poplar bogs (IMI), truffle soils (3451, 3454), soil of a mink farm (4230) and children's sandpits (1421). *B. piluliferum* has been observed down to a depth of 80 cm (2161) in soil and is intolerant of acid but very tolerant of alkaline soils (2741, 6328). Other known habitats include stems of *Urtica dioica*, hay (IMI), the rhizospheres of groundnut (2768), rice (2712) and wheat, paper products (IMI, 1875), and mouldy textiles (6328). The hyphae can grow chemotropically towards phosphate particles without dissolving them (3074). — It can serve as food for the mites *Pygmephorus mesembrinae* and *P. quadratus* (3104). It is susceptible to parasitism by *Pythium oligandrum in vitro* (1323).

The optimal temperature for growth is 25–30°C, and the maximum 40°C; the pH optimum is around 5·5 but it can still grow above pH 8·8 (2741). — Good decomposition of starch, pectin, xylan and CMC occurs and there is little variability in these abilities between isolates (654, 1432, 2741, 6076). When growing on straw, both cellulose and lignin are efficiently utilized and humic substances produced (2211). Cellophane is penetrated by broad fan-like "rooting branches"; the cellulases C_1 and C_x have been detected (1265, 4779, 5884, 5886, 6328). Chitin is decomposed (4264); in hairs, the hyphae penetrate by both boring hyphae and spreading fronds (1633). When growing on maple-wood strips, the fungus causes marked losses in both weight and tensile strength (2211); weight losses have also been measured after 12 weeks on *Populus tremuloides* and *P. grandidentata* wood (3767).

Botryotrichum Sacc. & March. *1885*

Botryotrichum piluliferum Sacc. & March. *1885* — Fig. 60.

? = *Sepedonium xylogenum* Sacc. *1882*
? = *Sepedonium niveum* Massee & Salm. *1902*
  = *Coccospora agricola* Goddard *1913*
? = *Botryotrichum keratinophilum* Kushwaha & Agrawal *1976*

Teleomorph: *Chaetomium piluliferum* J. Daniels *1961*

DESCRIPTIONS of the anamorph: Blochwitz (558), White and Downing (6328), Hammill (2240), and Downing (1450); teleomorph: Daniels (1265); conidiogenesis: Cole and Samson (1109). — Colonies spreading broadly and reaching 2·0–4·3 cm diam in one week at about 20°C with ± abundant whitish aerial mycelium but sometimes later becoming buff from the often abundant production of brown, rough-walled sterile setae, 2·8–3·9 μm wide (reminiscent of the ascomatal hairs); setae may be entirely absent in some isolates; reverse sometimes orange. Aleurioconidia produced on repeatedly racemosely branched hyaline conidiophores, globose, (10–)13–15(–21) μm diam, wall 3·0–3·5 μm thick, with the surface irregularly encrusted; phialoconidia commonly produced in chains. It has not yet been possible to test the mating behaviour as single-spore cultures rapidly loose their capacity to produce ascomata (1265). Ascomata are scarcely produced and mature after four weeks at 25°C; they are black, globose to subglobose, 280–560 × 220–480 μm; lateral and terminal hairs olive-brown, 500–1500 μm long, 4–6 μm wide, minutely roughened, slightly sinuous with ± circinate tips. Asci clavate. Ascospores ellipsoidal with faintly apiculate ends, with one germ pore, pale brown, 13–16 × 8–10·5 μm. — The teleomorph is close to *Chaetomium murorum* Corda *1837* which has narrower ascospores (13–16 × 7–8 μm), longer ascomatal hairs and no aleurioconidia. *Chaetomium piluliferoides* Udagawa & Horie *1975* has similar conidiophores with aleurioconidia 5–7·5 μm diam and produces abundant ascomata 200–240(–280) × 120–145 μm with rather short terminal hairs (200–250 μm) and fusiform ascospores 20–26 × 7–8(–9) μm. Hawksworth (2337) emphasized that the anamorph alone is not sufficient to recognize *Chaetomium piluliferum*, since comparable *Botryotrichum* states occur in three *Farrowia* and eight other *Chaetomium* species which need more comparative studies. Of these, *Ch. piluliferum* is probably the most common and has the most complex conidiophores and (sub)hyaline aleurioconidia with a thick and characteristically ornamented wall. The other described species *B. atrogriseum* van Beyma *1928* and *B. peruvianum* Matsush. *1975* have pigmented aleurioconidia of similar size.

    B. piluliferum was originally discovered in Belgium on rabbit dung; it has since been found on deer and goat excrements in Denmark (4917) and on field mouse dung in England (IMI). It has also been found in various soils, though it is generally rather rare in them. This species is difficult to isolate by dilution plates, and selective techniques such as buried cellophane (1265, 5884, 5886), soil plates with cellulose agar (4647, 4655) or hair baits (2091) are more

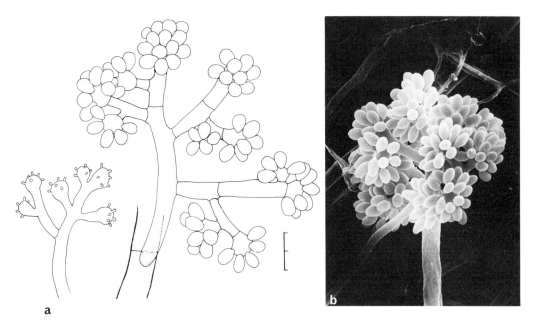

FIG. 61. *Botrytis cinerea*. a. Young conidiophores with beginning synchronous conidiation and half-mature conidia; b. a similar stage, SEM, × 800, orig. R. A. Samson.

Botrytis cinerea Pers. ex Nocca & Balb. *1821* — Fig. 61.

= *Sclerotium durum* Pers. ex St.–Am. *1821* (sclerotia)
= *Haplaria grisea* Link ex Leman *1821*
= *Phymatotrichum gemellum* Bonord. *1851*
= *Botrytis fuckeliana* Buchw. *1949*

For more synonyms see Hennebert (2400).

B. cinerea is a species complex connected to several distinct TELEOMORPHS including the plurivorous *Botryotinia fuckeliana* (de Bary) Whetzel *1945* (≡ *Peziza fuckeliana* de Bary *1869* ≡ *Sclerotinia fuckeliana* (de Bary) Fuckel *1870*), *Botryotinia convoluta* (Drayton) Whetzel *1945* (anamorph *Botrytis convoluta* Whetzel & Drayt. *1932*, on *Iris*), *Sclerotinia draytonii* Buddin & Wakef. *1946* (anamorph *Botrytis gladiolorum* Timmermans *1941*, on *Gladiolus*), *Botryotinia pelargonii* Røed *1949*, (anamorph *Botrytis pelargonii* Røed *1949*, on *Pelargonium*), *Botryotinia calthae* Henneb. & Elliott *1963* (anamorph *Botrytis calthae* Henneb. *1973*, on *Caltha palustris*), *Botryotinia ranunculi* Henneb. & Groves *1963* (anamorph *Botrytis ranunculi* Henneb. *1963*, on *Ranunculus septentrionalis*), and *Botryotinia ficariarum* Henneb. *1963* (anamorph *Botrytis ficariarum* Henneb. *1973*, on *Ficaria verna*).

DESCRIPTIONS: Gregory (2079), Groves and Loveland (2141), Hennebert (2397), Hennebert and Groves (2402), Menzinger (3757), Ellis (1603), and CMI Descriptions No. 431, *1974*; structure of sclerotia: Willetts (6371); conidiogenesis: Cole and Samson (1109). — Colonies spreading broadly, reaching 6·0 cm diam and more in ten days at 20°C on OA, at first hyaline but later becoming grey to greyish brown. Conidiophores arising irregularly, often in patches, without a basal swelling, 750 μm to over 2 mm long, brown below, smooth-walled, 16–30 μm wide, with an apical head of alternate branches. Conidia obovoid, usually with a protuberant hilum, pale brown, hydrophobic, smooth-walled (finely granulate under SEM; 2891), mostly 8–14 × 6–9 μm, length/width ratio 1·35–1·5(–1·7), liberated by hygroscopic movement of the shrinking conidiogenous cells. Micro-conidia as in numerous other species of the Sclerotiniaceae, globose, 2·5–3·0 μm diam. Sclerotia black, irregular in size and shape, to 15 mm diam, consisting of a densely packed medulla and a pseudoparenchymatous dark brown to black cortical layer of cells about 5–10 μm diam; these give rise to a tuberculate surface as viewed by SEM (6371). — Several similar *Botrytis* species with obovate conidia can be distinguished from *B. cinerea* as follows: *B. narcissicola* Kleb. *1906* (teleomorph *Botryotinia narcissicola* (Gregory *1941*) Buchw. *1949*) has slightly larger conidia 8–16 × 7·5–12 μm and small sclerotia 2–3 mm diam; *B. convoluta* Whetzel & Drayton *1932* has particularly large (to 16 × 18 mm), cerebriform sclerotia and grows on *Iris* species; *B. porri* (van Beyma) Buchw. *1949* (teleomorph *Botryotinia porri* (van Beyma *1927*) Whetzel *1945*) has slightly broader conidia 11–14 × 7–10 μm and particularly large, irregularly shaped sclerotia 10–40 mm diam; *B. ficariarum* Henneb. *1973* has creamy-white colonies, flat marginate small sclerotia, convex below, tardily blackening, and narrow conidia 10–15 × 5–7 μm, length/width ratio 2·0; *B. fabae* Sardiña *1929* has small sclerotia, usually 1–1·7 mm diam, and large conidia mostly 16–25 × 10–14 μm (5646); *B. pelargonii* Røed *1949* has rather short conidiophores 600–800 μm and sclerotia 1·5–7 mm diam, and grows on *Pelargonium* sp. — The other species of the *B. cinerea* complex can hardly be distinguished in the anamorphic state but only by details of sclerotial anatomy and ascoma morphology.

The colonial variability was ascribed by Hansen (2259) to the existence of a mycelial and a conidial basic type and used as an example of the "dual phenomenon". Menzinger (3759), however, showed that still greater possibilities for variaton exist in respect to conidial size, sclerotium formation, mycelial characters and the formation of micro-conidia. Segregation by repeated single-spore cultures led to the separation of six different types which remained constant; these recombined easily by heterokaryosis (3236, 3758). Hyphal cells and conidia are multinucleate and the conidial size is correlated with the number of nuclei included (3236); micro-conidia (spermatia) are always uninucleate. The species is heterothallic and shows bipolar compatibility (2141). — Ultrastructural studies show that the conidial wall is two-layered; the outer is an electron-dense, amorphous layer which ruptures during germination, irregular deformation of the mitochondria and elongation of ER strands occurring in the course of this process (763). Micro-bodies have been demonstrated in the hyphae (3691). — The major fatty acids of the mycelium are reported to be palmitic, oleic, linoleic and α-linolenic (2312).—DNA analysis gave 24·4% adenine, 25·0% guanine, 25·9% cytosine, 24·7% thymine or a GC content of 50·9% (3232).

B. cinerea has a worldwide distribution but occurs mainly in humid temperate and subtropical regions; it is very common in the phylloplane of plants but does not attack healthy leaves; it can be facultatively parasitic on a very wide range of plants, particularly of the Liliaceae and Asteraceae, causing blight or rot of leaves, flowers and fruits, the so-called grey mould, as it then covers the decayed tissue with conidiophores. Of economic importance are the fruit and leaf rots in grape, strawberries, cabbage, lettuce, etc., and damping-off and stem rot in lettuce and flax (430), seedling die-back or snow mould in forest nurseries and canker in woody plants (CMI Descriptions No. 431). In spite of some variation in virulence and host spectrum, it seems preferable not to distinguish formae speciales; 44 host plants are known in which *B. cinerea* may be seed-borne (4200). *B. cinerea* conidia are, however, typically air-borne and their terminal velocity in air has been determined as 0·22–0·45 cm sec^{-1} (2080).

They are liberated in particularly large numbers by rain-splash (2725).

B. cinerea is not regarded as a typically soil-borne fungus and is generally associated with plant residues in the superficial layers when in soil. A fluorescent antibody technique was used to trace this fungus on glass slides buried in soil (4627). Owing to its rapid growth, it can be isolated easily by various special techniques, such as screened immersion tubes and plates (982, 989). — Records of its occurrence include all of Europe, even the north of Scandinavia (1712, 1713), and also the USA, Canada, Israel (2762, 2764, 2772), Somalia (5048, 5049), Madagascar (4159), Jamaica (4886), Nepal (1826) and India (2854, 4700, 5622). It has been isolated relatively frequently from forest soil and litter, conifer swamps, and is also known from grassland, garden and arable soils, heathland, fen peat, silt, dunes and a saline sandy beach, salt marsh, mangrove mud, an ice cave, a uranium mine, a waste stabilization pond, sewage, and organic detritus in fresh water. Known soil types include chernozem, terra fusca, rankers and podzols. The individual reports are too numerous to be cited here. According to most of these, *B. cinerea* is particularly common in the upper soil layers, but has also been shown to be present down to 40–60 cm (3447). It has been isolated from various plant roots including cabbage (4451), sugar beet (4559), red clover (6493), alfalfa (3983), broad bean (6134), French beans (4451, 4452, 5759), strawberry (4133), oats (4443), wheat (3567), barley (4451), poplar (3452, 5898), *Salsola kali* (4665), and several Australian heathland plants (5819); it has been frequently reported from the rhizospheres of numerous plants but without a particular rhizosphere effect. *B. cinerea* can proliferate along lettuce plants down to the roots and spread from there to other root systems (4145). — Survival in soil is mainly facilitated by sclerotia, but both conidia and mycelia can survive for a considerable time. Hyphae on glass slides remained viable for 12 months in the range between vapour saturation and −70 bars water potential at 0°C without added nutrients (466); in the soil, survival for several weeks has been observed (4145). Conidia survive 2–6 months at 0°C, but only 1–3 months at 20°C and about −70 bars; they can survive in soil for several weeks, particularly at low temperatures and under dry conditions, but also in pure water (605, 2725). Sclerotia survive in dry soil for much longer than eight months, while in moist soil survival time can be less than one month (2725); the maximum temperature endured is 54°C for 14 days (2725). — Strains of *Bacillus, Pseudomonas* and *Chromobacterium* from dead lettuce leaves were found to be antagonistic to *B. cinerea* both *in vitro* and *in vivo, Bacillus* only above 15°C, the others in the range 4–25°C; an increase in pH to 7·8–8·4 *in vitro* may contribute to the inhibition; a mixed bacterial flora is even more efficient, causing lysis of the mycelium and germ tubes (4144). Various fungal genera and actinomycetes also are antagonistic, less so at 4°C than at 15–25°C, and their action is less dependent on pH (4145). *B. cinerea* is strongly inhibited *in vitro* by *Trichoderma harzianum* (6275). Its spread in sterile soil is much faster than in non-sterile soil (430, 2725). Prior colonization of plant tissues with *Cladosporium, Penicillium, Alternaria* or some bacteria prevents attack by *B. cinerea* (2725). The germination of conidia was inhibited *in vitro* by a sporostatic factor produced by various fungi (4882) but *B. cinerea* shows a comparatively low sensitivity to soil mycostasis (2559). Volatile substances from several fungi (1371), as well as acetaldehyde, ethyl acetate and others (54) can also inhibit germination. Bacteria of the phylloplane probably absorb the available nutrients in their polysaccharide slime sheath and thus prevent conidium germination (729, 5679); consequently applications of chloramphenicol and streptomycin improved germination on leaves (551). The sclerotia of *B. cinerea* can be parasitized by *Coniothyrium minitans, Rhizopus stolonifer, Gliocladium roseum, Trichoderma viride, Trichothecium roseum, Chrysosporium pannorum* and other fungi (1651, 2725).

Variations in growth and sporulation of *B. cinerea* must be considered with some reservation in view of the variability between different isolates. Conidium production is particularly variable, it can be stimulated by NH_4NO_3 (4472) or asparagin (2725) and light (4472), particularly near-UV (wave lengths 290–400 nm) (2468); sorbose, glycine or urea are inhibitory (2725). The temperature minimum for sporulation is 12°C in the field, the optimum 15°C (2725); superficial drying of the substrate also promotes sporulation; the minimum water potential is −70 bars (2093). — Sclerotia of the related *B. convoluta* have been induced to produce up to six crops of conidia at 5°C by exposing them to near-UV for 24 h, while the sclerotial weight decreased to 32% (2707). Near-UV can also induce sporulation in cultures pre-incubated in darkness (5726); photosensitivity is greatest after 4–5 days, cultures older than 10 days being unresponsive (2725). Blue light (wave lengths 380–525 nm) causes a reversible inhibition of photo-induced sporulation (5721, 5722, 5724, 5725, 5727), but near-UV or far-red (wave lengths >720 nm) reversed this effect (5723, 5725); these observations have been explained by the presence of a sporogenic substance, mycochrome, with an absorption maximum between 305 and 310 nm (2469, 5725) and possibly a second light receptor (5727) responsible for the far-red effect. Another more plausible explanation is to postulate an intermediate derivative of mycochrome which responds to far-red reactivation, the insertion of a dark period before reactivation reducing its effect (5724, 5725). Polarized light causes conidial germ tubes to grow in the plane of polarization (771); this phenomenon was explained by the assumption that a photoreceptor is localized in double-refracting dichroic structures of the hyphae (771). Conidiophores are positively phototropic, responding to near-UV or, most efficiently, to blue light of wave length 420 nm (2724). Illumination (including UV) also increased the free amino acid content in the mycelium (particularly tryptophan but not glutamic acid), but UV at 36 mW cm^{-2} for 30 min destroyed some protein (2725). — The germ tubes of densely sown conidia show a positive (2708) autotropism but a negative one when the medium is enriched with CO_2 (4883). Conidia do not germinate in distilled water but require 100 ppm of sugars, whilst amino acids or growth factors have no effect and do not increase germination when added to the sugars (1023); according to another report, biotin and thiamine (contained in a lentil extract) stimulated it (2725). Conidium germination can also be stimulated by substances diffusing from various pollen grains (646, 1025) and by cuticular waxes from conifer needles (5145); volatile exudates from the leaves or shoots of coniferous trees can both stimulate or inhibit germination, depending on the pretreatment of the leaf material (5146); phenolics from tomato leaf leachates are also inhibitory (1022). Conidia can germinate in the pH range 1·6–6·9 on mannitol, or pH 2·0–9·8 on sugar beet decoction, with a broad optimum between pH 3 and 7 (2725). Conidia can germinate between 5 and 32°C (5294); optimal germination has been reported at 15°C (5301) or 20°C, whilst subsequent growth was optimal at 30°C (2401, 5294). At −70 bars water potential 100% conidia germinate at 20°C and 80% at 5–15°C, at –140 bars 85% germinate at 20°C, and none at 5–15°C, while none germinate at all at −210 bars (2093, 2725). The lag phase of germination is shortest, at about 3 h, in rather young conidia (16-day-old cultures) but in extremely young or old and dry conidia it is 1–2 days; appressoria form at the tips of the germ tubes within 4–6 h at 20°C, but not at 30°C (5294). — Sclerotium formation has converse requirements to conidiogenesis, darkness and higher sugar concentrations being favourable (2725); three phases in their development have been distinguished: (a) initiation is best at high carbon concentrations, high nitrogen levels also being favourable, (b) development favoured by rich media with a high C/N ratio in moist and dark conditions, and (c) maturation and pigmentation after the nutrients are exhausted and, preferentially, at low temperatures. The pH optimum for sclerotium formation is 4, none are formed in the alkaline range, and the optimal temperature for their production is 11–13°C (2725) or 20°C (2401). Maturation may be favoured by an even lower pH of 2·5, ammonium oxalate (2869), a low temperature of 10–12°C or less and darkness (3145, 4472). Blue light has a suppressive effect whilst near-UV is not inhibitory if applied for less than 30 min (5726). The optimal temperature for germination of sclerotia is 27°C and the minimum is 3°C (3144). — The cardinal temperatures for growth are a minimum (−2–)5–12°C, an optimum (21–)22–25(–30)°C (2401) and a maximum of (28–)33–35°C (191, 2401, 5301); certain isolates are adapted to growth at low temperatures with a minimum of −2°C (734, 3153). The thermal death point (in apple juice) has been determined as 63°C for 25 min (3620). The pH range for growth extends from 2 to 8 (2725), the optimum being 3–5. The minimal water potential for vegetative growth is at −100 bars (2093). *B. cinerea* can grow with a limited O_2 supply, down to 1% or 1·4% (21, 1425, 1793), but the growth rate is then considerably reduced. A raised CO_2 concentration of 8% stimulates growth if 2% O_2 is supplied and only concentrations above 20% of CO_2 are inhibitory (2725); CO_2 fixation occurs (3106). — Growth (and sporulation) are favoured by media rich in nitrogen and phosphorus; a C:N:P ratio of 100:4:1 has been recommended for good growth and sporulation (1445). The enzymatic oxidation of glucose (hexose-monophosphate pathway) has been studied (1936) as has the role of glyoxalate transaminases in this process (1819). *B. cinerea* utilizes most sugars but not usually pentoses, mannose, lactose, raffinose, L-rhamnose, sorbose and inulin; it also grows on some glycosides, malic, tartaric and lactic acids, while citric and oxalic acids are poorly utilized and pyruvic, acetic and propionic not at all (2725, 3477). *B. cinerea* is one of the fungi with the greatest pectolytic activities (1432, 2250, 3414, 4237, 4779); the presence of endo- and exo-polygalacturonases (PG), pectin methylesterase and transeliminase have been demonstrated (1854, 2250, 2844, 5139, 5268, 6064); the efficiency and products of the PG have been studied in detail (657, 2752). Virulence and pectinase production of different isolates were not correlated (4472); the macerating effect of culture liquids could be increased by increasing the concentration of asparagine, but was reduced by adding glucose or sucrose (217). *B. cinerea* can grow on glucuronic or galacturonic acids as the sole C sources (4749). Amylolytic and proteolytic properties have been described (4545) and cellulase activities demonstrated on filter paper, cotton fabric (1425, 1828, 2718, 2844, 4184), wood cellulose and alkali jute (386), cellulose hydrate discs (3076) and CMC (1432, 2250, 2844). The degree of cellulose decomposition increases with raising N concentrations (4433). Araban (2843), arabinoxylan (mainly by the action of xylanase) (1753), cutin (3361) and usnic acid (322) can be decomposed. Tannin could be used as the sole C source at concentrations of up to 5% (1425); growth also occurs on gallic acid (1750), and polyphenol oxidase (laccase) is produced on several substrates (934, 1474). *B. cinerea* is rather osmotolerant and can grow with up to 2M sucrose (2725); increased osmotic values are required for sporulation. It is regarded as a non-staling fungus with unlimited mycelial growth, but inositol is accumulated in old cultures (2725) while most mannitol and all arabitol disappear in autolysing liquid cultures of 35–130 days age (3211). — *B. cinerea* can grow on a wide range of N sources, ammonium being less suitable than nitrate, asparagine and casein hydrolysate giving optimal growth (2725, 3477). Sulphate, persulphate, sulphite and sulphhydryl-S can all be utilized as sulphur sources (2725). A phosphatidase releases fatty acids and glycerylphosphoryl choline from lecithin. Zinc ions are essential for growth, and those of molybdenum, copper and iron all have a favourable influence (2725). With the possible exception of conidium germination, *B. cinerea* is prototrophic (1425). — Known metabolic products include oxalic, citric (5139), and glyoxylic acids, a low-molecular weight protein, urea, thiourea (1936, 5278), the antibacterial and fungistatic sesquiterpene botrydial (1714), and two phytotoxic polysaccharide fractions (83, 2870). The phytoalexin wyerone acid from *Vicia faba* can be metabolized (3585). — *B. cinerea* is highly sensitive to both SO_2 (2725) and ozone, the latter being inhibitory at concentrations above 0·5 ppm (3514). The development of fungicide resistance in *B. cinerea* has repeatedly been observed (2725); most remarkable is the appearance of isolates resistant to benzimidazole fungicides since *1971* (616, 1779) for which a one-gene mutation is responsible.

Broomella Sacc. *1883*

Broomella acuta Shoem. & E. Müll. *1963* — Fig. 62.

Anamorph: *Truncatella truncata* (Lév.) Stey. *1949*

   ≡ *Pestalotia truncata* Lév. *1846*
  = *Truncatella ramulosa* (van Beyma) Stey. *1949*
   ≡ *Pestalotia ramulosa* van Beyma *1933*

For numerous other synonyms see Guba (2148), and Steyaert (5573).

DESCRIPTIONS: Steyaert (5573), Guba (2148), Gams and Domsch (1887), and Matsushima (3680); teleomorph: Shoemaker and Müller (5305). — The melanconiaceous genus *Pestalotia* de Not. *1839* has been delimited in various ways. Guba (2148) used the number of terminal conidial setulae to separate *Monochaetia* (Sacc.) Allesch. *1902* and subsequently distinguished three sections within *Pestalotia* according to the number of conidial cells; Steyaert (5573), however, found the number of conidial cells to be a more reliable character and separated *Truncatella* Stey. *1949* (=*Pestalotia* sect. *Quadriloculatae* Guba *1929*) with two pigmented central cells from *Pestalotiopsis* Stey. *1949* with three and *Pestalotia* de Not. with four central pigmented cells, but included *Monochaetia* in *Pestalotiopsis*; his generic concepts have not been followed by most mycologists (e.g. 1473, 5305). Subsequently, Sutton (5650) has recognized not only Steyaert's genera but also *Seiridium* Nees ex Fr. *1821* emend. Sutton *1969* (a counterpart of *Pestalotiopsis* with 6-celled conidia) and *Monochaetia* because of its simple apical setulae and endogenous basal appendage which is narrower than the basal conidial frill.

Truncatella is thus characterized by acervuli (in culture appearing as sporodochia) with hyaline phialidic or annellidic conidiogenous cells and fusiform conidia with two pigmented central and two hyaline end cells bearing one to several terminal hyaline, filiform setulae and often a simple basal axial appendage. The variable species concepts are reflected by the numbers of 5 and 34 species recognized in this group by Steyaert (5573) and Guba (2148), respectively. Since these fungi are saprophytes or weak parasites and not host-specific, it seems likely that Steyaert's system is closer to reality.

The genus *Broomella* (=*Keissleria* Höhn. *1918*; type species *B. vitalbae* (Berk. & Br. *1859*) Sacc. *1883*) has been placed in the Amphisphaeriaceae Wint. and characterized by solitary or aggregate dark brown, short-beaked ascomata, unitunicate asci with a subapical chitinoid body (pulvillus) but no amyloid ring, 3-septate ascospores pigmented in the central cells, with a simple setula at either end. Anamorphs in *Truncatella* have one or several terminal conidial setulae (5305). — Three further *Broomella* species have been collected on *Clematis* species, the anamorphs of which can be distinguished as follows (5305): the type species *B. vitalbae* (in southern France) with pale ochraceous slender conidia, $35–45 \times 5–7\,\mu$m, and a single terminal setula; *B. excelsa* Shoem. & E. Müller *1963* (in Pakistan) with dark brown conidia, $17–22 \times 7–10\,\mu$m, and a single terminal setula; and *B. montaniensis* (Ell. & Everh. *1888*) E. Müller & Ahmad *1955* (in the USA) with warted brown conidia, $25–30 \times 8–10\,\mu$m, and 3–5

156

terminal setulae. — Another similar species is *Truncatella hartigii* (Tubeuf) Stey. *1949* (≡*Pestalotia hartigii* Tubeuf *1888*), causing stem infections in forest tree seedlings, which differs in having darker central conidial cells (5573).

Colonies of *Truncatella truncata* reach 5·3 cm diam in ten days at 20°C on OA. The conidia are 17–19(–21) × 6–8 µm, have rather pale brown, smooth-walled central cells and 1–3(–5) terminal setulae 7–30 µm long, arising at short distances beyond the conidial tip from the central axis. Hyaline, sickle-shaped, finely guttulate, 1-celled microconidia, 20–30 × 1·0–1·5 µm have also been observed in culture (5305). The teleomorph has been collected in southern France on stems of *Clematis flammula* and ascosporic isolates have proved the connection with the anamorph.

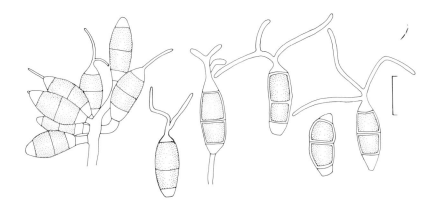

FIG. 62. *Broomella acuta*, attached and liberated conidia.

Species of *Pestalotia* sensu lato are known primarily as saprophytic inhabitants of plant substrates. Types with three dark-coloured central cells (*Pestalotiopsis*) are found especially in warm climates, while ones with two dark-coloured central cells (*Truncatella*) occur mostly in the temperate latitudes, *T. truncata* (*Broomella acuta*) being the commonest soil-borne species amongst them. — A worldwide distribution for *B. acuta* is indicated by reports from Alaska (3062), Austria (2530), Nepal (1827), New Zealand (5812), central Africa (3063), Brazil, India, Canada, the British Isles, Sweden and Cyprus (IMI). Soil isolates have been obtained from beech forest soil (4813), litter under *Abies grandis* (681), forest nurseries (6184), grassland (5812, 6182), arable soil (540, 918, 1889, 2161, 3450), a vineyard (4921) and estuarine mud (5316). This species is also common on plant substrates (IMI, 4917), including leaves of various grasses (191), seeds of *Avena fatua* (2961), and wood pulp (699). Occurrences in soil depths to 40 cm have been observed. Manuring has either no influence on its frequency or causes a slight reduction in it (2161). The fungus has been isolated from the rhizospheres of wheat (3567), dwarf beans (4452), *Calluna* (4814) and poplar (5898). It has also been reported from cultures of the collembolan *Hypogastrura* (1027) and can serve as food for the mite *Pygmephorus mesembrinae* (3104).

The optimal temperature for growth is 21°C but good growth occurs in the range 15–25°C, the maximum is 28°C and the minimum −3°C (191). Starch is degraded (1827), and starch or cellulose-containing substrates, as well as near-UV light, may stimulate sporulation which otherwise is often lost in culture. Decomposition of both xylan and cellulose is very good, although differences in this capacity occur between isolates (1432). *B. acuta* has been listed as one of the causal agents of soft rot in timber (1496). It exerts a specific antagonism to *Pythium ultimum* (1431), and produces the hexahydroisocoumarin derivative ramulosin (448, 5590) which has antimycotic properties (2374, 2427). At 10 ppm, ramulosin inhibits the germination of wheat and oats, at 100 ppm that of tomatoes and grasses, and at 1000 ppm that of corn, clover and *Sorghum* (2374). *B. acuta* can cause injuries to the roots of wheat, pea, rape (1430) and clover (6493).

Cephaliophora Thaxt. *1903*

Cephaliophora tropica Thaxt. *1903* — Fig. 63.

DESCRIPTIONS: Bainier (289), Wolf (6417), Agnihothrudu and Barua (42), Crook and Hindson (1224), Mayoral *et al*. (3695), Takada and Udagawa (5687), Subramanian (5626), Ellis (1603), and Matsushima (3680); cytological study: Visarathanonth (6098).

The Hyphomycete genus *Cephaliophora* (=*Cephalomyces* Bain. *1907*; type species *C. tropica* Thaxt., lectotype) is characterized by a subhyaline mycelium bearing numerous sessile or stalked ampulliform conidiogenous cells from which a number of blastoconidia are produced simultaneously; conidia one- to pluriseptate, hyaline to brown, clavate to cylindrical or occasionally lobed.

The two known species of this genus can be distinguished as follows: *C. tropica* has (1–)3–5(–8)-septate ellipsoidal to cylindrical conidia 14–21 μm wide, while *C. irregularis* Thaxt. *1903* (=*Cephalomyces nigricans* Bain. *1907*) has mostly only 1-septate, broadly clavate (or sometimes Y-shaped) conidia, 25–40 μm wide. The cells generally contain more than one nucleus, the average being 5–10; hyphal anastomosis has been observed; conidial heads originate from aerial hyphae showing knob-like protrusions into which most nuclei of the hyphal cell migrate (6098).

Colonies of *C. tropica* grow very fast with 12 mm radial daily increments at 25°C and 18 mm at 35°C, are deeply floccose, hyaline at first but later becoming brown and bearing abundant conidiophores. Chlamydospores arising as short lateral branches from aerial hyphae or as intercalary hyphal swellings, developing globose to broadly clavate conidiogenous ampullae 20–30 μm diam on which several (12–20) blastoconidia develop simultaneously from narrow denticles; short blunt denticles are left after conidium liberation. Conidia ellipsoidal to cylindrical, with broadly rounded apical cells (rarely with bilobed, Y-shaped apex) and soon emptied conical and apiculate basal cells, 3–5 (–8)-septate, hyaline but later becoming pale brown, smooth- and thick-walled, 28–45(–60) × 14–21 μm. Hyphal and conidial cells are multinucleate (5–10, up to 17 nuclei) (6098).

Both species of *Cephaliophora* are mainly coprophilous, and *C. tropica* is known principally from tropical and subtropical countries, including Australia (1224), Ghana, Indonesia, India (42), Jamaica, China, Liberia, Pakistan, and the Solomon Islands (1603), but is also known from the British Isles, Germany (1603), Hungary (5862) and Japan (3680, 5687). It was isolated originally from mongoose dung but has subsequently been obtained from deer (5862) and rabbit dung (6098), tobacco leaves on soil (6417), cocoa, shoe-leather (3695) and frequently the tea rhizosphere (42), *Piper betle* leaves (IMI) and silk worm excrements (5687).

At 25°C the growth rate is as great as at 37°C, although sporulation is less at the higher temperature (5687); the fastest growth was observed at 35°C but none occurs at 10° and 40°C (3695). D-fructose, D-glucose, D-galactose, D-mannose, cellobiose and D-xylose are all suitable C sources, whilst L-sorbose suppresses growth; starch and cellulose are not utilized (3695); all

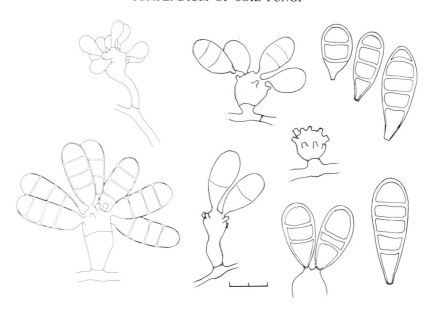

F₁G. 63. *Cephaliophora tropica*, conidiogenous cells with synchronous formation of blastoconidia (various isolates).

nine N sources tested (in the order ammonium nitrate and sulphate > sodium nitrate ~ urea and gelatin > glycine > neopeptone > L-asparagine > L-tryptophan) supported good growth; sporulation was best on sodium nitrate but poor on both ammonium salts and organic N sources apart from urea; the optimal pH is in the range 6·4–6·9, morphological alterations occur on acid media (6257).

Ceratocystis Ellis & Halst. *1890*

Ceratocystis fimbriata Ellis & Halst. *1890* — Fig. 64.

≡ *Ophiostoma fimbriatum* (Ellis & Halst.) Nannf. *1934*
≡ *Endoconidiophora fimbriata* (Ellis & Halst.) Davids. *1935*
= *Rostrella coffeae* Zimm. *1900*
= *Ceratocystis variospora* (Davids. *1944*) C. Moreau *1952*

Anamorph: *Chalara* (*Chalaropsis*) sp.

DESCRIPTIONS: teleomorph: Hunt (2612); anamorph: Nag Raj and Kendrick (4070), and CMI Descriptions No. 141, *1967*. — Species of *Ceratocystis* (type species *C. fimbriata*), mostly blue-staining wood-inhabiting fungi, are characterized by small globose perithecia with very long apical beaks, no paraphyses, and numerous evanescent globose asci with one-celled, colourless small spores of variable shape. Weijman and de Hoog (6256) presented convincing chemical evidence for the restriction of the genus to those species having phialoconidia of the *Chalara* type, species with branched conidiophores being referred to *Ophiostoma* Syd. *1919*.

C. *fimbriata* is characterized by ellipsoidal ascospores, flattened on one side, 4·5–8·0 × 3·5–6·5 μm, and surrounded by a sheath with a hat-shaped brim; this has been studied by TEM (5580, 6249). It is distinguished from the related *C. moniliformis* (Hedgc.) C. Moreau *1952* by the unornamented base of the perithecia (i.e. lacking setae) and the absence of terminal chlamydospores. Colonies spreading fairly rapidly, reaching about 4 cm diam in ten days at 24°C on OA or MEA, hyaline at first but later becoming light brown, loosely cottony. Conidium production beginning within two days and mature perithecia in fresh isolates (homothallic or heterothallic) forming within one week. Phialides hyaline, differentiated into a venter 14–33 μm long, and a collarette 16–37 μm long. Conidia either cylindrical with truncated ends, hyaline or pale brown, and usually 11–16(–25) × 3·5–5(–7) μm, or doliiform, slightly pigmented and 5–11 × 5–10 μm. Chlamydospores are produced singly and one-celled or in short chains, terminal, ellipsoidal to subglobose, brown, smooth and thick-walled, 10–20 × 7·5–13 μm. Isolates of different origins are often variable with respect to colony type, conidium production and pathogenicity, but are usually interfertile (6249). — The hyphal cell walls contain chitin, galactan and mannan (5493), but no cellulose (5443) or rhamnose (6256).

C. *fimbriata* has a worldwide distribution (2612, CMI Distribution Map No. 91, *1965*) and is very common in the tropics. As a parasite or saprophyte it occurs mostly on the bark and wood of many woody plants (2106, 4271, 6249) and various tubers (2612). Some strains cause a black rot of roots and stems of sweet potato (*Ipomoea batatas*), or cankers in stone fruit and other trees. Distribution occurs by wind, rain and insects (CMI Descriptions No. 141). It can be isolated selectively from insect vectors and stone fruit cankers by means of carrot disks (3907) but no direct isolations from soil appear to have been made. Several insects have been found to carry *Ceratocystis* canker (2457, 3906).

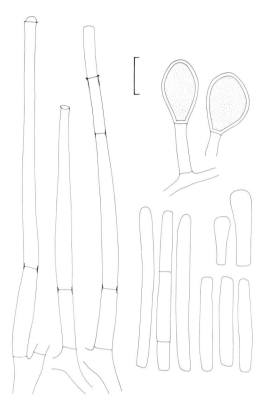

Fig. 64. *Ceratocystis fimbriata*, *Chalara*-type phialides with conidia and chlamydospores, CBS 103.40.

C. fimbriata requires thiamine for the production of perithecia and the optimum growth temperature lies in the range 24–27°C (6249). Ethylene is produced on PDA but a high degree of variability occurs between isolates (942). Other metabolic products include cinnamic acid, ipomeamarone, ipomeanin, batatic acid, furan-3-carboxylic acid (5278), propylacetate, and 8-hydroxy-6-methoxy-3-methylisocoumarin (3231). Pectolytic enzymes facilitate its penetration of plant tissues.

Chaetomidium (Zopf 1881) Sacc. 1882

Chaetomidium fimeti (Fuckel) Sacc. 1882 — Fig. 65.

≡ *Chaetomium fimeti* Fuckel *1861*
≡ *Thielavia fimeti* (Fuckel) Malloch & Cain *1973*
= *Chaetomidium subfimeti* Seth *1967*

DESCRIPTIONS: Bainier (290), Malone and Muskett (3556), Seth (5214), Malloch and Cain (3553), Udagawa and Furuya (5975), v. Arx (207), and Hawksworth and Wells (2340); ascoma development: Whiteside (6337). — *Chaetomidium* is a non-ostiolate counterpart of *Chaetomium*. It has been included in *Thielavia* by Malloch and Cain (3553) but separated again by von Arx (207) on account of its rather thick-walled, pseudoparenchymatous ascoma walls. Six species are now recognized.

Ch. fimeti has rather fast growing colonies, reaching 7·0 cm diam in ten days at 20°C on OA, with yellow-ochraceous, submerged mycelium, forming dense grey-brown layers of ascomata; odour spicy. *Acremonium*-like phialoconidia are present in some isolates. Homothallic. Ascomata 250–520 μm diam, (sub)globose, greyish green from the hairs. Hairs of two kinds: (a) dark brown, up to 3000 × 5–7 μm, slightly flexuous, almost smooth, arising from the ascoma base, (b) greyish-green, up to 330 × 3–5 μm, roughened with cupulate structures

FIG. 65. *Chaetomidium fimeti*. a. Perithecium with dark hairs arising from the base and paler hairs covering the entire surface, SEM, × 150; b. ascus and ascospores; CBS 827.71.

(2340), arising from all over the ascoma; both types unbranched. Asci clavate. Spores lemon-shaped, flattened, dark brown, (9–)11–13(–16) × 8–10(–13) × 7–8 μm. — *Chaetomidium subfimeti* Seth was separated by its smaller ascospores, 7·5–10 × 6·5–8·0 × 5·5–6·5 μm; the occurrence of intermediate isolates in Japan (5975) makes this distinction uncertain.

Very few data are available on the distribution of *Ch. fimeti* in soils. The fungus has been reported from arable soils in France (2161, 2162), California (IMI), the USSR (2871, 4703), rice fields (1519) and grassland soil (IMI) in India and soils in Japan (5975). *Ch. fimeti* has been found only in the uppermost 0–5 cm layer of soil (2161). It has also been isolated from garden compost (1425), wood buried in the soil (290), dung and plant remains (3113, 4548, 4917, 5215), birds' feathers and nests (2574, 2575), oat seeds (3556), leaves and paper (IMI). — It can serve as food for the mites *Pygmephorus mesembrinae* and *P. quadratus* (3104).

Cellulose decomposition is very good (1425); and maple-wood strips lose most of their tensile strength after three months (2211); it is autotrophic for growth substances. It also appears to have a low antibiotic activity against *Rhizoctonia solani* and a *Pythium* species (1425), but, conversely, the growth of the alga *Chlorella pyrenoidosa* is stimulated *in vitro*.

Chaetomium Kunze ex Fr. *1821*

= *Vanhallia* Marchand *1828*
= *Bolacotricha* Berk. & Br. *1851*
= *Bommerella* Marchal *1885*
= *Chaetomiotricha* Peyr. *1914*

Fam. Chaetomiaceae Wint. — Contributions to the MONOGRAPHIC TREATMENT: Bainier (290), Skolko and Groves (5402, 5403), Ames (124), Seth (5216), Hawksworth and Wells (2340), Millner (3825), Millner et al. (3827), and Dreyfuss (1469). — Ascomata mostly superficial, attached to the substrate by rhizoids, subglobose to vasiform, with lateral and terminal, septate hairs, which may be variously branched or contorted; with a mostly coarse ornamentation (2340). Ascoma wall consisting of several pseudoparenchymatous layers. Asci thin-walled, without apical structures, with a slender stalk, either cylindrical with uniseriate spores (20 species) or clavate with ± biseriate spores (most species), deliquescing ± simultaneously at an early age. Ascospores one-celled, pigmented, smooth-walled, of varying shape, mostly provided with one germ pore which is apical in all species treated here (2420, 3827), in most species binucleate. Most species are homothallic (5172). An *Acremonium*-like anamorph occurs in several species, particularly the heterothallic ones, with solitary aculeate phialides scattered over the aerial hyphae producing pyriform to dacryoid conidia in chains or heads; this state has little diagnostic value. Chlamydospores or aleurioconidia of *Botryotrichum* or *Humicola* types occur in some species (2337).

Within this genus 160–180 species have been recognized (2337, 3825) of which 56 have been described in recent years. The published descriptions are generally very incomplete with regard to ascoma wall structures, hair surface and ascospore details and rely too much on the variable characters of hair shape. Descriptions have also often been based on secondary collections and thus deviate from the original species concept. The identity of several species will require confirmation by the study of the type specimen; some studies in this direction were carried out by Hawksworth and Wells (2340) and Millner (3825). It should be noted that the types of most early epithets in the genus have not been examined since the time of Chivers (*1915*); the works of Ames (124) and Seth (5216) are not monographs in the accepted sense but rather uncritical compilations. Whilst most authors use the branching and coiling of the terminal hairs as the principal criteria for a first subdivision of the genus, Sörgel (5466) emphasized the reliability of ascospore features and suggested a subdivision into 13 species groups, which has been followed with some modifications by Dreyfuss (1469). The shape of the asci is generally a clear-cut criterion for a first subdivision of the genus, but in two species both clavate and cylindrical asci have been found in the same ascoma (3825). Wall structure is of potential value in the taxonomy of the genus (3825). The occurrence of morphologically similar homothallic and heterothallic isolates has in some cases been taken as criterion for species separation (1469, 5172); the interaction between self-fertile and self-sterile isolates has, as yet, only been studied for a few species (4006).

Chaetomium species are important as agents in the decay of cotton and other cellulosic man-made materials but have also been recognized as active agents of soft rot in wood and fruit rot in apple (3825). They also occur commonly on dung (3400, 3825) and straw. On the whole, they are important components in the decomposition of herbaceous and lignified plant matter reaching the soil. Six of the species treated here occur commonly on birds' feathers. mainly in the winter, particularly with eusynanthropic, terrestrial, polyphagous bird species and in nests of species which contain cellulosic or keratinic material with low or moderate moisture (2576).

Good sporulation is obtained in most species on substrates of a high C/N ratio, where the carbon source is provided in an insoluble form such as starch or cellulose and little soluble carbohydrate is added; suitable media are thus CMA (sometimes with 0·3% yeast extract), OA, PCA or 1–2% MEA with cellulose. Filter paper strips added to PCA prove particularly valuable. Very rich media encourage the production of aerial mycelium rendering perithecial hairs less amenable to examination under the dissecting microscope. Asci are studied in young and the perithecial hairs in ascomata at least two weeks old. Spores mounted in lactic acid must be heated to counteract shrinkage.

Key to the species treated and other selected soil-inhabiting species:

1   Asci cylindrical   **2**
         Asci clavate   **6**

2(1)   Asci 8-spored   **3**
         Asci 4-spored; terminal hairs loosely coiled and branched   *Ch. tetrasporum* (p. 182)

3(2)   Terminal hairs stiff and coiled, even in thickness; ascospores with an apical non-papillate germ pore; mesophilic   **4**
         Terminal hairs flexuose, dichotomously branched, not coiled, uneven in thickness; ascospores subglobose with an apical papillate germ pore, 7–9(–10) μm diam; thermophilic
                                     (*Ch. thermophilum*)

4(3)   Terminal hairs not or very rarely branched   **5**
         Terminal hairs with 2–3(–5) coils below but bearing numerous secondary paler-coloured much-branched arcuate to coiled hairs; ascospores 5·5–8 × 4–6 μm   (*Ch. medusarum*)

5(4)   Terminal hairs with numerous regular tight coils decreasing in diam towards the apex; ascospores 7·5–8·5 × 6·5–7 μm   (*Ch. brasiliense*)
         Terminal hairs with irregular coils of varying diam coiling in alternate directions; ascospores 9–12 × 6–9·5 μm   *Ch. crispatum* (p. 171)

6(1)   Aleurioconidia present   **7**
         Aleurioconidia absent   **8**

7(6)   Ascomata obpyriform, inconspicuous, with rudimentary terminal hairs forming a short tube; aleurioconidia hyaline to fuscous, formed on little-branched conidiophores, 7–10 μm diam; ascospores 7·5–9(–10) × 7–8·5 μm   cf. *Farrowia seminuda* (p. 301)
         Ascomata subglobose or globose, terminal hairs long and with circinate tips; aleurioconidia hyaline, formed on richly-branched conidiophores, 7–20 μm diam; ascospores 13–16 × 8–10·5 μm (cf. *Botryotrichum piluliferum*, p. 146)   (*Ch. piluliferum*)

8(6)   Terminal hairs frequently regularly dichotomously branched   **9**
         Terminal hairs unbranched or not regularly dichotomously branched   **11**

9(8)   Ascomata to 275 μm diam; ascospores not exceeding 8 μm in length   **10**
        Ascomata mainly 350–500 × 230–400 μm; ascospores lemon-shaped, 10–12·5 × 8–9·5 μm
                                     Ch. elatum (p. 172)

10(9)   Terminal hairs of two kinds, (a) unbranched long and straight, and (b) shorter dichotomously
        branched at acute angles; ascospores 5·5–6·5 × 3·5–5 μm   *Ch. funicola* (p. 174)
        Terminal hairs of one kind, dichotomously branched at obtuse angles, branches mainly strongly
        reflexed; ascospores 5–7 × 4–5·5 μm   *Ch. indicum* (p. 180)

11(8)   Terminal hairs undulate and flexuose   **12**
        Terminal hairs arcuate, coiled or with circinate tips   **15**

12(11)   Terminal hairs all unbranched; ascospores regular in shape   **13**
        Terminal hairs with occasional short lateral branches; ascospores very irregular in shape,
        13–25 × 7–16 μm; heterothallic   *Ch. nozdrenkoae* (p. 181)

13(12)   Ascomata ostiolate with the longest hairs arising around the ostiole   **14**
        Ascomata non-ostiolate with the longest hairs arising from the base of the ascomata; ascospores
        10–16 × 8–13 μm in face view   cf. *Chaetomidium fimeti* (p. 163)

14(13)   Ascospores broadly fusiform, with a subapical germ pore, 10–13 × 7–8 μm   (*Ch. jodhpurense*)
        Ascospores lemon-shaped, with an apical germ pore, 8·5–11 × 7–8·5 μm in face view
                                     Ch. globosum (p. 176)

15(11)   Terminal hairs erect and spreading to decumbent, not arcuate   **16**
        Terminal hairs strongly arcuately incurved forming a dense mass; ascospores 8–12·5 ×
        5–8·5 μm   (*Ch. gracile* = *Ch. arcuatum* = *Ch. erraticum*)

16(15)   Terminal hairs undulate or straight below with circinate tips   **17**
        Terminal hairs coiled from the base or only towards the tips   **18**

17(16)   Terminal hairs undulate below; ascospores with an apical germ pore, 13–16 × 7–8 μm; colony
        reverse white   (*Ch. murorum*)
        Terminal hairs straight below; ascospores with two apical germ pores, 7–10 × 5–6 μm; colony
        reverse red   (*Ch. trilaterale* = *Ch. cupreum*)

18(16)   Terminal hairs with numerous coils when fully developed   **19**
        Terminal hairs with 1–5 coils when fully developed   **20**

19(18)   Peridium of textura intricata; terminal hairs with 7–15(–20) coils decreasing in diam towards the
        tips, 3–5·5 μm thick at the base, not readily becoming detached from the ascomata en masse;
        ascospores brown, not distinctly angularly guttulate, 6–12 × 5·5–9 μm   (*Ch. spirale*)
        Peridium of textura angularis; terminal hairs with 3–12(–15) coils decreasing in diam towards the
        tips, 4–5·5 μm thick at the base, easily becoming detached from the ascomata en masse joined
        to the blackened apical cells of the ascomata; ascospores very pale olivaceous, generally
        distinctly angularly guttulate, 6·5–8·5 × 6·5–7·5 μm   *Ch. bostrychodes* (p. 168)

20(18)   Ascospores 5–6·5 × 4·5–5·5 μm; terminal hairs with 1–3 very uneven coils, coiling from the
        base   (*Ch. abuense*)
        Ascospores 8–10 × 7–8 μm; terminal hairs with 3–5 loose coils decreasing in diam towards the
        base, coiling starting some distance from the base   *Ch. cochliodes* (p. 169)

Chaetomium bostrychodes Zopf *1877* — Fig. 66.

= *Chaetomium formosum* Bain. *1910*
= *Chaetomium hyderabadense* Salam & Nusrath *1959*
= *Chaetomium guizotiae* Tilak & Reddy *1964*
= *Chaetomium biapiculatum* Lodha *1964*

DESCRIPTIONS: Bainier (290), Skolko and Groves (5403), Udagawa (5968), Ames (124), Hawksworth and Wells (2340), Millner (3825), Matsushima (3680), and Dreyfuss (1469); ascoma development: Ranga Rao (4725). — Colonies reaching 3·5 cm diam in ten days at 20°C on OA, without aerial mycelium, pigmentation and any specific odour. Homothallic. Ascomata subglobose to elongate-ovate, 135–280(–340) × 120–280 μm, wall dark brown, of textura angularis. Lateral hairs scanty, dark brown, straight; terminal hairs dark brown, ornamented with lacerate structures, 240–560 × 4·3–5·4 μm, straight below and with mostly 3–12 close or somewhat expanded coils of decreasing diameter above, sometimes branching in the distal part with markedly swollen branch bases. Asci clavate. Ascospores globose to subglobose, slightly flattened, at maturity obscurely pointed at both ends, pale olive-green to grey-blue, often with a large angular guttule, 6·5–8·5 × 6·5–7·5 μm, uninucleate. — *Ch. bostrychodes* is a member of a variable species group which also includes *Ch. caprinum* Bain. *1910*, *Ch. convolutum* Chivers *1912*, *Ch. microcephalum* L. Ames *1945* and *Ch. pachypodioides* L. Ames *1945* (1469) and the differences in ascoma and hair shapes are not correlated with physiological observations; intermediate strains regularly occur. Other similar species are the variable *Ch. spirale* Zopf *1881*, which has twice as many coils of decreasing diameter and larger ascospores 9–11 × 7–8 μm, and *Ch. cochliodes* (q.v.) with larger ascomata (270–370 × 200–310 μm) and ascospores (8–10 × 7–8 μm). — The nuclear divisions during ascus development have been studied and the haploid chromosome number was determined as 7 (4728). DNA analysis gave a GC content of 54% (5600).

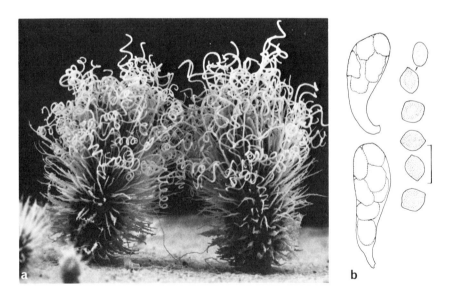

FIG. 66. *Chaetomium bostrychodes*. a. Perithecia, SEM, × 75; b. asci and ascospores; CBS 188.63.

Available data indicate a cosmopolitan distribution with the principal substrates being dung, seeds and soil. Soil isolations have been reported from the British Isles (167, 4655), France (2161), Belgium (4816), Germany (3042), Switzerland (566), Hungary (4223), Italy (270, 271), Greece (4647), Libya (6510), Tanzania (IMI), Central America (1697), Peru (2005), and India (1315, 1317, 1519, 2854, 4030, 4998). It is common on a wide variety of dead plant materials as is seen particularly in records from Malaysia, Pakistan, South Africa, Egypt, Zambia, Kenya, Hong Kong, Canada, and the USA (IMI). It has been isolated from forest soils (269, 270, 271, 272, 273, 2854, 4816), cultivated soils (1315, 1317, 1519, 2161, 3042), virgin soil (4030), slate slopes (566), sand dunes (4655), salt marsh (167), peaty soil (5559), and sewage sludge (IMI). In soil it has been found in layers between the surface and a depth of 20 cm (269, 272, 566, 1317, 4998). Other known habitats include the litter of salt marsh plants (167) and oak (3448), the rhizospheres of barley (3526) and several plants in an oak forest (4816), rotting paper (4928), birds' feathers (2574, 2575, 2576, 4652), and seeds of numerous plants (5376, 5403, IMI). However, particularly often colonized substrates seem to be the dung of dog (290), rabbit, (2279, 4439, 6085, IMI), goat, horse, rat (5376), antelope, ostrich, tortoise (IMI), deer, racoon and squirrel (4439).

The maximum temperature for growth is 33°C, for fructification 30–33°C; ascomata form within seven days at 27°C (1469). Ascospores germinate better in a 2% glucose or sucrose solution than in distilled water; Richard's medium with a pH 7 has been recommended for good growth (4928). A supply of vitamins is necessary for growth and fructification (1469). This fungus is able to decompose cellulose (2072, 4647, 4655, 6085) and moderately degrades the herbicide alachlor (5822). It produces nitrite when grown on amino acids as N source (479) and shows antagonistic properties against some common soil fungi (4031).

Chaetomium cochliodes Pall. 1910 — Fig. 67.

= Chaetomium flexuosum Pall. 1910
= Chaetomium lusitanicum M. Gomes 1953

DESCRIPTIONS: Skolko and Groves (5403), Udagawa (5968), Ames (124), Hawksworth and Wells (2340), Seth (5216), Dreyfuss (1469), and Udagawa and Takada (5984). — Colonies reaching over 9 cm diam in five days at 20°C on OA, with a greenish yellow submerged mycelium forming olivaceous crusts of ascomata, and with an odour suggestive of actinomycetes. Homothallic (702, 5171, 5172). Ascomata olive-brown, opaque, globose to subglobose, 270–370 × 200–310 μm. Lateral hairs dark brown below, paler above, straight or undulate, 5 μm wide near the base ; terminal hairs of two kinds: (a) dark brown, distinctly roughened with cupulate or globulate structures, 5–6 μm wide, straight in the lower half and with 3–5 loose coils above, (b) yellowish brown, more finely roughened, 3·0–3·5 μm wide, more irregularly twisted, with 3–5 coils. Asci clavate. Ascospores olive-brown, broadly ovate to lemon-shaped, 8–10 × 7–8 μm. — Ch. cochliodes is a very variable species but its further subdivision cannot be justified (1469). It is similar to Ch. angustum Chivers 1915, which has in addition straight terminal hairs, Ch. bostrychodes (q.v.) which has sometimes branched terminal hairs and smaller ascomata and spores, and Ch. globosum (q.v.) which has less differentiated undulate terminal hairs. — Vegetative hyphae are multinucleate while most cells of the ascogenous hyphae are uninucleate; the haploid number of chromosomes is estimated to be 7 (702). — The major sterol component (95%) of its hyphal lipids is ergosterol, and minor components of these are ergosta-5,7,9(11),22-tetraen-3β-ol and γ-linolenic acid (4968).

 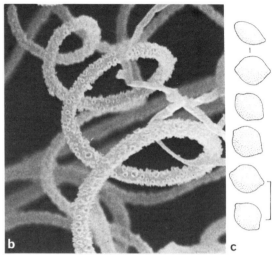

FIG. 67. *Chaetomium cochliodes.* a. Perithecia, SEM, $\times$ 90; b. detail of perithecial hair, SEM, $\times$ 1500; c. ascospores; CBS 814.73.

This species has a worldwide distribution (124, 5216); isolates from soil have been reported from Canada (234), the USA (2573, 4733), the British Isles (167, 745, 5559, IMI), Poland (1977), France (4162), Hungary (4223), Italy (5047), India (1315, 4997, IMI), Nepal (1826, 1827), Japan (2532, 3267, 5846), New Zealand (5812), Rhodesia, Tanzania (IMI), Egypt (4962), Kuwait (4000), Malawi (5313), Argentina (6445), Peru (2005), and Costa Rica (IMI). It has also been isolated from plant materials collected in New Guinea (5984), Swaziland, Kenya, Zambia, Pakistan, Sierra Leone, Malawi, South Africa, and Australia (IMI, 5216), and found on litter of salt marsh plants (167), leaves of tobacco (4214), crowns and roots of strawberry plants (2064), tomato roots (IMI), in the rhizosphere of *Coffea arabica* (5313) and the geocarposphere of groundnuts (2532), and on seeds of various crop plants (5403), including flax, peas, *Pinus patula* (IMI), and achenes of sycamore (1683), birds and birds' nests (2574, 2575, 2577), where it is relatively acidophilic and xerophilic (2576), dung of rat (3267), cow, horse (4439), antelope and rabbit, chitin strips buried in soil, caterpillars (IMI), decaying polypores (5984), paper, cardboard and wood (IMI). — It can be inhibited by an antibiotic from *Stilbella erythrocephala in vitro* (5362).

The maximum temperature for growth is 37°C, and that for sporulation 30°C; ascoma formation occurs within 7 8 days at 24°C (1469) and is optimal at 22°C, whereas the optimum for mycelial growth is 30°C (385) or lower (1827). Ascospores are released at a higher rate under conditions of high humidity than under dry conditions; in Britain a diurnal periodicity of release was observed with the maximum between 10.00 and 14.00 hours (2295). — *Ch. cochliodes* can decompose cotton fabric (2072) and cause soft-rot in wood (1496). For optimal growth Mn^{2+} is required (351) and nitrite is produced when grown on amino acids as sole N source (479). The fungus produces β-glucosidase (6159), 4,5-dihydroxy-3-methoxyphthalic acid (3231), orsellinic acid (5278), the phenolic metabolite cochliodinol which has antifungal and antibacterial properties, a fluorescent pigment (704, 705), the cytotoxic metabolites emodin, the chaetoglobosins A, B and C (6005), a chetomin-like substance (5954) and chetomin (725, 4969) which inhibits gram positive bacteria, to a lesser extent fungi and can be toxic to mammals (703) and HeLa cells. *Ch. cochliodes* can injure germinating pine seeds following wounding (1975).

Chaetomium crispatum (Fuckel) Fuckel *1870* — Fig. 68.

≡ *Sphaeria crispata* Fuckel *1867*
= *Chaetomium streptothrix* Quél. *1876*

DESCRIPTIONS: Bainier (290), Skolko and Groves (5403), Ames (124), Sörgel (5466), Seth (5216), and Hawksworth and Wells (2340). — Colonies reaching *c.* 5·0 cm diam in ten days at 20°C on OA, with much whitish or yellow-green aerial mycelium and scanty ascomata. Homothallic. Ascomata grey to grey-black, opaque, globose to subglobose, $200–280 \times 180–250 \mu$m. Lateral hairs numerous, light to dark brown, ± straight, 5–6 μm wide; terminal hairs dark olive-brown, 5–7 μm wide, ± coarsely roughened with annulate structures (by SEM) (2340), straight below and irregularly coiled above with several alternate loops. Asci cylindrical. Spores becoming dark bluish green, containing several refractive globules, lemon-shaped with broadly apiculate ends, $9–12 \times 6·0–9·5 \mu$m. — The *Chaetomium crispatum* group of Dreyfuss (1469) comprises all species with cylindrical asci; similar species with alternately coiled hairs are *Ch. contortum* Peck *1896* which has larger ascomata (to 330 μm diam), more coarsely roughened and densely looped hairs, and *Ch. simile* Massee & Salm. *1902* which has smaller ascomata ($185–230 \times 160–200 \mu$m) with less roughened hairs with longer arches. *Ch. tortile* Bain. *1909* is closest in its ascoma characters but has smaller ascospores $6–7 \times 4–5 \mu$m.

The few available data indicate a worldwide distribution (124, 5216), although this species has not been found frequently anywhere. Isolates have been reported from forest soils in Poland (269, 272, 273, 3138), salt marshes in the British Isles (413), soils in Italy (4183), citrus plantations in Libya (4085) and alkaline Usar soils (4030) and rice fields (IMI) in India. It has been found in the rhizosphere of barley (3526), on the dung of various animals (124, 4439, 4917, 5216, IMI), dead flies (5216), birds' feathers (2574, 2575), rotting potatoes (IMI), seaweeds (290) and seeds of various plants (3556, 5403, IMI). — It can serve as food for the mites *Pygmephorus mesembrinae* and *P. quadratus* (3104).

Cellulose in the form of filter paper is decomposed (1432), pectin utilized (4183), and in maple-wood strips a loss of tensile strength has been recorded (2211). Nitrite is produced when grown on amino acids as sole N source (479).

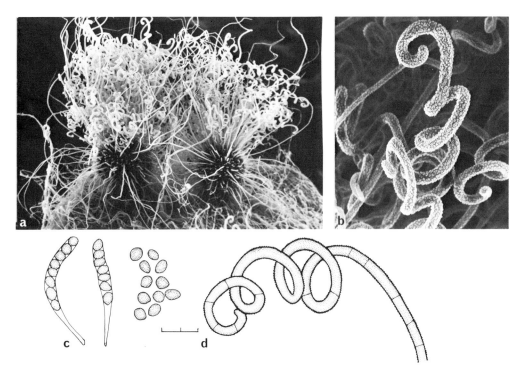

FIG. 68. *Chaetomium crispatum*. a. Two ascomata, SEM, × 75; b. detail of contorted hair, SEM, × 500; CBS 369.77; c. cylindrical asci and ascospores; d. perithecial hair.

Chaetomium elatum Kunze ex Steud. *1824* — Fig. 69.

= *Chaetomium atrum* Link *1824*
= *Chaetomium pannosum* Wallr. *1833*
= *Chaetomium lageniforme* Corda *1837*
= *Chaetomium graminis* Rabenh. *1851*
= *Chaetomium graminicola* Fuckel *1863*
= *Chaetomium velutinum* Ellis & Everh. *1885*
= *Chaetomium virgecephalum* L. Ames *1963*

DESCRIPTIONS: Bainier (290), Skolko and Groves (5402), Udagawa (5968), Takada (5685), Ames (124), Hawksworth and Wells (2340), Seth (5216), and Dreyfuss (1469); ascoma development: Moreau (3943). — Colonies reaching about 4 cm diam in five days at 20°C on OA, with yellow-green submerged and aerial mycelium on rich media; with an odour suggestive of actinomycetes. Mostly heterothallic (1811, 5171, 5213); single-spore isolates are interfertile only with conspecific isolates (5172) and otherwise produce sterile ascomata with unbranched hairs. *Acremonium*-like phialoconidia are abundantly produced. Typical ascomata are black, subglobose, to ovate, 350–500 × 230–400 μm. Lateral hairs dark olive-brown, 4–5 μm wide, unbranched; terminal hairs dark olive-brown, 5–7(–9) μm wide, mostly several times almost dichotomously branched in the upper part at angles of 90–120°, coarsely roughened with cupulate, annulate or globulate structures (by SEM) (2340). Asci clavate. Ascospores dark olivaceous-brown, flattened lemon-shaped, 10–12·5 × 8–9·5 µm in face view. — *Ch. virgecephalum* was distinguished from *Ch. elatum* by its self-fertility (1469, 5171) but as it was found to be interfertile with *Ch. elatum* isolates, this distinction is no longer justified (4006). — DNA analysis of *Ch. elatum* gave a GC content of 56·5% (5600).

FIG. 69. *Chaetomium elatum.* a. Two perithecia with dichotomously branched hairs, SEM, × 50; b. ascospores; CBS 154.59.

Together with *Ch. globosum*, this is the most widely distributed *Chaetomium* species but mainly in the temperate zone and rarely in the tropics. In soil it has rather rarely been found (124, 4245) and in general is most common on dead plant materials, particularly straw. — Reports from soil include ones from beech forest (273), soil under *Pinus maritima* in Yugoslavia (3798), spruce poles exposed to soil in a spruce stand (2833), light soils of citrus plantations in Israel (2764), dunes (4655) and desert soils in California (4733) and salt marshes in Kuwait (4000). It is also known on fallen leaves of salt marsh plants (167), decaying plant debris (2338, 6237), tomato roots (IMI), coarse fodder (4548), nests (2575, 2577) and feathers of various birds (2574, 2575, 4649, 4652), dung in Bulgaria (1684), Germany (5215) and Scotland (4931), and has been found rarely on seeds of oats (3556), cultivated grasses (3512), *Lolium*, leguminous and other plants (4120, 5402, IMI). This species is also recorded in sewage sludge and on wood and plywood (IMI). — *Ch. elatum* can serve as food for the mites *Pygmephorus mesembrinae* and *P. quadratus* (3104).

The maximum temperature for growth is 37°C and for sporulation only 27°C; ascomata are formed within 14–17 days at (18–)24°C (981, 1469) and their formation is favoured by a pH of 7·3 and darkness (981), not encouraged by Ca^{2+} ions (384) (unlike in *Ch. globosum*) and was found to be stimulated by leaf extracts of *Helleborus foetidus* (2195). — On media with asparagine or nitrate, only mycelium is produced (981) and highest yields were obtained with hydroxylamine and nitrite as N sources (479). The best C sources for mycelial growth were found to be cellulose and maltose and for fructification fructose (981) and peptone (385). For optimum growth Mn^{2+} is required (351). The enzymatic decomposition of CMC is very good, although marked differences occur between isolates (1432, 4224); reports concerning cellulase production are contradictory (386, 5386), and the activity seems to depend on the type of N source (2072). *Ch. elatum* is one of the organisms causing soft rot in timber (1496, 4191, 5072) and maple-wood strips are efficiently decomposed (2211). The formation of the anthraquinone derivative emodin (5278) has been reported and culture filtrates have a moderate antibacterial activity (6140).

Chaetomium funicola Cooke *1873* — Fig. 70.

= *Chaetomium setosum* Ellis & Everh. *1897*

DESCRIPTIONS: Skolko and Groves (5402), Udagawa (5968), Ames (124), Hawksworth and Wells (2340), Seth (5216), Millner (3825), Minoura *et al.* (3840), and Matsushima (3680); ascoma development: Cooke (1140). — Colonies reaching 3·0–3·5 cm diam in ten days at 20°C on OA, with scanty white aerial mycelium and a dense black layer of ascomata. Homothallic. Ascomata dark brown, globose to ellipsoidal, 135–240 × 135–180 μm, wall of textura angularis with a loose hyphal covering (textura intricata). Lateral hairs dark olive-brown with hyaline tips, to 4 μm wide, often rather sparse; terminal hairs dark olive-brown, of two kinds, (a) unbranched, long and straight, to 5 μm wide, gradually tapering to blunt tips, and (b) shorter, dichotomously branched at acute angles, tips rounded, 3–6 μm wide near the base; both types coarsely roughened with cupulate structures (2340). Asci clavate. Ascospores pale olive-green, becoming light brown, ovate to limoniform, the apical end with the germ pore more strongly pointed than the basal end, 5·5–6·5 × 3·5–5·0 μm, uninucleate (1140). — The similar *Ch. dolichotrichum* L. Ames *1945* has dichotomously branched terminal hairs with longer internodes and branching at wider angles besides much longer, straight, then once- or twice-forked hairs; young specimens of *Ch. dolichotrichum* can easily be mistaken for *Ch. funicola*. *Ch. indicum* Corda (q.v.) only has one type of dichotomously branching terminal hairs which have rigid ends.

Ch. funicola has a worldwide distribution (124, 5216) and occurs mainly on paper, timber and plant remains. Reports of its occurrence include Alaska (3062), Nova Scotia (IMI), the USA (1163, 1165, 2240, 2573), Germany (1424, 1736), Central America (1697), Brazil (3680), central Africa (3063), India (1519, 3865, 4030, 5376), Nepal (1826, 3840), Pakistan (3825), New Guinea, the Solomon Islands (5983), Japan (3267, 3680, 5846) and New Zealand (IMI). It has been reported from forest soils (2240, 2573, 3063, 3267), cultivated soils (540, 918, 4030, 6138, 6210), sandy soils (6182, 6184) and dunes (6414) and observed on litter of some salt-marsh plants (167), *Spartina alterniflora* (1958), and Scots pine (2344), decaying shoots of *Bothriochloa pertusa* (2955), graminaceous stems (4223), roots of strawberry (4133), cacao, cocos and tobacco plants (IMI), spruce poles exposed to spruce forest soil (2833), seeds of numerous cultivated plants (3512, 3556, 5402), in the rhizospheres of *Ammophila breviligulata* (6414), wilted pineapple plants (6210) and sugar cane (IMI), and on mycorrhiza of *Larix decidua* (1794). It has also been detected on birds' feathers (2574, 2575, 4649), and in birds' nests (2577) where it is comparatively xerophilic and often associated with *Ch. globosum* and *Ch. indicum* (2576). Other habitats recorded are dung of horse (4439) and rabbit, hay, coffee pulp, manihot chips, and canned strawberry products

FIG. 70. *Chaetomium funicola.* a. Perithecium with both unbranched and dichotomously branched apical hairs, SEM, × 200; b. detail of perithecial apex showing the thinner, dichotomous hairs, and ascospores, SEM, × 1000; c. detail of hairs with ornamentation, SEM, × 1100; CBS 166.49.

(IMI). — *Ch. funicola* is not susceptible to soil steaming (6183). It can serve as food for the mite *Glycyphagus destructor* (5380).

A minimum water potential of −110 bars is necessary for growth but optimal growth occurs at vapour saturation (3109). Dextrin, lactose, fructose (981), starch or $NaNO_3$ and $(NH_4)_2SO_4$ all stimulate fructification (385). High mycelial yields were obtained with hydroxylamine and nitrite as N sources (479). Fast fructification occurs at pH 5·5 but it also is possible to pH 8·2; the optimal temperature for this is 18–20°C (981). Good cellulose decomposition has been observed (386, 2072, 4224, 6326), particularly with ammonium salts as N sources (2072). *Ch. funicola* is one of the organims causing soft rot in timber (1494, 1496, 4191, 5072). — It produces the lactone colletodiol (3231) and also humic substances on cellulose-containing media (5240).

Chaetomium globosum Kunze ex Steud. *1824* — Fig. 71.

≡ *Chaetomium kunzeanum* Zopf *1881*
= *Chaetomium affine* Corda *1840*
= *Chaetomium setosum* Bain. *1910*
= *Chaetomidium barbatum* Traaen *1914*
= *Chaetomium subterraneum* Swift & Povah *1929*
= *Chaetomium japonicum* Saito & Okasaki *1939*

For many other synonyms see Seth (5216).

DESCRIPTIONS: Bainier (290), Skolko and Groves (5403), Ames (124), Sörgel (5466), Seth (5216), Lodha (3400), Hawksworth and Wells (2340), Millner (3825), Matsushima (3680), and Dreyfuss (1469); ascoma development: Chadefaud and Avellanas (936), and Whiteside (6335, 6336). — Colonies reaching 5·5 cm diam in ten days at 20°C on OA, with little aerial mycelium and a dense olivaceous or grey layer of ascomata; odour not pronounced.

FIG. 71. *Chaetomium globosum*. a. Perithecium with irregularly undulating hairs, SEM, × 150; b. detail of ostiole showing terminal hairs and ascospores, SEM, × 750; CBS 158.51.

Phialoconidia mostly absent. Homothallic. Ascomata dark brown to black, globose to subglobose, 225–350 × 200–350 μm, wall of textura intricata. Lateral hairs dark brown with paler tips, ± wavy, minutely roughened, to 3·5 μm wide; terminal hairs dark olive-brown with paler tips, wavy or very loosely coiled and intertwined, 3–4 μm wide throughout, roughened with ± cupulate or annulate structures (2340) also seen in TEM (4410). Asci clavate. Spores slightly reddish when very young, then pale greenish and finally dark olive-brown, flattened lemon-shaped, hardly apiculate, $8 \cdot 5–11 \cdot 0 \times 7 \cdot 0–8 \cdot 5 \times 6 \cdot 5–7 \cdot 5 \mu$m. — Five varieties have been distinguished (1469, 4223): var. *flavo-viride* Novák *1966* with olive-green and var. *griseum* Novák *1966* with mouse-grey ascomatal layers which otherwise both exactly match the above description; var. *ochraceoides* Dreyfuss *1976* has olive-green ascomatal mats, more densely undulate hairs and smaller ascospores, mostly $8–9 \times 6 \cdot 2–7 \cdot 0 \mu$m; var. *rectum* (Serg. *1961*) Dreyfuss *1976* has olive-green mats and straight hairs; and var. *arhizoides* Dreyfuss *1976* dark grey to black mats, rhizoids hardly developed and ascomata consequently easily detached from the substrate, and slightly larger ascospores, mostly $9 \cdot 3–10 \cdot 5 \times 8 \cdot 0–9 \cdot 0 \mu$m. It has not been possible to determine which variety is identical with the type of *Ch. globosum*. — *Ch. coarctatum* Serg. *1961*, described with broad, hardly apiculate ascospores $9 \cdot 5–11 \times 9–10 \mu$m, was considered as synonymous with *Ch. globosum* by Millner (3825) but retained separate by Dreyfuss (1469) because of the unusually broad spores. *Ch. olivaceum* Cooke & Ellis *1878*, repeatedly reported to be heterothallic but also listed as homothallic (5172), with ascospores of $8 \cdot 7–12 \cdot 0 \times 8 \cdot 7–10 \cdot 0 \mu$m, has now been restricted to large-spored homothallic isolates; similar heterothallic ones should be named *Ch. subaffine* Serg. *1961* (1469). *Ch. spirilliferum* Bain. *1910* ($16 \cdot 8 \times 8 \cdot 5 \mu$m), *Ch. undulatum* Bain. *1910* ($14–15 \times 8 \cdot 5 \mu$m) and *Ch. subglobosum* Serg. *1960* ($11 \cdot 5–13 \cdot 0 \times 10 \cdot 5–11 \cdot 5 \mu$m) all have been described as having larger ascospores but their delimitation is not yet clear. *Ch. kunzeanum* Zopf *1881* was explicitly introduced as a new name for *Ch. globosum* and is so typified by the type of the latter; the larger ascospores ($11–13 \times 8–9 \times 6–7 \mu$m) and the presence of phialoconidia, however, suggest that Zopf had *Ch. subaffine* before him. — TEM studies of meiosis (2561) show spindle fibres centrally attached to a spindle pole body. The behaviour of the nucleus during ascus development has been studied in detail (2563); the nuclear division is mitotic and the number of nuclei per cell varies between 3 and 10; n is either 5 or 6 (2564). — DNA analysis gave a GC content of $57 \cdot 5–58\%$ (5600). — Cell wall analysis yielded 69% polysaccharides, 29% protein, and 3% lipids; the polysaccharide fraction contained chitin and glucan with either $1,6\text{-}\beta\text{-}$ or $1,4\text{-}\beta$-linkages (3602). Mycelium contained $1 \cdot 4\%$ C_{14}-, $30 \cdot 6\%$ C_{16}-, $9 \cdot 6\%$ C_{18}-saturated and $10 \cdot 8\%$ C_{16}- and $9 \cdot 7\%$ C_{18}-monounsaturated fatty acids; the dominating amino acids were threonine ($14 \cdot 0\%$), glutamine ($12 \cdot 2\%$), asparagine ($10 \cdot 4\%$), proline ($9 \cdot 9\%$), glycine ($9 \cdot 8\%$), serine ($9 \cdot 2\%$) and alanine (8·5 mol % of total amino acids) (3232).

Ch. globosum is by far the commonest and most cosmopolitan *Chaetomium* species, especially on plant remains, seeds, compost, paper and other cellulosic substrates (124, 5216). A selective isolation medium containing 7% ethanol and 100 ppm streptomycin has been developed (1302). *Ch. globosum* has been observed very frequently in a very wide variety of soils, although its proportion in the total soil fungus population is relatively low. Reports are equally numerous for forest and cultivated soils and include some extreme habitats: an alpine soil with long snow coverage at an altitude of 2630 m (3976), an alpine pasture with *Nardus stricta* (3445), 3600 m in the Rocky Mountains (4711), soils of the sub-Himalayan tract (4998), salt marshes (8, 1958, 4000, 4001), sand dunes (4655), desert soils (2974, 4733, 4759), an open savannah (1559), highly saline estuarine sediments (655), marine sediments (2643), a vineyard soil with high copper content (4921), a conifer swamp (1039), and polluted streams (1154, 1166). It is sometimes found confined to the soil surface (2948) and mainly the litter layer, but is also known to penetrate to a depth of 60 cm (3447). Detailed reports describe its occurrence in the litter of salt-marsh plants (167), *Carex paniculata* (4644), Scots pine (269, 2344), beech (850), *Eugenia heyneana* (5513), *Nothofagus truncata* (4946) and other forest trees (2411, 3449). The minimum water potential for growth in the soil was found to be -85 bars (3109). It is not particularly sensitive to soil steaming (6183) and various soil fumigants (6287) but its frequency is apparently reduced by the application of several soil fungicides (6136). The hyphae can grow chemotropically towards phosphate granules but are incapable of dissolving them (3074). *Ch. globosum* has been found on residues of barley, cabbage, potato (3113), decaying *Pseudoscleropodium purum* (2969), leaves of *Brachypodium pinnatum* (2970, 2971), decaying grass clippings (2338), and the phyllosphere of various vascular plants (296), on sclerotia of *Sclerotinia sclerotiorum* (4699), in the rhizospheres of oats (986, 4443), barley (3526), beans (1407), groundnuts (2532), several sand dune plants (4371), *Salicornia* (4645), tea (41) and wilted pineapple plants (6210), on roots of broad beans (6134) and strawberry plants (2064, 4133), seeds and pods of groundnuts (346, 3220), and on various nuts and dried fruits (2258, 6252). It can also occur on grass compost as a late colonizer (3546), on straw compost after heating (1528), cotton tissue from conveyor belts (6163), decayed timber (1199, 3568), softwood and hardwood buried in the surface layer of soil (5237), pine wood exposed to seawater (831), wood pulp (3294), pellets, feathers and nests of free-living birds (2574, 2575, 4649, 4652, 6223) with a comparatively alkalophilic and hygrophilic behaviour (2576), dung of horse, squirrel (4439), goat (3400), rabbit (3400, 5215, 6085) and antelope (3839), material from a sewage treatment plant (1165), and rarely coarse fodder (4548), paper, plaster and books (IMI). This is the commonest seed-borne *Chaetomium* species found on numerous plant species (3172, 3556, 3670, 5009, 5376, 5403). — Ascospores can survive in a dried condition for > ten years (4361). Germination is inhibited by a mycostasis factor in soil (1407), by a sporostatic metabolite produced by *Fusarium oxysporum in vitro* (4882), and also by several *Streptomyces* species (2939); growth *in vitro* was inhibited by the presence of *Memnoniella echinata* (4); tannin extracts from freshly fallen oak leaf litter were strongly inhibitory (2282). Coating of corn seeds with spores of *Ch. globosum* was said to reduce both the *Fusarium* seedling blight (950) and storage microflora of kernels drastically (956). After soil or seed inoculation, successful control of *Cochliobolus victoriae* or *Monographella nivalis* has been reported (5955, 5956). — *Ch. globosum* can serve as food for the mites *Pygmephorus mesembrinae* and *P. quadratus* (3104).

Ascomata form best at 18–20°C (981) or, according to another report, at 24°C within ten days (1469); a low content of soluble sugars (<1·0% glucose, maltose, sucrose, cellobiose) in the medium (805) is a requirement for their formation; cornmeal agar (2563), maltose, fructose (981), soluble starch, cellulose, dextrin, sorbitol and mannitol promote ascoma formation and growth, with increasing concentrations; good N sources prove to be nitrates (981, 1495) and cyanamide (0·2%) but asparagine, urea, peptone and $(NH_4)_2SO_4$ are also utilized (385, 1495, 2868, 6078). Fructification is best at pH 7·3 (981). Calcium ions (10 ppm in a mineral nutrient solution) proved essential for ascoma formation but could be partially replaced by strontium or barium (384); the presence of CO_2 was also favourable (809). Good growth occurs with up to 40% CO_2 in the atmosphere. The onset of fructification is accompanied by the appearance of organic phosphates in the surrounding medium (810) and has also been induced by these compounds (806). Fungal extracts or the culture media of other fungi had a stimulating effect on sporulation, presumably through the production of organic phosphates and/or other unknown factors (807, 808, 964, 2868). Fructification is hardly stimulated at all by vitamins and jute extract (805), but abundant production occurred on *Melilotus albus* and *Dactylis glomerata* (981). Sporulation is greater in darkness than in light (1654). — The optimal temperature for the germination of ascospores is 24–28°C, the minimum 4–10°C or higher, and the maximum 38°C; the thermal death point for ascospores has been determined as 55–57°C for 10 min (959). At 50% salinity 10% spores were able to germinate at 25°C (832). The optimal temperature for growth lies in the range 16–25°C (981, 1494), possibly depending on the geographical origin of the isolate (1827), and the maximum temperature is 36–37°C (1469, 3826, 5874); the optimal pH range is 7·1–10·4 (6529); for optimal growth Mn^{2+} is required (349, 351). — Fructose as an intact molecule is absorbed only in traces but is broken down by extracellular enzymes prior to its uptake; the pH dependence of glucose and mannitol uptake, and glucokinase and fructokinase activities have been investigated (32, 6162); two to three times more mannitol is found in the mycelium after incubation on fructose or mannose than on glucose; total uptake of fructose was higher in 1% than in 0·5% solutions; apart from glucose no other free sugars accumulate in the mycelium; during uptake of the above three hexoses, a massive synthesis of both trehalose and glycogen takes place (31); the mycelium contains mannitol dehydrogenase. More mycelium is formed with acetate as the C source than with glucose (3329). The enzymes amylase, xylanase, mannanase, β-glucosidase and a rather weak pectinase have been investigated (825, 2917, 3471, 3472); xylan is broken down into equal parts of xylobiose and xylose (5475), the latter being utilized very well (2047); arabinoxylan is also degraded by xylanase (1753). — There are so many reports of its decomposition of cellulose that they cannot be cited here individually. The optimal temperature for cellulose degradation is 25–32°C (1495, 6163) and the optimal pH is given as 4·5 (5238); cellulase activity was found to increase with increasing concentrations of the nitrogen source (3303), and both cellulase and cellobiase activity were promoted by yeast extract (6129), biotin and to a lesser extent by Ca-pantothenate (38); ethyl malonate inhibits cellulase and cellobiase activity, partially at a concentration of 0·05% and completely at 0·5% (37, 39). — Usnic acid is degraded (322) and lignin sulphonate utilized to some extent as the sole C source (3260); in wood, *Ch. globosum* first attacks the polysaccharide components of the secondary cell wall, leaving large quantities of lignin behind (5073); there is scarcely any growth in the cell lumen, so that the fungus cannot be classed among the organisms causing brown rot (3741), but rather with the soft rot fungi (1199, 1494, 1496, 3323, 3768, 5072, 5073). There are reports of losses in weight and stability in wood fibre boards (3767, 3770) and in the wood of aspen, beech, birch, poplar, spruce, pine and other trees (125, 181, 1199, 1494, 3323, 3568, 3767, 4191). Wool is attacked only to a limited extent (6330); the fungus penetrates hairs by means of boring hyphae which then spread out internally (1633); the herbicide alachlor is metabolized (5822), cinerone is hydroxylated (5681) and urethan rubber can be degraded by the fungus (251). — A rennin-like proteinase complex has been found in the culture liquid (3049). Galactose, mannose (3640) and a galactobiosyl glucose (2047) were exuded in the culture solutions. Products include the anthraquinone derivatives chrysophanol (179) and emodin (5278), the cytotoxic cytochalasan metabolites chaetoglobosin A, B, C (5192), D, E and F (5193) and chetomin with toxicity against mammals and activity against gram positive bacteria (703, 4969), cochliodinol (704, 705), a bacteriolytic enzyme (*N*-acetylhexosaminidase) (2643) and, as part of a sporostatic principle, nonanoic acid (1905). When grown in an amino acid medium, nitrite is produced, especially at a C/N ratio of 5:1 (479). — Corn infested with *Ch. globosum* was found to be toxic to rats (1038). In a few cases an antagonistic activity against other fungi has been reported (569, 4031, 6205), but the *in vitro* effects are apparently smaller than those seen in other *Chaetomium* species (6140).

Chaetomium indicum Corda *1840* — Fig. 72.

= *Chaetomium melioloides* Cooke & Peck *1875*
= *Chaetomium setosum* Wint. *1887*

DESCRIPTIONS: Bainier (290), Skolko and Groves (5402), Udagawa (5968), Ames (124), Lodha (3400), Hawksworth and Wells (2340), Millner (3825), Minoura *et al.* (3840), and Matsushima (3680). — Colonies reaching 3·0 cm diam in ten days at 20°C on OA, usually without any aerial mycelium, ascomata forming a dense black layer. Homothallic. Ascomata dark brown to black, globose, wall of textura intricata, 110–180 μm diam. Lateral hairs dark brown, straight, simple, rather sparse; terminal hairs in a dense tuft, dark olive-brown; to 5·3 μm wide, dichotomously branched in the upper half, with branches at right angles or reflexed and roughened by tuberculate or slightly rugose structures (2340). Asci clavate. Ascospores olive-brown, flattened lemon-shaped 5·0–7·0 × 4·0–5·5(–6·0) μm. — The rather similar *Ch. erectum* Skolko & Groves *1948* has less branched, more rigid and smoother terminal hairs with easily collapsing tips; *Ch. reflexum* Skolko & Groves *1948* has strongly reflexed dichotomous branches with blunt tips (cf. also *Ch. funicola*). — DNA analysis gave a GC content of 58·5% (5600).

FIG. 72. *Chaetomium indicum*. a, b. Perithecia showing dichotomously branched, reflexed hairs, SEM, × 160 and × 250; c. ascospores; CBS 212.74.

The available data indicate a worldwide distribution (124, 5216) with a predominance in cultivated soils. It has been isolated from soils in the British Isles, Belgium (IMI), France (918), Germany (3042), Poland (1977), the USSR (2871), the USA (2482, 4313), Argentina (1827, 2012), Libya (4085), Chad (3415), India (1317, 2854, 4698) and Nepal (1826, 3840). Its habitats include forest soils under pine and oak (IMI), forest nurseries (1977, 2482), cultivated soils (3042) under wheat, sunflower (918), alfalfa, potatoes (2871) and citrus (4085), uncultivated soils (918), and wet prairie (4313). It has also been found on wheat roots (IMI), litter of salt marsh plants (167), pine needles, mouldy hay, straw, stored hops, timber

(IMI), beech stumps (3568), pulp and paper (4548, 6166), cow dung (3400, 3861), sewage sludge (IMI), composted municipal waste (3041), nests, feathers and pellets of free-living birds (2574, 2575, 2576, 2577), seeds of *Avena fatua* (2961), *Dactylis glomerata* (5940), other grasses (3512), barley (1757), oats, sorghum, rice, flax (IMI) and *Malus pumila* (4120), and rarely on coarse fodder (4548). *Ch. indicum* has been found to be associated with the ambrosia beetle, *Crossotarsus niponicus* (4102).

The optimal temperature for growth and fructification is in the range 25–35°C (385, 1827). Cellulose (1757, 2577, 3303, 4779, 5059) and starch (1827) are degraded, and cellulose degradation is favoured by yeast extract (6129); wood (1496) and keratinized substrates are attacked (1633). Nitrite is produced when grown on amino acids as sole C and N sources (479) and the production of 4-carboxy-2-oxo-3-phenyl-hept-3-enedioic acid has been reported (3231).

Chaetomium nozdrenkoae Serg. *1961* — Fig. 73.

= *Chaetomium difforme* W. Gams *1966*

DESCRIPTIONS: Sergejeva (5211), Gams (1878), Hawksworth and Wells (2340), and Seth (5216). — Colonies reaching 3·5 cm diam in five days at 20°C on OA, producing abundant white and often dense aerial mycelium, hyphae often particularly broad and densely septate; odour fruity. Heterothallic (1878, 5171, 5172). The type isolates of *Ch. nozdrenkoae* and *Ch. difforme* are interfertile. Producing abundant *Acremonium*-like phialoconidia. Ascomata dark-brown, subglobose to oblong, wall pseudoparenchymatous, 350–600 × 280–410 μm. Lateral hairs hyaline to pale brown, sometimes branched, 3·0–4·5 μm wide; terminal hairs pale to dark olive-brown with hyaline ends, finely to coarsely roughened with annulate structures (2340), undulate and irregularly branched, 2·5–4·0 μm wide. Asci clavate. Ascospores dark olive-brown, filled with numerous oil drops, of very irregular shape, ellipsoidal with ± broadly prominent ends, 13–25 × 7–16 μm, provided with two inconspicuous germ pores. — The rather similar *Chaetomium flavum* Omvik *1955* produces a yellow mycelium and has ascomata with coiled and apparently unbranched terminal hairs and similar ascospores (cf. also *Ch. tetrasporum* Hughes). *Ch. variosporum* Udagawa & Horie *1973* also has deformed ascospores with two germ pores but these are smaller, 10–12·5 × 6–10 μm, and the short terminal hairs are arcuate and incurved.

Ch. nozdrenkoae was originally isolated from fallow soil near Novosibirsk in the USSR (5211) and subsequently from washed soil particles of two wheatfield soils in Germany (1889) and other agricultural soils in the Netherlands (CBS). This species is apparently rather common in arable soils but, due to its heterothallic nature, is rarely recognized. It can serve as food for the mites *Pygmephorus mesembrinae* and *P. quadratus* (3104).

Pectolytic enzymes have been demonstrated after growing on Ca pectate; strong cellulose decomposition occurs on CMC, although there are differences in this ability between isolates (1432). It also causes very marked losses in weight and tensile strength on maple-wood strips, again with considerable variability between isolates (2211).

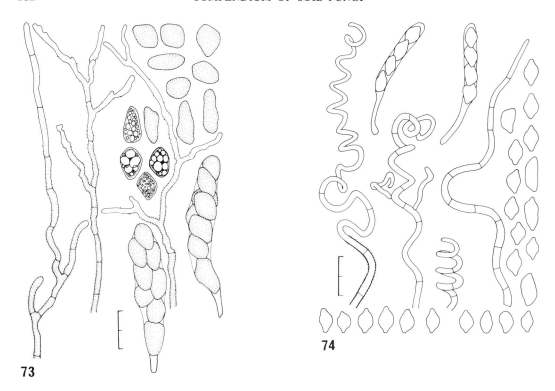

FIG. 73. *Chaetomium nozdrenkoae*, perithecial hairs, asci and ascospores, CBS 447.66 × 448.66.
FIG. 74. *Chaetomium tetrasporum*, perithecial hairs, cylindrical, 4-spored asci and ascospores, CBS 351.77.

Chaetomium tetrasporum Hughes *1946* — Fig. 74.

DESCRIPTIONS: Hughes (2593), Ames (124), Hawksworth and Wells (2340), Skolko and Groves (5403), Seth (5216), and Udagawa (5973). — Colonies reaching 4·8 cm diam in ten days at 20°C on OA, with a pale olivaceous aerial mycelium and greenish yellow reverse, producing ascomata scantily and often not at all. Homothallic. Phialoconidia scantily produced. Ascomata dark brown, globose to subglobose, 230–500 μm diam. Lateral hairs light brown, minutely roughened, straight or flexuose, 3–4 μm wide; terminal hairs light to dark brown, minutely roughened with tuberculate or cupulate structures (2340), with up to ten loose coils about 5 μm wide, often bearing several short lateral, undulate branches. Asci cylindrical with a slender stalk, 4-spored. Ascospores dark olivaceous-brown, irregularly umbonate, 10–14 × 7–9 μm. — Cf. also *Ch. nozdrenkoae*.

Ch. tetrasporum was originally collected in England on plant material in contact with rabbit dung. It has often been isolated from washed particles of wheatfield soils in Germany (1889) and soils under sugar beets or potatoes in the Netherlands (1614, 1616). Since similar isolates often remain sterile, it is likely, as in the case with *Ch. nozdrenkoae,* that this species is often unnoticed.

Ch. tetrasporum decomposes xylan and only slight differences in this capacity occur between isolates (1432). A weak antagonistic influence against *Gaeumannomyces graminis* has been demonstrated (1431), and the growth of *Chlorella pyrenoidosa in vitro* was stimulated by it (1430).

Chaetosphaeria Tul. *1863*

Chaetosphaeria vermicularioides (Sacc. & Roum.) W. Gams & Hol.-Jech. *1976* — Fig. 75.

   ≡ *Eriosphaeria vermicularioides* Sacc. & Roum. *1883*

ANAMORPHS: *Chloridium virescens* (Pers. ex Pers. *1822*) W. Gams & Hol.-Jech. *1976* var. *virescens*

   ≡ *Chloridium viride* Link ex Link *1824*

Chloridium virescens (Pers. ex Pers.) W. Gams & Hol.-Jech. var. *caudigerum* (Höhn.) W. Gams & Hol.-Jech. *1976*

   ≡ *Cirrhomyces caudigerus* Höhn. *1903*

Chloridium virescens (Pers. ex Pers.) W. Gams & Hol.-Jech. var. *chlamydosporum* (van Beyma) W. Gams & Hol.-Jech. *1976*

   ≡ *Bisporomyces chlamydosporis* van Beyma *1940*
   ≡ *Chloridium chlamydosporis* (van Beyma) Hughes *1958*

For further synonyms see Gams and Holubová-Jechová (1891).

DESCRIPTIONS: Mangenot (3568), Meyer (3789), Swart (5665), Ellis (1603), Hammill (2240), and Gams and Holubová-Jechová (1891). — The above three anamorphic taxa were relegated to varietal rank and connected with the teleomorph *Chaetosphaeria vermicularioides* in the monograph by Gams and Holubová-Jechová (1891). Common characters are unbranched pigmented conidiophores terminated by a shallow, rather wide collarette, within which several "conidiogenous loci" produce a sequence of conidia so that either a slimy head or a columnar cirrhus of conidia is formed. Percurrent elongation of the conidiophores commonly occurs. This mode of multiple conidium formation is an unusual modification of the phialide, but is also seen in *Gonytrichum* Nees ex Leman (q.v.) (5665) and *Codinaea* Maire *1937*; this has been illustrated by TEM in *C. virescens* var. *chlamydosporum* (2244). — Multiple conidium formation and the presence of globose or oval, pigmented chlamydospores serve to distinguish the anamorphs of this fungus from the other 12 species of *Chloridium* (1891). In culture, the colonies reach 1·2–2·5 cm diam in 14 days at 20–22°C. — The varieties can be distinguished with some degree of certainty on the natural substrate (decaying wood): var. *virescens* has long, bright green cirrhi of short conidia (2·7–3·8 × 2·0–2·5 μm), var. *caudigerum* whitish cirrhi of longer and more slender conidia (3·5–5·0 × 1·8–2·5 μm), and var. *chlamydosporum* exclusively heads of almost cylindrical conidia (3·0–5·0 × 1·7–2·7 μm). In culture, these differences are smoothened to a large extent and distinction is often hardly possible. If cirrhi occur in old cultures, var. *caudigerum* sometimes may be recognized amongst soil isolates, although pure cultures have so far mostly been referred to *Chloridium chlamydosporis*. The

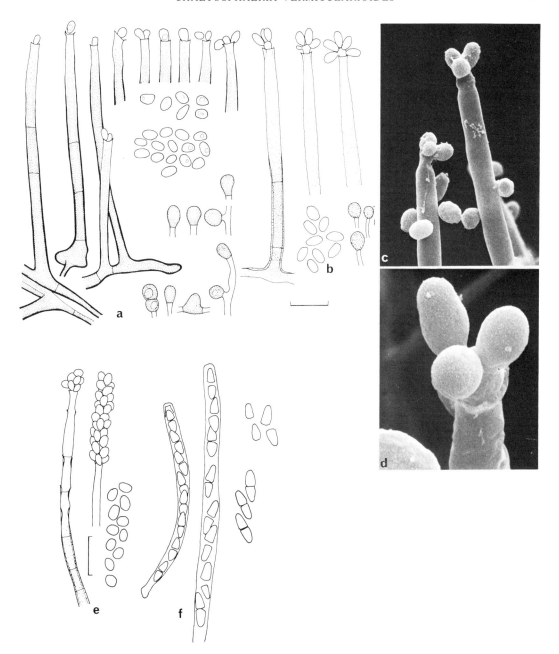

Fig. 75. *Chaetosphaeria vermicularioides*. a. Conidiophores, conidia and chlamydospores of *Chloridium virescens* var. *caudigerum*; b. var. *chlamydosporum* with strongly protuberant apex of the conidiogenous cell, CBS 526.73; c, d. details of conidiogenesis in *Chl. virescens var. virescens*, SEM, × 2000 and × 5000, CBS 919.73; e. conidiophores and conidial columns of the same variety, IMI 179577; f. asci and fragmenting ascospores, herb. CBS 675.74 (partly from 1891).

perithecia of the three anamorphic varieties appear to be indistinguishable; they are known exclusively from decaying wood.

This species has been observed very frequently on rotten wood and twigs of broad-leaved and sometimes coniferous trees, and also on *Pteridium, Urtica* and *Juncus* (1891, 3568, 5929). — A worldwide distribution is indicated by reports from Poland (272, 273), France (2963), the British Isles (989, 1376, 3234), the USA (1030, 1032, 1039, 2004, 2008, 2240, 2573), Canada (505), Japan (5846, 5929), Ceylon, Chile, Zaïre (1891), central Africa (3063) and Australia (5930). So far, most reports of soil isolates are known from forest soils including beech forest (272, 273), oak forest (989, 5811), a mixed *Salix-Populus* forest (2004, 2008), deciduous forests (2573), a conifer swamp (1039), white cedar forest (505), and a mesic conifer-hardwood forest (1030, 1032). It has also been found in a mediterranean brown soil (2963), peat bogs (1376) and peat soil under *Juncus* (3234), soil with steppe type vegetation (6347), an acid mine drainage stream in the USA (1166) and has been isolated from decaying bamboo stems (3063).

Suitable C sources include arabinose, xylose, glucose, fructose, galactose, mannose, maltose, sucrose, raffinose, starch, dextrin, inulin, and xylan and the fungus can grow on NH_4-tartrate and KNO_3 as N sources. For optimal growth, thiamine (or pyrimidine) and, in combination with thiamine, zinc ions (1 ppm) are required (4298). Moreover, the species is one of the organisms listed as causing soft-rot (1496) and weight loss in wood (4191). Polyphenoloxidase production is low (1891).

Chalara (Corda) Rabenh. *1844*
emend. Nag Raj & Kendrick *1976*

= *Cylindrocephalum* Bonord. *1851*
= *Thielaviopsis* Went *1893*
= *Chalaropsis* Peyr. *1916*
= *Hughesiella* Bat. & Vittal *1956*

Type species: *Chalara fusidioides* (Corda) Rabenh.

Teleomorphic genera: *Ceratocystis* Ellis & Halst., *Cryptendoxyla* Malloch & Cain

The largely dematiaceous genus *Chalara* is characterized by simple (rarely branched) conidiophores with terminal, integrated phialides. The phialides are mostly sharply differentiated into a basal venter of variable shape and an elongate, cylindrical collarette in which the conidia are produced in long basipetal chains. Conidia are mostly cylindrical (or obclavate or ellipsoidal) with truncated or somewhat rounded ends, hyaline or slightly pigmented, one-celled or septate. Characteristic pigmented chlamydospores occur in some species (those formerly distinguished as *Chalaropsis* or *Thielaviopsis*). Teleomorphs mostly belong to *Ceratocystis* Ellis & Halst. *1890* (*sensu stricto*) of the Microascaceae Luttrell ex Malloch and perithecia are sometimes produced in culture. *Chalara* with 70 species and six related genera have been monographed by Nag Raj and Kendrick (4070). Most species are saprophytes or wood blue-staining fungi, and only two soil-borne species are considered here; these both have characteristic chlamydospores and were only recently transferred to *Chalara* because of their phialidic form of sporulation.

Key to the species treated:

1   Chlamydospores in lateral or terminal clusters, cylindrical, 5–7-celled, fragmenting at maturity; conidia mostly 7·5–19 × 3–5 μm   *Ch. elegans* (p. 187)
       Chlamydospores 1-celled, ellipsoidal-subglobose, smooth and thick-walled, produced terminally and singly or rarely in short chains; conidia 7–23 × 2·5–4·0 μm.
       Chalara state of *Ceratocystis fimbriata* (p. 161)

Chalara elegans Nag Raj & Kendrick *1976* ("*1975*") — Fig. 76.

= *Thielaviopsis basicola* (Berk. & Br.) Ferr. *1912*
   ≡ *Torula basicola* Berk. & Br. *1850* (chlamydospores)
? = *Milowia nivea* Massee *1884*

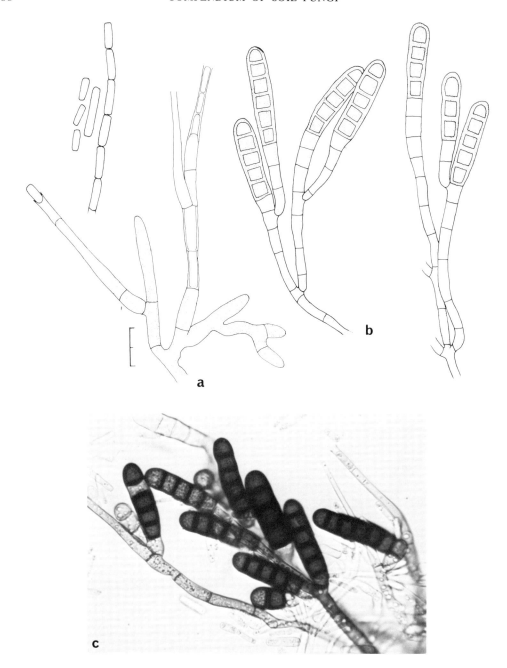

FIG. 76. *Chalara elegans*. a. Conidiophores with phialides and catenulate conidia; b. chlamydospores on sympodially proliferating conidiophores; CBS 430.74; c. cluster of chlamydospores, × 500.

DESCRIPTIONS: CMI Descriptions No. 170, *1968*, Ellis (1603), and Nag Raj and Kendrick (4070). — Colonies spreading, reaching 5·5 cm diam in ten days at 20°C on OA, appearing whitish or brown depending on the conidial type produced. Conidiophores basitonously branched a few times, 3–5-septate, with integrated terminal, slender, subhyaline phialides; conidia cohering in long chains, cylindrical, with truncated ends, 7·5–19 × 3–5 μm. Chlamydospores produced laterally or terminally in sympodially developing acropetal clusters (5904), consisting of a row of 5–7 cells, soon breaking apart in the manner of arthroconidia, dark brown, smooth-walled, cells 6·5–14 × 9–13 μm. A distinction between a grey (wild type) and a brown isolate proved to be based on unstable properties (2567). — *Ch. elegans* differs from the type species of *Thielaviopsis, Th. aethacetica* Went *1893* (=*Th. paradoxa* (de Seynes) Höhn. *1904*), the anamorph of *Ceratocystis paradoxa* (Dade) C. Moreau *1952*, in its pluricellular chlamydospores produced laterally in sympodial succession; in *C. paradoxa*, these are produced in basipetal chains as are the conidia and all intermediate stages occur. Since the epithet "*basicola*" refers to the chlamydospores only, the new epithet "*elegans*" was proposed for the phialidic form of sporulation by Nag Raj and Kendrick (4070). Because of this discrepancy with the type species of *Thielaviopsis*, and because the phialidic state gives more information about the natural affinity of this fungus, the classification in *Chalara* is preferred. — Cells of the vegetative hyphae, conidiophores, phialoconidia and chlamydospores are regularly uninucleate; anastomoses occur commonly, and n = 4 (2568). Plurinucleate conidia sometimes occur (2569). Chlamydospores are surrounded by a common, thin, two-layered wall, and a thicker, two-layered wall surrounds the individual cells; secession occurs along a thin electron-transparent central layer (1045). At the junction of the lateral and transverse walls, a weak zone is recognizable which breaks on germination to allow the germ tube to protrude laterally (5907). The protoplasts of the chlamydospore cells are connected through septal pores (1044). Cell walls of hyphae, phialoconidia, chlamydospores (1338) and conidiogenesis (2328) have all been studied by TEM and/or SEM.

Chalara elegans occurs in all continents from temperate to subtropical zones. It is a root pathogen of numerous plants but is particularly important on tobacco. Cultural characteristics, pathogenicity, and host range have all been reviewed (4753). A buried membrane-filter method has been devised to study its growth reactions *in situ* (23) and special agar media (3510, 4381, 5901), and bait techniques including carrot disks (2887, 6484), or other umbelliferous root tissues (3710) have been used for selective isolation from soil. — *Ch. elegans* occurs only rarely in soil analyses, however, but in California it was found in 197 of 402 samples. It appears in uncultivated and cultivated soils with about the same frequency (6485) and has been reported from steppe soils (4313), garden compost (1424), the rhizosphere of clover (3429), root surfaces of a very large number of plants, and the slime of a paper mill (5060). — The fungus survives in soil mainly by means of its chlamydospores (5903), but phialoconidia have been found to remain alive for 15 months in unsterilized moist clay loam but seem to be much less resistant in sandy soil or air-dry soils (5103) and they lysed within two weeks after being killed by UV light (1049). Phialoconidia showed high germination in both mull and mor soils and consequently survived at a low rate (633). Both types of propagules survive better at 9°C than at higher temperatures (3648) and the viability of phialoconidia (4405) and chlamydospores (1081) was found to be greatly reduced by a high moisture content of the soil (4405). Bacteria are able to perforate chlamydospores buried in soil (1080, 4276). This fungus was found to be inhibited *in vitro* by the phytoalexin α-tomatine (182), *Aspergillus fumigatus, Penicillium expansum, P. spinulosum, Talaromyces trachyspermus,* and other fungi (2753, 2754, 5440); a blue strain of a *Pseudomonas* stained both the hyphae and conidia blue and damaged the cytoplasm *in vitro* but not in soil (4468). — *Ch.*

elegans can serve as food for, and favour the multiplication of the nematode, *Aphelenchus avenae* (3579).

The germination of both phialoconidia and chlamydospores is stimulated by soil additives such as plant remains, extracts, cellulose and chitin when these are added to soil concurrently with the fungus (3350, 4385, 4388). Such organic additions counteract the mycostatic effects of soil on this species, but the response varies between isolates (27). The germination of phialoconidia is influenced less by the soil moisture than the availability of nutrients (3349). The phialoconidia germinate well on carrot extract *in vitro*, older conidia (≥36 days old) showing poorer germination; the temperature range for the germination of both phialoconidia and chlamydospores is 20–33°C, and the optimal pH range 4–8·5 (3668), the maximum germination of chlamydospores occurred at 25°C, but none at 35°C (24). Optimal germination of the chlamydospores was achieved after several months of maturation (4469), and their germination is also stimulated by alternating dry and moist conditions (3349). Suitable C sources for the germination of chlamydospores are D-glucose, sucrose, and D-mannose (3668). Chitinases (from different sources) are able to separate the chlamydospore cells (1043). — The influence of pH on mycelial growth is temperature-dependent, poor growth occurs at pH ≥7·0 (389) and the optimum lies in the range between 3·9–6·4 (3421); the optimal temperature for mycelial growth is 20–24°C (or higher) but no growth occurs at either 8°C or 37°C (2203, 3387, 3421). The thermal death point has been determined as 47·5°C for 30 min (613). — *Ch. elegans* grows well on sucrose as C source, together with asparagine (5554), as well as on L-aspartic acid, L-glutamic acid and L-arginine (5615) as N sources; the trace elements Fe, Zn, Cu, Mn and Mo (5554) and thiamine are also required, but riboflavin, although not essential, had a stimulatory effect (5615); growth and germination have high O_2 requirements (3045, 3046). This fungus shows proteolytic activity (3429), and on a bean root medium produces phosphatidases also found in infected root tissues (3437). The growth and reproduction of this species in soil is generally found to be correlated with its parasitic activity (390); pathogenicity has been observed within a wide range of soil-moisture regimes (388), but may be limited to the surface layers due to its O_2 requirement (3046). Attempts at biological control have been based on the susceptibility of germinated propagules to lysis; any additions with high C/N ratio favouring germination are potentially effective for biological control (24, 464), but the addition of nitrogen nullifies this effect (5462); the degree of control obtained in field situations, however, has been very limited (24, 5440).

Cheilymenia Boud. *1885*

Cheilymenia pulcherrima (Crouan) Boud. *1907* — Fig. 77.

≡ *Ascobolus pulcherrimus* Crouan *1858*
≡ *Lasiobolus pulcherrimus* (Crouan) Schröt. *1893*
≡ *Patella pulcherrima* (Crouan) Seaver *1928*

DESCRIPTIONS: Seaver (5170), Le Gal (3275), and Maas Geesteranus (3479); in culture: Woronin (6441); nomenclature: van Brummelen (755). — Colonies growing very rapidly, reaching 5 cm diam in four days at 20°C on MEA, with scanty, 6–16 μm wide submerged hyphae and no aerial mycelium. Chlamydospores mostly lateral but occasionally terminal, on short, irregularly curved, 8–10 μm wide, 1–3-celled stalks, ellipsoidal with a truncate base, pale brown, smooth- and thick-walled, 23–30 × 18–22 μm. By inoculation of identical or different isolates at two opposite points of a petri dish with 1% MEA, apothecia can be obtained in the contact zone. Ascogonia are vermiform lateral branches consisting of several swollen cells

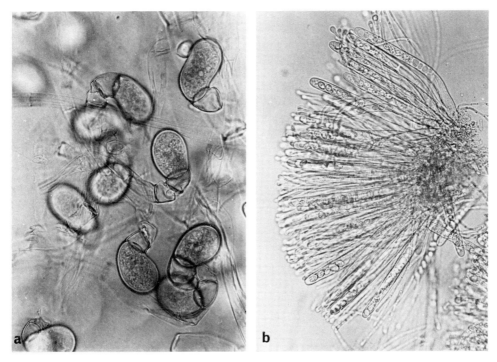

FIG. 77. *Cheilymenia pulcherrima*. a. Chlamydospores, × 500, b. squashed apothecium produced *in vitro*, × 200; CBS 864.68.

(6441). Apothecia 600–1000 μm diam, pale orange, surrounded by a series of pale brown, septate, acute, 200–250 μm long setae. Paraphyses branched near the base, frequently septate, slightly broadened at the tip. Asci operculate, cylindrical, 180–220 μm long, 8-spored; ascospores uniseriate, ellipsoidal, pale yellow, smooth-walled, on the natural substrate 21–30 × 12–18 μm (3275), *in vitro* often much smaller.

Ch. pulcherrima is unusual amongst the coprophilous Discomycetes in that it can easily be recognized in culture by its anamorph. The generic classification of this species is not yet satisfactorily resolved (755); *Lasiobolus* Sacc. emend. Kimbr. *1967* is not suitable because this is defined as having non-septate setae with a bulbous base and ascospores with de Bary bubbles.

Ch. pulcherrima has been isolated infrequently from washed soil particles of a wheatfield and other agricultural soils in Germany and the Netherlands (CBS) and once from beechwood soil in Poland (3138); otherwise it is known only from apothecia collected on cow and sheep dung. So far its distribution is documented for Europe, Nebraska and Bermuda (3136, 3479, 5170).

Chrysosporium Corda *1833*

= *Geomyces* Traaen *1914*
= *Glenosporella* Nannf. *1930*
= *Aleurisma* sensu Bisby *1944*

Type species: *Chrysosporium merdarium* (Link ex Grev.) Carm. (=*Chrysosporium corii* Corda *1833*)

Teleomorphic genera: *Pseudogymnoascus* Raillo, *Gymnoascus* Baran., *Arthroderma* Currey, *Ctenomyces* Eidam, *Aphanoascus* Zukal, and possibly other genera of Gymnoascaceae.

MONOGRAPHIC TREATMENTS: Carmichael (902), and Dominik (1418)*. — Vegetative hyphae and conidiogenous structures hyaline. "Aleurioconidia" hyaline or brightly coloured, one-celled, subglobose to pyriform or clavate, broadly attached to the supporting hyphae, borne singly at the tips of hyphae or lateral branches, or ± sessile on hyphae (solitary thalloconidia) or in short chains, mostly with alternate fragments aborting (alternate arthroconidia). Intercalary structures resembling conidia and larger and thick-walled cells able to act as chlamydospores are commonly produced.

Geomyces has been retained as a separate genus by Sigler and Carmichael (5322) on the basis of the erect and acutely branched conidiophores; this distinction is not accepted here as in freshly isolated cultures *Ch. merdarium* can show a similar tree-like type of branching. Two genera originally included in *Chrysosporium* by Carmichael (902) are now regarded as distinct: *Myceliophthora* Cost. *1894* (cf. *Thielavia heterothallica*) was removed by von Arx (205) because of its narrow conidiiferous denticles (blastoconidia); *Emmonsia* Cif. & Montem. *1959*, however, produces both similar "aleurioconidia" with a narrow base (which are better interpreted as solitary blastoconidia) and distinctive large, globose "adiaspores" which can appear in the lungs of rodents and man but are also formed on blood agar at 37°C. *Malbranchea* Sacc. *1882* (see 5322) produces alternating arthroconidia on undifferentiated hyphae which are not wider than the hypha which bear them, while in *Trichophyton* Malmst. (q.v.) microconidia are usually solitary, clavate to cylindrical, and commonly accompanied by larger, septate macroconidia.

Despite these segregations, *Chrysosporium* remains a heterogeneous assemblage of over ten species. Some species are cellulolytic soil fungi (e.g. *Ch. pannorum*, *Ch. pruinosum* and Ch. *tropicum*), whilst others (e.g. the anamorphs of *Arthroderma* and *Ctenomyces*) are keratinophilic. Of the nine species treated here, three are discussed under their *Arthroderma* teleomorphic names, and one is to be found under each of the names *Aphanoascus*, *Ctenomyces* and *Pseudogymnoascus*.

The colony habit and growth rates are described for cultures growing on a phytone-yeast extract (PYE) agar at 25°C (902), but similar results are obtained with MEA and OA.

*and van Oorschot *in* Stud. Mycol. **20**, *1980*.

Key to the species treated:

1 Conidia less than 4 μm diam **2**
 Conidia more than 4 μm diam **5**

2(1) Conidia ovoid-pyriform, smooth-walled or roughened, usually formed on short, tree-like conidiophores, 2–5 × 2–4 μm; cellulolytic species **3**
 Conidia clavate, smooth-walled, arising laterally from the hyphae and resembling microconidial *Trichophyton*; keratinophilic species **4**

3(2) Colonies ochraceous but with pink to vinaceous tints at least in patches, producing ascomata on OA in 2–3 weeks *Pseudogymnoascus roseus* (p. 676)
 Colonies white, yellow, orange, grey or ochraceous, never producing ascomata *Ch. pannorum* (p. 196)

4(2) Conidia borne singly on vegetative hyphae; homothallic, producing ascomata in 2–3 weeks *Arthroderma curreyi* (p. 67)
 Conidia arising singly or in dense groups on short undifferentiated pedicels or sessile; heterothallic *Arthroderma cuniculi* (p. 66)

5(1) Conidia 5–6 × 4–5 μm, globose to pyriform, smooth-walled to roughened, arising in short chains on branched or simple lateral hyphae; colonies with a yellow or green pigment; neither cellulolytic nor keratinophilic *Ch. merdarium* (p. 195)
 Conidia mostly larger; keratinophilic **6**

6(5) Conidia smooth-walled or slightly roughened, mostly solitary and on short lateral or terminal pedicels, pyriform with a truncate base, mostly 10–12 × 7–8 μm; colonies white to yellow *Aphanoascus fulvescens* (p. 44)
 Conidia conspicuously roughened; colonies buff to brown **7**

7(6) Colonies growing restrictedly (reaching less than 3·5 cm diam in 2 weeks on PYE at 25°C); conidia formed singly or in botryose groups on short and swollen lateral branches, subglobose to ovoid **8**
 Colonies growing rapidly (reaching 4·0–5·0 cm diam in 2 weeks on PYE at 25°C); conidia formed singly, pyriform, coarsely warted, mostly 13–15 × 7–10 μm; colonies buff with a white margin *Arthroderma tuberculatum* (p. 72)

8(7) Colonies reaching less than 1 cm diam in 2 weeks on PYE at 25°C, brown to olive with a white margin; conidia mostly 9–11 × 6·5–9·5 μm *Ch. asperatum* (p. 194)
 Colonies reaching up to 3·5 cm diam in 2 weeks on PYE at 25°C, buff with a white margin; conidia mostly 10–17 × 7–9 μm *Ctenomyces serratus* (p. 234)

Chrysosporium asperatum Carm. *1962* * — Fig. 78.

DESCRIPTION: Carmichael (902). — Colonies slow-growing, reaching less than 1·0 cm diam in two weeks on phytone-yeast extract agar at 25°C, flat, dense, dry, powdery, white, olive to brown, with white margin. Conidia arising directly on the hyphae or on either short or long lateral branches, frequently with 1–3 conidia forming on ampulliform swellings. Conidia subglobose to pyriform or clavate, thick-walled, hyaline to pale yellow, verrucose, 9–11 × 6·5–9·5 μm. — *Ch. asperatum* is very similar to the *Chrysosporium* state of *Ctenomyces serratus* (q.v.) but the conidia are shorter, the colonies are darker, and the rate of growth slower.

*Now *Myceliophora vellerea* (Sacc. & Speg.) van Oorschot (Stud. Mycol. **20**, 47. *1980*).

FIG. 78. *Chrysosporium asperatum*, attached and liberated conidia, CBS 479.76.

Ch. asperatum has been isolated from soil in Canada (305, 902), Czechoslovakia (3174), Siberia (5228), Afghanistan, Iran (2366), Chile (4569), Taiwan (6119), the Cook Islands in Polynesia (IMI), bird habitats in India (5046) and birds' feathers in West Siberia (5229).

Chrysosporium merdarium (Link ex Grev.) Carm. *1962* — Fig. 79.

≡ *Sporotrichum merdarium* Link ex Grev. *1823*
= *Chrysosporium corii* Corda *1833*
  ≡ *Sporotrichum corii* (Corda) Sacc. & Trav. *1911*
= *Blastomyces luteus* Cost. & Rolland *1888*

Teleomorph: *Gymnoascus uncinatus* Eidam *1880*

DESCRIPTIONS: anamorph: Carmichael (902), Sigler and Carmichael (5322), and Matsushima (3680); teleomorph: Benjamin (449), Samson (5023), and Orr *et al.* (4317). — Colonies growing moderately rapidly, reaching 3·0–4·5 cm diam in two weeks at 25°C on phytone-yeast extract agar, variable in colour and texture, downy or granular, often with a broad, glabrous or waxy marginal zone and a granular, intensely yellow centre which becomes green or reddish brown after 3–4 weeks; reverse yellow to olivaceous. After prolonged culture, isolates often degenerate and no longer show the characteristic catenulate arthroconidia and pigmentation, but mainly form intercalary chlamydospore-like structures. Conidia are typically formed on lateral branches as alternate arthroconidia, either subglobose to pyriform with a broadly flattened base or cuboid (when intercalary), smooth-walled to conspicuously roughened, mostly 5–6 × 4–5 μm. Distinct chlamydospores absent. — An uncommon var. *roseum* W. Gams & Domsch *1970* with pink, powdery colonies has been isolated from arable soils (1887). — Ascomata have been observed in culture in a few isolates only. In view of the variability of the anamorph, we prefer to use the *Chrysosporium* name if the teleomorph has not been observed.

Ch. merdarium was originally observed on the dung of various animals and leather (902), but has also been frequently isolated from soil and plant remains. It has been found in many parts of Europe and North America (902, 4315) and also in central Africa (3063) and Nepal (1826). It has been isolated from arable soils in Germany (1889) and the Netherlands (1614, 1616) and regularly appeared with a greater frequency after pea crops than after wheat or rape (1433).

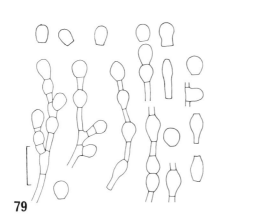 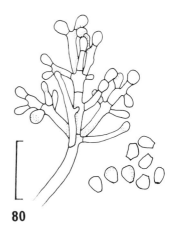

79   **80**

FIG. 79. *Chrysosporium merdarium*, alternating arthroconidia, CBS 225.74.
FIG. 80. *Chrysosporium pannorum*, tree-like branching of conidiophore with short chains of arthroconidia, CBS 226.74.

No growth occurs at 37°C (902). Both starch (1827) and pectin (1432) decomposition can be demonstrated but are slight in comparison with other fungi; *in vitro*, it stimulated the growth of *Chlorella pyrenoidosa* (1430).

Chrysosporium pannorum (Link) Hughes *1958* — Fig. 80.

≡ *Sporotrichum pannorum* Link *1824*
≡ *Geomyces pannorum* (Link) Sigler & Carm. *1976*
= *Corethropsis hominis* Vuill. *1913*
= *Geomyces vulgaris* Traaen *1914*
? = *Geomyces sulphureus* Traaen *1914*
? = *Geomyces auratus* Traaen *1914*
? = *Geomyces cretaceus* Traaen *1914*
? = *Sporotrichum carnis* Brooks & Hansf. *1923*
    ≡ *Aleurisma carnis* (Brooks & Hansf.) Bisby *1944*
= *Aleurisma lugdunense* Vuill. *1924*
= *Sporotrichum lipsiense* Benedek *1926*
? = *Sporotrichum cejpii* Fassatiová *1953*

DESCRIPTIONS: Traaen (5874), Carmichael (902), Fassatiová (1699), Taylor (5763), Sigler and Carmichael (5322), and Matsushima (3680); REVIEW: Williams and Pugh (6377). — Colonies usually slow-growing, reaching 1·0–2·0(–4·0) cm diam in two weeks on phytone-yeast extract agar at 25°C, floccose, granular or finely powdery but also sometimes smooth or with scattered tufts of aerial mycelium; the colour varies from white to yellow, grey, brown, olive, pink or lilac; reverse normally yellow but sometimes orange or brown with the pigment diffusing into the agar. Conidiophores arising from submerged or aerial hyphae, erect or standing out from mycelial tufts in all directions, to 50 μm long, bearing short chains of 2–4 alternate arthroconidia; arthroconidia cuneiform, pyriform to clavate, with a broadly truncate base when terminal but otherwise with both ends truncate; smooth to distinctly roughened, hyaline to greyish, mostly 2–5 × 2–4 μm. Distinct chlamydospores absent. — *Ch. pannorum* is microscopically indistinguishable from the anamorph of *Pseudogymnoascus roseus* Raillo (q.v.), which usually has pink to vinaceous colonies (at least in part) and produces ascomata on OA after three weeks. *Geomyces asperulatus* Sigler & Carmichael 1976 has longer conidial chains of 3–6 conidia, conidia which are thick-walled and slightly roughened, and colonies pale pink at first but later becoming a vivid mustard-yellow. — Although the correct nomenclature of this species has not yet been fully ascertained, it is now widely recognized that some of the numerous names listed as synonyms of it by Carmichael (902) represent different taxa; Williams and Pugh (6377) listed *Sporotrichum carnis*, *Geomyces cretaceus*, *G. vulgaris* and *S. cejpii* as distinct. It is likely that the great variability in the isolates obtained by Carmichael was at least partly due to the use of degenerated isolates and that colony pigmentation (viz. grey-brown/sulphur-yellow/golden-yellow/or white to buff) of Traaen's four *Geomyces* species which is correlated with smooth conidia in *G. auratus* and roughened ones in the other three species, is a reliable criterion for distinguishing fresh isolates. *Corethropsis hominis* and *S. carnis* are both described as having white colonies which become golden-yellow, *S. cejpii* is entomogenous and has compact white colonies and smooth-walled conidia. — Hyphae contain $C_{18:2}$ and $C_{16:0}$ as the most abundant fatty acids, besides $C_{18:3}$, $C_{18:1}$ and $C_{16:0}$ (1280).

Ch. pannorum is common in most soils (1376, 2008, 6377) preferentially in those with a neutral reaction (6184). It can be isolated particularly easily by dilution plates, soil plates with cellulose agar (6414), buried filter paper (5874) and cellophane (5886) or hair baits (2091). Its distributional range includes tundra soils in Canada and Alaska (1446, 3062), Spitsbergen (3064) and high altitudes (1826, 1876, 2005, 3445, 3538, 3976, 4569, 4711). With the exception of one report from Taiwan (6119), it has not been found in tropical and subtropical regions. It has been isolated especially frequently from various forest soils but is also known from grassland (159, 1700, 3445, 5812, 5846, 6182), soils with steppe type vegetation (4407), garden and arable soils (234, 1889, 3446, 3450), compost soil (1424), dunes (6414), saline soils (977, 3446), peat (1039, 3234, 5559), raw humus (989, 1876), activated sludge (1387), more frequently from soils to which cellulose (e.g. paper or paper pulp) has been added (4596) or from preservative-treated wood in ground contact (4193); it has also been found in slate rubble (566), children's sandpits (1421), caves (3453), and detritus in running fresh water (4429). It has been isolated from soil at depths down to 60 cm (3447). After a forest fire it was found to be slightly more frequent in the A_2 horizon of the burnt plot than in unburnt parts (6353). It was isolated regularly from a golf course after repeated applications of an organo-mercury fungicide (6378). Occurrences in the root region of *Pinus strobus* (1794), beans (4451), various forest plants (4814), peat bog plants (1376), *Andropogon scoparius* var. *septentrionalis* (6414) and wheat (5330) have also been reported. The fungus has been found with high frequency in a truffle soil (3454). It has proved common on the seeds of cultivated grasses (3512) and has also been isolated from bumble-bees (402, 5404), cultures of the collembola *Hypogastrura* (1027), earthworm casts (1429), and birds' nests (2577), feathers and pellets (2575). In one study it proved to be the most frequent fungus on leaf litter incubated at low temperature (2694) but was not able to survive under such conditions in the absence of other fungi (2695). — Growth *in vitro* was stopped by hyphal contacts with *Tuber melanosporum* (632).

The optimal temperature for growth is 18°C and the maximum 25–30°C (1377, 4596, 5874). *Ch. pannorum* can become adapted to low temperatures and grow below 0°C (740, 3153, 4596). — Xylan decomposition is very good (4192) and there is little variability in this activity between individual isolates (1432); starch and pectin are both utilized (1432, 1750, 1827, 4192), cellulose is decomposed in various substrates (4192, 6377), and mannanase is produced (4192); cellulose degradation increases with increasing concentration of the N source (4433, 4434). *Ch. pannorum* also causes a weight loss in birch and pine wood (4191); it is not keratinophilic but can form penetrating hyphae on hairs in pure culture (1633, 6122). — A significant selective antagonistic effect is exerted against *Pseudocercosporella herpotrichoides* (1431). *Ch. pannorum* slightly stimulated the growth of sugar beet in some greenhouse experiments (1427). Optimal growth occurs at the salt concentration of seawater, but a salt concentration three times higher is also tolerated (4596).

Cladorrhinum Sacc. & Marchal *1885*

Cladorrhinum foecundissimum Sacc. & Marchal *1885* — Fig. 81.

DESCRIPTIONS: von Arx and Gams (209), Gams and Domsch (1887), and Gams and Mouchacca (1892). — The hyphomycete genus *Cladorrhinum* (=*Bahupaathra* Subram. & Lodha *1964*) is characterized by the production of pustules of richly branching fertile hyphae which in each cell form a lateral (or terminal) collarette and release a number of small, one-celled hyaline phialoconidia which adhere in slimy balls. *Cladorrhinum* contains seven species (1892) amongst which the type species, *C. foecundissimum*, can easily be recognized by its yellow-green to pink colonies, globose to dacryoid conidia $2 \cdot 5 - 3 \cdot 5 \, \mu$m diam (which are ellipsoidal in the otherwise very similar anamorph of *Apiosordaria verruculosa* (C. N. Jensen) v. Arx & W. Gams (q.v.)), the absence of microsclerotia (unlike *C. bulbillosum* W. Gams & Mouch. *1980*) and the absence of dark brown, thick-walled hyphae (unlike *C. samala* (Subram. & Lodha) W. Gams & Mouch. *1980*). *C. foecundissimum* has fast-growing colonies with daily increments of 6–11 mm at 24°C and forms, particularly on OA, ochraceous or greyish pustules of slightly pigmented fertile hyphae with cells $4 - 12 \times 3 - 4 \cdot 5 \, \mu$m; collarettes sometimes to $4 \, \mu$m prominent and $1 - 2 \, \mu$m diam; conidia hyaline, globose to dacryoid, $2 \cdot 5 - 3 \cdot 0 \, \mu$m diam. Mating experiments aiming to obtain ascomata were unsuccessful (209).

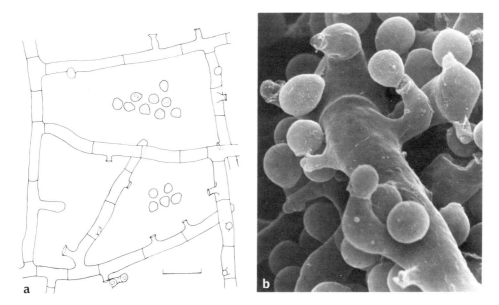

FIG. 81. *Cladorrhinum foecundissimum*. a. Rectangularly branching fertile hyphae with lateral collarettes and phialoconidia; b. details of conidiogenesis, SEM, × 4000, CBS 180.66.

Since the conidia are not capable of germination (209), the species has hardly ever been isolated from dilution plates. Its abundant occurrence on washed soil particles (1889) and buckwheat baits (209) must therefore be ascribed to isolations from clustered vegetative cells adhering to the particles. *C. foecundissimum* has been isolated from agricultural soils in Germany and the Netherlands, sometimes with great frequency, sometimes very rarely and at other times not at all (1889). Other reports include forest soils under beech in Poland (3138), under *Pinus-Chamaecyparis* in Japan (4971) and soil in New Guinea (6499). It has been found with significantly increased frequencies after growing rape than after wheat or peas (1433), and also after soil treatment with urea, liquid ammonia or ethylenediamine (4971). It has been found in earthworm casts (1429), was originally isolated from dung of wild boars and has subsequently been reported from horse dung (4917). — *C. foecundissimum* can serve as food for the mites *Pygmephorus mesembrinae* and *P. quadratus* (3104).

C. foecundissimum decomposes cellulose (1432, 2718) and can cause great weight losses in maplewood strips (2211). Injuries to peat roots have been observed *in vitro* (1430) and it has shown a marked antagonistic effect against *Rhizoctonia solani* (1431).

Cladosporium Link ex Fr. *1821*

= *Hormodendrum* Bonord. *1851*

Type species: *Cladosporium herbarum* (Pers.) Link ex Gray (lectotype)

Teleomorph: *Mycosphaerella* Johanson

Contributions to a MONOGRAPHIC TREATMENT: De Vries (6121), and Ellis (1603, 1604). — Colonies rather slow-growing, mostly olivaceous-brown to blackish brown, velvety or floccose, becoming powdery from the abundant conidia. Vegetative hyphae, conidiophores and conidia equally pigmented. Conidiophores ± distinct from the vegetative hyphae, erect, straight or flexuous, unbranched or with branches restricted to the apical region, in some species with geniculate sympodial elongation. Numerous branched acropetal conidial chains are formed from multiple conidiogenous loci either synchronously or in succession, with the lowest conidia often larger and septate (so-called "ramoconidia"), the upper ones one-celled, ellipsoidal or fusiform; blastoconidia forming on mostly 1–3 broadly conical denticles at the tip of the conidiophore, or subapically below a septum, or on the tip of previously formed conidia, with the delimiting septum usually thick-walled and darkened. Stromatic structures are commonly produced on host plants and *in vitro*, particularly on media containing ammonium sulphate. The racemosely branched conidiophores with acropleurogenous conidial chains arising at several levels are referred to as the *Hormodendrum* type, while geniculate and sympodially elongating conidiophores are termed the *Cladosporium* type (6121); these two types, although generally characteristic of a species, cannot always be sharply demarcated.

Some 500 species have already been described in *Cladosporium* and others continue to be so and a critical revision is overdue. De Vries (6121) redescribed nine species as typical members of the genus, and Ellis (1603, 1604) 15 and 28 species, respectively. For soil isolates, the arrangement of de Vries (6121) is generally used although the species distinguished there may require further subdivision. At present, the application of teleomorph names to *Cladosporium* conidial isolates is not justified. — Similar genera distinguished by some authors include *Hormoconis* v. Arx & de Vries *1973*, the anamorph of *Amorphotheca resinae* Parbery (q.v.), which has conidia without prominent scars arising from narrow denticles; *Fulvia* Cif. *1954* (type species *Fulvia fulva* (Cooke) Cif., pathogenic to tomatoes) which has unilateral swellings on the conidiophores and always simple branches; and *Zasmidium* Fr. *1849* emend. de Hoog *1977* (lectotype species *Zasmidium cellare* (Pers. ex Gray) Fr., correctly named *Rhinocladiella ellisii* D. Hawksw. *1977*), which has very long, sympodially elongating rhachides like *Rhinocladiella* with all structures verruculose, and forms microsclerotia.

Four ubiquitous and polyphagous species are treated here; some others are ± host-specific on senescent or dead plant material. *Cladosporium* species are amongst the most common air-borne fungi and thus have a worldwide distribution. The air spora is particularly rich in wet weather and water droplets contribute to the transportation of conidia (2293). A common feature of the species treated here is the 7-β-hydroxylation of progesterone (876).

For identification the most commonly used medium is 2% MEA or a glucose medium of pH 5·4 (6121); Petri dish cultures are best incubated at 18–20°C for about one week.

Key to the species treated:

1   Conidia with rounded ends, without prominent scars   cf. *Amorphotheca resinae* (p. 40)
     Conidia with apiculate ends and prominent scars   **2**

2(1) Conidia usually not exceeding 4·5 μm in width, smooth or slightly roughened; conidiophores not
      elongating sympodially   **3**
      Conidia usually exceeding 5·0 μm in width, distinctly verrucose; conidiophores commonly
      elongating sympodially   **4**

3(2) Most conidia (sub)globose, 3–4·5 μm diam, finely roughened   *C. sphaerospermum* (p. 209)
      Most conidia elongate, 3–7 × 2–7 μm, smooth-walled or almost so *C. cladosporioides* (p. 202)

4(2) One-celled conidia mostly 5·5–13 × 3·8–6·0 μm; somewhat larger 2–3-celled ones also pre-
      sent   *C. herbarum* (p. 204)
      Once-celled conidia mostly 7–15 × 5–9 μm; the commonly occurring 2–3-celled conidia consid-
      erably larger   *C. macrocarpum* (p. 208)

Cladosporium cladosporioides (Fres.) de Vries *1952* — Fig. 82.

    ≡ *Penicillium cladosporioides* Fres. *1850*
    ≡ *Hormodendrum cladosporioides* (Fres.) Sacc. *1880*
  = *Cladosporium hypophyllum* Fuckel *1870*

DESCRIPTIONS: De Vries (6121), Yamamoto (6467), Subramanian (5626), and Ellis (1603). — Colonies reaching 3·0–4·0 cm diam in ten days on MEA at 20°C; olivaceous-green to olivaceous-brown, velvety; reverse olivaceous-black. Conidiophores to 350 μm long but mostly much shorter, 2–6 μm wide, branching acropleurogenously and bearing numerous conidial chains arising below septa, but without swellings and sympodial elongations. Conidia ellipsoidal to lemon-shaped, mostly smooth-walled, rarely minutely verruculose, olivaceous-brown, one-celled 3–7(–11) × 2–4(–5) μm. — The most similar species to *C. cladosporioides* is *C. cucumerinum* Ell. & Arth. *1889*, which parasitizes Cucurbitaceae, and has woolly, pale grey-green colonies (cf. also *C. sphaerospermum*). Two similar species which may be mistaken for *C. cladosporioides* but do have nodose conidiophores are *C. tenuissimum* Cooke *1878* (cf. 1604) with very long conidiophores (to over 800 μm) which are apically sometimes unilaterally swollen, and smooth-walled or minutely verruculose conidia, 3–25 × 3–6 μm, and *C. oxysporum* Berk. & Curt. *1868* with conidiophores to about 500 μm, terminal and intercalary conidiiferous swellings, and ellipsoidal to lemon-shaped, smooth-walled conidia, 5–30 × 3–6 μm. — Conidia and hyphal cells of *C. cladosporioides* are uninucleate (6121). — DNA analysis gave a GC content of 49–50% (5600).

Fig. 82. *Cladosporium cladosporioides*, conidiophores and conidia, orig. J. Veerkamp.

Apart from *C. herbarum*, *C. cladosporioides* is the commonest species of its genus both on plant material and in soil. Reports of the occurrence of both species on leaf litter are extremely numerous and both are also very common inhabitants of the phylloplane and rank as widely distributed soil fungi (2741, 4422). *C. cladosporioides* appears frequently on dilution plates and is easily isolated by almost all other techniques (5812). After soil washing, which largely excludes the possibility of atmospheric infection, it has been mainly found to occur on organic soil particles (1614, 1888). The conidia are very commonly air-borne and their frequency has peaks in June–July and September–October in the British Isles (2293). — Its overall distribution ranges from frequent appearance in boreal-alpine regions to regular occurrences, though in small amounts, in tropical and subtropical zones. In the temperate zone this species has been isolated frequently from various forest as well as from grassland, garden and arable soils, dunes, salt marshes and saline beaches; there are too many reports to cite individually here. Furthermore, the fungus has been reported from caves, peat bogs, heaths and podzols. Its presence has been demonstrated in both clean (1166, 4429) and highly polluted streams (1154, 1155, 1157, 1166), a lake (4229), marine habitats (8, 4850), fields treated with sewage sludge (1163), activated sludge (1165, 1172, 1387), a waste stabilization pond (1166) and compost beds (3041); after the addition of sulphite liquor, its frequency in soil increased (1161). *C. cladosporioides* survives a drying out of the soil surface in summer (5000). There are reports of its isolation from the root surface of *Pinus strobus*, *Larix leptolepis* (1794), wheat (2023, 5330) and the poplar rhizosphere in truffle grounds (3452); in addition to this, it is particularly frequent on discoloured stored timber (812, 3210), wood pulp (6168), stored crops, seeds, birds' nests, feathers and pellets (2575, 2577), dry fodder (3258) and food products (509, 2572, 3153, 3930, 5205, 5980). — Only 25% of conidia introduced into the soil survived after 12 days incubation, but on colonized plant organs the survival rate was higher (4421). The mineralization of mycelia introduced into the soil takes place slowly, as is the case in many other fungi which also have pigmented hyphae (3633). — The mycelium can serve as food for the mites *Acarus gracilis*, *Aleuroglyphus ovatus*, *Glycyphagus destructor* and *Suidasia medanensis* (5380), and also for *Hypogastrura tullbergi* (3830), *Tenebrio angustus*, *Lathridius minutus*, *Microgramma arge* and *Lepinotus reticulatus* (5381).

Conidium production is considerably higher under moist than under dry conditions; it has been estimated to amount to $>2\cdot5 \times 10^4$ conidia mg^{-1} dry mycelium during nine weeks under wet conditions, and to $5\cdot7 \times 10^2$ under dry conditions (2294). In alternating light/darkness a distinct zonation of the cultures can be observed (1865). For growth on stored grain, vapour saturation is required (4493). Vegetative growth is possible in the temperature range 0–32°C, optimal development being at 20–28°C (1671, 4487); slight growth is still possible at −3°C (191) and even down to −10°C (509). — *C. cladosporioides* can grow on the following C sources: L-arabinose, cellobiose, D-galactose, D-glucose, D-xylose (1671), starch and pectin (1432, 1750, 3307). Arabinoxylan is degraded by the action of a xylanase (1753) but the cellulolytic activity of this species is low (654, 1168, 1671, 1750, 2718, 2741, 3307, 5386) and decreases with increasing concentration of the mineral N source (4433). Keratin is not utilized and if inoculated onto hairs, only their medulla is attacked (1633). Gallic and tannic acids are moderately utilized (1750); a number of phenyl carbonic acids can be assimilated; nitrite can be utilized as a N source just as well as nitrate (3842). In liquid cultures, DDT was found to stimulate growth (2488). On agar with lignin sulphonate as the sole C source, good growth also occurred. Laccase production has been observed (4894). On synthetic nutrient media free from growth substances, growth was only moderate (1425); vitamins $B_1 > B_6 > B_2$ stimulate growth, but a wide variability between isolates occurs in this respect (3838). The fungus produces ergosterol from *n*-alkanes; on *n*-undecane the yield was better than on glucose (3337). — Under favourable conditions, germinating conidia repel *Bacillus pumilus* and other bacteria from the barley phylloplane *in vitro* but not *in vivo* (1384). A mycostatic antibiotic, the isocoumarin derivative cladosporin, has been isolated (5166). The stimulation of corn seedlings by various *Cladosporium* species has been reported (4557). *C. cladosporioides* has been found to be associated with *Phialophora verrucosa* in one case of a corneal ulcer (4586) and the formation of a fungus ball in a human lung by *C. cladosporioides* has been reported in the USA (3194). — The fungus shows moderate osmotolerance (4001), is highly tolerant of phenyl-mercury compounds (2076) and also of γ-irradiation (3854); it was the most irradiation-resistant of five species tested, tolerating 100 krad with 10%, and 300 krad with >1% survival (4585). In a chronically irradiated soil its frequency increased up to an intensity of 820 R per day, but it was still present in plots receiving 2780 R per day (2009).

Cladosporium herbarum (Pers.) Link ex Gray *1821* — Fig. 83.

≡ *Dematium herbarum* Pers. ex Schleich. *1821*
= *Dematium graminum* Pers. *1822*
≡ *Cladosporium graminum* (Pers.) Link *1824*

= *Dematium epiphyllum* Pers. ex Hook. *1821*
  ≡ *Cladosporium epiphyllum* (Pers. ex Hook.) Nees *1817* ex Fr. *1832*

For numerous other synonyms see Hughes (2602).

Teleomorph: *Mycosphaerella tassiana* (de Not.) Johanson *1884*

  ≡ *Sphaerella tassiana* de Not. *1863*
= *Mycosphaerella tulasnei* (Jancz.) Lindau *1906*
  ≡ *Sphaerella tulasnei* Jancz. *1894*

DESCRIPTIONS: anamorph: De Vries (6121), Yamamoto (6467), Subramanian (5626), and Ellis (1603); teleomorph: von Arx (194, 195), and Barr (354). — Colonies reaching 3·0–3·7 cm diam in ten days on MEA at 20°C, olivaceous-green to olivaceous-brown, velvety; reverse olivaceous-black. Conidiophores to 250μm long and 3–6 μm wide, with terminal and intercalary swellings ($7–9 \mu$m diam) and geniculate elongations. Conidia ellipsoidal to cylindrical, with rounded ends, distinctly verruculose, golden brown, rather often 2- or more-celled, with scars somewhat protuberant, one-celled $5·5–13 \times 3·8–6·0 \mu$m. — The similar *C. murorum* Petr. *1941* has been distinguished from *C. herbarum* on account of its shorter terminal conidia (3–7 μm long) and also usually, the absence of septate conidia (4838); cf. also *C. cladosporioides*. — A time-lapse study of conidiogenesis has been carried out (2309). — Hyphal cells and conidia are generally 1-, or rarely, 2-nucleate (4812, 6121). During mitosis, eight dot-like chromosomes were reported (1208) but in other studies only five were noted (4812). DNA analysis gave a GC content of 55% (5600).

Together with *C. cladosporioides*, this is the most common *Cladosporium* species on dead organic matter (1603, 6121) and, in general, one of the most common fungi; it is one of the very early colonizers on dying and dead plant substrates, in particular leaves and stems of the most varied plant species including ferns, mosses, aquatic and desert plants. It was found to be the commonest species of all fungi on live leaves and dead material of many plants in Slapton (Devon) in a very moist climate (2338). The dry conidia are easily carried through the air and transported even over oceans (4351). In numerous investigations on the spore content of air, *Cladosporium* has been the dominant genus, with *C. herbarum* the commonest species of it (2587, 2629). A pronounced peak of occurrence was found in June–July in the British Isles (2293). In the summer and autumn, some 90% of the *Cladosporium* conidia from the air proved viable (1299). — Its worldwide occurrence in the very varied soils ranks as its third main habitat, and references to it in soil are extraordinarily numerous. *C. herbarum* sporulates freely and can be isolated easily by many techniques (5812), even hair baits (2091) and soil-washing (982, 1888). Its frequency is higher on organic than on mineral particles. — The numerous reports of this species extend from polar regions through temperate and mediterranean areas to subtropical and tropical countries. Reports from forest, grassland and arable soils are equally numerous. The following can be cited as extreme habitats: peat bogs (1376, 3856, 4180, 6520), conifer swamps (1039), lowland soils (2008, 3417), wet prairie soil (4313), soil under heath vegetation (5047), soil with tundra (1446) as well as with chaparral vegetation (1164), salt marsh (8, 4000, 4645, 4646) and estuarine sediments with high salinity (655, 4477), dunes (745, 3941, 4162), desert soils (100, 1512, 2974) and other very dry areas (1824), a guano cave (4312), a uranium mine (1703), activated sludge (1387), a river bed (3868), fresh- (3809, 3942, 4429) and seawater (4918), but not in highly polluted aquatic environments (1154). It is very frequent on decaying wood (2338, 3568, 3749) either as a surface colonizer (812) or a causal agent of "blue staining" (5061, 6572) and also on twigs of old trees of *Pseudotsuga menziesii* (5267). It has repeatedly been found on wood exposed to

Fig. 83. *Cladosporium herbarum*. a. Geniculately proliferating conidiophores and conidia, CBS 572.78; b. part of intact conidiophore, SEM, × 2000, orig. R. A. Samson.

soil (1956, 2833), and especially in wood pulp (4857, 5060, 6166, 6168). — It plays an important role as a colonizer of plant debris in litter though it also occurs at greater soil depths. A wide pH range from 4·4 (1713, 3417) to alkalinity (745, 3417, 3550, 3732, 5559) is tolerated, the optimum being given as pH 6 (1204). A certain prevalence in moist soils has occasionally been observed (2093); the minimum water potential was −155 bars for conidium germination (6436), −210 bars for vegetative growth and −140 bars for sporulation: the fungus can thus be regarded as rather xerophilic. It has frequently been observed in various rhizospheres, including those of *Funaria* (2860), ferns (2856, 2858), *Pinus laricio* (983), *P. maritima* (3798), onion (2159, 3733), wheat and oats (1614, 3567, 4443, 5330, 5825), barley (3870, 4451, 5514), rye grass (5815, 6133), virus-infected *Cynodon dactylon* more frequently than the virus-free grass (2855), flax (3870, 5826), alfalfa (3982, 3983, 5825), beans (4450, 4451), groundnuts (2768), clover (3429, 5815), potato (1614), tobacco (3431), coriander (1523), coffee (5313), *Trigonella foenum-graecum* (3733) with increasing occurrence after foliar spray with the antibiotic subamycin (2187), *Calluna vulgaris* (5220), *Trichodesma amplexicaule, Abutilon indicum* (3864), several sand dune (4371), peat bog (1376) and euphorbiaceous plants (3866), dying oak roots (963), roots of *Brassica campestris* (6030), sugar beet (4559), *Salicornia* (4645), *Halimione portulacoides* (1378) and other halophytes (3375). Very numerous reports exist from cereal grain (1752, 3207, 3556, 3584, 3989, 6121), especially wheat (1752, 2630, 2679, 2716, 3491, 4200, 4307, 4591, 5244), and also *Avena fatua* (2961), cultivated grasses (3512), groundnuts (2253, 2765, 3220), and castor beans (3172). It has been found to be associated with aecidiospores of *Cronartium comandrae* (4608), and has also been isolated from fir honey dew (651), the intestine of *Dendrolimus sibiricus* caterpillars (2848), pellets of the diplopod *Glomeris marginata* (4154), nests, feathers, and pellets of free-living birds (682, 2575, 2577, 3882, 4649, 4652, 6223), tunnels, storage places and nests of gerbils (5393), bee honeycombs (402), cultures of the collembolans *Hypogastrura* and *Onychiurus* (1027), liver, intestine, skin and forefoot of mould-infected frogs (1579), earthworm casts (2857), frescoes of a monastery (2666) and on an oil-painting (1626), very common on damp plaster (2338), frozen food products (509, 3153), fruit juices (3428) and butter (6070, 6071). — The germination of conidia in soil is reported as being unaffected (5144) or inhibited (4882) by sporostatic factors and also by aflatoxin (4806) and wortmannin (711) *in vitro*; the production of conidia was also found to be inhibited by rubratoxin B (4808) as was its growth *in vitro* by both *Chalara elegans* (2753) and oak leaf tannins (2282). — After steam sterilization recolonization of soil occurs rapidly (6184).

Release of the conidia in *C. herbarum* is correlated with rising temperatures, daylight, and declining relative humidity (4352), but conidium production is much higher under wet than under dry conditions; $1 \cdot 3 \times 10^4$ and $1 \cdot 1 \times 10^2$ conidia mg^{-1} dry weight mycelium have been estimated, respectively (2294). Data on the optimal growth temperatures vary between 18° and 28°C (191, 1204, 2939, 4487); the optimal temperature is lower for linear growth than mycelium production (2698), and the maximum lies between 28 and 32°C. Individual isolates can be adapted to low temperatures and growth of some is still possible at -6°C (191, 740, 2761). *C. herbarum* does not survive heat treatment of 50–60°C for 30 min (613) and has a thermal death point (in apple juice) at 63°C for 30 min (3620). — A nutrient medium for optimal fat production has been elaborated (5353). The relevance of dry weight determinations in growth studies has been discussed for *C. herbarum* as compared to the faster growing *Cunninghamella echinulata* (5006, 5007). Its decomposition of pectin is pronounced and varies little between isolates (1432, 1750, 1828, 3414, 3539); it is one of the pectolytic organisms involved in flax retting (6356). Its starch utilization is good (1425, 1750, 3307) and, according to numerous reports, cellulose decomposition occurs on the most varied substrates; different isolates differ in their optimum temperature (15–25°C) for cellulose decomposition (3618). Gallic (1750), tannic, benzoic, *p*-hydroxybenzoic, and 2,5-dihydroxybenzoic, protocatechuic, salicylic, anthranilic, phthalic acids and phenylacetate, can all be utilized as C sources (3842). With benzidine, a phenoloxidase can be demonstrated (934). Melanin formation has been demonstrated on gallic acid agar (6121). Autolysis of the mycelium yields a humic substance (4470). Relatively good growth occurs on both humic acid (1750) and lignin sulphonate (934, 1425). The degradation of epoxy- and dioctylsebacate plasticizers (5859), as well as urethane rubber (251), by *C. herbarum* has been reported. The sulphur-containing epicladosporic acid (3231), a-, β- and γ-hydroxybutyrates (4236) and an extracellular galactomannan (3893) are produced. Poor growth on NH_4^+ ions in nutrient solutions is attributed to a fall in the pH (2·7–3·8); NH_4^+ ions in low concentrations also suppress nitrate assimilation, possibly through a blocking of nitrate-nitrite reduction (3972); nitrite can be assimilated less efficiently than nitrate (3842). Growth is stimulated by vitamin $B_6 > B_1$ (3838), and the gibberellins A_3, A_4 and A_9 can be produced (5008). — Reduced O_2

supplies lead to reduced growth (1793, 6281), as do increases in the CO_2 content to 10% (1502). An adaptation to high salt concentrations has been observed (4477). — Mycelial extracts and culture filtrates were shown to contain indole-3-acetic acid, indole-3-acetonitrile and fructose-1,6-diphosphate (6034). — An endotoxin is formed by the fungus which, in its action (lesions on the mucous membrane of horses), resembles that of *Stachybotrys chartarum* (2050); mycelial extracts were found to be toxic to chicken embryos (1388). Toxic effects on warm-blooded animals have been recorded when severely infected cereals were used as fodder (2761, 6412). *C. herbarum* has been found to show a marked antagonistic effect on *Fusarium culmorum* (1431) and *Sporothrix schenckii* (569), but only a slight one against *Pythium "debaryanum"* (4516); otherwise, it is generally without any antibiotic effects (716, 2736). Spoilage of stored fruit is said to be possible but this proceeds slowly and is thus of only minor economic importance (2959); injuries have also been observed on stored tomatoes (2149). — The fungus has a relatively high tolerance for allyl alcohol as a fumigant (3991), UV (1512), and γ-rays (2786).

Cladosporium macrocarpum Preuss *1848* — Fig. 84.

DESCRIPTIONS: de Vries (6121), Subramanian (5626), Łacicova (3206), and Ellis (1603). — Colonies reaching 2·5–3·5 cm diam in ten days on MEA at 20°C, olive-green, often covered with a pale grey aerial mycelium. Conidiophores straight or flexuose, often geniculate and nodose by sympodial elongation, to 300 μm long and 4–8 μm wide, swellings (if present) 9–11 μm diam. Conidia usually in rather short chains, ellipsoidal, densely verruculose, mid-brown to olivaceous-brown, often 1–3-septate, one-celled mostly 7–17 × 5–8 μm. — *C. macrocarpum* is distinguished from *C. herbarum* by its larger and more verrucose conidia which adhere in shorter chains. — The connection with a *Mycosphaerella* teleomorph similar to *M. tassiana* (de Not.) Joh. *1884* has been reported (6121) but requires confirmation. — Hyphal cells and conidia are generally 1- or 2-nucleate (6121).

C. macrocarpum is common in temperate regions, particularly on dead herbaceous and woody plants (1603, 6121, IMI), though on the whole it is less common than *C. herbarum* and *C. cladosporioides*. — Records of its distribution include ones from Turkey, Libya, Egypt, Cyprus, France, Germany, the British Isles, Czechoslovakia, the USA, Tanzania, Kenya, South Africa, Iraq, Nepal, India and Australia. The few data on isolates from soils available refer to very diverse habitats, such as rock cliff on Surtsey (2406), soil with tundra type vegetation (1750), forest litter (2411), soil under *Pinus maritima* (3798), dunes (745), saline beach (4162), sediments immediately below a glacier (2530), and a uranium mine (1703). In the British Isles (Cardiff), it contributed 3–15% to a total catch of the air spora with the maximum in June and July (2293). There are reports on its occurrence in the rhizospheres of oats and barley (3006), on poles of spruce or pine exposed to forest soil with stands of beech and spruce (2833), seeds of cereals (3206, 3207, 4637, 5293) and *Dactylis glomerata* (5940), and feathers of free-living birds (2575).

Growth in mineral nutrient media is stimulated by yeast extract, particularly the vitamins B_1 and B_6 (3838). Conidium production was estimated to be about 4×10^3 conidia mg^{-1} dry mycelium under wet conditions and $3·4 \times 10^2$ under dry conditions, almost ten times lower than that of *C. cladosporioides* (2294). — Pectin and starch are degraded (1750); humic (1750), gallic and tannic acids are moderately utilized; a number of phenolic acids are assimilated; nitrite can serve as N source though less successfully than nitrate; proteolytic activity in *C. macrocarpum* has also been demonstrated (3842).

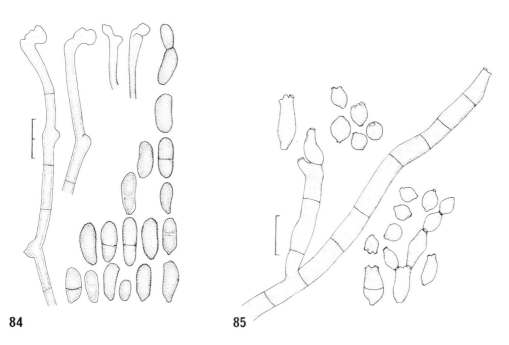

84   **85**

FIG. 84. *Cladosporium macrocarpum*, geniculately proliferating conidiophores and conidia.
FIG. 85. *Cladosporium sphaerospermum*, conidiophore and conidia, CBS 192.54.

Cladosporium sphaerospermum Penz. *1882* — Fig. 85.

DESCRIPTIONS: De Vries (6121), Yamamoto (6467), and Ellis (1603); conidiogenesis: Cole and Samson (1109). — Colonies reaching 1·5–2·0 cm diam in ten days on MEA at 20°C, olive-green to olivaceous-brown, velvety or densely powdery; reverse olivaceous-black. Conidiophores to $300\,\mu$m long but generally much shorter, $3–5\,\mu$m wide, with several conidial chains arising acropleurogenously below septa, not geniculate. Ramoconidia ± elongate and often septate, but most conidia (44–77%) (sub)globose, mid- to dark olive-brown, minutely verruculose, $3–4(–7)\,\mu$m diam. — The delimitation of this species from *C. cladosporioides* is not always sharp as in this species isolates with relatively short conidia occur.

C. sphaerospermum is a common and cosmopolitan species and occurs as a secondary invader on many different plants as well as in air, soil, food-stuffs, paint, and textiles and, occasionally, is isolated from man and animals (1603). — It has been observed in moorland soils in the British Isles (3234), at a recently deglaciated alpine site (2530), in forest (5048) and savannah type soils (5049) in Somalia, various soils in the Netherlands (1616), Turkey (4245), Libya (6510), Egypt (8, 3992), Pakistan (4855), Japan (5846) and on Tahiti (6084), air samples in the British Isles (Cardiff) where it accounted for 3–10% of the total catch with a peak in June–July (2293), and air of a cave in Piedmont/Italy (3453). It has been found in salt marshes (8), the eulittoral zone in Florida (4918) and a polluted stream (1166). A preference for substrates with a pH around neutrality is indicated (3453, 5048, 5049). It was found to be common on leaf litter of *Castanea sativa* (3448), *Betula alba* (3449), cotton (5009), and decaying stems of *Urtica dioica* (6463); it also occurs in the rhizosphere of broad beans (6511), on groundnut pods (1984), seeds of wheat, rice (2297, 3989), barley, *Sorghum*, and corn (57), pecans (2572), various foodstuffs (1303, 5980), in soil cultures of the earthworm *Pheretima californica* (1961), and in birds' nests (2577), feathers and pellets (2575).

Sporulation is estimated to reach $2·4 \times 10^4$ conidia mg^{-1} dry mycelium under wet, and $4·3 \times 10^2$ under dry conditions (2294). Soluble starch is well utilized and cellulase activity is low (3307). Growth is stimulated by yeast extract due to its supplementation with vitamins B$_6$ and B$_1$ (3838). A toxic action against chicken embryos has been demonstrated (1303).

Coccidioides Stiles *1896*

Coccidioides immitis Stiles in Rixford & Gilchrist *1896* — Fig. 86.

The hyphomycete genus *Coccidioides* is characterized by its dimorphism: a mycelial saprophytic phase with arthroconidia and a yeast-like parasitic phase with spherules (sporangia). A relationship with the Gymnoascaceae has been suggested.

DESCRIPTIONS: Baker *et al*. (295), Bogliolo and Arveira Neves (606), Sigler and Carmichael (5322) and many textbooks of medical mycology; general reviews: Huppert (2619) and Ajello (72); bibliography: Al-Doory and Pairon (91). — In the saprophytic phase colonies fast-growing, reaching 3·7 cm diam in 14 days at 25°C on 4% MEA, at first white, but later becoming brown; well developed also on PDA, Sabouraud-glucose agar, and particularly CMA; colony either flat and smooth or cottony. Aerial hyphae transformed into arthroconidia by a regularly spaced septation and subsequent degeneration of alternating cells, so that gaps are left between the conidia (like in *Malbranchea* Sacc. *1882* (5322)). Arthroconidia are highly variable, either cylindrical or barrel-shaped, thin-walled and 6–8(–15) × 2–4 μm, or thick-walled, chlamydospore-like and 3–6 μm long; not separating easily. Arthroconidia are 1–2-nucleate and vegetative hyphal cells 2–3-nucleate (3188). Spherical chlamydospores to 10(–20) μm diam also occurring. — Parasitic phase: Spherules 15– over 80 μm diam, wall approx. 2 μm thick, mostly smooth or with fine excrescences; numerous endospores formed by repeated cleavage and released by breakage of the wall, each 1–4-nucleate. For certain identification, the induction of spherules is essential and they are most easily demonstrated (a) on glucose broth with partially coagulated egg albumin under semi-aerobic conditions, (b) by addition of surface-active agents, or (c) by intraperitoneal injection into rodents. — The arthroconidial state has been classified in *Malbranchea* by Sigler and Carmichael (5322) and is similar to *M. dendritica* Sigler & Carm. *1976* which, however, has a denser arborescent, acute branching of the fertile hyphae and no yeast-like spherules. — Cell walls contain chitin, mannan and β-1,3-glucan (72).

C. immitis is the causal agent of a deep mycosis, coccidiomycosis, in both man and animals. This generally begins as a respiratory infection, noticeable as malaise or pneumonia; it may remain localized and even pass unnoticed (in over 50% of infected people) or spread and even become fatal. The vegetative phase is highly infectious and the greatest care is thus necessary in handling cultures. Unlike tuberculosis, person to person infection does not occur and neither does oral infection in predatory animals through their prey (72); buried carcasses may, however, infest the soil. — The peculiar distributional range of this fungus has mainly been established by coccidioidin skin tests of the populations. *C. immitis* is endemic to semiarid regions of California, Arizona and Texas and rare in other states of the USA (295); its occurrence has also been demonstrated in Italy (295), Central and South America as far south as Argentina (74). — The intradermal coccidioidin test is positive in people who have once been infected but, contrary to other serological techniques, does not give information on the current state of the disease. Cross reactions with histoplasmin occur, but the two pathogens do not occur in the same geographical areas (5669). — The fungus can be demonstrated in soil by

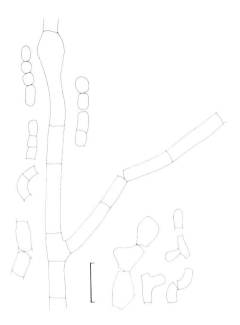

FIG. 86. *Coccidioides immitis*, fragmenting hypha and arthroconidia, CBS 711.73.

direct plating of a 1/10 dilution on 1·5% yeast extract agar with antibacterial antibiotics and cycloheximide, or by inoculating mice with this dilution and examining the internal organs of the animals after three weeks (5670). While the fungus thrives best in culture on rich media, in nature it grows in the poorest alkaline soils, possibly due to an inability to compete with other micro-organisms (1501). A connection with rodent excrements and an enrichment in front of their burrows has been observed (1549) and also near creosote bushes (*Larrea tridentata*) although this and some other plants do not support fungal growth (4846); an occurrence in a river bottom soil provides an exception (1501). — During the hot dry season, the upper soil layer is sterilized and *C. immitis* survives only at depths below 20 cm; during cooler periods it grows up through the soil and sporulates abundantly as the wet season begins; it grows easily in any sterilized soil and shows a certain prevalence at elevated salinity levels (2619). Effective antagonists include *Bacillus subtilis* and *Penicillium janthinellum* (1548); high salt concentrations are, however, tolerated more by *C. immitis* than by these antagonists (1548, 1576). The mycelial phase is better adapted to survival than the spherules (5476).

   C. immitis is not nutritionally exacting, and can even utilize ammonium lactate as its sole C and N source (2619). Some isolates do not sporulate with peptone in the medium (5670). Enzymes of the Embden-Meyerhof and pentose phosphate pathways and the Krebs cycle have been demonstrated, the latter to a lesser extent in the spherule phase than in the mycelium (72). Elastase activity has been reported (4849). Large inocula of arthroconidia suppress mycelial development; at low inoculum densities, 10% CO_2 in the atmosphere inhibits mycelial development (3407). Light has no influence on development (295) and the fungus can survive easily at 40°C. It is also highly tolerant of sodium borate (2619). Amongst several soil fungicides tested, 1-chloro-2-nitropropane is the most promising for the suppression of *C. immitis* locally, before earth-moving projects are commenced (72).

Cochliobolus Drechsler *1934*

= *Pseudocochliobolus* Tsuda *et al. 1977*

Type species: *Cochliobolus heterostrophus* (Drechsler) Drechsler

Anamorphic genera: *Bipolaris* Shoem., *Curvularia* Boedijn

Cochliobolus (Fam. Pleosporaceae Wint.) has been segregated from *Ophiobolus* Riess *1854* on account of its superficial, setose ascomata, long filiform and coiled ascospores, and thicker ascus walls. Asci open mostly by a circumscissile apical rupture (5302). — *Pseudocochliobolus* was segregated from *Cochliobolus* because of the columnar basal stroma which is absent in typical *Cochliobolus*, and less coiled ascospores. — Keys to the species: Luttrell and Rogerson (3456), Ammon (126), Ellis (1603) and Chidambaram *et al.* (1002) and Shoemaker (5304), sub *Drechslera*.

The four species treated here are *C. sativus*, *C. spicifer* (with *Bipolaris* anamorphs), *C. lunatus* and *C. geniculatus* (with *Curvularia* anamorphs); they are keyed out under their anamorphic genera.

Cochliobolus geniculatus R. Nelson *1964* — Fig. 87.

≡ *Pseudocochliobolus geniculatus* (R. Nelson) Tsuda *et al.* 1977

Anamorph: *Curvularia geniculata* (Tracy & Earle) Boedijn *1933*
≡ *Helminthosporium geniculatum* Tracy & Earle *1896*

DESCRIPTIONS: anamorph: Groves and Skolko (2143), Corbetta (1184), Ellis (1602, 1603), and Matsushima (3680); teleomorph: Nelson (4128). — Colonies reaching 3·5 cm diam in five days at 25°C on OA, dark brown, velvety or cottony, lacking stromata. Conidiophores arising singly or in groups, usually simple, straight or flexuose, occasionally geniculate, brown, paler near the apex, to 600 μm long (on the natural substrate, shorter in culture), 5–9 μm wide near the base. Poroconidia mostly 4-septate, usually curved or geniculate in the darkest and broadest central cell, the end cells subhyaline, 18–37(–48 on natural substrates) × 8–14 μm. — This species is heterothallic and ascomata often form after mating in columnar stromata on barley grains on Sachs' agar at 24°C under light; they then ripen within 35 days. Ascomata black, 490–940 μm high, with a distinct ostiolar beak; asci cylindrical to clavate, appearing unitunicate; ascospores ± coiled in the ascus, filiform, somewhat tapering towards the ends, hyaline, 6–16-septate, 160–270 × 4–7 μm.

Very many host plants are known for this species but these are mainly in tropical countries (1602); this fungus is, however, commonly reported from soil, for example in Jamaica (4886), the Bahamas (2006), India (2854, 2864, 4716), Canada (5828), the USA (2007, 2482, 4313), the USSR (2871), Brazil (6008), Peru (2005), Central America (1697), Italy (4538) and South Africa (1559). *C. geniculatus* has been isolated from arable soils under cotton and potatoes (2871), grassland (4538), a coconut grove (2006), willow and cottonwood lowland (2007), forests (3817) and forest nurseries (2482), dry prairie (1559, 4313), desert (4733), soils with different moisture regimes (1315) and estuarine sediments (655). Microhabitats known also include litter under chaparral vegetation (1164), palm leaves and angiosperm wood (3963), decaying shoots of *Bothriochloa pertusa* (2955), the rhizospheres of corn (6563), *Pennisetum typhoides* (3867), virus-infected *Cynodon dactylon* (2855), *Abutilon indicum* (3864) and *Cassia occidentalis* (6067), seeds of cultivated grasses (3512) and a termite mound (2857). *C. geniculatus* causes leaf spots in *Hamiltonia suaveolens* and *Clerodendrum infortunatum* (2864). Dry cultures have survived five years of storage without loss of germinability (5870). — It can serve as food for stored-product mites (5382).

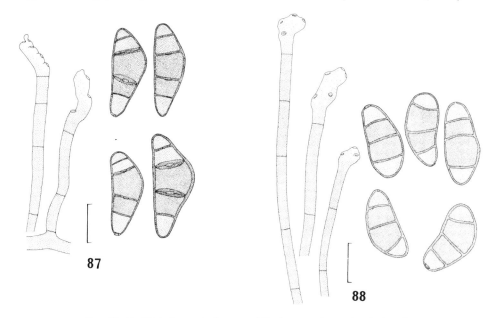

FIG. 87. *Cochliobolus geniculatus*, conidiophores and conidia, CBS 173.56.
FIG. 88. *Cochliobolus lunatus*, conidiophores and conidia, CBS 144.63.

With increasing temperatures (to 25°C) higher percentages of conidia germinate; percentage germination was found to be highest in 20-day-old cultures (2480, 2481); transfer from a humid to a dry atmosphere can initiate violent discharge of the conidia (3763). Colonization of *Plantago lanceolata* flowering stems in soil occurred in $O_2:CO_2$ mixtures between $1:17\cdot5(\%)$ to $21:0\cdot03(\%)$ (2096). — *C. geniculatus* can oxidize various Mn^{2+} salts (5828), produce a red pigment, cynodontin (1,4,5,8-tetrahydroxy-2-methylanthraquinone) (1563), hydroxylate progesterone in the 6β-, 11β- and 14α-positions (6562), and can grow on a large number of acids, alcohols and their esters used as plasticizers; it can also utilize *n*-butylitaconate, dibutylphthalate, and diethyleneglycol (473). — *C. geniculatus* has been isolated from a case of keratitis in man but was not regarded as the primary invader (4199), and also reported as being a causal agent of mycetomas (1624) and postoperative endocarditis (2897).

Cochliobolus lunatus R. Nelson & Haasis *1964* — Fig. 88.

≡ *Pseudocochliobolus lunatus* (R. Nelson & Haasis) Tsuda *et al. 1977*

Anamorph: *Curvularia lunata* (Wakker) Boedijn *1933*
   ≡ *Acrothecium lunatum* Wakker *1898*
   = *Helminthosporium curvulum* Sacc. *1916*
   = *Curvularia lycopersici* Tandon & Kakkar *1964*

DESCRIPTIONS: anamorph: Corbetta (1184), Groves and Skolko (2143), Ellis (1602, 1603), Benoit and Mathur (455), and Matsushima (3680); teleomorph: Nelson and Haasis (4130), and CMI Descriptions No. 474, *1975*. — Colonies reaching 6 cm diam in five days at 25°C on OA, dark brown, lanose or velvety. Conidiophores arising singly or in groups, simple or branched, straight or flexuose, sometimes geniculate, brown, paler near the apex, to 650 μm tall (on the natural substrate, shorter in culture), 5–9 μm wide near the base. Poroconidia 3-septate, curved in the third cell which is usually larger and often darker than the others, smooth-walled, 20–30 × 9–15 μm. — This species is heterothallic; ascomata often form after mating in columnar stromata on barley grains placed on Sachs' agar at 24°C; they then ripen within 20 days. Ascomata black, 410–700 μm high, with a distinct ostiolar beak; asci cylindrical or clavate, appearing unitunicate; spores coiled in the ascus, filiform, somewhat tapering towards the ends, hyaline, 6–15-septate, 130–270 × 3·8–6·5 μm (4130). — The tropical var. *aeria* (Batista *et al.*) M. B. Ellis *1966* differs from var. *lunata* (which usually has no stromata) by the abundant production of black stromata. The anamorph of *C. lunatus* resembles *Curvularia inaequalis* (Shear) Boedijn *1933*, which also occurs in the temperate zone, but that species usually has 4-septate conidia, 24–45 × 9–16 μm; *C. trifolii* (Kauffm.) Boedijn *1933*, which extends from the tropics to southern Europe, has similar conidia but with a prominent hilum. — DNA analysis of *C. lunatus* gave a GC content of 53% (5600).

C. lunatus is known from very numerous, mostly monocotyledonous host plants in many tropical countries but also from Canada, the British Isles, France and the Netherlands (1602); it is a facultative pathogen and may cause leaf spots and seedling blight (CMI Descriptions No. 474). It is the most commonly recorded *Curvularia* species on all parts of rice plants (455). — More than 50% of all reports of it are from cultivated soils in India where it occurs, for example, in rice fields (1519), fields with legumes (1317, 2164) and sugar cane (2861, 4610), grassland (3863, 4933), and further in forest soils (2854, 4995, 5000, 5512), uncultivated soil (4030), a pond bed (3868), estuaries (4477), mangrove mud (4700) and other marine habitats (468). Additional geographical areas from which it is known include high altitudes in Pamir (3850, 3856), Sri Lanka (4021), Bangladesh (2712), Panama (3963), Israel (2764, 2772), Iraq (92), Egypt (8), Libya (4084, 4085), Chad (3415), Somalia (5049), central Africa (3063), Zaïre (4159), and southern states in the USA (655, 1164, 1165, 3817, 4733, 4918). There appears to be no limitation to particular soil depths as it is frequently found in layers between 5 and over 30 cm (1317, 3865, 4610, 5349); its occurrence also seems to be independent of the water regime (655, 1315, 4021) but an increased frequency has been observed after NPK fertilization (2712). — In addition to the plant substrates listed above (1602), *C. lunatus* has been isolated from decaying leaves of bananas (3063), *Setaria glauca*, stems of sorghum and wheat (3219, 5314), litter of *Gymnosporia emarginata* (6108), shoots of *Bothriochloa pertusa* (2955), *Panicum* sp. (2679), cotton (5009) and leaves of *Polypodium* and *Cyclosorus* species (2859). It has also been found in the rhizospheres of barley (3526), rice (2712), corn (6563), *Euphorbia* species at all growth stages (3866), coriander (1523), sand dune plants (4371), *Dichanthium annulatum, Bothriochloa pertusa, Setaria glauca* (3271), *Pennisetum typhoideum* (3867), *Cynodon dactylon* (2855, 2862), *Cyperus triceps, C. pumilus* (2862), banana (2030) and sugar cane (2861, 4886). Further known microhabitats include sclerotia of *Sclerotinia sclerotiorum* (4699), seeds of groundnuts (1016, 2253, 2768, 3220), castor beans (3172), cotton (5009), rice (3584), barley, *Pennisetum typhoideum*, sorghum, wheat and corn (57, 3989), stored sweet potatoes (3066), earthworm casts, a termite mound (2857), birds (2575, 4652) and cotton fabrics (6014, 6326). It commonly becomes air-borne following the violent conidial discharge (3763) which reaches a maximum at about noon according to one report (6014). — It shows a moderate sensitivity to soil mycostasis (2559) and steam-sterilized soil is only recolonized slowly (3330). Sclerotial stromata may survive in soil for over two years and, following attacks by it, a period of three years without susceptible crops is recommended before *Gladiolus* can again be planted (CMI Descriptions No. 474). — *C. lunatus* can serve as food for the mite *Acarus siro* (5379).

Sporulation is significantly promoted by light (2788). For sporulation under submerged conditions, it requires corn steep liquor and 2% NaCl in a glucose-containing basal medium (6088). Optimal growth occurs at 24–30°C (CMI Descriptions No. 474), and at pH 3·5 (6014). — D-Glucose alone and in a mixture with D-fructose proved to be the most suitable C sources out of six monosaccharides tested (4373), but mannitol, cellobiose, lactose, maltose, sucrose, raffinose and starch are also utilized (6014). Thiamine, biotin and other vitamins improved growth and conidiation (5312). Growth *in vitro* was stimulated by root extracts from *Cassia tora* (5638). The degradation of cellulose (4651, 5234, 5388, 6326), keratin (1633), paddy and bajra straw (4611) is documented. *C. lunatus* grows well on rice husks where it produces cellulase and polygalacturonase (6089); pectic enzymes are produced adaptively (2944). Besides ammonia and nitrate (6014) numerous amino acids can serve as N sources with the exception of D,L-histidine (4372). Mn^{2+} salts are reported to be oxidized (5827, 5828). Progesterone is hydroxylated in the 6β-, $7a$-, 11β-, and $14a$-positions by constitutive enzymes (6562), and 1-dehydrocorticosterone-21-trimethylacetate is saponified (6014). The production of curvularin, orsellinic acid (3231) and the red pigment cynodontin (1,4,5,8-tetrahydroxy-2-methylanthraquinone) (1563) is reported. — *C. lunatus* can also be a causal agent of mycetoma in man (1624, 3520).

Cochliobolus sativus (Ito & Kuribayashi) Drechsler ex Dastur *1942* — Fig. 89.

≡ *Ophiobolus sativus* Ito & Kurib. *1929*

Anamorph: *Bipolaris sorokiniana* (Sacc.) Shoem. *1959*

≡ *Helminthosporium sorokinianum* Sacc. *1890*
≡ *Drechslera sorokiniana* (Sacc.) Subram. & Jain *1966*
= *Helminthosporium sativum* Pamm. *et al. 1910*
= *Helminthosporium californicum* Mackie & Baxton *1923*

DESCRIPTIONS: Luttrell (3455), with nomenclature of the anamorph; Shoemaker (5302), with complete developmental cycle; Malone and Muskett (3556), Ammon (126), Chidambaram *et al.* (1002), and Ellis (1603). — Colonies rather fast growing, reaching 5·5 cm diam in ten days at 20°C on OA, composed of dense masses of conidiophores, olivaceous-brown but later becoming black from abundant sporulation in the centre; reverse (olivaceous-)black. Co-nidiophores erect, straight or more commonly flexuose, bearing 1–6 conidia in sympodial succession; conidia (poroconidia) ellipsoidal, gradually tapering towards the ends, straight or slightly curved (particularly on the natural substrate), dark brown, smooth-walled, glossy, 6–9-distoseptate, mostly 60–100 × 18–23 μm. A SEM and TEM study of conidiogenesis has been carried out (1099). Ascomata are formed on boiled corn (5302) or barley grains on agar media of pH 4–6 (5832) after the mating of compatible isolates; the best results are obtained after incubation at 24°C for seven days and then at 20°C for 14 days (with or without light) (5832); at 24°C pycnidia develop on corn grains which form globose conidia (spermatia) 1·7 μm diam functioning in the fertilization of the trichogynes (5302). Asci bitunicate, cylindrical; ascospores vermiform, spirally coiled in the ascus, hyaline, multicellular at maturity, with 6–10 nuclei per cell, 160–360 × 6–9 μm, released in a slime drop. — The rather similar anamorph of *Cochliobolus setariae* (Ito & Kurib.) Drechsler ex Dastur *1942* (*Bipolaris setariae* (Sawada) Shoem. *1959* ≡ *Drechslera setariae* (Sawada) Subram. & Jain *1966*), mostly confined to *Setaria italica*, usually has straighter conidiophores and smaller conidia 45–100 × 10–15 μm; *C. victoriae* R. Nelson *1960* (anamorph *Bipolaris victoriae* (Meehan & Murphy) Shoem. *1959* ≡ *Drechslera victoriae* (Meehan & Murphy) Subram. & Jain *1966*) has ± pale golden-brown conidia which are narrower than those of *C. sativus*, measuring 40–120 × 12–19 μm. — Light-microscopic (2570) and TEM studies of intranuclear mitosis have been published (2571). In the conidiophores fascicles of microfibrils have been observed which may play a role in the migration of organelles (1098). — Vegetative hyphae are multinucleate (943). Heterokaryosis even in single conidia has been demonstrated with multiplely marked isolates (5830), and parasexual recombinations occur. A factor determining

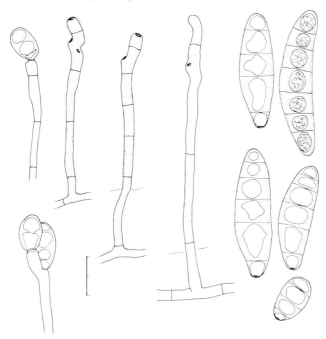

FIG. 89. *Cochliobolus sativus*, conidiophores and conidia, CBS 510.65.

albinism was found not to be linked with compatibility type factors (5832). Its virulence to cereals is controlled by at least two genes which are linked with colony colour; several factors blocking ascoma formation have been found (2543). — Cell wall hydrolysates are reported to contain 8·6% glucosamine, 8·3% galactosamine, 37·0% glucose, 2·0% mannose, and 2·0% galactose; the predominating amino acids in cell wall hydrolysates are alanine (11·2%), glutamine (10·3%), asparagine (9·9%), proline (9·0%), serine (9·0%), glycine (8·9%), valine (7·6%), leucine (7·5%), and threonine (7·1 mol % of total amino acids) (170). Lipids constitute about 10% of the dry weight of conidia; the most abundant carbohydrate is trehalose followed by mannitol (3667).

C. sativus is widespread on numerous plants causing black leaf-spots, seedling blights, foot and root rots and blotches mainly in grasses, including cereals (CMI Distribution Map 322, *1967*), and is very often seed-borne (1002). It has been studied most thoroughly in wheat in which it is the causal agent of a root-rot (e.g. 5013, 6065, 6066), particularly after sowing seed below 10 cm (3228), but also occurs in its rhizosphere (4086) and on seeds (2716). Wheat roots are colonized superficially *in vitro*, shoot growth is reduced, and seedling death often ensues (6305). — Reports of soil isolates are not numerous but include ones from Libya (4083, 4085), Iraq (92), Egypt (8, 3988, 3992, 4962), South Africa (4407), the USA (1041, 3817, 4733), Canada (1009, 1010), Australia (787), and India (2854, 5512). The relatively low spore densities in soils (8 to about 900 conidia g^{-1} soil) (1009, 1010) may explain its apparent absence when soils are analysed by conventional techniques. Low conidial densities can, however, be detected by flotation in a small amount of mineral oil with water where the conidia collect in the supernatant (3263); if this is then mixed with molten PDA and coated on microscope slides, the viability of conidia from soil can be determined (1008). In the field, it occurs more frequently on mature plants than on seedlings (1010) and the infection of wheat roots and subcrown internodes progresses gradually during the host's growing season (6065). On barley leaves (2679), *C. sativus* is less frequent than both *Drechslera graminea* (Rabenh. ex Schlecht.) Shoem. *1959* and *D. teres* (Sacc.) Shoem. *1959* (3511). It has also been found on the leaves and bark of *Citrus* (3988), in the rhizosphere or rhizoplane of barley, oats (3006), corn (6563), *Festuca sulcata* (3376), *Aristida coerulescens* (4083) and coriander (1523). It is frequently reported from seeds of wheat (2716, 3989, 5244), barley (2820, 2821, 3207), corn, sorghum (3989), cultivated grasses (3512) and many other plants (1002, 6407). — Cultures remained viable when stored under mineral oil for 15 years (691). Conidia remained viable after five years storage under dry conditions (1041), survival being favoured by low temperatures and humidities (3262, 3667). The fungus is able to survive extreme (−94°C to +23°C) diurnal fluctuations in temperature for at least 35 days (1242). An extra N source for survival is not required but if its addition speeds up the exhaustion of other substrates, then the survival rate will be shortened (1910). The median saprophytic survival period in soil at 20°C usually lasts longer than 16 weeks, but with nitrogen added, is reduced to 6–10 weeks (1913). On wheat seeds *C. sativus* was found to be able to survive about ten years (3492) although the number of viable conidia present was considerably reduced during the first few years of storage (4951). — A number of observations point to its being very susceptible to antagonistic action of bacteria (1930, 4272, 4277, 4280, 5329), actinomycetes (4731, 5570), and other fungi (1788, 5186); damage is mainly due to either direct parasitism or the action of lytic enzymes (1930); some specialized bacteria have even been found inside the conidial wall (4277, 4278) but the frequently observed perforation of conidia (4276) is caused by the action of the giant soil amoeba, *Arachnula impatiens* (4274). As a result of these susceptibilities, the fungus survives longest in sterilized soil (4731); conidia cannot germinate in non-sterile soil (4087, 4273) without the addition of nutrients (1041), the total rate of germination is given with 0–87% (1011) but growth is only slightly influenced by soil mycostasis (2559). *C. sativus* disappears gradually from buried cornmeal inoculum (5202) or from straw (126, 2997), this process being accelerated by high moisture levels (5181), or the addition of Mn, B and Zn ions (5624) or stable manure (574, 5182). It can grow into straw precolonized by *Penicillium* sp. or *Rhizopus stolonifer*, but not after pre-inoculation with *Aspergillus niger* or *Trichoderma viride* (5961). In general, the saprophytic colonization of non-sterile straw is slow (4087, 5331) and favoured by lower temperatures (about 10°C) (785); it is thus concluded that the fungus possesses a moderate competitive saprophytic ability (4734). Survival of conidia is low outside the rhizosphere of the host, and chlamydospore-like resting structures have been observed in soil (34). — Its conidial density in soil increases under wheat crops (3261, 5183) but is reduced after cropping rape, grass-clover or other legumes (5183); phosphate fertilization also can reduce it (6066). Root exudation activated by mineral fertilizers was found to increase root infection in wheat (1513). — *C. sativus* may serve as food for the mites *Tyrophagus putrescentiae* (5379), *Acarus gracilis*, *Aleuroglyphus ovatus*, and *Glycyphagus destructor* (5380). Means of biological control have been successfully applied, conidia on barley seeds combined with those of *Trichoderma viride* and *Gliocladium roseum* giving reduced infections (5962).

Optimal growth occurs at 27–28°C (126, 788, 4687) while the optimal temperature for conidium production is in the range 16–25°C, the minimum 10°C, and the maximum 29°C (1041). Light is both said to accelerate the production of conidia (2788) and to be without influence (1041). *In vitro* conidia are produced in large amounts on barley leaves (2929). These germinate on water agar or in distilled water to >90% in 12 h at 30–39°C, the optimum temperature for germ tube growth being 25–27°C (2481); percentage germination is increased by the addition of sucrose, maltose, D-galactose, D-fructose, D-mannose, D-glucose, L-arabinose (3667) or sterile soil extract (2998). The germinability of conidia increases with colony age (2481). — Dextrin was found to be more suitable for growth and sporulation than 18 other C sources tested (1069); D-arabinose is not very well utilized (3328). N requirements are relatively small (1912), a C/N ratio of 60 resulting in better growth than ratios of 10, 30 or 100, but the absolute concentration seems to be more important (C source 0·3–3·0%; N source 0·03–0·1%) (263). Better growth was obtained on ammonium than on nitrate salts but organic N sources were superior, asparagine, urea (5629), proline, valine, serine and histidine all supporting good growth; vitamins are not required (1068), but traces of Zn, Mn, and Fe ions all stimulate growth (4524). Good growth and sporulation was observed at glucose concentrations of 1–30 g l^{-1} and in the pH range 4–9 (2268, 4320) with longer conidia being formed at the lower concentrations and in the more acid pH range. D-Amino acids, which are suspected of playing a role in the production of phytotoxic metabolites, did not induce consistent reactions in growth, compared with their L-isomers, although the conidia formed were often shorter (2267). CO_2 concentrations of 10 and 20% lead to two maxima in dry weight yields (3482). Cellulose (500, 1912), pectin (4054), arabinoxylan (1753) and birch wood (4191) are all degraded, and Mn^{2+} salts oxidized (5828). Known metabolites include a red pigment, cynodontin (=sativin, 1,4,5,8-tetrahydroxy-2-methylanthraquinone) produced after photoactivation (1563, 5834), 9-hydroxyprehelminthosporol (3231), sterigmatocystin (4687), helminthosporol and helminthosporal (5278); (+)-epoxyfarnesol is more readily consumed than its (−)-enantiomer and is isomerized into a mixture of *trans-trans* and *cis-trans* isomers during fermentation (730, 5661); an autotoxic staling substance has been demonstrated (4425), as have phytotoxic compounds (1932, 3427) including victoxinine (4638). Culture filtrates have been shown to inhibit the development of wheat seedlings and cause a stunting of the roots in barley, oats, corn, sorghum and bajra (5624). A plant growth promoting substance, *cis*-sativenediol, has also been demonstrated in *C. sativus* (4235). The sesquiterpene fraction contained 95% (−)-sativene, 3% (−)-longifolene, a new seconlongifolane (1440) and two new isosativene derivatives (1441). Growth of the fungus in liquid culture in the presence of a number of amino acids was found to reduce its virulence to wheat seedlings (2556, 5829). — Its complete tolerance to benzimidazole and other systemic fungicides has led to increased infections by *C. sativus* after seed or foliar application of such fungicides (5069).

Cochliobolus spicifer R. Nelson *1964* — Fig. 90.

≡ *Pseudocochliobolus spicifer* (R. Nelson) Tsuda *et al. 1977*

Anamorph: *Bipolaris spicifera* (Bain.) Subram. *1971*

≡ *Brachycladium spiciferum* Bain. *1908*
≡ *Curvularia spicifera* (Bain.) Boedijn *1933*
≡ *Helminthosporium spiciferum* (Bain.) Nicot *1953*
≡ *Drechslera spicifera* (Bain.) v. Arx *1970*

DESCRIPTIONS: Nicot (4160), Nelson (4129) (with description of the teleomorph), Ellis (1603), Subramanian (5626), and Matsushima (3680). — Colonies reaching 3·0 cm diam in five days at 20°C on OA. Conidiophores flexuose, densely geniculate over long distances, with numerous conidial scars. Poroconidia oblong-ellipsoidal to almost cylindrical, golden-brown to dark olivaceous, smooth-walled, mostly 3-distoseptate, mostly 25–40 × 9–20 μm, hilum 2–3 μm wide, not prominent; germination bipolar (4160). A SEM and TEM study of the conidiophores and distoseptate conidia has been carried out (4799). This species is heterothallic (4129). Ascomata are produced after the mating of compatible isolates on barley grains on Sachs' agar at 24°C and in a pH range of 4·0–5·6 (with or without light) and ripen within 23 days (4129); ascomata black, 460–710 μm high, with an ostiolar beak; asci cylindrical-clavate, bitunicate but appearing unitunicate; spores filiform, ± tightly coiled in the asci, hyaline, 6–16-septate, 135–240 × 3·7 – 7·0 μm. — Two ascospore progenies from a particular cross were found to be compatible with both mating types although self-sterile; this dual mating capacity does not conform to any previously described sexual mechanism in ascomycetes (4131). — Similar species with oblong-ellipsoidal conidia are *Drechslera australiensis* (Bugnic.) Subram. & Jain ex M. B. Ellis *1971*, occurring on various plants in Australia, India and Africa, and with narrower conidia, 13–40 × 6–11 μm, with 3(–5) septa, and *D. dematioidea* (Bubák & Wróblewski) Subram. & Jain *1966*, found on grasses in Europe, North America and South Africa, with larger and usually 3–4-septate conidia, 20–70 × 10–16 μm. — The walls of the conidiophores, conidia and aerial hyphae of *C. spicifer* are principally two-layered: the inner hyaline layer can be hydrolysed but the outer pigmented layer which contains more chitin is more resistant (481). — Monoconidial isolates differ in sexual activity; a recessive allele blocks the formation of ascomata and another blocks ascus formation (6250).

C. spicifer is a very common and cosmopolitan species but appears to be most abundant in tropical and subtropical countries; it has been recorded from over 77 different plant species

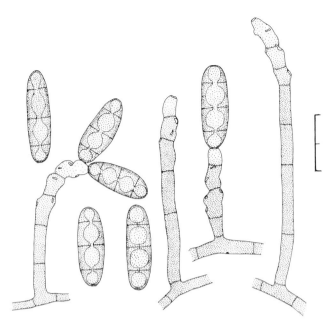

Fɪɢ. 90. *Cochliobolus spicifer*, conidiophores and conidia.

(1603) including 51 genera of grasses (4128), soil and air (1603). It may become economically important as the causal agent of cotton blight (1518) or leaf blight of tobacco (970, 971). — Occurrences in soil are reported from the USA (3817, 4733), South Africa (1559, 4407), Pakistan (4855), Kuwait (4000, 4001), Egypt (8) and, most frequently, India where it is known from peaty soil, loamy soil, Gangetic alluvium (1316), cultivated soil with *Crotalaria juncea, Cicer arietinum*, or *Cajanus cajan* (1317), and grassland (IMI). It has been isolated from sub-coleoptile lesions of the grass *Bouteloua gracilis* (4944) and mentioned as a seed contaminant on sycamore (1683), cotton (5009), onions (5467), wheat (3989, 5293), barley (57), corn, rice and sorghum (IMI, 3989). — The germination of conidia can be inhibited by soil mycostasis (5144); *in vitro* it was inhibited by emanations of *Ocimum basilicum* and *Origanum majorana* (35). Some lytic bacteria originating from *Tyrophagus* mite excrements can attack mechanically damaged hyphae, conidiophores and conidia and liberate melanin granules, thus contributing to humification (4801).

The conidia formed were larger on OA and PDA than on CMA, CzA and MEA, larger at 15° than at 28°C, and larger in 7- than in 14-day-old cultures, while light had no influence (4944). The optimal temperature for growth is between 30 and 33°C (788); the optimum pH 6, the minimum 4 and the maximum 9·2 (1518). — Good C sources for growth and sporulation are sucrose, ᴅ- and ʟ-xylose, ᴅ-galactose and maltose and, to a lesser extent, glucose and mannose, while ᴅ- and ʟ-sorbose, mannitol, raffinose, salicin, inulin and ʟ-rhamnose gave poor results; nitrates were the most suitable N sources, whilst ammonium salts reduced sporulation (1517, 2710, 4768); some organic nitrogen sources supported good growth but not sporulation (1517). *C. spicifer* can grow on media with 20–40% sucrose (4001). Macerating pectolytic enzymes are produced independently from the medium, most abundantly on glucose, followed by sucrose as carbon, and asparagine and sodium nitrate as nitrogen sources; the maximum yield is obtained at 20°C, followed by 15°C, and lowest amounts at 35–40°C (1518). *C. spicifer* has been found to cause a soft-rot of wood (1496). The production of the red pigment, cynodontin (1,4,5,8-tetrahydroxy-2-methylanthraquinone) is reported (1563). — Mycotic keratitis in man (6538) and subcutaneous mycoses in a cat and horses caused by *C. spicifer* have been described (2882, 4012).

Colletotrichum Corda *1832*

= *Vermicularia* Fr. *1849* [non *1825*]

Type species: *Colletotrichum dematium* (Pers. ex Fr.) Grove (=*Colletotrichum lineola* Corda)

Teleomorph: *Glomerella* v. Schrenk & Spauld.

REVISION of the genus: von Arx (196, 197). — Usually producing intraepidermal acervuli (which *in vitro* resemble sporodochia) with a rather dense layer of subulate or cylindrical, hyaline phialides supported by 1–3 short prismatic, hyaline or pigmented cells, sometimes interspersed with dark, septate, moderately pointed setae which may be long and arise from the basal stroma. Conidia cylindrical or falcate, 1-celled, hyaline, smooth-walled, aggregated in cream, orange, red or brownish slimy masses. A characteristic feature of the genus are the brown, rounded or lobed appressoria developed by germinating conidia; these are easily observed in water in hanging drop cultures.

Many congeneric species have been described in *Gloeosporium* Desm. & Mont. *1849* sensu Sacc. *1884*, this name being used if no setae are observed. This criterion is not, however, now regarded as generically significant and the name *Gloeosporium* s.s. placed as a synonym of *Marssonina* Magn. *1906* which has been suggested to be a *nomen conservandum* (197). About 20 *Colletotrichum* species are now accepted (196); most occur on the aerial parts of plants where they cause leaf spots and anthracnoses. Two species which are sometimes soil-borne are treated here.

Key to the species treated:

1   Conidia straight, cylindrical, with rounded ends, 10–21 × 4–6 μm *Glomerella cingulata* (p. 378)
     Conidia falcate, with a minutely truncate base, 18–30 × 3–5 μm
                                                                Colletotrichum dematium (p. 223)

Colletotrichum dematium (Pers. ex Fr.) Grove *1918* — Fig. 91.

≡ *Sphaeria dematium* Pers. ex Fr. *1821*
≡ *Vermicularia dematium* Fr. *1849*
= *Colletotrichum lineola* Corda *1832*

For further (about 80) synonyms see v. Arx (196).

DESCRIPTIONS: von Arx (196, 197), and Simmonds (5328). — Colonies reaching 4·5–4·8 cm diam in four days at about 24°C on MEA or OA (5328). Acervuli usually setose, 100–600 μm diam on the natural substrate. Conidia falcate with a minutely truncate base, hyaline, one-celled (often becoming 2-celled during germination), mostly 18–30 × 3–5 μm. — Several host-specific special forms have been distinguished: f.sp. *truncata* (Schw.) v. Arx *1957* on several legumes, f.sp. *spinaciae* (Ellis & Halst.) v. Arx *1957* on spinach, and f.sp. *circinans* (Berk.) v. Arx *1957* on onions. The only other species of the genus with falcate conidia described by von Arx (196) is *C. graminicola* (Ces.) G. W. Wils. *1914* (teleomorph *Glomerella tucumanensis* (Speg. *1896*) v. Arx & E. Müll. *1954)* which occurs on graminaceous hosts and has larger conidia, 22–40 × 5–7 μm. *C. trichellum* (Fr. ex Fr.) Duke *1928*, causing leaf spots on *Hedera helix*, has lobed appressoria and less curved conidia (5648).

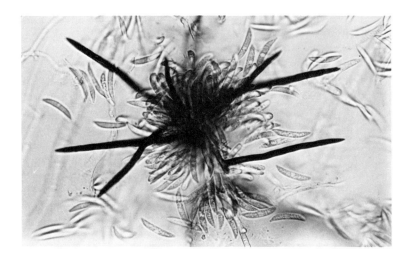

FIG. 91. *Colletotrichum dematium*, young acervulus with setae and phialoconidia, × 500.

C. dematium is a common saprophyte on dead plant substrates. Occasionally it becomes parasitic and causes fruit rots, leaf spots, anthracnose (196) and damping-off (429). The records at IMI contain over 100 plant species from which this fungus has been isolated; these originate from 32 countries, India, Pakistan and the British Isles predominating. In addition, *C. dematium* has been found in Germany, Cuba and Rhodesia. It has been isolated from seeds of *Capsicum annuum*, soya and beans (6407).

The optimal temperature for growth is 26–32°C (5328). *C. dematium* is partially hetero-trophic for biotin (1969) and good growth occurs with filter paper or xylan as sole C sources. Cholesterol is synthesized (4930).

Conidiobolus Bref. *1884*

Conidiobolus coronatus (Cost.) Batko *1964* — Fig. 92.

≡ *Boudierella coronata* Cost. *1897*
≡ *Delacroixia coronata* (Cost.) Sacc. & Syd. *1899*
≡ *Entomophthora coronata* (Cost.) Kevork. *1937*
≡ *Conidiobolus coronatus* (Cost.) Sriniv. & Thirum. *1964*
= *Conidiobolus villosus* G. W. Martin *1925*

DESCRIPTIONS: Martin (3623), Moreau and Moreau (3933), Frey (1835), Prasertphon (4617), Kjøller (3020), Krejzová (3129), and King (2987, 2989); nomenclature: Tyrrell and MacLeod (5966), and King (2988). — *Conidiobolus* (type species *C. utriculosus* Bref.) is the only genus of the Entomophthorales commonly isolated from soil. The genus has been surveyed by Srinivasan and Thirumalachar (5506) and monographed by King (2987, 2988, 2989). Twenty-seven species are now recognized.

The vegetative mycelium consists of ± short coenocytic hyphae (hyphal bodies) which advance at the tip while often withdrawing their contents at the other end and rapidly enter sporulation phases by developing positively phototropic unbranched conidiophores. Distinctions are made between (1) primary conidia which are ± globose, (2) replicative (secondary, tertiary, etc.) conidia which are globose or elongate, and are formed in all species, and (3) multiplicative conidia (micro-conidia) which occur only in some; all three types of conidia are discharged violently. Elongate, passively detached conidia also occur in some species. Zygospores arise through homothallic conjugation within the larger of two gametangia. Villose resting spores develop from globose conidia only in *C. coronatus*.

C. coronatus has very fast-growing colonies which reach 4·0–5·0 cm diam in three days at 21°C on Sabouraud-glucose agar (half strength) with 0·2% yeast extract or PDA, and appear fuzzy against a translucent background; older colonies often become pale brownish and powdery due to the presence of villose resting spores. Assimilative hyphae 4·5–17 μm wide, soon becoming septate. Conidiophores 31–210 × 9–25 μm, slightly distended centrally. Primary conidia mostly 20–46(–67) μm diam (average 32–44 μm) with 1–4 protruding papillae 5–13 × 3–14 μm on which the replicative conidia arise. Micro-conidia produced from primary conidia in clusters of 5–38 in some isolates only and after transfer to water agar, mostly 15–20 × 9–14 μm. Resting spores developing from primary conidia under relatively dry conditions by additional wall thickenings and especially by the secretion of villose appendages through the intact wall. Mycelial chlamydospores are usually absent and zygospores are unknown. — The forcible conidial discharge has been explained by a water-squirt mechanism (2648). The smaller conidia are shot to greater distances than the larger, primary conidia (2650) and thus have a better chance of becoming air-borne.

Conidiobolus is differentiated from *Entomophthora* Fres. *1856* not only by its usually saprophytic mode of life (not a reliable criterion), but also by the spherical to pyriform multinucleate primary conidia which form on unbranched conidiophores, and zygospores arising within one larger gametangium (2988). *C. coronatus* is the only species with villose resting spores.

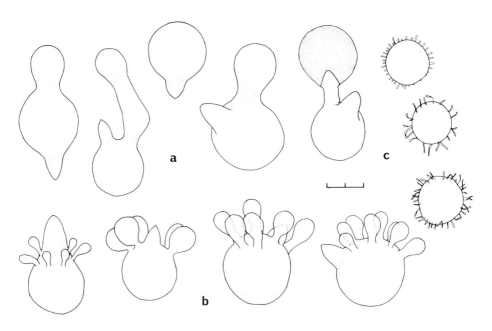

FIG. 92. *Conidiobolus coronatus*. a. Primary conidia; b. formation of multiplicative conidia; c. villose resting spores.

A SEM study of the conidial structures has been carried out (3664). In a TEM and cytochemical study of hyphae and conidia the localization of some hydrolytic enzymes in microbody-like organelles was demonstrated (1920). Cell wall hydrolysates contain the highest proportions of glucosamine, about equal amounts of D-glucose and D-mannose, and some D-glucuronic acid (2478), while cellulose and chitosan are absent. Hyphae contain 16–45% total lipids with the following (predominantly saturated) major fatty acids: $C_{12:0}$, $C_{13:0}$, $C_{14:0}$, $C_{16:0}$, $C_{18:1}$ and $C_{18:2}$, besides 18 other compounds (4040, 5964); the neutral lipid fraction contains mono-, di- and triglycerides, cholesterol (with its esters) and free fatty acids; the polar fraction is generally more strongly unsaturated than the neutral fraction and contains phospholipids with mainly sphingomyelin and some glycolipids with considerable amounts of cerebrosides (1325, 4041).

While species of *Entomophthora* are mostly insect and mite parasites, most species of *Conidiobolus* have been isolated from saprophytic habitats and all grow readily in axenic culture (2987). Five species have been isolated from insects, one from nematodes, and the most common species, *C. coronatus*, which also often occurs as an insect parasite, is also known as the causal agent of rhinoentomophthoromycosis in mammals including man (153, 1066, 3651, 4300, 4594, 4927, 6296). — This species is ubiquitous in moist leaf mould and other plant detritus (1462, 2987) from which it can be easily isolated due to the forcible discharge of the conidia: part of a sample simply is fastened to the lid of a petri dish over CMA with 0·5% peptone and the developing colonies isolated (1462, 2987). By this method this species has been found in 93% of the samples which yielded a *Conidiobolus* in the USA (2987); but is known from all continents (2987) including the Alaskan arctic (3062), mild temperate (3817) and tropical rain forests (1066, 1462, 1625), and soils in New Guinea and New Ireland (3020). It has also been isolated from the rhizosphere of banana (2030), rotten wood (3623), sewage (1166), and air in a fruit-bonded warehouse (3933). — The hyphae (but not the conidia) were able to survive in air-tight closed cultures for more than three years (158) but did not withstand a prolonged storage over silica gel at $-20°C$ (435).

The optimal temperature for growth on agar is $27(-33)°C$, the minimum 6°C, and the maximum 33°C (2217, 2218, 6490) or up to 40°C (1625); in liquid culture, optimal growth occurs below 18°C (2191); no growth but good survival has been found at 1°C (2218). High production of secondary conidia occurs in the pH range 8–10 in Tris buffer; these conidia showed a positive phototropism which was dependent on the duration and intensity of illumination (4355). Conidial discharge requires a high moisture level (2648) and some light although some discharge also occurs in the dark (852, 1547). Conidia germinate in the range 16–30°C (6489, 6490). Contrary to other Entomophthorales, 24% of the conidia can still germinate at -70 bars (6489), but loose their viability at -280 bars and lower water potentials after 6 h. Growth occurs in the pH range $3·5–8·0$ with a broad optimum at $5·5–7·0$ (6418). — Good growth and abundant sporulation occur on media containing coconut milk (1852). Glucose and trehalose are generally suitable C sources for *Conidiobolus* species, while galactose, starch and glycogen are poor (2191, 2987); further suitable substrates include fructose, mannose (1547, 4617), maltose, glycerol, oleate, stearate, palmitate, and casamino acids (2191). Complex N sources prove to be most suitable, but L-asparagine, ammonium salts and urea can be utilized to varying extents (2987, 2989) while nitrate cannot; other suitable compounds for growth are L-aspartic acid, glycine, L-alanine, L-serine and *N*-acetyl-D-glucosamine (99, 2191, 4616). A growth-promoting effect found with yeast extract is due to the nature of the N source and not growth factors (2191, 6418), but Mg^{2+} and Zn^{2+} can stimulate growth. Lipolytic, chitinolytic (adaptive, active in the pH range $3·2–9·5$) and proteolytic enzymes have been demonstrated (99, 1851). — Culture filtrates are toxic to the larvae of *Galleria mellonella* and adult *Musca autumnalis*; the toxin is heat-labile and could not be extracted with organic solvents (4616, 6491). *C. coronatus* is also pathogenic to the mosquitoes *Culex pipiens quinquefasciatus* and *Aedes taeniorhynchus* (the adults being most susceptible) (3420), the Guadeloupean parasol-ant *Acromyrmex octospinosus* (2935), the root maggots *Phorbia brassicae* and *Ph. platura* (3665), aphids (1547, 3128), and termites (2947, 6492). It has also been found associated with an *Onychiurus* species (1027) and the tick *Ixodes ricinus* (5018).

Coniothyrium Corda *1840* emend. Sacc. *1880*
[nomen conservandum]

= *Clisosporium* Fr. *1832* [nomen rejiciendum]
? = *Microsphaeropsis* Höhn. *1917*

Type species: *Coniothyrium palmarum* Corda *1840*

Coniothyrium fuckelii Sacc. has been proposed as neotype of *Coniothyrium* by various authors but is untenable under the present Code of nomenclature. — Contribution to a MONOGRAPHIC TREATMENT: Wollenweber and Hochapfel (6423); the following generic delimitation has been suggested by Sutton (5651): species with mostly two-celled, thick-walled brown and roughened conidia and conidiogenous cells mostly percurrently proliferating are regarded as *Coniothyrium* sensu stricto, whilst ones with one-celled, thin-walled and smooth (or roughened), pale or mid-brown conidia and phialidic conidiogenous cells form *Coniothyrium* sensu lato. — Pycnidia have a light-coloured or pigmented wall and a globose or irregularly folded cavity covered entirely by conidiogenous cells. Conidiogenous cells are hardly differentiated from the inner wall cells, broadly flask-shaped, with a narrow phialidic opening. Conidia are one- or exceptionally two-celled, light to dark brown, giving the pycnidia a dark appearance. — In *C. palmarum* the conidiogenous cells proliferate percurrently in a manner reminiscent of annellophores, and the conidia become one-septate, whilst in *C. fuckelii* this is not the case; therefore this and the other three species mentioned here might better be accommodated in a different genus.

Key to the species treated:

1   Conidia rod-shaped, mostly $5{\cdot}0–8{\cdot}0 \times 1{\cdot}5–2{\cdot}5\,\mu$m, slightly pigmented   *C. cerealis* (p. 229)
    Conidia broadly ellipsoidal, brown   **2**

2(1)   Conidia coarsely roughened, $4–6 \times 3{\cdot}5–4{\cdot}0\,\mu$m   *C. minitans* (p. 229)
    Conidia smooth-walled, $3{\cdot}5–4{\cdot}5 \times 2{\cdot}2–2{\cdot}7\,\mu$m   **3**

3(2)   Colonies producing much grey aerial mycelium and few scattered brown pycnidia; no yellow pigment produced   *Leptosphaeria coniothyrium* (p. 403)
    Colonies with little aerial mycelium and abundant pale pycnidia arranged in zones; usually producing yellow pigment in the agar, more strongly in some sectors than others
    C. sporulosum (p. 230)

Coniothyrium cerealis E. Müll. *1951*

DESCRIPTION: Müller in Zogg (6573). — Colonies slow-growing, reaching 3·0 cm diam in ten days at 20°C on OA, olivaceous due to the presence of woolly aerial mycelium, reverse green on OA, forming scattered pycnidia in older colonies on OA, lupin stems and similar substrates. Conidia rod-like, almost hyaline, becoming pale brown with age, 5·0–8·0(–11) × 1·5–2·5 μm (on the type specimen broader than in culture, 1·3–2·0 μm).

This species is thus far known from the original collection (ZT) on dead stems of wheat, rye and *Hordeum distichon* (6573), from arable soil under wheat in Germany where the identification was confirmed by examination of the type specimen (1889), agricultural soils in the Netherlands (CBS), and various grasses in Norway (191).

The optimal temperature for growth is 21°C, maximum 28°C, minimum −6°C (191). Pectin decomposition was found to be moderate and not very variable, but xylan decomposition was very good, with marked variability between isolates (1432). On maple-wood strips, great losses in weight and tensile strength occurred (2211). A strong pathogenicity on wheat roots has also been reported (6573).

Coniothyrium minitans W. A. Campbell *1947* — Fig. 93.

DESCRIPTION: Campbell (865). — Pycnidial wall dark; conidia roughened 4–6 × 3·5–4·0 μm. Cultural variability has been discussed by Turner and Tribe (5947). — TEM and SEM studies of the conidia show three wall layers, a vacuolated cytoplasm, and a distinctly verrucose surface (2804).

C. minitans has been reported from rice fields in India (IMI). Reports from Dutch soils (1614, 1617) prove to be based on misidentifications of *C. sporulosum*. This fungus is known almost exclusively from fungal sclerotia from which it has been reported in California (865), Hungary (6113), and the British Isles (5947). Other countries from which it is known include Finland, Germany, Poland, Scotland, South Africa (4407) and New Zealand (5947).

The optimal temperature for growth was found to be 20°C, the minimum at or below 10°C, and the maximum at 25°C, while no growth occurs at 30°C (865, 5947). It can infect sclerotia of *Botrytis cinerea*, *Sclerotinia sclerotiorum* (865, 1651, 2566, 6113), *S. trifoliorum* (6586, IMI) and others; a lytic action on hyphae of *S. sclerotiorum* was observed (2807), with endo- and exo-1,3-β-glucanases produced as lytic enzymes (2803). Pycnidia can be formed on and inside the sclerotia of *S. sclerotiorum* (1962). In soils enriched with *C. minitans*, up to 80% of the sclerotia of *S. trifoliorum* present became infected by *C. minitans* within 7–13 weeks (5885, 5946). In one month, pycnidia covered the surface of disintegrating sclerotia entirely (6113). *C. minitans* caused weight losses on wood of birch and pine (4191).

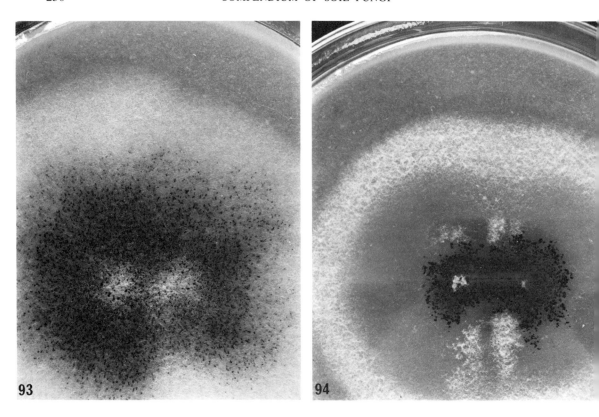

Fig. 93. *Coniothyrium minitans*, CBS 860.71.
Fig. 94. *Coniothyrium sporulosum*, CBS 303.77. Two-week-old cultures grown on OA in darkness, × 1·5.

Coniothyrium sporulosum (W. Gams & Domsch) van der Aa *1977* — Fig. 94.

≡ *Coniothyrium fuckelii* var. *sporulosum* W. Gams & Domsch *1970*

DESCRIPTION: Gams and Domsch (1887). — Colonies reaching 3·6 cm diam in ten days at 20°C on OA, producing little whitish aerial mycelium; pycnidia usually abundant, with pale walls, arranged in zones; yellow (later brown) pigment irregularly exuded into the agar, usually limited to some sectors of the colony and turning violet on addition of NaOH. Conidia indistinguishable from *C. fuckelii*, smooth-walled, 3·5–4·5 × 2·2–2·7 μm, visible through the transparent pycnidial wall.

C. *sporulosum* has frequently been isolated from agricultural soils (1889) and compost (1425) in Germany, and fields under potato, sugar beet, wheat and grasses in the Netherlands (1614). In the rhizosphere of potatoes it appeared more frequently after a wheat than a grass pre-crop (1617). It is also known from freshly harvested pecan nuts (2572).

Xylan (4192) and pectin are strongly decomposed, but differences in this capacity occur between isolates (1432); very good growth is obtained on lignin sulphonate (1425). *C. sporulosum* produces amylase, cellulase, and mannanase (4192) and causes high losses in both weight and tensile strength in maple-wood strips (2211) and other wood samples (4191). In pot experiments, a massive inoculation of the soil caused significant increases in the production of green matter in flax (1427).

Corynascus v. Arx *1973*

Corynascus sepedonium (Emmons) v. Arx *1973* — Fig. 95.

≡ *Thielavia sepedonium* Emmons *1932*
= *Thielavia sepedonium* var. *minor* Mehrotra & Bhattacharjee *1966*
= *Thielavia lutescens* Kamyschko *1965*

DESCRIPTIONS: Malloch and Cain *1973* (3553), Booth (638), Mouchacca (3995), v. Arx (207), Horie (2531), Minoura *et al.* (3840), and Matsushima (3679, 3680). For generic discussion see under *Thielavia*. — Colonies fast-growing, reaching 4·8 and 7·0 cm diam in six days on MEA at 25 and 37°C, respectively, becoming golden-yellow with the production of conidial masses. Ascomata globose, dark brown, lacking peridial hairs, (20–)50–120(–200) μm diam, wall at maturity composed of a layer of flattened, 4–8 μm wide, to 20 μm long cells, the surface exhibiting a characteristic reticulate engraving. Asci formed without croziers, obovate or nearly spherical, with a thin evanescent wall, 8-spored, 25–35 × 17–25 μm. Ascospores brown, smooth-walled, ellipsoidal to broadly fusiform, provided with a distinct germ pore at each end (11–)15–20 × 6–10 μm, length/width ratio about 1·8, uninucleate. Blastoconidia formed singly (or sometimes in short chains) on short denticles at the tips of small ampulliform conidiogenous cells, spherical, finely echinulate, nearly hyaline, thin-walled, 7–12 μm diam. — DNA analysis gave a GC content of 55·5% (5600). — *Th. sepedonium* var. *minor* falls within the range of this variable species. The three other species of *Corynascus* known all have smooth-walled conidia. *Achaetomiella* v. Arx *1970* is perhaps an ostiolate counterpart of *Corynascus*.

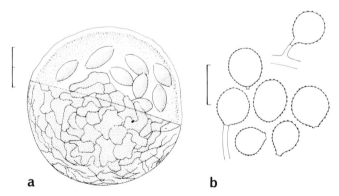

a   b

FIG. 95. *Corynascus sepedonium*. a. Partly opened ascoma with ascospores showing two germ pores; b. attached and liberated aleurioconidia; CBS 387.65.

C. sepedonium is a thermotolerant fungus with a worldwide distribution (3553, 3995). Its main substrates are dung and soil, but it is also known on human skin (3553). It has been reported from Egypt (3995), Senegal (207), Nigeria (638), Kenya (3553), South Africa (4407), Nepal (3840), India (638, 3732, 4030, 4477, 4698, 4700, 4727, 5046, 5512), Thailand (3680), northern Australia (638, 978), Japan (253, 2531, 3680), Guadalcanal (3679), Brazil (399), Argentina (CBS), the British Isles (160, 161, 163, 167, CBS), the Netherlands (3526, CBS), and Italy (1875, 3913). It occurs in uncultivated soils (4030, 4698), forest soil with *Eugenia* (5512), cultivated soil under sugar cane (3732), barley (3526) and rice (3913), grassland (161), savannah (4407), estuarine sediments (4477), a mangrove swamp (4700) and air (1666). *C. sepedonium* has been recorded as an early colonizer of the dead culms and leaves of grasses (161), from the litter of salt marsh plants (163, 167), wheat roots (4727), pea seeds (CBS), the rhizosphere (3526) and seeds (1757) of barley, birds' nests (168), chicken pens (5046), and paper products (1875, 6326).

The optimal temperature for growth is 25–35°C or more, the minimum 12°C, and the maximum 46°C (160, 2203, 3995). Suitable C sources include many mono-, di-, tri- and poly-saccharides and alcohols; it can grow on nitrate, ammonium and organic N compounds and no growth factors are required (2203). Arabinoxylan (1757), cellulose (4730, 6076) and CMC (1757) can all be degraded.

Ctenomyces Eidam *1880*

Ctenomyces serratus Eidam *1880* — Fig. 96.

Anamorph: *Chrysosporium* sp.
   ≡ *Chrysosporium serratum* (Eidam) Dominik *1967*

DESCRIPTIONS: Benjamin (449), Orr and Kuehn (4314); anamorph: Frey and Griffin (1834), and Carmichael (902), sub *Chrysosporium* sp. — *Ctenomyces* Eidam (Fam. Gymnoascaceae Baran.) is a monotypic genus. — Ascomata (gymnothecia) globose, orange-brown, about $100-350\,\mu$m diam excluding the appendages; peridium composed of a thin layer of densely interwoven, slender, pale orange, thin-walled hyphae, surrounded by a loose network of anastomosing, thick-walled, asperulate orange-brown hyphae, $3\cdot3-6\cdot0\,\mu$m wide, which give rise to numerous ctenoid appendages, mostly $100-150\,\mu$m long and consisting of 5–11 cells; each cell thick-walled ($3\,\mu$m), asperulate, growing out distally on one side to form a recurved, hook-like protuberance, $26-38\,\mu$m long. Asci subglobose, 8-spored; ascospores lenticular, asymmetrical, more convex on one side, pale orange in mass, $3\cdot3-3\cdot6 \times 2\cdot0-2\cdot6\,\mu$m. Colonies growing moderately rapidly, reaching $3\cdot5$ cm diam in two weeks on phytone-yeast extract agar at 25°C, flat, dense, dry, velvety to powdery, buff, with a white marginal zone; reverse brown centrally. Aleurioconidia obovate to globose, hyaline to buff in mass, verrucose, thick-walled, mostly $10-17 \times 7-9\,\mu$m, resembling those of *Chrysosporium asperatum* (q.v.) but slightly more elongate. SEM studies of conidia and peridial hyphae show them to have a verrucose ornamentation (6104). *C. serratus* is a heterothallic species, but mating is erratic (5189). — DNA analysis gave a GC content of $52\cdot2\%$ (5600). — Because of a mixture of different fungi in the original description, the name *Ctenomyces* has also often incorrectly been applied to *Arthroderma*-like fungi (4314).

   C. serratus is a worldwide keratinophilic fungus reported from the USA, Cuba, Panama (449, 4314), Argentina (6047), Easter Island (75), Australia, New Guinea, South Africa (4314), India (4314, 5046), Liberia (449, 4314), Tunisia (2721), Israel (1733), Italy (80, 893), Czechoslovakia (1013, 1014, 5338), Hungary, Germany, the British Isles (582, 4314), France (6106), and the Netherlands (CBS). It can be baited from soil most successfully and selectively with chicken feathers (6127), but also appears to prefer partridge and pheasant feathers (4346, 4653), other Galliformes, Passeriformes (4773) or other groups (2575, 4649, 4652) of birds. It is known from forest soil (893), grassland (162, 1013), garden and field soil (1733), saline habitats (1733, 4346, 4661), bird traps (4662), birds' nests (2575, 2577, 4650, 5046), frequently under starling roosts (79), and on dung (162, 4314). As a true soil inhabitant, the fungus is more successful than *Arthroderma curreyi* (q.v.) in colonizing keratinous substrates in soil (4648) and can also grow from these substrates into the soil.

   Suitable N sources include L-serine, L-proline, L-alanine, L-isoleucine, L-methionine, L-tyrosine and glycine (2541). — In artificial inoculations it caused skin lesions and inhibited hair growth in guinea pigs (2145).

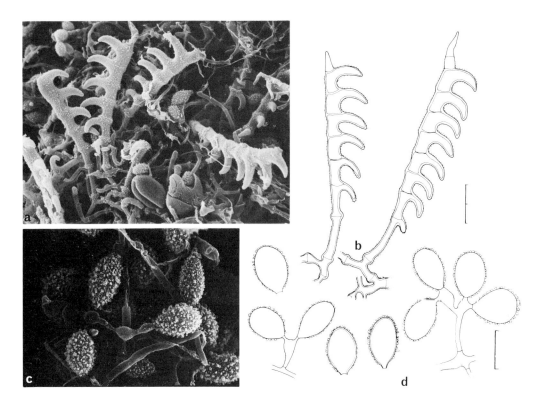

FIG. 96. *Ctenomyces serratus*. a. Peridial appendages, SEM, × 500; b. peridial appendages; c. *Chrysosporium*-type conidia, SEM, × 1000; d. conidia; CBS 187.61.

Cunninghamella Matr. *1903*

= *Actinocephalum* Saito *1905*
≡ *Saitomyces* Ricker *1906*
= *Muratella* Bain. & Sart. *1913*

MONOGRAPHIC TREATMENTS: Cutter (1253), Samson (5022), and Milko and Belyakova (3808). — Family Cunninghamellaceae Naumov. — Colonies very fast growing, white to grey; sporangiophores erect, ± verticillately branched or irregularly cymose, each branch ending in globose or pyriform vesicles from which several 1-spored, globose to oval, echinulate or smooth-walled sporangioles ("conidia") arise synchronously. Zygospores globose, dark brown, tuberculate, formed between equal suspensors; mostly heterothallic.

The only other genus now accepted in the Cunninghamellaceae (454) is *Thamnocephalis* Blakeslee *1905* distinguished by sterile, curved ends of the sporangiophores. — *Cunninghamella* species are mainly soil fungi of the mediterranean and subtropical zones; they are only rarely isolated in temperate regions. The genus now contains seven species; the two commonest treated here are both heterothallic and have globose to oval echinulate sporangioles.

Key to the species treated:

1   Mature colonies greyish due to the dark sporangioles; sporangiophore branches simple or verticillate; sporangioles smooth to short-echinulate, globose, 7–11 μm diam, occurring together with ovoid to ellipsoidal types, 9–13 × 6–10 μm   *C. elegans* (p. 239)

    Colonies remaining whitish to pale grey; sporangiophores verticillately or rather irregularly branched; sporangioles generally echinulate (rarely smooth-walled), 10–14 μm diam or 15–20 × 7–15 μm, giant sporangioles (14–28 μm) usually present in older colonies

                                               C. echinulata (p. 236)

Cunninghamella echinulata (Thaxt.) Thaxt. *1903* — Fig. 97.

  ≡ *Oedocephalum echinulatum* Thaxt. *1891*
= *Cunninghamella africana* Matr. *1903*
= *Actinocephalum japonicum* Saito *1905*
  ≡ *Cunninghamella japonica* (Saito) Pidopl. & Milko *1971*
  ≡ *Saitomyces japonicus* (Saito) Ricker *1906*

= *Muratella elegans* Bain. & Sart. *1913*
≡ *Cunninghamella bainieri* Naumov *1935*

DESCRIPTIONS: Cutter (1253), Nicot (4157), Raizada (4708), Samson (5022), Milko and Belyakova (3808), Zycha and Siepmann (6588), Milko (3807), and Watanabe (6212); sporogenesis: Cole and Samson (1109); TEM of sporangiole formation: Khan and Talbot (2953). — Colonies on MEA very fast growing, to 3 cm high, at first white but later becoming pale grey. Sporangiophores to 18 μm wide, at first with irregular but later with verticillate branches; vesicles globose to subglobose, terminal ones to 50 μm diam and lateral ones 15–30 μm diam. Sporangioles globose to subglobose and 10–14 μm diam (excluding the spines), or oval and 15–20 × 7–15 μm, some larger ones (14–28 μm diam) are borne on solitary vesicles in older colonies. Zygospores globose, brownish, with almost pointed, tuberculate projections, 30–50 μm diam. — Similar species are *C. blakesleeana* Lendner *1927* which has colonies that become yellowish and sporangiophores with simple or regularly verticillate branches, and *C. vesiculosa* Misra *1966* which has permanently white colonies and sporangioles with much longer spines (to 11 μm) and irregularly or cymosely branched sporangiophores with pyriform vesicles. — The spines consist of calcium oxalate dihydrate (weddellite) (2806). — A TEM analysis of sporogenesis showed a distinct outer sporangial wall and the spores are therefore interpreted as sporangioles (2953). During germination, a localized area of new wall material is deposited internally and this forms the wall of the germ tube (2952). Electron probe microanalysis showed a different distribution of elements in hyphal and spore walls (2805).

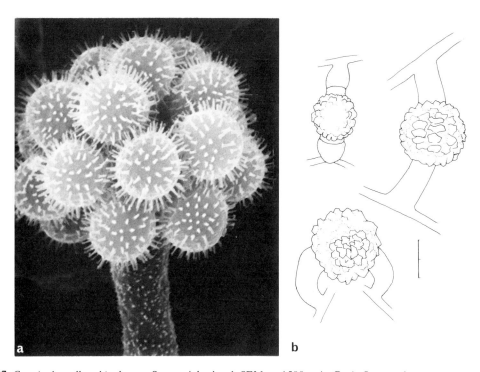

FIG. 97. *Cunninghamella echinulata*. a. Sporangiolar head, SEM, × 1500, orig. R. A. Samson; b. zygospores.

C. echinulata is a cosmopolitan species predominating in warmer zones. Numerous reports are available from India (too many to list individually here), Bangladesh (2712), Pakistan (61, 4855), Malaysia (6046), Singapore (3331), Indonesia (578), the Bismarck Archipelago (3020), Australia (977), Taiwan (2472, 6476), Central America (1697), Peru (2005), Argentina (1827), Jamaica (4886), the Bahamas (2006), South Africa (1555, 1558, 1559, 1827, 4407), equatorial West Africa (1420), central Africa (3063), Libya (4083, 4085), Egypt (4962), Israel (654, 2764, 4759), Syria (5392), Iraq (92), Spain (3417), France (918), Hungary (6551), Czechoslovakia (1702, 5022), the USSR (5298), Canada (5363), and the USA (1028, 1166, 2822). *C. echinulata* has been isolated from uncultivated (4030, 4698) and cultivated soils (3063, 3817, 3868, 4611, 4716, 4996), under legumes (1317), sugar cane (3732), wheat and beets (918), plantations with citrus (2764) and conifers (5363), forest soils (1420, 1525, 1558, 1702, 2482, 2854, 3817, 4716, 4995, 5512) after prescribed burning (2822), soil with steppe type vegetation (4407, 5298), desert soil (4733), sand dunes (4371), a waste pond (1166), and a cave (6551). Fertilization with NPK has been found to increase its frequency in soil (2712). Soil depth and pH seem to have little influence on its distribution. *C. echinulata* has been found on leaf litter of *Eucalyptus* (1558), decaying *Pseudoscleropodium purum* (2969), sclerotia of *Sclerotinia sclerotiorum* (4699), rotting wood (578), rotting straw of paddy and bajra (4611), nests, feathers and pellets of free-living birds (2575, 6223), and cotton fabrics (6326); it has been isolated from the roots of wheat (4727), cucumber (4726), *Setaria glauca, Dichanthium annulatum* (3272) and, with increasing frequencies, from those of *Trigonella foenum-graecum* after foliar spraying with the antibiotic subamycin (2187), also from the rhizospheres of *Artemisia herba-alba* (4084), *Aristida coerulescens* (4083), barley (3526), rice (2712), papaya (5345) and banana plants (2030), sugar cane (4886, 6212), tomato, carrot, kenaf (3733), various sand dune plants (4371), wilted pineapple plants (6210), and *Adiantum* (2856), and also the geocarposphere (2768) and kernels (2765) of groundnuts, cotton seeds (5009), and pecans (2258). — It can be inhibited by *Memnoniella echinata in vitro* (4) and parasitized by *Trichoderma viride* (1729).

Good growth occurs in the temperature range 17–44°C; the minimum is 6–7°C, and the maximum 40–45°C (517, 5022). Zygospores are formed between 25°C and 35°C, with an optimum of 30°C (5022). Sporangiospore germination and linear growth were found to be stimulated by indole-3-acetic acid (5006). The relevance of dry weight determinations in growth studies has been compared with those in *Cladosporium herbarum* (5007). — Growth on acetate is better than on D-glucose (3329). Endogenous respiration is inhibited by fatty acids of intermediate chain length (C_{10}–C_{14}) and suppressed (C_{10}–C_{12}) at high pH; the toxicity of short-chain fatty acids is correlated with increasing chain length and decreasing pH; with long-chain fatty acids, inhibition also increases with chain length; at pH 7·2–8·0 C_{14}–C_{18} and C_2–C_8 acids can be utilized as sole C sources except laurinic and capronic acids (3310); *n*-alkanes can be assimilated and accumulated in the mycelium (1940, 4498). Oxygen uptake by liquid cultures was found to be stimulated by adding 2% montmorillonite or kaolinite (5604). Starch is degraded (1827) and *C. echinulata* possesses a monooxygenase system capable of performing oxidative demethylations and hydroxylations (1730, 5824). The anthelminthic agent parbendazole can be transformed to its hydroxybutyl and carboxypropyl derivatives (6035). The gibberellins A_3, A_4 and A_9 were found to be produced in liquid culture (5008); tryptamine is oxidized to indole-3-acetic acid and tryptophol (1522). Epoxy plasticizers are degraded (5859). — Antibiotic actions against several pathogenic fungi (569), *Staphylococcus aureus* and *Salmonella typhosa* (3812) have been reported as well as an *in vitro* growth inhibition of various grass roots (3272). — Mycelial growth is still possible at $O_2:CO_2$-ratios of 1:18 (2096).

Cunninghamella elegans Lendner *1907* — Fig. 98.

= *Cunninghamella bertholletiae* Stadel *1911*
= *Cunninghamella japonica* (Saito) sensu Pidopl. & Milko *1971*

DESCRIPTIONS: Cutter (1253), Nicot (4157), Samson (5022), Zycha and Siepmann (6588), and Milko (3807); zygospore formation: Spalla (5488); sporogenesis: Cole and Samson (1109). — Colonies on MEA growing very rapidly, reaching 6·6 cm diam in three days on MEA at 24°C, to 4 cm high, at first white but later becoming rather dark grey and powdery with sporangiole development. Sporangiophores to $20\,\mu$m wide, with verticillate or solitary branches; vesicles subglobose to pyriform, the terminal ones to $40\,\mu$m and the lateral ones $10–30\,\mu$m diam. Sporangioles globose and $7–11\,\mu$m diam, or ellipsoidal and $9–13 \times 6–10\,\mu$m, smooth, verrucose or short-echinulate, hyaline singly but brownish in mass. Zygospores globose, brownish, with rather flat, tuberculate projections, $25–55\,\mu$m diam. — The only other species of the genus with grey colonies is *C. polymorpha* Pišpek *1929* which has more irregularly branched sporangiophores and smooth-walled sporangioles; cf. also under *C. echinulata*. — In mycelial extracts, glucose, fructose, malic, tartaric, aspartic and glutamic acids, serine, glycine, a-alanine and glutamine were found (5055). DNA analysis gave GC contents of 27·5–31% (5600).

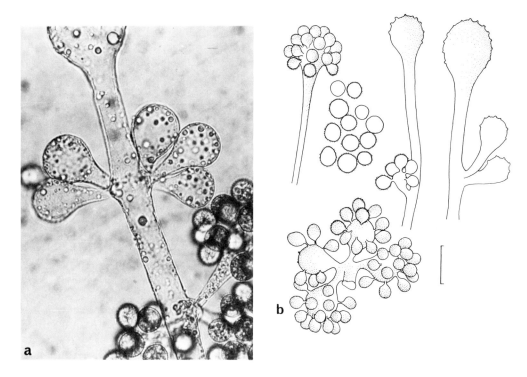

FIG. 98. *Cunninghamella elegans*. a. Sporangiophore with large terminal and five smaller verticillate vesicles, × 500; b. various stages of sporulation, CBS 481.66.

Isolates of *C. elegans* have been reported from Czechoslovakia (273, 1702, 1703), the Netherlands (5022), the British Isles (6184), Poland (650), the Ukraine (2948, 4474), Austria (3418), Italy (4538), Hungary (6551), Yugoslavia (2779, 4573, 5534), Spain (3417), Turkey (4245), Iraq (92), Israel (654, 2764, 2772, 4759), Libya (4085), Egypt (100), the Lake Chad area (3415), Zaïre (2966, 2967), central Africa (3063), the Ivory Coast (4159), South Africa (1555, 1556, 1558), Pakistan (4855), Nepal (1827), India (1315, 1519, 2854, 3431, 4030, 5000, 5349, 5512, 5622), New Guinea (3020), Indonesia (578), Taiwan (2472, 6476), Hong Kong (6516), Japan (3267), Australia (2091), the USA (655, 1028, 1166, 2482, 2573, 3817, 3965), Costa Rica and Panama (2030), Honduras (2031, 2035), the Bahamas (4312), Jamaica (4886) and Brazil (6007). Reports are particularly frequent from cultivated soils, but it has also been reported from uncultivated soils (4030), grassland (2573), a vineyard with a high copper content (4921) and rice fields (1519); there are numerous reports from forest soils, which are too many to list here individually. It has also been isolated from prairies (982, 4313), peat bogs (4474), salt marshes (164), dunes (3941), desert soils (100, 654, 2974), a uranium mine (1703), carst caves (3365), lake bottom mud in a carst cave (6551), estuarine sediments (655), and polluted water (1166). As with other members of the Mucorales, the upper soil layers are preferentially colonized (540, 2035, 6516). The numerous reports do not indicate any particular pH requirements, but occasionally a prevalence in acid soils has been found (4055). *C. elegans* grows particularly well in moist habitats (4313); no active growth occurs in soil at water potentials below −90 bars (3109). It has also been isolated from litter (2966) of various deciduous trees (650), *Eugenia heyneana* (5513), and *Eucalyptus maculata* (1558). Its occurrence on banana roots has been observed repeatedly (2030, 2031, 2035) but it is also known from roots of oat, barley (3004), sugar cane (4886), *Pinus taeda* (3159) and the rhizosphere of groundnuts (2768), tobacco (3431), *Rauwolfia canescens* (427), *Aristida coerulescens* (4083), virus-infected and healthy papaya plants (5345), and sugar cane (4029). Other known substrates include spoiled Brazil nuts (2515), groundnuts (3220), castor beans (3172), and sticks buried in soil (1956). — Sporangiole germination is inhibited by a sporostatic factor produced by *Fusarium oxysporum* (4878, 4882). Sporangioles were found to be relatively resistant to the lytic action of a streptomycete enzyme preparation *in vitro* (1049). *C. elegans* is inhibited *in vitro* by volatile products from leaves of *Ocimum basilicum* and *Origanum majorana* (35).

The optimal temperature for growth is *c.* 25°C (1827), and the maximum is 37°C for some isolates (5022) and to 50°C for others (2662). Zygospore formation occurs at 20–30°C (5022). — Good growth (in descending order) is obtained on glycogen, dextrin, lactose, sorbitol, and D-xylose (5053). Supplying phosphorus-starved mycelia with both sucrose and phosphorus induced maximum CO_2 production and the maximum yield in dry weight; all phosphorus fractions were detected in the mycelia with hexose mono- and diphosphates in equal portions (4076). The influence of colchicine and ascorbic acid on carbohydrate metabolism and mycelial components have been investigated (4074). *C. elegans* is autotrophic for thiamine (5133). Chitin synthetase has been found to be present, particularly in the microsomal fraction, but also in the cell wall, mitochondrial and soluble cytoplasmic mycelial fractions; maximum activity of this enzyme was observed during the late exponential growth phase (3923). Pectin (3127) and starch (1827) are both utilized; the proteolytic activity (2719, 2948) is optimal at pH 7 after two days growth at 37°C (3001). The relative activity of glyoxalate-transaminases has been determined; β-glucosidase is produced (6159), and tryptamine oxidized to indole-3-acetic acid and tryptophol (1522) and the indole alkaloid ajmalicine hydroxylated (427). Cellulose is not attacked, though hemicellulose can be utilized as a C source (3414); however, one report refers to an efficient decomposition of various forms of cellulose (4730). With KNO_3, L-asparagine or peptone as N sources, L-arabinose can be utilized (4075). *C. elegans* is able to assimilate and accumulate hydrocarbons from nutrient media; the composition of the mycelial lipids depends on the chain length of the assimilated alkanes, and the fat content in the hyphae is proportional to the chain length; phospholipids, mono-, di- and triglycerides, sterols, sterol esters, free fatty acids and hydrocarbons have been detected (1940); when grown on undecane, tridecane or pentadecane as sole C sources, both the uptake and the pool of intracellular amino acids were found to be decreased (2905, 3338). The proportion of the 15–16 free mycelial amino acids has been determined; L-glutamic and L-aspartic acids, as well as L-alanine, were found to dominate (2905). L-Leucine and L-asparagine are taken up preferentially from mixtures of amino acids (3240, 5054). Poor growth occurs on glycine, L-tyrosine, L-tryptophan and thiourea (5056). L-Glutamic acid is only assimilated in the presence of glucose, and the substitution of dodecane for D-glucose stimulates the accumulation of intracellular amino acids (3240). Suitable sulphur sources are K- and Na-bisulphate, K-bisulphite, K- and Na-metabisulphite, and $MgSO_4 \cdot 7\ H_2O$ (5057). Lipid synthesis was highest in media containing maltose, D-fructose, D-galactose, sucrose or D-mannose as C sources (431). The production of a sporostatic factor (nonanoic acid) by *C. elegans* has been reported (1905). — The growth of *Chalara elegans* can be inhibited *in vitro* by *Cunninghamella elegans* (2753). — Mycelial growth is still possible in O_2:CO_2 mixtures of 1:18 (2096). Two cases of human infections by *C. elegans* are known (3196).

Curvularia Boedijn *1933*

= *Malustela* Bat. & Lima *1960*

Type species: *Curvularia lunata* (Wakker) Boedijn *1933*

Teleomorph: *Cochliobolus* Drechsler *1934*

KEYS TO THE SPECIES: Corbetta (1184) and Ellis (1602, 1603), synoptic tables: Benoit and Mathur (455); REVIEW of the biology of some species: Upsher (6014). — Colonies form compact black stromata in many species. Conidiophores erect or ascendent, pigmented, geniculate from sympodial elongations, producing single conidia through conspicuous pores. Poroconidia simple, often curved, clavate, ellipsoidal, broadly fusiform, obovoid or pyriform, with three or more transverse septa, brown, with the end cells usually paler than the others; germination is bipolar; the hilum protrudes in some species. — A time-lapse study of the sympodial succession during conidiogenesis in *C. inaequalis* has been carried out by Kendrick and Cole (2926). — *Curvularia* comprises some 35 species of mostly subtropical and tropical facultative plant parasites; the two species treated here are ubiquitous on plant and other substrata, *C. lunata* extending farthest into the temperate zone. Both species treated here (under their teleomorphic names) have smooth-walled, curved, brown conidia with a non-protuberant hilum.

Key to the species treated:

1 Conidia predominantly 3-septate, the middle septum below the centre and the third cell strongly curved, tapering gradually towards the base, 20–30 × 9–15 μm   *Cochliobolus lunatus* (p. 215)
   Conidia predominantly 4-septate, the central cell distinctly geniculate, tapering gradually towards each end, 18–37 × 8–14 μm   *Cochliobolus geniculatus* (p. 213)

Cylindrocarpon Wollenw. *1913*
[nomen conservandum *1968*]

= *Fusidium* Link ex Fr. *1821* [nomen rejiciendum]
= *Allantospora* Wakker *1896* [nomen rejiciendum]
= *Moeszia* Bubák *1914*
= *Coleomyces* F. & V. Moreau *1937*

Types species: *Cylindrocarpon cylindroides* Wollenw.

Teleomorph: *Nectria* (Fr.) Fr.

MONOGRAPHIC TREATMENTS: Wollenweber (6422), Nicot (4158), and Booth (639). The hyphomycete genus *Cylindrocarpon* is characterized by rather fast-growing, hyaline or bright-coloured (whitish, beige, orange, brown or purple but not dematiaceous) colonies with velvety, felty or woolly aerial mycelium and diffuse or sporodochial sporulation. Conidiophores consisting of single phialides or repeatedly verticillately or almost penicillately branched structures, in some species also as complex sporodochia. Phialides cylindrical to subulate, with a single apical opening and a short collarette, producing hyaline, smooth-walled conidia which adhere in slimy masses; macro-conidia always present, ± cylindrical with rounded ends, straight or curved, with one to several transverse septa; 1-celled micro-conidia ± distinct from the macro-conidia are formed by many species. Chlamydospores present in some species, hyaline to brown, intercalary or terminal.

The similar *Fusarium* section *Martiella* Wollenw. is not sharply delimited from *Cylindrocarpon* but the conidia in that section generally have a more fusiform shape which is seen as a more gradual tapering of the apical cell and an asymmetrical protraction of the foot cell. *Cylindrocladium* Morgan (q.v.) has densely penicillate conidiophores from which long sterile processes arise. *Fusariella hughesii* Chabelska-Frydman *1964* has pale ochraceous cylindrical conidia adhering in irregular chains.

Cylindrocarpon has been subdivided into two sections: (a) sect. *Ditissima* Wollenw. *1917* without chlamydospores (correctly called sect. *Cylindrocarpon*) and (b) sect. *Chlamydospora* Wollenw. *1917* with chlamydospores. These sections have been further subdivided by Booth (639) according to the presence or absence of micro-conidia.

About 35 species are now known and can be roughly subdivided into two ecological groups: (a) ubiquitous soil-borne species with a particular affinity to plant roots (mainly sect. *Chlamydospora*) and which rarely become pathogenic, and (b) facultative plant parasites mainly found on woody hosts (mainly sect. *Ditissima*). — *Cylindrocarpon* colonies are grown on OA or other natural substrates (see under *Fusarium*), and conidia formed in sporodochia are preferred for their characterization, while samples from the colony margin are examined to assess the presence and importance of micro-conidia. When grown on MEA and other rich media, the shape of the macro conidia may vary considerably. Chlamydospore formation is judged by examination of at least two-week-old cultures.

Key to the soil-borne species:

1   Mycelial chlamydospores absent   **2**
    Mycelial chlamydospores present   **3**

2(1)   Colonies golden-yellow to brown; macro-conidia 5–7-septate, curved, 50–80 × 6–7·5 μm; micro-conidia very abundant, 4–9 × 1·5–3 μm   *Nectria coccinea* (p. 497)
    Colonies deeply violaceous; micro-conidia scanty, 6–14 × 3–5 μm; macro-conidia commonly 5-septate, mostly 55–60 × 6–7 μm
        (*Cylindrocarpon janthothele* var. *majus,* teleomorph: *Nectria mammoidea*)

3(1)   Colonies with a pungent, actinomycete-like odour; macro-conidia markedly curved, 30–70 × 5–8 μm   *C. olidum* (p. 247)
    Colonies without a pungent odour; macro-conidia straight or slightly curved   **4**

4(3)   Conidia mostly 1-septate, 15–35 × 2·5–4·5 μm, with an obliquely tapering apical cell
        C. magnusianum (p. 245)
    Conidia larger and often 3-septate   **5**

5(4)   Conidia mostly 1-septate, with an oblique and almost beaked apical cell, mostly 24–34 × 4–6 μm   *C. didymum* (p. 244)
    Conidia 1–3(–5)-septate, with a ± symmetrically rounded apical cell, mostly 20–40(–50) × 5–6·5(–7·5) μm   *Nectria radicicola* (p. 503)

Cylindrocarpon didymum (Hartig) Wollenw. *1926* — Fig. 99.

≡ *Fusisporium didymum* Hartig *1846*
≡ *Ramularia didyma* (Hartig) Wollenw. *1913*

DESCRIPTIONS: Wollenweber (6422), Nicot (4158), and Booth (639). — Colonies reaching 4·0 cm diam in seven days on PDA at about 20°C, whitish, beige to coffee-brown. Primary conidiophores long and slender, subsequent ones shorter, broader and more strongly branched. Phialides almost cylindrical, 18–23 × 2·5–3·5 μm. Conidia hardly separable into macro- and micro-conidial types 0–1(–2)-septate, oval to ellipsoidal or cylindrical, straight or slightly curved, the apical cell characteristically bent to one side and slightly beaked, mostly 24–34 × 4–6 μm. Chlamydospores terminal or intercalary, single or in chains, globose, smooth, hyaline at first but later becoming brown, 8·5–11 μm diam. — This species is close to *C. obtusisporum* (Cooke & Harkn. *1884*) Wollenw. *1926*, the anamorph of *Nectria tawa* Dingley *1951*, which has similarly bent apical cells but differs in the mostly 2–3-septate macro-conidia, 30–50 × 4·5–7·5 μm. *Nectria radicicola* (q.v.) can also have asymmetrical apical cells when grown on unsuitable media.

C. *didymum* is a rather uncommon species and the name has probably often been misapplied. It has been recorded from Europe, Egypt, Dominica, Canada (639), the USA (1032, 2004, 2008, 6350), Israel (2764), Pakistan, New Zealand, the West Indies (IMI), and Peru (2005). It has been isolated from forest soils under beech (272), *Salix* (2007) and mixed *Salix-Populus* communities (2004, 2008), conifer-hardwood stands (1032, 6350), a forest nursery (1978), conifer swamps (1039), alpine pastures (3445, 6082), agricultural soil (3450),

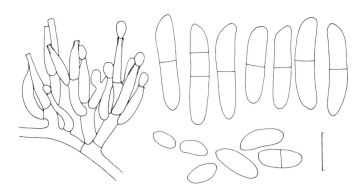

Fig. 99. *Cylindrocarpon didymum*, conidiophores, macro- and micro-conidia, CBS 159.34.

citrus plantations (2764), garden soil (4451), and light sandy soil (1614, 1616). *C. didymum* has been found frequently on plant roots including those of *Beta vulgaris*, *Rubus idaeus*, flax, potato (6422), barley, groundnuts (2768), *Brassica* (4451), poplar (5898), and ash (3156); a negative rhizosphere effect was found to occur in beans (4452). Further habitats reported include sticks buried in soil (1956), millipede droppings, and poultry feed (IMI). The fungus ranks as a very weak competitor compared with other *Cylindrocarpon* species (3682).

Optimal linear growth and good mycelium formation occur at 21°C and weak growth is obtained at 30°C (2698), although deviating results have also been reported (2939). Decomposition of xylan by *C. didymum* is very good and little variability in this capacity occurs between isolates (1432); cellulose is also degraded (1956) and birch wood eroded (4191). Antibiotic activity against various test fungi has been demonstrated (1431). — In a chronically γ-irradiated soil it was viable in plots receiving up to 500 R per day (2009).

Cylindrocarpon magnusianum (Sacc.) Wollenw. *1926* — Fig. 100.

≡ *Septocylindrium magnusianum* Sacc. *1878*
≡ *Ramularia magnusiana* (Sacc.) Lindau *1906*

Teleomorph: *Nectria ramulariae* (Wollenw.) E. Müll. *1962*

≡ *Neonectria ramulariae* Wollenw. *1917*

DESCRIPTIONS: Wollenweber (6422), Nicot (4158), and Booth (639). — Colonies reaching 4·4 cm diam in seven days on PDA at about 20°C; aerial mycelium whitish to cream, floccose, some isolates producing numerous reddish sclerotial bodies (but see below under *C. ehrenbergii*); reverse pale beige, or occasionally becoming reddish brown. Conidiophores ± branched or consisting of solitary phialides, in some isolates aggregated into discrete sporodochia which produce horn-like cream conidial masses several millimeters high. Phialides almost cylindrical, usually 10–12 × 3–4 μm. Conidia rather uniform in shape and size, slightly curved, the apical cell obliquely tapered, (0–)1(–3)-septate, 15–35 × 2·5–4·5 μm. Chlamydospores intercalary or terminal, hyaline at first, but later becoming brownish, smooth-walled or warted, ± globose, 9–12 μm diam. — The teleomorph (ascospores 12–15 × 3–4 μm) has apparently not been found since its original description, and, in view of the variability of the anamorph, we prefer to use the *Cylindrocarpon* name for soil isolates. — A very similar and possibly identical species is *Cylindrocarpon ehrenbergii* Wollenw. *1926* (≡ *Fusarium candidum* Ehrenb. *1818* ex Schlecht. *1824*) which is said to be the anamorph of *Neonectria caespitosa* (Fuckel) Wollenw. *1926* (≡ *Sphaerostilbe caespitosa* Fuckel *1873*) (ascospores 8–18 × 3–4·5 μm (6422)); the conidia of *C. ehrenbergii* are, however, slightly narrower, mostly 1-septate, 20–29 × 3–4 μm, and erumpent, greenish brown or red stromatic bodies are commonly present; *C. magnusianum*, on the other hand, is distinguished as having 3·5–4·5 μm wide conidia (6422), but these differences are apparently of little taxonomic significance. — *C. magnusianum* has narrower conidia than the other soil-borne *Cylindrocarpon* species and is easily recognized by the oblique apical cell of the conidia. The amount of sporodochium production and sclerotial bodies vary between isolates.

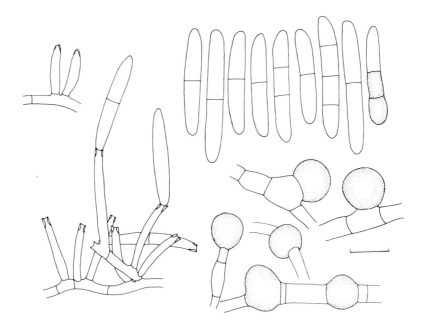

Fɪɢ. 100. *Cylindrocarpon magnusianum*, conidiophores, conidia and chlamydospores, CBS 182.36.

C. magnusianum is known from Europe, Egypt, and the USA from numerous host plants and soils (639); other reports are available from Norway (191), the USSR (2871), Canada (3959), Alaska (1750), the British Isles (850), France (2159), Germany (1889), Yugoslavia (3798), Italy (3450), and Nepal (1826, 1827). In the soils we have examined it is the second most comon species of the genus after *Nectria radicicola*. It has been isolated from forest soils under beech (850), aspen (3959), and *Pinus maritima* (3798), a forest nursery (IMI), agricultural soils (1889, 2159, 2871, 3450, 6294) and soils with tundra vegetation (1750). *C. magnusianum* has also been found as a colonizer of buried beech twigs and roots (850), on roots of onions (2159) and wheat (5330), and frost-damaged leaves of *Festuca pratensis* and *Phleum pratense* (191). In comparison with other species of the genus, it has a high competitive saprophytic ability (3682, 3683). Conidia rarely germinate in soil and tend to give rise to chlamydospores (3684).

Growth occurs in the temperature range $-2°–32°C$, the optimum being $19–24°C$ depending on the isolate (1188), the maximum $31°C$, and the minimum $-6°C$ (191). — Differences between isolates have also been reported in the ability to decompose xylan *in vitro* (1432). Starch, pectin and cellulose are also degraded (1750, 4433, 4434), birch wood eroded (4191), and gallic acid utilized (1750). *C. magnusianum* can produce the sideramines fusigen and fusigen B (3231). Antagonistic activity has been reported *in vitro* against *Gaeumannomyces graminis, Rhizoctonia solani, Pythium ultimum* (1431); the growth of *Chlorella pyrenoidosa* is also inhibited. — This species is one of the most important causal organisms of spring time root damage in alfalfa and sweet clover in Canada, but it is less virulent in the main growing season; cereal species tested are not susceptible to it (1188).

Cylindrocarpon olidum (Wollenw.) Wollenw. *1917* var. *olidum* — Fig. 101.

≡ *Ramularia olida* Wollenw. *1913*
? = *Cylindrocarpon curvatum* Hochapfel *1931*

DESCRIPTIONS: Wollenweber (6422), Nicot (4158), Gerlach (1942), and Booth (639). — Colonies rather slow-growing, reaching 3·5 cm diam in seven days on MEA at about 20°C,

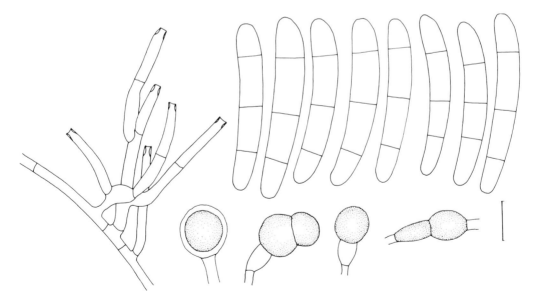

FIG. 101. *Cylindrocarpon olidum*, conidiophores, conidia and chlamydospores.

whitish, cream or beige, floccose to tufted; reverse amber to brown, with a penetrating earthy or actinomycete-like odour. Conidiophores branched but not aggregated in sporodochia. Phialides mostly 18–30 × 3–4 μm. Conidia distinctly curved with broadly rounded end cells, mostly 3–5-septate and 30–70 × 5–8 μm. Chlamydospores intercalary, formed singly or in chains, globose, smooth-walled, 10–15 μm diam. — Two varieties of this characteristic species have been distinguished: var. *suaveolens* Reink. *1936* found in soil from banana plantations in Central America has a caraway-like odour and slightly shorter conidia; and var. *crassum* Gerlach *1959*, pathogenic to various Cactaceae, has longer and broader conidia, mostly 53–80 × 6·5–10 μm, and a less penetrating odour.

C. olidum is known from Europe, Ghana, Canada (639), Colombia, Costa Rica, Guatemala, Honduras and Panama (4793), from numerous plants, particularly potato tubers (639), but is normally not pathogenic. It has also been reported from New Zealand (5812, 5813, 5814) and New Caledonia (IMI, 4793). *C. olidum* has been isolated repeatedly from washed soil particles of wheatfield soils (1889). In an alkaline forest soil, its greatest frequency was observed to be at depths of 25–40 cm (3683). There are also frequent reports from plant materials such as potato tubers, *Picea sitchensis* litter, *Pelargonium zonatum*, *Camellia sinensis*, *Asparagus officinalis*, and *Cocos nucifera* (IMI, 639, 1942). In comparison to other *Cylindrocarpon* species, *C. olidum* has a relatively high competitive saprophytic ability (3682, 3683).

Pectin is decomposed with little difference in this ability occurring between isolates (1432). Antagonistic activity against *Gaeumannomyces graminis* and *Rhizoctonia solani* has been observed *in vitro* (1431). In laboratory experiments, root damage occurred in wheat, peas and rape (1430).

Cylindrocladium Morgan *1892*

= *Candelospora* Hawley *1912*
= *Tetracytum* Vanderw. *1945*

Type species: *Cylindrocladium scoparium* Morgan

Teleomorph: *Calonectria* de Not.

MONOGRAPHIC TREATMENT: Boedijn and Reitsma (579). — Colonies spreading rapidly, light to dark brown in the centre, forming numerous chlamydospores in chains and often microsclerotial clusters. Conidiophores erect, in the upper part repeatedly forming solitary branches and

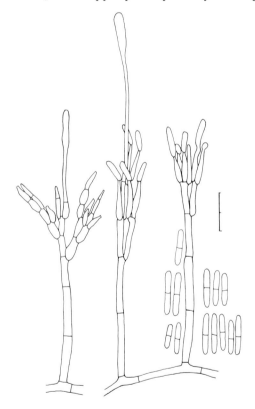

FIG. 102. *Cylindrocladium parvum*, penicillate conidiophores with cylindrical swollen sterile appendages, and conidia, CBS 874.68.

ultimately densely compacted penicillate groups of subulate or doliiform phialides; the main axis mostly a long, unbranched thread ending in a sterile swollen vesicle. Conidia usually aggregated in parallel bundles, cylindrical, with symmetrically rounded or almost truncated ends, straight or curved, 1- to many-septate.

Fungi with similar conidiophores but with numerous lateral sterile setae and mostly 1-celled conidia are placed in *Gliocephalotrichum* J. Ellis & Hesseltine *1962*. Since Boedijn and Reitsma's monograph which accepted seven species, the genus has been expanded markedly by about 15 species of which several are connected with *Calonectria* teleomorphs (see Rossman *in* Mycotaxon **8**: 485–558. *1979*).

Most *Cylindrocladium* species are polyphagous plant parasites, particularly in warmer zones, and there is a considerable phytopathological literature on their pathogenicity and control. — Many isolates do not sporulate on usual media; good results are obtained on PCA, hay decoction agar, CzA, and the latter with 20% sucrose (3964). Species differentiation by acrylamide gel electrophoresis seems to be promising (2617).

The two species treated here both have straight, generally 1-septate conidia.

Key to the species treated:

1   Conidia 15–21 × 2–3 μm   *C. parvum* (p. 250)
     Conidia 50–58 × 4–6 μm   *C. scoparium* (p. 251)

Cylindrocladium parvum P. J. Anderson *1919* — Fig. 102.

DESCRIPTIONS: Boedijn and Reitsma (579), Matsushima (3679, 3680), and Meyer (3789). — Colonies reaching 8 cm diam in seven days at 20–25°C, pale brown with white margins and several radially arranged darker fibres, becoming granular with the formation of sclerotial bodies. Conidiophores about 200 μm high, the ultimate branches bearing 2–4 phialides 11–14 × 2·0–2·5 μm; terminal thread 1·5–2·5 μm wide, ending in a club-shaped vesicle 25–36 × 4–5 μm. Conidia cylindrical, (0–)1-septate, 15–21 × 2–3 μm. — The similar *C. camelliae* Venkataramani & Venkata Ram *1961*, causing root rot on tea and *Acacia* species in India, Japan and Mauritius, has elongate–ellipsoidal to cylindrical vesicles, 10–30 × 4–5 μm, and narrower conidia, 8–16 × 1·8–2·2 μm. *Cylindrocarpon tenue* Bugn. *1939* lacks the sterile conidiophore processes and has conidia 16–20 × 1·5–2·5 μm, and *Cylindrocladium intermedium* Matsushima *1971* may be identical with this species.

C. parvum is perhaps the least pathogenic species of its genus and apparently has a worldwide distribution although it has not been frequently reported; records are available from the USA (139, 2616), Europe (CBS), the Congo basin (2966, 2967, 3063, 3790), Malawi (IMI), Mauritius (6590), Hawaii (3264), Brazil (IMI), New Caledonia (IMI, Papua-New Guinea (3679), Java (579), Australia (2091) and Japan (3680). *Geranium* leaf tissue has been used successfully as a bait (2616), and hair-baits buried in the soil are also colonized (2091). Known habitats include both forest (2616) and cultivated soils (CBS), a mangrove swamp (3264), and a chicken farm (2091). It has also been isolated from leaf litter (3063), stems and roots of roses (139), pine seedlings (IMI), tea roots (6590, IMI), *Viburnum* seedlings and strawberry plants (CBS). — It has been found to survive storage in water for ≥6 years (605).

C. *parvum* is able to decompose pectin (1432). Its growth was found to be stimulated *in vitro* by *Chlorella pyrenoidosa*.

Cylindrocladium scoparium Morgan *1892* — Fig. 103.

Teleomorph: *Calonectria* sp. (Sobers *1973)*

DESCRIPTIONS: Boedijn and Reitsma (579), Sobers and Seymour (5464), Morrison and French (3964), Alfieri *et al.* (98), Meyer (3789), CMI Descriptions No. 362, *1973*, Subramanian (5626), and Bertus (491). — Colonies reaching 3·0–4·2 cm diam in seven days at 25°C, reddish brown, radially striate, becoming granular from the abundant sclerotial bodies, margin whitish, reverse almost black in the centre. Conidiophores to 500 μm tall, ultimate branches bearing 2–4 doliiform to oval phialides, 7–12 × 3–4 μm; the main axis forming a 2·5–3·5 μm

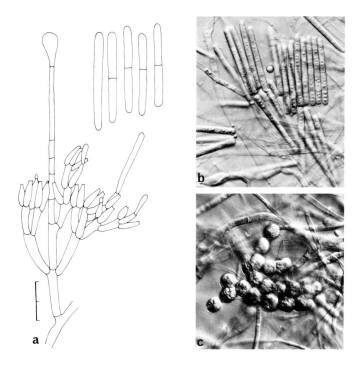

FIG. 103. *Cylindrocladium scoparium*. a. Penicillate conidiophore with sterile appendage, and conidia; b. conidia, NIC, × 400; c. young cluster of chlamydospores, NIC, × 400; CBS 217.54.

wide thread ending in a club-shaped, ellipsoidal or pyriform vesicle, 19–37 × 8–15 μm. Conidia cylindrical with rounded ends, straight, mostly one- (rarely 3-) septate, 50–58 × 4–6 μm. Microsclerotia of varying size are formed on several media; these are composed of chlamydospore-like cells. — Several similar species with straight, 1-septate conidia have been distinguished from *C. scoparium* in recent years because of minor differences in the terminal vesicle and in conidial size: *C. braziliense* (Bat. & Cif.) Peerally *1974* (≡*C. scoparium* var. *braziliense* Bat. & Cif. *1951*), pathogenic on *Eucalyptus* species, has smaller, clavate to ellipsoidal or spear-shaped vesicles 2·6–6·2 μm diam and conidia 24–38 × 2·0–2·8 μm. — *C. floridanum* Sobers & Seymour *1967* (5464), the anamorph of *Calonectria kyotensis* Terashita *Jan. 1968* (= *C. uniseptata* Gerlach *Apr. 1968* (1944) = *C. floridana* Sobers *1969*) has short and almost globose vesicles 8–20 μm diam, also produces lateral vesiculate stipes from secondary conidiophores, and has conidia 30–56 × 3·5–5·4 μm. The teleomorph of *Calonectria kyotensis* (homothallic, ascospores 18–50 × 5·3–6·8 μm, mostly 1-septate) has been obtained in cultures isolated from orchid roots on sterilized barley ears and alfalfa stems after three weeks and on several agar media after a longer time; this teleomorph was claimed to be connected with *C. scoparium* (1944), but subsequently Sobers (5463) was able to distinguish the anamorphs and revise the synonymy. *Calonectria kyotensis* also differs serologically from *C. scoparium* (4206); it has often been isolated from conifers and may account for most records of *C. scoparium* from such hosts. *C. ellipticum* Alfieri *et al. 1970* (98) has smaller ellipsoidal vesicles 9–20 × 7–9 μm and longer conidia 48–76 × 4·0–5·5 μm. *C. avesiculatum* Gill *et al. 1971* has appendages without inflated vesicles and conidia 51–78 × 3·5–4·7 μm. *C. clavatum* C. S. Hodges & Cardoso *May 1972* has clavate vesicles 35–60 × 4–5 μm and conidia 38–48 × 3·5–5·5 μm.

C. scoparium has a worldwide distribution and is known from North and South America, the West Indies, Europe, Africa, southern Asia, New Zealand (CMI Descriptions No. 362, 5626) and Australia (491). It is one of the most pathogenic and polyphagous species of its genus and may cause damping-off, seedling root rot and blight and leaf spots on a very wide range of hosts, particularly woody plants (773), but also beet, strawberry and watermelon. — Soil isolates have been obtained from forest nurseries (137, 773, 5799), forest soils under *Eucalyptus maculata* (1555), and clay deposits in carst caves (3364). — Conidia rapidly loose their viability under dry conditions (772) and it is the microsclerotia that are the main propagules responsible for its survival, spread and plant infection (772, 3345, 5799). These can survive in the temperature range −6°–45°C for at least several weeks (2615) and in non-sterile soil for over seven months (491). Their density in the soil has been determined by a combination of washing and wet-sieving (collecting on a 75 μm screen) with a selective agar medium containing Czapek ingredients with the addition of lactic acid (to render the pH 3·5) and 1000 ppm Tergitol NPX (nonyl phenyl polyethylene glycol) (5798). The recovery of a low inoculum from soil may be facilitated by adding 8% glucose to the soil 12 days before the analysis (2613). *C. scoparium* can also be baited with alfalfa roots (5798), leaves of *Azalea* (3344) or *Pelargonium cucullatum* buried for 4–7 days (2616). — Growth is inhibited by a *Lactarius* species (4438) and the phytoalexin (gluco-alkaloid) α-tomatine exuded by several *Solanum* species (182).

The optimal temperature for mycelial growth is in the range 25°–30°C, but at 37°C considerable growth still occurs (491). The production and germination of conidia requires a high humidity; isolates of *C. floridanum* from spruce seedlings showed optimal conidiation on CzA with 20% sucrose (772, 3964), but none on PDA or 2% MEA; however, an isolate of *C. floridanum* from peach and some *Azalea* isolates of *C. scoparium* also sporulated on the latter media (3964). — Microsclerotium production *in vitro* is optimal on a glucose-casein hydroly-sate agar at 24°–28°C, when light had no effect, and with a high C/N ratio to 100:1 (2613, 2614); a glucose-yeast extract-tyrosine agar (30·0 : 0·20 : 0·08 : 20·0 g l^{-1}) has been devised for production of microsclerotia (2615). — Galactose, glucose, maltose and mannose supported optimal growth in liquid media with several amino acids, peptone, ammonium nitrate or sulphate, potassium or sodium nitrate or urea as suitable N sources, but ammonium and nitrate ions inhibited the formation of microsclerotia (6230). A biologically active cyclotetrapeptide metabolite, Cyl-2, was found to be composed of L-pipecolic acid, L-isoleucine, D-*O*-methyl tyrosin and 2-amino-8-oxo-9,10-epoxydecanoic acid (2465, 2466).

Dendryphion Wallr. *1833*

Dendryphion nanum (Nees ex Gray) Hughes *1958* — Fig. 104.

≡ *Helmisporium nanum* Nees ex Gray *1821*
? = *Dactylium fumosum* Corda *1839*
   ≡ *Dendryphion fumosum* (Corda) Fr. *1849*
  = *Dendryphion laxum* Berk. & Br. *1851*

DESCRIPTIONS: Bainier (288), Subramanian (5623, 5626), Ellis (1603), Matsushima (3680), Shearer and Crane (5250), and Fungi Canadenses No. 147, *1979*. — The genus *Dendryphion* (type species *D. comosum* Wallr.) is characterized by well-differentiated, erect, darkly pigmented and apically branched conidiophores which form poroconidia through wide and darkened scars either singly or in acropetal chains with some sympodial elongation of the conidiogenous cells; the conidia are cylindrical with rounded ends, multiseptate and brown. — Whilst *D. comosum* occurs commonly on the dead stems of herbaceous plants, particularly *Urtica, D. nanum* is the only species that has repeatedly been isolated from soil.

Colonies of *D. nanum* reaching 3·5 cm diam in ten days at 20°C on OA. This species is characterized by moderate branching of the conidiophores which may become 80–300 μm long (on natural substrates, generally shorter in culture) and conidia which are almost cylindrical, slightly tapering at the ends, dark brown with much paler end cells, smooth-walled or verruculose, 5–11-septate, 45–90 × 10–12·5 μm.

D. nanum is common on dead herbaceous plants and also occurs on roots and cut tree stumps in Europe and North America (1603), India (5623), central Africa (3063) and Japan (3680). Isolations have been reported, for example, from *Heracleum sphondylium* (6462), rye grass, wheat, alfalfa, and beech (1603, 4917). It has only rarely been isolated from soils, such reports mainly being from cultivated ones (1616, 2161, 4451). — Conidial populations in soil can be estimated by the flotation technique developed for *Cochliobolus sativus* (1007) which has yielded up to 3500 conidia per gram soil; counts obtained with this technique were highest under rape, lower under wheat, and lowest in fallow fields; occurrence in Canada is consequently mainly confined to the rape-growing area (1007) but *D. nanum* is not a pathogen. There are reports of its occurrence on roots of other cereals, barley and cabbage (4451, 6030), on decaying *Urtica dioica* stems (6463), dead wood (3063), pine wood exposed in (diluted) seawater (831), balsa wood submerged in a river (5250) and seeds of cultivated grasses (3512). Conidia introduced into soil showed high survival rates (2499).

Decomposition of pectin and xylan is very good, with differences in capacity between individual isolates (1432). Antagonistic activity towards various test fungi has been demonstrated (1431).

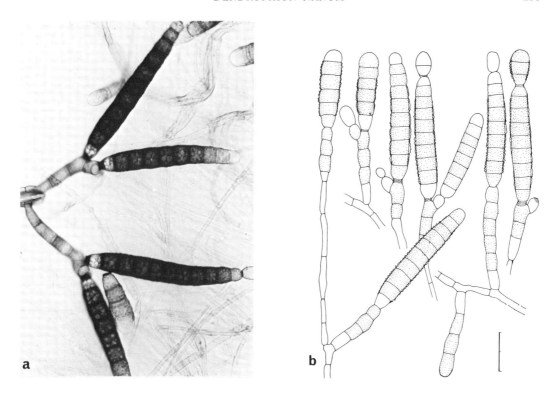

FIG. 104. *Dendryphion nanum.* a. Conidiophore and poroconidia, × 500; b. conidiogenesis.

Doratomyces   Corda *1829*

= *Stysanus* Corda *1837*
= *Echinobotryum* Corda *1831* (aleurioconidia)

Type species: *Doratomyces stemonitis* (Pers. ex Steud.) Morton & G. Sm. (≡*Doratomyces neesii* Corda)

MONOGRAPHIC TREATMENT: Morton and Smith (3974); morphogenetic studies: Breton (698). — Colonies rather slow growing, membranous to floccose, with hyaline or ± pigmented hyphae, giving rise to numerous, mostly pigmented synnemata (or coremia), which consist of a sterile stalk and a progressively elongating fertile portion which is cylindrical, aculeate or rounded in shape. Conidia formed on penicillate annellophores, catenate, globose to ovoid or mitriform, with a broadly truncate base, pigmented, smooth-walled or, in two species, warted. Simple conidiophores of the *Scopulariopsis* type sometimes present. Solitary, large, verrucose aleurioconidia (*Echinobotryum atrum* Corda) occur in addition to the synnematal conidia in *D. stemonitis*. — The synnema initials consist of a simple stalk (corresponding to a single conidiophore) which becomes septate and forms a fascicle of branches starting below the first septum; soon secondary branches arise at the base of the primary ones which grow downwards in a similarly fasciculate manner and anchor the synnema to the substratum by a rhizoid-like spreading, recalling vegetative hyphae. The synnema stalk is held together by a mucous substance and numerous anastomoses. The fertile upper portion is less densely aggregated and sporulation progresses upwards as long as the synnema can still elongate.

This genus, essentially a synnematal counterpart of *Scopulariopsis* Bain. (q.v.), is very similar to *Trichurus* Clem. & Shear (q.v.), which is distinguished primarily by its production of sterile setae in the fertile portion of the synnemata. There has been a debate as to whether the name *Doratomyces* or *Cephalotrichum* Link ex Fr. *1821* should be used for this genus; Link's genus originally comprised *C. rigescens* Link and *C. stemonitis* and was described as "*Trichiae affine genus*" and is typified (according to the protologue) by *C. rigescens* Link, which, although a *nomen dubium*, probably refers to a myxomycete. *Periconia stemonitis* Pers. was removed from *Cephalotrichum* by Link in *1825* as *Periconia subulata* Nees. — In *1837* Corda distinguished *Stysanus* from *Doratomyces* because of its catenate conidia which he apparently had not seen in *Doratomyces*.

Species of Doratomyces are common on decaying plant material, dung etc., but sometimes also occur in soil. In most species synnema development is particularly good on OA and other media with low sugar concentrations, and at temperatures of only about 12°C. In the monograph of Morton and Smith (3974), six species have been recognized, of which one (*D. phillipsii* (Berk. & Leighton) Morton & G. Sm. *1963*) has since been transferred to a monotypic genus (*Leightoniomyces* D. Hawksw. & Sutton *1977*), and three more have been described subsequently.

Key to the species treated:

1	Mature conidia smooth-walled or almost so	**2**
	Mature conidia roughened	**5**

2(1) Synnemata rarely exceeding 50 µm in length, with only a few branches growing downwards, bearing an almost flat-topped fertile head; conidia 4·5–8 × 3–4 µm   (*D. columnaris*)
     Synnemata much longer   **3**

3(2) Synnema stalks frequently consisting of a single hypha; fertile portions spreading out broadly; conidia with a rounded apex, smooth or faintly roughened, 5–7 × 3·5–4·5 µm
     D. purpureofuscus (p. 260)

     Synnema stalks always densely fasciculate, fertile portion slender, aculeate; conidia mostly with pointed apex, completely smooth-walled   **4**

4(3) Conidia 3–5 × 2–3 µm; aleurioconidia absent   *D. microsporus* (p. 257)
     Conidia 6–8·5 × 4·0–4·5 µm; ampulliform, strongly verrucose aleurioconidia, 8–12 × 6–8 µm, usually present   *D. stemonitis* (p. 262)

5(1) Conidia coarsely warted, with a rounded apex, 6–8·5 × 5–6 µm   *D. nanus* (p. 259)
     Conidia delicately roughened, with a pointed apex, 4·5–8·5 × 3·0–4·5 µm   (*D. asperulus*)

Doratomyces microsporus (Sacc.) Morton & G. Sm. *1963* — Fig. 105.

≡ *Stysanus microsporus* Sacc. *1878*

DESCRIPTIONS: Morton and Smith (3974), Matsushima (3680), and Ellis (1603). — Colonies reaching 1·9–2·4 cm diam in seven days at 24°C on MEA. Synnemata to 600 µm tall, with a slender ± cylindrical sporulating portion, ± pointed when young, more rounded when sporulating to the tip. Conidiogenous cells 4–9 × 3–4 µm, annellated zones 1·5–2·0 µm wide. Conidia ovoid to mitriform, with the tip ± pointed, smooth-walled, 3–5 × 2–3 µm. — *Stysanus medius* Sacc. *1881* was originally described as having conidia 5–6 × 3–3·5 µm and thus being intermediate between this species and *D. purpureofuscus* (q.v.); this taxon has remained doubtful.

D. microsporus has a worldwide distribution but does not usually occur with high frequencies (4647). It has been reported from various parts of the British Isles (163, 1376, 4429), the Netherlands (1616), Germany (1425, 1827), Poland (1421), Canada (IMI), and the USA (2573), to arid regions of Turkey (4245), Israel (654, 2764), Cyprus (IMI), India (1524, 3863, 4030, 4698, 4997), Nepal (1826), Japan (3680), New Guinea (6499), and Central America (1697). Characteristic substrates include rabbit dung (2279), cow dung (4758), garden compost (1425), composted manure (IMI), irrigated fields enriched with sewage (2467), children's sandpits (1421), organic detritus in fresh water (4429) and eutrophicated river sediments (1162). It has also been isolated from forest soils, especially the litter layer (269, 1558, 2411, 2573, 3138, 3447, 3680, 3817), salt-marshes (413), and heathland (5047). It is found preferentially in the upper soil layers but can occasionally occur down to 20 cm (3447). In Varanasi/India its frequency increased in the winter (3863). It has also been found on leaves of grasses and sedges, herbaceous stems (1603), oat straw (4916),

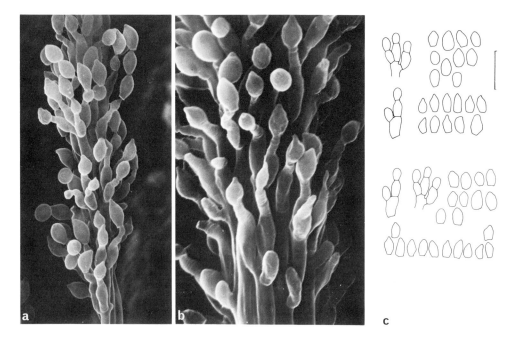

Fig. 105. *Doratomyces microsporus.* a. A small synnema, SEM, × 1250; b. detail showing annellated conidiogenous cells, SEM, × 2500; CBS 209.63; c. conidiogenous cells and conidia.

decaying roots of tomato (IMI), roots of alfalfa (3982, 3983) and oats (4443), in the rhizosphere of sugar beets (1614), on mouldy hay, dead wood (1603, IMI), and pine wood exposed to (diluted) seawater (831). It has been frequently found on oat seeds (3556), and seeds of cultivated grasses (3512). Growth *in vitro* can be inhibited by *Pholiota nameko* (3085).

The optimal temperature for growth is *c.* 25°C, regardless of the geographical origin of the isolate (1827). — *D. microsporus* can utilize starch (654, 1827) and decompose xylan efficiently, although there are pronounced differences in the latter ability between different isolates (1432). Cellulose decomposition has been demonstrated on cotton fabric and filter paper (654, 1425, 1432, 4647, 4655, 5386). Good growth occurs on lignin sulphonate (1425), in birch blocks (4191) and maple wood strips high losses in weight and tensile strength have been observed (2211) after three months incubation. Growth is possible in an atmosphere with reduced O_2 partial pressure (1425). — The development of wheat, pea and rape roots was significantly reduced by pure cultures of *D. microsporus in vitro* (1430).

Doratomyces nanus (Ehrenb. ex Link) Morton & G. Sm. *1963* — Fig. 106.

≡ *Periconia nana* Ehrenb. ex Link *1825*
≡ *Cephalotrichum nanum* (Ehrenb. ex Link) Hughes *1958*
= *Stysanus stemonitis* var. *fimetarius* Karst. *1887*
? = *Stysanus fimetarius* Massee & Salmon *1902*
= *Stysanus verrucosus* Oudem. *1903*

DESCRIPTIONS: Morton and Smith (3974), Breton (696), Ellis (1603), and Matsushima (3680). — Colonies reaching 1·7–2·0 cm diam in ten days on MEA at 18–23°C, or 2·0–2·5 cm in seven days at 24°C. Synnemata to 1–2 mm tall, particularly at low temperatures, with a ± cylindrical and apically tapering fertile portion. Conidiogenous cells 5–18 × 3·0–4·5 μm, annellated zones 1·7–2·3 μm wide. Conidia ovoid with a truncate base, apex ± rounded, distinctly verrucose, 6–8·5 × 5–6 μm. — Synnema development has been thoroughly studied (698). TEM analysis of the annellophores showed relatively short and inconspicuous annellations and conidial verrucosities appearing early in conidiogenesis (2245). — Various colony types can arise by sectoring: ones without synnemata, with small or thin synnemata together with numerous simple conidiophores, and ones lacking aerial mycelium and not sporulating except by solitary submerged aleurioconidia (698).

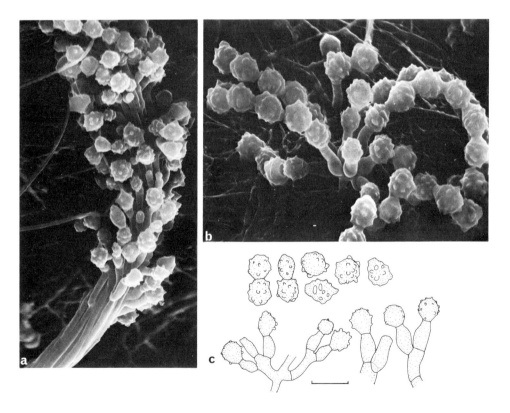

FIG. 106. *Doratomyces nanus*. a. Synnema, SEM, × 1250; b. solitary (*Scopulariopsis*-type) conidiophore with similar verrucose conidia, SEM, × 1500; c. conidiogenous cells and conidia; CBS 882.68.

This species is known mainly from decaying plant material, bark, and dung of deer, mouse, rabbit in the British Isles, the Netherlands and Finland (1603, 3974, 4757, IMI). It can be isolated with selective techniques for cellulolytic fungi (4655). The rather few reports from soil include arable soils in the Netherlands (1616), dunes in the British Isles (4655), mixed *Salix-Populus* forests in Wisconsin (2004, 2008), grassland soil in France (696), forest soil in central Africa (3063) and Japan, tundra in Alaska (3680), and water-logged organic soil in Canada (IMI). It has also been found on seeds of cultivated grasses (3512) and dead culms of *Juncus effusus* (IMI).

The optimal temperature for growth is 18°C, and the maximum 27°C; the highest density of synnemata appears at 23°C (695) but the longest (average 2350 μm) form at 8–12°C. Media poor in nutrients (1 g malt l^{-1}) favour synnema development (but not at 23°C), while mycelium production is promoted with increasing (5–10%) malt concentrations (694, 698). Griseofulvin (20 mg l^{-1}) reduces mycelial growth without stimulating synnema formation; light has no influence on their development. Synnemata do not respond to gravity but simply grow away from their substrate (698). — Its attack of maple-wood strips causes great losses in weight and tensile strength although there is a considerable variability between isolates (2211).

Doratomyces purpureofuscus (Fr.) Morton & G. Sm. *1963* — Fig. 107.

≡ *Aspergillus purpureofuscus* Fr. *1832*
≡ *Aspergillus purpureofuscus* Schw. *1832*
≡ *Stysanus purpureofuscus* (Fr.) Hughes *1953*
≡ *Cephalotrichum purpureofuscum* (Fr.) Hughes *1958*

For many other synonyms see Morton and Smith (3974).

DESCRIPTIONS: Morton and Smith (3974), Ellis (1603), and Matsushima (3680); TEM of conidiogenesis: Kiffer *et al.* (2965). — Colonies reaching 1–2 cm diam in two weeks at 18°C on MEA or 2·5–3·0 cm in seven days at 24°C. Synnemata usually 1–2 mm high, at 12°C often exceeding 3 mm, differentiated into ± long stipes and spherical to cylindrical fertile portions; stipes often retaining their original simple structure near the base. Conidiogenous cells 6–15 × 3·0–4·5 μm, with a pronounced annellated zone 1·8–2·8 μm wide. Conidia ovoid to oblong, truncate at the base, ± rounded at the apex, with the wall smooth or faintly roughened, sometimes with darker zones, 5–7 × 3·5–4·5 μm. — This species is very similar to *Trichurus spiralis* (q.v.) and may easily be confused with it when the sterile setae of the latter are poorly developed; the synnemata of *T. spiralis* usually have a thicker, fasciculate stipe and longer fertile region. — TEM observations of annellophore development show a *de novo* wall apposition with each new conidium formed comparable to that seen in phialidic conidiogenesis (2965). — Several colonial variants have been obtained by spontaneous segregation including ones with thin synnemata, no synnemata, and no sporulation at all, as well as an albino; an isolate with unusually tall synnemata obtained by using nitrosoguanidine permitted a detailed study of morphogenesis (697, 698). Synnema elongation proceeds three times slower than the growth of vegetative hyphae and the total synnema development lasts for 6–10 days (698). Excised synnemata transferred to a poor medium (0·2 g malt l^{-1}) soon cease to sporulate and

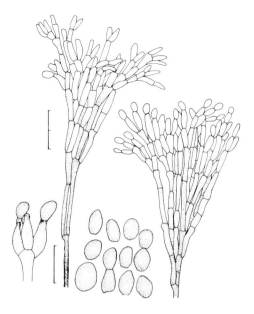

FIG. 107. *Doratomyces purpureofuscus*, reduced synnemata with simple stipes, conidiogenous cells and conidia, CBS 523.63.

start to produce secondary synnemata as long as the basal mycelium is not removed (697, 698).

D. purpureofuscus has not been isolated very frequently but its distribution seems to be worldwide. Data on soil isolations are available from Italy (3454), France (2965), the Netherlands (1614, 1616), Germany (IMI), the British Isles (1376, IMI), central Africa (3063), Egypt, Libya, India (IMI) and Japan (3680, 4972). Its occurrence on organic materials is documented from New Guinea, the Solomon Islands (5983), New Caledonia, Pakistan, Australia, Nigeria, and Sweden (IMI). Characteristic substrates include dead herbaceous stems, especially of *Heracleum* and *Brassica*, wood (3974) and dung (1603, 3680), for example from pig (5983), rabbits (6085), and horse (IMI). It has also been isolated from soil treated with urea (4972), composted municipal waste (IMI), mushroom compost (3974), sawdust buried in soil (2965), wooden baits exposed to river water (5249), and oat seeds (3556). It was found to be very abundant in one of six truffle (*Tuber magnatum*) soils in Italy (3454); its growth *in vitro* was stopped by hyphal contact with *Tuber melanosporum* (632).

At 23°C, besides 1–2 mm long synnemata, numerous superficial and submerged rudimentary conidiophores are formed; at 12°C with 0·2 g malt l^{-1} synnemata reach 3 mm on average in the wild type and 10 mm in a mutant strain. With 37 mg l^{-1} iodine-acetic acid or 20 ppm griseofulvin no solitary conidiophores are formed and synnematal development is stimulated. A mycelial extract was found to have morphogenic effects on this same species. With excised synnemata implanted in various positions on agar blocks a competition for nutrients has been observed within and between synnemata and the vegetative mycelium (698).

Doratomyces stemonitis (Pers. ex Steud.) Morton & G. Sm. *1963* — Fig. 108.

≡ *Isaria stemonitis* Pers. *1797*
≡ *Periconia stemonitis* (Pers.) Pers. *1801*
≡ *Cephalotrichum stemonitis* (Pers.) Link ex Steud. *1824*
≡ *Doratomyces neesii* Corda *1829*
≡ *Stysanus stemonitis* (Pers. ex Steud.) Corda *1837*
= *Echinobotryum atrum* Corda *1831* (aleurioconidia)

For numerous other synonyms see Morton and Smith (3974) and Hughes (2602).

DESCRIPTIONS: Morton and Smith (3974), Ellis (1603), Subramanian (5626), and Matsushima (3680); *Echinobotryum* state: Hennebert (2398). — Colonies reaching 1·7–2·4 cm diam in seven days on MEA at 24°C, at first grey but later becoming almost black; synnemata to 1200 μm tall, with ± cylindrical fertile portions. Conidiogenous cells 8–25 × 3·5–4·5 μm, annellated zones 2·7–3·3 μm wide. Conidia ovoid to mitriform, smooth-walled, 6·0–8·5 × 4·0–4·5 μm. Aleurioconidia of *Echinobotryum atrum* are often formed before the synnemata, in clusters on solitary hyphae or on the synnematal stipes, ampulliform with a ± pointed apex, verrucose, 8–12 × 6–8 μm. In isolates kept in culture for a long time, simple *Scopulariopsis*-like conidiophores may predominate.

D. stemonitis is the commonest species of its genus and has a worldwide distribution. In addition to occurrences on organic substrates such as wood, herbaceous stems and dung (1603), it has often been reported from soil. — Reports from soil are available from the British Isles (413, 1376), Germany (1889, 3042), Czechoslovakia (1703), Poland (1423), the USSR (2871), Austria (3418), Spain (3417), Turkey (4245), Israel (2768, 2777), South Africa (1559), Canada (505), India (41, 4610, 4716, 5622, 5626), Nepal (1826), China (3475), Japan (3680, 5846), New Zealand (5930), Chile (1824), and Brazil (4547). It is equally common in forest (269, 272, 273, 681, 1700, 3138, 3414), arable (1889, 2871, 2948, 3184, 3446, 3450, 4538, 4703, 6138) and other cultivated (2010, 2764, 2768, 3042, 3526, 3632, 4610, 4716) soils, but has also been found on slate slopes (566), in fen peat (5559), dunes (745, 4655), saline soils (3414, 3446), salt marsh (413), soils with steppe type vegetation (1559), polluted streams (1154, 1155, 1157, 1166), eutrophicated river sediments (1162), compost beds (3041), garden compost (1425), straw compost (953), and activated sludge (1166, 1172). An increased frequency has been found to occur after fertilization with NPK (2777) and a reduction after growing *Citrus* seedlings (3632). It has frequently been reported from samples of hardwood, various other plant remains, and dung (681, 1956, 2826, 3113, 3680, 3974, 4644, 4916, 5237, 5930), is a late colonizer of fallen beech leaves (2500), and has also frequently been isolated from the rhizosphere of corn (4553) and alfalfa (3982, 3983), roots of rape (6030), broad beans (6134), sugar beet (4559), and diseased tea plants (41). It can be present in the rhizospheres of various forest (4814) and steppe plants (3376), in coarse fodder (4548), birds' nests (2577), on oak stumps (4757), pine wood exposed to (diluted) seawater (831), and frozen meat (509). D. stemonitis is common on oat seed (3556). — The mycelium is relatively slowly decomposed in soil (3633). Its growth *in vitro* was found to be inhibited by a *Streptomyces* species (2939).

The production of conidia was found to be enhanced on media with elevated salt contents (seawater) (833). The optimal temperature for growth is 26°C (2939). — Xylan decomposition is very good, though differences in this capacity occur between individual isolates (1432). It can penetrate cellophane by "rooting branches" (5884, 5886), and hemicellulose (952) and

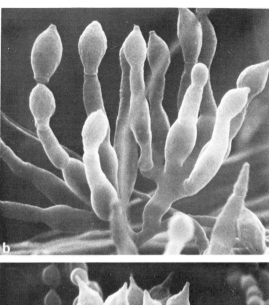

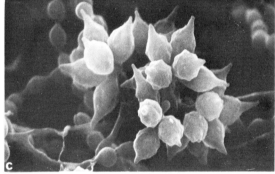

FIG. 108. *Doratomyces stemonitis.* a. A small synnema, SEM, × 800; b. solitary (*Scopulariopsis*-type) conidiophore, SEM, × c. 3000; c. *Echinobotryum*-type aleurioconidia, SEM, × 2000; CBS 180.35.

cellulose decomposition by it has been demonstrated (1956, 2741, 2948). Peptone, casein and tyrosine proved to be the best organic, and ammonium and nitrate suitable inorganic N sources for the production of cellulolytic enzymes; in addition, yeast extract mainly stimulated the formation of C_x while corn extract stimulated both C_1 and C_x enzymes (3383). Cyanamide can be degraded (2884). Good growth occurs on lignin sulphonate (1425), and maplewood strips (2211), beech and pine wood (2940, 5238) can be attacked. — *D. stemonitis* has been assumed to be the causal agent of "speck rot" on potatoes (4856).

Emericella Berk. & Br. *1857*

Anamorphs: *Aspergillus nidulans* group, *Aspergillus flavipes* group *pro parte*

Fam. Trichocomaceae Fischer. — For description see under *Aspergillus*. Of the 20 species treated by Raper and Fennell (4744), two are discussed here in addition to the recently described teleomorph of *Aspergillus niveus*.

Key to the species treated:

1   Ascospores orange-red, ridges prominent   **2**
        Ascospores faintly yellow, valves spiny, ridges very low   *E. nivea* (p. 269)

2(1)   Ascospore valves smooth; ridges mostly with entire margin and rib-like corrugations
                                                            E. nidulans (p. 264)
        Ascospore valves coarsely rugose, ridges entire but slightly corrugated   *E. rugulosa* (p. 270)

Emericella nidulans (Eidam) Vuill. *1927* — Fig. 109.

   ≡ *Sterigmatocystis nidulans* Eidam *1883*

Anamorph: *Aspergillus nidulans* (Eidam) Winter *1884*

DESCRIPTIONS: Raper and Fennell (4744), CMI Descriptions No. 93, *1966*, Subramanian (5626), and Hirayama and Udagawa (2464); GENETICAL REVIEW: Clutterbuck (1084). — Colonies reaching 5–6 cm diam in 14 days at 24–26°C on CzA, on MEA 6–7 cm in seven days, dark green from the conidia or more brownish from the ascomata; ascomata formed readily on CzA and MEA, more abundantly with urea or ammonium oxalate as sole N sources (4744), commonly 125–150 μm diam, surrounded by a yellowish layer of globose hülle cells; ascospores purple-red, lenticular, smooth-walled, with two equatorial crests 0·5–1·3 μm wide, rib-like corrugated, the body measuring 3·8–4·5 μm diam. — On the basis of ascospore variations, several varieties have been distinguished: var. *lata* (Thom & Raper *1939*) Subram. *1972* with crests 1·5–1·8 μm wide; var. *acristata* (Fennell & Raper *1955*) Subram. *1972* without crests; var. *dentata* (Sandhu & Sandhu *1963*) Subram. *1972* with the crests reduced to denticles; and var. *echinulata* (Fennell & Raper *1955*) Subram. *1972* with normal crests but echinulate valves. — SEM studies of ascospores have been carried out (3394). — Conidiophores arising from well differentiated small foot cells, smooth-walled, pigmented, mostly 75–100 µm long, with a hemispherical vesicle; conidial heads columnar; conidia globose, verrucose to rugose, somewhat wrinkled, 3–3·5 µm diam, uninucleate (6514). The vegetative hyphal cells and conidiophores are multinucleate (1082). The relative frequency of hyphae with two to eight nuclei varied with different C sources in the growth medium (2946). Nuclear division and chromosome cycle during karyokinesis have been studied (1083, 6255). Intranuclear mitosis in growing hyphae shows a central fibre and eight chromatic elements (4875). — SEM photographs of the conidia have been published (3395) and TEM illustrations show a thick, irregular wall, consisting of an electron-opaque outer layer and a thick transparent inner layer (6263). Conidiogenesis in wild-type and mutant strains has been compared (4283); the newest five or so conidia are hyaline at first, become pigmented later, and secede easily; the conidial surface shows a rodlet pattern in freeze-etch replicas (2422) and the rodlets are also visible in sections (1781); on germination, a new wall layer is formed inside, which gives rise to the germ tube while the outermost layer disintegrates (644); in dormant conidia three wall layers can be distinguished; the layer newly formed after hydration consists of an outer more opaque and an inner transparent fraction (1781). In both conidia and vegetative hyphae membraneous lomasomes have been described (6263, 6264). — Hülle cell initiation was studied by TEM and phase contrast microscopy, and these were seen to arise from delicate hyphae associated with ascomata and their morphogenesis recalls that of

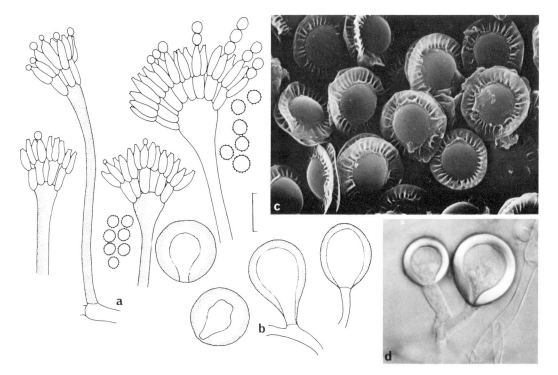

FIG. 109. *Emericella nidulans*. a. Conidial heads, b. hülle cells; c. ascospores, SEM, × 2500, orig. R. A. Samson; d. hülle cells, NIC, × 1000; CBS 124.59.

chlamydospores; the suspected ability of mature hülle cells to germinate has been confirmed (1606). — *E. nidulans* is one of the best known fungi genetically (1084, 1653, 4744). Its susceptibility to spontaneous mutations (6012) together with the frequent occurrence of aneuploidy (6013) lead to unstable phenotypic expression. *E. nidulans* is homothallic (4595, 4744) and heterokaryons are readily formed only with isolates of similar or identical genotypes (2121, 2756); variation in heterokaryon incompatibility is assumed to be not necessarily dependent on gene differences but rather cytoplasmic factors (2122). Analysis of 50 heterokaryotic hyphae gave a distribution of about 45% diploid and 55% haploid nuclei (1085). For genetic studies, crosses between mutants were induced (4595); ascomata from selfed cultures are smaller than those from hybrids (333). A technique has been devised to produce high proportions of ascomata (286), the production of which by various isolates was analysed quantitatively by a diallel crossing system; the numbers produced were found to be determined by genetic additive and possibly dominant epistatic effects (332). As a result from studies with mutants, a large number of genes has been identified and localized (4744); few loci are responsible for triggering the development of the conidial apparatus, seven affecting conidial colour (1082). Heterokaryotic conidial heads can show mixed conidial colours (4743). It was estimated that of about 1000 genes which affect conidiation, only 45 to 150 represent loci specifically involved in this process. — Zymograms of esterase and phosphatase indicated almost no intraspecific variation (1900). — DNA analysis gave a GC content of 51–54% (5600). — The polysaccharides of the hyphal cell wall consist of an alkali-soluble fraction (22%) of a-1,3-glucan and an insoluble fraction containing half chitin, and β-1,3- and a-1,4-glucan, some mannose and galactose (776, 6579). The main components of the fungal fat were oleic (40·3%), palmitic (20·9%), linoleic (17·0%), and stearic acids (15·9%) (5351).

Emericella nidulans is a typical soil fungus with a worldwide distribution. It has been recorded most frequently from India and other warm regions including Pakistan (61, 401, 4855), Bangladesh (2712), Nepal (1826, 3840), Israel (2764, 2768, 2772), Kuwait (4000, 4001), Syria (5392), Turkey (4245), Egypt (8, 3993, 3996, 4962), Tunisia (4055), Libya (6510), Somalia (5048, 5049), South Africa (1559), Australia (977, 978), Japan (2532), Tahiti (6084), Argentina (489, 2011, 2012, 6402, 6403, 6404, 6405), Peru (2005), California (1256), Florida (4918), South Carolina (6192) and Michigan (2010). Other records include ones from Canada (540), the British Isles (163, 166, 1376, 5559), France (2161, 3941, 4162), Germany (1424), Italy (270, 271), Poland (6520), Yugoslavia (5534), the Ukrainian SSR (4548, 5297), the Pamir region of the USSR (3850) and China (3475). It is known in various types of forest soils (4962, 4995, 5048), for example, tropical dry mixed (4716), under flood-plain hardwood stands (6192), *Eugenia* sp. (5512), beech (270, 271), and *Acacia* (2186) forest stands, uncultivated soils (540, 2179, 3868, 4030, 4855), grassland (3863, 4933), arable soils (44, 61, 4997), under wheat (2161), sugar cane (2861), rice (1519, 2712), cotton, potatoes (2871), citrus (2764), *Cicer arietinum*, lentil (2164), alfalfa (2871), and other legumes (1317), garden soil (61, 968, 2010, 2871, 4962) and garden compost (1424). Occasionally it is also found in soils with savannah type vegetation (1559, 5049), and recorded from desert soil (3996, 4733), sand beach (3941, 4162), peat (1376), a *Sphagnum* bog (6520), alkaline, peaty soil (5559), salt marshes (8, 4000) and sand dune plant debris (167, 4646), mangrove mud (4700), brackish soils (4477) and tide-washed coastal soils (1256, 4918) and coastal grassland debris (161), red sandy soil (poor in organic matter) and rich black clay soils (4540, 4736, 5349) and podzols (3414, 3416, 5534), soil treated with sewage sludge (1163), and a sewage stabilization pond (1166) and the cool parts of coal spoil tips (1665). It has been recorded in soil depths from 5–45 cm (3865, 3870, 5049) and most frequently in the range 20–30 cm (1317, 4995, 4996, 4998, 5048, 5349). It has been isolated from both acidic and alkaline soils with apparently no marked preference for either (270, 1559, 1713, 3729, 3732, 4245, 4698, 4700, 4736, 6510). It is active in the range between vapour saturation and -280 bars water potential to judge from hair bait colonization in soil (977); an increase in relative numbers occurred after soil fumigation (3631). — *E. nidulans* has been recorded in the phylloplane of *Adiantum* and *Polypodium* (2859) and as a colonizer of diverse plant debris including senescent culms and leaves of grasses (161, 166), *Ammophila* debris (163), leaf litter of *Eugenia heyneana* (5513), wheat straw compost (953), commonly on dried and rotting cotton leaves (5231), sclerotia of *Sclerotinia sclerotiorum* (4699), rarely on coarse fodder (4548), cotton fabrics (6326), plant debris in birds' nests (168, 5067), feathers and pellets of free-living birds (2575), earthworm casts (2857), goat (401) and rabbit (6085) dung, and fresh faeces of man (5943). It is also known in the rhizospheres of *Trifolium alexandrinum* (1963), groundnuts (2532, 2768, 3737, 4735), wheat (3737, 4727, 5514), barley (3526, 5514), *Dichanthium annulatum* (3271), flax (3870), coriander (1523) and several other crop plants (3737, 3870, 4726, 5345), steppe plants (3376), *Euphorbia thymifolia* (3866), ferns (2856, 2858), and *Funaria hygrometrica* (2860). Urea foliar sprays increased its rhizosphere population on wheat (45). *E. nidulans* is recorded also on stored seeds of oats, wheat, corn, sorghum (57, 1752, 3989, 4492), rice (2297, 2464) and cotton (5009, 6058), on nuts, dried fruits (6252) and groundnuts (5). This fungus is commonly air-borne but forms only a small percentage of the total air spora in England (1666, 2587, 3204), or of the *Aspergillus* air spora in Paris, Marseille, Lyon and London (3548). It was isolated five out of seven times from normal cow lungs (1399) and from many other domestic animals (63), and has been found associated with wild bees (*Apis*) in hives (402). — The fungus can serve as food for collembola (4007).

The optimum temperature for the production of ascomata lies between 20–30°C, in the pH range 6–9 (46, 4701), while the optimal temperature for conidiophore development is 36°C. The number of conidia per colony increased exponentially with a doubling time of $0\cdot17$–$0\cdot25$ h (254); good C sources for conidium production are glucose, sucrose, raffinose and mannitol, and good N sources peptone, $NaNO_3$, asparagine and urea; the suitability of various C/N combinations has been tested (46); the conditions for conidiation in submerged culture have been described (3641, 3642); the kinetics of germ tube emergence has been followed (285). Isolates from alkaline sites produced more ascomata at reduced water potentials and lower H^+ ion (pH 7–10) concentrations than ones from fertile soils (5782). Conidia germinate at temperatures between 12° and 37°C, and a minimum water potential at -270 bars (3873); they showed a 16 times higher UV resistance at -196°C than at 23°C (218). Optimal linear growth occurs at temperatures in the broad range 26–40°C (517, 4212), with a minimum at 11–13°C and a maximum at 47–48°C (517), with an optimum water potential between -20 and -40 bars (261, 3998); at -210 and -260 bars growth is possible (976, 5881). Fastest growth occurs with $0\cdot4$ M but half that speed with $1\cdot5$ M NaCl in the medium (3998). The thermal death point has been determined as 63°C for 25 min in apple juice (3620). *E. nidulans* tolerates low O_2 tensions (910, 1428) and a very wide pH range of 2–12, with (relatively) good growth between pH 5–11 (46, 4701) and an optimum between 6–7 (46, 3996). Maximum mycelial weight and fat production was at pH $3\cdot8$ (5355). — The apical branching in marginal vegetative hyphae was studied (5889) and the outer $37\,\mu$m wide zone of the colony found to represent the extension zone; exponential growth of the mycelium has a doubling time of $0\cdot95$ h; growth kinetic studies showed that the radial growth rate is proportional to the log of the initial glucose concentration; above 1% glucose the density of peripheral hyphae decreases with increasing glucose concentrations (5888). The doubling time for mycelial dry weight was $1\cdot45$ h on glucose, $2\cdot25$ h on L-arabinose and $3\cdot26$ h on $L(-)$xylose (2946). The maintenance requirement of glucose-limited chemostat cultures at 30°C was 0·029 g glucose per g of biomass (dry weight minus melanin) per hour (287, 911). Ascomata are not formed on manganese-deficient media which prevented the regular production of cyclic adenosine monophosphate (6585). — Ascomata appear on agar at 37°C after 72 h when all the glucose has been used and 1,3-a-glucanase mobilizes the reserve glucan from the cell walls. Sufficient formation of this reserve polymer depends on the initial glucose concentration in the medium and trace elements (B, Mo, Mn, Zn); 2-desoxyglucose inhibits 1,3-a-glucan formation and the corresponding enzyme activity; probably the same applies to manganese deficiency; ascomata formation is entirely dependent on the glucan-glucanase regulation system and is inversely related to conidium formation (i.e. at high sugar concentrations fewer conidia occur) (6581, 6582, 6583, 6584). After several conidial transfers the capacity to form ascomata may be lost due to the absence of some cytoplasmic factors (1222). — In developing conidiophores and growing conidial heads, high aryl sulphatase activity was detected (4810). — Conidia produced in submerged culture germinate in the absence of any inorganic salts but further growth requires $NaNO_3$, $MgSO_4$ and KH_2PO_4 (4999). In germinating conidia an acidic amino acid permease has been studied with L-glutamate, L-aspartate and L-cysteate as substrates (4876). After exhaustion of the C source in the medium, melanin formation begins and both a soluble and a more highly polymerized insoluble melanin fraction on the hyphal walls has been observed (4926). — Out of seven sugars tested in a combination experiment, fructose with either maltose or raffinose gave the highest dry weights (2624). Xylose, galactose (4212), ethanol, ethylamine (4353) and L- or D-malic acid (2052) are utilized. The poor utilization of a D-glucose/L-leucine substrate is thought to be caused by concomitant accumulation of isovaleric acid which acts as an inhibitor (1121). — E. nidulans is widely recognized as a good decomposer of starch and cellulose (932, 1425, 2718, 3618, 4775, 4779, 5076, 6326), as demonstrated on straw (4212, 4213, 4611), CMC, and cellulose powder. It can grow on tannin (1260), utilize arabinoxylan and produce 1,3-a-and 1,3-β-glucanases (6579, 6580). Phenoloxidases of the tyrosinase type are known (777, 3743) and a cholinesulphatase, the formation of which is de-repressed after the utilization of all available sulphur reserves (2625). The kinetics of arginase induction (1254) and the role of NADP L-glutamate dehydrogenase in ammonium regulation (2996) have been studied. The presence of low molecular weight sources of C, N or S repress extracellular and intracellular proteinase (1094). The best N sources for vegetative growth appear to be casein, peptone and sodium nitrate (4212), while methyl- and ethylamine can also be utilized; a possibly *de novo* enzyme synthesis of a nitrate reductase with nitrate as the sole N source was observed (1202). Two types of NADPH cytochrome-c oxidoreductases were distinguished, one of them seems to be closely associated with NADPH nitrate oxidoreductase (3485). E. nidulans grows well on L-tryptophan as sole N source without the formation of indole-3-acetic acid (1122) but it oxidizes tryptamine to indole-3-acetic acid and tryptophol (1522). Fat production on suitable C and N sources (1914, 4072) was optimized by the addition of phosphate (0·73%) and Mg^{2+} (0·5%) (1853); the lipid content was increased in cultures on synthetic media as compared to natural substrates (4642). E. nidulans can also utilize various hydrocarbons from fuel oil (4239) and grow on PVC (5148). Known products include gentiobiose and sophorose, synthesized from glucose (4744), an extracellular polysaccharide (3254), numerous amino acids (49), versicolorin (2680), averufin, emerin, emericellin (2681), nidurufin (238), nornidulin (=ustin), nidulin, dechloronornidulin, the anthraquinone pigment asperthecin in ascomata (3929, 4744, 5278), sterigmatocystin (1638, 2680), the siderochromes ferricrocin and ferrirhodin (3231, 6528), kojic acid (4458, 5278, 6388), oxalic acid (2730), riboflavin (1620), humic acid-like compounds (5555), orsellinic acid and a tris-dechloronornidulin, a precursor of nidulin (5317). Antibacterial antibiotics include the unsaturated δ-lactone asperline (175) and cordycepin (725, 2829). In addition to sterigmatocystin, a second mycotoxin of unknown structure termed nidulotoxin is reported (3231). A limited penicillin production occurs in most isolates (1489, 2514, 4744) and can be increased by the production of heterokaryons (2513). Staling substances were detected in the culture filtrates after 30 days (4425). Mutual inhibition was noticed between two strains cultured together on agar, where a 4 mm wide demarcation zone appeared (4462). — *E. nidulans* can inhibit the germination of cotton seed (6058) and cause a rot in apples (2581) and litchi fruits (4615) but is rarely pathogenic in man (4744) and then occurs mainly as a causal agent of mycetomas (1624) such as osteophilia (412), maduromycetoma (3519, 4155) and fibrosarcoma (3520). It can, however, cause pulmonary lesions in horse, ass and man and also aspergillosis in poultry. It is rarely also involved in bovine abortion or found associated with mastitis in cattle. Together with *Trichophyton rubrum*, it was observed in a case of onychomycosis and once also in one of mandibular periostitis (4744). It has also been recorded as a pathogen of adults or brood of wild bees (402). — Its growth is inhibited *in vitro* by garlic extract (5744).

Emericella nivea Wiley & Simmons 1973* — Fig. 110.

Anamorph: *Aspergillus niveus* Blochwitz 1929

DESCRIPTIONS: Raper and Fennell (4744), Subramanian (5626), and Wiley and Simmons (6361). — The anamorph belongs to the *Aspergillus flavipes* group. — Colonies reaching 2·5–3·0 cm diam in 14 days on CzA or MEA, white to pale lemon or pale luteous; exudate pale to brown. Ascomata found in few isolates, surrounded by yellow hülle cells; their wall consisting of one or several layers of thin-walled hyphae, light yellow. Ascospores lenticular with inconspicuous equatorial grooves and two very low equatorial crests; valves spiny, faintly yellow, 4·0–5·6 × 3·2–4·8 μm. Conidial heads small, white at first but becoming cream with age, radiating to loosely columnar; conidiophore wall hyaline, metulae covering the upper three-quarters of the hemispherical to ovate vesicle. Conidia globose, smooth-walled, 2·4–3·2 μm diam. Solitary or botryose aleurioconidia 3–5 μm diam occur (4598). — DNA analysis gave a GC content of 55% (5600).

E. nivea is most frequently reported from soils in India (1519, 2854, 4030, 4698, 4716, 4933, 4995, 4997, 5000, 5349), Thailand (6360) and Israel (654, 4759) but is also known from Somalia (5049), Czechoslovakia (1702), the USSR (2871, 5297), the USA (976) and Argentina (6404, 6405). It occurs in forest soils, cultivated land, grassland, soils with steppe type vegetation and under desert conditions. It was found in soil layers between 0–15 cm, occasionally with a very high frequency (5297, 5349). Available data on the soil pH indicate a prevalance under slightly alkaline conditions (5000, 5049). The fungus has been recorded in the rhizospheres of groundnuts (2768), *Cynodon dactylon* (2862) and barley (3526).

Hülle cell formation is suppressed by light (4045) but light promotes conidiation, the proliferation of phialides and the formation of sclerotia. The optimal temperature for linear

---

* Now *Fennellia nivea* (Wiley & Simmons) Samson 1979, because of the hyaline to pale yellow ascospores.

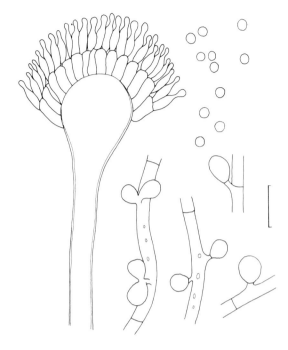

Fig. 110. *Emericella nivea*, conidial head with conidia, aleurioconidia formed in the vegetative mycelium, orig. A. C. Stolk.

growth was determined as both 36°C (654) and 41–42°C, the minimum 11–13°C and the maximum 47–48°C (517). — The degradation of cellulose (654, 3618, 4775), wool (4775) and gelatine (654), have been observed. Citrinin is produced (5278) and some antibacterial (4744) and antifungal (2945) activity reported. — Cereals contaminated with *E. nivea* have been found to be toxic to young ducklings (5158), and cerebral aspergilloma is produced in mice following intravenous inoculation (4599). — NaCl concentrations of to 20 and 25% in the nutrient media are tolerated (976, 5881).

Emericella rugulosa (Thom & Raper) C. R. Benjamin *1955* — Fig. 111.

Anamorph: *Aspergillus rugulosus* Thom & Raper *1939*

DESCRIPTIONS: Raper and Fennell (4744), Subramanian (5626), and Locci and Quaroni with SEM (3394, 3395). — The anamorph belongs to the *Aspergillus nidulans* group. — Colonies growing restrictedly, reaching 3·0–3·5 cm diam in 10–14 days at 24–26°C on CzA, with few green conidial heads and abundant purple-brown ascomata formed in several layers. Ascomata 225–350 μm diam, surrounded by dark brown, globose hülle cells; ascospores purple-red, lenticular, rugulose, with two sinuate equatorial crests, the body measuring 4·0–4·4 μm diam. Conidiophores smooth-walled, pale brownish, 50–80 μm long; conidial heads columnar; conidia globose, rugulose, 3–4 μm diam.

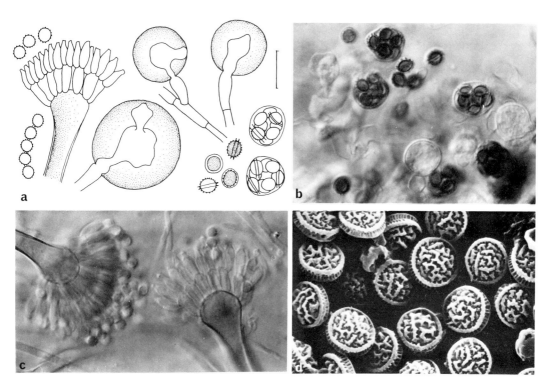

FIG. 111. *Emericella rugulosa.* a. Conidial head, hülle cells, asci and ascospores; b. young and mature asci, NIC, × 1000; c. conidial heads, NIC, × 1000; d. ascospores, SEM, × 2750, orig. R. A. Samson; CBS 133.60.

This fungus has been reported from soils in India (2179, 2854, 3737, 4700, IMI), Pakistan (401), Israel (2772), Egypt (IMI), Syria (5392), Iraq, South Africa (1559, IMI), the Sudan (IMI), Argentina (489, 6403, 6405), the USA (2482, 3817), the USSR (2871), France (242), and a coconut grove (2006) and a bat cave (4312) on the Bahamas. Other habitats include forest nurseries (2482), alkaline soils (1559), the eulittoral zone (4918) and mangrove mud (2006). It has also repeatedly been found in the rhizosphere of barley (3526) and wheat where its frequency increased after urea spraying (45) and is known from the rhizosphere of various other cultivated plants (3737). Further isolates have been obtained from groundnuts, *Guava* orchid, *Juglans regia*, hay in a compost heap (IMI), and chitin buried in a tropical soil (4264).

For optimal growth, a pH of 6·5 proved suitable. *E. rugulosa* degrades chitin (4264), cotton cellulose (3618) and produces C_x cellulase (4779). D-Arabinose cannot be utilized very well (3328). Known metabolites include asperthecin (5278), asperugin C (313), arugosin A, B and C (314, 316), aspertetronin A and B (315), shamixanthone and epishamixanthone (2682), emericellin (2681), but some further reports listed under *E. nidulans* certainly also apply to *E. rugulosa*. — The growth of *Pythium irregulare* (2945) and *Chalara elegans* (2753) was inhibited *in vitro*.

Emericellopsis van Beyma *1940*

= *Saturnomyces* Cain *1956*
= *Capsulotheca* Kamyschko *1960*

Type species: *Emericellopsis terricola* van Beyma

Anamorph: *Acremonium* Link ex Fr.

The genus *Emericellopsis* (order Eurotiales) has been reviewed several times, most recently by Gams (1882) who recognized six species; two additional species have been described subsequently by Belyakova (441).

Species of *Emericellopsis* are homothallic and form ascomata readily in culture. The cleistothecial wall is composed of several layers of hyaline, flattened cells. Asci are globose and contain 8 pigmented, more or less ellipsoidal spores with a characteristic ornamentation: an initially smooth, wide, gelatinous layer collapsing to form 3–6 longitudinal wings at maturity. — Species delimitation is somewhat problematic, since few characters are available and these show only a quantitative differentiation.

The most common species of the genus are *E. minima* and *E. terricola*, both of which have relatively small ascospores measuring $4 \cdot 7$–$6 \cdot 3 \times 2 \cdot 4$–$3 \cdot 8 \mu m$ and $6 \cdot 3$–$8 \cdot 5 \times 4 \cdot 1$–$5 \cdot 0 \mu m$, respectively. The width of the ascospore wings is variable and probably of little taxonomic value; in *E. terricola* the wings are characteristically split or laciniate, while in *E. minima* they are mostly simple. A convenient character to distinguish these two species is the relative size of the conidia: in *E. terricola* they are usually smaller than the ascospores, whilst in *E. minima* they are longer: $5 \cdot 0$–$8 \cdot 5 \times 1 \cdot 9$–$2 \cdot 5 \mu m$; only in the type strain of *E. salmosynnemata* Groskl. & Swift *1957* are they shorter. A not uncommon species is *E. glabra* (van Beyma) Backus & Orpurt *1961* distinguished by the ascospores measuring $8 \cdot 8$–$12 \cdot 1 \times 4 \cdot 9$–$7 \cdot 0 \mu m$ and having dark wings.

Emericellopsis minima Stolk *1955* — Fig. 112.

= *Saturnomyces humicola* Cain *1956*
≡ *Emericellopsis humicola* (Cain) Gilman *1956*
? = *Emericellopsis salmosynnemata* Groskl. & Swift *1957*
= *Emericellopsis microspora* Backus & Orpurt *1961*
= *Emericellopsis pusilla* Mathur *et al. 1963*

DESCRIPTIONS: Gams (1882), and Durrell (1509). — Colonies reaching $5 \cdot 0$ cm diam in ten days at 24°C on OA, pink, moist. Ascomata 30–$125(-300) \mu m$ diam, wall 6–$15 \mu m$ thick. Ascospores ellipsoidal, pale brown, surrounded by 4–6 hyaline wings, $4 \cdot 7$–$6 \cdot 3 \times 2 \cdot 4$–$3 \cdot 8 \mu m$.

272

Conidia of *Acremonium* type, ellipsoidal, $5 \cdot 5 – 8 \cdot 5 \times 1 \cdot 9 – 2 \cdot 5 \, \mu$m, in some isolates absent. — Durrell (1509) also regarded *E. terricola* as synonymous with *E. minima* while Belyakova (441) also kept *E. humicola* and *E. salmosynnemata* apart. Type strains of the four taxa placed as synonyms above were compared by Gams (1882) who found that sometimes they no longer showed the minor differences in spore size and wing diameter on which basis these species were originally distinguished. In addition many intermediate isolates were observed. — The hyphal cells, conidia and ascospores all proved to be uninucleate (456). — DNA analysis gave a GC content of $55 \cdot 5 – 56 \cdot 5 \%$ (2908).

E. minima has a worldwide distribution; it has been reported from soils in Italy (3913), France, the British Isles, the USA (1882, IMI), Canada (505), the USSR (2871), India (3526, 4477), Pakistan (401), Turkey (4245), Iraq, Bangladesh, Australia, Hong Kong, Mozambique (IMI), Zaïre (1882), Brazil (398), and Argentina (6445). Cultivated fields with barley (401, 3526), paddy (3913), potato (2871), fruit crops (IMI) and forest soils (505, IMI) are frequent sources in addition to reports from peat, low prairie (IMI), estuaries (4477) and a waste stabilization pond (1166). Characteristic substrates are apparently marine (and fresh water) mud sediments, as has been observed in Japan (5931, 5932) and Wyoming/USA (1294). — *E. minima* has also been found on plant remains of *Trifolium pratense* and *Sequoia*, in the rhizosphere of rice (IMI), and in banana root tissue (2031).

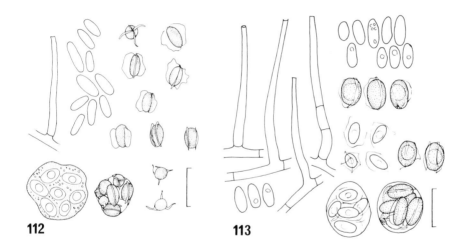

112   113

FIG. 112. *Emericellopsis minima*, phialide with conidia, young and mature asci and ascospores.
FIG. 113. *Emericellopsis terricola*, phialides with conidia, asci and ascospores, CBS 245.70.

E. minima is known to produce the cephalosporins N (2899) and C (1882), 6-aminopenicillic acid, the emericellopsins A and B (725), the polypeptide antibiotics zervamycin I and II (172, 174), *N*-acetyl-L-phenylalanyl-L-phenylalaninol (174) and the emerimycins II, III and IV (173). Utilization of amino compounds and fatty acids has been studied (1761, 1762).

Emericellopsis terricola van Beyma *1940* — Fig. 113.

= *Capsulotheca aspergilloides* Kamyschko *1960*

DESCRIPTIONS: Gams (1882), and Udagawa (5970). — Colonies reaching 4·7 cm diam in ten days at 24°C on OA, pink, moist. Ascomata 25–60(–120) μm diam. Ascospores ellipsoidal, pale brown, surrounded by 3–6 longitudinal hyaline wings, 6·3–8·5 × 4·1–5·0 μm. Conidia of *Acremonium*-type (not formed by all isolates), ellipsoidal, 4·7–8·0 × 2·1–3·2 μm. — The hyphal cells, conidia and ascospores are uninucleate (456). — The amino acid content of different isolates showed only slight differences (3476).

E. *terricola* has a worldwide distribution though it is generally rather rare. Besides its occurrence in the Netherlands (1614, 1616), the British Isles, Germany, the USSR, Italy, Canada (1882), the USA, India, and Pakistan, it has been reported from Zaïre (3790) and Central American banana soils (2031). Forests with beech (4813), *Salix-Populus* (2004) and white cedar (505), cultivated soils (1614, 1616, 1882, 3042, 3913), alpine pasture (3445), peat fields (4474), estuarine sediments of various degrees of salinity (655, 1882), organic detritus of fresh-water streams (3805, 4429), salt-marsh (167, 4646), greenhouse soils (1882) have all been reported as habitats. E. *terricola* has been isolated from bean roots though more frequently from control soil samples (4452), the rhizospheres of potato (1614, 1882), sugar beet and wheat (1614), roots of various forest plants on mor-type humus (4814), mycorrhizae of *Larix leptolepis* and *Quercus robur* (1794), provisions of the alkali bee *Nomia melanderi* (402) and is also found in the air (5970).

The temperature optimum is 30°C and the maximum 37°C while the pH minimum is 4·5 and the optimum pH 7·0 (1204) or 4·8–6·5 (508). Cleistothecia are abundantly produced on both cornmeal (2125) and oatmeal agar, particularly in fresh isolates. — Suitable C sources include glucose, sucrose and maltose; good growth occurs on inorganic nitrogen compounds, but supply of biotin is necessary (508). The production of proteolytic enzymes is very pronounced compared with other *Acremonium* species (4570). Xylan decomposition is very good, with little variability between isolates (1432). *Emericellopsis* species have been studied as producers of various antibiotics: 6-aminopenicillic acid, emericellopsin A and B (725), helvolic acid (5278) and cephalosporin N (=synnematin B) (508) have been demonstrated. Attempts to increase antibiotic production by heterokaryotization were unsuccessful (1689). — E. *terricola* is antiobiotically active against a series of human-pathogenic bacteria (2125, 4870) and *Rhizoctonia solani* for which it shows a high degree of specificity (1431). This fungus caused marked reductions in dry weight in wheat, rape and peas (1430) *in vitro*, but, according to other reports, it stimulates the growth of barley (293).

Emmonsiella Kwon-Chung *1972*

Emmonsiella capsulata Kwon-Chung *1972* — Fig. 114.

Anamorphs: *Histoplasma capsulatum* Darling *1906* var. *capsulatum, Histoplasma capsulatum* var. *duboisii* (Vanbreuseghem) Cif. *1960*

DESCRIPTIONS: ANAMORPH: Howell (2552), Berliner (474), and many textbooks of medical mycology; SEM study: Garrison and Lane (1919); reviews on ecology and medical aspects: Ajello *et al.* (76), and Mahvi (3528); discovery and description of the TELEOMORPH: Kwon-Chung (3190, 3191), with a TEM study by Glick and Kwon-Chung (2001).

Family Gymnoascaceae Baran. — The statement that the homothallic *Gymnoascus demon-breunii* Ajello & Cheng *1967* should be the teleomorph of *H. capsulatum* is erroneous, being based on a contaminated culture (3190, 3191).

E. capsulata is dimorphic with (a) a mycelial saprophytic phase growing at room tempera-ture on most agar media and particularly well on Sabouraud-glucose agar with KH_2PO_4, Sauton's medium and other synthetic media (135), and (b) a parasitic yeast-like phase formed both in internal infected animal organs and *in vitro* on brain-heart infusion agar with cystine or trypticase soya broth agar at 37°C. Conversion to the yeast-like phase is faster when the oxygen supply is reduced (3727). — Colonies reaching 1·8 cm diam in 14 days at 25°C on 4% MEA. Positive strains have a brown mycelium, are hardly converted to the yeast phase at 37°C on agar media, but give also positive infections in mice; negative strains are lighter brownish, have considerable aerial mycelium, and easily change to the yeast phase (3195). — In the saprophytic phase macroconidia are produced from aerial hyphae, on short stalks or sessile, they are globose, typically tuberculate by excrescences of the outermost wall layer as shown in TEM (1545), 7–15(–25) μm diam; near the agar surface smooth-walled conidia of a similar size and somewhat more irregular shape occur; micro-conidia similarly produced on aerial hyphae, smooth-walled, 2–6 μm diam. — Primary isolates tend to segregate into white sectors (A = albino type) with mainly smooth-walled conidia and more slowly growing pigmented sectors which bear abundant tuberculate macro-conidia (B = brown type) (474); these two types can be converted to indistinguishable yeast-like phases in continuous darkness (477) but they can still be distinguished by adding certain vital stains to the agar (476): on Nile blue A the brown type becomes bright blue, on bromecresol purple the albino bright yellow. — The yeast phase consists of subglobose cells with bipolar budding. Reversion to the mycelial phase is accompanied by the appearance of a rippled surface structure (1918). — *E. capsulata* is heterothallic. Ascomata are obtained after mating compatible isolates on alphacel-yeast extract, yeast extract or soil extract agars. The so far monotypic genus *Emmonsiella* is characterized by gymnothecia with (a) several radial, tight, cylindrical coils of 1·7–3·0 μm wide hyphae protruding in a stellate manner, and (b) sinuately branched, unconstricted, anastomosing peridial hyphae. Ascospores globose, smooth-walled, appearing finely warted under TEM (2001), 1·5–2·0 μm diam. — The related *Ajellomyces dermatitidis* McDonough & Lewis *1968*, the teleomorph of *Blastomyces dermatitidis* Gilchr. & Stokes

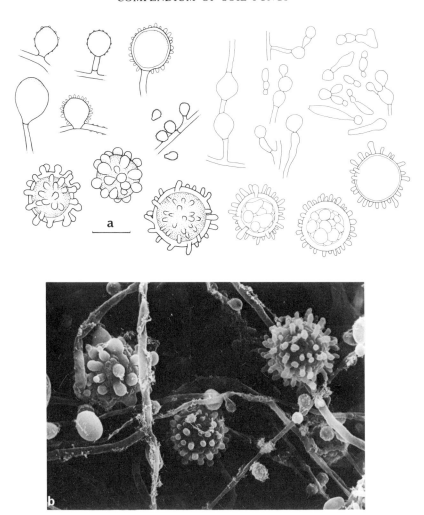

FIG. 114. *Emmonsiella capsulata.* a. Macro-conidia, micro-conidia, and yeast-like cells of CBS 213.53 and 137.72; b. macro- and micro-conidia, SEM, × 1500, orig. R. A. Samson.

1898, has thicker coiled peridial hyphae (2·3–7·5 μm) and a network of strongly constricted peridial hyphae (2001, 3190, 3192). — *Histoplasma duboisii* Vanbreuseghem *1952*, causal agent of African histoplasmosis, was distinguished by having larger cells in its yeast-like phase and the clinical symptoms; this was found to be compatible with *H. capsulatum* and is therefore now regarded as a variety of it (3193). — The cell walls of the yeast phase were found to contain 47% α-glucan, 31% β-glucan, 7·7% galactomannan, 11·5% chitin and 8·3% proteins; and those of the mycelial phase 18·8% β-glucan, 25% galactomannan, 26% chitin and 15% proteins (2877). According to another report, the mycelial phase contains about five times more mannan than the yeast phase (1416). Other investigators have reported 25% N-acetyl glucosamine, 21% glucose and 1·2% mannose (3232) and some galactose in the hydrolysate but no uronic acids (136); predominant amino acids are glycine (19%), glutamine

(13·5%), threonine (12%), serine (11%), valine, alanine and asparagine (each 7·9 mol % of the total amino acids); lipids contain 56% triglyceride, 16% phosphatidyl ethanolamine, 28% phosphatidyl choline (=lecithine); the cell walls of the mycelial phase contain 17·3% C_{16}- and 7·4% C_{18}-saturated, 1·3% C_{16}- and 33% C_{18}-mono-unsaturated fatty acids, while those of the yeast phase had 21% C_{16}- and 2% C_{18}-saturated, 3·3% C_{16}- and 66% C_{18}-mono-unsaturated fatty acids (3232). — Nuclear DNA of the yeast and mycelial phases gave GC contents of 47·5% and 46·9%, respectively (410). In the transition from the mycelial to the yeast phase, a sharp decrease in the total RNA content occurs; later stages are characterized by increased RNA synthesis (996).

Cultures of the highly infectious mycelial phase of this fungus, the causal agent of histoplasmosis in man and animals, must be handled with the utmost care. Infection occurs through the respiratory tract and may either pass unnoticed (in most of the infected persons), cause acute damage to the lungs, or spread and attack other organs. — E. capsulata is widespread in the New World (71, 1622), particularly in the eastern and central USA, and also known from India (5046), Peru (78), Puerto Rico (6536, 6537) and Panama (3038); it also occurs in Africa (3193) and warmer countries of Europe, especially Italy (3587) but is also known from Hungary, Romania, Poland and Norway (3586); the possibility of introductions from the Americas cannot be ruled out. — (+) And (−) strains have a different distribution: (−) strains have been isolated in Maryland, (+) strains in Washington D.C., and both types in Arkansas (3192). — The saprophytic phase is essentially a soil fungus whose presence has been mostly assessed by injecting soil suspensions (1/10) with antibiotics intraperitoneally into mice and examining their internal organs after eight weeks (69). A fractionated sieving and filtering technique was also devised to recover conidia (3038). Infected animals and skin tests with histoplasmin have given most information on the distribution (1544). A quick counterelectrophoretic technique has been developed for its recognition in blood serum (3025), and a semispecific fluorescent antibody staining technique has been described for it (2896). — Its frequent association with birds' roosts (682, 3190), bat tissues (6537) and bat guano (5046) has been investigated by many workers; chicken and oil birds (Steatornis caripensis) in particular promote infection (69) but E. capsulata does not attack these animals and is not excreted by them (69). Soils under birds' roosts positive for E. capsulata showed higher nitrogen, organic matter, phosphorus content and moisture holding capacity then negative ones; positive isolation was observed only if the roosts had been used for at least three years (6096). — Positive soils were characterized by the absence of montmorillonite-type clay minerals, E. capsulata never occurring together with these minerals (5603). — It was found to survive only about eight weeks in non-sterile soil and its low competitive saprophytic ability could not be enhanced by soil amendments, whilst a high moisture content facilitated survival (2032). Viable propagules were still detectable in soil cultures kept at 40°C for 15 weeks (71), but remained alive at 37°C over much longer periods in sterile soil; temperatures between −18°C and +37°C were endured for more than a year at high moisture levels but 37°C was not at low moisture. If a sterile soil surface is inoculated, the uppermost 5 cm become colonized, with sporulation restricted to the upper 1·5 cm (2028). — Dry sterile chicken manure and Seitz-filtered extracts of non-sterile manure inhibited growth, manure in soil decreased mycelial production but stimulated sporulation (2028). A soil pH range of 5–10 is generally suitable (3528) and in vitro this range has even been slightly expanded (475).

Formation of macro- and micro-conidia can be influenced by the medium: on Sabouraud-glucose with phosphate more macro-conidia are produced while on Sauton's medium more micro-conidia are formed (135). The optimum temperature for growth and sporulation is about 25°C (3528). Yeast-phase cells germinate best on casein hydrolysate at pH 7, but also germinate on a wide variety of carbohydrates but not on keto- and fatty acids (6488). The addition of citric acid to the basal medium at 30°C and 37°C induces the yeast phase (4566, 4567), while Ca and Mg ions can reverse the inhibitory effect of citric acid on mycelial growth; Zn ions stimulate conversion to the yeast phase at 37°C (4567). For maintenance of the yeast phase, a medium containing rabbit or horse blood, peptone, glucose, NaCl and cystine in a veal infusion is recommended (857). — Amongst a number of low molecular weight compounds tested, L-cysteine and L-cystine stimulated O_2 uptake most (1917). Ethylene production has also been reported (5278). Both mycelial and yeast phases have a strict O_2 requirement. Growth was improved by increasing the CO_2 concentration in a cysteine-glutamic acid-aspartic acid medium containing albumin (4564). The mycelial phase has no vitamin requirements at 25°C, while most yeast phase isolates tested required thiamine for their maximum growth at 37°C. All yeast phase isolates converted to the mycelial phase at 25°C after transfer to a minimal vitamin medium (4565). — Different isolates vary in their antigenic potential, and different serological tests are required for assessing the state of the disease in a patient; the skin test only indicates a past infection (5142, 5143). — Garlic extract is fungistatic at 0·2 ppm and fungicidal at 7 ppm (1780, 5744).

Epicoccum Link ex Schlecht. *1824*

Epicoccum purpurascens Ehrenb. ex Schlecht. *1824* — Fig. 115.

= *Epicoccum nigrum* Link ex Link *1825*

For numerous other synonyms see Schol-Schwarz (5128) and Subramanian (5626).

DESCRIPTIONS: Schol-Schwarz (5128), J. J. Ellis (1590), Malone and Muskett (3556), Ellis (1603), Subramanian (5626), Matsushima (3680), and Årsvoll (191); conidiogenesis: Cole and Samson (1109). — Colonies on MEA, OA, etc. fast-growing, reaching >6·0 cm diam in ten days at 20°C, lanose to felty, yellow, orange, red (particularly in light and at alkaline pH) or brown, sometimes also greenish; reverse similar, more intensely stained. Usually remaining sterile and producing no chlamydospores. Sporulation can be induced by exposure to "black" light, particularly on SEA, where growth is very thin and transparent; pulvinate sporodochia then appear in 8–10 days and are visible as black dots, 100–2000 μm diam. Drying of cultures on lupin stems before a transfer, or cool storage before returing to room temperature, may also induce sporulation (5128). Blastoconidia are formed singly on densely compacted, slightly pigmented, ± isodiametric conidiogenous cells. Conidia globose to pyriform, mostly 15–25 μm diam, with a funnel-shaped base and broad attachment scar, seceding with or without a protuberant basal cell; wall golden-brown to dark brown, verrucose, obscuring the septa which divide the conidia in various directions in up to 15 cells. — The conidia are actively discharged, often simultaneously from the whole sporodochium (2648, 3764); they germinate with numerous germ tubes (5128); ultrastructural studies showed mature dormant conidia to have two-layered walls; the germ tube forms by progressive evagination of the inner wall with a concomitant rupture of the outer wall (1586). TEM studies of the hyphae and conidia (757, 1492) showed that the hyphae are surrounded by a fibrous material which forms warts and may cause the hyphae to fuse in strands; the conidia are surrounded by a very thick, warted, pigmented outer layer. — The hyphal walls contain chitin and cellulose (3797). The pigment in the conidia has been characterized as indole melanin (1587).

Because of the different host plants and the variation in conidial sizes, many species have been described in *Epicoccum* but almost all were reduced to synonymy by Schol-Schwarz (5128). The closest fungus is *Phoma epicoccina* Punithalingam *et al. 1972* which differs from *E. purpurascens* only by the presence of sclerotia and pycnidia with short-cylindrical conidia 3–8 × 1·5–3 μm, and is known from grass seeds all over the world. Another species placed by some authors in *Epicoccum* is *E. andropogonis* (Ces.) Schol-Schwarz *1959* (≡*Cerebella andropogonis* Ces. *1851*) which forms extended, convolute smut-like sporodochia on many different grasses; in culture it is more greyish white to dark olivaceous, sometimes with a tinge of red; the conidia are constricted at the septa and constantly bear a conical or cylindrical basal cell.

Small black pustules of *E. purpurascens* are very frequently found on dead parts of numerous plants (5128) where the fungus is regarded as a secondary invader of damaged

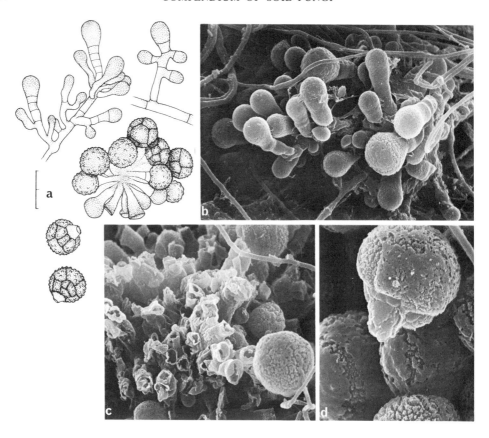

FIG. 115. *Epicoccum purpurascens*. a. Developmental stages of conidial pustules and two detached conidia; b. a young pustule, SEM, × 1500; c. pustule with conidiogenous cells after conidial liberation, SEM, × 1400; d. detached conidia, SEM, × 2500.

tissue (191), but it is also known on seeds (191, 3556), mouldy paper, textiles, insects, human skin and sputum, and very frequently in the air (2648, 3764, 5128). — The worldwide distribution of *E. purpurascens* is indicated by reports, for example, from the volcanic island of Surtsey (2406), the British Isles (1376, 1446, 1750, 2338, 3567), the Netherlands (1614, 1616, 5128), Italy (3538, 6082), Canada (505, 1446, 6352), the USA (1163, 1165, 1166, 1387, 2007, 4540, 5249), Egypt (8, 100, 3988), Somalia (5048, 5049), central Africa (3063), South Africa (1555, 1556, 1558, 1559), Pakistan (4855), India (1315, 3545, 4030, 4698, 5626), Nepal (1826), Bangladesh (2679), the Tuamotu Archipelago (6084), Japan (3680), Australia (2091), New Zealand (5815), and Honduras (2031). *E. purpurascens* has been isolated from the most varied soils, being especially frequent in forest soils (505, 1555, 1558, 5048, 6352) and on litter from deciduous trees (650, 1558, 2411, 3448, 3449, 3483, 4225) and conifers (681, 4445, 6080). It is also known from alpine soils (3976), including grassland (3445, 3538, 6082) and raw humus (1876); further in peaty soils (1376, 1564, 3234), salt-marshes (8, 163), heathland (5047, 5221), uncultivated (4698) and cultivated soils (540, 1614, 1616, 2482, 2764, 2816, 3545, 3567, 3817, 3988, 6186), saline sands (4162), desert soils (100, 4733), sand dunes (4540), soil under tundra (1446) and savannah vegetation

(1559, 5049), compost (1425, 3041), seawater (4918), more or less polluted fresh water (56, 1154, 1155, 1157, 1166, 4229, 4429, 5249), a waste pond, digested sewage (1165, 1166, 1172, 1387), fields treated with activated sludge (1163), and river sediments (1162). *E. purpurascens* is mainly found in the upper soil layers down to 20 cm (1564, 3447). It has been observed on twigs of *Pseudotsuga menziesii* (5267), the unrolled fronds (2013) and petioles (1821) of bracken, leaves of bananas (3063), peas (1373), sorghum, wheat (3219), *Panicum coloratum* (1560), *Cucurbita maxima, Areca catechu* (2679), *Hippophae rhamnoides* (3352, 3353), *Setaria glauca* (5233), *Typha latifolia* (4663), *Spartina alterniflora* (1959), culms of *Dactylis* (6237, 6238, 6243) and various other grasses (191, 2338), *Salsola kali* (4665), *Heracleum sphondylium* (6462), aging leaves of strawberry (2934), green and decaying leaves of cotton (5231), corn stalks (1590) and sheaths (3063), *Urtica dioica* (6463), and many others (5128), further on leaf litter of *Brachypodium pinnatum* (2970) and *Pseudoscleropodium purum* (2969). *E. purpurascens* has also been isolated from the roots of wheat (3567, 5330) and other cereal species (540, 3006), *Lolium perenne* (5815), *Stipa sareptana* (3376), strawberry (4132, 4133), rape (6030), broad beans (6511), beans (4451, 4452), and the rhizospheres of wheat (1614, 3567), potato, sugar beet (1614), and groundnuts (2768). It is also commonly found on seeds of *Poa pratensis* (1110), *Dactylis glomerata* (5940), *Avena fatua* (2961), corn (3989), barley (1751, 1752, 1753, 1754), oat (1752, 3556), pearl millet (3670), *Sorghum* (3989, 4120), wheat (1752, 1754, 2297, 3491, 3989, 4307, 4492, 4591, 5244), groundnuts (1016, 2765), sycamore (1683), *Pinus sylvestris* (6344), and pecans (2258, 2572), as well as on hay (2081, 4821), nests and pellets of free-living birds (2575), and cultures of the collembolan *Hypogastrura* (1027). It has been found in association with *Cronartium comandrae* on lodgepole pine (4608), on rotting wood (2338), pulp and paper (1875, 4857, 6168), cotton and textiles (3707). — Mycelium added to the soil is quickly decomposed (3633). Its growth *in vitro* is inhibited by tannins from oak leaves (2282). *E. purpurascens* can serve as food for the mite *Pygmephorus mesembrinae* (3104).

For conidium germination, at least −110 bars water potential is required (6243), optimal growth occurring at vapour saturation (4493). Conidiation is improved by light in the near UV range and alternating light/dark cycles (4036, 4132, 4592, 5106, 5664). Optimum temperature for growth is in the range 23°–28°C (1564, 4036), the minimum (−3–)4°C, and the maximum 45°C (191). The pH optimum is in the range 5·0–6·0; the strongest pigmentation is observed between pH 5·4 and 7·0, and no growth occurs at pH 2 or 10 (1564). — Sucrose supported better growth than glucose (1493), pectin is decomposed by both polygalacturonase and pectin methylesterase of *E. purpurascens* (2239); xylan decomposition is very good, with little difference in this capacity between isolates (1432); arabinoxylan is degraded by xylanase and cellulase action (1753); cellulose decomposition has been demonstrated (1425, 3934), and increases with the N source (4433); birch wood is eroded (4191); relatively good growth occurs on gallic acid (1750) and lignin sulphonate (3260). *E. purpurascens* is autotrophic for growth substances (1425). It produces considerable amounts of saturated and unsaturated fatty acids (1493), ergosterol (2088), the piperazine derivative epicorazine A (409), the antifungal substance flavipin with low phytotoxicity (320), and the flavipin-like pigment 3,4,5-trihydroxy-6-methylphthalaldehyde (1565). The striking pigments, the production of which is light- and temperature-dependent (1565, 2088), have been identified as β-carotene, γ-carotene, torularhodin and rhodoxanthin (2086). In liquid cultures metabolites resembling "humic acids" have been isolated (3636), the oxidation products of which consist of more aliphatic than aromatic compounds (5123, 5124). The production of orsellinic and cresorsellinic acids and their conversion into 20 different phenols has also been followed (2213); the presence of phenoloxidase has been demonstrated by the benzidine reaction (934). Growth was found to be stimulated in the presence of clay minerals (1738, 1739, 1740), while it was strongly retarded after a treatment at 60°C for 1 h (6344). — *E. purpurascens* reduces the pathogenicity of *Cochliobolus sativus* and has been shown to parasitize its mycelium (868). In experiments with pure cultures, the growth of barley (4919), *Pinus banksiana* and *Caragana arborescens* (6028) was inhibited. *E. purpurascens* is one of the causal organisms of leaf spot in oats (4015).

Eupenicillium Ludwig *1892*

= *Carpenteles* Langeron *1922*

Type species: *Eupenicillium crustaceum* Ludwig

Anamorphs: *Penicillium* sect. *Monoverticillata* and sect. *Asymmetrica* subsect. *Divaricata*

Family Trichocomaceae Fischer (=Eurotiaceae Clem. & Shear). The genus produces pseudoparenchymatous, firm, sclerotioid, non-ostiolate ascomata which ripen from the centre outwards; asci are formed singly or in chains.

The genus *Eupenicillium* was monographed by Scott (5159) who accepted 25 species. More recently a synoptic key for 36 *Eupenicillium* and 22 sclerotial *Penicillium* species was provided by Pitt (4579). Some of the species now placed here were described by Raper and Thom (4745) in the *Penicillium javanicum* (and *P. thomii*) series with simple (monoverticillate) conidiophores and in the *P. carpenteles* (and *P. raistrickii*) series with divaricate biverticillate conidiophores, but this delimitation is not sharp. In some cases, ascus formation has been discovered only later and such species were subsequently transferred to *Eupenicillium*.

Selective isolation of *Eupenicillium* species is facilitated by either soil steaming (6183) or pouring 60% ethanol over a soil sample for 6–8 min and subsequently diluting and heating the suspension to 60°C for 30 min (5159). For the observation of asci, colonies growing on OA slants for 2–6 weeks or longer are the most convenient. Conidium formation is usually increased if the sugar concentration in the medium is raised to 20% or more; maximum growth rate occurs at 28–30°C (5159). All species so far examined proved to be homothallic. All species treated here have asci arising singly as terminal or lateral swellings from the ascogenous hyphae, finally in clusters, ripening at random.

Key to the species treated:

1 Ascomata pseudoparenchymatous or slightly sclerotioid, generally ripening within two weeks on MEA or OA; ascospores without distinct equatorial ridges   **2**
   Ascomata definitely sclerotioid, generally not ripening within three weeks; ascospores with distinct equatorial ridges   **3**

2(1) Ascomata pseudoparenchymatous, cream or light tan; ascospores lenticular, finely echinulate, with a faint equatorial furrow, $3 \cdot 0$–$4 \cdot 0 \times 2 \cdot 5$–$3 \cdot 0\,\mu$m   *E. brefeldianum* (p. 284)
      Ascomata slightly sclerotioid, yellow or yellow-brown; ascospores lenticular, finely roughened with equatorial areas slightly flattened showing a trace of a furrow, $2 \cdot 5$–$3 \cdot 0 \times 2 \cdot 0$–$2 \cdot 5\,\mu$m

E. javanicum (p. 285)

3(1)   Colonies exuding abundant yellow-orange–red pigments; ascomata deep brown; ascospores with 2 prominent, widely separated, equatorial ridges, with ornamented valves, 5·5–6·5 (including the ridges) × 3·5–4·5 μm   *E. lapidosum* (p. 286)

        Colonies not producing such pigments; ascomata pale avellaneous or greyish; ascospores lenticular, with two prominent equatorial ridges and finely roughened valves, 2·5–3·0 × 2·0–2·5 μm   *E. shearii* (p. 288)

Eupenicillium brefeldianum (Dodge) Stolk & Scott *1967* — Fig. 116.

≡ *Carpenteles brefeldianum* (Dodge) Shear *1934*

Anamorph: *Penicillium brefeldianum* Dodge *1933* (but described inclusive of the teleomorph)

DESCRIPTIONS: Raper and Thom (4745), Stolk and Scott (5597), Scott (5159), and Udagawa and Takada (5982). — Colonies rather fast growing, reaching 3–4 cm diam in 14 days at 24°C on CzA. Ascomata 100–200 μm diam, ripening within one to two weeks. SEM of ascospores (5979) shows them to be spinulose over the entire surface. Conidiophores arising mostly from aerial hyphae, smooth-walled, monoverticillate, frequently with 1–2 branches inserted at different levels; conidia smooth-walled, subglobose to short ellipsoidal, 2·0–3·0 × 1·5–2·0 μm. — Acid hydrolysis of cell walls yielded glucose, galactose, glucosamine and oligosaccharides (4499). The content of long-chain fatty acids has been used to determine species relationships (1281). — DNA analysis gave a GC content of 51–53% (5600).

E. brefeldianum is apparently the commonest species of the genus with a worldwide distribution. Available data suggest a mainly tropical and subtropical distribution which includes Honduras (2035), Colombia (CBS), Argentina (489), Sri Lanka (4021), India (2853, 2854, 3583, 4698, 4995, 4996), Japan (2532, 6210), New Guinea and the Solomon Islands (5983, 5984), Egypt (3993), Israel (2768), Syria (5392), Somalia (5048), South Africa (5159), central Africa (3063), Cyprus, Tanzania, Kenya, Sierra Leone, Ivory Coast, Mauritius, Malaysia, Sarawak and Hong Kong (IMI). It has been isolated from cultivated (2035, 2768, 3526, 4996) as well as from forest soils (2853, 2854, 3063, 4995, 5048), uncultivated soil (4698), mountain pastures (4021), grassland, savannah (IMI), and an acid mine drainage stream (1166). Other reports include the rhizospheres of banana (2035), *Cassia occidentalis* (6067), and groundnuts (2532, 2768), groundnut seeds (2765), earthworm casts (2857), larvae of the lovebug, *Plecia nearctica* (3012), zebra dung, jute fabric exposed to soil, an injured seedling of *Piper nigrum* (IMI) and wilted pineapple plants (6210). — It survived a soil treatment with methyl bromide (6287).

No growth occurs at 5°C, but good growth is obtained at 37°C; very restricted growth occurs at −100 bars water potential (4578). — Cotton cellulose is degraded (3618). When grown on glucan, the fungus was found to produce endo-1,2-β-glucanase (4780). Known metabolites include brefeldin A and B, palitantin, frequentin (2204), the imidazole nucleoside bredinin with immunosuppressive activity (3899), griseofulvin and fulvic acid (5278). *E. brefeldianum* can grow on hydrocarbons from fuel oil (4239). It has been found to survive 70°C for 30 min in moist soil (615) or 80°C for 30 min in a soil-water suspension (614), but ascospores were killed at 80°C for 25 min in apple juice; cleistothecia can withstand 100°C for 30 min (5502).

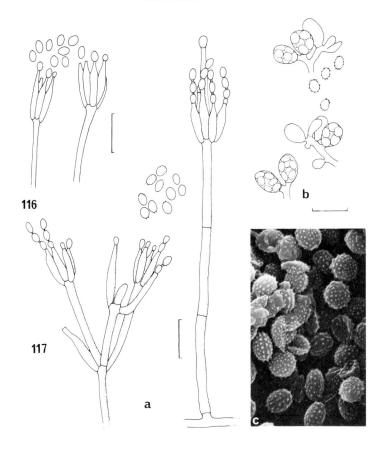

Fig. 116. *Eupenicillium brefeldianum*, penicilli and conidia, CBS 170.71.
Fig. 117. *Eupenicillium javanicum*. a. Penicilli; b. asci; c. ascospores, SEM, × 2600, orig. R. A. Samson; CBS 577.70.

Eupenicillium javanicum (van Beyma) Stolk & Scott *1967* — Fig. 117.

≡ *Carpenteles javanicum* (van Beyma) Shear *1934*

Anamorph: *Penicillium javanicum* van Beyma *1929* (but described inclusive of the teleomorph)

DESCRIPTIONS: Raper and Thom (4745), Stolk and Scott (5597), Scott (5159), Udagawa and Takada (5982), and Udagawa (5973). — Colonies growing fairly rapidly, reaching 3·5–4 cm diam in 12–14 days at 24°C on CzA. Ascomata 100–150 μm diam, ripening within two to three weeks, often deeply embedded in a sterile hyphal felt. SEM of ascospores (5979) shows them to be spinulose over the entire surface with some reticulate connections. Penicilli scanty, usually monoverticillate, with an occasional lateral branch, smooth-walled or slightly roughened, 2–6 phialides in a whorl; conidia ellipsoidal to pear-shaped, 2·3–3·0 × 1·5–2·0 μm, uninucleate. — Mechanism of nuclear division as in *P. expansum*; n = 4 chromosomes and one centriole (3198). — The content of long-chain fatty acids has been used to determine species relationships (1281). — DNA analysis gave a GC content of 51·5–53% (5600).

E. javanicum appears to have a worldwide distribution. It has been reported from the USA (655, 1040, 1166, 2482, 4226, 4313), Central America (1697), Colombia (CBS), Brazil (379), Argentina (2012), Jamaica (4886), Syria (5392), Turkey (4245), Spain (3417), Somalia (5048), Chad (3415), the Ivory Coast (4718), equatorial West Africa (1420), South Africa (1555, 3630, 5159), Sri Lanka (4021), India (1525, 3863, 4030, 4933), Papua-New Guinea (5983, 5984), and Japan (2532). Relatively numerous isolates come from grasslands (1555, 3863, 4021, 4933) and soils with steppe type vegetation (3417, 4313, 4718); it is also reported from forest (1040, 1525, 1555, 2482, 4226, 4718, 5048), cultivated (4886) and uncultivated soils (3630, 4030), an acid mine drainage stream (1166), and estuarine sediments (655). It has been found in the litter of *Eucalyptus maculata* (1558), the rhizospheres of groundnuts (2532), coriander (1523), sand dune plants (4371) and wilted pineapple plants (6210), also on the rhizoplane of sugar cane (4886) and tea (5159), on groundnuts (5159) and in nests of free-living birds (2575).

No growth occurs at 5°C, very good growth near 37°C, and very moderate growth at −100 bars water potential (4578). The fungus was found growing on hair baits on soil in the range between saturation and −280 bars (2097). — Cotton cellulose is degraded (3618). *E. javanicum* produces endo-1,3-β-glucanase when grown on laminarin, and 1,4-β-glucanase with cellulose as an inducing substrate (4778). Strong fat production together with citric acid from sugars has been known for a long time (663); media and conditions for optimal production have been described (1853, 1914). The formation of ketones from dodecane and tetradecane, and of esters from pentadecane was reported (4497).

Eupenicillium lapidosum Scott & Stolk *1967* — Fig. 118.

Anamorph: *Penicillium lapidosum* Raper & Fennell *1948*

DESCRIPTIONS: Raper and Thom (4745), Scott and Stolk (5160), Scott (5159), Udagawa and Horie (5978), and Tandon and Mehrotra (5734). Sclerotia were described by Raper and Thom (4745) in the monoverticillate *P. thomii* series. — Colonies fast growing, reaching 5–6 cm diam in 14 days at 24°C on CzA and MEA. Sclerotioid ascomata to 500 μm diam, surrounded by sterile yellow encrusted hyphae. SEM photographs of ascospores (5978, 5979) show, in addition to two large ridges, two smaller ones and some additional wrinkles on the valves giving a reticulate structure. Conidiophores scarce, appearing grey-green in the colony centre; arising mainly from aerial hyphae, typically divaricate, but also monoverticillate, stipe and metulae ± roughened; phialides usually 6–8 in a whorl; conidia ellipsoidal, smooth-walled, 2·2–3·0 × 1·7–2·2 μm. — The content of long-chain fatty acids has been used to determine species relationships (1281).

E. lapidosum has been reported from sand dunes (745), cultivated soil and air (5435) in the British Isles, forest soils in Belgium (4815), solonetz in the USSR (4677), heathland soil (3720) and other localities (977, 978) in Australia, soils in Papua-New Guinea (5978), and equatorial West Africa (1420), estuarine sediments (655) and other soils (682) in the USA, red sandy soil in South Africa (5159) and at high altitudes in Peru (2005). It is common in the succession of fungi on decaying leaves of *Eucalyptus regnans* (3480, 3483), *Fagus crenata* (4989, 4990), *Pseudoscleropodium purum* (2969), heathland plants (3720), and has been found in the rhizospheres of wheat (5435) and groundnuts (2768), in birds' nests (682) and in mull, moder and mor of forest soils (4815). It has also been isolated from frozen fruit cake (3153), canned blueberries (5159), corn seeds, and a rotten orange (5734).

The minimum temperature for growth is reported as 6–7°C, the optimum 26–28°C, and the maximum 41–42°C (517); other observations, however, indicate no growth at 37°C, and good growth at −100 bars water potential (4578). Conidia can germinate at 5°C. *E. lapidosum* can survive under anaerobic conditions and has an unusual heat resistance (5159), the thermal death point being as high as or higher than 70°C after an exposure for 30 min (613, 614, 615). In a chronically irradiated soil it did not occur at intensities of ⩾ 53 R per day (2009).

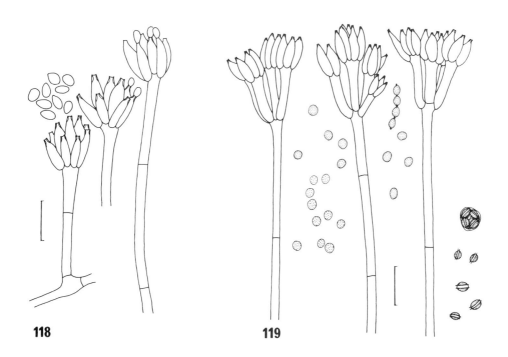

118   **119**

FIG. 118. *Eupenicillium lapidosum*, penicilli and conidia, CBS 279.39.
FIG. 119. *Eupenicillium shearii*, penicilli and conidia, ascus and ascospores, orig. J. Veerkamp.

Eupenicillium shearii Stolk & Scott *1967* — Fig. 119.

= *Penicillium asperum* (Shear *1934*) Raper & Thom *1949* sensu Raper and Thom [non *Carpenteles asperum* Shear *1934* = *Penicillium glaucum* Link sensu Brefeld *1874*]

Anamorph: *Penicillium shearii* Stolk & Scott *1967*

DESCRIPTIONS: Raper & Thom (4745), Stolk and Scott (5597), Scott (5159), and Udagawa and Takada (5982). — *Carpenteles asperum* Shear *1934* was claimed to be identical with the first observed teleomorph of *P. glaucum* Link sensu Brefeld *1874*, but the asci of *E. shearii* are borne singly and not in chains as illustrated by Brefeld for that species which is now referred to as *Eupenicillium crustaceum* Ludwig *1892* (5597). — Colonies reaching 2·5–3 cm diam in 14 days at 24°C. Ascomata to 500 μm diam, very hard; asci ripening after 4–5 weeks or more (or never produced). SEM photographs of ascospores show clearly the equatorial ridges and finely echinulate valves (5979). Ascospores at first uninucleate, later binucleate (3198). — Conidiophores arise mainly from the substrate and can reach 590 μm or more in length; penicilli divaricate and partly monoverticillate, all elements smooth-walled; phialides in whorls of 3–8; conidia ovoid to ellipsoidal, 2·2–3·0 × 2·0–2·5 μm, smooth-walled, uninucleate. — Nuclear divisions as in *Penicillium expansum,* n = 4 chromosomes and one centriole (3198).

E. shearii is widespread and of common occurrence in soil (5159). Distributional data are available from a number of tropical and subtropical areas which include forest soils in Honduras (5159), Colombia (CBS), Somalia (5048), central Africa (3063), South Africa (1555, 5159) and the Ivory Coast (4718), Japan (5982), New Guinea (5983), a rubber plantation in Sierra Leone (IMI), terra fusca in Austria (3365), arable soils under wheat and beet with an increase in frequency after NPK fertilization (2161), and a carst cave in southern Primor'e (3365). It has also been isolated from leaf litter of *Eucalyptus maculata* (1558) and cornmeal (5159).

No growth occurs at 5°C and almost none at 37°C, good growth is reported at −100 bars water activity (4578). The UV resistance of conidia was found to be higher at −196°C than at 23°C (218).

Eurotium Link ex Gray *1821*

Type species: *Eurotium herbariorum* (Wiggers) Link ex Gray *1821*

Anamorphs: *Aspergillus glaucus* Link ex Gray group

For description and introduction see under *Aspergillus*. MONOGRAPHIC TREATMENTS: Raper and Fennell (4744), and Blaser (554, 555). The genus now contains 19 species, three of which are discussed here; all three have lenticular ascospores of 6 μm diam or less (including the ridges).

Key to the species treated:

1   Ascospore valves roughened, with irregular ridges; conidia small (4·5–5·0 × 3·5–4·0 μm), echinulate; ascomata predominating in culture   *E. amstelodami* (p. 289)

       Ascospore valves smooth or finely roughened; conidia larger   **2**

2(1)   Ascospores with thin, flexuous ridges, crest-like, often interrupted and star-like; conidia mostly 4·5–5·5 μm long   *E. chevalieri* (p. 291)

       Ascospores without ridges or with low and rounded ridges and a broad and shallow furrow; conidia mostly 5–7·5 μm diam   *E. herbariorum* (p. 293)

Eurotium amstelodami Mangin *1909* — Fig. 120.

Anamorph: *Aspergillus amstelodami* (Mangin) Thom & Church *1926*

DESCRIPTIONS: Raper and Fennell (4744), Subramanian (5626), Blaser (555), and Hirayama and Udagawa (2464). — Colonies reaching 2·5–3·0 cm diam in 2–3 weeks on CzA, but 8–10 cm in two weeks on CzA with 20% sucrose, yellow to dull yellow-grey from ascomata and sterile hyphae; with addition of 20% sucrose producing abundant clustered bright yellow ascomata. Conidial heads deep olive-green. Ascospores with irregularly papillate valves and broad ridges, 4·7–5·0 × 3·6–3·8 μm. Conidia subglobose to ellipsoidal, with ends commonly flattened, finely papillate, mostly 4·5–5·0 × 3·5–4·0 μm, uninucleate (6514). — SEM studies of the life cycle (3392), ascospores (1073, 3394), and conidia (3395) have been carried out. — Using UV-induced mutants, 13 markers were mapped through meiotic and mitotic recombinations (3316). — In a numeric analysis of pyrolysis-gas-liquid chromatograms (6094) of conidia, a 53% average similarity was found between three isolates, i.e. slightly higher than with other species of the genus (6093). — Hyphae contain crystalline inclusions (3663). — DNA analysis gave a GC content of 52–55% (5600).

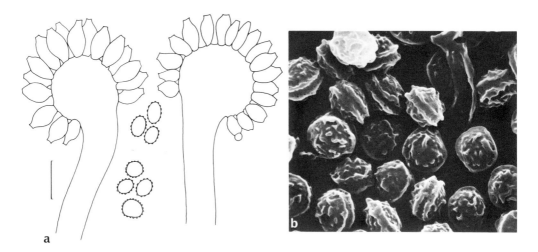

Fig. 120. *Eurotium amstelodami*. a. Conidial heads and conidia, CBS 520.65; b. ascospores, SEM, × 2500, orig. R. A. Samson.

The fungus is preferentially distributed in tropical and subtropical zones. Data are available from India (2179, 2854, 3729, 3732, 4030, 4477, 4698), Kuwait (4000, 4001), Pakistan (4855), Israel (2764, 2768, 2772), Syria (5392), Brazil (399), Peru (2005), Argentina (6403, 6404, 6405), South Africa (1415), Australia (978), the Tuamotu Archipelago (6084), Japan (2532), and Turkey (4245); it has also been reported from the USSR (2871), the USA (976, 1166, 2008, 3817), the British Isles (2587, 5559), and Italy (3452). It is frequently reported, although with low frequencies in some cases, from cultivated soils (2764, 2768, 2772, 2871, 3732, 3817), uncultivated soils (2179, 4030, 4698, 4855), also from a moist willow stand (2008), a teak forest (2854), alkaline peaty soil (5559), estuaries (4477), salt-marshes (4000), and the air (1166, 2587). — A relatively high number of isolates come from stored and/or decaying food products (4744), including frozen fruit cake (3153), fruit juices (5205), dough products (2067, 5980), milled rice in Pakistan and East Asia (2464), nuts and dried fruits (6252), and various other dried foodstuffs (4986, 5980). It was also found on seed samples of numerous plants including wheat (4306, 4492), *Avena fatua* (2961), corn (57), tomato (5899), groundnuts (1984, 3369), and sycamore (1683). It has been isolated from the rhizosphere of poplar (3452), barley, oat (3003), groundnuts (2768) and carrots (3733), decomposing wheat straw (953), dry fodder (3258), self-heated hay (4821), damaged flue-cured tobacco (6282), rabbit dung (6085), nests, feathers and pellets of free-living birds (2575), and optical glasses (2941). It has only occasionally been isolated from guts of the honey bee, *Apis mellifica* (1983).

The production of conidia is stimulated by light, formation of cleistothecia is inhibited by very high osmotic pressure and favoured by 35°C (1240). The optimal temperature and water potential for conidium germination and growth are 33°C and −100 bars (261), and the minimum water potential is ≥ − 350 bars (1192). The optimal temperature for vegetative growth is also 35°C (978) and the maximum 43–46°C (554); with 30–50% sucrose in the medium good growth and high production of antibiotics occur (107). — Of ten *Aspergillus* species tested, *E. amstelodami* was the best utilizer of glucose and sucrose (1018). L- Or

D-malic acid (2052) and also metaphosphate (4544) are utilized; β-D-xylopyranosidase is produced (4777) and xylan induces the production of endo-1,3-β-xylanase (3231). In the presence of high glucose concentrations, the corrosion of metals was observed (85). Known metabolites include riboflavin (1620), flavoglaucin, auroglaucin, physcion (=parietin), endocrocin, catenarin, and echinuline (4744, 5278). Flavoglaucin and echinuline are produced only concurrently with the formation of ascomata (107); the biosynthesis of echinuline from cyclo-L-alanyl-L-tryptophyl has been described (106, 5411). Other metabolites include cryptoechinuline A (887), the indole derivatives neoechinuline (339, 916) and neoechinuline D (1443), the antibiotics amodin A and B (725), with amodin B correlated to ascoma production and the peptide amodin A to that of conidia (1276). — Seed germination of wheat, peas and squash was found to be reduced by *E. amstelodami* (2276). The growth of barley seedlings was inhibited by *E. amstelodami* (4919). Corn contaminated with *E. amstelodami* was lethal to rabbits (4686) and toxic to mice (4986). — *E. amstelodami* has been isolated from aspergillosis in a pelican and from a cerebral abscess in a woman as well as from foot mycetoma (4744). — The fungus is able to grow on media containing 20–30% NaCl, 79% sucrose, or 33.3% $CaCl_2$ (976, 1475, 3166, 4697). — Ascospores can survive a 10 min exposure to 70°C at a rate of 3% (4580). — The growth *in vitro* was inhibited by garlic extract (5744).

Eurotium chevalieri Mangin *1909* — Fig. 121.

Anamorph: *Aspergillus chevalieri* (Mangin) Thom & Church *1926*

DESCRIPTIONS: Raper and Fennell (4744), Subramanian (5626), Blaser (555) and Hirayama and Udagawa (2464). — Colonies reaching 2.5–3.0 cm diam on CzA in 14 days, up to 8 cm on CzA with 20% sucrose at the optimal temperature of 30°C, grey-green to bluish grey in central areas from conidia. Ascomata abundant on CzA with 20% sucrose in a continuous yellow layer, covered with orange-red hyphae. Ascospores lens-shaped, with valves definitely roughened and with prominent equatorial crests 0.8–1.0 μm apart, with entire or interrupted, stellate margin, 4.6–5.0 × 3.4–3.8 μm. *E. intermedium* Blaser *1976* (anamorph *A. chevalieri* var. *intermedius* Thom & Raper *1941*) is intermediate between this species and *E. amstelodami* (q.v.) and has spinulose ascospore valves. SEM studies of ascospores (1073, 3394) and conidia (3395) have been carried out. Conidia ovate to ellipsoidal, mostly 4.5–5.5 μm long, with broadly papillate ornamentation; uninucleate (6514). Hyphae contain crystalline inclusions (3663). — In a numeric analysis of pyrolysis-gas-liquid chromatograms (6094) of conidia, a 45% average similarity was found between three isolates, i.e. slightly higher than with other isolates of the same group (6093).

E. chevalieri is mainly distributed in tropical and subtropical zones, including Israel (2764, 2768, 2772), Pakistan (4855), India (2179, 2851, 2854, 4698), Somalia (5048, 5049), Gambia (1984), Argentina (4119, 6404), Central America (1697), Japan (2532), but also in the USA (1166) and the British Isles (2587). It has been isolated from cultivated (2764, 2768, 2772) and uncultivated soils (2179, 4698), forest soils under teak (2851, 2854) and *Shorea* (2179), soils under steppe type vegetation (5048, 5049), estuaries (4477), and the air (1166, 2587). It has also been reported from grass litter (166), the rhizosphere (2532, 2768), seeds

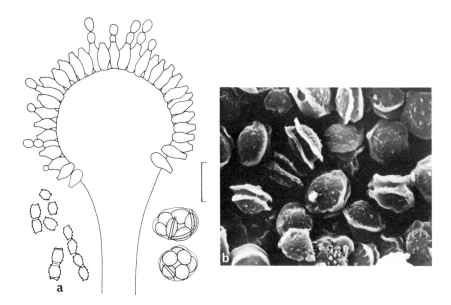

FIG. 121. *Eurotium chevalieri*. a. Conidial head, conidia and two asci, CBS 522.65; b. ascospores, CBS 129.54, SEM, × 2500, orig. R. A. Samson.

(3369) and pods (1984) of groundnuts, seeds of wheat (4492) and sorghum (3989), mouldy corn (4831) and pecans (2572, 5092), various dried or milled foodstuffs (4986, 5980), stored copra and cacao beans (1676), and cotton fabrics (6326). It has also been found in human maduromycosis (4119).

The formation of cleistothecia, with an optimum at 25°C, can be inhibited by very high osmotic pressure (1240) and intensive light irradiation (5750); the optimal growth temperature is in the range 30–35°C (978) with an optimal water potential at −100 bars for growth and conidium germination (261) and a minimum at > −350 bars (1192); the maximum temperatures are in the range 37–49°C (554). — Compared to other *Aspergillus* species tested, the utilization of glucose and sucrose is fair with a somewhat better utilization of sucrose (1018). Hydrocarbons from fuel oil are utilized (4239) and cinerone can be hydroxylated (5681). Cellulosic substrates or wood are not attacked (4775), but losses in tensile strength of fabrics have been reported (6326). Lipase is produced on coconut oil emulsion (1676). Known metabolites include L-alanyl-2-(1,1-dimethylallyl)-L-tryptophan anhydride (2235), flavoglaucin, auroglaucin, physcion (=parietin), physcion anthranol A and B, erythroglaucin, echinuline (4744, 5278), preechinuline (5584), gliotoxin (6366), and xanthocillin (3231). — The production of aflatoxin by *A. parasiticus* is drastically reduced when it is grown simultaneously with *E. chevalieri* on rice (618). — The germination and/or seedling growth of wheat, peas, squash, tomatoes (2276) and barley (293) was found to be inhibited; grain contaminated with *E. chevalieri* was toxic to young ducklings (4744, 5158) causing haemorrhagic syndrome; it also caused severe injuries in young calves (898, 1799). Some isolates are toxic to HeLa cells and mice (4986); *E. chevalieri* showed antagonistic activity against *Rhizoctonia bataticola* (1369). — It is highly osmotolerant and grows well on media containing up to 30% NaCl (976, 4697). A small number of ascospores (0·5%) can withstand a 10 min treatment of 80°C (4580).

Eurotium herbariorum (Wiggers) Link ex Gray *1821* — Fig. 122.

≡ *Mucor herbariorum* Wiggers *1780* ex Mérat *1821*
= *Eurotium repens* de Bary *1870*
= *Eurotium rubrum* Kœnig, Spieckermann & Bremer *1901*
= *Aspergillus sejunctus* Bain. & Sart. *1911*
= *Aspergillus umbrosus* Bain. & Sart. *1912*
= *Aspergillus mangini* Thom & Raper *1945*

Anamorph: *Aspergillus glaucus* Link ex Gray *1821*

The other above-mentioned epithets in *Aspergillus* have often been applied to the anamorph, but all of them were originally described including ascomata, and so strictly refer to the teleomorph.

DESCRIPTIONS: Raper and Fennell (4744), Hirayama and Udagawa (2464), Subramanian (5626), and Blaser (555). — Colonies growing restrictedly on CzA, reaching 1–2 cm diam in 21 days, but on CzA with 20% sucrose 5–7 cm in 14 days at 24–26°C, yellow-green to greenish grey due to the conidial heads, vegetative hyphae and reverse sometimes yellow-orange to red-brown; conidial heads pea-green, sparse or abundant. Ascomata abundantly produced on CzA with 20% sucrose or MEA with 40% sucrose, forming dense orange-yellow layers, surrounded by a loose network of yellow to orange-red hyphae. Asci in chains. Ascospores lenticular, valves smooth or slightly roughened, with a broad equatorial furrow with very low ridges but no wings, $5 \cdot 0$–$7 \cdot 0 \times 3 \cdot 8$–$4 \cdot 8 \,\mu$m. Conidia subglobose to ellipsoidal, densely covered with slender or thick spines, mostly $5 \cdot 0$–$7 \cdot 5 \,\mu$m diam, 1–2(–4)-nucleate (6514).

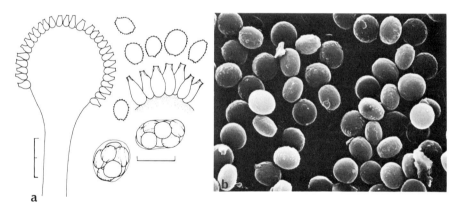

FIG. 122. *Eurotium herbariorum*. a. Conidial head with detail of phialides and conidia and two asci; b. ascospores, SEM, × 1500, orig. R. A. Samson; CBS 126.55.

For many decades, *E. herbariorum* and *A. glaucus* were considered as doubtful species until Malloch and Cain (3552) neotypified *E. herbariorum*. Blaser (555) extended this neotypification to include the anamorph *A. glaucus*. The neotype was originally named *E. herbariorum* ser. *minor* Mangin by Thom and Raper *1941* and later *Aspergillus mangini* Thom & Raper *1945* by Raper and Fennell (4744). Blaser (554) has shown that *E. rubrum*, *E. umbrosum* and *E. mangini* have overlapping ascospore dimensions and so reduced them to synonymy under the oldest epithet *herbariorum*. He still distinguished *E. repens* from this species by its smooth ascospores (or, if ornamented, then only provided with a low equatorial groove) and gave the ascospore dimensions of *E. repens* as $4·7–5·3 \times 3·2–4·4 \mu$m and of *E. herbariorum* as $5·2–6·8 \times 3·8–4·8 \mu$m. Since the presence of an equatorial groove is a variable character and the ascospore dimensions overlap strongly, *E. repens* is here treated as a synonym of *E. herbariorum*. — The anamorph of *E. repens* is often cited as either *Aspergillus repens* (Corda) Sacc. *1882* or *A. repens* de Bary *1870*. When de Bary described *E. repens* in *1870*, he deliberately did not name it *Aspergillus*, although this name was later ascribed to him by Fischer in *1890* and others. De Bary did not consider his fungus as synonymous with *A. glaucus* var. *repens* Corda *1842* and so in any case his *E. repens* cannot be considered as based on Corda's fungus (the type of which is now lost).

SEM studies of ascomata and ascospores (1073, 3394) and conidia (3395) have been carried out. The development of normal conidiophores is labile and proliferation of phialides frequently occurs due to environmental factors (5797). In a numerical analysis of pyrolysis-gas-liquid chromatograms of conidia (6094), 48% (*E. repens*) and 36% (*E. rubrum*) average similarities were found between three isolates of either group, i.e. usually not higher than with other *Eurotium* species (6093). The hyphae contain crystalline inclusions (3663). DNA analysis gave a GC content of 54·5% (*E. repens*) or 53–54% (*E. rubrum*) (5600).

The available data indicate a worldwide distribution with a preponderance in tropical and subtropical latitudes; reports of soil isolates include ones from India (969, 2179, 3545, 4048, 4371, 4611, 4698, 4700), Bangladesh (2712), Pakistan (401, 4855), Kuwait (4000, 4001), Japan (3267), Turkey (4245), Israel (2768), South Africa (1415), equatorial West Africa (1420), Australia (977, 978, 3720), Brazil (399), Argentina (6402, 6403, 6404, 6405), the southern USA (2573, 4733), Italy (5047), Germany (1424, 5316), Belgium (4816), the British Isles (2587, 2736), and the USSR (1482, 2871, 3003). It has been isolated from forest soils (1420, 2573, 2854, 3267, 4816), cultivated (2712, 2768, 2871, 3003, 3545, 4538) and uncultivated (401, 4698, 4855) soils, heathland (2736, 3720, 5047), peat (1259, 6520), an alpine soil with long snow coverage (3976), garden compost (1424), desert soil (4733), sand dunes (4371), estuarine silt (5316), salt marshes (4000), mangrove mud (4700), polluted streams (1482), off-shore ocean water (4918), frescoes in a monastery (2666), and air (2587). It is not uncommon on roots of sugar beet (4559), in the rhizospheres of barley, oats (3003, 3006), corn (4555), rice (2712), groundnuts (2768), and various plants in sand dunes (4371), steppe (3376), and oak forests (4816); it has also been recorded as an early colonist of decomposing straw (953), on damaged flue-cured tobacco (6282), achenes of sycamore (with increasing frequencies at increasing moisture and temperature) (1683), in silage (932), coarse fodder (3258, 4548), seeds (2765, 3369) and pods (346) of groundnuts, seeds of peas (2273), sorghum (795), wheat (1029, 3989, 4492), and corn (3645, 4831), where it is the most important invader when stored at moisture contents in the range 14–23% (795, 4306). Other known substrates include feathers and pellets of free-living birds (2575), pulp and paper (6166), leather and cotton fabrics (6326). As "*E. repens*" it has been reported to be a frequent contaminant of flour (3930), particularly milled rice in East Asia and the USA (2464), stored copra and cacao beans (1676), pecans (2258, 2572, 5092), spices and meat products (2201), fermented and dried food products (4986, 5980), and fruit juices (3428).

The optimal temperature for the formation of conidia and ascomata was found to be 25–30°C, the latter being inhibited by very high osmotic pressures (978, 1240); the temperature maximum is 37–40°C (554). The optimal temperature and water potential for ascospore germination and growth are 24°C and −100 bars (261), the minimum is as low as

−410 bars (1192). Colonization of hair baits in soil takes place at water potentials in the range −70 to −350 bars (977, 2095). — D-Glucose and sucrose are utilized almost equally well (1018). Cellophane is attacked (1877), but unprotected cellulose or woollen textiles are not (4775). When grown on ME broth with salicin, β-glucosidase is produced (6159). *E. herbariorum* produces a lipase on coconut oil emulsion (1676). Known metabolites include riboflavin (1620), flavoglaucin, auroglaucin, erythroglaucin, physcion, physcion anthranol A and B, echinuline and pteridine (4744, 5278), and apparently also aflatoxin B1 (3167). Hydrocarbons from fuel oil can be utilized as a sole carbon source (4239), and tryptamine is transformed to indole-3-acetic acid (1522). — The germination of infected wheat, peas, squash and tomato seeds is reduced (2276, 4306), and that of stored peas is strongly reduced by a heat-stable toxin (2273, 2275); the growth of barley seedlings can be inhibited by culture filtrates (293). *Venturia inaequalis* can be inhibited by *E. herbariorum in vitro* (6205). *E. herbariorum* is generally not pathogenic to animals, but has been recorded from an ear infection (4744). It is sometimes listed as a mycotoxic agent (3932) and extracts of infected corn proved to be toxic to mice (4831), while contaminated cereal products were so for young ducklings (5158). — *E. herbariorum* was found to withstand a 10 min heat treatment of 70°C with a survival rate of 2·3–3·2% (4580). It can grow with up to 30% NaCl in the medium (976, 4697).

Exophiala Carm. *1966*

= *Wangiella* McGinnis *1977*

Type species: *Exophiala salmonis* Carm.

Colonies growing ± slowly, dark olivaceous-brown to black, often slimy when young, but later becoming floccose or velvety. Vegetative hyphae pale brown, often forming long chains of swollen cells. Conidiophores mostly little differentiated; conidium formation taking place on terminal or lateral conidiogenous cells and, most characteristically, on indistinct lateral pegs arising just below the septa from intercalary cells of aerial hyphae but also from dislodged conidia or budding cells; conidiogenous pegs slightly protruding, phialidic with indistinct collarettes and, in some species, percurrently elongating during conidiogenesis (indistinct annellides or percurrent phialides). Conidia subhyaline, mostly 1-celled, elongate, with a slightly truncate hilum, showing a strong tendency to swell and form secondary conidia.

The genus *Exophiala* was monographed by de Hoog (2524), who distinguished it from *Rhinocladiella* Nannf. *1934* (see Fig. 123d) which has conidia produced on inconspicuous denticles in sympodial succession. But some other pleomorphic fungi which often also bear a few sympodially elongating *Rhinocladiella*-like conidiogenous cells are also considered under *Exophiala*. Even the type species, *Rhinocladiella atrovirens* Nannf. *1934*, was shown to be mostly, but not always, associated with *E. jeanselmei* var. *heteromorpha* (Nannf.) de Hoog *1977*. Cultures with predominant slimy budding cells are referred to as *Phaeococcomyces* de Hoog *1979* (≡ *Phaeococcus* de Hoog *1977*). *Exophiala* species have sometimes been referred to as *Phialophora* (q.v.); this genus, however, is clearly distinct by more broadly spreading, ± floccose colonies, ± flask-shaped phialides, mostly with a distinct, ± divergent collarette; pleurogenous sporulation (with openings not localized near the septa) occurs commonly only in the *Ph. hoffmannii*-group which, in addition, has much paler colonies than *Exophiala*.

In view of some conflicting views expressed in recent studies on *Exophiala* and related genera, some representative isolates were reexamined by R. A. Samson and W. Gams (unpublished data) which led to some modifications of de Hoog's scheme as shown in the following key. Less emphasis is given to the proportion of swollen and budding cells and the shape of the conidiogenous pegs, and the distinction between phialides and annellides is regarded as gradational and not warranting generic separation in this group; useful specific criteria were found in the maximal hyphal width and pigmentation, development of collarettes on the conidiogenous pegs and the size and shape of the smallest normal conidia.

The yeast-like budding cells can revert to true hyphae only when they have become thick-walled and accumulated sufficient reserve substances as seen in TEM (4331). A TEM study of annellide formation has been provided (2139). A fluorescent antibody technique did not give conclusive results for differentiation between species of "black yeasts" (5218).

Exophiala species are now known from human subcutaneous abscesses or chromomycosis, and from similar infections in fish; also from decaying wood, and, more rarely, from soil. The species are rather pleomorphic and variable and not sharply delimited. We suggest a new key

296

to the recognized species, but it is almost impossible to say whether the literature data compiled below really refer to any of these or *Rhinocladiella atrovirens*. Species determination has been standardized using cultures growing on OA for 5 and 15 days at about 20°C (2524), whilst in our recent study best differentiation was obtained on 2% MEA or cherry decoction agar after 12 days at 25°C.

Key to the recognized species:

1   Distinct erect conidiophores, at the base much darker than the vegetative hyphae, present
                                                                  (*E. spinifera*)
    Dark erect conidiophores absent   **2**

2(1)   Conidia often 2-celled, the smallest 8–12 × 2·5–3 μm   (*E. salmonis*)
       Conidia generally 1-celled, smaller   **3**

3(2)   Conidiogenous pegs with minute or rather conspicuous collarettes   **4**
       Conidigenous pegs without any collarette visible in the light microscope   **5**

4(3)   Conidiogenous cells often with multiple conidiogenous openings (polyphialidic) (see Fig. 124b); smallest conidia 3–4 × 1·5–2·0 μm; sclerotial bodies absent   (*E. dermatitidis*)
       Conidiogenous cells generally with single conidiogenous openings; smallest conidia 2·5–4 × 1·0–1·5 μm; sclerotial bodies often found in old cultures
                                     (*"Phialophora" heteromorpha* sensu Wang (6167))

5(3)   Vegetative hyphae to 3(–4) μm wide, rather dark brown; smallest conidia almost rod-shaped, 3·5–4 × 1·0–1·2 μm (see Fig. 124a)   (*Exophiala* state of *Dictyotrichiella mansonii*)
       Vegetative hyphae (without swellings) to 2·0(–3·0) μm wide, paler; smallest conidia obovoid to short-clavate and mostly broader   **6**

6(5)   Geniculate vegetative hyphae usually present (see Fig. 123b); smallest conidia 3·5–5 × 1·5–2·2 μm; chains of swollen cells present or absent: *E. jeanselmei*   **7**
       Geniculate vegetative hyphae absent; branching chains of swollen cells commonly present   **9**

7(6)   Sporulation almost exclusively pleurogenous   (*E. jeanselmei* var. *lecanii-corni*)
       Sporulation also often on discrete conidiogenous cells   **8**

8(7)   Conidiogenous pegs often arising from slender hyphal cells, but somewhat darkened, ± slender, terminal or lateral conidiogenous cells usually present *E. jeanselmei* var. *jeanselmei* (p. 298)
       Chains of swollen cells usually present and yeast-like cells abundant
                                (*E. jeanselmei* var. *heteromorpha*)

9(6)   Conidia ellipsoidal to obovoid, sometimes septate, the smallest 3–5 × 2–3 μm   (*E. pisciphila*)
       Conidia short-obovoid, the smallest 1·5–2·5 × 1·0–1·5 μm   (*E. moniliae*)

Exophiala jeanselmei (Langeron) McGinnis & Padhye *1977* — Fig. 123 a–c.

≡ *Torula jeanselmei* Langeron *1928*
≡ *Phialophora jeanselmei* (Langeron) Emmons *1928*
? = *Sporotrichum gougerotii* Matr. *1910*
   ≡ *Phialophora gougerotii* (Matr.) Borelli *1955* [sensu Borelli = *E. jeanselmei*, orig. possibly
      Sporothrix schenckii Hektoen & Perkins]
  = *Exophiala mansonii* (Cast.) de Hoog *1977* pro parte
    ≡ *Microsporum mansonii* Cast. *1905**
    ≡ *Rhinocladiella mansonii* (Cast.) Schol-Schwarz *1968* pro parte

For numerous synonyms see de Hoog (2524).

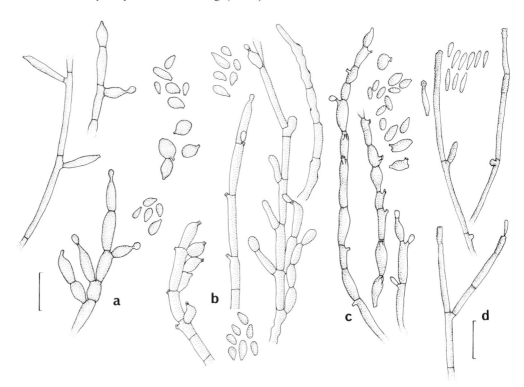

FIG. 123 a-c. *Exophiala jeanselmei.* Solitary conidiogenous cells and compound conidiophores with small primary conidia, swollen conidia with conidiogenous pegs, a. CBS 537.76, b. Kwon-Chung B-8639, c. CBS 526.76; — d. *Rhinocladiella atrovirens*, conidiophores with sympodially proliferating rhachis forming blastoconidia.

---

* Note: McGinnis (Mycopathologia **65:** 79–87. *1978*) recently provided evidence that the name *Microsporum mansonii* Cast. is of doubtful application since the neotype strain, CBS 158.58, cannot be responsible for the symptoms of tinea versicolor in man described by Castellani in *1905* for this fungus, but rather for those of tinea nigra described by Castellani in *1911*; thus a secondary isolate might have been taken for the type. Moreover, our recent studies (R. A. Samson and W. Gams, unpublished) showed that this isolate is different from the anamorph of *Dictyotrichiella mansonii* Schol-Schwarz *1968* (CBS 101.67, see our key), which was regarded as its teleomorph (2524, 5129), but much closer to *E. jeanselmei*. — Some of the information published for *E. mansonii* has therefore been incorporated in our text of *E. jeanselmei*.

DESCRIPTIONS: Mangenot (3568), Wang (6169), Nielsen *et al.* (4173), McGinnis and Padhye (3708), and de Hoog (2524); conidiogenesis: Cole and Samson (1109). — Colonies reaching 0·7–1·3 cm diam in 14 days at about 20°C, initially slightly slimy, soon becoming velvety or woolly, greenish grey to dark olivaceous-brown; reverse olivaceous-black. Vegetative hyphae brown, to 2·5 μm wide, smooth-walled, often showing a characteristic geniculation (see Fig. 123 b); swollen, moniliform hyphae present to various extent. The yeast-like budding cells are surrounded by slime capsules (3504). Conidia formed on lateral pegs of intercalary cells or on discrete slender, somewhat darkened conidiogenous cells, 6–15 × 2–3 μm, or on detached conidia; conidiophores of 2–4 cells occasionally formed; conidiogenous pegs 1–2(–3) μm long, slightly tapering, imperceptibly annellate; conidia hyaline, smooth-walled, clavate to obovate, with an inconspicuous basal hilum, the smallest 3·5–5 × 1·5–2·2 μm. — Different views subsist on the delimitation of *Trichosporium heteromorphum* Nannf. *1934*. While some dried authentic collections suggest that it is best characterized by phialidic openings with recognizable collarettes (6167), a living type culture and the original description do not give a clue to the presence of a collarette (2524), so that this taxon might only be a variety of *E. jeanselmei* (2524).

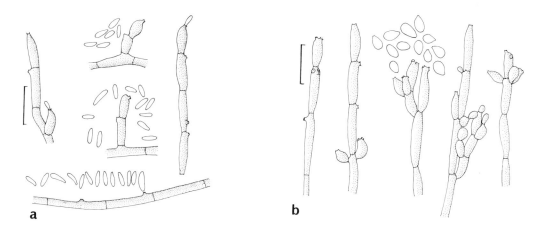

FIG. 124. a. *Exophiala* state of *Dictyotrichiella mansonii*, conidiophores and conidia, CBS 101.67; — b. *Exophiala dermatitidis*, conidiophores and conidia, CBS 207.35.

The occurrence of *E. jeanselmei* var. *jeanselmei* has been documented for France, the British Isles, the USA, Canada, the Netherlands, Sweden, Italy and India, from subcutaneous abscesses or mycetomas in man, coniferous and deciduous woods and polyvinylalcohol (2524, 4349); the var. *heteromorpha* in addition to the above areas is also known from Canada, Japan, Taiwan, China, Australia, Senegal and Argentina from similar substrates, with the addition of paper and paper pulp (6166), pine litter (2375), a stem of *Lunaria annua*, stream mud, a peatbog and soil (mostly recorded as *Rh. atrovirens*) (2524). Further reports under the epithet *mansonii* point to the cosmopolitan distribution of this variable species (505, 1164, 2573, 5390). *E. pisciphila* McGinnis & Ajello *1974* is also commonly soil-borne and may often have been confused with *E. jeanselmei* or other species. *E. "mansonii"* has also been reported from the rhizospheres of potatoes (1614) and wheat (2524), on decaying wood (2524, 5129), and *Populus tremula* (IMI). In potato fields, it showed a positive rhizosphere effect after a precrop of grass but not after wheat (1617).

Physiological investigations planned to differentiate between pathogenic and saprophytic isolates showed equally good growth of both on nitrate, nitrite and ammonium N, glucose, fructose, galactose, cellobiose, lactose, glycerol, starch, several other substrates and even plain agar (1837, 2524, 6170), while *E. "mansonii"* was found not to utilize ammonium salts or urea (2850). On ammonium as a N source, growth is yeast-like, while on peptone and yeast extract a dark pigment is exuded which is not produced on other media (1837, 1838). *E. jeanselmei* does not grow on hypoxanthine (3504). The pathogenic isolates grow at 37°C and poorly at 10°C, however, while saprophytic ones grow well at 10°C and not at 37°C (6170). The thermal death point of *E. "mansonii"* in soil has been determined as between 50°C and 60°C for 30 min (613). Fair growth occurs under anaerobic conditions, with requirements for thiamine and biotin (2850, 5680), and also on ethane, propane and *n*-butane as sole C sources (1298). Lignin-related compounds (*p*-hydroxybenzaldehyde, ferulic acid, syringaldehyde, vanillin) were shown to be utilized by the var. *heteromorpha* (2375). *E. "mansonii"* produces an extracellular polysaccharide composed of *N*-acetyl-D-glucosamine and -glucosamine-uronic acid residues (800).

Farrowia D. Hawksworth *1975*

Farrowia seminuda (L. Ames) D. Hawksworth *1975* — Fig. 125.

≡ *Chaetomium seminudum* L. Ames *1949*

DESCRIPTIONS: Skolko and Groves (5403), Ames (124), Hawksworth and Wells (2340), Hawksworth (2337), Seth (5216), Udagawa (5973), Minoura *et al.* (3840), and Matsushima (3680). — *Farrowia* has been described to replace *Chaetoceratostoma* Turconi & Maffei *1918* sensu auct., since the type species of that genus was found to belong to *Scopinella* Lév. *1847*. *Farrowia* is a member of the Chaetomiaceae Wint. and characterized by subglobose, brown to black ascomata raised on pedestal-like rhizoids, with lateral and terminal straight and unbranched hairs; terminal hairs arising synchronously from adjacent, elongated cells at the ascoma apex, forming a neck-like structure by fusion; hairs smooth. Asci unitunicate, clavate, short-stalked, 8-spored, rapidly deliquescing. Ascospores lemon-shaped, brown, smooth-walled, generally clearly uniguttulate, provided with one subapical germ pore. Aleurioconidia of *Botryotrichum* often abundant. Three species have been recognized (2337).

F. seminuda has fairly rapidly growing colonies, reaching $3 \cdot 0$–$3 \cdot 5$ cm diam in ten days at 20°C on OA, with little white flocculose aerial mycelium; reverse uncoloured or slightly orange. Homothallic. Ascomata obpyriform, mostly 150–180×75–100μm, yellowish brown, translucent when young. Lateral hairs mostly 70–$100 \times 2 \cdot 0$–$3 \cdot 5 \mu$m, pale brown to brown, smooth or slightly rugose near the base; terminal hairs rudimentary , pale brown to brown, 5–6μm wide and tapering to $2 \cdot 0$–$3 \cdot 5 \mu$m, forming a neck-like structure mostly 10–25μm, exceptionally to 75μm long. Asci clavate. Ascospores accumulating in a cirrhus at the ascoma apex, hyaline at first but becoming brown, subglobose to ellipsoidal, usually uniguttulate, biapiculate, $7 \cdot 5$–$9 \times 7 \cdot 0$–$8 \cdot 5 \times 4$–6μm. Aleurioconidia mostly abundant, formed on little-branched conidiophores, hyaline or slightly fuscous, smooth and thick-walled, ± globose, 7–10μm diam. — *F. seminuda* has the shortest terminal hairs in the genus *Farrowia*. *F. longicollea* (Krzem. & Badura *1954*) D. Hawksw. *1975* (=*Chaetoceratostoma longirostre* Farrow *1955*) has 40–70μm long hair tufts. *Chaetomium homopilatum* Omvik *1955* and some similar species have ornamented hairs and no fused terminal hairs (like other species of *Chaetomium*).

F. seminuda is a cosmopolitan, but rather uncommon species (2337, 5216), reported from Wisconsin (1040), Georgia (2240), Ohio (2573), and other southern states of the USA (2482), Canada, China (2337), Japan (3680, 5846), India (5375), Nepal (3840), New Zealand (5812), Brazil (6008), the Bahamas (2006), Trinidad (3831, 3832), Rhodesia (IMI), Angola, Israel (5171, 5172), Italy (270, 271), Poland (272, 273), France (2963) and Germany. Some of these reports may, however, be based on misdeterminations for some short-haired *Chaetomium* or *Achaetomiella* species. Known habitats include forest soils (270, 272, 273, 1040, 2482, 2963, 3680), grassland (5812), heavy clay under sugar cane (3831, 3832), brown soil (2963), leaf fragments in soil (IMI), mycorrhiza of *Suillus luteus* on

Pinus strobus (1794), roots of sugar cane (3832) and *Hevea* (IMI), rabbit dung (4439), seeds of tomato (5403) and groundnuts (IMI).

F. seminuda is able to degrade cellulose (2006). — In a chronically irradiated soil it did not survive intensities of >53 R per day (2009).

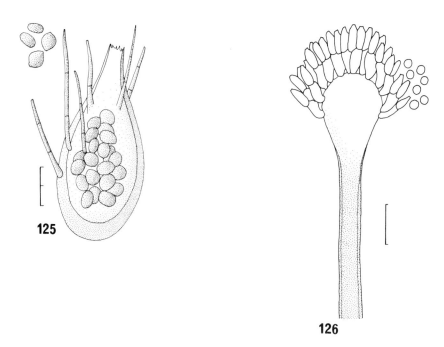

FIG. 125. *Farrowia seminuda*, perithecium and ascospores, CBS 549.64.
FIG. 126. *Fennellia flavipes*, conidial head of *Aspergillus flavipes*, CBS 587.65.

Fennellia Wiley & Simmons *1973*

Fennellia flavipes Wiley & Simmons *1973* — Fig. 126.

Anamorph: *Aspergillus flavipes* (Bain. & Sart.) Thom & Church *1926*
  ≡ *Sterigmatocystis flavipes* Bain. & Sart. *1911*

DESCRIPTIONS: Raper and Fennell (4744), Subramanian (5626), and Wiley and Simmons (6361). — Fam. Trichocomaceae Fischer. — Colonies growing rather slowly, reaching 3–5 cm diam in 10–14 days at 25°C on CzA and MEA, whitish from conidial masses, with brownish conidiophores shining through, reverse yellow-brown to red-brown. Some isolates form closely woven yellow masses of irregularly curved, thick-walled cells. Ascomata have been observed only in a few isolates from cellulosic material buried in Thailand forest soils (6360), and these were placed in the separate genus *Fennellia* (6361). Ascomata maturing inside the hyphal masses within three weeks, the wall consisting of several layers of yellow, thin-walled hyphae; asci mostly 4-spored; ascospores subglobose, with an inconspicuous equatorial groove, smooth-walled, hyaline to pale yellow, $6 \cdot 4$–$8 \cdot 0 \times 5 \cdot 6$–$6 \cdot 4$ μm. — For second species of *Fennellia* see under *Emericella nivea*. — The anamorph belongs to the *Aspergillus flavipes* group. Conidial heads sparse, loosely columnar, persistently white or becoming hazel. Conidiophores smooth-walled, pale yellow to light brown, $2 \cdot 4$–$3 \cdot 2$ μm diam, uninucleate (6514). Solitary or botryose aleurioconidia 4–6 μm diam occur (4598). — DNA analysis gave a GC content of 58% (5600).

F. *flavipes* has a worldwide distribution but with its main centres in tropical, subtropical and mediterranean zones, and is very well documented by numerous reports which cannot be listed in detail. In addition to its often dominant occurrence in cultivated and forest soils, it has been isolated from uncultivated soils (540, 2179, 4030, 4698), willow and cottonwood lowland (2007), soils with steppe type vegetation (1559, 1702, 4313, 4718, 5049), lateritic soils (399, 3417, 5000), sand dunes (4371), and loess (654, 4759). It has also been found in soils to an altitude of 3000 m (3850), in sewage (1165, 1166), fields treated with sewage (1165), a river bed (3868), brackish water (4477), salt marsh (4000), mangrove swamp (3264), an acid mine drainage stream and river water (1166), a bat cave (4312), carst caves (3365), and silt of an ice cave in the USSR (3367). The soil isolations of *F. flavipes* by far outnumber ones from other substrates. Other reported habitats include the rhizospheres of corn (4555), vegetable plants (3733), various sand dune plants (4371), and banana (2031, 2035), roots of strawberry (4133), rhizoids of *Anthoceros* sp. (2860), the ectomycorrhiza (with *Lactarius* sp.) of *Castanea sativa* (1794), truffle soils (3451), wheat seed (2097, 4492), cultures of the collembolan *Hypogastrura matura* (1027), and sheep dung (969). The fungus colonized hair baits exposed to soil in the range between vapour saturation to -140 bars water potential (977), and it can tolerate high soil temperatures to 50°C (5000, 5049); a neutral or slightly alkaline soil apparently favours its occurrence (654, 968, 1559, 3451, 4698, 5048, 5349, 6510). In a study of an Indian soil, *F. flavipes* was found to be a dominant species in the 15–30 cm layer (5349).

The formation of thick-walled sterile cells and aleurioconidia is favoured by darkness (4045). The cardinal temperatures for growth are given as: optimum 26–28°C or higher (1827), minimum 6–7°C or less and maximum 38–40°C. Faster growth was obtained at 40°C than at 25°C, but none occurred at 45°C (4598); the thermal death point has been determined as 63°C for 25 min in apple juice (3620). — Starch (1827), cellobiose, mannose and methyl-β-D-glucoside are suitable C sources (5387). It shows a moderate degradation of cellulose from wheat straw (4213), but vigorously attacks cotton cellulose (3618, 5388); optimal conditions for cellulose decomposition were found to be 29°C, pH 6, biotin 2 μg l^{-1} (5388), and Ca(CN)$_2$ as nitrogen source (6078). *F. flavipes* produces indole-3-acetic acid in liquid culture (294), i.a. from tryptamine (1522), utilizes *o*-, *m*- and *p*-xylene and various hydrocarbons from fuel oil (4239), transforms progesterone to testolactone (4744), can utilize the herbicide simazine as sole C source (2894) and degrade the herbicide atrazine (2892). Riboflavin (1620) and lipids are produced (1559). Other products include flavipin which has fungistatic properties (725, 4704), flavipucine (1741), 4,6-dihydroxy-2,3-dimethylbenzoic acid, orsellinic acid (3231) and a geodine-like metabolite (1336). — The growth of *Gaeumannomyces graminis* was inhibited by 80% by culture filtrates of *F. flavipes* (4744) and antimycotic activity was recorded against *Aspergillus fumigatus* and *Trichophyton mentagrophytes* (569). Culture filtrates were found to be toxic to the rice moth larvae *Corcyra cephalonica* (2732), and to show some antibiotic activity (6320); grain contaminated with it proved toxic to young ducklings (5158). Cerebral aspergilloma in mice appeared after its intravenous inoculation (4599). — *F. flavipes* is regarded as halophilic (4477), tolerating up to 40% sucrose (4001) or NaCl concentrations of 20% and above (1475, 5881), corresponding to $\geqslant -205$ bars water potential.

Fusarium Link ex Fr. *1821*

= *Fusisporium* Link *1809* ex Fr. *1821*
= *Selenosporium* Corda *1837*
= *Sporotrichella* Karst. *1887*

Type species: *F. sambucinum* Fuckel *1870* (= *Fusarium roseum* Link ex Gray *1821*, nomen ambiguum)

Teleomorphic genera: *Nectria* (Fr.) Fr., *Gibberella* Sacc., *Calonectria* de Not., *Plectosphaerella* Kleb.

The Hyphomycete genus *Fusarium* is characterized by usually fast growing, pale or bright-coloured colonies with a felty aerial mycelium and diffuse or sporodochial sporulation. The conidiophores are usually basitonously branched, if forming complex pustules termed sporodochia, if confluent pionnotes, or consisting in some species only of single phialides. The terminal branches are slender, slightly tapering phialides usually bearing one fertile opening (in some species several due to sympodial proliferation, i.e. "polyphialides"). Phialoconidia generally forming slimy masses, fusiform to sickle-shaped and 1- to many-septate, mostly differentiated into a ± beaked apical cell and a pedicellate basal cell. In addition to these macro-conidia, some species also produce one-celled, smaller, micro-conidia. Chlamydospores (terminal or intercalary) are characteristic of some species; but they are not visible in all isolates unless the cultures are very old or placed in sterile distilled water or grow on a poor agar medium. — Conidium ontogeny in *F. culmorum* has been studied by Marchant (3594) by TEM and SEM: the first formed conidium develops within a closed wall of the phialide tip which then bursts near the conidial base and sometimes remains visible as a cap on the conidium. Contrary to Marchant's interpretation, this development is regarded by us as typically phialidic.

Although perithecia are produced in culture in isolates of some species, the species discussed here are keyed out according to their anamorphs alone, as it is these which are most regularly encountered. Where the anamorph is well defined and no doubts persist about its connection with a particular teleomorph, the latter name has been used, while in others it is mentioned below the name of the anamorph.

The current taxonomy of *Fusarium* has been profoundly influenced by Wollenweber whose views are summarized in "Die Fusarien" (6424) where 142 species, varieties and forms are distributed amongst 16 sections. In view of the variability displayed by numerous isolates, some of these "species" were subsequently combined by Gordon (e.g. 2039), whose system was largely followed in the more recent monograph of Booth (640); Booth recognized 51 species and varieties in 12 sections. The school of Snyder and Hansen (5459, 5460, 5461) adopted a broader approach accepting only nine species but designated (incorrectly) some of the "species" recognized by other workers as "cultivars". Since information based on their broad species concept (in particular with *Fusarium "roseum"*) must mostly be regarded as not reproducible, we do not report it here. Joffe (2770) adopted a different system with 33 species and 14 varieties. Gerlach (1945, 1947, 1948) maintains a somewhat more differentiated system than Booth, and details of it have been worked out in the monographs of the sections *Sporotrichiella* by Seemüller (5174) and *Liseola* by Nirenberg (4194). A laboratory guide by Booth (641) was issued in *1977*, a survey of *Fusarium* species in the tropics (1947) and an illustrated atlas to the species recognized by Gerlach was published in *1981* (1948). For standard descriptions, we generally refer to these publications.

Most species of *Fusarium* are soil fungi with cosmopolitan distributions and are active in decomposing cellulosic plant substrates; some species are plant parasites, causing root and stem rot, vascular wilt, fruit rot or ear diseases. In some species, host-specific pathogenic strains are distinguished as *formae speciales* which otherwise do not differ from saprophytic strains of the same species. The available information on pathogenicity and *formae speciales* has not been included in this compilation. Pathogenicity to man is rare, but many species cause storage rots and are important toxin producers.

Most *Fusarium* species are best identified from cultures grown on OA or potato-sucrose agar (PSA) slants in daylight where sporulation can occur in less than one week or more than one month, depending on the species; sporulation can be promoted by "black light" illumination. In critical cases, cultures need to be grown on at least six different media (1941) so that the variability can be assessed and the most suitable medium selected for microscopic study. Conidial drawings and measurements in most publications are based on water mounts, as in lactic acid or lactophenol the conidia appear narrower.

The genus has been subdivided into 16 groups by Wollenweber and Reinking (6424) or 12 sections by Booth (640). Although some of the sections combine species with related teleomorphs, they do rather little to facilitate identification. We therefore refrain from keying them out.

Identification of species is difficult with the available dichotomous keys, because of the variability between isolates, and because not all features required are always well developed. To overcome this difficulty, a synoptic survey is presented here with a list of characters and the numbers of the species showing them. Another synoptic key to tropical *Fusarium* species has been constructed by Gerlach (1947). In a group of alternative characters, usually only the less common ones are mentioned. Numbers of species with a certain character not always present are indicated between parenthesis.

Synoptic key to the species treated:

Colonies slow growing (less than 3·0 cm diam in 4 days at 25°C on OA or PSA): 2, 7, (13), 14, 16, 25, (26)
Colonies moist, ± floccose or granulose: 5, 9, (12), 16, (17), 22
Colonies moist, smooth, slimy from pionnotes
    Colonies cream: 9, 12, 14, 22, 25
    Colonies greenish: 9, 12, 22
    Colonies orange: 2, 7, (14), (16)
Colonies floccose
    Colonies whitish: (7), (13), 16, (22), 27
    Colonies cream: (8), (13), (17), 19, (22), 24
    Colonies ochraceous: 8, (10), (13), (19), 21, 24, (27)

Colonies pink: 3, (15), (16), (17), 18, (20), (21), 23, 24
Colonies red: 1, (3), 4, 6, (10), 11, 20, 23, 26
Colonies vinaceous to violet: 5, 15, 17, (26)
Colonies with discrete sporodochia: 1, 3, (5), 6, (8), 9, (11), (12), 13, (15), 16, 17, 19, (20), (22), (23), (26)

Polyphialides generally present: (3), 4, (15), 21, 23
Monophialides short (less than 12 μm) and not in compact whorls: 16, 17, (18), 19
Phialides bearing micro-conidia very long (>30 μm): 5, 9, 12, 22

Micro-conidia abundant
  pear-shaped to lemon-shaped: (15), 21, 23, 26
  almost globose: 18, (23)
  club-shaped with truncate base: 15
  ellipsoidal or slightly curved: (2), 5, 9, 12, 17, 19, 22, (27)
  fusiform: 4

Macro-conidia with a non-pedicellate basal cell
  mostly two-celled, not exceeding 25 μm long: 7, (16), 25
  two-celled, exceeding 25 μm long, curved: 2
  more-celled: 14, (16)
Macro-conidia with a pedicellate basal cell
  foot-cell pronounced, apex beaked: 1, 3, 6, 8, 13, 15, 17, 18, 20, 23, 26
  foot-cell less pronounced: 5, 12, 19, 21, 22, 24
  central part straight, cylindrical: 5, 13, 19, 21
  whole conidium moderately curved: 3, 4, 6, 8, 9, 10, 11, 12, 15, 17, 18, 20, 21, 22, 23, 24, 26, 27
  whole conidium strongly curved: 1, (8)
Macro-conidia usually less than 4·5 μm diam: 1, 2, 3, 4, 7, (10), (12), 13, 15, 16, (17), 25, (26)
Macro-conidia often exceeding 6 μm diam: 6, 9
Macro-conidia usually less than 35 μm long: 2, (4), 7, (10), 16, (18), (19), (21), 25
Macro-conidia often exceeding 50 μm long: 3, (8), 9, 11, 12, (15)

Chlamydospores mostly terminal: 5, 9, 12, 17, 19, (20), 22, 27
Chlamydospores only intercalary, mostly in chains: 1, 4, (5), 6, 7, 8, 10, (11), (12), 13, (14), 20, (21), 23,
   24, 26
Chlamydospores absent: 2, 3, (11), 15, 16, 18, 25

Sclerotial plectenchymatous bodies abundant
  cream to ochraceous: (3), (8), 10, (13), 17, 19, 20, 26
  red: 10, (20), 23, 26
  often blue: 1, (5), 13, 15, 17, (22)

List of the species treated:

ANAMORPHS	TELEOMORPHS [species treated under their anamorph name placed in square brackets]
(1) *Fusarium acuminatum*	— *Gibberella acuminata*
(2) *F. aquaeductuum*	— [*Nectria purtonii*]
(3) *F. avenaceum*	— *Gibberella avenacea*
(4) *F. chlamydosporum*	
(5) (*F. coeruleum*)	

(6)	*F. culmorum*	
(7)	*F. dimerum*	
(8)	*F. equiseti*	— *Gibberella intricans*
(9)	*F. eumartii*	— *Nectria haematococca*
(10)	*F. flocciferum*	— [*?Gibberella heterochroma*]
(11)	*F. graminearum*	— *Gibberella zeae*
(12)	(*F. javanicum*)	— (*Nectria ipomoeae*)
(13)	*F. lateritium*	— *Gibberella baccata*
(14)	*F. merismoides*	
(15)	*F. moniliforme*	— [*Gibberella fujikuroi*]
(16)	*Gerlachia nivalis*	— *Monographella nivalis*
(17)	*F. oxysporum*	
(18)	*F. poae*	
(19)	*F. redolens*	
(20)	*F. sambucinum*	— *Gibberella pulicaris*
(21)	*F. semitectum*	
(22)	*F. solani*	— [*Nectria haematococca* var. *brevicona*]
(23)	*F. sporotrichioides*	
(24)	(*F. sulphureum*)	— (*Gibberella cyanogena*)
(25)	*F. tabacinum*	— *Plectosphaerella cucumerina*
(26)	*F. tricinctum*	
(27)	*F. ventricosum*	— *Nectria ventricosa*

Fusarium aquaeductuum (Radlk. & Rabenh.) Lagerh. *1891* — Fig. *127*.

≡ *Selenosporium aquaeductuum* Radlk. & Rabenh. *1863*

Teleomorph: *Nectria purtonii* (Grev.) Berk. *1860*

≡ *Sphaeria purtonii* Grev. *1828*
= *Nectria applanata* Fuckel *1871*
? = *Nectria episphaeria* (Tode ex Sprengel) Fr. *1849* var. *coronata* Wollenw. *1931*

DESCRIPTIONS: Wollenweber and Reinking (6424), Booth (640), Joffe (2770); teleomorph: Booth (636), and Samuels (5032). — Section *Eupionnotes* Wollenw. (*Episphaeria* sensu C. Booth). — Colonies slow growing, reaching 5 mm diam in 4 days at 25°C on PSA, covered with orange slime from the sporodochial or pionnotal conidia or partly covered with a white floccose mycelium; reverse yellow-brown due to diffusing pigment. Conidia slender sickle-shaped, moderately to strongly curved, mostly indistinctly 1-septate with the basal cell hardly pedicellate, 14–35 × 2–3(–3·5) μm. Chlamydospores absent. — *F. aquaeductuum* var. *aquaeductuum* is distinguished from the more common var. *medium* Wollenw. *1931*, the anamorph of *Nectria episphaeria* (Tode ex Sprengel) Fr. *1849*, which often has ellipsoidal to clavate micro-conidia and longer, 1–3-septate macro-conidia 15–55 × 2·5–3·5 μm. The acospores of both teleomorphs measure 8–11 × 3·5–4·5 μm, but the perithecia of *N. purtonii* have a thicker (26–55 μm), gelatinous wall. — The majority of the hyphal cells are uninucleate but up to five nuclei have been observed (2496).

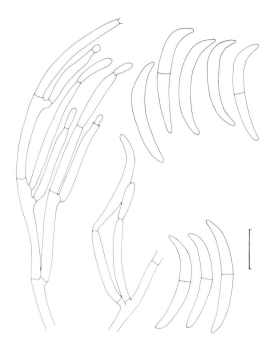

F<small>IG</small>. 127. *Fusarium aquaeductuum*, conidiophore, phialides and conidia, CBS 268.53.

F. aquaeductuum with its varieties is one of the most common fungi in all kinds of flowing water and water pipes (1946, 6424, 6425) but is also frequent in the bleeding sap of trees and on dead twigs; it is commonly associated with other fungi. The teleomorph is often found on sphaeriaceous fungi as is *N. epalsphaeria*. It has been observed in most cases on dead wood, in the British Isles, the Netherlands, Canada, New Zealand, Japan, Pakistan, India, Rhodesia and Uganda (IMI). In the USA (1213), the British Isles (2338) and the USSR (3809, 4548) in fresh water; in the USA it is also known in sewage sludge (1165, 1166, 1167), soils treated with sewage sludge (1163), and polluted water (1166), in Canada in the slime of a paper mill (IMI), in Norway in river water polluted with spent sulphite liquor (5548), in India in a mangrove swamp (4700), and in Ireland on cut peat (1376). Other habitats include sticks exposed to soil (1956), and the fungi *Diatrype disciformis* (4583), *Fomes* spp. and *Xylaria polymorpha* (IMI). — It was found to be strongly inhibited by *Streptomyces* species (2939).

The optimal growth temperature lies in the range 20–28°C (5548). Cellulose is degraded (1956), particularly well at a C/N ratio of 15:1 with ammonium sulphate as the N source (5675). It grows well on organic N sources, does not require B-vitamins, and is adapted to a wide pH range of 4–9 (4363). Several polyphenols are utilized (378). The following pigments have been identified: γ-carotene, lycopene, poly-*cis*-lycopene, neurosporaxanthin, and rhodoxanthin (2087, 4751). The synthesis of carotenoids was found to be light-induced (3225) at wavelengths <520 nm (4750) or induced in the presence of mercuribenzoate in the dark (5790).

Fusarium chlamydosporum Wollenw. & Reink. *1925* — Fig. 128.

≡ *Fusarium sporotrichioides* Sherb. var. *chlamydosporum* (Wollenw. & Reink.) Joffe *1974*
= *Fusarium fusarioides* (Frag. & Cif.) C. Booth *1971*
  ≡ *Dactylium fusarioides* Frag. & Cif. *1927*
= *Fusarium tricinctum* (Corda) Sacc. emend. Sny. & Hans. *1945* pro parte

DESCRIPTIONS: Wollenweber and Reinking (6424), Booth (640), Seemüller (5174), and Joffe (2770). — Section *Sporotrichiella* Wollenw. (*Arthrosporiella* sensu C. Booth). — The identity of *F. chlamydosporum* is well established by Wollenweber's descriptions and type material preserved at Berlin and there is therefore no reason to reject this name as *nomen confusum* (640). — Colonies rather fast growing, reaching 4·5–5·7 cm diam in four days at *c.* 20°C on MEA or OA; aerial mycelium abundantly developed, intensely pink, red or vinaceous, rarely whitish, ochraceous or brownish; agar surface intensely red-vinaceous or brown-red. Sporodochia producing macro-conidia on normal phialides rarely occur under "black light"; sporodochia orange, flesh or ochraceous. Conidiophores scattered over the aerial mycelium, richly branched; phialides with numerous sympodial proliferations (denticles) bearing one micro-conidium on each opening (i.e. strictly polyblastic); micro-conidia accumulating in dry heads, fusiform or elongate, 8·5–10(–15) × 2·5–3·0(–3·7) μm; macro-conidia rarely produced and appearing only in sporodochia, 3(–5) septate, slightly curved, 30–38 × 3·0–4·5 μm. Chlamydospores numerous, intercalary, often roughened, 7–17 μm diam; smaller hyphal swellings also occur. Sclerotial bodies generally absent. — The hyphal cells mostly have one nucleus, but up to five have been observed (2496).

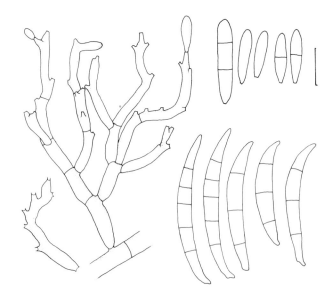

FIG. 128. *Fusarium chlamydosporum*, conidiophores with numerous sympodial proliferations, micro-conidia and macro-conidia, CBS 698.74.

This fungus has not been isolated often in soil analyses and can be regarded as a relatively rare soil inhabitant (2045). It is known from soils in Turkey (4183), Iraq (92), Nepal (1826),

Namibia (1827) and India where it occurs in forest soils (1525, 2854, 4716, 5512), cultivated soils (1519, 3868, 6059), soils with widely different water regimes (1315), grassland (3865, 4933), and associated with sorghum ear-moulds (5314). It is also known from cultivated soils in Australia (945) and Pakistan (61), was frequently encountered in Columbia (4795) and has been found in the Sonoran desert of the southern USA (4733), in estuaries in Yugoslavia (4850), and in a creek in Ohio/USA (1166). It can occur in the rhizospheres of coriander (1523), *Cassia tora* (521, 6067) and various sand dune plants (4371), on roots (945) and culms of wheat (2043), on *Arundo donax* (CBS), birch wood exposed to soil (5061), seeds of soya, flax (6407), rice (4929), *Sorghum vulgare* (57) and *Pennisetum typhoides* (3670), and pecan nuts (2572). — Its growth was inhibited by *Memnoniella echinata in vitro* (4).

The minimum temperature for growth is 5°C, the optimum 27·5°C, and the maximum 37·5°C (5174). Compared with eight other *Fusarium* species, *F. chlamydosporum* gave the highest yield on 0·1% glycine as N source and a moderate yield on 3% raffinose as C source (2623). It can cause wilting of bean seedlings *in vitro* (2775).

Fusarium culmorum (W. G. Sm.) Sacc. *1895* — Fig. 129.

≡ *Fusisporium culmorum* W. G. Sm. *1884*
= *Fusarium culmorum* var. *cerealis* (Cooke) Wollenw. *1931*
 ≡ *Fusarium cerealis* Cooke *1878*
= *Fusarium roseum* Link ex Gray emend. Sny. & Hans. *1945* pro parte

DESCRIPTIONS: Wollenweber and Reinking (6424), Gordon (2039), Gerlach and Ershad (1949), Booth (640), Joffe (2770), and CMI Descriptions No. 26, *1964*; the macroscopic features are described by Colhoun and Park (1119). — Section *Fusarium* (= *Discolor* group). — Colonies very fast growing, reaching 8·5 cm diam in four days at 25°C on PSA or OA, intensely red to red-brown or purple at the agar surface, with whitish, ochraceous, to red aerial mycelium which is difficult to wet (unlike *Gibberella zeae*, q.v. (1119)). Sporodochia abundantly produced in 1–2 weeks, often confluent to pionnotes, with red-brown conidial slime. Conidiophores abundantly branched with short and wide phialides. Conidia rather broad, falcate, with a short subterminally slightly constricted and almost pointed apical cell, and a distinctly pedicellate basal cell, rather thick-walled and slightly pigmented, mostly 5-septate and 30–50 × 5·0–7·5 μm, cells mostly slightly longer than wide. Chlamydospores not very common in the hyphae, mostly intercalary, solitary or in chains, globose to oval, brownish, 9–14 μm diam; chlamydospores also often formed from conidial cells. The ability to produce chlamydospores is probably the reason that this species is isolated in greater quantities from washed soil particles than many other fungi, particularly from organic particles (1888, 1889). — Investigations have been made on the development of fine structures during conidium formation (3594), the conidia themselves (17, 3201, 3591, 3592), the development of endoconidial chlamydospores (870), the distribution of nuclei in the conidia and hyphal cells (514, 943, 2496), and the migration of nuclei from the conidial cell during germination (4668). The nuclei contain 8 chromosomes (641). — The main component of conidial walls is chitin, while an outer mucilaginous layer consists of xylan (3591). In hydrolysed conidial walls

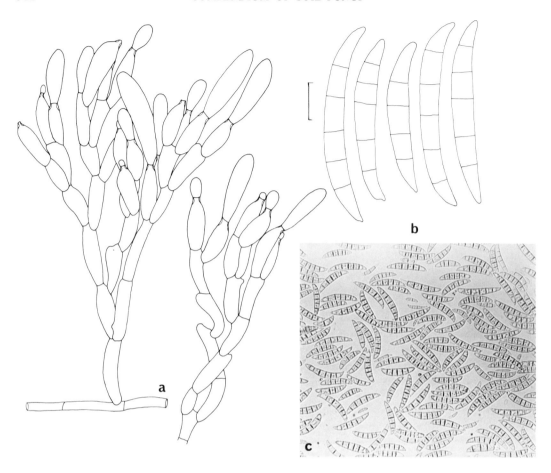

FIG. 129. *Fusarium culmorum*. a. Densely branching conidiophores of beginning sporodochium formation; b. conidia; CBS 251.52, orig. J. Veerkamp; c. conidia, × 500.

the following sugars were found: mainly galactose, glucose and mannose (3201), with lesser amounts of arabinose and xylose, in addition to the amino acids glutamic acid, glycine, alanine, histidine, proline, phenylalanine, valine and leucine (2049, 3201, 3591). Lipids account for 8·6% of the mycelial dry weight; neutral lipids are represented by mono-, di- and triglycerides, free fatty acids (C_{16}-saturated and C_{18}-unsaturated acids), free sterols, sterol esters and a number of phospholipids (3201, 4205).

Having an extraordinarily wide range of host plants, in particular seeds, *F. culmorum* is a common cosmopolitan species although with some slight preference for temperate latitudes (1949, 2045). In the tropics, it seems to be less frequent in comparison with other *Fusarium* species. — It causes pre-emergence seedling blight, root and foot rot and head blight in cereals, brown patch of turf, foot rot of asparagus, carnation, leek and pea, and storage rot of apple, potato, sugar beet and others (CMI Descriptions No. 26). Selective isolation from soil-exposed kaolin pellets containing vanillic acid has been reported (2801). A fluorescent antibody technique for direct observation in soil micro-habitats was described (2539, 3595). — It is rarely reported from forest (1713, 5363), forest steppe (3362) or forest nursery soils (1978, 6184) but, on the other hand, frequently from grassland (159, 163, 165, 1555, 3131, 3790, 4948, 5813, 5814) and especially arable soils, including a particularly thorough study of its behaviour under Canadian conditions (2041). It has also been found in garden soil (4451, 6289), dunes (3941, 4162, 4655), salt marshes (1379, 4645), peat (1375, 1376, 1438, 4474, 5559), polder soils (4658), fresh water (1166, 4429), seawater in the eulittoral zone (4918), and in a guano cave in the Bahamas (4312). It was isolated with a high frequency from washed organic soil particles (1614). It occurs mainly in the uppermost soil layers (159, 6184) and apparently more frequently in neutral or slightly alkaline soils (159, 745, 5031, 5559). It is sensitive to a soil treatment with formaldehyde (1655), but increased initially up to 31 days after soil treatments with the fungicides captan, dicloran, milcol and triarimol (6136), and effectively colonized plant material with high residues of the herbicide diquat (6367). It proved destructive to wheat in sandy, dry soils in the USA, but was rarely so in moist soils with a high organic content (5865). In one study, *F. culmorum* occurred with significantly greater frequencies after growing wheat than after peas or rape (1433). This fungus can hardly be eliminated effectively by crop rotation measures; cultivation of oats seems to favour its survival (3261). Only a limited number of examples will be given here of the numerous reports of its isolation from various substrates: amongst two flax cultivars, the fungus occurred less frequently on the variety resistant to *F. oxysporum* f.sp. *lini* (5826); a clear rhizosphere effect was shown in beans (4451, 4452), and it is further reported from the rhizospheres of *Lolium perenne* (5815, 6132, 6133), wheat (1614, 1752, 2538, 3567), various grasses (1614, 3131, 3376) and groundnuts (2768); on oats (986, 3002, 3006, 4443) the frequency increased in the rhizosphere with the age of the plant (4446). It has also been found on live and dead roots of barley (3002, 4450, 4451), roots and culm bases of rye and wheat (945, 1119, 3567, 4136, 5330), red clover (6493), and corn (4553, 4556). *F. culmorum* is commonly isolated from seeds of oats (3556), barley (1753, 2820, 6407), wheat (2038, 2039, 4307, 5293), other cereals (2039), *Avena fatua* (2961), cultivated grasses (3512) and other plants (2040, 6407). Wheat grains occasionally show a higher infection at the time of harvest than after storage (4492). Maximum incidence in soil in Wales occurred in autumn and late spring; it is sensitive to warm and dry as well as cold and wet soils (2115, 2116). Although it is one of the first colonizers of wheat straw, it can later be supplanted by *Penicillium* species (953, 3423, 4966); according to some other reports (753, 3213), however, straw colonized by *F. culmorum* is only slightly colonized by other organisms. — Survival in straw is favoured by a high nutrient content (822, 1909) and dead straw under some conditions is less intensely colonized than are live culms (1137). It can survive a storage in water for ≥ 7 years (605). — In general *F. culmorum* can be ranked as a soil inhabitant with a high competitive saprophytic ability (820, 821, 1431, 3423, 4734). An antagonistic influence by *Streptomyces albidoflavus* is demonstrable under conditions of nutrition and sorption that favour both growth and production of the antagonist (5399, 5400, 5401); hyphae can also be lysed by *Streptomyces* species (2800). The use of inhibiting organisms for soil decontamination has been attempted in artificial cultures, sterile soils, and composted oat straw (998, 2115, 3213, 5960). Root exudates of onion inhibit the growth of *F. culmorum in vitro* (1072). — The fungus can serve as food for collembola and woodlice (4007).

The optimal temperature for growth is 25°C, the maximum 31°C, and the minimum 0°C (191, CMI Descriptions No. 26). At high densities (10^5/ml), conidia show self-inhibition which can be overcome by sucrose or wheat root exudates (4301). The formation of chlamydospores *in vitro* was enhanced by high nitrogen and sucrose concentrations (870). During germination, glucosamine is the main component of glucose metabolism (3596). Optimal growth on agar occurs at pH values between 4·8 and 8·0 (1375, 4914, 6532); the minimum water potential is at −130 bars (2093); colonization of hair baits occurred at water potentials between −40 and vapour saturation (2095). — In contrast to many other fungi, L-arabinose is utilized well (3328) and mannitol and lactic acid can also be used as C sources (4914). Good decomposition of pectin and cellulose in various forms is evidenced by numerous reports; the degradation of arabinoxylan takes place more via xylanase than by cellulase activity (1753); the optimum temperature for cellulose decomposition is 25°C, and increasing activity also occurs with rising O_2 partial pressure (6163). Nitrates are better N sources for cellulose decomposition than are ammonium salts, while sucrose and glucose (2%) inhibit cellulase activity (6062). Gluten proved a good N source (4914) and L-leucine is utilized as a sole N source without accumulation of isovaleric acid (1121); with L-tryptophan as N source, indole-3-lactic acid appears as a metabolic product (1122). Utilization of sodium glutamate is better than potassium or ammonium nitrate and ammonium sulphate; N uptake decreases with the age of the mycelium (5010). Lignin sulphonate is utilized relatively well (1425, 3260). Effective degradation of the herbicide simazine *in vitro* is also reported (5616). *F. culmorum* is osmotolerant (3934), autotrophic for growth substances (1425), and can grow in an atmosphere with 3–7% CO_2 quite normally (3440); good development is also possible under reduced O_2 partial pressure (1425). It was found to produce extracellular phosphatases weakly (498) and also L-asparaginase (4094). A morphogenic substance causing vacuolization in the hyphae of other fungi was found in ageing cultures (4435) and a staled culture filtrate inhibited the growth of *Talaromyces flavus* (1526). The formation of droplets during hyphal exudation was found to be closely related to colony ageing (3724). Humic substances are formed after autolysis of the hyphae (4470). Other metabolic products that have been isolated and described include the pigments fuscofusarin (3231), rubrofusarin, aurofusarin and culmorin (216, 3951, 5277), the sesquiterpenoids cyclonerodiol and cyclonerotriol (1668, 2261), and a peptide-like substance culmomarasmin (3015) acting as wilt toxin (3224). — In spite of its high potential pathogenicity for a wide range of hosts, *F. culmorum* has also been isolated very frequently from healthy plants; in some cases no loss in yield has been found even at high isolation frequencies (5031). There are numerous studies showing injuries in wheat, particularly at high inoculum densities, high temperatures (1118) and low soil humidities (453, 1119, 1430, 4954, 4966, 6573). Seed germination in wild oats is reduced by this fungus when soil humidity is low, but this effect diminishes with increased temperature (2962). — It showed antagonistic activity against *Chalara elegans* (2753). Extracts of the mycelium cause lethal skin lesions in rabbits (1747), and with guinea pigs as test animals the production of an emetic material was demonstrated (4630).

Fusarium dimerum Penzig *1882* — Fig. 130.

≡ *Fusarium aquaeductuum* (Radlk. & Rabenh.) Lagerh. var. *dimerum* (Penzig) Raillo *1950*
= *Fusarium episphaeria* (Tode) Sny. & Hans. *1945* pro parte (nom. illeg., Art. 59)

DESCRIPTIONS: Wollenweber and Reinking (6424), Gordon (2039), Booth (640), and Joffe (2770). — Section *Eupionnotes* Wollenw. (*Episphaeria* sensu C. Booth). — Colonies slow-growing, reaching 2·7 cm diam in four days at 25°C on PSA or OA, surface usually orange to deep apricot from conidial slime; aerial mycelium sometimes floccose, whitish. Conidiophores loosely branched, with short, swollen phialides, 10–18 × 4–5 μm. Conidia falcate-lunate, mostly 1–(3)-septate in or somewhat above the centre, with the basal cell hardly pedicellate, 15–25 × 2·5–4·0 μm. Chlamydospores intercalary, single or in chains, hyaline, smooth-walled, 8–12 μm diam. — Several varieties distinguished by Wollenweber and Reinking (6424) are no longer regarded as sufficiently distinct (except var. *violaceum* Wollenw. *1931* with smaller, often non-septate conidia and dark violet sclerotial stromata). *F. dimerum* may resemble *Monographella nivalis* (q.v.) which, however, sporulates only under light and has shorter phialides and usually somewhat wider (3–5 μm) and often 3-septate conidia and no chlamydospores. *Plectosphaerella cucumerina* (Lindf.) W. Gams (q.v.) has less pigmented colonies and much straighter, shorter conidia. — The majority of hyphal cells contain one nucleus, but up to five have been observed (2496).

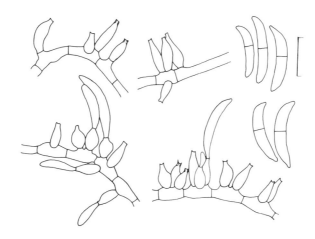

Fig. 130. *Fusarium dimerum*, densely aggregated phialides and conidia, CBS 187.79.

Fusarium dimerum is a typical soil fungus. It has been reported from Europe, Africa, America, Asia and Australia but, compared with some other *Fusarium* species, less frequently (2045); in Central and South America it is apparently more common (4795, 4796). It has been found in very different soils: uncultivated (4698, 4855), forest (3549, 4180), grassland (4180, 4948), arable soils (1048, 1614, 1616, 1966, 2039, 2712, 2768, 4179, 6059), light citrus soils (2764), fen (3413), rendzina (3414, 3418), limestone, sandstone and marl soils (3127). It does not withstand high salt concentrations in soil (92) but has been found in a mangrove swamp (3264) and a sewage treatment plant (1165). It was isolated with fluctuating but generally low frequencies from washed soil particles in two wheat-field soils (1889) and was reported as common in garden compost (1424). A very long list of plant substrates colonized by *F. dimerum* has been compiled (6425); it appears to be rare on wheat seeds (2039), and has been isolated from cotton seeds (5009), the rhizospheres of groundnuts (2768), rice (2712) and sand-dune plants (4371). — It has been isolated from a case of keratomycosis (6540).

There is evidence of pectin decomposition (3127, 3414), utilization of xylan with differences in capacity between individual isolates (1432), and cellulose decomposition at pH 6·5 as tested on filter paper and CMC (1432, 6060). Phenol lignin supports the growth of *F. dimerum* but none occurs on calcium lignin sulphonate (1744).

Fusarium flocciferum Corda *1828* — Fig. 131.

Teleomorph: ? *Gibberella heterochroma* Wollenw. *1917*

DESCRIPTIONS: Wollenweber and Reinking (6424), Gerlach and Ershad (1949), and Booth (640). — Section *Fusarium* (= *Discolor* group). — Colonies fast-growing, reaching 4 cm diam in four days at 25°C, grey-ochraceous to brown, aerial mycelium felty and with carmine patches. Scarcely sporulating, sporodochia rarely produced. Conidiophores loosely branched, each branch bearing 2–3 phialides, 16–22 × 2–5 μm. Conidia slender fusiform, slightly curved, with pointed apical and slightly pedicellate basal cells, 3–5-septate and 22–35(–48) × 3·5–5·0 μm. Chlamydospores abundantly produced, mostly in intercalary chains or clusters, commonly red-brown, cells 8–11 μm diam. — This species is close to *Gibberella zeae* (Schw.) Petch (q.v.), which often sporulates equally poorly, but has a stronger red pigmentation of the colonies, more commonly 5-septate conidia measuring 41–60 × 4·5–5·5 μm, and usually no chlamydospores. The mycelium of *F. flocciferum*, in contrast to that of *Gibberella zeae* (1119), is not easily wetted with water.

F. flocciferum occurs mainly on fruits, stems, twigs, roots and bulbs of various plants in Europe and North America (6424). It has been isolated from washed soil particles from two different agricultural soils in Germany (1889); in this case there was a regular and significantly

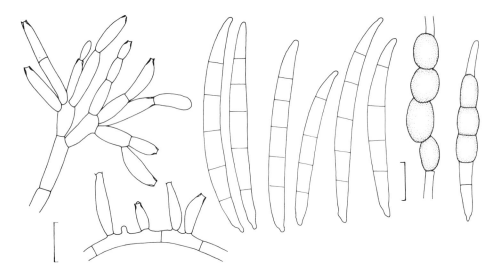

FIG. 131. *Fusarium flocciferum*, conidiophores, conidia, and intercalary chlamydospores formed in a hypha and a conidium, CBS 792.70.

more frequent occurrence after growing wheat than after peas and rape (1433). Apart from this, and the reports of Wollenweber and Reinking (6425), there are very few data on soil isolates and distribution available; these include garden soil (6289), soils in the Netherlands (1614, 1616), Poland (1423) and Turkey (4245). *F. flocciferum* is known to cause disorders in beds of cultivated mushrooms and to be the agent of black rot in potatoes in the northwestern states of the USA (640). It has also been isolated from dead corn stalks in Iran (1949).

Xylan is rapidly decomposed by *F. flocciferum* with little variation between different isolates, and cellulose decomposition has been assessed on filter paper (1432). *F. flocciferum* showed antagonistic activity towards *Gaeumannomyces graminis* and *Rhizoctonia solani in vitro* (1431).

Fusarium merismoides Corda *1838* var. *merismoides* — Fig. 132.

= *Fusarium episphaeria* (Tode) Fr. emend. Sny. & Hans. *1945* pro parte (nom. illeg., Art. 59)

DESCRIPTIONS: Wollenweber and Reinking (6424), Gordon (2039), Booth (640), and Joffe (2770). — Section *Eupionnotes* Wollenw. (*Episphaeria* C. Booth). — Colonies slow-growing, reaching 0·9 cm diam in four days at 25°C, mostly without any distinct aerial mycelium,

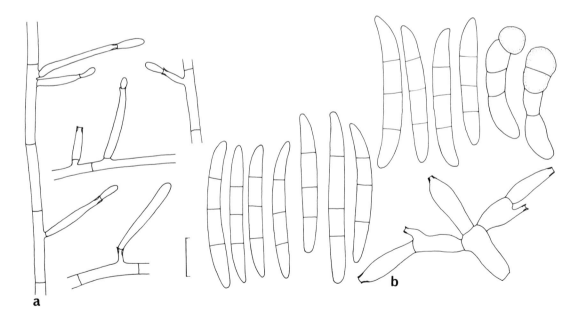

FIG. 132. *Fusarium merismoides*. a. Phialides and conidia of var. *merismoides*, CBS 186.34; b. conidiophore, conidia and conidia with chlamydospores of var. *chlamydosporale*, CBS 179.31.

quickly becoming slimy from pionnotal conidia, cream to peach or orange. Conidiophores consisting of simple lateral phialides or a basal cell with a cluster of phialides; phialides 12–20 × 3–5 μm, sometimes percurrently or sympodially proliferating. Macro-conidia fusiform, little curved, with rounded apical and hardly differentiated basal cells, 3(–4)-septate, 30–45 × 3·5–5·0 μm. Chlamydospores sometimes appearing in old cultures, mostly intercalary and forming chains, hyaline, smooth-walled, 8–12 μm diam. — Two varieties distinguished by Wollenweber have been synonymized by various authors but are kept separate by Gerlach (1945, 1948). The var. *crassum* Wollenw. *1931,* which is known only from water (1946), may be distinguished by the larger (3-septate mostly 33–60 × 4–6 μm), blunt and sometimes also 5-septate conidia; the var. *chlamydosporale* Wollenw. *1931* also has larger (3-septate 27–53 × 3·3–4·8 μm) and slightly more uncinate conidia and more abundant chlamydospores; var. *acetilereum* Tubaki *et al. 1976* has vinaceous colonies and almost symmetrically tapering and curved conidia, while in var. *violaceum* Gerlach *1978* they taper very gradually and are less curved in the basal half than in the upper part and the colonies are more violet.

Unlike most other *Fusarium* species, *F. merismoides* is primarily a soil fungus and there are no reliable reports of its association with any plant disease. — *F. merismoides* is a cosmopolite and occurs in soils of temperate and tropical regions (2045), particularly in slightly acid to alkaline soils (3127). It is easily isolated from dilution plates and by soil washing. *F. merismoides* is clearly less frequent in warmer regions, but there are reports from Israel (2762, 2764), Syria (277, 5392), Iraq (92), the northern Sahara (2974), equatorial West Africa (1420) and India (4030). It has only rarely been isolated from forest soils (1713, 2004, 3127, 3549) but more frequently from grassland (277, 1702) and arable soils (277, 1614, 1616, 2786, 2871, 2948, 3982, 4180, 4556), possibly with an increased incidence after alfalfa (1330). In Canadian soils it formed 1% of all the *Fusarium* isolates (2041). It also occurs in sludge (640), estuarine silt (5316), polluted and fresh water (3942), and water pipes (1946). *F. merismoides* has frequently been shown to be present in rabbit dung (2279), children's sandpits (1421) and garden compost (1424). It has been isolated with high frequencies from washed particles of two arable soils (1889). Further reports include the phyllosphere of vascular plants (296), pine roots (with lower frequencies than in surrounding soil) (5776), accumulations in the rhizosphere of red clover (3429) and on roots of alfalfa (3982, 3983), beans (4452), pears (5499), corn (4556), wheat (3567), *Stipa lessingiana* and *Festuca sulcata* (3376). It was rarely found on pecans (2572) and in wheat grains and is not generally regarded as a contaminant of seeds (2041).

Sporulation under submerged conditions occurs on sucrose-containing media (6088). *F. merismoides* decomposes pectin, although less intensely than many other fungi (1432), and cellulose decomposition is slight or absent (1425, 4184). Good growth (in decreasing order) occurs on glycerol, glucose, mannitol, ethanol, 2-butyne-1,4-diol, 1,2-ethanediol, 1,3-propanediol, 1,3- and 1,4-butanediol, also on 1,6-hexanediol, 3-butyne-1-ol, methanol (3895), and on calcium lignin sulphonate (1744). Organic acids are used in the following preferential order: fumarate, acrylate, acetate, lactate, propionate, succinate and β-hydroxypropionate (3895). From 2-butyne-1,4-diol the following metabolites are derived: acetylene dicarboxylic acid and its esters with 2-butyne-1,4-diol (3896), and 2,4,6-triketosuberic and 2,4,6,8-tetraketosebacic acids (3897); 2-alkyne-1-ol dehydrogenase formation is induced (3898); this feature is a peculiarity of var. *acetilereum* Tubaki *et al.* Antagonistic activity is shown towards *Rhizoctonia solani, Phytophthora cactorum, Fusarium culmorum* and a *Pythium* species (1425).

Fusarium moniliforme Sheld. *1904* — Fig. 133.

Teleomorphs: *Gibberella fujikuroi* (Saw.) Wollenw. *1931*

≡ *Lisea fujikuroi* Saw. *1917*

Gibberella moniliformis Winel. *1924*

DESCRIPTIONS: Wollenweber and Reinking (6424), Booth (640), and Joffe (2770), but most detailed: Nirenberg (4194).

The section *Liseola* Wollenw. *et al.* is characterized by densely branched conidiophores, abundant straight, clavate to ellipsoidal micro-conidia which often cohere in long chains, and often sympodially proliferating phialides (polyphialides), slender falcate macro-conidia, and absence of chlamydospores. Colonies are pale flesh to vinaceous shades. — The three species and several varieties placed by Wollenweber and Reinking (6424) in sect. *Liseola* have in recent years generally been regarded as representing a single species with one teleomorph, *Gibberella fujikuroi* (640, CMI Descriptions No. 22, *1964*). — In the monograph by Nirenberg (4194) the shape of the micro-conidia and the sympodial proliferation of the phialides were used in addition to the macro-conidial characters to distinguish eight species and three varieties; these taxa are ecologically different and three are connected with different teleomorphs in two species and one variety: *F. verticillioides* (Sacc.) Nirenberg *1976* with *G. moniliformis* Winel. *1924*; *F. fujikuroi* Nirenb. *1976* with *Gibberella fujikuroi* (Saw.) Wollenw. *1931* var. *fujikuroi*; and *F. sacchari* var. *subglutinans* (Wollenw. & Reink.) Nirenb. *1976* with *G. fujikuroi* (Saw.) Wollenw. var. *subglutinans* Edwards *1933*. — *F. verticillioides* (Sacc. *1881*) Nirenb. *1976* is an earlier epithet for *F. moniliforme* sensu stricto and its use will prevent confusion with *F. moniliforme* sensu lato which we maintain, however, for the present compilation.

The most frequent of these taxa are characterized and distinguished as follows:

1   Phialides with a single opening; micro-conidia clavate, mostly $7–10 \times 2\cdot5–3\cdot2\,\mu$m, forming long chains; macro-conidia 3–5-septate, $31–58 \times 2\cdot7–3\cdot6\,\mu$m   *F. verticillioides*

     Phialides proliferating sympodially   **2**

2(1)   Micro-conidia clavate or pyriform (with straight to concave outline)   **3**

     Micro-conidia narrowly ovoid (with convex outline)   **4**

3(2)   Micro-conidia forming long chains on all media, $7–9(–11) \times 2\cdot2–3\cdot2\ (–7\cdot5)\,\mu$m; macro-conidia rarely produced, almost straight, $27–58 \times 3\cdot3–4\cdot5\,\mu$m   *F. proliferatum*

     Micro-conidia usually forming slimy heads or rarely short chains, $8–10 \times 2\cdot5–3\cdot3\,\mu$m; macro-conidia commonly produced, 3–5-septate, $33–60 \times 3\cdot0–4\cdot0\,\mu$m   *F. fujikuroi*

4(1)   Macro-conidia rarely and sporodochia hardly produced; micro-conidia usually one-celled, $7–11 \times 2\cdot5–3\cdot0\,\mu$m   *F. sacchari* var. *sacchari*

     Macro-conidia easily produced in sporodochia, 3(–5)-septate, $27–54 \times 3\cdot5–4\cdot5\,\mu$m; conidia in the aerial mycelium often 1–3-septate   *F. sacchari* var. *subglutinans*

The subsequent compilation of information concerning *F. moniliforme* could not be ascribed to any of these species with certainty, but may mostly apply to *F. verticillioides*. — The majority of conidia and hyphal cells are uninucleate, but up to 5 nuclei have been observed (514, 2496), the nuclei contain 7 chromosomes (641, 4669). — The mycelium of an isolate has been found to contain virus particles (674). — Hyphal wall hydrolysates were found to contain glucose, galactose and mannose (3640). Alanine, lysine, γ-aminobutyric, glutamic and

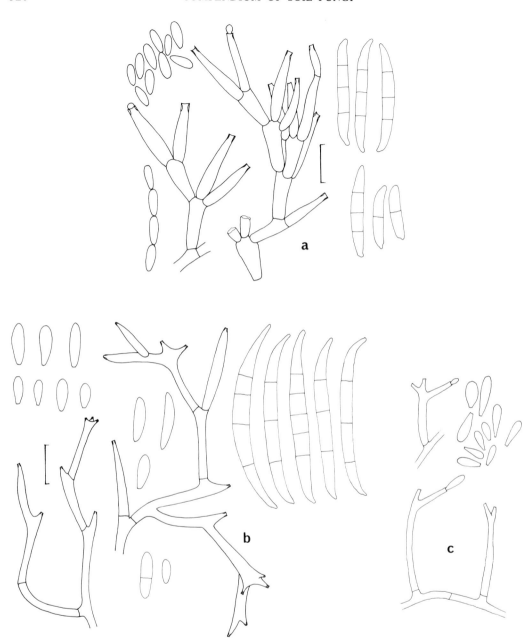

FIG. 133. *Fusarium "moniliforme"*. — a. *F. verticillioides*, conidiophores, catenulate micro-conidia and macro-conidia, CBS 218.76; b. *F. sacchari* var. *subglutinans*, sympodially proliferating conidiophores, clavate-ellipsoidal micro-condidia and macro-conidia, CBS 215.76, orig. J. Veerkamp; c. *F. proliferatum*, sympodially proliferating conidiophores and clavate micro-conidia, CBS 480.77.

aspartic acids are the main components of mycelial protein (518, 5897). — Isozyme patterns of malate dehydrogenase, peroxidase, acid phosphatase, glucokinase and acetylesterase were found to be distinctly different from those of other Fusaria (4767). — DNA analysis gave a GC content of 50–51% (5600).

F. verticillioides is common on sugar cane, corn (57, 3645, 4194, 4796), rice, banana, asparagus, cotton, etc., where it causes a seedling and stem rot; it is polyphagous and usually weakly virulent. It causes curly top ("pokkah boeng") in sugar cane and cob rot in corn. *F. proliferatum* is common on various plants in the tropical and temperate zones (particularly in greenhouses) and may play a role in the foot rot of rice and fruit rot in apples and pears. *F. fujikuroi* is the causal agent of the Bakanae disease on rice, and also occurs rarely on sugar cane; it is the most active producer of gibberellin and occurs mainly in eastern Asia and Australia. *F. sacchari* var. *sacchari* specifically occurs on sugar cane in which it causes red stem rot and wilt; probably also on banana. *F. sacchari* var. *subglutinans* is common on various host plants, particularly corn, millet, and rice; this variety is polyphagous and weakly virulent (4194). All other information compiled here is simply for "*F. moniliforme*".

A selective isolation technique based on culture media with malachite green, PCNB and dicrysticin has been described (5235). — It has been isolated from soils in Central America (2042, 4795, 4796), India (1315, 1317, 1519, 2854, 2861, 4698), Jamaica (4886), Japan (5846), Hong Kong (1021), Peru (2005), South Africa (1555, 4407), Israel (2762, 2764, 2776), Egypt (8), Libya (4085), Italy (4538), Czechoslovakia (1702), the USSR (3364, 4365, 4474, 4556), the USA (2007, 2573, 6253), Canada (2041), and rarely from Iceland (3089). It has frequently been isolated from various deciduous forest soils (3127, 3362, 3817, 4537, 6253), in willow and poplar lowland (2004, 2007, 3549), savannah (4407), grassland soils (1555, 2573, 4538), arable soils (1317, 1632, 2041, 2776, 2861, 3817, 3868, 4886), paddy fields (3913), a vineyard with copper accumulation (4921), and citrus soils (2764, 4085), and also reported from the Atlantic coast of France (4178), peat fields (4474), fen (4365), caves (3364, 4137), a uranium mine (1703), growing stalactites of a cave in Indiana/USA (2315), alkaline Usar soils in India (4698), and salt marshes (8). — It has regularly been isolated from roots of banana (2030), sugar cane (2861), sugar beet (4559), broad beans (6511), alfalfa and red clover (909), rice stems (2679), and more rarely roots and stems of wheat (2492, 2538, 3219). It has also been observed in the rhizospheres of oats (6202), wheat (3095), corn (4556), sugar cane (2861), banana plants (2030), *Ocimum basilicum, Origanum majorana,* beans (35), groundnuts (2768), red clover (3429) and various sand dune plants (4371), and on the rhizoids of a *Riccia* and *Anthoceros* species (2860). Isolates have been obtained from litter and the phylloplane of *Panicum coloratum* (1560), groundnut pods (346, 1984), pecans (2258, 2572), sycamore achenes (1683), bananas (6158), some subtropical fruits (2769), seeds of wheat (2038, 3989, 4307, 5293, 6407), corn (3989, 6407), oats, rice, green pepper, flax, beans, groundnuts (6407), soya (4975, 6407), cotton (5009), and *Pennisetum typhoides* (3670). Higher isolation counts occur in soils after cropping soya bean than after corn (6379). Other habitats reported include the dung of geckos (5366), wild bees (402), larvae and adults of the picnic beetle, *Glischrochilus quadrisignatus* (6399), feathers of free-living birds (2575), activated sludge (1387), a waste stabilization pond where it was rare (1160), other waters (1946), wood pulp (6168), and some sticks exposed to river water were colonized by *F. moniliforme* (5249). — *F. moniliforme* is the causal agent of stem rot in corn (1515, 3072) and sugar cane, and also affects the growth of cotton seedlings (6058). Growth *in vitro* is retarded by 6-methoxybenzoxazolinone (6341), a compound present in ether extracts of corn plants (6340). — Several bacteria (4272), particularly a *Pseudomonas* species (5296, 6002) and strains of an *Azotobacter* (3216) showed strong antagonistic properties towards this fungus. — It can survive at low temperatures in the form of thickened cells (8–12 μm) (4241, 4242) and was still found viable after 8 months incubation in soil (4243). — Germination is inhibited *in vitro* by volatiles from leaves of *Origanum majorana* and, to a lesser extent, *Ocimum basilicum* (35). Growth *in vitro* is inhibited by *Memnoniella echinata* (4), and also by cochliodinol, an antibiotic produced by some *Chaetomium* species (704). — *F. moniliforme* can serve as food for the springtail *Hypogastrura tullbergi* (3830) and the mites *Acotyledon redikorzevi* (5379), *Aleuroglyphus ovatus* and *Suidasia medanensis* (5380).

The optimal growth temperature is 22·5–27·5°C, the maximum 32–37°C and the minimum 2·5–5·0°C (4194); it has also been reported that little growth occurs at 6°C and none at >30°C (2776, 5599). The minimum humidity for vegetative growth is at −180 bars water potential (2093), with conidium germination occurring between vapour saturation and −140 bars (2776); the optimal osmotic potential lies at about −10 bars but no growth occurs at −150 bars (6395). Sorghum grain was found to be colonized at a moisture content of 25% (795) and shelled corn at 18·4% (3069). — Good utilization of the following C sources has been reported: glucose, mannose, galactose, fructose, rhamnose, arabinose, xylose, ribose, sucrose, maltose, starch, sorbitol, mannitol, lactate, and gluconate (1000); moderate growth is obtained with raffinose and glycine as sole C and N sources (2623). Growth phases in fermenter cultures and under varying nutrient conditions have been thoroughly studied (652); yeast-like growth is observed in a number of liquid media at 25°C under agitation (2958). The production of mycelium is stimulated significantly by iron and zinc ions (5052), and thiamin, biotin or inositol (4614); for maximal growth 0·09% phosphorus and 0·013% sulphur are required in the medium (2727). Sporulation is promoted by light (12:12 h light:dark at 20–23°C) and the production of macro-conidia may be induced by "black light" (4194). Sporulation under submerged conditions occurs on sucrose-containing media (6088). *F. moniliforme* tolerates 10% CO_2 (4029) but 20% CO_2 in the atmosphere reduces growth *in vitro* by 15% (1504). Amylase and inulase activities are found to be localized in the microsomal fraction (4746); the optimal conditions for amylase production (i.e. inducers, pH, phosphate concentration) have been investigated (2271). There are reports of the production of galactose oxidase (1894), 1,6-β-glucanhydrolase (5279), dextranase (5335, 5336), pectinases (884, 2184, 5367, 5368, 6049, 6089), glutamic acid decarboxylase (2270), acid phosphatases (6501), penicillin V-acylase (6040), ribonuclease (4295) and phenol oxidase (934). Several polyphenols are utilized (378). A cellulolytic enzyme was also shown to be present in corn ears inoculated with this fungus (1791); *F. moniliforme* causes a marked loss in tensile strength in textiles (3618, 6326), utilizes cotton cellulose (3618), filter paper and bacterial cellulose with the strongest activity at pH 5·5 (6060). It does not hydrolyse fructosan from *Aspergillus sydowii* but converts those of *Acetobacter* sp. and *Bacillus subtilis-pumilus* very well (3404). It can also attack wool (6330). A proteinase complex resembling rennin was found in liquid cultures (3049). Degradation of the herbicide simazine (2894), N-dealkylation of the allylamino group in atrazine (2892) and of the acaricide chlorphenamidin (2781) have been demonstrated *in vitro*. Cortisol is converted to its dehydro derivative (1610). The hydrocarbons decane, undecane, hexadecane and paraffin are oxidized (627, 3112). In nutrient solutions originally devoid of amino acids, alanine and glutamic acid together with traces of glycine, lysine, threonine and valine were found (895) as well as glucuronic acid, glucose, galactose, and mannose (3640); a substance with morphogenic properties on other fungi occurs in ageing cultures (4435). *F. moniliforme* produces fusaric acid in varying quantities (512, 607, 6061), fusarinolic acid (1830), indole-3-acetic acid (294, 6115) and also gibberellic acids and their precursors (237, 415, 2812, 5295) and derivatives (396). The biosynthesis of gibberellins was said to be promoted by light (3774). Thiamine, pyridoxine, nicotinic acid and biotin are synthesized; ammonium salts were superior for supporting this synthesis (5247). Other metabolites include malonic acid, jasmonoylisoleucine, β-*epi*-(–)-manoyloxide, orsellinic acid (3231), the benzoxanthentrione pigments bikaverin and norbikaverin (3019), fusigen (= fusarinine C), fusigen B (= fusarinine B), zearalenone (3231), 5-hydroxy-methylfuran-2-carboxylic acid, phenylethyl alcohol, phthalic acid and demethyl phthalate (5278), cyclonerodiol (1228), fujenal, fujenoic acid, (–)kaurene, (–)kauranol, a number of kaurenolides (395, 1226, 2357, 6469), 16α-hydroxy-(–)-kauran-19-al, 16α-hydroxy-(–)-kauran-19-oic acid, $5a$,$8a$-epidioxy-$5a$-ergosta-6,9,(11),22-triene-3β-ol-12-one, cerevisterol, 4-oxododecanedioic acid (5209), Δ^2-anhydro-mevalonic acid lactone in iron-deficient media (2918), $4b\beta$,7-dihydroxy-1-methyl-8-methylene-gibba-1,3,4a(10a)-triene-10-one (1227), *trans*-geranylgeraniol, copalyl pyrophosphate, and farnesyl triphosphate (5251). — A suppression of root growth has been observed in wheat (713) but not in oats. The release of scopoletine from oat roots was, however, induced (3637) and the growth of barley, according to the reports available, partly stimulated (293) or partly retarded (4629). The fungus suppresses the growth of *Chalara elegans* (2753), *Gaeumannomyces graminis* and *Rhizoctonia bataticola in vitro* (1369). The production of substances toxic to young ducklings (5158), chicken embryos (1388), cockerels (1116), mammals (733) or compounds with emetic properties (4630) has been reported. Cases of keratitis in man (2808) and limb paralysis in pigs and poultry (3932) caused by *F. moniliforme* have been reported. — It is able to grow well under anaerobic conditions (1244) and tolerate >15% NaCl in the medium (5881), corresponding to about −145 bars osmotic potential.

Fusarium oxysporum Schlecht. *1824* emend. Sny. & Hans. *1940* pro maxima parte — Fig. 134.

= *Fusarium bulbigenum* Cooke & Massee *1887*
= *Fusarium vasinfectum* Atk. *1892*
= *Fusarium tracheiphilum* E. F. Smith *1899*
= *Fusarium dianthi* Prill. & Delacr. *1899*
= *Fusarium lini* Bolley *1902*
= *Fusarium orthoceras* Appel & Wollenw. *1910*
= *Fusarium conglutinans* Wollenw. *1913*
= *Fusarium angustum* Sherb. *1915*
= *Fusarium bostrycoides* Wollenw. & Reink. *1925*

DESCRIPTIONS: Wollenweber and Reinking (6424), Gordon (2039), Gerlach and Ershad (1949), Booth (640), Joffe (2770), and CMI Descriptions No. 28, *1964* and No's. 211–220, *1970*. — Section *Elegans* Wollenw. — Colonies fast-growing, reaching 4·5(–6·5) cm diam in four days at 25°C; aerial mycelium sparse to abundant and floccose, becoming felted, white or peach, but usually with a purple or violet tinge, more intense at the stromatic agar surface. Some isolates have a characteristic aromatic odour suggesting lilac, some produce sporodochia with an orange slime of macro-conidia. Micro-conidia generally abundant, mostly borne on short (often reduced), simple, lateral phialides or from sparsely branched conidiophores, never forming chains, mostly 0-septate, ellipsoidal to cylindrical, straight or often curved, 5–12 × 2·3–3·5 µm. Macro-conidia fusiform, moderately curved, pointed at both ends, basal cells pedicellate, 3(–5)-septate, (20–)27–46(–60) × 3·0–4·5(–5·0) µm. Chlamydospores terminal or intercalary in hyphae, often also in conidia, hyaline, smooth-walled or roughened, 5–15 μm diam. Sclerotial pustules present in some isolates, pale to green or deep violet.

F. oxysporum is one of the most variable *Fusarium* species. The nine species and numerous varieties distinguished in sect. *Elegans* by Wollenweber and Reinking (6424) were combined into one by Snyder and Hansen (5459) and this concept has been generally accepted, apart from the fact that *F. redolens* Wollenw. (q.v.) is now mostly treated as distinct (1943). — Chlamydospores, which are essential for survival in the soil are not readily formed

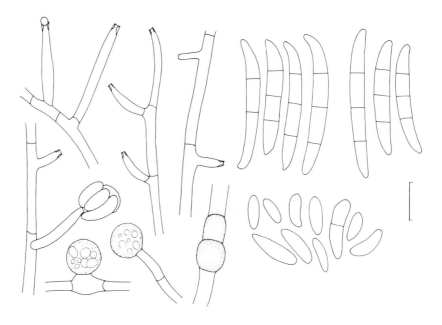

Fig. 134. *Fusarium oxysporum,* short phialides, macro- and micro-conidia, terminal and intercalary chlamydospores, CBS 242.59.

by all isolates, but may be induced by transferring agar squares to distilled water, soil extract or to a two-salt medium (640, 5571); their formation is also favoured by traces of a C source (4682), a low C/N ratio (897), or even the herbicide trifluralin (5738). — The phytopathological literature on more than 100 pathotypes (640, 2046) has not been considered in this compilation. — The teleomorph is still unknown (2044), despite assertion to the contrary (827) and the formation of hyphal anastomoses as a possible mechanism of genetic recombination has been investigated (2495). — In TEM and light-microscopic studies the time course of mitosis and the ultrastructure of the mitotic nucleus have been investigated (65). The majority of cells are uni- or binucleate, but up to five nuclei have been observed (397, 514, 2496); individual nuclei contain 12 chromosomes (641, 4669). Light intensity and concentration of the N source largely influence the colony habit (826). Single conidium isolations can be segregated in an unstable orange type and a stable purple type which showed a number of distinct characteristics (3256). — The fine structure of hyphae (2109) and chlamydospores (2110, 2111, 5571) were investigated by TEM. Numerous large micro-bodies have been observed in hyphae within host tissues but not in agar-grown hyphae (6292). — Serological methods (5774) and amino acid composition analyses (5897) have been used to distinguish between *Fusarium* species, including *F. oxysporum*. Isoenzyme patterns of different formae speciales with numerous isolates were found to differ both in electrophoretic mobility and intensity; the following enzymes have been studied: malate dehydrogenase, peroxidase, phosphatases, glucokinase and esterases (3792, 4767). — Extracts of the mycelium contained the sterols ergosterol, ergosterol peroxide, cerevisterol and, as the principal fatty acids, palmitic, stearic, oleic, and linoleic acids (5537). — DNA analysis gave a GC content of 50·5% (5600).

Fusarium oxysporum has a worldwide distribution, mostly as a soil saprophyte in a wide range of soils and on numerous host plants (640, 2045) in which specialized pathotypes may cause vascular wilt or damping-off. The species is isolated easily by all usual techniques. In order to measure low inoculum densities in soil, selective techniques have, however, been developed (667, 3083, 3084, 4384, 4424, 4426, 4924, 5373, 6440); a medium containing oxgall and TCNB in V8-juice agar proved most efficient for this (668). — Examples of the very numerous reports on its distribution in soils include ones from the arctic zone (1169), the Alps (3445, 3976, 6082), the Himalaya (4995), central Africa (3063), Hawaii (3264), India, North America and Europe. It has very frequently been found in grassland (2041, 3445) and soils with steppe type vegetation (1559, 4313, 4407, 5049, 6347, 6398), heathland (5047), or tundra (1169). In comparison to its very frequent occurrence in cultivated soils, this fungus seems to be less common in forest soils, although it has been isolated from samples taken under beech (1702, 4537), hornbeam (1702), aspen (3959), willow-poplar communities (2004, 2008, 3549), acacia (2186, 5048), teak (2854), white cedar (505) and pine (681, 3447). It is also known from uncultivated alkaline soils (4698), dunes, especially in the alkaline pH range (745), truffle grounds (3451), peat (3530, 4474), marshland (413), salt-marshes (8), mangrove mud (3264), estuaries (4477, 5316), and other coastal habitats (4918), irrigated (3497) and sewage-treated (1163) soil, and activated sludge (1387). — Although the viable spore content diminishes with soil depth (2161, 6182, 6184), *F. oxysporum* has still been found frequently at 30 cm (1317, 4995, 4996, 5349) and even at 50 cm or below (2948, 5812). It requires high humidities (977, 2095) and is very tolerant with respect to pH (6182), although it apparently is more common in neutral to slightly acid soils (159, 505, 3871, 4538, 5049, 5349, 5755, 6082). In contrast to *Cylindrocarpon destructans,* the characteristic habitats reported for *F. oxysporum* are warm, humid, occasionally slightly acid soils (5755). With a hyphal isolation technique, an increased frequency after ploughing was shown (6186). In plots treated with farmyard manure, it has been isolated more frequently than from those treated with other manures (2762), especially the unmanured plots (2895). In one of two soils investigated, the frequency of *F. oxysporum* increased after growing peas (1433). This was also the predominant *Fusarium* species associated with monoculture of soya beans (6203). The proximity of roots of radish, tomato, and lettuce causes its chlamydospores to germinate (2704); extracts from *Pinus ponderosa* leaf litter were found to stimulate germination of chlamydospores, but then their germination was not normal and the viability was reduced (5865, 5866). On pine and larch roots dilution counts were higher than in surrounding soils in some studies (5776) but the reverse in others (2101). Occurrence in the rhizosphere has been reported from cucumber (4726), clover (909, 3429), wheat (3567), beans associated with *Origanum majorana* or *Ocimum basilicum* (35), *Cicer arietinum* (377), *Aristida coerulescens* (4083), *Cassia occidentalis* (6067), *Coffea arabica* (5313), *Carica papaya* (5345), grasses (3088), wheat (2538, 3567, 4727, 5514), barley (3870), oats (6202), rice (2712), *Brassica* (4451, 4452), poplar (3452), *Picea glauca* (6026), and *Polypodium* sp. (2856). On roots it has been detectable down to 35 cm (3002). Reports of its isolation from roots are extraordinarily numerous and cannot be cited here individually. In general, living roots were colonized more intensely than dead barley roots (4450), but colonization has also been shown to increase in both dead broad bean roots (6134) and ageing barley roots (4444). It is an early colonizer of buried roots (1965), and dead roots are a suitable substrate for survival (5607). However, dead substrates are not, as a rule, colonized if they are already occupied by other fungi (4423). The fungus has been observed infrequently on seeds of *Avena fatua* (2961), cereals (2039, 2538, 3584, 3989, 6495), cotton (6058), flax (3517), groundnuts (1984, 2765, 2768), soya, peas, beans, flax (6407), pecans (2258), stored olives (309), bananas (6158), onion bulbs, potatoes (2538), sweet potatoes (3066) and other foodstuffs (1303). It has also been found on larvae of the picnic beetle, *Glischrochilus quadrisignatus* (6399), feathers of free-living birds (2575), wooden furniture (3707) sticks or poles exposed to soil (1956, 2833), and decaying leaf litter of *Pinus* species (681), *Eucalyptus maculata* (1558), *Polypodium* sp., *Dryopteris* sp. (2859), other plant debris (2705) and sclerotia of *Sclerotinia sclerotiorum* (4699).

High survival rates of at least one year in naturally infested soil or up to 17 years in soil tubes in pure culture have been reported (3716). It can survive a storage in water for ≥7 years (605). Pathogenicity tends to be lost after repeated transfers on agar media but was maintained after two years storage at −18°C (4044). Both the conidia and the mycelium soon form chlamydospores, in which activity is immediately restored on the introduction of fresh organic substrates (4423). *F. oxysporum* possesses a higher competitive saprophytic ability than other root parasites (4961, 5755). Dispersal in the soil is probably brought about by conidia transported by water movements (4423). *F. oxysporum* can be antagonized *in vitro* by *Streptomyces* (997, 998) and *Trichoderma* species (1343, 1344), *Zygorrhynchus moelleri*, *Paecilomyces lilacinus*, *Penicillium janthinellum*, *Gliocladium roseum* (5755) and a *Lactarius* species (4438). *Bacillus cereus*, *B. megaterium*, *Actinomyces* and *Pseudomonas* species were able to lyse both living and dead mycelium (1049, 5755); a high proportion of the mycelium is also lysed by glucanase and chitinase (4499). — The fungus is inhibited *in vitro* by coch-liodinol, an antibiotic metabolite of *Chaetomium* species (704). *F. oxysporum* showed but slight sensitivity to avenacin, a phytoalexin of oat roots (4287), but was strongly inhibited by α-tomatine from *Solanum* species (182); *F. oxysporum* is also inhibited to varying degrees by soil mycostatic factors (1407, 6284). Decline phenomena have been observed with pathogenic strains after repeated cultivation of host plants in suppressive soils but not with saprophytic strains (5865). With wheat as a precrop, the population density in the rhizosphere of vegetable plants was decreased, with broad beans and tomatoes it was increased (492). — The nematode *Aphelenchus avenae* was found to control the fungus in soil (3037), whereas soil inoculation with antagonistic *Bacillus* species was not successful; the addition of organic substrates (5208), especially chitinous (828, 2951, 3885) or ones enriched with streptomycetes (3143), sup-pressed *F. oxysporum* considerably. Inoculation of seed material with an unidentified *Acremonium* species reduced wilt in tomatoes (507). — *F. oxysporum* can serve as food for the mite *Pygmephorus mesembrinae* (3104).

This fast-growing fungus has an optimal temperature for growth between 25 and 30°C (2361, 4294, 5755, 6521), a maximum at or below 37°C, and a minimum above 5°C (826, 1204); the thermal death point was determined as between 57·5 and 60°C for 30 min in soil (613). On agar, a wide range from pH 2·2–9·0 is tolerated (2741, 2784, 3096, 5755) with an optimum at pH 7·7 (4294). The minimum water potential lies between −125 and −155 bars, semi-optimal growth occurring at −40 bars (2093). — Good C sources for sporulation include starch and mannitol (999, 4294). For optimal sporulation, 20–25°C with 12:12 h alternating light and dark are recommended (6521); mercury lamps proved suitable for inducing macro-conidium formation (665) which also occurs under submerged conditions on sucrose-containing media (6088). Readily utilized C sources include succinic and lactic acids, glycerol, mannitol, cellobiose (4914), starch, maltose (999), and sucrose (4294); lactose is more suitable than xylose (1153); on raffinose and glycine mycelial yields are low compared with other *Fusarium* species (2623). Increased respiration and growth rates occur on glucose and galactose, a large increase of O_2 uptake on L-proline and L-cysteic acid, but an inhibition on DL-arginine, tyrosine, DL-methionine and other amino acids (5478). Alanine (4979), asparagine, aspartic acid (999, 1153, 4294) and gluten are suitable N sources. Invertase does not appear as an extracellular enzyme in the medium but is localized on the cytoplasmic membrane (4978); amylacetate and inulin can both be utilized at low concentrations (1153); amylase and inulase activity was found to be highest in the microsomal fraction of cell homogenates (4746). Pectin decomposition (1432, 3539, 4294, 6049, 6437) takes place by the action of polygalacturonases, depolymerase and pectin methylesterase (1748, 4354, 4392), summarized as "phytolysin" (2937); numerous workers have reported good decomposition of cellulose supplied as filter paper, cotton fibres, bacterial cellulose or CMC, and the formation of β-glucosidase (377, 6159), dextranase (5335), 1,2-β-D-glucanase (4780) and L-asparaginase (4094). *F. oxysporum* grows relatively well on lignin sulphonate (3260) or *p*-hydroxybenzaldehyde, ferulic acid and vanillin as well as on gallic acid and tannin (544, 1153, 1206, 1748, 4914). Benzoic acid, cinnamic acid, flavonoids and isoflavones are also utilized, and *o*-coumaric acid is metabolized to 4-hydroxycoumarine (377). No utilization of chitin occurs (2706); hairs are penetrated by boring hyphae which then spread internally (1633). *F. oxysporum* grows on various plasticizers (diesters of aliphatic 2-basic acids) (473), hydrocarbons from fuel oil (3384, 4239), and *in vitro* degrades atrazine by N-dealkylation of the alkylamino-group (2892), simazine (2158, 2894), metobromuron (5957), various carbanilinate herbicides (2893) and DDT (1635). It solubilizes Ca phosphate (5210), reduces PCNB (4100) and Fe^{3+} ions (4330), and growth is likely to be influenced by bicarbonate ions rather than by CO_2 dissolved in the medium (3482). *F. oxysporum* is tolerant of both high CO_2 concentrations (775, 1244, 1502, 2511, 3481, 3791, 5680, 5853) and salt concentrations up to 18% (309, 4477), corresponding to an osmotic potential of -180 bars. In PDA containing 5% NaCl, growth is reduced but sporulation not (1991). Potassium deficiency is said to retard growth, and phosphorus deficiency to interfere with sporulation (3096, 5850). Ca^{2+} ions are apparently important for protein synthesis (5850), though, according to other reports (5554), they are not required in nutrient media. The addition of zinc and iron ions is essential for growth and copper is stimulatory (4914, 5052, 6508). *F. oxysporum* is autotrophic for growth substances, however (1153), and it can grow without O_2 (612) provided that yeast extract, MnO_2, nitrate, selenite or Fe^{3+} ions are present (2176). In a closed culture system including $^{35}SO_2$, the uptake of labelled sulphur within the fungal colony was demonstrated (1210). — Synthesis of carotenoids occurs in light, while naphthoquinones are formed mainly on substrates with a C/N ratio of $\geqslant 20/1$ (897). Extracellular phosphomonoesterase (498), pyruvic, α-ketoglutaric (4978), xylonic and arabinoic acids (5278) are released into the culture medium as well as a red pigment identified as bikaverin (= lycopersin) (701, 1191, 1327). *F. oxysporum* was found to convert cortisol to its dehydro-derivative (1610). In ageing cultures compounds arise which have morphogenic and sporostatic effects on other fungi (4425, 4428, 4435, 4881, 4882); most of the inhibitors are volatile and include aldehydes and alcohols (4878); one was identified as nonanoic acid (1905). The following metabolic products are known from *F. oxysporum*: fusarenone, nivalenol diacetate (5993), oxysporin, enniatin A and B, lycomarasmin and bostrycoidin (512, 843). The production of fusaric acid is favoured by glucose (2%) as C source, whilst pectin, polypectate and CMC are far less suitable (1301). — The fungus was found to parasitize oospores of *Phytophthora cactorum* (5454), antagonize a number of fungi *in vitro* (2753, 4427, 6001) and to inhibit the development of the nematode *Heterodera schachtii* in sugar beets (2823). With high infestation of *F. oxysporum*, the penetration of *Pratylenchus penetrans* into alfalfa roots is made easier (1542, 1543). On the other hand, the fungus can be ingested by nematodes (*Aphelenchus*), springtails (*Folsomia fimetaria*) and woodlice (*Oniscus asellus*) (3579, 4007). In addition to the extensive range of host plants in which the fungus occurs as a specialized wilt pathogen, injurious influences have been observed on cereal species, including barley (4919) and wheat (1430, 6573); on the other hand, no primary pathogenicity to wheat was observed under experimental conditions (2789, 6201). *F. oxysporum* has often been isolated as the causal agent of damping-off in cultivated mushrooms (6431). A few isolates with the ability of producing growth-promoting substances have been reported (519, 4553). *F. oxysporum* was found to destroy both brood and honeycombs of wild bees (402) and isolates with toxic influence on brine shrimps and chicken embryos have been found (1303). — Mostly under other names (e.g. *Hyalopus onychophilus* (Vuill.) Aschieri, *Cephalosporium keratoplasticum* Morikawa), Fusaria of sect. *Elegans* have become known as human pathogens causing onychomycoses (1633, 1634, 4949).

Fusarium poae (Peck) Wollenw. *1913* — Fig. 135.

   ≡ *Sporotrichum poae* Peck *1902*
   = *Fusarium tricinctum* (Corda) Sacc. emend. Sny. & Hans. *1945* pro parte

DESCRIPTIONS: Wollenweber and Reinking (6424), Gordon (2039), Seemüller (5174), Booth (640), Joffe (2770) and CMI Descriptions No. 308, *1971*. — Section *Sporotrichiella* Wollenw. — Colonies fast growing, reaching 5·6–8·8 cm diam in four days at 20–27°C on MEA or OA, aerial mycelium loosely cottony, white or pink, at the agar surface sometimes red but less so than in other species of the section; odour characteristically fruity (suggesting peaches). Conidiophores ± richly branched but usually not in sporodochia, with rather short and broadly ending, non-proliferating phialides, 6–18 × 3 μm. Micro-conidia predominantly produced, globose to broadly ellipsoidal or pyriform, with a broadly apiculate base, 6–10 × 5·5–7·5 μm, usually one-celled. Macro-conidia rather sparsely produced, slightly curved, generally 2–3-septate and 18–38 × 3·8–7·0 μm, more rarely 5-septate and to 56 μm long. Chlamydospores not produced, but thickened, swollen hyphal portions may function as chlamydospores. — Similar subglobose micro-conidia occur in *F. sporotrichioides* var. *minus* Wollenw. *1935*, but in that taxon the phialides proliferate sympodially and chlamydospores are present. — Hyphal cells are mostly 2-nucleate but rarely up to 8-nucleate (2496). Nuclei contain 6 chromosomes (641).

   F. poae is common on gramineous and herbaceous plants, mainly in temperate regions (640). Soil isolations have been reported from Germany (3127, 4180), the Ukrainian SSR (4556), Iraq (92), dry and wet pastures in the British Isles (165) and sand dunes in France (3941). In Canada, it represented less than 1% of the total soil *Fusarium* isolates (2041) and was found in low frequencies in coniferous soil of Newfoundland (5363). In India, it has been found in both cultivated (1317, 4996) and uncultivated soils (4030) and in sal (1525), teak (2854) and Himalayan forests (4995). It also seems to be common in polluted and non-polluted aqueous environments (1166, 1946, 4850, 5060) and has been repeatedly

F‍IG. 135. *Fusarium poae,* aggregated and loosely scattered conidiophores, subglobose micro-conidia, and macro-conidia, CBS 446.67, orig. J. Veerkamp.

isolated from cereal (2039) and other seeds (2040), particularly wheat (1752, 2437, 2716, 6407), barley (1752, 2820, 3207), oats (6407), corn (3781), various common grasses (3512), and *Avena fatua* (2961), and onion bulbs (2538). *F. poae* is also known from the rhizosphere of corn (4556), roots of wheat (945), alfalfa (2538), and red clover (6493) but is regarded as non-pathogenic to cereal plants (1118). On the rhizoplane of *Trigonella foenum-graecum* its frequency increased following foliar application of the antibiotic subamycin (2187). — It may occur at soil depths of ⩾30 cm (1317, 1525). It survived a soil treatment with ethylene dibromide (6287). — The fungus is inhibited by low concentrations of cochliodinol *in vitro* (704). — On corn and other cereals, *F. poae* is commonly associated with the mite *Siteroptes graminum* which feeds on the fungus and distributes it (5174).

The optimal temperature for growth is 22·5–27·5°C (2771, 3187), the maximum 32–33°C and the minimum 2·5(–9)°C (3187, 5174). Suitable C sources are maltose, starch and lactose (3187); arabinoxylan is degraded by the action of a xylanase rather than cellulases (1735). Iron and zinc ions are essential for optimum growth (5052). — Like *F. tricinctum* (q.v.) it produces the mycotoxins neosolaniol, T–2, HT–2 (2778, 5992, 5993) and acetyl T–2 toxins (3107); toxin production is optimal at pH 5·6, in darkness, and at low (2771) or changing temperatures (2761). *F. poae* was found to be toxic to Guinea pigs (4630). Toxic isolates survive and grow at −2°C.

Fusarium redolens Wollenw. *1913* — Fig. 136.

≡ *Fusarium oxysporum* var. *redolens* Gordon *1952*
= *Fusarium oxysporum* Schlecht. *1824* emend. Sny. & Hans. *1940* pro parte

DESCRIPTIONS: Wollenweber and Reinking (6424), Gordon (2039), Booth (640), Joffe (2770), Hantschke (2262), CMI Descriptions No. 27, *1964*, and most detailed by Gerlach (1943). — Section *Elegans* Wollenw. — Colonies rather fast-growing, reaching 4·5 cm diam in four days at 25°C on PSA or OA, usually less strongly pigmented than *F. oxysporum*, discolouring the agar pale beige to reddish brown (never violet); an aromatic odour suggesting lilac mostly present. Micro-conidia slightly larger than in *F. oxysporum*, 7·5–15 × 2·5–4·2 µm. Macro-conidia often formed in pale (rarely orange) sporodochia, wider in the upper third than in the middle with scarcely pointed apical cells (somewhat resembling those of *F. solani*), distinctly 3(–5)-septate, 25–38(–50) × 4·0–5·5 µm. Chlamydospores arising in the mycelium and the conidia, terminal and intercalary, mostly smooth-walled, 6–14 µm diam. Sclerotial bodies cream to brownish, or more rarely greenish. — Hyphal cells are mostly uninucleate, but up to 6 nuclei have been observed (2496); the nuclei contain 8 chromosomes (4669).

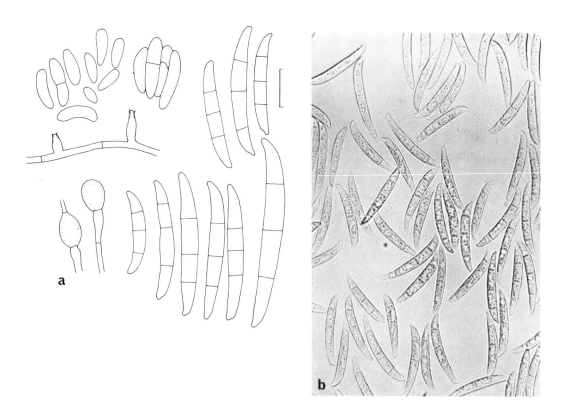

FIG. 136. *Fusarium redolens*. a. Short phialides, micro-conidia, macro-conidia and chlamydospores, CBS 128.73; b. macro-conidia, × 500.

The proportion of *F. redolens* among the Fusaria of sect. *Elegans* is variable. It has been recorded on numerous host plants, particularly peas, spinach and carnations, where it can cause wilt or damping-off in seedlings, and sometimes a cortical rot. — It is widespread in the temperate zone and also in tropical regions (640, 1951, 2043, 2045, 2262). *F. redolens* is commonly soil-borne, and in soils cultivated with cereals in Canada it is very common, making up 35% of all Fusaria (2041). In soils permanently cropped with wheat in Grignon/France it was also common (918) but, to the contrary, in corresponding soils in Germany it was extremely rare. It has also been isolated from soils in Israel (2762), Iraq (92), India, Hungary and the British Isles (IMI). There are further reports from wheat roots (2492), the rhizosphere of castor plants (IMI), roots of pear trees (5499), remains of cabbage and potato plants (3113), rarely cereal seeds (2039) and cysts of the nematode *Globodera rostochiensis* (3090).

Some fungitoxic effects of *F. redolens* have been described (4139). The hyphae grow chemotropically towards phosphate granules without dissolving them (3074). No injury to wheat was found to occur in greenhouse experiments (2789).

Fusarium semitectum Berk. & Rav. *1875* var. *semitectum* — Fig. 137.

≡ *Pseudofusarium semitectum* (Berk. & Rav.) Mats. *1975*

DESCRIPTIONS: Wollenweber and Reinking (6424), Gordon (2039), Gerlach and Ershad (1949), Booth (640), and Joffe (2770). — Section *Arthrosporiella* Sherb. — Colonies fast-growing, reaching about 6 cm diam in four days at 25°C on PSA or OA. Aerial mycelium floccose, at first whitish but later becoming avellaneous to buff-brown; at the agar surface and in reverse becoming peach but never red. Sporodochia absent. Conidiophores scattered in the aerial mycelium, loosely branched; phialides slender, cylindrical, 19–24 × 2–4 μm, often sympodially proliferating, producing one conidium at each opening. Conidia fusiform, almost straight, with slightly bent, beaked apical and conical, hardly apiculate, basal cells, 3–5-septate, variable in size, 17–40(–50) × 3·0–4·5 μm. Micro-conidia absent. Chlamydospores often sparse, globose, intercalary, pale, formed singly or in chains, 5–10 μm diam. — *F. semitectum* var. *majus* Wollenw. *1931* is distinguished by 3–7-septate conidia to 55 × 6·5 μm. — *F. camptoceras* Wollenw. & Reinking *1925*, which also has similar polyphialides, is distinguished by its even more heterogeneous, more strongly curved and wider conidia, 5-septate 32–37 × 4·5–6·0 μm. — Hyphal and conidial cells are normally uninucleate, but up to 5 nuclei may occur in the hyphal cells (514, 2496).

F. semitectum is extremely common, particularly in tropical and subtropical countries and mainly occurs on plant tissues. It causes a storage rot in various fruits and potatoes, but is seldom pathogenic to living plants although it may cause damping-off in tomato seedlings (640). It is often imported to Europe with tropical fruits (6424). — Occurrences in soil have been reported from both tropical and temperate regions (2045), including Canada where it was particularly rare (2041), the USA (4733), the USSR (2871, 4548, 4556), Poland (1421), Germany (3127, 4180, 5316), Hungary (2538), Australia (945), India (4371, 4698), Sri Lanka (4021), Pakistan (4855), Kuwait (4001), South Africa (1555, 1558), Syria (5392), Israel (2764, 2768, 2772), Libya (6510), Colombia (4795) and Trinidad where it appeared to

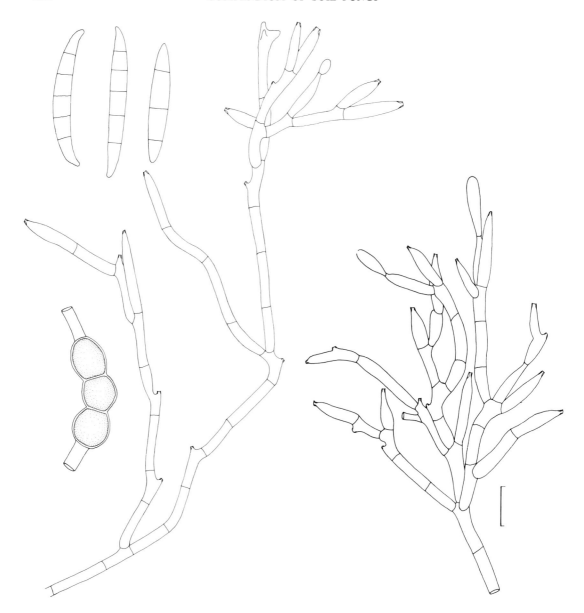

Fig. 137. *Fusarium semitectum*, sympodially proliferating conidiophores, conidia and intercalary chlamydospore, orig.
J. Veerkamp.

be one of the most frequent *Fusarium* species (2042). It has been found in both cultivated
(2764, 2768, 2772) and uncultivated (4698) soils, forest soil (1555), dry and mesic grassland
(4021), sand dunes (4371), children's sandpits (1421), saline soil (4001), desert soil (4733),
estuarine silt (5316), and running water (1166). *F. semitectum* is also known from frost-
damaged leaves of various grasses (191), leaf litter of *Eucalyptus maculata* (1558), stems of

sorghum (3219), roots of wheat (945) and clover (3429), the rhizospheres of wheat (2538, 4558), barley, oat (3002), corn (4556), groundnuts (2768), alfalfa, peppermint (2538) and various sand dune plants (4371), seeds of barley (6407), wheat (2038, 2039, 2716), oats (3097, 6407), sorghum (57), *Pennisetum typhoides* (3670), common grasses (3512), pods and seeds of groundnuts (1984, 2765), soya (4975, 6407), peas (2040, 6407), beans, green pepper, flax (6407), onions (5467), pecans (2572), various parts of cotton plants (5009), avocado fruits (2769), stored sweet potatoes (3066) and, most commonly, bananas (4227, 6158). It has also been isolated from hay (4821), coarse fodder (4548) and cotton fabrics (6326). — Its growth was inhibited by *Memnoniella echinata in vitro* (4).

The optimal temperature for growth is 25°C, the minimum 3°(or 15°C), and the maximum 34–37°C, but no growth occurs at 40°C (191, 2846). Good growth is obtained between pH 4·5 and 7·5, with an optimum of 6·0; optimal pH for production of conidia lies between pH 5·0 and 6·5 (2845). — This fungus can degrade starch, glycogen, inulin (4746) and cellulose (53, 6060, 6326). Suitable C sources include glucose, mannose, galactose, fructose, sorbose, rhamnose, arabinose, xylose, ribose, sucrose, maltose, sorbitol, mannitol, lactate, gluconate, acetic, gluconic and lactic acids; L- and D-amino acids are equally well utilized (1000). Compared with other *Fusarium* species, relatively poor growth on raffinose and glycine as sole C and N sources is obtained (2623). Various polyphenols can be utilized (378). *F. semitectum* produces enzymes with specificities for *N*-phenoxyacetyl amino acids and phenoxymethyl penicillin (penicillin amidase) (406) and high yields of an extracellular phosphomonoesterase (498). One strain was detected out of 44 different fungal isolates, which produced an estrogen-like substance (933) which was subsequently identified as zearalenol and 8′-hydroxyzearalenone (5583). — *F. semitectum* was found to parasitize conidiophores and conidia of *Cercospora mori* (4747). — It can grow on media with 3–10% sucrose (4001).

Fusarium solani (Mart.) Sacc. *1881* — Fig. 138.

≡ *Fusisporium solani* Mart. *1842*
≡ *Fusarium solani* (Mart.) Appel & Wollenw. *1910*
= *Fusarium solani* (Mart.) Sacc. emend. Sny. & Hans. *1941* pro parte

Teleomorph: *Nectria haematococca* var. *brevicona* (Wollenw.) Gerlach *1980*

≡ *Hypomyces haematococcus* var. *breviconus* Wollenw. *1930*

DESCRIPTIONS: Wollenweber and Reinking (6424), Gerlach and Ershad (1949), and Joffe (2770). Other authors take *F. solani* in the broader sense of Snyder and Hansen (5460), to include *F. javanicum* Koord. and *F. eumartii* Carp. (640, 2039, CMI Descriptions No. 29, *1964*). — Section *Martiella* Wollenw. — Colonies rather fast growing, reaching 3·2 cm diam in four days at 25°C on PSA or OA, green to bluish brown at the agar surface, with little aerial mycelium but abundant cream to buff conidial slime formed in sporodochia or pionnotes.

Micro-conidia usually abundant, produced on elongate, sometimes verticillate, conidiophores, 8–16 × 2–4(–5) μm. Macro-conidia produced in variable quantities on shorter, branched conidiophores which soon form sporodochia, usually moderately curved, with short, blunt apical and indistinctly pedicellate basal cells, mostly indistinctly 3-septate, 28–42 × 4–6μm (if 5-septate then to 65 μm long). Chlamydospores commonly produced singly or in pairs, in terminal, lateral, or more rarely intercalary positions, hyaline, smooth- or rough-walled, 6–10 μm diam. Chlamydospores are often formed from macro-conidial and micro-conidial cells under adverse conditions.

The three species and several varieties of section *Martiella* and one species of sect. *Ventricosum* distinguished by Wollenweber and Reinking (6424) were combined into the single species *F. solani* by Snyder and Hansen (5460). Gordon (2039) and Gerlach (1945) separated out *F. coeruleum* (Lib.) ex Sacc. *1886* as a good species, while Booth (640) treated it as a variety of *F. solani*. This fungus is distinguished from *F. solani* by the deep blue-purple colonies and macro-conidia with an almost cylindrical central part; it is a well-known parasite of stored potatoes in which it causes a dry rot. *Nectria ventricosa* C. Booth (q.v.) (anamorph *Fusarium ventricosum* Appel & Wollenw.) is distinguished from *F. solani* by its whitish colonies which usually do not form sporodochia and have no distinct micro-conidia, whilst *Nectria haematococca* Berk. & Br. *1873* (anamorph *Fusarium eumartii* Carpenter *1915*) has mostly 5-septate macro-conidia 52–68 × 5·2–7 μm, and *Nectria ipomoeae* Halst. *1892* (anamorph *Fusarium javanicum* Koord. *1907*) has them 3–5-septate, 28–56 × 4–5·3μm. *Nectria haematococca* var. *brevicona* is hardly separable from var. *haematococca* in the perithecia but only in the anamorph.

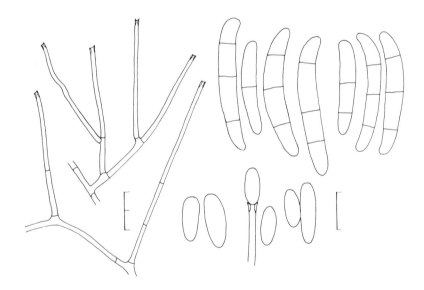

Fɪɢ. 138. *Fusarium solani,* long, little-branched conidiophores, macro- and microconidia.

TEM and SEM studies of macro-conidia (5780) show filamentous processes by which the conidia can be agglutinated; conidial septa are provided with a pore and Woronin bodies. Wall formation during chlamydospore formation from conidia has also been studied by TEM (4279). — Hyphal and conidial cells are regularly uninucleate (514), but up to three nuclei were found in hyphal cells (2496); the chromosome number is n = 8 (4669) or 4–5 (397). — *F. solani* can be separated from species of other sections by serological double diffusion tests (5484). Amongst the amino acids present in the mycelium, alanine predominated; the mycelium also contains large amounts of δ-amino-*n*-butyric (3081), palmitic, oleic, linoleic and linolenic acids (2312). — DNA analysis gave a GC content of 50·5% (5600).

Because of varying species concepts, the reports on the distribution and physiology compiled below must be interpreted with some reservations. Wollenweber and Reinking (6424, 6425) found *F. solani* s.s. on numerous plants, mainly in temperate regions, while other reports speak of particularly common occurrences in the tropics; this latter observation seems to apply more to *Nectria haematococca*. — More than 15 formae speciales have been described which are specialized pathogens of particular host plants. *F. solani* is usually somewhat less frequent than *F. oxysporum*. Various selective isolation methods have been described (667, 1434, 4108, 4384); an improved method is based on the use of soil particles 40–200 μm diam plated on complex media including antibiotics and TCNB (4924). — There is only one report on its occurrence in alpine habitats (3445), and a few isolates from Iceland (3089), but, on the other hand, there are numerous reports from soils in the tropics and subtropics including Israel (2762, 2764, 2768, 2772), Syria (277, 5392), Egypt (8, 4962), Libya (4083, 4084, 4085, 6510), the Sudan (4222), Madagascar and the Ivory Coast (4159), Zaïre (3790), South Africa (1559, 3630, 4407), Namibia (1827), Pakistan (61, 4855), India (41, 1315, 1966, 2186, 3545, 4477, 4716, 4995, 4996, 5622, 6059), Nepal (1826), Sri Lanka (4021), Singapore (3332, 3334), Malaysia (3333), Hong Kong (1048), Japan (5846), Hawaii (3264), Jamaica (4886), in Central America particularly frequently (1697, 2031, 4794, 4796), Trinidad (2042), the Bahamas (2006), Peru (2005) and Colombia (4795). From the temperate zone there are also very numerous reports from grassland and arable soils which cannot be cited here individually. The distribution of *F. solani* has been thoroughly investigated in Canadian soils (2041). It was found to be relatively rare on washed particles in soils under wheat in Germany (1889). This species has been isolated more rarely from forest than from agricultural soils. It has also been detected in heathland soil (5047), various citrus soils (2764, 3632, 4085), a coconut grove (2006), paddy fields (1966, 3333), dunes (3941, 4162, 4655), desert soil (4083), salt-marsh (8, 4646) and other saline soils (3446), mangrove swamp (3264), estuaries (4477), seawater (468, 4918), fresh water (1166, 2338, 3809, 3942, 4429), water pipes (1946), polluted streams (1154, 1155, 1157, 1166), river sediments (1162), waste stabilization ponds (1160, 1166) and sewage treatment plants (1165, 1166). The available data for soil depths show considerable divergence; in spite of a marked diminution in viable spore counts with depth, occurrences down to 80 cm (159, 2161, 2948, 3630) have been documented. The pH value of soils manured with NPK and farmyard manure (2161) had little influence on its occurrence, but a relative increase following soil fumigation was observed (3631); in general, however, *F. solani* has a relatively high sensitivity to soil fungicides (1426). It was found associated with monoculture of soya beans (6203) where it was isolated from stems, roots and residues (4240); it occurred with higher frequencies after growing corn (6379) or peas (1433) than after other crops, although counts were lower in soils under corn than under prairie (6398). It is known in the rhizospheres of various plants; accumulations being observed particularly with legumes, including alfalfa (909, 2538, 2816), beans (35, 4451, 4452), clover (909, 2816, 3429, 5448, 5815, 6493) and groundnuts (2768). The germination of chlamydospores, with tropism of the germ tubes towards roots, has been observed with radish, tomato (2704), peas, wheat and lettuce (1134, 2703). *F. solani* has been isolated from the roots of numerous cultivated and wild plants, too numerous to be named here individually. It is rare in cereal grains (2039, 4307), but isolates have been obtained from castor seeds (3172) and various other seeds (2040, 5009, 6407), and pods and seeds of groundnuts (346, 1984, 2765, 2768) where it possibly acts synergistically with *Pythium myriotylum* in causing pod rot (1818). It has also been found on the sclerotia of *Sclerotium cepivorum* (3990) and in truffle soils (3451). — *F. solani* was also isolated from larvae and adults of the picnic beetle, *Glischrochilus quadrisignatus* (6399) and found to be the dominant fungal symbiont of the ambrosia beetle, *Xyleborus ferrugineus* (297) and a facultative parasite of the tick *Ixodes ricinus* (5018). It can be used as food by the nematode *Aphelenchus avenae* (3579, 3580), springtails and woodlice (4007). It also occurred in nests of gerbils (5393) and in cultures of earthworms (*Allolobophora caliginosa*) particularly frequently (1961). — It is one of the first colonizers of cotton root baits exposed to soil (1965) and of composted grass (3546); colonization of hair baits occurs only at vapour saturation (2095). It is not uncommon on wood and has been isolated from poles exposed to soil (1956, 2833) and boards in a bath-house (5016). Wood of *Pinus ponderosa* buried in soil was first colonized by *F. solani* which was later superseded by *Trichoderma viride* (3769). — The hyphae can be parasitized by *Trichoderma* species (5186) and in various soils a negative correlation between it and the occurrence of *T. viride* exists (2763). Streptomycetes (997, 998), *Bacillus subtilis* (2393) and a *Lactarius* species (4438) showed antagonistic activity towards *F. solani in vitro*. The mycelium is rapidly decomposed in soil (3633); lysis of the hyphae by chitinase and various 1,3-β-glucanases from a *Streptomyces* species has been investigated (5408). — *F. solani* survives as chlamydospores in soil and conidia added to the soil are quickly transformed into chlamydospores (4279) although this process is prevented in soils which are suppressive to *F. solani* (5865). Chlamydospores formed at a low pH (with addition of ammonium chloride) favour survival (5104). Spent coffee grounds were found to first stimulate the germination of chlamydospores in soil, but then the germ tubes are lysed (26). Organic substrates with high C/N ratios led to the suppression of *F. solani* f.sp. *phaseoli* on beans (301, 3685, 5462, 5867). The germination of conidia was found to be inhibited by a sporostatic factor produced by *F. oxysporum* (4882), cochliodinol (704), and mycostatic effects (6284) but, in general, it has a medium sensitivity to soil mycostasis (2559). After soil amendment with chitin, a growth-inhibiting substance was extracted from soil in *n*-butanol (5453). After soil treatment with the fungicide captan, increased dilution counts of *F. solani* were obtained (6026). Conidia survived 36 months storage on silica gel at−20°C (435). Mycelium can survive a storage in water for ≥7 years (605).

The optimal temperature for linear growth is 27–31°C (999, 5942, 6532), possibly depending on the geographical origin of the isolate (1827); good growth occurs at 37°C (2154). The response of *F. solani* to fluctuating temperatures has been analysed (6254). The pH optimum *in vitro* is at 7·8 (1574) but with a wide range of tolerance on either side (3096, 6532); the minimum water potential permitting growth is at −140 bars (2093) or, according to another report, −70 bars (3109). In contrast to numerous other *Fusaria*, sporodochia, which easily pass into pionnotes, are produced in complete darkness (6521); more abundant and more typical macro-conidia were, however, found to occur when grown under near-UV light intermittent with darkness (5106). — The formation of chlamydospores is controlled by exogenous C and N sources in conjunction with conidium densities (2100, 5104), enhanced by bacterial metabolites (4712), and induced on low pH media (3·2–4·0); no chlamydospores are formed under anaerobic conditions (1091). The system responsible for the formation of chlamydospores develops early in the normal germination process but is triggered by environmental conditions (1090). At high densities, chlamydospores require both C and N sources for germination, while at low densities, an exogenous C source is sufficient (2100), 2 ng glucose-C per chlamydospore being determined as the minimum requirement

(2102). — Sugars in the culture solution are assimilated in the order mannose > glucose ~ arabinose ~ rhamnose > galactose ~ fructose > sorbose (5736); other suitable C sources for growth are ribose, xylose (5448), raffinose (2623), mannitol, starch and maltose (999, 1827). On pectin-containing substrates pectolytic enzymes are produced (1432, 2948, 3127, 3414, 6437). The decomposition of xylan is good, although differences occur between isolates (1432). Cellulose decomposition has been demonstrated on filter paper, cotton fabrics, CMC, and bacterial cellulose (53, 1432, 2006, 2741, 2948, 4184, 6163, 6326); the pH optimum for cellulose decomposition is 6·5 (6060) and the temperature optimum 30°C (6163). Keratin in hair is attacked by boring hyphae which spread out internally (1633). Growth on lignin sulphonate (1425, 3260) or phenol lignin (1744) is considered good to moderate. Soft rot due to *F. solani* has been observed in wood (1496, 3767, 4191); characteristic features of this are the penetration of the cell walls, entry through pits, destruction of the torus and penetration into the vessel walls (3121). Polyvinyl acetate and polyvinyl alcohol, and dihexyladipate are degraded (5859); protocatechuic acid is metabolized by f.sp. *phaseoli* (3312); the herbicide propanil can also be used as sole C source but the metabolite dichloroaniline formed in this process was found to be toxic and to retard further propanil decomposition (3226); propanil induces the production of an acylamidase (3227); the pesticides chlorphenamidine, dicryl, propachlor, propham, pentanochlor (2893), bromophos (5558) and PCNB (935) are also degraded *in vitro*. The phytoalexin phaseollin is metabolized to 1a-hydroxy phaseollone by the f.sp. *phaseoli* (2436). Cortisol can be converted into its dehydro-derivative by this species (1610). It is able to degrade 1-naphthol (611) and grow on a large number of *n*-paraffins, monohydric alcohols and fatty acid esters (3123), and to reduce Fe^{3+} to Fe^{2+} (4330). A potassium deficiency is said to retard growth, and one of phosphorus to suspend sporulation; nitrates are better N sources than ammonium salts (3096). The growth of young cultures (in the presence of 4% glucose) is promoted by thiamine at concentrations from 10 μg l^{-1} upwards (5942). *F. solani* tolerates CO_2 concentrations to 20% without any significant deleterious effects on its growth (1502, 1504) and grows well in atmospheres with reduced O_2 partial pressure (612, 1425, 5680, 6163). It produces acid and alkaline phosphomonoesterases (498), L-asparaginase (4094), and an estrogenic metabolite (F2 = zearalenone) (3858). In ageing cultures, the production of ergosterol (3082) and a substance with morphogenic properties has been demonstrated (4435). Among the phytotoxic metabolic products (naphthazarins), fusarubin (4939) deserves special mention; javanicin, marticin and isomarticin have been found only in small quantities (2936), and bostrycoidin (a β-aza-anthraquinone derivative) is a less common metabolite (190); an iron deficiency induces the production of Δ^2-anhydromevalonic acid lactone (2918); the pigment solanione has been isolated and identified (6265); and the toxic metabolites solaniol (2683), neosolaniol, T–2 toxin, HT–2 toxin (cf. *F. tricinctum*), and diacetoxyscirpenol (5993) have been found. — *F. solani* is listed as one capable of causing mycetomas (1624), and it has repeatedly been isolated from human keratitis and corneal ulcers (2154, 2808, 3836, 6539). It caused a lethal mycosis in the southern pine beetle, *Dendroctonus frontalis* (3915), can infect larvae of *Nomia* bees (402), and was also reported as a wound parasite causing lethal epidemics in the crayfish *Thelohania* sp. (6087). — *F. solani* inhibits the growth of *Chalara elegans* (2753) *in vitro* and has also been identified as a mycoparasite of *Phycomyces blakesleanus*, *Aspergillus clavatus* and *Mucor hiemalis* (5668). There are reports of growth stimulation in barley by its culture filtrates (293), but seed germination of *Avena fatua* was inhibited by it (2961).

Fusarium sporotrichioides Sherb. *1915* var. *sporotrichioides* — Fig. 139.

≡ *Fusarium sporotrichiella* Bilai *1955* var. *sporotrichioides* (Sherb.) Bilai *1955*
= *Sporotrichella rosea* Karst. *1887* [non *Fusarium roseum* Link ex Gray *1821*]
= *Fusarium tricinctum* (Corda) Sacc. emend. Sny. & Hans. *1945* pro parte

DESCRIPTIONS: Wollenweber and Reinking (6424), Gordon (2039), Seemüller (5174), Booth (640) and Joffe (2770). — Section *Sporotrichiella* Wollenw. (*Arthrosporiella* sensu C. Booth). — Colonies fast-growing, reaching 8·0–8·8 cm diam in four days at 25–30°C on MEA or OA; aerial mycelium abundant, cottony, at first whitish but later becoming yellow, pink, red to purple or brownish. Red-brown sporodochia may be abundant or absent. Conidiophores ± richly branched, bearing phialides 5–18 ×3·5–5 μm which often proliferate sympodially. Micro-conidia irregular in shape, subglobose and 5–7 μm diam, or pyriform to fusiform, in

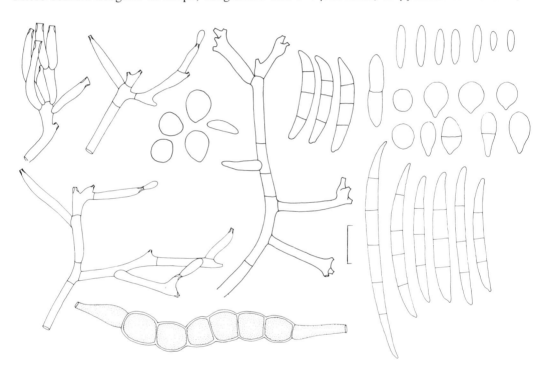

FIG. 139. *Fusarium sporotrichioides*, conidiophores with sympodial proliferation, narrow-ellipsoidal and pyriform micro-conidia, macro-conidia, and intercalary chlamydospores; various CBS strains, partly orig. J. Veerkamp.

variable proportions, 6–11(–16) × 3–4 μm, 0(–2)-septate. Macro-conidia scattered in the aerial mycelium or formed in sporodochia, moderately curved, 3–5-septate, 29–46 × 3·7–5·3 μm. Chlamydospores numerous, intercalary (or rarely terminal), often in chains or clusters, ochraceous, 7–15 μm diam. Sclerotial plectenchymatous bodies commonly produced, particularly on rice or wheat grains, dark red to brown. — *F. sporotrichioides* var. *minus* Wollenw. *1930* is distinguished by the absence of both sclerotial bodies and sporodochia, and the somewhat wider conidia. — Hyphal cells are mostly 3-nucleate but may be up to 8-nucleate (2496).

On account of frequently inaccurate determinations, only few reports can be cited from the literature on the ecology of this species. There is a long list of host plants, especially relating to its occurrence on cereal crops (oats, corn) and pome fruits (6425). — *F. sporotrichioides* is distributed in the temperate and tropical zones (2045). In contrast to *F. poae* (q.v.), which is a much more specialized species of corn and other cereal seeds, *F. sporotrichioides* is generally a soil fungus (2041, 5174) and has been reported from Iraq (92), equatorial West Africa (1420), India (6059), Australia, the USA (IMI), Canada (2041) and the USSR (2871). It has also been found on wheat grains (2038, 2039, 2538, 3097, 4492), cucumber seeds (2040), alfalfa roots, fodder mixtures (2538), and feathers of free-living birds (2575).

The optimal temperature for growth is 22·5–27·5(–30)°C (2771, 3187), the maximum 35°C and the minimum 2·5–7°C (3187, 5174). The humidity minimum for vegetative growth is at −170 bars, but half the optimal growth occurs at −55 bars water potential (2093). Toxic isolates were found to grow at −2°C (2761). Conidia are equally well produced in either continuous darkness or under "black light"/dark sequences (5106). — Suitable C sources are maltose, starch and raffinose (3187). The fungus shows cellulolytic activity (4730, 6289). Vanillic, *p*-hydroxybenzoic, ferulic acids and naringenin are readily degraded (378). — Like *F. tricinctum*, it produces the potent mycotoxins neosolaniol, T-2 toxin, and HT-2 toxin (2778, 5992). Toxin production was highest when grown at low (5–8°C) temperatures, pH 5·6 and in darkness (2771). — It inhibits the growth of *Chalara elegans in vitro* (2753); it has highly toxic effects on rodents when they are fed with contaminated cereals. Toxic aleukia caused by eating infected cereals (which were also infected by *F. poae* and other fungi) has led to many deaths in the Örenburg district and other areas of the USSR (6412).

Fusarium tricinctum (Corda) Sacc. *1886* — Fig. 140.

≡ *Selenosporium tricinctum* Corda *1838*
≡ *Fusarium sporotrichiella* Bilai var. *tricinctum* (Corda) Bilai *1955*
= *Fusarium citriforme* Jamalainen *1943*
= *Fusarium sporotrichioides* Sherb. *1915* sensu Gordon *1952*

DESCRIPTIONS: Wollenweber and Reinking (6424), Seemüller (5174), and Booth (640). — Section *Sporotrichiella* Wollenw. — Colonies reaching 3·2–4·0 cm diam in four days at 20–25°C on MEA or OA (5174). Aerial mycelium forming a compact cushion, red to vinaceous or purple, partly also white or ochraceous, agar surface in similar, dark red shades. Sporodochia not always produced and generally appearing later (after 6–8 weeks, particularly under "black light"), orange to flesh, on a stromatic or sclerotial base. Conidiophores ± richly branched, bearing slender phialides, 10–30 × 2–3 μm, not proliferating sympodially. Micro-conidia scattered on the aerial mycelium as a cream powder, mostly lemon-shaped, sometimes pyriform, ellipsoidal or fusiform. 8–11(–14) × (2·0–)4·5–7·5 μm, rarely two-celled. Macro-conidia mostly produced only in sporodochia, moderately curved, 3(–5)-septate,

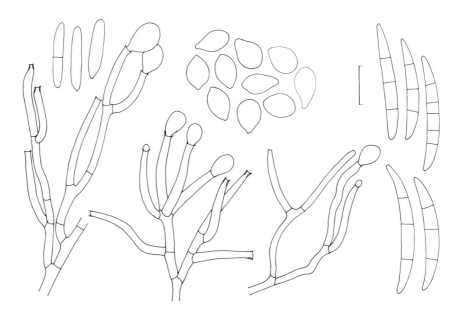

Fig. 140. *Fusarium tricinctum*, conidiophores, cylindrical and pyriform-limoniform micro-conidia and macro-conidia, CBS 253.50.

$24-50 \times 3\cdot2-4\cdot5$ μm. Chlamydospores not common, intercalary, in chains or clusters, becoming brown, $7-18\,\mu$m diam. Sclerotial bodies commonly produced, cream, ochraceous, red or blackish. — Isozyme patterns, distinct from those of other *Fusarium* species studied, were found in malate dehydrogenase, peroxidase, acid phosphatase, glucokinase and acetylesterase (4767).

Occurrence of this fungus in cultivated soil, forest soils, prairie and grassland has been reported from Germany, Switzerland, Sweden, Finland, Scotland, the USSR (5174), Iceland (3088, 3089), the USA (3817, 4313, 6201, 6203, 6253, 6398), Ireland (IMI), Italy (3445, 4538), Spain (3446, 3447), Somalia (5048), and India (2164, 6059). It has been isolated from cereals and various grasses, red clover, carnation (5174), pecans (2258), infrequently from pods and seeds of soya beans (4240), and various other substrates also in Germany, Czechoslovakia, Finland and Australia (IMI).

The optimal temperature for growth is $22\cdot5-23°$C, the maximum $31-32\cdot5°$C, and the minimum $0-10°$C (3187, 5174). — Suitable C sources include maltose, starch and lactose (3187). *F. tricinctum* has become well known for its production of toxic metabolites which include $4\beta,15$-diacetoxy-12,13-epoxy-Δ^9-trichothecen-3α-ol (= diacetoxyscirpenol) (319, 5993) with phytotoxic properties and also able to cause oedema and intradermal haemorrhages in rat skin (1981), and $4\beta,15$-diacetoxy-8α-(3-methylbutyryloxy)-12,13-epoxy-Δ^9-trichothecen-3α-ol (= T-2 toxin (317, 318, 319, 790, 5992), the latter apparently with distinct differences between isolates (2778); $3\alpha,4\beta$-dihydroxy-15-acetoxy-8α(3-methylbutyryloxy) 12,13-epoxy-Δ^9-trichothcccnc (= HT-2 toxin) (319, 790, 5992), neosolaniol (5992), 4-acetamido-4-hydroxy-2-butenoic acid-γ-lactone (1595, 6486) and the estrogenic metabolite 6-(10-hydroxy-6-oxo-*trans*-1-undecyl)-β-resorcylic acid lactone (= F-2, RAL, zearalenone) (851, 2684, 3858) involved in the estrogenic syndrome in swine. It may cause wilting in cotton plants *in vitro* (2775). — It should be noted, however, that most reports on *F. tricinctum* sensu Snyder and Hansen do not in fact refer to this species but to others of the same section.

Gaeumannomyces v. Arx & Olivier 1952

Gaeumannomyces graminis (Sacc.) v. Arx & Olivier *1952* — Fig. 141.

   ≡ *Ophiobolus graminis* (Sacc. *1875*) Sacc. *1881*
   = *Linocarpon cariceti* (Berk. & Br.) sensu Petrak *1952* [non *Sphaeria cariceti* Berk. & Br. *1861*]
   = *Ophiobolus oryzinus* Sacc. *1916*

DESCRIPTIONS: Walker (6147), CMI Descriptions No. 381–383, *1973*, and Fungi Canadenses No. 37, *1974*; review: Walker (6148). — On account of its unitunicate asci, this species belongs to the Diaporthaceae Höhn. in the Sphaeriales and thus has been excluded from *Ophiobolus* Riess *1854* with bitunicate asci (Pleosporaceae Wint.) and placed in the separate genus *Gaeumannomyces* (212). Petrak (4535) combined this genus with *Linocarpon* Syd. *1917* and regarded *G. graminis* as conspecific with *L. cariceti* (Berk. & Br.) Petrak *1952*, but his view has been rejected by Schrantz (5135), Eriksson (1647), Skou (5405) and Walker (6147), since *Linocarpon* has applanate perithecia and an apical papilla, while *Gaeumannomyces* has spherical perithecia elongated into an apical beak and a plectenchymatous wall.

   The dark olivaceous-grey cultures, reaching 4·5 cm diam and more in ten days at 20°C on MEA, which at first are sterile, can be recognized as *Gaeumannomyces* by virtue of their rapid growth, ± sparse aerial mycelium, and differentiation into broad runner hyphae and more delicate mycelial hyphae. Sickle-shaped *Phialophora*-like micro-conidia similar to *Ph. radicicola* Cain *1952* (see below) are formed freely. Perithecia appear in illuminated Petri dish cultures on OA and some other low-sugar media after several weeks; *G. graminis* is homothallic (6148).

   Comparative studies between isolates assigned to *Ph. radicicola* and *G. graminis* by Deacon (1321), considering morphological, physiological and pathological criteria, proved that the two fungi are distinct species. In addition, *Cephalosporium maydis* Samra *et al. 1963* also corresponds to the micro-conidial form (1882). Walker (6147), after reexamining all pertinent type collections, recognized three varieties of *G. graminis*: var. *graminis,* causing crown sheath rot on rice and also growing on numerous other grasses; characterized by 3–5-septate ascospores 80–100 × 2·5–3·0 µm, lobed brown appressoria on the host or artificially infected wheat coleoptiles; particularly in tropical and subtropical countries. — Var. *tritici* J. Walker *1972*, causing take-all on wheat, but also on barley, rye, corn and surviving on numerous grasses; has ascospores similar to var. *graminis*, but simple appressoria; distribution shown in CMI Map No. 334, *1972*; especially in temperate zones of all continents. — Var. *avenae* (E. M. Turner) Dennis *1960*, causing take-all on oats, barley and wheat and many grasses (patch disease); ascospores several-septate, 100–125 × 2·5–3·5 µm, and simple appressoria; widespread in temperate regions of Europe and N-America, but also in Australasia and Oceania. — The anamorphs have been disposed as follows (6430): contradictory previous observations on conidium germinability are explained by the joint occurrence in *G. graminis* of sickle-shaped, slender conidia (3–7 × 1–2 µm) which do not germinate, and larger and straighter conidia (av. 10 × 3 µm) which do germinate; the latter develop after 2–4 days at 25°C while the former take 2–3 weeks.

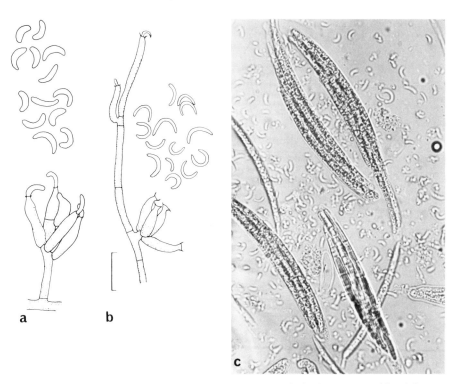

Fig. 141. *Gaeumannomyces graminis*. a. Phialides and sickle-shaped *Phialophora*-type conidia of *G. graminis*; b. the same of *Phialophora radicicola* var. *radicicola*, CBS 350.77; c. asci, ascospores and conidia from a *G. graminis* perithecium produced *in vitro*, × 500.

Phialophora radicicola Cain was isolated from browned roots of corn in Canada and has only sickle-shaped conidia measuring 5–9 × 0·7–1·5 μm, which do not germinate. *Ph. radicicola* sensu Scott (5168) = *Ph. radicicola* var. *graminicola* Deacon *1974* (probably the anamorph of *Gaeumannomyces cylindrosporus* Hornby *et al. 1977*) was isolated from grasses in the British Isles and has slightly curved conidia, av. 4·5 × 2·0 μm and slightly lobed appressoria. Colonies grow half as fast as in the remaining varieties; sickle-shaped, non-germinating conidia are absent. From these observations and the shorter ascospores (55–75 × 4–5 μm) of the teleomorph observed on insterile grass roots it is concluded that this fungus is specifically distinct.

The cells of the hyphae *of G. graminis* are predominantly uninucleate (943). — Ethanol extracts of the mycelium are reported to contain glucose and trehalose as the only soluble carbohydrates (4684). The mycelium of some isolates exhibiting reduced pathogenicity was found to contain virus particles (674, 4755, 5496, 6148). — An immunoelectrophoretic analysis gave a slightly higher number of precipitin bands within the vars. *tritici* and *avenae* than between them and few cross reactions with other fungi (2505).

The take-all fungus of wheat is one of the most important pathogenic fungi on roots and bases of culms in cereals. An exhaustive review of recent work on its taxonomy, pathogenicity and ecology has been published (6148). Most of the information compiled below refers to var. *tritici*. There are only few reports of its isolation from root-free soil; isolation from soil suspensions is usually unsuccessful (6186). A fluorescent-antibody staining technique has been described to distinguish *G. graminis* from other fungi in the rhizosphere and on roots of wheat and oats (1020). From washed soil particles of a wheatfield it was isolated only infrequently (1889) and particularly from organic particles (1888, 2536). — Ascospores are forcibly discharged from the perithecia. In dry conditions no ascospores could be detected in the air 50 cm above infested stubbles, but after two hours rain, the number released reached 3700 spores m^{-3} air (2083). Infection with an ascospore suspension was successful on sterile exposed wheat roots (736, 6300) but not in non-sterile soil (737). — *G. graminis* is generally regarded as a weak competitor when growing saprophytically (820, 821, 4734); in this context the C/N ratio of the substrate plays an important role. Organic manuring and growing non-susceptible crop plants reduces the infection potential in the soil, to judge from numerous reports in the phytopathological literature, whereas organic nitrogen sources tend to increase take-all. The disease is usually more severe in light soils and at high pH values (i.e. increased by liming), particularly if the rhizosphere-pH is maintained at or above 7 (influence of the N-source). High moisture content of the soil also enhances development of the disease (6148). A high ammonium:nitrate ratio in the rhizosphere is sometimes found to reduce disease (e.g. $(NH_4)_2SO_4$ application in spring) (747, 5415). — Various grasses can act as susceptible intermediate hosts (738); its explosive appearance on wheat in the first years following polder reclamation in the Netherlands is ascribed to occurrence of ascospore infections on reeds and other grasses that appeared as initial colonizers (1952). Survival is considerable in stubble and even without susceptible crops, the fungus remained viable for one to over five years; survival may be drastically reduced by competing soil micro-organisms (6148). The fungus survives without loss of pathogenicity for ten months on straw (834) or roots (746). The survival capacity declines with increasing resistance of the host grass (1906), age and degradation of host residues (2537), increasing soil acidity (3217), soil depth (947) and temperature (6576); a temperature change of $-94°C$ for 19·5 h and $23 ± 2°C$ for 4·5 h had no adverse effect (1242). — After inoculation with a virulent monospore isolate, perithecia were produced on the living culm base, whilst less pathogenic isolates fruited only on moribund stubble (1319). The soil may be colonized from roots only slightly or not at all (1716, 3423, 4022). In non-sterile soil the mycelium labelled with ^{32}P extended over the roots at a rate of 5–6 mm per week (4885); in sterile soil, a spread of 35 mm in ten days has been recorded (3217). A chemotropic response towards the roots of hosts occurs from a distance of 5–8 mm in soil (6148). Inoculum viability did not decline noticeably in irradiated, reinoculated soil within 24 weeks but in non-sterile soil samples it did for 33% (948). Survival is also considerably prolonged by large amounts of available nitrogen (212, 944, 946, 947, 948, 1318, 1911, 5167, 6148, 6310), inhibition by the microflora of root colonization depends on the preceding crops (6574, 6575); crop rotation is thus an important control measure against take-all. The rhizosphere microflora of non-host plants was found to be more efficient in suppressing the infestation than that of susceptible cereals (6577). — After continuous cropping of susceptible cereals for five years or more, the initially strong infestation of the soil declines due to an accumulation of some antagonistic agents (1952, 6577). In pure culture several fungi have proved antagonistic to *G. graminis*, for example *Trichoderma viride, Fusarium culmorum, Myrothecium verrucaria,* and *Paecilomyces lilacinus* (731, 3217, 4590, 5409). Strong antagonistic action by actinomycetes caused a significant suppression of the parasite in pot cultures after artificial infection (1552). This mechanism does not, however, provide a satisfactory explanation of the reduction of infestation by green manuring (1553, 2127). Var. *avenae* has been found to be inhibited *in vitro* by a methoxyhydroquinone glucoside occurring in oats (4286) and the triterpene glycosides avenacin and aescin (4287). A specific antagonistic effect on virulent isolates was shown to occur in polder soils in the Netherlands (1952), and this appears to be responsible for the rapid decline seen in the first two years following wheat monoculture; strains of *Pseudomonas fluorescens* may be involved in this form of specific antagonism (1139). The occurrence of the antagonistic *Pseudomonas putida* in the rhizosphere can be stimulated by adding ammonium (299). — In a systematic combination experiment *in vitro* with numerous soil fungi, *G. graminis* was less susceptible to inhibition than *Rhizoctonia solani*, but was more easily affected than *Pythium ultimum*, *Fusarium culmorum* and *Pseudocercosporella herpotrichoides* (1431). Hyperparasitism on hyphae of *G. graminis* by *Trichoderma viride* (5409) and *Didymella exitialis* (5315) has been observed; lysis of the mycelium in the soil has been observed on Cholodny plates (6406) and examined more recently by SEM studies of colonized wheat roots (4925). A volatile heat-stable chemical factor found in some soils is able to induce autolysis of the mycelium and is regenerated during this process (5391). Soil inoculation or enrichment with antagonistic micro-organisms reduced the attack on wheat in pot experiments (4022, 5039, 5409), but not under field conditions. Five or more additions of mycelium to soil at frequent intervals initiated a decline of *G. graminis* after several weeks and a pronounced accumulation of antagonistic bacteria and actinomycetes (6578). Some soils in eastern Washington are so strongly suppressive, that they are used at a 1% ratio to render other soils suppressive after fumigation (5292). — The use of *Phialophora radicicola* var. *graminicola* as an agent of biological control has been considered (1322). This fungus, which is often abundant in soil following grazing, spreads over wheat roots without causing any vascular discoloration and prevents infection by *G. graminis* var. *tritici* and var. *avenae* in turf (1320).

Initiation of perithecia requires illumination (737, 3244, 6309), the range of 390–450 nm being effective (6302). In general, high C source concentrations inhibit perithecium formation but increase vegetative growth; maximum numbers of perithecia are produced on agar media containing 1% glucose and 0·2% asparagine (6301). *G. graminis* var. *graminis* grows well at 30°C and faster at 25°C than var. *tritici* (6147). The temperature range for perithecium production is 13–24°C with an optimum of 20°C (6302). The addition of mixed populations of bacteria (3283) considerably stimulates the formation of perithecia as does flooding the mycelium with sterile or non-sterile water for 48 h (1989). — For linear growth, a temperature range of 20–25°C and pH of 5 or more are given as optimal (6191, 6370). With a minimum for vegetative growth at −15 bars water potential, this fungus is very hygrophilic (2093); water potentials of −1·5 bars have been reported to enhance its growth, of −20 to −25 bars to reduce it, and of −50 bars to stop it (1138, 1991). — Optimal growth occurs on the C sources glucose, sucrose, trehalose, dextrin, starch, cellobiose, xylose, fructose and maltose (1988, 4684); D-mannose and L-arabinose are also utilized well, but D-mannitol, D-arabinose and inulin are not utilized (1987, 3328, 4684, 6191). The disaccharides fructose and lactose were hydrolysed before being taken up (4684). The minimum glucose concentration for vegetative growth is 0·2%, the optimum 0·75–4%, and the maximum tolerated 10%, which corresponds to approximately −15 bars osmotic potential. Pectinmethylesterase and cellulase production is less in var. *tritici* than in var. *avenae*, and relatively high N concentrations are required for cellulolysis (6148). The most suitable N sources are arginine, asparagine, glycine and ammonium tartrate. Ammonium salts are more suitable than nitrates (1988), but inorganic N compounds are inferior to organic ones (1786, 6191); a synthetic medium for optium growth has been described (1988). Lysine and threonine (4×10^{-3}M) inhibited growth *in vitro* while cystine and cysteine are said to inhibit growth of var. *avenae* even at 10^{-6}M but to stimulate var. *tritici* (5944, 5945, 6148). Thiamine can be synthesized from its moieties by the fungus (1786, 1988, 6191); there is a strict heterotrophy for biotin (6324); 100μg l^{-1} of biotin and 10 mg l^{-1} of thiamine are regarded as sufficient; other vitamins are without effect (1987). Low O_2 partial pressures are tolerated, but 10% CO_2 in air inhibited growth by about 40% and 30% CO_2 by about 80% (1502); there are reports of an even higher CO_2 sensitivity (5422).

Gelasinospora Dowding *1933* emend. Cailleux *1971*

Gelasinospora retispora Cain *1950* — Fig. 142.

= *Gelasinospora reticulispora* (Greis & Greis-Dengler) C. & M. Moreau *1951*
≡ *Rosellinia reticulispora* Greis & Greis-Dengler *1940* (nomen inval., Art. 36)

The sordariaceous genus *Gelasinospora* (type species *G. tetrasperma* Dowding) is characterized by black, smooth perithecia with or without an ostiole (i.e. inclusive of *Anixiella* Cain *1961*), stalked cylindrical asci with persistent or evanescent walls and a ± conspicuous apical ring, and blackening, thick-walled ascospores with a characteristic foveolate or reticulate sculpture. Micro-conidia of the *Cladorrhinum* type are occasionally present.

   Contributions to a monograph of the genus: Cain (839) and Cailleux (837); cytotaxonomy of three species: Sun *et al.* (5644). The genus now contains about 20, mostly coprophilous, species, the majority of which are known from warmer regions.

   DESCRIPTIONS: Cain (839), Udagawa and Takada (5983), Matsushima (3679, 3680), and Udagawa and Takada (5984). — *G. retispora* is characterized by fast-growing colonies, pyriform black perithecia 700–1000 × 400–600 μm, covered with some inconspicuous short hyaline hairs. Asci cylindrical, 250–300 × 20–40 μm, stalked, persistent, with a distinct apical ring, 8-spored. Ascospores uniseriate, ellipsoidal, 28–33(–37) × 14–17(–20) μm, wall with a smooth surface and an underlying thickened network in the pigmented epispore layer (observable best with concentrated NaOCl), and two opposite germ pores which in mature spores are more transparent than the wall ornamentation (3747). — *G. retispora* is the only species of the genus with this kind of spore ornamentation (characteristic of the section *Reticulocostulae* Cailleux), while the most similar *G. cerealis* has narrower light spots in the epispore which correspond to indentations in the endospore, and broader ascospores, 30–35 × 23–27 μm. — Non-ostiolate ascomata have been observed in a normally ostiolate isolate where the hyphae reach the colony edge in light or darkness. Morphological changes during perithecial development after a localized photoinduction, starting from a coiled initial, have been followed microscopically (2660).

   G. retispora has only rarely been reported from plant material, but a worldwide distribution is indicated. It was originally isolated from *Nothofagus* wood in Chile, then from seeds of *Beta vulgaris* in the Netherlands, apple twigs in Quebec (839), *Agropyron pungens* leaves (167) and *Picea sitchensis* in the British Isles, *Camellia sinensis* in Kenya (IMI) and *Pyrus sikkimensis* in India (5508). — It has rarely been isolated from soil, including soils under beech wood in Belgium, red pine in Canada (IMI), forest soils in central Africa (3063), sand dunes (745) and soils under *Calluna* heath in the British Isles (5221), soils in New Guinea (3679, 5983, 5984) and Japan (2532, 3680). Other reports include the rhizospheres of groundnuts (2532) and various forest plants (4814) and rabbit excrements (5221). Its apparent absence in northern Europe has been noted (3441), but *G. retispora* recurred once in a wheatfield soil in Germany. — For development in soil, water potentials of between −6·5 and −10 bars have been found to be optimal (5226).

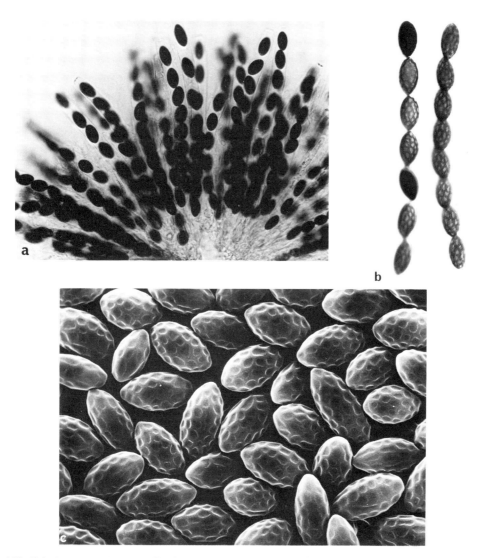

FIG. 142. *Gelasinospora retispora*. a. Crushed perithecium showing radiating asci, × 200, CBS 868.68; b. two asci, NIC, × 400; c. ascospores, SEM, × 750, orig. R. A. Samson; CBS 218.58.

Perithecium formation is induced by light after a preceding dark phase (1843, 2656, 2657). A brief exposure of the hyphae to white, near UV or blue light during the initial growth phase suppresses perithecium formation (2658); photosensitivity begins at a colony age of at least 30 h and is highest after 72 h, but after more than 48 h darkness fruiting is delayed accordingly. The time of illumination required for optimal perithecium formation decreases with the increasing dark phase from 10 000 to 100 sec, and the most effective wavelengths are shifted from near UV to blue (2657). Monochromatic light was most effective at wavelengths of 450–460 nm, with secondary peaks at 370, 420 and 480 nm; above 520 nm, no effect was observed (2659). At 37°C rapid colony growth but hardly any fruiting is observed (5984).

Geotrichum Link ex Leman *1821*

Geotrichum candidum Link ex Leman *1821* — Fig. 143.

    = *Oidium lactis* Fres. *1850*
      ≡ *Oospora lactis* (Fres.) Sacc. *1886*
      ≡ *Endomyces lactis* (Fres.) Windisch *1951*

For numerous other synonyms see Carmichael (901) and Morenz (3945).

Teleomorphs: *Dipodascus geotrichum* (E. Butler & L. J. Petersen) v. Arx *1978*

      ≡ *Endomyces geotrichum* E. Butler & L. J. Petersen *1972*
      ≡ *Galactomyces geotrichum* (E. Butler & L. J. Petersen) Redhead & Malloch *1977*
        Dipodascus australiensis v. Arx & Barker *1978*

DESCRIPTIONS: TELEOMORPHS: Butler and Petersen (818), v. Arx (201, 208), and Redhead and Malloch (4769); ANAMORPH: Morenz (3945, 3946), Caretta (891), Butler (814), Carmichael (901), and Pelhate (4494); REVIEWS on biology: Windisch (6400), Morenz (3945, 3947), and Butler *et al.* (819); CONIDIOGENESIS in cinematography: Cole and Kendrick (1107), TEM and SEM: Cole (1101, 1102).

The genus *Dipodascus* Lagerh. *1892* (=*Galactomyces* Redh. & Malloch *1977*), family Dipodascaceae Gäumann, comprises Hemiascomycetes with arthroconidia and naked, globose to elongate asci arising from conjugating gametangial cells, and ascospores which are ellipsoidal to subglobose, and either ornamented or surrounded by a sheath. This genus has been expanded by v. Arx (208) to cover species with one to numerous spores per ascus, while the genus *Endomyces* Rees *1870* (type species *E. decipiens* Rees; fam. Endomycetaceae Schröt. *1893* = Ascoideaceae Schröt. *1894*) has been restricted to three species which form asci directly on the hyphae without copulation and have hat-shaped ascospores and blastoconidia (4769). *Dipodascus* as now conceived contains eight species.

D. geotrichum is homothallic but mostly self-sterile, has globose asci containing a single, thick-walled, ornamented ascospore, 7–10 × 6–9 μm, as also shown in TEM (3126). The colonies grow somewhat faster (5–7 mm daily extension at 24°C) than most conidial isolates (4–5 mm). *D. australiensis* forms asci with 1–20 ascospores, 3–6 × 2·5–4 μm, surrounded by a gelatinous sheath. Colonies grow slightly more slowly (3·5–4 mm daily increment) than other, conidial isolates. This species is therefore taken as a proof of the close relationship between *D. geotrichum* and the other many-spored species of the genus (208).

The form-genus *Geotrichum* (type species *G. candidum*) is characterized by creeping, mostly submerged, septate hyphae; the septa are perforated by micro-pores arranged in a ring (2300, 2301, 3126, 6385). When the hyphae become air-borne, they soon fragment into arthroconidia which remain cylindrical or become barrel-shaped or ellipsoidal. Blastoconidia are sometimes formed laterally on the hyphae (regarded by Windisch (6400) as the asci of his "*Endomyces lactis*").

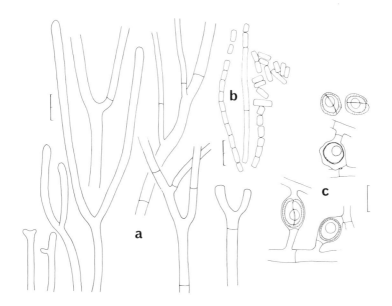

FIG. 143. *Geotrichum candidum*. a. Dichotomous branching of marginal hyphae, CBS 187.67; b. conidial chains; c. *Dipodascus geotrichum*, asci and ascospores formed between CBS 774.71 and 775.71.

G. candidum has fast-growing colonies, reaching 5·0–6·0 cm diam in 5 days at 25°C on Sabouraud-glucose or other agars, white, butyrous or membranous; odour often fruity. Advancing hyphae dichotomously branched, 7–11 μm wide. Conidial chains mostly aerial, erect or decumbent; conidia mostly 6–12(–20) × 3–6(–9) μm; (1–)2(–4)-nucleate (901).

The genus *Geotrichum* with ascomycete relationship can be distinguished from the rather similar genera *Trichosporon* Behrend (q.v.) and *Protendomycopsis* Windisch *1965* with basidiomycete relationship by ascending arthroconidial chains and absent or scanty blasto-conidia. The only other known equally fast-growing species of *Dipodascus* is *D. reessii* (van der Walt) v. Arx *1978* which has irregularly branched, 10–14 μm wide advancing hyphae and 1-spored asci; its ascospores are ellipsoidal to subglobose, thick-walled, and 6–7 × 4·5–6 μm. Other *Geotrichum* (about six are now recognized) and *Dipodascus* species grow less than 3 mm per day at 24°C and have no dichotomously branched advancing hyphae.

A serological comparison between various *Geotrichum* species has been carried out (3945). In a TEM study of the hyphae, a distinction was made between the apical cells and non-apical cells, the former containing numerous vesicles (5547); double-membrane-bounded vesicles are numerous close to the cell wall and associated with invaginations of the plasmalemma (5545). During conidium germination, a new wall layer forms inside the three layers which make up the mature conidial wall (5546). — DNA analysis gave a GC content of 40·5% (3232).

D. geotrichum is known from soil in Puerto Rico (818), gas oil in Germany and paper pulp in France (CBS); *D. australiensis* from rotting cladodes of *Opuntia inermis* in Queensland (208). — *G. candidum* is an extremely common fungus with a worldwide distribution, occurring mainly in soil, air, water, sewage, various plant substrates, such as mouldy hay, straw, corn (4494) and other cereals, and beet, baker's dough, bread, husks of fermentations, more rarely in breweries, and very frequently on milk and milk products (814, 3945, 5985).

As a causal agent of sour rot of citrus fruits it has a considerable importance (819, 3525). It also plays a role as causal agent of a tomato fruit rot (814), particularly after the fruits have been stored at 0–5°C, and can also attack ripe fruits of muskmelon, squash, and cucumber (814). It is commonly isolated from human faeces (in 25–30% of the samples), sputum and skin, but less frequently from animals (3945). It can cause infections called geotrichosis which are either of an endogenous origin and affect oral, bronchial and bronchopulmonary epithelia, or of an exogenous origin affecting the skin; allergic reactions to *G. candidum* have also been reported (901, 1745, 3342, 3945, 3947).

A selective isolation method is based on the tolerance of *G. candidum* to novobiocin and CO_2 (817). Soil isolates have been reported from Canada (505), the USA (655, 1039, 1163, 1387, 4733, 4918) including Alaska (1171), Poland (272, 1421, 1423, 4230), the British Isles (2923), Germany (3042), Austria (3418), the Ukrainian SSR (4474, 4548), Italy (3445, 3453, 4183), Turkey (4245), Israel (2764), South Africa (4407), central Africa (3063), India (1519, 2854, 3863, 4477, 4996, 5512), Japan (3267), New Guinea (3020), Hawaii (3014), Honduras (2031, 2035), Brazil (398), and Peru (2005). It has been found in cultivated soils (3042, 4996), under paddy (1519), banana (2031, 2035), grape (2719), and citrus (2764); in forest soils (2030, 2923), for example, under beech (272), teak (2854), white cedar (505), larch (3267), and *Eugenia* (5512). Other known habitats include grassland (3863), savannah (4407), an alpine pasture (3445), sand dunes (1423), desert soil (4733), rendzinas (3414), river (3809) and estuarine sediments (655, 4477), other marine habitats (468, 3014, 4918), fresh water (3942), recently deglaciated soil (1171), caves (3365, 3453), children's sandpits (1421), timber in a copper mine (2664), peat bogs (with high frequencies) (1039, 4474), pine litter (2344, 2345, 6080), grass roots (331), hazel mycorrhiza (1794), and the rhizospheres of banana (2031, 2035), groundnuts (2768), various steppe plants (3376), and *Polypodium* sp. (2856). It has also been found in sewage and activated sludge (814, 1165, 1167, 1170, 1387, 2923, 5928), composted municipal waste (3041), a waste stabilization pond (1166), on rotting paper (4548), cotton, textiles, furniture (3707), and, particularly, in paper mills (699, 1672, 5060). Other known substrates include rice seeds (4106) and polished rice (5980), frozen fruit cake (3153), fruit juices (5205), bananas (6158), *Drosophila melanogaster*, bees (814), birds' roosts (682, 2575), skin scrapings of domestic animals (2155), chicken coops, bat caves, cattle yards (2156), dung of rabbits (6085), piglets and dogs (63), and rarely coarse fodder (4548). It can be isolated from soils enriched with aromatic compounds related to lignin (2375). — Cell walls are degraded by the action of different enzymes of *Streptomyces satsumaensis* (5319), and its growth *in vitro* can be stopped by hyphal contacts with *Tuber melanosporum* (632).

The conditions for arthroconidium formation have been described (6400); good sporulation is dependent on the balance between C and N sources (4436). Conidia can germinate without any external energy sources, but show the phenomenon of self-inhibition at high densities (4437) and a negative autotropism during germination depending on the O_2 supply (4877). The optimal temperature for growth was found to be 25–27°C (1673) in isolates from plants, 30–31°C in isolates from animals (1612), and the maximum 35–38°C (3945, 5985). The thermal death point in soil is in the range 49–54·5°C for 15 min (6454) or 30 min (6453) or, according to other sources, 63°C for 25 min in apple juice (3620). — Good growth is supported by D-glucose, D-mannose, D-xylose, L-sorbose, D-fructose, D-galactose, sucrose, D-mannitol, D-sorbitol, ethanol, and glycerol (1166, 1673, 5344, 5985), but the sugars are not fermented (3945); citric acid, arabinose, cellobiose, maltose, rhamnose, raffinose (818, 3945, 6400) and cellulose (1673) are not utilized. The relationships between a lag phase in the utilization of C sources, specific growth rate, and the width of the peripheral growth zone have been analysed (847). In chemostat cultures the maintenance coefficient was 0·13 mmoles for glucose and 0·61 mmoles O_2 consumed per g dry weight per h; conidium formation was correlated with a low growth rate (4884). Suitable N sources include L-asparagine, urea, ammonium tartrate and sulphate, peptone, and glycine; L-histidine and L-tryptophan are weekly utilized, and nitrate is not assimilated (1166, 1673, 3945, 6400). *p*-Hydroxybenzoic acid, ferulic acid, syringaldehyde, and vanillin can be degraded (2375). The addition of vitamins is necessary for growth (3946). The production of glucose, galactose, mannose and ribose has been reported (3640), and the biosynthesis of the alkaloids, agroclavine, elymo-clavine, lysergic acid, and ergosine has been followed in different growth stages (1608, 1609). Strong lipase production was observed on rice bran, olive oil, oleic and linoleic acids (2696), or, according to other reports, exclusively on oils or fatty acids (2744, 5913). During the germination of the arthroconidia, polygalacturonase is liberated in the presence of pectin and sodium polypectate (334). An NADP-specific glutamate dehydrogenase is synthesized (and its formation repressed by NH_4Cl and urea) (338); a tyrosine-requiring mutant has been shown to produce phenylalanine hydroxylase, an enzyme which converts L-phenylalanine to tyrosine (3704); a mannitol dehydrogenase has been purified from cell-free extracts, which catalyses the oxidation of D-mannitol, D-sorbitol, D-arabitol and xylitol (3321). The formation of ornithine transcarbamylase is induced by an external supply of urea, and inhibited by arginine (337); the activity of NH_4- and glutamine-dependent carbamylphosphate synthesis in the presence of urea was also studied during conidium germination (336). — Mycelial extracts were found to be toxic to chicken embryos (1388). — *G. candidum* showed fair or good growth with a reduced supply of dissolved O_2 in the agar medium (679, 3814) or under anaerobic conditions (1244, 5680). With reduced O_2 conditions, hyphal cells are longer and usually unbranched, while their lateral branches arise in the vicinity of any localized O_2 supply (4879). A linear decrease in growth is found with increasing CO_2 concentrations beyond 1% (6280). — In a chronically irradiated soil *G. candidum* still occurred at intensities of up to 53 R per day (2009).

Gibberella Sacc. *1877*

Type species: *Gibberella pulicaris* (Fr. ex Fr.) Sacc.

Anamorph: *Fusarium* Link ex Fr.

Fam. Hypocreaceae de Not. — Perithecia ± superficial, often gregarious, with or without a basal stroma, fleshy, dark blue or violet; paraphyses absent. Asci cylindrical, unitunicate, with an undifferentiated apex. Ascospores ellipsoidal to fusiform, mostly 4-celled, hyaline to subhyaline. — *Gibberella* is delimited against *Calonectria* de Not. *1867* by its dark perithecia and against *Nectria* (Fr.) Fr. also by its many-celled ascospores.

The species of this genus have usually been treated in the literature dealing with *Fusarium* (q.v.), and the six species treated here are keyed out under their *Fusarium* anamorphs.

Gibberella acuminata Wollenw. *1943* — Fig. 144.

= *Gibberella saubinetii* (Dur. & Mont.) Sacc. f. *dahliae* Sacc. *1880*
= *Gibberella acuminata* C. Booth *1971*

Anamorph: *Fusarium acuminatum* Ell. & Everh. *1895*

≡ *Fusarium scirpi* Lamb. & Fautr. var. *accuminatum* (Ell. & Everh.) Wollenw. *1931*
≡ *Fusarium gibbosum* Appel & Wollenw. emend. Bilai var. *acuminatum* (Ell. & Everh.) Bilai *1955*
= *Fusarium roseum* Link emend. Sny. & Hans. *1945* pro parte

DESCRIPTIONS: Wollenweber and Reinking (6424), Gordon (2039), Booth (640), Gerlach and Ershad (1949), and Joffe (2770). — Anamorph belonging to *Fusarium* section *Gibbosum* Wollenw. — Colonies fast growing, reaching 4·5 cm diam in four days at 25°C on PSA or OA, strongly carmine-red, aerial mycelium ochraceous to brownish, floccose, with reddish brown sporodochial stromata. Conidia slender sickle-shaped, strongly curved, with pedicellate basal cell, 3–5-septate, 24–54 × 3·0–4·5 μm. Chlamydospores mostly intercalary and in chains, brownish, 7–20 μm diam. Ascospores of *G. acuminata* 22–26 × 5–6 μm (smaller than in *G. intricans* Wollenw., q.v.). *F. acuminatum* was considered worthy of specific rank by Gordon (2039) on account of its red colonies and the distinct teleomorph.

The distribution of *G. acuminata* is worldwide on a wide variety of plants (2043, 2045), not only in the temperate zone, but also in tropical countries (640, 2045, 6425). In Canada it constituted 3% of all *Fusarium* soil isolates (2041). Soil isolates are known from Australia (977, 978), Hong Kong (1048), Kuwait (4000, 4001), India, Sri Lanka, Pakistan, Nigeria,

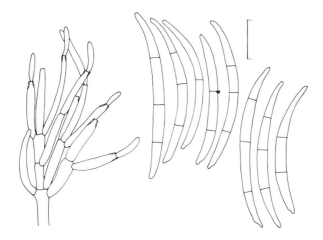

Fig. 144. *Gibberella acuminata*, conidiophore and conidia, CBS 680.74.

Libya (6510), Turkey (3473, 4245), the British Isles (640), and the Ukranian SSR (4556). It is also known from a wide range of plants and micro-environments in the USA (1165, 1166), Finland (6493), Northern Ireland (4429), Hungary (2538), Denmark, Germany, Israel, the Sudan, Egypt, Kenya, Bangladesh and Thailand (640). It has been found in soils under cereals (2041) and often appears as a secondary colonizer of injured roots, for example those of cotton (1965), alfalfa and clover (6493, IMI) and is also recorded in the rhizospheres of corn (4556), wheat and alfalfa (2538), on seeds of cereals (2038, 2039, 2538), carrot (IMI) and those of other plants (2043), pods of groundnuts (346), on organic detritus in fresh water (4429), a waste stabilization pond (1166), saline soil (4000, 4001), and a sewage treatment plant (1165).

G. acuminata is not affected by CO_2 partial pressures deviating from normal (3481). The growth of *Fusarium solani*, *Nectria radicicola* and *Pythium periplocum* can be inhibited by *G. acuminata in vitro* (6029). It is able to grow on media containing 20–40% sucrose (4001), which represents an osmotic potential of about −20 to −60 bars.

Gibberella avenacea R. J. Cook *1967* — Fig. 145.

Anamorph: *Fusarium avenaceum* (Fr.) Sacc. *1886*

≡ *Fusisporium avenaceum* Fr. *1832*
= *Fusarium roseum* Link emend. Sny. & Hans. *1945* pro parte

DESCRIPTIONS: Wollenweber and Reinking (6424), Gordon (2039), Schneider (5118), Cook (1135), Booth (640), Joffe (2770), and CMI Descriptions No. 25, *1964*; summary of the macroscopic features: Colhoun and Park (1119). The variability has been investigated repeatedly (1189, 5118). — Anamorph belonging to *Fusarium* section *Roseum* Wollenw.

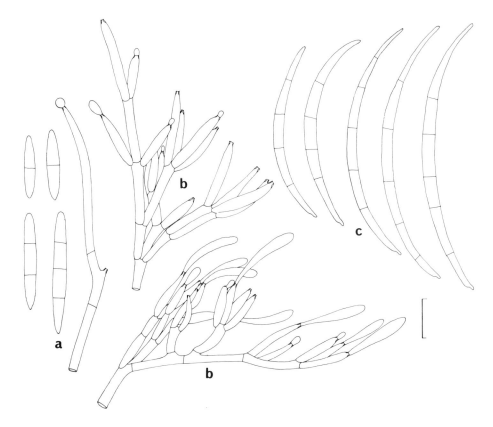

Fɪɢ. 145. *Gibberella avenacea*. a. Sympodially proliferating conidiophore (exceptionally observed) in young colony and primary conidia of irregular shape; b. aggregated conidiophores of beginning sporodochia; c. sporodochial conidia; CBS 121.73, orig. J. Veerkamp.

(*Arthrosporiella* sensu C. Booth). — Colonies fast growing, reaching 5·4 cm diam in four days at 25°C on PSA or OA; aerial mycelium rose-red fringed with white, reddish brown from below; in old cultures flecked with orange sporodochia. Conidiophores have been described as being initially scattered throughout the aerial mycelium, and sometimes bearing polyphialides with shorter conidia of irregular shapes (640). The observation of polyphialides may apply to *F. arthrosporioides* Sherb. *1915* but it could not be verified for *F. avenaceum* (1948). Sporodochia containing regularly falcate, long and slender conidia with an elongated apical and distinctly pedicellate basal cell, 3(–5)-septate, 30–66(–80) × 3·0–4·5 µm. Chlamydospores absent. — Most authors recognized only one species, *F. avenaceum*, with such slender sporodochial conidia, whereas Gerlach (1945) preferred to distinguish *F. graminum* Corda *1837* with straighter conidia, 30–55 × 2·5–4·5 µm (listed by Booth under *F. heterosporum* Nees ex Fr. *1832* in section *Discolor*) and possibly *F. detonianum* Sacc. *1886* with longer, up to 9-septate conidia, 66–97 × 3–4 µm. The last species treated in section *Roseum* (6424), *F. arthrosporioides* Sherb. (originally the type species of sect. *Arthrosporiella* Sherb. *1915*), with shorter conidia, 22–55 × 4–6 µm, is placed by Booth (640) in section *Gibbosum* although chlamydospores are absent and polyphialides have originally been described. — Hyphal cells of *G. avenacea* often contain between 1 and 4 nuclei but up to 5–10 have been observed (397,

943, 2496), particularly in the peripheral tip cells; cells of the macro-conidia were found to contain one nucleus each (514) with n=6 (4669). Atypical anastomoses with isolates of other *Fusarium* species have been observed (5118). The production of sporodochia is promoted by UV-irradiation (5106).

G. avenacea is a cosmopolitan weak parasite of numerous plants, in particular cereals, where it may become severely pathogenic on roots of overwintering crops; but it is generally less frequent than *F. culmorum* and *Gibberella zeae* (640). It is very widely distributed in the temperate zone but also occurs in the tropics and subtropics (2045). It can be isolated easily from soil suspensions and from washed, particularly organic, soil particles (1888, 1889) as well as by other techniques. In general it is found more rarely in soils than on plant material of very different kinds. — Its frequent occurrence in grassland and arable soils cannot be documented here by individual citations. It has also been isolated from garden soil (4716, 6289) and found rarely in hardwood forest (1555, 2854, 3127, 3549, 5512) and moorland soils (3234). *G. avenacea* mainly occurs in the uppermost soil layers penetrated by roots (5812). It occurs in the soil significantly more frequently after ploughing-in oat plants (1330) or growing wheat than under peas or rape (1433); it was frequent in permanent wheat soils (918). There are reports from the rhizospheres of alfalfa (3982, 3983), clover (5815), onion (2159), *Lolium perenne* (5815) and other grasses (3368), corn (4556), oats (3002, 4443), barley (3002, 3006), wheat (1614, 3095, 4305), several sand-dune plants (4371), and ferns (2858), and the rhizoids of *Funaria hygrometrica* (2860), and the rhizoplanes of red clover (6493), alfalfa (2538), wheat (945), and *Amaranthus oleraceus* (2679). Furthermore, it is common on seeds of oat and barley (1751, 1752, 1754, 2039, 2820, 3556), wheat (2038, 2039, 2437, 4307, 4492), *Avena fatua* (2961), cultivated grasses (3512) and *Cyamopsis tetragonoloba* (5350), and known on germinating seeds of clover (4120) and many other plants (2040). It was found on frost-damaged leaves of various grasses (191) and in the leaf litter of *Eucalyptus maculata* (1558) and has been isolated from the sclerotia of *Sclerotinia borealis* and *Botrytis cinerea* (3531), and water (1946). — While true chlamydospores do not occur, *G. avenacea* forms a type of resting spore within macro-conidia and hyphae when in contact with soil; these survival units may survive for 18 or more months (2269). It survived in soil cultures at 5°C for eight years, and at room temperature for three years (1189). It may cause serious losses in cereals due to its role as causal agent of pre-emergence and seedling blight in cooler climates (see CMI Descriptions No. 25). It possesses a high competitive saprophytic ability (4734) but this ability may vary between isolates (1796). — In artificial culture, its growth was suppressed by both *Streptomyces rimosus* (998) and cochliodinol (704).

The optimal temperature for growth is 25°C, the maximum 31°C, and the minimum −3°C (191). This fungus still grows even at a water potential of −130 bars; half of its optimal growth occurs at −40 bars (2093). The pH optimum for different isolates ranges between 5·4 and 6·7 (1796, 5118). — Glucose, mannose, galactose, fructose, sorbose, rhamnose, arabinose, xylose, ribose, sucrose, maltose, starch, sorbitol, mannitol, lactate, gluconate are all listed as suitable C sources; D- and L-amino acids are equally well assimilated (1000). Varying capacities for decomposing pectin, xylan and cellulose are reported (1432, 3618, 6289) and the influence of temperature, pH and nitrogen on cellulose decomposition has been studied (1795); arabinoxylan is degraded by xylanase and cellulase (1753). *G. avenacea* grows relatively well in nutrient solutions with 5% lignin sulphonate as the sole C source (3260); ferulic and *p*-hydroxybenzoic acids can be readily utilized (378); several hydrocarbons from fuel oil can also be utilized (4239). On potassium-deficient nutrient media, growth is reduced; phosphorus deficiency reduced sporulation, and nitrates were utilized better than ammonium salts (3096). Biotin increases growth, but mutants autotrophic for biotin have also been observed (4861). Enniatin B has been described as a metabolic product of *G. avenacea* and this has a low antibiotic activity; its production is dependent on the N source (5839). An antibiotic activity towards *Chalara elegans* (2753), *Rhizoctonia solani, Gaeumannomyces graminis* and a *Pythium* species has been demonstrated *in vitro* (1425). — In laboratory experiments, *G. avenacea* proved highly pathogenic to wheat, peas and rape (1430); on the other hand, greenhouse infection tests had only slight success with wheat (2789) and infection increased with low soil humidity (1118, 1119, 1120). This fungus can be pathogenic to clover (5815) but, in contrast, corn seedlings growing on sand are said to be stimulated by both the living fungus and by culture filtrates (4557). Ingestion with food causes vomiting in mammals (733).

Gibberella baccata (Wallr.) Sacc. *1883* — Fig. 146.

≡ *Sphaeria baccata* Wallr. *1833*
≡ *Gibberella pulicaris* (Fr.) Sacc. subsp. *baccata* (Wallr.) Sacc. *1878*
= *Gibberella lateritium* (Nees) Sny. & Hans. *1945* (nomen illeg., Art. 59)

Anamorph: *Fusarium lateritium* Nees ex Link *1824*

= *Fusarium lateritium* Nees ex Link emend. Sny. & Hans. *1945* pro parte

DESCRIPTIONS: Wollenweber and Reinking (6424), Gordon (2039), Booth (640), Gerlach and Ershad (1949), Joffe (2770), and CMI Descriptions No. 310, *1971*. — Anamorph belonging to *Fusarium* section *Lateritium* Wollenw. — Colonies comparatively slow-growing, reaching 2·8 cm diam in four days at 25°C on PSA, covered with floccose to felted, yellow to reddish brown aerial mycelium; stromatic agar surface and sclerotial bodies pink or tan to violet or blue-black. Sporodochia usually abundantly formed on the sclerotia, producing a deep orange conidial slime. Conidiophores abundantly branched, phialides doliiform to cylindrical, 10–30 × 2·5–4·0 μm, sometimes percurrently proliferating. Conidia almost straight in the central part, with beaked apical and distinctly pedicellate basal cells, 3–5(–7)-septate, 22–50(–75) ×3–4·5(–5·0) μm. Chlamydospores sparse in the mycelium, sometimes formed from conidial cells, 7–8 μm diam. Perithecia sometimes produced on the stromatic pustules in agar cultures, but more commonly on sterilized wheat straw after 3–4 weeks at 25°C in light; homothallic and heterothallic isolates are known to occur. Perithecia appearing black, with a slightly verrucose peridium. Ascospores smooth, hyaline, 1–3-septate, 13–18 × 5–8μm. — Two varieties of *F. lateritium* with larger conidia have been distinguished: var. *majus* (Wollenw. *1917*) Wollenw. *1931* with mostly 5-septate conidia, 46–64 × 3·5–4·7 μm and var. *longum* Wollenw. *1931* with 5-septate conidia 52–69 ×3·9–4·9 μm (6424). Amongst the other species of the section, *F. stilboides* Wollenw. *1924*, the anamorph of *Gibberella stilboides* Gordon ex C. Booth *1971*, has carmine red colonies and mostly 5-septate conidia 48–73 × 3·5–5·0μm and *F. xylarioides* Stey. *1948* produces numerous sharply curved micro-conidia; it is retained in sect. *Elegans* by Gerlach (1948) and so is *F. udum* Butler *1910* with abundant oval to allantoid micro-conidia. — Hyphal and conidial cells of *G. baccata* are usually uninucleate, but the latter may be up to 5-nucleate (514, 2496).

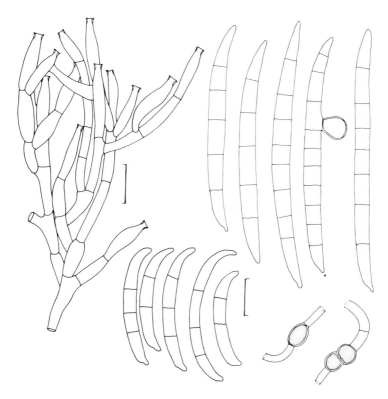

FIG. 146. *Gibberella baccata*, conidiophore, conidia and chlamydospores (of *Fusarium lateritium* var. *longum*), CBS 737.74, partly orig. J. Veerkamp.

The distribution of this species is worldwide on woody hosts (640, 2043, 2045, CMI Descriptions No. 310) such as *Buxus, Citrus, Juglans, Pinus, Coffea, Olea* and many others, but it also occurs on cereals and ornamental shrubs; it is responsible for die-back, bud and twig blight or canker. Three *formae speciales* occurring on cereals, mulberry, pines and a variety *buxi* C. Booth *1971* on *Buxus* leaves have been distinguished. — It is only relatively infrequently isolated from seeds and soil, but soil isolates are known from both tropical and temperate regions (2045) including Bangladesh (2712), India, Australia (IMI), Hong Kong (IMI, 1048), Hawaii (3264), the Bahamas (2006), Egypt (4962), Syria (277, 5392), Turkey (4245), Spain (3446), Italy (3445), France (2161, 4178), Germany (3127), the British Isles (4429), the USA from the Sonoran desert (4733), Georgia (3817), and Oklahoma (1632), and the USSR (2871, 3364, 4548, 4556). *G. baccata* occurs in cultivated (277, 1048, 2712, 3446, 3817, 4178, 4962) as well as in forest soils (1713), under grassland and prairie type vegetation (277, 1632, 3445), in chalk soil (4152), chernozem (4556), clay of a carst cave (3364), mangrove swamps (3264, IMI) and unpolluted streams (4429). It has been isolated from the rhizospheres of corn (4556), rice (2712), cabbage and potato plants (3113), litter of *Betula alba* (3449), leaves of vascular plants (296), stumps of *Fraxinus* (3568), and various other plant remains (4548). It is also known from *Suillus luteus* mycorrhiza on *Pinus strobus* (1794), seeds of cereals (2039, 6407), stored olives (309), other foodstuffs (1303), citrus fruit (1873, 2042, 2769) and cotton fabrics (6326). It was inhibited by tannin from oak leaf litter *in vitro* (2282). — It can serve as food for the nematode *Aphelenchus avenae (3579)*.

The optimal temperature for the germination of conidia is 25°C (1873). — Pectin, cellulose, and urea are all utilized (3127, 3618, 6060, 6326) and under iron-deficient nutrient conditions Δ^2-anhydromevalonic acid lactone is produced (2918). — The mycotoxin diacetoxyscirpenol, produced from this species, was found to have larvicidal properties (1112), and the related T-2 toxin (790, 3858) and zearalenone (2684) have also been isolated from it. A toxic principle causing death of brine shrimps and chicken embryos (1303, 1388) and toxic leucopenia in man has been described (1815). Further metabolic products include baccatine A, a substance with phytotoxic (1855) and antibiotic properties against both bacteria (725) and other fungi (2151). An antagonistic influence on the vascular pathogen *Eutypa armeniacae* has also been reported (913).

Gibberella intricans Wollenw. *1931* — Fig. 147.

Anamorph: *Fusarium equiseti* (Corda) Sacc. *1886* sensu Gordon *1952*

    ≡ *Selenosporium equiseti* Corda *1838*
  = *Fusarium scirpi* Lamb. & Fautr. *1894*
  = *Fusarium gibbosum* Appel & Wollenw. *1910*
  = *Fusarium roseum* Link emend. Sny. & Hans. *1945* pro parte

DESCRIPTIONS: Wollenweber and Reinking (6424), Gordon (2039), Gerlach and Ershad (1949), Booth (640), and Joffe (2770). — Anamorph belonging to *Fusarium* section *Gibbosum* Wollenw. — Colonies fast-growing, reaching 5·8 cm diam in four days at 25°C on PSA or OA, with abundant peach to buff, later brown, but never red, aerial mycelium and similarly coloured agar surface; reverse initially peach but becoming dark brown later. Conidia often sparse, sometimes abundant in yellowish to salmon sporodochia. Conidiophores densely branched, bearing obclavate phialides, 12–17 × 3–4 μm, with a single apical opening. Conidia of variable shape, 3–5-septate, with distinctly pedicellate basal cells, either little curved and with a blunt apical cell, 31–47 × 4·0–5·0 μm (responding to the description of *F. equiseti* var. *bullatum* (Sherb.) Wollenw. *1931*) or strongly bent mainly in the central part with strongly elongate apical and basal cells, to 66 μm long, and more commonly forming sporodochia (as described for *F. scirpi*, with *F. equiseti* var. *equiseti* as an intermediate). Chlamydospores abundantly produced in intercalary chains, ochraceous, thick-walled, verrucose, 7–9 μm diam. This species can usually be recognized by its ochre coloration and coarsely warted, catenulate, ochraceous chlamydospores even in the sterile state, but sporulation can be promoted by near-UV irradiation. Perithecia are not usually formed in culture. — Gordon (2039) has combined the species and varieties of section *Gibbosum* without any red pigmentation as one species (*F. equiseti*) because no clear line could be drawn between these different types although the extreme types differ very strongly indeed. — The similar *F. semitectum* Berk. & Rav. (q.v.) can be distinguished by the presence of polyphialides and straighter, less pedicellate conidia. — Hyphal cells of *F. equiseti* contain mostly 2(–5) nuclei (2496); a single nucleus was found in conidial cells (514) which on division gives rise to 8 chromatin strands (641, 4669).

    G. intricans has been isolated from numerous plants and fruits in temperate to subtropical

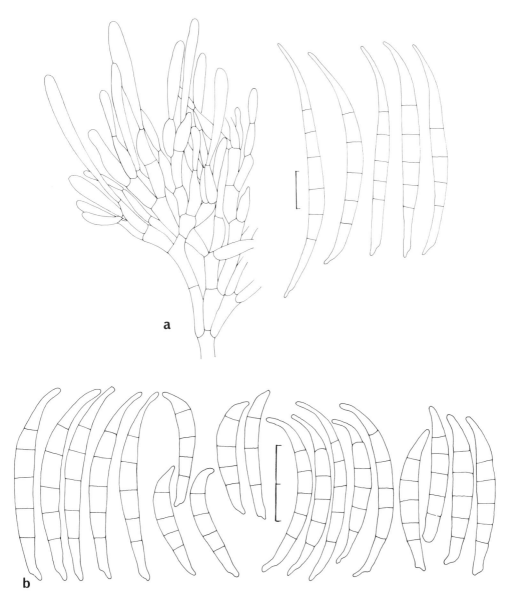

F<small>IG</small>. 147. *Gibberella intricans.* a. Aggregated conidiophore of beginning sporodochium and conidia of tropical isolate, orig. J. Veerkamp; b. conidia of various isolates.

zones, in particular cereals, where it may cause stem and root rot (640). — It is a cosmopolitan soil fungus (2045), principally occurring in arable soils. Its distribution extends from Alaska (6138) to the tropics and subtropics, where it ranks as one of the most frequent species of the genus. It is reported from Israel (2762, 2764, 2768, 2772), Syria (5392), Libya (4083, 6510), the Sudan (4222), Zaïre (3790), equatorial West Africa (1420), South Africa (1559, 4407), Iraq (92), Kuwait (4000, 4001), Pakistan (61, 4855), India (41, 5314, 5622, 6059), and particularly frequently in Central and South America (2042, 4547, 4795, 4796). It has been found rarely in hardwood forests (3549), frequently in nursery soils (6184), soils under savannah type vegetation (4407), but most commonly in grassland soils (159, 977, 1614, 3127, 4179) and in arable soil by numerous, not cited authors. In Canada, *F. equiseti* is one of the most frequent *Fusarium* species (2041). It has proved to be common in salt marshes (4000) and polluted waters, for example, a waste stabilization pond in Ohio/USA (1166), a sewage treatment plant (1165) and children's sandpits (1421). It has been isolated from washed organic soil particles more frequently than mineral ones and dilution plates (1614) and preferentially from soil layers close to the surface (2161). It occurred frequently at pH values close to 7 (159, 6184) and at high soil moistures (977), and is tolerant of high CO_2 concentrations in soil (2096). It has been isolated regularly from plant roots of the most varied origin, including those of cabbage (4451), alfalfa (3983), beans (4451), cotton (5009), pears (5499), tea (41), *Artemisia herba-alba* (4084), onion (2159), some halophytes (3375), reported particularly often from cereals such as oats (3002), barley (3002, 3006, 4444), corn (4553, 4556), wheat (945, 2491, 2949, 3095, 4136, 4558, 5330), rice (3584), and various grasses (191), and also in the rhizospheres of groundnuts (2768), grasses, wheat (1614, 2538), barley, alfalfa (2538), *Coffea arabica* (5313), *Aristida coerulescens* (4083) and ageing roots of various steppe grasses in Kazakhstan (3376). It is also frequently associated with sorghum ear-moulds in India (5314) and isolated from cereal grains in Canada (2039), especially those of wheat (2038, 4307); but also of barley (57), pearl millet (3670), onions (5467), and groundnuts (2768) and found on coarse fodder (4548) and salted olives (309). It has been isolated from cysts of the nematode *Globodera rostochiensis* (3090), earthworm casts (1429) and sticks exposed to soil (1956). — Conidium germination is inhibited in soil by mycostatic factors (6284) while *in vitro* the fungus was inhibited by *Streptomyces* species (2939).

The optimal temperature for growth is 21°C, the maximum 28°C, and the minimum −3°C observed in Norway (191), but in another study the optimum was given as 30°C (2939); the minimum water potential is at −110 bars and the optimum at vapour saturation (977, 3109). Growth on media with 20–40% sucrose is reported (4001); this corresponding to osmotic potentials between −20 and −60 bars. Variability has been found with respect to light requirements for sporulation (5106). Sporulation under submerged conditions occurs on sucrose-containing media (6088). — Moderate growth is obtained on raffinose and glycine as C and N sources, respectively (2623). Pectin decomposition is good but variations of this capacity occur in different isolates (1432); the microsomal fraction shows amylase and inulase activity (4746). Cellulose utilization (filter paper, cotton fabric, CMC) is good but also variable (932, 1425, 1432, 1956, 6289, 6326), with the optimum pH at 5·5 for this (6060). Growth on 5% lignin sulphonate is moderate (3260), *n*-paraffins and monohydric alcohols are utilized for growth (3123), and *G. intricans* is autotrophic for growth substances (1425). It produces an extracellular phosphomonoesterase (498), cyclo-L-prolylglycine (6323), and (±)-2-acetamido-2,5-dihydro-5-ketofuran (6322). Amongst numerous toxic metabolic products (790, 2130, 2131), diacetoxyscirpenol, together with enniatin, is the most important (1308, 1309), but the influence of this compound on bacteria and other fungi is slight (712). — In pure culture the fungus causes injury to barley (4919), wheat and peas (1430). *F. equiseti* s.str. is reported to be hardly pathogenic to *Agropyron cristatum* and *Bromus inermis*, whereas isolates corresponding to *F. scirpi* were pathogenic (5413). Stimulation of growth has been observed on both corn seedlings (4553) and tomatoes (15). When the fungus has accumulated in fodder cereals, it can be toxic to warm-blooded animals (6412); toxic effects on experimental animals have been described (733, 1815).

Gibberella pulicaris (Fr. ex Fr.) Sacc. *1877* — Fig. 148.

≡ *Sphaeria pulicaris* Fr. ex Fr. *1823*

Anamorph: *Fusarium sambucinum* Fuckel *1870* var. *sambucinum*

= *Fusarium roseum* Link ex Gray *1821* (nomen ambiguum)
= *Fusarium sarcochroum* (Desm. *1850*) Sacc. *1879* fide Booth (640)
= *Fusarium discolor* Appel & Wollenw. *1910*
= *Fusarium roseum* Link emend. Sny. & Hans. *1945* pro parte

Amongst several substrates mentioned by Link (*1809*) for *F. roseum*, Gray in *1821* cited only stems of a malvaceous plant as the substrate in the validating text and thus determined the application of the name (cf. Wollenweber, Fus. autogr. del., No. 311, *1916*) although two other of Link's specimens are extant which bear different fungi (6424). The correct name of the anamorph would thus be *Fusarium roseum* Link ex Gray, had this name not been used by Snyder and Hansen (5461) in a different sense, thus rendering it ambiguous.

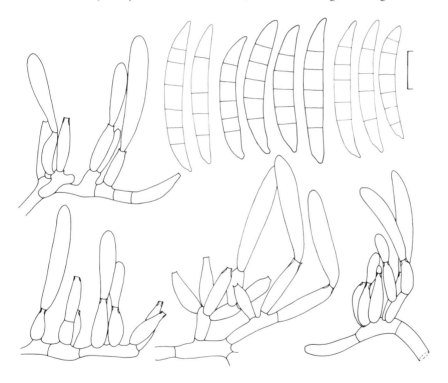

Fɪɢ. 148. *Gibberella pulicaris*, conidiophores and conidia, CBS 136.24, partly orig. J. Veerkamp.

DESCRIPTIONS: Wollenweber and Reinking (6424), Gordon (2039), Booth (640), Joffe (2770), and CMI Descriptions No. 385, *1973*. — Anamorph belonging to *Fusarium* section *Fusarium* (= *Discolor* group). — Colonies fast-growing, reaching 5·2 cm diam in four days at 25°C on PSA or OA, rose, becoming blood-red particularly at the agar surface (more bluish on neutral or slightly alkaline media); aerial mycelium white to rose; sporodochia present in most strains, covered with an orange conidial slime, solitary conidiophores also present. Conidiophores repeatedly branched, phialides cylindrical to doliiform, $16–22 \times 4–5 \mu$m. Macroconidia fusiform, curved, with short, pointed apical and distinctly pedicellate basal cells, 3–5-septate, the cells usually not longer than wide, hyaline $30–45(–55) \times 4·0–5·5 \mu$m. Chlamydospores sparse, intercalary, terminal or lateral, often also developing from conidial cells, $6–11 \mu$m diam. Dark brown sclerotial bodies usually present. Perithecia known mainly from woody substrates, but also obtained on sterilized straw after four weeks at 25°C by mating compatible strains. Ascospores 3-septate, slightly constricted at the septa, smooth-walled, hyaline, $20–28(–38) \times 6–9(–10) \mu$m. — The varieties and forms distinguished by Wollenweber and Reinking (6424) are generally no longer recognized, with the exceptions of *F. sambucinum* var. *coeruleum* Wollenw. *1917* with shorter conidia (av. 30 μm long), more reddish purple pigmentation and dark blue sclerotia, and *Fusarium sulphureum* Schlecht. *1824* ($\equiv$ *F. sambucinum* f. 6 Wollenw. *1931*), the anamorph of *Gibberella cyanogena* (Desm.) Sacc. *1883* (= *G. saubinetii* (Mont.) Sacc. *1879*), which does not develop any red pigmentation but is mainly yellow to peach, later becoming brown, and has conidia with more gradually tapering apical and less pedicellate basal cells. *F. culmorum* (W. G. Sm.) Sacc. (q.v.) may be confused with *G. pulicaris* but has wider (to 7 μm) and slightly pigmented conidia. *Fusarium sarcochroum* (Desm.) Sacc. *1879* sensu Wollenw. is a different species of sect. *Lateritium* whilst the type of *F. sarcochroum* was shown to represent *F. sambucinum* by Booth (640). — Hyphal cells are mostly 3–4(up to 8)-nucleate (2496), and the conidial cells uninucleate (514). — DNA analysis gave a GC content of 51·5% (5600).

G. pulicaris is common in the temperate to mediterranean regions of the northern hemisphere, and possibly not native in the southern hemisphere (CMI Descriptions No. 385) but it also occurs in the tropics (1947). It is associated with various plant diseases including root rots, especially in cereals, but also in flax, lupin, strawberry and tomato amongst others, storage rot in potatoes, fruit rot in pome fruits and cucurbits (6424), and canker and die-back in woody plants and also hops (CMI Descriptions No. 385; 640, 2043, 2045, 6425). — A wide distribution in soil is documented (2045) with reports ranging from the White Sea (3368), sediments below a glacier (2530), peat fields in the Ukrainian SSR (4474), a waste stabilization pond in Ohio/USA (1166) to soils in mediterranean and (sub)tropical areas in Turkey (4245), Syria (5392), Israel (2764), equatorial West Africa (1420), South Africa (1559), Pakistan (4855) and India (41). Reports are numerous, particularly for grassland and arable soils, though *G. pulicaris*, in contrast to other *Fusarium* species, occurs fairly regularly also in forest soils (3362, 4180), forest nurseries (6184), but also sand dunes (1423, 3941) and children's sandpits (1421). Data on soil depth are not uniform and range from observations of occurrence in the surface (6182) to the subsoil layers (5812). *G. pulicaris* occurs in soils of a wide pH range, from neutral to alkaline (159, 3871, 6182). The optimum temperature for the colonization of bean roots is given as 20°C (5758). Manuring with PK has been said to increase its frequency in the soil, whilst NK and NP reduced it (2163). Observations on its occurrence on plant roots are numerous; particularly mentioned here are various steppe grasses (3376), oats and barley (3002, 3006, 4450, 4902), corn (4556), wheat (945, 2492), sugar beet (4559), garlic (2159), tea (41), alfalfa (3983), red clover (3429, 6493), broad beans (5757) and beans (4451, 4452, 4902, 5758, 5759), where the hyphae are confined to the surface. In wheat, the hyphae penetrate deep into the cortical parenchyma and give impression of a kind of mycorrhiza (4444). *G. pulicaris* has frequently been found in the rhizospheres of wheat (3095, 4558) and groundnuts (2768), and several authors have noted its sparse occurrence in cereal grains (2038, 2039, 5293); it has also been isolated from seeds and pods of groundnuts (1984,

2768), cysts of the nematode *Globodera rostochiensis* (3090), tunnels and storage places of gerbils (5393); it appeared frequently in earthworm (*Pheretima californica*) cultures (1961), and sometimes in water (1160, 1946). — Forty-four per cent of chloroform-fumigated mycelium was mineralized within ten days at 22°C (133). — *G. pulicaris* can serve as food for the mites *Pygmephorus mesembrinae* and *P. quadratus* (3104).

G. pulicaris has a pectinase activity with some variation occurring between individual isolates (1432), and cellulose decomposition observed on filter paper (1425, 4184, 6289), which is most efficient at pH 5·5 (6060). No growth substances are required (1425). *G. pulicaris* forms diacetoxyscirpenol (1787) and the enniatins A and B (239, 5278), the maximum yield of enniatins appearing in ageing cultures at 20°C on lactose or glycerol and with tryptone as N source (240). Other metabolites include γ-carotene, lycopene, rhodoxanthin (2087), and an extracellular phosphomonoesterase (498). — *G. pulicaris* inhibits the growth of *Chalara elegans* (2753), *Rhizoctonia solani, Gaeumannomyces graminis* (1425), and some soil bacteria (3127). Germination and shoot growth were promoted in corn (4553) and inhibited in barley (293) by *G. pulicaris*. It can be one of the causes of toxic leucopenia in mammals (1815).

Gibberella zeae (Schw.) Petch *1936* — Fig. 149.

≡ *Sphaeria zeae* Schw. *1822*
= *Gibberella saubinetii* (Mont. *1846*) Sacc. *1879* pro parte, sensu Wollenweber [*G. saubinetii* fide C. Booth = *Gibberella cyanogena* (Desm.) Sacc. *1883*, teleomorph of *F. sulphureum* Schlecht. *1824*]

Anamorph: *Fusarium graminearum* Schwabe *1838*

= *Fusarium roseum* Link *1809* emend. Sny. & Hans. *1945* pro parte

DESCRIPTIONS: Wollenweber and Reinking (6424), Gordon (2039), Booth (640), Joffe (2770), and CMI Descriptions No. 382, *1973*. — Anamorph belonging to *Fusarium* section *Fusarium* (= *Discolor* group). — Colonies very fast growing, reaching 9 cm diam in four days at 25°C, greyish rose to livid-red to crimson, often becoming vinaceous with a brown tinge, aerial mycelium floccose, somewhat lighter coloured and becoming brown, easily wetted (1119). Sporulation often scarce, but the formation of sporodochia can be induced by near-UV irradiation. Densely branched conidiophores occurring besides solitary phialides; phialides doliiform, 10–14 × 3·5–4·5 μm, sometimes percurrently but not sympodially proliferating. Conidia slender falcate, moderately curved, with pointed and curved apical and pedicellate basal cells, mostly 5–6-septate and 41–60(–80) × 4·0–5·5 μm. Chlamydospores scarce and often completely absent, mostly intercalary and in chains, 10–12 μm diam. — For species differentiation compare also *F. flocciferum*. — The teleomorph is commonly associated with the anamorph on cereal hosts. Perithecia dark blue, 140–250 μm diam, tuberculate; ascospores 3-septate, 18–27 × 3·0–4·0(–5·0) μm. This species is normally homothallic and some isolates produce perithecia in culture on natural substrates such as wheat straw or carnation leaf tissue (5909). — Conidial cells are uninucleate, whilst vegetative hyphal cells may be plurinucleate (514, 943) with n=6 rod-shaped chromatin strands (4669).

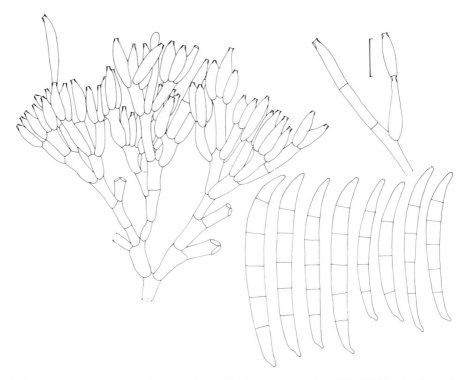

FIG. 149. *Gibberella zeae*, sporodochial and simple conidiophores and conidia, CBS 389.62, orig. J. Veerkamp.

G. zeae is mainly known as a colonizer and sometimes a parasite of graminaceous and other plant hosts from all continents (6424, 6425). It may cause pre- and post-emergence blight, root and foot rot, crown rot, culm decay and cob or head rot in corn, wheat, barley, oats (640, 5140), sorghum (3237) and die-back in carnation (4125). — Reports of isolations from soil are available from tropical and temperate regions (2045), for example, the British Isles (163, 1447), the Netherlands (1614, 1616), Poland (1423), Hungary (2538), Austria (3413), the Ukrainian SSR (4556), and South Africa (1415, 1559). It is common on seeds of cereals (2039, 2538), particularly barley (2820) and corn (3781), and is also known on seeds of common grasses (3512). Other habitats include debris of *Ammophila* sp. (163), organic material in unpolluted streams (4429), pods of groundnuts (1984), cabbage and barley residues (3113), wheat roots (945), the rhizospheres of barley, oats (3002), rice (2712), corn (4556) and alfalfa (2538), rarely coarse fodder (4548), and the tick *Ixodes ricinus* (5018). — Inhibition by soil bacteria (4272) and also by a-tomatine (182) has been demonstrated. Survival on infested wheat straw is drastically reduced at high temperatures and a broad range of soil humidities (786). The phenomenon of induced "suppressiveness" of a soil has been observed with *G. zeae* (5865).

The optimal temperature for growth is about 25°C (640, 788). Good growth *in vitro* occurs between pH 5 and 8 with an optimum at pH 6·7–7·2 (640). Macro-conidia also form in shake cultures on media containing sucrose (6088) or CMC, mineral salts and yeast extract (883) or acetate at pH 8 (885). A method for uniform production of ascomata has been described

(5908); from 15°C up to 29°C ascoma formation increases with rising temperatures; ascospore discharge is favoured by low temperatures with no discharge above 26°C; production of ascomata is best at low intensities of near-UV light without dark period; 5–6-day-old cultures were found to be most photosensitive with respect to ascoma initiation (5909). — Several polyphenols are utilized (378). Proteolytic enzymes are produced abundantly when grown on glycerol, lactate or starch (4640) as C sources and on NH_4Cl as N source (4641). *G. zeae* produces pectinmethylesterase, cellulase (884), zearalenone, a metabolite with estrogenic action (3857, 6018) and several zearalenone derivatives (623, 5544), zeagenin, aureofusarin (3231), fusarenon-X, nivalenol (5993), T-2 toxin (see under *Fusarium tricinctum*) (790), rubrofusarin and ravenelin (5278). Arabinoxylan is degraded by both xylanase and cellulase (1753). — CO_2/O_2 ratios up to 20:0 cause little or no reductions in growth (3481).

Gilmaniella Barron *1964*

Gilmaniella humicola Barron *1964* — Fig. 150.

– *Adhogamina ruchira* Subram. & Lodha *1964*

DESCRIPTIONS: Barron (361), and Loquet (3411). — The genus *Gilmaniella* resembles *Humicola* Traaen (q.v.), but the conidia are generally terminal or lateral, often in botryose clusters, and provided with a distinct germ pore. The solitary blastoconidia are attached to the conidiogenous cells by a broad base and do not secede easily. In *G. humicola* they are mainly 7–10 μm diam, whilst in *G. macrospora* Moustafa *1975* they are 14–18 μm diam and in *G. subornata* Morinaga *et al. 1978* 9–16(–20) μm diam and spinulose below a smooth outermost layer. Colonies of *G. humicola* reach 6·5 cm diam in ten days at 20°C on MEA.

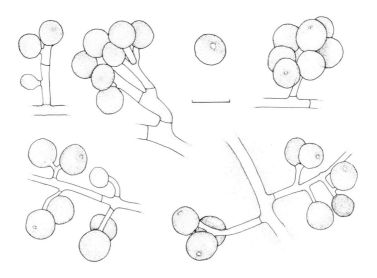

FIG. 150. *Gilmaniella humicola*, clusters of aleurioconidia, CBS 660.74.

G. humicola is a very representative example of the heat-tolerant fungi (613, 614). Heat treatment of soil for 30 min at 70°C not only eliminates most other fungi, but also stimulates germination of the conidia of this species. With this technique the fungus has been isolated from 65% of the greenhouse, 45% of agricultural, and 4% of forest and heath soil samples tested in the Netherlands; the fungus is primarily coprophilous, however, and is introduced into arable soils with manure (615). It is also one of the characteristic moulds of heated mushroom compost and greenhouse soil (615). It has been isolated from burnt and unburnt chaparral soils in California/USA at soil depths between 15 and 30 cm (1164), fumigated soil in England, and saline mud under *Halimione portulacoides* in France (3411). There are further soil isolations recorded from central Africa (3063), Namibia (1827), Egypt, Canada (IMI), India, Japan (36, 2532, 5846) and the Solomon Islands (5983). It has been found in the rhizospheres of wheat (427) and groundnuts (2532), on rotting roots, on *Berberis,* tomatoes, potatoes, beet seedlings, wilting *Clematis* plants, pea plants, *Olea europaea*, various plant debris, horse dung (IMI, 3113) and feathers of free-living birds (2575).

Good growth occurs at 22°C (4974) with an optimum of 45°C or less (1827), and a maximum of 55°C (613, 614, 4329); this fungus survived a heat treatment of soil at 70°C for 30 min (615). — Utilization of glucose, galactose, fructose, sorbitol, dulcitol, xylose, maltose, lactose, sucrose, trehalose, cellobiose, raffinose, inulin, starch, glycerol, and gluconate is reported. It has a high cellulase activity (36) and is an excellent decomposer of bark. Corrosion and attack of birch wood has been observed (4191). Ferulic acid is utilized (544). Suitable N sources are ammonium salts, urea, and amino-N (4329). Contrary to some *Humicola* species, *G. humicola* produces a mixture of two monohydroxy derivatives from progesterone (877). — It was found to cause superficial necrotic lesions in potato tubers (4974).

Gliocladium Corda *1840*
(incl. *Clonostachys* Corda *1839*)

Type species: *Gliocladium penicillioides* Corda

Teleomorphic genera; *Nectria* (Fr.) Fr., *Hypocrea* Fr., *Nectriopsis* Maire (*Hypomyces* sensu Samuels), *Roumegueriella* Speg. (= *Lilliputia* Boud. & Pat.)

Contributions to a MONOGRAPHIC TREATMENT: Pinkerton (4568), Raper and Thom (4745), Pidoplichko (4548), and Morquer *et al.* (3958).

The hyphomycete genus *Gliocladium* is often characterized as a counterpart of *Penicillium* with slimy conidia. The differences between these two genera are, however, more fundamental. The most characteristic features are ± densely penicillate conidiophores which bear slimy, one-celled hyaline or brightly pigmented, smooth-walled conidia in heads or columns. Besides the penicillate conidiophores, primary, simply verticillate conidiophores may be present which is the most characteristic feature of *Clonostachys* in addition to the conidial columns (1884). If the conidia cohere in chains, they are ± imbricate and devoid of connectives (contrary to *Penicillium)*. The conidia are often slightly curved with a somewhat truncate base; in isolates with predominantly *Verticillium*-like primary conidiophores, this feature may be useful for their separation from *Verticillium* itself, in which the conidia are usually symmetrical. Another similar genus is *Sarocladium* W. Gams & D. Hawksw. (cf. discussion under *Verticillium*); *Myrothecium* Tode ex Fr. (q.v.) is more than an aggregated *Gliocladium* with dark green conidia since its sporodochia are sharply delimited.

The genus *Gliocladium* is still not satisfactorily delimited and the most satisfactory species concept within it not yet determined. Here we confine ourselves to recognition of some apparently well-known and rather broadly defined species but would stress that neither the species concepts nor their nomenclature are finally determined. — *Gliocladium* species are equally common on a wide range of plant remains and in soil.

Key to the species treated:

1   Colonies spreading broadly (*Trichoderma*-like); conidia forming green slimy masses; phialides broadly ampulliform   *G. virens* (p. 375)
         Colonies growing more restrictedly and not *Trichoderma*-like; conidia forming whitish, pink or green masses; phialides ± aculeate   **2**

2(1)   All conidiophores of a similar, densely penicillate type; conidiophore stipes ± roughened; conidia symmetrical   **3**
         Conidiophores differentiated into *Verticillium*-like primary and penicillate secondary conidiophores; conidiophore stipes usually smooth-walled; conidia mostly asymmetrical   **4**

3(2)	Conidial masses dark green, paler singly; conidia 3–4 × 1·7–2·7 μm	*G. viride* (p. 376)
	Conidial masses yellow to orange; conidia 3·5–8 × 1·5–3·0 μm	(*G. penicillioides*)
4(2)	Primary conidiophores with convergent phialides	**5**
	Primary conidiophores with divergent phialides	**6**
5(4)	Conidial masses cream	(*G. solani*)
	Conidial masses dark green	(*G. nigrovirens*)
6(4)	Conidia turning green after 6–10 days	*G. catenulatum* (p. 369)
	Conidia in pink shades	**7**
7(6)	Conidia cylindrical, (5–)6–8(–11) × 1·5–3 μm	(*Clonostachys compactiuscula*)
	Conidia asymmetrically navicular, 5–7·5 × 2·0–3·5 μm	*G. roseum* (p. 371)

Gliocladium catenulatum Gilm. & Abbott *1927* — Fig. 151.

= *Gliocladium varians* Pidopl. *1931*
= *Gliocladium roseum* var. *viride* Rall *1965*

DESCRIPTIONS: Gilman and Abbott (1985), Raper and Thom (4745), Morquer *et al.* (3958), and Subramanian (5626). — Colonies spreading rather broadly, reaching 2·5–3·5 cm diam in ten days at 20°C on MEA. Generally as *G. roseum* (q.v.), but distinguished by the conidial areas turning pale to olive-green after 7–10 days; reverse colourless, yellow or intensely salmon (under light). Conidiophores of both *Verticillium*-like and densely penicillate types, the latter with long conidial columns; stipes to 125 μm long, appearing warted when observed directly, but smooth in liquid mounts. Conidia as in *G. roseum*, 4–7·5 × 3–4(–4·5) μm. Hyaline chlamydospores may be present, terminal and intercalary, 7–10 μm diam. — According to Raper and Thom (4745), *Gliocladium* isolates with light-green conidia are combined in this one species; *G. catenulatum* can, however, be distinguished from the similar but darker *G. nigrovirens* van Beyma *1931* which has appressed phialides even on the primary conidiophores.

The reports of *G. catenulatum* are not very frequent; nevertheless, they point to a worldwide distribution in the most varied soils coming mainly from Europe, North America, and Australia. Reports from warmer zones include Egypt (8, 3993), Turkey (4245, IMI), Cyprus, Pakistan, Kenya, Swaziland, Rhodesia (IMI), Iraq (92, IMI), Zaïre (3790), South Africa (4407), Nepal (1827), India (4030, 4996, IMI), Borneo (IMI), Central America (1697), Argentina (1827), and Hong Kong (IMI). It has been isolated from forest soils (272, 681, 850, 1978, 2068, 3138, 3817), willow-cottonwood lowlands (2004, 2007, 2008, 3549), grassland (6182), savannah (4407), salt marshes (8) and arable soils (very numerous reports). It was found to colonize sticks (1956) or twigs (850) buried in the soil (1956), and has also been found on boards of a bathhouse (5016), damp plywood (IMI), leaf litter (269, 681, 2411), roots of broad beans (6134), groundnut pods (346), *Iris* bulbs, potato tubers and a number of wild and cultivated tropical plants (IMI). Other substrates reported include fen peat (5559), running water (1154, 1155, 1166), children's sandpits (1421), soils treated with sewage sludge (1163) and sewage from a treatment plant (1165). It has been found, though with diminishing

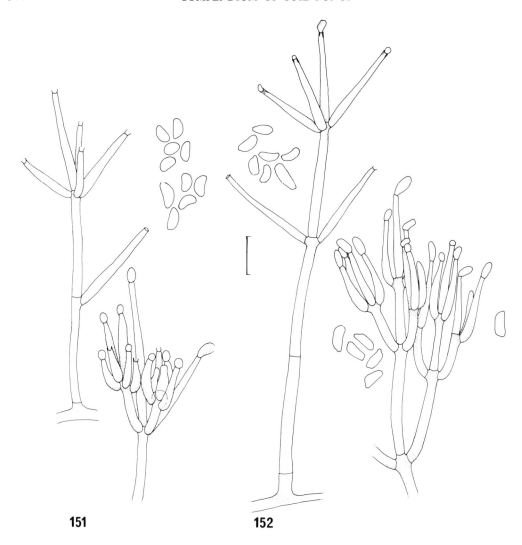

151 **152**

FIG. 151. *Gliocladium catenulatum*, primary, *Verticillium*-type and secondary, penicillate conidiophores and conidia, CBS 227.48.

FIG. 152. *Gliocladium roseum*, primary, *Verticillium*-type and secondary, penicillate conidiophores and conidia.

frequencies, down to soil depths of 80 cm (1823, 2161). In neutral to alkaline soils (pH 6·5 to 8·0) it occurred in over 50% of the samples examined (5559). It has also been isolated from soils with a high copper content (2922, 4921). A significantly higher frequency was found in a soil treated with the insecticide mirex (2796). Hair baits were colonized only at vapour saturation (977). A reduced frequency in the soil was observed after growing corn (1330). It has been isolated from the root regions of barley, wheat (IMI, 3567), groundnuts (2768), beans and cabbage, though without any noticeable rhizosphere effects (4451, 4452), mouldy hay (IMI), and feathers and pellets of free-living birds (2575).

The optimal temperature for growth is (25–)26–28°C (3958), regardless of the geographical origin of the isolate (1827). *G. catenulatum* can grow in the pH range 3–8·2, and has the optimum at 5·6 (3958). — The most suitable C source is sucrose, followed by D-fructose, D-glucose, maltose, some sugar alcohols, starch and cellulose; the optimal concentration of the C source was determined as 0·4–0·8%, and the optimal C/N ratio as 10 (3958); β-glucosidase (5266), amylase and cellulase (1827, 3618, 3958, 5059) activities in *G. catenulatum* have been demonstrated. Erosion and weight losses in birch and pine wood are documented (4191). *G. catenulatum* can decompose pectin, though less actively than some other fungi tested (1432). Peptone and various amino acids were found to support better growth than nitrate (3958); milk clotting and gelatine liquefaction proceed rapidly (3958). Though it has been isolated from buried chitin (1422), hardly any chitinase production was observed *in vitro* (2068). In the presence of keratin chips, Mg^{2+}, and $PO_4{}^{3-}$ ions, large quantities of $NH_4MgPO_4 \cdot 6H_2O$ were precipitated (2880). — A very strong inhibition was observed against seedling roots of peas (5266), wheat, and rape comparable with the action of true root parasites (1430).

Gliocladium roseum Bain. *1907* — Fig. 152.

= *Acrostalagmus roseus* Bain. *1905*
= *Isaria clonostachoides* Pritchard & Porte *1922*
= *Verticillium foexii* van Beyma *1928*
= *Gliocladium verticilloides* Pidopl. *1930*
= *Gliocladium cholodnyi* Pidopl. *1931*
= *Gliocladium aureum* Rader *1948*
? = *Verticillium intertextum* Isaac & Davies *1955*

Synonyms partly fide Isaac (2671) and Raper and Thom (4745).

Teleomorph: *Nectria ochroleuca* (Schw.) Berk. *1875*

= *Nectria pallida* Ell. & Everh. *1894*
= *Nectria gliocladioides* Smalley & Hans. *1957*

For numerous other synonyms see Samuels (5032).

DESCRIPTIONS: anamorph: Pinkerton (4568), Raper and Thom (4745), Isaac (2671), Subramanian (5626), Hanlin (2252), and Pidoplichko (4548); morphogenesis: Morquer *et al.* (3958); teleomorph: Udagawa and Horie (5977), Smalley and Hansen (5414), Dingley (1402), and Samuels (5032); GENERAL REVIEW: Pugh and Dickinson (4657). — Colonies spreading rather broadly, reaching 2·4–3·5 cm diam in ten days at 20°C on MEA, granular to felty, white, pink or salmon; reverse yellow or salmon (under light). Conidiophores erect on aerial or submerged hyphae or hyphal ropes, and of two kinds: (a) primary conidiophores with divergent, *Verticillium*-like branches, 100–200 μm tall, phialides in whorls of 3–4, 17–30 × 3 μm, conidial heads discrete; and (b) secondary conidiophores, formed particularly in the colony centre, densely penicillate, stipes 45–125 μm long, phialides in closely appressed whorls of 4–7, 10–18(–22) × 2–3 μm, conidia cohering in imbricate columns or irregular slimy masses. Conidia from both types of conidiophores elongate, slightly asymmetrical, apex obliquely rounded, base slightly protruding, hyaline or pinkish in mass, smooth-walled, $(3 \cdot 2-) 5-7(-8 \cdot 4) \times 3-4(-4 \cdot 8) \mu m$. Hyaline chlamydospore-like swollen cells, $6-9 \mu m$ diam, and sclerotium-like hyaline hyphal knots, $30-100 \mu m$ diam, may occur in old cultures. — The teleomorph is only rarely formed in culture by homothallic isolates or in heterothallic ones after mating (5414, CBS). Perithecia globose to pyriform, solitary or in dense clusters on a stroma, yellow to orange, collapsing when dry, smooth-walled, $200-310 \mu m$ diam, wall $35-45 \mu m$ thick. Asci clavate, thin-walled, mostly $50-68 \times 5-9 \mu m$. Ascospores ellipsoidal, hyaline, smooth-walled or finely echinulate, $10-12 \times 3-4 \mu m$.

G. roseum may be regarded as a collective species which shows a great deal of variation between isolates in the colony habit and proportion of the two types of conidiophores; the microscopic features are, however, quite constant. This fungus has to be classified in *Clonostachys* (1884) but as the oldest epithet is not yet determined and its circumscription is still vague, the commonly used name *G. roseum* is retained here. — Isolates only forming conidia should not be referred to *N. ochroleuca*. *V. intertextum* Isaac & Davies is very similar but lacks penicillate conidiophores. *Gliocladium solani* (Harting) Petch *1944*, the anamorph of *Nectria solani* Reinke & Berth. *1879* (636), has phialides which are appressed even on the primary conidiophores and do not turn pink under light. *Gliocladium vermoesenii* (Biourge) Thom *1930*, reported to differ from *G. roseum* mainly in the more intensely pink colonies (4745), has conidia cohering end-to-end in chains with minute connectives and thus is better referred to as *Penicillium vermoesenii* Biourge *1923*. — A number of closely related teleomorphic species with *Gliocladium* conidia belong to the *Nectria ochroleuca* group (636) which includes *N. stenospora* Berk. & Br. *1875* with minutely verrucose ascospores, $(9-) 10-14 \times 4-5 \mu m$; *N. kowhai* Dingley *1956* with hyaline verrucose ascospores, $12-20 \times 5-6 \mu m$; *N. egmontensis* Dingley *1956* with echinulate ascospores $14-18 \times 3-4 \mu m$; *N. byssicola* Berk. & Br. *1873* with perithecial walls covered with large white warts and smooth or spinulose ascospores, $11-14 \times 4-5 \mu m$ (all three with usually more compacted, *Dendrodochium*-like conidiophores); *N. solani* Reinke & Berth. *1879* with ascospores $10-14 \times 3 \cdot 5-5 \mu m$ and cream colonies (see above); *N. pallidula* Cooke *1888* with smooth or spinulose ascospores, $13-17 \times 4-5 \mu m$, and either a *Gliocladium* (636) or possibly a *Fusisporella* (5032) anamorph; *N. aureo-fulva* Cooke & Ellis *1878* with smooth or spinulose ascospores, $12-16 \times 4-5 \mu m$; and *N. apocyni* Peck *1873* with larger, finely spinulose ascospores, $18-22 \times 5-6 \mu m$; the last two species have *Dendrodochium*-like compacted conidiophores.

G. roseum is a common soil fungus and colonist of rotting plants with a worldwide distribution. The *Nectria ochroleuca* group is the most frequently encountered group of *Nectria* species in tropical and subtropical areas (5032); it is known from many parts of the USA, Bermuda, Cuba, Puerto Rico, Jamaica, Dominica, Colombia, Venezuela, Brazil, Argentina (5032), New Zealand (1402), Czechoslovakia, South Africa (CBS) and Japan (5977). It is mostly found on bark, and less frequently on herbaceous tissue, fleshy fruits of many plants and fungi (5032). — On account of its copious production of phialoconidia, *G. roseum* is easy to isolate by dilution plates and many other techniques, including selective methods designed for cellulolytic (1379, 4647) or keratinolytic fungi (1422). It has been isolated selectively from kaolin tablets with a-conidendrol buried in the soil (2801). — Its overall distribution reaches from East Siberia (3652) and northern Scotland (163), to Turkey (4245), Iraq (92), Syria (277, 5392), Israel (2764, 2768, 2772), Libya (6510), Egypt (3993), the Sudan (4222), and other parts of Africa (1420, 1559, 4159, 4718, 4719), India (41, 1524, 3863, 3865, 5000, 5512), Nepal (1826), Central America (1697), Argentina (1827), Jamaica (4886), the Bahamas (2006), Singapore (3331), Japan (2532, 3267, 5846), Australia, Tasmania and New

Zealand (977, 2705, 5930). Its frequent occurrence has been observed in various mixed deciduous forests (1040, 2004, 2007, 2008, 2573, 3549, 4225), it has also been found in aspen stands (3959), beech (272, 3138), oak (1700), teak (2854), and *Casuarina* (2006) forests. Very numerous reports are also available from arable soils and grassland. It is found more rarely in coniferous forests (269, 505, 681, 2344, 3267) and other reports include truffle soils (3451), heathland soil (3720, 5047, 5811, 5819), peaty soil (163, 1376, 5559), salt marsh (1379, 4646, 4657), estuarine sediments of high salinity (655), saline soils (3446), dunes (745, 1423, 4655, 4657), soils with and without prescribed burning (2822), soil with steppe type vegetation (4718, 4719, 6347), desert soils (1512), river sediments and river water (1162, 1166, 4429). It seems to be absent from running streams with a high degree of pollution by sewage (1154, 1155, 1157), but has also been found in a waste stabilization pond, a sewage treatment plant (1166), activated sludge (1387), and soils treated with sewage sludge (1163). It appears in the eulittoral zone in seawater (4918) and has been isolated from freshly reclaimed polder soils (4658). It occurs in the surface layers of soil (2161, 2948, 6352), but has also been isolated from deeper layers (159, 6184). It is prevalent in neutral to alkaline soil and is much less frequent in podzols and peat (4657). It is very common in manure of pH 7·5 (4657). Its occurrence in calcareous soils has repeatedly been reported (3451, 4152, 4657). Seasonal fluctuations have been established in Wisconsin/USA with the minimum occurring in autumn and the maximum in January (2004). According to some reports (2163, 2895), *G. roseum* is stimulated by NPK fertilizers in various combinations, but according to others it proved indifferent to NPK and farmyard manure (2161). It is relatively sensitive to soil treatments with formaldehyde (1655) but occurred with high frequencies in soils treated with the fungicides captan, quintozene, thiram, benomyl, dicloran, milcol, and triarimol (6135, 6136). It occurs in compost beds (3041), and vineyard soils with high copper contents (4921). Its frequency was found to be increased after growing soya beans and peas, but decreased after oats (1433, 6379). — In general, the germination of the conidia can be inhibited in the soil (1407, 2703), but in mull or mor humus almost all conidia germinate; they are not lysed in mull where 100% survive two weeks (633). In chitin-amended soil a growth-inhibiting principle could be extracted by *n*-butanol (5453). Conidia can be induced to germinate in the vicinity of roots of peas, radish, tomatoes and lettuce (2704). Germination was found to be inhibited by exudates from onions *in vitro* (1072). — In addition to sitka spruce, some salt-marsh plants (4657), cabbage (4451), tea (41), alfalfa (5825), clover (5815), peas (5560), broad beans (5757, 6134), and French beans have been especially thoroughly investigated for root colonization (1406, 4451, 4452, 5758, 5759); *G. roseum* penetrates only a few cells deep into the root cortex; it has also been found on roots of beech, *Iris* (4657), cotton (5009), sugar cane (4886), oats (4443, 5825), live and dead roots of barley (3006, 4450, 4451), *Lolium perenne* (5815, 6133) and wheat (2816, 3567, 5330, 5825) and in the rhizospheres of poplar (3452), groundnuts (2532, 2768), sugar cane (2861), *Coffea arabica* (5313), *Cassia occidentalis* (6067) and some peat bog plants (165). In most cases senescent roots are more strongly colonized than young ones, except in *Lolium perenne* where young white roots supported the strongest development (6133); with *Halimione portulacoides* R/S ratios between 1/7 and 2/3 were recorded (4657). It has often been found to be present on coarse fodder (4548), in birds' nests (2577), on feathers and pellets (2575), in wood pulp (3294), on filter paper exposed in a river (4432), boards in a bath-house (5016), wood used in a copper mine (2664), aged *Ulex* stems (2338), and sticks exposed to soil (1956, 5237), cotton material of a conveyor belting (6163), organic debris in soil (2705), and rabbit dung (2338). It has also been isolated from bananas (6158), seeds of *Avena fatua* (2961), oats, barley (3556), wheat (1752, 4492) and *Cyamopsis tetragonaloba* (5350). — It can colonize hair baits in soil at water potentials between saturation and −40 bars (977, 2095).

Good growth occurs between 20° and 35°C; the temperature minimum is 4–8°C, and the optimum 25–28°C; the maximum for sporulation is 29°C (2671, 3958); the pH optimum is 6·4–8·0 or 6·1–6·6, but the pH range 3·2–10·5 is tolerated (3958). On PDA containing 5% NaCl growth is inhibited (1991). Pink coloration of the colonies occurs mainly in light (2671). — The most suitable of many C sources tested was found to be maltose followed by glucose and sucrose, but, according to other reports, D-glucose and D-fructose were as suitable as maltose and better than acetate (3329, 3958). Other suitable substrates are glycerol, starch and glycogen (3958); tannin up to 4·6% is tolerated and utilized (2706). The optimal concentration of C sources was determined as 1·2%, and the optimal C/N ratio as 21–22 (3958). Chitin (2706, 5926), pectin (2948, 3127, 3414, 5926) and xylan (1432) are all well utilized with relatively little variation in this capacity between isolates. Cellulose (textiles, CMC) is severely attacked (1432, 2006, 2706, 3618, 3934, 5059, 5926, 6326); the optimal pH for this action is 5·4 (5238) and the degree of cellulose decomposition achieved depends on the concentration of the N source (4433). The insecticide carbaryl is hydrolysed to 1-naphthyl N-hydroxymethylcarbamate and other metabolites (610, 3380). Trimethylarsine is produced from various As compounds (1207). Nitrate is a very suitable N source (2671), but much less satisfactory than L-aspartic and L-glutamic acids, in addition to L-alanine, L-leucine, L-phenylalanine, L-tyrosine, L-arginine, peptone and urea (3958). Caseine and gelatine can also be utilized (1425, 2948, 5926), and synthetic dipeptides degraded (3699). A proteinase complex similar to rennin was shown to be present in liquid cultures (3049) and milk-clotting enzymes have been observed (3958). There are investigations with purified proteinases from *G. roseum* (3013). Sinigrin (4774) and ferulic acid (544) are utilized. No particular growth substances are required (1425). *G. roseum* grows chemotropically towards phosphate granules without dissolving them (3074). Orsellinic acid decarboxylase (4534), orcinol, 1,3-dihydroxy-4,5-dimethylbenzene (4531), 1,5-dihydroxy-2,3-dimethylbenzene, 2-hydroxy-6-methylbenzoic acid (3231), and humic substances (5240) have been isolated from culture filtrates. The pigments aurantiogliocladin (and its dihydro-derivative), rubrogliocladin and gliorosein have often also been shown to be present and their structures elucidated (4340, 4533, 6099); some further benzoquinone and toluquinone derivatives have also become known (3231, 4533, 5572). A slight gibberellin activity has been reported (237). —*G. roseum* is the best known destructive mycoparasite: hyphae of *Ceratocystis fimbriata*, *Trichothecium roseum*, *Thamnidium elegans* (350), *Botrytis aclada* (6149), and *Verticillium dahliae* (2065) are all actively parasitized by entwining, killing and penetration (350, 5282). *G. roseum* has also repeatedly been shown to be a parasite of the sclerotia of various fungi (1652, 2886, 3531, 6149) and inhibitory effects have been observed towards *Pythium "debaryanum"* (4516, 6029), *Chalara elegans* (5440), *Aureobasidium pullulans*, *Gibberella acuminata*, *Fusarium solani*, *Aphanomyces euteiches* (6029), *Rhizoctonia solani* (4658), a number of cellulolytic saprophytic fungi (3544) and *Bacillus subtilis* (3599). On inoculating *G. roseum* and *Cochliobolus sativus* together onto barley seed, *C. sativus* was suppressed (5962); *Verticillium* wilt in alfalfa was also found to be efficiently reduced by *G. roseum* (232, 236, 920). Infection experiments with potato, tomato, salsify, eggplant, *Antirrhinum*, cucumber, strawberry, lupin, carnation and tobacco did not reveal any pathogenicity (2671) but it can attack weakened plant tissues, at least in clover (1734, 2975, 2976, 4322), peas (1430), alfalfa (233), *Cyamopsis tetragonoloba* (5350), and beans (2579). — Mycelial growth is more sensitive to γ-irradiation than is the germination of conidia (5748). $CO_2:O_2$ mixtures are tolerated up to a proportion of 18:1 (2096).

Gliocladium virens Miller, Giddens & Foster *1957* — Fig. 153.

Teleomorphs: cf. *Hypocrea sublutea* Doi *1971*, cf. *Hypocrea psychrophila* E. Müller, Aebi & J. Webster *1972*

DESCRIPTIONS: Webster and Lomas (6245), Miller *et al.* (3817), and Komatsu (3086). — This fungus has often been misidentified as *Trichoderma viride* (6245). — Colonies very fast-growing, reaching 5·8 cm diam in five days at 20°C on OA; the habit suggesting a *Trichoderma* but with the phialides appressed and bearing one large drop of green conidia on each whorl. Conidia short-ellipsoidal, smooth-walled, rather large, mostly 4·5–6 × 3·5–4 μm.

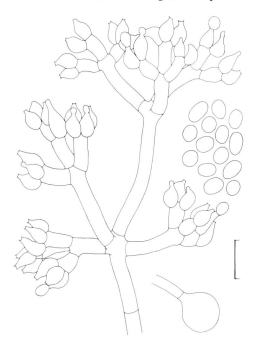

FIG. 153. *Gliocladium virens*, conidiophore, conidia and chlamydospore, CBS 512.66.

G. virens is a common soil fungus regularly received for identification at CBS. Published reports stem from soils in the British Isles (1343, IMI), forest soils in Japan (3086), Ohio (2573), Georgia (2240, 3817) and other southern states of the USA (2482), and Zaïre (2967). It has been found to predominate in a peat bog under *Sphagnum recurvum* (CBS). It has been isolated from sticks exposed to soil (1956), healthy and diseased strawberry roots in Illinois/USA (4133), and logs used in the cultivation of *Lentinus edodes* in Japan (3956).

Optimal growth occurs at 25–32°C (3086). This fungus produces the antifungal antibiotics gliotoxin, viridin (120), and non-volatile antibiotics active against *Heterobasidion annosum* and other fungi; *Lentinus edodes* is found to be inhibited *in vitro* (3086). — A high tolerance to CO_2 has been demonstrated (2096). On media containing 5% NaCl, *G. virens* shows reduced growth and sporulation (1991).

Gliocladium viride Matr. *1893* — Fig. 154.

= *Gliocladium deliquescens* Sopp *1912*

Teleomorph: cf. *Hypocrea lutea* (Tode ex Fr. *1822*) Petch *1937*, fide Doi *1966*; cf. *Hypocrea gelatinosa* (Tode ex Fr. *1822*) Fr. *1849*

DESCRIPTIONS: Raper and Thom (4745), and Komatsu (3086). — Mycelial growth slower than in *Trichoderma* species, filling a Petri dish in eight days at 20°C. Conidiophores uniformly erect, densely penicillate and bearing dark green masses of conidia. Conidia short-ellipsoidal, smooth-walled, 3–4 × 1·7–2·7 μm. Chlamydospores hyaline, sometimes in chains, 6–10 × 4 8 μm. — *G. viride* closely resembles the anamorph of *Hypocrea gelatinosa*, but that has less compact and more irregularly branched conidiophores and conidia 3·7–4·7 × 2·8–3·0 μm (6240). — SEM studies of conidiophores have been carried out (2302).

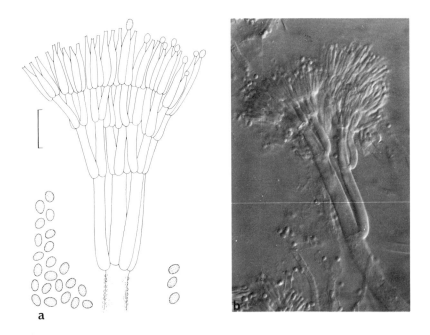

FIG. 154. *Gliocladium viride*. a. Penicillate conidiophore and conidia; b. densely penicillate conidiophore, NIC, × 500.

From the available data it can be concluded that this fungus is not restricted to particular geographical regions. It was originally reported in France and Norway, but has subsequently been found in the British Isles (1376, 2923, 4429), Czechoslovakia (2577), Poland (272), Germany (3041), the Netherlands (1614, 1616), Romania (2664), Syria (5392), Canada (IMI), the USA (1032, 1040, 1166, 1167, 2482, 2573), Japan (1414), India (1519, 5000), Hawaii (296, 3264) and the Bismarck Archipelago (3020). There seems to be a majority of records for forest soils (272, 1032, 1040, 2482, 2923, 5000), but it has also been found in grassland (2573), paddy soil (1519), a mangrove swamp (3264), fresh water (4429), a waste stabilization pond and an acid mine drainage stream (1166), sewage sludge (1167), composted urban waste (3041), cut peat (1376), litter of poplar (IMI), leaf surfaces (296), pine poles exposed to soil (2833), timber used in a copper mine (2664), paper (1875), the rhizosphere of groundnuts (2768), the rhizoplanes of bean, barley and cabbage (4451), and birds' nests (2577), feathers and pellets (2575).

The temperature minimum for growth is 6°C, the optimum 25–35°C, and the maximum 36–37·5°C (3958). Conidia can germinate between −70 bars water potential and saturation (3086). For mycelial growth, water potentials between −30 bars and vapour saturation are required (3109). It can grow in the pH range 2–9 with an optimum at 5·7–6, but sporulates well at pH 3·5 (3958). — It can grow on 0·2% formaldehyde, 7% methanol, and 5% sodium formate as sole C sources (4991). The most suitable C source is sucrose, followed by D-fructose, D-glucose, maltose, some alcohols and starch; the optimal concentration of the C source was given as 1–1·5% and the optimal C/N ratio as 42 (3958). Acetate is equally well utilized as glucose (3329). The fungus degrades cellulose (3568, 3618, 6076) and corrodes the wood of birch and pine (3568, 4191). Various amino acids and peptone are superior N sources compared with nitrate; proteolytic enzymes are present (3958). — The growth of wood-degrading fungi is reported to be inhibited by *G. viride* (2847), of which *Lentinus edodes* is inhibited *in vitro* by coiling hyphae of *G. viride*; lethal actions are pronounced at pH 3·3–4·1 (3086); it is also a lethal parasite on *Pholiota nameko* (3085).

Glomerella Spauld. & v. Schrenk *1903*

Glomerella cingulata (Stonem.) Spauld. & v. Schrenk *1903* — Fig. 155.

≡ *Gnomoniopsis cingulata* Stonem. *1898*

For further (about 120) synonyms see v. Arx and Müller (210).

Anamorph: *Colletotrichum gloeosporioides* Penzig *1882*

= *Gloeosporium fructigenum* Berk. *1856*

For further (over 600) synonyms see v. Arx (196).

The epithet *gloeosporioides* has been retained for the anamorph as the best known one not based on a particular host name, although it is not the oldest available (196).

DESCRIPTIONS: both states: von Arx (196); anamorph: von Arx (197), Simmonds (5328), and Stephan (5561); teleomorph: Dennis (1348). — Fam. Polystigmataceae Höhn. ex Nannf. — Colonies growing with a daily radial increment of 5–7 mm at 24°C and reaching 6·0–6·2 cm diam in four days at their optimum temperature 27–29°C on MEA or OA (5328), grey, with variable amounts of lanose aerial mycelium. Conidia appearing scantily on solitary phialides but normally in light or brown sporodochia (corresponding to the acervuli developed on the natural substrate); a basal stromatic cushion is covered with a dense layer of cylindrical, slightly tapering phialides to 20 μm long; sometimes these are interspersed with dark brown, tapering, blunt, septate setae. Conidia cylindrical with a rounded apex and a slightly truncated base, hyaline, filled with granular cytoplasm, 10–21 × 4–6 μm, forming orange-red slimy masses, germinating by irregularly rounded, brown appressoria. — Ascomata mainly confined to the natural substrate, where they are submerged with slightly prominent papillae, but also formed by some isolates *in vitro* where their production can be induced by UV-irradiation and the mating of compatible isolates (1792, 3277, 5567, 6316). Ascomata 85–300 μm diam, dark brown; asci few in number, clavate, thin-walled, with truncate, thin apex and biseriate ascospores; ascospores ellipsoidal to fusiform, hyaline, smooth-walled, mostly 10–21 × 4–6 μm.

Both the teleomorph and anamorph are highly variable and can be segregated into numerous physiologically, biologically or genetically differentiated types (196, 5561, 6316). Stable criteria are, however, the shape and width (but not length) of the conidia and the orange-red conidial slime (5561) (cf. introduction under *Colletotrichum*). — Some morphologically almost indistinguishable but host-specific taxa have been retained: *C. lindemuthianum* (Sacc. & Magn.) Briosi & Cav. *1889*, causing anthracnose on *Phaseolus*, and *C. orbiculare* (Berk. & Mont.) v. Arx *1957*, causing leaf spots on Cucurbitaceae, both have a slower growth rate (2–3 mm daily increment at 24°C) and almost black colonies. *C. trifolii* Bain & Essary *1906*, causing anthracnose on legumes, has slow-growing colonies and shorter conidia, 12–18 × 4·5–6 μm. *C. musae* (Berk. & Curt.) v. Arx *1957*, growing on *Musa* and other tropical fruits, has a higher growth rate (8–10 mm daily increment) and the same shorter

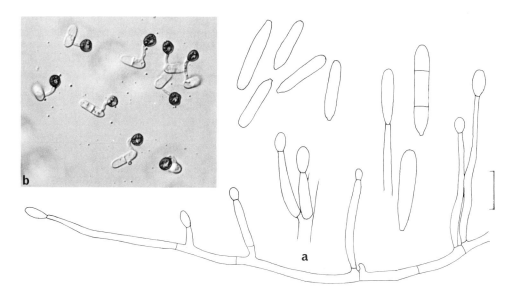

Fig. 155. *Glomerella cingulata*. a. Phialides and conidia; b. germinating conidia with appressoria, CBS 234.49, NIC, × 500.

conidial dimensions. It was shown by discriminant analysis that *C. coffeanum* Noack *1901*, causal agent of coffee berry disease in various African countries, can be distinguished from *G. cingulata* on account of its very slow growth (to 2 mm daily increment), more truncate conidial ends and very long phialides (25–60 µm) (2454). — Another soil-borne species, commonly found on potato tubers as a causal agent of dry rot, is *C. coccodes* (Wallr.) Hughes *1958* (= *C. atramentarium* (Berk. & Br.) Taubenh. *1916*) which produces black stromatic sclerotia 200–600 µm diam and conidia similar to those of *G. cingulata*. The other accepted host-specific species with straight conidia have ones longer than in *G. cingulata*.

Ascospore isolates often segregate into fast-growing, pale grey ('+') and slower, darker colonies which form numerous ascoma initials ('−'); the combination of both results in fertile ascomata (196, 6316). Environmental factors, such as the C/N ratio or pH, influence the numbers of nuclei in the hyphae which reach 5 to 35 in hyphal tips and to 9 in cells of older hyphae; the conidia are generally uninucleate (5562). Evidence for heterokaryosis has been presented by experiments with UV-derived auxotrophic mutants (5563) and anastomosis occurring in every stage of development (5562). Other genetic studies of various segregates and their sporulation behaviour have been reviewed (196).

The anamorph of *G. cingulata* (on many host species only the anamorph, *Colletotrichum gloeosporioides,* is known) is the commonest species of *Colletotrichum* and has a worldwide distribution on a very wide range of host plants (196). It often causes no damage but can cause fruit rots and stem anthracnoses (in both temperate and tropical countries) or leaf spots (mainly in tropical countries and greenhouses). Infection may be systemic but latent for a long time (196). There are numerous records from herbaceous plants (IMI, 6208) and fruit rots (1348, 4615, 6158), including the important one of *Coffea arabica* (2455, 2456). *G. cingulata* is one of the early colonizers of ageing leaves of sugar cane (2583); it has also been isolated from litter of *Gymnosporia emarginata* (6108) and the phyllospheres of *Phoenix sylvestris* and

Caesalpinia pulcherrima (2679) in Bangladesh and groundnuts in Gambia (1984). — Reports of its occurrence in soil are scarce; they include ones from South Africa (4407), New Guinea, New Zealand and India (IMI). It has also been found in a waste stabilization pond (1166). *G. cingulata* is not very sensitive to soil mycostasis (2559). It can survive dry periods of at least a year without loss of viability (6297) and storage in water for seven (or more) years (605). In both mull and mor humus, conidia germinated and lysed almost completely within two weeks (633). — In Malaysia its conidial densities in the air reached a peak at 23 h on dry days, but fell during rain (6207).

Good conidium production occurs in the pH range 5–6·5, on thick layers of carrot juice agar, temperatures between 20° and 29°C, and water potentials above −140 bars (411, 3822, 4675, 6297); reports on the influence of light on conidium production are contradictory, both a minimum of 12 h light (3822) and continuous light have been said to favour conidium production (6206), which ceased after two days of darkness; according to another report, conidiation and formation of acervuli is lowest in continuous light (6297). Suitable C sources for sporulation include arabinose, galactose and xylose; the best N sources are peptone, casamino acids and KNO_3 (4675). Conidium production was also found to be encouraged by media containing unsaturated fatty acids, for example oleic and linoleic; saturated fatty acids had no such effect (523). UV-irradiation stimulates ascoma production (1792, 3277, 5567) but young mycelia are easily killed by UV-irradiation (5567). — Conidium germination is optimal between 21·5 and 30°C (6206), but conidia were still observed to germinate at 5°C after one month (734); germination requires vapour saturation and is drastically reduced at water potentials below −15 bars (6206). A self-inhibitor of protein synthesis is effective in densely sown conidia (3356). — On cornmeal agar the optimum temperature for growth is 25–26·5°C, the maximum 30°C and the minimum 5°C; on natural substrates (apple), however, no growth was observed below 10°C (734, 6206); optimal growth occurs at pH 5·5 (4675). On potato-sucrose media, high conidial densities led to yeast-like development of cultures at 21°C; this became mycelial at temperatures >25°C and when exposed to light (3355). — Mycelial growth is possible without vitamins but is enhanced by thiamine, biotin, riboflavin, inositol, folic acid and ascorbic acid (5310). Of 13 carbon sources tested, fructose gave the best growth followed by sucrose and glucose, the best N sources were asparagine and nitrates (4675). In *G. cingulata* the production of glucose oxidase has been described (1894); oligosaccharides can be utilized (3218) by either hydrolysis (except for lactose) (965) or by transglycosidation (2883); maltotriose is produced from maltose (965). *G. cingulata* produces much pectolytic and little cellulolytic enzymes both on papaya fruits and *in vitro* (3902), and also arabanase on araban (3231). It produces *p*-diphenol oxidase (6154) and can grow on salicylic or 2,3-dihydroxybenzoic acids (801) and on plasticizers and related organic compounds, for example 2,5-dimethyl-3-hexene-2,5-diol and mixed *n*-octyl-, *n*-decylphthalate (473). The mycelium was found to contain cerebroside sulphate (144), triglycerides and phosphoglycerides (143), and the phytotoxin aspergillomarasmine A is produced in batch cultures (670). *G. cingulata* is able to transform steroids by dehydrogenation (1196). Less than 10% of pentachloronitrobenzene is degraded in six days (935). Autolysis occurring in nutrient-deficient media is ascribed to chitinase and β-D-glucosidase activities (3058).

Gongronella Ribaldi *1952*

Gongronella butleri (Lendner) Peyronel & Dal Vesco *1955* — Fig. 156.

≡ *Absidia butleri* Lendner *1926*
≡ *Gongronella butleri* (Lendner) Picci *1955*
≡ *Mortierella butleri* (Lendner) Chalabuda *1967*
= *Gongronella urceolifera* Ribaldi *1952*
= *Mucor vesiculosus* G. Sm. *1957*

DESCRIPTIONS: Hesseltine and Ellis (2429), Hammill (2240), and Watanabe (6212). — The genus *Gongronella* belongs to the Mucoraceae Dumort. — Colonies relatively slow growing, resembling *Absidia*, but with stolons and rhizoids rather poorly differentiated; sporangiophores erect and mostly branched. Sporangia globose, with a reduced columella, supported by a conspicuous, more or less globose apophysis. Spores oval, mostly flattened on one side,

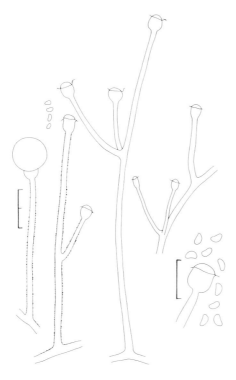

FIG. 156. *Gongronella butleri*. Sporangiophores with intact and emptied sporangia showing apophysate columellae, detail of apophysate columella and sporangiospores, CBS 969.73.

$2 \cdot 2$–$4 \cdot 5 \times 1 \cdot 6$–$2 \cdot 5\,\mu$m; in the only other species of the genus, *G. lacrispora* Hesselt. & J. Ellis *1961*, they are dacryoid to pyriform. Heterothallic; zygospores spiny as in *Mucor*, lacking finger-like appendages. Globose chlamydospores, 4–6 μm diam are commonly formed.

This fungus has a worldwide distribution with apparently higher frequencies in (sub)tropical regions. It is known from Zaïre (4159), equatorial West Africa (1420), Ghana, Nigeria, Uganda, New Caledonia, Sri Lanka, Singapore (IMI), India (1315, 2854, 3863, 3865, 4997, 5000, 5058, 5512, 5513), Malaysia (6046, IMI), Madagascar (4159), South Africa (1555, 1556, 1558, 1559, 4407), Mauritius, Trinidad (IMI), Brazil (6009), New Zealand (2705, 5814), Okinawa (6210), the USA (1040, 1330, 2008, 2240, 2385, 2482, 2573, 3817, 4313) including Hawaii (569), Italy (3450, 3913, 4538, 5047), Canada (234), the Ukrainian SSR (4474, 5298), and the British Isles (5432). It is mainly found in forest soils including forest nurseries (2482), moist *Salix-Populus* stands (2008), *Patharia* hillside forest (5000), *Eugenia* forest (5512), *Eucalyptus* (1555, 1558), and other forest types (2240, 2385, 2573, 2854, 3790, 3817, 4995). It is also isolated frequently from arable soils (234, 1420, 3450, 4538, 4996, 4997), under corn (1330), paddy (3913) and sugar cane (6009), grasslands (164, 1555, 2705, 3863, 3865, 4538), soils with steppe type vegetation (1559, 4313, 4407, 5298), *Calluna* heath (5047), peat fields (4474), marshes (164), wasteland, ant hills, garden soil (5432), and palm plantations (4159, 6046). The pH of the recorded soil habitats ranges between 6 and 7. *G. butleri* usually occurs in the surface layer of the soil (6009), but may be still abundant at depths of 30 cm or more (1556, 1558, 3865, 4995, 4996, 5000, 6046). It was observed at low frequencies in acid mine drainage streams and waste stabilization ponds (1166). There are reports of isolations from roots of *Pinus taeda* (3159), *Cocos*, *Hevea* sp., and *Imperata cylindrica* (IMI), rhizoids of *Riccia* sp. (2860), the rhizospheres of barley, oats (3004), wilted pineapple plants (6210) and *Dacrydium araucarioides* (IMI), underground parts of sugar cane (6212), and *Eugenia heyneana* leaf litter (5513).

Growth and zygospore formation are optimal at 25°C. The indole alkaloids corynanthine and ajmalicine are hydroxylated *in vitro* (438).

Gonytrichum Nees ex Leman *1821*

Gonytrichum macrocladum (Sacc.) Hughes *1951* — Fig. 157.

≡ *Chaetopsis macroclada* Sacc. *1877*
≡ *Mesobotrys macroclada* (Sacc.) Sacc. *1880*

The genus *Gonytrichum* is characterized by pigmented, erect, simple or branched conidio-phores bearing whorls of phialides on short basal cells, the so-called collar hyphae. The conidia are formed within a broad collarette from several conidiogenous loci in succession (as in *Chloridium*) (5665).

DESCRIPTIONS: Hughes (2594), Swart (5665), and Gams and Holubová-Jechová (1891). — The genus contains four species (1891) of which both *G. macrocladum* and *G. chlamydosporium* Barron & Bhatt *1967* have a simple conidiophore axis and several whorls of phialides in the median part; *G. macrocladum* often has sterile setae in the upper part of the conidiophores and slightly pigmented conidia $3 \cdot 5$–$5 \cdot 8 \times 2 \cdot 0$–$3 \cdot 6 \,\mu$m, while *G. chlamydosporium* has no sterile setae and hyaline conidia 2–$4 \cdot 5 \times 1 \cdot 0$–$2 \cdot 0 \,\mu$m and abundant thick-walled chlamydospores in rows or clusters. Colonies of *G. macrocladum* grow very slowly, reaching $1 \cdot 8$ cm diam in 10–14 days at 20–22°C on MEA. — In young cultures or on unsuitable media, simple conidiophores suggesting a species of *Chloridium* are formed.

Gonytrichum species are fairly common on decaying wood; *G. macrocladum* is the one most commonly isolated from soil but it never occurs at high frequencies. — A worldwide

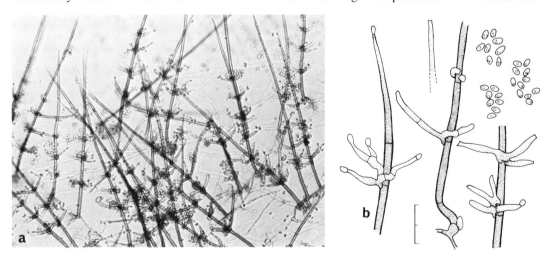

FIG. 157. *Gonytrichum macrocladum.* a. Conidiophores × 200; b. details of conidiophores with "collar hyphae" and conidia, CBS 875.68.

distribution with some preference for warmer zones is evidenced by reports from central Europe, Spain (3417), Italy (1874), Israel (2764, 2772), Zaïre (3790), central Africa (3063), South Africa (1559), Sierra Leone (IMI), India (1519, IMI), Japan (2532, 5846, 5935), the USA (2007, 2008, 2240, 2573), Canada (IMI, 505), New Zealand, Jamaica (4886, IMI), and Central America (1697, 2031, 2035). It has been reported from *Salix-Populus* communities (2004, 2007, 2008), deciduous forests (2573), a white cedar forest (505), grassland (IMI), and especially arable soils (2772, 2816, 3063, 4886, 6563), soil with steppe type vegetation (1559, 6347), a vineyard soil with high copper content (4921), citrus soils (2764) and banana plantations (2031, 2035). It is mainly found in the surface soil layers (2035, 3790); it also occurs on wood submerged in water (4955, 5249), rotten oak wood and bark, dead *Castanea sativa* branches (IMI), sterilized *Castanopsis* and *Quercus* leaves exposed to soil (5935), and *Acacia karroo* litter (IMI). Additional reports include the rhizospheres of groundnuts (2532, 2768) and sugar cane (4886, IMI), roots of corn (6563), oats, barley (3006) and banana (2035), and isolates from birds' roosts (682), feathers (2575), and fruit juices (5205).

G. macrocladum inhibits the growth of *Pholiota nameko* at 25°C but is overgrown by this fungus at 12°C (3085).

Gymnoascus Baran. *1872*

Gymnoascus reessii Baran. *1872* — Fig. 158.

= *Gymnoascus corniculatus* Orr & Plunkett *1963*

DESCRIPTIONS: Kuehn (3150), Takada (5685), Orr *et al.* (4317), and Samson (5023). — Colonies grow rather slowly, reaching 2·7 cm diam in 10 days at 20°C on OA. Ascoma initials consisting of swollen antheridia and coiled ascogonia produced as outgrowths of vegetative hyphae. Peridial hyphae yellow to ochraceous, smooth to roughened, bearing short, branched or unbranched, thick-walled appendages to 25 μm long. Ascospores oblate, yellowish, smooth-walled, 2·5–4·2 × 1·5–3·0 μm. Conidia absent. — The other three recognized species of the genus have longer peridial appendages. — SEM studies show the ascospores to have wart-like surfaces (6105). — DNA analysis gave a GC content of 54·5–55% (5600).

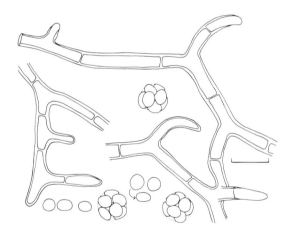

FIG. 158. *Gymnoascus reessii*, peridial appendages, asci and ascospores, CBS 497.64.

The distribution of this species is presumably worldwide as reports extend from central Africa (3063) to Greenland (4317) and also include the USA (4315, 6350, 6414), Chile (1824), Peru (2005), Israel (2764), Italy (271), Germany (5215), the British Isles, Australia (IMI), Bangladesh (2712), Pakistan (401), Japan (5846, 5970), New Guinea and the Solomon Islands (5983). It is observed mostly on dung of sheep, horse, donkey, cow, goat, rat and other rodents, fox, dog, deer, lizard, man, etc. (401, 3401, 4317, 4916, 4917, 5215, 6085), but has also been isolated from the most varied soils, including those of deciduous and coniferous forests (270, 271, 272, 273, 989, 3063, 3138, 6350), open oak forest (1700), light citrus soils (2764), desert and other arid soils (654, 1824, 2974, 4733, 4759), slate slopes (566), chernozem (4703), relatively coarse loamy sands (6350), salt-marsh (162, 167), and sand dunes (6414). It has also been isolated from the ectomycorrhiza of *Corylus avellana* (1794), the root region of rice (2712) and *Andropogon scoparius* var. *septentrionalis* (6414), rotting hay (4548), diseased pupae of wild bees (402), the intestine of *Dendrolimus sibiricus* caterpillars (2848), and nests, feathers and droppings of free-living birds (2575). It can serve as food for the mites *Pygmephorus mesembrinae* and *P. quadratus* (3104).

The utilization of nutrients by eight isolates of *G. reessii* was comprehensively studied (3154) and it was shown that glycine is particularly suitable as a N source, peptone, casamino acids, and nitrates also being utilized; maltose is suitable to most isolates (3154). Biotin and pyridoxin are both required for optimal growth (1970).

Harzia Cost. *1888*

Harzia acremonioides (Harz) Cost. *1888* — Fig. 159.

≡ *Monosporium acremonioides* Harz *1871*
= *Acremoniella atra* (Corda) Sacc. *1886*
≡ *Acremonium atrum* Corda *1837*

DESCRIPTIONS: Groves and Skolko (2144), and Ellis (1603). In a review of the nomenclature, Holubová-Jechová (2517) pointed out that the well-known generic name *Acremoniella* Sacc. *1886* is illegitimate and has to be replaced by the later name *Harzia* Cost. Since Corda's epithet "*atrum*" implies a black fungus his name is doubtfully referable to the present species and so it is preferable to use the later but well characterized epithet "*acremonioides* Harz".

The genus *Harzia* is characterized by a hyaline spreading mycelium and cymosely branching hyaline conidiophores bearing solitary brown thick-walled blastoconidia with narrow bases. In addition, a phialidic second anamorph with apically inflated, *Aspergillus*-like conidiophores occurs, suggestive of an affinity with *Melanospora* Corda *1837*.

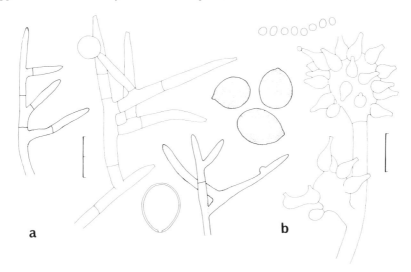

a   b

FIG. 159. *Harzia acremonioides*. a. Sympodially branching conidiophores and blastoconidia; b. conidial head with phialides and micro-conidia; CBS 100.61.

Harzia contains three accepted species: *H. acremonioides* with colonies reaching 3·3 cm diam in five days at 20°C on MEA and almost smooth-walled obovoid conidia, 20–30 × 15–20 μm, *H. verrucosa* (Togn.) Hol.–Jech. *1974* with verrucose, globose to subglobose conidia, 16–23 μm diam, and *H. velata* (Onions & D. Jones) Hol.–Jech. *1974* with globose conidia, 12–18 µm diam, surrounded by a fine membrane standing out 1·5–2·5 µm and sometimes becoming wrinkled; the conidia of *H. velata* secede less easily than in the other two species and it is hyperparasitic on various hyphomycetes.

H. acremonioides is by far the most common species of its genus and has been found in all climatic regions of the world (1603). It occurs in the soil but should not be regarded as primarily a soil fungus. — There are reports from coniferous forests in Hungary (1712) and Japan (3267), carst caves in Yugoslavia (3365) and the USSR (3364), peat bogs in Ireland (1376), and bogs in a coniferous forest in Sweden (IMI). It has further been isolated from soils in the Netherlands (1614, 1616), the British Isles, Germany, the USA, Canada, Australia, New Zealand, Papua, Rhodesia, Mozambique, Sierra Leone and Kenya (1603). The main recorded substrates are the seeds of different plants where it is often found in association with *Alternaria alternata* (2144). It has been isolated from seeds of clover (6283), grass (3512), peas, timothy, wheat, oats, cotton, radish and sorghum (1603), from rotting stems and/or leaves of *Betula alba* (3449), *Salsola kali* (4665), gramineaceous plants (4548), tomato, clover, beans, radish, corn cobs, seedlings of *Nicotiana* sp. and *Elaeis guineensis*, branches of *Acer campestre*, bamboo, *Sambucus* sp., oak logs and timber. It has also been reported as a saprophyte on the cultivated plants beet, alfalfa, ryegrass and *Chrysanthemum cinerariifolium*, and the wild plants *Pteridium aquilinum*, *Heracleum sphondylium*, *Rumex acetosella*, *Scrophularia nodosa* and *Urtica* sp.; it has also been found on sclerotia of *Sclerotinia sclerotiorum*, in hay, on the dung of opossum and kangaroo (1603), in birds' nests (2577), on feathers (2575), and on frescoes of a monastery (2666).

Heteroconium Petr. *1949*

Heteroconium chaetospira (Grove) M. B. Ellis *1976* — Fig. 160.

≡ *Septocylindrium chaetospira* Grove *1886*
≡ *Septonema chaetospira* (Grove) Hughes *1952*

DESCRIPTIONS: Hughes (2598), Ellis (1604), and Matsushima (3680). — Colonies slow-growing, reaching 8–10 mm diam in 10 days at 20°C, olivaceous-brown, powdery, reverse black. Vegetative hyphae and conidiophores pale olivaceous; conidiophores little differentiated, to $50\,\mu$m tall, 3–$4\,\mu$m wide. Conidia formed in single acropetal chains at the tips of the conidiophores, fusiform, usually slightly curved, pale brown, completely smooth-walled, with (0–)1–3(–4) septa, 20–35×3–$4\,\mu$m.

Ellis (1604) transferred the commonly encountered species formerly classified in *Septonema* Corda *1837* (2598) to *Heteroconium* in which the conidiophores are, although erect and

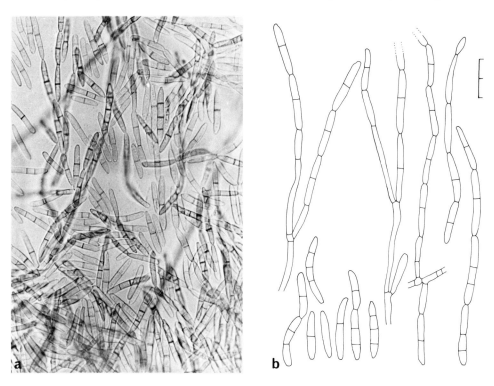

FIG. 160. *Heteroconium chaetospira*. a. Conidial mass, × 500; b. conidium formation in acropetal chains; CBS 514.63.

mostly many-celled (macronematous), much less differentiated than in *Septonema*; conidial chains arise singly in *Heteroconium* and in groups of 2–3 in *Septonema*. While *Heteroconium* species sporulate profusely in culture, *Septonema secedens* Corda *1837* generally remains sterile. Both genera differ from *Cladosporium* Link ex Fr. (q.v.) by the regularly septate blastoconidia which are not provided with protuberant thickened scars. — *Septonema chaetospira* var. *pini* Bourchier *1961* is an equally common taxon and distinguished by mostly two-celled, straight conidia, 12–23 × 2–3 μm. Many more, sometimes soil-borne species of this genus have not yet properly been named and delimited.

H. chaetospira is mainly known from rotten wood of various deciduous trees, but also from millipede droppings in the British Isles (1604), Japan (3680) and the USSR (5252). It has been found in wheatfield soils in Germany and has also repeatedly been isolated from arable soils in the Netherlands (CBS). Further known habitats are roots of *Ammophila arenaria* in sand dunes (IMI) and organic detritus in fresh water (4429).

H. chaetospira produces laccase (2516). — In a chronically irradiated pine-oak forest soil it still occurred at intensities of ⩾1250 R per day (2009).

Hormiactis Preuss *1851* sensu Sacc. & Marchal *1885*

Hormiactis candida* Höhn. *1924 — Fig. 161.

Original DESCRIPTION only: v. Höhnel (2489). — The genus *Hormiactis* is characterized by acropetal branching chains of hyaline, fusiform, mostly bicellular blastoconidia borne on undifferentiated erect hyphae. It is distinguished from *Cladosporium* by the absence of pigmented structures and a differentiated conidiophore stipe and more regular, mostly bicellular conidia.

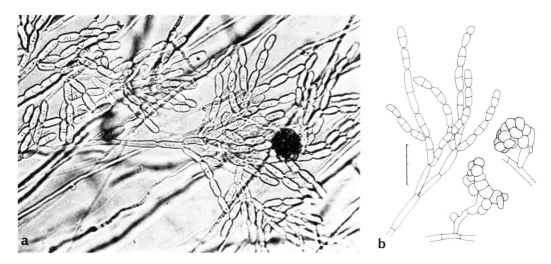

FIG. 161. *Hormiactis candida*. a. Conidiophore with acropetal conidial chains, CBS 252.62, × 500; b. conidiophore and clusters of chlamydospores.

In *H. candida* the conidia measure 10–19 × (2–)3–4 μm, and brownish irregularly shaped microsclerotia, to 40 μm diam and with cells 4–7 μm diam appear in three-week-old cultures. Colonies grow rather slowly, reaching 2·8 cm diam in ten days at 20°C on OA. — The only other recognized species of the genus, *H. fimicola* Sacc. & March. *1885,* which is common on mushroom compost and various kinds of dung, has conidia 17–22 × 3–5 μm. — Of the original species placed in the genus, *H. alba* Preuss *1851* has two-celled phialoconidia, while *H. fusca* Preuss (≡ *Hormiactella fusca* (Preuss) Sacc. *1886*) is close to *Septonema* Corda *1837* and has both pigmented hyphae and similar blastoconidia. A thorough revision of *Hormiactis* and related genera is required.

H. candida was originally described from a decayed fir twig with a *Tubercularia* on it in the Wienerwald/Austria. Our own observations have shown it to be commonly well developed on

Urtica stems. Very few data are available on its occurrence in soil; it has been isolated from leaf litter in the Botanical Garden in Turin (270, 3448), muck soil of a white cedar forest in Canada (505) and litter of *Pinus nigra* var. *laricio* in the British Isles (983). It has rarely also been found in garden compost (1425) and on washed particles from two wheatfield soils (1889) and was isolated in abundance from urban soil in Oegstgeest/Netherlands (CBS).

When growing on straw, it was found to utilize lignin and cellulose and to produce humic substances; after three months culture on maple-wood strips it caused high losses in both weight and tensile strength, but isolates can vary markedly in this capacity (2211).

Humicola Traaen *1914*

? = *Monotospora* Corda *1837* sensu auct. (nomen ambiguum)
 = *Melanogone* Wollenw. & Richter *1934*

Type species: *Humicola fuscoatra* Traaen (lectotype)

KEY TO THE SPECIES: Fassatiová (1701); discussion of the nomenclature: Mason (3661), and Fassatiová (1698).

The hyphomycete genus *Humicola* has been characterized since its first description by the possession of two types of conidia: (a) conspicuous, brown, solitary aleurioconidia which are globose to elongated, smooth-walled but surrounded by a slimy melanizing sheath, formed laterally, terminally or sometimes intercalarily (then called chlamydospores); and (b) hyaline, small, obovoid phialoconidia produced on awl-shaped *Acremonium*-like phialides scattered over the mycelium which may be entirely absent in some isolates. — The similar *Thermomyces* Tsikl. (q.v.) is separated by its stalked and mostly ornamented aleurioconidia. *Staphylotrichum* J. Meyer & Nicot (q.v.) can easily be mistaken for *Humicola* when not sporulating typically, but that genus has a characteristic golden-brown cottony mycelium. — Several species added to *Humicola* do not form phialoconidia at all, rendering the genus heterogeneous. We consider that it is preferable to exclude species with pigmented hyphae or catenulate chlamydospores from it. The following taxa with different types of second anamorphs should also be omitted from *Humicola*: *Sporothrix inflata* de Hoog *1974* (=*Humicola dimorphospora* Roxon & Jong *1974*), *Phialophora mutabilis* (van Beyma) Schol-Schwarz *1970*, *Sagenomella humicola* (Onions & Barron) W. Gams *1978*, *Verticillium nigrescens* Pethybr. (q.v.), and *Scolecobasidium terreum* Abbott (q.v.) (=*Humicola minima* Fassatiová *1967*).

Humicola species are widely distributed in neutral or alkaline soils (6329). They are infrequently isolated by dilution and soil plate techniques but more frequently detected by selective methods using cellulosic substrates (1551, 2114, 5874, 5884, 5886). About six species are recognized in this genus.

Key to the most important species:

1	Mesophilic species, no growth at $\geqslant 37°C$	**2**
	Thermophilic species, optimal growth at $40–45°C$	**3**

2(1) Colonies light grey above, reverse somewhat darker, in old cultures becoming black, not zonate; aleurioconidia $<10\,\mu$m diam   *H. fuscoatra* var. *fuscoatra* (p. 394)
  Colonies olivaceous above and in reverse, later turning black, often zonate; aleurioconidia generally $>10\,\mu$m diam   *H. grisea* var. *grisea* (p. 395)

3(1) Aleurioconidia dark brown, globose or oval, 8–16 μm diam, lateral or intercalary, often

catenulate; colonies light brown, later turning dull black   (*H. grisea* var. *thermoidea*)
Aleurioconidia yellow-brown, globose or elongate, usually 7–12·5 μm diam, commonly forming
intercalary chains; colonies at first white, later greyish brown   (*H. insolens*)

Humicola fuscoatra Traaen *1914* var. *fuscoatra* — Fig. 162.

DESCRIPTIONS: Traaen (5874), Fassatiová (1701), and Ellis (1603). — Colonies reaching
2·0–2·5 cm diam in ten days at 20°C on OA, greyish white, cottony, to 3 mm high, reverse
evenly greyish black. Aleurioconidia 7–10(–12) μm diam. Phialoconidia present in most
isolates, obovoid, 3–5 × 1·5–2·0 μm. Hyphal cells and aleurioconidia plurinucleate (5–7
nuclei), phialoconidia uninucleate (485). — The var. *longispora* Fassatiová *1967* differs in
having elongate and predominantly intercalary aleurioconidia, 11–17 × 7–10 μm, and no
phialoconidia. *H. insolens* Cooney & Emerson *1964* (1174) is a somewhat similar thermophilic species which often produces intercalary ± catenulate chlamydospores and no
phialoconidia. — Different values have been given for the GC content of the DNA of *H.
fuscoatra*, 52·5 (5600) or 32·5 (3295), but this appears to be very variable in morphologically
hardly distinguishable isolates; it is generally around 30% (486).

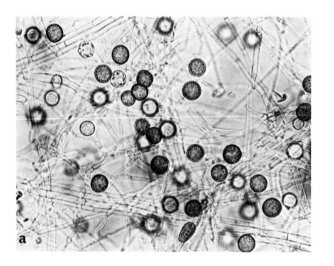
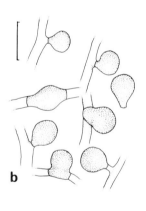

FIG. 162. *Humicola fuscoatra*. a. Aleurioconidia, × 500; b. formation of aleurioconidia, CBS 516.67.

A worldwide distribution is indicated by its occurrence in the Netherlands (4658), Norway
(5874), Sweden (5465), Poland (272), Czechoslovakia (1702), Greece (4647), Italy (3454,
3913), Spain (3446), France (3451), Turkey (4245), Pakistan (61, 4855), Iraq (92), Egypt (8,
3992), Libya (4083), equatorial West Africa (1420), South Africa (1559, 3630, 4407), India
(1315, 1317, 2186, 3865, 3868, 4651, 4698, 4716, 4933), Chile (4569), and Central America
(1697). It has been isolated from beech-wood soil (272, 1702), soils under spruce (5465) and
Acacia mollissima (2186), tropical mixed forest (4716), forest steppe (1700), open savannah

(1559, 4407), grassland (159, 1632, 3865, 4933), arable soils (61, 1317, 3446, 3450, 3868, 4538, 4716, 4788), polder soils (4658), paddy fields (3913), fruit orchards (3630, 3632), saline soils (8, 3446), sandy and clay soils (3992), desert (4083) and Usar soils (4698). It has been observed at soil depths down to 60 cm (3630) and mainly in soils with a neutral reaction (159, 3451). Rapid recolonization after fumigation of citrus soils occurs (3629). It has been isolated from the rhizospheres of wheat (3567), *Aristida coerulescens* (4083) and wilted pineapple plants (6210), leaves of vascular plants (296), soils in association with *Tuber magnatum* (3451, 3452, 3454), and feathers of free-living birds (2575). — It can serve as food for the mites *Pygmephorus mesembrinae* and *P. quadratus* (3104).

The temperature optimum for mycelial growth is 20°C and the maximum 26°C (5874); production of aleurioconidia is stimulated by near-UV light (4132). — *H. fuscoatra* is well known to be an efficient cellulose decomposer (1432, 5386, 6326) and thus can be isolated selectively (4647). Cellulase C_1 is produced in greater quantities than C_x (4779); 1,3-β-D-glucanase (laminarinase) was shown to be present in culture filtrates (4778). *H. fuscoatra* hydrolyses both fructosan from *Aspergillus sydowii* and inulin well, but attacks fructosan from *Acetobacter acetigenum* and *Bacillus subtilis* var. *pumilus* less efficiently (3404). *H. fuscoatra* can utilize humic and fulvic acids and lignosulphonate, and produces some laccase (934). Leucine aminopeptidase, peroxidase and both acid and alkaline phosphatases are formed, but these enzymes were not observed in *H. grisea* (3925). Progesterone is transformed to testololactone (877). — Antagonistic activity against *Bacillus subtilis* (3599) and parasitism of oospores of *Phytophthora cactorum* (5454) have been observed.

Humicola grisea Traaen *1914* var. *grisea* — Fig. 163.

= *Basisporium gallarum* Molliard *1902* sensu Dale *1912* [non Molliard]
= *Monotospora daleae* Mason *1933*
= *Melanogone puccinioides* Wollenw. & Richter *1934*

DESCRIPTIONS: Traaen (5874), Mason (sub *Monotospora daleae*) (3660), White and Downing (6329), Tribe (5886), Fassatiová (1701), and Ellis (1603); conidiogenesis: Cole and Samson (1109); additional data on nomenclature: Mason (3661), and White and Downing (6329). — Colonies reaching 4·0 cm diam in ten days at 20°C on OA; young colonies on MEA and similar media are characteristically olivaceous, lighter towards the margin, darker in reverse and often zonate, the fluffy aerial mycelium lighter, often zonate, later becoming olivaceous-black. Aleurioconidia dark brown, (9–)12–17 μm diam, surrounded with a vaguely delimited pigmented sheath which contains melanized granules when examined under TEM (487). Phialoconidia present in most isolates, adhering in chains or slimy heads, obovate, 3–3·5 × 1·5–2·0 μm. The phialoconidia suggest an affinity with *Chaetomium* but no teleomorph was obtained in our numerous mating experiments. — Hyphal cells and aleurioconidia are plurinucleate (8–16 nuclei), while phialoconidia are always uninucleate (485). — TEM of intercalary chlamydospores showed a secondary wall layer, while aleurioconidia had a uniform, but outwardly melanized thick wall in which a germ pore has been demonstrated (2112). — The aleurioconidia contain numerous globules of β-hydroxybutyrate (487). The pigment has been determined as indole-melanin (1587). The

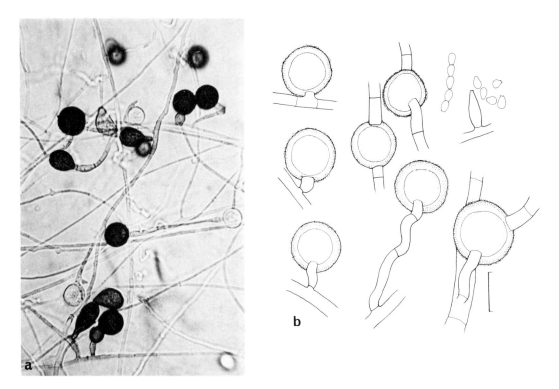

FIG. 163. *Humicola grisea*. a. Aleurioconidia, × 500; b. aleurioconidia and phialide with phialoconidia, CBS 217.34.

mycelium contains 15·3% C_{16}-, 1·5% C_{18}-saturated and 30·9% C_{18}-monounsaturated fatty acids (3232). — DNA analyses gave differing GC contents of 36·3% for a mesophilic (3295) and 45·5% for a thermophilic strain (5600). — *H. grisea* var. *thermoidea* Cooney & Emerson *1964* (1174) is a somewhat similar thermophilic fungus (temperature optimum 35–40°C) with grey to black (never olivaceous) colonies and dark brown aleurioconidia 8–16 μm diam and no phialoconidia.

H. grisea has often been isolated by selective techniques using cellulosic substrates, by immersion techniques (5812, 5813), and hair baits (2091). Reports range from Norway (5874), the USA (6329), Canada (505, 1750) and high altitudes in Peru (2005), Nepal (1826) and Himalaya (4994), to mediterranean (8, 2768, 3992, 4245) and tropical (1519, 1697, 2006, 4159, 5048) areas; it has also been found in Brazil (398), Japan (3267), Tasmania and New Zealand (5930). There are too many reports from various forest and arable soils to cite individually here; it is also known from grassland (163, 1551, 5812, 5814, 6082, 6182), soils with tundra (1750) or savannah type vegetation (1700, 5049), dunes (6414), saline (4000, 4001) and desert soils (4733), carst caves (3365), coal spoil tips (1665), salt-marshes (8), estuarine sediments with rather high salinities (655), with low frequencies in heathland soils (989, 3720, 5811, 5819), also from seawater (4918), wood exposed to soil (5237) or river water (5249), a sewage purification pond (1166), river water only when sewage pollution is low (1154, 1155, 1157), soils treated with activated sewage sludge (1163, 2467), a sewage treatment plant (1165), composted urban waste, compost beds (3041), as an early colonizer in grass compost (3546), from a copper-enriched vineyard soil (4921) and air (5784). *H. grisea* occurs mainly in the upper soil layers (6184) but also down to the B_2 horizon in grassland and arable soils (540). In the rhizosphere of beans its frequency was lower than in the surrounding soil (4451, 4452) but higher than on the rhizoplane (4453); in young plants, however, an increasing rhizosphere effect was noted (5801); further reports are available from the root region of poplar (5898), *Fraxinus* (3570) and various other trees (3142), *Ammophila* (6414), barley (4451), *Lolium perenne* (5815), wheat (2023), corn (6563), dying broad beans (6134), groundnuts (2768), euphorbiaceous plants (3866) and wilted pineapple plants (6210). It has been isolated from seeds of sorghum and wheat (3491, 3989), *Avena fatua* (2961) and cultivated grasses (3512), and is also known from decaying sheaths and blades of *Setaria glauca* (5233), faeces of the ape *Anthropopithecus* in India (5366), and nests and feathers of free-living birds (2575). — *H. grisea* spreads slowly in steam-sterilized soil (6184); only 25% of the aleürioconidia introduced into non-sterile soil survived 12 days incubation, but they survived better when grown on plant material (4421). *H. grisea* is inhibited *in vitro* by *Streptomyces griseus* and other *Streptomyces* species (2939). It can serve as food for the mites *Pygmephorus mesembrinae* and *P. quadratus* (3104).

Optimal growth occurs at 25°C (1827, 5874) and pH 7·5–8·0 (3871); like *H. fuscoatra*, this fungus is decidedly alkaliphilic. The production of aleurioconidia is stimulated by near-UV light (4132); they did not survive 80°C for 5 min (4257). — Good C sources are glucose, sucrose, fructose, inulin, starch and xylan, but glycerol, maltose, and mannitol are poorly utilized (348, 4192, 5874). Mannanase is produced (4192); pectinase activity occurs (1750) but is slight compared to other fungi (1432). Its marked capacity for cellulose decomposition has been described repeatedly, which increases with the concentration of the N source (4433), and has a pH optimum at 7 (5238); only slight differences in this capacity were found between isolates (1432, 6326, 6329). Chitin is also decomposed (1425, 4264) and keratin attacked by boring and spreading hyphae (1633). Colonization of wood caused losses in weight and tensile strength (2211, 2940, 3770, 4191). — Mn^{2+}, Zn^{2+} and Ca^{2+} ions promote mycelial development, synergistically in combinations; at concentrations exceeding 0·2 mM, Mn^{2+} inhibits growth (4297). *H. grisea* can utilize humic and fulvic acids and lignosulphonate; phenol oxidase is produced on pyrocatechol (934); it is heterotrophic for growth substances (1425). — Antagonistic actions towards *Stemphylium sarciniforme* (4422), *Rhizoctonia solani* (1425) and some human-pathogenic fungi (569) are documented. The production of an antibiotic identical with radicicol has been demonstrated (6591). Progesterone is transformed to testololactone (877). Root and shoot weights in wheat, rape and pea seedlings were significantly reduced *in vitro* (1427, 1430). Mycelial extracts showed toxic effects against brine shrimps (1304).

Hypomyces (Fr.) Tul. *1860*

Hypomyces chrysospermus Tul. *1860* — Fig. 164.

≡ *Apiocrea chrysosperma* (Tul.) Sydow *1920*

Anamorph: *Sepedonium chrysospermum* (Bull.) Link ex Fr. *1832*

≡ *Mucor chrysospermus* Bull. ex Purton *1821*

DESCRIPTION of cultures: Arnold (186). — The *Hypomyces* teleomorph is known only from decaying agarics and boleti: the yellow-brown ascomata are submerged in a hyphal mat; asci cylindrical with an apical thickening and the ascospores fusiform, biapiculate, with a septum in the lower third (*Apiocrea* Syd. *1920* has been distinguished from *Hypomyces* on this character) and longitudinally striate walls. — The anamorph is readily obtained in culture, reaching 5·0 cm diam in ten days at 20°C on MEA. It is mainly characterized by the abundant globose, yellow, solitary blastic conidia (aleurioconidia) borne on branched conidiophores; the conidia themselves are densely verrucose or spiny, have a thick and distinctly layered wall, and measure 15–25 μm diam. In young colonies, a few smooth-walled, ellipsoidal to cylindrical conidia are produced on long and slender phialides. The pigmentation of the aleurioconidia and the presence of phialoconidia distinguish *Sepedonium* from the similar genus *Histoplasma* Darling (cf. *Emmonsiella*). — Mannose and glucose were the only sugars known in the cell walls other than *N*-acetylglucosamine, and the lipids were found to consist of phospholipids, di- and triglycerides, sterols, sterol esters and free fatty acids (1417).

H. chrysospermus is extremely common in the temperate zone as a parasite of Boletales fruit-bodies mainly in the genera *Boletus* (sensu lato), *Suillus, Gomphidius, Paxillus* and also of *Scleroderma* and *Serpula lacrymans* (IMI, 1538, 3130) from which it can easily pass into the soil. — This fungus has also been reported from cultivated soils in India (1315, 1519) and Israel (2764), heathland in Australia (3720), the Sonoran desert of the USA (4733), salt-marshes in Egypt (8), forest soils in Georgia/USA (3817), the British Isles (IMI) and Austria (1828), compost soil in Germany (1424), children's sandpits in Poland (1421), and soils in Brazil (6008), Hawaii (569) and Canada (IMI). It is also known from bat habitats in India (5046) and feathers, nests and pellets of free-living birds in Czechoslovakia (2575).

The growth on glucose and $NaNO_3$ by *H. chrysospermus* is dependent on the presence of biotin (3238). Pectin and cellulose can be degraded by this fungus (3414). The structure of an antifungal metabolite, the azaphilone chrysodine (1079), and the biosynthesis of the yellow tropolone pigment sepedonin have been analysed (5699, 6447); labelled acetate and formiate can be incorporated as precursors suggesting the polyketomethylene biosynthetic pathway of the latter metabolite (5698).

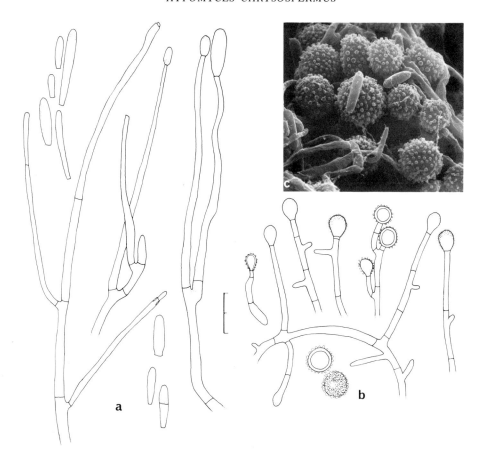

FIG. 164. *Hypomyces chrysospermus*. a. Conidiophores with phialides and phialoconidia of young colony; b. conidiophores with aleurioconidia; c. both kinds of conidia, SEM, × 1000, orig. R. A. Samson; CBS 140.23.

Idriella P. E. Nelson & Wilhelm *1956*

Idriella lunata P. E. Nelson & Wilhelm *1956* — Fig. 165.

? − *Idriella couratarii* Ram *1970*

DESCRIPTIONS: Nelson and Wilhelm (4126), Barron (364), Ellis (1603), Nicot and Mouchacca (4171), Nicot and Charpentié (4168), and Matsushima (3680).

The genus *Idriella* (= *Chloridiella* Arnaud *1953*, nom. inval., Art. 36) was originally described as monotypic, but several species have recently been added to it. Its circumscription was emended by Nicot and Mouchacca (4171) to cover hyphomycetes with hyaline or pigmented hyphae, erect subulate, pigmented conidiophores consisting of a single conidiogenous cell or several cells with an apical conidiogenous cell which forms a somewhat geniculate and inconspicuously denticulate conidiiferous rhachis in the uppermost part. Blastoconidia are formed in sympodial succession, arranged in heads or ears, mostly one-celled, falcate, hyaline, smooth-walled. Brown, pluricellular *Trichocladium*-like chlamydospores are generally present.

I. lunata forms effuse, rather slow-growing, brownish colonies on PCA, reaching 2·8–3·0 cm diam in ten days at 20°C on MEA. Conidiogenous cells usually formed singly directly on aerial hyphae, pale brown, somewhat darker than the subtending hyphae, 13–26(–35) × 3–4(–5) μm, tapering to 1–2 μm above. Conidia produced in dry heads (sometimes also compacted by slime), falcate, simple, 7–13(–20) × 2–2·5(–4) μm. Chlamydospores golden brown, 1-4-septate, smooth-walled, 10–25 × 7–9 μm. Colonies on media other than PCA tend to remain sterile. — Ten species have now been described in this genus: the closest to *I. lunata* are *I. vandalurensis* Vittal *1970* with septate conidiophores and 2-celled conidia 12–23 × 2–3 μm; *Chloridiella leucopoda* (Bonord.) ss. Arnaud *1953* (nom. inval., Art. 43; ≡*Camptoum leucopodum* Bonord. *1851*) with longer septate conidiophores and conidia similar to those of *I. lunata* (4168), probably identical to *I. variabilis* Matsushima *1971*; the anamorph of *Hymenoscyphus caudatus* (Karst.) Dennis *1964* with loose synnemata, 2–3-septate conidiophores 25–30 × 4–5 μm and simple conidia (3–4-septate at germination) to 25 μm long; and *I. desertorum* Nicot & Mouchacca *1972* usually with single conidiogenous cells to 55(–69) μm long, distinct conidiiferous denticles, rather clavate conidia 6–11 × 1·7–3·2 μm, and chlamydospores mostly one- to two-celled and 4·3–7·7 μm diam.

I. lunata was originally described as the causal agent of a strawberry root rot in California. It has been isolated on various occasions in Canada from glasshouse, arable and forest soils (364, 505, IMI), and has also been found in soils in New Caledonia (IMI). In addition to its rare occurrence in arable soils under wheat in Germany, the fungus was also shown to be present in a soil under *Pinus sylvestris* in Austria and garden soil in the Netherlands (CBS) and in the surface soil of managed prairie stands (6347). Apart from strawberries (4133), the species has been observed on dead leaves of oaks, *Pasania edulis*, and *Rhododendron metternichii* in Japan (3680), and on roots of pepper, tomato, carrot, lettuce and *Potentilla* in California, without causing any damage to these plants (4126).

I. lunata sporulates only on media with low sugar contents (4126).

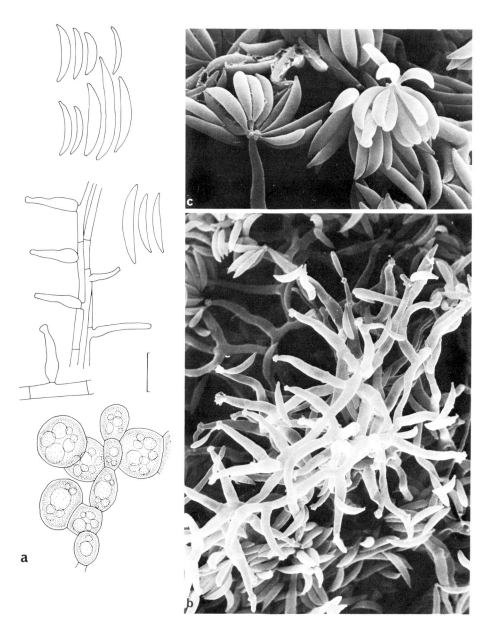

FIG. 165. *Idriella lunata*. a. Conidiogenous cells, conidia and chlamydospores; b. densely aggregated conidiogenous cells with apical clusters of conidial denticles, SEM, × 1500; c. detail of conidial heads, SEM, × 2750; CBS 404.78.

Khuskia H. Hudson *1963*

Khuskia oryzae H. Hudson *1963*

Anamorph: *Nigrospora oryzae* (Berk. & Br.) Petch *1924*

≡ *Monotospora oryzae* Berk. & Br. *1873*
= *Basisporium gallarum* Moll. *1902*
= *Glenospora elasticae* Koord. *1907*
= *Nigrospora gorlenkoana* Novobr. *1972*

DESCRIPTIONS: teleomorph: Hudson (2584); anamorph: Mason (3659, 3660), Ellis (1603), Potlaichuk (4605), Subramanian (5626), and Matsushima (3680); conidial discharge mechanism: Webster (6236).

The monotypic genus *Khuskia* (fam. Amphisphaeriaceae Wint.) is only known from the natural substrate of *Vetiveria zizanioides* and *Panicum maximum*. Perithecia form in elongate clusters of 1–7 below the epidermis but are later erumpent, with papillate ostioles, to 250 μm diam. Asci clavate, short-stalked, unitunicate, without any apical differentiation. Ascospores hyaline, somewhat allantoid, with the distal end more rounded than the basal, 16–21 × 5–7 μm; usually becoming unequally two-celled on germination.

Colonies spreading broadly, reaching 5·5–8·0 cm diam at 20°C in five days on MEA, grey, cottony; reverse olivaceous-black. Conidia of the *Nigrospora* type (q.v.) (11–)12–14(–20) μm diam. — DNA analysis gave a GC content of 42–62% (5600).

Kh. oryzae is very common in tropical countries and also occurs sometimes in temperate regions (1603). The principal hosts are monocotyledonous plants, such as rice (leaves, glumes), banana (fruit), coconut, corn (kernels), wheat, rye, barley, and sugar cane (3207, 3584, 3659, 3781, 3989, 4548, 6407); the different *Nigrospora* species are not restricted to any of these plants. *Kh. oryzae* has also been isolated from forest soil in central Africa (3063), the rhizosphere of wilted pineapple plants (6210), pecans (2258, 2572) and strawberry plants (4133) and became known as the causal agent of an apple fruit rot in Allahabad/India (2954). *Kh. oryzae* is a suitable food for mites (3208).

Optimal germination of conidia occurs at 40°C on D-maltose- or D-glucose-containing media; no germination occurs at 60°C (5045). Dextrin and L-arabinose were found to favour mycelial growth most and acetate and citrate least. Starch is also utilized (1827, 6519) and, with urea as N source, its growth is stimulated by biotin (104). *Kh. oryzae* produces griseofulvin (726, 1849). — Phytotoxic metabolites occur *in vitro* and their production was found to depend on the provision of a suitable N source, such as asparagine, tryptophan or tyrosine (4103). Corn contaminated with *Kh. oryzae* is toxic when fed to rats (5230).

Leptosphaeria Ces. & de Not. *1863*

Type species: *Leptosphaeria doliolum* (Pers. ex Hook.) Ces. & de Not.

Anamorphic genera: *Phoma* Sacc., *Coniothyrium* Corda emend. Sacc., *Camarosporium* Schulz., *Hendersonia* Sacc., *Rhabdospora* (Sacc.) Sacc., *Stagonospora* (Sacc.) Sacc.

Fam. Pleosporaceae Wint. — Ascomata scattered or aggregated, partly immersed in the host tissue at least at the beginning, ± broadly conical, ostiolate, smooth-walled (lacking hairs or setae); wall black, carbonaceous, rather thick, composed of large, globose, thick-walled cells. Asci bitunicate, clavate to subcylindrical, with a short stalk, often with a refringent apical cap, mostly 8-spored; ascospores fusiform, cylindrical or filiform, with three to many transverse septa, yellow-brown to almost hyaline, mostly lacking a gelatinous sheath.

Of over 700 species described in *Leptosphaeria*, about 50 can be recognized in this genus; those occurring in Switzerland have been monographed by Müller (4003); for generic synonymy and further references cf. v. Arx and Müller (211). *Leptosphaeria* species are usually saprophytic, occasionally parasitic on dicotyledonous hosts.

The two species treated here are keyed out under their *Coniothyrium* and *Phoma* anamorph genera.

Leptosphaeria coniothyrium (Fuckel) Sacc. *1875*

≡ *Sphaeria coniothyrium* Fuckel *1870*

Anamorph: *Coniothyrium fuckelii* Sacc. *1875*

DESCRIPTIONS: teleomorph: Müller (4003); anamorph: Wollenweber and Hochapfel (6423). — Colonies reaching 2·6–2·8 cm diam in seven days on OA at 20°C, grey-brown due to the aerial mycelium; pycnidia sparsely produced, brown to black. Conidia short-ellipsoidal, light brown, smooth-walled, 3·5–4·5 × 2·2–2·7 µm. Conidiogenesis is as in *Phoma* (2811); mature conidia have a thick, inner, electron-transparent and a thin, outer, electron-dense wall layer. — The connection between teleomorph and anamorph has been known since the work of Saccardo in *1875*, but was first demonstrated in culture experiments by Koch (3065). Because of difficulties in species delimitation (cf. under *Coniothyrium sporulosum*), the reports listed below concerning the distribution of this fungus must be treated with reservation.

Both the teleomorph and anamorph are known from dead stems of woody and herbaceous plants (1963, 3568, 4003, CMI Distribution Map No. 185, *1967*), in particular of the Rosaceae, for example *Rubus,* where it can cause a die-back. — Its distribution is worldwide, including Canada (234), the USA (1166, 1171, 2573), many countries of Europe, also Israel (2764, 2768), Libya (6510), Kuwait (4001), Somalia (5048), Zaïre (3790), central Africa (3063), South Africa (1415), India (4977), Pakistan (1963), New Zealand (2705) and Peru (2005). According to some authors, it is particularly frequent in forest soils (272, 3549, 3790) but there are also numerous reports from agricultural soils (234, 540, 1616, 1889, 2768, 2871, 3006, 3450, 3817, 3913, 4703) and grasslands (2573, 3856, 6138); furthermore, it has been isolated from soil under steppe vegetation (1559), an alpine pasture (3445), a forest nursery (1978, 6184), saline soil (4001), estuarine sediments (655), polluted streams (1157, 1166) and wood pulp (6168). *L. coniothyrium* has been isolated mainly from the upper soil layers (6184), although it penetrated down to the C_1-horizon in a recently deglaciated soil in Alaska (1171). The optimum pH for *L. coniothyrium* in the soil is above 6·5 (2895). It has been isolated from the rhizospheres of wheat (3567), oats (4443), alfalfa (3982), groundnuts (2768), and peas, with or without an accumulation effect (5825). Other known substrates include the roots of poplar (5898), *Lolium perenne* and clover in New Zealand (5815), litter of *Panicum coloratum* (1560), decaying stems of *Pteridium aquilinum* (1821), sticks exposed to soil (1956), stubble residues (3856) and pecans (2572). — The fungus can serve as food for the springtail *Folsomia fimetaria* (4007). It has been found to be inhibited by *Streptomyces* species (2939), *Chalara elegans* (2753), and by hyphal contact with *Tuber melanosporum in vitro* (632). It survived storage in water for at least seven years (605).

The optimal temperature for growth is about 26°C and no growth occurs below a water potential of −20 bars (2939). *L. coniothyrium* is able to oxidize Mn^{2+} salts (5151, 5828) with optimal activity at pH 6·5 (6054). — It is entirely heterotrophic for thiamine and partially so for both biotin and inositol (2994). The decomposition of pectin, xylan (1432) and cellulose in the form of filter paper (2706) has been reported, 4·6% tannin is tolerated in the medium (2706) and the production of a β-glucosidase able to split several heteroglucosides has been described (6159). — A slight antagonistic activity against *Colletotrichum coccodes, C. linicola, Rhizoctonia solani, Sclerotinia sclerotiorum, Aspergillus niger* and *Botrytis aclada* has been observed (6111).

Leptosphaeria maculans (Desm.) Ces. & de Not. *1863* — Fig. 166.

≡ *Sphaeria maculans* Desm. *1846*

Anamorph: *Phoma lingam* (Tode ex Fr.) Desm. *1849*

≡ *Sphaeria lingam* Tode ex Fr. *1823*
≡ *Plenodomus lingam* (Tode ex Fr.) Höhn. *1911*
= *Plenodomus rabenhorstii* Preuss *1851*
= *Phoma brassicae* (Thüm.) Sacc. *1884*

For further synonyms see Boerema and van Kesteren (602).

DESCRIPTIONS: Pound (4606), Boerema and van Kesteren (602), Boerema (592), and CMI Descriptions No. 331, *1972.*—Colonies reaching 3·7 cm diam in ten days at 20°C on OA, light to dark grey due to the aerial mycelium, pycnidia scattered and of two distinct

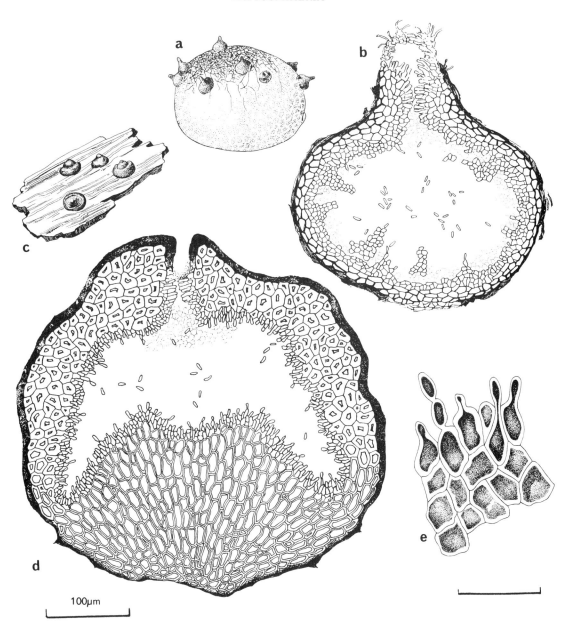

Fig. 166. *Leptosphaeria maculans.* a. Normal *Phoma*-type pycnidia on cabbage seed (× 16); b. the same in vertical section; c. sclerenchymatous type of pycnidia on cabbage stem (× 16); d. the same in vertical section; e. conidiogenous cells; orig. G. H. Boerema (from 602, reproduced with permission of the Editors of Persoonia); left bar applies to Figures b and d).

phenotypes: either with mainly pseudoparenchymatous walls or thicker and pseudoscleren-chymatous walls; pycnidia brown, opening at a late stage of development with one or several papillate necks. Conidia formed as in other *Phoma* species; they are hyaline, ellipsoidal, 4–6 × 1·5–2·0(–3) μm. TEM studies of conidiogenesis have been carried out (593). — The sclerenchymatous pycnidia were used in the characterization of the genus *Plenodomus* Preuss *1851* which is now regarded only as a section of *Phoma* Sacc. (q.v.) (593). The production of pycnidia from ascospore isolates has been described by Smith and Sutton (5437).

L. maculans is known to be the causal agent of leaf spots, lesions, canker and discoloration on stalks (black-leg), pods and tubers of *Brassica* species in many areas but is most prevalent in temperate regions (see CMI Distribution Map No. 73, *1969*). A technique for determining its inoculum density in the soil has been described (1972). — With the exception of occurrences in calcareous grassland (4152) and the rhizosphere of sand dune plants (4371), it has not been reported in soil analyses and thus is not generally listed as soil-borne (1439) but rather seed-borne (602). In wheatfields in Germany it has been detected only twice.

No growth occurs at pH 3·5. The utilization of starch has been found to be weak (2119) and the production of a β-glucosidase described (6159). — It is inhibited *in vitro* by *Trichoderma viride* (6011).

Mammaria Ces. 1854

Mammaria echinobotryoides Ces. *1854* — Fig. 167.

≡ *Trichosporium echinobotryoides* (Ces.) Sacc. *1886*
= *Echinobotryum laeve* Sacc. *1877*
= *Acrotheca solaris* Sacc. *1883*

DESCRIPTIONS: Hughes (2601), Hennebert (2398), Barron (364), Sugiyama (5634), Ellis (1603), and Matsushima (3680). — The monotypic genus *Mammaria* is characterized by brown, lanose mycelia consisting of pale-brown hyphae with darker septa, and clusters of up to 45 aleurioconidia formed in sympodial succession on scarcely differentiated and densely septate conidiophores; aleurioconidia brown, smooth-walled, provided with a longitudinal germ slit. Colonies reaching 1·5 cm diam in seven days at 28°C, isabelline to olivaceous-brown, fluffy; reverse fuscous-black. Aleurioconidia mostly sessile, ellipsoidal to pyriform,

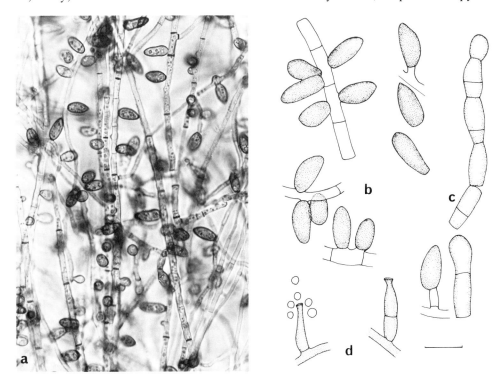

FIG. 167. *Mammaria echinobotryoides*. a. Aleurioconidia, × 500; b. aleurioconidia; c. hypha fragmenting like arthroconidia; d. phialides formed in old culture.

slightly apiculate at the tip, broadly flattened at the basal scar which is $2 \cdot 0$–$4 \cdot 5\,\mu$m diam, $8 \cdot 5$–$17 \cdot 5 \times 4 \cdot 5$–$8 \cdot 5\,\mu$m. Some slightly swollen cells of the vegetative hyphae may secede like arthroconidia. In ageing cultures, a *Phialophora* state very similar to *Ph. cyclaminis* van Beyma (q.v.) is sometimes formed which has globose conidia $1 \cdot 5$–$1 \cdot 8\,\mu$m diam (2398).

This species is known principally from rotting plant material in the temperate latitudes of Europe, North America (2398) and Japan (3680). *M. echinobotryoides* has rather rarely been isolated from soil but is known from forest soils in France (2963), Canada (505) and central Africa (3063), garden soil in England (4452), wheatfield soils in Germany, and other cultivated soils in Japan (5634), the Netherlands (1616) and Belgium, salt-marsh mud in the British Isles (IMI), and peat soils in Canada (364). Isolations from fresh water and/or organic substrates in fresh water have been reported from Northern Ireland (4429), the USA (5249) and the Indian Himalayas (5634). It has been isolated from the rhizospheres of bean plants (4452) and groundnuts (2532), and decaying wood (3063).

Aleurioconidia require a C source for germination; starch, lactose and sodium citrate proved not to be suitable for this purpose. Arthroconidia, however, did not show any specific substrate requirements for germination while phialoconidia completely failed to germinate on numerous media tested (4430). — *M. echinobotryoides* has a temperature minimum of $\leqslant 10°$C and does not grow at $37°$C (5634). — When growing on straw, it can utilize lignin and cellulose, forming humic acid-like substances; on maple-wood strips, it can cause considerable losses of weight and tensile strength (2211).

Mariannaea   Arnaud ex Samson *1974*

Mariannaea elegans (Corda) Samson *1974* — Fig. 168.

≡ *Penicillium elegans* Corda *1838*
≡ *Paecilomyces elegans* (Corda) Mason & Hughes *1951*

DESCRIPTIONS: Brown and Smith (743), Hughes (2596) and Samson (5024). — The genus *Mariannaea* has been segregated from *Paecilomyces* Bain. (q.v.) because of its almost aculeate phialides and imbricate conidial chains. — Arnaud regarded his *Mariannaea elegans* as distinct from *Penicillium elegans* Corda, but his material and illustration agree perfectly with Corda's type material (5024).

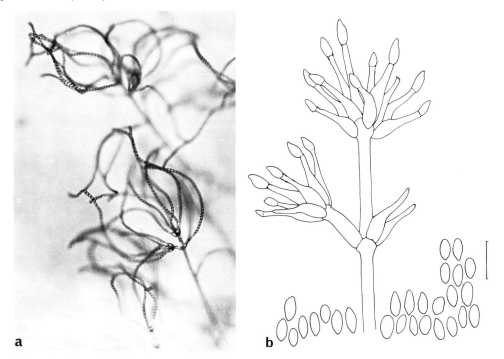

FIG. 168. *Mariannaea elegans*. a. Low power view showing imbricate conidial chains; b. verticillate conidiophore and conidia.

Colonies reach 5–7 cm diam in 14 days at 25°C on MEA. The slender awl-shaped phialides have a symmetrical opening through which the young conidia are pushed out obliquely; subsequently they develop an asymmetrical ellipsoidal to fusiform shape, 4–6 × 1·5–2·5 μm, and cohere in imbricate chains. Catenulate chlamydospores are present. SEM photographs of conidiophores and conidia have been published (5024).

Penicillium elegans was originally described from coniferous bark; isolates from bark invariably form a brown reverse in culture, whereas soil isolates generally have a red-purple reverse, grow faster and have no tendency to fasciculate growth, this latter type having been distinguished as var. *punicea* Samson *1974* (5024). A second species of this genus, *Mariannaea camptospora* Samson *1974*, has been isolated both from wood and forest soil.

M. elegans var. *elegans* is very common on decaying coniferous bark and wood but also occurs in forest soil; it has been recorded from Germany, the Netherlands, France, the British Isles (2596), Italy, Poland, Canada, the USA, and South Africa; the var. *punicea* is recorded from soils in Germany, the Netherlands, the USA, Canada, and Zaïre (5024). In the subsequent records, no distinction between the two varieties could be made. Besides a wide distribution in the temperate latitudes, it has been documented for Chad (3415), South Africa (1555, 1556), Zaïre (2966, 2967, 3790), central Africa (3063), Brazil (399), Chile (1824), Japan (2532, 4989, 5846), and Nepal (1826). It has mainly been isolated from forest soils, among which beech woods have been mentioned in particular (272, 273, 3138, 4989), but also under *Eucalyptus maculata* (1558), white cedar (505) and pines (269, 681, 6352), especially in the litter layer (681, 2411, 3448). In Zaïre it is found in both shaded cultivated and forest soils (3790); it is also known in prairie and grassland soils (982, 1555, 2573, 4313). It has been observed in calcareous soils (4152), parachernozem (3414), running water (5249) with a low degree of pollution (1154, 4429) but also in a waste stabilization pond, an acid mine drainage stream (1166) and on fields treated with digested sewage sludge (1163). In arable soils, it has been isolated from the rhizospheres of groundnuts (2532), oats (4443) and wheat (3567). *M. elegans* has further been isolated from pine chips after storage (3140), beech wood veneers inoculated with soil (5238), stained birch wood (5061), boards in a sauna (5016), wood in a copper mine (2664), wood baits exposed in soil (1956, 5237), and the fruitbodies of *Pholiota nameko* (3085). —The conidia of *M. elegans* can serve as food for amoebae (2348).

Good growth occurs at 30°C, traces even at 50°C, the optimum pH range is 5·5–8·0; sporulation is best between pH 6·0 and 6·5; below pH 5·0 sporulation and growth are reduced. — Cellulose decomposition (3414, 4191, 5926) is relatively strong, but no exoenzymes were shown in the medium (656). Pectinase, chitinase and proteinase activities have been demonstrated (5926). *M. elegans* is somewhat more thermophilic than other common cellulose decomposers (5024). The var. *elegans* was shown to produce an electrophoretically uniform humic acid (3200). It produces the antiviral steroid antibiotic paecilomycerol (2889) without inhibition of fungi and bacteria. It grows chemotropically towards phosphate granules without dissolving them (3074). — *M. elegans* inhibits the mycelial growth of *Pholiota nameko* at 25°C and at 12°C it is overgrown by *Ph. nameko*; it does not affect the fruitbody development in *Ph. nameko* (3085).

Memnoniella Höhn. *1923*

Memnoniella echinata (Riv.) Galloway *1933* — Fig. 169.

≡ *Penicillium echinatum* Riv. *1873*
≡ *Stachybotrys echinata* (Riv.) G. Sm. *1962*
= *Memnoniella aterrima* Höhn. *1923*

DESCRIPTIONS: Ellis (1603), Matsushima (3679, 3680), Subramanian (5626), and Jong and Davis (2815); discussion of the nomenclature: White *et al.* (6332).

The genus *Memnoniella* is very close to *Stachybotrys* but is distinguished by the catenate arrangement of its conidia. It has been monographed in the same publications as *Stachybotrys* (q.v.). Smith (5434) did not consider the difference between the two genera to be sufficient to warrant their separation, but most recent authors, though generally with some hesitation, retain them as distinct. Contrary to *Stachybotrys*, in *Memnoniella* new conidia arise in basipetal succession before the previous ones are mature as evidenced by TEM studies (863). — Five species have been recognized and *M. echinata* is distinguished from the others by its relatively small, slightly flattened, and coarsely roughened conidia.

Colonies reaching 1·3 cm diam in five days on OA at 24°C, becoming blackish granulate with the production of conidia; reverse yellowish brown to brownish grey. Conidiophores 70–90 μm long and 3–5 μm wide, hyaline at first, later becoming olivaceous and sometimes

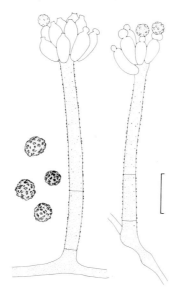

FIG. 169. *Memnoniella echinata*, conidiophores and conidia, CBS 304.54.

minutely roughened. Phialides obovate to ellipsoidal, pale olivaceous, $7-10 \times 3-4\,\mu$m. Conidia cohering in long chains, globose but slightly compressed at the ends, at first hyaline and smooth-walled but at maturity becoming dark olive-grey, coarsely warted, and $3-6\,\mu$m diam. — The similar *M. subsimplex* (Cooke *1883*) Deighton *1960* has coarsely warted conidia $6-9\,\mu$m diam; *M. zingiberis* Vasant Rao *1962* has perfectly globose and warted conidia $(4\cdot5-)6\cdot0-6\cdot5(-7\cdot0)\,\mu$m diam; and *M. levispora* Subram. *1954* has smooth-walled conidia $3-7$ μm diam.

M. echinata is a cosmopolitan fungus, and its occurrence has been documented in Denmark (2745), the British Isles, Austria, and Italy, although it is mainly found in tropical countries (1603, 2815). It is known from Papua-New Guinea (3679), Japan (3680), Honduras (2031, 2035), India (2854, 3865, 4716), the USA (2482, 3817) and Israel (4758). It has been isolated from forest soil (2923), grassland (3445, 3865) and cultivated soils (3817). Other known substrates and micro-habitats are very varied and include the rhizospheres of banana (2030, 2031, 2035), tobacco (3231), pines (3572), and several sand dune plants (4371), decaying sheaths and blades of *Setaria glauca* (5233, 5234), stems of *Salvadora persica* (1963), cotton leaves (5231), litter of *Atlantia monophylla* (6108), submerged parts of bamboo (4), birds (4652), and cotton textiles (4106, 4186, 4758, 6163, 6326, 6332).

Good growth occurs in the range $15-30°$C, and the maximum temperature for conidium germination is $37°$C (2815). — It showed good growth on methyl-β-D-glucoside, cellobiose, mannose (5387) and even with calcium cyanamide as sole N source (6078). Cellulose is strongly degraded (53, 631, 2815, 4779, 5234, 6129, 6326), optimally at $30°$C (6163). It has also been found to degrade mannan and carob gum (4781) and to attack hardboard (3770) and woollen fabrics (6330). *M. echinata* can produce acetic acid (4510). — Inhibitory effects on numerous soil fungi have been observed *in vitro* (4).

Metarrhizium Sorok. *1883*

Metarrhizium anisopliae (Metschn.) Sorok. *1883* — Fig. 170.

≡ *Entomophthora anisopliae* Metschn. *1879*
= *Isaria destructor* Metschn. *1880*
≡ *Oospora destructor* (Metschn.) Delacr. *1893*
= *Metarrhizium album* Petch *1931*
= *Metarrhizium brunneum* Petch *1935*
= *Myrothecium commune* Pidopl. & Kiril. *1969*
= *Metarrhizium velutinum* Borowska *et al. 1970*

For further synonyms see Tulloch (5939).

In spite of the numerous arguments to the contrary (5939), we retain the spelling *Metarrhizium* and regard it as a necessary linguistic correction of the original spelling (Art. 73).

DESCRIPTIONS: Tulloch (5939), Latch (3233), and Veen (6052).—*Metarrhizium* is distinguished from the otherwise similar *Myrothecium* Tode ex Fr. (q.v.) by its dry subhyaline phialoconidia compacted into regular chains and columns; the column formation is due only to the aggregation of the elongate conidia themselves, unlike some otherwise rather similar *Penicillium* species where the conidiophores are aggregated (synnemata). — In *Metarrhizium* two species are now recognized: *M. flavoviride* W. Gams & Rozsypal *1973* with ellipsoidal conidia, 7–9(–11) × 4·5–5·5 μm and yellow-green to olivaceous-buff colonies, and *M. anisopliae* with cylindrical conidia and colonies in various darker shades of green, or sometimes sepia or isabelline; colonies rather slow growing, reaching 2·0 cm diam in ten days at 20°C on OA. *M. anisopliae* has two varieties, the more common, small-spored var. *anisopliae* with conidia (3·5–)5·0–8·0(–9·0) × 2·5–3·5(–4·5) μm and var. *major* (Johnson) Tulloch *1976* with conidia 10·0–14·0(–18·0) μm long. Serological differences were found between isolates of these two varieties (1693). — The phialide neck possesses a thickened, electron-transparent wall similar to that seen in most phialidic genera; the blown-out conidia are delimited at the base by a completely closed septum which thickens during maturation and provides the basis for conidium coherence to judge from TEM studies (2242). — The conidia are uninucleate (5831, 6051) and contain numerous lipid and protein droplets together with multivesicular bodies near the delimiting septum (2242). Further TEM work on the conidia, germ tubes and their ability to penetrate insects has been carried out (6524, 6525, 6526, 6527). — Schaerffenberg (5080) claimed having observed a teleomorph which might belong to the Hypocreaceae, but this observation has not been confirmed. Nuclear migration by means of anastomoses (5831) and the combination of auxotrophic mutants (5833) have given evidence for its heterokaryosis. — The major fatty acids of neutral and polar lipids in the mycelium were found to be palmitic, oleic and linoleic, and with smaller amounts stearic and palmitoleic (5965).

M. anisopliae is one of the most important insect pathogens with a high degree of

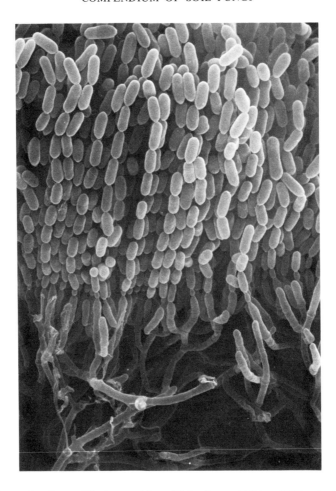

Fig. 170. *Metarrhizium anisopliae*, conidiophores with conidial columns, SEM, × 1000, orig. R. A. Samson.

specialization, occurring almost entirely in two families of Coleoptera, the Elateridae and Curculionidae (3257). A list of 204 host insect species has been compiled (6052) and a selective medium for its isolation from insects described (6053). — On insects, it has a worldwide distribution and in soil it has been reported from, for example, Nepal (1826), New Zealand, New Caledonia (IMI), the Bahamas (2006), the USA (2240, 2482, 2822, 3817), Canada (505), Northern Ireland (4429), Italy (488), Turkey (4245), the USSR (IMI), and Zaïre (3790). Its non-insect habitats include cysts of the nematodes *Heterodera schachtii* and *Globodera rostochiensis* (3090), corn fields (3490) and forest soils (364, 2573) in Canada, banana soils in Honduras (2031), bean- and cornfields in Brazil, grassland soils in New Zealand (IMI), wheatfields in Germany irregularly, forest soils after burning (2822), muck soil (505), organic detritus in unpolluted streams (4429), river sediments (IMI), a mangrove swamp (3264), lead mine spoil heaps (IMI), birds' roosts (682) and healthy strawberry roots (4133). — *M. anisopliae* is not able to grow out from dead chitinous substrates into adjacent non-sterile soil (2582). Low temperatures (8°C), high or very low humidities, CO_2 accumulation or O_2 deficiency favour the survival of isolated conidia which in this way were able to survive for up to 455 days; under other conditions they had lost much of their capacity to germinate after 56 days (1076, 6164). The germination of the conidia of *M. anisopliae* in soil appears to be strongly inhibited by the soil microflora (6164); an *Aeromonas* species was found to suppress the growth of *M. anisopliae in vitro* (6490).

The production of conidia requires water potentials above -100 bars, but below 10°C and above 35°C no sporulation occurs (6164); the optimal germination of conidia, however, is obtained at vapour saturation (6052), and the optimal temperature for this is 25–30°C; no germination occurs at water potentials of less than -160 bars; the thermal death point for conidia has been determined as 49°C for 10 min in water (6164, 6490). *M. anisopliae* has an optimum growth temperature of about 25°C (1827) and can grow *in vitro* in the pH range 3·3–8·5. — The growth on various N and C sources has been investigated and glucose found to be utilized better than acetate (3329). In shake cultures, the mycelium becomes transformed into a kind of blastospores (6052). Degradation studies on starch (1827) with positive results, and on cellulose with both negative (386, 3325, 6326) and positive (1425) results have been carried out. Wool (6330) and chitin are decomposed: the latter moderately (2706) to strongly (2582, 3325, 5019); fat (optimally at pH 6·7) and glycogen decomposition and a slight proteinase capacity have also been observed (3329). *M. anisopliae* is autotrophic for growth substances (1425, 6052) and weakly antagonistic towards *Pythium* sp. and *Phytophthora cactorum* (1425). — On account of the high humidity requirements, infection of the host insects occurs mainly orally; artificial infection has been successful in *Nemocestes incomptus* (1211), *Bombyx mori* and *Oryctes rhinoceros* (4715). In a *Pyrilla* species the penetration of the integument by chitinolysis and the subsequent spread of the fungus through the blood system, fat bodies, glandular tissues, digestive tract and nervous tissues of the host have been followed (2709). *M. anisopliae* produces a substance that paralyses the larvae of *Galleria mellonella* and *Bombyx mori* (4863, 4864), and the cytochalasins C and D (= zygosporin A) (522) which are highly toxic to mouse fibroblasts in culture (93, 914). Other toxic metabolites include the cyclodepsipeptides destruxin A, B, C and D, L-prolyl-L-leucine anhydride, L-prolyl-L-valine anhydride and desmethyl destruxin B (3068, 4865, 5654, 5655, 5656, 5720).

Microascus Zukal *1885*

= *Fairmania* Sacc. *1906*
= *Peristomium* Lechm. *1913*
= *Nephrospora* Loub. *1923*

Type species: *Microascus longirostris* Zukal

Anamorph: *Scopulariopsis* Bain.

MONOGRAPHIC TREATMENTS: Barron *et al.* (366), Morton and Smith (3974), Udagawa (5972), and von Arx with most recent key (206). — Fam. Microascaceae Luttrell ex Malloch. — Aerial mycelium consisting of pigmented hyphae. Ascomata superficial or partly immersed, carbonaceous, glabrous or setose, with a well-differentiated and sometimes cylindrical (or neck-like) ostiole, extruding the spores in a long cirrhus. Asci ovate, 8-spored, evanescent. Ascospores small, one-celled, asymmetrical, often reniform, heart-shaped or triangular, dextrinoid when young, provided with one small germ pore. — Thirteen species have been recognized (206). Von Arx (204) segregated *Pithoascus* v. Arx *1973* from *Microascus* on account of its often non-ostiolate or inconspicuously ostiolate ascomata and the absence of any *Scopulariopsis* conidia but scanty *Scopulariopsis* conidia have been found since in *P. intermedius* (Emmons & Dodge) v. Arx *1973*. — *M. trigonosporus* can easily be distinguished from all other species of the genus by its small, triangular ascospores (4–5·5 μm in longest diam) with broadly rounded ends.

Microascus trigonosporus Emmons & Dodge *1931* — Fig. 171.

Anamorph: *Scopulariopsis trigonospora* Emmons & Dodge *1931*

DESCRIPTIONS: Barron *et al.* (366), Morton and Smith (3974), Udagawa (5969), and Matsushima (3680); ascoma development: Corlett (1186). — Colonies reaching 9–17 mm diam in seven days on MEA at 24°C, at first pale buff but soon turning mouse-grey; reverse greyish but later becoming dark grey to almost black. Ascoma initials consisting of short chromophilic side branches of vegetative hyphae which become curled and 3–4-celled. Ascomata carbonaceous, glabrous or covered with scattered smooth-walled hairs, base spherical, 125–200 μm diam, neck ± cylindrical, usually to 75(−250) μm long. Asci ovate, 10–22 × 7–8 μm, catenate. Ascospores triangular in plane view, with 3 indented sides and broadly rounded ends, red-brown in mass, 4–5·5 × 3–4 μm, uninucleate. *Scopulariopsis* state producing mostly solitary, rarely verticillate, annellophores on fasciculate aerial hyphae, conidiogenous cells 10–15 × 2–3 μm. Conidia globose to subglobose or ovoid, narrowly truncate at the base and

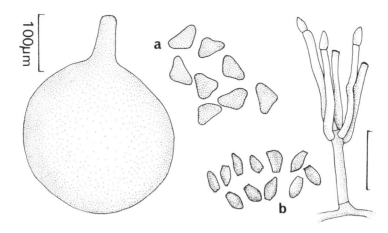

Fɪɢ. 171. *Microascus trigonosporus.* a. Perithecium and ascospores, CBS 601.67; b. *Scopulariopsis*-type conidiophore and conidia, CBS 366.65.

rounded or papillate at the tip, 4·0–4·5 × 3–4 μm. — The var. *macrosporus* Orr *1961* is distinguished by ascospores 5·5–7·5 × 5–6·5 μm and conidia 5–7 × 4–5 μm.

M. trigonosporus has been frequently isolated from soil in the USA and Canada and also El Salvador (366). It has been reported from soils in Brazil (399), Pakistan (2621), cultivated soil under cotton in the USSR (2871), the Sonoran desert in the USA (4733), the Negev desert in Israel (654), salt-marshes in Kuwait (4000), coastal sediments in Norway (5085), a slightly alkaline truffle soil in France (3451) and loess in Israel (4759). Other substrates include dead wood of *Abies balsamea* (366), seeds of various cereals (CBS), sorghum and onion (366), milled rice from Taiwan and Burma (5969), freshly harvested pecans (2572), a mouse (366), chicken manure (3680), rat dung (366) and a skin lesion on a human foot (366).

The optimal temperature for growth and sporulation is 30°C; at 18° and 37°C a quarter of the optimal growth occurs and sporulation is scanty; no growth occurs at 5°C (3974, 5085); 4% salt content (in seawater) favours growth (5085). *M. trigonosporus* has been shown to degrade cellulose (5086, 6076).

Microdochium Syd. *1924*

Microdochium bolleyi (Sprague) de Hoog & Hermanides–Nijhof *1977* — Fig. 172.

≡ *Gloeosporium bolleyi* Sprague *1948*
≡ *Aureobasidium bolleyi* (Sprague) v. Arx *1957*

DESCRIPTIONS: Sprague (5497), de Hoog and Hermanides–Nijhof (2526), and von Arx (197).

The genus *Microdochium* (type species *M. phragmitis* Syd.) has been redescribed and emended by Sutton *et al.* (5652) to cover four species growing on living and dead leaves of various plants. Sporodochia erumpent, subepidermal; conidiogenous cells discrete, little differentiated, ampulliform, producing a few apical conidiiferous denticles; blastoconidia fusiform or ± falcate, continuous or 1-septate, hyaline, smooth-walled. — After some further additions to this genus, *Gloeosporium bolleyi* has been transferred to it as the eighth species on account of its sympodial conidiogenesis (2526).

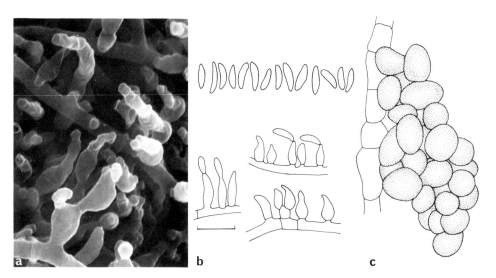

FIG. 172. *Microdochium bolleyi*. a. Conidiogenous cells, SEM, × *c*. 2500; b. conidiogenous cells and conidia; c. chlamydospore cluster.

Colonies spreading broadly, reaching 3·2–3·8 cm diam in seven days at about 20°C, moist, smooth, mostly without aerial mycelium, pink, red-brown to black, releasing an orange pigment into the agar, later darkening due to the formation of clustered chlamydospores with cells 6–9 μm diam. Conidia formed on lateral projections of intercalary cells or more frequently on discrete conidiogenous cells which arise singly or in sporodochial clusters, either directly on creeping hyphae or on globose subtending cells, ampulliform, forming apical clusters of conidia in an inconspicuously sympodial order; on the short, irregularly shaped rhachides the conidial scars can hardly be traced by light microscopy and SEM (Fig. 170c). Conidia crescent-shaped, one-celled, hyaline, thin-walled, normally 5·5–8·5 × 1·6–2·2 μm.

This species differs from other species of *Microdochium* (except *M. phragmitis* Syd.) in its much faster growth rate and abundant sporulation on only moderately rich media like 2% MEA, OA, CMA or PCA. There is no tendency to form discrete stromatic pustules as in *Colletotrichum graminicola* (Ces.) G. W. Wils. *1914* with which this fungus has sometimes been confused (5497). Some species of *Aureobasidium* Viala & Boyer (q.v.) formerly placed in *Kabatiella* Bubák *1923* have similarly falcate but much larger blastoconidia produced synchronously on minute denticles (2412).

M. bolleyi can be isolated equally easily from dilution plates and washed soil particles. The scarcity of references to it in the literature must be attributable to difficulties in identification, as in our experience this is a very common fungus in agricultural soils. It has been found regularly on the roots of 121 graminaceous species examined in various areas of the USA and its host range appears to be unlimited (5497). — From washed, predominantly organic particles, it has also been isolated with high frequencies (1614, 1888, 1889), and it is one of the most frequently isolated fungi in soils in the Netherlands (1614, 1616). It has often been shown to be present on both healthy and diseased internodes and roots of wheat throughout its growing season (2492, 4791, 5236) but is particularly abundant on young and very old parts of the plant (2491, 2492). Its percentage in the total soil fungus population was regularly significantly higher after growing wheat than after peas or rape (1433). A strong accumulation in the rhizosphere of grass monocultures (1614, 1615), and in the rhizospheres of potato, sugar beet and wheat has been observed (1614). Other substrates recorded include aerated hay (4821) and preservative-treated wood (4193). — Its development in soil is easily suppressed by bacteria (5497). Forty-two per cent of chloroform-fumigated mycelium was mineralized within ten days at 22°C (133). *M. bolleyi* can serve as food for the mites *Pygmephorus mesembrinae* and *P. quadratus* (3104).

Pectin, xylan and CMC are intensely decomposed by *M. bolleyi*, but different isolates vary in their efficiency to do this (1432). *M. bolleyi* is generally not pathogenic, but can nevertheless provoke seedling diseases in cereals, and occasionally also in alfalfa and gourd, at least in infection experiments, again with variations in behaviour between individual isolates (5497). Root injuries, with reductions in dry weight, have been observed in peas and wheat *in vitro* (1430), and barley roots were stunted by *M. bolleyi* in sandy soils in northern Scotland (CBS).

Microsporum Gruby *1843*

Type species: *Microsporum audouinii* Gruby

Teleomorph: *Nannizzia* Stockd.

GENERIC DEFINITION: Ajello (73); KEY TO THE SPECIES: Caprilli *et al.* (886). — Colonies hyaline, producing both micro- and macro-conidia; macro-conidia borne on conspicuous stalk cells, often arranged in branched conidiophores, echinulate, multiseptate, variable in form: fusiform to obovate, thin- or thick-walled, rarely formed in some species; micro-conidia pyriform to obovate. Hyphal and conidial cells are both multinucleate (886). Conidiogenesis is regarded by some authors as holoblastic (1862) and by others as solitary thallic (1109). A time-lapse study showed that the distal segment of the fertile hyphal element is converted into a conidium, although the septum which delimits the conidium at its base is often formed after conidial enlargement (1109). A short cylindrical aborting cell below the conidium allows the rhexolytic discharge of the conidium.

The separation of this genus from *Trichophyton* Malmst. (q.v.) is generally based on the conidial roughness, although the presence of distinct conidiophore cells also seem to provide a useful criterion. Fourteen species are now recognized (73). Those of the *M. gypseum* complex can become pathogenic in mammals, whereas the other species treated here are mainly geophilic saprophytes. The techniques of study are the same as in *Arthroderma* (q.v.). The most common species can be distinguished by the following key; four are treated here under their *Nannizzia* teleomorph names. *N. gypsea* and *N. incurvata* can hardly be distinguished on the basis of their anamorphs only.

Key to the species treated:

1   Macro-conidia thick-walled ($2–4 \mu$m)   **2**
       Macro-conidia thin-walled ($1·0–1·5 \mu$m), fusiform, with broadly truncate base   **3**

2(1)   Macro-conidia cylindrical with conical end cells, mostly $59–62 \times 10–11 \mu$m; colony reverse sometimes red   *Nannizzia grubyia* (p. 490)
      Macro-conidia fusiform, $40–75 \times 15 \mu$m; colony reverse deep purplish red
                (*Nannizzia cajetani*, anamorph *M. cookei*)

3(1)   Macro-conidia $35–50 \times 8–11 \mu$m; colonies reaching $6·0–7·0$ cm diam in 14 days at 25°C, with a dense downy to granular aerial mycelium; reverse straw, usually with a trace of primrose
                Nannizzia fulva (p. 488)
      Macro-conidia $35–53 \times 10–14 \mu$m (*M. gypseum* complex)   **4**

4(3)   Colonies reaching over $8·5$ cm diam in 14 days at 25°C, somewhat radiating, with a finely granular aerial mycelium; reverse straw with a trace of primrose   *Nannizzia gypsea* (p. 491)
      Colonies reaching $5·0–6·0$ cm diam in 14 days at 25°C, strongly radiating, with very thin aerial mycelium; reverse deep ochraceous-amber   *Nannizzia incurvata* (p. 495)

Minimedusa Weresub & LeClair *1971*

Minimedusa polyspora (Hotson) Weresub & LeClair *1971* — Fig. 173.

≡ *Papulaspora polyspora* Hotson *1912*

DESCRIPTIONS: Hotson (2544), and Weresub and LeClair (6290). — The genus *Minimedusa* was segregated from *Papulaspora* Preuss (q.v.) because of its unusual development of the bulbils from a group of lateral, 2–4 μm wide, branches and the presence of clamp-connections, mainly in the advancing zone. The bulbils have a rounded surface and pseudoparenchymatous context of textura angularis.

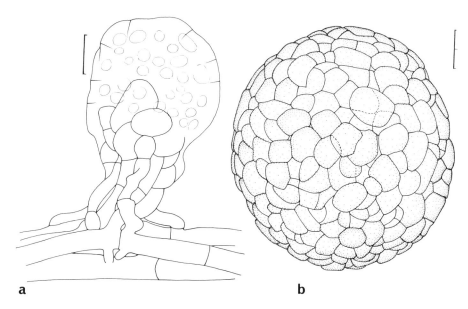

a   b

FIG. 173. *Minimedusa polyspora*. a. Bulbil initial arising from a bundle of lateral hyphae; b. mature bulbil; CBS 880.68.

   M. polyspora has a scanty hyaline, procumbent mycelium, reaching 3·0–4·0 cm diam in five days at 20°C on MEA; clamp-connections are occasionally present in the peripheral zone. Bulbils originating from the aggregation of lateral aerial hyphae, eventually forming a dense powdery mass, globose, smooth, reddish brown, 90–130 μm diam, with 150–200 polygonal cells visible in surface view, surface cells mostly 10–20 μm diam, not differentiated from the central cells. When grown on wood above a liquid medium the peripheral cells may round themselves off to form individual spores (2544). — The similar *Myriococcum praecox* Fr.

1823 (=*Papulaspora byssina* Hotson *1917*) hardly produces any aerial mycelium, has bulbils consisting of smaller (7–12 μm diam) strictly polygonal cells, and no clamp-connections.

M. polyspora was originally isolated from straw, old paper and cotton flowers in Cuba (2544). Since then it has been reported from oat grain in Ireland (3556), and very rarely observed in arable soils under wheat in Germany (CBS). It is also known from agricultural soils in the Netherlands and Denmark, soil in North Carolina and mushroom substrate kept at 30°C in the Netherlands (CBS). It has frequently been observed on damaged spots of stored sugar beets (J. W. Veenbaas-Rijks, pers. comm.).

An enzyme with the capacity to oxidize dehydromatricarianol has been found in culture filtrates of *M. polyspora* (2479).

Monacrosporium Oudem. *1885*

Monacrosporium psychrophilum (Drechsler) R. C. Cooke & Dickinson *1965* — Fig. 174.

≡ *Dactylaria psychrophila* Drechsler *1944*
≡ *Genicularia psychrophila* (Drechsler) Rifai *1968*
≡ *Geniculifera psychrophila* (Drechsler) Rifai *1975*

Subramanian (5625) and Cooke and Dickinson (1150) redescribed *Monacrosporium* (=*Golovinia* Mekhtieva *1967*) as usually forming simple conidiophores with ± fusoid hyaline conidia with a large central cell; this genus then accommodates most nematophagous species formerly placed in *Dactylella* Grove *1884*. KEY TO THE SPECIES: Cooke and Godfrey (1151).

DESCRIPTIONS: Drechsler (1460), Duddington (1479), and Shepherd (5253). — Colonies reaching 4·0 cm diam in ten days at 20°C on MEA. Conidiophores 150–500 μm tall, simple or with 1–4 short apical branches formed by sympodial proliferation, subapical conidia sometimes sessile. Conidia rather broadly attached to the conidiophore tip, fusiform with the largest width above the middle and a slightly rounded tip, with 1(–2) septa near the tip and usually 2 near the base, 46–71 × 21–29 μm. Trapping nematodes by means of adhesive network. — *M. psychrophilum* is distinguished from its closest relatives, *M. eudermatum* (Drechsler) Subram. *1963* by the slightly narrower and apically more pointed conidia, and from *M. reticulatum* (Peach) R. C. Cooke & Dickinson *1965* by the slightly broader conidia and occasional branching of the conidiophores. Rifai (4841) transferred *M. psychrophilum* to *Genicularia* Rifai & R. C. Cooke *1966* emend. Rifai *1968* (correctly *Geniculifera* Rifai *1975*, non *Genicularia* Rouss. ex Desv. *1808*) in which he also included species with pluriseptate conidia. Since a generic separation based on monoblastic or occasionally sympodially proliferating polyblastic conidiophores in the nematophagous fungi seems to have little taxonomic value, we have retained this species in *Monacrosporium* here.

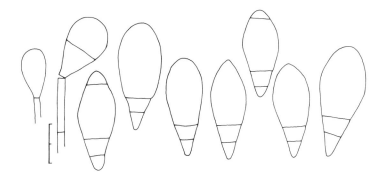

FIG. 174. *Monacrosporium psychrophilum*, attached and liberated conidia, CBS 548.63.

The fungus was originally isolated from old potato plants in the USA (1460) but has also been isolated, albeit rarely, from agricultural soils in the British Isles (1481) and Germany. It is also occasionally isolated from mosses (1479, 1480).

Good growth and sporulation are obtained on cornmeal or potato-carrot agars, sporulation being particularly profuse in pure culture (1479). The optimal temperature for conidium formation and the development of adhesive networks lies in the range 15–20°C, whilst at 28–32°C no adhesive networks are formed (1460); this facet is expressed in the name of the species. As in *Arthrobotrys* species, rapid growth, high competitive saprophytic ability and the formation of adhesive networks are all related to a low efficiency in nematode control (1143).

Monascus van Tiegh. *1884*

Monascus ruber van Tiegh. *1884* — Fig. 175.

= *Physomyces heterosporus* Harz *1890*
? = *Monascus purpureus* Went *1895*
= *Allescheria gayoni* (Cost.) Sacc. & Syd. *1899*
= *Backusia terricola* Thirum. *et al. 1964*

Anamorph: *Basipetospora rubra* Cole & Kendrick *1968*

DESCRIPTIONS: Takada (5685) and Moreau (3931); cytology: Young (6507); conidiogenesis: Cole and Kendrick (1104).

The genus *Monascus* has an isolated taxonomic position (fam. Monascaceae Schröt.) and is characterized by stalked, non-ostiolate ascomata (cleistothecia) developing from a terminal vesicle and which become surrounded by hyphae growing out from the base and forming a 4μm thick wall. The ascomata contain a limited number of globose, soon evanescent asci, which are no longer recognizable at maturity when the ascospores fill the cavity.

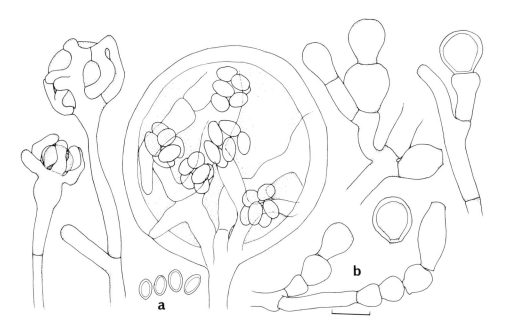

FIG. 175. *Monascus ruber*. a. Various stages of ascoma development, with mature ascospores; b. conidium formation of *Basipetospora* type.

The colonies of *M. ruber* are fast-growing, reaching 7·5–8·5 cm diam in seven days at 25–30°C on MEA or PDA (1104); they are thinly felty, usually in various shades of greyish orange to intense red with more intensely coloured reverse. Ascomata profusely produced, ± hyaline to brownish, 20–70 μm diam. Ascospores ellipsoidal, hyaline or slightly orange, thick- and smooth-walled, (4·0–)5–6(–7) × (3–)4–5 μm. Anamorph reminiscent of *Scopulariopsis*, with undifferentiated conidiogenous cells, but during conidiogenesis the conidiogenous cell is gradually shortened by the incorporation of small portions of its wall into the lower half of the conidial wall (1104). Conidia mostly in short or rather long chains, subglobose to pyriform with a very broadly truncate base, (4·5–)6–12(–16) × 4·5–10(–14) μm, thin-walled, hyaline or pale brownish, sometimes thick-walled and larger, in short chains or solitary, and functioning as chlamydospores.

The species delimitation in *Monascus* is not yet sufficiently elucidated and the genus is badly in need of a monographic revision. While some authors regard differences in colony pigmentation and conidial size as gradual and insufficient for species distinction, Young (6506) recognized four species in addition to *M. ruber*: *M. purpureus* Went *1895* with globose to subglobose ascospores, 6 × 5 μm and purple cultures; *M. barkeri* Dangeard *1903* with ascospores 8 × 4 μm, violet cultures and a temperature maximum of 30°C; *M. mucoroides* van Tiegh. *1884* with globose ascospores 8 μm diam and white colonies; and *M. olei* Piedallu *1910* characterized by some physiological differences. Most other epithets appearing in the literature have not been validly published and are probably synonyms of *M. ruber*. The osmophilic *Xeromyces bisporus* Fraser *1953* (≡*Monascus bisporus* (Fraser) v. Arx *1970*) with white colonies and almost sessile ascomata, 2-spored asci and lunate ascospores 9–11 × 4–5 μm is quite distinct and best retained in *Xeromyces* Fraser *1953*.

M. ruber is commonly found on starch-containing foods kept at higher temperatures and has a worldwide distribution. *M. ruber* and *M. purpureus* were originally described from slices of cooked potatoes in France and discolored rice in eastern Asia, respectively; subsequent reports under both names come from a starch factory in South Africa, mouldy corn silage in the USA, Australia (1104) and France (3931), milled rice in Japan (5685), soil in India (1104) and alluvial grassland in the British Isles (160). The fungus is also known to occur in Canada, Israel, Ethiopia, Nigeria and Swaziland (IMI). Other substrates recorded include the seeds of oat, soya and sorghum, tobacco, *Theobroma cacao*, and palm kernels (IMI).

Growth is possible in the temperature range 18–40°C with the optimum for different isolates varying between 30° and 37°C; the optimum for conidiation was found to be 37–40°C, and that for ascoma formation 25–37°C (160, 3559, 6506). — Good growth is possible in 1% lactic acid and up to 3·5% is tolerated; in SO_2-water ascomata form up to 0·1% concentration (6506). Optimal growth with little pigmentation occurs on organic N sources; nitrate stimulates conidiation and ascoma formation, while ammonium is inhibitory (890). Sexual reproduction has also been observed in submerged cultures (889). With yeast extract or nitrate as nitrogen sources (pH 6·5) red pigments are formed, whilst with ammonium or NH_4NO_3 (pH 2·5) orange ones form; the orange pigments are named monascorubin and rubropunctatin (890), and additional yellow pigments are monascin and ankaflavin; their structures have been elucidated (3335). The amount of pigment produced by *M. ruber* is less than in isolates attributed to some other *Monascus* species; the red pigments are often used for food colouring in Asia. Pigment production is optimal at 25°C with pH 6·5, glucose or ethanol as C sources, and certain organic N sources (6503). — A hot air treatment for short periods at 96°C is survived (6506). — A metabolite of *M. ruber* proved to be highly toxic to chicken embryos (2723).

Monocillium Saksena *1955*

Monocillium mucidum W. Gams *1971* — Fig. 176.

DESCRIPTION: Gams (1882). — The genus *Monocillium* (teleomorph *Niesslia* Auersw. *1869*) is distinguished from *Acremonium* (q.v.) by the basal thickening of the phialide walls and from *Torulomyces* Lembke (q.v.) by disconnected conidial chains or heads. — Colonies of *M. mucidum* spread relatively fast, reaching 2·0–3·0 cm diam in ten days at 20°C; they are pink, moist, and lack aerial mycelium. Phialides arising from the substrate, gradually tapering in the upper, thin-walled part, 15–25(–40) × 1·5–2·0 μm. Conidia forming slimy masses, obovoid, 2·8–3·9 × 1·8–2·2 μm.

M. *mucidum* appears to be widely distributed in cultivated soils in Europe (1882, CBS) and is also known from soil in a coconut grove in the Bahamas (2006) and under grassland in Japan (5844, 5846).

Decomposition of xylan is very good, there being little variation in activity between different isolates; cellulose decomposition in filter paper (1432) and severe injury to pea roots *in vitro* have been recorded (1430). The production of cephalosporin C has been reported (3957).

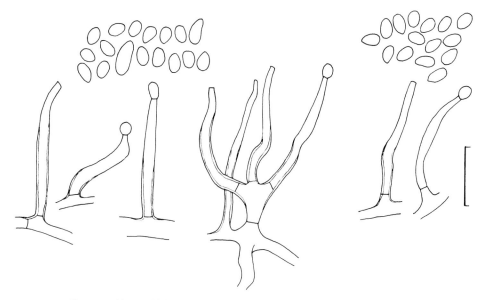

FIG. 176. *Monocillium mucidum*, phialides with wall-thickening in the lower part and conidia, CBS 404.66.

Monographella Petr. *1924*

Monographella nivalis (Schaffn.) E. Müll. *1977* — Fig. 177.

≡ *Calonectria nivalis* Schaffn. *1913*
≡ *Griphosphaeria nivalis* (Schaffn.) E. Müll. & v. Arx *1955*
≡ *Micronectriella nivalis* (Schaffn.) C. Booth *1971*
= *Sphaerulina divergens* Rehm *1913*
  ≡ *Monographella divergens* (Rehm) Petr. *1924*
= *Calonectria graminicola* (Berk. & Br.) sensu Wollenw. *1913* [non *Nectria graminicola* Berk. & Br. *1858*]

Anamorph: *Gerlachia nivalis* (Ces. ex Sacc.) W. Gams & E. Müll. *1980*

  ≡ *Fusarium nivale* Ces. ex Sacc. *1892*
= *Fusarium nivale* (Fr.) Sorauer *1901*
  ≡ *Lanosa nivalis* Fr. *1825* pro parte

DESCRIPTIONS: anamorph: Gordon (2039), Joffe (2770), Gams and Müller, *1980**; including the teleomorph: Wollenweber and Reinking (6424), Booth (640), Müller and v. Arx (4005), Müller (4004), and CMI Descriptions No. 309, *1971*. — The anamorph was previously placed in *Fusarium* section *Arachnites* Wollenw. but is regarded as generically distinct because of its annellidic conidiogenesis. — Colonies slow-growing, reaching slightly more than 1 cm diam in four days at 25°C, white to pale peach or apricot, aerial mycelium ± floccose or felted. Sporulating sparsely in the aerial mycelium or in pale orange sporodochia which appear after 10–14 days only under daylight or near-UV (5035). Conidiogenous cells short, doliiform to obpyriform, 7–9 × 2·5–3 µm, showing an apical annellated zone under SEM (Fig. 177b-d). Conidia broadly falcate, with almost equal ends, the base only slightly flattened, 1(–3)-septate, 10–18(–27) × 2·5–3·5 µm. Chlamydospores absent. — *G. nivalis* var. *major* (Wollenw. *1931*) W. Gams & E. Müll. *1980* (teleomorph *Monographella nivalis* var. *neglecta* (Krampe) W. Gams & E. Müll. *1980*) is distinguished by mostly 3-septate conidia to 35 µm long. Well-sporulating cultures of *M. nivalis* are similar to *F. dimerum*, which, however, has more asymmetrical, generally 1-septate conidia and sporulates abundantly also in darkness. — *M. nivalis* is not congeneric with *Calonectria* de Not. *1867*, *Plectosphaerella* Kleb. (q.v.), *Micronectriella* Höhn. *1906* (211) or *Griphosphaeria* Höhn. *1918* (= *Discostroma* Clem. *1909*), but *Monographella* (= *Griphosphaerella* Petr. *1927*), fam. Amphisphaeriaceae Wint., is the appropriate genus (4004).

   M. nivalis is homothallic and produces perithecia on sterilized straw (rarely on agar media) after 10–15 days at 18°C (640). Perithecia in culture are superficial, at first white to pink, later becoming greyish black; on the natural substrate, e.g. winter wheat in Oregon (1136), they are immersed in the leaf sheath. Ascospores hyaline, fusiform, 1–3-septate, 10–17 × 3·5–4·5 µm. — Hyphal cells are mostly uninucleate, but up to 5 nuclei have been observed in them (2496).

---

* *Neth. J. Pl. Path.* **86,** 45–53.

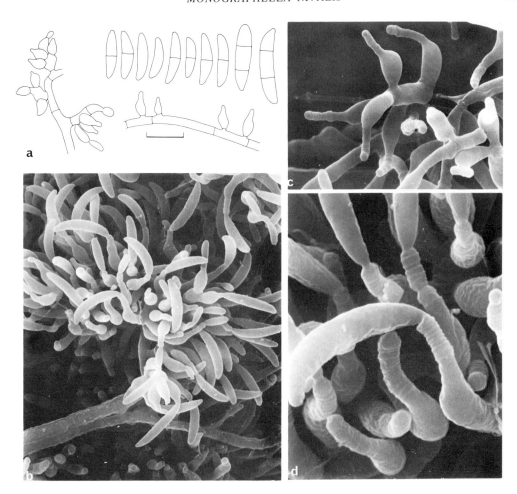

Fig. 177. *Monographella nivalis*. a. Conidiophores and conidia, CBS 146.68; b. sporodochium, SEM, c. × 1000; c, d. annellated conidogenous cells, SEM, c. × 2250, d. × 5000; b and d. CBS 162.77, c. CBS 319.78.

M. nivalis is a serious pathogen of cereals and grasses, especially in cold to temperate regions (2045) of the northern hemisphere, but also in Australia and New Zealand, and may cause total losses of winter-sown wheat and rye. Tropical records are subjected to doubt (CMI Descriptions No. 309). Losses in cereals are particularly severe in areas with a thick and long-lasting snow cover where the fungus can thrive at temperatures near freezing point. *M. nivalis* is spread mainly through seeds but also by growth through soil. — Soil isolates have been reported from grassland in Iceland (5321), tall-grass prairie and an abandoned field in Oklahoma/USA (1632), the Sonoran desert in the USA (4733), cultivated soil under oat in Ireland (1447), cultivated soil and wet meadows and other soils of different geological origin in Germany (3127, 4180), an alpine pasture with *Nardus stricta* (3445) in Italy, soils in Iraq (92), Peru (2005) and the USSR (2948). Isolates have also been reported from India in uncultivated (4030) and forest soils (2854, 5000), cultivated soils (2712, 3817, 3868) with lentils and gram (2164), grassland (3863, 4933), and grass plots with *Desmostachya bipinnata* and *Dichanthium annulatum* (3865). *M. nivalis* has frequently been found in the rhizospheres of barley (5514), rice (2712), and the ferns *Polypodium, Cyclosorus, Dryopteris* and *Adiantum* (2856); it is also known from the root regions of *Pennisetum typhoideum* (3867), *Abutilon indicum* (3864), *Carica papaya* (5345), euphorbiaceous plants (3866), *Setaria glauca, Dichanthium annulatum* (3272) and oats (4754). Other habitats include frost-damaged leaves of various grasses (191), the leaves of vascular plants (296), decaying leaves of *Polypodium* and *Cyclosorus* (2859), wooden posts in soil (1744), seeds of barley, oat and wheat (1752, 2045, 2820), pecans (2572), and organic material in fresh water (4429). — This fungus has relatively good saprophytic ability and can survive in soil in the absence of host plants without the formation of any resting structures and grows well over the surface and through soil (643, 5756); spread in soil (643) and survival in plant debris (4754) is favoured by low soil temperatures; high N doses increased infection by it in grassland whereas balanced NPK fertilization gave a disease-free turf (CMI Descriptions No. 309).

M. nivalis has a temperature optimum of 20–21°C, but good growth occurs at 0°C, the maximum is 28°C and the minimum −5°C (191, 1118, 2713). — Urea is readily used as a N source (3127). *M. nivalis* can degrade pectin (2239, 3127), tannin, cellulose, hemicellulose, *p*-hydroxybenzaldehyde, straw (2165, 3127, 3618), phenol lignin (1744), Na-humate if other C sources are also supplied (5673), and the fungicides benzoxazolinone and 6-methoxybenzoxazolinone (6130). It produces an extracellular NAD-nucleosidase (5538), cyclo-L-prolylglycine (6323), the estrogenic compound zearalenone (3858) and a number of toxic metabolites such as 4-acetamido-4-hydroxy-2-butenoic acid-γ-lactone (789, 6486, 6487), the sesquiterpene nivalenol with cytotoxic properties (1842, 4985), fusarenone (1842, 3953, 5752, 5993) (=nivalenol monoacetate) which causes degradation of polyribosomes from mouse fibroblasts (4258) and inhibits protein synthesis (2425), nivalenol diacetate (1842, 5751), HT–2 toxin (6487), and a substance which induces vomiting in guinea pigs (4630). — The growth of *Chalara elegans* was inhibited *in vitro* (2753). — Carbon from labelled KCN was incorporated into alanine (108). — This fungus is highly sensitive to γ-irradiation (5748).

Mortierella Coemans *1863*

= *Haplosporangium* Thaxter *1914*
= *Azygozygum* Chesters *1933*
= *Naumoviella* Novotelnova *1950*

The genus *Mortierella* has been placed in a separate family Mortierellaceae A. Fischer of the Mucorales because of several unusual features. All structures are generally more delicate than in the Mucoraceae and the sporangia have no columella and are split into numerous, few, or single spores; they are consequently called sporangioles by some authors, especially if they contain only one or two spores. The mycelium is mostly dichotomously branched and has a characteristic garlic-like odour.

The genus *Mortierella* was monographed in *1941* by Linnemann (3359) and again in *1970* by Zycha *et al.* (6588), who considered it to comprise 83 species grouped into 11 sections. Turner published some critical studies of selected groups (5948, 5949) but the keys recently proposed by Russian workers (2950, 3807) made little contribution to our knowledge of the genus. In the detailed treatments published, some of the criteria used for the characterization of sporangiophore branching were found to be vague or unreliable, and Gams (1881) thus proposed an alternative key to ten sections and subsequently (1883) one to 73 recognized species.

Mortierella can be subdivided into two main parts: the group around *M. isabellina* (subgen. *Micromucor* W. Gams *1977*) has rather slow-growing, velvety colonies mostly with pigmented sporangia whose best development is seen on 2% MEA. This group has been monographed by Turner (5948), whose arrangement is adopted here. Because of the presence of a small columella, *M. ramanniana* has often been placed in *Mucor* but in fact neither genus is quite appropriate and even the true family is uncertain because zygospores are unknown. The remaining *Mortierella* species (subgen. *Mortierella*) have a thin, spreading mycelium with hyaline sporangia and mostly a garlic-like odour. For this group, soil-extract agar (1598) and potato-carrot agar are the most suitable media for observing sporulation, while on MEA (as recommended by Linnemann) too much aerial mycelium is formed which tends to obscure the sporangiophores. Observation under the compound microscope of the whole undisturbed colony (in Petri dishes) or squares of agar on which a drop of water and a cover slip are gently placed, give the clearest views of sporangiophore branching. Many species form chlamydospores in the agar and as these may also be supported by short aerial stalks and have an ornamented wall, they are then usually termed stylospores ("Stielgemmen"). Recent investigations have, however, shown that all possible intermediates occur between aerial and submerged, ornamented and smooth chlamydospores; this terminologic distinction is thus of little practical value. Moreover, the term stylospores has also been applied to one-spored sporangia (3359, 6588) and so has caused much confusion. In addition to the differences in wall structure, an important criterion for distinguishing between the two types is that sporangiophores taper upwards whilst chlamydospore stalks are of a constant narrow width and usually much shorter.

Zygospores of *Mortierella* were initially reported to be invested and to form small fruiting-bodies similar to those of *Endogone* Link ex Fr. *1823*. Since Gams and Williams (1893) reported naked zygospores with unequal suspensors for the heterothallic *Mortierella parvispora*, a similar development has been shown in 16 further species, amongst which only *M. epigama* W. Gams & Domsch *1972* and *M. chlamydospora* (Chesters) van der Plaats-Niterink *1976* are homothallic. The invested type of zygospore is apparently restricted to species of sections *Simplex* and *Mortierella*. For inducing zygospore formation, potato-carrot, cornmeal, "Pablum" (or "Bambix") baby-food, or hempseed agars are recommended; temperatures must not exceed 18°C for some species and the mating partners must always be inoculated close to each other.

Species of *Mortierella* are amongst the commonest soil fungi and, as they grow rather fast and often sporulate freely, they can easily be isolated by the conventional techniques. They probably play a rather minor role in soil metabolism but are generally capable of decomposing chitin (2706).

Keys to the sections and species of *Mortierella*:

1   Colonies velvety, not exceeding 3 mm in height; sporangia mostly ochraceous or vinaceous
                                                                      Subgen. *Micromucor* (p. 433)
       Aerial mycelium consisting of longer ascendent and prostrate hyphae, white, cottony; mostly
            with a garlic-like odour (Subgen. *Mortierella*)   **2**

2(1)   Sporangiophores always unbranched   **3**
       Sporangiophores branched (at least sometimes)   **5**

3(2)   Sporangiophores usually exceeding $200\,\mu$m in length   (Sect. *Simplex*)
       Sporangiophores less than $150\,\mu$m in length   **4**

4(3)   Sporangia — at least partly — many-spored; sporangiophores with distinctly widening base (cf.
            also *M. horticola* in Sect. *Stylospora*   Sect. *Alpina* (p. 433)
       Sporangia 1-spored; very slender sporangiophores arising in dense rows from the aerial hyphae
                                                                              (Sect. *Schmuckeri*)

5(2)   Sporangiophores racemosely branched with a thick main stem and thin, short branches   **6**
       Branching in another way   **7**

6(5)   Branches arising above the middle of the sporangiophore   Sect. *Mortierella* (p. 452)
       (only *M. polycephala* treated; if sporangiophores shorter than $100\,\mu$m with strongly swollen
            base, cf. *M. bisporalis*)
       Branches arising in the uppermost part of the sporangiophore in clusters from an inflated region
                                                                          (Sect. *Actinomortierella*)

7(5)   Branches arising mainly in the lower part of the sporangiophore (basitonous)   **8**
       Branches arising in the middle or upper part of the sporangiophore (mesotonous or acrotonous)
                                                                                                  9

8(7)   Sporangia containing many or at least several smooth or ornamented spores
                                                                       Sect. *Hygrophila* (p. 433)
       Sporangia 1-spored, often ornamented   Sect. *Stylospora* (p. 434)

9(7) Sporangia many-spored; sporangiophores often bent upwards above an ascendent basal part and with a minute columella Sect. *Spinosa* (p. 434)

Sporangia 1- or 2-spored; sporangiophores short, with broad base, strongly tapered in the central part and arising in dense rows from the aerial hyphae

(Sect. *Haplosporangium*) *M. bisporalis* (p. 438)

Subgen. *Micromucor* W. Gams

1 Sporangia 1-spored, hyaline; colonies white *M. nana* (p. 449)

Sporangia many-spored **2**

2(1) Colonies in shades of ochraceous-grey; spores slightly angular, 2–3 μm diam; small chlamydospores scarcely produced *M. isabellina* (p. 447)

Colonies pink, russet or lilac **3**

3(2) Chlamydospores small and scarce or absent **4**

Chlamydospores filled with lipid droplets, abundantly produced; sporangiophores always with a small distinct columella **5**

4(3) Sporangiophores slightly widened below the sporangium, sporangial wall mostly remaining as a large collarette; small columella often present; spores angular, 2–3 μm diam (*M. longicollis*)

Sporangiophores not widened below the sporangium, without a collarette; columella hardly developed; spores angular, 3–4 μm diam *M. vinacea* (p. 458)

5(3) Spores angular; colonies somewhat brownish *M. ramanniana* var. *angulispora* (p. 456)

Spores rounded; colonies ochre-red to vinaceous **6**

6(5) Spores ellipsoidal; fungus requiring thiamine *M. ramanniana* var. *ramanniana* (p. 454)

Spores globose; fungus not requiring thiamine (*M. ramanniana* var. *autotrophica*)

Subgen. *Mortierella*

Section *Alpina* Linnem.

1 Sporangia always 1-spored, globose, finely echinulate cf. *M. horticola* (p. 444)

Sporangia, at least partly, many-spored and spores of many-spored sporangia elongate **2**

2(1) Fimbriate chlamydospores, 20–60(–120) μm diam, present (*M. alliacea*)

Only small, indistinct, smooth-walled chlamydospores occasionally present *M. alpina* (p. 435)

Section *Hygrophila* Linnem. emend. W. Gams

1 Sporangiophores not more than 120 μm long; spores more or less globose **2**

Sporangiophores longer or spores ellipsoidal-cylindrical **4**

2(1) Sporangiophores often with verticillate branches; sporangia few-spored, spores 4·5–12 μm diam, finely wrinkled   cf. *M. verticillata* (p. 457)
       (if spores larger than 12 μm diam, cf. *M. hyalina*)
       Sporangiophores with branches arising at different levels; sporangia many-spored   **3**

3(2) Spores 4–7(–10) μm diam; chlamydospores absent   *M. minutissima* (p. 448)
       Spores 2–3 μm diam; chlamydospores numerous, 20–100(–300) μm diam   (*M. macrocystis*)

4(1) Spores globose to subglobose   **5**
       Spores distinctly elongate   **6**

5(4) Sporangiophores with apophysis-like inflation; spores smooth-walled, 8–12 μm diam; chlamydospores often in groups   (*M. beljakovae*)
       Sporangiophores without any inflation; spores minutely roughened, 8–25 μm diam; chlamydospores formed solitarily   *M. hyalina* (p. 446)

6(4) Sporangiophores 2–3 mm long, 18–20 μm wide at the base; spores 5–10 × 3–5 μm; chlamydospores absent   *M. bainieri* (p. 437)
       Sporangiophores shorter and/or more slender   **7**

7(6) Homothallic species with numerous naked zygospores produced on MEA; spores fusiform with rounded ends, 9–14 × 3–6 μm; chlamydospores absent   (*M. epigama*)
       Heterothallic species; spores ellipsoidal to cylindrical, 6–13(–16) × 3·5–7·0 μm; chlamydospores 10–18 μm diam numerous   *M. elongata* (p. 439)

Section *Stylospora* Linnem.

1   Sporangiophores always unbranched; sporangia 7–12 μm diam, minutely spinulose
                                                                                 M. horticola (p. 444)
       Sporangiophores basitonously branched   **2**

2(1) Sporangia with reticulate walls   (*M. stylospora*)
       Sporangia with spinulose or smooth walls   **3**

3(2) Sporangiophores 5–7 μm wide at the base, strongly tapering in the middle part to 1·0–1·8 μm at the tip; sporangioles echinulate, 8–18 μm diam   (*M. lignicola*)
       Sporangiophores more slender, tapering gradually to 1·5–3 μm at the tip   **4**

4(3) Sporangia smooth-walled, 10–25 μm diam   (*M. zonata*)
       Sporangia finely ornamented, smaller   **5**

5(4) Sporangia finely spinulose, always one-spored, with the outermost layer firmly attached to the spore   *M. humilis* (p. 444)
       Sporangia often few-spored, spore wall irregularly wrinkled with a loose outer layer
                                                                               M. verticillata (p. 457)

Section *Spinosa* Linnem.

1   Spores globose to subglobose; chlamydospores small and smooth or absent   **2**
       Spores ellipsoidal to cylindrical; chlamydospores provided with irregular appendages   **3**

2(1)   Spores not exceeding 4 μm diam   *M. parvispora* (p. 451)
       Spores larger, 6–12(–14) μm diam   *M. gamsii* (p. 442)

3(1)   Sporangia leaving a pronounced collarette; spores with a double membrane *M. wolfii* (p. 459)
       Sporangia not leaving a collarette, but with a trace of a columella; spores with a single membrane
                                                                             M. exigua (p. 441)

Mortierella alpina Peyronel *1913* — Fig. 178.

= *Mortierella renispora* Dixon-Stewart *1932*
= *Mortierella thaxteri* Björling *1936* (nomen inval., Art. 36)
= *Mortierella monospora* Linnem. *1936* (nomen inval., Art. 36)
= *Mortierella acuminata* Linnem. *1936* (nomen inval., Art. 36)

DESCRIPTIONS: Linnemann (3359, 6588), Björling (542), and Turner and Pugh (5949).— Section *Alpina* Linnem. — Colonies reaching 2·5 cm diam in five days at 20°C on MEA. Sporangiophores simple, short (to 120 μm long), with a widened and often irregularly swollen base;

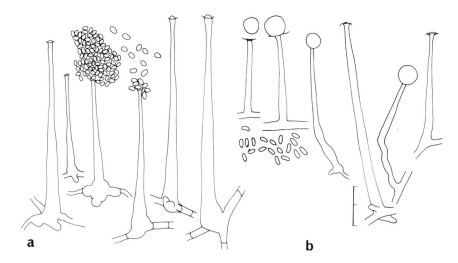

FIG. 178. *Mortierella alpina*. a. Sporangiophores and sporangiospores of isolate with predominant many-spored sporangia; b. isolate with deciduous sporangia, CBS 219.35.

sporangia at least partly containing numerous ellipsoidal spores, 3–4 × 2 μm. — Turner and Pugh (5949) and Khalabuda (2949) noted that entire sporangia may fall off without separating into single spores; this phenomenon formerly served for the distinction of *M. acuminata* and *M. monospora* but since authentic cultures (CBS 219.35 and CBS 250.53) of these taxa also form many-spored sporangia, *M. acuminata* and *M. monospora* are now regarded as synonyms of *M. alpina* (1881). All isolates so far studied also produced at least some deciduous sporangia as described for *M. thaxteri*, whilst in *M. alpina* they were said to be absent. Some variation is observed in these deciduous sporangia; they may be globose or ellipsoidal, smooth-walled or verrucose, and sometimes be subdivided a few times. Invested zygospores were described for *M. renispora* (1410), while Kuhlman (3161) discovered the formation of naked zygospores in the aerial mycelium of *M. alpina* after mating compatible isolates (with oval verrucose deciduous sporangia). For the rest, no significant difference was observed between the type culture of *M. renispora* (CBS 210.32, with globose and smooth-walled deciduous sporangia) and other cultures of *M. alpina*. The scarce and atypical chlamydospores are not regarded as significant. Hyphal knots commonly occurring in the mycelium of this and other *Mortierella* species, considered to be zygospores by Linnemann (3359), do not contain zygospores as already shown by Björling (542).

The species of *Mortierella* can be divided fairly sharply into two ecological groups according to their preference for particular pH ranges (3360). *M. alpina*, probably the commonest species of the genus, together with *M. gamsii*, *M. exigua* and *M. minutissima*, belongs to the inhabitants of neutral and/or alkaline habitats (271, 3360, 3451, 3497, 4152, 6182). *M. alpina* is easily isolated by dilution plates, immersion tubes and plates, and from washed soil particles; after washing it was isolated more frequently from organic than from mineral soil particles (1614, 1888); buried chitin is also suitable for isolation (2068). — The overall distribution, which cannot be documented here in detail, embraces Alaska (3062), Spitsbergen (3064), Lapland (3360), Antarctica (3235), alpine soil with long snow coverage (3976) and pasture (3445), and sediments below a glacier (2530). The species is not restricted to alpine habitats and has, although not very frequently, also been reported for warmer zones such as the southern steppe, forest steppe and the Poles'e in the Ukraine (2950), rice fields in Italy (3913), highland sites in Mexico (3360) and Peru (2005), further in Brazil (6007), the Ivory Coast (4719), South Africa (4407), India (3738, 4610) and Singapore (3331). According to our own observations, this is also the most common *Mortierella* species in mediterranean soils in Spain, southern France, Sardinia and Corsica. *M. alpina* is frequent under all types of vegetation: spruce (3975), pine (307, 2068, 2344, 3447, 3798, 4445), Douglas fir (6588), larch (3267), beech (270, 271, 272, 273, 1631, 3414, 3538, 4537) and mixed forest (2004, 2007, 2008, 2950, 6350). It is further known in forest nursery soils (6021, 6184), forest soil after 20 years of prescribed burning (2822), grassland (159, 164, 982, 3497, 3871, 5812, 5813, 5814, 6182), soil with steppe type vegetation (4407, 4719, 6347), and very common in arable soils; it is also the most common *Mortierella* in salt-marsh soils (4645, 4646, 5949), estuarine sediments, even with high salinities (655), further in moorland (163, 1376, 3119, 3234, 3414), heathland (989, 2736, 5047, 5811), dunes (745), desert (4733), saline soils in Spain (3446) and caves in the Caucasus (3364) and in Piedmont/Italy (3453). *M. alpina* was said to be associated with the occurrence of *Tuber magnatum* (3451, 3452, 3454) and to have antibiotic effects in truffle soil (1488). It penetrates to great soil depths (159, 540, 2161, 3447, 6184) and the recorded maximum is 135 cm (164). The spore content in the soil can even increase with depth if the pH value rises (6182); the minimum pH required has been given as 6·5 (3359), 6·2 (3360) and 6·0 (3871, 6182), but the species occurred also occasionally in British podzols on sites where no iron-pan is formed (3721); it also occurs very rarely in alpine raw-humus soils (1876). It occurs preferentially in soils which are very humid (164). While manuring with PK, NP and NPK is said to favour it slightly (2163), farmyard manure plus mineral fertilizers does not (2467). It is reduced in numbers after application of soil fungicides (1426). In contrast to other Mucorales, *M. alpina* has been shown to be present in sewage-irrigated fields (2161), and also isolated from the root regions of *Juniperus* (3324), *Salicornia* (4645), potato (1614), clover (5815), grasses (1614, 3331, 5815), and wheat (1614, 2949, 2950), and also earth-worm casts (1429). — There is a report on growth inhibition by sterile root-inhabiting fungi (4770). It appears to be intolerant of high copper levels in soil (2922); formaldehyde-treated soil is slowly recolonized (6184) and there is no growth on dry substrates (5226). This species has also been isolated from liver lesions of a calf in New Zealand (5441).—In contrast with *M. ramanniana* var. *angulispora*, it can easily be parasitized by *Dimargaris verticillata* (2817, 3560).

Of the floccose *Mortierella* species, *M. alpina* possesses a relatively low growth rate. It can become adapted to low temperatures and is able to show some growth even at 0°C (6588) or 1°C (3235); growth was slightly better at 20°C than at 15 or 24°C (2817). Fresh isolates sporulate freely, especially on soil extract agar, while during preservation on rich media this capacity can soon be lost. — It grows well on media containing peptone, L-glutamic acid or glycine as N source (2817). Chitin utilization is very good (1425, 2068). Hexadecane and solid paraffins are utilized (3810). *M. alpina* is autotrophic for growth substances (1425). The production of mycoferritin (675), ethanol and acetic acid has been reported (4509) as well as a large amount of a tetra-unsaturated C_{20} fatty acid, which has tentatively been identified as 5,8,11,14-eicosatetraenoic acid (2312).—Antagonistic activity *in vitro* is displayed towards *Rhizoctonia solani* and *Gaeumannomyces graminis* (1425, 3494). The culture filtrate proved toxic to wheat seedlings (3582) and on agar media with a low N content, it may promote growth of aseptically grown *Buddleya davidii* and *Senecio squalidus* (5569). — It can also grow on media with 4% NaCl, corresponding to −35 bars osmotic potential (5949).

Mortierella bainieri Cost. *1889* — Fig. 179.

DESCRIPTION: Linnemann (3359, 6588); zygospores: Kuhlman (3160). — Section *Hygrophila* Linnem. — Colonies reaching 6 cm diam in five days on MEA at 20°C; characterized by particularly large and abundantly branched sporangiophores, to more than $1000\,\mu$m tall, spores 6–10(–13) × 3–5(–6) μm, and the absence of chlamydospores. A few zygospores were obtained by Kuhlman (3160) after mating compatible isolates on "Pablum" and hempseed agars and three weeks incubation at 26°C. — The similar *M. elongata* (q.v.) differs in its shorter sporangiophores and the presence of chlamydospores.

M. bainieri is most commonly observed on decaying agarics, especially those of the genus *Amanita* (6588, IMI), but has also frequently been reported from forest nurseries in southern states of the USA (2482), arctic soils in Alaska (3062), mixed hardwoods in the Appalachians (2385) and in stands of beech in Italy (270, 271) and Poland (272, 273). This fungus is also known in pine forests in Poland (269), rarely in cultivated soils in the USA (3817), dry and waterlogged pastures (164, 165), heathland (5221) and also in marshes, swamps and/or peat bogs in the British Isles (163, 164, 1376) and Poland (6520) and in fresh water in France (3942). Its distribution appears to be independent of soil depth and it can reach the mineral horizon and depths to 105 cm (272, 273). It has also been isolated from *Pinus taeda* roots (3159). — On cultivated *Agaricus bisporus* it causes the "shaggy stipe" disease (1778), whose occurrence is apparently enhanced by the application of the fungicide benomyl to control other moulds (1777).

On a glucose-containing medium, the production of ethanol and acetic and oxalic acids has been observed (4509). Hexadecane and solid paraffins are moderately utilized (3810).

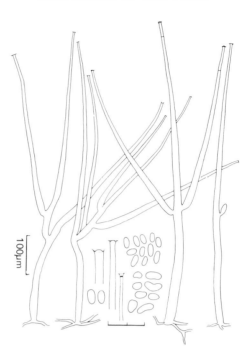

Fig. 179. *Mortierella bainieri*, sporangiophores with details of sporangiophore tips and sporangiospores, CBS 442.68.

Mortierella bisporalis (Thaxt.) Björling *1936* — Fig. 180.

≡ *Haplosporangium bisporale* Thaxt. *1914*
= *Haplosporangium decipiens* Thaxt. *1914*
≡ *Mortierella decipiens* (Thaxt.) Björling *1936*

DESCRIPTIONS: Thaxter (5788), Linnemann (6588), Nicot (4161), and Vallier (6037). — Section *Haplosporangium* (Thaxt.) W. Gams. — Colonies reaching 4·2 cm diam in five days on MEA at 20°C. Sporangiophores arising in dense rows along the aerial hyphae, 50–100 μm long, 4–9 μm wide at the base but strongly tapering towards the tip. Sporangia 1–2(–3)-spored; spores 8–13 μm diam, with finely verrucose walls. Björling (542) and Gams (1881) discussed the inclusion of the genus *Haplosporangium* Thaxt. in *Mortierella*. — *M. bisporalis* is traditionally distinguished from the single-spored *M. decipiens* by its bisporous sporangia, and Thaxter (5788) separated *M. decipiens* on the basis of its nodding (recurved) rather than straight sporangiophores. Our isolates with 1-spored and 2-spored sporangia were identical in having nodding sporangia as well as in all other respects and so we regard these taxa as conspecific. — *M. bisporalis* grows sparingly in pure culture and can only be maintained with difficulty over a number of transfers. A good medium for vegetative growth and abundant sporulation proved to be PDA with rat dung (2424). — A TEM study by Dijkstra (1394) of a 1-spored isolate showed the sporangial (sporangiole) nature of the spores (contrary to Dijkstra's interpretation); these are covered by a thin sheath, the sporangial wall; the thick verrucose spore wall is 2-layered, and the germ tube wall is continuous with the inner spore wall. — DNA analysis gave a GC content of 52% (5600).

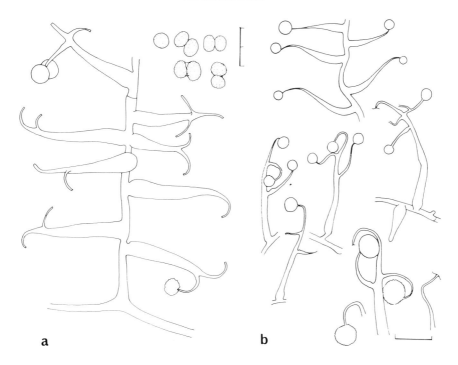

FIG. 180. *Mortierella bisporalis*. a. Isolate with 2-spored, b. isolate with only 1-spored sporangioles, CBS 873.68.

This species, first found in the USA on dung of various animals (2424, 5788), has been isolated only rarely since then: from sand dunes in the British Isles (745), desert soils south of Algiers (2974), soils in an oak and beech forest in Czechoslovakia (1700, 1702), beech litter in the Turin Botanical Garden (270, 271), more frequently from beech forests in Poland (272, 273, 3138) and Italy (271), and old forest stands in the USA (2385). It is also known from arable soil and earthworm casts in Germany (1429). Slide-traps proved a particularly suitable isolation technique (5221).

Mortierella elongata Linnem. *1941* — Fig. 181.

= *Mortierella rishikesha* B. S. Mehrotra & B. R. Mehrotra *1964*

DESCRIPTIONS: Linnemann (3359, 6588), and Watanabe (6212); zygospores: Gams *et al.* (1886). — Section *Hygrophila* Linnem. — Colonies reaching to 10 cm diam in five days on MEA at 20°C. Sporangiophores to more than 400 µm long, basitonously branched, 5–8(–12) µm wide at the base and gradually tapering towards the tip. Spores ellipsoidal to short cylindrical, 6–13(–16) × 4–6(–7) µm. Small, ± elongate chlamydospores, 10–18 µm diam, abundantly produced. Zygospore formation in this heterothallic species has been obtained by Gams *et al.* (1886) on hay extract, SAB sucrose and potato-carrot agars, and by Kuhlman (3160) on hempseed and "Pablum" agars. — Kuhlman's isolates, however, did not mate with those of Gams *et al.* and differed in having a wider sporangiophore tip (6–11 μm as compared with 1·5–3·5 μm in *M. elongata*) and were found to be homothallic and so they are regarded as a distinct species, *M. kuhlmanii* W. Gams *1977*. The similar homothallic *M. epigama* W. Gams & Domsch 1972, which produces numerous zygospores on 2% MEA but no chlamydospores, has slightly fusiform spores. — *M. rishikesha*, isolated from forest soil in India, is reported as having chlamydospores with appendages like *M. exigua,* but the type strain (CBS 652.68) hardly differed from *M. elongata* and produced a few zygospores with CBS strain 122.71 of *M. elongata.*

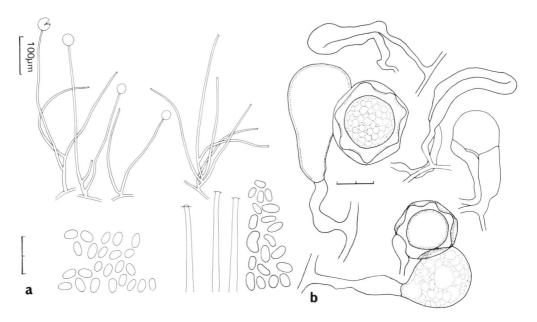

FIG. 181. *Mortierella elongata.* a. Sporangiophores with details of sporangiophore tips and sporangiospores; b. various stages of zygospore formation.

M. elongata is one of the most widely distributed *Mortierella* species, especially in neutral soils. It ranges from Alaska (3062), Swedish Lapland (3360), Poland (273), the British Isles (164, 982, 989, 1375, 1438), the Netherlands (1614, 1616), Germany (1888, 1889), Austria (2530), and the USSR (2949, 2950), to soils in Italy (3450, 3913, 6588), Turkey (4245), the USA (2385, 6350), highland in Mexico (3360), New Zealand (5812), and Japan (3267). It has been isolated from mixed forest soils (2385) and those under pine (989, 5813), larch (3267), beech (273, 6588), forest nurseries (6350), forest steppe (2950), heathland (5811), grassland (164, 982, 5812, 5813), arable soils (1616, 2949, 3450, 3913), under wheat mainly from organic particles (1614, 1888, 1889), peat (1375, 1376) with increased frequencies after cultivation (1438), and alpine sediments below glaciers (2530). As to maximum soil depth, its occurrence down to 135 cm is recorded (164). *M. elongata* is also known in the rhizospheres of poplar (3452), potato (1614, 1617), wheat and other crops (1614, 2950), roots of *Trifolium repens* and *Lolium perenne* (5815), various forest plants

(3159, 4814), hardwood stumps, *Carthamus tinctorius* (IMI), underground parts of sugar cane (6212) and on *Pteridium* petioles (1821). — Thirty-one per cent of chloroform-fumigated mycelium was mineralized within ten days at 22°C (133).

The optimum temperature for zygospore production lies between 20° and 26°C (1886) and that for growth is 25°C (1375). — Pectin decomposition is relatively weak compared with many other fungi (1432). Chitin decomposition occurs (2706, 6057) and hexadecane, solid paraffins and other hydrocarbons from petroleum are all efficiently utilized (3810).

Mortierella exigua Linnem. *1941* — Fig. 182.

= *Mortierella indica* B. S. Mehrotra *1960*
= *Mortierella sterilis* B. S. Mehrotra & B. R. Mehrotra *1964*
= *Mortierella spinosa* Linnem. sensu Milko *1974*

DESCRIPTIONS: Linnemann (3359, 6588), and Milko (3807). — Section *Spinosa* Linnem. — Colonies reaching 9 cm diam in five days on MEA at 20°C. Sporangiophores at least partly to 600–1000 μm tall, with pronounced acrotonous branching, 6–20 μm wide at the base,

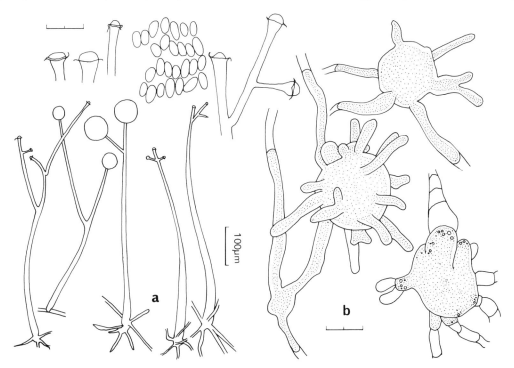

FIG. 182. *Mortierella exigua*. a. Sporangiophores with details of sporangiophore tips showing small columellae and sporangiospores; b. amoeba-like chlamydospores.

tapering to 3–6(–9) μm at the tip. Spores cylindrical, 6–10 × 3–5·5(–7) μm. Chlamydospores more or less globose, 15–45 μm, giving rise to a number of irregularly radiating hyphae which are partly filled with plasma on maturity. — This interpretation of the species, which was inadequately described by Linnemann (3359), and of which no type strain is preserved, is inferred from the observation that all isolates matching the description of the spores and the "amoeba-like" chlamydospores examined eventually produce large sporangiophores with acrotonous branching on suitable media: an isolate from M. Turner of this type was also confirmed by Linnemann as *M. exigua*. — The description of *M. indica* (no authentic culture existing) agrees perfectly with the present concept of this species. An original culture of *M. sterilis* (CBS 655.68), isolated from soil in India, still produces numerous sterile branches in the sporangiophores as described, but otherwise agrees in all respects with *M. exigua*. — *M. exigua* can be distinguished from *M. gamsii* by the cylindrical spores and by the larger chlamydospores which have irregular appendages, the characteristic large sporangiophores arising relatively late.

This species mainly occurs in deciduous and coniferous forests on calcareous soils (3360) and extends from Alaska (3062) to highlands (2700 m) in Mexico (3360), but is otherwise known only in the temperate zone. — The reported habitats include beech forest on mull in Poland and Denmark (273, 2745), grassland in the British Isles and New Zealand (164, 5812), arable soils in the Netherlands and Germany (1614, 1616, 1889), garden compost in Germany (1425) and heathland in the British Isles (5811). It has been isolated from acidic alpine raw-humus soils very rarely but then occurs in an atypical form (1876). The pH range has been given as 5·3–7·4 (3360). It occurred with reduced frequencies after fungicide application to soil (1426). *M. exigua* has been demonstrated in the rhizospheres of grasses (1614, 5815), clover (5815), potato, wheat (1614) and ericaceous plants on peat bogs (6416). A slightly increased frequency in soil after a wheat crop has been observed (1433).

Good C sources include inulin, fructose, mannitol and galactose, while trehalose, melibiose, dulcitol, glycerol, malic and tartaric acids are utilized poorly (5053). From a mixture of four amino acids, leucine, asparagine, alanine and glycine (5054) were taken up preferentially in that order and no growth was obtained on L-arginine and L-histidine (5056). Chitin is attacked very strongly (1425, 2706). Sulphate, persulphate and sulphite proved to be utilizable S sources (administered as 32 mg S·l^{-1} (5057). An analysis of mycelial extracts gave nine different free amino acids and, while no sucrose was found, glucose, fructose, malic and succinic acids were present (5055).

Mortierella gamsii Milko *1974* — Fig. 183.

= *Mortierella candelabrum* van Tieghem & Le-Monnier *1873* sensu Gams *1970*
= *Mortierella spinosa* Linnem. *1936* (nomen inval., Art. 36)
? = *Mortierella mutabilis* Linnem. *1941*

DESCRIPTIONS: Linnemann (3359, 6588), Turner and Pugh (5949), and Milko (3807); zygospores: Kuhlman (3161).—Section *Spinosa* Linnem.—Colonies reaching 8 cm diam in five days on MEA at 20°C. Sporangiophores 400–800 μm long, 10–>20 μm wide at the base, often bent upwards, mesotonously to acrotonously branched. The specific epithet *"spinosa"* indicates the presence of a small columella-like projection at the sporangiophore tip. Spores globose to slightly ellipsoidal, 6–12(–14) μm diam. Small chlamydospores, to 20 μm diam, are regularly produced. Zygospores were obtained on hempseed agar at 20°C (3161). — *M. candelabrum* was originally described as having chlamydospores and not stylospores and may have been based either on this species or on *M. hyalina*; the original illustration of *M. candelabrum* shows mesotonously or almost basitonously branched sporangiophores and this epithet has been subsequently interpreted in at least three different ways and caused much confusion. We have consequently refrained from our previous interpretation (1887) and accept the unequivocal name *M. gamsii*, regarding *M. candelabrum* as of doubtful application. The type culture of *M. mutabilis* and some similar isolates show a particularly irregular pattern of the sporangiophore branching, but for the rest agree with *M. gamsii*. CBS 308.52 (*M. mutabilis*) produced a few zygospores with CBS 552.73 (*M. gamsii*) on PCA. Milko (3807) found it necessary to redescribe the species, because he had confused *M. spinosa* with *M. exigua*, and thus created an acceptable epithet for the not validly published *M. spinosa*.

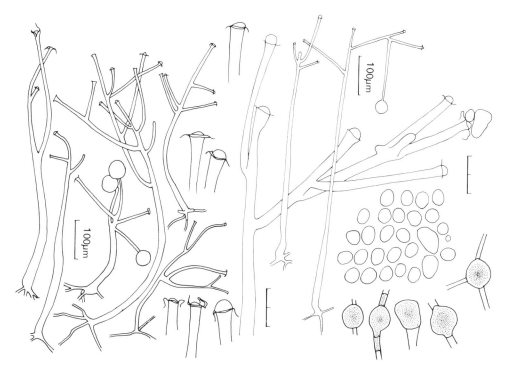

FIG. 183. *Mortierella gamsii*. Variation in sporangiophore branching with details of sporangiophore tips showing small columellae, sporangiospores and chlamydospores.

M. gamsii is one of the most widely distributed *Mortierella* species and occurs mainly in forest soils. Reports of it range from Greenland (4174), Spitsbergen (3064), Murmansk (4703) and the Buhara region of the USSR (2871), Finland (5663), Denmark (2745), Poland (272, 273), Germany (1631, 1889), Austria (1876), the Netherlands (1614), the British Isles

(159, 164, 745, 4646, 5949) to Mexican highlands (3360), Brazil (6007), India (282), the USA (2004, 2008) and New Zealand (5812, 5814). It is frequently noted in hardwood forests on calcareous (3360) and black soil (1631); consequently there are numerous reports from beech forests (272, 273, 2745, 5813), mixed hardwood forests (2004, 2008) but fewer from coniferous ones (5663). It has, however, been shown to be present in various central European podzols (3416), raw humus (1876), and also occurs in grassland (159, 164, 982, 1614, 5812, 5814), arable soils (2871), more rarely in salt-marsh under *Salicornia* (4646, 5949) and dunes (745). *M. gamsii* has been isolated very irregularly from washed particles of two wheatfield soils (1889). The pH range is given as 4·2–7·8 (6588) and it occurs at all soil depths, the maximum observed being 135 cm (164), although it is apparently more often found in the upper horizons. It has also been demonstrated on roots of *Lolium perenne*, clover (5815) and *Pinus taeda* (3159), in the rhizospheres of grasses and wheat (1614), on *Pteridium aquilinum* (1821), particularly frequently on polypores of the most varied kinds (4916, and personal observation), and on organic detritus in fresh water (4429). — Forty-one per cent of chloroform-fumigated mycelium was mineralized within ten days at 22°C (133).

This species is known to be a chitin decomposer (2706). On a glucose-containing medium the production of ethanol and acetic acid has been described (4509). Hexadecane and solid paraffins are efficiently utilized (3810). It does not require thiamine (4860).

Mortierella horticola Linnem. *1941* — Fig. 184.

DESCRIPTION: Linnemann (3359, 6588). — Section *Stylospora* Linnem. — Colonies reaching 5·5 cm diam in five days on MEA at 20°C. Sporangiophores unbranched, 40–100(–170) μm long, with 1-spored spinulose sporangia, 6–11(–13) μm diam. Unlike the otherwise similar *M. alpina*, the base of the sporangiophores is not differentiated.

This species was named a "garden-dweller" (3359) because of its occurrence in a garden bed with forest plants and in a calcareous forest soil in Germany. Since then it has been reported from a cultivated black soil in Germany (1631) and roots of *Lolium perenne* and *Trifolium repens* in New Zealand (5815). It is also found very rarely in arable soil in Germany, in Ecuador it has been isolated from soil in cacao plantations, in Japan (3267) and in Sweden (CBS) it has been found in forest soils.

The ability of *M. horticola* to decompose chitin has been demonstrated (2706).

Mortierella humilis Linnem. *1936* ex W. Gams *1977* — Fig. 185.

DESCRIPTIONS: Linnemann (3359, 6588), and Gams (1883); zygospores: Chien *et al.* (1003). — Section *Stylospora* Linnem. — Colonies reaching to 7 cm diam in five days on

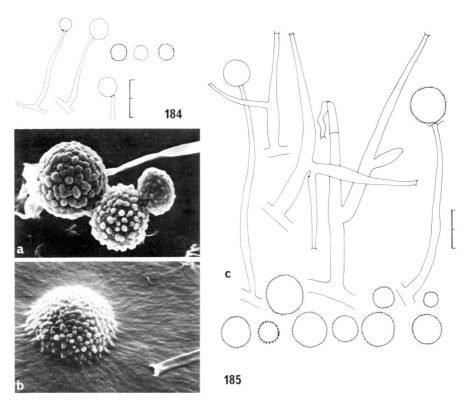

Fig. 184. *Mortierella horticola*, unbranched sporangiophores with 1-spored sporangia, CBS 869.68.
Fig. 185. *Mortierella humilis*. a, b. Warted sporangiospores, SEM, a. × 2100, b. × 4100, orig. K. M. Old;
c. sporangiophores with liberated sporangiospores.

MEA at 20°C. Sporangiophores basitonously branched, 50–200 µm long. Sporangia consistently 1-spored, 6–15 µm diam, finely verrucose by light microscopy and SEM (1003, Fig. 185a, b), with the outer wall firmly adhering to the spore. This is a heterothallic species forming naked zygospores on PCA, CMA and other media between 5° and 22°C (1003). —*M. humilis* is perfectly interfertile with *M. verticillata* (q.v.) which is distinguished by the usually more-spored sporangia and wrinkled surface of the spores.

M. humilis is a very frequent species, especially in forest soils, grassland soils and compost (6588). It is particularly easy to isolate with immersion plates and tubes (982, 989, 5813) but has also been observed on hair baits (2091). — Its overall distribution extends from Swedish Lapland (3360), Denmark (2745), the British Isles (159, 163, 164, 1376, 5811), the Netherlands (1614, 1616), Germany (1889, 3786), Austria (1876), the USSR (3807), Italy (3445, 3976), highlands in Mexico (3360), the USA (1039, 2004), New Zealand (5812), and Japan (3267). *M. humilis* occurs in coniferous forest soils (3267, 3786, 5813), under beech on mull and mor (2745), frequently in oak forests (989, 5811, 5813), *Salix-Populus* mixed forest (2004), a conifer swamp (1039), grassland (159, 164, 165, 982, 3445, 5812, 6182), arable soil (1616, 1889), heathland (989, 5811), alpine raw humus (1876), and marsh and moorland soils (163, 1376, 3360). It is able to penetrate to great soil depths (159, 3267) and the maximum recorded is 135 cm (164). *M. humilis* occurs preferentially in acidic soils and, according to observations made so far, has a pH range of 3·7–6·4 (3359, 3360, 3871, 6182, 6588); very humid soils are also preferentially colonized (1039). There are also reports of its occurrence in the root regions of *Trifolium repens, Lolium perenne* (5815) and *Avena* (4443) and in cultures of a collembolan species (1027).

M. humilis (misidentified as *M. verticillata*) was found to decompose chitin (2706) and to utilize hexadecane, solid paraffins and high-molecular alcohols, while *M. verticillata* does not (3810). In a chronically irradiated soil, it did not occur at intensities of ≥120 R per day (2009).

Mortierella hyalina (Harz) W. Gams *1970* — Fig. 186.

≡ *Hydrophora hyalina* Harz *1871*
= *Mortierella candelabrum* var. *minor* Grove *1891*
= *Mortierella hygrophila* Linnem. *1936* (nomen inval., Art. 36)

DESCRIPTION: Linnemann (3359, 6588); synonymy: Gams and Domsch (1887). — Section *Hygrophila* Linnem. — Colonies reaching 7–8 cm diam in five days on MEA at 20°C. Sporangiophores abundantly basitonously branched, 200–>1000 μm long. Spores more or less globose, large, variable in size but mostly more than 10 μm diam, with minutely verrucose walls. Small chlamydospores are also abundantly produced. — Isolates having smaller sporangiophores and few-spored sporangia (e.g. *M. hygrophila* var. *minuta* Linnem. *1941*) can be distinguished from *M. verticillata* (q.v.) mainly on the basis of spore size.

M. hyalina has a worldwide distribution in forested and cultivated soils in the temperate zone, particularly in compost-enriched soil, but also on decaying macromycete fruit-bodies (6588). *M. hyalina* has been reported from Swedish Lapland (3360), the British Isles (159, 163, 164, 745, 1598, 2068, 4445, 5221), Germany (1889, 2467), the USSR (2950), Corsica (6588), Nepal (1826), India (3738) and New Zealand (5813, 5815). It is common in forest soils, particularly under pine (983, 1598, 2068, 4445) and also occurs in soils of forest nurseries (3908) and forest steppe (2950), arable soils (1889), heathland (5221), moorland (163), grassland (159, 164, 982) and dunes (745); pH values between 4 and 5 are preferred (6588). It has been isolated more frequently from organic than from inorganic soil particles (1888). It occurs primarily in the upper soil horizons (983, 2068), but has been reported to depths of 135 cm (164). Following steam sterilization of soil, it could no longer be isolated but was present as a recolonist 25 months later (3908). Thirty-eight per cent of chloroform-fumigated mycelium was mineralized within ten days at 22°C (133). *M. hyalina* regularly is found in sewage-treated land, especially those areas suffering from irrigation sickness (2467) and has also been isolated from hay and straw (4548). This species has also been demonstrated on frost-damaged leaves of various grasses (191), roots of *Trifolium repens* (5815), wheat and other crops (2950), deer dung (3738), and found to be the causal agent of a tracheomycosis in a buzzard (2490).

The optimal temperature for growth is 21°C, the maximum 28°C, the minimum −3°C (191). — When cultured on glucose with nitrate as the N source and without any growth factors, no growth occurs, but after adding ammonium ions there was no dependence on the addition of particular amino acids or the vitamins B_1, B_2, pyridoxin, calcium pantothenate, nicotinic, ascorbic or *p*-aminobenzoic acids, biotin or inositol; peptone is the best N source and yeast extract (0·2%) also promotes growth; in addition to glucose, maltose is utilized well, but arabinose proved to be less suitable (5809). Pectin decomposition can be demonstrated but is low in comparison with most other fungi tested (1432). Selective isolation from buried chitin indicates that *M. hyalina* has some chitinolytic activity (2068), and it is also able to utilize hexadecane, solid paraffins and diesel fuel extract (3810).

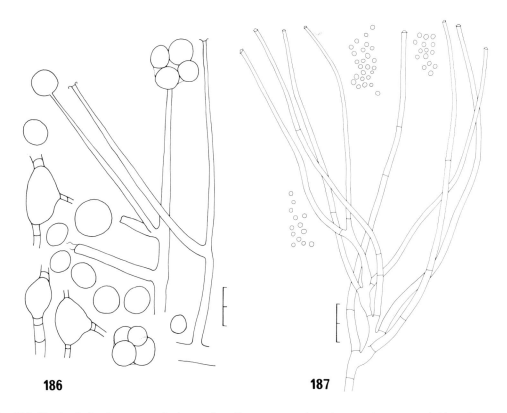

186   **187**

FIG. 186. *Mortierella hyalina,* sporangiophores of small type, sporangia and sporangiospores, and chlamydospores.
FIG. 187. *Mortierella isabellina*, branched sporangiophore and sporangiospores.

Mortierella isabellina Oudem. *1902* — Fig. 187.

? = *Mortierella humicola* Oudem. *1902*
   = *Mortierella atrogrisea* van Beyma *1929*
   = *Mortierella isabellina* var. *ramifica* Dixon-Stewart *1932*
? = *Mortierella fusca* Wolf *1954*
   = *Mortierella isabellina* var. *atra* Khalabuda & Zhdanova *1957*

DESCRIPTIONS and discussion of species variability: Linnemann (3359, 6588), and Turner (5948). — Subgenus *Micromucor* W. Gams. — Colonies reaching 1·7 cm diam in five days on MEA at 20°C, velvety, drab-grey to fuscous from the sporangia; sporangiophores 150–300 μm long. Spores slightly angular, 2–3 μm diam. Small chlamydospores sparingly produced. It is uncertain whether *M. fusca* is conspecific with *M. isabellina* as its branches are said to be shorter and the spores 4–5 μm diam. — DNA analysis gave a GC content of 50% (5600).

M. isabellina is common in forest and heathland soils of the cool temperate zones. There are reports from, for example, Denmark (2745), the British Isles (745, 1376, 2502, 2736, 5221, 5222, 5559, 6182, 6383), Poland (273, 1978, 6520), Czechoslovakia (1703), Italy (3975), the USSR (3807), Canada (3959, 3962, 5363), the USA (1032, 1039, 2008, 2482, 2573, 3817, 6350, 6351), including arctic soils in Alaska (3062), Nepal (1826), Australia (1410, 3721), and Antarctica (IMI). Exceptions are reports from the Canary Islands (1827), Brazil (6007) and South Africa (1555). It has been isolated most commonly from forest soils (2573, 3817) including ones with alder and conifers (6351), spruce (3975), larch (3267), beech (273, 2745), aspen (3959), wet *Salix-Populus* communities (2008), dry and mesic conifer-hardwood communities (1032, 6350), a boreal mixed forest (2008), also in a forest soil after 20 years of prescribed burning in the USA (2822) and forest nurseries (1978). It has also been recorded from peat bogs and conifer swamps (1039, 6520), podzol (3414), peat-podzol transitional soils (2502), an alkaline peaty soil (5559), a copper swamp (2922), vineyards enriched with copper (4921), grassland (1555, 6182), heathland (2736, 3721, 5220, 5221, 5222), acid sand dunes (745), cultivated soils rarely (6007), an acid mine drainage stream (1166), and a uranium mine (1703). *M. isabellina* has been isolated from soil depths of 30–45 cm (5220). It has also been reported from the phylloplane of vascular plants (296), frequently from litter of Scots pine (2344), spruce (3786), and *Eucalyptus maculata* (1558), from willow stumps (3568), decaying *Pteridium aquilinum* (1821), the rhizospheres of groundnuts (2768), barley and oat (3004), roots of *Pinus taeda* (3159), *P. sylvestris* (3572), *P. nigra var. laricio* with frequencies increasing with root age (4448), and several ericaceous plants (5220, 6416), decaying wood and bark (6588), and faecal pellets of the diplopod *Glomeris marginata* (4154).

Optimum growth temperature is at 25°C (1827); in soil *M. isabellina* can grow actively only at water potentials between −0·2 and −0·5 bars (5226). — Isolates regarded as var. *atrogrisea* utilized hexadecane and solid paraffins, while var. *isabellina* did not (3810). An ATP-dependent citrate lyase is produced by various species of the subgen. *Micromucor* (230) and 2-deacylusnic acid isolated as a metabolite (3484). — In a chronically irradiated soil there was a sharp drop in numbers of viable cells at intensities of 500 R per day but the fungus was still present at 1400 R per day (2009). After heating soil for 30 min to 70°C it could still be isolated amongst the heat-resistant fungi (615).

Mortierella minutissima van Tiegh. *1876* — Fig. 188.

DESCRIPTIONS: Linnemann (3359, 6588), and Turner and Pugh (5949); zygospores: Kuhlman (3160). — Section *Hygrophila* Linnem. emend. W. Gams. — Colonies reaching 6 cm diam in five days on MEA at 20°C. Sporangiophores short, 50–180 μm long but usually less than

100 μm, basitonously branched. Spores globose, less than 8 μm diam. A few zygospores were obtained (3160) after the mating of compatible isolates on "Pablum" or hemp seed agars at 18°C. — DNA analysis gave a GC content of 49% (5600).

M. minutissima is a very widely distributed species, particularly in moderately acid forest soils (3360, 6588), extending from Alaska (3062), Spitsbergen (3064), Sweden (5465), the USSR (2950, 3807), where it occurred in the southern steppe, more commonly in forest steppe, and also in the Poles'e of the Ukraine (2950), across alpine habitats (1876, 3445, 3976) and to high altitudes (2500 m) in Mexico (3360), also in the British Isles (159, 307, 745, 989, 4445, 4646, 5559, 5811, 5949), the Netherlands (1616), Denmark (2745), Germany (2467), Austria (3414), the USA (2004, 2008, 2573), Australia (6182), and New Zealand (5812, 5814). It has been reported from beech (2745), oak (5811), spruce (5465), pine (307, 989, 4445) and mixed hardwood forests (2004, 2007, 2008), heathland soils (989, 5811), grassland (159, 1614, 5812, 5814, 6182), riverside soils (3414), fen peat (5559), more rarely salt-marshes (4646, 5949), sand dunes (745), arable soils (1616) and sewage-treated land (2467). The pH range, previously given as 6·0–7·8 (3359), can now be extended to 3·8–8·6 on the basis of subsequent observations (3871, 6182). *M. minutissima* has been shown to be present at particularly high frequencies in habitats with high lime content (4152). Where the pH value increases with soil depth, a higher concentration of the species has been found in the deeper soil layers (4445, 6182, 6184) but otherwise it is most frequent in horizons rich in organic matter (2008, 3445, 3976). In the southern steppe soils of the Ukraine it occurred only with irrigation (2950). After serial soil washings it could be isolated from both organic and mineral particles (307). This species has been observed on various decaying plant remains (1821, 3113, 4548) and roots of *Pinus taeda* (3159), clover, *Lolium perenne* (5815) and oaks (4443), birds' roosts (682), polypores (6588), decaying agarics and extruded sap of elm (2950).

M. minutissima can be selectively isolated from buried chitin (2068) which it can decompose (2706); hexadecane is also moderately utilized (3810). The production of a substance with antibiotic activity against *Mycobacterium phlei* has been reported (3812).

Mortierella nana Linnem. *1941* — Fig. 189.

= *Mortierella alba* Mańka & Gierczak *1961*

DESCRIPTIONS: Linnemann (3359, 6588), and Turner (5948). — Subgenus *Micromucor* W. Gams. — Colonies slow-growing, reaching 0·7–1·4 cm diam in five days on MEA at 20°C, velvety, to 4 mm deep, pure white. Sporangiophores 30–80 μm long, with short, ± verticillate branches arising close together from a slightly swollen node. Sporangia consistently 1-spored, 5–10(–13) μm diam, smooth-walled. Chlamydospores absent. — The discovery of *M. roseo-nana* W. Gams & Gleeson *1977* which has reddish 1-spored sporangia points to the affinity of this species with the other species of the subgenus *Micromucor*.

The available data indicate a wide distribution in cool temperate zones including dry and mesic conifer-hardwood communities in Wisconsin (1032, 6350), deciduous and evergreen

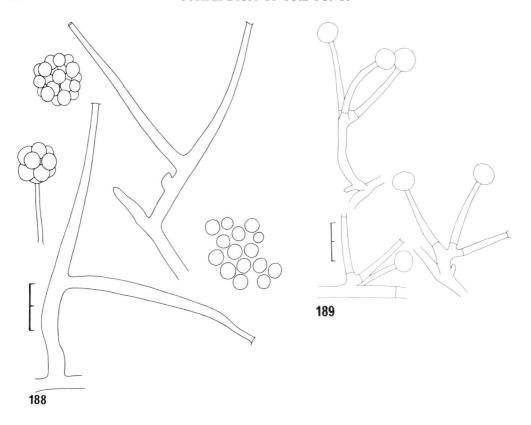

FIG. 188. *Mortierella minutissima*, sporangiophores, sporangia and sporangiospores.
FIG. 189. *Mortierella nana*, sporangiophores with 1-spored sporangia, CBS 444.68.

forests in Ohio (2573), boreal forests (3962, 5363) and stands dominated by aspen (3959) in Canada, a larch forest in Japan (3267), spruce forest in Sweden (5465), various forest soils in Germany (6588) and a sandy loam forest soil in Belgium where it was the dominant fungus (6285). Other records include a mull type forest soil under beech in Denmark (2745), forest nurseries in Poland (1978), grey sandy forest soil under deciduous forests in the Ukraine (2950), grassland in the British Isles (164) and New Zealand (5812), soils in Pakistan (3473), heathland soil in Australia (3721), podzols in the British Isles (989), cultivated brown soil (1631) and soils treated with sewage in Germany (2467). It has also been isolated from the rhizospheres of barley and oat (3004), roots of *Pinus taeda* (3159), *P. sylvestris*, beech, *Pseudotsuga* (6588), a macromycete fruit-body (3807) and earthworm casts (1429).

This fungus was found to liberate N from urea-formaldehyde fertilizers (4010) and to stimulate fruit-body formation in *Psilocybe panaeoliformis* (2901). In a chronically irradiated soil, a sharp drop in numbers of viable cells occurred at 53 R per day, no cells survived at 230 R per day (2009).

Mortierella parvispora Linnem. *1941* — Fig. 190.

= *Mortierella gracilis* Linnem. *1941*

DESCRIPTIONS: Linnemann (3359, 6588), and Turner and Pugh (5949); zygospores: Gams and Williams (1893), and Kuhlman (3160). — Section *Spinosa* Linnem. — Colonies reaching 3·5–4 cm diam in five days on MEA at 20°C. Sporangiophores (particularly on soil extract agar and PCA) tall, 150–400(–600) μm long and strongly tapering towards the tip from an often curved, 5–12 μm wide base, mesotonously to acrotonously branched, commonly forming a minute columella. Spores globose, 2–3 μm diam. Chlamydospores absent. — Naked zygo-spores were discovered after mating compatible isolates on SAB glucose, acidified Czapek-Dox with yeast, soil extract, PDA, cornmeal, CMC (1893), hempseed and "Pablum" agars (3160) and are not produced above 18°C. — Sporangiophores are very variable in length and width, but morphologically different isolates are usually interfertile. *M. gracilis* has a rather diffuse branching of generally shorter sporangiophores but as the short columellae and the spores are identical to those in *M. parvispora* and the type strain produced zygospores with a (–) strain of *M. parvispora*, these are regarded as conspecific. — The hyphal walls contain a fucose polysaccharide besides glucan and chitin and are particularly resistant to enzymatic lysis (4499). — DNA analysis gave a GC content of 50·5% (5600).

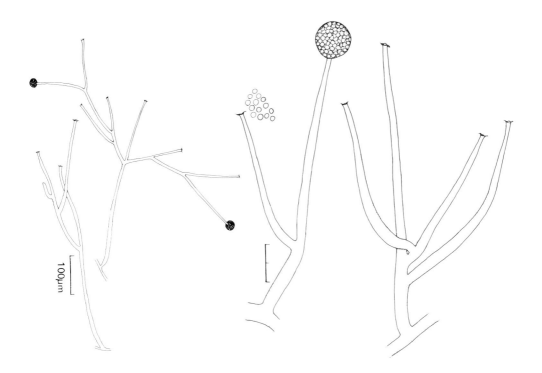

FIG. 190. *Mortierella parvispora,* sporangiophores at different magnifications, with sporangia and sporangiospores, CBS 316.61.

With the exception of one report from Brazil (6007), this species seems to be restricted to the cool temperate zones. Records include arctic (1750) and antarctic areas (3062, 3235), Sweden (5465), numerous reports from the British Isles (e.g. 1375, 1376, 1438), the Netherlands (1614, 1616), Germany (6588), Austria (2530), Poland (273), the USA (1032, 1039, 2573) and New Zealand (5812). The majority of sites are acid forest soils with pines (307, 1893, 2923, 2950, 6383), beech (273), larch (3267), spruce (5465), dry to mesic conifer-hardwood communities (1032), conifer swamps (1039), deciduous and evergreen forests (2573), dry or wet grassland (159, 163, 164, 1614, 5812), peat and moorland (163, 1375, 1376, 1438), heathland (5221, 5222), podzol (989), acid sand dunes (745), salt-marsh (4646, 5949), an inland-marsh (164), and a recently deglaciated alpine soil (2530). It occurs only rarely in cultivated soils (1616, 4646). *M. parvispora* was found to be closely associated with organic particles in both soil (307) and fresh water (4429) and was also reported on roots of heath plants (5220) and *Pinus nigra* var. *laricio* (4448), and in the rhizospheres of grasses and wheat (1614). The pH range in soil has been given as 3·4–5·2 (6588); *M. parvispora* has been found down to soil depths of 30–45 cm (5220).

The optimal growth temperature is 20°C (1375), and the optimal water potential is at −0·2 bars (5226). Maltose, starch and pectin are utilized (1375, 1750) and trimethylamine is demethylated to dimethylamine (255) by *M. parvispora* which also utilizes hexadecane and solid paraffins weakly (3810).

Mortierella polycephala Coemans *1863* — Fig. 191.

= *Mortierella crystallina* Harz *1871*
= *Mortierella lemonnieri* Vuill. *1918*
= *Mortierella canina* Dauphin *1908*
= *Mortierella raphani* Dauphin *1908*
= *Mortierella vantieghemii* H. Bachmann *1900*

DESCRIPTIONS: Linnemann (3359, 6588) and Mehrotra and Mehrotra (3738); discussion of synonymy: Turner and Pugh (5949). — Section *Mortierella*. — Colonies reaching 7·5 cm diam in five days on MEA at 20°C. Sporangiophores often arising in groups, typically racemosely branched, 180–400 μm long. Spores smooth-walled, globose, 8–12 μm diam. Numerous chlamydospores ("stylospores", "Stielgemmen") borne on short slender stalks in the aerial mycelium, slightly flattened, covered with blunt warts, 15–22 μm diam; less ornamented and smaller chlamydospores commonly occur in the agar. The repeated branching of the sporangiophores is not recognized as a satisfactory criterion to distinguish *M. polycephala* from further species (5949). — Zygospores were described by van Tieghem and Le Monnier in *1873* and Dauphin in *1908* as being invested by a thick layer of hyphae and formed in homothallic cultures, but in recent decades it has not been possible to repeat this observation, although a weak mating reaction was observed by Chien (1003) with an isolate of his *M. indohii* which does not produce sporangiophores and has globose, strongly spiny chlamydospores.

M. polycephala was originally observed on mouse, rat and bat excrements and there are

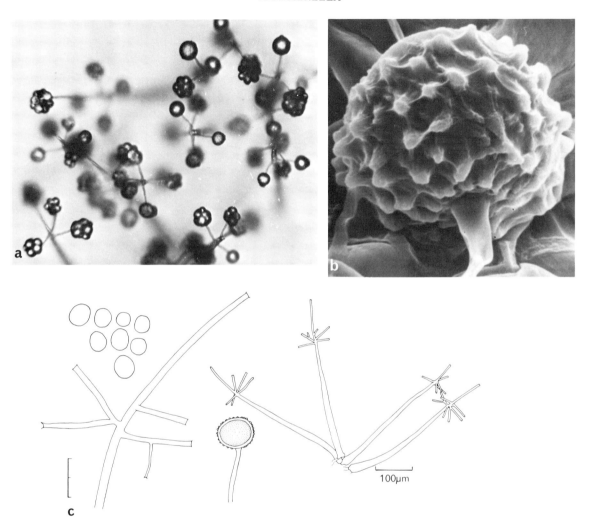

FIG. 191. *Mortierella polycephala*. a. Sporangiophores seen from above showing racemose branching, × 200; b. chlamydospore (stylospore), SEM, × 3200; c. sporangiophores with detail of the tip, sporangiospores and stalked chlamydospore.

several reports of occurrences on decaying polypores (6588). — In soil analyses *M. polycephala* is generally found only rarely, but has a wide distribution. Most records are from Europe, including the Buhara region (2871) and the southern and central Poles'e of the USSR, in grey sandy forest soils after irrigation and manure application (2950), and in Germany where it was frequently found in garden compost (1424). It is also known from Brazil (6007), Indonesia (578) and India (3735). This species occurs, apparently without particular prevalence, in beech (272, 273, 2779) and spruce forests (3786, 4574, 4916), grassland (164, 982, 5814), *Calluna* heath (989, 5811) and arable soils (918, 1889), and is also recorded from slate slopes in Switzerland (566), salt-marsh (164, 4646, 5949), coastal sand

(4178) and Mexican desert soils (4733). It appears to be restricted to the uppermost soil layers (164). There are also reports of its isolation from alfalfa roots (3983), the rhizospheres of clover (3429), wheat, tomato and cabbage (2950), various plant remains and mouldy millet (4548).

The decomposition of chitin by this species is very good; it is heterotrophic for growth substances (1425) and able to liquefy gelatin (513). The production of ethanol and acetic and oxalic acids when grown on a glucose-containing medium has been demonstrated (4509). Hexadecane and solid paraffins are only weakly utilized (3810). When *M. polycephala* is artificially enriched to a high level in the soil, the green-matter production of rape in some pot experiments diminished significantly (1427). — It has also been reported as the causal agent of a lung mycosis (5132) and mycotic abortion in cattle (6004).

Mortierella ramanniana (Möller) Linnem. *1941* var. *ramanniana* — Fig. 192 a.

≡ *Mucor ramannianus* Möller *1903*

DESCRIPTIONS: Linnemann (3359, 6588), Turner (5948), and most detailed in Evans (1656). — Subgenus *Micromucor* W. Gams. — This species was first described from forest soil in Denmark by Möller (3901) as *Mucor ramannianus* but, in spite of the presence of a small columella, it was transferred to *Mortierella* by Linnemann (3359) on account of its close resemblance to *M. isabellina* (q.v.). — Colonies reaching 2·4–2·7 cm diam in five days on MEA at 20°C, velvety, 2–3 mm deep, vinaceous-brown from the colour of the sporangia. Sporangiophores 300–700 μm long, simple or branched; after dehiscence of the sporangium wall, a small but definite columella is visible. Spores oval to ellipsoidal, 2·5–5·5 × 1·5–3 μm. Chlamydospores abundant, to 100 μm diam, thick-walled, round or of irregular shape, with dense and oily contents, abundantly produced. Septation of mycelia occurs soon after germination of sporangiospores (1735). — The separation of isolates into var. *autotrophica* E. H. Evans *1971* and var. *ramanniana* (1656) is not only based on the thiamine (thiazole) independence of the former, but also on the pale congo-pink colonies and globose spores (2–3 μm diam) as opposed to the darker colonies and ellipsoidal spores of var. *ramanniana*. While var. *ramanniana* is characteristically associated with living roots and other plant material, var. *autotrophica* is a soil fungus most common in podzol and other acid sandy soils, particularly in the deeper layers (1657). The distinction of these varieties cannot yet be carried out consistently in the data compiled below. — Esterase profiles proved to be of no value for taxonomical purposes in subgen. *Micromucor* (4485). — DNA analysis gave a GC content of 49% (5600). — The analysis of cell walls yielded chitin, chitosan and other polysaccharides; the main component of the sporangiospore walls was found to be glucan (2800). Fatty acids (myristic, palmitic, palmitoleic, stearic, oleic, linoleic, linolenic) in the sporangiospores are found to be more highly saturated than those in the mycelium (5643).

M. ramanniana is one of the most widespread *Mortierella* species. The very numerous reports, which cannot be listed individually here, show a clear preference for cold and temperate zones: records range from Alaska (1750), northern Canada (3959, 3962, 5363), Murmansk (4703), East Siberia (3652), the Amur region (432), Norway (2206) and Sweden

(5465) to Poland (269, 272), Denmark (2745), Ireland (1376, 3918), Great Britain (with very numerous reports), Germany (825, 1712), Austria (1828) and the Poles'e in the Ukraine (2950). Its distribution extends into mediterranean countries, such as Yugoslavia (5534), northern parts of Italy (270, 271, 3538, 4537, 5047, 6082), Spain (3447), Teneriffe and Madeira (6588), and it is also found in the USA (1028, 1032, 1039, 2573, 6444), Australia (3020, 3720, 3721), Nepal (1826), Japan (3267), high elevations (>4000 m) in Peru (2005) and Mexico (3360). Some notable exceptions from this pattern are findings in Brazil (6007), South Africa (1555), Taiwan (2472) and Singapore (3331). — As pointed out in an earlier review (5223), *M. ramanniana* can be regarded as a typical inhabitant of forest soils; examples include an upland boreal forest (3962), deciduous forests (825, 2573) under beech (270, 271, 272, 1702, 2745, 3538, 4537), oak, hornbeam (1702), aspen (3959), and *Eucalyptus maculata* (1555), dry to mesic conifer-hardwood stands (2385), and old hardwood stands (2385), coniferous forests (5363) under pine (269, 1598, 1637, 2923, 3447, 6381, 6383), Douglas fir (6444), spruce (432, 5465), and larch (3267), forest nurseries (2482) and forest soil after prescribed burning (2822). It is also characteristic of podzols (989, 2502, 3414, 3416, 3652, 5534), particularly those with a well-defined pan (3721), also in terra fusca and loess (3414), tundra (1750), heathland soils (2736, 3720, 5047, 5221, 5222, 5223) and peat or peat bogs (432, 1039, 1376, 3119, 3918, 4474, 5559, 6082). It is further found in the humus layer (273, 1028, 2923, 4537) of mull and mor soils (2745, 4815), particularly in the deeper horizons (272, 2736, 3267, 3720, 5220, 5223, 6381, 6383). Spores can be transported by water down to 10 cm and more, but this kind of movement is unlikely to determine its distribution in soil (2410). Isolations have also been made from acid sand dunes (745) and in general there is a distinct prevalence in acid habitats (269, 861, 1555, 1713, 3331, 3360, 5223); the pH range of its habitats is given as 3·0–5·4 (1712, 5363, 6588). *M. ramanniana* has been frequently isolated from litter including that of Scots pine (2343, 2344), Norway spruce (3786), *Picea sitchensis, Fagus crenata* (4989, 4990), oak, hornbeam, hazel (4817), *Eucalyptus regnans* (3483), *E. maculata* (1558), *Pteridium aquilinum* (1821), and *Pseudoscleropodium purum* (2969). Other reports include the phylloplane of *Brachypodium pinnatum* (2971) and *Carex paniculata* (4644), roots of Ericaceae (2950, 5220, 5223, 6416), beech (2272), spruce, pine and *Cryptomeria* (1657, 3159). It has also been found in the rhizospheres of oats, barley (3004), beans, cabbage (4451) and some forest plants (3142, 4816); a negative rhizosphere effect observed with Australian heathland plants (3142) should be ascribed to the var. *autotrophica* (1657). *M. ramanniana* has further been isolated from cultures of the collembolan *Hypogastrura* (1027), pellets of the diplopod *Glomeris marginata* (4154), dung of rabbits (6085) and feathers of free-living birds (2575). — The hyphae can be lysed by *Streptomyces* sp. (2800), and the response to general soil mycostasis has been found to be very high (2559).

The optimal growth temperature is between 15° and 25°C (1827). The optimal pH for growth and sporulation is reported to lie between 3 and 4 (5223). — D-Arabinose can moderately be utilized (3328), citrate and urea cannot serve as sole C sources (229). An ATP-dependent citrate lyase activity has been reported (230). *M. ramanniana* is well known as a test organism for thiamine or its thiazole moiety (except the var. *autotrophica*), but it is independent of biotin and accumulates that in the medium (1328). The utilization of starch (1750, 1827), cellulose derivatives and pectin has been described (1750, 2736, 3414, 3538); the degradation of usnic acid has been reported (322). Ethanol, acetic, oxalic (4509), pyruvic and α-oxoglutaric acids (4639) are produced on glucose-containing media. The antibiotic ramycin (1392), identical to fusidic acid from *Acremonium fusidioides* (q.v.), active against gram positive bacteria has been found in 2 out of 16 isolates of *M. ramanniana* (1393). The pigment β-carotene is produced after exhaustion of nitrogen in the growth medium

(229). — In greenhouse experiments, the growth of flax and tomato plants has been inhibited (1427). — Little growth of *M. ramanniana* occurs with 10% CO_2 in the atmosphere (784). In a chronically irradiated soil it was still present at intensities of 1400 R per day (2009). After heating soil samples to 70°C (not 80°), this species was isolated amongst the "heat-resistant fungi" (615).

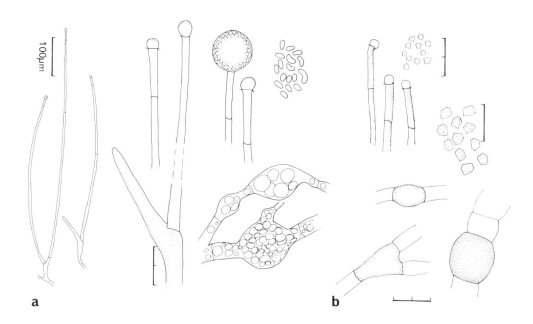

FIG. 192 a. *Mortierella ramanniana* var. *ramanniana*, sporangiophores with detail of tips with columellae and sporangiospores, and chlamydospores, CBS 219.47. — b. *Mortierella ramanniana* var. *angulispora*, sporangiophore tips, sporangiospores and chlamydospores, CBS 603.68.

Mortierella ramanniana var. *angulispora* (Naumov) Linnem. *1941* — Fig. 192 b.

≡ *Mucor angulisporus* Naumov *1935*

This variety differs from var. *ramanniana* mainly in the angular spores (as in *M. vinacea*). Colonies reaching 1·6 cm diam in five days on MEA at 20°C. The pigmentation, columella and chlamydospores are almost indistinguishable from those of var. *ramanniana*. — DNA analysis gave a GC content of 46% (5600).

According to the available data, this fungus has a worldwide distribution: it has been isolated from soils in Alaska (1171, 1750, 3062), Swedish Lapland (3360), Germany (6588), high altitudes in Austria (1876), Italy (3538, 3976), Romania (4379), Mexico (3360) and Peru (2005), and also from Canada (3962, 6352), the USA (1039, 1163, 6351), the USSR (2950), Australia (IMI), Japan (3267), and Brazil (6007, IMI). It has been reported from mixed forests (3962, 6351), coniferous forests (3267, 6352), conifer swamps (1039), beech forests (4537), soils with tundra vegetation (1750), recently deglaciated soil (1171), arable soil under sugar cane (6007, IMI), soils treated with sewage sludge (1163), and an acid mine drainage stream (1166). The pH range of its habitats is given as 4·1–5·5 (6588). It has been isolated from the rhizospheres of barley and oats (3004) and roots of *Pinus sylvestris* (3572). — It can be parasitized by *Dimargaris verticillata* (2817) only if thiamine is supplied.

M. ramanniana var. *angulispora* utilizes pectin and starch (1750); it is either thiamine-independent (1328) and completely auxo-autotrophic (1329) or only slightly stimulated in its growth by this substance (1656). It has been found to accumulate biotin in the medium (1328).

Mortierella verticillata Linnem. *1941* — Fig. 193.

= *Mortierella marburgensis* Linnem. *1936* (nomen inval., Art. 36)
= *Haplosporangium fasciculatum* Nicot *1957*

Nomenclature according to Gams (1883)

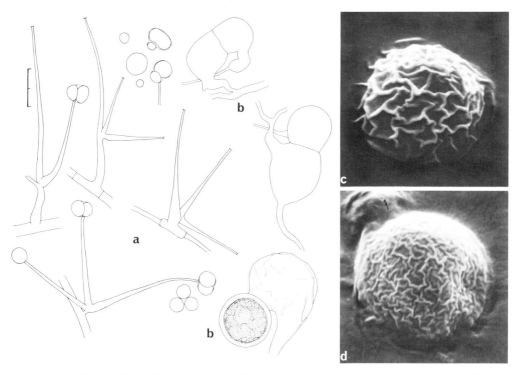

Fig. 193. *Mortierella verticillata*. a. Sporangiophores with few-spored sporangia and sporangiospores; b. various stages of zygospore formation, CBS 130.66 × 131.66; c,d. sporangiospores showing wrinkled surface, SEM, × 5000, CBS 380.66, orig. K. M. Old.

DESCRIPTION: Linnemann (3359, 6588); zygospores: Williams *et al.* (6382), and Chien *et al.* (1003). — Section *Stylospora* Linnem. — Colonies reaching 7·2 cm diam in five days on MEA at 20°C. Sporangiophores more or less verticillately branched, arising with or without a short stalk from the aerial hyphae, 50–120 μm long, with plurisporous or 1-spored sporangia, 5–10 μm diam. A wrinkled spore surface seems to be due to the loosely attached sporangium wall as seen in SEM (1003, Fig. 193c, d). — In the original isolate of *M. verticillata*, a few 2-spored sporangia were observed (1003) thus leaving no doubt about the conspecificity with *M. marburgensis*. *M. verticillata* is heterothallic and forms thick-walled, naked zygospores on 1% chitin agar or OA at 14–23(–25)°C (6382) as well as on hempseed and "Pablum" agars (3160). It is interfertile with *M. humilis* (1003) and is therefore placed in the section *Stylospora* (1883) in spite of the mostly several (up to 10) spores in each sporangium.

M. verticillata is rather common in forest soils in the temperate zone (2385, 2573, 4445, 6588). It has been recorded from the Alaskan arctic (3062), northern and central Europe (3360, 3416), the British Isles (745, 4445, 5220, 5221, 5222), Poland (6520), the USSR (2950, 3807), the USA (2385), Mexico (3360) and Italy (6082). Besides its occurrence in forest soils it is known from heathland soils (5220, 5221, 5222), forest steppe and steppe (2950), peat bogs (6520), podzols (3416) and sand dunes (745). *M. verticillata* has been found on litter of pine (2344) and spruce (3786), the rhizosphere of wheat in a crop rotation (2950), roots of *Pinus taeda* (3159) and *P. nigra* var. *laricio*, and heath plants (5220). It has been grouped as an acidophilous fungus (6082). By the soil washing technique it was exclusively isolated from organic soil particles in England (307) and has also been detected in organic detritus in fresh water (4429).

Degradation of chitin has been reported (2068).

Mortierella vinacea Dixon-Stewart *1932* — Fig. 194.

DESCRIPTIONS: Dixon-Stewart (1410), Linnemann (3359, 6588), and Turner (5948). — Subgenus *Micromucor* W. Gams. — Colonies reaching 1·5–1·7 cm diam in five days on MEA at 20°C, velvety, whitish to pink to russet-vinaceous depending on the density of the sporangia. Sporangiophores *c.* 100(–200) μm long, usually with whorled or corymbose branching, almost without a columella. Spores isodiametrical, sharply angular, 3–5(–8) μm diam. Small chlamydospores, *c.*10 μm diam, sparingly produced. — This species is distinguished from the rather similar *M. ramanniana* var. *angulispora* by the weaker pigmentation of the colonies, less pronounced columellae and the absence of large lipid-rich chlamydospores.

M. vinacea is a very widely distributed species, above all in acidic forest and cultivated soils (3360, 3817, 6588) and is reported for Alaska (1750, 3324), Scandinavia (3360, 3416, 5465), the USSR (2950, 3807), alpine sites in the Rocky Mountains (4711) and the Alps (1876), high altitudes (to 3700 m) in Mexico (3360), and other soils in the USA (2573), Canada (3959, 5363), Turkey (4245), Australia (1410) and Japan (5856). Reports from forest soils relate to deciduous (273, 989, 2573, 2745, 3959, 5811), coniferous (505, 989, 4445, 5363, 5465, 5813) and especially mixed forests (1030, 1032, 1204, 2004, 2008, 3817, 6073, 6350), forest nurseries (1978, 6184), and forest soils after 20 years of prescribed burning (2822). Its occurrence in podzol soils has been observed repeatedly (3416, 3571, 3871, 3962) but *M. vinacea* also occurs in peat bogs (1039, 3360, 6520), under *Calluna* heath (5811), Australian bush and heath (1410, 3720), tundra type vegetation (1750), acidic grassland (6182), in arable soils (1410, 1631, 2482) and on washed soil particles of a wheatfield (1889). The recorded tolerated pH range of soils extends from 3·4 to 6·0(–6·5) (3871, 5363, 6182, 6588) and the optimum has been indicated as pH 4·5 (1204). Considerable soil depths down to the C-horizon can be colonized (3720, 4445). Seasonal variations in occurrence were found to occur in Wisconsin, with the lowest frequency in April (1030). *M. vinacea* is often detected in the root regions of various plants, including *Pinus densiflora* (penetrating the roots in a manner reminiscent of a mycorrhiza and also penetrating carpophores of *Armillaria matsutake*) (5856), *Pinus taeda* (3159), spruce (3573), poplar (5898), barley (4450), cabbage (4451) as well as some Australian heathland plants (5819). A detailed investigation into its occurrence on beans showed there to be no accumulation in the rhizosphere (4451, 4452, 4453), the hyphae either growing only externally on the roots (5759) or being entirely absent from the root surface (4453). This fungus has been isolated mainly from mineral particles of the rhizosphere of dwarf beans (5801), but can also be obtained from organic ones (307). — An increased occurrence was encountered with lower humidities (5758), although the optimum water potential in soil is reported as –0·2 bar (5226); rapid recolonization of the soil by *M. vinacea* following steam sterilization has been observed (6184).

The temperature optimum for growth has been given as 30°C and the maximum as 37°C (1204). — Starch and pectin are utilized by this species (1750), which is autotrophic for thiazole (5948). ATP-dependent citrate lyase activity is moderate (230), a-galactosidase has been isolated from *M. vinacea* in a crystalline form (5657) and is used to hydrolyse raffinose in beet molasses (3059). Cellulose is readily degraded in the presence of high levels of a N source (4434). Hexadecane and solid paraffins are weakly utilized (3810). The influence of the culture medium on fat synthesis has been investigated (987). Protoplasts of *M. vinacea* were obtained from the mycelium with lytic enzymes from a *Streptomyces* sp. (4484). *M. vinacea in vitro* is able to inhibit the growth of *Cochliobolus sativus* and *Rhizoctonia solani* (3571). — After heating soil samples to 70°C (not 80°C), this species was isolated amongst the heat-resistant fungi (615). In a chronically irradiated soil no viable cells were found at intensities of ⩾53 R per day (2009).

Mortierella wolfii B.S. Mehrotra & Baijal *1963* — Fig. 195.

DESCRIPTIONS: Mehrotra and Baijal (3735), Evans (1664), Di Menna *et al.* (1399), Linnemann (6588), and CMI Descriptions No. 530, *1977*. — Section *Spinosa* Linnem. — Colonies reaching 1·6 cm diam in five days at 20°C, but 6 cm diam in three days at 30°C on MEA. Sporangiophores 80–250(–400) μm tall, 6–15(to >20) μm wide at the base, with a compact cluster of short acrotonous branches. On dehiscence of the sporangia, a conspicuous flaring collarette, mostly without a trace of a columella, is formed. Spores short-cylindrical, with a double membrane, 6–10 × 3–5 μm. Chlamydospores in some isolates abundant, to 35 μm diam, with numerous amoeba-like appendages similar to those of *M. exigua* (q.v.). — Because of this latter feature and the acrotonous branching, we regard this species, together with *M. pulchella* Linnem. *1941*, as small representatives of the section *Spinosa*.

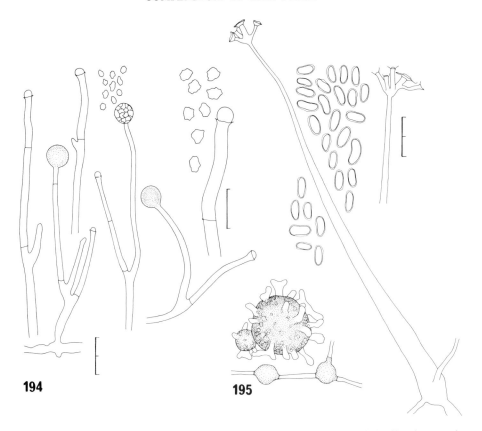

FIG. 194. *Mortierella vinacea*, sporangiophores with sporangia and small columellae, with detail and sporangiospores, CBS 561.63.
FIG. 195. *Mortierella wolfii*, sporangiophores with acrotonous branching, sporangiospores, and amoeba-like and smooth chlamydospores, CBS 612.70.

M. wolfii was first isolated from sandy soils in Rajasthan/India (3735) and has regularly been found in overheated, spoiled silage of pH 8–9, where it occurs as a secondary invader but does not survive the heating period (247). It has also been isolated from coal spoil tips with temperatures above 30°C (1664, 1665) and rotten hay in New Zealand (CBS).

This is the only *Mortierella* species that grows well at 40–42°C (1665), more restrictedly at 20°C (1399) and has a maximum growth temperature of 48°C (1665). — Because of this temperature adaptation, it is the most pathogenic species of the genus to mammals. Apart from *Absidia corymbifera* and *Aspergillus fumigatus*, *M. wolfii* is the most important causal agent of mycotic abortion, pneumonia and systemic mycosis in cattle in New Zealand (912), but has also been observed once in England (2265). Spores injected intravenously into cows caused the same disease and proved lethal to mice and rabbits (1399) but not sheep (1185, 1399). — From the washed mycelia a nephro-endotoxin was isolated (1290), the toxic fraction of which contained 60% lipid and 39% protein; the proline-rich protein is the active principle in its mycotic behaviour (1289).

Mucor Mich. ex St.–Am. *1821*

= *Chlamydomucor* Bref. *1889*

Type species: *Mucor mucedo* Mich. ex St.–Am.

Contributions to a MONOGRAPHIC TREATMENT: Schipper (5097, 5098, 5099, 5101), and Scholer (5125). A comprehensive key to the recognized species is given by Schipper (5101). A survey of the physiology of the sexual state is provided by Van den Ende and Stegwee (1628) and one of differentiation given by Gooday (2026).

The fam. Mucoraceae Dumort. is characterized by the occurrence of exclusively many-spored sporangia with well developed columellae, and no branches ending in sterile spines. Of the 18 accepted genera in the family, *Mucor* is generally regarded as the most primitive. Colonies are very fast growing (filling a Petri dish in 3–5 days at 20°C) and often several cm high, white to yellow and later often becoming dark grey with the development of the sporangia. Sporangiophores forming a dense mat, erect, without basal rhizoids, ± sympodially or sometimes racemosely (not dichotomously) branched, bearing terminal many-spored sporangia without apophyses, with a large columella; wall soon deliquescing or persistent and rupturing at maturity, often covered with spines of calcium oxalate; sporangiospores hyaline, greyish or brownish, smooth-walled or finely ornamented, globose to ellipsoidal. Hyaline chlamydospores present in some species. Submerged hyphae may form yeast-like cells which are often called oidia. Zygospores produced in the aerial mycelium between compatible mating types, dark brown, with characteristic ± stellate warts (5102); suspensors equal; mostly heterothallic (physiologically dioecious), rarely homothallic.

The closest genus to *Mucor* is *Parasitella* Bain. *1903*, a facultative hyperparasite of other Mucoraceae, which has similar sporangiophores but finger-like processes on the zygospore suspensors. *Actinomucor* Schostak. (q.v.) and *Rhizomucor* Lucet & Cost. (q.v.) are both distinguished by the presence of rhizoids and stolons; the former has whitish tall colonies and is mesophilic, whilst the latter has dark grey low colonies and is thermophilic. See also *Amylomyces* Calmette *1889* which is mentioned under *Rhizopus*.

Most published arrangements of the species of *Mucor* are unsatisfactory as they do not conform with observations on sexual compatibility. In contributions to a more modern system, isolates that are interfertile and produce normal zygospores with each other have been regarded as conspecific even if they differ morphologically (5097, 5099); in the latter case they have been distinguished only as formae (although a somewhat higher rank might be justified in some cases). Forty-nine taxa are thus recognized (5101). The elucidation of the genetic background of speciation would require the analysis of germinating zygospores. Care must be taken not to confuse interspecific azygospores with zygospores (2026). Consequently, most of the data compiled here have to be treated with considerable reservation since misidentifications might be very numerous. An analysis of eleven homologous enzymes in 31 isolates was used to assess their genetic distance and yielded a dendrogram very different from the present key (5606); it does not seem to provide an insight into taxonomic relationships.

As in all the Mucorales, the cell walls consist mainly of chitin. *Mucor* species commonly occur on organic matter, particularly dung lying on the ground, but do not generally penetrate to great soil depths. They are regarded as true coprophilous fungi and the sporangiospores can survive the intestinal passage of rabbits (6241). Sporangiospores are usually distributed by rain splash, with the exception of *M. plumbeus* which is commonly air-borne.

Colonies are grown in Petri dishes (often 5 cm high or more) on 4% MEA, whey agar, cherry decoction agar or PDA at defined temperatures, mostly in darkness. The maintenance of certain temperatures and the composition of the media are crucial conditions for obtaining zygospores; these conditions are different for each species.

Key to the species treated:

1   Stolons and rhizoids present; thermophilic; colonies grey to grey-brown, less than 5 mm high
cf. *Rhizomucor pusillus* (p. 700)
Stolons and rhizoids absent; not thermophilic; colonies often more than 5 mm high   **2**

2(1)   Sporangiophores often over 20 μm wide and sporangia often over 80 μm diam, with spinulose walls; sporangia often produced at different levels, sporangiophores either tall and little branched or very short and much branched   **3**
Sporangiophores generally less than 20 μm wide and sporangia less than 80 μm diam, with spinulose or smooth walls; sporangia usually not produced in distinct levels   **7**

3(2)   Sporangiospores spherical and 3·5–7 μm diam when grown at 20°C, ellipsoidal and 5·5–8 × 3–4 μm at 10°C; columellae subglobose (at 20°C) to ellipsoidal or slightly pyriform (mostly at 10°C)   (*M. strictus*)
Sporangiospores ellipsoidal (irrespective of the temperature); columellae obovoid, ellipsoidal or commonly pyriform   **4**

4(3)   Tall sporangiophores usually unbranched above; colonies grown in darkness with a second level of short, much branched sporangiophores; columellae obovoid, ellipsoidal to pyriform   **5**
Tall sporangiophores usually sympodially branched; with or without a second level of short sporangiophores; columellae generally obovoid   **6**

5(4)   Colonies growing rapidly on cherry decoction agar or acid MEA; sporangiospores mostly not exceeding 10 μm in length   *M. piriformis* (p. 474)
Colonies growing more restrictedly on cherry decoction agar or acid MEA than on neutral media; sporangiospores often exceeding 10 μm in length   *M. mucedo* (p. 472)

6(4)   Colonies grown in darkness pale to dark grey; a layer of short sporangiophores commonly present; sporangiospores 4·5–8 × 3·0–5·5 μm   (*M. saturninus*)
Colonies grown in darkness olive-buff; a layer of short sporangiophores rarely occurring (if present producing larger sporangiospores than the tall sporangiophores); sporangiospores 7–12 × 4–6·5 μm   (*M. flavus*)

7(2)   Columellae usually with apical projections; sporangial wall spinulose, rupturing at maturity; sporangiospores globose, brownish, punctate   *M. plumbeus* (p. 475)
Columellae without apical projections; sporangial wall spinulose or smooth-walled, rupturing or deliquescing; sporangiospores ellipsoidal or globose, hyaline to greyish, smooth-walled   **8**

8(7)   Sporangiophores repeatedly, mostly sympodially branched, to 20 μm wide; columellae ellipsoidal, cylindrical to pyriform (rarely globose); sporangiospores subglobose to ellipsoidal   **9**
Sporangiophores unbranched or scarcely sympodially branched, to 12(–15) μm wide; sporangial wall ± smooth, deliquescing; columellae globose or sometimes also oval; sporangiospores generally not globose   **15**

9(8) Sporangiophores branched in a sympodial fashion; chlamydospores generally absent in the sporangiophores; columellae ellipsoidal, obovoid or globose; sporangial wall slightly spinulose, mostly deliquescing; sporangiospores short-ellipsoidal (l/w 1·3–1·6) to subglobose, hyaline, about 4–6 μm diam **10**

Sporangiophores branched in a mixed sympodial and monopodial fashion; chlamydospores often abundant in the sporangiophores; sporangial wall spinulose, breaking at maturity; columellae ellipsoidal with a truncate base to pyriform, rarely subglobose; sporangiospores short-ellipsoidal or subglobose (l/w 1·1–1·3), greyish and to 8–10 μm diam **13**

10 (9) Colonies greyish; sporangiophores to 7(–10) μm wide; sporangia black **11**
Colonies brownish; sporangiophores to 15 μm wide; sporangia brownish to brownish grey **12**

11(10) Sporangiospores subglobose, 3·5–5·5(–6) μm diam; sporangial wall persistent, breaking at maturity (*M. circinelloides* f. *janssenii*)
Sporangiospores ellipsoidal, 4·5–6·0(–8) × 3·8–4·5 μm; sporangial wall deliquescing
M. circinelloides f. *griseo-cyanus* (p. 466)

12(10) Columellae globose; sporangiospores ellipsoidal or irregularly shaped, 5·5–17·5 × 3·5–12·5 μm
(*M. circinelloides* f. *lusitanicus*)
Columellae obovoid; sporangiospores ellipsoidal, 4·4–6·8 × 3·7–4·7 μm
M. circinelloides f. *circinelloides* (p. 463)

13 (9) Sporangiospores mostly (sub)globose, 4–8 μm diam   *M. racemosus* f. *sphaerosporus* (p. 480)
Sporangiospores broadly ellipsoidal, 5·5–8·5(–9·5) × 4–7 μm **14**

14(13) Columellae obovoid to broadly pyriform, to over 50 μm high; monopodial branches short (to 200 μm)   *M. racemosus* f. *racemosus* (p. 477)
Columellae subglobose, obovoid or ellipsoidal, to 20 μm high; monopodial branches often long (to *c.* 400 μm)   (*M. racemosus* f. *chibinensis*)

15 (8) Sympodial branches arising a short distance below the sporangia; sporangia blackish brown; sporangiospores short-cylindrical, 3·5–5·2 × 2·5–3·7 μm   *M. hiemalis* f. *silvaticus* (p. 470)
Sympodial branches arising a long distance below the sporangia; sporangia yellowish to dark brown; sporangiospores mostly ellipsoidal **16**

16(15) Sporangia yellowish; columellae globose; sporangiospores narrowly ellipsoidal (l/w 2·6–3·0), 3·3–9·5 × 1·4–4·0 μm   (*M. hiemalis* f. *luteus*)
Sporangia brownish to dark brown; sporangiospores somewhat broader **17**

17(16) Sporangiospores cylindrical-ellipsoidal, 4·5–6·8 × 2·5–3·5 μm   *M. hiemalis* f. *corticola* (p. 470)
Sporangiospores ellipsoidal, sometimes flattened on one side, (4–)5·5–8·5 × 2·5–5·5 μm **18**

18(17) Homothallic   (*M. genevensis*)
Heterothallic   *M. hiemalis* f. *hiemalis* (p. 466)

Mucor circinelloides van Tiegh. *1875* f. *circinelloides* — Fig. 196 a.

= *Mucor ambiguus* Vuill. *1886*
= *Mucor alternans* van Tiegh. *1887*
= *Mucor javanicus* Wehmer *1900*
= *Mucor prainii* Chodat & Nechitch *1904*
= *Mucor griseo-roseus* Linnem. *1936*

DESCRIPTIONS: Hagem (2206), Schipper (5099), and Raizada (4708); zygospores: Schipper (5094). — Colonies on MEA at 20°C 6–20 mm high, in the dark pale grey-olivaceous but in the light olivaceous-buff with an apricot reverse. Sporangiophores to 17 μm wide, differentiated into two levels (mainly in the dark): tall ones repeatedly sympodially branched with long and short branches, infrequently circinate; short sporangiophores profusely branched, often with circinate branches and slightly encrusted walls. Sporangia at first whitish to yellowish but later becoming brownish grey, to 80(–100) μm diam, wall slightly spinulose, deliquescent on tall, but persistent on short sporangiophores. Columellae either large and ovoid to ellipsoidal or small and ± globose, dilute brownish grey, to 50 μm diam. Sporangiospores ellipsoidal, 4·5–6·5(–9) × 3·5–5·0 μm. Chlamydospores formed in and on the substrate, few in number. Heterothallic. Zygospores formed on whey agar at 25°C, mostly close to the substrate, reddish brown to dark brown, to 100 μm diam, with stellate spines to 7 μm high; suspensors slightly unequal, very short (5102). — Three morphologically distinct but interfertile forms have been distinguished (cf. key): f. *lusitanicus* (Bruderlein) Schipper *1976* (≡ *M. lusitanicus* Bruderlein *1916* = *M. griseo-lilacinus* Povah *1917*), f. *griseo-cyanus* (Hagem) Schipper *1976* (q.v.), and f. *janssenii* (Lendner) Schipper *1976* (≡ *M. janssenii* Lendner *1908* = *M. tenellus* Ling Young *1930* ≡ *Circinella tenella* (Ling Young) Zycha *1935* = *M. kursanovii* Milko & Beljakova *1967*). — Sporangiophore morphogenesis was compared with that of *Rhizopus oryzae*; in contact with the substrate, sporangiophores regenerate vegetative hyphae (1926). — DNA analysis gave a GC content of 43% (5600). Mycelial wall hydrolysates were found to contain 24·3% hexosamine (3232) and relatively large amounts of polyphosphate (1390). — Hyphae, oidia and sporangiospores have been studied by TEM (5703, 5704, 5705).

This fungus was found to occur in most climatic zones including top soil of permafrost areas in Alaska (3021), soils in Canada (6352), the USA (2740, 5671), Manchuria (5099), the USSR (2871, 4474), Finland (5099), Poland (1421, 6520), Czechoslovakia (1702), Germany, the Netherlands (5099), Denmark (2741), the British Isles (413, 861), France (2161, 5099), Yugoslavia (4573), Turkey (4245, 5099), Egypt (8, 4962), Israel (2764), Iraq (92), Kuwait (4001), Tunisia (4055), South Africa (1555, 1558, 4407, 5099), Pakistan (4855), Nepal (1826), India (3730, 4371, 4933, 5058), Indonesia (5099), Japan (3267, 5099), and Central America (1697). It has been isolated from forest soils (2741) under pines (6352), larch (3267), oak (1702), and *Eucalyptus maculata* (1555, 1558), cultivated and uncultivated soils (2161, 2740, 2764, 4962, 5671), grassland (4933), soils with steppe (1702) or savannah type vegetation (4407), garden soil (3730), peat bogs (4474, 6520), children's sandpits (1421), salt-marsh (8, 413, 4001), and sand dunes (861, 4371). Its frequency in the soil is reduced by the application of captan, TMTD and allylalcohol (1426). It has frequently been isolated from waste materials as in manure pits (968), composted urban waste (3041), garden compost (1424), chicken manure (578), dung of some arctic animals (4174), hawk pellets (6223), but is also known from the rhizospheres of barley, oats (3004) and various sand dune plants (4371), wheat seeds (2716, 4492), various decaying plant materials (3113) and moist hay (4548); it has also been found in the air in Norway (2206) and Taiwan (3377), and in river water (3809).

Growth and sporulation occur in the range 5–30°C; at 37°C growth is very poor and at 5°C only short sporangiophores are formed (517, 5099); optimal mycelium production in liquid culture occurs at 25°C (3736). — Thiamine is not required (4860), and starch and protein are degraded (1827). It can grow on and utilize 0·25% tannic acid (3052). The production of lipids (3736), and extracellular lipase (2444, 5309), phosphatase (5309), riboflavin-α-glucoside (5662), and D-glucose, D-galactose, and D-mannose as constituents of intra- and extracellular polysaccharides (2049, 3640) have been reported. *M. circinelloides* f. *circinelloides* can convert DDT into dechlorinated water-soluble metabolites (134, 2827) and inhibit the growth of *Chalara elegans in vitro* (2753). — This fungus is considered to play a role in phycomycoses of man (1624), cattle, pigs and poultry (3932) and is also reported to spread in internal organs of frogs causing their death (1817). — *M. circinelloides* f. *circinelloides* tolerates 3–10% sucrose (4001) or even 15% NaCl (5881) in the medium. Under anaerobic conditions it assumes a yeast-like growth (3319).

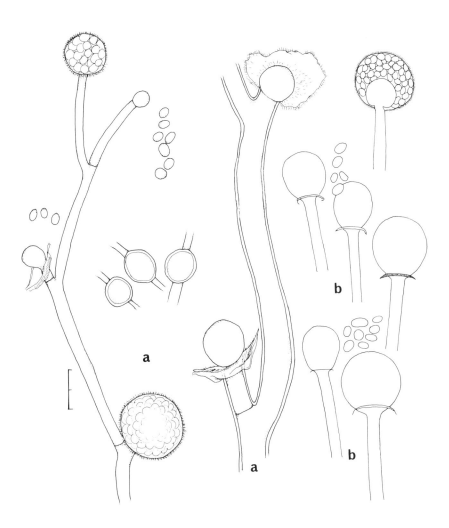

FIG. 196 a. *Mucor circinelloides* f. *circinelloides*, sporangiophores with sporangia and columellae, sporangiospores and chlamydospores. — b. *M. circinelloides* f. *griseo-cyanus*, narrower sporangiophores with an intact sporangium, columellae and sporangiospores, CBS 198.29.

Mucor circinelloides f. *griseo-cyanus* (Hagem) Schipper *1976* — Fig. 196 b.

≡ *Mucor griseo-cyanus* Hagem *1908*

DESCRIPTIONS: Hagem (2206), and Schipper (5095, 5099); zygospores; Schipper (5094). — Colonies on MEA at 20°C 6–13 mm high, pale neutral grey in the dark (in the light olive-buff with yellow reverse). Sporangiophores to 6(–10) μm wide, sympodially or monopodially branched. Sporangia at first pale yellowish, later becoming brownish black to black, to 60 μm diam (rarely larger), wall slightly spinulose, deliquescing. Columellae either large and ovoid or small and globose, greyish, to 35 μm diam. Sporangiospores irregular in shape and size, oval or broadly oval, a few globose, 4·5–6·0(–8·0) × 3·8–4·5 μm, smooth-walled. Chlamydospores and "oidia" occurring in the submerged mycelium. Heterothallic. Zygospores formed on PDA at 25°C, to 80 μm diam, with stellate projections to 5(–7) μm high (5102). — *Mucor janssenii* Lendner *1908* (cf. key) was once regarded as a form of *M. griseo-cyanus* (5095) but both taxa have subsequently been subsumed under *M. circinelloides* (5099). — DNA analysis gave a GC content of 44% (5600).

M. *circinelloides* f. *griseo-cyanus* has been reported from Norway (2206), Ireland (1376), Poland (269, 272, 273, 1421), Yugoslavia (4573), Italy (3913), Portugal (IMI), Egypt (4962), Iraq (92), Uganda (IMI), South Africa (5095), and the USA (2573, 5095, 5671). It has been isolated from deciduous and coniferous forests (269, 272, 273, 2573), cultivated and uncultivated soils (4962, 5671), rice fields (3913), peat (1376, 3119), children's sandpits (1421), and an ant hill (IMI). It has also been found in the rhizospheres of barley and oats (3004), stored wheat (4492), nests of free-living birds (2575), and the air (IMI).

Optimal growth occurs in the temperature range 20–30°C, slow growth at 10–15°C (minimum 5°C) and none at 37°C (5095). — It also sporulates in submerged culture on malt extract medium (6088). — It shows a weak phosphatase activity (5309). Several steroids are hydroxylated (1644) and aflatoxin B_1 can be transformed to a blue-fluorescent compound (1359). — Culture filtrates of this fungus were found to have a stimulatory influence on the growth and reproduction of tomatoes (15).

Mucor hiemalis Wehmer *1903* f. *hiemalis* — Fig. 197 a.

? = *Mucor subtilissimus* Oudem. *1898*
? = *Mucor adventitius* Oudem. *1902*
  = *Mucor lausannensis* Lendner *1908*
  = *Mucor vallesiacus* Lendner *1918*

DESCRIPTIONS: Hagem (2206), Christenberry (1028), and Schipper (5097); zygospores: Schipper (5094), and Spalla (5489). — Colonies on MEA at 20°C to 15 mm high, buff, reverse (pale) luteous in the light (greyish in darkness). Sporangiophores to 12(–14) μm wide, slightly sympodially branched; sporangia at first yellowish, later becoming dark brown, to 70 μm diam, with deliquescing walls; columellae globose when young, later ± ellipsoidal, to 38 × 30 μm, sometimes with yellowish contents; sporangiospores ellipsoidal, sometimes flattened at one side, $5 \cdot 5–8 \cdot 5 \times 2 \cdot 5–5 \cdot 5 \mu$m. Oidia present in submerged hyphae. Chlamydospores absent. Heterothallic. Zygospores formed on MEA or PDA at 15°C throughout the aerial mycelium but mainly near the agar surface, to $70(–100) \mu$m diam, blackish brown, covered with stellate spines to $3 \cdot 5–4 \mu$m high (5102), suspensors ± equal. Azygospores have been observed in some diploid heterozygous single-spore isolates (1929). Septation of mycelia occurs soon after germination of sporangiospores (1735). — *M. hiemalis* is a highly variable species. Eight morphologically slightly different strains (including ones called *M. subtilissimus*) have been differentiated in Japan (4065, 4066). Three interfertile forms have been distinguished nomenclaturally (cf. key) from f. *hiemalis*: f. *silvaticus* (Hagem) Schipper (q.v.), f. *luteus* Schipper *1978* (≡ *M. luteus* Linnem. *1936*, nomen inval., Art. 36) and f. *corticola* (Hagem) Schipper (q.v.). The last two names listed above amongst the synonyms also refer to fungi slightly different from f. *hiemalis*. *M. genevensis* Lendner *1908* is very close to f. *hiemalis* but constantly homothallic. — The sporangial size was found to be temperature-dependent, reaching the maximum in the range 22–30°C (5789). A cytological study has been presented (4874) and mitosis has been followed in light and electron microscopy (3701). Changes in the distribution of nucleic acids, nuclei, mitochondria and reserve substances during initiation and development of zygophores and zygospores have also been observed (2334). — DNA analysis gave a GC content of $42 \cdot 5–43 \cdot 5\%$ (5600). — Fatty acids occur in a more saturated state in sporangiospores than in the mycelium; myristic, palmitic, palmitoleic, stearic, oleic, and linolenic acids have been identified (5643). In spite of considerable variation, the isozyme patterns determined for six enzymes in 30 isolates coincided with the present arrangement of several forms within *M. hiemalis* (2326).

M. hiemalis f. *hiemalis* is one of the commonest soil fungi and the most frequent representative of the Mucorales. On account of its rapid growth, it can be isolated easily by immersion tubes and immersion plate methods (982, 5220, 5812), from quite large soil crumbs (6138), buried chitin strips (2068) and washed soil particles (1888, 4449, 4452). — *M. hiemalis* f. *hiemalis* has a worldwide distribution; it extends from arctic and alpine soils to the tropics, reports being too numerous to cite individually here. It occurs in the most varied soils and forms of vegetation, including grassland, arable, and forest soils (especially the litter layer); in a pinewood soil it was exclusively isolated from washed organic particles (307). Other known and more extreme habitats include red lava stones on Surtsey (2406), estuarine sediments (655, 4477), even at high salinities (5316), salt-marsh (4646, 4987), peat bogs (1376, 3119, 3234, 3414, 4180, 6520), heathland (989, 3720, 5047, 5222, 5819), mountain forests and Páramo soil in Columbia at 3100–3900 m altitude (CBS), conifer swamps (1039), a copper swamp (2922), children's sandpits (1421), sand dunes (745, 3941, 6414), brown coal (4474), slate slopes (566), various podzol soils (1876, 3413, 3414, 3416, 3962, 6383), unpolluted (3809, 3868, 4429) or polluted (1154, 1157, 1162, 1166) streams, sewage sludge (1165, 1167, 1387), and fields treated with sewage sludge (1163). Increased frequencies have been reported after mineral fertilization (3498) and also after repeated cropping with rape (1433). After a forest fire its presence was higher in the H horizon of the burnt site in comparison to the unburnt (6353). *M. hiemalis* f. *hiemalis* is one of the least substrate-specific coprophilous Mucoraceae; reports include the dung of rabbits (2279, 6085), mice and cows (3730), feathers, nests and pellets of free-living birds (682, 2338, 2572, 6223), and pellets of the diplopod *Glomeris marginata* (4154). It is found in soils with a very wide pH range, though it appears to occur preferentially in neutral or slightly alkaline soils (159, 745, 3730, 4055, 6510). As with most Mucoraceae, it occurs predominantly in the uppermost soil layers, but occasionally reaches greater depths (1171, 3267), sometimes down to 100 cm (164). Its occurrence is apparently not limited to dry habitats, as it has been found as frequently in desert as in wet meadow soils (164, 4313). Seasonal variations, with a minimum in January, have been observed in Wisconsin (2004). — *M. hiemalis* f. *hiemalis* can stabilize soil aggregates (224, 225), but the mycelium is degraded in soil at a fast rate (3787). Forty-six per cent of chloroform-fumigated mycelium was mineralized within ten days at 22°C (133). Its frequency in the soil is reduced by application of captan, TMTD, nabam, vapam and allylalcohol (1426). — Its rapid growth allows the fungus to colonize fresh substrates rapidly, though it is soon superseded by other micro-organisms. This applies particularly to the colonization of roots, where considerable rhizosphere effects have been noted in young stages. *M. hiemalis* f. *hiemalis* has been observed in the root region of various trees (3142) including spruce (3573), *Pinus taeda* (3159), poplar (3452, 5898), and *Fraxinus* (3156), but also flax (5826), alfalfa (3982, 3983), French beans (4451, 4452), broad beans (6134), clover (3429, 5815), banana (2030), sugar cane (2861, 4886), various heathland plants (5220, 5819), forest herbs (4814), ferns (2856, 2858), moss (2860), *Ammophila breviligulata* (6414), wheat (4727), oats, barley (3004, 4443), *Lolium perenne* (2611, 5815), rice (1282), corn (4553) and some vegetables (3730, 3733). It has been isolated from stored wheat grain (4492), cotton seeds (5009), groundnut pods (3220), pecans (2572), stored sweet potatoes (3066), castor beans (3172), and corn silage (1092). The list of dead substrates colonized is very extensive and does not indicate any particular preferences. — Like most *Mucor* species, it can be parasitized by numerous obligate mycoparasites of the Piptocephalidaceae and Dimargaritaceae; while most species of *Piptocephalis* usually cause little harm to their host, *Syncephalis* species can considerably damage it (450); there are detailed studies on the parasites *Dimargaris verticillata* (3560) and *Syncephalis californica* (2618). — The fungus was inhibited *in vitro* by tannins (1206, 2282) and by the antagonistic action of *Trichoderma koningii*, *T. viride* (1344) and *Stilbella erythrocephala* (5362). The penetration of hyphae by some hyperparasitic fungi, for example *Fusarium solani*, can cause the formation of callosities in the sporangiophores (5668).

Growth and sporulation are good in the range 5–25°C; the optimum is 15–25°C, and no growth occurs at 30°C (832, 1204, 5097, 5789). On saline media, the production of zygospores and sporangiospores and germination of sporangiospores is restricted (832, 833); the percentage of germination of zygospores is in general low, after 30–90 days mostly <1%; 90% gave rise to (+)-cultures which showed more vigour than (−)-cultures (1928). Zygospore formation can be inhibited by a streptomycete metabolite (3011). The germination of sporangiospores takes place in the presence of oxygen (6436). Sporulation *in vitro* was found to be inhibited by aflatoxin (4806). The thermal death point has been determined as 63°C for 30 min in apple juice (3620); certain isolates are cryophilic and capable of growing at or below 0°C (2761, 3235, 5789). *M. hiemalis* f. *hiemalis* can grow *in vitro* at pH values between 2·5 and over 8·8 (2741). The sporangiophores are positively phototropic and attain a greater length in light than in darkness (1645). — Pectin (at pH 3·0–6·7) can be utilized as a C source with little variability in this ability occurring between isolates (1432, 3127, 3414, 4690). Urea can also be used as the sole C source; hemicelluloses (3414), starch (1827) and chitin (1425, 2068) are also decomposed. An endo-1,6-β-glucanase has been demonstrated (3894). Contrary to other species of its genus, nitrate cannot be utilized by the forms of *M. hiemalis* (4065). For optimal growth in liquid culture, the concentration of NH_4NO_3 was found to have particular importance (1808). No thiamine is generally required (5133) but its addition can stimulate growth (4639) and some isolates were found to be thiamine-dependent (4066). Relatively high activities of proteinase and phosphatase were observed (3049, 5309) as well as an amino acid oxidase (3467). *M. hiemalis* f. *hiemalis* has been used for investigations of β-carotene synthesis (2123, 5278). It produces the alkaloid ergosine (1609). Glucuronic acid, fucose, galactose and mannose have been found as components of extracellular polysaccharides

(3640). Ethylene (3465, 3466, 3467, 3468, 3469), ethanol, and acetic (4509), pyruvic and 2-oxoglutaric acids (4639) are produced. The herbicide paraquat is accumulated in the mycelium *in vitro* (5446). — *M. hiemalis* f. *hiemalis* has been found amongst the hyperparasites of sclerotia of *Sclerotinia borealis, S. sclerotiorum, S. trifoliorum,* and *Claviceps purpurea* (1652, 2886, 3531). It is reported to parasitize Lepidoptera, Coleoptera and Diptera species (6588), and to have antibacterial (1092) and antifungal (2753) activities. It can cause injuries to pine seeds on wounding (1975), and *in vitro* in pea seedlings (1430). Corn seedlings (4553) and barley shoots (3005) are said to be stimulated by *M. hiemalis* f. *hiemalis* and so is the sporulation of *Pilobolus kleinii* (5361). — The fungus tolerates $CO_2:O_2$ mixtures up to a ratio of 18:1 (2096), while growth can still be observed at 100% CO_2 (5602).

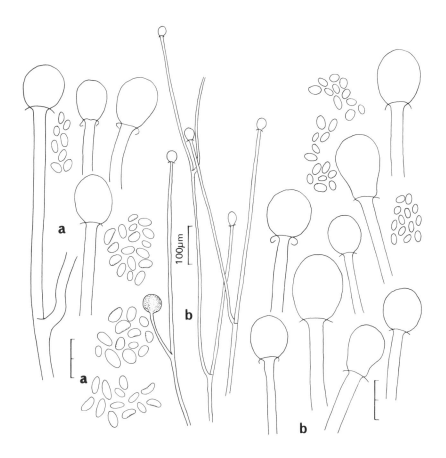

FIG. 197 a. *Mucor hiemalis* f. *hiemalis*, sporangiophore tips with columellae and sporangiospore samples from various isolates growing on different media. — b. *M. hiemalis* f. *corticola*, sporangiophores, sporangiophore tips with columellae and sporangiospores, CBS 534.78.

Mucor hiemalis f. *corticola* (Hagem) Schipper *1973* — Fig. 197 b.

≡ *Mucor corticolus* Hagem *1910*

DESCRIPTION: Schipper (5097); zygospores: Schipper (5094). — Characters as in f. *hiemalis*, but colonies to 18 mm high, rosy buff; sporangia dark brown, to 60 μm diam; columellae ellipsoidal, to 44 × 29 μm; sporangiospores mostly cylindrical-ellipsoidal, 4·5–7(–8) × 2·7–3·5 μm. Zygospores formed on PDA at 20°C, throughout the aerial mycelium, black, to 95(–110) μm diam, covered with irregularly stellate projections, 3·5–4(–7) μm high (5102). Azygospores also often present.

This form has been reported less commonly, and mainly from the temperate zone, including Norway, Denmark, the Netherlands (5097), the British Isles (1376, 2338), Poland (269, 272, 273), the USSR (3652, 5097), Germany (1889), Czechoslovakia (1702), Austria (1828, 5097), and the USA (2573); other reports come from Israel (4759), Taiwan (2472), Japan (3267), and Mexico (3360). It has been isolated from forest soils (1828, 2573) under pine (269), larch (3267) and beech (272, 273), soils under steppe type vegetation (1702), and peat bogs (1376). It was preferentially isolated from ranker and chernozem (3414) and from the litter and humus layers in forest soils (269, 272, 273, 3416). It has been found on *Peziza repanda* (2338), roots of clover (3429) and broad beans (6134), hay after self-heating (4821), horse dung (3377) and in river water (3809).

Growth and sporulation are good in the temperature range 5–25°C; at 30°C no sporulation, and at 37°C no growth occur (5097). — Pectin and urea are utilized (3414).

Mucor hiemalis f. *silvaticus* (Hagem) Schipper *1973* — Fig. 198.

≡ *Mucor silvaticus* Hagem *1908*

DESCRIPTIONS: Hagem (2206), and Schipper (5097); zygospores: Schipper (5094). — Characters as in f. *hiemalis* but colonies pale olivaceous-grey in both light and darkness; sporangiophores 7–8(–15) μm wide, with long sympodial branches originating a short distance below the previous sporangium; sporangia blackish brown, columellae globose, rarely ellipsoidal, to 38(–49) μm diam; sporangiospores cylindrical-oblong, 3·5–5·2 × 2·6–3·5 μm. Zygospores formed in greatest numbers on cherry decoction agar at 10°C or 15°C (2206), occurring throughout the aerial mycelium, to 70 μm diam, covered with irregularly stellate projections (5102). Azygospores generally also present in addition to the normal zygospores.

In common with *M. hiemalis* f. *hiemalis*, f. *silvaticus* can be considered a true soil inhabitant (4421). According to the available data, this form seems to occur most frequently in regions with a cool or temperate climate: Greenland (4174), northern Scotland (163), Norway (2206), Denmark (2745, 5097), Poland (272, 273, 1421), the USSR (2948, 4474, 5097), Germany (5097), Austria (3413, 4179), the USA (1032, 1039, 2385), Canada (3962), China (3475), and high altitudes in Peru (2005); but it has also been found in Yugoslavia (4573), Italy (3913, 4538), South Africa (1555, 1556), Japan (3267), and India (4736, 5058). It is mainly found in

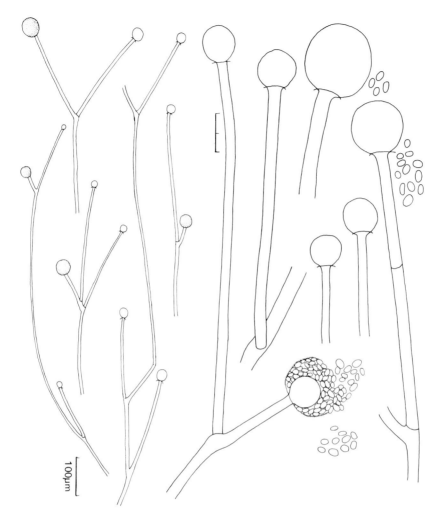

Fig. 198. *Mucor hiemalis* f. *silvaticus*, sporangiophores with long and some relatively short branches, sporangiophore tips with sporangium and columellae, and sporangiospores, mostly CBS 508.66.

soils rich in organic substrates such as forest soils (1032, 2206, 2948, 3267, 4179), the L- and H-layers of beech forests (272, 273, 2745), under *Eucalyptus* (1555), in peat (3119, 4474), heathland (5047), conifer swamps (1039), and grassland (163, 1555); it has rarely been isolated from cultivated soils (2948, 3913, 4538), found in children's sandpits (1421), the air (2206), and river water (3809). Other known substrates include brown coal (4474), roots of *Pinus taeda* (3159), decaying stems of *Pteridium aquilinum* (1821), litter of pine (3801) and *Eucalyptus maculata* (1558), corn silage (1092), and dung of various animals in the arctic (4174) and in Japan (3267). — It can serve as food *in vitro* for the psocopteran *Lepinotus reticulatus* (5381), and a number of mites (5379, 5380).

Growth and sporulation occur in the temperature range 5–25°C; at 30°C growth is restricted, and at 37°C is nil (5097). — *M. hiemalis* f. *silvaticus* has a relatively high lipase production (5309) and shows proteolytic activity (2948). Some antibacterial activity has been reported (1092).

Mucor mucedo Mich. ex St.–Am. *1821* — Fig. 199.

= *Mucor griseo-ochraceus* Naumov *1915*
= *Mucor murorum* Naumov *1915*
= *Mucor coprophilus* Povah *1917*

DESCRIPTION: Schipper (5098); zygospores: Schipper (5094); review of differentiation and life cycle: Gooday (2026). — Colonies on MEA at 20°C to 25(–50) mm high, smoke-grey in the dark, yellowish brown in the light; odour aromatic; in darkness differentiated into a sparse layer of tall, usually simple, to 40 μm wide sporangiophores and a denser layer of short, repeatedly branched sporangiophores (these latter absent in light). Sporangia on tall sporangiophores to 250 μm diam, brownish grey at maturity, wall spinulose, deliquescing; columellae obovoid to ellipsoidal, to 160 × 125 μm, sometimes with reddish brown contents; sporangia on short sporangiophores somewhat smaller, slightly darker, with persistent walls. Sporangiospores thick-walled, with granular contents, greyish, cylindrical-ellipsoidal to broadly ellipsoidal, 10·5–13·5 × 5·5–7·5 μm, to subglobose, 8–9 μm diam. Chlamydospores absent. Heterothallic. Zygospores formed on MEA at 15°C near the agar surface, black, to 250 μm diam, with irregular warts to 15 μm high (5102); suspensors ± equal, ± swollen, yellowish to brownish. — The closest species is *M. piriformis* Fischer (q.v.) which grows better on cherry decoction agar or MEA with pH 4 and has more regularly ellipsoidal sporangiospores (with no tendency to become cylindrical). — The phases of zygospore development have been described, zygospore walls contain mainly chitin and chitosan with black pigments and sporopollenin (2026, 2027), while the sporangiospore walls consist mainly of glucan; hyphal cell walls contained L-fucose, D-galactose and D-gluconic acid at a ratio of 3·0:1·0:2·1 (3890). More saturated fatty acids are contained in spores than in mycelia, while the total amount is smaller in spores; identified fatty acids include myristic, palmitic, palmitoleic, stearic, oleic, linoleic and linolenic acids (5643). — DNA analysis gave a GC content of 29·5% (5600).

Contrary to some literature reports, *M. mucedo* as presently understood (the reports compiled here have to be considered with considerable reservation) is relatively rare, but apparently has a worldwide distribution. — It has been reported from Alaska (1171), the British Isles (861, 2338, 4429), Norway (2206), Finland (1712, 5663), Poland (273), France (242), the Netherlands, Germany, the USSR, Japan (5098), the USA (660, 2385, 2573, 3817), Spain (3417), Algeria (2972), Libya (4084), Egypt (4962), equatorial West Africa (1420), South Africa (1415), India (4997, 5058, 5512), Nepal (1826), Indonesia (578), Taiwan (2472, 6476), and Australia (3720). In general it is more frequently found on dung (horse, cow, kangaroo, dog, ostrich, angora goat, rabbit, field mouse, and bat) than in soil (1028, 2206, 2338, 3358, 3882, 6085, 6588, IMI). The majority of soil isolates come from various forest soils, mainly hardwood (272, 273, 2385, 2573, 3138, 4573, 5512), in conjunction with leaf litter (269, 3448, 3449, 3720, 5513). It has also been found in forest nurseries (6184),

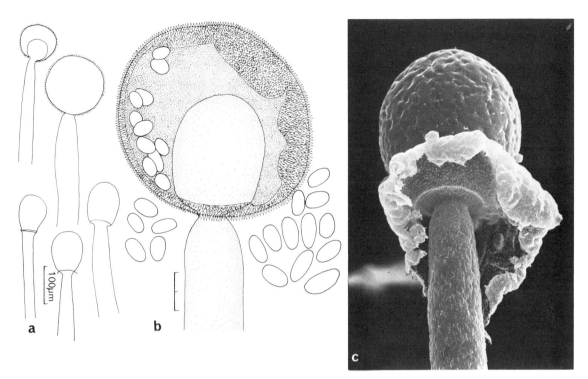

FIG. 199. *Mucor mucedo.* a. Sporangiophore tips with some (relatively small) sporangia and columellae; b. dehiscing sporangium with sporangiospores; CBS 228.29; c. sporangium after dehiscence, SEM, × 700, orig. R. A. Samson.

soils with steppe type vegetation (2972), heathland soils (3720), arable and garden soils (1420, 2740, 3817, 4962), vineyards (242), citrus plantations (2764, 4085), and various other cultivated soils (1028, 2762, 4759, 4788, 4997), terra rossa (3417), and a granite outcrop (660). The pH range in soil is given as 4·2–7 (1713). *M. mucedo* has also been isolated from the rhizospheres of *Aristida coerulescens* (4083) and *Juncus acutiflorus* (IMI), organic detritus in fresh water (4429), wooden planks in a bathhouse (5016), wood used in a copper mine (2664), stored wheat (4492), pellets of free-living birds (2575), and mummified honey-bees (402).

Growth and sporulation occur in the range 5–25°C, with a minimum at or below 0°C, the optimum at 22°C, and no growth at 30°C; at 15°C and below it forms short, recurved branches in the sporangiophores and has somewhat larger spores (5098, 5789). — *M. mucedo* is the most commonly used test organism in studies of sexual processes in Mucorales (2026). Trisporic acids B and C (identical to those in *Blakeslea trispora* and many other Mucorales) act on both (+) and (−) mating types and induce progametangium formation (780, 1628, 1629). Zygotropism is induced by volatile factors similar to trisporic acid precursors (1628, 3780), which are found to consist of 4-hydroxy methyltrisporate in the (+), and trisporins in the (−) mating type (4185, 6295); further factors determining zygospore formation were analysed using induced mutants (6456). Sexual vigour is lost by frequent transfers, but loss can be prevented by cold storage or changing temperatures (20:3°C) (1628). The production of sexual hormones depends on a sufficient supply of O_2 and occurs only at a certain age of the mycelium (2025). — Sporangiophore formation is stimulated in blue or white light (3285). The thermal death point was determined as 63°C for 25 min in apple juice (3620). Zygospore formation can be inhibited by a streptomycete metabolite (3011). — Growth in liquid culture can be supported by volatile compounds derived from germinating seeds of beans, cabbage, corn, cucumber and peas (5091). Dicarboxy-amino acids are suitable N sources; fatty acids can be utilized; an extracellular amylase is formed adaptively with optimum production at pH 5; protein is weakly degraded (5564). Pyruvic, 2-oxoglutaric (4639), acetic and oxalic acids, ethanol and ethylene are produced (2639, 4509); also (amongst others) an extracellular heteroglycan composed of fucose, mannose, glucose, galactose, N-acetyl glucosamine and N-acetyl galactosamine (3892). Lipase activity can be observed in resting cultures, with a pH optimum at 6·4 and a temperature optimum at 38°C (5564). *M. mucedo* is autotrophic for thiamine (5133). It is said to produce precursors of humic substances (4470). Experiments *in vitro* have revealed the formation of water-soluble metabolites when grown with DDT (134).

Mucor piriformis Fischer *1892* — Fig. 200.

= *Mucor wosnessenskii* Schostak. *1898*
= *Mucor albo-ater* Naumov *1915*
= *Mucor albo-ater* var. *sphaerosporus* Naumov *1935*

DESCRIPTION: Schipper (5098); zygospores: Schipper (5094). — Colonies on MEA at 20°C to 45(–70) mm high, whitish to olivaceous-buff, odour aromatic; in the dark differentiated into tall and short sporangiophores, both to 40(–70) μm wide, the taller with short lateral branches and the shorter sympodially branched. Sporangia blackish, to 300(–350) μm diam, wall spinulose, deliquescing; columellae either large and obovoid, or small and cylindrical-ellipsoidal, pyriform or subglobose, to 170(–190) × 150 μm, sometimes with brownish content; sporangiospores greyish, ellipsoidal, 7–9·5 × 4–7 μm, intermixed with some sub-globose to globose, more strongly pigmented ones, originating from short sporangiophores. Chlamydospores absent. Heterothallic. Zygospores formed on cherry decoction agar at 15°C, near the agar surface, black, to 210(–240) μm diam, covered with irregular warts (5102), suspensors ± equal, yellowish to brownish. — *M. piriformis* is regarded as a variable species and not subdivided into forms (5098). It is close to *M. mucedo* (q.v.) which has more yellowish brown and brighter colonies and sporangia and more frequently narrow-cylindrical columellae; interspecific matings do not yield fully developed zygospores (5098). *M. saturninus* Hagem *1910* and *M. flavus* Bain. *1903* both have more strongly sympodially branched tall sporangiophores and ovoid columellae, but colonies of the former are ± pale smoke-grey and of the latter olive-buff. — The submerged oidia have been examined by TEM (5703, 5704).

M. piriformis has been reported rather rarely and is obviously limited to the temperate zone. Its occurrence in soil is known from the British Isles (861), Poland (273), the USSR (3184, 5393), Germany, Austria, the Netherlands and France (5098), but also from Taiwan (2472) and mountain forests in Columbia at 3100 m (CBS). It has been isolated from forest soils (273, 3184), cultivated soil (3184), sand dunes (861), beech leaf litter (3538), rotten wood and sawdust of *Fraxinus* and beech (3568). A principal substrate is, however, the dung of rabbits (6588) and cows (1028); it has also been found in nests of gerbils (5393), on millet grain, rotting pears in cold storage (3408), and in river water (3809). — *M. piriformis* is relatively resistant to the avenacin-related triterpene glycoside aescin (4287).

Growth and sporulation occur in the range 5–20°C; the optimum is in the range 10–15°C or higher (3408); at 25°C growth is slow and none occurs at 30°C; at 10–15°C larger sporangia are formed than at 20°C (5098). — Pectin is utilized (3538) and ethanol, acetic and oxalic acids are produced (4509). — Cereals severely contaminated by *M. piriformis* proved toxic to warm-blooded animals (6412).

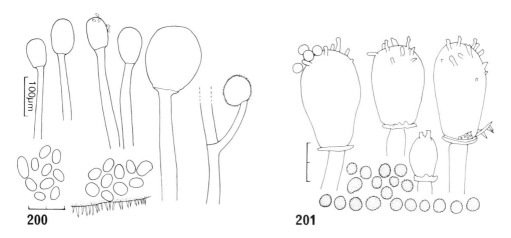

FIG. 200. *Mucor piriformis*, sporangiophore tips with columellae, detail of sporangial wall and sporangiospores, CBS 528.68.
FIG. 201. *Mucor plumbeus*, columellae with irregular projections and spinulose sporangiospores.

Mucor plumbeus Bonord. *1864* — Fig. 201.

= *Mucor spinosus* van Tiegh. *1876*
  ≡ *Mucor plumbeus* var. *spinosus* (van Tiegh.) Zycha *1935*
= *Mucor spinescens* Lendner *1908*
= *Mucor adriaticus* Pišpek *1929*
= *Mucor brunneus* Naumov *1935*
= *Mucor brunneogriseus* Sarbhoy *1968*

DESCRIPTIONS: Zycha *et al.* (6588), Hagem (2206), Mehrotra *et al.* (3739), and Schipper (5099). — This is one of the most characteristic and generally correctly identified species of *Mucor*. Colonies on MEA at 20°C 2–20 mm high in both darkness and light, ± dark grey or pale olive-grey. Sporangiophores branched both sympodially and monopodially, to 21 μm wide, constricted immediately below the sporangium, walls slightly encrusted. Sporangia at first hyaline, later becoming dark brownish grey, wall distinctly spinulose with crystals of calcium oxalate (2806), rupturing at maturity. Columellae pyriform obovoid, ellipsoidal to cylindrical-ellipsoidal, to 50(–72) × 25(–50) μm, brown or grey, usually provided with one to several irregular projections to 5 μm long. Sporangiospores globose (a few ellipsoidal or irregularly shaped), (5–)7–8(–10) μm diam, yellowish brown, densely verruculose.

Chlamydospores sometimes present in the sporangiophores but not abundant. Heterothallic. Zygospores formed on MEA at 24°C, throughout the aerial mycelium, dark brown, to 85 μm diam, covered with short irregularly stellate spines to 3 μm high (5102). — Another species with verruculose spores is *M. fuscus* Bain. *1903* which has smooth to slightly denticulate columellae, sporangiophores sympodially branched at short distances below the sporangia, and distantly verruculose sporangiospores, 8–10 μm diam.

M. plumbeus is a very common species and has a worldwide distribution. Contrary to other species of its genus, it commonly becomes air-borne (1028, 2206, 2653, 6588) due to the release of dry spores at sporangial dehiscence. — It has been reported from the far north in Murmansk (4703), and temperate zone countries including Poland (1421), Denmark, the Netherlands, Germany, Austria, Switzerland, France, Canada (5099), Italy (3453), Hungary (6551), and Spain (3417); there are also reports from Algeria (2972), Somalia (5049), the East African mountains and the Sahara margins (3415), the Congo basin (4159, 5099), Madagascar (4159), South Africa (1559), California/USA (4733), highlands of Mexico (3360), Brazil (4547), the Philippines (4823), Indonesia (578), Australia (5819), New Zealand (5812), India (5058, 5622), and Nepal (1826). Particularly numerous are isolations from the most varied forest soils (these reports are too many to cite individually here), and also cultivated soils (918, 2159, 2740, 2871, 4538, 4788, 4823, 6138), grassland (5812, 6082), orchards (3539) and citrus plantations (2764), soils with steppe type vegetation (2972, 3417, 5049), desert soils (4733), heathland and peaty soils (2741, 4474), marsh (164) and estuarine sediments (5316), river sediments (1162) with strong sewage pollution (1157, 1166), sewage sludge (1165, 1166), fields treated with sewage sludge (1163), slate slopes (566), caves (3453, 6551), children's sandpits (1421), and frescoes of a monastery (2666). A very wide pH range in soil has repeatedly been observed (4055) which extends particularly far (pH 8) into the alkaline range (861). *M. plumbeus* has been reported from the rhizospheres or roots of alfalfa (3982), oats and barley (3004), *Holcus mollis* (4814), various Australian heathland plants (5819), and *Pinus sylvestris* (3572). It has frequently been isolated from various plant remains (650, 3113, 4548, 5389), hay (4821), dung of various animals (1028, 6223, 6588), feathers of free-living birds (2575), the tick *Ixodes ricinus* (5018), fresh water (1166, 4429), wood pulp (6168), beech bark (3568), sticks exposed to soil (1956), wood used in a copper mine (2664), stored seeds of wheat (3491, 4492) and oat (3556), and pecans (2572). — *M. plumbeus* has been isolated from sclerotia of *Sclerotinia borealis* and *Claviceps purpurea* (3531). Its hyphae were reported to be invaded by *Trichoderma viride* (1511) and the obligate hyperparasite *Syncephalis californica* (2618).

Germinating spores are negatively autotropic (4883); sporostatic factors originating from culture filtrates, one of them identified as nonanoic acid (1905), cause a self-inhibition of spore germination (4882). Sporulation is suppressed by rubratoxin B, a metabolite from *Penicillium rubrum* and *P. purpurogenum* (4808). Good growth and sporulation occur in the temperature range 5–20°C; the minimum temperature for growth is 4–5°C and the maximum 35–37°C (4369); at 5–10°C the turf is short and at 28–30°C sporulation is poor (5099). The minimum water potential for vegetative growth is −100 bars (2093). — Pectin (3127, 3414, 3539), and more especially chitin (2068), can be degraded. *M. plumbeus* can utilize D-mannose, sucrose, D-sorbitol, and citric acid as sole C sources, and NH_4Cl, L-histidine and urea as sole N sources, but not methionine as the sole sulphur source (3739); lignosulphonate, and humic and fulvic acids were found to be utilized to some extent by *M. plumbeus* (934). Ageing cultures contain a morphogenic factor which inhibits hyphal growth and promotes autolysis (4425). The fungus produces pyruvic and 2-oxoglutaric acids (4639) and a phenol oxidase (934). Extracellular and intracellular polysaccharides contain polymers of fucose, glucose, galactose and mannose in addition to glucuronic acid (only extracellular) and glucosamine (only intracellular) (2800, 3640). Sporangium formation in *Pilobolus kleinii* was found to be stimulated (5361) in dual cultures by *M. plumbeus.* — *M. plumbeus* proved very sensitive to soil treatments with CS_2 (5001) and formaldehyde (1655). Its growth *in vitro* is strongly inhibited by garlic extract (5744).

Mucor racemosus Fres. *1850* f. *racemosus* — Fig. 202 a.

≡ *Chlamydomucor racemosus* (Fres.) Bref. *1889*
= *Mucor dimorphosporus* Lendner *1908*
= *Mucor christianensis* Hagem *1910*
= *Mucor varians* Pišpek *1929* [non Povah *1917*]

DESCRIPTIONS: Schipper (5095, 5099), Hagem (2206), Zycha *et al.* (6588), Raizada (4708), and CMI Descriptions No. 529, *1977*; zygospores: Schipper (5094). — Colonies on MEA at 20°C to 15(–40) mm high, buff to smoke-grey in the dark and smoke-grey in the light. Sporangiophores to 14(–17) μm wide, some tall and some short but hardly differentiated into layers, walls slightly encrusted, with sympodial and monopodial branches, the latter short, perpendicular and somewhat recurved. Sporangia at first hyaline, later becoming brownish, to 80(–90) μm diam, wall spinulose, rupturing at maturity. Columellae ovoid, oval, or cylindrical-oval, to 55(–64) × 37(–40) μm, pale brownish. Sporangiospores broad-oval to subglobose, 5·5–8·5(–10) × 4–7 μm, greyish. Chlamydospores numerous in the sporangiophores (even in the columellae), barrel-shaped to subglobose, slightly yellowish, to 18 μm long. Heterothallic. Zygospores formed on prune or cherry decoction agar at 5–10°C, occurring throughout the aerial mycelium, reddish brown to brown, to 90(–110) μm diam, covered with irregularly stellate spines about 5 μm tall (5102), suspensors equal, to 35 μm long. — Two morphological forms, interfertile with *M. racemosus*, have been distinguished (cf. key): f. *sphaerosporus* (Hagem) Schipper (q.v.) and f. *chibinensis* (Neophytova) Schipper *1976*. — More saturated fatty acids occur in the spores than in the mycelium but the total content in the spores is lower; myristic, palmitic, palmitoleic, oleic, linoleic, and linolenic acids were identified (5643).

This species was one of the first soil fungi ever to have been discovered and was already isolated in *1886* (22). — This species and *M. hiemalis* f. *hiemalis* are the commonest species of the genus and have a worldwide distribution (2741, 6588). In central Europe, *M. racemosus* has even been named as the most frequent member of the genus (2779). It has been reported from Alaska (6138), Greenland (4174), the USSR (3652, 4703), Finland, Norway (1712, 2206, 5663), and other parts of Europe (the reports are too numerous to cite individually). Other reports are available from Tunisia (4055), the Sahara (2973, 2974, 3415), Libya (4084), Egypt (8), Israel (2762), Iraq (92), Somalia (5048), Zaïre (3415), South Africa (1559), India (41, 1966, 2854, 3868, 4995, 4996, 4997, 4998, 5058, 5622), Sri Lanka (4021), Indonesia (578), Taiwan (2472, 6476), Brazil (398, 6007), and Chile (1824). Records from forest soils are rather more numerous than ones from cultivated soils. It appears equally often in a great variety of soils (2762, 3413, 3414, 3416, 3418) and vegetation types, including peat bogs (861, 3119, 4474), heathland (5047), soils under wet or dry grassland (164, 165, 3417, 4021, 4180, 6182, 6294), and steppe (1559, 4084); some extreme habitats noted are slate slopes (566), sand dunes (861, 1477), salt-marshes (8, 413, 861, 4646, 4987), saline soils (3446), caves (3364, 3365, 6551), a uranium mine (1703), sewage sludge (1166), fields treated with sewage sludge (1163), and also running fresh water (1166, 3809, 4429) with varying degrees of pollution (1154, 1157, 1166, 1482). In the soils studied, pH values between

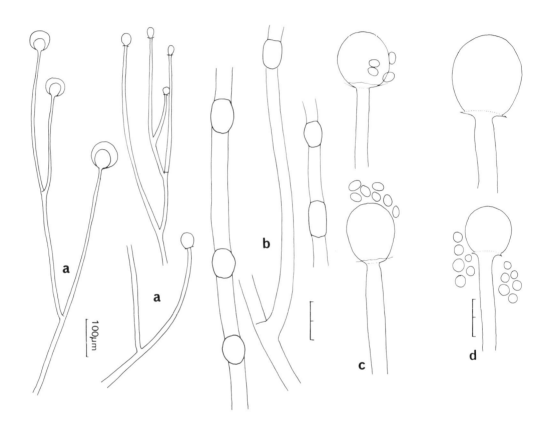

FIG. 202 a–c. *Mucor racemosus* f. *racemosus*. a. Branching sporangiophores with intact sporangia and columellae; b. detail of chlamydospore formation in the sporangiophores; c. sporangiophore tips with columellae and sporangiospores; CBS 248.35. — d. *M. racemosus* f. *sphaerosporus*, sporangiophore tips with columellae and sporangiospores, CBS 115.08.

4·4 (163, 1713) and 8 (861, 3417) are reported, while *in vitro*, the range tolerated may extend from pH 2·0 to 8·5 (4369). Although, in general, the uppermost soil layers are preferentially colonized (1823, 1966, 2948, 4996, 4998), the species has also been found at 40 cm (2779) and, in isolated cases, down to 100 cm (164). The frequency of *M. racemosus* was found to be strongly enhanced by fertilization with NPK and the application of farmyard manure (2762), but reduced after the application of the soil fungicides allylalcohol, captan, nabam, and TMTD (1426). *M. racemosus* is primarily a soil fungus but has nevertheless often been found on dung (1028, 6588), nests, feathers and droppings of free-living birds (682, 2575), plant remains (1821, 2338, 2344, 3113, 4548), sclerotia of *Sclerotinia sclerotiorum* (4699), and in fruit juice (3428). It also occurs in the air (2206, 6588), in earthworm casts (1429), and on dead adult bees and in honeycombs (402). Further habitats recorded include stored seeds of *Pinus sylvestris* (6344), seeds of wheat (2716, 3491, 4492), barley (3207), and rice (2297), tomato (5899), groundnuts (2765, 2768), and pecans (2572). It has been isolated from the rhizospheres of *Polypodium* and *Adiantum* (2856), oats and barley (3004, 4443), *Stipa sareptana* (3376), alfalfa (3982), beans (4451), groundnuts (2768), tea (41), flax (5826), various Solanaceae (3362) and forest trees (3142), and also the roots of *Pinus taeda* (3159). — It was consumed in a feeding experiment by the collembolan *Folsomia fimetaria* (4007). *M. racemosus* f. *racemosus* can be parasitized by the obligate hyperparasites *Dimargaris verticillata* (3560) and *Syncephalis californica* (2618).

White or blue light is required for the normal production of sporangia (3285). Zygospore formation can be inhibited by a streptomycete metabolite (3011). The mean doubling time of the mycelium in submerged culture was determined as $7 \cdot 2$ h (5890). The minimum water potential for vegetative growth is -110 bars, and for sporulation -70 bars. Growth and sporulation are possible in the temperature range 5–30°C, the optimum is in the range 22–25°C, the minimum -4°C, and the maximum 33°C; no growth occurs at 37°C (4369, 4759, 5095, 5099, 5789); the thermal death point in fruit juice was determined as 63°C for 25 min (3620); toxigenic isolates are said to be psychrophilic (2761). In nitrogen- and carbon-starved cultures chlamydospores are produced (5891). — Under anaerobic conditions, as in many *Mucor* species, a budding mycelium ("oidia") is formed (6436). Following a shift from anaerobic to aerobic conditions, the hyphal growth is strongly self-inhibited by the formation of cyclic adenosine $3',5'$-monophosphate (3229), but in air the budding habit is induced by this metabolite (4478). — Pectin is rapidly decomposed (3127, 3414, 4238); the activity of glyoxylate-transaminase has been determined (1819); signs of utilization of CMC have been reported (2736) and hemicellulose found to be attacked (1828). Proteinase (1846, 2948, 3414), lipase (1676) and polyphenoloxidase (934) activities have also been demonstrated; conflicting data exist on the utilization of lignin sulphonate (934, 3260); epoxy plasticizers were found to be utilized as C sources (5859), and the herbicides simazine and atrazine as sole N sources (3799); tryptamine is oxidized to indole-3-acetic acid (1522); phenylamide herbicides can be metabolized by oxidative transformations (5824) and the insecticide carbaryl is hydroxylated (610). *M. racemosus* f. *racemosus* is autotrophic for growth substances (1425) and produces riboflavin-α-glucoside (5662). Potassium is released from orthoclase and oligoclase (4008) and nitrogen from urea-formaldehyde fertilizers (4010). With an excess of phosphate in the medium, the fungus accumulates 5–6% P (dry-weight basis) in the mycelium (2714). Pyruvic, 2-oxoglutaric (4639), citric and succinic acids (4009), fucose, mannose, galactose and glucuronic acid (3639) have been identified in culture filtrates. Besides the morphogenic cyclic adenosine $3',5'$-monophosphate (4478), cyclic guanosine $3',5'$-monophosphate was found, the largest amounts in germinating spores (4310). — The production of a substance with antibiotic activity against various test bacteria has been reported (3812, 4853). Mycelial extracts have a protistocidal activity against *Paramecium caudatum* (1456) and are also toxic to rats (1388). *M. racemosus* f. *racemosus* is regarded as one of the causes of mastitis in cattle and it has often been isolated from rabbit lungs (63). The growth of oats and barley seedlings was found to be inhibited by both mycelium and culture filtrates (3005). Spore germination of the moss *Tetraphis pellucida* was stimulated (6019).

Mucor racemosus f. *sphaerosporus* (Hagem) Schipper *1970* — Fig. 202 d.

≡ *Mucor sphaerosporus* Hagem *1908*
? = *Mucor globosus* Fischer *1892*
= *Mucor globosus* var. *intermedius* Sacc. *1913*

DESCRIPTIONS: Schipper (5095, 5099), Hagem (2206); zygospores: Schipper (5094). — Characters as in f. *racemosus*, but colonies more clearly two-layered, with a dense layer 1–2 mm high, and a sparse layer 15–25(–40) mm high, brownish grey to light grey-olive in the dark and greyish olive in the light; the short sporangiophores mainly sympodially branched. Sporangiospores globose, 4–8 μm diam, infrequently broadly oval or irregularly shaped, smooth-walled to slightly punctate (at low temperatures more roughened and more frequently oval). Heterothallic. Zygospores formed on cherry decoction agar at 10°C throughout the aerial mycelium but mainly in the lower part, to 90 μm diam, suspensors shorter than in f. *racemosus* (to 15 μm long). — For delimitation of the forms of *M. racemosus* see under f. *racemosus*. *M. globosus* Fischer was said to have no chlamydospores and larger sporangia and its identity is consequently doubtful. — DNA analysis gave a GC content of 38%; the mycelium is reported to contain 2·1% C_{12}-, 7·6% C_{16}-, 6·9% C_{18}- saturated and 7·7% C_{16}- and 25·8% C_{18}-monounsaturated fatty acids (3232).

The relatively few records of this taxon indicate a worldwide distribution which includes Norway (2206, 5099), the British Isles (413, 1376, 5099), Poland (269, 272, 1421), Czechoslovakia (1700, 2575), Yugoslavia (4573), the USSR (4474), the Netherlands, Germany (5099), Austria (1828, 3413, 3418), Italy (3445), the USA (682, 1166, 1632, 4733), Chile (1824), Japan (5099), Indonesia (578), India (4995), Bangladesh (2712), Pakistan (4855), and Iraq (92). It seems to be common in forest soils (269, 272, 1700, 1828, 4995), but is also known in peat (1376, 3119, 4474), alpine pastures (3445), salt-marsh (413), desert soil (4733), soils under steppe type vegetation (1632), a chalk soil (1259), podzol, rendzina, chernozem, terra fusca, and loess (3413, 3414, 3416, 3418). Fertilization with NPK increased its frequency in the rhizosphere of rice (2712). It has also been reported from the rhizospheres of *Stipa lessingiana, S. sareptana,* and *Artemisia frigida* (3376), dying roots of broad beans (6134), various decaying plant remains, pine litter (681), moist chaff (4548), sewage (1166), wood, rabbit dung (IMI), birds' roosts (682, 2575), children's sandpits (1421), and air (2206). — Spore germination in soil can be inhibited by mycostasis (6284). The fungus can be parasitized by the obligate hyperparasite *Dimargaris verticillata* (3560), and serve as food for a number of mites (5379) and other micro-arthropods (3830, 5380, 5381).

Growth and sporulation are possible in the temperature range 5–25°C, the best growth occurs at 15–18°C, poor growth at 25°C, and none has been observed at 28°C (2206, 5095, 5099). — Pectin, hemicelluloses and urea can all be utilized (3414).

Myrothecium Tode ex Fr. *1821*

= *Myxormia* Berk. & Br. *1850*
= *Myrotheciella* Speg. *1910*
= *Exotrichum* Syd. *1914*
= *Starkeyomyces* Agnihothrudu *1956*

Type species: *Myrothecium inundatum* Tode ex Gray (lectotype)

Teleomorph: *Nectria* (Fr.) Fr. (known in two species)

MONOGRAPHIC TREATMENT: Tulloch (5938); key also in Ellis (1603).

Conspicuous sporodochia occurring on both plant material and in culture, cupulate, pustulate or stalked, synnema-like, formed of densely compacted conidiophores arising from a ± developed stroma and bearing a mass of slimy, green to black conidia. Sporodochia surrounded by differentiated, hyaline, often curling marginal hairs; hyaline or darkened setae present in some species. Conidiophores hyaline or slightly olivaceous, repeatedly branching with several branches arising from each node, the ultimate branches being the phialides. Phialides compacted in a dense parallel layer, cylindrical with conically tapering tips and usually undifferentiated collarettes. Conidia 1-celled, subhyaline but dark olivaceous in mass. — Thirteen species are now recognized (5938).

Myrothecium species, particularly *M. verrucaria*, are well known as cellulolytic saprophytes, *M. inundatum* showing the least cellulolytic activity. They are also well known producers of antibiotics and toxins and *M. roridum* is also a plant parasite, causing leaf spots in various plants.

Colonies are grown on PDA, or better OA or CMA, under daylight for at least ten days; plant substrates like lupin stems (5938) or clover plants (1235) may further stimulate fructification. — The three species treated here all have pustulate or cupulate sporodochia without setae, phialides less than 20 μm long, and conidia generally less than 10 μm long.

Key to the species treated:

1	Conidia longitudinally striate, fusiform, 6·5–14 × 2·5–4·5 μm	*M. cinctum* (p. 482)
	Conidia not striate, not exceeding 8 μm in length	**2**
2(1)	Conidia rod-shaped or narrowly ellipsoidal, 5·5–7 × 1·5–2·0 μm	*M. roridum* (p. 483)
	Conidia fusiform, 6·5–8 × 2·0–3·5 μm	*M. verrucaria* (p. 485)

Myrothecium cinctum (Corda) Sacc. *1886* — Fig. 203.

≡ *Fusarium cinctum* Corda *1842*
= *Myrothecium striatisporum* Preston *1948*
= *Myrothecium brachysporum* Nicot *1961*
= *Myrothecium ucrainicum* Pidopl. *1969*
= *Myrothecium longistriatisporum* Matsushima *1971*

DESCRIPTIONS: Tulloch (5938), Ellis (1603), and Matsushima (3680). — Colonies attaining
5·5–7·0 cm diam in 14 days at 25°C on PDA, initially white to rosy buff; reverse rosy buff.
Sporodochia 300–600 μm diam on the natural substrate, sometimes stalked, synnematous, to
400 μm high, olivaceous-black. Phialides 2–3(–4) in a whorl, 10–20 × 2·0–2·5 μm. Conidia
broadly or narrowly fusiform, the apical end pointed, the basal end truncate and sometimes
darkened, with up to eight longitudinal or oblique striae visible in one plane,
6·5–14 × 2·5–4·5 μm.

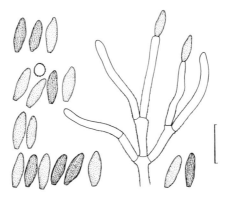

FIG. 203. *Myrothecium cinctum*, conidiophore with striate conidia, CBS 277.48.

The cosmopolitan distribution of *M. cinctum* has been documented with collections from
Czechoslovakia, Italy, France, the Netherlands, Germany, the British Isles, the USSR, Iran,
Pakistan, India, Indonesia, Malaysia, Australia, New Caledonia, New Zealand, Papua-New
Guinea, Egypt, Algeria, Sierra Leone, Uganda, Togo, Ghana, South Africa, the USA and
Canada (5938). It is mostly found on dead plant material, particularly of grasses (5938). Soil
isolates have been reported from agricultural soils (1614, 1616, 3063, 3450), wet and dry
mesic prairie soils (4313), forest soil (3063), savannah with *Acacia karroo* (4407), and a highly
alkaline soil (4698). *M. cinctum* has also been found in the rhizosphere of *Trifolium
alexandrinum* (1963) and as a secondary colonizer of decaying sheaths and blades of *Setaria
glauca* (5233). — Its growth *in vitro* was found to be inhibited by *Memnoniella echinata*
(4).

M. cinctum is able to decompose cellulose (5938) and produce coprogen B (3231).

Myrothecium roridum Tode ex Steudel *1824* — Fig. 204.

= *Gliocladium nigrum* F. & V. Moreau *1941*

For other synonyms see Tulloch (5938).

DESCRIPTIONS: Tulloch (5938), Preston (4632, 4633), Ellis (1603), Subramanian (5626), Matsushima (3680), and CMI Descriptions No. 253, *1970*. — Colonies reaching 4·0–6·0 cm diam in 14 days at 25°C on PDA, mycelium white to rosy buff; reverse rosy buff or yellow. Sporodochia 60–750 μm diam, olivaceous–black, flattened or convex, often formed in concentric zones. Phialides 11–16 × 1·5–2·0 μm. Conidia rod-shaped or narrowly ellipsoidal, mostly with both ends rounded but the basal one slightly more truncate, 5·5–7·0 × 1·5–2·0 μm.

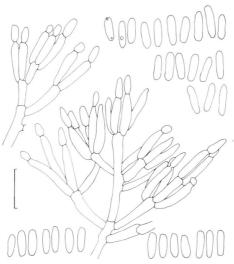

FIG. 204. *Myrothecium roridum*, conidiophores and conidia.

The cosmopolitan distribution of this species has been documented with collections from Germany, the British Isles, France, Denmark, Egypt, Mauritius, Sudan, the Yemen, Nigeria, Tanzania, Sierra Leone, Kenya, Ghana, Zambia, Turkey, India, Sri Lanka, Pakistan, Burma, Thailand, Malaysia, the Solomon Islands, Indonesia, New Zealand, China, the USA, Venezuela, Trinidad, Mexico, and Argentina (5938). *M. roridum* has been found on numerous host plants and on cellulosic materials (4633, 5938). Soil isolations have been reported from Israel (654, 2764, 2768, 2772, 4759), Libya (6510), Egypt (8), Kuwait (4001), Nepal (1826), Somalia (5048), Zaïre (3790), equatorial West Africa (1420), South Africa (4407), Japan (5846), Canada (505), the USSR (2871) and the Netherlands (1616). It has been found in forest soils (505, 2004, 2007, 2008, 2030, 2854, 3447), forest nursery soils (2482), soils with steppe type vegetation (1700, 4407), *Calluna* heath (5047), grassland (3863, 3865), very frequently in cultivated soils, sandy soils (4651), arid soils (654), slate slopes (566), dunes (745), salt marshes (8, 4001) and coastal sediments (4477). The pH range of the soil habitats is broad, although its occurrence in dunes was found to be confined to alkaline sites (745). In sewage-irrigated fields, the fungus was found with particular frequencies in those showing irrigation sickness (2467), but in running water (4429) it has been found only when the degree of pollution was low (1154). Plant substrates recorded include decaying shoots of *Setaria glauca* (5233, 5234) and *Bothriochloa pertusa* (2955), leaves of ferns at early stages of decay (2859), leaves of palms (3963), *Atlantia monophylla* (6108), *Casearia tomentosa* (4976) and cotton (5009), fruits of *Onobrychis viciifolia*, birch catkins (4917), and seeds of soya (4975), *Dahlia* and *Nasturtium* (5820); it has also been found on birds' feathers (4652). The distribution on grasses in New Zealand seems not to be affected by the season (1400) but in India its frequency increased during the rainy period (3863). It has been observed on roots of both tomatoes (1531) and banana (2030), and in the rhizospheres of groundnuts (2768), *Carica papaya* (5345), and several euphorbiaceous (3866) and sand dune plants (4371); in that of *Trigonella foenum-graecum* it decreased after foliar application of the antibiotic subamycin (2187).

The optimal temperature for growth on agar is 25–27°C (1721, 6017); growth is reduced at 35°C when cultivated in the dark (1234). *M. roridum* tolerates a wide pH range, but growth is reduced at pH 2·8 and 9·2 (6017). Isolates from saline substrates were found to grow best on seawater agar and to be adapted to high salt concentrations (4001, 4477). — Growth and sporulation were optimal with D-glucose, followed by raffinose as C sources, while the most suitable N sources were found to be glutamic acid, nitrates and urea (972). The decomposition of starch (1827), CMC (1432) and native cellulose has been reported (6326). The chromanes myrochromanol and myrochromanon (5718) and also ethylene are produced (2641), and various substances with antibiotic activities occur as metabolic products: wortmannin (4519), coprogen B (3231), myrothecin (3067, 4138), particularly the sesquiterpenes roridin A, B, D and E, trichodermol (= roridin C), the verrucarins A, B, H and J (586, 2205, 5719), 2-dehydroverrucarin A (6587) and necrocitin which induces wilt symptoms in tomatoes (4475). In plant tests this fungus proved to be the most active antibiont of 39 soil fungi investigated (4139). An antibiotically active component of undefined nature ("glutinosin") has previously been shown to be present in nutrient solutions in which *M. roridum* had grown (718). — This is the only species of *Myrothecium* that can become pathogenic to plants, commonly causing necrotic lesions or shot-holes on leaves and also occurring on petioles, stems and fruits (e.g. of tomato). It can also cause considerable losses in cotton, coffee, *Antirrhinum* and *Viola* (CMI Descriptions No. 253). It has further been reported to be pathogenic to *Abelmoschus esculentus, Phaseolus radiatus* (5800), potato, lupin (739), red clover (1234), *Gardenia* (356, 1721), wheat, peas, rape (1430), *Gloxinia* (3374), cotton, cucurbits (4043), *Solanum melongena* (4111) and water chestnut (4593). Germinating conidia produce a metabolite similar to necrocitin which facilitates the penetration of germinating conidia into red clover tissues (1235). Components of culture filtrates caused two types of cytotoxicity in cell cultures, one of which was fatal to sheep and calves (3966).

Myrothecium verrucaria (Alb. & Schw.) Ditm. ex Steudel *1824* — Fig. 205.

≡ *Peziza verrucaria* Alb. & Schw. *1805*
= *Gliocladium fimbriatum* Gilm. & Abbott *1927*
= *Metarrhizium glutinosum* Pope *1949*
= *Starkeyomyces koorchalomoides* Agnihothrudu *1956*

DESCRIPTIONS: Tulloch (5938), Ellis (1603), Preston (4632), Marasas *et al.* (3590), Matsushima (3680), and White and Downing (6327). — Colonies reaching 4·0–5·0 cm diam in 14 days at 25°C on PDA, mycelium white to rosy buff, forming diffuse or coalescent olivaceous to black sporodochia; reverse rosy buff. Phialides 3–6 in a whorl, 10·5–14·5 × 1·5–2·0 μm. Conidia

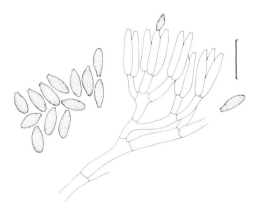

FIG. 205. *Myrothecium verrucaria*, conidiophore and conidia, CBS 187.46.

broadly fusiform, the apical end pointed and the basal truncate, bearing an apical, funnel-shaped appendage (visible in ammoniacal erythrosin (5938) or other stains (5804) or under phase contrast), 6·5–8 × 2·0–3·5 μm. — A similar soil-borne fungus with less differentiated sporodochia and irregular but discrete conidial chains and no appendages is the anamorph of *Nectria pityrodes* Mont. *1842*. — Galactose and mannose have been shown in both cell hydrolysates and culture solutions (3640). In the conidia, trehalose accumulates as a reserve material (3564); for the transport of this, a constitutive and an inducible transport mechanism exist (3564).

The worldwide distribution of this species has been documented by collections including ones from Denmark, Cyprus, Turkey, Egypt, the Sudan, Kenya, Tanzania, Gambia, Rhodesia, Ghana, Nigeria, Nepal, China, Australia, Malaysia, Indonesia, Papua-New Guinea, New Zealand and Venezuela (5938). It has been found on various plant material, cellulosic matter and in soil (5938) and isolated with selective methods for cellulolytic fungi (4647). Soil isolates have been reported from Canada (505), the USA (1163, 1165, 1166, 1330, 6351), Poland

(1421, 3498), the USSR (2871), the British Isles (6182), the Netherlands (1616), France (242), Spain (3417), Italy (488), Israel (654, 2762, 2768, 2772, 4759) where it was sometimes very frequent (2764), Egypt (8), the Ivory Coast (4159), central Africa (3063), Zaïre (2966, 2967, 3790), South Africa (3590, 4407), Iraq (92), Kuwait (4001), Pakistan (2621, 4855), Japan (3680), India (1317, 1524, 3865, 4736, 4933), Jamaica (4886) and the Bahamas (2006). It has been demonstrated in both mixed and hardwood forests (1040, 2004, 2008, 3417, 4225, 6351), grassland (1524, 3498, 3865, 4933, 6182), often in soils with steppe type vegetation (1700, 4407), dunes (6414), saline soil (4001), desert soils (654, 1512, 4733), children's sandpits (1421), various cultivated soils where its incidence was found to be increased by growing alfalfa (1330) or N fertilization (3498), citrus soils (2764, 3630, 3632) where it survived steaming (3631), in estuarine sediments (655), running and still water with varying degrees of sewage pollution (1154, 1155, 1157, 1166), sewage sludge (1165, 1167) and fields treated with sludge (1163), seawater off Florida (4918), and the air in Wisconsin (2004). A wide variety of plant substrates can be colonized by *M. verrucaria* (5938). Accumulations have been observed in the rhizospheres of *Ammophila arenaria* (163), *A. breviligulata, Andropogon scoparius* var. *septentrionalis* (6414), *Acacia karroo* (3590), alfalfa (1330), groundnut (2768), sugar cane (4886), wilted pineapple plants (6210), *Bothriochloa pertusa* (3271), oats and barley (3006); it has been isolated from roots of broad beans (6134), seeds of wheat, sorghum (3989), *Capsicum annuum*, peas (6407), soya and rice (5820), stored sweet potatoes (3066), wood exposed to running water (5249), and wood used in a copper mine (2664). — Tetrahydrostilbene contained in the wood of *Maclura pomifera* (Moraceae) arrests the growth of the fungus *in vitro* (347); its growth *in vitro* is also inhibited by α-tomatine (182). — The fungus seems to be a weak competitor in soil, as mycelium of it introduced into the soil is easily decomposed (2128, 3633). Growth in soil was found to be moderately influenced by mycostasis (2559).

The production of conidia was found to be independent of light (1234, 6518) but two rhythms regulating sporulation were observed: an endogenous one of 56 h persisting up to two months under constant conditions and a 24 h rhythm controlled by a daily photocycle (2746). Good sporulation has been found to occur on maltose-containing media (6519), and also on pea, bean and flax seed (2128). Germination of densely sown conidia is self-inhibitory through the secretion of growth substances, but this inhibition can be removed by the addition of biotin (3562). — The optimal temperature for growth is 27°C, the minimum 12°C, and the fungus still grows at 35°C (1234) but not at 37°C (2128); it does not survive at $\geqslant 50$°C for 30 min (613). — Dextrin and L-arabinose support its growth best, starch was less favourable, and acetate and citrate least so (6519). Sucrose is metabolized by a non-hydrolytic system, although invertase is also present; melezitose and turanose are also assimilated without hydrolysis (3561). It can grow on media with 20–40% sucrose (4001). The relationship between respiration and mycelial growth phases, CO_2 production, pH, and assimilation of sugars, has been investigated (1274). Pectinase activity has been reported (2128, 4779) and xylan is hydrolysed, although xylose has not been detected as a decomposition product (1921, 3092). Methyl-β-glucoside, mannose and cellobiose are all utilized (5387) and 1,3-β-D-glucanase is released (4778). — *M. verrucaria* is one of the most potent cellulose decomposers known (386, 1168, 2128, 3092, 3093, 4597, 5068, 6010, 6326). Several enzymes must be involved in this process (2118, 5196) to explain the contradictions in the literature concerning the optimum pH and temperature ranges given for this process (2071, 3092, 5068, 5206). The enzyme C_x has been shown to be present in larger amounts than C_1 (4779). The course of cellulose decomposition by this species has been followed on cotton fibres with the aid of cell-free preparations (2224). As products of cellulose decomposition, cellotriose, cellobiose and glucose have been identified in the presence of cellobiase and transglucosidase (3093); the presence of yeast extract ($>10\,\mu g\,ml^{-1}$) (6129) and sugars (1717, 5388) suppresses cellulase production but organic solvents such as phenylethyl alcohol were found to stimulate its production (5610). C_x is produced on glucose- or glycerol-containing media when the soluble C source becomes limiting (2608). Its erosion of birch wood has been observed *in vitro* (4191). — Nitrate is reduced in liquid media (3969) and calcium cyanamide found to be a better N source than $NaNO_3$ (6078); the production of a cyanamide hydratase has been demonstrated (5611). The requirements for N, P, Mg and organic acids for growth, reproduction and enzyme production have been thoroughly investigated (937, 1275, 3303, 5388, 6519) as this fungus is one of the standard organisms used in textile degradation tests. Plasticizers (diesters of aliphatic dicarbonic acids) can be attacked (473) but wool is only weakly degraded (6330). There are numerous reports of the production of the sesquiterpenes verrucarin A, B, C, D, E, F, G and J, the roridins A, B, D, E and H (586, 587, 2205, 5719, 5876, 5877), muconomycin A (6109), coprogen B (3231), gliotoxin (424, 2785) and the antifungal glutinosin (709, 722). An antiviral antibiotic, (−)rugulosin, produced by *M. verrucaria* can inhibit phage multiplication without affecting either the free phage or host bacterium (4095). — Antibiotic activity of the fungus (without isolation of the active principle) towards various test organisms has been repeatedly reported (569, 718, 868, 2735). *M. verrucaria* can act as a mycelial parasite of *Cochliobolus sativus* (868) and has been described as a facultative parasite of peach roots (6288), although on artificial infection after soil steaming no deleterious effects on citrus plants were observed (3635). — The production of three substances causing skin lesions has been reported (671, 2205). A cytotoxin present in culture filtrates can cause death in both sheep and calves where the symptoms are similar to those observed in lambs fed with roridin A and verrucarin A (3966); verrucarin A is one of the most potent cytostatica known (4941). — The fungus is highly sensitive to γ-irradiation (2786).

Nannizzia Stockd. *1961*

Type species: *Nannizzia incurvata* Stockd.

Anamorph: *Microsporum* Gruby

Fam. Gymnoascaceae Baran. — Ascoma initials consisting of a clavate antheridium surrounded by a spiralled ascogonium. Ascomata (gymnothecia) globose, white to yellowish; peridium consisting of a network of hyaline, septate hyphae with verticillate or single branches, cells moderately thick-walled, densely asperulate, ± symmetrically and moderately constricted, without protuberances at the ends; bearing numerous free ends or slender, straight or spiral appendages or macro-conidia. Asci and ascospores as in *Arthroderma*. Eight heterothallic species are known which are all keratinophilic (see introduction to *Arthroderma*); some of these sometimes occur as dermatophytes in man and various other mammals.

Key to the species treated:

1   Macro-conidia thick-walled (2–4 μm), ± cylindrical; peridial hyphae dichotomously and divergently branched, the cells with one slight constriction   *N. grubyia* (p. 490)
   Macro-conidia thin-walled (1·0–1·5 μm), fusiform; peridial hyphae verticillately branched, the cells with 1–3 constrictions (*Microsporum gypseum* complex)   **2**

2(1)   Peridial hyphae with up to 5 verticillate, convergent branches, often bearing straight slender appendages   *N. incurvata* (p. 495)
   Peridial hyphae with up to 4 verticillate, divergent arcuate branches   **3**

3(2)   Ascomata 500–1250 μm diam, formed mainly on hair above non-sterile soil; peridial hyphae bearing numerous spiral appendages terminally and laterally   *N. fulva* (p. 488)
   Ascomata 300–750 μm diam, also formed on hair above sterile soil; peridial hyphae bearing spiralled and straight appendages mostly terminally   *N. gypsea* (p. 491)

Nannizzia fulva Stockd. *1963* — Fig. 206.

Anamorph: *Microsporum gypseum* complex, *Microsporum fulvum* Uriburu *1909*

DESCRIPTION: Stockdale (5587). — A heterothallic species which produces ascomata on hairs, particularly when in contact with non-sterile soil. Ascomata pale buff, 500–1250 μm diam (excluding appendages); peridial hyphae densely verrucose, bearing up to four branches in whorls, straight or divergent; cells 10–25 × 4–7 μm, with 1–3 symmetrical constrictions; bearing terminally, and often also laterally, spiralled or, more rarely, straight appendages or macro-conidia. Ascospores smooth-walled, lenticular, 2·5–4·0 × 1·5–2·0 μm. Colonies reaching 6·0–7·0 cm diam in 14 days at 25°C on 2% MEA, dense, downy to granular, pale buff, not strongly radiating; reverse straw, usually with a trace of primrose in the centre. Macro-conidia moderately thick-walled (0·8–1·2 μm), verrucose, predominantly cylindrical with a rounded apex or clavate, 25–58 × 7·5–12 μm (mostly 35–50 × 8–11 μm), rarely bearing a whip-like appendage, supported by comparatively long (10–100(–300) × 2·0–3·5 μm), little branched conidiophores; micro-conidia clavate, 3·3–8·3 × 1·7–3·3 μm, smooth-walled or slightly

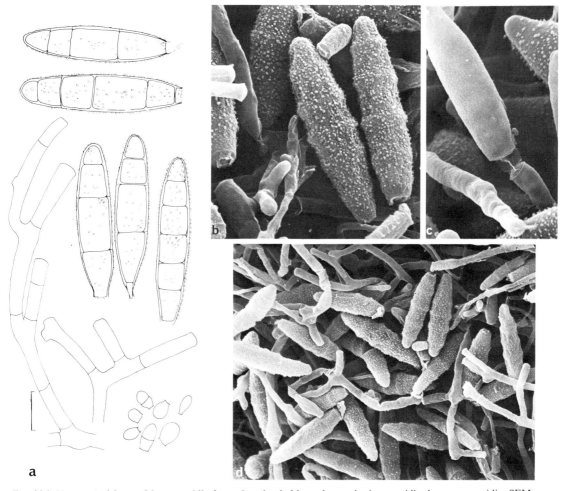

FIG. 206. *Nannizzia fulva*. a. Macro-conidia formed on hyphal branches and micro-conidia; b. macro-conidia, SEM, × 1500; c. a young macro-conidium liberated at an emptied basal cell, SEM, × 2000; d. macro-conidia and tips of the subtending hyphal branches, SEM, × 750; CBS 167.64.

roughened; a SEM study of the macro-conidia showed vesicles in addition to very fine papillae (6103) and no differences between *N. fulva, N. gypsea* and *N. incurvata* (4561).

N. fulva has usually not been distinguished from other species of the *M. gypseum* complex. Its reliable identification is documented for the British Isles, the USA, Hungary, Yugoslavia, Argentina (5587), and Taiwan (6119) from soil and skin lesions in humans. — It is known to produce an elastase (4849).

Nannizzia grubyia Georg, Ajello, Friedman & Brinkman *1962* — Fig. 207.

Anamorph: *Microsporum vanbreuseghemii* Georg *et al. 1962*

DESCRIPTIONS: Georg. *et al.* (1937), Palsson (4368), Caprilli *et al.* (886), and Dvořák and Otčenášek (1521). — A heterothallic species which forms ascomata on soil-hair plates (70, 1937, 4347). Ascomata at first white but becoming buff at maturity, 150–600 μm diam (excluding the appendages); peridial hyphae hyaline, uncinately, dichotomously branched, the cells gradually and slightly constricted in the centre, very frequently bearing spiralled appendages. Asci globose, 8-spored; ascospores light yellow in mass, smooth-walled, oblate, $3\cdot0 \times 2\cdot4\,\mu$m. Colonies fast-growing, reaching $7\cdot0$ cm diam in 14 days at 24°C on 2% MEA, and $8\cdot0$ cm at the optimum temperature of 30°C; flat, powdery or downy, cream to yellow or pink, reverse yellow or ± red. Macro-conidia numerous, borne on lateral or terminal stalk cells, cylindro-fusiform, with thick ($2\cdot0$–$2\cdot5\,\mu$m) and echinulate walls, mostly 7–10-septate, 59–62×10–$11\,\mu$m. Micro-conidia usually numerous, borne singly and laterally, pyriform to obovate, about $9 \times 4\,\mu$m. — The macro-conidia are similar to those of *Arthroderma uncinatum* (1938) but distinctly roughened and supported by stalk cells. Macro-conidia contain an average number of about four nuclei per cell (2367). Further cytological and morphological data are given by Caprilli *et al.* (886). — Based on disc electrophoretic protein patterns, *M. vanbreuseghemii* was found to belong to one cohesive group also including *M. cookei, M. canis, M. gypseum, M. distortum, M. ferrugineum, M. nanum* and *M. audouinii* (3223). Pyrolysis-gas-liquid chromatography of several *Microsporum* species has not yielded reproducible results suitable for use in classification (903).

By employing the hair bait method, this fungus has been recovered from soils in the USA (68), Canada (68, 305), New Zealand (3614), Australia (977), Argentina (6047), Brazil (4897), Norway (3351), Sweden (4368), Finland (5015), Poland (3530), Germany (580, 582, 584, 4536, 5521), France (1197), Italy (893), Hungary (2541), Romania (118), Nigeria (77), Tunisia (2721) and India (5217); it has also been found in stables in Bulgaria (306). *N. grubyia* is not recognized as an important cause of dermatophytosis, but reports of it include isolations from hair or skin lesions in opossum (3613), a Malabar squirrel (1938) and dogs (68, 1937).

Ascoma formation has been observed not only on soil and hair medium (70, 4347) but also on non-keratinous, modified alphacel medium with oatmeal as its most essential component (6270). — The germination of conidia and subsequent growth of mycelium is inhibited by light, particularly at short wavelengths (761). Using human hair, the principle inducing ascospore formation was identified as α-keratinose (6354). In colonizing hair, the species is

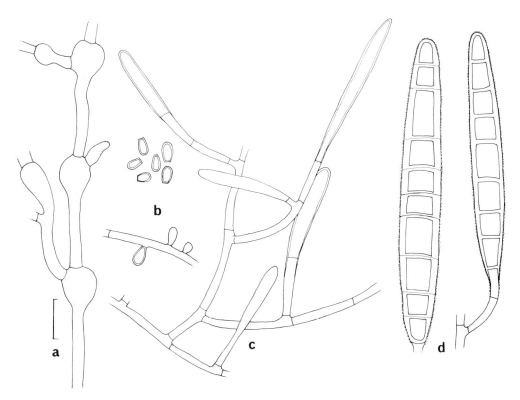

FIG. 207. *Nannizzia grubyia.* a. Racquet hyphae; b. micro-conidia; c. young still attached macro-condia; d. mature macro-conidia; CBS 243.66.

restricted to humidity at vapour saturation (977). — Compared with other keratinophilic fungi, this species does not belong to the very active keratin degraders (30% within 77 days) (585). It has been shown to be able to utilize cow foot oil and wool wax (581). — An experimental infection of guinea pigs with *N. grubyia* was successful (1938).

Nannizzia gypsea (Nann.) Stockd. *1963* — Fig. 208.

≡ *Gymnoascus gypseus* Nann. *1927*

Anamorph: *Microsporum gypseum* (Bodin) Guiart & Grigorakis *1928* pro parte

≡ *Achorion gypseum* Bodin *1907*

For 14 further synonyms see Ajello (67).

DESCRIPTIONS: Griffin (2092), Stockdale (5587), and Dvořák and Otčenášek (1521); co-nidiogenesis: Cole and Samson (1109). — Heterothallic. Ascomata formed on soil-hair plates with and without sterilization, pale buff, 300–750(–900) μm diam (excluding the appendages); peridial hyphae densely verrucose, bearing up to four divergent, verticillate branches; cells 8–20 × 4–7 μm, with 1–3 symmetrical constrictions; bearing straight or spiralled appendages mostly terminally. Ascospores smooth-walled, lenticular, 2·5–4·0 × 1·5–2·0 μm. Colonies reaching more than 8·6 cm diam in 14 days on 2% MEA at 25°C, thin, finely granular, pale buff, radiating; reverse straw with a trace of primrose, sometimes rosy-vinaceous in the centre. Macro-conidia moderately thick-walled (0·8–1·2 μm), verrucose, ellipsoidal to fusiform, to 5-septate, 25–58 × 8·5–15 μm, mostly 35–50 × 10–12·5 μm, often with a whip-like appendage to 30 × 1·0–1·5 μm; conidiophores 10–80 × 2·5–6 μm, repeatedly branched. Micro-conidia clavate, smooth-walled or slightly roughened, 3·5–8·5 × 1·7–3·3 μm. Hyaline chlamydospores rarely produced (4636). A SEM study of macro-conidial surfaces showed very dense spherical outgrowths in a somewhat serial arrangement (6103). — It is not certain whether this species or *N. incurvata* is the teleomorph of the original *Microsporum gypseum sensu stricto* (5587), but unspecified reports on the anamorph have been compiled under this species, although they may equally apply to *N. incurvata* (q.v.). Of 58 clinical isolates of *Microsporum gypseum* collected in the USA, 50 were found to be *N. gypsea*, eight *N. incurvata*, and none *N. fulva* (6269). — SEM/TEM studies of macro-conidium development (82, 4561) revealed an attachment to the conidiophore by a rigid circular disk-like septum and an electron-dense outer layer continuous with the warty structures; the role of lysosomes (4357) and changes in both the mycelial wall and conidial coat glycoproteins (4360) during different stages of macro-conidium germination have been investigated. The ultrastructure of ascosporogenesis has been studied by TEM (2453). — Studies on the sexual recombinations between mutants and wild type strains showed that aconidial mutants may be produced by one

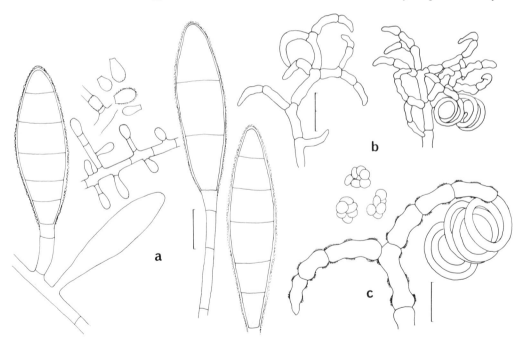

Fig. 208. *Nannizzia gypsea*. a. Macro- and micro-conidia; b. peridial hyphae with partly spiralled appendages; c. detail of appendage with three asci; CBS 170.64 × 171.64; d. *Microsporum gypseum* showing the natural arrangement of macro-conidia, SEM, × c. 850, orig. R. A. Samson.

or more gene mutations (6267). Although the three species of the *N. gypseum* complex are intersterile, they were found to give moderate sexual reactions (formation of ascoma initials) with opposite mating types of *Arthroderma simii* Stockd. *et al. 1965* (5588). Proof is given for the existence of a parasexual cycle (3287). — The protein profiles of *M. gypseum, M. canis, Trichophyton mentagrophytes, T. tonsurans, T. rubrum* and *Epidermophyton floccosum* showed no significant intraspecific variation and each species was characterized by its own profile (1900). — Four chromosomal structures have been distinguished in the vegetative nucleus of *M. gypseum* (6268); macro-conidia contain a total of 39–55 nuclei (3922) and 5–8 nuclei per cell (886, 2367). The pleomorphic cultural patterns seen in this species were demonstrated to be due to true mutations (6267). — Hyphal cell wall hydrolysates contained 31·2% N-acetyl-glucosamine, 0·25% galactosamine, 45·9% neutral sugars, 36·6% glucose, 10·3% mannose, and traces of galactose; the predominating amino acids are threonine (12·3%), serine (10·8%), glycine (10·8%), proline (9·6%), phenylalanine (8·1%), glutamine (7·2%) and alanine (6·9 mol % of the total amino acids) (3232). According to another report, hydrolysates of purified cell walls contained mannose and glucose in the proportion 1:1·4 (4202). The fatty acids palmitic, stearic, oleic and linoleic occur in the mycelium (6409). — Pyrolysis-gas-liquid chromatography has not proved suitable for taxonomical purposes in this complex (903).

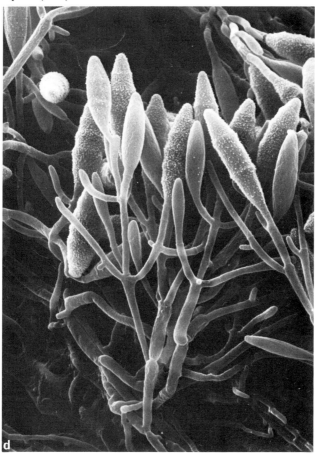

Reliable reports, including induction of the teleomorph, exist from Easter Island (75), Australia, New Zealand, the USA and the British Isles, from skin lesions and soil (5587). — The following data refer to the *Microsporum gypseum* complex, as it is commonly isolated from soil by the hair bait technique. This is a saprophytic soil inhabitant with a worldwide distribution (4327); it can, however, become infectious and cause dermatophytoses in man and animals (1624). Records of its occurrence in soil include ones from New Zealand (3613, 3614), New South-Wales (2092), tropical Queensland (4636), Polynesia (3614), Hawaii (3014), Taiwan (6119), Chile (6544, 6545), Colombia (4896), Peru (78), Brazil (4488, 4897), Argentina (6047), the Bahamas (6117), New York, Texas, Georgia and Tennessee (67, 88, 6269), Scotland (1312), Sweden (4368), the USSR (6533, 6534), Poland (3530), Czechoslovakia (1013, 1014, 5338), Germany (3743, 4536, 5521), Belgium (79), France (1197, 6106), Hungary (2541), Romania (118), Italy (80, 403, 886, 893), Spain (1216), Turkey, Iran, Afghanistan, Pakistan (2366), Tunisia (2721), Israel (1733), Egypt (89, 5766), the Sudan, Ethiopia (5766), Nigeria (77), Guinea (1423), equatorial West Africa (1420), Kenya, South Africa (89, 90), Iran (103), India (2644, 4345, 4346, 4932) and Hong Kong (5951). It has been isolated with higher frequencies from cultivated but unmanured soils than from uncultivated ones (5951), frequently from garden soil (893, 1733, 2721, 4345), soils from or close to animal and human dwellings (306, 893, 1420, 2644, 4345, 4536, 4932, 6533), grassland (1013, 1014), frequently from soils of marmot burrows (403), caves (88, 1216) and even uninhabited mountainous regions (3742). Other habitats include beach sands (3014), marine muds (4346), a sewage treatment plant (5337), and silage (6533). *M. gypseum* has been commonly recorded from soil under starling roosts, feathers and pellets of free-living birds (4346, 4650), the hair and skin of monkeys, dogs, mice, rats (2155, 3754) and other small mammals (4324); in the tropics, it also occurs on woollen fabrics (6326). It has been recognized as the causal agent of dermatomycoses in cattle and man in Brazil (2540), Uruguay (3502), Zaïre (6039), Czechoslovakia (1012), France (4504) and Germany (4840).

Soil-hair agar plates (4347) and alphacel medium (6270) are suitable for the induction of the teleomorph (4347); no ascomata are produced in matings on a modified Sabouraud medium (5695). — *M. gypseum* grows rapidly in culture and the typical multiseptate macro-conidia are produced at a very early stage (67). Physiological and biochemical changes in germinating macro-conidia have been thoroughly studied (1395, 4356, 4358). No growth occurs below pH 5 (3175). — L-Serine proved to be an excellent N source; L-alanine, L-valine, L-isoleucine, L-aspartic and L-glutamic acids, L-phenylalanine, L-tyrosine and L-proline are also utilized (2541, 5347). In the presence of glucose, cystine was intensively metabolized with the release of sulphite, sulphate and S-sulphocysteine into the medium (3176, 3178). Cystine, S-sulphocysteine, cysteic acid as well as oligo- and polypeptides have also been found in the course of keratin decomposition (3177, 3179). *M. gypseum* showed amylase (6570), amidase, aminopeptidase (cytosol), carboxypeptidase (1263), elastase (6269) and high lipolytic activities (583, 4201). Free amino acids are released during the germination of the macro-conidia; in addition, the medium used in germination has been found to contain a proteinase capable of hydrolysing conidial walls (3281) and having exo- as well as endokeratinase activity (4359). The fungitoxic antibiotic nystatin (879) and the fungicide nitrofungin (880) can be degraded by adapted strains. *M. gypseum* produces ferricrocin (3231) and two antibiotically active antimetabolites (1743). Ammonium-magnesium phosphate is precipitated when grown on a keratin substrate supplemented by KH_2PO_4 and $MgSO_4 \cdot 7H_2O$ (2880); phosphate stimulates keratin degradation (3175). — Growth *in vitro* can be inhibited by garlic extract (5744). *N. gypsea* tolerates 10% NaCl in the medium, no growth occurs at 12% NaCl (2875). Hyphae are sensitive to UV irradiation and are killed by five days exposure to $<14\,000\,\mu\mathrm{W} \cdot \sec^{-1} \cdot \mathrm{cm}^{-2}$; conidia last for six days at $56\,000\,\mu\mathrm{W}$ (1001).

Nannizzia incurvata Stockd. *1961* — Fig. 209.

Anamorph: *Microsporum gypseum* (Bodin) Guiart & Grigorakis *1928* pro parte

DESCRIPTIONS: Stockdale (5586, 5587), and Hejtmánková and Hejtmánek (2368); ascoma development: Kwon Chung (3189). — A heterothallic species. Ascomata mostly 350–500 μm diam (excluding the appendages), white at first but becoming yellowish buff at maturity. Peridial hyphae bearing up to five often convergent verticillate branches, otherwise as in *N. gypsea*. Ascospores smooth-walled, lenticular, 2·5–3·6 × 1·5–2·0 μm. Colonies reaching 5·0–6·0 cm diam in 14 days on 2% MEA at 25°C, growth mostly submerged and strongly radiating, with a thin, pale buff surface growth; reverse deep ochraceous-amber, occasionally in zones, and with rosy-vinaceous traces. Macro-conidia as in *N. gypsea*, mostly 40–53 × 11–14 μm, often with a whip-like appendage. — Heterokaryosis has been demonstrated between morphological and biochemical mutants (3286). Heterokaryotic constitution cannot be transferred by micro-conidia which are uninucleate; in non-septate macro-conidia the numbers of nuclei range from 2 to 21; in septate macro-conidia most nuclei are located in the central cells and the upper part (3288); the number of chromosomes was estimated to be n = 7 (2370). More rapidly growing sectors which arise from recombinant mutants have been shown to be diploid (3289). Somatic karyokinesis has been demonstrated (2369, 2370). Parthenogenetic production of fertile cleistothecia was observed under the influence of culture

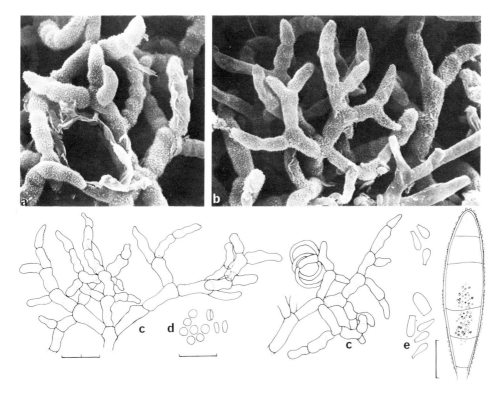

FIG. 209. *Nannizzia incurvata*. a, b. Peridial appendages, SEM, × 1000; c. peridial appendages; d. ascospores; e. micro-conidia and a macro-conidium; CBS 173.64 × M 139.

filtrates from eight *Chrysosporium* species; the conidia derived however, differed morphologically from the parent strain (442). The pyrolysis-gas-liquid-chromatographic patterns of whole cells of *N. incurvata* have been compared with those of other Gymnoascaceae (903, 5188). — SEM studies of conidiogenesis and surface structure of macro-conidia showed no difference between this species and *N. gypsea* or *N. fulva* (4561).

Soil isolates of *N. incurvata* are known from Germany (582, 584), Czechoslovakia (1013), the Netherlands (CBS) and Iran (103); it has also been reported from free-living birds (2575) and the fur of the field mouse *Microtus arvalis* (762).

Ascomata are produced after mating on hair or chicken feathers above soil, and also on a modified Sabouraud medium (5695), alphacel cellulose (6270), MEA, OA, "pablum" cereal or rabbit dung agar (1313, 4347, 5586). The maximum temperature for ascoma production was found to be 30°C and continuous light was inhibitory; the tolerated pH range of *N. incurvata* is relatively wide, good growth occurring in the range pH 5–8; the optimal temperature for growth is in the range 22–30°C, the minimum 4°C, and the maximum 34°C (1313).

Nectria (Fr.) Fr. *1849*

≡ *Hypocrea* sect. *Nectria* Fr. *1825*
= *Sphaerostilbe* Tul. *1861*
= *Dialonectria* (Sacc. *1883*) Cooke *1884*
= *Lasionectria* (Sacc. *1883*) Cooke *1884*
= *Calostilbe* Sacc. & P. Sydow *1902*
= *Nectriopsis* Maire *1911*
  ≡ *Hyphonectria* (Sacc. *1883*) Petch *1937*
= *Neonectria* Wollenw. *1917*
= *Neuronectria* Munk *1957*

Type species: *Nectria cinnabarina* (Tode ex Fr.) Fr. (lectotype)

Anamorphic genera: *Tubercularia* Tode ex Fr., *Acremonium* Link ex Fr. (see *A. butyri*), *Cylindrocarpon* Wollenw. (q.v.), *Fusarium* Link ex Fr. (q.v.), *Verticillium* Nees ex Link, *Stilbella* Lindau, *Myrothecium* Tode ex Fr., and others.

Contributions to a MONOGRAPHIC TREATMENT: Booth (636) for the British Isles, and Samuels (5032) for species formerly placed in subgenus *Hyphonectria*. — The genus *Nectria* is placed by some authors in its own family, *Nectriaceae* Martin, and by others in the *Hypocreaceae* de Not. — Ascomata (perithecia) ± superficial, often gregarious, with or without a basal stroma, fleshy, cream, orange, red, purple or violet, glabrous or setose. Paraphyses absent. Asci cylindrical, unitunicate, with an undifferentiated apex. Ascospores ellipsoidal or naviculate with ends ± rounded, equally 2-celled, smooth, minutely spinulose or striate, hyaline, yellow or brownish.

The structure of the ascoma wall is now regarded as the primary criterion for subdivision of the genus into several groups (5032). Over 650 names have been published in *Nectria,* of which 32 have been treated by Booth (636) for the British Isles, and 35 formerly placed in subgenus *Hyphonectria* by Samuels (5032); the majority of epithets in the genus thus await a critical revision. The species treated here are keyed out under their anamorphic genera.

Nectria coccinea (Pers. ex Fr.) Fr. *1849* — Fig. 210.

≡ *Sphaeria coccinea* Pers. ex Fr. *1823*

Anamorph: *Cylindrocarpon candidum* (Link ex Gray) Wollenw. *1926*

≡ *Fusidium candidum* Link ex Gray *1821*

DESCRIPTIONS: anamorph: Wollenweber (6422), Nicot (4158), Booth (639), and CMI Descriptions No. 532, *1977*; teleomorph: Booth (636). — Colonies reaching 4 cm diam in two weeks on PDA at about 20°C, golden-yellow to brownish, floccose, felted or fibrose, with reddish brown stromatic pustules; reverse brownish. Micro-conidia abundant, ellipsoidal to cylindrical, 4–9 × 1·5–3·0 μm, arising from little-branched conidiophores; macro-conidia tardily produced, regularly curved, cylindrical with the end cells tapering slightly towards the rounded ends, mostly 4–7-septate, 50–80 × 5·0–7·5 μm. Chlamydospores absent. — Ascomata commonly form on beech and other deciduous wood (4442) and are also obtained in culture after mating compatible isolates (639, 4441) on sterilized beech twigs on MEA after seven weeks. Ascomata aggregated on erumpent stromata, bright red, ovate to subglobose with pointed ostiolar papillae, smooth-walled, 250–350 μm diam. Ascospores ellipsoidal, hyaline, smooth to slightly verrucose, 12–15 × 5–6(–7) μm. — The four varieties *crassum,*

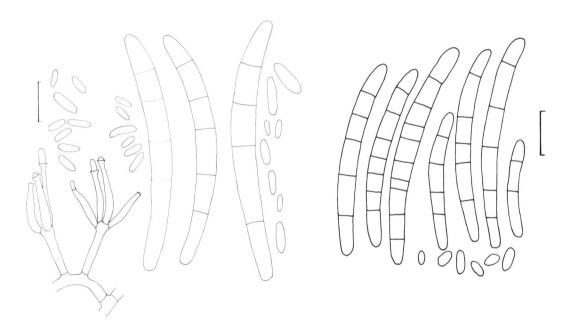

Fig. 210. *Nectria coccinea,* conidiophores forming micro-conidia, and macro-conidia.

majus, medium and *minus* Wollenw. *1928* have not been regarded as sufficiently distinct to merit recognition, but *Cylindrocarpon faginatum* C. Booth *1966*, anamorph of *N. coccinea* var. *faginata* Lohman *et al. 1943*, with larger macro-conidia 60–110 × 6–7 μm, has been retained (639). — Other similar species include *Cylindrocarpon album* (Sacc.) Wollenw. *1917*, the anamorph of *Nectria punicea* (Schmidt ex Fr.) Fr. ex Rabenh. *1858*, a weak pathogen of various trees, with 5–7-septate conidia 44–80 × 5–7·5 μm, but up to 9-septate ones 120 μm long; *Cylindrocarpon heteronema* (Berk. & Br.) Wollenw. *1928* (= *C. mali* (Allesch.) Wollenw. *1928*), the anamorph of *Nectria galligena* Bres. *1901*, causing cankers on fruit and other trees, has almost straight conidia, mostly 4–5-septate, and 45–65 × 4–7 μm.

N. coccinea is very frequently found on twigs and wood of numerous host plants, but is most common on *Fagus.* It also occurs on 15 other deciduous tree genera, *Taxus* (636), on dung and in soil (639). Contrary to *Cylindrocarpon willkommii* (Lindau) Wollenw. *1928*, the anamorph of *Nectria ditissima* Tul. *1865*, which causes beech canker, this fungus is associated with a bark disease occurring in succession to the scale insect *Cryptococcus fagi* (3557, 4441). — *N. coccinea* is known from Europe, North America and New Zealand (639); reports are available from France (3557), the British Isles (2338, 4440), Germany (6572), Denmark, the Netherlands, Sweden, Czechoslovakia, India (IMI), Canada (234) and Peru (2005). It has been isolated from cultivated soil under red clover (234), chitin strips buried in sand dunes (IMI), herbivore dung (4439) including horse and rabbit dung (6157), and decaying roots of *Manihot utilissima* (1568). — The fungus ranks as a weak competitor in comparison with other *Cylindrocarpon* species (3682, 3683). It was found to be colonized and inhibited by the mycoparasite *Nematogonium ferrugineum in vitro* (5284).

The highest growth rate is reached in the temperature range 22–27°C, but significant growth occurs at 5°C and little at 2°C (4441). — From culture filtrates and mycelial extracts the metabolites chloronectrin, the antibiotically active ascochlorin, and two of the latter's derivatives have been isolated (94, 5728).

Nectria haematococca Berk. & Br. *1873* var. *haematococca* — Fig. 211.

≡ *Hypomyces haematococcus* (Berk. & Br.) Wollenw. *1926*
= *Nectria citri* P. Henn. *1909*
= *Nectria asperata* Rehm *1909*

Anamorph: *Fusarium eumartii* Carp. *1915*

≡ *Fusarium solani* var. *eumartii* (Carp.) Wollenw. *1931*

DESCRIPTIONS: Wollenweber and Reinking (6424), Gerlach and Ershad (1949), and Joffe (2770); teleomorph: Booth (637), Samuels (5032), and Hanlin (2254), the latter with SEM pictures of ascospores. — Anamorph belonging to *Fusarium* section *Martiella* Wollenw. — Colonies rather fast growing, reaching 7·5 cm diam in ten days at 24°C on OA, macroscopically resembling those of *F. solani*, i.e. usually with green tints in the agar surface and cream to buff conidial slime in sporodochia or pionnotes. Micro-conidia produced on long, slender phialides, not abundant, 4–14 × 2·5–5·0 μm. Macro-conidia produced on shorter, flask-shaped phialides, abundant, falcate, moderately curved, with bluntly beaked apical and pedicellate basal cells, mainly 5–7-septate (but to 8–9-septate), (36–)50–85 × 5–7(–8) μm. Chlamydospores as in *F. solani*. This species has homothallic and heterothallic strains and perithecia are often produced in culture. Perithecia coarsely tuberculate and frequently found on the natural substrate in the tropics (637, 640). Ascospores two-celled, densely striate, 11–16 × 4·5–6·0 μm.

F. eumartii is often synonymized with *F. solani*, as was suggested by Snyder and Hansen (5460) but the common occurrence of multiseptate broad macro-conidia appears to justify its separation from typical *F. solani* with generally 3–5-septate and much shorter conidia. It is also regarded by many authors as synonymous with *F. javanicum* Koord. *1907* (teleomorph *Nectria ipomoeae* Halst. *1892*), but this synonymy has not been accepted by Gerlach (1945, 1948, 1949) because the macro-conidia are larger, commonly 6–8-septate and more robust, whilst those of *F. javanicum* are given as commonly 5–6-septate and 46–60 × 4·5–5·5 μm. — Whilst

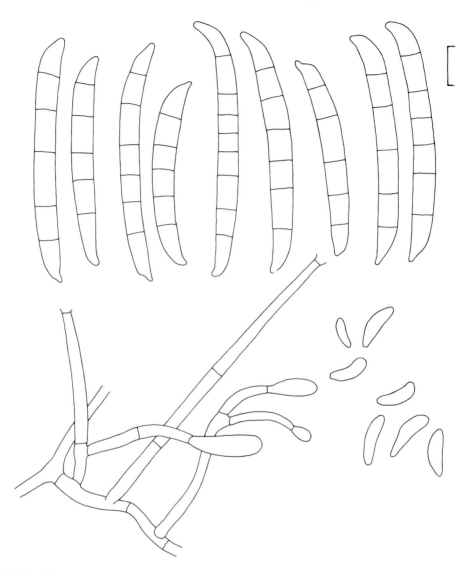

FIG. 211. *Nectria haematococca*, macro-conidia, condiophores and micro-conidia, CBS 225.58.

in *F. solani* numerous host-specific forms are known, the connection of a pathogenic f.sp. *pisi* Reichle *et al. 1964* with this teleomorph has been proven unequivocally only once (4790). — While the teleomorphs of *N. haematococca* var. *haematococca*, *N. haematococca* var. *brevicona* and *N. ipomoeae* can hardly be distinguished, *N. ventricosa* C. Booth (q.v.) differs from *N. haematococca* by its smooth-walled perithecia and verrucose ascospores, 14–17 × 7–10 μm (637). The conclusion that *N. haematococca* and *Hypomyces solani* Reinke & Berth. 1879 (≡ *N. ventricosa*) should be synonymous (4683) was based on a misidentification. — The majority of conidial cells are uninucleate, but three- or plurinucleate cells occur in hyphae (514, 2496).

N. haematococca is a common parasite of potato tubers in North America (where it is best known under the name *F. eumartii*) and on various other plants, particularly in the tropics of all continents, e.g. *Coffea, Theobroma, Hevea, Ficus, Manihot* and many others (6424, 6425, IMI). — This fungus has only rather rarely been reported in soil isolation studies as distinct from *F. solani*: in Italy in rice (3913), Syria (277, 5392), Israel (2773, 2774), Honduras (4794), India in dry tropical forests soils (4716) and cultivated soils (4996), in Czechoslovakia in soils with steppe type vegetation (1702) and in a uranium mine (1703). It is also known from the rhizospheres of clover (3429) and wheat (4558) and sugar beet roots (4559); in Germany commonly from various plant roots (mainly orchids) in warm greenhouses (W. Gerlach, pers. comm.).

The optimal temperature for growth is 24°C and the optimal pH 7; growth is better in darkness than in light when the fungus is kept above the optimum growth temperature, i.e. at about 32°C (2774). — Perithecia are formed (in suitable isolates) on most of the usual media, but preferentially on natural substrates, after 2–4 weeks; light is not essential for perithecium formation (4683) but, according to other reports, perithecium formation does respond favourably to light in the near-UV range (1243). The production of perithecia has been found to be inhibited by morphactin (2274). — The growth of this species is supported by most of the usual carbohydrates, including sorbose, rhamnose, arabinose, xylose, ribose, starch, sorbitol, mannitol, lactate and gluconate (1000). *N. haematococca* forms L-asparaginase (4094) and degrades both cellulose (6060, 6326) and birch wood (3568); woollen fabrics are also attacked (6330). Known metabolic products include fusarubin, marticin, isomarticin (2936), hydroxyjavanicin and javanicin (188) which has antibiotic properties. — The growth of *Chalara elegans* was inhibited *in vitro* (2753). Mycelial extracts caused toxic reactions on the skin of rabbits (2774).

Nectria inventa Pethybr. *1919* — Fig. 212.

? = *Sphaeria erythrella* Wallr. *1833*

Anamorph: *Verticillium tenerum* (Nees ex Pers.) Link *1824*: Fr. *1832*

    ≡ *Verticillium tenerum* Nees *1816* (deval. name)
    ≡ *Botrytis tenera* Nees ex Pers. *1822*
  = *Sporotrichum lateritium* Ehrenb. *1818* ex Link *1824*
    ≡ *Verticillium lateritium* (Ehrenb. ex Link) Rabenh. *1844*
  = *Sporotrichum luteo-album* Link *1809* ex Link *1824*
    ≡ *Verticillium luteo-album* (Link ex Link) Subram. *1971*
  = *Sporotrichum cinnamomeum* Wallr. *1833*
  = *Verticillium ochrorubrum* Desm. *1834*
  = *Verticillium affine* Corda *1837*
  = *Acrostalagmus cinnabarinus* Corda *1838*
    ≡ *Verticillium cinnabarinum* (Corda) Reinke & Berth. *1879*
  = *Verticillium rufum* Rabenh. *1844*
  = *Verticillium ruberrimum* Bonord. *1851*
  = *Acrostalagmus fulvus* Berk. & Br. *1875*

Nomenclature according to Hughes (2596) and Subramanian (5626).

DESCRIPTIONS: teleomorph: Pethybridge (4527); anamorph: Hughes (2596), Subramanian (5626), and Matsushima (3680). — Colonies spreading broadly, reaching 3·5–4·0 cm diam in ten days at 20°C on MEA, thinly velvety, brick, dull orange to orange-brown due to the pigmented conidiophores and conidia. Conidiophores densely crowded, erect, pale orange-brown, 200–370 μm tall, repeatedly verticillately branched in one to several orders. Phialides slender, flask-shaped, hyaline, 12–23 × 2–4 μm. Conidia oval, pale reddish brown in mass, 3·5–5·0 × 2·0–2·5 μm. Chlamydospores, microsclerotia and resting mycelium absent. — The teleomorph was found on old potato tubers: perithecia dark red, the upper part covered with short stiff hairs; ascospores ellipsoidal, two-celled, hyaline, 9–10 × 4–5 μm. — The identity of *V. tenerum* was confirmed by the discovery of the type specimen in the Munich herbarium (M). The connection with the teleomorph has been confirmed only once by single ascospore isolates (4527). Deviating sectors commonly occur in agar cultures (2674).

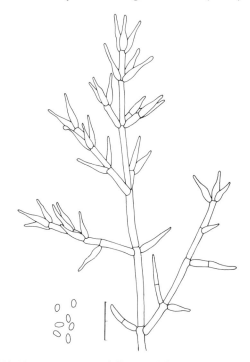

FIG. 212. *Nectria inventa*, verticillate conidiophore and phialoconidia.

N. inventa is a very common cosmopolitan fungus, mainly growing on various organic remains such as herbaceous stalks and twigs of various shrubs and trees, exposed wood, leaves and fruits, slime fluxes, potato, beet root etc., cloth, paper, dung, milk and soil (2596). It can be isolated successfully from soil by a cellulose agar technique (1379). Because of its striking dull orange to orange-brown coloration, it has often been reported in soil analyses, though it seems, in spite of its wide distribution, not to be a major component of the total soil mycoflora. — Reports from soil include ones from the Netherlands (1614, 1616), the British Isles (163, 1379, 1598, 5559), Canada (505, 6352), the USA (1163, 2008, 2010), the USSR (1482, 2871), Poland (1421), Austria (1823), Germany (1424, 1889, 5316), Italy (270, 3452), France (918, 2161, 3941, 4162), Spain (3417), Turkey (4245), Somalia (5048, 5049), the Sahara (3415), South Africa (1559, 4407), India (3865, 4698, 5000), Sri Lanka (2596), Nepal (1826), Japan (2532, 3680), Hawaii (569) and Australia (2596). It is known from uncultivated (4698), various forest (270, 505, 569, 1598, 2008, 3417, 5048, 6352) and arable soils (569, 918, 1614, 1616, 1889, 2161), grassland (163, 3417, 3865), salt-marsh (413, 1379, 4646), an alkaline peaty soil (5559), cut peat (1376), sand dunes (3941, 4162, 4655), heathland (3720), soils with steppe type vegetation (1559, 4407), garden compost (1424), and soils treated with sewage sludge (1163). Other known habitats include running and still water (4429) with varying degrees of pollution (1154, 1165, 1166, 1482), river sediments (1162), estuarine silt (5316), and sticks exposed to seawater (4546). *V. tenerum* can penetrate only to moderate soil depths (2161) and its frequency is scarcely influenced by soil pH. No particular accumulation on plant roots is known, but there are reports from the root of sugar beet (4559), the rhizospheres of wheat (5330), various steppe grasses (3376), groundnuts (2532), oak (963), poplar (3452), hazel (1794), and various other forest plants (4814). It has been isolated from pine litter (681), seeds of wheat (2716, 4492), oats (3556) and *Avena fatua* (2961), germinating leguminous seeds (4120), cotton fibres in conveyor belts (6163), nests, feathers and pellets of free-living birds (2575), and aeciospores of *Cronartium comandrae* (4608).

The optimal temperature for growth is in the range 15–25°C (1827). Conidiation is stimulated by violet, blue, or blue-green light (3619). — Pectin decomposition has been demonstrated (3127, 3414, 5926), as has the utilization of starch (1425, 1827, 5926) and cellulose in various forms (1425, 2718, 2878, 3127, 3414, 4655, 5926); in cellulose decomposition little variability between several isolates occurred (1432); the fungus can utilize urea as C source (3127, 3414). Chitin is also decomposed (1425, 4264, 5926) and the growth on lignin sulphonate is classed as very good (1425). *N. inventa* causes considerable losses in both weight and tensile strength when growing on maple-wood strips (2211). It produces the piperazine derivative $11a$, $11'a$-dihydroxychaetocin, which has antibacterial properties (2325), and also biotin (3833). It is autotrophic for growth substances. — It displays an antagonistic activity *in vitro* towards *Phytophthora cactorum* (1425), *Aspergillus fumigatus* (569) and a weak activity towards *Gaeumannomyces graminis* (1425) and it was shown to invade hyphae of *Alternaria brassicae* (5917); it was found to inhibit the growth of barley seedlings (4919), but sometimes to stimulate that of citrus seedlings (3635). Water-soluble pigments causing wilt symptoms in cotton were isolated from the mycelium and culture liquid, amongst which a compound ferrilaterin was characterized as a Fe(III)-trihydroxamate complex (980). Toxicity to warm-blooded animals has been shown using infested cereals (6412). — *N. inventa* was isolated from keratomycosis in one of 88 cases (1992).

Nectria radicicola Gerlach & L. Nilsson *1963* — Fig. 213.

Anamorph: *Cylindrocarpon destructans* (Zinssm.) Scholten *1968*

   ≡ *Ramularia destructans* Zinssm. *1918*
  = *Allantospora radicicola* Wakker *1896*
  = *Moeszia cylindroides* Bubák *1914* sensu Tubaki *1958*
  = *Cylindrocarpon radicicola* Wollenw. *1926*

DESCRIPTIONS: Wollenweber (6422), Nicot (4158), Booth (639), Gerlach (1941), Scholten

(5131), Gerlach and Ershad (1949), and Matsushima (3680); teleomorph: Gerlach and Nilsson (1950). — Colonies reaching 10–12 mm diam in seven days on PDA at about 20°C; aerial mycelium floccose to felted, whitish, beige to pale brown or reddish brown; reverse beige to deep reddish brown; odour often suggesting toilet soap. Conidiophores repeatedly branched and bearing subulate phialides, 18–35 μm long, 2·5–3 μm wide when bearing micro-conidia, to 4·5 μm when bearing macro-conidia; sporodochia with cream to beige conidial slime commonly produced. Macro-conidia cylindrical in the central part, straight or slightly curved, the apical cell symmetrically rounded in sporodochial conidia but in others slightly asymmetrical, mostly 1–3-septate, 20–40(–50) × 5–6·5(–7·5) μm; micro-conidia little differentiated from the macro-conidia, 6–10 × 3·5–4 μm. Chlamydospores abundantly produced, intercalary or terminal, singly or in chains, smooth or warted, mostly somewhat

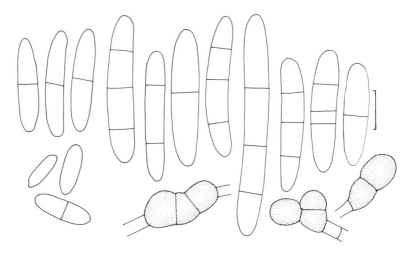

Fig. 213. *Nectria radicicola,* macro-conidia, micro-conidia, and chlamydospores, CBS 158.31.

pigmented, 9–16 μm diam. — The teleomorph has been found on cyclamen tubers (1950) and on various tree ferns in a greenhouse (6224). Ascomata scattered, red to red-brown, subglobose, 150–320 μm diam; ascospores hyaline, smooth-walled, mostly 10–13 × 3–3·5 μm. The teleomorph cannot be distinguished from that of some other similar species without a knowledge of the conidia. — The var. *crassum* (Wollenw.) C. Booth *1966,* known from Europe and Jamaica, has 8–11 μm wide conidia. Atypically developed conidia of *N. radicicola* may be mistaken for *C. didymum* (Hartig) Wollenw. (q.v.) or *C. obtusisporum* (Cooke & Harkn.) Wollenw. *1926.* Another similar species without mycelial chlamydospores is *C. lucidum* C. Booth *1966.*

N. radicicola is the commonest species of *Cylindrocarpon* and has been frequently reported from soil, but is particularly common on roots and other underground parts of a large number of woody and herbaceous plants. — It is known from Europe, North America, New Zealand, Australia, Borneo, Tanzania, Nigeria (639) and Iran (1949), but reports of its occurrence in subtropical and tropical areas are relatively rare. This species is isolated fairly regularly by means of dilution plate and other techniques (2091, 5813). Isolates are known from various forest soils, particularly nurseries and litter. It has often been recorded with high frequencies in grassland, garden and arable soils, peat and heathland; individual reports are too numerous to be listed here. In flowing water, the species was found only when the degree of pollution was low (1154). It can penetrate to considerable soil depths (to 100 cm) (159, 5812), but in one forest soil its highest frequency was in the lower half of the A-horizon (6182). The available data show that the wide pH range 3·1–8·0 is tolerated but neutral or slightly acid (159, 4523, 5559, 5758, 5814, 5815, 6182) and moist (3109, 4313, 5758, 6182) habitats are preferred. Its frequency in soil was found to be reduced by farmyard manure (2764) but increased by N fertilization (3498). After a forest fire, it reached a higher frequency than in unburnt soil (6353). An alternation of this common species with *Fusarium oxysporum* in different soils has sometimes been observed and can be explained by the fact that *N. radicicola* occurs particularly in cool, moist soils with low acidity, but is absent in soils with seasonal desiccation (5755, 5815). The fungus was isolated with significantly greater frequencies in wheat fields after growing wheat than after either pea or rape crops (1433), while in potato fields its frequency was higher after wheat than after barley (1617). It has been isolated with higher frequencies from washed organic than from mineral particles (1888, 4449). It is further reported from sticks buried in soil (2833). — *N. radicicola* is well known as a colonizer of the root surface of numerous plants, including spruce (3573), *Pinus strobus* and *Larix leptolepis* (1794, 3683), *Fraxinus* (3141, 3156), poplar (5898) and various other forest trees (3142, 4814), *Halimione portulacoides* (1378), pear (5499), strawberry (3683), flax (5826), cabbage (4451), bean (4451, 4452, 4453, 5759, 5801), lupins (4400), clover (5815, 6493), broad beans (5757), onion (2159), rye grass (5815), *Elytrigia repens* (4449), oats (4443), barley (4450, 4451), and wheat (3567, 4523, 4727, 5330). It has also been isolated from the rhizosphere of *Pinus nigra* var. *laricio* (4448), various peat bog plants (1376), wheat (1614, 3567), groundnuts (2768), sugar beet (1614), onion (4447), the mycorrhizae of *Tuber magnatum* with poplar (3452), *Suillus luteus* with *Pinus strobus* and of *S. grevillei* with *Larix* species (1794). Garlic root extracts were found to inhibit its growth (1072) whereas pea roots had no chemotropic influence, at least on conidial germ tubes (3684). Optimum colonization of hair baits in soil occurred between −40 bars water potential and saturation (2095). — The saprophytic competitive ability of *N. radicicola* is low compared with that of *Fusarium oxysporum* and *F. culmorum* (3683, 5755). It can survive storage in water for ⩾6 years (605). The micro-conidia scarcely germinate in soil and have a marked tendency to form chlamydo-spores accompanied by a lysis of both conidia and mycelium, in which process coryneform bacteria are mainly involved (3684). Its growth *in vitro* is also inhibited by *Trichoderma viride* (6029), *Zygorrhynchus moelleri*, *Paecilomyces lilacinus*, *Penicillium janthinellum* and espe-cially *Gliocladium roseum* (5755). *N. radicicola* occurred in cysts of the nematodes *Globodera rostochiensis*, *Heterodera avenae* and *H. schachtii* (3090) and with particular frequency in culture vessels of an earthworm (*Allolobophora caliginosa*) (1961).

The optimal temperature for growth is 20–21°C (4914, 5755) or lower (1827), and the maximum 37°C; the thermal death point has been determined as 47·5–50°C for 30 min (613). The optimal pH for growth *in vitro* is 4 (5755). — Sucrose, glycerol, and succinic and lactic acids are suitable C sources. Gluten and also uracil, which is not attacked by many other fungi, can serve as N sources (4914). In shake cultures, glutamic acid, proline, phenylalanine and leucine were also utilized well, whereas histidine, aspartic acid, threonine and valine were poorly utilized (228). Though there was great variation between individual isolates, pectin, xylan, cellulose, CMC (1432, 2706), and starch (1827) were also found to be utilized well. Growth was also relatively good on *p*-hydroxybenzaldehyde, ferulic acid and vanillin (4914). Up to 4·6% tannin is tolerated in the culture medium (2706). *N. radicicola* is prototrophic (1425, 4914). Calcium phosphate is solubilized (5210). Twelve amino acids were found to be secreted into the medium (228) and fusaric acid and Δ^2-anhydromevalonic acid lactone produced (2918). — An antagonistic activity, but with variability between isolates, has been observed against numerous saprophytic and phytopathogenic fungi (1431, 2868, 6325). *N. radicicola* produces the mycostatic antibiotics radicicolin and monorden (= radicicol) (1663, 3231, 3860, 6029, 6592) in addition to the phytotoxic necrolide (= brefeldin A, cyanein, decumbin) (1658). — *N. radicicola* is regarded as a weak parasite of higher plants but has gained importance in the ornamentals cyclamen (host-specific isolates), *Begonia, Ficus, Hydrangea, Saintpaulia* (1941, 5131), and rhododendron stocks (6225), besides pine seedlings, oak (2285, 3047), alfalfa, sweet clover (1188), *Lupinus arboreus* (3683) and *Panax ginseng* (1050). It is also responsible for a storage rot in carrots (1215) and potatoes (639), and severe injuries on seedlings of *Pinus banksiana, Caragana arborescens* (6028) and, *in vitro*, on pea plants have been observed (1430).

Nectria ventricosa C. Booth *1971* — Fig. 214.

≡ *Hypomyces solani* Reinke & Berth. *1879* [non *Nectria solani* Reinke & Berth. *1879*] — [non *Hypomyces solani sensu* Snyder & Hansen *1941*]
≡ *Hyponectria solani* (Reinke & Berth.) Petch *1937*
≡ *Nectriopsis solani* (Reinke & Berth.) C. Booth emend. *1959*

Anamorph: *Fusarium ventricosum* Appel & Wollenw. *1910*

≡ *Fusarium solani* var. *ventricosum* (Appel & Wollenw.) Joffe *1974*
? = *Fusarium argillaceum* (Fr. *1832*) Sacc. *1886*
= *Fusarium cuneiforme* Sherb. *1915*

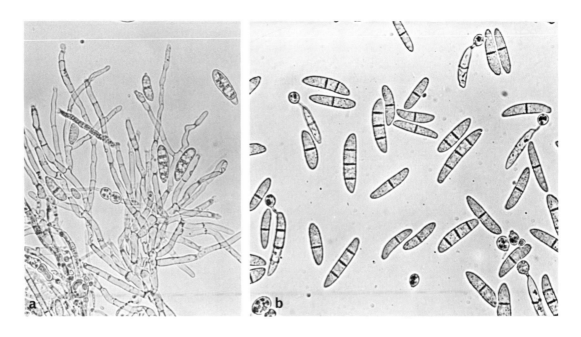

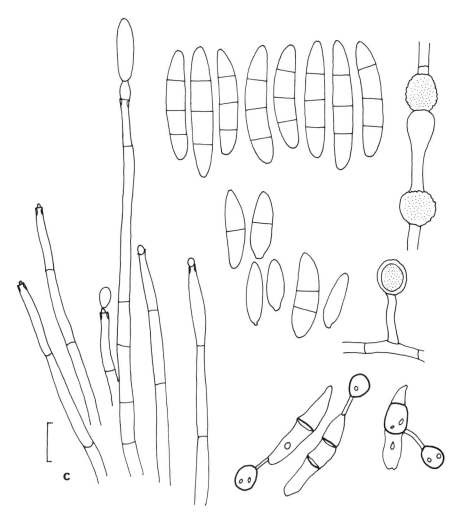

FIG. 214. *Nectria ventricosa*. a. Conidiophores; b. conidia, partly with chlamydospore formation, × 500 (orig. Inst. f. Mikrobiologie, Biologische Bundesanstalt, Berlin–Dahlem); c. conidiophores, conidia and chlamydospore formation in hyphae and conidia (orig. W. Gerlach); isolate 8743.

DESCRIPTIONS: Booth (637, 640).—Anamorph belonging to *Fusarium* section *Martiella* Wollenw. or *Ventricosum* Wollenw. — *F. ventricosum* differs from the other species of section *Martiella* in the absence of sporodochia and distinct micro-conidia and has therefore often been placed in a separate section *Ventricosum*; the rather broad and blunt macro-conidia and terminal and lateral chlamydospores, however, agree with those of sect. *Martiella*. — Colonies reaching 3 cm diam in four days at 25°C on PSA or OA, white to light grey, with some floccose aerial mycelium, later developing a yellow-brown discoloration under light. Conidiophores branching near the base, the long branches repeatedly septate, producing both macro- and micro-conidia, not aggregating in sporodochia. Conidia 0–3-septate, slightly curved, widest above the middle, with a blunt tip, gradually tapering (cuneiform) towards the hardly pedicellate base, 24–48 × 5–8 µm when 3-septate. Chlamydospores abundant, formed singly or in groups of 2–3, terminal or lateral on short stalks, globose, hyaline, warted or smooth-walled, 8–12 µm diam. — The genus *Nectriopsis* Maire *1911*, originally distinguished by perithecia seated on or immersed in a subiculum, is no longer considered as distinct from *Nectria*. Perithecia are known from decaying potatoes only and differ from those of the *N. haematococca* complex in the smooth peridium and warted, light brown ascospores, 14–20 × 7–10 µm. Wollenweber and Reinking's (6424) description of striate ascospores clearly refers to a member of the *N. haematococca* complex; their concept of *Hypomyces solani* was also followed by Snyder and Hansen (5460) and has caused much confusion.

Nectria ventricosa occurs mainly on decaying discarded potato tubers, but does not seem to play a role as a storage rot (640, 3113). Other plant substrates include beets, melons, groundnuts (637, 6425, IMI). — It is occasionally isolated from soil, for example, arable soils in Germany and the Netherlands with low frequencies (CBS), garden compost (1424), and various soils with beech stands and meadow (3127). It has also been reported from the French Atlantic coast (4178), Syria (5392), estuaries and fresh and seawater in Yugoslavia (4850).

Neocosmospora E. F. Sm. *1899*

Neocosmospora vasinfecta E. F. Sm. *1899* — Fig. 215.

= *Neocosmospora vasinfecta* var. *major* Rama Rao *1963*

DESCRIPTIONS: Doguet (1412), Udagawa (5971), Hammill (2240), and Minoura *et al.* (3840). — The genus *Neocosmospora* belongs to the Sphaeriales, family Hypocreaceae de Not., but also resembles the Melanosporaceae Bessey in some respects (4900, 5971). — It is characterized by light orange-brown, superficial, ± translucent, aparaphysate perithecia with cylindrical asci containing subglobose, pale orange-brown to almost hyaline one-celled ascospores. Anamorph *Cephalosporium*-like, but as the colonies grow fast and the hyphae are wide, it recalls the micro-conidial states of *Fusarium* rather than *Acremonium*.

Five species are known, all from soil in warm climates: *N. vasinfecta* E. F. Sm. with subglobose to ellipsoidal ascospores with reticulate walls, 12–16 × 9–10·5 μm, colonies reaching 2·8 cm diam in five days at 25°C on OA; *N. africana* v. Arx *1955* with smaller and almost smooth ascospores (reticulate only when young but similar to *N. vasinfecta* with SEM), 9–10·5 × 8·5–10 μm; *N. parva* Mahoney *1976* with all structures smaller than in the previous species and ascospores 9·5–10 × 5–5·5 μm; *N. ornamentata* Freitas Barbosa *1965* with ellipsoidal to ovoid ascospores with an irregular undulate epispore to 2 μm wide, 8–17 × 7–11 μm, and *N. striata* Udagawa & Horie *1975* with ascospores covered with 8–10 transverse crests, 7–10 × 4–6 μm. — All species produce thick-walled chlamydospores which permit prolonged survival in soil. — Ultrastructure of the ascus wall (3555, 6197) and ascosporogenesis (4782) have been studied.

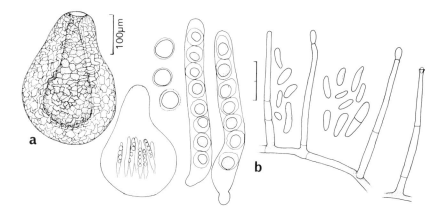

FIG. 215. *Neocosmospora vasinfecta*. a. Perithecia, asci and ascospores; b. conidiophores and conidia; CBS 237.55.

N. vasinfecta is by far the commonest species of the genus (2938, 5971). Without exception, this species has thus far been isolated only from soils in tropical or subtropical regions: the very numerous reports include ones from the southern USA, India, Israel (2764, 2768, 2772), Nepal (3840), Syria (5392), the Congo basin (4159), Chad (3415), South Africa (1415, 1555, 1556, 1559, 4407), Namibia (1827), Argentina (489), Brazil (6008), Peru (2005), Japan (2532, 5971) and the Solomon Islands (5686). *N. vasinfecta* is known from uncultivated soils (4030, 4698), forest soils (1555, 2186, 2240, 2482, 2854, 3817, 4995, 5512, 5841), cultivated soils (3817) under rice (1519), groundnuts (2768, 2772), legumes (1317) and sugar cane (3732), citrus (2764) and coffee (4159) plantations, grassland (1555, 3415, 3863, 3865, 4933), soil under steppe type vegetation (1559, 4407, 5841); it has also been found in an estuarine sediment (655). It has been isolated from the litter of *Eucalyptus maculata* (1558), the rhizospheres of groundnut (2532, 2768, 4735), tobacco (3431), barley (3526) and other plants (5971), roots of cotton (5009), the seeds of groundnuts (3220, 3702) and from chitin buried in tropical soil (4264). — It was found to be inhibited by a sporostatic factor produced by *Fusarium oxysporum* (4882), but it also produces an inhibitor of this type (1905).

Perithecia are easily formed in pure culture on OA, CMA and other media; perithecium formation is said to be favoured by Mn^{2+} ions (349, 351). — The fungus apparently requires Zn^{2+} ions, the uptake mechanism of which has been studied in detail (4467). Cl^- ions are accumulated in the cells against a concentration gradient (3811). D-Arabinose can be utilized (3328), starch (1827), chitin (4264), and straw of both rice and *Pennisetum typhoideum* are decomposed (4611). The phytotoxic metabolites fusarubin, novarubin, marticinine, nor-javanicin and javanicin (2938) are produced. — *N. vasinfecta* is a well-known pathogen which causes root- and fruit-rots and seedling damping-off in Malvaceae, legumes, Piperaceae, Cucurbitaceae etc., including pepper, groundnuts, soya, beans, coconuts, *Albizzia*, *Crotalaria* spp. and others (2938, 4197, 5426, 5783, 5971). It damages pea plants and cress seedlings by excretion of phytotoxins identified as naphthazarin derivatives which also have antibacterial and antifungal activities. It also causes stem browning in soya, bean, and cowpea, but does not appear to invade the vascular tissues (4542).

Neosartorya Malloch & Cain *1972*

Neosartorya fischeri (Wehmer) Malloch & Cain *1972* — Fig. 216.

= *Sartorya fumigata* Vuill. *1927* (sensu Benjamin *1955*)

Anamorph: *Aspergillus fischeri* Wehmer *1907* (described including the teleomorph)

DESCRIPTIONS: Raper and Fennell (4744), Subramanian (5626), Udagawa (5967), and Udagawa and Takada (5984). — Fam. Trichocomaceae Fischer. — The anamorph belongs to the *Aspergillus fumigatus* group. — Colonies reaching 4–6 cm diam in 14 days at 25°C on CzA; conidial masses in pale blue-green shades, but usually not abundant; conidia subglobose to ellipsoidal, slightly roughened, $2\cdot0$–$2\cdot5\,\mu$m diam, or to $3\cdot0\,\mu$m long. Ascomata are abundantly produced on CzA and MEA (without any increased sugar content) at 25°C, and more sparsely at 37°C. The ascomata are white to cream and contain ascospores with distinct equatorial crests, measuring $7\cdot0 \times 4\cdot0\,\mu$m; the crests are often fused and so conceal the underlying furrow. The valves have a network of ridges in var. *fischeri*; three other varieties have been distinguished mainly on the basis of their ascospores: var. *glabra* (Fennell & Raper) Malloch & Cain *1972* with prominent equatorial crests and smooth valves, var. *spinosa* (Raper & Fennell) Malloch & Cain *1972* with spiny valves, and var. *verrucosa* (Udagawa & Kawasaki) Malloch & Cain *1972* with valves with irregular warts. — SEM investigations of the conidia (3395) and ascospores (3394) have been reported. The rodlet pattern on the conidial surface of var. *fischeri* as seen in freeze-etch replicas is very prominent with no underlying surface showing (2422). — DNA analysis gave a GC content of $54\cdot5\%$ (5600).

N. fischeri is thermotolerant and its selective isolation is facilitated by soil steaming (614, 6183). — Data on its occurrence are available from India (2179, 2854, 3732, 4030, 4698, 4995), Pakistan (401), Bangladesh (2712), Australia (978, 3720), Japan (2532, 5846, 5981), the Solomon Islands (5984), New Guinea (5983), Brazil (398, 399), Argentina (6404), the Bahamas (2006), Jamaica (4886), South Africa (1555), Uganda (3839), Zaïre (3790, 4159), Israel (2772), Turkey (4245), the USSR (2871, 5297), the British Isles (5559, 6184), and the USA (1166). Isolations have been obtained from uncultivated soils (4030, 4698), a Himalayan forest (4995), forest soil with *Eucalyptus maculata* (1555), teak forest (2854), forest nurseries (6184), palm plantations (2006, 4159), cultivated soil under sugar cane (3732, 4886), rice (2712), cotton, potatoes (287), and groundnuts (2772), heathland soil (3720), alkaline peaty soil (5559), and an acid mine drainage stream (1166). It has also been frequently isolated from the rhizospheres of barley (3526), rice (2712), banana (2035), and groundnuts (2768), the litter of *Eucalyptus maculata* (1558), chitin buried in tropical soil (4264), milled rice (5967), leather (6326), and paper products (1875). *N. fischeri* is a rapid recolonizer of steam-sterilized soil (6184) and burnt peat soil (615). — It was inhibited by *Memnoniella echinata in vitro* (4).

N. fischeri grows well in the temperature range 26–45°C, with a minimum of 11–13°C and a

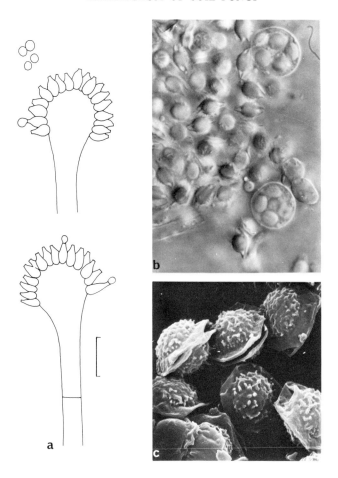

Fig. 216. *Neosartorya fischeri.* a. Conidial heads and conidia; b. asci and ascospores, NIC, × 1000; CBS 865.70; c. ascospores, SEM, × 3000, orig. R. A. Samson, CBS 111.51.

maximum of 51–52°C (517, 3729). Ascospore germination is initiated by exposure to 60–65°C for 30 min (6187). — Potassium pyruvate is a good C source (6124). Hydrocarbons from fuel oil are utilized (4239); cellulosic substrates and wool are attacked (2006, 4775). 1,2-β-Glucanase is produced on the corresponding glucan (4780), and alkali-soluble glucans are present in the mycelium (2049). On an organic medium, more lipid was produced than on a synthetic one (4642). Other metabolites reported include terrein (5278) and, by the var. *spinosa*, ferricrocin, fusigen (1381) and the chlorine-containing fumigachlorin (3231). — *N. fischeri* can grow on media containing 0·2% sinigrin (4774) or 15% NaCl (4697).

Neotestudina Segretain & Destombes *1961*

Neotestudina rosatii Segretain & Destombes *1961* — Fig. 217.

≡ *Zopfia rosatii* (Segretain & Destombes) D. Hawksw. & C. Booth *1974*
= *Pseudophaeotrichum sudanense* Aue, E. Müll. & Stoll *1969*
= *Pseudodelitschia corianderi* Kapoor *et al. 1975*

DESCRIPTIONS: Segretain and Destombes (5176), Segretain *et al.* (5178), Aue *et al.* (241), v. Arx (199), and Hawksworth and Booth (2339).

The family Zopfiaceae Arnaud (incl. Testudinaceae v. Arx) is characterized by dark, cleistothecium-like ascomata arising from parenchymatous initials in the aerial mycelium with the peridial wall often divided into plates of radiating cells (cephalothecoid); asci bitunicate but becoming thin-walled at maturity, obovate to almost spherical; ascospores two-celled, pigmented, occasionally with germ pores. *Neotestudina* has smooth-walled ascospores without prominent germ pores and which are not particularly darkened at the septum.

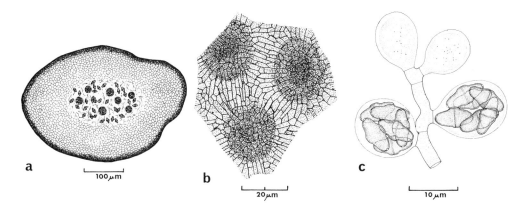

FIG. 217. *Neotestudina rosatii*. a. Vertical section of young ascoma; b. surface view of peridium; c. asci with young ascospores; orig. D. L. Hawksworth (from 2339).

N. rosatii has rather slow-growing colonies, reaching 2·5 cm diam in 14 days on PDA at 20°C, but growing considerably faster at 33°C, initially whitish but later becoming greyish black to brownish black, somewhat felted, reverse black. Ascomata black, 300–450 μm diam, with cephalothecoid and very thick peridium; asci 16–24 × 10–16 μm; ascospores of very variable shape, bicampanulate or biconical, ± constricted at the septum, with a rather thick, brown wall, 9–12 × 5–8 μm. — The other two species recognized in the genus by v. Arx and Müller (211), *N. bilgramii* (D. Hawksw. *et al.*) v. Arx & E. Müll. *1975* and *N. cunninghamii*

(D. Hawksw. & C. Booth) v. Arx & E. Müll. *1975*, have slightly larger ascospores with four paler longitudinal furrows and at the septum less constricted ascospores in more cylindrical asci, respectively; *N. bilgramii* has been renamed *Ulospora bilgramii* (D. Hawksw. *et al.*) D. Hawksw., Malloch & Sivanesan *1979*. Both these species are not known to be pathogenic to man.

N. rosatii was originally isolated from severe white-grained mycetomas in man in Somalia and other African countries (4908, 5176, 5178) and Australia (2339) but is also known from groundnut, coriander, *Vigna sinensis* and soil in Africa, Australia and India (241, 2339). It has repeatedly been isolated from dry soil in Senegal (5177).

Optimal growth occurs in the range 30–37°C (5176) or at 33°C, with the minimum at 12°C and the maximum at 41°C (241); ascomata develop in the range 18–36°C but mature only up to 34°C; light does not stimulate fructification (241). — Vitamins are not required but slightly stimulate growth with ammonium ions as the N source; nitrate and L-asparagine are less suitable N sources. With cellulose instead of glucose, fructification was stimulated on all nitrogen sources tested, whilst on glucose ascomata appeared only with KNO_3 or L-asparagine, particularly at reduced sugar concentrations ($\leq 0 \cdot 2\%$) (241). Maltose and xylose are less suitable C sources (241) and on lactose growth is poor (5176).

Nigrospora Zimmerm. *1902*

= *Basisporium* Moll. *1902*
= *Dichotomella* Sacc. *1914*

Type species: *Nigrospora panici* Zimmerm.

Teleomorph: *Khuskia* H. Hudson

MONOGRAPHIC TREATMENTS: Mason (3659), Potlaĭchuk (4605), Hudson (2584), and Ellis (1603). — Colonies consisting of hyaline or slightly pigmented hyphae. Blastoconidia are formed on broad, flask-shaped conidiophores which arise as short lateral branches from prostrate vegetative hyphae; conidia solitary, prolate, black, opaque, smooth-walled, provided with an equatorial germ slit. The penultimate conidiophore cell is extended as a fine tapering process through the ultimate cell so that it almost reaches the conidium, and the bursting of the penultimate cell produces a jet of fluid which forcibly discharges the conidium (2648, 6236).

Five species have been recognized (1603) which differ only in the diameter of the conidia: *Khuskia oryzae* (q.v.), (11–)12–14(–20) μm; *N. sphaerica* (q.v.), (14–)16–18(–22) μm; *N. sacchari* (Speg.) Mason *1927*, (16·5–)20–22(–24) μm; *N. panici* Zimmerm. *1902* (= *N. arundinacea* (Cooke & Massee) Potlaĭchuk *1952*), 25–30 μm; and *N. padwickii* Prasad *et al.* *1960*, 33–42 μm. In addition *N. musae* McLennan & Hoëtte *1933* may be distinguished, the type strain of which has conidia 9–12·5 μm diam. — *Kh. oryzae* and *N. sphaerica* are often difficult to distinguish in culture because the latter tends to form smaller conidia than on the natural substrate (2584, 3660). However, the combination of both taxa under the older but rather uncertain name *maydis* as *N. maydis* (Garov.) Jechová *1963* is not recommended (2584).

Nigrospora sphaerica (Sacc.) Mason *1927* — Fig. 218.

≡ *Trichosporium sphaericum* Sacc. *1882*
= *Epicoccum levisporum* Pat. *1893*
= *Coniosporium extremorum* Syd. *1913*

DESCRIPTIONS: Mason (3659, 3660), Ellis (1603), Potlaĭchuk (4605), Subramanian (5626), and Matsushima (3680); conidiogenesis: Cole and Samson (1109). — Colonies reaching 5·5–8·0 cm in five days at 20°C on MEA; conidia (14–)16–18(–22) μm diam. Other characteristics as in *Khuskia oryzae* (q.v.).

N. sphaerica prefers warmer regions. It has been reported from India (1315, 2854, 3732,

FIG. 218. *Nigrospora sphaerica*, various stages of conidium formation.

3865, 4030, 4700, 5000), the Bahamas (4312), Central America (1697), southern states of the USA (2482), Ohio (1165, 1166), Pennsylvania (1163), Australia (3720), central Africa (3063), Israel (4758, 4759), and Italy (4538). — It has been isolated from forest soils (2854, 3063), forest nurseries (2482), uncultivated (4030) and cultivated soils under sugar cane (3732) and cotton (5009), grassland (3865, 4538), heathland (3720) and sand (4758, 4759). Other known habitats include bat caves with guano deposits (4312), a mangrove swamp (4700), a sewage treatment plant (1165), a waste pond (1166), and soils treated with sludge (1163). It is common on seeds of corn (3069), *Dactylis glomerata* (5940), groundnuts (2253), castor beans (3172), and pecans (2572). It has been isolated from the rhizosphere of banana (2030), ageing leaves of sugar cane (2583), decaying *Setaria glauca* (5233), sorghum (3219), banana fruits (6158), leaves of various other plants (296), feathers of free-living birds (2575), and a fungus-contaminated oil painting (1626). — *N. sphaerica* is a very suitable food for insects (5381) and mites (5379).

N. sphaerica decomposes cellulose (52). It produces the diterpene derivative aphidicolin which has antiviral properties (5536), triglycerides, and free fatty acids which can act as attractants for the flour beetle, *Tribolium confusum* (5535). Mycelial extracts were found to be toxic to rats (1388).

Oidiodendron Robak *1932*

= *Stephanosporium* Dal Vesco *1961*

Type species: *Oidiodendron tenuissimum* (Peck) Hughes
   (= *Oidiodendron fuscum* Robak) (lectotype)

Teleomorphic genera: *Byssoascus* v. Arx, *Myxotrichum* Kunze, *Toxotrichum* Orr & Kuehn

Contributions to a MONOGRAPHIC TREATMENT: Robak (4857), Barron (360), and Ellis (1603, 1604).

Colonies slow-growing, usually not exceeding 2 cm diam in ten days on MEA at 20–25°C, of variable structure within a species, smooth, wrinkled or tufted; vegetative hyphae ± hyaline but often with a strongly pigmented exudate. Conidiophores erect, distinctly pigmented, arising from submerged or aerial hyphae and often also from ascendent mycelial tufts. Conidiophores apically ± verticillately branched by the formation of a succession of several short or rather long branching chains of arthroconidia; conidial chains may also develop from undifferentiated hyphae; between the arthroconidia often abortive portions remain which after dehiscence form a fringe or short appendage. Conidia maturing in basipetal succession, one-celled, globose, elongate, cylindrical or barrel-shaped, smooth-walled or ornamented, hyaline or pigmented. Pigmented chlamydospores present in two species.

The closest genus to *Oidiodendron* is *Chrysosporium* Corda (q.v.) which differs in having unpigmented conidiophores. — The enterarthric (or enterothallic) conidiogenesis has been studied in detail for *O. truncatum* by time-lapse cinematography (1107) and SEM (1101).

Oidiodendron species are common inhabitants of the humus layers in forest soils but also occur on decaying wood and bark where they can often form macroscopically visible patches. — Descriptions are usually based on colonies growing on (non-neutralized) MEA or PDA, but an acid cherry decoction agar has proved to be particularly suitable for inducing typical sporulation; streak cultures of conidial inoculum also favours sporulation.

Since the available keys by Barron (360) and Ellis (1603, 1604) do not cover some common soil-borne species, a key to some additional species which are not treated in the text is provided here: excluded are the species with known teleomorph which have paler, little differentiated conidiophores and usually produce ascomata *in vitro*.

Key to the species (excluding those with known teleomorphs):

1	Conidia lens-shaped, provided with a thick-walled ring, 3·5–5·5 × 2–3 µm	*O. cerealis* (p. 518)
	Conidia without a thick-walled ring	**2**

2(1)   Colonies exuding a dark red or brown pigment into the agar; conidia ellipsoidal,
        2–4 × 1·5–2·5 μm   **3**
        Colonies not exuding a dark pigment into the agar   **4**

3(2)   Colonies exuding a red pigment; conidia pale greyish green, smooth-walled
                                                                    O. rhodogenum (p. 522)
        Colonies exuding a dark brown pigment; conidia brown, roughened   *O. tenuissimum* (p. 522)

4(2)   Pigmented chlamydospores present; conidiophores not exceeding 50 μm in length   **5**
        Pigmented chlamydospores absent; conidiophores commonly exceeding 50 μm in length **6**

5(4)   Conidia cylindrical or barrel-shaped, 2–6 × 1·2–2·0 μm   (*O. chlamydosporicum*)
        Conidia subglobose to short-cylindrical, 2·5–4(–5) × 2·0–2·5 μm   (*O. scytaloides*)

6(4)   Conidia regularly subglobose or globose, thick-walled, with a fine wrinkled ornamentation,
        2·7–5·0 μm diam   (*O. periconioides*)
        Conidia not regularly rounded, smooth-walled or verruculose   **7**

7(6)   Colonies purplish black; conidia ± globose, with numerous warts, 2·5–4 μm diam, or to 5·5 μm
        long   *O. echinulatum* (p. 519)
        Colonies not purplish black   **8**

8(7)   Colonies yellowish grey or yellowish green   **9**
        Colonies darker grey or brownish   **10**

9(8)   Conidiophores 150–250 μm long; conidial chains often undulate   (*O. citrinum*)
        Conidiophores to 125 μm long; conidial chains straight   (*O. flavum*)

10(8)   Conidia cylindrical or barrel-shaped   **11**
        Conidia (sub)globose to short-cylindrical   **13**

11(10) Conidiophores mostly 250–350 μm long; conidia 2·5–4·0 × 2·0–2·5 μm, with slightly rounded
        ends, often forming curled chains   (*O. majus*)
        Conidiophores mostly 100–150 μm long; conidia with sharply truncate ends   **12**

12(11) Conidia olivaceous-brown, with darker frills at the ends, 4·0–6·5 × 1·5–3·0 μm
                                                                    O. truncatum (p. 524)
        Conidia hyaline, with inconspicuous frills, 3·0–3·5 × 1·5–2·0 μm   (*O. pilicola*)

13(10) Conidia roughened, with a dark brown outer wall, globose to subglobose, 2–4 × 1·5–2·5 μm
                                                                    O. tenuissimum (p. 522)
        Conidia smooth-walled to finely roughened, dilute grey-green, ovoid to short-cylindrical,
        2·0–3·6 × 1·6–2·0 μm   *O. griseum* (p. 521)

Oidiodendron cerealis (Thüm.) Barron *1962* — Fig. 219.

≡ *Sporotrichum cerealis* Thüm. *1880*
≡ *Trichosporium cerealis* (Thüm.) Sacc. *1886*
≡ *Stephanosporium cerealis* (Thüm.) Swart *1965*
= *Trichosporium olivatrum* Sacc. *1882*
= *Oidiodendron nigrum* Robak *1932*
= *Haplographium fuligineum* van Beyma *1933*
= *Stephanosporium atrum* Dal Vesco *1961*

DESCRIPTIONS: Robak (4857), Melin and Nannfeldt (3749), Wang (6168), Barron (358, 360),
Ellis (1603), and Swart (5667). — Colonies reaching 1·0–1·5 cm diam in ten days at 20–25°C

on PDA, violaceous to dark violaceous-grey and finally purple-black; reverse purple-black, without a diffusing pigment. Conidiophores 10–20 μm long and about 2·5 μm wide, sometimes lacking. Conidia lens-shaped, smooth-walled, brown, surrounded by a thick-walled ring which may be oriented longitudinally, transversely or obliquely in the chain, 3·5–5·5 × 2·0–3·0 μm. —The thickened ring surrounding the conidia has been taken as a criterion for separating *Stephanosporium* Dal Vesco *1961* from *Oidiodendron* (e.g. 1603). The development of this wall-thickening has also been studied by TEM (4800).

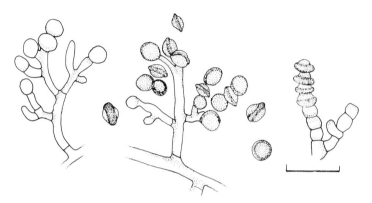

Fig. 219. *Oidiodendron cerealis*, conidiophores and conidia.

The distribution of this species appears to be worldwide including not only the temperate zones of Europe and North America but also Pakistan and the Philippines (1603). It has been isolated from acid sand dunes (745) and salt-marsh mud in the British Isles (745), soils in Spitsbergen (3064), a cedar-birch bog in Canada (358), savannah with *Acacia karroo* in South Africa (4407), the Sonoran desert in the USA (4733) and soils in Peru (2005). It has also been found on the bark of *Larix decidua*, the rhizoplane of *Ammophila* (IMI) and in the rhizospheres of *Stipa lessingiana, Festuca sulcata* and *Linosyris tatarica* (3376), on seeds of cultivated grasses (3512), stumps of beech (3568), wooden boards of a bath-house (5016), pulp and paper (4857, 6166), and in the air (IMI).

Optimal growth occurs with 50–100% O_2 in the atmosphere, and below 50% it is proportional to the O_2 concentration (6565). The indole melanin present in hyphae and conidia (5300) protects the fungus against desiccation, but to a lesser extent than in *Cladosporium* and *Ulocladium* species (6567), and contributes to a γ- and UV-ray resistance (5300, 6566). *O. cerealis* is able to decompose cellulose (4857).

Oidiodendron echinulatum Barron *1962* — Fig. 220.

DESCRIPTIONS: Barron (360), Ellis (1604), and Tokumasu (5845). — Colonies reaching 1 cm diam in ten days at 25°C on PDA, violaceous to purple and finally purplish black; reverse purplish black with the pigment diffusing into the agar; a violaceous exudate is often also present. Conidiophores 50–100(–150) μm long and 2·5 μm wide, brown, but often sporulating without differentiated conidiophores. Conidia globose or rarely ovoid, coarsely roughened with numerous warts, thick-walled, at first hyaline but becoming dark brown, 2·5–4·0 μm diam or to 5·5 μm long.

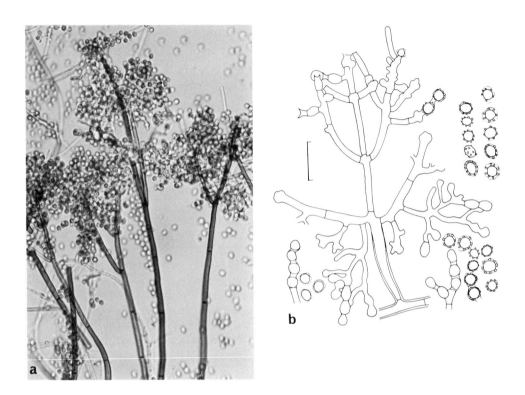

FIG. 220. *Oidiodendron echinulatum*. a. Group of conidiophores, × 500, CBS 113.65; b. details of conidium formation.

O. echinulatum was originally isolated in Ontario/Canada from peaty soil (360, 505) and has subsequently been found in a conifer swamp in Wisconsin (1039), humus and mineral soil of deciduous and coniferous forests and grassland in Japan (5846, 5847), agricultural soils in Germany, beech forest soil in Denmark (CBS), wheat rhizosphere in Ireland (3567), remains of cabbage and potato plants in the USSR (3113) and on rabbit dung in Japan (5845). — It can serve as food for the mite *Oppiella nova* (2960).

Xylan is decomposed very vigorously and cellulose can also be utilized (1432, 4191). — In a chronically irradiated soil, the frequency of *O. echinulatum* declined sharply at intensities of 230 R per day (2009).

Oidiodendron griseum Robak *1934* — Fig. 221.

DESCRIPTIONS: Robak in Melin and Nannfeld (3749), Barron (360), Tokumasu (5845), Wang (6168), and Ellis (1603). — Colonies reaching 1·0–1·5 cm diam in 14 days at 25°C on PDA, at first greyish, then olive-grey and finally olive-brown; reverse grey and later olivaceous-black, without a diffusing pigment; aerial hyphae comparatively strongly pigmented. Conidiophores mostly 90–100 μm long and 1·2–2·0 μm wide. Conidia dilute grey-green, almost smooth-walled, 2·0–3·6 × 1·6–2·0 μm.— This species differs from the rather similar *O. tenuissimum* (q.v.) in the pale and almost smooth walls of the conidia.

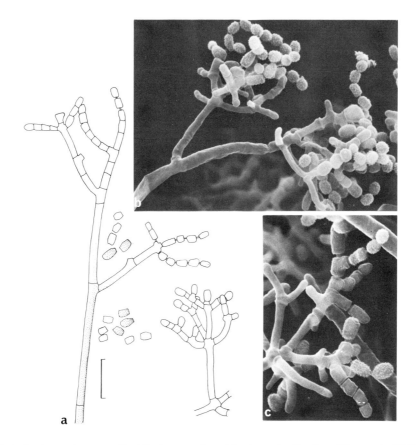

FIG. 221. *Oidiodendron griseum*. a. Conidiophores and conidia; b, c. details of conidiogenesis, SEM, b. × 2000, c. × 2500; CBS 701.74.

O. griseum was originally isolated from the air in Sweden and subsequently from cedar and spruce bogs in Ontario/Canada (360). There are now other records from the USA (1040, 6168), Britain (IMI) and Ireland (1376, 1438), France (2963), the USSR (1637), Japan (5845, 5846, 5847), central Africa (3063), New Zealand (5812) and Trinidad (IMI). It has been found in wet-mesic deciduous forests (1040), soils under pine (2963, 5363, 5845, 5846), white fir (2963), a forest nursery (IMI), peat bogs (1376, 1438), heathland (2736), grassland

(5812), and cultivated soil under cacao (IMI). It has also been isolated from garden compost (1424), pine litter (2923), rotten wood (IMI), decaying petioles of *Pteridium aquilinum* (1821), rarely in humus and decaying mosses in bogs (5847), the rhizosphere of peat bog plants (1376), as a recolonizer of steam-sterilized soil (3908), from hay (IMI), paper (360), wood pulp (6168) and air (3749, IMI). — The germination of conidia is inhibited in cut-away peat (1436).

Optimal growth occurs at 30°C. The best C sources for growth proved to be fructose and arabinose (1375), but both starch and cellulose are also degraded (1425, 2736, 3568) and birch wood eroded (3568, 4191).

Oidiodendron rhodogenum Robak *1932* — Fig. 222.

DESCRIPTIONS: Robak (4857), Barron (360), Kiffer *et al.* (2964), and Ellis (1604). — Colonies reaching 0·5–1 cm diam in ten days at 18–20°C, at first whitish but later becoming grey-green or grey-brown; reverse red-brown with a red pigment diffusing into the agar. Conidiophores 80–175(–210) μm long. Conidia short cylindrical, almost smooth-walled to finely roughened, pale greyish green, 2·0–4·0 × 1·5–2·2 μm.

This species is slightly less common than others of its genus. It has been reported from various substrates, mainly from Europe (1604), but also from the USA (1040). It was originally isolated from wood pulp in Norway (4857) and has since been obtained, using cellulosic baits, from pine forest soils in France (2963, 2964) and the British Isles (IMI), and from soils of a wet-mesic deciduous forest in Wisconsin/USA (1040). It has been found on decaying litter of the moss *Pseudoscleropodium purum* (2969), on lichens (5428), straw, rotten wood, and in the air of a cork factory (IMI).

Oidiodendron tenuissimum (Peck) Hughes *1958* — Fig. 223.

≡ *Periconia tenuissima* Peck *1893*
= *Oidiodendron fuscum* Robak *1932*

DESCRIPTIONS: Robak (4857), Barron (360), Tokumasu (5845), and Ellis (1603). — Colonies reaching 1·5 cm diam in 10–14 days at 25°C on PDA, in various grey-brown shades; reverse grey-brown to dark brown, with or without a brown pigment diffusing into the agar. Conidiophores 100–200(–300) μm long in the unbranched part, 1·5–3·0 μm wide, smooth-walled or roughened. Conidia globose to ovoid, outer wall dark and distinctly roughened, 2–4 × 1·5–2·5 μm, frequently separated by long connectives. — Three groups of this variable taxon have been distinguished by Barron (360): (a) fuscous-brown colonies without a pigment diffusing into the agar; (b) violaceous-grey colonies with a dark brown pigment diffusing into the agar (corresponding with the description of *O. fuscum* Robak); and (c) light to dark grey

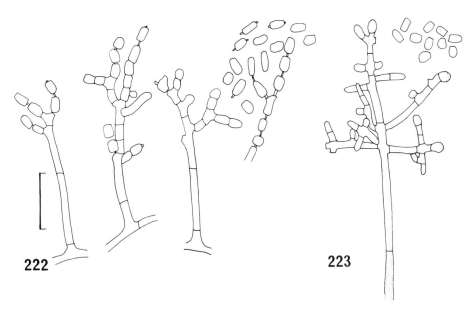

FIG. 222. *Oidiodendron rhodogenum*, short conidiophores and conidia, CBS 849.68.
FIG. 223. *Oidiodendron tenuissimum*, conidiophore and conidia, CBS 920.73.

colonies with little pigment diffusing into the agar. — *O. tenuissimum* is connected to *O. griseum* (q.v.) by intermediate isolates but the latter usually has lighter and less roughened conidia. *O. periconioides* Morrall *1968* differs in having regularly rounded, brown, thick-walled, very regularly ornamented conidia, $2 \cdot 7$–$5 \cdot 0 \,\mu$m diam.

O. tenuissimum has commonly been isolated from various soils, wood and polypores in Canada (360) and wood pulp in Norway (4857). Other records are available from Poland (3497), the British Isles (159, 2338, 3234, 4429, 4445, 5428, 6182, 6381, 6383), the Netherlands (IMI), France (2963), the USA (1039), Japan (5845, 5846, 5847), Peru (2005), Nigeria and Portugal (IMI). Preferred habitats are forest soils with pine (2963, 4445, 5363, 5428, 5845, 6381, 6383), but it is also known from soils in mixed forests (3962), forest nurseries (IMI), open bogs or conifer swamps (1039), moorland and peat bogs (3234), grassland (159, 3497, 5845, 5847, 6182) and rarely in arable soils (IMI). It has been isolated from rotting *Castanea* cupules (2338), litter of *Abies grandis*, *Picea sitchensis*, Scot's pine (2343, 2344) and *Pteridium aquilinum*, bark and cork, non-mycorrhizal beech roots (IMI), the rhizoplane of *Pinus nigra* var. *laricio* (4448), and organic and mineral samples of forest soils (307, 5847).

It has been found to grow better at 15°C than at 25°C (1377). — *O. tenuissimum* can cause weight losses and erosion of birch wood (4191). Known metabolic products include fuscin and dihydrofuscin (5278); fuscin has a high affinity to mitochondrial SH-groups and interferes with respiration (6090). — In a chronically irradiated soil the frequency of *O. tenuissimum* decreased at 120 R per day (2009).

Oidiodendron truncatum Barron *1962* — Fig. 224.

DESCRIPTIONS: Barron (360), Ellis (1604), and Tokumasu (5845). — Colonies reaching 1 cm diam in ten days at 25°C on PDA, olive-grey to olive-brown; reverse brown; a little reddish brown pigment sometimes diffusing into the agar. Conidiophores 100–150(–200) μm long and 2·5–3·5 μm wide. Conidia barrel-shaped with both ends broadly truncate and often bearing darkened frills, with a smooth, gelatinous wall, but becoming roughened with age, brown, 4·0–6·5 × 1·5–3·0 μm. — The similar *O. pilicola* Y. Kobayasi *1969* has almost cylindrical or barrel-shaped, hyaline conidia 3·0–3·5 × 1·5–2·0 μm. The anamorph of *Byssoascus striatisporus* (Barron & C. Booth *1966*) v. Arx *1971* has conidiophores to 100 μm long which are pigmented only in the lower third and the colonies produce thick crusts of white, confluent ascomata with striate fusiform ascospores. — Time-lapse (1107) and SEM studies (1101) of conidiogenesis have been carried out; the orientation of the rodlet pattern of the conidial surface (freeze-etched replicas) reflects the extension of the wall during conidiogenesis (1100).

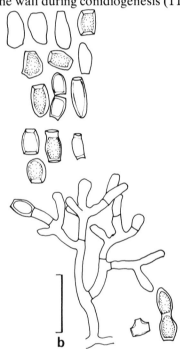

FIG. 224. *Oidiodendron truncatum*. a. Upper part of conidiophore, × 500; b. a small conidiophore and conidia, CBS 115.65.

O. *truncatum* was originally isolated in Canada from soil of a mixed forest near Guelph and from peat soil near Toronto (360). It has also now become known from aspen stands (3959) and conifer plantations (5363) in Canada, open bogs and conifer swamps in Wisconsin/USA (1039), soils in Japan (5845) and Italy, grassland in New Zealand, garden soil in the British Isles (1604, IMI), and the rhizospheres of bean (4452) and wheat (3567). It has occurred rarely in wheatfield soils in Germany.

O. *truncatum* showed an antibiotic activity against *Candida albicans* and several bacteria tested *in vitro* (3600). In a chronically irradiated soil, it survived up to 53 R per day (2009).

Paecilomyces Bain. *1907*

= *Spicaria* auct. [non Harting *1846*]

Type species: *Paecilomyces variotii* Bain.

Teleomorphic genera: *Cordyceps* (Fr.) Link, *Byssochlamys* Westl., *Thermoascus* Miehe, *Talaromyces* C. R. Benjamin

MONOGRAPHIC TREATMENT: Brown and Smith (743), and Samson (5024). — *Paecilomyces* is related to *Penicillium*, and the techniques used for its study are the same as for that genus, although CzA is better replaced by OA. *Paecilomyces* is distinguished from *Penicillium* mainly by the divergent phialides which consist of a swollen base tapering into a rather long and slender neck; colonies are never typically green. The subsequent treatment follows Samson's monograph, which also contains valuable information on distribution and physiology of the species.

   The genus *Paecilomyces* is subdivided into two sections: *Paecilomyces* which includes nine species, in addition to the type species, *P. variotii*, and the anamorphs of some *Byssochlamys*, *Thermoascus* and *Talaromyces* species, and the rather heterogeneous section *Isarioidea* with 22 species. The genus *Mariannaea* Arnaud ex Samson (q.v.) has been segregated from *Paecilomyces* because of its *Verticillium*-like conidiophores, awl-shaped phialides and imbricate conidial chains. Samson did, however, include *Penicillium lilacinum* Thom in *Paecilomyces* as it is close to *P. marquandii*; these two species are somewhat intermediate between the two genera and are reminiscent of species of the *Penicillium janthinellum* series in particular.

Key to the species treated:

1   Colonies yellow-brown; odour sweet aromatic; usually thermotolerant or thermophilic *(Sect. Paecilomyces)*
       Ascomata absent: conidiophores repeatedly branched; conidia of different sizes; more or less ellipsoidal; chlamydospores present   *P. variotii* (p. 534)
     Colonies of other bright colours; odour not sweet aromatic; mesophilic (Sect. *Isarioidea*)   **2**

2(1)   Colonies vinaceous to violet   **3**
       Colonies other colours   **4**

3(2)   Conidiophore stalks pigmented, rough-walled; reverse uncoloured or purple; aleurioconidia absent   *P. lilacinus* (p. 530)
       Conidiophore stalks hyaline, smooth-walled; reverse yellow; aleurioconidia usually present
              P. marquandii (p. 533)

4(2)   Conidia finely echinulate; colonies white to pink, thin, powdery, synnemata absent; reverse
       usually dark green   *P. carneus* (p. 526)
       Conidia smooth-walled; colonies usually deep and woolly, forming synnemata on insects and
       rarely in culture; colony reverse not green   **5**

5(4)   Colonies usually pink; conidia fusiform to cylindrical, 3–4 × 1–2 μm *P. fumosoroseus* (p. 529)

       Colonies white to luteous; conidia ellipsoidal to fusiform, less than 3·5 μm long
                                                                      P. farinosus (p. 527)

Paecilomyces carneus (Duché & Heim) A. H. S. Brown & G. Sm. *1957* — Fig. 225.

      ≡ *Spicaria carnea* Duché & Heim *1931*
    = *Paecilomyces austriacus* Szilvinyi *1941*
    = *Spicaria carnosa* J. H. Miller, Giddens & Foster *1957*
  ? = *Spicaria decumbens* Oudem. *1902*

For some other synonyms see Samson (5024).

DESCRIPTIONS: Brown and Smith (743), Samson (5024), and Fungi Canadenses No. 152, *1979*. — *P. carneus* can easily be recognized by its slowly growing pale pink colonies, reaching 1·4–2·2 cm diam in 14 days at 25°C on MEA, often dark green reverse, slender phialides with a very narrow neck, and rough-walled conidia, 3–4 × 2–2·5 μm.

Although there are rather few published reports, this is regarded as a frequent soil fungus (743). It has been reported from France, the Netherlands (1614, 1616, 5024), Australia, New Zealand (2705, 5930), Afghanistan, Pakistan (3473), Nepal (1826), Japan (5024, 5846), Canada (234, 505, 3959), equatorial West Africa (1420) and Zaïre in forests and shaded cultivated soils (3790). It has frequently been isolated from conifer swamps (1039), evergreen and deciduous forests in Ohio (2573), and humid mixed forests in Wisconsin/USA (1030, 1032, 1040, 2004, 2007, 2008, 4225, 6350) where it has a maximum incidence after summer precipitation, in stands with aspen (3959), from podzols with rather low frequencies (3962) but particularly in less acid deeper layers (307, 2068, 4445). It is also known from grassland (159, 2573, 2705), under steppe vegetation (6347), arable (1614, 1616, 1889, 2482, 3817) and garden soils (4451), and dunes (1477). It seems to be less frequently found in highly acid than in chalk soils (4152), but is most frequent in soils with a high organic content (307, 2008, 2705) and in finer textured sandy loams (6350). In cultivated soils, an increasing frequency after repeated wheat crops has been observed (1433, 1617). It has been isolated from the rhizosphere of grasses (1614), and from wheat (1614, 3567), barley and bean roots (4451, 4452, 5801). It is also found on rabbit dung (6085), wood pulp and paper (4857, 6168). — Forty-five percent of chloroform-fumigated mycelium was mineralized within ten days at 22°C (133).

P. carneus grows very slowly, but sporulates rapidly and freely. Cellulose is modestly degraded (2706, 6085), and CMC (1432) and chitin are well utilized (2068, 2706). *P. carneus* grows on *n*-paraffin and other non-sugar C sources and produces on these substrates the

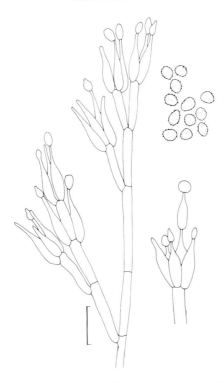

Fɪɢ. 225. *Paecilomyces carneus*, conidiophore, whorl of phialides and conidia.

β-lactam antibiotics deacetoxycephalosporin C, cephalosporin C, and deacetylcephalosporin C (3017).—*P. carneus* belongs to the few fungi which showed, *in vitro*, marked specific antagonistic effects towards *Pseudocercosporella herpotrichoides* (1431). In laboratory experiments, it caused some injury to pea shoots and roots (1430).

Paecilomyces farinosus (Holm ex Gray) A. H. S. Brown & G. Sm. *1957* — Fig. 226.

≡ *Corynoides farinosa* Holm ex Gray *1821*
≡ *Isaria farinosa* (Holm ex Gray) Fr. *1832*
= *Penicillium alboaurantium* G. Smith *1957*

For many other synonyms see Samson (5024).

Teleomorph: *Cordyceps memorabilis* Ces. *1861*

Dᴇsᴄʀɪᴘᴛɪᴏɴs: Samson (5024), Skou (5404), and Fungi Canadenses No. 153, *1979*; teleomorph: Pacioni and Frizzi (4338). — Colonies growing moderately fast, reaching 4–6 cm diam in 14 days at 25°C on MEA, deep and powdery, remaining white or turning bright yellow, sometimes with conspicuous yellow synnemata; conidiophores usually erect,

100–300 μm tall, bearing several whorls of flask-shaped phialides. Conidia ellipsoidal to fusiform, 2·0–3·0 × 1·0–1·8 μm. On insects the fungus produces white to yellow synnemata, up to 1·2(–3) cm long, which are often found on the soil surface. — A similar species is the tropical entomogenous *Paecilomyces amoeneroseus* (P. Henn.) Samson *1974* with intensely pink pigmentation and almost globose conidia. — Analysis of the fungal lipids showed palmitic, oleic and linoleic acids to be the major components with smaller amounts of stearic, palmitoleic and *a*-linolenic acids (5965).

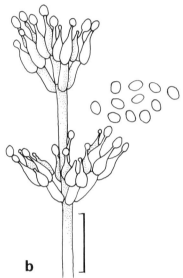

FIG. 226. *Paecilomyces farinosus*. a. Conidiophore, SEM, × 2850, orig. R. A. Samson; b. conidiophore and conidia.

P. farinosus is a ubiquitous insect parasite, equally common in temperate and tropical zones, with an unlimited host range (3257, 5026). When it is isolated from soil, the possibility cannot be ruled out, that the original substrate was an insect (5024) and its synnemata are commonly found arising from insects and emerging through the soil (CBS). It has been recorded from soils in South Africa (4407), Nepal (1826), Japan (5846), Brazil (399), Czechoslovakia (1702), and the USA (1039, 2007, 2008, 6351). Forest soils are clearly the preferred habitat including pure stands of alder and conifers (6351), *Populus-Salix* floodplain communities (2008), pine stands (6352), an oak forest (1702), a conifer swamp and open bogs (1039); it has been isolated from an open savannah with *Acacia karroo* (4407), leaf litter of birch (2411), poplar (6101) and pine (2923), roots of strawberry (4132, 4133), rhizosphere of peat bog plants (1376), aecidiospores of *Cronartium comandrae* (4608), feathers and pellets of free-living birds (2575), hair baits exposed to soil (1422), sewage sludge (1165), an acid mine drainage stream (1166) and organic detritus in fresh water (4429). — When stored at 8°C under dry and dark conditions *in vitro*, 90% of the conidia were found viable after 200 days (1076). — The fungus is sensitive to mycostatic factors in soil (1075). The synnemata can be parasitized by *Melanospora parasitica* and *Tritirachium isariae* (5024).

Good sporulation occurs, amongst others, on soil extract agar, particularly under "black" light (4132) and good growth is obtained at 27°C (291). Synnema initiation and development is light-sensitive (1095, 5083). — Glucose, sucrose and starch support good growth up to concentrations of 10% (1095). Suitable N sources include glutamic and aspartic acids, ammonium oxalate, citrate and tartrate (156). Weight loss of beech wood after incubation with *P. farinosus* has been reported (1199). — *P. farinosus* has been used for biological control experiments with various insects (4023). Its pathogenicity to bees (5863), bumble bees (5404), *Leptinotarsa decemlineata* (291) and *Pyrausta nubilalis* (5864) has received particular attention.

Paecilomyces fumosoroseus (Wize) A. H. S. Brown & G. Sm. *1957* — Fig. 227.

≡ *Isaria fumosorosea* Wize *1904*
= *Spicaria aphodii* Vuill. *1910*

For some other synonyms see Samson (5024).

DESCRIPTIONS: Brown and Smith (743), Samson (5024), and Fungi Canadenses No. 154, *1979*. — Colonies rather slow growing, reaching 4 cm diam in 14 days at 25°C on MEA, raised, floccose, pink; sometimes producing synnemata on agar; erect conidiophores bearing several compact whorls of phialides which have a strongly inflated base; conidia cylindrical to fusiform, 3–4 × 1–2 μm. — A similar tropical, entomogenous species is *P. javanicus* (Friederichs & Bally) A. H. S. Brown & G. Sm. *1957* with white to cream colonies, more slender phialides, and conidia 4–7·5 × 1·4–1·7 μm.

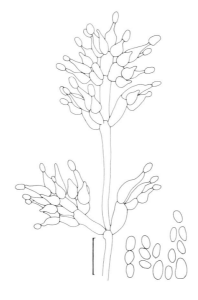

FIG. 227. *Paecilomyces fumosoroseus*, conidiophore and conidia.

P. fumosoroseus is known mainly as an insect parasite with a wide host range and a worldwide distribution from Europe to Africa and Asia (5024, 5026). Soil isolates are known from the Netherlands (5024), Germany (CBS), Canada (505, 6352), and Brazil (399). Known habitats are, for example, forest soils (2573) with white cedar (505) and pine (6352), rarely in arable soils (5024) and fen peat (5559). — This species, together with the preceding one, *Metarrhizium anisopliae* and *Beauveria* species, has been used in the biological control of insects (4023). It has also been isolated from nests of free-living birds (2575) and lung tissue of giant tortoises (1939).

The depsipeptide beauvericin is also produced by *P. fumosoroseus* (478).

Paecilomyces lilacinus (Thom) Samson *1974* — Fig. 228.

≡ *Penicillium lilacinum* Thom *1910*
= *Spicaria violacea* Petch *1932* [non Abbott *1926*]

For some other synonyms see Samson (5024).

DESCRIPTIONS: Raper and Thom (4745), Samson (5024), and Fungi Canadenses No. 156, *1979*. — Colonies reaching 5–7 cm diam in 14 days at 25°C on MEA, in vinaceous shades, with the reverse uncoloured or vinaceous; conidiophores erect, 400–600 µm tall, mostly arising solitarily from the horizontal mycelium, rarely synnematous, stalks 3–4 µm wide, yellow to purple, rough-walled, with densely clustered metulae and phialides. Conidia ellipsoidal to fusiform, smooth-walled to slightly roughened, 2·5–3·0 × 2·0–2·2 µm. Aleurioconidia absent. The species may produce mononematous conidiophores or synnemata on insects. — The species is most similar to *P. marquandii* (q.v.), and intergrading forms exist. — The main components of the fungal lipids are oleic (38·6%), palmitic (32·3%), linoleic (13·4%), and stearic acids (9·4%) (5351, 5352). — DNA analysis gave a GC content of 59–61% (5600).

P. lilacinus is a typically soil-borne fungus; its occurrence on insects in tropical regions is of secondary importance (5024). This species has been reported from numerous parts of the world, but seems to be most frequent in warmer regions. — There are numerous reports from India, and others from Pakistan (60, 4855), Nepal (1826), Sri Lanka (4021), Malaysia (6046), tropical Australia (978), Hong Kong (1021), Hawaii (3264), Japan (2532, 3337), Jamaica (4886), Honduras (2031), Chile (1824), Argentina (1827), South Africa (1559, 3630, 4407), Madagascar (4159), the Ivory Coast (4159, 4718), Ghana (5026), central Africa (3063), equatorial West Africa (1420), Somalia (5048, 5049), Libya (6510), Egypt (3993), Israel (654, 2764, 2772, 4759), Turkey (4245), Italy (270, 271, 3450, 3454, 3538, 5047), Spain (3417, 3446, 3447), and Yugoslavia (5534). It has also been reported from France (242, 918, 2161, 3451), Austria (1823, 3413, 3418, 5678), Germany (5316), Poland (272, 1423, 3530, 4137), the Netherlands (1614, 1616), the British Isles (6182), the USSR (2871, 3363, 3850, 4365, 4548), Oklahoma (1632), Wyoming (1293), Wisconsin (1040, 2007, 2008, 4313), California (1164), North Carolina (655), Ohio (1166), Colorado (3259), Pennsylvania (1163), various southern states of the USA (13, 2482), Canada (540, 6352), and recently deglaciated soil in southeast Alaska (1171). It has been isolated from uncultivated soils (540,

3259, 3630, 4030, 4855), coniferous and deciduous forest soils (too numerous to list here individually), cultivated soils (1420, 1423, 1616, 3450), under cotton, potato (2871), sunflower, rape, beet, wheat (918, 1614, 2161, 3446), corn (4716), paddy (1519), alfalfa (2871), groundnuts (2532, 2768, 2772), sugar cane (3732, 4886), vineyards (242) with high copper content (4921), plantations of citrus (2764), coffee (4159), banana (2031) and palms (6046), grassland (1614, 4021, 6182), soils with steppe (1632, 3417, 4313, 4407, 5049) and chaparral vegetation (1164), heathland (5047), peat bogs (3530, 4365), garden compost (1424), desert soil (654), sand dunes (1423), caves (4137), a mine (5871), jungle (6046), a mangrove swamp (3264), estuarine sediments (655, 4477, 5316), oil fields (3337) and soils treated with oil waste (3384) or sewage sludge (1163). Occurrence in loess (4759), podzol (5534), rendzina, chernozem, and solonchak (3413, 3414, 3418) has also been described. It can be found in soils down to 40 cm and more (2161, 4759, 5000); neutral to slightly alkaline soils seem to be preferred (1259, 1559, 3417, 3450, 3451, 3732, 5048, 5049, 6182, 6510). Hair baits in soil are colonized at water potentials from saturation to −140 bars (977); in isolation experiments, it has been found on chitin buried in soil (4264) and it predominated on washed organic soil particles (1021); higher frequencies were obtained by soil dilution than by soil plate methods (4452). An increased occurrence has been reported after precrops of alfalfa or oat (2816) and also by NPK fertilization (2161, 2162, 2163); *P. lilacinus* belongs to the re-colonizing mycoflora of steam-treated soil (3908), and has been found with high frequencies in soils treated with the fungicides captan, quintozene, thiram and benomyl (6135). The colonization of bean roots by *P. lilacinus* in particular has received considerable attention (4451, 4453, 5758), but this species is also known from roots of wheat (5330), barley (4450, 4451), sugar cane (4886), cabbage (4451), and banana (2030, 2035). It has been reported from the rhizospheres of groundnuts, pineapple, banana, tomato, flax, and some steppe plants (16, 2035, 2532, 2768, 3376, 5826, 6210), and also from those of oak (3454) and poplar (3452) in truffle grounds. Other substrates include leaf litter of *Abies grandis* (681) and *Pinus radiata* (6080), seeds of cultivated grasses (3512), pecans (2572), stored sweet potatoes (3066), coarse fodder (4548), corn silages (1092), roosts, nests and feathers of free-living birds (682, 2575), and lungs of small wild mammals (2153). — Germination of conidia is inhibited by soil mycostatic factors (1408). — The mites *Acarus siro* and *Tyrophagus putrescentiae* have been observed feeding on the mycelium (5383).

P. lilacinus has a wide (8–38°C) temperature range (1491) regardless of the geographical origin of the isolates (1827); optimal growth occurs in the range 26–30°C and none at 5°C (654, 4578); the optimum pH is 6·5, its tolerance range is pH 2–10. It can still grow at osmotic potentials of $c.$−270 bars (5881) but grows at $c.$−90 bars with reduced vigour (4578). Optimal sporulation has been observed on media with 1% NaCl, optimal growth at 3%, and none at 5% (3773). Best growth occurs with as little as $0·2 O_2$ partial pressure (6565). — When grown on racemic mixtures, it shows a preference for L-tartaric acid (2052). Suitable C sources are sucrose, mannitol, glucose and maltose; good N sources are $NaNO_3$, NH_4Cl, NH_4NO_3, yeast extract and peptone; based on nutrient studies, the lipid production of *P. lilacinus* has been optimized (1491, 4072, 4541). Sugar uptake is stimulated by the addition of riboflavin and nicotinic acid (4071) and respiration inhibited in the presence of montmorillonite in the medium (5604). Metaphosphate is utilized (4544) and K^+ ions are liberated *in vitro* from orthoklase and oligoklase (4008) by the production of organic acids (4009). Isolates from silkworms yielded oxalic, dipicolinic, and succinic acids and some unidentified amino acids, as well as large amounts of D-mannitol (5024). Starch is utilized (6570) and the degradation of pentosans (xylan) via pentosanases, of glucans (pustulan, dextran) via 1,6-α-, 1,3-β- and 1,6-β-glucanases described (749, 3231, 5335). Chitin is also degraded (1425,

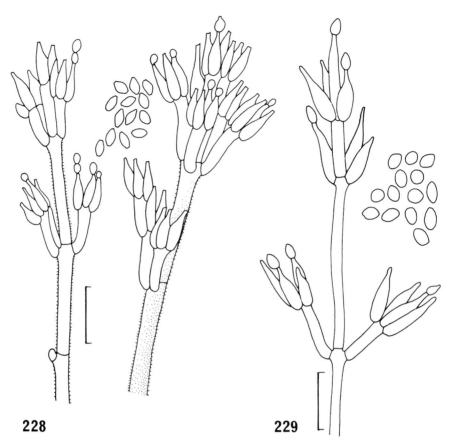

228   229

FIG: 228. *Paecilomyces lilacinus*, conidiophores and conidia.
FIG. 229. *Paecilomyces marquandii*, conidiophore and conidia.

4264). *P. lilacinus* is strongly proteolytic (141, 654, 1425, 2719), decomposes keratin (3530) to 20% in 63 days (585). It hydroxylates cinerone to cinerolone (5681). It can grow on sinigrin (4774), tannins from horse chestnut (3315), tannic acid (3052), and hydrocarbons from fuel oil (3384, 4239), and on *n*-alkanes produces ergosterol and organic acids (3337). — It has some antagonistic activity against bacteria (515, 3599) and fungi (516, 716); the peptide antibiotics leucinostatin, effective against some gram positive bacteria and a wide range of fungi (171), and lilacinin have been described (5024). *P. lilacinus* colonizes sclerotia of *Sclerotinia sclerotiorum* (2886) and *S. borealis* (3531). It was found to produce indole-3-acetic acid (294, 6115) and stimulate the growth of barley seedlings (293); growth of flax was inhibited in greenhouse experiments (1427). — A case of endophthalmitis after a glaucoma operation was found to be caused by *P. lilacinus* (3540).

Paecilomyces marquandii (Massee) Hughes *1951* — Fig. 229.

≡ *Verticillium marquandii* Massee *1898*
= *Spicaria violacea* Abbott *1926*

DESCRIPTIONS: Hughes (2596), Brown and Smith (743), Samson (5024), and Fungi Canadenses No. 157, *1979*; conidiogenesis: Cole and Samson (1109). — Colonies reaching 5–7 cm diam in 14 days at 25°C on MEA, pale vinaceous to violet, aerial mycelium floccose; reverse usually bright yellow to orange-yellow. Conidiophores erect, 50–300 μm tall, 2·5–3 μm wide, arising from the substrate mycelium or from the aerial mycelium, smooth-walled, hyaline, bearing loose whorls of branches and phialides. Otherwise as *P. lilacinus*. Globose thin-walled aleurioconidia, 3·5 μm diam, occur in the submerged mycelium.

P. marquandii is apparently an exclusively soil-borne species not known to be pathogenic to insects; according to our experiences, it is more common than *P. lilacinus*, whereas the latter (as *Penicillium lilacinum*) still ranges amongst the most frequently recognized *Penicillium* species in the soil. The species distinction may not always have been correct. — *P. marquandii* has a worldwide distribution from temperate to tropical latitudes. Thus, it has been isolated from soils in the Netherlands (1614, 1616), Austria (3418), Czechoslovakia (1702), the USSR (2871, 4554), the USA (reports too numerous to list here individually), Canada (505, 1067, 3959), Spain (3417), Turkey (4245), Israel (654, 2764), Syria (5392), Zaïre (3790), central Africa (3063), the Ivory Coast (4718, 4719), South Africa (1555, 1559, 4407), India (1519, 4030, 4698), Pakistan (3473), Nepal (1826), Jamaica (4886), the Bahamas (2006, 4312), Brazil (6008), Central America (1697), New Zealand (2705), and Japan (5846). It has been found in very varied soil types (505, 3414, 3418), in forest soils (1702) under aspen (3959), alder, pure conifer stands (6351) and *Eucalyptus maculata* (1555), evergreen and deciduous forests (2573), mixed hardwood in Wisconsin (1030, 1032, 1040, 2004, 2008, 3549, 4225) where this species occurs in areas with medium humidity under ash (1204) and with high humus accumulation (5880), grassland (1614, 3417, 4313, 6182), particularly in the upper layer (0–10 cm) (1067), soils with steppe type vegetation (6347), arable and other cultivated soils (too numerous to list here individually) down to a depth of 40 cm (2161) and is accumulated near the underlying chalky boulder clay when the top-soil is acidic (6182). *P. marquandii* is also reported from fields treated with sewage sludge (1163), in sewage sludge itself (1165), streams with a lower degree of pollution (1154, 1157), river sediments (1162), estuarine silt (5316), sand dunes (745), carst caves (3365), and bat guano (4312). The pH range is very wide, with a clear preference for higher values (745, 3451, 4152, 5239, 6182); the optimum is pH 5–6 and the fungus is described as occurring preferentially under dry conditions (1032, 1204, 3417, 4313). *P. marquandii* has been isolated from pine litter (3798), pine humus (681), peat (5559), truffle grounds (3451, 3454), roots of strawberry (4133), the rhizosphere of corn (4553, 4554), wheat (1614, 3567), grasses (1614, 5815), *Beta vulgaris* (5024), and sugar cane (4886). In the rhizosphere of *Lupinus angustifolius* a marked accumulation was noted, in contrast with *P. lilacinus* (4400). It was found on aecidiospores of *Cronartium comandrae* (4608), on fruit-bodies of *Hygrophorus* (5024), and in birds' roosts (682), feathers and pellets (2575). — Conidium germination in soil is inhibited by a mycostatic factor (2703); germination was observed in the vicinity of roots of peas and radish (2704).

The optimal temperature for growth is 25°C, at 5°C growth is very limited, the maximum is 30°C and at 37°C no growth occurs (1204, 4578). Optimal growth occurs at water potentials of

−45 bars (5316). Dextran (from *Leuconostoc mesenteroides*) is decomposed by an extracellular enzyme with the formation of isomaltose (5911); *P. marquandii* utilizes starch (1827), gelatine (654) and chitin (2706); cellulose decomposition is absent or is very slight in comparison to other fungi (3618). It can utilize nitrite as a N source (10); in the presence of keratin chips, PO_4^{3-} and Mg^{2+} ions, large quantities of $NH_4MgPO_4 \cdot 6H_2O$ crystals are formed (2880). — Growth is inhibited at >3% (v/v) CO_2 (784); the fungus is sensitive to CS_2 (3987); O_2 uptake was reduced by saturated C_8- to C_{11}- fatty acids as the sole C source, but favoured by compounds with shorter or longer chains (2783). — Antagonistic activity towards *Rhizoctonia solani* (3495) and other fungi (569) has been reported. Stimulatory effects were observed with corn plants (4553), while growth of barley seedlings remained unaffected (293).

Paecilomyces variotii Bain. *1907* — Fig. 230.

= *Penicillium divaricatum* Thom *1910*
   ≡ *Spicaria divaricata* (Thom) Gilman & Abbot *1929*
= *Penicillium aureo-cinnamomeum* Biourge *1923*
= *Paecilomyces indicus* Rai, Tewary & Mukerji *1969*

For many other synonyms see Samson (5024).

DESCRIPTIONS: Brown and Smith (743), Samson (5024), and Fungi Canadenses No. 151. *1979*. — This species is very variable in the shape and size of the phialides and conidia; numerous synonyms were compiled by Samson who adopted a rather wide species concept because of this variability. — Colonies spreading broadly, reaching 8 cm diam in 14 days at 25°C on MEA, powdery, isabelline, olivaceous, darkening with age; odour sweet aromatic. Conidiophores repeatedly verticillate, with rather slender, flask-shaped phialides. Conidia hyaline to yellow, ellipsoidal, of very unequal size within one colony, $3 \cdot 2$–$5 \cdot 0 \times 2 \cdot 0$–$4 \cdot 0\,\mu$m (to $15 \times 5\,\mu$m); chlamydospores usually present, borne singly or in short chains, intercalary and terminal, ± globose, brown, 4–8 μm diam. — This species is similar to the anamorph of *Thermoascus crustaceus* (Apinis & Chesters) Stolk *1965* which can be distinguished besides by the ascomata, by a higher temperature optimum (40°C) and by the absence of chlamydospores. — DNA analysis gave a GC content of $50 \cdot 5$–51% (5600).

The geographical distribution of this thermotolerant fungus (168) shows an obvious concentration in warmer climates and arid regions (5024). It has been frequently reported from India, also from Kuwait (4000, 4001), Pakistan (61), Nepal (1826), Japan (2532, 3267), Polynesia (6084), southern states of the USA (1164, 2482, 2573, 3817, 4918), Central America (1697, 2031), Brazil (398, 399), Egypt (8, 3993, 4962), equatorial West Africa (1420), and Italy (270, 271); it has also been isolated from soils in southeastern Alaska (1171), the USA (1039, 1166, 2822, 6351), Poland (1421, 1422), and the British Isles (166, 1376, 1666, 3157). It is common in forest soils (270, 271, 2482, 2573, 2854, 3817, 4716, 4995, 6351), also after prescribed burning (2822), further in conifer swamps (1039), soil with chaparral vegetation (1164), grassland (166, 3863, 3865, 4933), cultivated (1420, 3545, 3817, 3868, 4716, 4962, 4996) and uncultivated soils (4030, 4698), saline soils (8, 4000, 4001), polluted or non-polluted water (1166, 3809), seawater (4918), wood exposed to seawater (4546), and air (1166, 1666, 3204, 5024). *P. variotii* has repeatedly been recorded in

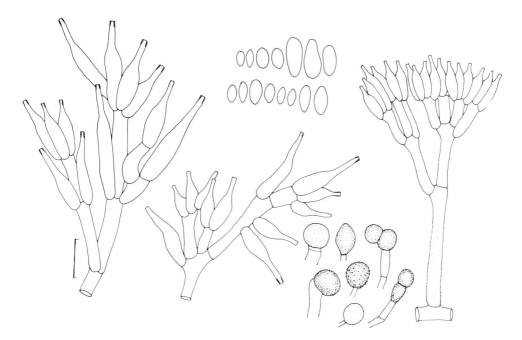

Fig. 230. *Paecilomyces variotii*, various types of conidiophores, conidia, and chlamydospores, orig. R. A. Samson (from 5024).

soil depths down to over 30 cm (3865, 4996, 5000). Steam-sterilized soil is rapidly recolonized (3330). Isolations have been made from palm leaves (3963), Scots pine leaf litter (2344, 2345), peat (1376, 3157), hay (5024), oakwood (5024, 5237), discoloured wood (5283), wood pulp and slime accumulations in paper mills (3294, 5060), sewage treatment plants (1165), composted urban wastes (3041, 3051), decomposing straw (4611) and other self-heating substrates (5024), birds' nests, feathers and pellets (168, 2575), hair baits in soil (1422, 1423), children's sandpits (1421), earthworm casts (2857), and dung of arctic animals (4174). *P. variotii* has been found in the rhizospheres of *Bothriochloa pertusa* (3271), *Trichodesma amplexicaule* (3864), groundnuts (2532), wheat (5514), banana (2030, 2031), onion, carrot and cabbage (3733), peat bog plants (1376), on roots of sugar beet (4559), seeds of cotton (5009), rice (2297), wheat (2297, 2716, 3989, 5293), corn, sorghum (3989), pods of groundnuts (3702), pecans (2572), spoiled olives in storage (309) and numerous other plants and plant parts (4548, 5024). It is a common agent in the deterioration of food-stuffs (5024, 5980) and can apparently develop on some most unusual substrates including optical lenses (2941), leather, various chemical solutions, photographic paper, synthetic rubber, creosoted wood, mouldy cigars and ink (5024).

P. variotii is moderately thermophilic with a growth optimum between 25 and 35°C (1827), in some isolates 40°C; it can still grow at 60°C (2662) and tolerates high temperatures during composting processes (3041) and fruit juice pasteurization (3428, 3620). It can grow at osmotic potentials of $c.-260$ bars (309, 4001, 5881). — Out of seven sugars tested in a combination experiment, galactose with either sucrose or maltose gave the highest yields in dry weight (2624). *P. variotii* deteriorates jute fibre and paper (5024), causes soft rot of timber, and stains oakwood yellow (5072); it was found to grow on 0·2% formaldehyde, 5% methanol and 5% sodium formate as sole C source (4991), utilize a great number of different plasticizers (473) and grow on PVC (5148) and kerosene (4935); corrosion of metals is reported in media which favour acid production (85). On media with laminarin as a substrate, it produces 1,3-β-D-glucanase (4778) and with mannan, mannose, glucose and starch 1,4-β-mannanase (4781). It was found to be a good producer of citric (5858) and gluconic acids (2147); humus-like substances were formed when grown on starch (5240); caprylate is converted into 2-heptanone (1933). Metabolites include the sideramin ferrirubin (6528), indole-3-acetic acid, fusigen (294, 5024, 6115), the antifungal antibiotic variotin (5706, 5733) and a penicillin-like substance (799). — *P. variotii* is said to stimulate the growth of barley seedlings (293). Contaminated grain proved to be toxic to young ducklings (5158). — As the causal agent of paecilomycosis it has been isolated from a generalized infection in a dog, an aborted calf, mycotic stomach ulcer in a calf, kidney infection of a horse, from brain and lung of a turkey and a rat caecum (5024); it has also rarely been isolated from man in cases of endocarditis and infection of the lacrymal sacs (5943); toxicosis from moulded feed has been observed in rabbits, swine and chicken; isolates from infected organs proved to be toxic (5024).

Papulaspora Preuss *1851*

Type species: *Papulaspora sepedonioides* Preuss

Contributions to a MONOGRAPHIC TREATMENT: Hotson (2544, 2545), and Weresub and LeClair (6290).

The genus *Papulaspora* was interpreted rather widely by Hotson (2544, 2545) as consisting of sterile mycelia with various kinds of "bulbils"; the use of the term bulbils has subsequently been restricted by Weresub and LeClair (6290) to homogeneous pseudoparenchymatous bodies occurring only in the basidiomycetous genera *Burgoa* Goid. *1938* and *Minimedusa* Weresub & LeClair *1971*, while "papulaspores", thallodic propagules differentiated almost from the inception into central and sheathing cells, occur amongst the mycelia of some species of *Melanospora* Corda (e.g. *M. fallax* Zukal *1889*) and some probably related hyphomycetes. In *Papulaspora* species, phialidic states resembling that of *Harzia* Cost. (q.v.) are often also found. About 20 species now remain in *Papulaspora* (6290).

Comparable bulbillose genera are *Minimedusa* Weresub & LeClair (q.v.) in which the bulbils arise from the aggregation of several lateral hyphae and clamp-connections rarely occur in peripheral hyphae, and *Myriococcum* Fr. *1823* (type species *M. praecox* Fr. = *Papulaspora byssina* Hotson *1917*) which has dense masses of small-celled bulbils 100–250 μm diam arising from one to several lateral branches twining around the main branch.

The ecology of *Papulaspora* species is relatively little known; most have been isolated from dung, decaying plant material or soil, and some from *Gladiolus* and *Crocus* corms and *Helianthus* tubers. Some species are mycoparasites.

The two species treated here both produce abundant yellowish to brown papulaspores which develop from intercalary primordia.

Key to the species treated:

1   Papulaspores mostly submerged, mostly 88–150 μm diam; central cells 1–6, 28–55 μm diam; surface cells moderately protruding   *P. immersa* (p. 538)
     Papulaspores mostly in the aerial mycelium, mostly 140–170 μm diam; central cells numerous, 10–20 μm diam; surface cells markedly protruding   *P. irregularis* (p. 539)

Papulaspora immersa Hotson *1912* — Fig. 231.

DESCRIPTIONS: Hotson (2544), and Nicot (4156). — Colonies reaching 8·0–9·0 cm diam in five days at 20°C on MEA; aerial mycelium scanty. Papulaspores originating from intercalary cells, pale brownish yellow, irregular in outline, 88–150(–260) μm diam, often submerged in the agar; showing 50–70(–100) moderately protruding cells in surface view; central cells 1–6, comparatively large, 28–55 μm diam, angular, darker than the peripheral ones. Hyphal cells multinucleate. Phialidic state unknown.

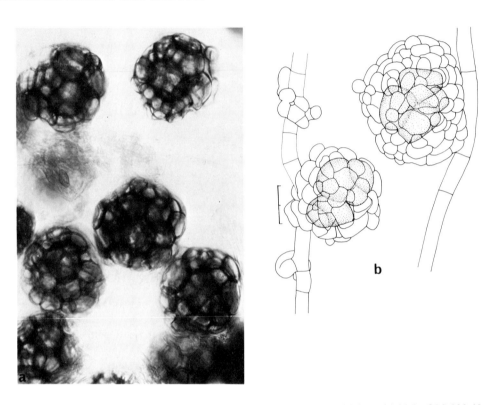

FIG. 231. *Papulaspora immersa*. a. Papulaspores, × 500; b. two papulaspores with lateral initials; CBS 890.68.

P. *immersa* was originally isolated from horse, dog and rabbit dung in the USA and Canada (2544). It has been found subsequently in a garden soil in France (4156), and agricultural soil in the Netherlands (CBS) and on decomposing cellulose in Israel (4758). In arable soils under wheat in Germany it was the commonest *Papulaspora* species found (1889).

P. *immersa* is able to decompose pectin (1432).

Papulaspora irregularis Hotson *1912*

DESCRIPTIONS: Hotson (2544), and Watanabe (6213). — Mycelium white, ± procumbent, reaching over 5·0 cm diam in five days at 20°C on MEA. Papulaspores originating from intercalary cells in the aerial mycelium, at first hyaline, later becoming pale straw-coloured, ± globose with prominent peripheral cells, 140–170 μm diam, or irregular in outline and 250–300 μm diam; showing 90–150 cells in surface view; central cells numerous, less than 20 μm diam, angular with slightly thinner walls than the peripheral hyphae. Hyphal cells multi-nucleate. Phialidic state unknown.

P. irregularis was originally isolated from rat dung in the USA. It has since been isolated once from sugar cane ratoons in Taiwan (6213), infrequently from greenhouse soil in the Netherlands (CBS), and very rarely from wheatfield soils in Germany (1889).

Penicillium Link ex Fr. *1821*

= *Coremium* Link ex Fr. *1821*
= *Citromyces* Wehmer *1893*

Type species: *Penicillium expansum* Link ex Gray (5029)

Teleomorphic genera (fam. Trichocomaceac Fischer = Eurotiaceae Clem. & Shear): *Eupenicillium* Ludwig, *Talaromyces* C. R. Benjamin, *Trichocoma* Jungh., *Hamigera* Stolk & Samson

MONOGRAPHIC TREATMENT: Raper and Thom (4745) with a supplement by Kulik (3164), and Pidoplichko (4549); subsection *Fasciculata* and related species: Samson *et al.* (5029).

The Hyphomycete genus *Penicillium* is characterized by macronematous penicillate conidiophores, the ultimate branches of which are verticillate phialides (often called sterigmata) while the penultimate branches are termed metulae; the conidiophore stipe has often been referred to as the conidiophore; the phialides form basipetal chains of dry conidia which usually form their surface ornamentation in a maturation phase after complete separation from the conidiogenous plasma; at that stage the conidial apices are often differentiated into connectives. — The process of conidium formation was studied by time-lapse cinematography in *Penicillium corylophilum* (1105), but the same sequence of events can be expected in any species of the genus; the constant length of the phialide wall during conidium formation becomes evident. TEM studies of conidiogenesis in various species show a progressive internal thickening of the phialide wall as increasing numbers of conidia are formed.

Species which form ascomata have been retained in *Penicillium* for a long time (4745) but are now correctly assigned to *Talaromyces* (q.v.) (with quickly ripening ascomata with a loose hyphal covering) and *Eupenicillium* (q.v.) (with sclerotial bodies which form asci in the centre, sometimes at a very late stage and thus were often not found in young cultures), and some other genera.

Raper and Thom (4745) in *1949* accepted 137 species including several teleomorphs; 113 species published until *1968* were compiled by Kulik (3164) and provided with an auxiliary key; since then approximately 60 further species (including teleomorphs) have been described. Pidoplichko (4549) provided a key to about 235 species including many described by Russian authors. In an important recent study of type cultures by Pitt (4578), growth at 5°C, 37°C, and at 25°C with a water potential of −100 bars (medium with 25% glycerol) was used for distinguishing species. Raper and Thom's (4745) keys are mainly based on conidiophore ramification and colony structure; the latter criterion in particular often gives artificial subdivisions, and some of the series distinguished by these authors are consequently not recognized here. Samson *et al.* (5029) tried to modify the whole system in a revision of the subsection *Fasciculata* by placing the main emphasis on phialide structure and adopting modern terminology for conidiogenesis. They use the terms "one-stage branched" (e.g. Fig. 232, 248) and "two-stage branched" (e.g. Fig. 233, 235, 242) to replace biverticillate and biverticillate with additional branches. The shape of the phialides is one of the most important criteria in *Penicillium* taxonomy; they are mostly differentiated into a cylindrical venter and a tapered neck, but this tapering can be very pronounced (*P. janthinellum* series) or it can be gradual, rendering the phialides lanceolate (section *Biverticillata-Symmetrica*, e.g. Fig. 245, 260, 261). A considerable variability, particularly of the macroscopic characters, between isolates of one species is now recognized and several species previously distinguished have been merged into one. The alternative key to the 40 species and varieties discussed here is based on the arrangement of Samson *et al*. (5029) and their personal suggestions; this key is preliminary in nature and should eventually be superseded by an up-to-date revision of the whole genus.

For standard descriptions we generally refer to Raper and Thom (4745) and Samson *et al*. (5029). The techniques used by these authors consist in growing the fungus in glass Petri dishes with CzA or 2% MEA; cultures are inoculated at three points; during inoculation and incubation, Petri dishes are kept in inverted position and incubation is at room temperature or at 25°C.

Species of *Aspergillus* and *Penicillium* are ubiquitous saprophytes, whose conidia are easily distributed through the atmosphere. Whereas *Aspergillus* occurs primarily in warmer regions and on heat-generating substrates, *Penicillium* species predominate in soils of the temperate regions. In soil analyses by the dilution plate technique, *Penicillium* species are detected with particular frequency. In contrast, their frequencies are considerably reduced when the soil washing technique is used: the total number of species present, however, in general, remains unaltered. Very little is yet known of interactions between *Penicillium* species and other soil fungi. As a rule, *Penicillium* species appear to be detectable at greater soil depths than species of other genera, and are only slightly concentrated in rhizospheres. Because it is known that they frequently produce antibiotic metabolites, *Penicillium* is, with regard to biochemistry, one of the most thoroughly investigated genera of fungi. All strains thus far tested have been able to solubilize metaphosphates and utilize them as their phosphorus source (4544); many species have been found to contain mycoviruses (674).

Key to the species of *Penicillium* and related genera treated:

1	Cleistothecia or sclerotia present	**2**
	Cleistothecia or sclerotia absent	**4**
2(1)	Cleistothecia soft, covered by a few to several layers of loose hyphae, ripening within 1–2 weeks	cf. *Talaromyces* (p. 752)
	Cleistothecia firm, of pseudoparenchymatous sclerotioid structure or only sterile sclerotia present	**3**
3(2)	Sclerotia on all media abundantly produced, flesh to pink, not embedded amongst orange-red hyphae, never developing asci; conidiophores simple, bearing loose conidial columns	*P. thomii* (p. 602)
	Sclerotia sooner or later developing asci in the centre; conidiophores simple, one-stage branched or two-stage branched, divaricate	cf. *Eupenicillium* (p. 283)
4(1)	Conidiophores strictly unbranched, exceptionally with an additional branch	**5**
	Conidiophores at least one-stage branched	**12**
5(4)	Colonies spreading (at least 3·0 cm diam in 12 days)	**6**
	Colonies restricted (not exceeding 2·5 cm diam in 12 days)	**8**

6(5)   Conidia ellipsoidal, conspicuously roughened, sometimes with spiral bands, in tangled chains;
       reverse dull peach to flesh   *P. lividum* (p. 585)
       Conidia globose, in long compact columns   **7**

7(6)   Conidia smooth-walled or finely roughened; conidiophores arising regularly from the sub-
       stratum, forming a dense layer, often developing from aerial hyphae in central areas; reverse
       uncoloured or yellow-orange to brown   *P. frequentans* (p. 568)
       Conidia distinctly roughened; conidiophores often arising from aerial hyphae, occasionally also
       from the substratum; reverse uncoloured or pinkish to purplish   *P. spinulosum* (p. 597)

8(5)   Conidiophores generally arising from the substratum, to 100 μm tall; conidia in compact
       columns   *P. implicatum* (p. 577)
       Conidiophores arising mostly from aerial hyphae, somewhat shorter; conidia in loose columns or
       in tangled chains   **9**

9(8)   Colonies grey or brownish; conidiophores usually not exceeding 25 μm, phialides with a distinct,
       thin conidium-bearing neck; conidia coarsely rough-walled, in short chains with often less
       than 10 conidia   **10**
       Colonies grey-green or blue-green; conidiophores about 50–100 μm; phialides with less strongly
       constricted tip; conidia smooth-walled or finely roughened, in loose columns   **11**

10(9) Colonies thin, velvety; phialides 5·5–7 × 2·5–3·5 μm; conidia 4·5–5·5 μm diam
                                                                         P. sacculum (p. 595)
       Colonies floccose; phialides mostly 5 × 1 μm; conidia mostly 2·0–2·5 μm diam
                                                                         P. restrictum (p. 591)

11(9) Colonies loose-textured, growing restrictedly on CzA, more rapidly on MEA, comparatively
       thin, dull blue-green; conidia 2·0–2·5 μm long, smooth-walled; conidiophores borne at
       successive nodes upon prostrately spreading hyphae; reverse uncoloured
                                                                         P. decumbens (p. 561)
       Colonies growing very restrictedly on CzA and MEA, blue-green; reverse uncoloured or light
       vinaceous-grey; conidiophores arising from densely entwined aerial hyphae, vegetative
       mycelium also present; conidia 2·5–3·0 μm long, smooth or rough-walled
                                                                         P. fellutanum (p. 566)

12(4) Phialides lanceolate, in compact whorls on metulae, usually without additional branches (sect.
       Biverticillata-Symmetrica); colonies often intermixed with yellow or red vegetative hyphae;
       reverse bright coloured; conidial chains tangled   **39**
       Phialides not lanceolate, usually with a short but distinct, or a long and conspicuous neck, rarely
       without a neck; additional branches often present   **13**

13(12) Phialides less then 6 μm long, short cylindrical, usually without a distinct neck; conidiophores
       two- to four-stage branched with divergent branches; colonies grey, loosely synnematous
                                                                         P. griseofulvum (p. 573)
       Phialides longer than 6 μm, with distinct neck   **14**

14(13) Phialides 15–28 μm long, cylindrical, with an abruptly tapering neck; conidia ellipsoidal to
       cylindrical, commonly 5–8 × 4–6 μm and larger; colonies olivaceous on MEA, very restricted
       and thin on CzA; causing green citrus rot   *P. digitatum* (p. 562)
       Not combining the above characters   **15**

15(14) Phialides cylindrical, tapering abruptly to a conspicuous, fairly long narrow conidium-bearing
       neck; conidial chains tangled; conidia ellipsoidal or sometimes subglobose in older cultures **16**
       Phialides not as above   **17**

16(15) Colonies with glaucous conidial areas, often intermixed with bright-coloured vegetative
       mycelium; reverse usually red, yellow or purple, occasionally colourless; conidiophores
       smooth-walled or roughened, irregularly divergently branched; conidia smooth-walled or
       finely roughened, 3·0–3·5 μm long   *P. janthinellum* (p. 580)

Colonies usually pale blue-green; reverse colourless or slightly yellow; conidiophores coarsely roughened; penicilli commonly consisting of a terminal verticil of divergent metulae; branches occasionally present; conidia finely echinulate, $2\cdot5$–$3\cdot0\,\mu$m long

P. simplicissimum (p. 597)

17(15) Metulae and/or branches divergent; colonies grey-green, grey, olive-grey or dark brown; conidia slightly roughened or echinulate; phialides flask-shaped, tapering abruptly into a rather short neck **18**

Not combining the above characters **20**

18(17) Conidiophores arising as short branches from long trailing aerial hyphae, occasionally longer when arising from the substrate; metulae conspicuously divergent; conidia subglobose to globose **19**

Conidiophores longer; metulae less conspicuously divergent; conidia subglobose to ellipsoidal, ornamented with distinct spiral bands, $2\cdot5$–$4\cdot0 \times 2\cdot5$–$3\cdot0\,\mu$m *P. daleae* (p. 560)

19(18) Colonies grey-green, floccose; reverse orange to rich brown, occasionally uncoloured, conidiophores coarsely roughened; conidia at first short ellipsoidal, becoming globose, finely roughened, 2–$2\cdot5\,\mu$m diam *P. canescens* (p. 549)

Colonies grey to dark olive-grey; reverse deep orange; conidiophores smooth-walled; conidia subglobose to globose, echinulate, 3–$3\cdot5\,\mu$m diam *P. nigricans* (p. 586)

20(17) Conidiophores strongly synnematous; synnemata consisting of a compact white or pink, up to $1(-2)$ cm long stalk and a distinct club-like green conidial head *P. claviforme* (p. 557)

Conidiophores mononematous or loosely synnematous, synnemata never showing a clear differentiation into a stalk and conidial head **21**

21(20) Conidiophores usually consisting of a stipe bearing one apical verticil of metulae with phialides (one-stage branched); branches occasionally present and often not appearing as integral part of the terminal penicillus **22**

Conidiophores usually consisting of a stipe bearing a well-defined apical penicillus which is composed of branches, metulae and phialides (two-stage branched) **29**

22(21) Metulae 2–3 in the penicillus, slightly appressed; conidia strongly ellipsoidal, $4\cdot5$–$6\cdot5 \times 3$–$4\,\mu$m, forming crusts and giving the colonies a silky appearance; reverse often pinkish

P. oxalicum (p. 588)

Conidia smaller, not forming crusts **23**

23(22) Reverse on CzA and MEA yellowish brown to dark yellow-green, colour diffusing into the agar; colonies on CzA somewhat restricted; conidiophore stipes usually rough-walled **24**

Not combining the above characters **25**

24(23) Colonies yellow-green; branches absent; conidia ellipsoidal, smooth- or occasionally rough-walled, $3\cdot5$–$4 \times 2\cdot2$–$3\cdot0\,\mu$m *P. herquei* (p. 575)

Colonies greyish blue; branches sometimes present (depending on the isolate), arising fairly low on the conidiophores; conidia globose, conspicuously roughened, 3–$3\cdot5\,\mu$m diam

P. atrovenetum (p. 545)

25(23) Branches rarely present; conidia smooth-walled, forming columns **26**

Conidiophores usually with several branches; conidia globose, $2\cdot0$–$2\cdot5\,\mu$m diam, rough-walled, in loose columns or tangled chains, ± grey-green **28**

26(25) Colonies on MEA spreading rather broadly; metulae (2–3) often of unequal length; reverse green or almost black, usually with darker sectors, but agar uncoloured

P. corylophilum (p. 558)

Colonies on MEA growing more restrictedly; metulae of equal length; reverse bright yellow, yellowish or uncoloured **27**

27(26) Conidial areas blue-green to grey-green; reverse usually bright yellow *P. citrinum* (p. 555)

Conidial areas dull blue-green to yellow-green; reverse uncoloured or yellowish

P. steckii (p. 599)

28(25) Conidial chains tangled; conidiophores smooth-walled *P. waksmanii* (p. 610)
Conidial chains in loose columns; stipes smooth- to rough-walled *P. jensenii* (p. 583)

29(21) Conidiophores mononematous; odour absent, undefinable or fruity **30**
Conidiophores usually more or less synnematous, rarely mononematous; odour in most species earthy or pungent, strong **33**

30(29) Branches and metulae divergent **31**
Branches and metulae convergent and strongly appressed **32**

31(30) Stipes usually rough-walled; conidia globose, rough-walled, 2·5–3·0 μm; colonies glaucous-grey, lanose; reverse colourless or yellowish; colonies reaching 2·5–3 cm diam in 14 days
P. lanosum (p. 584)
Stipes smooth-walled; conidia globose to ellipsoidal, smooth-walled, 3–4 × 2·8–3·8 μm; colonies blue-green or grey-green, velvety; reverse intensely yellow (sometimes uncoloured); colonies reaching 2·5–4 cm diam in ten days on CzA, on MEA growing even faster
P. chrysogenum (p. 550)

32(30) Conidiophore elements swollen; stipe 300–500 μm long and usually exceeding 4 μm in width slightly roughened *P. brevicompactum* (p. 546)
Conidiophore elements not swollen; not exceeding 4 μm diam, stipe smooth-walled, 100–250 μm long and 3·5–4·0 μm wide *P. stoloniferum* (p. 600)

33(29) Conidiophores smooth-walled on CzA and MEA **34**
Conidiophores rough-walled **35**

34(33) Colonies fast growing, reaching 4–5 cm diam after 10 days; sometimes strongly synnematous; odour reminiscent of apples; conidia subglobose or ellipsoidal; causing rot of pomaceous fruits *P. expansum* (p. 563)
Colonies growing restrictedly, reaching 2–2·5 cm diam after 12 days; odour aromatic perfume-like; conidia ellipsoidal to cylindrical, causing blue citrus rot *P. italicum* (p. 579)

35(33) Synnemata often forming distinct granulose zones; reverse yellow to orange-brown; odour fruity combined with a pungent component; conidia ellipsoidal, 3·0–3·5 × 2·0–2·5 μm
P. granulatum (p. 572)
Synnemata less distinct and not or indistinctly zonate; reverse uncoloured or yellowish, sometimes later orange-brown or purplish; conidia globose to subglobose, 3·5–4·0 μm diam
P. verrucosum **36**

36(35) Colonies growing restrictedly, reaching 2·5–3·5 cm diam after 14 days, bright yellow-green
P. verrucosum var. *verrucosum* (p. 604)
Colonies reaching 4 to >5 cm diam in 14 days, of other colours **37**

37(36) Conidial areas blue-green, often becoming yellow-green or light grey-green with age; in fresh isolates, showing a yellow to ochraceous centre *P. verrucosum* var. *cyclopium* (p. 606)
Conidial areas dark yellow-green, without a bluish tint **38**

38(37) Conidiophores in fresh isolates predominantly loosely synnematous; yellow mycelium and orange-brown exudate usually present *P. verrucosum* var. *corymbiferum* (p. 606)
Conidiophores predominantly mononematous; yellow mycelium absent, exudate absent or colourless *P. verrucosum* var. *melanochlorum* (p. 609)

39(12) Conidia smooth-walled or delicately rough-walled **40**
Conidia conspicuously rough-walled, apiculate **43**

40(39) Colonies usually spreading broadly (rarely restricted), mostly reaching 4·5–5·5 cm diam in 14 days, typically funiculose or tufted, mostly with yellow vegetative mycelium; reverse usually red; conidiophores often greenish; conidia usually delicately rough-walled, subglobose to ellipsoidal, apiculate; in some isolates producing dark sclerotia *P. funiculosum* (p. 570)
Colonies growing somewhat more restrictedly, less funiculose, rather floccose; conidia perfectly smooth-walled **41**

41(40) Conidia strongly ellipsoidal to fusiform, variable in size, the larger ones slightly roughened, the smaller ones smooth-walled; colonies light yellow to grey-green, partly covered with yellow, encrusted hyphae; reverse red-orange, yellow-orange or greenish brown, without diffusion into the agar   *P. variabile* (p. 603)

Conidia of more uniform size, perfectly smooth-walled   **42**

42(41) Colonies developing an intense red or purple-red pigmentation diffusing into the agar; conidia subglobose to ellipsoidal, usually not apiculate   *P. rubrum* (p. 593)

Colonies not developing an intense red pigmentation, showing areas of orange or yellow mycelium; conidia dark green, thick-walled, ± apiculate   *P. islandicum* (p. 577)

43(39) Colonies on CzA growing very restrictedly, reaching 1·0–1·5 cm diam in 14 days, tough, felty, white to flesh, never developing an intense red pigmentation; conidia ellipsoidal, conspicuously roughened   *P. rugulosum* (p. 594)

Colonies on CzA growing somewhat faster, reaching 2·0–3·0 cm diam after 14 days, dark blue to yellow-green, usually partly covered with yellow, encrusted aerial hyphae, developing an intense red pigmentation diffusing into the agar; conidia subglobose to ellipsoidal, thick-walled, roughened, with roughening sometimes arranged in transverse bands

      P. purpurogenum (p. 590)

Penicillium atrovenetum G. Sm. *1956* — Fig. 232.

DESCRIPTION: Smith (5431). — *P. herquei* series or rather *P. citrinum* series. — Colonies growing somewhat restrictedly, reaching 2·0 cm diam in ten days at 20°C on MEA, with bright bluish green, later deep greyish blue conidial areas; reverse intensely yellowish brown, later dark green to blackish, with the pigment diffusing into the agar. Conidiophores partly rough-walled; penicilli typically with an additional divergent branch; metulae somewhat divergent. Conidia globose, coarsely warty, 3·0–3·5 μm diam. — Species identification still presents problems and some reports thus have to be accepted with caution. — The fatty acid analysis (6043) gave 0·4% C_{14}-, 13·8% C_{16}-, 3·6% C_{18}- and 1·9% C_{20}-saturated acids, and 2·7% C_{16}- and 30·9% C_{18}-mono-unsaturated acids (3232).

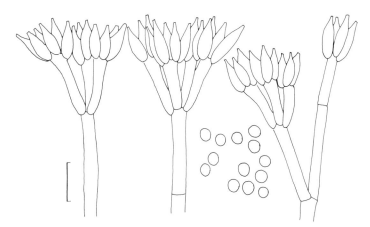

FIG. 232. *Penicillium atrovenetum*, penicilli and conidia, CBS 241.56, orig. R. A. Samson.

P. atrovenetum was originally isolated from agricultural soils and dunes in England but subsequently found rather commonly in wheatfield soils in Germany (1889). It has rarely been found in sandy soils of high salinity in France (4162), soils in Egypt and South Africa (IMI), and on dead petioles of *Pteridium aquilinum* (1821). — Dead conidia were lysed in soil within seven days (1049).

Swelling of the conidia prior to germination requires glucose, phosphate and oxygen; for germ-tube formation the presence of a N source is also required (2053, 2055). Pectin decomposition is strong, but with differences in this capacity occurring between isolates (1432). The formation of nitrite but not of nitrate has been detected in media containing peptone or NH_4^+ ions (3622); when grown on nitrite, the formation of nitrous oxide has been observed (6502). The pigment atrovenetin (371, 4122), β-nitropropionic acid (4707, 5245) and an enzyme system oxidizing β-nitropropionic acid (2146) are produced. — In a chronically irradiated soil the fungus survived in the range 15–120 R per day (2009).

Penicillium brevicompactum Dierckx *1901* — Fig. 233.

DESCRIPTION: Raper and Thom (4745). — Members of the *P. brevicompactum* series are

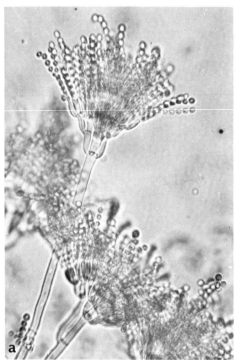
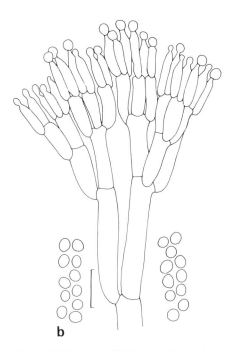

FIG. 233. *Penicillium brevicompactum.* a. Compacted penicilli with conidial chains, × 500; b. penicillus and conidia; c. penicilli, SEM, × 1600, orig. R. A. Samson.

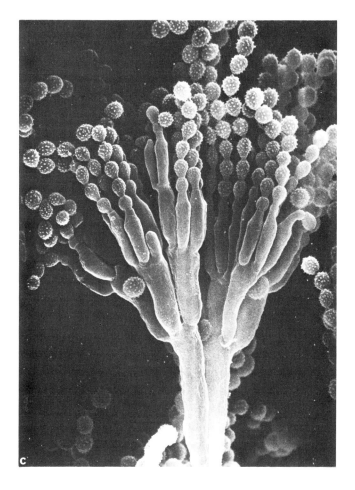

characterized by (mostly) rather slow-growing, greyish green colonies, uncoloured or brownish reverse, and long conidiophores with two-stage (sometimes three-stage) branched, very compressed penicilli. Colonies reaching 1·5–2 cm in 14 days at 24°C on CzA and MEA. Conidiophore stipes 300–500 μm long and usually more than 4 μm wide. Metulae and branches often swollen. Conidia globose to subglobose, slightly roughened, mostly 3·5–4·0 μm diam. — The similar *P. paxilli* Bain. *1907* lacks additional branches and might also be accommodated in the *P. citrinum* series close to *P. herquei* (q.v.); cf. also *P. stoloniferum.* — The rodlet pattern of the conidial surface as seen in freeze-etched replicas is prominent and slightly interlocked; surfaces are smooth except for localized raised areas (2421). — Some isolates contain virus particles (2527, 5043, 6433); a possible correlation with the production of mycophenolic acid as an antiviral metabolite has been suggested (1358). — DNA analysis gave a GC content of 51–53% (5600).

P. brevicompactum is a cosmopolitan species, especially in soil, but never particularly frequent. Reports include ones from the White Sea (3368), the USSR (3365, 3367, 3652, 3871, 4474), more northerly latitudes of Asia in the Amur region (432), as well as in

Tien-Shan, the Altai (5647) and Pamir (3850), Czechoslovakia (1700, 1703), the British Isles (163, 1376, 5220, 5222), Wisconsin/USA (1030, 1032, 1040, 2004, 2007, 2008, 3549, 4225, 4226), Italy (3453), Turkey (4245), Israel (654, 2762, 2764, 4759), Syria (5392), Egypt (8, 3993, 3997), Iraq (92), Kuwait (4000), the Ivory Coast (4159), Guinea (3852), South Africa (1559), Australia (3720), India (1519), Japan (2532, 5846) and Brazil (4547). It has been isolated from very varied soil types, such as podzol, grey forest soil and chernozem (2871, 3652, 3871). Reports are somewhat less frequent for cultivated (e.g. 2762, 2768) than for forest soils. It is common under humid mixed hardwood (1030, 1032, 1040, 2004, 2007, 2008, 3549, 4225, 4226), beech (270, 271, 272, 273, 3538), oak (1700), but also various coniferous forests (681, 3447, 3975, 5465) and in conifer swamps (1039). It has also been reported from peat bogs (1376, 4474), surface layers of peat (1437), peaty soils (163), heathland soils (3720, 5220, 5222), grassland (159, 163, 2573, 3363), soils with taiga (3366) and steppe type vegetation (1559), dunes (745), salt-marshes (8, 4000), and desert soils (654, 3997, 4733); it is rarely isolated from garden compost (1424). The following have been recognized as further habitats: caves (3453, 4137), carst (3365) and ice caves (3367) and a uranium mine (1703); streams (1166, 4429) with only moderate degrees of sewage pollution (1154, 1157) but also activated sludge (1387). It has been isolated from pine litter (681), a decaying *Boletus* sp. and an old fruit-body of *Calvatia excipuliformis* (2338), aecidiospores of *Cronartium comandrae* (4608), the rhizospheres of wheat (3567), groundnuts (2532, 2768), some peat bog plants (1376), heathland plants with high frequencies (5819) and rarely halophytes (3375), and roots of oak (5394), poplar (3452) and various steppe grasses (3376); it has been isolated from stored wheat grain (2097, 3989, 4492), organo-mercury-treated barley and wheat seeds (2077), seeds of groundnut (2765), *Pennisetum typhoideum* (57), cultivated grasses (3512) and corn (3875), corn silages (1092), sycamore achenes (1683), pecans (2572, 5092), wood pulp and paper (6168), wood exposed to seawater (4546), rabbit dung (6085), nests and feathers of free-living birds (2575), brown coal (4474), stored apple, grape and tomato fruits (342), fruit juices (5205) and frozen fruit cake (3153). It has been found in accidental wound infections on caterpillars of *Malacosoma neustria* (4024). It belongs to the *Penicillium* species that often occur in the deeper soil layers (432, 3447) and is stimulated by fertilizing with N, K and NK (2895), but intolerant of high salinity (92).

Optimum conidium germination occurs at 23°C; the temperature range for mycelial growth is −2 to +30°C with an optimum at 23°C (3876) and no growth at 37°C; good growth occurs at a water potential of −100 bars (4578). Growth on hair baits in soil was still observed at −200 bars (2095). — Positive evidence has been obtained for the ability to decompose starch, tannin, protein (1425), cellulose (1425, 3618, 6085), pectin and xylan (1432, 3538). This fungus can utilize hydrocarbons from fuel oil (4239) and is also reported to grow on various plastic polymers (5148, 5859). Organic acids are secreted in such quantities that they can lead to metal corrosion (85). From the culture filtrate ribonuclease was isolated (2642). The known metabolites from *P. brevicompactum* include compactose (5278), arabinose, ribose, a glucan, citric and traces of other acids, gluconic acid (2014), 2-oxo-D-gluconic acid, D-glucuronic acid, the sesquiterpenes pebrolide, 1-deoxypebrolide and desacetylpebrolide (3231), 3,5-dihydroxyphthalic acid (4333), 2-carboxy-3,5-dihydroxy benzyl methylketone, 2-carboxy-3,5-dihydroxy phenyl acetyl carbinol (4334), the toxin mycophenolic acid (419, 536, 860, 1087, 1088, 6448) with antibiotic (14, 1782), cytotoxic (6005), phytotoxic (6448) and antiviral (1358) properties, mycophenolic acid diol lactone, ethyl mycophenolate, mychromenic acid (3231), ergosterol palmitate, *meso*-erythritol (5278), the piperazine derivatives brevianamide A to F (524). — *P. brevicompactum* shows a pronounced inhibitory activity towards *Rhizoctonia solani* (1425, 1431) and also *Bacillus cereus* var. *mycoides*, *B. subtilis*,

Staphylococcus aureus, Streptomyces griseus, and *Azotobacter chroococcum* (1092, 3852); it is also antagonistic towards *Fusarium sporotrichioides, Neurospora sitophila* and *Escherichia coli* (5394); it has been isolated from sclerotia of *Sclerotium cepivorum* (3990). It inhibits the growth of pea, vetch and wheat *in vitro* (3852). On stored *Gladiolus* bulbs it may cause deterioration and subsequent growth inhibition (5878). It proved toxic when it is accumulated in cereals (1815, 6412). — It can grow anaerobically in an atmosphere of N_2, but requires biotin and thiamine under these conditions (5680).

Penicillium canescens Sopp *1912* — Fig. 234.

DESCRIPTIONS: Raper and Thom (4745), and Mehrotra and Tandon (3740). — The *P. canescens* series is close to the *P. nigricans* series, but the colonies are more grey-green than greyish brown (conidia hyaline to greyish under the microscope). Colonies growing somewhat restrictedly, reaching 3·5–4 cm diam in 10–13 days at 25°C on CzA, growth slightly slower on MEA; reverse orange, becoming rich brown with age. Conidiophores conspicuously roughened, metulae strongly divergent. Conidia at first ovate, later becoming globose, smooth-walled or slightly roughened, 2·0–2·5 μm diam, forming loose columns.

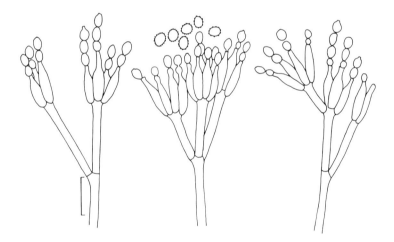

FIG. 234. *Penicillium canescens,* penicilli and conidia, CBS 288.53.

A worldwide distribution is documented by observations from the British Isles (163, 2502, 6182), Sweden (1446), the USSR (3652, 3871, 4474), Czechoslovakia (1700), Germany (825), Austria (3414, 3418), Poland (1978), France (242), Italy (3976, 5047, 6082), Wisconsin (2004, 2007, 2008, 4225) and Oregon/USA (6351), Canada (3959), soils to above 4000 m in Peru (2005) and Pamir (3850); also in desert soil in Egypt (3996, 3997) and other soils in Turkey (4245), Syria (277), Iraq (92), Libya (6510), central Africa (3063), South Africa (1559, 4407), India (3740), Japan (5846), Australia (977, 3720) and Tasmania (5930).

It has been isolated from solonetz (4677), grey forest soil, chernozem (3871), gley, pseudogley, ranker, terra fusca, podzol, and semipodzol (825, 3414, 3416, 3418, 3652); it is found to be frequent in mixed hardwood forests (2004, 2007, 2008, 4225), mixed alder-conifer stands (6351) and vegetations dominated by aspen (3959), forest nurseries (1978), calcareous forest soils (3451, 4152), humus under *Pinus ponderosa* (681), truffle soils (3451), grassland (6182), heathland soils (3720, 5047), peat fields (4474), tundra (1446), taiga (3871) and steppe soils (1700), dunes (4162), vineyards (242), and, more frequently, arable soils (918, 977, 2159, 2161, 2762, 3450); it has also been found in an acid mine drainage stream (1166). Its frequency was found to be increased by NP fertilization (2163). It has been observed on leaf and needle litter (650, 681), on wood sticks exposed to soil (1956) or seawater (4546), also in stored wheat grain (4492) and wheat seeds in soil (2097). The maximum frequency was recorded at a soil depth of 5 cm but it has also been found down to 20–50 cm (825, 2161). It has been isolated from the rhizospheres of clover (3429), onion (2159), tomato (16), and Australian heathland plants (5819); also from the phylloplane of vascular plants (296), storage places of gerbils (5393), and feathers of free-living birds (2575). — It has been found to be associated with cultures of the collembolan *Hypogastrura macgillivrayi* (1027) and is suitable as food for the mites *Acarus siro* and *Tyrophagus putrescentiae* (5383).

Little growth occurs at 5°C and 37°C, good growth still at a water potential of −100 bars (4578), optimal growth at *c*. −18 bars (0·4 M NaCl), 50% growth reduction at *c*. −95 bars, no growth at –150 bars (3998, 6395). Optimal growth was obtained at 0·013 atm O_2 partial pressure (1428). — Cellulose fibres do not appear to be attacked (3618), cellulase production could, however, be induced by jute cellulose and to a lesser extent by filter paper (386); methylcellulose can be attacked (2878). *P. canescens* can grow on nitrite as the sole N source (10). The metabolic product canescin possesses a moderate antimycotic action (720). — The culture filtrate was found to inhibit bacteria (515) and fungi (516) but stimulated growth in barley seedlings (293) and sprouts of tomato seedlings, whereas roots were inhibited (5779). — Indole-3-acetic acid has been described as a metabolic product of *P. canescens* (294, 6115).

Penicillium chrysogenum Thom *1910* — Fig. 235.

= *Penicillium notatum* Westling *1911*

For some other synonyms see Samson *et al.* (5027, 5029).

DESCRIPTIONS: Raper and Thom (4745), and Samson *et al.* (5029). — *Penicillium chrysogenum*, the only species recognized in its series, is characterized by thin, spreading colonies, reaching 2·5–4 cm diam in 14 days at 25°C on CzA, velvety, grass-green to bluish green with pale to bright yellow, sometimes vinaceous reverse and yellow exudate; conidiophores two- to three-stage branched at rather wide angles. — The four species distinguished by Raper and Thom (4745) in the *P. chrysogenum* series have been combined by Samson *et al.* (5029) to one because the conidial shape varies within single cultures, and the type cultures of *P. notatum* and *P. chrysogenum* proved to be indistinguishable. — Conidia

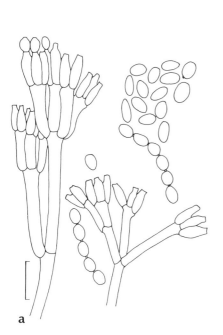

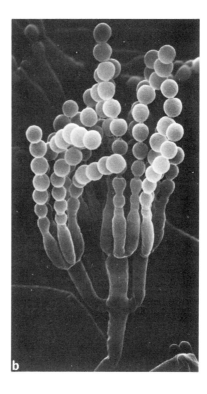

FIG. 235. *Penicillium chrysogenum*. a. Penicilli and conidia, CBS 196.46; b. penicillus with conidial chains, CBS 306.48, SEM, × 2250, orig. R. A. Samson.

smooth-walled, ellipsoidal, $3 \cdot 0$–$4 \cdot 0 \times 2 \cdot 8$–$3 \cdot 5 \mu$m diam ("*P. chrysogenum*" morph) or globose to subglobose, $3 \cdot 0$–$3 \cdot 3 \mu$m diam ("*P. notatum*" morph). Odour often fruity, suggesting apples or pineapples, sometimes undefinable. — Nuclear division is similar to that in *P. expansum*: n = 4 or 5, depending on whether one of the staining bodies is regarded as a centriole or a chromosome (1209); conidia are uninucleate whilst the vegetative cells are multinucleate. — Conidium germination (on sucrose-asparagine broth) is accompanied by an increase in total N, protein and nucleic acid contents which decrease again during germ tube formation; protein content again increases subsequently to a constant level; the lipid content decreases during germination (3624). — The rodlet pattern of the conidium surface as seen in freeze-etched replicas is not interlacing and has few rodlets in bundles; their surfaces are thus relatively smooth (2421). — Resting conidia show four wall layers, the outermost of which dissolves during germination whilst the inner, together with a newly formed layer, gives rise to the germ tube; the latter subsequently exudes an electron-opaque outer layer (3627, 3700). Some lomasomes, spherosomes and two nucleoli per cell are present (3700). — In shaken liquid cultures, conidia germinate after 4–12 h to form thin, short-lived, and subsequently wider hyphae which give rise to a second generation of mycelium (1478). — Protoplasts were obtained from mycelium by combinations of cellulase with snail gut juice or with the culture filtrates from *Streptomyces graminofaciens* (148) or *S. venezuelae* (1709). — A standard penicillin strain NRRL 1249 B21 segregated into light and dark types (5044). Several investigations on the genetics have been reviewed (312). Mutants have often been induced in order to increase the penicillin yield. Heterokaryon formation and parasexuality have been studied and rather stable diploids with larger conidia found. During haploidization, genes can be recombined by exchange of whole chromosomes, but crossing-over occurs only very rarely. Mutational loci were arranged in three haploidization groups. Virus particles have been detected in some isolates (328, 1229, 2510, 4107, 5043). — Purified cell wall hydrolysates contained 34–43% glucose, 4–12% galactose, 1–8% mannose, 18–19·5% glucosamine, 1–2% rhamnose, 2·1% xylose, and traces of galacturonic acid (3232), although mannan was not detected in the purified walls by X-ray diffraction analysis (3124). The main wall constituents are, however, a glucan and chitin (169). The dominating amino acids of the mycelium are serine (12·8%), glycine (11·4%), alanine (10·7%), asparagine (9·9%), threonine (9·7%), proline (9·4%), glutamine (9·1 mol % of the total amino acids) (3232). The conidial walls contain more amino acids than the hyphal walls, histidine, arginine, and lysine being found mainly here; galactose, glucose, mannose and traces of rhamnose were found in the hydrolysate of resting conidia (in that order); glucosamine is found in walls of both resting and swollen conidia, while galactosamine appears in germinating conidia and mycelium (3626). The mycelium was found to contain 23·8% C_{16}-, 9·0% C_{18}-, 4·5% C_{20}- and traces of C_{12}- and C_{14}-saturated fatty acids, 3·0% C_{16}- and 4·7% C_{18}-mono-unsaturated acids when grown on a glucose medium (3232); conidia contained 10·5% C_{16}-, 1·0% C_{17}-, 2·8% C_{18}-saturated fatty acids, 0·7% C_{16}- and 30·5% C_{18}-mono-unsaturated acids, and 53·3 and 0·8% C_{18}- di- and tri-unsaturated acids, respectively (1277). The content of long-chain fatty acids has been used to study species relationships (1281). — DNA analysis gave a GC content of 51–54·5% (5600).

Penicillium chrysogenum is the best known penicillin-producing fungus and is very widely distributed as a soil fungus, occurring in the temperate zones in forest, grassland and arable soils with comparable frequencies. — Its overall distribution embraces glacial ice in Alaska (1171), soils in the Amur region (432) and numerous other locations in the USSR, Spitsbergen (3064), Finland (3801), Sweden (1446), Poland (1423, 4137), Czechoslovakia (1703), the British Isles (745, 1376), the USA (1166, 1632, 2573), Italy (3445, 3538), Spain (3446), France (242, 2161), Turkey (4245), Syria (277, 5392), Israel (654, 2762, 2764, 2768, 4759), Egypt (8, 100, 3988, 3993, 3996, 3997, 4962), Libya (6510), Iraq (92), Kuwait (4000, 4001), Somalia (5048, 5049), Ivory Coast (4159, 4718), central Africa (3063), equatorial West Africa (1420), South Africa (1559, 3630), China (3475), Japan (3267, 5846), Pakistan (60), India (1317, 2164, 2853, 2854, 3732, 3863, 3868, 4698, 4716, 4997, 5512), Polynesia (6084), Australia (977, 978, 3720), the Bahamas (2006), Brazil (399), and Peru (2005). It occurs in the most varied soil types including podzol, chernozem, terra fusca, chestnut-brown soil (277, 3416, 3871, 4378, 5534), and also solonchak (977) and solonetz (277, 4677). Examples for its occurrence in forest soils are teak (2853, 2854), *Eugenia* (5512) and *Casuarina* (2006) forests, deciduous forests (2573) with beech (272, 273, 1702, 3138, 3538), aspen (3959), and oak (989, 1700, 5811), and coniferous forests (681, 3267, 3975), mainly with pine stands (3447, 3798). *P. chrysogenum* has also been isolated from uncultivated soils (4698), grassland (1632, 3363, 3445), and very frequently from cultivated soils. It is known from vineyards (242, 2719) and soils of citrus plantations (2764, 3988), further in soil under *Calluna* heath (3720, 5047, 5811), soils with steppe type (1700), prairie, tundra and taiga vegetations (1446, 1559, 3366), *Sphagnum* bogs (1376, 6520), peat fields (4474, 6571) and fens (4365, 6571). It is also known from saline soils (8, 3446, 4000, 4001, 4162, 4759), sand dunes (745, 1423, 1477, 6414), desert soil (100, 3996, 3997), caves (1423, 3364, 3367, 4137), a uranium mine (1703), river sediments (1162), streams, sometimes with a high degree of pollution (1154, 1155, 1157, 1166) and treated sewage

(1165). In irrigated fields its frequency was reduced after applying large quantities of sewage (2467). It seems to tolerate a cyanamide treatment (0·2%) of the soil (3050) and occurs down to great soil depth (3975). Hair baits are colonized in soil at water potentials of up to −140 bars (977). *P. chrysogenum* has been isolated from decaying cabbage and barley plants (3113), *Pseudoscleropodium purum* (2969), dead *Pteridium* petioles (1821), *Carex paniculata* litter (4644), *Citrus* leaves and bark (3988), needle and leaf litter of pine (681, 2344, 6080), oak (2411), and *Eugenia heyneana* (5513); it has also been found in the rhizospheres of ferns (2856, 2858), spruce (3573), oak (5394), hazel (1794), alfalfa (3982), clover (3429), pea (5825), broad beans (6511), groundnuts (2768), onion (2159, 3733), flax (5826), barley (3870), tomato (16), *Carica papaya* (5345), *Aristida coerulescens* (4083), *Trichodesma amplexicaule* (3864), *Pennisetum typhoideum* (3867), *Setaria glauca* (3271), *Ammophila breviligulata, Andropogon scoparius* var. *septentrionalis* (6414), various steppe grasses (3376), peat bog plants (1376), Australian heathland plants (5819), euphorbiaceous plants (3866), and halophytes (3375). It is also found on stored seeds of barley (3207), wheat (2716, 3491, 3989, 4492), rice (3181) and other cereals (57, 3491, 3875, 3989, 4492), wheat seeds in soil (2097) including those treated with organo-mercury seed-dressings (2077), seeds of *Poa pratensis* (1110), *Pennisetum typhoideum* (57), tomato (5899), flax (3517) and groundnuts (2765). Additional habitats include damp hay (4548), nests of free-living birds (2575) and gerbils (5393), dead adult bees, honeycombs (402) and rarely the intestinal content of bees (1983), the alimentary canal of lizards (5943), pulp and paper (6168), stored olives (309, 4545) and grapes (342), nuts and dried fruits (2572, 5092, 6252), flour-based foodstuffs (5980), frozen fruit cake (3153) and fruit juices (5205). It is also known from frescoes of a monastery (2666), and optical lenses (2941). Though conidia germinate on glass slides exposed to the soil without any additional nutrients, the germ tubes are lysed after a few days (1006). — The fungus can serve as food for the mites *Acarus siro, Tyrophagus putrescentiae* (5383), and *Pygmephorus mesembrinae* (3104).

Maximum production of conidia was found to require 0·038 g glucose g^{-1} dry matter of mycelium h^{-1}, while the maintenance of metabolism requires 0·022 g glucose (4845). For sporulation in submerged culture, iron (1979) and calcium ions (1805) are required; lowering the N source increased sporulation (2199). During conidiogenesis, the tips of the phialides contain a high concentration of SH-groups in the cell wall (4576). The initial sporulation time is inversely proportional to the logarithm of inoculum load (2200). Best germination of conidia at 20°C and 30°C, and partial germination in seawater (832) was found. Germination of conidia requires external C sources (glucose, galactose, xylose, arabinose) and N sources but no growth factors (3625). Germinating conidia possess β-glycosidase and split scilliroside (5598). — *P. chrysogenum* grows at 5°C but not at 37°C (4578), with the optimum at 23°C (3876); growth has been observed at water activities between saturation and −350 bars (977, 978, 3876, 4001, 4578, 5881). Aeration of pure cultures with increased CO_2 concentration promoted mycelial growth (341). The mean doubling time in submerged culture was calculated to be 4·3 h (5890). *P. chrysogenum* possesses all enzymes of the Embden-Meyerhof-Parnas and hexose-monophosphate glycolytic pathways (3470, 5323, 6097). The relative activity of glyoxylate transaminase has been determined (1819) and the production of numerous enzymes has been described, including mevalonic acid dehydrogenase (146), 6-phosphogluconate dehydrogenase (3541, 3816), glutathione reductase (6438), alkaline and acid phosphodiesterase and phosphomonoesterase, ribonuclease (497), catalase (2865) and sulphate adenyl transferase (5959). Synthesis of glucose oxidase is promoted by NO_3^- ions, whereas NH_4^+ ions are inhibitory, maintenance of pH 3·0–4·5 is favourable (2051). Of 24 sugars tested, sucrose, glucose, galactose and mannose as well as glycerol were oxidized with particular intensity (6097). Amylase activity has been demonstrated with the optimum occurring at 45°C and pH 4·8–5·2 (625, 626, 4545); there are data on production of pectinases (2655, 3432, 3539, 4545, 4779, 5688); suitable sources for production of polygalacturonase (PG) and pectin methylesterase are pectin, pectic acid, D- and L-galacturonic acid and also mucic acid; NH_4^+ salts are better N sources for pectinase production than nitrates or amino acids (4539). *P. chrysogenum* proved to be a relatively weak PG producer when compared with numerous other fungi (4237), optimum PG production occurring at pH 3·3 (6049). Cellulolytic activity on (carboxy-)methyl cellulose (2736, 2878, 4779) is ascribed to three enzyme components (4528) but native cellulose is not attacked (386, 3618, 6326); xylan decomposition has, however, been described (1432). The fungus forms proteolytic enzymes (756, 2450, 2719, 3049, 4545); diploid strains exhibit a higher production of β-galactosidase, alkaline proteinase and glucose oxidase than the parent strains (4463). Phenylacetic acid can be completely oxidized (2476) and tryptamine is oxidized to tryptophol (1522); D-tryptophan is partly, and L-tryptophan completely, utilized (5150). In the initial phases of growth, glutamic acid appears in the nutrient medium and later also valine; proteins occur only in small amounts (3500); after complete assimilation of the available inorganic N, considerable quantities of organic N compounds are also excreted (3968), particularly cystine, glutamic acid and alanine (4739) as well as the peptide δ-(α-aminoadipyl)-cysteinyl valine (5278). Hairs are not decomposed but colonized in the medulla (1633). A lysophospholipase is produced (1681, 2903); in replacement cultures, high amounts of fat (up to 14·8% of mycelium dry weight) are produced, and a suitable medium for fat production has been described (4073); the amount of lipids produced (4642) is low when grown on pentoses (D-arabinose, D-xylose, D-ribose) compared to high yields obtained on hexoses (1405); lauric acid is oxidized to undecanone-2 (3139), "lard oil" and oleinic acid are oxidized (4901); hydrocarbons from fuel oil are utilized (4239); plasticizers (diesters of aliphatic 2-basic acids) and plastic polymers are also attacked (473, 5148); very effective degradation of the herbicide simazine was observed *in vitro* (5616); the pesticides dicryl, propanil, propham and solan are degraded to the aniline moiety by *P. chrysogenum* (2893). Nitrogen is liberated from urea-formaldehyde fertilizer (4010); selenite and tellurite are reduced (4577); the uptake of sulphate is followed by the formation of 3-phospho-adenosin-5-phosphosulphate and choline-*O*-sulphate (5175); metaphosphate is utilized (4544). Optimum ratios for cations are given with K/Ca = 1:3, K/Mg = 2:1, Ca/Mg = 9:2 (95). — The following metabolites have been found: citric acid (3074), α-ketoglutaric acid (2477), gluconic acid (534, 3693), α-methyl-*n*-butyric acid, 2-decene-1,10-dioic acid, 2-pyruvyl aminobenzamide, indole-3-acetic acid (3231), riboflavin-*a*-glucoside and riboflavin oligosaccharides (5662), the flavour compounds 2-octen-1-ol, 3-octanol, 3-methylbutanol (2867), stearyl alcohol, sorbicillin, xanthocillin, and pyrocalciferol (5278). Galactose, mannose (3640), arabinose, xylose, rhamnose, ribose (2049) and galactosyl lactose (5278) have been shown in the culture solution. Peptidogalactomannans of the cell walls are active antigens (1895), as is the quinazolinone chrysogine (2448). The following can be mentioned as antibiotically active metabolic products: penicillin (512, 1086, 1763, 4228, with voluminous subsequent literature), negapillin (5290), 6-aminopenicillanic acid (3016), and the sideramines coprogen (6528), fusigen (147, 1380) and fusigen B (3231). Further products are chrysogenin (1086), cerebrin, fungisterol (5278) and dethiobiotin (3231). — When tested against five plant-pathogenic fungi, *P. chrysogenum* showed a specific strong inhibitory action on *Pythium ultimum* (1431); other authors have observed an antagonistic activity against *Candida albicans, Fusarium oxysporum* (4304), *Fusarium sporotrichioides* and *Neurospora sitophila* (5394). In culture filtrates, a substance, noxiversin, with antiviral activities has been detected (1141); an antiviral agent and interferon induction have been investigated (329). Mycelial extracts were found to be toxic to chicken embryos (1388). — *P. chrysogenum* has been isolated from sclerotia of *Sclerotium cepivorum* and shows an antagonistic activity against this fungus (3990). Penicillin was shown to be produced in inoculated sterile soil and on corn seeds (2449). Soil in which the fungus has been accumulated led to a significant stimulation of green matter production in flax (1427). The germination of corn seeds and seedling growth were stimulated in the presence of *P. chrysogenum* (4553), but culture filtrates were found to inhibit the growth of barley seedlings *in vitro* (293). — The fungus is sensitive to γ-rays, with only 0·5% conidia surviving 100 krd h^{-1} (4585). In a chronically irradiated soil, *P. chrysogenum* appeared with increasing numbers of viable cells up to 230 R per day; no viable cells were found at 1400 R day (2009). It exhibits a rather high tolerance to soil fumigants (viz. formaldehyde, CS_2, and allylalcohol) (3991).

Penicillium citrinum Thom *1910* — Fig. 236.

DESCRIPTION: Raper and Thom (4745). — Members of the *P. citrinum* series are characterized by smooth-walled conidiophore stipes bearing divergent whorls of metulae usually without further branches. Conidia globose to subglobose, smooth-walled, not exceeding 3 μm diam. Colonies of *P. citrinum* grow restrictedly, reaching 2·0–2·5 cm diam in 10–14 days at 24°C on CzA, up to 2·0 cm on MEA, blue-green, reverse bright yellow with the same pigmentation diffusing into the agar.

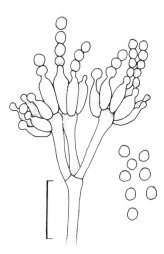

FIG. 236. *Penicillium citrinum*, penicillus and conidia.

P. citrinum is a cosmopolitan and frequent fungus. In addition to numerous other reports, there are observations from Murmansk (4703) and the Buhara region of the USSR (2871), Scotland (163), Sweden (1446), caves in Poland (4137), alpine sites (1876, 2530, 3538,

3976), Turkey (4183), Syria (5392), Egypt (8, 3993, 3997, 4962), Israel (654, 2764, 2768, 4759), Somalia (5048, 5049), the Ivory Coast (4159), central Africa (3063), South Africa (1555, 1556, 4407), Zaïre (3790), Pakistan (61, 4855), India (1519, 1525, 2186, 4698, 4736, 4997), Nepal (1826), Sri Lanka (4021), Tahiti, Polynesia (6084), Australia (977), Japan (2532, 3267, 5846), China (3475), Central America (1697, 2031), Brazil (379, 398, 399, 4547), Chile (1824) and Peru (2005). It occurs in a particularly wide range of soil types, including chestnut soils, takyr (3871), rendzina, parachernozem, chernozem, terra fusca (3414, 3418), and podzol (2923, 3416, 5534). It is very widely distributed (individual citations omitted here) in forest, grassland and arable soils; also in heathlands (2736, 5047), savannahs (4407), tundra, taiga (3871), and uncultivated soils (4698), desert soil (3997), a saline beach (4162), salt-marshes (8) and seawater (4918). — It is relatively sensitive to a soil treatment with formaldehyde (1655). *P. citrinum* is predominantly isolated from the surface soil layers (3975). The minimum water potential for colonization of hair baits and vegetative growth was found to be −350 bars (977, 978, 2093, 2939, 3876). It has been isolated from washed mineral as well as organic soil particles (307). It has been reported from roots of sugar beets (4559), the rhizospheres of coriander (1523), beans (1369), groundnuts (2532, 2768), *Astragalus paucijugus* (5416), poplar (3452), various forest plants (4814) and halophytes (3375) as well as leaves of vascular plants (296). It has been isolated from plant remains of the most varied kind, including cabbage and potato plants (3113), coarse fodder and hay (4548), leaf litter of deciduous (650) and coniferous trees (681, 2344, 2923, 6080), litter of *Carex paniculata* (4644), petioles of *Pteridium aquilinum* (1821), sclerotia of *Sclerotinia sclerotiorum* (4699), wood pulp and paper (6168), sticks exposed to soil (1956) and cut peat (1376). It is also found in fresh and/or stored wheat grain (3491, 3989), rice (3181), sorghum (3989), corn (57, 3875, 3989), cotton (5009), pecans (2572, 5092), groundnut seeds and/or pods (346, 2765, 3702), stored grapes (342), and sweet potatoes (3066). It has also been isolated from nests and feathers of free-living birds (2575, 5067), tunnels of gerbils (5393), diseased *Nomia* bee larvae (402), dried cereals and legumes (5980), flour-based food-stuffs (2067, 5980), frozen fruit cake (3153) and fruit juices (5205).

Conidia germinate in the temperature range 12–37°C; the minimum water potential is −270 bars (3873, 3876). Conidia germinate on the β-glucoside salicin but not on the related aglucon (6159). Germination is inhibited by volatile compounds from *Ocimum basilicum* but stimulated by those of *Origanum majorana* (35). — Optimum temperature for growth lies in the range 26–30°C (517, 654, 3876), the minimum 5–7°C and little growth occurs at 37°C; good growth is possible at a water potential of −100 bars (4578). — *P. citrinum* is an intensive decomposer of pectin (3539, 4779) and starch (654) but the utilization of cellulose is only partially positive (386, 654, 2736, 3618) and, with native cellulose, sometimes negative (3618, 6326). Nitrite (10) and metaphosphate (4544) are utilized. *P. citrinum* possesses a chymosin-like enzyme complex (6, 7, 3049) and produces an acid carboxypeptidase (6498) and an RNA-degrading phosphodiesterase (3180). Lignosulphonate, humic and fulvic acids (934), polyurethane substrates (251) and the medulla of hairs are attacked (1633); plasticizers (473, 5859) and hydrocarbons from fuel oil can be used as C source (4239). Very effective degradation of the herbicide simazine has been reported (5616). The fungus is autotrophic for growth substances (1425). An extracellular polysaccharide was determined as 1,5-β-linked malonogalactan (3075). Gluconic (2147, 3694) and kojic acids (4458) are produced; also orcinol, protocatechuic acid, orsellinic acid and other hydroxybenzenes (1249, 3231), citrinin (2434, 4887, 5152) and several related metabolites (1250); citrinin, which has fungistatic and hyphae-narrowing properties (4880, 6075), inhibits the germination and growth of wheat, mustard and red clover (6448) and suppresses the growth-promoting properties of indole-3-

acetic acid on *Avena* coleoptiles (2697). Further metabolites are the sideramine coprogen with low antibiotic activity (6528), the isocoumarin sclerotinin A (3231), and the flavour compound 2-octen-1-ol (2867). — In plate tests, strong antagonistic activities towards *Rhizoctonia bataticola* (1369), *Gaeumannomyces graminis, Pythium* sp. (1425), *Sclerotinia sclerotiorum* (4699), *Staphylococcus aureus* and *Salmonella typhi* (2736) have been observed. *P. citrinum* inhibits the growth of corn plants in liquid culture and reduces the absorption of strontium (1832). The culture filtrate had no effect on barley (293), and growth of the fungus is stated to be stimulated by oat and inhibited by corn plants in aseptic mixed cultures (1536). — It was found to be associated with mycotic keratitis in man (2154). — Development is possible in an atmosphere low in O_2 (1425). It is able to grow on 5% tannin *in vitro* and shows considerable tolerance to CS_2 (5001).

Penicillium claviforme Bain. *1905* — Fig. 237.

DESCRIPTIONS: Raper and Thom (4745), and Samson *et al.* (5029). — *P. claviforme* occupies an isolated position in the genus and is characterized by distinct synnemata (coremia) up to 2 cm tall with sterile white or pink stipes and grey-green conidial heads; few solitary conidiophores are formed; reverse ± brown. — Colonies reaching 2·5–3 cm diam in 12–14 days at 24°C on CzA and MEA. Conidiophores in the synnemata two- to four-stage branched with the stipes often interlaced. Conidia ellipsoidal, smooth-walled, 4·0–4·5 × 3·0–3·5 μm. — The only other species placed in the *P. claviforme* series, *P. clavigerum* Demelius *1923*, has synnemata without a distinct sterile stipe. Other species with large synnemata have lanceolate phialides of the *Symmetrica*-type. — The ultrastructure of conidiogenesis has been investigated (1771). Nuclear division in the phialides and nuclear migration into the conidia have been studied by TEM (6522). Conidia are uninucleate (6523); the rodlet pattern of the conidial surface is slightly interlacing, and surfaces are relatively smooth (2421). — DNA analysis gave a GC content of 50–52% (5600).

P. claviforme seems to be a ubiquitous species, but does not appear to occur with high frequencies anywhere. Data are available from soils in the British Isles (413, 5559), Poland (1421), the USSR (2871, 4474), Austria (1828, 1876), the USA (2482, 3817, 5841) including Alaska (1750), Canada (6352), Australia (3720), Chile (1824), Israel (2772), Syria (5392), and Turkey (4183); it has also been found in Colombia (CBS), Cyprus, Guinea, and India (IMI). It has been isolated from forest (1828, 2482, 3817, 6352) and cultivated soils (2772, 2871), soils under steppe type (5841) and tundra (1750) vegetation, heathland (3720), peaty soil (4474, 5559), raw humus (1876), salt-marsh (413) and children's sandpits (1421). It has also been reported from litter of *Pinus monticola* (681), remains of potato plants (3113), *Pteridium aquilinum* petioles (1821), moist hay (4548), *Avena fatua* seeds (2961), and various tropical plants including *Citrus nobilis, Ananas comosus* and *Musa sapientum*, and dung of birds, kangaroo and opossum (IMI).

Virtually no growth occurs at either 5°C or 37°C and little at a water potential of −100 bars (4578). The formation of synnemata requires Mn^{2+} ions in the growth medium (5835), it is induced by both light in the short-wave range (350 and 450–460 nm) and temperature changes

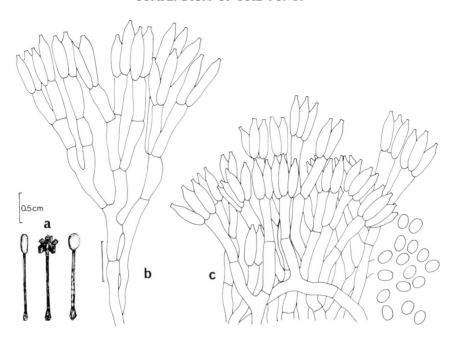

Fig. 237. *Penicillium claviforme*. a. Synnemata with conidial masses, sometimes splitting in columns; b, c. aggregated conidiophores and conidia; orig. R. A. Samson (from 5029).

in mutants of *P. claviforme* (1690, 1691, 4571). — Spontaneous lysis of conidia and deficient sporulation seen in mutant strains are ascribed to abnormal nuclear behaviour (6523). Terminal hyphae burst when flooded with hypotonic solutions or rapidly heated (370). — Germinating conidia produce β-glucosidase on salicin-containing media (6159). Several acid ribonucleases are liberated into the growth medium (278, 496). An optimum growth medium has been described as containing $1 \cdot 25\%$ each of glucose and fructose, $0 \cdot 5\%$ asparagine and 1 ppm thiamine (4572); the presence of an amino acid in the growth medium seems to be essential (5835). Cellulose (2718, 3414, 5076), pectin (4183) and starch (1750) are degraded, glucose, galactose and mannose being constituents of extra- and intracellular polysaccharides (3640). *P. claviforme* can grow on gallic (1750) and tannic acids (1260, 3052). Peptidogalactomannans of the cell wall are antigenically active (1895). The fungus produces $1,3$-β-glucanases (3231) and also the toxic patulin (=claviformin) (939, 5278) which has antibacterial activity. — An *in vitro* activity against *Venturia inaequalis* has been reported (6205). In one study, *P. claviforme* was the most efficient antagonist of 118 fungi and bacteria against *Heterobasidion annosum* (4820).

Penicillium corylophilum Dierckx *1901* — Fig. 238.

DESCRIPTION: Raper and Thom (4745). — *P. citrinum* series. — Colonies on MEA rather fast growing, reaching $2 \cdot 5$–3 cm diam after 12 days at 25°C on CzA, reverse usually dark green, often with darker sectors but not staining the agar. Conidiophores with metulae of unequal length. Conidia subglobose to ellipsoidal, smooth-walled, usually $2·5–3·0\,\mu$m in length. — Conidiogenesis has been studied by Cole and Kendrick (1105). — DNA analysis gave a GC content of 52% (5600).

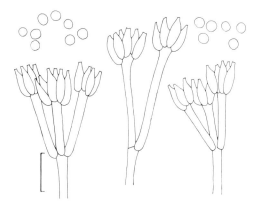

FIG. 238. *Penicillium corylophilum*, penicilli and conidia, CBS 312.48, orig. R. A. Samson.

P. corylophilum evidently occurs less frequently in cool–temperate regions than in warmer climates. It has been recorded from Spain (3447), Italy (3445, 3913, 3975, 4538, 5047, 6082), Turkey (4245), Syria (5392), Israel (654, 2764, 4759), Egypt (8, 3989, 3993, 3997), the Ivory Coast (4718), Somalia (5048), India (1519, 4610), Nepal (1826), Australia (977, 978), Japan (3267), Hawaii (569, 3264), Peru (2005), the USA (655, 1387, 2482, 3817), the USSR (432, 3363, 3365, 3850, 4474), and Germany (1424). It has been found in forest soils (3267, 3447, 3538, 4718, 5048), forest nurseries (2482), cultivated soils with citrus (2764), paddy rice (1519, 3913) and sugar cane (4610), garden compost (1424), grassland (3363, 3445, 4538), open bogs (6082), peat fields (4474), cut and uncut peat (1376, 1437), soils under steppe type (4718) and heathland vegetation (5047), desert soil (3997), fresh water (4229), acid mine water (5343), sewage (1166, 1387). a mangrove swamp (3264), salt-marshes (8), estuarine sediments (655), and caves (3365, 4137). — It has also been isolated from leaf litter of pine (681, 6080), *Carex paniculata* (4644), aecidiospores of *Cronartium comandrae* (4608), the rhizosphere of clover (3429), grains of wheat, sorghum and barley (3989, 4038), pecans (2258, 2572), sticks buried in soil (1956), cotton, textiles (3707), dead adult bees (*Apis*) and honeycombs (402), frozen fruit cakes (3153) and other foodstuffs (1303).

Little growth occurs at 5°C and none at 37°C, but optimum growth at 35°C has also been reported (1827); the thermal death point has been determined at 60°C for 30 min (613). For vegetative growth, a water potential between −70 and −120 bars is required (977, 4578). — *P. corylophilum* degrades starch (1827) and grows on media with 5% tannin (1425) or on substrates with hexadecane, dodecane, octadecane, cyclohexane, toluene, benzene or kerosene as sole C source in the presence of yeast extract and at pH 2 (5343). It produces D-xylonic acid (3231), ethylene (2641) and a viridicatin-related metabolite (4500); nitrite can be used as an N source (10), and tryptamine is oxidized to indole-3-acetic acid and tryptophol (1522). — Some antibacterial activity (515) has also been demonstrated. *P. corylophilum* proved to be toxic to brine shrimps and chicken embryos (1303).

Penicillium daleae Zaleski *1927* — Fig. 239.

DESCRIPTION: Raper and Thom (4745). — *P. daleae* has been placed in the *P. janthinellum* series by Raper and Thom, but is perhaps better referred to the *P. nigricans* series (5596). — Colonies reaching 4–4·5 cm diam in 12 days at 24°C on CzA, 8 cm on MEA, greyish brown, reverse uncoloured to brown. Conidia ellipsoidal to subglobose, with coarse roughening in winding bands, 2·5–4 × 2·5–3 μm. — DNA analysis gave a GC content of 52·5% (5600).

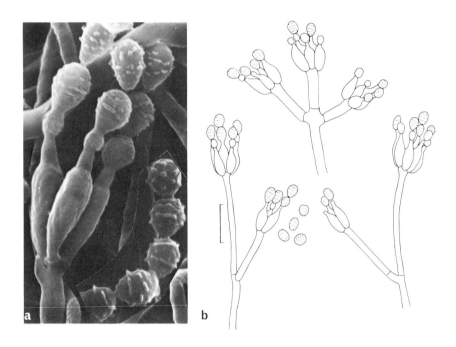

FIG. 239. *Penicillium daleae*. a. Phialides and conidia, SEM, × 5000; b. divaricate penicilli and conidia with spiral ornamentation; CBS 211.28.

P. daleae was originally reported from conifer and beech forest soils in Poland (6531), later from soils in the British Isles, Canada and France (IMI), North Carolina (655), deciduous forests in Ohio (2573), and Wisconsin/USA (1034, 6351), the Arctic (1750), the USSR (1637, 3652, 5393), Denmark (CBS), Sweden (5465) and Irak (92); it is also known from paddy fields in India (1519), birch litter (IMI), the rhizosphere of clover (3429), and tunnels of gerbils (5393). — It can be consumed by the springtail, *Hypogastrura tullbergi* (3830).

Little growth occurs at 5°C, none at 37°C, and moderate growth at a water potential of −100 bars (1578). Starch and pectin are degraded (1750). Production of kojic acid (6388) and peptidases (3231, 6497, 6498) has been reported.

Penicillium decumbens Thom *1910* — Fig. 240.

DESCRIPTION: Raper and Thom (4745). — The *P. decumbens* series is characterized by short, simple conidiophores arising from creeping aerial hyphae. — Colonies usually growing restrictedly, reaching 2·0–3·0 cm diam in 12–14 days at 25°C on CzA, velvety to floccose, dull blue-green; reverse colourless or slightly greenish; conidiophores 50–100 μm long; conidia forming loose columns, ellipsoidal to subglobose, smooth-walled, appearing green under the microscope, 2·0–2·5 μm long. — DNA analysis gave a GC content of 52% (5600).

P. decumbens has a worldwide distribution with most data available from temperate zones, including the USSR (432, 1637, 2948, 4677, 6571), the USA (655, 1166, 2007, 2482, 2822, 3817), Canada (2922, 6352), Sweden (1446), the British Isles (413, 745), Ireland (1376), France (4162), Poland (1978), Czechoslovakia (1700, 1703) and Austria (3413); it has also been recorded from Italy (3913, 4537, 4538, 5047), Turkey (4245), Egypt (3993, 3997, 4962), Libya (6510), Israel (2768), Iraq (92), South Africa (1559, 3630), Florida/USA (4918), Peru (2005), Hawaii (3264) and India (1519, 5622). It has repeatedly been isolated from uncultivated (3630) and forest soils (432, 1637, 2007, 2482, 2948, 3817, 4537, 6352), also after prescribed burning (2822), a forest nursery (1978), cultivated soils (1519, 2768, 3913, 4538), peaty soils and peat bogs (432, 1376, 3414, 4474, 6571), heathland (5047), soils with tundra (1446) or steppe type vegetation (1559, 1700), desert soil (3997), sand dunes

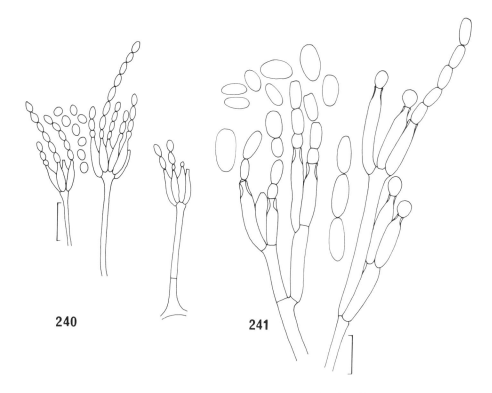

240   241

FIG. 240. *Penicillium decumbens*, monoverticillate penicilli and conidia, CBS 258.33.
FIG. 241. *Penicillium digitatum*, penicilli and conidia, CBS 319.48.

(745), salt-marsh (413), fens (3413), a copper swamp (2922), a mangrove swamp (3264), estuarine sediments (655), coastal areas (4162, 4918), chalk soil (4152), podzols, gley, pseudogley (432, 3414, 3416), solonetz (4677), polluted water (1166) and a uranium mine (1703). — It has been reported as common on leaf litter of *Abies grandis* (681), *Pinus sylvestris* (2344) and *P. radiata* (6080), isolated with equal frequencies from mineral and organic soil particles (307), found on the roots of *Pinus nigra* var. *laricio* (4448), in the rhizospheres of various sand dune (6414), steppe (3376) and peat bog plants (1376), clover (3429), and groundnuts (2768); on seeds of barley, corn, *Pennisetum typhoideum*, *Sorghum vulgare* (57), groundnuts (2765, 2772) and pecans (2258), and is also known to occur on feathers of free-living birds (2575), in frozen fruit cake (3153) and fruit juice (5205). — Germination of conidia in soil is inhibited by soil mycostasis (5144).

Growth is very limited at 5°C, good at 37°C, and is possible at a water potential of −100 bars (4578). On glass-microbead media, a decrease in metabolic activity occurs with decreasing bead sizes. *P. decumbens* tolerates a wide range of growth temperatures and pH conditions (5760). — It utilizes cellulose (2948, 3414), grows on nitrite as a N source (10), produces an acid carboxypeptidase (6497, 6498), degrades the herbicide atrazine by *N*-alkylation (2892), and reduces steroid A ring double bonds (3824). The major odorous constituents of *P. decumbens* have been identified as the sesquiterpene thujopsene, 3-octanone, nerolidol, 1-octen-3-ol, and phenylethylalcohol (2216). This fungus produces 2,6-dipicolinic acid (5278) and decumbin, a metabolite toxic to rats and goldfish (3231, 5378). — *P. decumbens* inhibits germinating wheat seeds (5378) and antifungal activity has been reported (516, 569).

Penicillium digitatum (Pers. ex St.-Am.) Sacc. *1882* — Fig. 241.

≡ *Monilia digitata* Pers. ex St.-Am. *1821*

DESCRIPTIONS: Raper and Thom (4745), and CMI Descriptions No. 96, *1966*. — The *P. digitatum* series with its large cylindrical conidia occupies an isolated position in the genus. The somewhat divergent, irregularly branched penicilli recall those of *Paecilomyces*. — Colonies reaching 1 cm diam in 10–14 days at 24°C on CzA, 6–8 cm on MEA, mostly greenish brown, reverse uncoloured or dull tan; growth very poor and restricted on CzA. Conidia typically cylindrical (partly subglobose) with a broadly truncate base and evenly rounded tip, smooth-walled, commonly 3.5–5.0×3.0–$3.5\,\mu$m, later much bigger and mostly 6–8×4–$6\,\mu$m. It produces an odour of decaying citrus fruits *in vitro*. — The rodlet pattern of the conidium surface as seen in freeze-etched replicas is interlacing and characteristically slightly raised with some areas of the underlying surface showing; the surfaces are relatively rough (2421). — Nuclear division is as in *P. expansum* (3198) with n = 5; vegetative cells are multinucleate while the conidia are mostly binucleate. Heterokaryosis could not be demonstrated (5614). — Mycelial hydrolysates contain 45.4% glucose, 5.7% glucosamine, 3.8% galactose and traces of rhamnose and xylose (3232). — DNA analysis gave a GC content of 52% (5600).

The principal substrates of *P. digitatum* are citrus fruits on which it causes green rot; it is rather a soil-invader than a true soil-inhabitant (4421). — *P. digitatum* has a worldwide distribution showing a preference for warmer climates as indicated by its occurrences in Italy (3453), Libya (4085), Egypt (3997), South Africa (1555), India (967, 2854, 4030, 5349), Bangladesh (2712), Australia (3720), Oklahoma (1632), Wisconsin (1039), North Carolina (655), the British Isles (745, 2338, 3918, 4429), and the USSR (2948). It has been isolated from forest soils under *Eucalyptus maculata* (1555) and teak (2853, 2854), uncultivated soils (4030), cultivated and abandoned land (1632, 2948), citrus plantations (4085), heathland (3720), peat bogs (3918), sand dunes (745), desert soil (3997), a cave in Piedmont/Italy (3453), conifer swamps (1039), estuarine sediments (655) and fresh water (4429). It is also known from composted municipal waste (3041), litter of beech (3538) and *Eucalyptus maculata* (1558), *Spartina townsendii* (5389), coarse fodder (4548), the rhizospheres of ferns (2858), clover (3429), rice (2712), and *Artemisia herba-alba* (4084), seeds of corn (3875), and fruit juices (5205).

Conidiation of *P. digitatum* cultures can successfully be synchronized on media lacking vitamins, in particular *p*-aminobenzoic acid; concentration of C and N sources also controls synchronization (6548), but has no influence on the conidiation time, while glutamic acid as sole N source can considerably accelerate conidiation (6549). Germination of conidia is stimulated by calcium (1305). — Temperature requirements for growth fall in the broad range 17–42°C with a minimum of 6–7°C and a maximum of either 37°C (4578) or 47–48°C (517); pH optimum is 3–6 with the maximum 7·8 (3612); and little growth occurs at a water potential of –100 bars (4578). — Nitrite can be used as a nitrogen source (10). Peptone, urea, asparagine and ammonium are all good N sources although nitrate is not. Most sugars can be utilized but not inulin or soluble starch; citric and malic acids are the only suitable organic acids; and, although growth factors are not required, these may promote growth (1719). The fungus produces a staling substance with morphogenic properties (4425). Conidia were found to contain D-mannitol, *meso*-erythritol, glycerol, a,a-trehalose, and *meso*-inositol (1539); pectin is degraded (1899) and production of polygalacturonases has been confirmed in virulent and avirulent strains (1898), pectate lyase and arabanase were demonstrated (802); when grown on glucose or pectin, it produces a-L-arabinofuranosidase (1097); the glycosides scilliroside and desacetyldigilanide are cleaved (5598); *P. digitatum* also produces an alkaline phosphatase (4459), xylanases (1097) and has cellulolytic (1899, 2948, 3538) and proteolytic (3429) properties. Known metabolites include ethylene under any cultural conditions (531, 1720, 6509) and 2,6-dipicolinic acid (5278). — Heat treatment of conidia for 5 min at 50°C is survived by 30% of the conidia, and γ-irradiation with 100 krd by 0·8% (343). — In mixed infections with *P. italicum* on citrus fruits, *P. digitatum* tends to predominate (1993); virulence towards citrus fruits is lost in some mutants but restored by diploidization (463).

Penicillium expansum Link ex Gray *1821* — Fig. 242.

= *Penicillium glaucum* Link *1809* pro parte

For some other synonyms see Raper and Thom (4745), and Samson *et al.* (5029).

DESCRIPTIONS: Raper and Thom (4745), Samson *et al.* (5029), and CMI Descriptions No. 27,

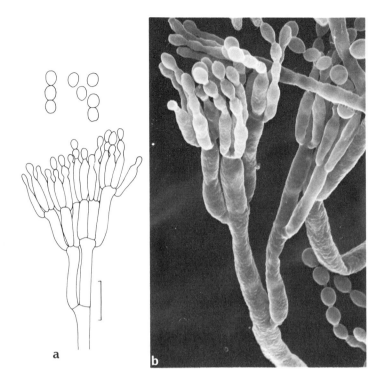

Fig. 242. *Penicillium expansum*. a. Penicillus and conidia; b. penicillus, SEM, × 2000, CBS 146.45.

1966. — Colonies fast growing, reaching 4–5 cm diam in 14 days at 24°C on CzA and MEA. Conidiophores in fresh isolates typically loosely synnematous (rarely distinctly synnematous), giving the colony a zonate appearance, light green, reverse colourless to yellow-brown. Conidiophores usually smooth-walled on CzA, finely roughened on MEA, to 400 (–700) μm long; penicilli two- to three-stage branched with numerous, usually somewhat appressed metulae. Phialide tip thin-walled, the first five or so conidia adhering to the phialide are usually weakly delimited and hyaline. Conidia subglobose to ellipsoidal, smooth-walled, 3·0–3·5 μm diam. Odour aromatic, fruity, suggesting apples. The capacity to cause a rapid apple rot has been used as a taxonomic character. Synnema formation can be stimulated by 20 ppm of the fungicide botran in the medium (3872). — Vegetative cells are mostly 5–10-nucleate, while conidia are mostly uninucleate; time-lapse cinematography showed that a single conidium can be formed within 2 hours. Young conidia are cylindrical and become inflated before the separating septum is laid down (3198). — The occurrence of a parasexual cycle has been demonstrated (362); seven genes were resolved in two linkage groups. Nuclear division has been thoroughly studied; chromosomes are arranged in a row along a nuclear filament and divide longitudinally; n = 4 + 1 centriole in a wild type, n = 10 in artificial diploids (3197); the chromophilic part is ring-shaped in resting nuclei. Vegetative nuclei contain a central body of RNA, a single diffuse chromophilic structure in a non-condensed state, and a bipartite spindle pole body which acts as a centre for the distribution of chromosome material during division and as a motion centre during migration. Fresh conidia contain chromosomes in a condensed, spiralized state but they become despiralized on activation. — Dominating amino acids contained in the conidia are cystine (10·2%), asparagine (9·9%), glutamine (8·9%), serine (8·7%), threonine (8·4%), glycine (8·0%), and alanine (7·9 mol % of the total amino acids) (3232). No surface lipids have been found in the cell wall (4834).

P. expansum ranks as one of the commonest *Penicillium* species on fruit and various rotting substrates and, in addition, as a widely distributed soil fungus. It is easily isolated from soil dilution plates, with Thornton's screened immersion plates (5812), and from washed soil particles (1889). It is also not infrequently detected in the atmosphere. — The overall distribution of *P. expansum* ranges from Greenland (4174), arctic tundra (1169, 1446, 3871), and alpine pastures (4179), to the subtropics, including Israel (2764), Egypt (4962), the Sahara (2974) and Pakistan (60, 4855). There are especially numerous reports, which cannot be mentioned individually here, from various forest and arable soils; it has also been isolated from fallow soil (918), grassland (163, 1700, 4304, 4538, 5812), heathland (3720, 5047), peat fields (4474, 5559), salt-marsh (413), saline beach (4162), soils with high salinity (3259, 3414, 3446), cliffs along the French Atlantic coast (4178), high altitudes in Piedmont/Italy (3453) and rendzina (3418) and solonetz (4677) type soils. It has often been isolated from polluted streams and river sediments (1154, 1155, 1157, 1162, 1166), but also obtained from a sewage treatment plant (1165) and organic detritus in fresh water (4429). It was found to predominate in the tilled layer of soils (2161, 2948, 4554), but is able to penetrate into deeper layers (432, 3720). Isolations are known from litter of pine (681, 2923), beech (3538), the root region of ash (3156), poplar (3452), Australian heathland plants (5819), corn (4554), groundnut (2768), clover (3429), and various steppe grasses (3376), sticks exposed to soil (1956) or seawater (4546), damp hay, nests and feathers of free-living birds (2575, 4649), tunnels of gerbils (5393), stored apples (645), grain of wheat (3491, 4492), corn (3069, 3875) and barley (4038), seeds of cultivated grasses (3512), castor beans (3172), tomato (5899), and pods of groundnuts (3220); it has also been isolated from dead adult bees (*Apis, Nomia*), hives and combs (402) and from various food products including pecans (2258, 2572, 5092), cold-stored meat (740) and meat products (3844), frozen fruit cake (3153) and fruit juices (3428, 5205). — Growth is inhibited *in vitro* by cochliodinol, an antibiotic metabolite of *Chaetomium* spp. (704); conidiogenesis is inhibited by rubratoxin B, a metabolite obtained from *Penicillium purpurogenum* and *P. rubrum* (4808).

The optimum temperature for the germination of conidia is reported as 23–30°C (3876), but some can germinate after one month's incubation at 0°C (734). In contrast to many other fungi, germination of conidia in *P. expansum* is little affected by aflatoxin (4806). Some individual isolates are adapted to very low temperatures and still grow satisfactorily at 0°C (740, 3153). The temperature minimum for vegetative growth is given as -3°C (4369), while the optimum for linear growth on agar is 23–26°C and for mycelium production 27°C (2698, 2939, 3876). The thermal death point in apple juice is reported as 63°C for 23 min (3620). Lowest tolerated water potential is for growth at -250 bars, for sporulation at -200 to -210 bars (3876, 4369). The growth rate is not significantly influenced by elevated O_2 partial pressures up to 0·2 MPa; the actual O_2 requirements are very low, growth inhibition occurring at c. 0·56 mg l^{-1} H_2O (3814). *P. expansum* did not grow in 70% CO_2 atmospheres and showed reduced growth at 10% CO_2 (2020). — The fungus is listed as resistant to phenylmercuric acetate (2077). It sporulates freely; in alternating light and darkness, growth is zonate (1654). — The high capacity for pectin decomposition in connection with apple rot (647, 1432, 1899, 3127, 5503) has given rise to numerous investigations from which the following results are cited: NH_4NO_3 is a better N source for production of polygalacturonase than

NaNO$_3$ or asparagine (2183, 2185); pectin lyase is produced on media with pectin and polypectate as substrates (5486); the production of pectin lyase, polygalacturonase and endo-β-1,4-glucanase is repressed by arabinose, mannose, galactose, glucose, sucrose, raffinose and galacturonic and glutamic acids (5487). The temperature optimum for production of pectolytic enzymes is 20–21°C (5505); galacturonic acids of varying degrees of polymerization occur as intermediate products (657, 2752, 4337); a plant tissue-macerating factor, phytolysin, has been described, which also degrades pectin and araban (2937). — There are numerous observations on the decomposition of cellulose (e.g. 1425, 3618, 5076); the production of cellulases in culture media is dependent on the nutritional conditions (1899); CMC is hydrolysed with a low degree of variability between different isolates (1432); and amylase (1425), xylanases (1432), β-mannosidase (4781), carboxypeptidase (6498) and proteinases (3429) can be produced. — Tannin-containing extracts from tea plants are decomposed to simple catechols and gallic acid which can then serve as C sources (6048); tannin concentrations up to 10% are tolerated (3127). The mycodextran nigeran (1452), gluconic acid (5278) and ethylene (2639) have been shown to be among the metabolic products; in ageing cultures an unknown substance accumulates which acts morphogenically on hyphae of other fungi (4435). Peptidogalactomannans, components of the cell walls, are antigenically active (1895). *P. expansum* produces the antibiotics citrinin, with activity against acid-fast gram positive bacteria (512), penicillic acid (3844), and, above all, patulin (= expansin), with high activity against numerous fungi and bacteria (152, 5440), high phytotoxicity (352, 3030) and mammalian toxicity (3231). — Patulin is produced in apples (706, 760, 6394), and on PD broth more than on apples by all of 77 isolates tested at a wide range of temperatures (5477). Since this fungus accumulates in the root region of apple trees following the excretion of phenolic substances from the roots, a connection has been assumed to exist between its phytotoxicity and "soil sickness" (603). *P. expansum* displays an antagonistic action *in vitro* towards the green alga *Chlorella pyrenoidosa*, various bacteria (515, 4304, 4503), saprophytic fungi, human-pathogenic fungi (516, 1431, 5394), and *Rhizoctonia solani* and *Alternaria alternata* involved in damping-off of tomato and other plants; biological control has not, however, been achieved (568). *P. expansum* is also toxic to *Tenebrio molitor* larvae (4809). It has also been found to be associated with mycotic keratitis in man (2154). Live cultures in a laboratory experiment inhibited root growth of oats, wheat, corn, grasses, alfalfa, rape and peas (1427, 1430, 3430, 3637, 4554); a root rot in sugar beets has also been associated with *P. expansum* (4548). — Since this fungus is one of the most common contaminants in mould-fermented sausages and other meat products, it may be a dangerous source of toxins, although patulin has not been demonstrated in these substrates (115, 3282, 3844).

Penicillium fellutanum Biourge *1923* — Fig. 243.

DESCRIPTION: Raper and Thom (4745). — *P. decumbens* series. — Colonies reaching 2·0–2·5 cm diam in 12–14 days at 25°C on CzA, bluish green, covered with closely woven aerial hyphae. Conidiophores 50–100 μm long. Conidia forming loose columns, ellipsoidal to subglobose, thick-walled, smooth or finely roughened, 2·5–3·0 μm long. — DNA analysis gave a GC content of 52·5–53·5% (5600).

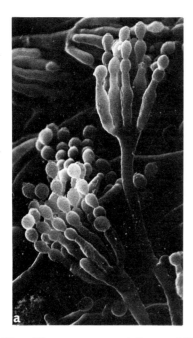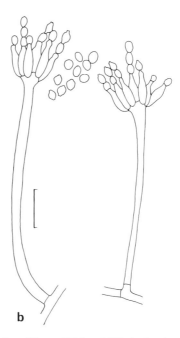

FIG. 243. *Penicillium fellutanum*, monoverticillate penicilli and conidia, a. SEM, × 2000; b. drawing; CBS 330.59.

P. fellutanum is a cosmopolitan species occurring in a wide range of soils. Data are available from the British Isles (159, 2736, 5221, 5222, 6182), Ireland (1376), the USSR (432, 4474, 4548), Belgium (4815), Austria (1823, 1828, 3413, 3418), Spain (3417), Italy (3913, 4538, 6082), Turkey (4245), Egypt (3997), Israel (2768), equatorial West Africa (1420), the USA (655, 660, 1165, 1166), Central America (1697), Peru (2005), India (967, 1519, 4371) and China (3475). Isolations come from forest soils (432, 1828, 3417, 3965, 4538), grassland (159, 3417, 6082, 6182), cultivated land with paddy rice (1519, 3913), corn, alfalfa (4538) and groundnuts (2768); it is also known from heathland (2736, 3720, 5221), peaty soil and bogs (432, 1376, 4474), desert soil (3997), and sand dunes (4371); it has been found in podzols, pseudogley, gley, chernozem, rendzina, terra fusca and loess (432, 3413, 3414, 3416, 3418), on a granite outcrop (660), in a sewage treatment plant (1165, 1166), estuarine sediments (655) and air (967). — It has been recorded from leaf litter of bean plants (4548), *Abies grandis* (681) and *Pinus radiata* (6080); the rhizospheres of groundnuts (2768) and sand dune plants (4371), on pecans (2572) and groundnut seeds (2772), casts of the earthworm *Pheretima californica* (1961), nests of free-living birds (2575), frozen fruit cake (3153) and in fruit juices (5205). — Hair baits in soil are colonized at a water potential of −85 bars (2095). *P. fellutanum* is also suitable as food for the springtail *Folsomia fimetaria* (4007).

The optimum temperature for vegetative growth is 26–28°C, with the minimum 6–7°C and the maximum 41–42°C or less (517, 4578); the optimum water potential is at −40 bars (or higher), the minimum at −280 bars (1192). — It utilizes nitrite and metaphosphate (4544), degrades CMC (2736), decomposes paddy and *Pennisetum typhoideum* straw (4611), and can utilize hydrocarbons from fuel oil (4239). Metabolites include citrinin (512), spinulosin, γ-methyltetronic acid (5278), dehydrocarolic and carlosic acids (678). — Culture filtrates are reported to have some antifungal activity (516).

Penicillium frequentans Westling *1911* — Fig. 244.

DESCRIPTION: Raper and Thom (4745).—*P. frequentans* series.—Colonies reaching 5–6 cm diam in 12–14 days at 25°C on CzA, dark green, velvety. Reverse variable in colour, typically yellow-brown (if red, cf. *P. purpurescens* (Sopp) Raper & Thom *1949*). Conidiophores 100–200 μm long; conidia in long, compact columns, globose to subglobose, 3·0–3·5 μm diam, smooth-walled or finely roughened. — DNA analysis gave a GC content of 51% (5600).

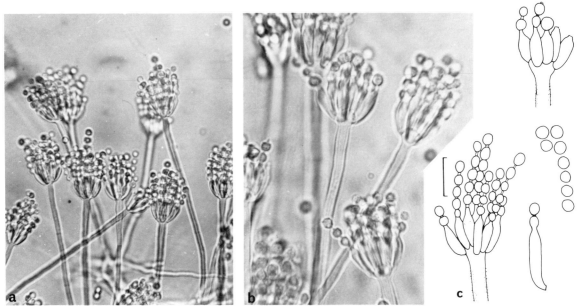

FIG. 244. *Penicillium frequentans*, monoverticillate penicilli with conidial columns, a. × 500, b. × 1000; c. detail of penicilli and conidia.

As the specific epithet implies, this is a very frequent species, particularly in acid forest soils (3817). It has a worldwide distribution; in view of the vast number of reports, only those from unusual or otherwise rarely mentioned sites are cited here: White Sea (3368), arctic tundra (1750), soils near Murmansk (4703), northern and central parts of the USSR (4677, 6571), the Amur region (432), the Turkmenian SSR (4304), Buhara (2871), Krasnodar (3364) and Zhitomir region (4474), Ukraine (1482), alpine sediments on the foot of glaciers (2530), alpine raw humus in Austria (1876), various soils in Romania (4378), Spain (3417), Syria (5392), Egypt (8), Libya (6510), Somalia (5048, 5049), equatorial West Africa (1420), Guinea (3852), central Africa (3063), Zaïre (3790), Ivory Coast (4718), South Africa (1559, 4407), Chile (1824), Brazil (379, 398, 399, 4547), Tahiti (6084), the Bahamas (2006), Hawaii (3264), Japan (3267), Tasmania and New Zealand (5930). It has been isolated from almost all soil types (3414), though particularly podzol and grey forest soils (3871, 4378), and even solonetz (4677), also rendzina, terra fusca (3418), loess (4759) and desert soils (654, 2974, 3997, 4733). Amongst numerous forest soils, it is mentioned especially frequently under oak (650, 989, 1700, 5811) and beech (272, 2719, 3538, 4537); in a comparison of different humus types, *P. frequentans* was found sometimes only in mull (4815) but sometimes predominantly in mor (4814). There are numerous reports from peaty soils (163, 1376, 2502, 3234, 3918, 4365, 4474, 6520, 6571), cut peat (1376, 1437), brown coal (4474), and from neutral fens (5559). It is often found under *Calluna* (989, 2736, 5047, 5811) and in other heaths (5819), under grassland (2573), tundra (1446, 3871) or steppe type vegetation (1559, 1700, 4407). In arable and other cultivated soils it tends to be rare, with the exception of reports from the USSR (2871), Poland (1423), Italy (4538), India (4996, 4997) and the Philippines (4823). It is reportedly rare in garden compost (1424) but has been found on sewage-treated land (2467), in a sewage treatment plant (1165), polluted water (1166, 1482), organic detritus in fresh water (4429), fresh water (2338), estuarine and seawater (4850), salt-marshes and a mangrove swamp (3264). It is also frequently mentioned as an inhabitant of sediments in caves: an ice cave (3367) and carst caves (3364, 3365) in the USSR, caves in Poland (4137), and a uranium mine in Czechoslovakia (1703). *P. frequentans* penetrates to medium or greater soil depths (3720, 3975). The pH range is given as 3·8–4·4 (3871, 5363, 6182), but it is nevertheless also found in chalk soils (4152). Its occurrence in wheat fields was enhanced by organic and mineral fertilizers (4198). It has repeatedly been found in the litter of pine (269, 2343, 2344, 2923, 3798, 6080), other conifers (2343, 4984) and deciduous trees (650, 2411, 3483), on remains of barley and cabbage plants (3113), in the phylloplane of vascular plants (296), sticks exposed to seawater (4546). It has also been found in the rhizospheres of various forest plants (4814), including ferns (2856, 2858, 5510), spruce (3573) and oak (5394), heathland plants (5819), *Lolium perenne* (5815), peat bog plants (1376), legumes (5416), and other cultivated plants (3429, 3737, 6067). It is frequent in stored wheat grain (2716, 3491, 3989, 4492), fresh and stored corn (3875, 3989), sorghum (3989), and seeds of tomato (5899) and groundnut (2765); it has been isolated from decaying Brazil nuts (2515), stored grapes and strawberries (342), dried fruits and nuts (2572, 5092, 6252), frozen fruit cake (3153), fruit juices (5205), and flour-based foodstuffs (5980); further on wood pulp and paper (6168), rarely on coarse fodder (4548), in birds' roosts and feathers (682, 2575, 2577), tunnels of gerbils (5393), rabbit dung (6085), the alimentary canal of a frog (5943), the intestine of *Dendrolimus sibiricus* caterpillars (2848), and bee hives (402). Numerous isolates were obtained from the gut of worker honeybees (1983). *P. frequentans* has also been found on optical glasses (2941, 4833). — It shows a moderate sensitivity to soil mycostasis (2559).

P. frequentans can grow at 5°C but not at 37°C, and at a water potential of −100 bars (4578). The optimum temperature for growth is reported to be 23°C, and for germination of conidia 23–30°C and −200 bars (3876); an osmotic potential of −210 bars (20% NaCl) is well tolerated (976), and this fungus is regarded as osmophilic (3934, 4369). Hair baits in soil were colonized at minimum water potentials of *c.* −200 bars (977, 2095). The thermal death point in apple juice is reported as 63°C for 25 min (3620). Germination does not require an exogenous energy source but the formation of germ tubes does (6496). — Intensive pectin (3539) and starch (1750) decomposition has been demonstrated; polygalacturonase and pectin methylesterase (101) are produced and cellulose in various forms is utilized (1425, 2736, 2878, 3538, 3618, 5079, 6085) although native cellulose is not (3618). Peptidogalactomannans of the cell wall are antigenically active (1895). Proteolytic activity is very strong; an acid carboxypeptidase is produced (6497, 6498); good growth occurs on 5–10% tannin or tannic acid (1425, 6083). *P. frequentans* attacks plasticizers (473) and plastic polymers (5148). It can utilize hydrocarbons from fuel oil (4239) and grow on lignosulphonate (934). Fulvic and humic acids are decomposed to some extent (934, 3582) by hydrolysis of ether bonds (4471); fulvic acid (at 1%) proved toxic to *P. frequentans* but it seems to be transformed into humic acid (3671); the utilization of humates, however, was reported as good (3672). The herbicide paraquat can be degraded (5447) and accumulated in the mycelium (5446), and atrazine, prometryne and simazine are said to be used as N and C sources (3799). Potassium is released from orthoclase and oligoclase (4008), metaphosphate is utilized (4544). Metabolic products include succinic, tartaric, citric and oxalic acids (1355, 4009), a-, β- and γ-hydroxybutyrates (4236), the hydroxybenzophenone sulochrin, the hydroxydiphenyl ether asterric acid, isomers of the grisan bisdechlorogeodin, the anthraquinones questinol, questin and emodin (3527, 5579), the amino acid hadacidin (2830), ergosterol palmitate (5278), the nephrotoxic citromycetin (2135, 2433), and the antifungal palitantin (512, 1245). Frequentin (1246), closely related to palitantin, was shown to be present on seeds of mustard, pea and wheat inoculated with *P. frequentans* (6451). — *P. frequentans* inhibits *Streptomyces griseus* (3852), *Erwinia carotovora* (2956), *Staphylococcus aureus* (4304), *Beauveria bassiana, B. brongniartii* (2848) and *Fusarium sporotrichioides* (5394), but the antibiotic activity of different isolates is not uniform (2736). *P. frequentans* has often been described as a parasite of the sclerotia of various fungi (1652, 2886, 3531, 3990). Seed treatment with *P. frequentans* suppressed the development of *Pythium* infections in mustard or beet seedlings (3378, 6450). Artificial infection after soil steaming did not affect the growth of citrus plants (3635). — The fungus is involved in suberosis, a respiratory disease in workers in cork industries which is accompanied by appearance of precipitins for *P. frequentans* (248).

Penicillium funiculosum Thom *1910* — Fig. 245.

= *Penicillium purpurogenum* Stoll var. *rubrisclerotium* Thom *1915*

DESCRIPTION: Raper and Thom (4745). — The distinction of series in the section *Biverticillata-Symmetrica* is of little value. *P. funiculosum* is an extremely variable species of this subsection, generally with broadly spreading colonies (5595), reaching 4·5–5·5 cm diam in 12–14 days at 24°C on CzA, growing faster on MEA. Sporulating areas yellow-green. The funiculose or tufted habit of the colonies is not always well developed. Odour earthy, or sometimes aromatic. Yellow vegetative mycelium often present on the colony surface. Reverse pink to deep red or orange-brown, occasionally almost black. Conidiophore stipes smooth or nearly so, usually greenish brown. Conidia ellipsoidal to subglobose, apiculate, delicately roughened, rarely smooth-walled, 2·5–3·5 × 2·0–2·5 μm. In fresh isolates, reddish brown to dark brown sclerotia, 200–300 μm diam, are often produced in concentric zones.—*P. funiculosum* differs from *P. purpurogenum* Stoll (q.v.) in the greenish brown conidiophores and penicilli and the smaller, less ornamented conidia. The similar *P. rubrum* Stoll (q.v.) has a more velvety growth and more constant and diffusing pigmentation, whilst the slower growth rate reported is not a very reliable criterion. — Some isolates contained virus particles (2509). — DNA analysis gave a GC content of 49·5–50% (5600).

P. funiculosum is very common in all climatic zones with the possible exception of extremely cold habitats and seems to prefer acid substrates. — Distributional data are available from the British Isles (745, 1375, 1376, 1438, 5559), the Netherlands (1614, 1616), Poland (269, 272), from solonetz, podzol, gley, brown soil and other soil types in the USSR (432, 774, 961, 1637, 2871, 3366, 3652, 4474, 4677, 6571), Germany (2467), France (242, 2161), Italy (3450, 4538, 5047), Spain (3446), Turkey (4245), Syria (5392), Israel (2764, 2768, 2772), Iraq (92), Egypt (8, 3993, 3997), Libya (4083, 6510), South Africa (1555, 4407), Zaïre (4159), central Africa (3063), the Ivory Coast (4718, 4719), Canada (6352), various states of the USA, Central America (1697), the Bahamas (2006), Brazil (379), Peru (2005), Pakistan (61), very numerous reports from India, also Sri Lanka (4021), Japan (3267, 5846), Hawaii (3264) and Australia (3721). — It has frequently been isolated from forest soils (2482, 2573, 3063, 3817,

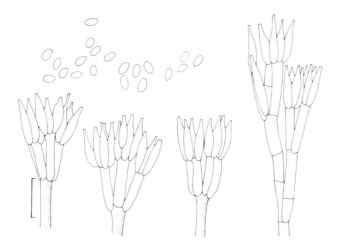

FIG. 245. *Penicillium funiculosum*, penicilli and conidia, CBS 884.72, orig. R. A. Samson.

4226, 5000, 6351), under pine (269, 1637, 6352), larch (3267), teak (2853, 2854), beech (272), dry to mesic conifer-hardwood communities (1032), wet-mesic deciduous forests (1040), stands with *Eucalyptus maculata* (1555), *Acacia mollissima* (2186) and *Casuarina* (2006), and tropical (4716) and Himalayan (4995) forests, soil in South Carolina/USA after prescribed burning (2822), further uncultivated soils (3259, 3446, 4030, 4698), rarely in grasslands (1614, 4021, 4933), more often in cultivated soil (1614, 1616, 2871, 3450, 3652, 3817, 4996) with wheat, beet (2161), paddy'rice (774, 1519), cotton (13), citrus (2764), coffee (4159), alfalfa, corn (1330, 4538, 4716), groundnuts (2768, 2772) and lentils (2164), soils under steppe type vegetation (4313, 4718, 4719, 6347), very frequently in heathland (3721, 5047), peaty soils and peat bogs (961, 1375, 1376, 1438, 4474, 5559, 6571), marshy taiga (3366), sand dunes (745, 6414), desert soil (3997, 4083, 4733), in alpine zones (4711), lowland *Salix* communities (2008), open bogs and conifer swamps (1039), estuarine sediments (655), salt-marshes (8) and a mangrove swamp (3264). *P. funiculosum* is reduced in numbers in sewage-treated soils (1163, 2467), but is also known from a sewage treatment plant (1165), activated sludge (1172), polluted waters and air (1166). — Soil depths of 30 cm and more are not unusual (272, 4995, 5000, 5349). The fungus has been frequently isolated from leaf litter of pine (681, 2343, 2344, 2345), fir (2343, 3567, 4984), spruce (2343) and *Eucalyptus maculata* (1558); from the phylloplane of vascular plants (296), corms of *Gladiolus* (5369), ectomycorrhizae of *Suillus grevillei* with *Larix leptolepis* (1794), sclerotia of *Sclerotinia sclerotiorum* (4699), the rhizospheres of wheat (3567), sugar cane (2861), banana (2030), groundnut (1897, 2768), other crop plants (3737), *Cynodon dactylon* (2855), *Festuca lessingiana, F. sulcata* (3376), euphorbiaceous plants (3866), sand dune plants (6414), peat bog plants (1376), Norway maple (919), *Aristida coerulescens* (4083), and roots of strawberries (4133). It is also found on seeds of corn (3875, 3989), wheat (2297, 3989), barley (6274), rice (3181, 5967), sorghum (3989), cultivated grasses (3512), tomato (5899), pecans (2572) and seeds and pods of groundnuts (1016, 1984, 2765, 2772, 3702, 4602). Other recorded substrates include nests, feathers and pellets of free-living birds (2575), brown coal (4474), optical lenses (2941), frozen fruit cake (3153), worker pupae of bees (4631), and the lungs of small wild mammals in India (2153). — The mites *Caloglyphus redikorzevi* (5379), *Acarus gracilis* and *Glycyphagus destructor* (5380) can feed on this fungus.

The optimal growth temperature is in the range 25–28°C, the minimum being reported as 17–19°C, with no growth at 5°C, and the maximum as 41–42°C (517, 1375, 4578); the thermal death point is at *c.* 70°C for 30 min (613); little growth occurs at a water potential of −100 bars (4578). — Suitable C sources include glucose, sucrose, maltose, starch, galactose (919, 1375) while good N sources are glycine (1375), asparagine, $NaNO_3$ and NH_4NO_3 (919). This fungus utilizes fructosans (3404), cellobiose (4778), dextran (938, 5911) with high yield of dextranase (3100, 5335, 5632), xylan (3231) and 1,2-β-glucan (4780). On salicin β-glucosidase is produced (as in numerous other fungi) (6159) and it grows on sinigrin (4774). An exo-1,3-a-glucanase (2298) and β-mannosidase on mannan, mannose, glucose and starch (4781) are produced. The cellulolytic activity (2006, 5076) depends largely on the availability of N (3303). *P. funiculosum* is also able to degrade *Sphagnum* moss (962), aspen and birch wood (4191) and utilize ferulic acid (544). An acid carboxypeptidase is produced (6497, 6498); hydrocarbons from fuel oil are utilized (4239). The following metabolic products have been found: malonic acid, cholesterol, the viridin-related 11-desacetoxy-wortmannin (2202), islandicin (3231), 2,6-dipicolinic acid (5278), the ribonucleoprotein helenin which has antiviral activity (3317, 5306, 5307) and is apparently identical with double-stranded RNA of viral origin (330, 5496), funicone, orsellinic acid, mitorubrin, mitorubrinol, mitorubrinic acid (3393) and the flavour compound 3-methylbutanol (2867). — *P. funiculosum* has but a weak antibiotic activity against bacteria (515) and fungi (516, 569, 4699), culture filtrates were reported to contain indole-3-acetic acid (294) and to inhibit the growth of barley seedlings (293) and germination of tomato, radish and eggplant seeds (919). — In a chronically irradiated soil *P. funiculosum* appeared with increasing numbers up to intensities of 120 R per day, at 820 R per day some viable cells were still present (2009).

Penicillium granulatum Bain. *1905* — Fig. 246.

DESCRIPTIONS: Raper and Thom (4745), Udagawa (5967), and Samson *et al.* (5029). — Colonies reaching 2–3 cm diam in seven days at 24°C on CzA, 1·5–2 cm on MEA; conidiophores commonly loosely synnematous, giving the colony a coarsely granulate zonate appearance, blue-green. Reverse yellow to orange-brown. Odour aromatic fruity combined with the pungent component of *P. verrucosum*. Conidiophores two- to three-stage branched, usually conspicuously rough-walled. Conidia strongly ellipsoidal, later sometimes subglobose, smooth-walled, mostly 3·0–3·5 µm long. — DNA analysis gave a GC content of 49·5% (5600).

This is one of the rarer *Penicillium* species but nevertheless has a worldwide distribution, especially in temperate latitudes, but is also known from Peru (2005), Japan (5846), Egypt (3997), Israel (2762, 2764, 2768), Syria (5392) and Australia (3720). It has been isolated from tundra, podzol, grey forest soil (3871), chernozem (3414), and both mor and mull (4814). It has been reported from beech forests (272, 3138), other deciduous forests (2573), humus under fir and pine (681), pine litter (269, 2344), mixed hardwood forests (1030, 1204, 2004, 2008, 4225), heathland soils (3720), *Sphagnum* bogs (6520), dunes, saline beach (4162), caves (3453) and cultivated soils (e.g. 2762, 2768). It is most frequent in soils of pH 5 (1204). In mixed forests it showed its highest density in dry to mesic habitats. — It has been isolated from roots of sugar beet (4559), mycorrhizae and dying roots of oak (963, 1794), the rhizospheres of groundnuts (2768) and various forest plants (4814), and is also known from stored wheat grain (4492), milled rice (5967), stored eggplant (342), honeycombs (*Apis*) (402) and the intestine of *Dendrolimus sibiricus* caterpillars (2848).

The temperature optimum is 25°C with little growth occurring at 5°C and the maximum 30°C (1204, 4578); it grows at water potentials of −100 bars (4578). Xylan decomposition is very good (1432), cellulose is utilized (5076), and it has a high tolerance to tannin (1204, 1206, 1260, 3052). It produces viridicatin (1058), a viridicatin-related alkaloid-like metabolite (4500) and a tremorgenic toxin (1059), and shows a strong antagonistic activity towards *Beauveria bassiana*, *B. brongniartii* (2848), *Rhizoctonia solani*, *Gaeumannomyces graminis Pythium ultimum*, *Fusarium culmorum* and *Pseudocercosporella herpotrichoides* (1431). A mycelial homogenate of *P. granulatum* inhibited the growth of tomato roots (5779).

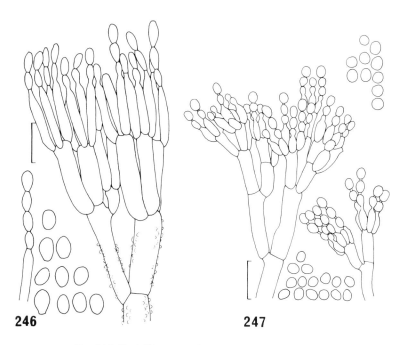

246   **247**

FIG. 246. *Penicillium granulatum*, penicillus and conidia.
FIG. 247. *Penicillium griseofulvum*, penicilli and conidia.

Penicillium griseofulvum Dierckx *1901* — Fig. 247.

= *Penicillium patulum* Bain. *1906*
= *Penicillium urticae* Bain. *1907*

DESCRIPTIONS: Raper and Thom (4745), Mehrotra and Tandon (3740), and Samson *et al.* (5029). — Colonies growing restrictedly, reaching 1·8–2 cm diam in seven days at 24°C on CzA, 2·0–2·4 cm on MEA, light greyish green, reverse yellow to orange-brown. Odour fragrant in some isolates but not pronounced in others. Conidiophores three- to four-stage branched, the branching rather divergent; phialides less than 6·5 μm long, with very short necks, soon evanescent. Conidia ellipsoidal, smooth-walled, 2·5–3·5 × 2·2–2·5 μm. — The conidial nucleus divides at the time of formation of the first germ tube, one nucleus remaining in the conidium to give rise to nuclei for one or two further germ tubes (1769). During germination, a third wall layer which is continuous with the germ tube wall is formed in the conidia (1772). — Constituents of hydrolysed cell walls include 12·3% glucosamine, 0·5% galactosamine (3232), galactose, arabinose, xylose and rhamnose (2049). Conidia contain 18·6% C_{16}-, traces of C_{17}-, and 6·3% C_{18}-saturated fatty acids, traces of $C_{16:1}$-, 13·6% $C_{18:1}$-, 59·2% $C_{18:2}$- and 2·3% $C_{18:3}$-unsaturated acids (1277). The content of long-chain fatty acids has been used to study species relationships (1281).

P. griseofulvum has a worldwide distribution, though it is usually not very frequent. It has been reported from sediments below glaciers in Alaska (1171), soils in the British Isles (2338), Germany (1889), the USSR (4304, 4554), France (918), Spain (3446), Israel (2764, 2768), Somalia (978, 5049), Pakistan (60), India (3740), Australia (977, 3720), Brazil (379) and Peru (2005). It is rather rare in forest soils, particularly under conifers (681, 3362, 3447), but known from heathland soils (3720), under shrub (2160) and steppe type vegetation (5049), desert soils (1512), arable soils (918, 1616, 1889, 2161, 2162, 2768), particularly under corn (4554) and in a wheatfield soil in Australia where it was one of the most abundant *Penicillium* species (6186), also frequent in stubble-mulched soil after wheat cropping (4218); it has also been reported from citrus soils (2764), saline soils (3446), a uranium mine (1703), caves (3453, 4137), and sewage (1154). There are numerous reports of it reaching soil depths of 40 cm and more. Hair baits were still colonized at low water potentials of −150 bars (977). After ploughing under of plant remains, increased frequencies can occur and the associated irregular plant growth has been attributed to patulin production (4215, 4216). The antibiotic patulin is formed in detectable quantities in sterilized soil, especially after addition of organic substrates (2126, 4216). *P. griseofulvum* also occurs on pecans (2572), stored grain of wheat (3491, 3989), corn (3875), and rarely barley (6274). It has been isolated from the rhizospheres of potato (1614), tomato (16), groundnut (2768) and corn, but not in adjacent root-free soil (4554). Increased incidence in the wheat rhizosphere (1614) has been observed following previous crops of winter wheat or alfalfa (2816). It is also reported on vole (2338) and rabbit (6085) dung, tunnels of gerbils (5393), feathers of free-living birds (2575), and refrigerated dough products (2067, 3153).

Conidia germinate in the temperature range 4–37°C, optimally at 23–30°C; the minimum water potential is −190 bars; mycelial growth occurs in the range 4–35°C with the optimum at 23°C, no growth occurring at 37°C (3873, 3876, 4578). — Germination requires glucose (1770, 3971); galactose and meliobiose also induce germination of conidia but do not support growth, while the reverse was found to apply for fructose and lactose (1776). Conidium formation can be induced by Ca^{2+} ions (in a glucose-nitrate medium), by Cu^{2+} ions at later growth stages, and transfer to a N-free medium (3971). The uptake of L-leucine is pH-dependent, 98% of it being found as free amino acid in the mycelium (6319); the amino acid pool is not restricted to hyphal tips (459); organic N compounds are released into mineral media (3968). Extra- and intracellular polysaccharides are composed of glucose, galactose and mannose (3640). Peptidogalactomannans of the cell wall are antigenically active (1895). Cellulose decomposition (6085) has been shown on textiles (3618) and with CMC (1432). Lignosulphonate, fulvic and humic acids (934) and sinigrin (4774) are utilized and β-glucosidase produced on scilliroside (5598). Numerous metabolic products have been determined including ethylene (2641), patulin (4217, 4219), desoxypatulinic acid (5164), triacetic acid lactone (5156), mycelianamide, several hydroxy benzenes, including pyrogallol, toluhydroquinone, gentisyl alcohol, hydroxybenzene aldehydes, ketones and carboxylic acids including gentisic acid, protocatechuic acid, 6-formyl- and 6-methylsalicylic acid and orsellinic acid, gentisyl aldehyde, the griseophenones A, B, C (3231), gentisylquinone (1630), and the antifungal antibiotic griseofulvin (3457, 3722, 4336) and other grisans related to it. Further metabolites reported are griseoxanthone C (3231), coprogen (6528), mycophenolic (5278), shikimic (3231) and kojic acids (4458). The toxic and antibiotic metabolite patulin (syn. expansin, clavacin, claviformin) (152, 528, 5601) inhibits both gram positive and gram negative bacteria and also phytopathogenic fungi; the biosynthesis of patulin has been investigated (5156, 5157), particularly the conversion of 6-methyl-salicylic acid into patulin (1803) and the role played in this by the key enzyme 3-hydroxybenzyl-alcohol dehydrogenase (1802). — *P. griseofulvum, in vitro,* is one of the most active antibionts known against a range of pathogenic soil fungi (1431) and bacteria (520, 4304, 4503). Soil inoculation against *Rhizoctonia* was not successful (6434), but, on the other hand, *P. griseofulvum* reduced plant growth (428); in pure culture the fungus was stimulated in the presence of corn seedlings (1536) and simultaneously led to injuries in corn (4553) ascribed to patulin production and its effect as an antagonist of indole-3-acetic acid (4215). Griseofulvin is held responsible for growth inhibition of mustard, red clover and wheat seedlings (6448). This antibiotic has become very important in the oral treatment of dermatomycoses and is produced on a large scale by commercially selected strains of *P. griseofulvum* and *P. nigricans. P. griseofulvum* is very toxic to chicken and rats when accumulated in cereals (733, 5158).

Penicillium herquei Bain. & Sart. *1912* — Fig. 248.

DESCRIPTION: Raper and Thom (4745). — Smith (5431) transferred the *P. herquei* series from the *Biverticillata-Symmetrica* to the *Asymmetrica* because the phialides have an abruptly tapering neck. The conidiophores are strictly two-stage branched (because of their compact structures resembling the *P. brevicompactum* series), with very dense metulae and a coarsely roughened stipe; colonies grow rather restrictedly, reaching 2–3 cm diam in 14 days at 24°C on CzA, 6–7 cm on MEA, the abundant aerial mycelium is yellow and the conidial areas are yellow-green to light olivaceous-green, reverse dark yellow-green. Conidia in more or less tangled chains, ellipsoidal, almost smooth-walled but in some isolates rough-walled, $3 \cdot 5 – 4 \cdot 0 \times 2 \cdot 2 – 3 \cdot 0 \, \mu$m. — The rodlet pattern of the conidial surface as seen in freeze-etched replicas is slightly interlacing with some raised areas, the individual surfaces varying from relatively smooth to rough (2421). — DNA analysis gave a GC content of 48·5% (5600).

Reports on the distribution are relatively rare and identification is not always certain. A worldwide distribution is indicated by the following records: recently deglaciated soils in Alaska (1171), soils in the USA (1039, 1040, 1164, 1166, 2007, 2482, 3817, 4226, 4733), the USSR (2871, 3366, 4474, 6571), Australia (978), Hong Kong (1021), Sri Lanka (4021), India (4996), Peru (2005), South Africa (1559), Somalia (5048), Zaïre (3790), Egypt (4962) and Italy (3913). It has been found in forest soils (1021, 3790, 3817, 5048) after prescribed burning (1164), forest nurseries (2482), *Salix-Populus* communities (2007), flood-plain communities with maple and ash (4226), wet-mesic deciduous forests (1040), grassland (4021), open savannah (1559), taiga (3366), low and high moor type habitats (4474, 6571), open bogs (1039), cultivated soil (2871, 3817, 3913, 4996), desert (4733), salt-marsh (4962), caves (3364, 3365, 3367), and polluted soil and water (1166). It has been isolated more frequently from organic than mineral soil particles (1021), and also from the rhizosphere of crop plants (3737), the geocarposphere of groundnuts (2768), leaves of vascular plants (296), freshly fallen leaves of *Fagus crenata* (4989), cotton fabrics (3707, 6326), various foodstuffs (1303), dried cereals and legumes (5980). — Forty-eight per cent of chloroform-fumigated mycelium was mineralized in soil within ten days at 22°C (133). — The fungus may serve as food for mites (5383).

P. herquei hardly grows at 5°C and there is no growth at 37°C; good growth occurs at –100 bars water potential (4578), and some growth even at *c.* −260 bars (25% NaCl) (5881). Conidia are not produced in continuous darkness or at high light intensities and the green mycelial pigment is produced only in daylight (4837). — It degrades cotton fabrics (6326), can grow on tannin as C source (1206) and produces an acid carboxypeptidase (6498). Nitrite can be used as an N source (10). Growth *in vitro* was found to be slightly stimulated by root extracts and exudates of *Crotalaria medicaginea* (5638). The following metabolites closely related to atrovenetin have been described for this species: herqueichrysin, herqueinone, isoherqueinone, norherqueinone and deoxyherqueinone (371, 741, 1839, 4104, 4105, 5340); further metabolites are parietin (= physcion) (5278), rubratoxin B (3231) and herquein, the latter with antibacterial activity (799). Antimycotic (569) and mycotoxic properties have been described (569, 1303).

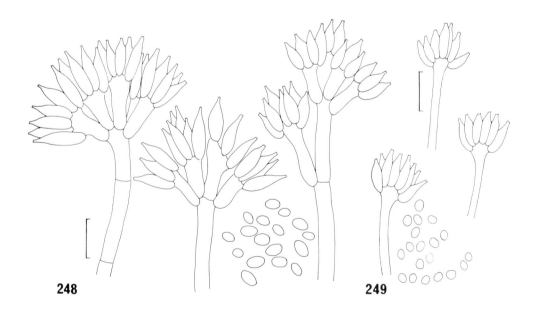

248   **249**

FIG. 248. *Penicillium herquei,* penicilli and conidia, CBS 336.48, orig. R. A. Samson.
FIG. 249. *Penicillium implicatum,* monoverticillate penicilli and conidia, orig. R. A. Samson.

Penicillium implicatum Biourge *1923* — Fig. 249.

DESCRIPTION: Raper and Thom (4745). — Colonies growing restrictedly, reaching 1·5–2·0 cm diam in 14 days at 25°C on CzA, bluish green with yellow-orange to deep brown reverse and evenly coloured agar. Conidiophores to 100μm long. Conidia forming loose columns, ellipsoidal to subglobose, smooth-walled or nearly so, 2·0–3·0 × 2·0–2·5 μm. — DNA analysis gave a GC content of 54% (5600).

The worldwide reports of this fungus include ones from Alaska on recently deglaciated soil (1171), the USSR (432, 3365, 3366, 4474, 6571), Ireland (3918), Sweden (1446), Poland (4137, 6520), Czechoslovakia (1700, 1703), Italy (270, 271, 3453, 3976, 5047), Spain (3446), Turkey (4183), equatorial West Africa (1420), the USA (655, 1166, 1387, 4313, 4733, 4918), Canada (1446), Japan (5846), Peru (2005) and Brazil (399). It has been found in cultivated soil under wheat (3446), forest soil under beech (270, 271), podzols (989) and brown forest soils (432), heathland (5047), peat bogs and fens (3918, 4474, 6571), soils under tundra (1446, 3366) or steppe vegetation (1700, 4313), desert soil (4733), an alpine soil with long snow coverage (3976), caves (3365, 3453, 4137), a uranium mine (1703), the eulittoral range of the ocean in Florida (4918), estuarine sediments (655), activated sludge (1387), sewage and polluted water (1166). It has also been isolated from leaf litter of pine (2344), decaying *Eucalyptus regnans* (3483), corn seeds (3875), birds' roosts (682), the intestine and integuments of *Dendrolimus sibiricus* caterpillars (2848), leather (6326), and frozen fruit cake (3153). — *P. implicatum* is strongly inhibited by a metabolite from *Lactarius* sp. (4438).

No growth occurs at 5°C and limited growth is obtained at 37°C and −100 bars water potential (4578). — *P. implicatum* can use tannin as C source (1206). Known metabolites are the nephrotoxic citrinin (4587), sclerotiorin and 7-*epi*-sclerotiorin (5278). — This fungus is said to have a low antibacterial activity (515, 2956) and inhibits, *in vitro*, the growth of *Ophiostoma minus* and *O. ips* (355). — In a chronically irradiated soil the fungus did not occur at intensities of ⩾ 15 R per day (2009).

Penicillium islandicum Sopp *1912* — Fig. 250.

DESCRIPTION: Raper and Thom (4745). — *P. islandicum* is close to *P. rubrum* (q.v.) but the reverse is orange-brown to red without any diffusion of the pigment into the agar. Colonies growing somewhat less restrictedly, reaching 2·5–3·5 cm diam after 14 days, consisting of a felt of orange to red encrusted mycelium from which funiculose hyphae arise which bear the short conidiophores. Conidial areas dark green. Conidia ellipsoidal, smooth and thick-walled, ± apiculate, 3·0–3·5 × 2·5–3·0 μm.

P. islandicum has a worldwide distribution, ranging from the original description from Iceland, the Amur region (432), and solonetz soils (4677) in the USSR, to observations in the USA (1032, 1039, 1166, 2007, 2008), Australia (3720), India (1519, 3732, 4716), central Africa (3063), Israel (2764), Italy (311, 4538, 5047) and Spain (3446). Forest soils in particular are mentioned as habitats including podzols (432), soils under *Populus* (2007,

2008), conifer-hardwood stands (1032) and tropical vegetation (4716); amongst cultivated soils it has been reported from paddy rice fields (1519), under alfalfa and corn (4538), sugar cane (3732), and citrus (2764); it has also been isolated from heathland (3720, 5047), open bogs (1039) and an acid mine drainage stream (1166). It has been reported from the rhizosphere of groundnuts (2768), and seeds of stored rice, wheat (2297) and corn (57), pecans (2572, 5092), flour-type foodstuffs (5980) and feathers of free-living birds (2575).

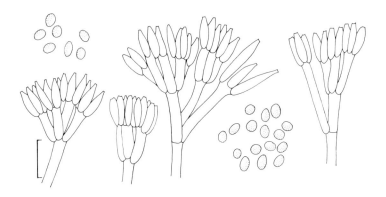

FIG. 250. *Penicillium islandicum*, penicilli and conidia, CBS 189.68, orig. R. A. Samson.

Optimal growth occurs at 31°C, with a maximum of 38°C and a minimum of 10°C, little growth is obtained at −100 bars water potential and the minimum for growth is −240 bars (261, 4578). — Degradation of glucan via a 1,2-β-glucanase (4780) has been described and β-D-xylopyranosidase can be produced (4777). The extra- and intracellular production of glucose, galactose and mannose has been studied (3640), as has the production of the polysaccharide islandinic acid (1535). Known metabolites are malonic acid, 3-hydroxyphthalic acid, erythroskyrin (3231), the anthraquinones chrysophanol, islandicin (2549), emodin, catenarin, endocrocin (3231), flavoskyrin (2550, 5207), 5,6-dihydro- and 5,6,7,8-tetrahydrocatenarin, rugulosin, dianhydrorugulosin, roseoskyrin, auroskyrin, iridoskyrin, rhodoislandin A and B, skyrin, punicoskyrin, aurantioskyrin, dicatenarin, skyrinol, oxyskyrin, rubroskyrin, luteoskyrin (664, 3231, 5276, 5697, 5991) with hepatotoxic properties (3888, 6015, 6016), islanditoxin (3654) and a toxic chlorine-containing peptide (3231). — Cereal grain contaminated with *P. islandicum* has been found to be toxic to young ducklings (5158), chicken and mice (4671). It is one of the most toxic fungi in food which, as the causal agent of "yellow rice", has been responsible for many cases of acute and chronic liver lesions, particularly cirrhoses and primary carcinoma, in Japan (3929, 6412). The toxic substances are mainly the pigments luteoskyrin and islandicin.

Penicillium italicum Wehmer *1894* var. *italicum* — Fig. 251.

DESCRIPTIONS: Raper and Thom (4745), CMI Descriptions No. 99, *1966*, and Samson *et al.* (5029). — Colonies growing restrictedly, reaching 2·0–2·5 cm diam in 10–14 days at 24°C on CzA, 5–6 cm on MEA, grey-green; reverse pale grey to yellow-brown. A faster growing avellaneous variety has recently been found in Israel (5029). Odour sweet, suggesting perfume. Conidiophores usually loosely synnematous, particularly in the colony margins, smooth-walled, two-stage branched with appressed penicilli. Phialides cylindrical, slender, with a short but distinct neck. Conidia ellipsoidal to cylindrical, smooth-walled, 4–5 × 2·5–3·5 μm. Sclerotia may be present in fresh isolates. — Nuclear divisions occur as in *P. expansum* (3198); the conidia are uninucleate but vegetative cells are multinucleate. A parasexual cycle has been demonstrated (5614). — Hydrolysates of the mycelial wall contained 51·6% glucose, 9·0% glucosamine, 3·8% galactose and traces of rhamnose and xylose (3232). — DNA analysis gave a GC content of 51·5% (5600).

The principal substrate of *P. italicum* are all kinds of citrus fruits in which it causes blue rot. — The majority of the distributional data are from locations in warmer climates: Egypt (8, 3988, 3993), Libya (4085), Israel, Cyprus, Ghana, Zambia, Rhodesia, South Africa and New Zealand (IMI); exceptions are reports from an alpine beech forest (3538), caves in Italy (3453), and isolations from the British Isles (IMI) and Austria (3418). It has been isolated

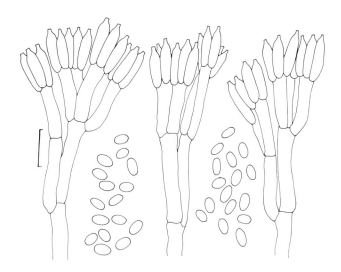

FIG. 251. *Penicillium italicum*, penicilli and conidia, orig. R. A. Samson (from 5029).

from soil in citrus plantations (3988, 4085), wet grassland (4180), rendzina, terra fusca (3414, 3418), salt-marshes (8), organic detritus in fresh water (4429), litter of conifers and deciduous trees, rye grass seeds, yams, hay (IMI), tomato fruits (342) and fruit juices (5205).

Growth occurs at 5°C but not at 37°C; good growth was obtained at −100 bars water potential (4578). Optimum germination of conidia occurs in 1 mmol cm⁻³ glucose solution (4988). Optimum growth occurs in the pH range 2·9–6·5 with a maximum tolerated pH of 7·9 (3612). — A rapid consumption of glucose and L-leucine (as sole N source) and the accumulation of isovaleric acid in the medium has been measured (1121). Invertase is localized within the hyphal cells (3734). Cellulose and pectin are degraded (1494, 3414, 4184), and the production of polygalacturonases has been investigated on different substrates (1898); *P. italicum* can grow on 0·25% tannic acid (3052); nitrite can be used as an N source (10) and L-tryptophan is utilized (1122); proteinases have also been reported (3231). Abscisic acid was found as a metabolite when *P. italicum* was grown on orange tissue (4938); the formation of diketopiperazines has been described (5162). — Growth of *Trichoderma harzianum* and *T. koningii* was inhibited *in vitro* (1526). Virulence to attack citrus fruits can be lost in some mutants but restored by diploidization (463).

Penicillium janthinellum Biourge *1923* — Fig. 252.

DESCRIPTION: Raper and Thom (4745). — The *P. janthinellum* series is well characterized amongst the subsection *Divaricata* by phialides with an abruptly and strongly tapering conidiogenous tip. Pigmentation of the reverse and roughening of the conidia and co-nidiophores all are highly variable within the series; therefore, the delimitation of species is often uncertain. — *P. daleae* Zal. (q.v.), classified by Raper and Thom (4745) in the *P. janthinellum* series, is better placed in the *P. nigricans* series, while amongst the remaining species only *P. simplicissimum* (Oudem.) Thom (q.v.) is clearly distinguished, not only by its coarsely roughened conidiophores but also the denser ramification. — Colonies spreading broadly, reaching 5–7 cm diam in ten days at 24°C on CzA and MEA, pale grey to greenish grey, reverse variously coloured but mostly in bright yellow-green or red colours. Conidiophores with few divergent branches, smooth-walled or slightly roughened. Conidia strongly ellipsoidal, sometimes later becoming subglobose, often with apiculate ends, more or less rough-walled, mostly 3·0–3·5 μm long. — DNA analysis gave a GC content of 53·5–54% (5600).

Species of the *P. janthinellum*-series are amongst the commonest soil fungi (4745) and the overall distribution is illustrated here only by some selected examples: Siberian (3652), arctic (1446, 3368) and alpine habitats (1876, 3538) and high altitudes in Tian-Shan and the Altai (5647), throughout the temperate and mediterranean zones and into the warmest zones, including arid zones of the Negev (654), Libya (6510), Egypt (3997), Syria (277), Madagascar (4159), central Africa (3063), Zaïre (3790), Ivory Coast (4718), Pakistan (60, 4855), India (1519, 2178, 2186, 4995), Nepal (1826), Sri Lanka (4021), Japan (3267, 5846), Hawaii (3264), Peru (2005), Central America (1697) and Brazil (398, 399, 4547). It has been regarded as a pioneer species in an oak-maple community in Wisconsin (5880) and is common in many types of forest soils: under *Fagus* (3538, 4537) and *Nothofagus* (4947), other deciduous forests (2573), pines (6352), larch (3267), white cedar (505), dry to mesic conifer-hardwood (1032), *Salix-Populus* lowland communities (2007), floodplain communities with maple, elm and ash (4226), mixed alder-conifer stands (6351) and *Acacia* (2178, 2186). It often occurs in soils rich in organic matter (2008), such as peat (163, 432, 4474) and grassland (277, 3363, 3497, 3498, 4021), and is frequent in heathland (3721, 5047) and under tundra (1446), steppe (1702, 4718) or chaparral vegetation (1164). The *P. janthinellum*-series has also been isolated from caves (4137) and a uranium mine (1703). It has been shown to occur at soil depths between 0·5 and 100 cm (163, 3267, 4759, 4995). Its distribution covers a wide pH range although in grassland slightly acid soils are preferred (6182) as are the more acid saline soils (3871); nevertheless, this series has also been isolated from lime soils (4152) and alkaline sand dunes (745). It occurs in very varied humidity ranges (4313) and spread in soil on hair baits occurred from saturation to −140 bars water potential (977, 2095). *P. janthinellum* has also been found in a mangrove swamp (3264), fields treated with sewage sludge (1163), polluted waters (1166), river sediments and streams (1162). A seasonal decline in January has been reported in Wisconsin (2004). In soils under corn it was more common than in fields with alfalfa or sugar beet (1330). Manuring with N, P, PK, NPK and farmyard manure generally raised the viable spore content in soils (2762, 2895, 3498); in a podzol it was reduced by liming and dung but raised by $(NH_4)_2SO_4$ (3849). *P. janthinellum* is tolerant of copper accumulation in the soil (4921), but fumigation with methylbromide reduces the fungus drastically (1268, 6026). Forty-four per cent of chloroform-fumigated mycelium was mineralized in soil within ten days at 22°C (133). After growing peas repeatedly, *P. janthinellum* was shown to be significantly more frequent than after wheat and rape (1433) and it appeared in higher numbers after barley crops than after wheat (1617). It has been isolated from litter of aspen, poplar (6101), *Pinus taeda* (2344) and *P. radiata* (6080), the phylloplane of *Acacia* and *Metrosideros* (296), plant roots on various occasions (but without clear rhizosphere effects), roots of *Picea abies* (3573), *Pinus strobus* (1794), flax (5826), tomato (16), alfalfa, peas (5825), dying broad beans (6134), various herbaceous plants on mull, moder and mor (4814), various Australian heathland plants (5819) and halophytes (3375), roots of *Lolium perenne* (5815), oat, wheat (5825), corn (1330, 4553, 4554) and various steppe grasses (3376). It has also been found on sticks buried in soil (1956), in seawater (4546), on wood pulp and paper (3294, 6168), wooden furniture (3707) and seeds of *Pinus sylvestris* (6344), wheat (3491), corn (3875) and groundnuts (2765, 4602), in birds' roosts (682), feathers and pellets (2575), and on food products (3153, 5205, 5980). — It may serve as food for the springtail *Hypogastrura tullbergi* (3830).

No growth occurs at 5°C, the optimal temperature is 25°C, and the maximum either 30°C (1204) or >37°C; good growth occurs at −100 bars water potential (4578). Utilization of starch (654, 655, 1750), pectin (655, 1432, 1750, 3538, 3539) and cellulose in various forms (386, 655, 1425, 2948, 3538, 3618, 5059) has been reported. Three xylanases were purified from cultures on wheat bran koji (5700). *P. janthinellum* can grow on humic (1750) and ferulic acids (544). It is also able to degrade tannins (1204, 1206, 3315) and hydrocarbons from fuel oil (4239), and the pesticides atrazine (2892), chlorphenamidine, chlorpropham, dicryl, propanil, propachlor, propham and solan (2893). Nitrite can be used as an N source (10). The production of ribonuclease and proteinases (1971), the composition and enzymatic properties of peptidases (2498, 5246, 6497, 6498), in particular of the non-specific, SH-dependent penicillocarboxypeptidase-S (2813), have been investigated. Strong acid production occurs on carbohydrate-rich substrates. A fungitoxic metabolic product, janthinellin, has been isolated from the mycelium (4589) and griseofulvin production is also reported (3231). — *P. janthinellum* is antagonistic towards various bacteria (515, 4304, 4553, 5394), the fungi *Chalara elegans* (2753), *Rhizoctonia solani*, *Gaeumannomyces graminis*, *Pythium ultimum*,

Neosartorya fischeri and *Mucor plumbeus* (1431, 1526, 6024), and human pathogens (569, 1548). The culture filtrate had a toxic effect on wheat (3849, 3851), and the growth of both corn (4553, 4557) and tomato seedlings (5779) were inhibited by it. — With >3% CO_2 in the atmosphere growth is inhibited (784); *P. janthinellum* does not survive a heat-treatment of 60°C for 1 h (6344). It proved to be resistant to phenylmercuric acetate (2077).

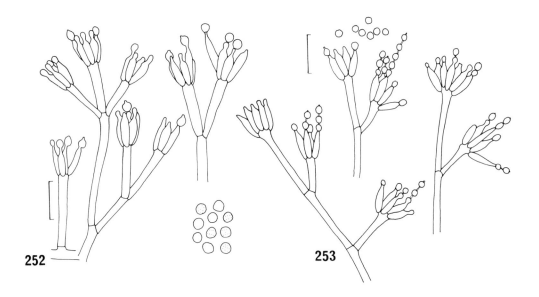

FIG. 252. *Penicillium janthinellum*, divaricate penicilli and conidia.
FIG. 253. *Penicillium jensenii,* divaricate penicilli and conidia, CBS 342.48.

Penicillium jensenii Zaleski *1927* — Fig. 253.

DESCRIPTIONS: Raper and Thom (4745), and Mehrotra and Tandon (3740). — *Penicillium canescens* series. — Colonies reaching 2·5–3·0 cm diam in 12–14 days at 24°C on CzA and MEA, strongly folded or wrinkled, greyish green, reverse uncoloured or dull peach. Conidiophore stipes smooth to delicately rough-walled, often bearing several branches, penicilli divaricate. Phialides with well-defined conidium-bearing necks; conidia globose to subglobose, delicately roughened, 2·0–2·5 µm diam. — *P. jensenii* is related to *P. waksmanii* Zal. (q.v.) which has more tangled conidial chains and also to *P. lanosum* Westl. (q.v.) which has a different, lanose growth habit and more diffusely branched penicilli. — DNA analysis gave a GC content of 50·5% (5600).

P. jensenii is one of the less frequent *Penicillium* species, and published reports have to be treated with caution as misidentifications are frequent. Its overall distribution includes Ireland (1446), Scotland (163, 3918), the Netherlands (1614), Czechoslovakia (1703), Poland (6531), Italy (3913), Israel (654, 2762, 2764, 2768, 4759), different regions in the USSR (432, 3365, 4474, 4677), the USA (1166, 4918, 6351), India (3740), Australia (2091), New Zealand (5812), South Africa (1555, 1556), central Africa (3063) and Honduras (2035). It is known from beech (272, 4537) and spruce forests (3975), mixed alder-conifer and pure alder stands (6351), soils of a forest nursery (6184), heathland (719), tundra (1446), peat fields (4474), grassland (1555, 5812), arable soils (2482, 2762, 2768, 2786, 4554), paddy rice fields (3913), caves (3365) and a uranium mine (1703). Its frequency may be increased by applications of N and NPK (2895). In soils under wheat in Germany, it was found to be very rare but it is known from sewage-irrigated fields near Berlin (2467). It has also been reported from estuarine sediments of various salinities (655), polluted streams (1154, 1157) and sea-water in the eulittoral zone (4918). It has been observed close to the soil surface (2035) and also at greater depths (3975, 4554). It has been isolated from the rhizospheres of corn (4553, 4554), *Lolium perenne* (5815), groundnuts (2768), clover (3429), tomato (16), banana (2035) and oak (5394). Other substrates include pine litter (681), wood pulp (6168), stored wheat grain (3491) and groundnut seeds (2765), olives (4545), corn silages (1092) and frozen fruit cake (3153). As a producer of gliotoxin, this species was, together with *P. nigricans*, found to be responsible for the fungitoxicity of a heathland soil (719). — It can also serve as food for the woodlouse *Oniscus asellus* (4007).

Little growth occurs at 5°C and none at 37°C; good growth occurs at −100 bars water potential (4578), and −240 bars are still tolerated when incubated at 10°C; no germination is possible at water potentials less than −350 bars (6344). — Decomposition of pectin (4545, 4779, 5688), starch, gelatine (654, 4545) and cellulose (3618) has been demonstrated; lipolytic activity has been observed (4545). *P. jensenii* produces gentisic (3231), succinic, citric, tartaric and oxalic acids (4009); respiration is stimulated by the presence of montmorillonite in the culture medium (5604); it can utilize N from a urea-formaldehyde fertilizer (4010), and nitrite as an N source (10); coumarin (439) as well as *o*-coumaric acid can act as sole C source, in the latter case with the formation of 4-hydroxycoumarin and dicoumarol (439, 5500); it produces gliotoxin (719) and citrinin (5278). — It is reported to inhibit *Staphylococcus aureus*, *Candida albicans* (3599), *Botrytis aclada* (719), *Rhizoctonia solani* (1431) and *Aspergillus fumigatus in vitro* (569), although other reports indicate that it is inactive against fungi (516); it is, however, antagonistic to a number of bacteria (515, 1092, 4550, 5394). It has been isolated from sclerotia of *Sclerotium cepivorum* (3990). Effects of stimulation and inhibition have been shown of both corn (4557) and tomato seedlings (5779).

Penicillium lanosum Westling *1911* — Fig. 254.

DESCRIPTIONS: Raper and Thom (4745), and Samson *et al.* (5029). — Colonies reaching 2 cm diam in seven days at 24°C on CzA and MEA, floccose, cottony, at first white but later becoming pale greyish green, reverse colourless to drab. Conidiophores arising from submerged or aerial hyphae, smooth-walled or delicately roughened, two-stage branched. Conidia globose to subglobose, finely granular, 2·5–3 μm diam. — Similar conidiophores are found in the fasciculate *P. hordei* Stolk *1969*. *P. lanosum* is probably also related to *P. jensenii* (q.v.), both species having similar slender phialides with well-defined, short but thin, conidium-bearing necks. Branches in *P. lanosum* often arise low down the conidiophore and so do not appear as an integral part of the terminal penicillus; thus they approximate to the branching type seen in *P. jensenii*.

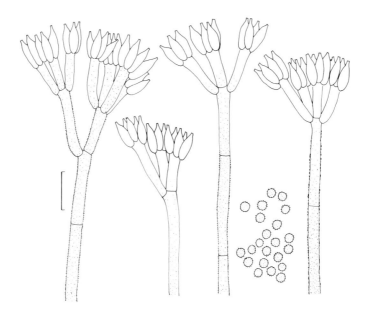

FIG. 254. *Penicillium lanosum*, penicilli and conidia, orig. R. A. Samson (from 5029).

This fungus probably has a worldwide distribution and has been reported from Spitsbergen (3064), Alaska (1171, 3062), Canada (3959, 3962, 5363), the USA (1166, 3817), the British Isles (2736), Poland (273, 4137), the USSR (2871, 4548, 6571), Czechoslovakia (1700, 1702, 1703), Austria (1876, 5678), France (4162, 4921), Italy (3913, 4538, 5047), Spain (3446), Egypt (8, 100), Israel (2764, 2768, 4759), the Negev (654), Iraq (92), Kuwait (4000), South Africa (4407), Zaïre (4159), Central America (1697), Peru (2005), Brazil (379, 399), the Bahamas (4312) and Japan (5846). It occurs equally in forest (273, 1700, 3959, 3962, 5363) and cultivated soils (2764, 2768, 3446, 3913, 4538), also in grassland (4538), even vineyards with high copper content (4921), under steppe vegetation (1700, 1702), heathland (2736, 5047), peat soil (1376, 4180), fens (6571), sand dunes (4162), salt-marshes (4000), recently deglaciated soil (1171), sandstone (3127), desert soil (100), loess (4759), caves

(4137, 4312), a uranium mine (1703), activated sludge (1387), and creeks (1166). — It was the most common fungus on decomposing pine litter (2344, 3801) and other plant remains (4548); also isolated from the rhizospheres of *Saxifraga aizoides* (2530), groundnuts (2768) and pineapple plants (6210), roots of heathland plants (5220), seeds of wheat (3989) and groundnuts (2765, 2768, 2772), pecans (2258, 2572), foodstuffs (1303, 5980) and the integuments of *Dendrolimus sibiricus* caterpillars (2848).

P. lanosum grows at 5°C, but not at 37°C, and some growth is seen at –100 bars water potential (4578). — Cellulose and CMC are degraded (2736, 5076); *P. lanosum* utilizes hydrocarbons from fuel oil (4239) and is known to oxidize decane, undecane, hexadecane and (food) paraffin (3112). — Toxicity to brine shrimps and chicken embryos (1303) and an antibiotic action against *Staphylococcus aureus* (4503), *Beauveria bassiana* and *B. brongniartii* (2848) has been reported; sclerotia of *Sclerotium cepivorum* can be colonized by *P. lanosum* (3990).

Penicillium lividum Westling *1911* — Fig. 255.

DESCRIPTION: Raper and Thom (4745). — *P. lividum* series. — Colonies reaching 3·0–3·5 cm diam after 10–12 days at 25°C on CzA, greyish blue-green, reverse dull peach to flesh on CzA, orange on MEA; conidiophores to 400–600 μm long; phialides more divergent than in the *P. frequentans* series. Conidia coherent in tangled chains, ellipsoidal, later subglobose, clearly roughened, showing spiral bands, 3·0–4·0 × 2·6–3·0 μm. — DNA analysis gave a GC content of 51% (5600).

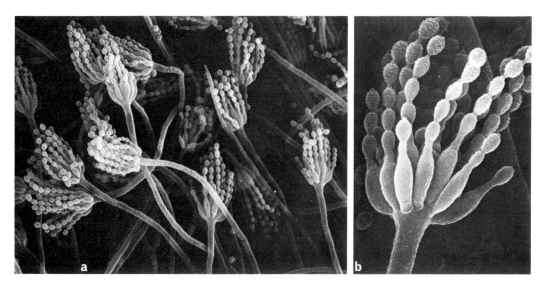

FIG. 255. *Penicillium lividum*, monoverticillate penicilli and ornamented conidia, SEM, a. × 750, b. × 2000; CBS 287.65.

P. lividum is a relatively rare species with a distribution mainly in northern latitudes: arctic tundra (1750), sites near Murmansk (4703) and other places in the USSR (4474, 4548, 4677), Scotland (2344), Ireland (3918), Poland (650), Canada (505), commonly in alpine raw-humus soils (1876), forest steppes near Perm (3362), very rare in Israel (2764), also in Texas/USA (3965), Oregon/USA (6351), Japan (4971), and Peru (2005). It has been isolated from forest soils (6138, 6351), particularly under beech (650), white cedar (505), mixed *Pinus-Chamaecyparis* (4971), spruce (1828) and pine (307), above all in peat bogs (3414, 3918, 4474, 6138) and fens (3413), and rarely dunes (745), arable soils, plant remains and hay (4548). It has also been found in the rhizosphere of tomato (16), on the roots of *Pinus nigra* var. *laricio* (4448) and feathers of free-living birds (2575). — It has been found with higher frequencies in soils treated with hippuric acid (4971).

Growth is very limited at 5°C and none occurs at 37°C; it is able to grow at −100 bars water potential (4578). — Starch, pectin (1750, 5688) and cellulose decomposition have been demonstrated (1828, 2736); an acid carboxypeptidase is produced (6497, 6498); nitrite can be used as a N source (10); lauric acid is oxidized to undecanone-2 (3139). *P. lividum* produces citrinin (4587) with bacteriostatic effects on various organisms (512). — In plate tests inhibitory effects upon *Candida albicans* (2736), *Botrytis cinerea*, *Aureobasidium lini*, *Verticillium tenerum* and *Fusarium oxysporum* have been observed (2956). Culture filtrates were found toxic to wheat (3582). — In a chronically irradiated soil it was present up to intensities of 820 R per day (2009).

Penicillium nigricans Bain. ex Thom *1930* — Fig. 256.

DESCRIPTION: Raper and Thom (4745); conidiogenesis: Cole and Samson (1109). — The members of the *P. nigricans* series are distinguished amongst the *Divaricata* in having phialides with little tapering tips and strongly pigmented conidia which appear brown under the microscope. — Colonies reaching 2·5–3·0 cm diam after 10–12 days at 25°C on CzA and other media, fairly deeply felty, light grey to dark olive-grey; reverse yellow to orange or ferruginous. Conidiophores often arise as branches from long trailing hyphae; metulae strongly divergent. Conidia globose, coarsely warted, 3·0–3·5 μm diam. — The rodlet pattern of the conidial surface has been studied in freeze-etched replicas (2421). — Mycelial proteins have a relatively high content of glutamic acid, tyrosine and proline (3034). Linoleic acid contributes more than 60% of the fatty acids in the mycelium (3035).

P. nigricans is a very common soil fungus although there are relatively few reports from warmer zones including ones from Italy (3538, 3913, 5047), Israel (2762, 2764, 2768), Syria (277, 5392), Egypt (8, 3993), Peru (2005), Brazil (379, 399), the Bahamas (2006), Japan (5846), tropical Australia (978), and Indian forests (1525, 2186, 2853, 2854, 5000). It is extraordinarily widespread in the temperate zone in all kinds of forest, grassland and arable soils; it is also known from heathland (989, 2736, 3720, 3721, 5047, 5220, 5811), peat soils (163, 1376, 1438, 2502, 3850, 3918, 5559), dunes (745, 4162), burnt and unburnt soils with chaparral vegetation (1164), soils after prescribed burning (2822), carst caves (3364, 3365), alkaline soils (4030, 5349), mangrove mud (4700), rice fields (1519), estuarine sediments of medium salinity (655), salt-marshes (8), polluted streams (1154, 1157, 1166), sewage-treated

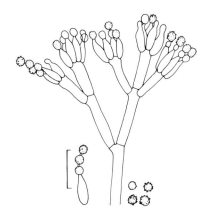

FIG. 256. *Penicillium nigricans*, divaricate penicillus, phialide and conidia.

land (2467), river sediments and river water (1162). The viable spore content generally decreases with soil depth, although an accumulation at 10–15 cm has also been reported (163); in isolated cases very great soil depths have been reached (159, 1525, 3630, 3720, 6182, 6184). It has been reported from soils of a wide pH range (3871, 6182) but the optimum is given as pH 5–6 (399, 1204). It has been found in numerous soil types (3414); it is equally widely distributed in mull, moder and mor (4815) but also known in chalk soils (4152). It was more frequently isolated from washed mineral than organic soil particles (1614, 1888). It occurs at a wide range of humidities (977, 2095, 4313), but more frequently in medium and dry habitats (1204, 4313) and is resistant to summer drying (5000). Steam- and formaldehyde-sterilized soil is recolonized by *P. nigricans* at a slow rate (6184), but it has a relatively high competitive saprophytic ability (2791) and is not much affected by soil mycostasis (633). The viable spore content was found to be reduced by both lime and farmyard manure but increased by N fertilization (1438, 3498, 3849), it also shows a high tolerance for various soil fungicides (1426, 6135, 6136). It was reduced after repeated cultivation of corn (1330) but increased in numbers by wheat cultivation (1433). — *P. nigricans* has been isolated from the litter of Scots pine (2344), remains of barley, cabbage and potato plants (3113), the phylloplane of vascular plants (296), roots of broad bean (6134), the rhizospheres of the fern *Struthiopteris* (2858), citrus (3630), sunflower (4305), oak (4816, 5394), hazel (1794), various forest herbs (4814), Australian heathland plants (5819), peat bog plants (1376), coriander (1523), clover (3429), groundnut (2768), cucumber, rice (4726), grasses (1614, 5815), barley (5514), wheat (1614, 2949, 3567, 4727) and corn (4554). It is also known from garden compost (1424), birds' feathers, roosts and nests (682, 2575), rabbit dung (6085), fruit juices (5205), frozen fruit cake (3153), pecans (2572), wheat and sorghum grain (3989, 4492).

The optimum temperature for growth is 25°C, with little growth occurring at 5°C, a maximum of 33°C, and no growth at 37°C (517, 1204, 4578). — Good decomposition of starch (1425) and pectin, with variability between different isolates, has been demonstrated (1432, 3539). There is a series of positive reports for cellulose fibre and CMC decomposition (1425, 1432, 2006, 2736, 2948, 3618, 5386); it can also utilize hydrocarbons from fuel oil (3384, 4239). 1,3-β-D-Glucanase is formed (4778). It can grow on 5% tannin solution (1206, 1425) and on coumarin as well as on *o*-coumaric acid with hydroxylation to 4-

hydroxycoumarin (439). The formation of succinic, citric, tartaric and oxalic acids occurs in liquid culture (4009). Production of the piperazine derivatives cyclo-di-(L-phenylalanyl) and mycelianamide (3231), *meso*-erythritol, dechlorogriseofulvin (5278), and of the antibiotic griseofulvin in pure culture (708, 3457, 3458), sterile soil (6449) and also possibly in non-sterile soil after extremely high rates of inoculation (723) has been demonstrated. Good N sources for griseofulvin production *in vitro* are $NaNO_3$ and also most amino acids (3034). — The research on griseofulvin has been reviewed (707): it has strong antagonistic action against various bacteria (4550, 5394), human-pathogenic fungi (516, 569) and also *Neurospora sitophila, Fusarium sporotrichioides* (5394), *Phytophthora cactorum, Rhizoctonia solani, Gaeumannomyces graminis, Pythium ultimum, Fusarium culmorum, Pseudocercosporella herpotrichoides* (1425, 1431), *Candida albicans* (2736) and *Botrytis aclada* (716). Soil inoculation is said to be effective against *Chalara elegans* (3635) while inoculation of mustard seed material is reported to protect it against *Pythium* attack (6450). Griseofulvin possesses strong phytotoxic activity (723) and, in particular, causes root stunting in wheat (3849, 3851, 5591). Barley (293) and oat seedlings were injured in soil in which the fungus was accumulated to a high degree (723). Injuries on pea roots were observed *in vitro* (1430); corn seedlings were partly stimulated, partly inhibited (4553). A mycotoxic action has been reported (1303). — *P. nigricans* has a relatively low tolerance to CO_2; in an atmosphere of >5% CO_2 growth was considerably inhibited (784, 1502). Both conidia and mycelium are relatively resistant to γ-radiation (5748). In a chronically irradiated soil there was a significant drop in numbers of viable cells at 120 R per day; no conidia survived at 1400 R per day (2009).

Penicillium oxalicum Currie & Thom *1915* — Fig. 257.

DESCRIPTION: Raper and Thom (4745). — *P. oxalicum* series. Colonies spreading, reaching 3·5–5·0 cm diam in ten days at 24°C on CzA, faster on MEA, dull green, reverse uncoloured or pink. Conidiophores typically two-stage branched. Phialides cylindrical, tip distinctly tapering. Conidia often forming deep crusts which appear silky, strongly ellipsoidal, smooth-walled, 4·5–6·5 × 3–4 μm. — The closely related *P. atramentosum* Thom *1910* is distinguished by its maroon reverse. — DNA analysis gave a GC content of 53% (5600).

P. oxalicum is a cosmopolitan species, although the available data indicate a preferential occurrence in warmer climates: Pakistan (60, 4855), India (967, 2853, 2854, 3732, 4371, 5622), Hawaii (3264), Japan (2532, 4987), China (3475), southern states of the USA (660, 2482, 3817, 4918), but also North Carolina (655), Oregon (6351), and Illinois (5671), and in addition the Bahamas (4312), Honduras (2031), Brazil (399), Iraq (92), Israel (654, 4759), Egypt (8, 3993, 3997), Italy (3453, 3539) and Spain (3446, 3447). Exceptions from this pattern are reports from Alaska (1171), high altitudes in Peru (2005), Northern Ireland (4429), Ireland (3918), Poland (6520), and the USSR (2871, 3364, 4474, 4548, 4677, 6571), including soils in the Pamir (3850). *P. oxalicum* has been isolated from a multitude of soils and habitats, including forest soils (2482, 3817, 6351) under beech (3538) and teak (2853, 2854), cultivated soils (3817) with wheat (3446), sugar cane (3732), cotton (2871), banana plantations (2031) and orchards (3539); a possible reduction in numbers after growing alfalfa is indicated (1330). It is also reported from peat bogs, peat fields (3850, 3918, 4474, 6520),

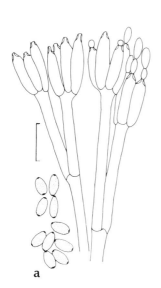
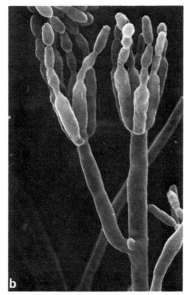

Fig. 257. *Penicillium oxalicum*. a. Penicillus and conidia; b. penicillus, SEM, × 1500; CBS 347.68.

fens (6571), dunes (4371), salt-marsh (8, 4987), desert soil (654, 3997), caves (3364, 3453) with guano deposits (4312), a mangrove swamp (3264) and other marine habitats (655, 4918, 5316), standing and running polluted waters (1166, 4429), and a sewage treatment plant (1165). It has been found on a granite outcrop (660), in loess (4759), chalk soil (4152), podzols (3414, 3416) and solonetz (4677). — There are only limited data on its distribution in other habitats or substrates which include the rhizospheres of barley (3526), *Brassica nigra* (3870), *Trifolium alexandrinum* (1963), groundnut (2532), banana plants (2030), ferns (2858) and sand dune plants (4371), roots and leaves of cotton (5009) and seeds of corn (3069, 3070, 3875, 3989) and wheat (3989, 4492), pecans (5092), hay, coarse fodder, plant remains (4548), nests and feathers of free-living birds (2575), and brown coal (4474); it has also been reported as frequent in the slime of paper mills (5060) and on tan liquors (1260).

Optimal growth occurs at 20–25°C with little at 5°C and good growth at 37°C, also at −100 bars water potential (30, 4578); conidia require about 23°C and may tolerate −200 bars for germination (3876). On NaCl-containing media an osmotic potential of −90 bars (10% NaCl) was found unsuitable for both growth and germination (4987). — The concentration of respiratory enzymes increases during germination (4296). Glucose oxidase activity has been described (1894). *P. oxalicum* can grow on nitrite as a N source (10). Cellulases (386), pectinases (2239, 3538, 3539) and an acid carboxypeptidase are known to be produced (6497, 6498). The conditions for lipid production at different temperatures have been investigated (3736). On media with high sugar contents particularly large amounts of oxalic acid accumulate (1241). On cornmeal both the hepatotoxic secalonic acid D and oxaline are produced (5574). — *P. oxalicum* has some antibiotic activity against fungi (716) and bacteria (1760); contaminated grain was found to be toxic to young ducklings (5158); phytotoxic properties have also been described (6542). Seedlings growing from infected corn develop seedling blight, and injured corn ears develop a rot for which *P. oxalicum* is mainly responsible (3070, 3071).

Penicillium purpurogenum Stoll *1904* — Fig. 258.

DESCRIPTION: Raper and Thom (4745). — Section *Biverticillata-Symmetrica.* — Colonies growing restrictedly, reaching 1·5–2·5 cm diam in 14 days at 25°C on CzA (more on MEA); sporulating areas deep yellow-green, reverse deep red to dark reddish purple with the pigment diffusing into the medium. On MEA sometimes with aromatic odour suggesting apples or walnuts. Conidiophore stipes hyaline, conspicuously encrusted. Conidia ellipsoidal at first but later becoming subglobose, apiculate, irregularly roughened, thick-walled, sometimes showing distinct transverse bands, 3·0–3·5 × 2·5–3·0 µm. — DNA analysis gave a GC content of 54·5% (5600).

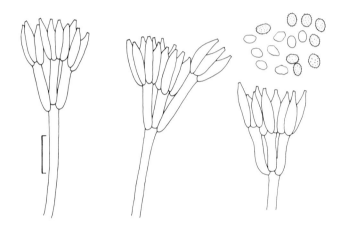

FIG. 258. *Penicillium purpurogenum*, penicilli and conidia, CBS 257.37, orig. R. A. Samson.

P. purpurogenum has a worldwide distribution as indicated by reports from Alaska (1171, 1750), Ireland (4182), the USSR (628, 1637, 3365, 3652, 4548, 4677), Czechoslovakia (1700), Poland (269, 272, 273, 4137), Germany (825), Austria (3413, 3418), France (242), Italy (270, 271, 3539, 4538, 6082), Turkey (4245), Syria (5392), Israel (2764, 2768), Egypt (8, 100, 3993, 3997), Canada (540), the USA (1039, 1166, 1330, 2482, 2573, 3817, 4313, 6351), the Bahamas (2006), Honduras (2031), Brazil (399), central Africa (3063), Zaïre (2967), Pakistan (60, 61), India (3732, 4995), Japan (2532), Hawaii (3264), Polynesia (5983) and Australia (978, 3720). It has been isolated from uncultivated (540), recently deglaciated (1171) and forest soils (825, 1700, 2482, 2573, 3063, 3652, 3817, 6351) including soils under beech (270, 271, 272, 273) and pine (269, 6352), conifer swamps (1039), *Sphagnum* bogs (6082), cultivated land (3652, 3817) under rice (1519), alfalfa, corn (4538), with higher incidence under corn than under alfalfa or sugar beet (1330); further found in soils under groundnuts (2768) and sugar cane (3732), vineyards (242), citrus (2764), banana (2031) and coconut (2006) plantations, orchards (3539), grassland (4538), mesic soils with steppe type vegetation (4313), heathland (3720), and caves (3365, 4137). It is commonly isolated from podzols (825, 3413, 3416, 3652), and also reported from rendzina, terra fusca, pseudogley, gley (3414, 3418), solonetz (4677), sandy soil (4184), desert soil (100, 3997), polluted water (1166) and a mangrove swamp (3264). It has been isolated from litter of barley, cabbage, potato (3113), bean plants (4548), deciduous trees (650), *Abies firma* (4984), and the moss *Pseudoscleropodium purum* (2969); also from sticks exposed to seawater (4546), the rhizospheres of banana (2035), groundnut (2532, 2768), tomato (16), several other crop (3737) and steppe plants (3376), stored seeds of wheat (2716, 3989), corn (3875) and groundnuts (2765, 2772), strawberries (342), dried nuts and fruits (2572, 6252), nests and feathers of free-living birds (2575), leather (6326), and optical lenses (2941).

Optimum temperature for growth is given as 30°C (3876) and it is still able to grow at 37°C but not 5°C; *in vitro* hardly any growth occurs at -100 bars water potential (4578) although hair baits in soil were colonized at -70 to -210 bars (977). Blue light is inhibitory to growth but stimulates conidium production (5750). — *P. purpurogenum* is regarded as an acidophilic fungus (1825). Cells were found to contain appreciable amounts of free D-amino acids (506); nitrite can be used as a N source (10). Cellulose, pectin and starch are degraded (1750, 2006, 4184, 5076) and amylase production has been studied in detail (3284); it produces riboflavin-a-glucoside (5662), an acid proteinase (3231), and carboxypeptidase (6498), grows on tannins (3052, 3315) and degrades the herbicide simazine (2894). When grown on tryptophan, a red pigment is produced (6209); other known pigments are the binaphtho-quinones purpurogenone (4869) and deoxypurpurogenone (4868), and purpuride (2995). Further products are kojic acid (4458), gluconic acid (2147), glauconic acid (6358), rubratoxin A and B (4113) and possibly related phytotoxins (628) as well as rugulovasin A and B (3231). Culture filtrates were found to contain indole-3-acetic acid (294), to have insecticidal properties (with variability between isolates), to inhibit the germination of vetch, pea, mustard, *Avena fatua*, and wheat (629) and to inhibit the growth of both wheat (3851, 3853, 3855) and barley (293) seedlings; a phytotoxin similar to rubratoxin B has been extracted, which proved to be effective in biotests with wheat seedlings and other plants (628). — *P. purpurogenum* is relatively tolerant of short-term soil-steaming (6183). In a chronically irradiated soil it appeared with increasing numbers at 120 R per day, while no viable cells were found at 820 R per day (2009).

Penicillium restrictum Gilman & Abbott *1927* — Fig. 259.

DESCRIPTION: Raper and Thom (4745). — *P. restrictum* series, which comprises the slowest-growing species of Sect. *Monoverticillata*. – Colonies reaching 2·0–2·5 cm diam in 12–14 days at 25°C on various media, brownish grey with much aerial mycelium and rather few conidiophores; reverse yellow to peach or drab. Conidiophores arising from creeping aerial hyphae, short (to 25 μm long), mostly monoverticillate, rarely irregularly branched, bearing few small phialides with very thin conidium-bearing necks. Conidia in short chains, globose, conspicuously roughened, 2·0–2·5(–3·0) μm diam.

P. restrictum is a common soil fungus (4745) with a worldwide distribution. Records include ones from the Swedish (1446) and Siberian (3871) tundra, alpine pastures (3445), high mountains in Tian Shan and the Altai (5647), soils near Peking (3475), from Ireland (1376), Czechoslovakia (1702), Germany (1433, 1889), Yugoslavia (4850), Italy (271, 3913, 4537), France (918), Spain (3446), Turkey (4245), Israel (2762, 2764), Syria (5392), the east African mountains (3415), South Africa (4407), India (1519), Bangladesh (2712), Australia

FIG. 259. *Penicillium restrictum*, monoverticillate short conidiophores and conidia, CBS 367.48.

(978, 5930), Japan (3267), Honduras (2031), Peru (2005), Brazil (4547), Georgia/USA (3817) and Canada (3959, 6352). Among the numerous reports, ones from various forest and arable soils predominate. In Wisconsin/USA both mesic and dry prairies and hardwood forest have been given as typical habitats (569, 1204, 2008, 4313, 6347). *P. restrictum* is more rarely found in grassland (1700), savannah (4407), heathland (3720, 5220), podzols (432, 3414, 3416), solonetz (4677), fen (3413), peat bogs (1376, 3414, 6520), peat fields (4474), dunes (745), sand blows (4540), desert (4733) and saline soils (3446). *P. restrictum* extends to great depths, to the C horizon or down to 50 cm (540, 3447, 3720, 4554). The distribution of this species seems to be largely independent of soil pH; it was found to be the dominating fungus one year after soil fumigation with monomethyldithiocarbamate, ethylenedibromide or D-D (3631). A stimulation by NPK manuring has been reported (2712, 2764, 2777). It has been regularly and significantly more frequently isolated after growing peas than after wheat or rape (1433). It has also been isolated from a polluted creek (1166), seawater, estuarine and fresh water (4850). Other reports include the litter of pine species (2344, 2923, 6080), and remains of barley, cabbage and potato plants (3113). The rhizospheres of peat bog plants (1376), Australian heathland plants (5819), steppe plants in Kazakhstan (3376), clover (3429), alfalfa (3982), flax (5826), rice (2712), wheat (2949, 5330), corn (4554), mycorrhiza of *Pinus strobus* (1794) and truffle soils (3451). It has also been found associated with aecidiospores of *Cronartium comandrae* (4608) and isolated from feathers of free-living birds (2575).

The optimal pH is 5, the optimal temperature 26–30°C; very limited growth occurs at 5°C and the maximum for growth is 37–42°C (517, 1204, 4578). Colonization of hair baits occurs in soil at –40 to –200 bars (2095).—Nitrite can be used as a N source (10). Growth appears to be inhibited by a CO_2 concentration of $>3\%$ in the atmosphere (784). — *P. restrictum* produces gliotoxin (3231, 5041) and is known to be antagonistic towards bacteria (520, 3414), particularly *Staphylococcus aureus, Bacillus subtilis, B. cereus* var. *mycoides, Proteus vulgaris* (515) and *Erwinia carotovora* (2956). Above average inhibition effects by morphologically slightly deviating isolates on *Rhizoctonia solani, Gaeumannomyces graminis, Pythium ultimum, Fusarium culmorum* and *Pseudocercosporella herpotrichoides* have been reported (1431), although other workers had found no antifungal effects (516). Growth of barley was not influenced by the culture filtrate (293). — *P. restrictum* is relatively resistant to γ-radiation (5748).

Penicillium rubrum Stoll *1904* — Fig. 260.

DESCRIPTION: Raper and Thom (4745). — Section *Biverticillata-Symmetrica.* — Colonies reaching 4–5 cm diam in 14 days, conidial areas yellow-green to grey-green, reverse deep red, the pigmentation diffusing into the agar. Conidiophore stipes hyaline, with smooth or slightly granulose walls. Conidia ellipsoidal (or in some isolates subglobose), smooth-walled, usually not apiculate, $2\cdot2$–$3\cdot5 \times 2\cdot0$–$2\cdot5\,\mu$m.

The worldwide distribution of *P. rubrum* embraces localities in the British Isles (745, 2736), Poland (269, 272, 273, 1421), Czechoslovakia (1700), the USSR (2948, 4677), France (918, 2963), Italy (3445, 3453, 3538, 3913, 3976, 4537, 4538, 5047), Syria (5392), Israel (2764, 2768, 2777), the USA (13, 1039, 1166, 2482, 2573, 3817, 4918), Peru (2005), Brazil (379), Central America (1697), Tahiti (6084), Australia (3719, 3720), New Zealand (5812), Somalia (5048, 5049), equatorial West Africa (1420), central Africa (3063), Zaïre (2967), the Ivory Coast (4718, 4719), India (3583, 4716, 4995) and Pakistan (60). Reports from forest soils are particularly frequent (1700, 2482, 2573, 2948, 2963, 3063, 3817, 4716, 4995, 5048), especially under beech (272, 273, 3538, 4537) or pine (269). It is also known from grassland (2573, 3445, 5812), cultivated soils (1420, 3817) under rape (918), alfalfa, corn (13, 4538, 4716), paddy rice (3913), soya (13), groundnut (2768), citrus (2764, 2777), but also soils under steppe type vegetation (1700, 4718, 4719, 5049, 6347), heathland (2736, 3720, 5047), conifer swamps (1039), sand dunes (745), caves (3453), an alpine soil with long snow coverage (3976), lowland (3414), podzol (3719) and solonetz soils (4677). The frequency of *P. rubrum* in soil is reported to be stimulated by the application of NP and NPK fertilizers (2777). Other reports include isolates from seawater of the eulittoral zone (4918), polluted fresh water (1166), sewage sludge (1165, 1167), garden compost (1424), children's sandpits (1421), soil around human dwellings (1420), the rhizospheres of groundnuts (2768), *Trifolium alexandrinum* (1963) and other cultivated plants (3737), seeds of wheat (4492) and tomato (5899), seeds and pods of groundnuts (1984, 2772), pecans (2572), grapes (342) and frozen fruit cake (3153).

The optimum temperature for growth is 26–28°C, none occurring at 5°C, but with good growth at 37°C, and a maximum at 41–42°C; little growth occurs at −100 bars (517, 4578). Blue light is inhibitory to growth but stimulates conidium production (5750). — This species is able to degrade cellulose (2736, 2948), it grows on tannin (1425), utilizes nitrite as an N source (10), produces levanase (3412) and an acid proteinase (3231), and can utilize hydrocarbons from fuel oil (4239). Metabolites reported include the pigments (−)-mitorubrin and (−)-mitorubrinol (770, 1019), phoenicin (799), tetrahydrophoenicin, ergosta-4,6,8(14),22-tetraen-3-one (3231), kojic acid (4458) and the rubratoxins A and B (1618, 3985, 3986, 5869). — Toxicity of *P. rubrum* to swine (792, 6391, 6413) and ducklings (5158) has been frequently reported. It has an antibiotic activity against both bacteria (516) and fungi (3719); rubratoxin B interferes with wall synthesis in various fungi, causes hyphal deformations in *Rhizopus stolonifer, Aspergillus flavus* and *A. niger*, leading to plasmoptysis in the latter, and also inhibits sporulation (4808). Growth of rape plants was slightly stimulated in greenhouse experiments (1427). — In chronically irradiated soils ranging from 53 to 2780 R per day, there was a constant decline in numbers of propagules (2009).

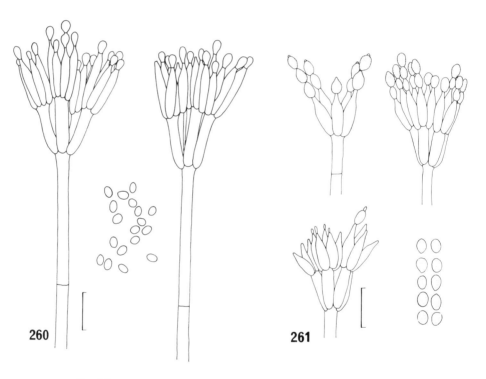

FIG. 260. *Penicillium rubrum,* penicilli and conidia, orig. J. Veerkamp.
FIG. 261. *Penicillium rugulosum,* penicilli and conidia.

Penicillium rugulosum Thom *1910* — Fig. 261.

DESCRIPTIONS: Raper and Thom (4745), and Udagawa and Takada (5984). — Section *Biverticillata-Symmetrica.* — Colonies growing restrictedly, reaching 2·5–3·0 cm diam in 14 days at 25°C on MEA, 1·0–2·5 cm on CzA; conidial areas yellow-green to dark green, reverse variously coloured, yellow to orange-brown, but never red. Conidiophore stipes smooth-walled, penicilli usually symmetrically branched, sometimes irregular. Conidia ellipsoidal, apiculate, conspicuously roughened, 3·0–3·5 × 2·5–3·0 μm. — The rodlet pattern of the conidial surface is interlacing and often raised to give the distinctly roughened surface (2421). — DNA analysis gave a GC content of 50–55% (5600).

P. rugulosum is one of the most widely distributed species of the *Biverticillata-Symmetrica*, though considerably rarer than many asymmetric species. — In addition to the temperate latitudes, with records from Canada (505), the USA (1166, 1330, 2573, 6351), Ireland (1376), the Netherlands (1614, 1616), France (242), Poland (1421, 1422) and the USSR (774, 2871, 4474, 4677, 6571), it has also been observed in warmer regions including Yugoslavia (4850, 5534), Italy (3538, 3913), Turkey (4245), Spain (3417), Israel (2764), Iraq (92), Egypt (3997), Zaïre (3790), Polynesia (5983), Australia (3720), India (1519), Bangladesh (2712), Sri Lanka (4021), the Bahamas (2006), and Peru (2005). Reported habitats include forest soils (505, 1700, 2573, 3538, 3549, 3790, 3965, 5678, 6351), heathland (3720, 5047), grassland (1700, 3417, 3790, 4021), arable soils (1519, 1614, 2871, 3450, 3913, 6138), paddy fields (774), vineyards (242), citrus plantations (2764), cut peat (1376), peat fields (4474), fen (6571), solonetz soils (4677), desert soil (3997), children's sandpits (1421), fresh and seawater, estuaries (4850, 5316) and polluted streams (1157, 1166). It has been observed at soil depths to 100 cm (3720). The frequency of *P. rugulosum* is reported to be increased by NPK fertilizers (2712) and growing corn (1330). It has also been found on remains of cabbage and potato plants (3113), in the rhizospheres of rice (2712), wheat, oats, alfalfa and peas (3982, 5825), and on the root surface in wheat (2023). This species is also frequent on damp hay and straw (4548), seeds of wheat (3491) and rice (3181), stored apples (342), in flour-based foodstuffs (5980), and reported as associated with aecidiospores of *Cronartium comandrae* (4608); also found on rabbit dung (6085), cotton, textiles, wooden furniture (3707), feathers of free-living birds (2575) and brown coal (4474).

Little growth occurs at 5°C and none at 37°C; moderate growth is reported at −100 bars water potential, and the minimum for vegetative growth is −280 bars and for sporulation −210 bars (2093, 4369, 4578, 5984). Optimal growth occurs at osmotic potentials of −20 to −50(−90) bars on media containing NaCl (5316).—*P. rugulosum* can decompose various forms of cellulose (2006, 4433, 5076, 6085) but not cellulose fibres (3618), degrade xylan (3231), utilize thiourea (3230) or nitrite (10) as sole N sources, degrade the pesticides atrazine (2892), chlorphenamidine, dicryl, propanil, propham and solan (2893), and grow on polyethylene plasticizers (5148). Metabolites include the antibacterial and antimycotic substances rugulosin (664, 3061, 5994), which has hepatotoxic properties, tardin (512), spiculisporic acid (5278), skyrin, luteoskyrin (cf. *P. islandicum*) (664, 692) and the toxic rugulovasins A and B (3231). — It showed a pronounced inhibitory effect against *Pseudocercosporella herpotrichoides* (1431) and some bacteria *in vitro* (515). Hyperparasitism on *Aspergillus niger* has been observed (4904).—Growth was reported to be little affected by 10% CO_2 (784), but this observation was not confirmed by other studies (4029).

Penicillium sacculum Dale *1926* — Fig. 262.

≡ *Eladia saccula* (Dale) G. Sm. *1961*
= *Scopulariopsis verticillioides* Kamyschko *1961*

DESCRIPTION: Smith (5433), who regarded this species as related to *Scopulariopsis*, but conidiogenesis is strictly phialidic. This species can be placed in the *P. restrictum* series on the basis of the phialidic structure and the short conidial chains. — Colonies reaching 2·2 cm diam in ten days at 20°C on MEA, thinly velvety, grey-green, often with white sectors, reverse uncoloured or greyish. Conidiophores short, monoverticillate or irregularly branched; phialides inflated, 5·5–6 × 3·0–3·5 μm; conidia forming short chains, globose, coarsely ornamented, 5–6 μm diam.

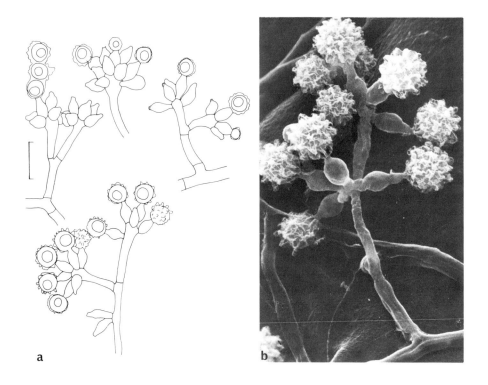

FIG. 262. *Penicillium sacculum*. a. Penicilli and conidia; b. penicillus, SEM, × 2000, CBS 449.69.

This species has been isolated only rarely and by few authors. Originally it was reported from mountain peat in Scotland (1259), but has subsequently been found in other locations of the British Isles (989, 3567), the Netherlands, Belgium (CBS), Austria (1876), Spain (5433), Germany, Denmark (CBS), the USSR (5252), Canada (505), Peru (IMI), central Africa (3063) and India (427). Known habitats are alpine raw humus (1876), forest soils (3063) under white cedar (505) and pine (989, 5252), agricultural soil (CBS), heathland (989), lead mine spoil heaps (IMI), wheat rhizosphere (3567) and potato roots (CBS). — The conidia of *P. sacculum* can be induced to germinate by heating (6187).

Penicillium simplicissimum (Oudem.) Thom *1930* — Fig. 263.

≡ *Spicaria simplicissima* Oudem. *1903*

DESCRIPTION: Raper and Thom (4745). — *P. simplicissimum* is distinguished from other species of the *P. janthinellum* series by the denser ramification of the conidiophores, the conspicuously roughened stipe, branches and conidia, and the uncoloured to yellow reverse. — Colonies reaching 4·0–4·5 cm diam in 12–14 days at 24°C on CzA, 5–9 cm on MEA. Penicilli commonly consist of more or less well-defined terminal verticils of 2–4 divergent metulae bearing phialides. Conidia subglobose, finely echinulate, 2·5–3·0 μm long. — DNA analysis gave a GC content of 52% (5600).

A worldwide distribution of *P. simplicissimum* is evidenced by reports from Alaska (1750), the USSR (2948, 3365), Czechoslovakia (1700, 1702, 1703), the Netherlands (1614, 1616), the British Isles (745), France (2161), Austria (5678), Romania (4378), Italy (4538, 5047, 6082), the USA (660, 1166, 2007, 2008, 2482, 3817, 4313, 4918, 6351), Peru (2005), Brazil (398, 399), Somalia (5048), central Africa (3063), South Africa (1555, 4407), India (2178, 2186, 3732), Nepal (1826), Hong Kong (1021), Japan (5846) and Hawaii (296, 3264). It is known from forest soils (1021, 2482, 2948, 3063, 3817, 5048, 6351), floodplain communities with *Salix* and *Populus* (2007, 2008), stands with *Acacia* (2186), grassland (1555, 4538, 6082), rarely cultivated soils (1614, 1616, 2161, 3732, 3817), frequently soils under steppe vegetation (1700, 1702, 4313, 4407), heathland (5047), acid sand dunes (745), carst caves (3365), a uranium mine (1703), a granite outcrop (660), alpine podzols (4378), the eulittoral range (4918), and a mangrove swamp (3264). The reported soil pH comprises a range of 5·8–7·4 (4538, 5047, 5048). This species has occurred in potato fields at higher frequencies after barley crops than after wheat (1617) and it has been preferentially isolated from washed mineral particles (1021). It is also known from the phylloplane of vascular plants (296), the rhizospheres of banana (2030), clover (3429), *Stipa lessingiana, S. sareptana* and *Festuca sulcata* (3376), and sticks buried in soil (1956).

Limited growth occurs at 5°C, good growth at both 37°C and −100 bars water potential (4578). — *P. simplicissimum* can utilize pectin, starch (1750, 1827), cellulose (2948, 3618) and tannin (1206) and produces an acid carboxypeptidase (6498). It can grow on humic acid (1750) and utilize nitrite as the sole N source (10). Rock powder is reported to be solubilized by this species with the liberation of Ti^{4+} (5327), Si^{4+}, Al^{3+}, Fe^{3+}, Mg^{2+} ions (5326). — In the presence of *P. simplicissimum*, growth of axenically grown *Tagetes erecta* was significantly stimulated and root exudation was reduced in the early stages of development (2238).

Penicillium spinulosum Thom *1910* — Fig. 264.

DESCRIPTION: Raper and Thom (4745). — *P. frequentans* series. — Similar to *P. frequentans* (q.v.) but differing in the somewhat more floccose colonies with some white aerial mycelium, and the usually less coloured reverse. Colonies reaching 4·5–5·5 cm diam in 12–14 days at 24°C on CzA, 6–7 cm on MEA. Conidiophores occasionally form an additional branch; conidia more strongly roughened than in *P. frequentans,* 3·0–5·5 μm diam. — Acid hydrolysates of the cell walls contained glucose, galactose, glucosamine, and oligosaccharides (4499).

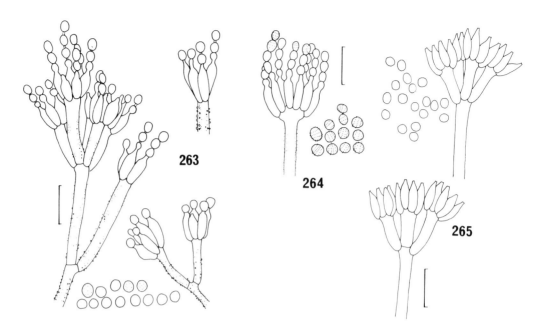

FIG. 263. *Penicillium simplicissimum*, divaricate penicilli and conidia.
FIG. 264. *Penicillium spinulosum*, monoverticillate penicillus and conidia.
FIG. 265. *Penicillium steckii,* penicilli and conidia, CBS 260.55, orig. R. A. Samson.

P. spinulosum is a very common species with a worldwide distribution principally in forest soils. It extends from sediments of recently glaciated areas in Alaska (1171) and soils near the White Sea (3368), East Siberia (3652) and the Amur region (432), through the temperate latitudes with records from the British Isles (1376, 1438, 3234, 4429, 5221, 5222, 6381), the Netherlands (1614, 1616), Czechoslovakia (1702, 1703), the USSR (3364, 3365, 3366, 3367, 4365, 4474, 4677, 6571), the USA (1032, 2007, 2008, 2922, 3817, 6351), Canada (5363), Italy (3538), Egypt (3993), the Chad region and the Congo basin (3415), South Africa (4407), Pakistan (4855), Bangladesh (2712), India (1519, 3330, 4736), Australia (977, 2091), New Zealand (5930), Japan (3267) and Brazil (4547). In the USSR it has been reported mainly from podzols, grey forest soil, chernozem (432, 3871) and solonetz (4677). It has also been recorded for coniferous forests (432, 3267, 3975, 5363, 6351) and conifer swamps (1039) and frequently encountered under pine species (1637, 3447, 3798, 6381) where it is especially abundant in the litter layer (269, 2344, 2923, 2925, 6383), although still relatively well represented in deeper layers (3447, 4407, 6383). It has also been isolated from conifer-hardwood forests (1030, 1032), stands with beech (272, 2719), (mixed) oak (1702, 5811), *Salix, Populus* (2004, 2007, 2008), and maple-elm-ash floodplains (4225). In deciduous forests it has also been observed in the litter layer (2411, 3538). *P. spinulosum* occurs in chalk soils (4152), in tundra and taiga (3366) and soil under steppe vegetation (1702, 4407) but is particularly frequent in heathland soils (989, 3720, 5220, 5221, 5222, 5811) and peat (432, 1376, 1438, 3234, 3918, 4365, 4474), even fens (5559), and also in dunes (745), deserts (4733), caves (3364, 3365, 3367) and a uranium mine (1703). There are no reports for grassland soils and it is not found in arable soils with any great frequency (1614, 1616, 2161, 4554). Other reports include rice fields (1519, 2712, 3913), soil treated with sewage (2467), sewage itself (1166), estuarine sediments, even those of high salinity (655), polluted streams (1154, 1155, 4429) and a copper swamp (2922). In forest nursery soils, *P. spinulosum* was found to be one of the first recolonizers after steaming or formalin treatment (3908), although the colonization process is slow (6184). Data for occurrence in the rhizosphere are available for Australian heathland plants (5819), peat bog plants (1376), oak (5394), rice (2712) and clover (3429). It is also found in stored wheat grain (4492), on damp hay (4548), in the phylloplane of vascular plants (296), on litter of *Carex paniculata* (4644), *Nothofagus truncata* (4946, 4947), dead petioles of *Pteridium aquilinum* (1821), decaying *Pseudoscleropodium purum* (2969), *Spartina townsendii* (5389) and decaying pine chips (3140). It is also frequently found on rabbit dung (2279), and reported from flour-based foodstuffs (3930, 5980), frozen fruit cake (3153), fruit juices (5205) and leather (6326). — It can serve as food for the woodlouse *Oniscus asellus* (4007).

The minimum temperature for growth is 6–7°C, the maximum 41–42°C or less, and the optimum 26–28°C (517, 4578). Growth still occurs at −210 bars (2093) or, according to another report, at −280 bars (4369); hair baits in soil are colonized between saturation and −180 bars water potential (977, 2095, 4493). — Pectin (2362, 3538, 5688) and cellulose can be decomposed (2362, 3618, 4184, 5076), the enzyme C_x is formed abundantly but C_1 is absent (4779) and so native cellulose is not attacked (2363, 6326). *P. spinulosum* has been isolated from buried chitin strips (2069) and is one of the few fungi able to decompose cutin (2362, 2363, 3361) and *Sphagnum* moss (962). Nitrite can be used as a N source (10). The production of an acid carboxypeptidase (6498) and a β-glucosidase on salicin (6159) has been established. *P. spinulosum* is able to utilize ferulic acid (544), hydrocarbons from fuel oil (4239), and can even grow on tannin (1206) and plasticizers of polyethylene sheets (5148). Metabolic products detected include spinulosin (151, 533, 4335) which is active against *Chalara elegans* and some other fungi (5440), fumigatin, phyllostine (3231), citromycetin (5278), citric acid (534), small quantities of orsellinic acid, orcinol (4532), tri- and tetra-hydroxytoluene (3231) and undecanone-2 (3139). Strong lipid synthesis occurs on succinic, fumaric and malic acids (5291), glucose, xylose, sucrose, and inulin (1914) as C sources, and NH_4Cl or NH_4NO_3 as N sources in the presence of phosphate and magnesium (1914). — Culture filtrates are found to be inhibitory to various bacteria (515, 2956, 5394) and fungi (2736, 2956, 5394), *Pseudocercosporella herpotrichoides* being very strongly inhibited in plate tests (1431). *P. spinulosum* also occurs as a sclerotium parasite on *Sclerotinia* species (1652). Occasional isolations from keratomycoses have been reported (1992). In one study, the growth of corn seedlings was stimulated by culture filtrates (4557). — Growth is inhibited at >3% CO_2 (784). In a chronically irradiated soil it survived at intensities of ⩾1250 R per day (2009).

Penicillium steckii Zaleski *1927* — Fig. 265.

DESCRIPTION: Raper and Thom (4745). —*P. citrinum* series. —*P. steckii* is very close to *P. citrinum*, but the reverse is uncoloured or dull yellow. Colonies reaching 2 cm diam in 10–12 days at 24°C on CzA and MEA. Conidia globose to subglobose, smooth-walled or delicately roughened, 2·0–2·5 μm diam.

P. steckii evidently has a worldwide distribution and is reported from the arctic tundra

(1750), various regions of the USSR (432, 1637, 2871, 4474, 4677), Czechoslovakia (1700), Poland (272, 6531), the Netherlands (1614), Romania (4378), many times from the USA, also Peru (2005), Brazil (398, 399), Italy (3453, 3538), Syria (5392), Israel (654, 2764, 2768, 4759), Egypt (3993, 3997), Iraq (92), central Africa (3063), South Africa (1555, 1556), Pakistan (60), India (1519, 2186, 2853, 2854, 4698, 4997) and Japan (2532, 3267). It is known from uncultivated (4698) and forest soils (1637, 1700, 2482, 2573, 3063, 3817) under beech (272, 3538), teak (2853, 2854), larch (3267), pine, spruce, oak, birch (432), *Salix-Populus* communities (2007, 2008), dry-mesic conifer-hardwood forests (1032, 6351), wet-mesic deciduous forests (1040), conifer swamps (1039), under *Acacia melanoxylon* (2186), grassland (1555, 1614) or steppe vegetation (1700), cultivated soils (1614, 3817, 4997), under rice (1519), cotton, potatoes, alfalfa (2871), groundnut (2768) and citrus (2764), peat (1376, 4474), desert soils (654, 3997, 4733), caves (3453), solonetz (4677), rendzina (3414), alpine podzol (4378), chalk soil (4152), loess (4759), activated sludge (1387), an acid mine drainage stream (1166), and estuarine sediments (655). It has been isolated from the leaves of vascular plants (296), litter of pine (681, 2344), sclerotia of *Sclerotinia sclerotiorum* (4699), roots of banana (2035) and pines (3572), the rhizospheres of various crop plants (3737) and steppe plants (3376), and from pods (1984), seeds (2765) and the rhizosphere (2532, 2768) of groundnut. Other reports include seeds of wheat (3989) and corn (3875), pecans (2572, 5092), stored grape fruits (342), other foodstuffs (1303, 3153), hawk pellets (6223), and cotton fabric (6326). — *P. steckii* is eaten by the springtail *Folsomia fimetaria* (4007).

No growth occurs at both 5°C and 37°C, but good growth is reported at −100 bars water potential (4578). — Nitrite can serve as a N source (10). Starch, pectin and cellulose can be degraded (1750, 3538, 6326) and it can grow on plasticizers in polyethylene and PVC (5148), dihexyladipate and epoxyplasticizers (5859). It produces curvularin (5278), has toxic effects on mammals (1815), rats, chicken embryos and brine shrimps (1303, 1388), a slight antibacterial activity (515) and inhibits the growth of *Sclerotinia sclerotiorum in vitro* (4699). — In a chronically irradiated soil, the fungus did not occur at intensities of >120 R per day (2009).

Penicillium stoloniferum Thom *1910* — Fig. 266.

DESCRIPTION: Raper and Thom (4745). — *P. brevicompactum* series. — *P. stoloniferum* is close to *P. brevicompactum* Dierckx (q.v.) and linked with it by the occurrence of intermediate isolates. All the conidiogenous structures are more slender then in *P. brevicompactum*, the stipe usually not exceeding 4 μm in width. Colonies reaching 3–4 cm diam in 14 days at 24°C on CzA or MEA. Conidia ellipsoidal but later becoming globose to subglobose or pyriform, finely roughened, 2·5–3·5 μm diam. — Some isolates contain virus particles (676, 2509, 2527, 5043); a virus-infected isolate contained considerably higher amounts (20·1%) of galactosamine than virus-free isolates (0·5–1·1%), the corresponding data for glucosamine were 18·5 and 15·2%, respectively (3232). Isolates producing mycophenolic acid did not contain virus-like particles whilst some non-producing isolates did (1358). Heterokaryosis proved to be an effective means of virus transmission (3320).

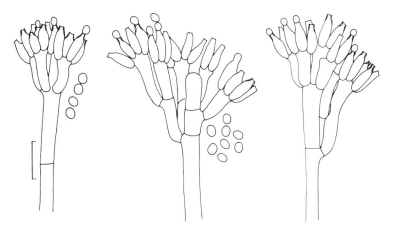

Fɪɢ. 266. *Penicillium stoloniferum*, penicilli and conidia, CBS 227.28.

This apparently rather rare fungus has been reported from the Arctic (1750), the USSR (1637, 3652, 4365, 4474), Ireland (3918), Poland (272, 1421, 1978), Austria (1876), France (2161), Italy (3453, 4538), Syria (5392), Israel (2768), Iraq (92), Egypt (3993), the USA (1032, 1039, 1387, 2573, 4313, 4733, 5671, 6347), the Bahamas (2006), and New Zealand (5812). It is known from forest soils under beech (272), *Casuarina* (2006), dry-mesic conifer-hardwood forests (1032), deciduous forests (2573) and forest nurseries (1978), cultivated soils under wheat (2161), alfalfa, corn (4538) and groundnuts (2768), grassland soils (5812) where it has a high frequency after irrigation (4948), open bogs (1039), peat bogs and peat fields (1259, 3918, 4365, 4474), wet-mesic soils with steppe vegetation (4313, 6347), podzols (3652), chalk soils (4152), desert soil (4733), caves (3453), children's sandpits (1421), and activated sludge (1387). It has been isolated from pine litter (2344), the rhizospheres of *Saxifraga aizoides* (2530), groundnut (2768), tomato (16) and wheat (4727), wheat seeds (4492), stored grape fruits (342), decaying mushrooms, coarse fodder (4548), rabbit dung (6085), sewage (1166) and dead adult bees (*Apis*) (402).

Little growth occurs at 5°C and none at 37°C, but growth is reported at −100 bars water potential (4578). — Starch and pectin can be degraded (1750, 2239), and it is able to utilize nitrite as a N source and grow on dihexyladipate (5859). *P. stoloniferum* produces mycophenolic acid (5278), which has some antiviral activity (1358), and an interferon-inducing principle, statolon (1596), which was identified as virus-RNA (330, 1361, 5496). This species is also said to have some antibacterial activity (515). — Destructive infection of inflorescences of *Poinsettia pulcherrima* in greenhouses has been repeatedly observed in Switzerland (567).

Penicillium thomii Maire *1917* — Fig. 267.

DESCRIPTIONS: Raper and Thom (4745), and Udagawa (5967). — The *Penicillium thomii* series is close to the *P. frequentans* series but differs primarily in the presence of sclerotia. — Colonies reaching 3·5–4·0 cm diam in 12 days at 25°C on CzA, producing numerous hard pink sclerotia (not in clusters) to 300–350 μm diam. Conidiophores often scarce in fresh isolates, but more abundant on hay decoction and PCA, delicately echinulate; reverse pale yellow to pinkish brown. Conidia in compact columns, ellipsoidal, 3·0–3·5 μm long, smooth-walled. — Fatty acid analysis of conidia gave the following results: C_{16} (25·4%), $C_{16:1}$ (3·0%), C_{17} (trace), C_{18} (7·3%), $C_{18:1}$ (21·8%), $C_{18:2}$ (38·1%), and $C_{18:3}$ (4·5%) (1277). The content of long-chain fatty acids has been used to determine species relationships (1281).

According to the available data, *P. thomii* is more widely distributed in temperate than in tropical regions. It has been reported from Canada (540), Alaska (1171), very frequently from other states of the USA and the British Isles, Sweden (1446), Belgium (4816), France (3451), Italy (3452, 4537, 5047, 6082), Spain (3446, 3447), Austria (1876, 3418, 5678), Poland (6520), Czechoslovakia (1700), the USSR (432, 961, 1637, 3366, 3652, 4677, 6571), Romania (4378), Chile (1824), Peru (2005), Brazil (399), South Africa (4407), the Ivory Coast (4718), Zaïre (4159), Somalia (5048, 5049), Libya (6510), Israel (2768), Iraq (92), Australia (977, 978, 3720), New Zealand (5812), India (1519) and Japan (3267, 5846). Isolations have been made from uncultivated soils (540, 3652), recently deglaciated soil (1171), in very numerous cases from forest soils, grassland (163, 2573, 5812), rarely from cultivated land (1519, 2768, 3446), from soil with tundra (1750, 3366) and steppe vegetation (1559, 1700, 4313, 4407, 5049), heathlands (2736, 3720, 5047, 5221, 5222), very frequently from peat bogs, from acid sand dunes (745), but also typical chalk soils (4152), truffle soil (3451, 3452), desert soil (4733), conifer swamps (1039), a copper swamp (2922), podzols (432, 989, 3416, 3652, 4378), solonetz (4677), gley, terra fusca, loess (3414), seawater in the eulittoral zone (4918), estuarine sediments (655), polluted water (1166) and a sewage treatment plant (1165). — It has also been found on decaying petioles of *Pteridium aquilinum* (1821), cotton bolls (5009), litter of *Carex paniculata* (4644), coniferous (681) and deciduous trees (2411), wood of birch, maple, and beech (1295), roots of heath plants (5220), in the rhizospheres of oat (5825), groundnut (2768), tomato (16), peat bog plants (1376) and steppe plants (3376), on brown coal (4474), feathers of free-living birds (2575), slime of a paper mill (5060), stored rice (5967), frozen fruit cake (3153) and fruit juices (5205). — This fungus has been isolated from the integuments of *Dendrolimus sibiricus* caterpillars (2848) and can serve as food for the mites *Acarus siro* and *Tyrophagus putrescentiae* (5383). — Growth is inhibited by oak leaf tannins (2282), but, according to another source, it can also grow on 0·3% gallotannins (3315).

Little growth occurs at 5°C and none at 37°C, good growth is reported at −100 bars water potential (4578). Hair baits in soil can be colonized between −70 bars and −210 bars (977). *P. thomii* has been found able to survive a heat treatment of soil at 70°C but not 80°C for 30 min (614). — Pectin is utilized (1750), cellulose is degraded (2736, 5059), nitrite can serve as a N source (10), and hydrocarbons from fuel oil can be utilized (4239). Hadacidin (= asymmetrin), *N*-formyl-hydroxyaminoacetic acid (6543), and penicillic acid (2888) can be produced, and antibiotic activity against bacteria and fungi (569, 2956, 3599) has been observed. Inhibition of the growth of bean plants can also occur (6542). — In a chronically irradiated soil *P. thomii* appeared with increasing numbers at intensities of 120 R per day; no viable cells were counted at 2780 R per day (2009).

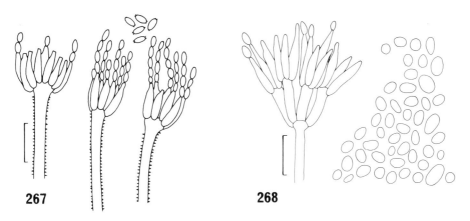

FIG. 267. *Penicillium thomii*, monoverticillate penicilli and conidial columns, CBS 111.66.
FIG. 268. *Penicillium variabile*, penicillus and conidia.

Penicillium variabile Sopp *1912* — Fig. 268.

DESCRIPTION: Raper and Thom (4745). — Section *Biverticillata-Symmetrica*. — Isolates vary in growth rate, mostly reaching 3·0–4·0 cm diam in 14 days at 25°C on CzA. Colonies often covered with sterile yellow aerial mycelium, reverse yellow to orange or greenish brown. Conidia ellipsoidal to fusiform, of different size, smooth or irregularly roughened, especially the larger conidia, mostly 3·0–3·5 × 2·0–2·5 μm, sometimes much larger. — DNA analysis gave a GC content of 50% (5600).

P. *variabile* is widely distributed and records include recently deglaciated soil in Alaska (1171), soils in Ireland (1376, 3918), Northern Ireland (4429), Belgium (4816), France (242, 3451), Italy (3913), the USSR (432, 4304, 4474, 4677), Canada (6352), the USA (1039, 1166, 1387, 3817, 4226, 4313, 4918, 6351, 6414), Syria (5392), Egypt (3993, 3997), Somalia (5048, 5049), the Ivory Coast (4159, 4719), central Africa (3063), Namibia (1827), India (1519, 2853, 2854, 3545, 3732, 4716, 4995, 4997), Pakistan (60), Nepal (1826), Australia (977, 978), Brazil (379, 399), Peru (2005) and Hawaii (3264). In the USSR, tundra, grey forest soils and solonetz are reported as the main areas of distribution (3871, 4677). Records from forest and cultivated soils are roughly equally numerous; examples include stands with beech (272) and teak (2853, 2854), oak forests on mull, moder and mor humus (1700, 4814, 4815, 4816), mixed hardwood forests (1204, 4225), coniferous forests (269, 432, 3362, 3975, 6351, 6352), conifer swamps (1039), Himalayan forests (4995), grassland (1204), cultivated soils (918, 2159, 2161, 3446, 3450, 4538, 4997) with citrus (2764), rice (1519, 3545, 3913), corn (4716) or sugar cane (3732), vineyards (242, 2719), soil with steppe vegetation (4313, 4719), peat soils (1376, 3918, 4474), sand dunes (6414), deserts (3997, 4733), carst caves (3365), heathland (5047), fen (5559), river sediments and water (1162, 4429) even with high pollution (1157, 1166), a mangrove swamp (3264) and seawater of the eulittoral zone (4918).

P. variabile is sometimes found only in the uppermost soil layers (2035, 4995, 5048), but has been reported down to about 40 cm (2161, 3975). In one study of soils under mixed hardwood forests, the greatest frequency occurred at pH 5 (1204). *P. variabile* contributes to the stabilization of soil aggregates (2280). It is also known from decaying stems of *Spartina townsendii* (5389), the late stages of composting *Diplachne fusca* (3546), the rhizospheres of wheat, oats and alfalfa (5825), clover (3429), beans (1369), tomato (16), banana (2035), oak (5394), *Ammophila breviligulata* (6414), the mycorrhiza of hazel (1794), truffle soils (3451), soil around groundnut pods (2768), seeds of wheat, sorghum, corn (3875, 3989) and *Pinus sylvestris* (6344), pecans (2572), sticks exposed to soil (1956), rabbit dung (6085), activated sludge (1387), sewage (1165), wood pulp and paper (6168), fruit juices (5205), and optical lenses (2941).—*P. variabile* shows a relatively high sensitivity to soil mycostasis (2559).

The optimum temperature for growth is 25°C or higher, depending on the geographical origin of the isolates (1827); no growth occurs at 5°C and little growth at both 37°C and −100 bars water potential (1204, 3876, 4578). No growth occurred *in vitro* at either pH 1·6–1·8 or 10·1–11·1 (2784). Hair baits in soil are colonized between saturation and −280 bars water potential (977). — There is evidence for degradation of starch (1827), *a*-cellulose, CMC and methyl cellulose (2878, 6286); the action of purified cellulase preparations (3091, 4364) and the degradation of crystalline and amorphous forms of cellulose have been investigated (123); native cotton cellulose is not attacked (3618). Proteolytic activity and tannin tolerance are high (1206, 1425, 1827); nitrite can be used as a N source (10). *P. variabile* grows on polyvinylacetate (5859), produces ochratoxin A (1057), skyrin and rugulosin (5278), and ferrirubin which either promotes (6589) or slightly inhibits the growth of *Staphylococcus aureus* (6528). — Antagonistic activity of *P. variabile* towards various fungi (1369, 1425, 1431) and bacteria (515, 4304, 5394) has been reported. Cereals contaminated with this fungus proved toxic for chickens (5158). The mycelial homogenate has been found to inhibit the growth of tomato seedlings, while the culture filtrate stimulated it (5779). — In a chronically irradiated soil it showed a drop in numbers of viable cells at 53 R per day, no cells surviving at 1400 R per day (2009). *P. variabile* did not survive a heat treatment of 60°C for 1 h (6344).

Penicillium verrucosum Dierckx *1901* var. *verrucosum*

= *Penicillium viridicatum* Westling *1911*
= *Penicillium lanoso-viride* Thom *1930*

DESCRIPTIONS: Raper and Thom (4745), and Samson *et al.* (5029). — Several "species" of the series *P. ochraceum*, *P. viridicatum*, *P. cyclopium*, *P. terrestre* and partly *P. expansum* and *P. granulatum* have been combined in an expanded species concept by Samson *et al.* (5029), for which the oldest available name is *P. verrucosum* Dierckx. The use of this hitherto almost forgotten name will avoid confusion between former and present species concepts. Some flocccose isolates (e.g. *P. terrestre* Jensen *1912, P. lanoso-viride* Thom *1930*) differ slightly in colony habit, but not in microscopic morphology. Some strains, hitherto distinguished on small differences in colony colour which are difficult to verify in practice, are now regarded as seven varieties of *P. verrucosum* (5029). All show (on MEA more than on CzA) the same roughened, two-(to three-)stage branched conidiophores, which are usually loosely synnematous in fresh isolates. — Colonies reaching (2·0–) 2·5–3·5 cm diam in 14 days at 25°C on CzA, on MEA sometimes faster. Penicilli moderately appressed. Conidia at first ellipsoidal, later becoming globose to subglobose, smooth-walled, 3·0–4·0 μm diam. — *P. verrucosum* can be distinguished from the similar *P. expansum* Link ex Gray (q.v.) by a lower degree of ramification of the penicilli and by a more pronounced wall thickening at the phialide tip. Furthermore it does not cause a rapid rot of pomaceous fruits as *P. expansum* does. — The variety *verrucosum* is recognized by bright yellow-green, usually rather slow-growing colonies and yellowish or brownish reverse. — The content of long-chain fatty acids has been used to determine species relationships (1281). Pyrolysis-gas-liquid chromatography of conidia proved not to give conclusive results for taxonomical purposes. — DNA analysis (of *P. viridicatum*) gave a GC content of 54% (5600).

Penicillium verrucosum var. *verrucosum* has been recorded from soils in the Buhara (2871) and Amur regions (432), the Pamir (3850), the Ukraine (2948), and ice caves (3367) of the USSR, also from the British Isles (745, 5221), Czechoslovakia (1700), Yugoslavia (5534), Italy (3453, 3538, 5047), Turkey (4245), Syria (277, 5392), Egypt (3993, 3997), Peru (2005) and Australia (3720). Observations come from forest soils (432, 1700, 2948, 3538), soils with steppe vegetation (1700), wet meadows (2948), heathland (3720, 5047, 5221), sand dunes (745), desert soil (3997), caves (3453), podzols (432, 5534), peaty soil (1259), solonetz (277, 4677) and sewage (1166). — It has been isolated from remains of barley, cabbage and potato plants (3113), litter of *Pinus monticola* (681) and deciduous trees (650), the rhizospheres of tomato (16), *Ammodendron conollyi* (5416), *Stipa lessingiana, S. sareptana* and *Festuca sulcata* (3376), the rhizoplane of heath plants (5220), sticks buried in soil (1956), tunnels of gerbils (5393), feathers of free-living birds (2575), seeds of corn (3875), wheat (3989, 4492), rice (3181) and groundnuts (1016), coarse fodder (4548), flour-based foodstuffs (5980), stored grape fruits and melons (342), frozen fruit cake (3153), fruit juices (5205) and mould-fermented sausages (3844). It is inhibited by volatile metabolites from bacteria (3920, 3921, 3922).

Conidia germinate at temperatures between 4 and 37°C, minimum water potential at −270 bars (3873). The four varieties of *P. verrucosum* treated can grow within a temperature range of (−2 or) + 4°C to 30 or 35°C, with the optimum between 21 and 23°C; there is usually no growth at 37°C (3153, 3876, 4578). The thermal death point for var. *verrucosum* was found to be 63°C for 30 min in apple juice (3620). — Pectin and cellulose are degraded (3538). Metabolic products include the flavour compounds 2-octen-1-ol and 1-octanol (2867), and ochratoxin A (5163, 5165, 6145), the production of which on wheat and barley is favoured at −100 to −140 bars water potential (2296); further brevianamide A (4858, 6393), citrinin (5165), penicillic acid (3844), ergosterol (6145), ergosteryl palmitate, *meso*-erythritol, mannitol, viridicatic acid (539), viridicatol, viridicatin (5278) which is weakly antibiotic against *Mycobacterium tuberculosis* (1238), xanthomegnin, viomellein, rubrosulphin, viopurpurin (5515), 3-*O*-methylviridicatin (3231), cyclopenin, cyclopenol, and intermediates in the formation of viridicatin and viridicatol (3426). — Over 80% of the isolates from sausages were potentially toxigenic on laboratory media (3844). Contaminated barley has proved to be toxic to rats (3134) and contaminated rice to mice (768), nephrotoxicosis has been observed in pigs (3932). — *P. verrucosum* var. *verrucosum* was found sensitive to β-irradiation, sensitized strains being less tolerant to higher incubation temperatures (4027); the sensitivity is also age-dependent (3547).

Penicillium verrucosum var. *corymbiferum* (Westling) Samson, Stolk & Hadlok 1976

≡ *Penicillium corymbiferum* Westling *1911*

Colonies reaching 4·5–5·0 cm diam in 8–10 days at 24°C on CzA, 6–7 cm on MEA, yellow-green, coarsely granulate from synnematous conidiophores, yellow aerial mycelium and orange-brown exudate usually present; reverse of similar colour. Conidia usually only 3·0–3·3 μm diam. — Ultrastructure of conidium ontogeny in this variety has been studied by Fletcher (1771) who demonstrated a progressive wall thickening in the phialide neck. — Pyrolysis-gas-liquid chromatography of conidia proved to be not conclusive for taxonomic purposes (3168).

The available data indicate a worldwide distribution of this variety including the British Isles, the Netherlands, Canada (IMI), the USA (655, 2573), Czechoslovakia (1702), the USSR (2872, 3365), Poland (1423, 6520), France (2161), Spain (3417), Turkey, Pakistan, central and South Africa (IMI, 3063). It has been isolated from cultivated (2161, 2872), forest (1702, 2573) and floodplain soils (3417), podzols (3416), gley, chernozem (3414), peat bogs (6520), sand dunes (1423), estuarine sediments (655) and carst caves (3365). Besides occurrences in the rhizospheres of wheat and barley, it was found in those of tomato (16), on rhizomes of *Iris* (4960, 5121), and is known from bulbs of tulip, *Lachenalia* and *Narcissus, Gladiolus* corms, moist hay (4548), flour-based foodstuffs (5980), silage, *Phyllanthus emblica* fruits (3740) and air (IMI).

Temperature relationships are the same as in var. *verrucosum* (4578). *P. verrucosum* var. *corymbiferum* was found to grow on and utilize sinigrin (4774). Alliin lyase was isolated from it (1505); the formation of an antibiotic active against *Staphylococcus aureus* has been demonstrated (4503).

Penicillium verrucosum var. *cyclopium* (Westling) Samson, Stolk & Hadlok 1976 — Fig. 269.

≡ *Penicillium cyclopium* Westling *1911*
= *Penicillium puberulum* Bain. *1907*
= *Penicillium palitans* Westling *1911*
= *Penicillium solitum* Westling *1911*
= *Penicillium casei* Staub *1911*
= *Penicillium terrestre* Jensen *1912*
= *Penicillium aurantiovirens* Biourge *1923*
= *Penicillium olivino-viride* Biourge *1923*
= *Penicillium crustosum* Thom *1930*

For some other synonyms see Samson *et al.* (5029).

Colonies reaching 4·5–5 cm diam in 14 days at 24°C on CzA or MEA, sometimes only 3–4 cm (*P. puberulum*), blue-green, in fresh isolates often fading to light ochraceous in the colony

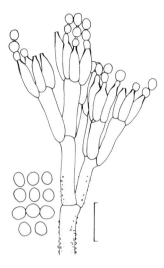

FIG. 269. *Penicillium verrucosum* var. *cyclopium*, penicillus and conidia.

centre; reverse uncoloured, yellow, orange or brown. Conidiophores usually finely roughened, in some isolates smooth-walled on CzA but always roughened on MEA. — *P. puberulum* falls within the range of variability of *P. verrucosum* var. *cyclopium*, although synnematous conidiophores may be absent even with the addition of botran to the medium (3872). — The rodlet pattern of the conidial surfaces shows interlacing loose fascicles with relatively smooth surface (2421). — Pyrolysis-gas-liquid chromatography of the conidia proved to be not conclusive for taxonomic purposes (3168). — DNA analysis gave a GC content of 51–52% for *P. cyclopium* and 49–52% for *P. puberulum* (5600).

P. verrucosum var. *cyclopium* is one of the most widely distributed *Penicillium* taxa. An immunofluorescence technique to detect mycelium of this fungus is available (6200). — It has very frequently been found in Europe and many parts of the USSR; reports from Spitsbergen (3064), the White Sea (3368) and East Siberia (3652) indicate a certain preference for cool-temperate zones. Nevertheless, it has also been found in Syria (5392), Egypt (3993, 3997), Libya (6510), Kuwait (4000, 4001), Pakistan (60, 61), Somalia (5049), South Africa (1559), the Ivory Coast (4159, 4718), India (4698), Peru (2005), China (3475), Japan (3267), Australia and New Zealand (3720, 5812, 5930). It occurs to an equal extent in grey forest soil, chernozem, chestnut brown soil, solonetz (3414, 4677), takyr (3871), rendzina and podzol (3414, 3416, 4378). Reports from various forest soils are somewhat more numerous than those from arable soils; it is also frequent in peat bogs (1376, 6082), different types of peat fields (4474, 6571), heathland (5047, 5220), grassland (2573), tundra, taiga (1446), savannah (1559, 5049), steppes (1700, 1702, 3362, 4554), desert soil (3997, 4733), dunes (745), children's sandpits (1421), a uranium mine (1703), caves (3364, 3453), on growing stalactites (2315) and other calcareous substrates (4152). It has also been found in saline soil (4001), fresh (2338, 4429) and polluted water (1154, 1155, 1166, 1482), estuarine sediments with high salinity (655) and even seawater in the eulittoral zone (4918). Its incidence is largely independent of soil pH (745), which *in vitro* can range from 2 to 10 (4369, 5895), and it has been observed down to 50 cm depth (3447, 4554). Its frequency in the soil is said to be reduced by applications of calcium fertilizers and dung, but increased by ammonium sulphate

(3849). There are detailed investigations of its occurrence in pine litter in particular (269, 2344, 2345, 2923, 6080), and also observations of it on litter of *Abies grandis* (681), remains of barley, cabbage and potato plants (3113), decaying petioles of *Pteridium aquilinum* and in "die-back" stands of *Spartina townsendii* (5389). It has been isolated from the rhizospheres of wheat (3567), corn (4553, 4554), various steppe grasses (3376), alfalfa (3982), clover (3429), broad bean, cotton (6511), tomato (16), peat bog plants (1376), oak (5394) and various halophytes (3375). It is also frequent in stored seeds of *Pinus sylvestris* (6344), wheat (2297, 3989, 4492, 5895), barley (4038, 6199, 6274), corn (3875, 3989) and rice (2297, 3181), pecans (2572, 5092), decaying groundnuts (2483), stored apple, pear, grape, strawberry, melon, green pepper and tomato (342), hay, straw (4548), preservative-treated wood in ground contact (4193), wood pulp (6168), garden compost (1424), sewage (1166), dung of angora goat (3882) and rabbits (6085), tunnels, storage places and nests of gerbils (5393), nests and feathers of free-living birds (2575), dead adult bees (*Apis*), larvae, pupae and combs (402, 4631); further substrates recorded include fermented food (5980), frozen fruit cake (3153), fruit juices (5205), refrigerated dough products (2067) and flour (3930). — Germination of conidia *in vitro* is inhibited by volatile substances from *Ocimum basilicum* and *Origanum majorana* (35). *P. verrucosum* var. *cyclopium* can serve as food for several mites (5379, 5380, 5383).

Conidia germinate at temperatures between 4° and 30°C and a minimum water potential of −270 bars (3873, 3876). Cardinal temperatures for growth are the same as for var. *verrucosum* (4578); the thermal death point, determined in apple juice, was 62°C for 20 min (3620). *P. verrucosum* var. *cyclopium* grows well at −100 bars water potential (4578), the minimum is in the range −230 to −270 bars and the optimum at −30 bars (261, 2093, 6344); it is rather osmophilic (1991, 3934) and acidophilic (5895). — Optimized nutrient media for sporulation and growth have been described (5895). Starch can be utilized (655, 3944, 5896) and is superior to glucose as a substrate for vegetative growth; galactose, fructose, maltose, raffinose, and sucrose can also be utilized (5894). Tests for pectinase were positive (655, 1432, 2239, 3539) within the pH range 3–7 (4690); xylan utilization is very pronounced (1432, 3231). Decomposition of filter paper is good (1425, 3414) but other observations on decomposition of cellulose fibres had negative results (655, 3618); CMC breakdown is moderate as compared to other fungi (2878). Production of proteinase (3429, 5896) and lipase (3944) has been reported. *P. verrucosum* var. *cyclopium* can grow on 0·2% sinigrin (4774) and on 5% tannin solution as the sole C source and good development is possible in atmospheres deficient in O_2 (1425), but the fungus requires biotin and thiamine under anaerobic conditions (5680). It produces enough organic acids to be able to corrode metals (85). Lignosulphonate, humic and fulvic acids are utilized (934), epoxy and dihexyladipate plasticizers were found to be degraded (5859), and thus the fungus can grow on polyethylene (5148). *P. verrucosum* var. *cyclopium* was able to use the herbicides simazine, atrazine and prometryne (3799); metaphosphate can also be utilized (4544). Metabolic products which have been identified include orsellinic, cyclopaldic and cyclopolic acids (3231) which have more or less pronounced antifungal properties (537), the unsaturated dihydroxyaldehyde palitantin (535), the anthraquinones citreorosein (= ω-hydroxy emodin), emodic acid (150, 3231), viridicatin, viridicatol (512, 677, 1058, 3426), 3-*O*-methylviridicatin (245), 2,3-dihydroxy-4-phenylquinoline (2190), the weakly antibacterial cyclopenin (677) and cyclopenol; further α- and β-cyclopiazonic acid (2518, 2519, 2521, 5077) and derivatives of it, and penicillic acid (116, 460, 532, 6408) with considerable antibiotic properties (1934). — Both cyclopiazonic acid (4670, 6005) and penicillic acid (3231) have acute toxic effects on mammals, and it can be assumed that these mycotoxins are the causal agents of liver and kidney lesions in mice fed with contaminated corn (899). — Further products include ergosterol (6482) and the tremorgenic metabolites tremortin A and B (1059, 2546, 6392) and tremorgen (3231). — As *P. puberulum* and as *P. aurantiovirens*, this fungus is further known to produce the tropolones puberulic acid and puberulonic acid (3231, 6005), a mycotoxin of unknown structure (6390), and poly-(L)-malic acid which acts as a proteinase inhibitor (5289). Reports of the production of aflatoxins (1303, 2483) have not been confirmed (1360, 6389). For *P. terrestre* the production of terrestric acid, a 4-hydroxy-γ-lactone, was described (3231). — Production of fungitoxic substances by *P. verrucosum* var. *cyclopium* occurs also in non-sterile soil without addition of organic material (3853). Antagonistic activity *in vitro* towards numerous fungi and bacteria has been demonstrated (515, 516, 1425, 2753, 2754, 2872, 4304, 4550, 5394). It produces lesions on wounded bulb scales of *Scilla nutans* (3487, 5346). Culture filtrates had a toxic effect on wheat (3849, 3851, 3855) and barley (293) but stimulated corn (4553) and spruce seedlings (3588). Injurious effects could be demonstrated in a reduced nitrogen content in peas and vetches, but not wheat (3855). — The fungus is highly sensitive to γ-rays (4585). Isolations from seeds treated with phenyl mercuric acetate indicate a high mercury tolerance (2077).

Penicillium verrucosum var. *melanochlorum* Samson, Stolk & Hadlok *1976*

= *Penicillium palitans* Westling *1911* sensu Raper & Thom *1949*

Colonies reaching 3–4 cm diam in 14 days at 25°C on CzA, 4–5 cm on MEA, dark green. Mononematous conidiophores predominate.

This variety is not very widely distributed and appears to be adapted to cool-temperate climates; it has been reported from Spitsbergen (3064), a lagoon on the volcanic island of Surtsey (2406), the USSR (3365, 3367), the USA (1039, 1040, 2008), the British Isles (2736), France (2161), Italy (3538, 5047), Egypt (3993), Iraq (92), Peru (2005) and New Zealand (5812). Known habitats include wet-mesic deciduous forests (1040), floodplain communities with willow (2008), beech forest in the Alps (3538), a conifer swamp (1039), grassland (5812), heathland (5047), arable soil (2161), caves (3365, 3367, 3453), and polluted streams (1166). It has been reported from the rhizospheres of clover (3429) and tomato (16), seeds of corn (3069, 3875), pecans (5092), sticks buried in soil (1956), frozen fruit cake (3153) and adult bees (402).

Growth temperature requirements are the same as for var. *verrucosum* (4578). The optimum temperature for germination of conidia in var. *melanochlorum* is 23°C (3876). At 2°C it can grow at −240 bars water potential (6344). — Cellulose (1956, 2736) and pectin (3538) can be degraded. Known metabolites include palitantin, frequentin (1245), viridicatin (1058), penicillic acid (3231), ochratoxin A, citrinin (5165), tremortin (2546) and/or a tremorgenic toxin (1054, 1059). Contaminated corn was found to be toxic to cattle (86). — *P. verrucosum* var. *melanochlorum* was found to be relatively resistant to phenyl mercuric acetate (2077).

Penicillium waksmanii Zaleski *1927* — Fig. 270.

DESCRIPTION: Raper and Thom (4745), who placed this species in the *Ramigena* series, but it resembles also species of the *P. citrinum* series and also shows some affinity with *P. jensenii* Zaleski (q.v.). — Colonies growing rather restrictedly, reaching 1·5–2·0 cm diam in ten days at 24°C on CzA, 2·5–3·0 cm on MEA, pale blue-green, reverse pale pinkish. Smooth-walled branches arising low on the conidiophores and thus not appearing as an integral part on the terminal penicillus. Conidia globose to subglobose, delicately roughened, 2·0–2·5 μm diam. — The content of long-chain fatty acids has been used to determine species relationships (1281).

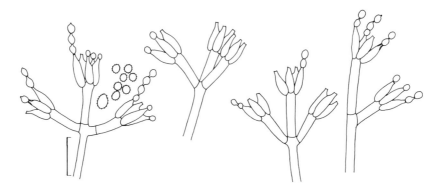

FIG. 270. *Penicillium waksmanii*, divaricate penicilli and conidia, CBS 586.70.

P. waksmanii has a worldwide distribution, although some reports raise doubts about the correctness of the identifications. — It has been reported from the North of Scotland (163), the Netherlands (1614, 1616), Czechoslovakia (1700, 1703), Poland (1421, 1978), the USSR (4474, 6571), France (242), Spain (3446), Italy (3450, 3913, 6082), Syria (5392), the USA (660, 1040, 1166, 2482, 2573, 3817, 4733, 4918), New Zealand (4947, 5812), Australia (3720), Peru (2005), Brazil (399), the Bahamas (2006), Somalia (5048), South Africa (1559, 4407), India (1519) and Japan (3267). — It has been isolated from forest soils (1040, 1700, 2006, 2573, 3267, 3817, 4537, 4947, 5048), forest nurseries (1978, 2482), grassland (163, 5812), cultivated soils (242, 1519, 1614, 1616, 2006, 3446, 3450, 3817, 3913), heathland (3720), peat fields (4474), *Sphagnum* bogs (6082), soils under steppe type vegetation (1559, 1700, 4407), desert soils (4733), a uranium mine (1703), solonetz (6571), rendzina (3414), a granite outcrop (660), a mine (5871), children's sandpits (1421), seawater in the eulittoral zone (4918), and an acid mine drainage stream (1166). It has also been found on leaf litter (6080), pine roots (3572), in the rhizospheres of wheat (3567), wilted pineapple plants (6210) and steppe plants (3376), stored wheat seeds (4492), fermented food, flour-based foodstuffs (5980), frozen fruit cake (3153), fruit juices (5205), and nests, feathers and pellets of free-living birds (2575). — Its frequency in grassland is said to be increased by N fertilization (3498); methylbromide-treated soil is only slowly recolonized (1268).

The optimum temperature for growth is in the range 26–28°C, the minimum 6–7°C, and no growth occurs at 37°C (517, 4578); it can grow at −100 bars water potential (4578). — Degradation of parathion to aminoparathion (4738) and antibiotic activity against both bacteria and fungi have been demonstrated (515, 569, 2956, 2753, 2754). *P. waksmanii* grows on larvae of honey-bees infected by *Bacillus larvae* (American foulbrood) and effectively reduces the bacterial infestation (5952). — In a chronically irradiated soil the fungus was still found viable at intensities of 2780 R per day (2009).

Periconia Tode ex Fr. *1821*

Periconia macrospinosa Lefebvre & A. G. Johnson *1949* — Fig. 271.

DESCRIPTIONS: Lefebvre *et al.* (3274), Mason and Ellis (3662), and Ellis (1603).

The Hyphomycete genus *Periconia* (= *Sporocybe* Fr. *1825* = *Sporodum* Corda *1836* = *Trichocephalum* Cost. *1888*) has been monographed for the British Isles by Mason and Ellis (3662), and for India by Raghuveer Rao and Dev Rao (4691). The type species *P. lichenoides* Tode ex Mérat *1821* is now regarded as a doubtful species, and *P. byssoides* Pers. ex Schw. *1822* is perhaps most representative of the genus.

Periconia is characterized by hyaline and partly pigmented vegetative hyphae. Conidiophores erect, mostly several cells long, brown, bearing an apical or subapical head of dense branches with numerous short, usually monoblastic, conidiogenous cells which form short, sometimes branched chains of blastoconidia. Conidia ± globose, brown, rough-walled, formed in acropetal chains but maturing in basipetal sequence (3662). Similar conidia are also often produced without differentiated conidiophore stipes. — The superficially similar genus *Periconiella* Sacc. *1885* has conidiogenous cells with sympodial cicatrized rhachides. — Mason and Ellis (3662) recognized 15 species in *Periconia* for the British Isles; Ellis (1603, 1604) described 27 species, and Raghuveer Rao and Dev Rao (4691) keyed out 28 species for India.

Colonies of *P. macrospinosa* grow moderately fast, reaching 3·2–4·7 cm diam in ten days at 20°C on MEA; vegetative hyphae hyaline to pale brown, 3–6 μm wide; mat dark brown, compactly cottony; reverse often vinaceous. Conidiophores dark brown, to 420 μm long but usually much shorter, 7–12 μm wide at the base, bearing an irregular apical cluster of fertile branches. Conidia 18–35 μm diam (including the spines), dark brown, coarsely echinulate with spines 2–7 μm long, supported by several smaller, paler, verruculose cells, 5–9 μm diam. *P. macrospinosa* is a very distinctive species distinguished by the long and broad spines on the large conidia.

P. macrospinosa is by far the most commonly observed soil-borne species of its genus in the temperate zone, while other species are more restricted to plant substrates, particularly in tropical countries. — Records of *P. macrospinosa* include ones from the British Isles (1603, 3662), the Netherlands, Germany (CBS), the Ukrainian SSR (6563), Italy (4538), the USA (2004), Canada (505, 6352), central Africa (3063), Jamaica (IMI), Peru (2005), Japan (2634, 3267, 5846), Hong Kong (IMI), New Guinea (6499), Australia (6186), New Zealand (5815), India (427) and Iraq (1603). It is most frequently found in arable soils (4538, 6563), but is also known in fallow soil (IMI), forest soils under *Thuja occidentalis* (505), larch (3267), pine (6352), beech (4813) and mixed hardwood forests (2004, 2008), and grassland (6182). Manuring has been reported to reduce its frequency and it can be found down to 20 cm in the soil (2161). *P. macrospinosa* was originally isolated from roots of *Sorghum vulgare* in Kansas (3274) but does not appear to be pathogenic to any graminaceous plant. It has been most frequently reported from the rhizosphere and roots of various plants, including grasses (5815, IMI), corn (4553), wheat, especially from the seminal and coronal parts (2492, 3662), oats,

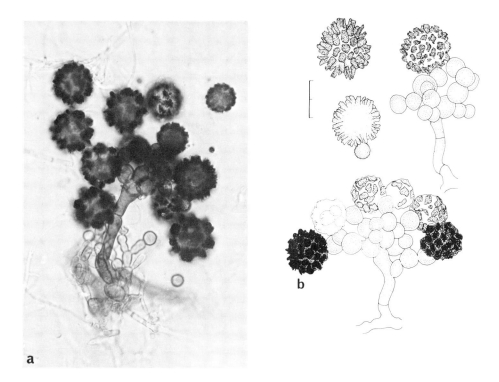

FIG. 271. *Periconia macrospinosa*. a. Conidiophore and conidia, × 500; b. conidiogenesis, young and mature conidia; CBS 876.68.

barley (3006), clover (5815), and forest plants in moder humus (4814). It has also been isolated from stems and leaves of cotton, wheat (427), *Prunus, Chenopodium album* (1603), and strawberry (4133), and seeds of groundnuts (2253) and other legumes (2634).

P. macrospinosa decomposes xylan very strongly though differences in this ability occur between isolates (1432); chitin is also degraded (2706). It is able to oxidize Mn^{2+} ions (5828). Known metabolites include a number of compounds containing chlorine (1980, 2504).

Petriellidium Malloch *1970*

Petriellidium boydii (Shear) Malloch *1970* — Fig. 272.

≡ *Allescheria boydii* Shear *1922*

Anamorph: (a) solitary conidiophores: *Scedosporium apiospermum* (Sacc.) Sacc. ex Castell. & Chalmers *1919*
≡ *Monosporium apiospermum* Sacc. *1911*
= *Cephalosporium boydii* Shear *1922*
= *Indiella americana* Delamare & Gatti *1929*
(b) synnematous conidiophores: *Graphium* species
≡ *Dendrostilbella boydii* Shear *1922*
? = *Graphium eumorphum* (Sacc.) Sacc. *1886*
≡ *Sporocybe eumorpha* Sacc. *1882*

DESCRIPTIONS: Emmons (1621), and Emmons *et al.* (1624); generic diagnosis: Malloch (3551); generic revision: v. Arx (203). — *Petriellidium* belongs to the family Microascaceae Luttrell ex Malloch which is characterized by usually less hairy perithecia than in the Chaetomiaceae, and reddish brown ascospores which are dextrinoid when young, and provided with 1 or 2 germ pores. — *Petriellidium* has fairly rapidly spreading colonies reaching about 5·0 cm diam in five days at 25°C on OA. Ascomata arising from coiled ascogonia, non-ostiolate, globose, thin-walled and semi-translucent; peridium 1–2 cells thick. Asci (sub)globose, unitunicate, 8-spored; ascospores ellipsoidal, orange to copper-coloured, smooth-walled, with two germ pores. Conidia pyriform, pale grey, formed in dry heads on solitary conidiogenous cells or as solitary aleurioconidia, sometimes on pigmented *Graphium*-like synnemata.

Petriellidium is a non-ostiolate counterpart of *Petriella* Curzi *1930* which has similar dimorphic, solitary and synnematous conidiophores, although the latter usually predominate. — *Allescheria* Sacc. & Syd. *1899* (type species *A. gayoni* (Cost.) Sacc. & Syd.) is probably a synonym of *Monascus* van Tiegh. (q.v.); *Monosporium* Bonord. *1851* is a nomen illegitimum (2602), and there is no generic name available for the anamorph apart from *Scedosporium* which was only incidentally mentioned by Saccardo *1911* but subsequently validated by Castellani and Chalmers in *1919* and fully described by Brumpt in *1922*. The synnematous anamorph is not clearly distinct from that found in various *Petriella* species (although it is usually less dominant) and produces conidia on irregularly annellate conidiogenous cells similar to those occurring in the solitary conidiophores (Fig. 272); this mode of conidiogenesis is unusual and suggests a sympodial elongation (under the light microscope) because each conidium is pushed aside after its delimitation, thus forming divergent heads. — The other non-ostiolate genus of the Microascaceae, *Kernia* Nieuwl. *1916*, has slower-growing colonies and much thicker, black ascomatal peridia, which are frequently also setose.

P. boydii has relatively fast-growing colonies, reaching 4·0–5·0 cm diam in five days at 25°C on OA, at first whitish to greyish, later becoming grey-brown, floccose to lanose. Homothallic.

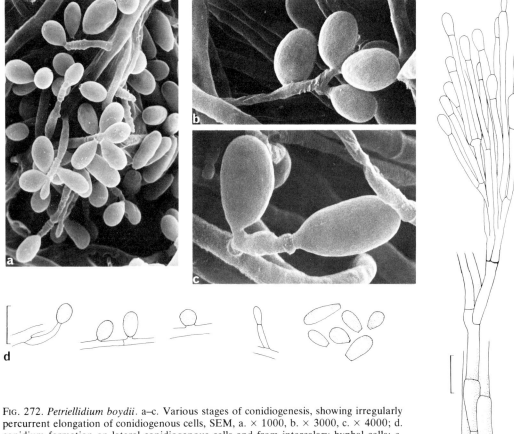

FIG. 272. *Petriellidium boydii.* a–c. Various stages of conidiogenesis, showing irregularly percurrent elongation of conidiogenous cells, SEM, a. × 1000, b. × 3000, c. × 4000; d. conidium formation on lateral conidiogenous cells and from intercalary hyphal cells; e. synnematous sporulation.

Ascomata spherical, mostly submerged but also developed in the aerial mycelium on acidified Sabouraud-glucose agar (1621), 140–200 μm or more diam, wall 4–6 μm thick, composed of irregularly interwoven brown hyphae, sometimes covered with loose brown thick-walled hyphae. Asci 12–18 × 9–13 μm, evanescent; ascospores ellipsoidal, symmetrical or slightly flattened, orange-brown, with two germ pores, 6–6·5(–7) × 3·5–4·0 μm, length/width ratio 1·5–1·6. Conidia of two types: (a) *Scedosporium* type, broadly clavate to ovoid, rounded above, truncate at the base, grey-brown, 6–14 × 3·5–6 μm; and (b) *Graphium* type, narrowly clavate, hyaline, 5–7 × 2–3 μm. — The other five *Petriellidium* species recognized (203) differ primarily from *P. boydii* in the ascospore dimensions; the closest species are *P. angustum* Malloch & Cain *1972* which has narrower ascospores, 6–7 × 3·0–3·8 μm, length/width 1·8, and only solitary conidiophores with almost cylindrical conidia 5–10 × 3–4·5 μm, and *P. fusoideum* v. Arx *1973*, which has more fusiform, pale straw-coloured ascospores of the same dimensions as *P. boydii* and very similar conidia of both types. — Apart from *Graphium fructicola* El. & Em. Marchal *1921*, *G. eumorphum* has the largest conidia described in its genus; however, this cannot reliably be distinguished from some *Petriella* species without available cultures. — A white type has been obtained from *P. boydii* by mutation, which produces less conidia, more aerial hyphae, and no ascomata (1570). — Hyphal structures have been studied with TEM (2215). — A standardized antigen production using mild extraction yielded mainly a-D-glucans and glucoproteins (6569). — DNA analysis gave a GC content of 53·5% (5600).

P. boydii has a worldwide distribution, particularly in soil and is well known as causal agent of mycetoma and other deep-seated mycoses (cf. below); but it had not been known to cause mycetomas or other infections in some tropical countries although it occurs there commonly (4350); the first case of mycetoma in India due to this fungus was reported only in *1973* (5786). It can be isolated from soils by various techniques including hair baiting (5015, 5766). — *P. boydii* has also been reported from soils in Finland (5015), the British Isles (IMI), Germany (3041), Egypt, Sudan, Ethiopia (5766), Algeria, Tunisia, Argentina, Brazil, Paraguay, Trinidad, the Virgin Islands (66), the USA (2573) including Hawaii (3014), Canada (1624), Australia (IMI), and India (427, 2156, 4350, 4476, 4477); in the state of Maharashtra/ India 45% of mainly non-acidic soil samples contained the fungus (4350); unusually high frequencies occurred in soils under *Hevea* in Sri Lanka (CBS). Known substrates include chicken pens, soil of a zoologic garden, cultivated soil, estuarine sediments (468, 2156, 3014, 4476, 4477), polluted water (1170), coastal tide-washed areas (1256), a mangrove swamp (IMI), a sewage treatment plant (1165), and composted municipal waste (3041). Other known habitats include marine algae (1256), silage, and diesel fuel (IMI). — Isolations from man or animals have been reported from Canada (304, 1621), the USA (1648, 4114), the British Isles, the Netherlands, Italy, Rhodesia, Sri Lanka (IMI) and India (2153).

Good growth occurs in the temperature range 18–40°C (160), the optimum is 30°C, the minimum between 15 and 20°C, and the maximum between 40 and 45°C; the optimum pH is 7·0–7·6, but *P. boydii* is able to grow in the wide pH range 3·6–10·8 (6419). — Best growth is obtained on D-xylose, D-fructose or trehalose, and fair growth on D-glucose, D-galactose, D-mannose, sucrose, dextrin, soluble starch, some organic acids and alcohols; the best N sources are glycine, DL-valine, L-glutamic acid, L-tryptophan, and NaNO$_3$ (6419). Proteinase (3231), amylase, cellulase, xylanase, and mannanase (4192) are produced. *P. boydii* causes considerable losses in weight in wood of birch and pine (4191). It is halophilic (4477) and can grow at salt concentrations up to 3·4% (3008). Growth on 1·6% suspended defatted chips of cow horn (a-keratin) as sole C and N source led to only 30% decomposition in 6 months (H. Ziegler, pers. commun.). — *P. boydii* is well known as the causal agent of a foot mycetoma with white to yellow grains (1624), pulmonary mycosis (183), mycotic abortion in cattle (437), systemically spreading infections of the central nervous system (1801), and mycotic infections of the external auditory canal and the cornea (1624, 4114, 4949).

Peziza L. ex St.-Am. *1821*

Peziza ostracoderma Korf *1961* — Fig. 273.

≡ *Plicaria fulva* R. Schneider *1954* [non *Peziza fulva* Mich. ex Pers. *1822*]

Anamorph: *Chromelosporium fulvum* (Link) McGinty, Hennebert & Korf *1975*

    ≡ *Botrytis fulva* Link *1824*
    ≡ *Sporotrichum fulvum* (Link) Fr. *1832*
  = *Chromelosporium ollare* (Pers.) Hennebert *1973* (nomen illeg., Art. 13)
    ≡ *Dematium ollare* Pers. *1801* (devalidated name)
  = *Botrytis luteo-brunnea* Krzem. & Badura *1955*
  = *Mycotypha dichotoma* Wolf *1955*

Misapplications:

Ostracoderma epigaeum (Link) Hellmers *1965* [*Botrytis epigaea* Link *1824* = *Chromelosporium terrestre* (Fr.) M. B. Ellis *1976*]
Botrytis crystallina (Bonord.) Sacc. *1886* [*Polyactis crystallina* Bonord. *1864* = *Chromelosporium coerulescens* (Bonord.) Hennebert *1973*]
Botrytis spectabilis Harz *1871* [*Clonostachys spectabilis* (Harz) Oud. & Sacc. *1886* = *Chromelosporium ochraceum* Corda *1833*]

Chromelosporium Corda *1833* (= *Hyphelia* Fr. *1849*, non Fr. *1825*) is characterized by colonies which are effuse, velvety, or tufted, and white to rose, ochraceous or brown (2400). Conidiophores erect, solitary or caespitose, stipes hyaline or ochraceous, forming dichotomously branching fertile "brushes" at the apex; branches ± long, cylindrical, straight, hyaline, simultaneously producing numerous blastoconidia on short denticles. Conidia globose, subglobose or ovate, hyaline to pale coloured, smooth-walled or ± ornamented. — *Ostracoderma* Fr. *1825*, a name often incorrectly used for this fungus, has closed fruit-bodies (2400). Ten species of *Chromelosporium* are known, of which four are certainly connected with teleomorphs in *Peziza* (2400).

    DESCRIPTIONS: Schneider (5117), Rieth (4839), Fergus (1722), Hellmers (2371), Barron (364), Pudelko (4643), and Ellis (1604); conidiogenesis: Hughes and Bisalputra (2592) and Cole and Samson (1109). MONOGRAPHIC TREATMENT: Hennebert and Korf (2403). — Colonies reaching over 9 cm diam in five days at 20°C on MEA. Conidiophores 300–1000 μm high, stipe to 600 × 8–17 μm, hyaline to fulvous, 1–5 times dichotomously branched, ending in slightly inflated, clavate conidiogenous cells. Conidia produced singly on denticles 3×1 μm, napiform, thin- and smooth-walled, fulvous, 5–13 μm diam. The botryose blastoconidia are produced synchronously on the conidiogenous cells and their wall is in continuous connection with that of the conidiogenous cell; at maturity a septum is layed down which facilitates secession (2592). — The teleomorph of *P. ostracoderma* can easily be observed on peat soil, steam-sterilized soil, and pots in greenhouses (2403). The apothecia are at first cupulate but soon flatten out, 5–15(–30) mm diam, with a fulvous to dark brown hymenium, paler

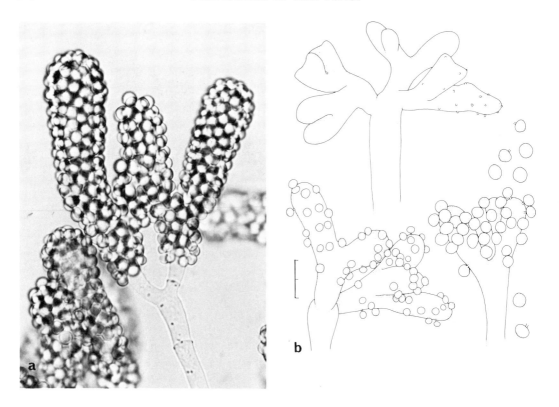

FIG. 273. *Peziza ostracoderma*. a. *Chromelosporium*-type conidiophore with dichotomous branching, × 500; b. various stages of conidiogenesis, CBS 202.68.

excipulum, pruinose and sometimes with a violaceous cast. Ascospores ellipsoidal, covered with a fine cyanophilic reticulum, 11–13·5 × 6–8(–10) μm. — The apothecia of *P. atrovinosa* Cooke & Gerard *1875*, a species which has sometimes been confused with *P. ostracoderma*, are 2–8 cm diam, vinaceous-brown to purplish black, and produce ascospores having large irregular reticulations and warts; their dimensions are as in *P. ostracoderma*. — The anamorph of *P. ostracoderma* resembles two other *Chromelosporium* species which can also be found on soil surfaces: *Ch. ochraceum* Corda *1833* with mononematous conidiophores, and *Ch. terrestre* (Fr.) M. B. Ellis *1976* with partly synnematous conidiophores; both these two species have verruculose conidia 4–6 μm diam. *Ch. fulvum* also resembles *Dichobotrys abundans* Henneb. *1973*, the anamorph of *Trichophaea abundans* (Karst.) Boud. *1907* which also occurs on steamed soil and burnt wood; this species has ochraceous colonies with dichotomously branched or coral-like conidiophores bearing short, terminally swollen branches; conidia subglobose, pedicellate, smooth-walled, oblate, 8–11 × 7–9 μm. — Apothecia of certain isolates of *P. ostracoderma* contained spherical virus-like particles (1382, 1383).

P. ostracoderma is commonly referred to as the "peat mould" as it frequently occurs in both its anamorph and teleomorph on peaty soil, steam-sterilized soil and pots, walls in greenhouses, and mushroom beds (then called "brown mould") in both Europe and North America (1722, 2403, 4839, 5117). It is less commonly found in forest soils, for example, under beech

(3138) and *Thuja* (IMI), but is also known from the Sonoran desert in the southern USA (4733), garden compost in Germany (1424), dung, an infected ear of a pig, air (2403) and human sputum (3723).

Optimal conidium production occurs on sterilized soil at pH 4·9–5·4 (5117). The conidia can germinate on a wide variety of media and in the pH range 3–8; the optimal temperature for germination is 20–33°C, the minimum 3°C, and the maximum 38–42°C (1726). The optimum pH for growth lies between pH 6·6 and 7·4 (5117). Both starch and cellulose are degraded (1425) and ethylene can be produced (2641).

Phialophora Medlar *1915*

= *Cadophora* Lagerb. & Melin *1927*
= *Margarinomyces* Laxa *1930*
= *Lecythophora* Nannf. *1934*

Type species: *Phialophora verrucosa* Medlar

Teleomorphic genera: *Pyrenopeziza* Fuckel, *Mollisia* (Fr.) Karst., *Ascocoryne* Groves & Wilson, *Coniochaeta* (Sacc.) Massee, *Gaeumannomyces* v. Arx & Olivier

Contributions to a monographic treatment: Schol-Schwarz (5130), Wang (6168), Cole and Kendrick (1108), Sivasithamparam (5390), and Ellis (1604).

Colonies mostly rather slow-growing, usually olivaceous-black but sometimes hyaline to pink or becoming brown with age, with or without aerial mycelium. Conidiogenesis phialidic; phialides arising solitarily from vegetative hyphae or on branched conidiophores, often in clusters; reduced phialides often arising from intercalary cells without a basal septum (adelophialides) in some species; phialides mostly flask-shaped or cylindrical, usually with a distinct collarette; they may proliferate percurrently through the collarette (common in old cultures) or sympodially below the collarette. Conidia one-celled, globose to ellipsoidal or allantoid, ± hyaline, aggregated in slimy heads or chains in some lignicolous species (1891). — In species with a conspicuous collarette, the first-formed conidium is fully differentiated inside the closed phialide apex (sometimes even two or more conidia) before the wall bursts apically. In *Ph. bubakii* (Laxa) Schol-Schwarz *1970* (formerly placed in *Margarinomyces* Laxa (495, 1108)), the collarette is very short but may be more conspicuous in older colonies; intermediate forms exist and make the distinction of *Margarinomyces* impractical.

Hyaline or pink *Phialophora* isolates may resemble *Acremonium* Link ex Fr. (q.v.), but they usually produce abundant adelophialides in superficial hyphae, whilst in *Acremonium* these occur only on submerged hyphae and the collarettes are less pronounced. *Cladorrhinum* Sacc. & Marchal (q.v.) has more broadly spreading colonies with sporulation limited to superficial pustules consisting of fertile hyphae in which most cells form a sessile collarette (pleurophialides). *Myrioconium* Syd. *1912*, the micro-conidial state of numerous Sclerotiniaceae, has broadly spreading, hyaline or centrally pigmented colonies consisting of broad hyphae, and sporulates mostly in sporodochium-like dense clusters of hyaline phialides which have very conspicuous collarettes and form subglobose conidia. — *Phialophora*, as circumscribed by Schol-Schwarz (5130), is a rather heterogeneous assemblage of anamorphs of unrelated ascomycetes. These anamorphs can roughly be grouped according to their teleomorph relationships (Table II), but this grouping covers only a fraction of the known approximately 35 anamorphic taxa; nine additional species with catenulate conidia have recently been described by Gams (1891). A considerable number of other species still require description, but species delimitation is particularly difficult in this genus; it must be appreciated that *Ph. fastigiata, Ph. malorum* and *Ph. hoffmannii*, for example, are species groups rather than clearly defined taxa.

TABLE II. Teleomorph connections within the genus *Phialophora*

Teleomorphic genus	Characters of the *Phialophora* state
Mollisia — Pyrenopeziza	Olivaceous-black colonies with aerial mycelium
Ascocoryne	Purplish brown to violet slimy colonies with little aerial mycelium and a strong tendency to sporodochium formation
Gaeumannomyces	Broadly spreading, olivaceous-brown colonies, differentiated into pigmented, broad runner hyphae and paler, thinner, lateral hyphae; conidia mostly elongate and curved
Coniochaeta	Colonies slimy, cream to orange or sometimes with purplish brown tinges in the centre; adelophialides predominant (*Ph. hoffmannii* group)

Phialophora species are very common on decaying wood in which they may cause a blue stain (3749) and are also frequent in wood pulp (6168); they may become soil-borne secondarily. Five species are vascular plant parasites of which only one, *Ph. cinerescens*, is treated here as it is also a root parasite (cf. also *Gaeumannomyces*). *Ph. verrucosa* and *Ph. americana* have sometimes been found as causal agents of chromoblastomycosis in man (comparable with that produced by *Exophiala* species). — Isolation, maintenance and examination usually pose few problems, but for standardization streak cultures on 2% MEA and OA incubated at about 20°C are recommended.

Key to the more important species which may be soil-borne:

| 1 | At least some conidia globose or subglobose | **2** |
| | Globose or subglobose conidia absent | **3** |

2(1) Conidia all regularly globose to dacryoid, hyaline, $1 \cdot 5$–$2 \cdot 5\,\mu$m diam   *Ph. cyclaminis* (p. 625)
 Conidia dimorphic: (a) globose, brown, $2 \cdot 5$–$3 \cdot 5\,\mu$m diam, and (b) ellipsoidal, hyaline, 2–$6 \times 1 \cdot 0$–$3 \cdot 0\,\mu$m   *Ph. richardsiae* (p. 628)

3(1) Colonies lacking aerial mycelium, often producing adelophialides, cream to orange, sometimes darkening in the centre with age; with or without chlamydospores   **4**
 Colonies usually producing some aerial mycelium and mostly distinct phialides, grey or olivaceous-black; without chlamydospores   **6**

4(3) Colonies remaining pink to orange   (*Ph. hoffmannii* group)
 Colonies darkening in the centre   **5**

5(4) Colonies darkening due to abundant chlamydospores (mostly 5–$8 \times 3 \cdot 5$–$5 \cdot 5\,\mu$m, terminal or intercalary)   (*Ph. mutabilis* group)
 Colonies darkening due to pigmented swollen hyphal cells   (*Ph. lignicola*)

6(3)   Colonies spreading broadly (filling a Petri dish in two weeks or less), with differentiated darker
       and broader "runner hyphae"; conidia mostly sickle-shaped   *cf. Gaeumannomyces* (p. 342)
       Colonies growing more restrictedly, without runner hyphae; conidia mostly straight   **7**

7(6)   Colonies greyish; phialides arranged in dense clusters, conspicuously darkening at the short
       collarettes; conidia ± pigmented   *Ph. cinerescens* (p. 624)
       Colonies mostly rather olivaceous or dark brown; phialides less densely aggregated; conidia
       hyaline   **8**

8(7)   Collarette forming a distinct cup, darker than the rest of the phialide   **9**
       Collarette less pronounced, not darker than the rest of the phialide   **10**

9(8)   Collarette with divergent cup, not longer than wide   (*Ph. verrucosa*)
       Collarette with elongate cylindrical base, often much longer than wide
                                                                          Ph. americana (p. 622)

10(8) Colonies dark olivaceous-brown or reddish brown in early stages; hyphae and phialides
       pigmented and usually irregularly encrusted; conidia ovoid to ellipsoidal, usually
       3–6 × 2–3 μm   *Ph. fastigiata* (p. 626)
       Colonies pale olivaceous at first, becoming darker after two or more weeks; hyphae hyaline or
       slightly greenish and hardly encrusted; phialides hyaline; conidia ellipsoidal,
       4·5–7 × 2·5–3·0 μm   *Ph. malorum* (p. 627)

Phialophora americana (Nannf.) Hughes *1958* — Fig. 274.

≡ *Cadophora americana* Nannf. *1934*

DESCRIPTIONS: Melin and Nannfeldt (3749), Mangenot (3568), Moreau (3928), Schol-Schwarz
(5130), Udagawa and Takada (5984), Cole and Kendrick (1108), Wang (6168), Sivasithamparam (5390), and Ellis (1604) (mostly as *Ph. verrucosa* Medlar); conidiogenesis: Cole and
Samson (1109). — Colonies on MEA reaching 0·8–1·0 cm diam in ten days at 20°C and to
14 mm at 30°C, floccose to woolly, olivaceous-grey at first but later becoming black; reverse
olivaceous-black; vegetative hyphae brown, smooth-walled, sometimes with moniliform
swellings. Phialides arising directly from the hyphae or on short lateral branches in diverging
groups of 2–3, often also reduced to sessile collarettes, broadly ampulliform, pale brown or
olivaceous-brown, strongly constricted below the collarette, 8–26 × 3–5 μm; collarette conspicuous, vase-shaped, darker brown than the rest of the phialide, 3–5 μm long and
2·5–3 μm wide at the flaring margin; percurrent proliferation is often observed. Conidia
ovoid to ellipsoidal with a slightly apiculate base, hyaline but becoming brownish with age,
3–5 × 1·5–3·0 μm. Thick-walled typical chlamydospores absent. — *Ph. americana* is a very
distinctive species by virtue of its dark vase-shaped collarettes. *Ph. americana* has usually been
synonymized with *Ph. verrucosa* Medlar *1915*, but Hughes (2602) and Zweibel and Wang*
regarded the two species as distinct because of the much shorter, hardly vase-shaped collarette in
the latter. Human-pathogenic isolates are found amongst both and growth and sporulation are
better at 30 than at 20°C. Amongst the cultures kept at CBS only one represents *Ph.
verrucosa*; thus we assume that most reports compiled here refer to *Ph. americana*. — In
immunodiffusion and immunoelectrophoresis tests, *Ph. americana* ("*verrucosa*") shared 5 out
of 8 antigens with *Cladosporium carrionii* and 2 out of 8 with *Rhinocladiella pedrosoi*

* Scient. Publ. PanAm. Health Org. **356**: 91–100. *1978*.

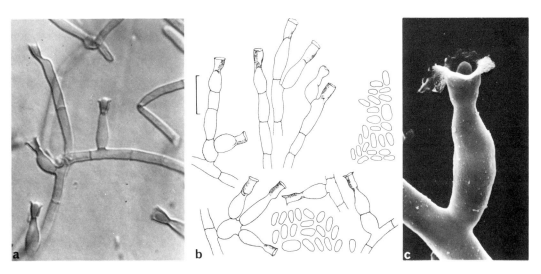

Fig. 274. *Phialophora americana*. a. Phialides with darker collarettes, NIC, × 1000, CBS 400.67; b. conidiophores and conidia; c. a single phialide, SEM, × 5000, orig. R. A. Samson.

(1177). — The septa show simple pores and Woronin bodies (1178). — Cell wall hydrolysates were found to contain 36% glucose, 8·3% mannose, 5·9% glucosamine, and traces of galactose (5672). — DNA analysis gave a GC content of 51% (5600).

Ph. americana is widely distributed in Europe and North America on wood, wood products, in soil and in air (1604). A selective isolation technique developed for human-pathogenic fungi and based on fractionated filtering facilitated its discovery at low inoculum densities in soil (3038). — The fungus has been isolated from soil in Spitsbergen (3064), cultivated soil under red clover in Canada (234), soil (3038) and palm leaves (3963) in Panama, wheat rhizosphere in Australia (5390), bark in New Guinea (5984), forest soil and plant debris, abandoned wasps' nests, birds' nests, and bird droppings in different locations in Uruguay (1960), ash stumps (3568), wooden boards of a bath-house (5016), and wood pulp and paper (6166). — Growth *in vitro* can be inhibited by *Pholiota nameko* (3085).

Manganese (Mn^{2+}) is required for conidium formation, with synergistic effects of Zn^{2+}, Fe^{2+} and Cu^{2+}. Ammonium salts and nitrates are suitable N sources. In man, sclerotic cells are formed and similar structures can also be induced on serum- or blood-containing agar media at 37°C; the absence of Mn^{2+} ions or a low pH (<3) stimulates their formation (4803). In one isolate abortive fruit-bodies have been observed; of 19 carbon and 44 nitrogen sources tested, L-arabinose, D-mannose, D-fructose, maltose and D-sorbitol with glutamic acid and numerous amino acids in combination with glucose gave the most luxurious growth (6298). *Ph. americana* caused very slight weight losses in birch wood and did not decompose cotton cellulose (3568). Hyphae grown in submerged culture contain 18% (dry wt) alkali-extractable proteinaceous melanins and exude another protein-carbohydrate pigment into the medium (4804). — *Ph. americana* (and *Ph. verrucosa*) have been reported from chromoblastomycosis and corneal ulcers of man in the USA and South America (759, 1623, 1624, 3502, 4098, 4586).

Phialophora cinerescens (Wollenw.) van Beyma *1940* — Fig. 275.

≡ *Verticillium cinerescens* Wollenw. *1930*

DESCRIPTIONS: van Beyma (495), Hantschke (2262), Moreau (3928), Schol-Schwarz (5130), Burge and Isaak (783), Ellis (1604), Tirilly and Moreau (5838), and CMI Descriptions No. 503, *1976.* — Colonies slow-growing, reaching 1·2–1·7 cm diam in ten days at 20°C on MEA, at first greyish but later becoming brownish grey, fluffy, woolly, zonate, reverse brownish grey to olivaceous-black; hyphae hyaline to pale brownish. Phialides usually formed in dense clusters on very short stalk cells, ± convergently bent, slender ampulliform, with a distinct, slightly flaring collarette, becoming brownish with age, phialide apex and collarette slightly darker than the rest of the phialide, 8–12 × 2·5–3·5 μm. Conidia slightly pigmented from their initiation and becoming greenish with age, ellipsoidal to cylindrical, slightly apiculate at the base, 3–5(–7) × 2·5–3·5 μm. Chlamydospores absent, but chlamydospore-like swollen hyphal cells occur. — Hyphal cells and conidia are usually uninucleate (2496); diploid conidia occur rarely under natural conditions but more frequently with actinomycin D treatment (5837).

Ph. cinerescens is known almost exclusively from its role as the most important causal agent of carnation wilt into which a considerable amount of phytopathological work has been conducted (e.g. 2262, 3928, 5838, and CMI Descriptions No. 503). Besides the ornamental carnation, numerous other Caryophyllaceae can also be attacked (4506). *Ph. cinerescens* is known from various European countries, including France (3935), Germany (2262), Denmark, the British Isles, Poland and Italy, but also the USA, New Zealand and the USSR (5130, CMI Descriptions No. 503). — A similar fungus isolated from tracheomycosis in *Helianthus annuus* but not pathogenic to carnations, is now regarded as *Ph. asteris* (Dawson) Burge & Isaac *1974* f.sp. *helianthi* Tirilly & C. Moreau *1976* (5838). There is one report of isolation from soil, under olive trees and cork oaks in Sardinia (488). In non-sterile soil, the

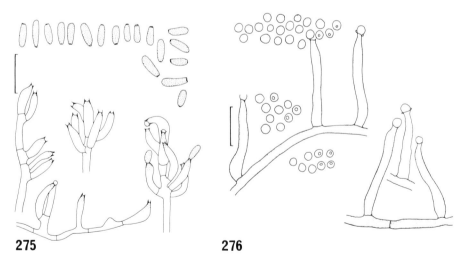

275   **276**

FIG. 275. *Phialophora cinerescens*, conidiophores and conidia, CBS 418.50.
FIG. 276. *Phialophora cyclaminis*, phialides and conidia, CBS 166.42.

fungus can survive as pigmented, thick-walled phialoconidia for over two years, but not with high moisture and raised temperatures (4187).

The optimal temperature for the induction of regular growth and sporulation is 11°C (3936, 5836); rapid changes in temperature may also stimulate sporulation; the optimum for linear growth is about 15–20°C (CMI Descriptions No. 503), and the infection of plants is more rapid at 22°C than at 15–20°C; conidium germination also has an optimum at 22°C (3937); the absolute maximum for growth is 33°C. At 22–28°C growth is irregular with a tendency to sector formation; subcultures from sectors are no longer pathogenic to carnation and antagonistic to *Pythium splendens* (5836), although the pathogenicity is not normally lost after prolonged cultivation. — Glucose and sucrose concentrations of up to 20% give increasing dry weights, but sporulation is reduced considerably above 5% glucose and less so with sucrose (3938). On Czapek-Dox agar better sporulation is obtained at higher temperatures than on 1% or 2% MEA; conidium size is inversely correlated with temperature (3939). No degradation of indolyl-3-acetic acid is observed but the activity of auxinoxidases from other sources may be enhanced (1476). *Ph. cinerescens* can cause erosion in birch wood (4191). On mineral media with sucrose it produces relatively large amounts of glutamic acid, alanine and phenylalanine (4151).

Phialophora cyclaminis van Beyma *1942* — Fig. 276.

DESCRIPTIONS: Schol-Schwarz (5130), Sivasithamparam (5390), and Ellis (1604). — Colonies reaching 3·0–3·5 cm diam in ten days at 20°C on MEA, woolly, dark olivaceous-grey; reverse olivaceous-black, hyphae pigmented, with darker septa. Phialides formed singly or on sparsely branched conidiophores, slender flask-shaped to cylindrical, brown, with a distinct, flared collarette, 15–30 × 3–4 μm. Conidia globose to dacryoid, hyaline, 1·5–2·5 μm diam. Chlamydospores absent. — In its mycelial characters, this species resembles the *Phialophora* state of *Mammaria echinobotryoides* Ces. (q.v.). — Some isolates with globose conidia but distinguished by shorter, swollen phialides and clusters of swollen cells commonly isolated from agricultural soils in Germany and the Netherlands have been described by Gams and Domsch (1887) as *Phialophora cyclaminis* groups A and B.

Ph. cyclaminis was originally isolated from *Cyclamen* tubers in the Netherlands, and has subsequently been obtained from wood in Italy, cocos fibres buried in cultivated soils, aquarium water in the Netherlands (5130, CBS), *Sida rhombifolia* and dung in India (521, 1604, CBS), and leaf-spots on *Cymbidium* in New Zealand (CBS). It has been isolated rarely from washed soil particles from wheatfields in Germany (1889), other arable soils in the Netherlands (1614, 1616), a wheat rhizosphere in Australia (5390), and soil in Zaïre (3790).

Ph. cyclaminis can utilize lignin and cellulose when growing on straw. It caused great losses in weight and tensile strength in maple-wood strips, with slight differences between individual isolates (2211), and was also found to attack aspen and birch wood (4191).

Phialophora fastigiata (Lagerb. & Melin) Conant *1937* — Fig. 277.

≡ *Cadophora fastigiata* Lagerb. & Melin *1927*

DESCRIPTIONS: Melin and Nannfeldt (3749), van Beyma (495), Hughes (2596), Mangenot (3569), Schol-Schwarz (5130), Wang (6168), Ellis (1603), Cole and Kendrik (1108), and Sivasithamparam (5390). — Colonies reaching 2·3–2·5 cm diam in ten days at 20°C on MEA, olivaceous-brown or reddish brown, floccose, frequently producing mycelial strands with a ± hyaline margin, reverse dark brown to black. Vegetative hyphae generally brownish and distinctly encrusted. Phialides borne singly or in densely convergent clusters on lateral branches, flask-shaped, light brown, with a conspicuous, funnel-shaped collarette, 6–15(–24) × 2·2–3(–4·2) μm; slightly constricted below the collarette to 1·0–1·8 μm wide. Conidia obovoid to ellipsoidal, usually apiculate at the base, hyaline, becoming light brown with age, 4·0–5·5(–6·7) × 2·0–3·0(–3·8) μm. Chlamydospores absent. — DNA analysis gave a GC content of 49·5% (5600). — *Ph. fastigiata* is a variable and poorly delimited species; the similar *Ph. melinii* (Nannf.) Conant *1936* grows somewhat faster (3·0 cm diam in ten days at 20°C on MEA), has more aerial mycelium, less conspicuous collarettes with inwardly curved margins and phialides more commonly borne singly (1108). *Ph. malorum* (q.v.) has generally less pigmented and more olivaceous colonies with less aerial mycelium, smooth-walled hyphae, and hyaline phialides with less conspicuous collarettes and slightly larger conidia.

Ph. fastigiata is the only *Phialophora* species that is fairly regularly specifically mentioned as a soil fungus in the literature and has been reported not only throughout Europe and North America (1603, 5130), but also from Uganda, the USSR, New Zealand, and Australia (IMI). It was originally observed on wood, in which it causes a blue discoloration (3209, 3210, 3749, 5061, 6572), and has often been found on rotten hardwood (3568), preservative-treated wood (4193), and also on stored pine wood chips (3140), sticks exposed to soil (1956, 2833), wooden boards of a bath-house (5016) and wood stumps (4757, IMI); wood pulp and slime from paper mills (1053, 1672, 1673, 5060) are preferential substrates. The first reports of this species were from countries rich in wood: Sweden (3210, 3749), Norway (4857), Canada (699, 1672) and (more rarely) from New York State (6168). There are only a few reports of isolations from soil, including the arctic tundra (1750), Spitsbergen (3064), sand on the volcanic island of Surtsey (2406), sediments below a glacier in Alaska (1171), arable soils in Germany (1889) and the Netherlands (1614, 1616), forest soils in Canada (505, 6101), pine forest podzols (307, 2068) and moorland soils in the British Isles (3234), conifer swamps (1039), and soils with steppe vegetation (6347). It has been isolated with equal frequencies from organic and mineral particles of washed forest soil samples (307). In an extremely acid podzol profile, *Ph. fastigiata* was found in the less acid C-horizon (2068). In arable soils, it occurred at low and variable frequencies (1889). In addition, it has been isolated from water and air in Sweden (3749), estuarine sediments in Germany (5316) and polluted running water in the USA (1154, 1166). — Isolations are also known from wheat, *Vicia faba*, *Mercurialis perennis*, *Aster* (IMI), wheat rhizosphere (5390), the rhizoplane of *Pinus nigra* var. *laricio* (4448) and seeds of cultivated grasses (3512). — The fungus was found to be inhibited by a *Streptomyces* species *in vitro* (2939).

Cellulose and pectin are degraded and growth on gallic acid occurs (1750). Amylase, xylanase, cellulase and mannanase are produced (4192). Wood of aspen, birch and pine is attacked and eroded (4191). — In a chronically γ-irradiated soil it did not occur at intensities of >6 R per day (2009).

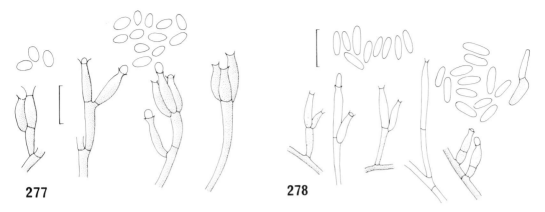

FIG. 277. *Phialophora fastigiata*, conidiophores and conidia, CBS 226.30.
FIG. 278. *Phialophora malorum*, conidiophores and conidia, CBS 260.32.

Phialophora malorum (Kidd & Beaum.) McColloch *1942* — Fig. 278.

≡ *Sporotrichum malorum* Kidd & Beaum. *1924*
= *Torula heroderae* Korab *1929*
  ≡ *Phialophora heroderae* (Korab) van Beyma *1942*
= *Phialophora luteo-olivacea* van Beyma *1940*
= *Phialophora atra* van Beyma *1942*
= *Phialophora goidanichii* Delitala *1952*

DESCRIPTIONS: McColloch (3698), Moreau (3928) (as *Ph. atra*), Schol-Schwarz (5130), Udagawa and Takada (5984), and Sivasithamparam (5390). — Colonies reaching 2·0–2·3 cm diam in ten days at 20°C on MEA, olivaceous-black but very variable between isolates (3698), with little aerial mycelium, centrally floccose, and with a radial striation apparent in the submerged mycelium; reverse olivaceous-black; vegetative hyphae pale olivaceous, mostly smooth-walled. Phialides arising singly or in clusters on stalk-cells, slender flask-shaped, often somewhat curved towards the conidiophore axis, hyaline, $6–12·5 \times 1·5–2·5 \mu$m. Conidia ellipsoidal to cylindrical with a slightly apiculate base, $4–7 \times 1·5–2·5(–3·0) \mu$m. Chlamydospores absent. — For species delimitation see under *Ph. fastigiata*.

Ph. malorum is almost as common as *Ph. fastigiata*, although less frequently reported. It is known from numerous European countries, the USA, Australia (5390) and New Guinea (5984). Isolates are known from wood, soil, various chemicals, waste water, stalactites, and air (5130, CBS). It has been isolated from sand on the new volcanic island of Surtsey (2406), the arctic tundra (1750), forest and forest steppe soils (1700), wheat rhizosphere (5390), preservative-treated wood (4193), cankerous apple wood and, most commonly, rotting apples (3698). — It can parasitize nematode cysts (CBS).

The optimal pH for growth is in the range 4·4–6·0. *Ph. malorum* is able to oxidize tannic and gallic acids (3698), degrade cotton cellulose and starch (1750, 3618), and grow on gallic and humic acids (1750). No growth occurs at 37°C (5984).

Phialophora richardsiae (Nannf.) Conant *1937* — Fig. 279.

≡ *Cadophora richardsiae* Nannf. apud Melin & Nannf. *1934*
= *Cadophora brunnescens* Davidson *1935*
  ≡ *Phialophora brunnescens* (Davidson) Conant *1937*
= *Phialophora caliciformis* G. Sm. *1962*

DESCRIPTIONS: Melin and Nannfeldt (3749), van Beyma (495), Brewer (699), Moreau (3928), Schol-Schwarz (5130), Wang (6168), Nicot (4164), Nicot and Caillat (4167), Cole and Kendrick (1108), and Ellis (1604); SEM and TEM studies of conidiogenesis: Oláh and Reisinger (4270). — Colonies reaching 2·0–2·5 cm diam in ten days at 20°C on MEA, floccose or woolly, grey-brown, fulvous or olivaceous-brown; reverse brown to black; vegetative hyphae ± brown. Phialides arising singly or in sparsely branching divergent conidiophores. Phialides and conidia of two kinds: primary phialides pale brown with inconspicuous cupulate collarettes, 8–16 × 1·5–3·0 μm, secondary phialides bearing very conspicuous, dark brown, broadly flaring collarettes, 8–25× 2·5–4·2 μm; conidia from the primary phialides ellipsoidal, hyaline, 2–6 × 1·0–3·0 μm, from secondary phialides (sub)glob-ose or dacryoid, brown, 2·5–3·5 μm diam, both types occurring in very variable proportions. Chlamydospores absent. — *Ph. richardsiae* is one of the most distinct, although rather variable, species in *Phialophora*. — The walls of hyaline conidia consist almost exclusively of polysaccharides, while in pigmented ones they have a thick melanized layer and contain much protein (4270).

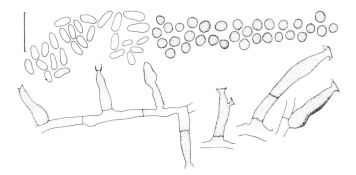

FIG. 279. *Phialophora richardsiae*, phialides and two kinds of phialoconidia, CBS 270.33.

Ph. richardsiae is a rather uncommon species but has a worldwide distribution. It has been reported from wood and ground wood pulp in Sweden (3749), the USA (6166) and central Africa (4164), the slime from paper mills in Canada (699) and fermented corn dough in Mexico (6003). Soil isolates are known from Tahiti (6084), Brazil (400), the USA (1039), the Netherlands (1614, 1616), Germany (CBS), and India (1604, 4933). Known habitats and substrates include open bogs and conifer swamps (1039), arable soils (1614, 1616) and grassland (4933), a sewage treatment plant and waste stabilization ponds, wax (CBS), plastics (4167), and kerncls of *Elaeis guineensis* (IMI). — Cultures stored at −30°C survived for over two years (3444).

The optimal temperature for growth is 25–30°C, the minimum 10–15°C, and the maximum <35°C. Suitable C sources are D-mannose, D-fructose, D-xylose and cellobiose, but no growth occurs on D-arabinose; good N sources are nitrate, L-proline, and L-alanine but not L-cysteine-hydrochloride (700). Cellulose is degraded (3303), and wood attacked and eroded (1496, 3442, 3443, 4191). — *Ph. richardsiae* has been associated on man with a subcutaneous cystic granuloma (1623, 5149) and further isolated from a prostate gland (5130).

Phoma Sacc. *1880* [nomen conservandum]

= *Plenodomus* Preuss *1851*
= *Leptophoma* Höhn. *1915*
= *Bakerophoma* Died. *1916*
= *Sclerophomella* Höhn. *1918*
= *Polyopeus* Horne *1920*
= *Deuterophoma* Petri *1929*
= *Peyronellaea* Goid. ex Togliani *1952*

Type species: *Phoma herbarum* Westend.

GENERIC DEFINITION: Boerema and Bollen (593). — Pycnidia separate, not aggregated on a stroma and with a single ostiole or occasionally confluent and multiostiolate; wall thin, with a pigmented surface, sometimes with a sclerotioid base (sect. *Plenodomus* (Preuss) v. Arx). Conidiogenous cells hardly differentiated from the inner wall cells, with a narrow phialidic opening, producing large numbers of slimy, mostly one-celled, ellipsoidal to cylindrical, hyaline conidia. — Conidium formation in *Phoma* was first described as a monopolar repetitive budding but subsequently recognized as being normally phialidic because each new conidium is delimited at the base by a three-layered septum (593). If septation occurs in the conidia, the whole wall is involved in septum formation (secondary septation). Pigmented chlamydospores are formed in some species either in simple chains or aggregated in "dictyochlamydospores" resembling the conidia of *Alternaria* or *Stemphylium* (formerly distinguished as *Peyronellaea*).

Phoma is by far the largest and most widely distributed genus of the Sphaeropsidales. Whilst many species have previously been included in *Phyllosticta* because of their growth on leaves, in *Ascochyta* because of their septate conidia, or in *Pyrenochaeta* because of setose pycnidia, these genera can now be delimited more precisely on details of conidium formation, with the consequence that the soil-borne species with hyaline conidia can almost exclusively be accommodated in *Phoma*.

Cultures are best grown on OA and 4% MEA (based on malt syrup) in Petri dishes, centrally inoculated by an agar disk, and then kept at 20–22°C in darkness for seven days followed by seven days alternating light/darkness. For colour reactions, a drop of conc. NaOH is applied at the margin of colonies growing on MEA or OA. — The identification of *Phoma* species requires well standardized techniques of culture and the simultaneous inspection of several characters, since single characters are usually not specific. Therefore Boerema and Dorenbosch (594, 1439) and Boerema (592) have produced only synoptic tables for identification. The following dichotomous key is based on the data of Dorenbosch (1439) but should ideally be used in conjunction with her synoptic table.

Key to the species treated:

1   Pigmented chlamydospores present   **2**
        Pigmented chlamydospores absent (pale sclerotium-like bodies sometimes present)   **5**

2(1)   Chlamydospores in single rows only, loosely catenulate, without longitudinal septa   **3**
        Chlamydospores in compact chains, often with longitudinal septa (dictyochlamydospores)   **4**

3(2)   Chlamydospores elongate, 4·5(–10) μm diam; conidia always one-celled; not producing crystals
        on MEA   *Ph. eupyrena* (p. 632)
        Chlamydospores ± globose, 10–14(–22) μm diam; conidia partly two-celled; often producing
        crystals on MEA   *Ph. medicaginis* var. *pinodella* (p. 640)

4(2)   Catenulate dictyochlamydospores typically present; fresh isolates usually with little aerial
        mycelium   *Ph. glomerata* (p. 636)
        Chlamydospores rarely with longitudinal septa, and if so, not in long chains; colonies mostly with
        abundant olivaceous aerial mycelium   *Ph. pomorum* (p. 642)

5(1)   Pycnidia pseudosclerenchymatous, the wall becoming red with iodine (sect. *Plenodomus*)   **6**
        Pycnidia not pseudosclerenchymatous, wall not becoming red with iodine   **7**

6(5)   Pycnidia depressed hemispherical with a rather sharply delimited ostiolar beak; colonies
        reaching 2·5–3·0 cm diam in 7 days at 20°C on OA; mycelial mats greyish, often with pinkish
        discoloration of the medium; not host-specific   (*Leptosphaeria doliolum*)
        Pycnidia flask-shaped with gradually emerging ostiolar beak; colonies reaching 1·5–5·0 cm diam
        in 7 days at 20°C on OA; mycelial mat whitish, greenish or yellowish brown; mostly associated
        with crucifers   *Leptosphaeria maculans* (p. 404)

7(5)   Mature pycnidia with dense dark short setae around the ostiole   *Ph. leveillei* (p. 639)
        Pycnidia not or sparsely setose   **8**

8(7)   Conidia small (not exceeding 5 μm in length)   **9**
        Conidia commonly exceeding 5 μm in length   **10**

9(8)   Colonies reaching 1·5–3 cm diam in 7 days on OA, with greyish green aerial mat, producing
        intense yellow-green pigment in the reverse, particularly on OA; conidia mostly 2·5–4·0 ×
        2·0 μm, hardly guttulate   (*Ph. fimeti*)
        Colonies reaching 5–7 cm diam in 7 days on OA, with little aerial mycelium, sometimes with buff
        to apricot reverse; conidia mostly 3·5–4·0 × 2·0 μm, distinctly (1–)2-guttulate
                                                  (*Ph. putaminum*)

10(8)   Colonies slow-growing, reaching 2–4 cm diam in 7 days on OA, with grey felty aerial mycelium
        and an orange-red reverse; pseudosclerotial bodies often present; conidia rod-shaped,
        3·5–6·5 × 1·5-2·5 μm   *Ph. chrysanthemicola* (p. 632)
        Colonies usually faster growing, reverse uncoloured or darker red, but then with little aerial
        mycelium; pseudosclerotia absent; conidia usually wider   **11**

11(10) Colonies with little aerial mycelium, often producing a red pigment in the agar (MEA), which
        turns violet-blue on addition of NaOH; conidia always one-celled   *Ph. herbarum* (p. 636)
        Colonies with abundant floccose, whitish to olivaceous-grey aerial mycelium, not producing a red
        pigment in the agar, but turning green and then red on addition of NaOH (on MEA); conidia
        sometimes two-celled   *Ph. exigua* (p. 634)

Phoma chrysanthemicola Hollós *1907*

DESCRIPTIONS: Hawkins *et al.* (2336), Dorenbosch (1439), and Schneider and Plate (5122). — Colonies rather slow-growing, reaching (2–)6–8 cm diam in seven days at 20–22°C on OA, with pale to dark grey aerial mycelium and the agar characteristically discoloured orange-red (fading with addition of NaOH). Pycnidia sparingly produced, often hairy; conidia rod-shaped, $3\cdot5$–$6\cdot5 \times 1\cdot5$–$2\cdot5\,\mu$m. Pseudosclerotial bodies often present. — *Ph. chrysanthemicola* is not to be confused with *Ph. chrysanthemi* Vogl. *1902* which causes flower blight in chrysanthemums. *Ph. chrysanthemicola* f.sp. *chrysanthemicola* R. Schneider & Boerema *1975* is regarded as a specialized form exclusively pathogenic to *Chrysanthemum indicum* (5119), whilst other soil-borne isolates are not.

Ph. chrysanthemicola was originally described from dead stems of *Chrysanthemum indicum* but later re-described as a pathogen of various *Chrysanthemum* species in the British Isles (2336, 4489), the Netherlands (1439), Germany (5122) and India (5511). In the Netherlands it has also been isolated from *Achillea millefolium* and other wild plants (1439, 5119) and in Belgium from stunted roots of *Scorzonera* (659). In Germany (1889) and the Netherlands (1614) it has been found in cultivated soil, and in Australia in the rhizosphere of wheat (IMI).

The optimal growth temperature lies between 22 and 28°C, the minimum is 2°C, and the maximum 35°C. Pycnidia develop best in direct sunlight or under near-UV at 300–400 nm (5122); mineral media with cellulose, glycerol, sorbitol or mannitol are suitable for pycnidium production (659). — Pectinases and xylanases are produced, with little difference in capacity between individual isolates (1432). *Ph. chrysanthemicola* causes great losses in both weight and tensile strength when growing on maple-wood strips (2211). — The growth *in vitro* of the alga *Chlorella pyrenoidosa* was found to be stimulated by this fungus (1430).

Phoma eupyrena Sacc. *1879* — Fig. 280.

DESCRIPTIONS: Wollenweber (6421), Dennis (1346), Boerema and van Kesteren (601), Kranz (3120), Dorenbosch (1439), and Boerema (592). — Colonies reaching $4\cdot5$ cm diam in seven days at 20–22°C on OA, at first dark green but soon becoming black from the chlamydospores which start to be produced after one week, chlamydospores arising singly or in chains, consisting of elongate cells, mostly 4–5(–10)μm diam. Conidia ellipsoidal, $3\cdot5$–$6 \times 1\cdot5$–$3\cdot0\,\mu$m. — Wollenweber (6421) was the first author to describe the chlamydospores in culture and the species has since then often been referred to as "sensu Wollenweber"; as similar chlamydospores were present also on the type specimen (1439) this qualification is not necessary. — *Ph. eupyrena* might be confused with *Ph. pomorum* (q.v.) which does not always form dictyochlamydospores and often only in old cultures; in that species the individual chlamydospore cells are generally globose (595).

In spite of few literature reports, *Ph. eupyrena* can be regarded as a very frequent soil fungus, as in arable soils in both Germany and the Netherlands it occurs with high frequencies (1614, 1616, 1889, 4791) but it is also known from the USA (1387), the British Isles, India and Malaysia (4698, IMI). The presence of chlamydospores seems to favour its isolation from

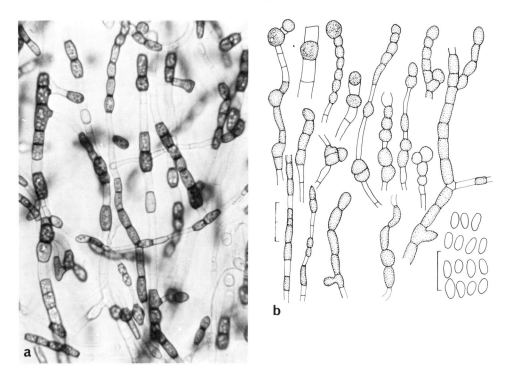

FIG. 280. *Phoma eupyrena*. a. Chlamydospore chains, × 500; b. chlamydospore chains and pycnoconidia.

washed soil particles, whilst the conidia account for high numbers obtained in dilution plate studies. — *Ph. eupyrena* prefers the upper soil layers (6184), has been reported from sandy forest nursery soils (3908, 6184), and has been isolated somewhat more frequently after growing rape than after wheat or peas (1433). It was first described as growing in association with potatoes (3542, 6421) but has also been isolated frequently from soils where potatoes had never previously been grown (1889). It is also known from Usar soils (4698) and grassland (IMI) and was the first *Phoma* species to be discovered on the volcanic island of Surtsey (592). Thirty-eight per cent of chloroform-fumigated mycelium was mineralized within ten days at 22°C (133). In soils sterilized by steam and formalin, it was found among the first colonizers (3908, 6184). Other substrates include the rhizospheres of potato, wheat and grasses (1614), frost-damaged leaves of various grasses (191), seedlings of various weed plants (2247), *Zingiber officinale*, *Rubus idaeus* and *Salvia splendens* (IMI), various other plant remains (601), stained maple wood (6572), activated sludge (1387), and earthworm casts, of which it may be a frequent inhabitant (1429).

The optimal temperature for growth is about 20–21°C, the maximum 28–32°C, the minimum −3°C and the pH optimum 4 (191, 3120); pycnidia form mainly below 20°C. The intensity of chlamydospore formation varies from isolate to isolate and is dependent on the sugar content of the medium (3542). *Ph. eupyrena* shows maximal growth on media with 1% NaCl (1428). — Pectin and xylan are decomposed rapidly; the capacity of different isolates is, however, variable (1432). Ferulic acid is utilized (544) and birch wood is attacked (4191). The development of the alga *Chlorella pyrenoidosa* is promoted *in vitro* (1430) by this fungus which also causes an economically unimportant dry rot in potato tubers (3120, 6421); it is mostly regarded as a harmless secondary invader (601, 1346, 3542).

Phoma exigua Desm. *1849* var. *exigua* — Fig. 281.

= *Phyllosticta destructiva* Desm. *1847* [non *Phoma destructiva* Plowright *1881*]
= *Ascochyta phaseolorum* Sacc. *1878*
= *Phoma solanicola* Prill. & Delacr. *1890*
= *Ascochyta pirina* Pegl. *1894*
= *Phoma tuberosa* Melhus *et al. 1961*

For about 100 synonyms see van der Aa and van Kesteren (1), Boerema and Dorenbosch (594), and Boerema (592).

DESCRIPTIONS: Dennis (1346), Malcolmson (3542), Boerema and van Kesteren (601), Maas (3478), CMI Descriptions No. 81, *1966*, Boerema and Höweler (600), Dorenbosch (1439), van der Aa and van Kesteren (1), Boerema and Dorenbosch (594), and Boerema (592); TEM study of conidiogenesis: Boerema and Bollen (593). — Colonies on MEA growing slightly faster than on OA, reaching (2–)6–8 cm diam in seven days at 20–22°C on OA, variable in habit, whitish to olivaceous-grey, irregularly lobed. On application of a drop of NaOH, a blue-green reaction changing to reddish brown coloration occurs (particularly on MEA). Conidia oblong, mostly $4.0–8.5 \times 2.0–3.0\,\mu$m, sometimes two-celled. — Whilst the var. *exigua*, characterized by an irregular growth pattern, is the most frequent *Phoma* species on many kinds of plant remains (594), several more host-specific varieties have been distinguished: var. *linicola* (Naum. & Vass.) Maas *1965* (≡ *Ascochyta linicola* Naum. & Vass. *1926*) which causes footrot in flax and has more uniform, slow-growing colonies; var. *foveata* (Foister) Boerema *1967* (≡ *Phoma foveata* Foister *1940*) which causes "gangrene" in potatoes and forms an abundant yellow-brown pigment (red in alkaline media) on MEA (particularly with the addition of 25 ppm thiophanate methyl to the medium) and similar crystals consisting of pachybasin and three other anthraquinones; var. *sambuci-nigrae* (Sacc.) Boerema & Höweler *1967* (≡ *Phoma herbarum* f. *sambuci-nigrae* Sacc. *1884*) which causes leaf spots in *Sambucus nigra* and forms uniform, faster growing colonies; and var. *inoxydabilis* Boerema & Vegh apud Vegh *et al. 1974* which causes stem blight in *Vinca minor* and does not give any NaOH reaction. — The formation of pycnidia can be stimulated by a chemical factor from a complementary strain when they are grown together (5232). — All following data refer to var. *exigua* alone (unless otherwise stated).

This fungus has been reported from Germany (1889), the Netherlands (1614, 1616), Sweden (1446), Libya (6510), and India (5840). It has been isolated from more than 200 genera of phanerogams (594). On dying plant substrates in the close proximity of soil, this is the commonest pycnidial fungus found, but it also has a wide distribution in the soil itself. Soil reports include grassland (5840), arable soils (1616, 1889) under potato and wheat (1614), soils with tundra vegetation (1446), an alkaline sandy soil (6510) and the rhizospheres of potato and wheat (1614). It has also been isolated by means of strip-baits from the air above a potato field (1640) and from cysts of the nematode *Globodera rostochiensis* (3090).

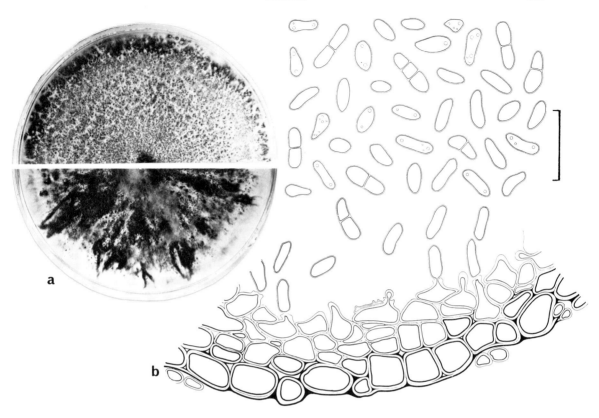

FIG. 281. *Phoma exigua*. a. Two different two-week-old cultures on OA, orig. G. H. Boerema (from 594); b. section through pycnidium showing conidiogenous cells and conidia, orig. G. H. Boerema (from 600, reproduced with permission of the Editors of Persoonia).

The optimal temperature for the germination of conidia is 25°C, and the maximum 28°C, while optimal growth occurs at 22–26°C (3405), the maximum at about 30°C and the minimum at 4°C; the pH optimum lies between pH 4 and 6 depending on the isolate (3120). At high concentrations conidia do not germinate because of self-inhibition; heating to 40–42°C for 10 min reduces the percentage germination markedly in both var. *exigua* and var. *foveata* (1641). — Pectin and xylan are both decomposed (1432). Metabolites identified include an antibiotic substance "E" active against both gram positive and negative bacteria as well as *Fusarium coeruleum* and *Phytophthora infestans* (3406), and the cytochalasins A and B (= phomine) with phytotoxic properties (669, 5161). — Antagonistic activity is also exhibited against *Gaeumannomyces graminis*, *Rhizoctonia solani* and *Pythium ultimum* (1431). The germination of flax seed was inhibited by two phytotoxins produced by var. *linicola* (6171); var. *foveata* is a true parasite on all parts of the potato plant (601, 1346, 1639, 3120), its occurrence being favoured by high soil moisture levels (3543), and var. *exigua* can cause considerable losses in dry weight in roots of wheat, pea and rape seedlings *in vitro* (1430), and var. *inoxydabilis* (cited as var. *exigua*) was found to be implicated in die-back of stems and wilting of *Vinca minor* (4473).

Phoma glomerata (Corda) Wollenw. & Hochapfel *1936* — Fig. 282.

≡ *Coniothyrium glomeratum* Corda *1840*
≡ *Peyronellaea glomerata* (Corda) Goid. ex ̀ ̀ oghani *1952*
= *Phoma radicis-vaccinii* Ternetz *1907*
= *Phoma alternariacea* F. T. Brooks & Searle *1921*
= *Phoma hominis* Agostini & Tredici *1937*

For further synonyms see Boerema *et al.* (595, 596), Boerema and Dorenbosch (594).

DESCRIPTIONS: Boerema and van Kesteren (601), Sutton (5649), Boerema *et al.* (595, 598), CMI Descriptions No. 134, *1967*, Dorenbosch (1439), Boerema and Dorenbosch (594), and Boerema (592). — Colonies fast-growing, reaching (3·5–)5–7·5 cm diam in seven days at 20–22°C on OA, producing mostly little aerial mycelium (although sectors with woolly olivaceous aerial mycelium may occur) and abundant pycnidia, olivaceous buff to dull green, darkening from the production of dictyochlamydospores. Pycnidia mostly regularly globose. Conidia ellipsoidal, mostly 6–7·5 × 3·0–3·5 μm, becoming light olivaceous. Dictyochlamydospores commonly arising in unbranched or branched chains of 2–20, resembling the conidia of *Alternaria*, usually 18–20 × 12–30 μm.

Ph. glomerata is a ubiquitous fungus, frequently found on dead plant material (595, 2338, 4409), and numerous host plants have been listed (594). Its occurrence in soil has been reviewed (595); it is particularly frequently reported from cultivated soils under groundnuts (2768, 2772, IMI), a citrus plantation in Israel (2764), leguminous fields in India (1317, IMI), high altitudes in Peru (2005), a sandy beach ridge in Canada (IMI), estuarine sediments in North Carolina/USA (655) and also from children's sandpits (1421) and frescoes of a monastery (2666). It colonizes roots (595), the rhizosphere of groundnuts (2768), and has been reported from onion seeds in India (5467) and rice seeds in Ghana (1273), Nigeria and South Africa (597). In lenticels of potatoes, it is regarded as a secondary invader (601). In wheat, it can cause severe leaf spotting after a wet period of 24 h (2542); it was found to be associated with a leaf spot disease in apples without being the causal agent (3486).

Cotton cellulose is degraded by *Phoma glomerata* (3618), and its enzyme C_x, which is produced on cellulose, sucrose and cellobiose, has an optimal activity at 30°C and in a broad pH range of 5·0–7·6 (1580). *Ph. glomerata* causes a soft-rot in wood (1496) and decomposition of birch wood has been observed (4191). — Its pathogenicity to man (granuloma of the foot, mycoses of the hand and genital tract, otomycosis, rhinitis, etc., and allergenic effects) has been described in several cases (595).

Phoma herbarum Westend. *1852*

= *Phoma urticae* S. Schulz. & Sacc. *1869*
= *Phoma oleracea* Sacc. *1882*
= *Phoma violacea* (Bertel *1904*) Eveleigh *1961*
= *Phoma pigmentivora* Massee *1911*
= *Phoma hibernica* Grimes, O'Connor & Cummins *1932*
= *Phoma lignicola* Rennerfelt *1937*

DESCRIPTIONS: Grimes *et al.* (2119), Dennis (1346), Eveleigh (1670), Sutton (5649), and

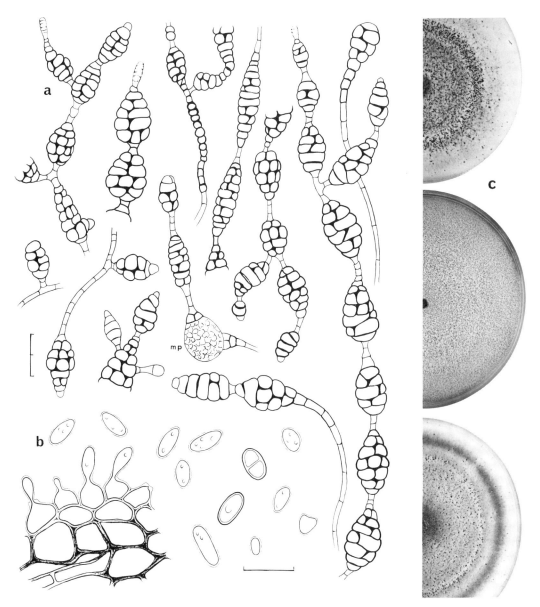

FIG. 282. *Phoma glomerata*. a. Chains of dictyochlamydospores (mp = micropycnidium developing from a dictyo-chlamydospore); b. conidiogenous cells and conidia, orig. G. H. Boerema (from 595, reproduced with permission of the Editors of Persoonia); c. three different two-week-old cultures on OA, orig. G. H. Boerema (from 594).

Boerema (590, 592); TEM study of conidiogenesis: Boerema and Bollen (593). — Colonies reaching 3·5–4·5 cm diam in seven days at 20–22°C on OA. Aerial mycelium usually sparse, pycnidia abundant and producing orange conidial slime (sectors with more olivaceous-brown aerial mycelium may occur); a reddish pigment is usually produced in the agar which, on the application of NaOH, turns blue. Conidia ellipsoidal, mostly one-celled and 4·5–5 × 2·0–2·5 μm. No dictyochlamydospores are produced. — Several varieties described of this species proved to be either indistinguishable from *Ph. herbarum* or to belong to different species (591).

Ph. herbarum has a worldwide distribution and is known from the most varied substrates including herbaceous and woody plants, soil and water (590, 601, 1670, 2338). In soil analyses, it is the *Phoma* species that is most frequently determined with certainty; its frequency of occurrence, however, in our experience, is lower than that of *Ph. eupyrena.* — In addition to numerous reports from the temperate zone, there are many from subtropical areas including Iraq (92), Egypt (8), Israel (2764, 2768, 2772, 4759) and India (1524, 2854, 3863, 3865, 4030, 4698, 5000). Known habitats include forest (2854, 3549, 3798), uncultivated (4030, 4698) and cultivated soils (1614, 1616, 2768, 4152), grassland (3865), soils with tundra vegetation (1750), sand dunes (6414), garden compost (1424), fresh (4229) and polluted water (1166), eutrophicated river sediments (1162), salt-marshes (8), estuarine sediments of all degrees of salinity (655), activated sludge (1387), fields treated with sewage sludge (1163) and a sewage sludge treatment plant (1165). *Ph. herbarum* tolerates the seasonal drying of the soil (5000). It has been isolated from decaying stems of *Bothriochloa pertusa* (2955) and *Urtica dioica* (6463), and the rhizospheres of *Ammophila breviligulata, Andropogon scoparius* (6414), groundnut (2768), *Setaria glauca* (3271), potato (1614) and wheat (IMI); also found on pecans (2258, 2572), hay (4821), stained wood of pine (812) and birch (5061), and slime of a paper mill (5060). — It can serve as food for the mite *Pygmephorus mesembrinae* (3104).

The rate of linear growth is highly variable but is somewhat reduced by constant light; the formation of the violet pigment, however, is stimulated by light, whilst the culture becomes black in the dark. *Ph. herbarum* grows in the pH range 3–>8, with optima at both 4·5 and 7·5; the optimal temperature for pycnidium production is at 18–21°C and for vegetative growth 22–25°C, while the maximum is at 35°C. The thermal death point has been determined as 75°C for 30 min in soil (613). — Fair growth is obtained in an O_2-free atmosphere but biotin and thiamine are required under these conditions (5680); ten days exposure to 10 atm of pure O_2 were lethal (848). Conidia germinate at water potentials to −210 bars (1670). — *Ph. herbarum* utilizes CMC and xylan well (1432), and also starch, pectin (1750), and a large number of other carbohydrates and fatty acids (1669); weak growth is obtained on gallic acid (1750). *Ph. herbarum* is regarded as one of the causal organisms of a soft-rot in wood (1496). It is autotrophic for growth substances (1669). Known metabolites include the anthraquinones helminthosporin and cynodontin (3231), three isomerous chromanones (1857) and a light-induced sporogenic substance (5893). — *Ph. herbarum* inhibits the growth of the alga *Chlorella pyrenoidosa in vitro.* It proves to be pathogenic to *Avena fatua* seedlings after artificial infection (2961, 2962), although otherwise it is not a primary pathogen. It was once isolated from a lesion on a human leg but it is not known whether *Ph. herbarum* was the primary pathogen in that case (303).

Phoma leveillei Boerema & Bollen *1975* — Fig. 283.

≡ *Pyrenochaeta acicola* (Lév.) Sacc. *1884*
≡ *Vermicularia acicola* Lév. *1848* [non *Phoma acicola* (Lév.) Sacc. *1881*]
= *Pyrenochaeta spinaciae* Verona & Negru *1966*

DESCRIPTIONS: Gams and Domsch (1887), Dorenbosch (1439), and Boerema and Bollen (593); TEM study of conidiogenesis and nomenclature: Boerema and Bollen (593). — Colonies slow-growing, reaching 2–3 cm diam in seven days at 20–22°C on OA, grey with a light olivaceous tint, velvety. Pycnidia abundantly produced in the centre of the colonies, olivaceous-brown, darker around the ostiole and covered with short (30–70 μm long), thick-walled, dark brown setae; conidia ellipsoidal to cylindrical, mostly 3–7 × 1–3 μm. — The soil-borne *Ph. leveillei* differs from the similarly setose root parasite of onions, *Phoma terrestris* Hansen *1929* (≡ *Pyrenochaeta terrestris* (Hansen) Gorenz *et al. 1948*), in its slower growth and the absence of any red pigmentation of the colonies. On the basis of conidiogenesis, *Ph. leveillei* belongs to *Phoma*, while the setose pycnidia are similar to those of *Pyrenochaeta* de Not. *1849*; that genus, however, has elongate, branched conidiophores with several lateral and terminal phialidic openings.

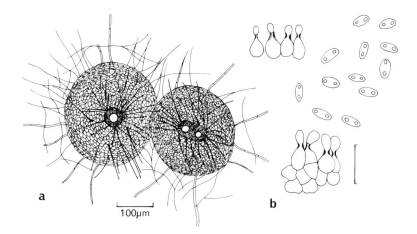

a

100µm

b

FIG. 283. *Phoma leveillei*. a. Setose pycnidia seen from above; b. conidiogenous cells and conidia; orig. J. Veerkamp.

Ph. leveillei, originally described as *Vermicularia acicola* from fallen pine needles in the Vosges, has since been isolated fairly frequently from washed particles of two arable soils in Germany (1889), and found to be abundant on mineral particles of agricultural soils in the Netherlands (1614); in the latter case its frequency was found to be increased by monocultures of grass or wheat (1615). In the USA it has been isolated from soil of the Sonoran desert (4733), and in Canada from a sandy beach ridge (IMI). Other habitats include the rhizospheres of potato, wheat and grasses (1614), *Fragaria vesca*, leaves and seedlings of pine, wood of *Ribes* and *Populus*, stem bases of *Callistephus* and *Campanula*, and decomposing wood (1439).

Xylan, pectin and cellulose are all readily decomposed and there is little difference in these abilities between isolates (1432); cellulose decomposition is slow and only at low concentration of the N source (4433, 4434). *Ph. leveillei* causes considerable losses in weight and tensile strength when grown on maple-wood strips (2211).

Phoma medicaginis Malbr. & Roum. *1886* var. *pinodella* (L. K. Jones) Boerema *1965* — Fig. 284.

≡ *Ascochyta pinodella* L. K. Jones *1927*
= *Ascochyta imperfecta* Peck *1912*
= *Phoma trifolii* E. M. Johnson & Valleau *1933*

DESCRIPTIONS: Boerema *et al.* (599), Dorenbosch (1439), and CMI Descriptions No. 518, *1976*. — Colonies reaching 4·0–6·5 cm diam in seven days at 20–22°C on OA, very variable in their production of mycelium and pycnidia, flat, with light grey to black aerial mycelium. Pycnidia usually arising in radial sectors but sometimes also irregularly scattered. Conidia ellipsoidal, 4–9 × 2–4 μm, sometimes two-celled. Crystals forming on MEA or OA; after some weeks these are macroscopically visible. Chlamydospores consisting of rows of dark, swollen cells, 10–14(–22) μm diam, particularly frequent in dark sectors, abundantly produced after one week. — *Ph. medicaginis* var. *pinodella* is a weak parasite of pea, red clover and other leguminous plants; it is differentiated from the similar var. *medicaginis*, causal agent of black stem and footrot in alfalfa, which produces macroscopically uniform colonies with a grey, gradually darkening, felty mat of aerial mycelium, regularly scattered pycnidia, and fewer crystals. *Ph. medicaginis* is distinguished from *Ph. eupyrena* by the larger and less numerous chlamydospores. — Heterokaryosis and parasexuality have been observed (5036).

Ph. medicaginis var. *pinodella* has been isolated from arable soils in both Germany (1889) and the Netherlands (1439, 1614, 1616). On washed soil particles, it was found to occur with a significantly greater frequency after growing peas than after wheat or rape (1433). It has also been isolated from seeds and stems of peas (6407, IMI), clover roots (6493) and the rhizosphere of potato plants (1614); in one study, it was isolated more frequently from corn than pea roots (4940). — *Ph. medicaginis* var. *pinodella* is inhibited *in vitro* by metabolites from the coprophilous *Stilbella erythrocephala* (5362).

Optimal pycnidium formation in var. *medicaginis* occurred at 30°C and on galactose, maltose, lactose, cellobiose, starch, inulin and pectin as C sources, and with nitrate or L-isoleucine as N sources; conidia are, however, produced most abundantly at 20°C and on cellobiose or lactose as C sources; ammonium salts, L-histidine and L-leucine are all able to stimulate chlamydospore formation (1052). Conidia are self-inhibiting when germinating in high density on dry agar surfaces (1051). — Filter paper (4748) and CMC are rapidly attacked, but in the latter case there are considerable differences in capacity between individual isolates (1432). O- (3,13)-diacetyl-gibberellic acid is deacetylated (5116). — Pronounced antagonistic activity towards *Pythium ultimum* has been observed (1431). Severe injuries to pea roots have been found in laboratory experiments (1430). Var. *medicaginis* produces a non-specific

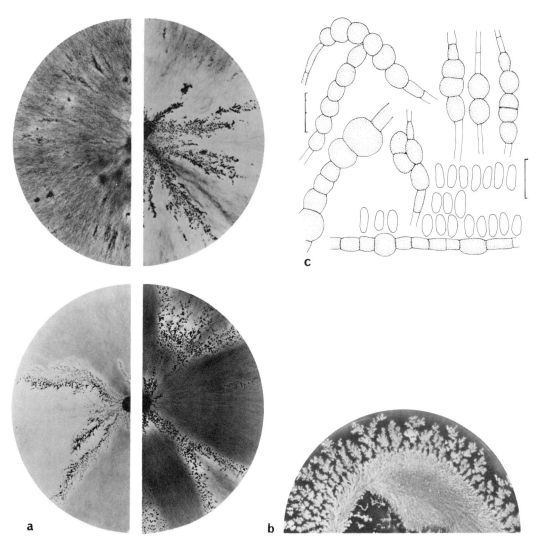

FIG. 284. *Phoma medicaginis* var. *pinodella*. a. Four different three-week-old cultures on cherry decoction agar; b. culture on MEA showing radiating crystals, orig. G. H. Boerema (from 599, reproduced with permission Neth. Soc. Pl. Path.); c. chlamydospore chains and pycnoconidia.

inhibitor of seed germination in rape and flax (3215), but otherwise this fungus is only a weak seed- and soil-borne parasite of leguminous plants, becoming injurious only under exceptional conditions (599).

Phoma pomorum Thüm. *1879* — Fig. 285.

> = *Phoma prunicola* (Opitz) Wollenw. & Hochapfel *1936*
>> ≡ *Depazea prunicola* Opitz *1852* [non *Phoma prunicola* Schw. *1832*]
>> ≡ *Peyronellaea prunicola* (Opitz) Goid. *1946*
> = *Phyllosticta pyrina* Sacc. *1878*
> = *Phyllosticta cydoniicola* Allesch. *1897*
> = *Peyronellaea nicotiae* Leduc *1958*

For further synonyms and nomenclature see Boerema *et al.* (595–598), and Boerema and Dorenbosch (594).

DESCRIPTIONS: Sutton (5649), CMI Descriptions No. 135, *1967*, Ristanović (4851), Dorenbosch (1439), Boerema and Dorenbosch (594), and Boerema *et al.* (595, 598); nomenclature: Boerema *et al.* (597); TEM studies of conidiogenesis: Boerema and Bollen (593), and Jones (2811). — Colonies moderately fast growing, reaching 4–6(–7·5) cm diam in seven days on OA, mostly with abundant olivaceous aerial mycelium and pycnidia scattered throughout. Pycnidia often with a furrowed wall and distinct ostiole. Conidia ellipsoidal, mostly 5–7 × 2·0–3·4 μm, eventually becoming light olivaceous. Chlamydospores solitary, mostly in single rows, 8–10 μm diam, sometimes also with longitudinal septa and forming dictyo-chlamydospores; the latter usually formed singly and resembling the conidia of *Stemphylium*.

This soil-borne fungus has been found on a wide variety of dead and diseased plant material (595, 596, 1439); a list of the host plants has been compiled (594). It has been reported from the arctic (1750), the USA (2573), the Netherlands, Nigeria (IMI), Turkey (4245) and India (4698). *Ph. pomorum* is perhaps best known as the causal agent of leaf spots on *Prunus* and *Pyrus* species in Europe (595, 1439, 4851). It has also been found on *Fagopyrum tataricum* and in the rhizosphere of groundnuts (IMI).

The optimal temperature for germination and growth is 18–21°C; light has no influence and the pH range of 3–10 is tolerated with an optimum at 7–8 for growth and at both pH 6·0–6·5 and 9–10 for sporulation (4851). — Pectin, starch and cellulose are slowly degraded by this species while inulin is not utilized; moderate growth occurs on both gallic and tannic acids (1750, 4852).

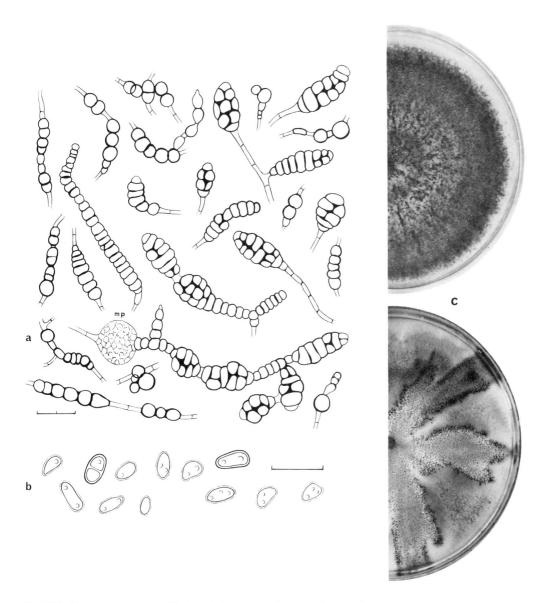

FIG. 285. *Phoma pomorum*. a. Chains of dictyochlamydospores (mp = micropycnidium developing from a dictyo-chlamydospore); b. conidia; orig. G. H. Boerema (from 595, reproduced with permission of the Editors of Persoonia); c. two different two-week-old cultures on OA, orig. G. H. Boerema (from 594).

Phymatotrichopsis Henneb. *1973*

Phymatotrichopsis omnivora (Duggar) Henneb. *1973* — Fig. 286.

≡ *Phymatotrichum omnivorum* Duggar *1916*
≡ *Ozonium omnivorum* Shear *1907* (mycelial strands)

The true teleomorph is unknown, despite several reports to the contrary which have mentioned species such as *Hydnum omnivorum* Shear *1925*, *Sistotrema brinkmannii* (Bres.) J. Erikss. *1948* (≡ *Trechispora brinkmannii* (Bres.) Roberts & Jackson *1943*) (323, 563, 4899).

DESCRIPTIONS: Neal *et al.* (4116), and Bloss (563); generic diagnosis: Hennebert (2400); sclerotia: King and Loomis (2986).—Colonies very fast growing *in vitro*, with 1 cm daily increment on MEA at 25°C, spiderweb-like, floccose, matted or compacted, yellow-ochraceous to brown, hyphae often funiculose (mycelial strands) and forming aerial cruciate, hair-like setae. Conidiophores borne laterally on the hyphae, simple or branched, moniliform, with the terminal and subterminal cells inflated, globose to ellipsoidal, $20–28 \times 15–20\,\mu$m, producing numerous blastoconidia ± simultaneously. Conidia globose to ovate, smooth- and thin-walled, with a rather broad hilum, $4·8–5·5\,\mu$m diam, or $6–8 \times 5–6\,\mu$m. Sparsely sporulating *in vitro*. Sclerotia readily produced from hyphal strands on sterile soil, round to ovoid, 1–2 mm diam, often aggregated into clusters of about 10 mm diam, at first light yellow but later becoming reddish brown, consisting of closely packed, thin-walled intermixed cells of large and small sizes and a thin cover of brown, moderately thicker-walled cells. — *Phymatotrichopsis* is a monotypic genus and all other species hitherto placed in *Phymatotrichum* have been referred elsewhere (2400). Reports of clamp connections (323, 563) are probably based on cultures contaminated with the commonly air-borne *Sistotrema brinkmannii*. — The major phosphatides in mycelia, conidia and sclerotia are phosphatidyl choline, phosphatidyl ethanolamine and lysophosphatidyl choline; the most abundant unsaturated fatty acids were linoleic and oleic, while the predominant saturated one was found to be palmitic (2170). The lipid composition in the mycelium and sclerotia differ with 47·9% and 27·1% for polar lipids, 1·6 and 17·5% for monoglycerides, and 14·3 and 6·7% for diglycerides, respectively (2173).

Ph. omnivora is a common causal agent of root rot and wilt on cotton, particularly in the southern and southeastern USA and northern Mexico (563, 2986, 4116, 5248), but it occurs on over 2000 dicotyledonous plants (563, 5753). Interestingly, citrus does not support its growth (2986). Cotton seedlings are highly resistant to attack, notwithstanding any saprophytic micro-organisms present or the carbon content of the roots (1530); but the plants are most susceptible when fruiting, due to a reduced carbon content of the root bark (1530); this is also the period of highest CO_2 concentration in the upper soil layers which is conducive to infection (3463). — *Ph. omnivora* is prevalent in alkaline, calcareous montmorillonite clays with a pH about 8·2 (3463) but does not occur in either more alkaline clay soils with a high sodium content or acid soils of pH 4·7 (3459), where little sclerotium and strand formation occurs and the fungus cannot survive. Growth and pathogenicity are limited in

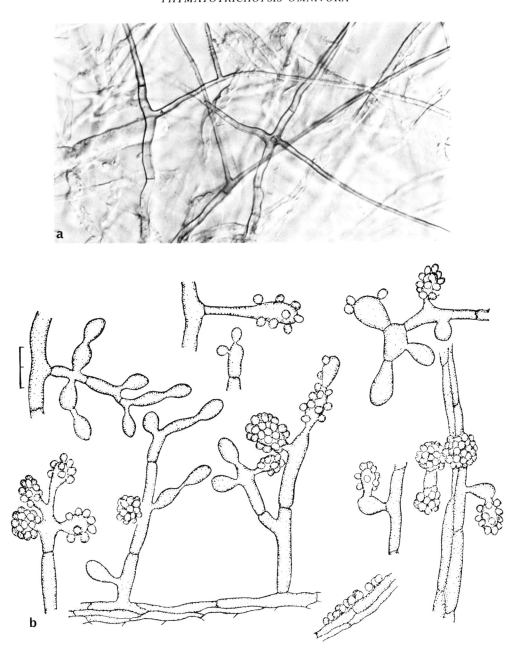

FIG. 286. *Phymatotrichopsis omnivora*. a. Thick-walled cruciate hyphae formed in the aerial mycelium, NIC, × 500, CBS 183.79; b. types of conidiophores and conidium formation, from Duggar (Ann. Mo. bot. Gd. 3: 16, 1916, reproduced with the Editors' permission).

manured soils (2985) and fertilization can reduce losses by enhancing plant vigour and accelerating crop maturation (2818). *Ph. omnivora* may occur in soil independently of plant roots in both cultivated and fallow soil; strands are detectable in cotton field soil six weeks before any symptoms in the plants themselves become evident (1529). — Sclerotia are the most endurable resting stage of this species; these can be quantitatively recovered by sieving soil with 1 mm mesh sieves (3463). The greatest numbers of sclerotia have been found in lawn and arable soils in the depth range 30–60(–90) cm but they can occur down to 244 cm (2984, 2986, 3459). Conidium formation has been observed along barrier trenches down to 180 cm (2986). — In the surface layers of soil, the fungus is rapidly killed by heat and drought in the summer months (4898). Mycelium enclosed in glass fibre bags and introduced into the rhizosphere of cotton was dead after four months but it can survive in naturally infected roots for nine months at 10°C (6317). Hyphal strands obtained from soil cultures in the laboratory were still viable after eight months but in fallow field soils they can survive for one to several years (2984, 4116). Sclerotia on cotton roots in slightly moist sand survived and were found to germinate normally after over 2·5 years (2984). Populations of *Ph. omnivora* can be drastically reduced by cropping pea or barley or adding pea plants to the soil; this may be due to a premature germination of the resting stages of the fungus which then become vulnerable to antagonistic micro-organisms (973). An increased carbon content (sugars and polysaccharides) of cotton roots (induced by removing branches and bolls) can check the infection, and recovery occurs by an overgrowth of new root bark; this process is accompanied by a strong increase of *Pseudomonas* species in the rhizosphere which may antagonize *Ph. omnivora*. On corn roots, antagonistic phenomena are more pronounced and the fungus cannot colonize these at all unless the soil is sterilized (1530).

Light is a prerequisite for conidium production from the time a yellow pigment forms which may act as a photoreceptor (324). Conidium formation can be enhanced on media containing fatty acids or sterols (324); they also germinate rapidly on these media. — Strand formation is reported to be markedly less on agar media than on soil and fails on liquid media (564); strands can form on various types of agar media, unless K_2HPO_4 exceeds 1·4%, or glucose, asparagine, peptone and $MgSO_4·7H_2O$ are added (564). — *Ph. omnivora* has an unusually high CO_2 tolerance (2452) and on agar media growth still occurs at 74% CO_2 (3463); at 100% CO_2 or N_2 no growth occurs, but the fungus is still alive after eight days exposure to such conditions (4115). Molecular nitrogen is taken up to a limited extent and ethylene produced (3464). With 5% CO_2, more glucose is metabolized through the EMP pathway than is the case in air (2166). CO_2 accumulation in soil enhances both growth and sclerotium formation (3462, 3463). In non-sterile soil columns good linear growth (1·6 cm per day) occurred in the range 0·5–5% CO_2 but not with air or 50% CO_2, the latter causing a decrease of the pH to 6·8 and a twenty-fold increase in free calcium ions which may themselves be inhibitory (3463). It has been concluded that HCO_3^- ions are essential to this fungus and the assimilation of CO_2 by starved mycelium has been observed; conversely, soil aeration by deep ploughing gives some measure of control (3459, 3463). — The optimal temperature for growth and sclerotium and hyphal strand formation is (25–)27·5(–30)°C (3460, 3461, 4898), little growth occurs below 20°C, none at 10°C, good growth still at 35°C, and the maximum for growth is 37°C (4898). Sclerotium production is favoured in non-sterile soil cultures by the introduction of a fresh inoculum (from diseased roots), by a high moisture content, and also by temperatures in the range 21–32°C (1261). Sclerotia develop their full pigmentation only at higher temperatures (4898). — The catalase content of sclerotia reaches a maximum after a post-ripening period and then gradually declines; a decline in germinability correlates with that of catalase (2983). Sclerotia require a short dry period prior to germination in soil

(563). — The pH range for growth *in vitro* is 3–7 (3460). — Glycogen is a suitable C source (563, 1677, 2168, 6317) but growth can also be supported by D-glucose, D-fructose, and D-mannose (553), D-galactose, D-maltose, sucrose, lactose, xylose, D-mannitol, inulin, dextrin, starch and cellulose of different kinds (3914). The utilization of C sources was found to be variable depending on whether a solid or liquid medium was used. At reduced O_2 partial pressures, growth is retarded more strongly in liquid than on agar media (1677, 3914). Substrates metabolized initially are highly reduced compounds (lipids) and *Ph. omnivora* does not respond to exogenous oxidized substrates (carbohydrates) unless endogenous reserves are depleted (3460). Both organic and inorganic N sources are equally well utilized (1677). The uptake of inorganic N sources is determined by the balance of other ions in the solution. Nitrite is tolerated and can be utilized (5714, 5715), and ammonium can have an inhibitory influence if its uptake causes a marked fall in the pH (552). An optimal balance of mineral salts in solution was found to be 8 mM K_2HPO_4, 3 mM $MgSO_4 \cdot 7H_2O$, 2 mM KCl, and 2 ppm Fe, Mn, and Zn ions (5714). *Ph. omnivora* produces proteinases (3231); enzymes of the Embden-Meyerhof-Parnas pathway and hexose-monophosphate pathway have been demonstrated (2167). — A phytotoxin is produced in its infections of plant roots which kills the roots prior to penetration by the fungus (6222).

Phytophthora de Bary *1876*

= *Pythiomorpha* Petersen *1909*

Type species: *Phytophthora infestans* (Mont.) de Bary

MONOGRAPHIC TREATMENT: Tucker (5936); keys to the species; Waterhouse (6216)*; compilation of all published diagnoses: Waterhouse (6219); review of development: Zentmyer and Erwin (6557); studies on mating behaviour: Haasis and Nelson (2197).

Fam. Pythiaceae Schröt. — Colonies comparatively slow-growing, in most species not exceeding 4·0 cm diam in four days at 25°C on CMA and other media. Oogonia developing as in *Pythium*; antheridium stalk usually arising at a considerable distance from the oogonium, thus making attempts to distinguish monoclinous and diclinous meaningless; the antheridial cell is termed paragynous when it is applied to the side of the oogonium (mostly near its base), and amphigynous when the oogonial hypha first pierces through the antheridium before forming the oogonium. Sporangia are either produced singly or in a sympodial succession on mostly unbranched hyphae; internal proliferation often also occurs; zoospores are liberated through an apical opening where the wall is often thickened and projecting to form a papilla; otherwise sporangia may germinate directly by a germ tube and often form a secondary sporangium. Thin-walled hyphal swellings, not delimited from the hyphae by septa, occur in some species; in others they are thick-walled, delimited by a septum and termed chlamydospores.

Genetic (1584) and cytological (688) studies have demonstrated that meiosis occurs in the gametangia and that the vegetative mycelium and sporangia are diploid (1864).

Phytophthora species are not normally soil fungi, but rather highly specialized plant parasites, more specialized than in *Pythium*, and more often attacking above-ground plant parts where they can form aerial, sometimes wind-borne, sporangia. The importance of terrestrial populations has been reviewed by Schmitthenner (5113), and their distribution by Tucker (5937).

Most techniques of maintenance and identification are the same as listed under *Pythium*. The formation of sporangia is generally dependent on water; *Ph. cactorum* produces sporangia also on CMA, OA, bean agar or water agar. Petri's solution or diluted soil extracts or pond water are generally suitable. Oogonia are produced on various agar media, including V8-juice agar or a synthetic medium containing glucose, asparagine, thiamine and β-sitosterol (4826). Mating of A1 × A2 types in heterothallic species on CMA or OA is often used to obtain oogonia, but these may also be induced by contact with strains of a different species of opposite mating type. Oogonium production can also be induced by mechanical damage or the influence of H_2O_2, diethyl ether, *Trichoderma viride* (4787) or root extracts (6554).

Phytophthora species generally grow more slowly than those of *Pythium* and are therefore less easily isolated by selective media. Baiting with buried avocados, pears, tomatoes, strawberries or unripe apples (3910) was used successfully before the advent of selective inhibitors of other fungi which have been listed under *Pythium* here (2387). The addition of

---

* and Newhook, F. J. *et al. in* Mycol. Pap. **143**. *1978.*

gallic acid, as well as pimaricin, has proved particularly suitable in increasing the selectivity (1783). Baiting with lupin radicles (4619) or conifer needles is still valuable for the detection of low inoculum levels (1262). The success of selective isolation is mainly determined by the germinability of the propagules (oogonia not being recovered) (1262). Amongst several antibacterial antibiotics tested, aureomycin and terramycin inhibited growth of *Phytophthora* considerably, streptomycin moderately, and chloramphenicol and erythromycin not at all (5153).

Amongst the species compiled by Waterhouse (6219), some 44 species are to be recognized and six species and one variety have been described since then. Only two rather ubiquituous and polyphagous species are treated here.

Key to the species treated:

1   Sporangia readily produced on agar media in sympodial succession, with an apical, hemispherical, prominent wall thickening; oogonia and antheridia readily produced; thick-walled, globose chlamydospores sometimes present   *Ph. cactorum* (p. 649)

      Sporangia produced only upon transfer to water, in sympodial succession or with internal proliferation, with an inconspicuous apical wall thickening; oogonia and antheridia rare in single cultures but inducible by combination with opposite mating types of *Ph. cryptogea* or several other species of *Phytophthora*; thin-walled, globose hyphal swellings in terminal and lateral clusters are commonly produced   *Ph. cinnamomi* (p. 653)

Phytophthora cactorum (Leb. & Cohn) Schröt. *1886* — Fig. 287.

   ≡ *Peronospora cactorum* Leb. & Cohn *1870*
   ≡ *Phytophthora omnivora* de Bary *1881*
 = *Phytophthora fagi* (Hartig) Hartig *1879*
   ≡ *Peronospora fagi* Hartig *1876*
 = *Phytophthora paeoniae* Cooper & Porter *1928*

DESCRIPTIONS: Blackwell (547), Stamps (5524), Waterhouse (6216, 6219), Ershad (1649), and CMI Descriptions No. 111, *1966*. — Sporangia readily produced on various agar media in sympodial succession, 36–50(–55) × 28–35(–40) μm, separating with part of the pedicel (about 4 μm); sporangial apex with a hemispherical, prominent, thickened papilla, not proliferating, leaving a narrow pore on dehiscence. Homothallic; oogonia and antheridia readily produced; oogonia mostly 27–33 μm diam, oospores aplerotic, 20–26 μm diam; antheridia paragynous, attached near the base of the oogonium. Chlamydospores thick-walled (1·0–1·5 μm), globose, 19–33(–55) μm diam, occurring sometimes, particularly under the influence of fluctuating temperatures. — In the similar *Ph. nicotianae* van Breda de Haan *1896* (incl. *Ph. parasitica* Dastur *1913*), chlamydospores are much more common, whilst oogonia appear rarely and usually only after mating. *Ph. citricola* Sawada *1927*, considered the same as *Ph. cactorum* var. *applanata* Chester *1932* by some authors (5155), has a less

FIG. 287. *Phytophthora cactorum*, sporangia and oogonia with paragynous antheridia, CBS 268.55.

prominent apical papilla but is physiologically and pathologically similar to *Ph. cactorum*. — Mutations have been induced by X-ray irradiation of >4·5 krd (767) and *N*-methyl-*N*'-nitro-*N*-nitrosoguanidine (1584). — *Ph. cactorum* is one of the most clearly defined *Phytophthora* species but a considerable variability has been demonstrated to occur within it (767, 5524); isolates differing in growth rate have been derived from one isolate in the sexual phase but not from sporangia (3499), while in another study the zoospores from one sporangium gave rise to different colony types which also differed in their fertility (5524). — Acrylamide-gel-electrophoresis of the proteins of three *Phytophthora* species showed that the *Ph. cactorum* isolates belong to one well-defined group (1982, 2222). — DNA analysis gave a GC content of 53·5% (5600).

Ph. cactorum has a worldwide distribution, mainly in temperate zones. It is known as a root, stem and die-back pathogen of some 160 host plants, but is particularly important in the Rosaceae (4175, 4729), and does not usually occur in soil analyses. It has been isolated selectively by burying apple (3712, 5154), lupin rootlets and conifer needles (1262) or safflower seedling baits (326), glass fibre strips (3276), media containing antibiotics (1540, 5450), benomyl, Na taurocholate (3658, 3714) or gallic acid (1783), and on selective media even by dilution plates (3784, 5452). — Reports of its occurrence are available from all over Europe, the USA (4153), Canada (3712), Iran (1649), the southern parts of Africa, Australia, New Zealand, Venezuela, and Argentina, as reviewed by Meyer (3784). It has been isolated from a great number of plants or fruits, including apple, pear, plum (e.g. 97, 689, 5087, 5088) where it causes a collar rot on the stem and a fruit rot, also from strawberry with crown and fruit rot (5088), and various forest tree seedlings (3784). Isolates from different host plants have been found to differ in their virulence to other plants, but no biotypes have been separated as special forms (689, CMI Descriptions No. 111). Wet and poorly drained soils are reported to favour infection (3784). Soils around both healthy and diseased apple trees are usually infested (326) to depths of 25 cm, exceptionally 45 cm (689), and it is concluded that the fungus is able to remain there in a saprophytic phase. The propagules that can be isolated from soil are mainly sporangia and chlamydospores (3785). Because of a great sensitivity of these structures to frost (3785, 5458) and drought, the fungus cannot be isolated at mean air temperatures below 8°C (5225) and it almost disappears from soil in December to January in Europe (689). Contrary to sporangia which are formed freely in soil, the oospores which develop only in host tissues (3784, 3785) are far more resistant and generally do not germinate immediately but only after a period of frost; it is these that are responsible for the further survival of the population; they can germinate after a resting period by the formation of germ tubes or a sporangium. The longest recorded period of survival in natural, host-free soil of medium moisture is two years (327), and at a water potential of -0.5 bars more than 300 days (5458); *in vitro* it survived a storage in water for $\geqslant 7$ years (605). Sporangia are generally regarded as the most important means of distribution (1996). Sporangia and zoospores originating from infected fruits left in water can reach the crowns of fruit trees through irrigation water (689). The mycelium does not grow out from sporangia in soil without supplementary nutrients, but zoospores can be released (1997) at temperatures of 10–30°C (3785); the mycelial growth can extend over 2–3 mm in non-sterile, and 20–30 mm in sterile, soil (3785). At -0.3 bars water potential and 29°C, 80% of the mycelium was lysed within two days, at 15°C only 10%, and at 4°C most of it remained viable for 30 days; mycelium does not survive in air-dried soil (5458). In other studies mycelium was found to survive for 6–10 days in non-sterile soil (3784, 3785) but was completely lysed within two weeks (1997). Zoospores remain motile in soil between 15 min and 3 h, depending on the soil moisture level, and can remain viable for up to 32 days (3784); survival of the fungus is generally favoured by low temperatures and low soil moisture (326, 3713), but may be reduced by urea (3713). — The fate of sporangia and mycelia has been studied after marking with a fluorescent brightener (a stilbene compound) (1995, 1996): sporangia spread evenly over a model soil penetrated to only 2.5 mm (1995); growth from mycelium or zoospores in sterile soil is found to be stimulated by the addition of glucose, but then no sporulation occurs; in non-sterile soil, sporangium formation is reported to be correlated with the soil moisture content (1996). At 15°C sporangia germinate indirectly by the formation of zoospores, but at 25°C mostly directly with the production of germ tubes and often also a secondary sporangium (1996, 3785), also in the temperature range 24–30°C and on an agar medium containing mineral salts, DL-threonine, thiamine and sucrose, 80% of the added sporangia germinated directly and $\leqslant 1\%$ released zoospores (5452). — Oospores formed in field soils germinated after at least 4 weeks rest at 10–25°C (3785). Otherwise their germination requires a longer resting period including some frost, or treatment with glucuronidase and sulphatase (327) or $CaCl_2$ (1582), all of which can speed up the maturation process. Light is required for oospore germination, with optimal efficiency at wave lengths of either 400–450 or about 700 nm, suggesting that a phytochrome acts as receptor; oospores can consequently germinate only at the soil surface (888). The optimum temperature for oospore germination is either 16°C (689) or 20°C (327), with none occurring at 28°C (the optimum for mycelial growth); $CaCl_2$ at $\geqslant 25$ ppm reduces germination (327). At germination, the oospore wall is partly dissolved from the inside and a germ tube protrudes which often forms a sporangium a short distance away (3784). — The application of castor-bean meal reduced infections of fallen apples, but hardly affected the *Phytophthora* population in the soil (689). On infected strawberry plants transplanted to disinfected soil, the fungus can spread more easily than in non-sterile soil (3909). Both these observations have been attributed to antibiotic effects exerted by soil micro-organisms; both bacteria and actinomycetes have been repeatedly found to show antibiotic activity *in vitro* against *Ph. cactorum* (3784). Oospores inoculated into test soils were found to be parasitized by various hyphomycetes, bacteria and chytridiomycetes (5454). From apple bark, stimulatory, inhibitory (phloridzin) and absolutely inhibitory proteinaceous fractions have been extracted (1925).

The optimal temperature for mycelial growth is (20–)25–26(–28)°C, the minimum $\leqslant 2\cdot 5$°C, and the maximum 30–32·5°C (1649, 5153, 5155, 6216), while the optimal pH is 4·5–5·0 (854, 855). — Sporangia form, for example, on semi-solid (0·2% agar) lima-bean agar at various concentrations (5109) and can be promoted by nutrient deficiency or addition of sterols (1997, 2388). No sporangia were found to occur on mycelia buried in soil at 4°C and −0·1 to −3·0 bars water potential for eight days; most sporangia appeared at 10–29°C and −0·1 bars water potential (5458). No sporangia appear under reduced O_2 partial pressures (3881). — Zoospores show a positive chemotactic response towards host roots (3784) and a wide variety of chemicals, mostly known as components of root exudates (2957). Glucose additions to soil induce the germination of sporangia which is, however, inhibited by asparagine (5458). — Oospores are abundantly produced on rape-seed agar (5066), media supplemented with sterols (1581, 1997, 2278, 2388, 3251, 5071, 5318), or with wheat germ oil when asparagine is used as an N source (3027). Sterols (cholesterol, ergosterol, cholestanol) are not so much incorporated into but adsorbed by the hyphae (1583, 5318). Cultures grown with β-sitosterol show an increase in cell mass and also in the specific activities of aldolase, glutamic acid transaminase and glucose-6-phosphate dehydrogenase (846). L-Cystine proved to be the amino acid most efficient in stimulating oospore production, but also suitable are DL-valine, L-aspartic acid, L-asparagine, L-glutamic acid, creatine, DL-citrulline, DL-serine, L-histidine, L-tyrosine, and L-threonine (3253). Optimum O_2/CO_2 gas mixtures for oospore formation are in the range 20:0–1:0–1:5 (%/%), higher CO_2 proportions inhibiting oospore formation (3881); optimal growth occurs at 21% O_2 and reduced growth at lower O_2 concentrations, the lowest for visible growth being 0·2% (1203). Equal numbers of oospores are found to be produced in both darkness and under red light, while either white or blue light gave only a minimal production (2277). Pure oospore preparations can be obtained by treating the mycelium for seven days with a mixture of cellulase and hemicellulase (5451). — For vegetative growth, the best C source out of ten tested was sucrose (1046); both soya oil and cholesterol were found to be superior to glucose in this respect (2388). Growth of the mycelium is also stimulated by wheat germ oil (3027), and thiamine is reported to be required for growth (2413, 4860). Suitable N sources are L-arginine, L-histidine, DL-alanine, L-asparagine, L-aspartic acid, L-glutamine, L-glutamic acid, glycine, L-proline, L-serine, L-threonine, L-phenylalanine, casein hydrolysate, $Ca(NO_3)_2$, and NH_4NO_3 (855, 1046, 3252); growth is stimulated by $CaCl_2$ (1582). *Ph. cactorum* hydrolyses starch very rapidly (2470). — It is sensitive to streptomycine which is accumulated in its hyphae more strongly than in those of fungi insensitive to this antibiotic (6112).

Phytophthora cinnamomi Rands *1922* — Fig. 288.

DESCRIPTIONS: Rands (4724), Tucker (5936), Waterhouse (6216, 6219), Ho and Zentmyer (2471), and CMI Descriptions No. 113, *1966*. — Sporangia produced in sympodial succession or with internal proliferations, mostly $57-67 \times 33-39 \mu$m, with an inconspicuous apical thickening, leaving a wide (over 8μm diam) opening on dehiscence. Sporangium formation can best be induced by non-sterile soil extracts or specific bacteria; under axenic conditions only a thorough washing with a salt solution has proved successful (6557). *Ph. cinnamomi* is heterothallic but oogonium formation has also been observed in single cultures, for example on oat grains submerged in water for 2–3 months (4934); oogonia and antheridia can be induced by mating with opposite strains of *Ph. cinnamomi* (1863, 4621), *Ph. cryptogea* or other *Phytophthora* species; the compatibility types A1 and A2 are not correlated with the formation of either antheridia or oogonia (1864, 5071), but A1 isolates usually grow more slowly (3881) and produce sporangia and hyphal swellings more readily (5255). Oogonia $32-40(-58) \mu$m diam, smooth-walled; oospores almost plerotic, wall 2μm thick; antheridia amphigynous. Most characteristic of this species are thin-walled, globose hyphal swellings (often termed "chlamydospores"), about 40μm diam and abundantly produced in terminal or lateral clusters; TEM studies of these hyphal swellings showed them to be very similar to sporangia in having an unstratified wall, but to differ mainly by the absence of an exit pore for zoospore release, flagella, flagellar vacuoles and cleavage vesicles (2372); before germination a new layer is formed inside, which gives rise to the germ tube (6219). — Similar but less abundant and not botryose hyphal swellings occur also in *Ph. cambivora* Petri *1917* which has verrucose oogonial walls and produces sexual organs in combination with *Ph. nicotianae*. — The morphological variability of *Ph. cinnamomi* is rather low (975, 5254). Both fast-growing (0·6–1·1 cm increment per day at 25°C) and slow-growing (less than 0·6 cm per day) isolates of compatibility type A2 have been observed (5254), depending on their origin. — Zoospores are mostly uninucleate (5254), the chromosome number of the diploid state being determined as 9–10 (688). Interspecific combinations with *Ph. nicotianae* sometimes gave progeny from germinated oospores showing recombinant characters, but this observation could also have arisen by self-fertilization (572). — Different isolates showed identical or similar protein patterns, three different esterase, and identical peroxidase patterns

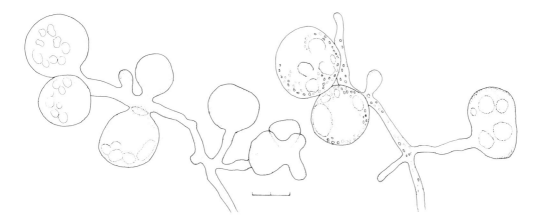

FIG. 288. *Phytophthora cinnamomi*, hyphal swellings.

(1064, 2222); *Ph. cinnamomi* has been distinguished serologically from related species (3778). — The cell wall hydrolysates have been found to contain 88% glucose (from cellulose), 0·3% glucosamine, 0·6% mannose, and traces of galactosamine and ribose; the dominating amino acids of the mycelium are threonine (14·8%), glutamine (11·1%), serine (9·8%), alanine (9·0%), and asparagine (9·0 mol% of the total amino acids) (3232). — DNA analyses of different isolates gave GC contents in the range 49–57% (5600).

Ph. cinnamomi has a worldwide distribution, occurs on over 900 host plants, but is most frequent in tropical and subtropical zones (1212, 3775). — The use of seeds or roots of *Lupinus angustifolius* as bait facilitates both its isolation and differentiation from other *Phytophthora* species (974, 1954, 1955, 4619, 4620); other useful baits include conifer needles in water above soil (1262), rooted pineapple crowns or heart leaves, and avocado fruits (128, 3028, 6558). Selective media with suitable antibiotics (1540, 2387, 3162, 3658, 5905, 5906) have also been described for its isolation from soil. — Isolates have been obtained from soils in Australia (728, 1306, 3608, 3610, 4619, 4620, 4622), Indonesia (4724), New Zealand, Malaysia, the Fiji Islands, Papua-New Guinea, Hawaii (128, 426, 3029, 3032, 3033), and the USA (2198, 2381, 2385, 3817, 3845, 4153, 4915). It is further known from Honduras, Puerto Rico, Mexico, Peru, Chile, Argentina, South Africa, Rhodesia, Portugal, Spain, France, and the British Isles (CMI Descriptions No. 113). *Ph. cinnamomi* is claimed not to be indigenous in Southeast- and West-Australia but to have been introduced there in the course of logging (4923, 6306). — The compatibility type A2 is much more widespread than A1 which was originally known only from Hawaii (1863), but subsequently found in both continental USA (2198, 3881) and Australia (951, 5255). — *Ph. cinnamomi* has been found as a pathogen, for example, in the root region or on roots of *Cinnamomum* (4724), *Eucalyptus* (3610, 3611, 6308), *Erica* and *Rhododendron* (2501), avocado, peach (3845), *Vaccinium corymbosum* (3804), lupin (3000), pineapple (128), in forest nursery plants (2380, 4328), such as *Pseudotsuga* (3158) and *Pinus* species (1306, 3000, 4915), and also on *Nothofagus* sp. (6304) and in old forest stands (2385). Isolates from different host plants differ in their virulence to other plant species (2501, CMI Descriptions No. 113), but no special forms have been described. Roots of *Pinus echinata* in mycorrhizal associations with *Pisolithus arhizus*, *Thelephora terrestris* or *Cenococcum geophilum* are protected from invasions by *Ph. cinnamomi* (3656) unless they are simultaneously attacked by nematodes (340). Direct observations of mycelia in soil using immunofluorescence techniques permitted hyphae on root surfaces and in roots to be recognized but not those in soil itself (3537). — Zoospore movement is the normal means of dispersal while the mycelium shows a limited (6561) or no growth at all in non-sterile soil (3158). In newly infested areas the fungus can spread at 6·6 m per year downhill (much less quickly uphill) and attack numerous plants, thus changing the whole vegetation, as has occurred in W. Australia (6307). It can also be distributed with road-making gravel and by tracked vehicles (6303, 6306). — *Ph. cinnamomi* has been recovered with higher frequencies from the upper 15 cm of soil than from deeper layers, and also more often from wet than dry soils (4620), especially if a subsurface clay layer prevents drainage (3610). Low temperatures and dry soils were found to be correlated with a decrease in *Ph. cinnamomi* populations in Australia (6307); here a minimum density was observed in June (winter) and a maximum in December–March (summer) in sites with a high infection risk (3610). In dry soils *Ph. cinnamomi* was found not to be able to survive for more than two months (3158); its survival in soil is also shortened by temperatures of 36°C (2460). It can, nevertheless, survive in soil for six years or more in the absence of host plants (6557, CMI Descriptions No. 113) if the moisture content is sufficiently high (5542, 6561). It can survive storage in water for ≥7 years *in vitro* (605). *Ph. cinnamomi* does not normally survive winters in Germany and the Netherlands unless it is artificially inoculated at depths below 10 cm. It can survive there at $+1°C$ and -55 bars water potential for over three months, but at $-2°C$ for less than 20 days (5542). — Hyphal swellings are more persistent in soil than sporangia and zoospores; after four weeks their viability was found to decrease by 5 and 83% in moist and waterlogged soils, respectively (2627, 3033). The production of hyphal swellings can also be temporarily reduced in the presence of either *Trichoderma harzianum* or *T. polysporum* (2920). Cholesterol stimulates the formation of hyphal swellings which, under suitable conditions, germinate by germ tubes (6557); their germination is stimulated by the addition of glucose and asparagine (3848). The formation of oospores and/or hyphal swellings within or outside infected plant tissues are the usual means of survival (3847). Oospore formation has been observed after the exposure of colonized glass fibre tape to non-sterile soil (3847) with an optimal production at O_2 concentrations of 1–5% without CO_2 (3881); it also occurs over a wide range of soil moisture regimes (4785) and under monochromatic light in the range 320–1350 nm, although decreasing with increasing ($\geqslant 100\,\mu$W cm^{-2}) light intensities (4828). Oospore formation in single A2 (but not A1) isolates can be stimulated by avocado root extract, *Trichoderma viride* (685, 3984, 4785, 4786), *T. koningii* (4623) and some other *Trichoderma* species both *in vitro* on sterol-containing media (686) and in soil (687); the latter effect has been ascribed to volatile antibiotics released by *Trichoderma* (686) and interpreted as a defence mechanism (687). Oospore germination requires light in blue or far red wavelengths (4787, 4827). — Sporangium formation is low at 14°C, and high at 22–28°C and pH values in the range pH 4–7 (975); their function is much more dependent on sufficient moisture (water potential not below $-0·4$ bars) than that of hyphal swellings, but is insensitive to the nutritional state and pH of the soil (4785). Sporangia are formed only under normal CO_2 and O_2 concentrations or without CO_2 (3881), but mycelial growth was not inhibited by 20% CO_2 (1504). In submerged soil, zoospores can be released from sporangia and dispersed by both rain splash and run-off water (3033). Zoospores can also be released from colonized *Cicer arietinum* substrates on transfer to non-sterile soil (974). — The inoculum potential was reduced considerably in soil amended with cotton waste, wheat chaff, and soya-bean meal (1986). Chicken manure, alfalfa meal, feather-meal mix, seaweed, and aqueous extracts of treated soils prevented zoospore release (5902). — Numerous basidiomycetes show mycostatic effects against *Ph. cinnamomi in vitro* (4618). Its hyphae can be attacked rapidly in soil by *Trichoderma viride* with subsequent lysis (687, 4785). In TEM studies, bacteria have also been seen to cause alterations in the outer cell wall of sporangia (727). — In S. Australia some infertile soils with excessive moisture were found to be conducive to *Ph. cinnamomi* infection, whilst in fertile rain forest areas in N.E. Australia suppressive soils occur in avocado groves; suppressiveness can be induced, for example, by organic mulching and liming, and soils thus treated do not contain bacteria which induce sporangium production, while the bacterial flora is richer than in the conducive soils and contains antagonistic bacteria and actinomycetes; suppressiveness can be destroyed by steaming at 100°C for 30 min, but not at 49°C (728). In Hawaii a similar distinction between conducive and suppressive areas has been made (3032, 3033), although there the fungus could be isolated from both. The suppressiveness of the soil seems to be directly correlated to the organic matter content and to its microbial activity (4135).

Optimal growth occurs at temperatures of (20–)24–28(–32·5)°C, the minimum is 5–9(–16)°C, and the maximum (30–)32–34(–36)°C (1863, 2460, 3029, 4752, 5254, 6216, 6559). Optimum pH *in vitro* is 5 (854, 855), no growth occurring at a water potential of or below -50 bars (6395), and 50% growth reduction between -4 and -16 bars, depending on the type and the toxicity of the osmotica used (5565); isolates of the mating type A2 had a wider temperature optimum between 22·5–32·5°C than type A1 (5254). — Starch is hydro-lysed rather rapidly (2470). The amounts of the enzymes amylase, β-glucosidase and phosphatase varied strongly among four isolates tested (2919). Growth *in vitro* is supported by casein hydrolysate, DL-alanine, L-arginine, L-asparagine, NH_4^+ and NO_3^- ions (855, 4752); thiamine is required (4752, 4860). The formation of sporangia in water can be enhanced by non-sterile soil extracts (260), by certain bacterial isolates (*Bacillus, Pseudomonas, Flavobac-terium* and *Chromobacterium*) (727, 728, 6556), and also axenically by repeated washings of mycelial discs in salt solutions (2628) or with millipore-filtered soil leachates (727, 3657). Sporangium formation is reduced by continuous light (4828, 6560) but favoured (on washed mycelia) by fluorescent light; the presence of Ca^{2+}, Mg^{2+}, and Fe^{2+} ions is essential (979), while a small dosage of only 0·2 ppm Cu^{2+} is inhibitory (2928). The positive chemotaxis of zoospores and the chemotropism of germ tubes towards a wide range of chemicals (111, 2230, 2957, 3656) and to host (avocado) roots (6555) are functional in plant infection. At con-centrations above 37μmol H^+, 628 μmol K^+, 5040 μmol NH_4^+, and 6480 μmol l^{-1} Na^+ (given as chlorides), the chemotaxis was reversed (110); if exposed to solutions of higher concentrations, zoospores immediately encyst (110). Increasing ethanol concentrations dis-turb the reversion of chemotaxis (112). — *Ph. cinnamomi* can synthesize both geraniol and farnesol, pyrophosphates and squalene but, like other Pythiaceae, not squalene-2,3-epoxide and sterols (2056). Mycolaminaran from *Ph. cinnamomi* and other species caused wilting in cacao and tomato at 0·01–0·5 mg ml^{-1} (2913).

Pithomyces Berk. & Br. *1873*

Pithomyces chartarum (Berk. & Curt.) M. B. Ellis *1960* — Fig. 289.

≡ *Sporidesmium chartarum* Berk. & Curt. *1874*
= *Sporidesmium echinulatum* Speg. *1879*
  ≡ *Scheleobrachea echinulata* (Speg.) Hughes *1958*
= *Sporidesmium bakeri* Syd. *1914*
= *Pithomyces maydicus* (Sacc.) M. B. Ellis *1960*

DESCRIPTIONS: Hughes (2600), Ellis (1600, 1603), Dingley (1404), Matsushima (3680), and CMI Descriptions No. 540, *1977*.

The genus *Pithomyces* (type species *P. flavus* Berk. & Br.; syn. *Scheleobrachea* Hughes *1958*) was revised by Ellis (1600), who later recognized 15 species (1603, 1604). — Solitary blastoconidia arising from short cylindrical lateral pegs of undifferentiated aerial hyphae, broadly attached and liberated through the fracture of the peg, a part of which remains attached to the base of the conidium (Fig. 289d). Conidia ellipsoidal, clavate, lemon-shaped, obovoid or (ob)pyriform, wall pale to dark brown, smooth or verruculose, with 0–13 transverse and often one or more longitudinal septa.

P. chartarum has rather fast-growing colonies, covering the surface of a 9 cm Petri dish in eight days at 24°C on PDA. Colonies at first hyaline, later becoming dark grey, usually zonate and often forming differentiated sectors. Vegetative hyphae of three types: (a) (sub)hyaline, 2–4·5 μm wide, sparsely septate, often becoming verrucose; (b) dark brown, thin-walled, 5–8 μm wide, densely septate, smooth-walled; and (c) similar to the latter but echinulate. Fertile hyphae subhyaline to pale olive, 2–5 μm wide; conidiiferous pegs 2·5–10 × 2·0–3·5 μm. Conidia broadly ellipsoidal, with 3(–4) transverse and mostly two longitudinal septa, often somewhat constricted at the septa, brown to dark brown, verruculose, 18–29 × 10–17 μm. — *P. chartarum* is a variable species and its morphology is greatly influenced by both temperature and the nature of the medium (1404). — *P. maydicus* was formerly distinguished by paler brown conidia with mostly two transverse and sometimes one longitudinal septum, 12–20 × 6–12 μm (1600), but reported to be conspecific by Dingley (1404). *P. sacchari* (Speg.) M. B. Ellis *1960* differs in the darker brown conidia which are variable in shape, ± clavate, sometimes curved, with 3 transverse and 1–2 oblique or longitudinal septa, 12–25 × 5–15 μm. — The number of nuclei in the hyphae and conidiophores is variable (1404). — The hyphal cell walls are reported to contain polymers of 9·5% glucosamine and 39% neutral sugars, while those of conidia contain 3·6% hexosamine and 45% neutral sugars (3232).

Sporidesmium chartarum was originally isolated from decayed paper in the USA, and *S. bakeri* from banana leaves in the Philippines (2600). *P. chartarum* is generally common on dead leaves and stems of many plants (1600); it has been reported from South Africa (3589), Ghana (2600), central Africa (3063), Nigeria, Kenya (IMI), Rhodesia, Malawi, the Sudan, Sierra Leone, Jamaica, Mauritius, Malaysia, New Zealand, Australia (1398, 1600, 2600),

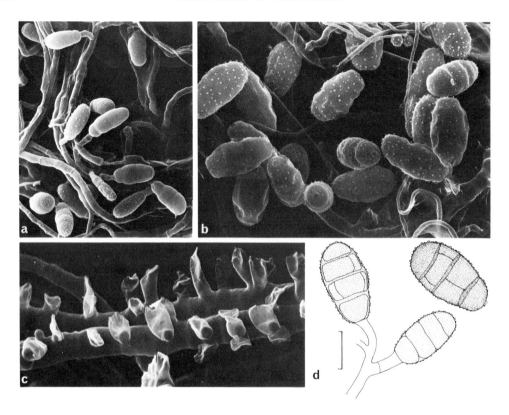

FIG. 289. *Pithomyces chartarum*. a. Young colony with attached conidia, SEM, × 750; b. older conidia, partly detached, SEM, × 1000; c. conidiogenous cells showing denticles after conidial dehiscence, SEM, × 2000; d. attached and liberated conidia; CBS 679.71.

Brazil (398), the USA (4918, 5682), Czechoslovakia (2575), and the British Isles (2082, 3205). Soil isolates are known from Japan (3680), India (IMI), banana plantations in Honduras (2031), grassland in Australia (1746), and rarely agricultural soils in the Netherlands (CBS). Records from dead plant substrates (732, 5233, 5234) include leaves of grasses (1401, 1404, 2082, 5682), *Panicum coloratum* (1560), corn (3063), *Bridelia ferruginea* (2600), *Cassia* spp. (4977), palms (3963), bananas (3063), and various other plants (296, 2600), strawberry roots (4133), seeds of alfalfa, sorghum (IMI) and wheat (5293), and sticks exposed in river- (5249) or seawater (4918). It has also been obtained from air over grassland, forests (2082, 3205) and banana plantations (1600), feathers of free-living birds (2575), and paper (1600). — Conidia did not survive a passage through the digestive tract of the earthworm *Lumbricus rubellus* (2933). *P. chartarum* was found to be inhibited *in vitro*, by both *Bacillus subtilis* (1214) and cochliodinol, a metabolite of *Chaetomium* spp. (704).

The production of conidia and vegetative growth are both good at 24°C; conidia do not germinate at water potentials at or below –140 bars, and in the field the strongest conidiation occurs after rain (732). The production of conidia is stimulated *in vitro* by near-UV light (4132). — *P. chartarum* produces several cyclodepsipeptides including a pithomycolide-fraction (4696) and the sporidesmolides I, II, III (3231), and IV (2692), and the piperazine derivatives sporidesmin, sporidesmin B, C, D, E, and F (2484, 3231). Sporidesmin production was found to be independent of the rate of mycelial growth (1435); however, some isolates not producing this compound are reported (5989); lower temperatures resulted in a higher sporidesmin content of conidia (1307); the sporidesmins are highly potent mycotoxins, mainly responsible for the mammalian toxicity of *P. chartarum*; the ingestion of contaminated grass causes facial eczema in sheep and cattle (733, 4505, 5816) involving photosensitation of the skin, liver lesions (3967) and raised phylloerythrin concentrations in the blood (1065); the sporidesmins are also phytotoxic (6442). *P. chartarum* is also said to be involved in a glume blotch disease of rice (2327).

Plectosphaerella Kleb. *1930*

Plectosphaerella cucumerina (Lindf.) W. Gams apud Domsch & Gams *1972* — Fig. 290.

> ≡ *Venturia cucumerina* Lindf. *1919*
> = *Plectosphaerella cucumeris* Kleb. *1930*
>> ≡ *Micronectriella cucumeris* (Kleb.) C. Booth *1971*
> = *Nectria septomyxa* Wollenw. *1926*
>> ≡ *Sphaerella solani* Ellis & Everh. *1893* sensu Wollenw. [non Ellis & Everh.; non *Nectria solani* Reinke & Berth. *1879*]

Anamorph: *Fusarium tabacinum* (van Beyma) W. Gams *1968*

> ≡ *Cephalosporium tabacinum* van Beyma *1933*
> = *Cephalosporiopsis imperfecta* F. & V. Moreau *1941*
> = *Cephalosporium ciferrii* Verona *1939*
> = *Septomyxa affinis* (Sherb.) Wollenw. *1930*
>> ≡ *Fusarium affine* Fautr. & Lamb. *1896* sensu Sherb. *1915* [non Fautr. & Lamb.]

For further data on nomenclature see Gams and Gerlagh (1890).

DESCRIPTIONS: Gams and Gerlagh (1890), Booth (640), Tubaki and Ito (5933), and Joffe (2770). — The neglected name *Venturia cucumerina* was reinstated by Elbakyan (1575). The setae around the perithecium neck, described by Lindfors have not, however, been found by later authors but otherwise there is no doubt about the identity of his fungus with *P. cucumeris* Kleb. — The teleomorph has been found only rarely, but perithecia can form in two weeks in suitable isolates on various media (1890). Perithecia slender flask-shaped with a dark brown rounded base and a pale slender neck, $200–300 \times 90–110 \mu$m. Asci cylindrical, thin-walled, lacking an apical differentiation, $50–65 \times 6 \cdot 5–7 \cdot 5 \mu$m, 8-spored; ascospores ± distichous, ellipsoidal, two-celled, hyaline, finely warted, $10 \cdot 8–12 \cdot 0 \times 2 \cdot 8–3 \cdot 0 \mu$m. — This fungus is not congeneric with *Monographella nivalis* (Schaffnit) E. Müll. (q.v.), as suggested by Booth (640), which has amyloid asci. The genus *Micronectriella* Höhn. *1906* was shown to be synonymous with *Sphaerulina* Sacc. *1878* (211).

Anamorph belonging to *Fusarium* sect. *Eupionnotes* Wollenw. Colonies reaching $4 \cdot 0–5 \cdot 0$ cm diam in ten days on MEA or OA at 20°C, cream or pinkish, plane and moist from conidial slime, aerial mycelium generally absent or scanty. Phialides solitary or in whorls, $12–30 \times 3 \cdot 0–4 \cdot 5 \mu$m at the slightly inflated base. The characteristic one-septate, multiguttulate conidia, slightly curved and pointed at the ends, $8 \cdot 2–13 \cdot 5 \times 2 \cdot 2–3 \cdot 0 \mu$m, are easily seen on OA, but less clearly recognizable on other media; this is probably why this species has relatively seldom been correctly determined.

P. cucumerina is a very common fungus both in the soil and on decaying plant material. In addition to reports from the temperate zone of Europe and North America (1890), it has been isolated in Cuba, Australia, New Zealand, Malaysia, Mauritius, Tanzania, Malawi and Egypt

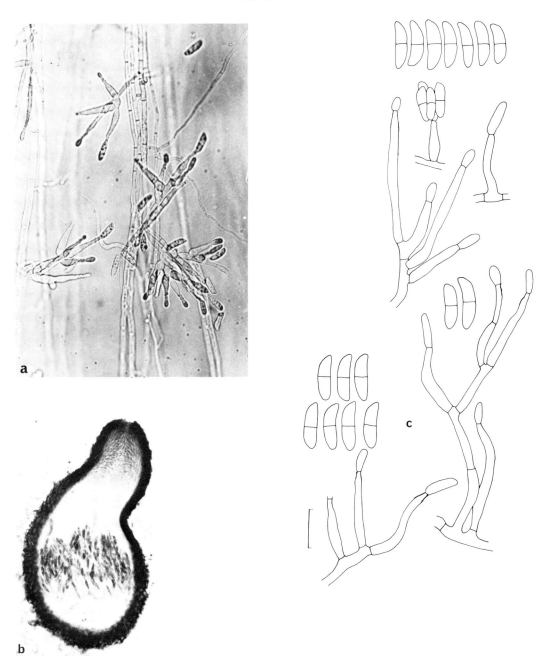

FIG. 290. *Plectosphaerella cucumerina*. a. Conidiophores arising from hyphal bundle, × 500; b. vertical section through perithecium (stained in cotton-blue), × 250; c. conidiophores and phialoconidia.

(IMI). Reports in the literature relate almost exclusively to arable soils (540, 918, 1614, 1616, 1617, 1889, 2159, 2161, 3490, 4658) or occasionally also to sand dunes and saline beaches (1890, 3941). The fungus occurs preferentially in depths of 5–10 cm and almost disappears at 40 cm (2161). Its frequency is slightly increased by manuring: NPK>N>NK>unmanured, or N>P>K (2162, 2163) but not further influenced by NPK together with farmyard manure (2161). *P. cucumerina* was isolated with significantly greater frequencies after growing rape than after wheat or peas; in the polder areas of the Netherlands, it was less frequent under permanent wheat than under crop rotations (1433) and less frequent after barley than after wheat in the rhizosphere of potatoes (1616, 1617). Strains forming perithecia accumulate in greenhouse soil in which cucumbers are grown regularly, with steaming between crops (G. J. Bollen, pers. comm.). There are reports of isolations from the rhizospheres of potato, sugar beet, grasses, wheat (1614), flax, with particular frequency from a variety resistant to *Fusarium* (5826), garlic (2159), *Ammophila arenaria*, mustard, speedwell, broad bean, tobacco, pansies (1890), also on vole dung (2338), in the gut content of an adult wheat bulb fly and in the air of a barley field (IMI). It is sometimes found on aerial parts of plants including potato, tomato, tobacco, cucumber, melon, phlox, lettuce and *Canna* (1890).

In a respiration test, the following C sources were utilized well: oxalacetate, acetate, D(+)-glucose, D(+)-xylose, D(−)-fructose and ethanol; whereas citrate, ketoglutarate and gluconate relatively poorly (5356). Pectin and CMC can be attacked strongly, but native cellulose hardly at all (1432, 6326, 6356). Decomposition of xylan is generally very good, although some variation occurs between isolates (1432). Several studies were carried out on steroid transformations by *P. cucumerina* (9, 5357). It can tolerate up to 2% methanol, ethanol or dimethylformamide (5356). — It has been found to produce necrotic spots on potato stalks (6420), cause a fruit rot in young cucumbers in the greenhouse (3024), and an intense wilting in beets (grown for seed production), additional to that caused by *Verticillium dahliae* (5490). Attempts to infect potatoes or cucumbers with soil isolates were unsuccessful (1890), but it is reported as the cause of a wilt of cucumber in the USSR (1575); we observed no pathogenicity on wheat or rape *in vitro*, but injuries occurred on pea roots (1430). Damage to tobacco and pansies is also known (1890). A high soil temperature and high soil moisture were found to favour seedling diseases in several weed plants (2247). The development of *Chlorella pyrenoidosa* was found to be stimulated in the presence of *P. cucumerina* (1430).

Pleospora Rabenh. ex Ces. & de Not. *1863*

Pleospora herbarum (Fr. ex. Fr.) Rabenh. *1857* — Fig. 291.

≡ *Sphaeria herbarum* Fr. ex. Fr. *1823*

Anamorph: *Stemphylium botryosum* Wallr. *1833*

DESCRIPTIONS: Groves and Skolko (2142), Malone and Muskett (3556), Webster and Lucas (6246), Simmons (5332, 5333), Ellis (1603), Verona and Firpi (6074), Matsushima (3680), and CMI Descriptions No. 150, *1967*.

The anamorph is the type species of the genus *Stemphylium* Wallr. *1833* which has been monographed by Neergaard (4118) and subsequently carefully documented (5332) and studied in connection with its *Pleospora* teleomorphs by Simmons (5333). The conidiophores are pigmented, erect, with a swollen, darker apical region, and give rise to a single poroconidium. After conidial discharge, proliferation may occur percurrently through the scar of the previous conidium and the same process be repeated several times. Conidia ellipsoidal or ovoid, muriform, with one major central and several secondary transverse, and also several longitudinal septa. — Cultivation on natural media (e.g. V8-juice agar, OA or PCA) and incubation under alternating temperatures (5120) or near-UV light (3242) can improve both conidiation and ascoma formation.

This is also the type species of *Pleospora* (nomen cons.) which has simple black asco-stromata, bitunicate cylindrical asci interspersed with paraphyses, and muriform, pigmented ascospores. In *P. herbarum* the ascostromata mature in 2–12 months *in vitro* at low temperatures, are dark brown, 100–500 μm diam, and provided with a short apical beak; ascospores oblong with obtusely rounded ends, more flattened at the base, constricted at one or three of the initial transverse septa, yellow-brown, at maturity mostly with seven transverse and 1–3 longitudinal septa in each section, 26–50 × 10–20 μm, mostly 30–40 × 14–19 μm (6246).

Colonies extensive, reaching 6·0–8·0 cm diam in nine days on MEA at the optimal temperature of 22–24°C, consisting of hyaline to brown hyphae, forming small pulvinate stromata from which numerous conidiophores may arise. Conidiophores straight or bent, usually simple, 20–72 μm long, and 4–6 μm wide, darker brown and roughened at the apex which is swollen to 7–9 μm diam. Conidia subglobose, ovoid or subdoliiform, constricted at the median and to a lesser extent at the other 2–3 septa, pale to deep olivaceous-brown, (20–)24–38 × (12–)15–26 μm, minutely warted in SEM (1187). A TEM study showed complete wall continuity between the conidiogenous cell and the young conidium (holoblastic conidiogenesis) (906). During conidiogenesis the conidiogenous cell undergoes senescence within two days and ultimately dies as evidenced by TEM studies (907). — In degenerating cultures, the conidia are less easily detached and may resemble chlamydospores but the conidiophore swelling remains clearly recognizable. — Similar species with subglobose conidia are *Stemphylium sarciniforme* (Cav.) Wiltsh. *1938* which has smooth-walled and less con-stricted conidia, and *S. globuliferum* (Vestergr.) Simmons *1969* which has smaller, echinulate

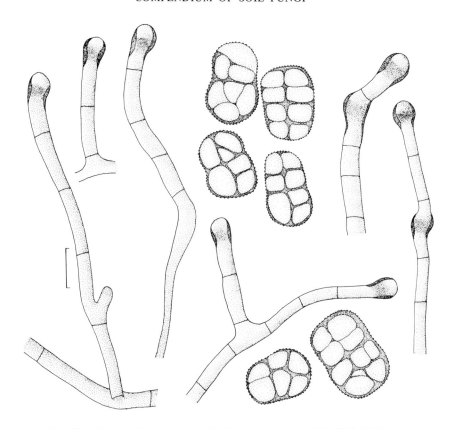

FIG. 291. *Pleospora herbarum*, conidiophores and poroconidia, CBS 406.76.

conidia, 28–30 × 25–28 μm. — Somatic division of the nuclei in vegetative hyphae was found to be strictly mitotic; chromosome counts at metaphase gave n = 6 (6195). Hyphal cells are mostly uninucleate (6195) and all conidial nuclei derive from one present in the terminal cell of the conidiophore (6196). — The mycelium of one isolate was found to contain virus particles (674).

P. herbarum is a cosmopolitan species, very common in temperate and subtropical regions, where it occurs on a wide range of host plants (2066, CMI Descriptions No. 150); it causes leaf spots or net blotch, particularly on alfalfa, vegetable plants and ornamentals. Physiologically specialized formae speciales occur. — It has been reported, for example, from the USSR (2871, 3850, 3856, 4703), Finland (6493), the British Isles (159, 167, 2923), the Netherlands (1614), Germany (5120), Austria (1828, 3418), Yugoslavia (4507), France (2161), Spain (3417), Egypt (8, 3988, 3992), Syria (277), Libya (4085, 6510), Iraq (92), Zaïre (4159), Somalia (5049), Chile (1824, 4569), Central America (1697), Australia (3720), Japan (3267), and the USA (1166, 3077, 4733, 6351). Soil isolates have been reported from conifer forests (2923, 3267, 6351), fertilized forest stands (1828), grassland (159, 277, 1614), arable land under wheat and beet (2161), citrus (3988, 4085) and coffee (4159) plantations, soils with steppe type vegetation (5049), heathland (3720), peat fields (3850, 4474), saline soils (8,

167, 3077), podzols (3856), pseudogley, rendzina, and terra fusca (3413, 3414, 3418), terra rossa (3417), desert soils (4733), as well as polluted fresh water (1166). *P. herbarum* has been isolated from the leaf litter of deciduous trees (650), bark and leaves of citrus (3988), various other leaves and stems (296, 2338), cereal stubble (3856), coarse fodder (with low frequencies) (4548), cut peat (1376), roots of young bean plants (4451), sugar beets (4559) and red clover (6493), and the rhizospheres of *Astragalus paucijugus* (5416), groundnuts (2768) and various steppe plants (3376). As a typically seed-borne fungus it has been reported from tomato (5899), flax (3517), wheat and barley (1752, 2630, 3989, 5293), *Dactylis glomerata* (5940), and other cultivated grasses (3512). — It can serve as food for the springtail *Hypogastrura tullbergi* (3830), the mite *Acarus gracilis* (5380), the coleopteran *Tenebrio angustus* and the psocopteran *Lepinotus reticulatus* (5381). — Its growth is inhibited by the phytoalexin α-tomatine (182).

The number and maturity of ascostromata formed depend on the concentration of the C source; ascostromata are produced in a pH range of 5·9–7·2. In conidiogenesis there is an inductive phase stimulated by near-UV light at wavelengths of 238 and 313 nm (3243, 3244, 3246, 5892, 5893) and temperatures of 5–10°C for $\geqslant 8$ days (3246); there is also a terminal phase when light is inhibitory at wavelengths between 240 and 650 nm (3245). The optimal temperature for conidium germination is 25°C, the minimum 2·3°C, and the maximum 35°C (4507). The conidia are violently discharged after a lowering of the relative humidity in the presence of light (3247). — The optimal temperature for mycelial growth is 22–25(–27)°C, the minimum 2°C and the maximum 30°C (5120); the thermal death point has been determined as 60°C for 30 min (613). A pH range of 3–9 is tolerated; diurnal temperature changes can lead to zonal growth but light appears to have no influence on zonation (1597). — *P. herbarum* has low nutrient requirements but on poor media produces sparce growth, few ascomata and abundant conidia; maltose seems to be the most suitable C source (1597). Cellulose is degraded (3414). — The phytoalexins sativan, pisatin, phaseollin, medicarpin and maackiain can be degraded *in vitro* and *in vivo* to the corresponding isoflavan derivatives (2355, 2356, 2447, 5557). Maackiain is readily converted to dihydro-maackiain which is subsequently further degraded (2446). — Culture filtrates are said to stimulate the development of tomato seedlings (15). A phytotoxin, termed pleosporin and exuded in liquid culture was found to cause a wilt in tomato (4844). Another phytotoxic chromone glucoside, stemphylin, which is toxic to lettuce leaves, has been isolated (335). — *P. herbarum* does not survive γ-irradiation with 625 krd (3854).

Preussia   Fuckel *1866*

= *Perisporium* Corda *1838* [non Fr. *1829*]
= *Sporormiella* Ellis & Everh. *1892*
= *Ohleriella* Earle *1902*
= *Honoratia* Cif. *et al. 1962*
= *Sporormiopsis* Breton & Faurel *1964*

Type species: *Preussia funiculata* (Preuss) Fuckel

Anamorph (where known): *Phoma*-like

MONOGRAPHIC TREATMENT of the non-ostiolate species: Cain (841); ostiolate species (*Sporormiella*): Ahmed and Cain (59); soil-borne species: v. Arx and Storm (214). — Fam. Sporormiaceae Munk. — *Preussia* species form rather slow-growing greyish, woolly colonies, often with dark red reverse. Ascomata immersed or semi-immersed, scattered or aggregated, dark brown to black, smooth or covered with hairs or scales in the upper part, ostiolate (sometimes markedly beaked) or non-ostiolate. Pseudoparaphyses filiform, disappearing at maturity. Asci fasciculate, clavate to cylindrical, short-stalked or elongated, inconspicuously bitunicate, 8-spored. Ascospores 3- to many-septate, dark brown at maturity, each cell with a ± longitudinal or oblique germ slit, mostly surrounded by a gelatinous sheath, the cells usually separating readily at maturity.

The genus comprises about 12 non-ostiolate and 78 ostiolate, mostly coprophilous, species. Those with ostiolate ascomata have been placed in the genus *Sporormiella* Ellis & Everh. by Ahmed and Cain (59). In soil-borne species, both ostiolate and non-ostiolate ascomata often occur which render this subdivision impractical (204, 214). In culture, fructification can usually be induced by using OA or CMA and near-UV irradiation.

A related ostiolate genus is *Sporormia* de Not. *1845* which has many-septate ascospores without germ slits. *Westerdykella* Stolk *1955* has non-ostiolate ascomata with 4-celled ascospores which fragment into their part cells before maturation and give the impression of many-spored asci; smooth-spored species of *Westerdykella* are often classified in *Preussia* or *Pycnidiophora* Clum *1955*, but their spores have no germ slit. — All *Preussia* species treated here have 4-celled ascospores.

Key to the species treated:

<table>
<tr><td>1</td><td>All ascospores forming a single fascicle within the broad ascus</td><td>*P. fleischhakii* (p. 668)</td></tr>
<tr><td></td><td>Ascospores forming at different levels in the ascus</td><td>**2**</td></tr>
</table>

2(1)   Ascospores obliquely septate   *P. funiculata* (p. 669)
       Ascospores transversely septate   **3**

3(2)   Ascoma wall semitranslucent; ascospores 28–32(–34) × 5–6 μm, pale to dark brown, gelatinous
         sheath narrow   (*P. minima*)
       Ascoma wall opaque; ascospores 28–35 × 5–7 μm, dark brown, apically broadly rounded,
         gelatinous sheath generally absent   *P. aemulans* (p. 667)

Preussia aemulans (Rehm) v. Arx *1973* — Fig. 292.

≡ *Ohleria aemulans* Rehm *1912*
≡ *Sporormia aemulans* (Rehm) v. Arx *1967*
= *Sporormia aemulans* var. *ostiolata* v. Arx *1967*

DESCRIPTION: v. Arx and Storm (214). — Colonies growing with a daily increment of
1·7–2·1 mm, reaching 2·6–3·2 cm diam in ten days at 24°C on OA, floccose, grey; reverse
often vinaceous. Homothallic. Ascomata superficial, globose, somewhat depressed, black,
glabrous, generally non-ostiolate (sometimes ostiolate when growing, for example, on lupin
stems: "var. *ostiolata*") and provided with a papilla where the wall is thinner, 260–540 μm
diam. Asci clavate to cylindrical, stalk 15–25 μm long, 8-spored, 110–140 × 12–15 μm.
Ascospores 3–4-seriate, 4-celled, moderately constricted at the septa, not easily separating
into single cells, dark brown, thick-walled, 28–35 × 5–7 μm, end-cells slightly tapering and
broadly rounded, 7–10·5 μm long, gelatinous sheath absent. — Anamorph *Phoma*-like,
pycnidia 55–115 μm diam, conidia subglobose, 2·5–3 × 1·5–2 μm. — *Preussia leporina*
(Niessl) v. Arx *1973* (≡ *Sporormia leporina* Niessl *1878* ≡ *Sporormiella leporina* (Niessl)
Ahmed & Cain *1972*) is an allied coprophilous species, but this has flask-shaped ascomata and
slightly narrower (5·5–6·5 μm wide), paler and apically pointed ascospores with a wide
gelatinous sheath. *Preussia terricola* Cain *1961*, originally isolated from soil in Honduras, has
ascomata embedded in a brown hyphal mat and ascospores deeply constricted at the septa,
26–32 × 5–5·5 μm, and without a gelatinous sheath. Another similar, very common but
mainly coprophilous species is *Preussia minima* (Auersw.) v. Arx *1973* (≡ *Sporormia minima*
Auersw. *1868*) which has semitranslucent non-ostiolate ascomata and slightly smaller ascospores, 28–32(–34) × 5–6 μm.

   P. aemulans has only been very rarely reported but is now known from Germany, the
Netherlands, Switzerland, Canada, and Mozambique (214, CBS). Most isolates are from soil,
including mangrove soil on Inhaca Island (214), but cultivated soils are more usual; *P.
aemulans* occurred fairly regularly in wheatfield soils in Germany (1889). Other known
substrates include *Peltigera polydactyla* growing on the soil surface, wood in a cooling tower,
and buds of an apple tree (CBS).

   P. aemulans can utilize lignin and cellulose in decomposing straw, and cause great losses in
both weight and tensile strength in maple-wood strips (2211). — It displays a marked specific
antagonistic activity against *Rhizoctonia solani* (1431) and encourages the growth of *Chlorella
pyrenoidosa in vitro*; it has also been reported to inhibit the growth of pea roots (1430).

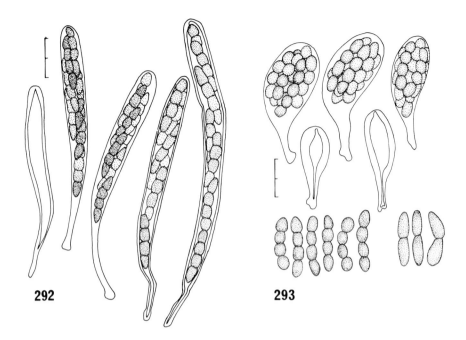

FIG. 292. *Preussia aemulans,* young and mature asci, CBS 120.66.
FIG. 293. *Preussia fleischhakii,* young and mature asci and ascospores, CBS 126.66.

Preussia fleischhakii (Auersw.) Cain *1961* — Fig. 293.

≡ *Sporormia fleischhakii* Auersw. *1866*
≡ *Fleischhakia laevis* Auersw. *1869*
? = *Preussia vulgaris* (Corda) Cain *1961*
≡ *Perisporium vulgare* Corda *1838*
= *Sporormia fasciculata* C. N. Jensen *1912*
= *Sporormia montana* Peyron. *1916*
= *Sporormia petasoniformis* C. Moreau *1953*

DESCRIPTIONS: Cain (841), v. Arx and Storm (214), and Truszkowska *et al.* (5900). — Colonies moderately fast growing, with a daily increment of 0·9–1·3 mm, reaching 3·0 cm diam in ten days at 24°C on OA, aerial mycelium woolly, whitish to grey or pink; reverse grey at first but often becoming pink to vinaceous. Ascomata superficial, subglobose, slightly depressed from above, provided with a papilla but generally non-ostiolate, black, glabrous, shining, 300–450(–510) μm diam; peridium composed of almost isodiametric angular cells, 5–8 μm diam. Asci subglobose to broadly clavate, 43–60 × 20–28 μm, stalk very short, 10–14 μm. Ascospores in a single fascicle, 4-celled, deeply constricted at the septa, cells easily separating, 26–35 × 5·5–7·5 μm, end-cells broadly rounded, 8–10 μm long; gelatinous sheath not observed. — Anamorph *Phoma*-like, rarely formed, indistinguishable from that of *P. aemulans.* — *P. vulgaris* was treated as a distinct species by Cain (841) because of its narrower ascospores (26–31 × 5 μm) and longer stalked asci (20 μm), but was regarded as doubtfully synonymous with *P. fleischhakii* by v. Arx and Storm (214).

The not too frequent records of *P. fleischhakii* indicate that it may be restricted to the cool temperate zones. Reports of soil isolates are available from the British Isles (6182, 6383, IMI), Canada (505), the Netherlands (1614, 1616), Belgium (CBS), Poland (5900), Germany (1425, 6383), and Italy (CBS). The first observations of this species were mainly on dung and rotten wood (841) but there are also reports from forest soils (6383), e.g. under white cedar (505) and hardwood trees (1712), sandy grassland (6182), arable soil (1616, 4703, 5900), infrequently from wheatfield soils (CBS), also garden compost (1425), the rhizospheres of wheat (3567) and potato (1614), litter of birch and hazel (2411), *Spartina* leaves (167), dead petioles of *Pteridium aquilinum* (1821), and human nail scrapings (CBS). — The isolation frequencies increased following soil fumigation with CS_2 (3987).

P. fleischhakii exhibits proteolytic activity and grows very well on lignin sulphonate (1425); lignin and cellulose in decomposing straw are utilized, and slight losses in both weight and tensile strength in maple-wood strips occur (2211). No growth-substances are required (1425). A strong antagonistic activity has been observed against *Rhizoctonia solani*, *Gaeumannomyces graminis* (1425), and *Fusarium culmorum* (1431), little against *Staphylococcus aureus*, but more against *Bacillus subtilis* and *Candida albicans* (3599). The thermal death point was determined in soil as 60°C for 30 min (613).

Preussia funiculata (Preuss) Fuckel *1866* — Fig. 294.

≡ *Perisporium funiculatum* Preuss *1851*

DESCRIPTIONS: Cain (841), v. Arx and Storm (214), and Udagawa and Furuya (5974); ascoma development and nuclear division: Kowalski (3115). — Colonies rather fast growing with a

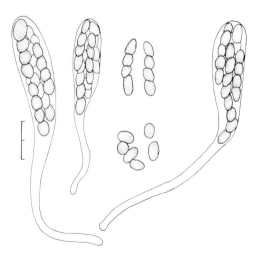

FIG. 294. *Preussia funiculata*, asci and ascospores, partly fragmenting, CBS 127.66.

daily increment of 1·7–2 mm, reaching 3·0–4·0 cm diam in ten days at 24°C on OA; aerial mycelium whitish, reverse becoming red. Ascomata superficial, black, glabrous, shining, depressed globose, mostly non-ostiolate, 200–400(–550) μm diam; peridium thin, membranaceous to coriaceous, surface consisting of thick-walled, angular cells, 3·5–10 μm diam. Asci broadly clavate, 90–120(–200) × 16–20(–28) μm including the long slender stalk (to 65(–140) μm). Ascospores obliquely 3–4 seriate, dark olivaceous-brown at maturity, 4-celled, the septa oblique, deeply constricted at the septa and easily separating into single cells, (24–)28–32(–38) × (5–)6–7(–7·5) μm, end-cells 8–11 μm long, gradually tapering towards the rounded end. Anamorph unknown. — The oblique ascospores septa are highly specific for this species.

P. funiculata is mainly a coprophilous species and is known from Europe (including the USSR), Canada (841), Japan (5974) and Senegal (CBS). It has been observed on dog, rabbit, horse, porcupine (841), and goat (5974) dung, but is also reported from pine litter (269), beech forest soil (272, 273, 3138), wheatfield soil (CBS), decaying wood, reeds, straw, jute, tomato seeds (841), dead leaves of *Agropyron junceum* and *Spartina* (167), and overwintering hay (4548). — It has been isolated from soil after heating to 60°C (CBS).

Pseudeurotium van Beyma *1937*

= *Levispora* Routien *1957*

Type species: *Pseudeurotium zonatum* van Beyma *1937*

Contribution to a monographic treatment: Booth (638); review: Malloch and Cain (3553). — The family Pseudeurotiaceae Malloch & Cain has been separated from the Trichocomaceae Fischer because of the presence of dark, closed ascoma walls and the absence of catenulate phialoconidia.

Pseudeurotium is characterized by glabrous, globose ascomata which are covered with one layer of polygonal dark brown cells without any pre-formed sutures. The initials consist of coiled hyphae (5592). Asci produced as side branches from ascogenous hyphae, globose to ellipsoidal, 4–8-spored. Ascospores globose to ellipsoidal, one-celled, becoming olive-brown at maturity, without germ pores. — Anamorph, when present, *Sporothrix*-like, with irregularly swollen conidiogeneous cells. — A good medium for inducing ascoma formation is OA. Of the seven species distinguished by Booth (638) perhaps only the two species treated here are distinct members of this genus (3553) but the group is in need of a detailed monographic survey. — Close genera are *Nigrosabulum* Malloch & Cain *1970* which has permanently hyaline ascospores and an *Acremonium* anamorph, *Fragosphaeria* Shear *1923* which has ascomata breaking open along cephalothecoid sutures and reniform ascospores, and *Ephemeroascus* van Emden *1973* which has ascomata with a stromatic wall and a *Verticillium* anamorph. Smooth-spored species of *Westerdykella* Stolk *1955* may resemble *Pseudeurotium* but in that genus, the many-spored asci arise from the early fragmentation of septate ascospores in bitunicate asci.

Key to the species treated:

1	Ascospores globose, 3–4·5 μm diam	*P. zonatum* (p. 672)
	Ascospores ellipsoidal, 4·5–6 × 3·5–4·5 μm	*P. ovale* (p. 671)

Pseudeurotium ovale Stolk *1955* — Fig. 295.

DESCRIPTIONS: Stolk (5592), Booth (638), and Udagawa (5973). — Colonies growing slowly, reaching 3·2 cm diam in ten days at 24°C on OA, initially white, floccose, becoming grey with the development of ascomata, ± zonate; reverse initially white to yellow-green, later becoming olive-buff. Ascomata developing in the aerial and submerged mycelium, globose, dark brown to black, 90–180 μm diam. Asci globose to ellipsoidal, 7·5–9 × 6·5–8 μm. Ascospores hyaline, becoming olive-brown, ellipsoidal, 5·5–6 × 3·5–4 μm. Conidiogenous cells and conidia as in *P. zonatum*.

P. ovale was originally isolated from cysts of the nematode *Globodera rostochiensis* collected on Jersey, in the Channel Islands. It has later been found in soils in the British Isles (4429, IMI), Germany (CBS), the Netherlands (1614, 1616), Pakistan, the Ivory Coast (IMI), Hong Kong (1021), and Japan (2532, 5973), and from a variety of substrates in Malaysia, Sabah, India, and Trinidad (IMI). In a forest podzol, it was isolated preferentially from organic substrates (1021). It has been found in the rhizospheres of groundnut (2532) and *Hevea*, on dung of deer (401), rabbit and sheep (IMI), and organic detritus in fresh water (4429). It has rarely been isolated from wheatfield soils (CBS). — Its isolation frequencies were increased after a CS_2 treatment of the soil (3987).

P. ovale does not grow at 37°C (5973). It shows antagonistic activity towards *Pythium "debaryanum"*, *Pseudocercosporella herpotrichoides* and *Gaeumannomyces graminis* (1431). Known metabolites include an antibiotically active sesquiterpene, ovalicin, which has an immunosuppressive activity (624, 873, 5320, 5729), and pseurotin (557).

Pseudeurotium zonatum van Beyma *1937* — Fig. 296.

= *Levispora terricola* Routien *1957*
= *Pseudeurotium bakeri* C. Booth *1961*

DESCRIPTIONS: Stolk (5592), Booth (638), and Udagawa (5973). — Colonies slow-growing, reaching 3·0 cm diam in ten days at 24°C on OA, at first whitish, floccose, becoming grey with the development of ascomata; reverse darker grey from ascomata. Ascomata globose, glabrous, 100–230 μm diam, wall composed of angular isodiametric or elongate cells in surface view, mostly 4–6 μm diam. Asci globose, pyriform to ellipsoidal, 7–9 × 6–6·5 μm, mostly 8-spored. Ascospores globose, remaining hyaline for a long time, but eventually becoming olive-brown when fully mature, 3·0–4·5 μm diam. Conidiogenous cells 7–20 × 2–3 μm, cylindrical to slightly flask-shaped, with an inconspicuously denticulate tip, bearing a few ellipsoidal conidia, 4–6 × 2·5–3·5(–4·5) μm. — DNA analysis gave a GC content of 57% (5600). — A similar species is *P. desertorum* Mouchacca *1971* distinguished by its larger globose ascospores 5·2–7·2 μm diam.

P. zonatum is the commonest species of its genus. It has been isolated many times from soil although on the whole it is not a very common fungus. Soil isolates are now known from the British Isles (159, 2338, IMI), the Netherlands (1614), Belgium (4813, 4814, 4815), Germany (1424, 1889), the USSR (4474), Austria (1876), Hungary (2541), Italy (3913), the USA (1387, 2573, 4225), Canada (505), New Zealand (IMI), Nepal (1826), Pakistan (401), Japan (2532, 5973), Colombia (CBS), and central Africa (3063). Known habitats include forest soils under white cedar (505), beech (4813), and oak, mostly with mor and more rarely

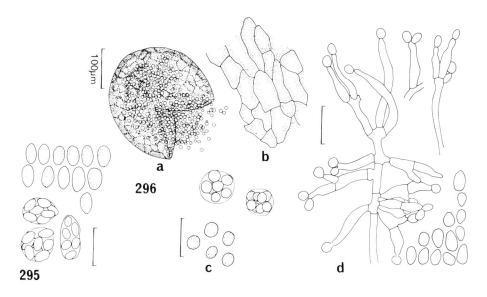

FIG. 295. *Pseudeurotium ovale*, three asci and ascospores.
FIG. 296. *Pseudeurotium zonatum*. a. Crushed ascoma with spreading ascospores; b. portion of wall in surface view; c. two asci and ascospores; d. conidiophores and sympodially produced blastoconidia.

acid mull type humus (4814, 4815), *Acer-Ulmus-Fraxinus* communities of flood plains (4225), alluvial grassland (159), cultivated soils under wheat (1889), potato or sugar beet (1614), ricefield soils (3913), garden soil (2541) and compost (1424), peat fields (4474), at high frequencies in subalpine raw-humus soils (1876), cow dung (2338), activated sludge (1387), silage and timber of a warship submerged in the Baltic Sea (CBS), litter of salt-marsh plants (167), the geocarposphere of groundnut (2532), the rhizospheres of sugar beet (1614) and various forest plants (4814), and bean roots (4452). — It was inhibited *in vitro* by *Pseudo-cercosporella herpotrichoides* (1431). *P. zonatum* has been isolated selectively from wood pulp with a nutrient medium containing *o*-phenylphenol (699).

P. zonatum has a temperature maximum near 37°C (5973) and thus ranks as mesophilic (160). Xylan decomposition is very good (4192) with variations in this capacity occurring between isolates (1432); birch and pine wood are degraded (4191), cellulose (filter paper) is decomposed (1425) and cellulase, amylase and mannanase are produced (4192). Good growth is possible in an atmosphere poor in O_2 (1425). — It can inhibit the growth of *Chlorella pyrenoidosa in vitro,* and in pot experiments, it caused a significant reduction of fresh weight in rape seedlings (1427).

Pseudobotrytis Krzem. & Badura *1954*

Pseudobotrytis terrestris (Timonin) Subramanian *1956* — Fig. 297.

≡ *Spicularia terrestris* Timonin *1940*
≡ *Umbellula terrestris* (Timonin) Morris *1955*
= *Pseudobotrytis fusca* Krzem. & Badura *1954*

DESCRIPTIONS: Subramanian (5626), Ellis (1603), Matsushima (3679, 3680), Hammill (2240), Tubaki (5929), Barron (364), and Meyer (3789); conidiogenesis: Cole and Samson (1109).

Colonies growing slowly, reaching 2·5–3·5 cm diam in ten days on MEA at 20°C, brown or greyish brown. Conidiophores erect, consisting of an unbranched ± straight stipe, brown and 4–5 μm wide in the lower part, giving rise to 6–12(–18) umbellate, almost hyaline conidiogenous cells; conidiogenous cells 14–23 × 2–3 μm, somewhat tapering towards the tip and ending in an expanded (4–6 μm diam) denticulate portion, producing dry heads of conidia in sympodial succession and gradually becoming swollen during this process (2243). Blastoconidia ellipsoidal with apiculate base to clavate, 1-septate, pale brown, 6·5–9 × 2·8–3·5 μm. Rows of dark brown, globose or ellipsoidal chlamydospores, 6–7 × 3–4 μm or 3–6 μm diam, are sometimes present. — The hyphomycete genus *Pseudobotrytis* (= *Umbellula* Morris *1955*) remained monotypic for some years, but a second species, *P. bisbyi* Timonin *1961*, is known from soil in Honduras and has non-septate conidia of a similar size and these produce only one germ tube on germination.

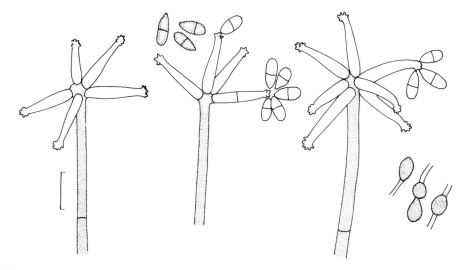

FIG. 297. *Pseudobotrytis terrestris*, conidiophores, sympodially produced blastoconidia and chlamydospores, CBS 104.54.

P. terrestris is a rather uncommon, although cosmopolitan, soil fungus reported from Europe, North America, New Zealand, Jamaica (1603), Panama (2034), Zaïre (3789), India (5626), Japan (3680, 5846, 5929), and New Guinea (3679). Known habitats include tundra soils (3680), cultivated soil under sugar cane (4886), forest soils (2240), arable soil under alfalfa (IMI), and decaying stems and leaves of various plants (273, 1603, 2034, 3680, 5626) including *Brassica*, sugar cane and beech (IMI).

Pseudogymnoascus Raillo *1929*

Pseudogymnoascus roseus Raillo *1929* — Fig. 298.

≡ *Gymnoascus roseus* (Raillo) Apinis *1964*
= *Pseudogymnoascus vinaceus* Raillo *1929*
= *Gymnoascus rhousiogongylinus* Wener & Cain *1970*

Anamorph: *Geomyces vinaceus* Dal Vesco *1957*
= *Chrysosporium pannorum* (Link) Hughes *1958* pro parte (q.v.)

DESCRIPTIONS: Apinis (162), and Samson (5023). — Fam. Gymnoascaceae Baran. — Colonies reaching 2·0–3·0 cm diam in 14 days at 25°C on OA. Ascoma initials consisting of coiled ascogonia without a recognizable antheridium. Peridium consisting of a loose network of interwoven smooth to roughened, 2·2–3·5 μm wide hyphae, at first hyaline but later becoming yellow-brown to reddish; appendages simple, short, blunt, straight, thin-walled, to 15 μm long. Ascospores ellipsoidal to fusiform, smooth-walled, yellow to pinkish, 3·0–4·0 × 2·0–2·5 μm. Conidia *Chrysosporium*-like, globose to ellipsoidal with a truncate base, hyaline, smooth-walled, 2·0–4·5 × 1·5–2·5 μm. — The genus *Pseudogymnoascus* was originally distinguished from *Gymnoascus* by the absence of hooked appendages on the peridium; since however, short and thin-walled appendages are present, Apinis (162) combined the genera. On account of the ascoma initials and ascospore shape, however, Samson (5023) separated them again and his view is followed here. — DNA analysis gave a GC content of 50·5% (5600).

P. roseus has a worldwide distribution and has been reported from the British Isles (167, 4429), Poland (1977, 3497), the USSR (2871, 3982), Italy (3445, 3450, 3453, 4538, 6082), India (5622), Nepal (1826), the Philippines (4823), New Zealand (5812, 5813), the USA (1039, 1040, 4313, 6350) and Central America (1697). It has been isolated from various forest soils (989, 1039, 1040, 4703, 5813, 6350), a forest nursery soil (1977), arable soils (2871, 3450, 3982, 4538), under paddy (4823), grassland (3497, 4313, 5812, 6082), sand dunes (1377), leaf litter in a salt-marsh (167), the rhizospheres of alfalfa (3982, 3983) and bean (4452), and the mycorrhiza of *Pinus strobus* (1794). Extreme habitats discovered are an alpine pasture (3445) and caves (3453). A preference for relatively humid habitats has been observed (4313). Other reports include compost made of urban waste (3041), organic detritus in fresh water (4429), and nests and pellets of free-living birds (2575).

Xylan decomposition is very good and of constant intensity in various isolates (1432). Birch wood is corroded by *P. roseus* (4191), which also inhibits the growth *in vitro* of both *Cochliobolus sativus* and *Rhizoctonia solani* (3571).

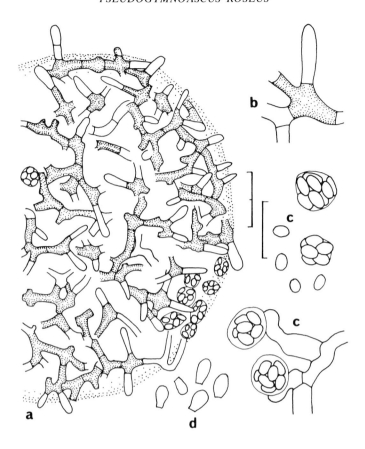

FIG. 298. *Pseudogymnoascus roseus.* a. Half ascoma in surface view showing peridial hyphae with short appendages; b. detail of appendages; c. asci and ascospores; d. *Chrysosporium*-type conidia (right bar gives magnification for b – d).

Pythium Pringsheim *1858* [nomen conservandum *1974*]

[non *Pythium* Nees *1823*, nomen rejiciendum]
  = *Artotrogus* Mont. *1845*
  = *Nematosporangium* Schröter *1897*
  = *Sphaerosporangium* Sparrow *1931*

Type species: *Pythium monospermum* Pringsheim (lectotype)

Fam. Pythiaceae Schröt. — Colonies ± fast-growing, mostly colourless, hyphae to 10 μm wide (average 5 μm) and initially unseptate, ± racemosely branched. Sexual propagation is by oogonia which are formed in terminal or intercalary positions; the oogonia contain one oospore in most species; the oospores may be thin- or thick-walled (inspissate) and may fill the oogonium (plerotic) or not (aplerotic). Fertilization is by one or several antheridia which may originate from the oogonial stalk at various distances from the oogonium (monoclinous) or from a different hypha (diclinous) which may arise from a different mycelium (heterothallic or, better, dioecious species); sometimes the part of the oogonial stalk adjacent to the oogonium forms the antheridium (hypogynous). In some species the oogonia have a characteristic spiny sculpture. The ultrastructure of oogonial development has been described by McKeen (3717). — Asexual propagation occurs by zoospores formed in two steps. Sporangia differentiate on the mycelium and may be thread-like (nematosporangiate), irregularly swollen, or globose (sphaerosporangiate). Under suitable conditions the sporangia produce an evacuation tube with a terminal vesicle in which the zoospores differentiate and from which they are liberated in a short time. Many species form globose, sporangium-like structures which never produce zoospores and therefore have been called hyphal swellings (sometimes also "hyphal bodies"); their distinction from zoosporangia can be difficult. — Isolates which do not form oogonia are suspected to belong to heterothallic species, but even after mating with numerous tester strains many fail to form oogonia and cannot be positively identified (4581).

The fine structure of the vegetative cells has been studied by TEM (1176, 2137, 2331). Mycelial analyses showed high contents of α-alanine and the characteristic occurrences of asparagine and glutamine (2458). Acrylamide-gel electrophoresis of component proteins has been applied for species differentiation (2018).

Heterothallic species are differentiated into male and female mating types which are capable of producing either antheridia or oogonia only upon contact with each other. Contrary to other fungi, the oomycete mycelium is mostly regarded as diploid whilst the haploid phase is limited to the gamete nuclei.

Reviews on the ecology and significance of *Pythium* populations in soil are provided by Schmitthenner (5113) and Hendrix and Campbell (2382). While the nematosporangiate species occur preferentially in aquatic or semiaquatic habitats, the sphaerosporangiate species and particularly those with hyphal swellings, are better adapted to life in soil. Destructive plant root parasites are found in both groups and the host-plant relationships so far known have been compiled by Rangaswami (4729). Because of their great sensitivity to antagonistic and other environmental influences, *Pythium* species do not usually cause as much damage as their density in the soil might suggest. The infection of plant roots requires high soil moistures and reduced oxygen tensions (5113). *Pythium* species are generally more frequent in arable and other cultivated soils than in forest soils, and more common in loam than sandy soils (6027).

Pythium species have been isolated comparatively rarely in general soil fungal analyses, even from washed soil particles, since many species of this genus are sensitive to the antibacterial antibiotics commonly used in such techniques. Previously, *Pythium* species were isolated from detritus and plant fragments simply by placing them on water agar because of their rapid growth. Subsequently many selective isolation techniques have been devised (1434, 2382, 3436) based on the incorporation of polyene antibiotics like mycostatin (100 units ml^{-1}), pimaricin (50–100 ppm) or endomycin (5 ppm), or the fungicide benomyl into the agar medium. A quantitative isolation technique was devised by Schmitthenner (5111) who plated soil crumbs on agar plates, but the highly selective media also allow the adaptation of Warcup soil plates or dilution plates (2387, 3436, 6027) if relatively high amounts of soil are used. No single technique is, however, equally suitable for all *Pythium* species (2382). Baiting on buried filter paper discs is also very efficient if they are soaked in nutrients with pimaricin (4873). The addition of oatmeal to soil increases the success of detection using cucumber seedlings as test plants (666). Because of their rapid growth and propagation, a quick change of dominance and species composition in certain localities has been observed, rendering quantitative determinations of uncertain value (4582).

For the maintenance of cultures CMA, OA and PCA are generally used. Sexual and asexual structures are best seen on transparent agar media such as PCA, chopped carrots in water agar, or Schmitthenner's medium (5111); PCA and hempseed agar are generally used for mating experiments. The addition of cholesterol, β-sitosterol (0·01 g l^{-1}), sunflower oil or wheat germ oil (0·5 g l^{-1}) may greatly enhance oogonium formation (2382, 6217). Thiamine is generally not required in the species treated here, except *P. oligandrum* (4836).

The most widely used monograph is that of Middleton (3800). Waterhouse provided a key to the recognized species (6217) and compiled original descriptions of all species described until *1968* (6218). Recently several heterothallic species have become known which require mating before a reliable identification is possible (4581). Waterhouse recognized 90 species, since which time 14 additional species have been described.

The name most commonly encountered in the literature is *Pythium debaryanum* Hesse *1874* but this is a nomen ambiguum based on a mixture of different organisms (6218). Middleton (3800) interpreted this species differently from Hesse but did not delimit it sufficiently clearly and so caused further confusion. It is now recognized that this name has been most commonly applied to *P. irregulare*, *P. sylvaticum* and to a lesser extent perhaps *P. ultimum* (2382, 2384). As a result of these variations in usage, this name has to be considered as doubtful.

Key to the species treated and some related ones:

1	Oogonial wall with numerous spiny or blunt projections	**2**
	Oogonial wall smooth or with few irregular projections	**4**
2(1)	Oospores ± plerotic or almost filling the oogonium; projections mostly 4·5 μm long, blunt	*P. mamillatum* (p. 686)
	Oospores definitely aplerotic	**3**

3(2) Projections in a dense pattern, acute, 3–7 μm long; antheridia 0–1 per oogonium, mostly diclinous; sporangia consisting of interconnected globose elements, 24–45 μm diam
P. oligandrum (p. 688)
Projections of variable density and length, blunt, cylindrical; antheridia monoclinous, 1–2(–4) per oogonium; sporangia ± globose, 10–27 μm diam *P. irregulare* (p. 684)

4(1) Hyphal swellings catenulate without intercalary hyphal fragments; heterothallic species **5**
Hyphal swellings if present not catenulate; mostly homothallic species **6**

5(4) Sporangia not normally formed; hyphal swellings deciduous, 18–24 μm diam
P. intermedium (p. 683)
Sporangia commonly produced as irregularly swollen filaments; hyphal swellings 10–20 μm diam, not deciduous (*P. catenulatum*)

6(4) Sporangia filamentous, scarcely differing from the vegetative hyphae in appearance; oospores aplerotic, frequently with 4–5 antheridia; hyphal swellings absent (*P. dissotocum*)
Sporangia ± differentiated from the vegetative hyphae **7**

7(6) Sporangia lobulate, irregularly swollen, with or without branching **8**
Sporangia or hyphal swellings discrete, spherical or elongate (only non-proliferating species treated) **10**

8(7) Antheridia usually monoclinous, often intercalary, broadly barrel-shaped or cylindrical; oogonial stalk not curved towards the antheridium (as in *P. deliense*); oospores aplerotic; mainly in the tropics *P. aphanidermatum* (p. 681)
Antheridia never intercalary, but sometimes hypogynous, more slender; oospores plerotic; mainly in cooler climates **9**

9(8) Oogonia 16–30 μm diam, with 1–6 antheridia; antheridia arising remotely from the oogonium (*P. graminicola*)
Oogonia 12–19 μm diam; antheridia arising at 5–10 μm from the oogonium
P. torulosum (p. 693)

10(7) Heterothallic species **11**
Homothallic species **12**

11(10) Hyphal swellings 24–30(–40) μm diam; antheridia 2–4 per oogonium *P. sylvaticum* (p. 692)
Hyphal swellings 15–21(–25) μm diam; antheridia up to 8 per oogonium (*P. heterothallicum*)
(If after mating with tester strains no oogonia are formed, no species determination is possible).

12(10) Oospores plerotic, 12–18 μm diam, oogonia often intercalary; colonies slow growing (linear growth 0·7 cm per day at 20°C on CMA) *P. rostratum* (p. 690)
Oospores aplerotic; colonies faster growing **13**

13(12) Antheridia very short stalked or sessile **14**
Antheridia distinctly stalked **15**

14(13) Oogonia mostly terminal, 11–23 μm diam, oospore wall 2 μm or more thick; antheridia 1–2 per oogonium, sessile or sometimes diclinous *P. ultimum* (p. 694)
Oogonia often intercalary and in chains, 11–28 μm diam; oospore wall 1·0–1·5 μm thick; antheridia 1–5 per oogonium, mono- or diclinous, stalked or sessile
P. paroecandrum (p. 689)

15(13) Oospore wall less than 1·5 μm thick; oogonium wall often with irregular, blunt, cylindrical projections; antheridium stalk variable in length, sometimes branched, antheridial cell apically attached to the oogonium *P. irregulare* (p. 684)
Oospore wall 2 μm thick; oogonium wall always smooth; antheridium stalk constantly several times the length of the cell, unbranched; antheridial cell bell-shaped and broadly applied to the oogonium (*P. vexans*)

Pythium aphanidermatum (Edson) Fitzp. *1923* — Fig. 299.

≡ *Rheosporangium aphanidermatum* Edson *1915*
= *Pythium aphanidermatum* var. *hawaiiense* Sideris *1931*

DESCRIPTIONS: Middleton (3800), Drechsler (1465), Hickman (2442), Waterhouse (6218), Takahashi *et al.* (5693), CMI Descriptions No. 36, *1964*, and Watanabe (6211). — Colonies growing with a daily increment of 2·5–3·0 cm at 25°C on CMA. Sporangia inflated filamentous, readily producing zoospores on transfer to water at 20–30°C. Oogonia mostly terminal, spherical, 22–27 μm diam; oospores aplerotic, 17–19 μm diam, moderately thick-walled; antheridia usually monoclinous, intercalary or terminal, 1–2 per oogonium, barrel- or dome-shaped or cylindrical, 10–14 × 9–11 μm. Oospore diameter was found to decrease somewhat after repeated subculturing for 30 months (2383). — In addition to *P. aphanidermatum*, intercalary antheridia occur in two related species: *P. butleri* Subramaniam *1919* with oogonia 16–33 μm diam, straight stalks and mostly diclinous antheridia; and *P. deliense* Meurs *1934* (incl. *P. indicum* Balakrishnan *1948*) with oogonia 12–31 μm diam and stalks characteristically bending towards the antheridium (1465). — The tips of actively growing hyphae do not contain dividing nuclei; during anaphase achromatic structures resembling spindle fibres with defined poles have been observed (5212). The mycelium contains relatively high amounts of a-alanine, and some proline (2458). — The ultrastructure of swarming and encysted zoospores has been examined (3060).

P. aphanidermatum is a plurivorous soil-borne plant parasite with a mainly tropical distribution (4729, CMI Descriptions No. 36) but is found in all continents. Reports are available, for example, from the southern USA (794, 2381, 2382, 4153, 5529), Lebanon and Syria (62), Iran (1704), Iraq (92), Zaïre (4752), South Africa (1415), India (4716, 5512), Japan (5690, 5693), Taiwan (6211), the Philippines (2640) and Australia (782). Its occurrence in central and western Europe is generally limited to greenhouses (CBS) with the exception of reports from Austria and Czechoslovakia (3800); it has been rarely isolated in Canada (6021). — Besides several selective techniques for *Pythium* in general (325, 793, 794, 3436, 5692), a baiting technique selective for *P. aphanidermatum* using streptomycin-pimaricin-soaked potato cubes (2461), and a dilution technique with 100 ppm pimaricin, 200 ppm streptomycin sulphate, 150 ppm rose bengal and 5 ppm benomyl in CMA, have been devised (794). Best isolation results were obtained with an incubation at 38–40°C and a pH above 6 (3436). — *P. aphanidermatum* has been reported from forest soil with *Eugenia* (5512), dry as well as tropical mixed forest and cultivated soils (325, 4716), the rhizosphere of *Cynodon* sp. (2386), cotton roots (IMI), small grains, corn, sugar cane (2382, 6211), sugar beet (1704, 5693), pineapple, beans (2473), tomatoes (2977) and many other host plants (CMI Descriptions No. 36). It causes root damage in numerous plants mainly at higher temperature (2382). It occurs with lower frequency in alkaline soils (4709). Nitrogen fertilization reduced the *P. aphanidermatum* population and damping-off in tomatoes (4710), whilst soil treatments with benomyl and related fungicides increased its incidence in cowpea (6380). Naturally infested soil from various sites contained 10–250 oospores per gram (794). Amendment of soil with various chitinous materials significantly increased the *P. aphanidermatum* populations in non-sterile soil (5374). — In artificially inoculated soil, survival was higher at pH 6 than at pH 8, and higher at 20°C than at 30°C (4709). Oospores survive in soil better at lower (4°C) than at higher (40°C) temperatures (5529, 5530) but this fungus does not survive at low temperatures for long periods (62). Encysted zoospores remain viable in soil for seven days but cannot grow out saprophytically (2382). Hyphae are quickly degraded upon

Fig. 299. *Pythium aphanidermatum*. a. Irregularly swollen sporangia; b. oogonia with saccate antheridia; CBS 258.30.

contact with soil, particularly by certain actinomycetes (3185); *in vitro*, it can be strongly inhibited by *Trichoderma harzianum* (1682). The presence of L-amino acids shortens the longevity of cultures *in vitro* (3080).

According to numerous reports, the minimum growth temperature is 8–9°C, the optimum 34–36°C and the maximum <46°C. Linear growth can continue indefinitely in long tubes with PDA and 20% V-8 juice agar media at about 3·0 cm/day at 24°C but stops after three weeks in the same liquid media (3031). Good growth occurs at pH 9, optimal growth at 6, and none at 4 (3436, 5690). — The production of oospores can be stimulated by cholesterol (258, 2388), and Schmitthenner's synthetic medium with the addition of 10 ppm cholesterol has been recommended (25). Sonication of aqueous suspensions of mycelial fragments containing oospores can be used as a method for obtaining mycelium-free oospores (5070). Optimum conditions for oospore germination are exposure to light (when kept on media with $CaCl_2$) (5114), temperatures of 30–35°C and pH 7·0–7·4; older oospores germinate more readily than younger ones (25, 258, 5530) and do so directly in asparagine-amended soil with moisture levels between saturation and −15 bars, the optimum being between −0·01 and −0·1 bars (5526); percentage germination was increased (5531) or remained unaffected (5070) by treatment with a commercial snail enzyme, or passing of oospores through live water snails, and was also promoted by selective media containing casein, gallic acid, yeast extract and thiamine (1784, 5531). Maltose, starch, sucrose and fructose supported >90% germination (5525, 5531). Oospore germination was strongly reduced by freezing of aqueous oospore suspensions (5070). — Sporangium formation can be induced on hemp-seed, wheat endosperm or rice floating in water at pH 5·4 (2057). Zoospores were found to be attracted by root hairs and the elongation zone of young cotton roots and also by glutamic acid (5492). Oxygen uptake was low during the motile stage and encystment period of the zoospores but increased considerably as the cysts germinated; zoospores can remain motile for a long period without an external nutrient supply (955). Germination of zoospores was found to be stimulated by pea root exudates, D-alanine, L-glutamine, L-histidine, DL-homoserine, L-proline, lactic acid, tartaric acid, and D-fructose (954). — In growth tests with different C sources, *P. aphanidermatum* performed better with soya oil or cholesterol than with glucose (2388); casein hydrolysate proved to be the most suitable N source (3117, 3118). Isolates vary significantly in both pathogenicity and physiological behaviour (2140). A macerating enzyme is produced on cucumber (5950) and potato (4771) tissues, but no correlation was found between pectic enzyme production and pathogenicity (4771). Endo- and exopolygalacturon-ase, pectin methylesterase and pectin methyl transeliminase were detected and could be partly suppressed by some fungicides (1351). Cellulose is degraded, and both nitrite and nitrate can be used as N sources (4752). — Of the antibiotics nystatin (1810), streptomycin and pimaricin 100 ppm are tolerated (2459) but, contrary to other Pythiaceae, not gallic acid (3436).

Pythium intermedium de Bary *1881* — Fig. 300.

DESCRIPTIONS: Middleton (3800), Waterhouse (6218), van der Plaats-Niterink (4581), and CMI Descriptions No. 40, *1964.* — Colonies growing with a daily increment of 2·5 cm at 25°C on CMA. Hyphal swellings spherical or sometimes pyriform, 18–20(–25) μm diam, in deciduous chains of up to 13 cells. Zoospores have been described once but are not normally formed on transfer to water. This species is heterothallic (4581); oogonia are formed sparsely in dual cultures after about six days at exactly 20°C (4581), intercalary or terminal, 18–27 μm diam; oospores aplerotic, 13–22 μm diam, wall 1–2 μm thick, 1(–2) per oogonium; antheridia 1–7 per oogonium, with long stalks, often branched near the oogonium. — Protein patterns have been determined by starch-gel and acrylamide-gel electrophoresis (2018) and found to differ from those of other *Pythium* species.

P. intermedium has a worldwide distribution but predominates in the temperate zone (62, 4581) and is regarded as a typical soil inhabitant. Numerous host plants from different families have also been listed (4581, 4729, CMI Descriptions No. 40), in which it sometimes causes damping-off of seedlings, foot rot and root rot. It has rarely been isolated by dilution plates using non-selective media but often with immersion tubes (982) and other special techniques (e.g. 62, 1433). — It has been reported from cultivated soils (982, 1433), wet grassland (164), forest nurseries (6021, 6185), old forest stands (2385) and also an urban soil in Iraq (92). In the Netherlands (4581), it is one of the commonest *Pythium* species and has been isolated from numerous host plants, for example, flax roots; it has also been found with gradually increasing frequencies from newly reclaimed polder soils (4582). The distribution appears to

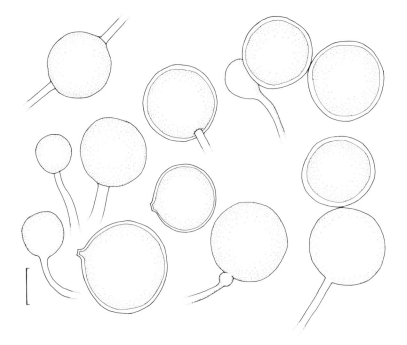

FIG. 300. *Pythium intermedium*, development of catenulate hyphal swellings, CBS 224.68.

be to a large extent independent of soil pH (6185). *P. intermedium* penetrates only to a depth of *c.* 25 cm in soil (164). It is known as a common saprophyte on grass roots in the spring in the USA (5498), and has also been reported from the rhizosphere of *Cynodon* sp. (2386). After repeated croppings of wheat, its frequency was significantly lower than after peas or rape (1433).

The optimum temperature for growth is 25–28°C, the minimum 1°C, and the maximum 34°C (3800), but lower cardinal temperatures have also been reported (4581). — Growth of roots of rape, wheat and peas was inhibited by a pure culture of *P. intermedium* in laboratory experiments (1430).

Pythium irregulare Buisman *1927* — Fig. 301.

= *Pythium debaryanum* sensu Drechsler *1953*
= *Pythium polymorphon* Sideris *1932*

DESCRIPTIONS: Middleton (3800), Waterhouse (6218), and Biesbrock and Hendrix (510). — Colonies growing with a daily increment of 2·5–3·0 cm at 25°C on CMA. Sporangia (and hyphal swellings) mostly globose or obovate, terminal or intercalary, $10-27\,\mu$m diam; zoospores sometimes produced in water at 5–15°C. Oogonia globose to limoniform, $(10-)16-18(-28)\,\mu$m, usually intercalary, irregular in outline, smooth or with irregularly cylindrical and blunt projections; oospores mostly aplerotic, $8-25\,\mu$m and mostly $14-16\,\mu$m diam, wall $1\cdot0-1\cdot5\,\mu$m thick. Antheridia mostly monoclinous, sometimes diclinous, 1–2(–4) per oogonium, antheridial cell short, clavate, stalk long, often falcate (particularly on rich media). After repeated subculturing for 30 months, the sizes of hyphal swellings and oogonia were found to increase somewhat (2383). — Isolates with predominantly smooth oogonia have often been identified as *Pythium debaryanum* sensu Middleton (e.g. 1464). — Starch-gel electrophoresis gave protein patterns distinct from several other *Pythium* species (1063). Oleic and palmitic acids were the principal acids of the polar lipid and free fatty acid fractions (501).

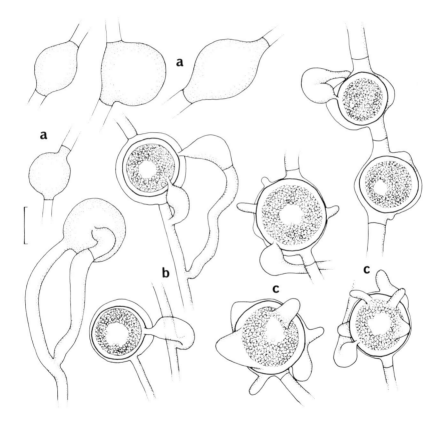

FIG. 301. *Pythium irregulare*. a. Hyphal swellings; b. smooth oogonia of CBS 287.31; c. oogonia with irregular projections of CBS 250.28.

P. irregulare is both one of the most widespread and most pathogenic species of its genus in the temperate zone (62, 4729). The numerous reports of *P. "debaryanum"* have not been incorporated in the present compilation, because of uncertainties to which species they may refer. — Media with antibiotics (62) or gallic acid (3436) and temperatures of 20–25°C proved useful for its selective isolation from soil. In a comparative study of 15 selected media the one containing 5 ppm pimaricin, 300 ppm vancomycin, 10 ppm rose bengal and 100 ppm PCNB gave the highest recovery (3846). In the Netherlands, *P. irregulare* occurred equally frequently in alluvial meadows and agricultural soils under flax, but more rarely in polder and forest soils as well as on various plants (4582). Other reports include dry grassland, marsh and swamps in the British Isles (164) and cultivated soils in New Zealand (4873). Most reports, however, are from diseased plants: as a causal agent of damping-off in forest (mostly pine) nurseries in the southern USA (2380) and southern Australia (1306, 6020, 6027), and also in pine forests (6027) and old forest stands in the USA (2385); further hosts include seedlings of peach (511), *Eucalyptus* (3609) and *Salix* (5495), also causing a stem canker in the latter as well as numerous other plants in New Zealand (4871) and South Africa (1415). *P. irregulare* has further been isolated from pineapple plantations in Hawaii (3029), cotton roots and rhizosphere (525, 4905) where it causes damage particularly in combinations with nematodes, subterranean clover in Australia at low temperatures and in wet soils (5609), tomato (4872), various cacti in greenhouses in Germany (3133), carrots in connection with rusty root disease (353, 3802) and beans (4560). — In soil (without amendments) *P. irregulare* was more sensitive than *in vitro* to griseofulvin, actidione and aureomycin (6025). Leachates from a nursery soil had a strong inhibitory influence on *P. irregulare* particularly when samples were taken in late summer or autumn (6024). In containers with milled pine bark and sand (50 : 50) and *Ilex crenata* plants, *P. irregulare* multiplied more profusely than on pure pine bark; on both substrates, the population declined markedly after seven weeks and the fungus penetrated (with decreasing frequency) down to 17 cm (2152). Survival and growth were usually favoured by higher CO_2 concentrations in the presence of normal O_2 concentrations, whilst a reduced O_2 concentration alone had no comparable effect (1902).

The production of hyphal swellings and oogonia is stimulated by Ca^{2+} ions (6479). Hyphal swellings do not germinate in autoclaved tap water but they do so in the immediate vicinity of roots (511). The club-shaped appressoria are optimally developed *in vitro* on cellophane discs between 15° and 25°C, at pH 6–7 on media containing glucose or rhamnose, organic N sources and biotin (47). For growth in soil, vapour saturation is required (3109). It can grow in the temperature range 5–40°C with an optimum at 25–30°C (511, 874, 5767), minimum at 6°C, and maximum at 36·5°C (5767); the thermal death point has been determined as 52·5°C for 30 min (613); pH had no distinct influence on mycelial growth (874). — At higher temperatures, a reduced sugar content was found to occur in the mycelia, while the synthesis of proteins and lipids increased (874). Polygalacturonase and polymethylgalacturonase activity has been described (4134). A highly active thermolabile phytotoxin damaging mainly Chenopodiaceae is produced (1994, 3638). Growth is stimulated by vegetable oils and cholesterol (96, 5105). — In contrast to many other fungi, *P. irregulare* tolerates relatively high concentrations of the triterpene glycoside aescin (4287). — The growth of *Rhizoctonia solani*, *Aphanomyces euteiches* and several other *Pythium* spp. was found to be stimulated by a growth factor produced by *P. irregulare* (6477).

Pythium mamillatum Meurs *1928* — Fig. 302.

DESCRIPTIONS: Middleton (3800), Hickman (2441), Barton (374), Waterhouse (6218), and CMI Descriptions No. 117, *1966*. — Colonies growing with a daily increment of 2·5–2·8 cm at

25°C on CMA. Sporangia and hyphal swellings ± globose, 14–21(–27) µm diam, terminal or intercalary; zoospores readily produced on transfer to water at 15–20°C. Oogonia abundant, globose, 13–19 µm diam (excluding the projections), usually terminal, with numerous obtuse-conical projections, 2·5–6 µm long; oospores ± plerotic, wall 0·8–1·6 µm thick; antheridia monoclinous, usually single, arising close to the oogonium.

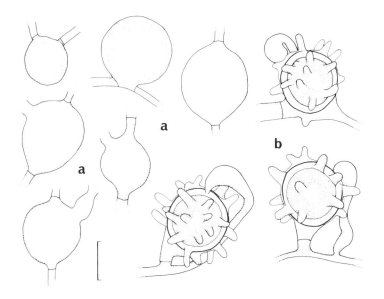

F$_{IG}$. 302. *Pythium mamillatum*; a. Sporangia, partly showing discharge tube; b. oogonia with projections; CBS 251.28.

P. mamillatum is a somewhat less common species reported from all continents (CMI Descriptions No. 117, CBS); numerous host plants from different families have been listed (4729, CMI Descriptions No. 117) in which it may cause damping-off or root rot. It has been isolated with selective media and incubation at 20–25°C from various soils in Maryland and New Jersey/USA (2381, 3436); it has been reported from the British Isles (374, 2441) but not below pH 5·4; in the Netherlands it has been isolated repeatedly from alluvial meadows and various plant roots (4582); it is commonly found in forest nurseries in Australia (6020, 6027), with higher frequencies in sandy than in loamy soils (6027), and less frequently in forest nurseries in Canada (6021); on *Eucalyptus* seedlings in Australia it was present but hardly pathogenic (3609). It is also found in tropical dry forest and cultivated soils in India (4716) and on sugar cane in Taiwan (6211). — Growth in soil is inhibited by the fungicide fenaminosulf (48), *in vitro* by antibiotics produced by *Penicillium expansum* or by 100 ppm streptomycin (2459, CMI Descriptions No. 117). Longevity is reduced *in vitro* by the presence of L-amino acids (3080).

The minimum temperature for growth is *c.* 0°C, the optimum 28–30°C, and the maximum 35°C (CMI Descriptions No. 117). Oospores were readily formed in cultivated but not woodland soils due to the low pH (3·7) of the latter (374); oospores germinate only in the vicinity of roots (e.g. of turnip seedlings), and an increase in germination was found to occur with exudates of such roots (373). The production of oogonia and oospores was not affected by vegetable oils or sterols (96). Appressoria are produced *in vitro* on cellophane discs (47). It shows no thiamine requirement in a glucose-asparagine medium (4836). Saprophytic colonization of wood substrates in soil was greatest at 10°C, at high water content, at a pH range 5·5–7·0 and in soil with a fine texture (375). — *P. mamillatum* is known to inhibit the growth of a number of soil fungi *in vitro*, for example, *Mucor plumbeus, Pythium "debaryanum", P. ultimum, Emericella nidulans, Aspergillus niger, Fusarium coeruleum, Zygorrhynchus moelleri, Aureobasidium pullulans,* and *Rhizopus stolonifer* (4427). — A high tolerance of gallic acid can facilitate its selective isolation (3436).

Pythium oligandrum Drechsler *1930* — Fig. 303.

DESCRIPTIONS: Middleton (3800), Drechsler (1461), Chesters and Hickman (984), Waterhouse (6218), Watanabe (6211), and CMI Descriptions No. 119, *1966*. — Colonies very fast-growing, with a daily increment of 2·7–3·0 cm at 25°C on CMA, very thin and, due to the evanescent hyphae, hardly visible after one week or more. Sporangia terminal or mostly intercalary, sometimes in irregular complexes of 2–5 elements, subglobose, 25–45 µm diam; zoospores formed on transfer to water at 15–18°C. Oogonia terminal or intercalary, subglobose, 17–35 µm diam (excluding the projections), with numerous (mostly 35–75) conical, pointed projections 3–7 µm long; oospores somewhat aplerotic, 15–30 µm diam, wall about 1·5 µm thick. Antheridia absent in many cases, the oogonia developing parthenogenetically; when present the antheridia are mostly diclinous. — *P. oligandrum* is similar to *P. acanthicum* Drechsler *1930* and *P. periplocum* Drechsler *1930*; the former has plerotic oospores, oogonia with somewhat shorter (to 5 µm long) and rounded projections, regularly 1(–2) usually monoclinous antheridia, sporangia similar to those of *P. oligandrum* but usually not catenulate, and even thinner colonies without any aerial mycelium; *P. periplocum* has aplerotic oospores, oogonia with 2–4 µm long, rounded projections, 1–4 antheridia and irregularly swollen sporangia; the aerial mycelium is equally scanty.

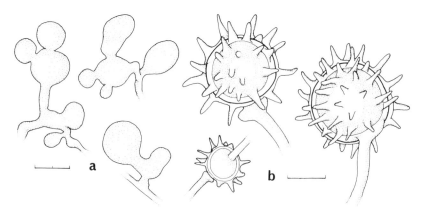

FIG. 303. *Pythium oligandrum*. a. Irregularly catenulate sporangia; b. spinulose oogonia lacking antheridia; CBS 382.34.

P. oligandrum is a common soil-borne species reported from all continents, predominantly in the temperate zone, and numerous host plants (4729, CMI Descriptions No. 119) on which it may cause damping-off, stem and root rots, and also fruit rots in cucumbers and water melon (1461, CMI Descriptions No. 119). It was originally isolated from pea roots to which it can cause considerable damage, but has subsequently also been found on numerous other plants (1461, 3029, 3800, 5767). Selective isolation of *P. oligandrum* on media containing inhibitors of other fungi (62, 1433, 5111) or by baiting techniques is easy. — It has frequently been isolated from soils in the British Isles, Egypt (IMI), forest nurseries in Canada (6021), and more rarely from *Pinus radiata* plantations in Australia (1306, 6027). In the Netherlands it is very common in polder soils, alluvial meadows, and on various plant roots, but on flax roots it was less common than in the surrounding soil (4582). — Together with certain cellulolytic fungi, it grows well on cellophane film, though only with species of a moderate competitive ability; with *Stachybotrys chartarum* and *Trichoderma viride* it grows poorly, and with *Rhizoctonia solani* not at all. *P. oligandrum* obviously profits from the cellulolytic activity of the partner and has been regarded as "parasitic sugar fungus" (5887). — *P. oligandrum* is also an aggressive mycoparasite which can attack *Gaeumannomyces graminis* var. *tritici*, *Botryotrichum piluliferum*, *Pythium ultimum*, *Monographella nivalis* and *Agaricus* sp. by means of unmodified hyphae overgrowing their colonies and reducing their growth; *Fusarium culmorum* and *Rhizoctonia solani* were less susceptible and some other basidiomycetes suppressed *P. oligandrum* by hyphal interference (1323, 1461).

The optimum temperature for growth *in vitro* is between 25 and 32°C, the minimum 10–11°C, and the maximum 37–38°C (3800, 5767). The oospores germinate in distilled water after a period of ageing (1461). — *P. oligandrum* can utilize (with the addition of thiamine) the following N sources: D-arginine, L-aspartic and D-glutamic acids, glycine, DL-alanine, DL-isoleucine and L-proline; it is heterotrophic for pyrimidine (3291, 4836).

Pythium paroecandrum Drechsler *1930* — Fig. 304.

DESCRIPTIONS: Drechsler (1459), Middleton (3800), and Waterhouse (6218). — Colonies growing half as fast as those of *P. ultimum* and *P. irregulare*, with a daily increment of 1·5 cm at 24°C on CMA. Sporangia subglobose to prolate, typically intercalary, 12–33 μm diam; zoospores abundantly produced on transfer to water at 15–20°C. Oogonia mostly intercalary, subglobose, often including a portion of the supporting hypha on one or both ends, (11–)18–25(–28) μm diam; oospores aplerotic, 10–22 μm diam, wall 1·0–1·5 μm thick; antheridia mono- or diclinous, 1–5 per oogonium, almost sessile or stalked. The formation of appressoria has been studied on cellophane discs placed on agar cultures (47).

P. paroecandrum seems to be a cosmopolitan, although infrequently reported species, occurring in temperate and warmer zones but scarcely in the tropics (62, 4729). It has been successfully isolated from soil at temperatures of 20–25°C by means of selective media based on gallic acid (3436) or antibiotics (62). It was originally isolated from roots of *Allium vineale* and *A. cepa* and has subsequently been observed on *Impatiens pallida*, *Sanguinaria canadensis*, *Aloë* spp. (3800), *Phaseolus vulgaris*, *Melilotus*, *Lupinus*, *Chrysanthemum* (IMI), carrots (3802), diseased alfalfa (4729, 5112), *Picea sitchensis* (2098), and some Cactaceae (931). *P.*

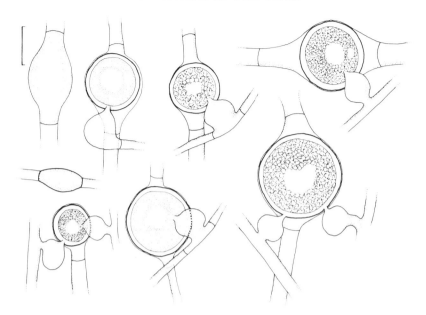

FIG. 304. *Pythium paroecandrum*, sporangia and oogonia with laterally formed diclinous antheridia; CBS 157.69.

paroecandrum is scarcely or not at all pathogenic, although sugar beet, tomato and pea have been found to be somewhat susceptible (1332). It has been isolated from the rhizosphere of *Cynodon* sp. (2386) and uncommonly from forest nurseries in Australia (6027, CBS) but was also isolated from soils in old forest stands in the USA (2385). In the Netherlands it is quite common in polder soils, alluvial meadow soils, on flax and hyacinth roots, and on water plants (1616, 4582).

The optimal temperature for growth *in vitro* is 25–28°C, the minimum 7°C and the maximum 37°C (3800). — Suitable C sources include D-maltose and L-asparagine (3752). Pretreated, but not native cellulose can be decomposed (5754). Growth is stimulated by cholesterol (5105), which also proved to be a better C source than glucose (2388). Cholesterol supports better growth in the presence of Ca^{2+} ions (3752); it is rapidly taken up, incorporated into the plasma membrane, and translocated through the hyphae (1331); its application enhances the pathogenicity of *P. paroecandrum* in pea, soya and alfalfa but suppresses fungal penetration in sugar beet and tomato (1332). — Streptomycin is accumulated in the mycelium to ten times the concentration in the ambient medium (6112).

Pythium rostratum Butler *1907* — Fig. 305.

= *Pythium diameson* Sideris *1932*

DESCRIPTIONS: Middleton (3800), and Waterhouse (6218). — Colonies growing more slowly than in any other *Pythium* species, with a daily increment of 0·7–0·9 cm at 25°C on CMA, with a dense and regular margin. Sporangia terminal or intercalary, globose to oval, 23–34 μm diam (considerably larger than the oogonia); zoospores are produced on transfer to water at 5°C. Oogonia globose to subglobose, mostly intercalary, 13–29 μm diam; oospores plerotic, with a moderately thick wall; antheridia sometimes hypogynous, mostly monoclinous, 1(–2) per oogonium, often reduced to a lateral swelling adjacent to the oogonium. The formation of appressoria has been studied on cellophane discs placed on agar cultures (47). — Acrylamide-gel electrophoresis gave a protein pattern distinct from other *Pythium* species tested (2018).

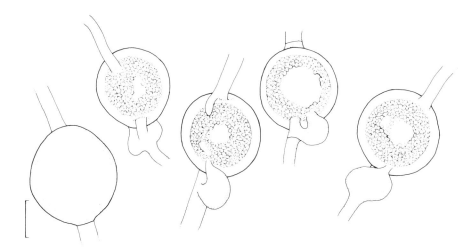

Fig. 305. *Pythium rostratum*, a sporangium and oogonia with plerotic oospores and almost sessile monoclinous antheridia; CBS 172.68.

P. rostratum has been isolated more rarely than many other species of the genus, probably because of its slow growth rate, but it nevertheless seems to be cosmopolitan, predominating in the temperate zones. — A considerable number of host plants are known (3800, 4729) and these include various cereals. This fungus is generally not, however, pathogenic (3029, 4729, 4872). Together with other *Pythium* species, it can be isolated from soil with the aid of selective media (62). It is known from soils in Iceland (2787), Canada (6021), the continental USA where it is fairly common (2381), the south-eastern USA (866), Hawaii (4729), soils in South Africa (4729), the Lebanon, Syria (62), Australia (6027, IMI), New Zealand (4871, 4872), the British Isles (6185), Germany (62, 1433), and the Netherlands (4582). It has been isolated from forest soils (866) with *Eucalyptus* (IMI), forest nurseries (6021, 6027, 6185), arable soil (62, 1433), commonly from polder soils, alluvial meadows, more rarely from various plants, including flax, and also pond water (4582). Crop rotation can apparently exert a significant influence (5110). It has also been found in the rhizosphere of *Cynodon* sp. (2386), and isolated from the effluent of a sewage treatment plant (3779). *P. rostratum* is mainly found in the surface layers of the soil (159) and its incidence diminishes to a depth of 60 cm (164). It is mainly found in soils with pH values around neutral (159). — High atmospheric humidity is required for active growth; the least tolerated water potentials are <–20 bars, with optimal development occurring at vapour saturation (3109); nevertheless, it can withstand long periods of desiccation (164) and is the only *Pythium* species found in some particularly dry habitats (4582).

Optimal growth occurs at 25°C, with a minimum of 4°C and a maximum of 31°C (3800). — *P. rostratum* is a relatively good cellulose decomposer after suitable pretreatments of the substrate (5754).

Pythium sylvaticum W. A. Campbell & Hendrix *1967* — Fig. 306.

= *Pythium debaryanum* var. *pelargonii* H. Braun *1925*

DESCRIPTIONS: Hendrix and Campbell (2384), Pratt and Green (4624), and van der Plaats-Niterink (4581). — Colonies are rather fast-growing, with a daily increment of 3·0 cm at 25–28°C on CMA. This is essentially a heterothallic species although homothallic isolates occur infrequently; in old cultures of female isolates oogonia are often formed near the inoculation point (4581); different isolates vary in sexual strength (4624). Sexual response was greatest at 26°C in darkness; lima bean agar (4624, 4626) supplemented with neomycin and penicillin (4625) proved suitable for mating in addition to CMA, PCA (867, 4581) and hemp-seed agar (4377, 4624). Oogonia forming in intercalary or terminal positions, globose, (11–)13–27(–36) μm diam; oospores aplerotic, (9–)10–24(–30) μm diam, wall 1–2 μm thick; antheridia diclinous or rarely monoclinous, 1–4(–6) per oogonium, stalk usually bifurcate near the oogonium. Hyphal swellings abundant, mostly terminal, sometimes intercalary, ± globose, 11–30 μm diam; zoospores unknown. — *P. sylvaticum* resembles *P. heterothallicum* Campbell & Hendrix *1968* which has hyphal swellings of 15–21(–25) μm diam, oogonia with up to 8 antheridia with frequently branched stalks and aerial mycelium forming zonate or rosette patterns on CMA; the two heterothallic species are not interfertile (4581). — The postulated synonymy of *P. debaryanum* sensu Middleton with *P. sylvaticum* (2384) is open to doubt since in the latter the antheridia are more irregularly applied to the oogonia whilst in the former they were illustrated as being more distinct and contacting the oogonia only with the tip as in *P. irregulare* (A. J. van der Plaats-Niterink, pers. comm.).

Because of its heterothallic nature, *P. sylvaticum* has been recognized only since 1967 and its distribution is still imperfectly known. Where it has been recognized, it proves to be one of the commonest *Pythium* species in temperate zones. It was first isolated from soils in various regions of the southern USA (867), where it is common in forest nursery soils with pine seedlings, in which it causes damping-off (2380, 4624), and in old forest stands (2385). It is also common in forest nurseries in Canada (6023). *P. sylvaticum* has also been reported from arable soils in Germany and the Netherlands (1614, 1616); in the latter country it has also been isolated frequently from polder soils, alluvial meadows, forest soils, and roots of various plants including (with high frequencies) flax (4582). Some preliminary observations suggest that it is rather common throughout Europe (4581). — It has been isolated from the rhizosphere of *Cynodon* sp. (2386), diseased strawberries (4624), sweetgum (1737), apple seedlings in which it caused a root rot (4034), and has been found to affect the growth of various other crop plants (559, 1887, 3802).

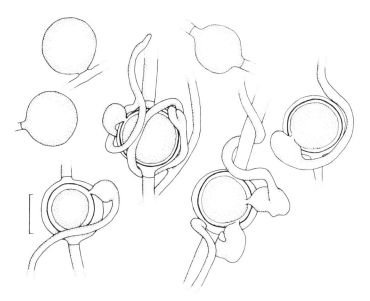

Fɪɢ. 306. *Pythium sylvaticum*, hyphal swellings and oogonia with entwining antheridia formed between CBS 225.68 and 232.68.

The optimal temperature for growth and oospore production is 25–30°C, the minimum 0°C, and the maximum >35°C (867, 2558, 4581). — Oospore production in mating experiments was stimulated by cholesterol (4626) or structurally and stereochemically related sterols (1004) which affect the female part and are rapidly taken up from the medium (2389). *P. sylvaticum* produces polygalacturonase; cellulase activity has been detected in culture filtrates (4134); the fungus produces a growth-regulating substance (560) which might be identical to indole-3-acetic acid (4603).

Pythium torulosum Coker & Patterson *1927* — Fig. 307.

DESCRIPTIONS: Middleton (3800), Kouyeas and Kouyeas (3111), and Waterhouse (6218). — Colonies growing with a daily increment of 1·3 cm at 25°C on CMA. Sporangia consisting of few to numerous irregularly branched and inflated elements; zoospores easily formed on transfer to water at about 15°C. Oogonia terminal, on short lateral branches, or intercalary, globose to subglobose, 12–19 µm diam; oospores plerotic, 11–15 µm diam, with a moderately thin wall (0·8–2·0 µm); antheridia monoclinous, 1–2 per oogonium, short-stalked, soon becoming evanescent. — The most similar species, *P. dissimile* Vaartaja *1965*, has oogonia 9–26 µm diam with no or inconspicuous hypogynous antheridia; *P. vanterpoolii* V. & H. Kouyeas *1963* has oogonia 14–22 µm diam, thick-walled oospores (2–3(–4) µm), and sporangia consisting of catenulate, subglobose elements; *P. graminicola* Subramaniam *1928* (and some other species) have much larger oogonia, 16–29(–40) µm diam and with 1–6 antheridia each.

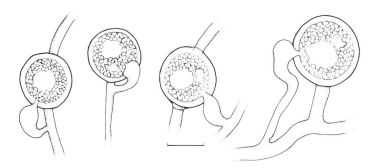

Fig. 307. *Pythium torulosum*, oogonia with plerotic oospores and antheridia, CBS 405.67.

P. torulosum is rather infrequently reported although a cosmopolitan species equally distributed in temperate and subtropical zones. It can easily be isolated from soil on selective media (62) or on filter paper bait discs impregnated with pimaricin (4873). — It is widespread in the continental USA (2381, 2385) and Hawaii (3029) and has also been reported from agricultural soils in Germany (62), Italy (4729), forest nurseries in Canada (6021), and grassland soils in New Zealand (4873) and the British Isles (159, 164). In the Netherlands it is rather common in polder soils and alluvial meadows, as it is also in Czechoslovakia (CBS). It has been found mainly in the upper soil layers (159) but has been detected down to 100 cm (164). *P. torulosum* has been isolated preferentially from soils with almost neutral pH values (159). It was originally described from mosses in water and has subsequently been isolated from roots of barley, wheat, various grasses, spruce and pine (3800), and also detected on roots of some other plants (5767). Pathogenicity has been proved to wheat (3800) and (mildly) to tomato (4872). Further known habitats include droppings of birds which contribute to the long-range dispersal of the fungus (5808), fresh water (3800), and bog water and moss cushions in Iceland (2787).

Optimal temperatures between 25 and 30°C have been reported for growth *in vitro* (3800), with a minimum of 6–7°C, and a maximum of 35–38°C (5767). The minimum water potential for growth is −30 bars, optimal growth occurring at saturation (3109). — Several inorganic S sources are equally well suited, regardless of the oxidation state of the sulphur; the following compounds are also utilized: DL-methionine, thioacetamide, L-cysteine, and sodium thioglycolate. A ratio of 1 : 1 : 4 of K_2HPO_4, $MgSO_4 \cdot 7H_2O$ and KNO_3 was found to be favourable (1806). Survival *in vitro* is reduced on media containing L-amino acids (3080).

Pythium ultimum Trow *1901* var. *ultimum* — Fig. 308.

DESCRIPTIONS: Middleton (3800), Waterhouse (6218), Kouyeas (3108), and Grouet (2129). — Colonies growing with a daily increment of 3·0 cm at 25°C on CMA. Hyphal swellings and sporangia mostly terminal and globose, 12–28 μm diam. Zoospores sometimes formed on transfer to water at 5°C; these can also be liberated from well ripened oospores transferred to water at 10°C (1463). Oogonia mostly terminal, globose, 19–23 μm diam; oospores aplerotic, 15–18 μm diam, with a 2·0–2·5 μm thick wall; antheridia mostly monoclinous but sometimes diclinous, almost sessile or short-stalked, 1(–2) per oogonium, broadly swollen. The ultrastructure of oogonium development has been studied by TEM (3593). The size of hyphal swellings, and sometimes oogonia, was found to decline after 30 months of repeated subculturing (2383). Club-shaped appressoria were formed on cellophane discs placed on agar media (47). — *P. ultimum* var. *sporangiiferum* Drechsler *1960* was found to release zoospores at higher temperatures than var. *ultimum* (1466). The similar *P. violae* Chesters & Hickman *1944* can be distinguished from *P. ultimum* by its production of 3–5 antheridia per oogonium with longer stalks and thinner oospore walls, oogonia 24–30 μm diam, and hyphal swellings mostly 22–30 μm. — TEM studies of *P. ultimum* hyphae showed dictyosomes associated with the nuclear envelope (2138) which were particularly frequent in the subapical zone of the hyphae which elongated by the incorporation of vesicles in the plasmalemma (2137). — The amino acid content of the mycelium has been analysed (2458); hyphal walls contain only traces of glucosamine (1389). The hyphae contain 3–48% lipids, depending upon the stage of development, which consist mainly of triglycerides besides free fatty acids, phosphatides and monoglycerides (672). Acrylamide-gel electrophoresis gave a protein pattern indistinguishable from that of *P. "debaryanum"* and *P. irregulare* (2018), but there are also controversial reports (1063).

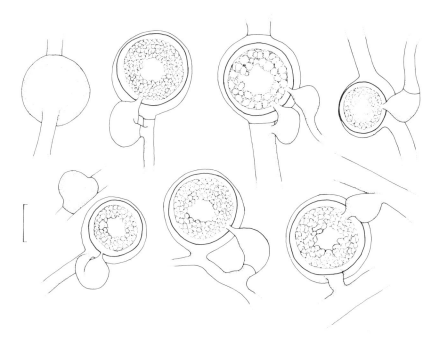

Fig. 308. *Pythium ultimum*, hyphal swellings and oogonia with monoclinous and diclinous antheridia.

P. ultimum is probably one of the commonest species of the genus and can more frequently be isolated in temperate than in tropical countries. Long lists of hosts have been compiled for this phytopathogenic species (3800, 4729) but these can be extended by numerous subsequent reports in the phytopathological literature. Selective media for its isolation, together with other *Pythium* species, have been described which incorporate antibiotics (62, 2459, 5527, 5692) and gallic acid (3436). In a comparative study of 15 selected media, cornmeal agar with 5(–100) ppm pimaricin, 300 ppm vancomycin, 10 ppm rose bengal and 100 ppm PCNB gave the highest recovery (3846). — Soil isolates have been reported, for example, from forest nurseries in Australia (6020, 6027), New Zealand (380, 4871), the British Isles (2098, 6184, 6185), Canada (6021) and Iceland (2787); grassland (982) and arable land (62, 1433, 3817, 5111), particularly after growing beans and peas but rarely after oats, and with seasonal variations (3116, 5110); garden soil (374, 931), pear orchards in California where it was the commonest *Pythium* species (4153), and alkaline soils (6185). In chitin-amended soils, an increased frequency has been recorded (781) as has one on chitin-coated sugar beet seeds, due to some soluble nutrients present, although its growth is not promoted by purified chitin (1990). Similarly, green manuring greatly stimulated the multiplication of *P. ultimum* on lettuce, but this effect was reversed after three weeks (6226). In Britain, the greatest incidence in grassland soils was found to occur in winter (6182). Soil water content appears to have less influence on its distribution than pore size which must be greater than the diameter of the oogonia (2094). At relatively high moisture levels preferentially oogonia, and at lower ones more hyphal swellings are produced (284). Two mechanisms for survival have been described: (a) under reduced nutrient supply by contraction of the protoplasm in the parent hyphal swellings, and (b) under good nutrient conditions by formation of new terminal and intercalary hyphal swellings (5528). *P. ultimum* can survive a storage in water for seven years (605). Soils infested with *P. ultimum* and stored for two years at −18°C maintained their pathogenic potential (4044). — The oogonia can be parasitized by several *Dactylella* species (1467, 1468). The fungus cannot, in contrast to other pathogenic soil fungi tested, serve as food for the nematode *Aphelenchus avenae* (3580); on the other hand, injuries on sugar beets were cumulated by *Heterodera schachtii* and additive with those caused by *P. aphanidermatum* (6334). A cumulative effect can occur also between *P. ultimum* and *Phytophthora megasperma* on soya plants, whereas *Glomus mosseae* and *Rhizobium* reduced the damage to them (1024). — Experiments *in vitro* show a relatively high sensitivity to antagonistic fungi and streptomycetes (1180, 1431, 5370, 6325). The fungus has been found in the rhizosphere of various plants, particularly legumes (4871); an accumulation of antagonists around the roots of resistant plants has been observed; the root extract of resistant plants themselves has been said to have an inhibitory effect (102). Root exudates of *Pinus resinosa*, however, stimulated growth and germ tube development (50). Soil leachates from a nursery had an inhibitory influence on *P. ultimum*, particularly when samples were taken in late summer and autumn (6024). Several fractions from extracts of clay and sandy soils inhibited mycelial growth whilst others were able to stimulate it and counteract the former (6022). Seed treatment with *Trichoderma viride* and *Penicillium frequentans* gave good results in controlling *P. ultimum* in beet (3378); its growth *in vitro* was also inhibited by volatile metabolites from *Trichoderma koningii* and *T. viride* (1344), and also by *Bacillus subtilis* (2393). The mycelium is lysed in non-sterile soil within 1–2 weeks leaving the oospores free (258). Hyphal swellings do not usually germinate in soil due to mycostasis; the inhibition can be counteracted by the addition of glucose, fructose, maltose and sucrose, polysaccharides, or various amino acids and yeast extract; mixtures of these C and N sources are especially effective (51).

   The optimal temperature for the germination of hyphal swellings lies in the range 15–25°C; in the absence of organic compounds, the germ tubes rapidly lyse in soil (51). Fructification *in vitro* is promoted by plant extracts (2982), thiamine is said to be required for sexual reproduction (4752), and cholesterol stimulated oogonium formation optimally at 10–20°C

(258). The influence of Ca^{2+} ions on the formation of oogonia has been investigated (6479, 6512). Thick-walled, dormant oospores become thin-walled in water-saturated soil after 1–10 weeks and germinate readily on nutrient media or in water; thin-walled oospores are less resistant to drought and can infect French beans rapidly (3435). — Optimal growth occurs in the range 24–28°C (5371), the minimum temperature for growth is 1°C, and the maximum a little above 30°C, and 37–38°C in var. *sporangiiferum* (A. J. van der Plaats-Niterink, pers. comm.). Isolates from Puerto Rico did not grow below 12°C (121); the thermal death point has been determined as 49·5–50°C for 30 min in soil (613). The most favourable pH range is 5·0–6·5 (389, 1806). — Pectin degradation is quite pronounced (1351, 3919, 4771), with only slight differences between individual isolates (1432); chitin is degraded (781), and C_x cellulases (3919) or endocellulase (1351) are produced. *P. ultimum* can efficiently utilize glucose, fructose, starch, sucrose and inulin. Growth is stimulated by cholesterol (5105) but inhibited by the alkaloid a-tomatine (182). Nitrite is not used (4752) and nitrate-N and ammonium-N are poor N sources; L-aspartic acid, L-asparagine, L-glutamine, L-alanine, L-proline, casein hydrolysate, and peptone are preferentially utilized (2335, 3117, 3118). Inorganic S sources tested proved equally suitable, regardless of the oxidation state of the sulphur; the following organic S sources can also be utilized: DL-methionine, thioacetamide, L-cysteine, sodium thioglycolate, DL-cysteine; a very unsuitable source is DL-ethionine; a favourable (though not obligate) ratio for good growth is $K_2HPO_4 : MgSO_4 \cdot 7H_2O : KNO_3$ in the proportions 0·25 : 0·25 : 2·0 (1806). Mitochondrial preparations displayed succinoxidase and DPNH oxidase activities; enzymes of the Krebs cycle could not, however, be demonstrated (5854). Streptomycin absorption is as in *P. paroecandrum* (q.v.) apart from its higher streptomycin tolerance (2459, 6112). Mycelial mats can accumulate considerable amounts of dieldrin, DDT and PCNB (3057). Chloramphenicol was found to inhibit growth, but not O_2 uptake which was inhibited by antimycin A (5424). The phytopathological literature provides numerous reports on the influence of low temperature, high soil humidity, soil pH and fungus/nematode associations on the infection of higher plants. Good compatibility with cellulolytic fungi has been noted (5887). The growth of a number of rhizosphere fungi is stimulated by a metabolite of *P. ultimum* and three other *Pythium* species (6477). This fungus has been regarded as the vesicular-arbuscular endophyte of *Allium* (2335). Production of a growth-promoting substance is thought to be responsible for a stimulatory effect noted on aseptically grown *Buddleya davidii* and *Senecio squalidus* (5569). — *P. ultimum* tolerates high CO_2 concentrations up to 18 times that of O_2 (2096), but was also found inactive at high $CO_2 : O_2$ ratios (3481). The fungus is highly sensitive to both steam and formalin soil treatments (6185, 6364).

Ramichloridium Stahel ex de Hoog *1977*

Ramichloridium schulzeri (Sacc.) de Hoog *1977* — Fig. 309.

≡ *Psilobotrys schulzeri* Sacc. *1884*
≡ *Rhinocladiella schulzeri* (Sacc.) Matsushima *1975*
= *Acrotheca acuta* Grove *1916*
  ≡ *Pleurophragmium acutum* (Grove) M. B. Ellis *1976*
= *Rhinotrichum multisporum* Doguet *1953* (nomen inval., Art. 36)

DESCRIPTIONS: Hughes (2595), Ellis (1604), Gams and Domsch (1887), Matsushima (3680), and de Hoog (2524).

The Hyphomycete genus *Ramichloridium* (= *Pleurophragmium* Cost. *1888* sensu M. B. Ellis *1976* pro parte; type species *Ramichloridium apiculatum* (Miller *et al. 1957*) de Hoog *1977*) is characterized by erect, mostly unbranched conspicuously differentiated conidiophores, darker than the hyaline or pale vegetative hyphae, with several thin septa; conidiogenous cells terminally integrated, lighter brown, the terminal rhachis with scattered conidiiferous denticles. Conidia hyaline or almost so, mostly one-celled, subglobose to fusiform. The thirteen known species have been monographed by de Hoog (2524).

Colonies of *R. schulzeri* spreading, reaching 2·6–3·8(–5·7) cm diam in 14 days at about 20°C, pale orange to light brown or slightly olivaceous; reverse pink to orange. Conidiophores often to 250 μm tall, 3–6-celled, reddish or greyish brown. Conidia slightly clavate with acuminate bases, smooth-walled or finely warted, subhyaline, (6·3–)7·5–9(–10·5) × 3–4 μm. — Three varieties have been distinguished: var. *flexuosum* de Hoog *1977*, described from Surinam soils, differs from var. *schulzeri* in having slightly wider conidia (4–5 μm) with more rounded bases and a prominent hilum; and var. *tritici* (M. B. Ellis) de Hoog *1977* (≡ *Pleurophragmium tritici* M. B. Ellis *1976*) described from wheat stems in Ireland, has frequently septate conidia.

R. schulzeri has been observed on dead stems of *Urtica dioica* infected with *Leptosphaeria acuta* (2595), wheat in Australia, *Chloris* seed in Rhodesia, *Pieris* leaves in Japan, apple leaves in Switzerland (2524), a hornbeam trunk in Italy, and a dead stem in Madras India (5626); it was found to be associated with a strawberry root rot in Illinois (4133); isolates have been recorded from cultivated soils under wheat in Germany (1889), *Salix* communities in Wisconsin (2004, 2008), and other soils in the Netherlands, the USA, Canada, Zaïre, Madagascar and South Africa (2524).

R. schulzeri causes considerable losses in weight and tensile strength of thin maple-wood strips with differences in this ability between individual isolates (2211).

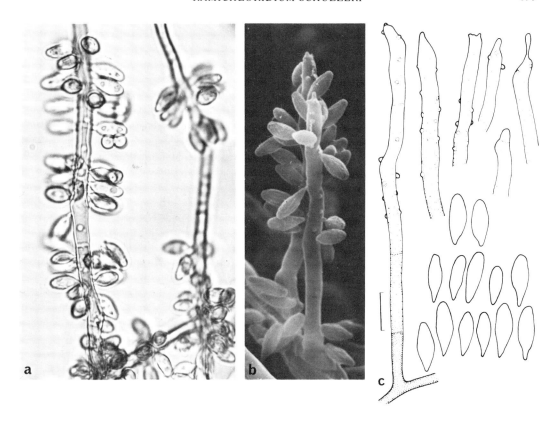

FIG. 309. *Ramichloridium schulzeri*. a. Conidiiferous rhachis, × 1000; b. conidiophore with some attached conidia, SEM, × 1200, CBS 304.73, orig. R. A. Samson; c. conidiophores with conidiiferous denticles (from 1887, reproduced with permission J. Cramer).

Rhizomucor Lucet & Cost. ex Vuill. *1931*

Rhizomucor pusillus (Lindt) Schipper *1978* — Fig. 310.

≡ *Mucor pusillus* Lindt *1886*
= *Mucor septatus* Bezold & Siebenmann *1889*
  ≡ *Rhizomucor septatus* (Bezold & Siebenmann) Lucet & Cost. *1901*
= *Rhizomucor parasiticus* Lucet & Cost. *1899*
  ≡ *Rhizopus parasiticus* (Lucet & Cost.) Lendner *1908*
= *Mucor buntingii* Lendner *1930*
= *Mucor hagemii* Naumov *1935*

DESCRIPTIONS: Lucet and Costantin (3425), Scholer (5125), Cooney and Emerson (1174), Mehrotra *et al.* (3739), Schipper (5101), and CMI Descriptions No. 527, *1977*; zygospores: Schipper (5094). — The genus *Rhizomucor* is distinguished from *Mucor* by the presence of rhizoids at the base of the sporangiophores and the thermophilic behaviour of its three species. — *Rh. pusillus* has 2–3 mm high colonies, at first white, but later becoming deep neutral grey to greyish sepia. Sporangiophores about $10\,\mu$m wide, sympodially and, in the upper part, often umbellately branched, colourless to yellow-brown; sporangia 50–$80\,\mu$m diam, spinulose, rupturing at maturity; columellae subglobose to slightly elongate, often to $60\,\mu$m high, 15–$35\,\mu$m diam; sporangiospores globose to subglobose, hyaline, smooth-walled, 3–$5\,\mu$m diam. Chlamydospores absent. Heterothallic. Zygospores form between compatible isolates on MEA or YpSs agar at 30 or 40°C and throughout the aerial mycelium; they are *Mucor*-like, reddish brown to black, 45–$63(-70)\,\mu$m diam, covered with flat, irregular warts (5102). — *Rh. pusillus* has often been confused with the equally thermophilic, thinly growing and equally common *Rh. miehei* (Cooney & Emerson) Schipper *1978* (≡ *Mucor miehei* Cooney & Emerson *1964*) which is regularly homothallic, while in *Rh. pusillus* homothallic isolates have only exceptionally been found (5094, 5125); its colonies are a pale dirty grey-brown to beige-brown, the sporangiophores are loosely sympodially (not umbellately) branched, and no growth occurs at 22°C and very slow growth at 25–30°C. *Rh. pusillus*, in contrast to *Rh. miehei*, also assimilates sucrose, melezitose and a-methyl-D-glucose and does not require an external thiamine supply (5125). — Azygospores could be induced in interspecific matings with *Absidia corymbifera* but not with *Mucor* species (5100). — Morphological and ultrastructural changes have been observed at the maximal growth temperature of 55°C (1589). — The mycelium is reported to contain 21·2% hexosamine, 23·5% C_{16}-, 2·9% C_{18}- and 0·8% C_{20}-saturated and 1·2% C_{16}- and 59·4% C_{18}-mono-unsaturated fatty acids (3232). — DNA analysis gave a GC content of 48% (5600).

Rh. pusillus has been reported as a thermophilic fungus from all over the world, including the British Isles (163, 166, 1551, 1664, 1665, 1666, 2588), Czechoslovakia (1703), Chad (3415), South Africa (1557), Indonesia (578, 2358), India (3730, 3739, 3868, 5058), Japan (252) and the USA (1174, 5747). It has mainly been found on composting and fermenting

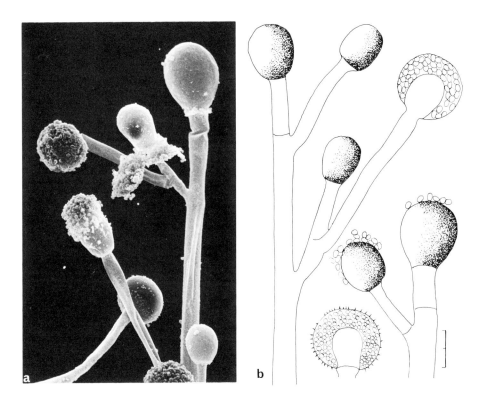

Fig. 310. *Rhizomucor pusillus.* a. Sporangiophore with diffluent sporangia and columellae, SEM, × 380; b. sporangiophores with sporangia and columellae; CBS 354.68.

substrates, for example: garden compost (1557), municipal wastes (2874, 3041, 3051), cultivated mushroom beds (1723, 1727), horse dung (1174), composted wheat straw (953) and *Diplachne* grass (3546), guayule (1174), leaf mould (1174), self-heated hay (4548, 4821), self-heated (252) or rotting (4263) corn, seeds of cacao (1174, 5138), barley, oat and wheat (1751, 1752, 1757, 4492), groundnuts (1984), pecans (2572), but has also been recorded from nests, feathers and pellets of birds (168, 2575, 3843, 5067) and some other animals (4174) and from air (1174, 1666, 2206, 2588, 3204, 5784). Known soil habitats include grassland (1551), coal spoil tips (1664, 1665), peat (3157), and a uranium mine (1703).

Growth is slow at 20°C, moderate at 22°C, and optimal up to 50°C; at 55°C only sterile mycelium is produced, and at 57°C no growth occurs. In contrast, *Rh. miehei* still grows moderately well at 57°C (1174). *Rh. pusillus* has been found to survive a heat treatment of 68°C for 45 min (1727). It grows and sporulates, and sporangiospores germinate well, when incubated in sun-heated soil of temperate latitudes (2699). A minimum of 0·2% oxygen is required for mycelial growth, and 0·7% for sporulation (1350). — Both *Rhizomucor* species can rapidly utilize numerous C sources, including D-xylose, D-ribose, L-arabinose, D-mannose, D-glucose, D-fructose, D-glucosamine, trehalose, melibiose, cellobiose, lactose, raffinose, dextrin, pectin, starch, arabinoxylan, glycogen, arbutin, salicin, D-sorbitol, D-mannitol, glycerol, pyruvic, succinic, stearic, oleic, palmitic and myristic acids, but not maltose, cellulose and ethanol; some other substrates are moderately attacked (29, 1349, 1724, 1757, 1758, 4321, 5126, 5741). 1,4-β-Xylosidase (1758) and endo-1,4-β-glucanase (5471) are produced. Suitable N sources are NH_4Cl, nitrate, nitrite, L-histidine, L-cystine, adenine, caseine, peptone, glycyl-glycine, and urea (3739, 5640); the most suitable amino acids (administered singly) are glycine, L-alanine, L-proline and L-aspartic acid, no growth occurs on glutathion and cytosine (5640); both potassium sulphate and sodium sulphite are suitable S sources (3739). Lipase (5309, 5473), phosphatase (5309), a strong milk-clotting enzyme (176), acid proteinase (5472) and an alcohol dehydrogenase (1999) are produced; small amounts of 2-oxoglutaric acid (4639), acetic acid and ethanol also occur in liquid culture (4509); ergosterol, 22-dihydroergosterol, 24-methylenedihydrolanosterol, and obtusifoliol are produced in the presence of methionine or mevalonic acid (3761). More unsaturated fatty acids are synthesized at 25° than at 48°C (5643). — *Rh. pusillus* is known as the causal agent of a lung mycosis and other internal infections in various animals, including horse, dog, cow (also causing abortion and mastitis), pig, rabbit and duck (404, 1174, 1399, 1814, 3054). Mycelial extracts contain mycotoxins effective against brine shrimps and chicken embryos (1304).

Rhizopus Ehrenb. *1821*

Type species: *Rhizopus stolonifer* (Ehrenb. ex Link) Lind

Contributions to a MONOGRAPHIC TREATMENT: Inui *et al.* (2661) recognized 13 species groups and one variety, mainly based on physiological studies; Scholer (5125, 5127) studied the potentially pathogenic species, of which five were recognized. Numerical taxonomy: Dabinett and Wellman (1255).

Fam. Mucoraceae Dumort. — Colonies very fast-growing, forming ± hyaline aerial hyphae, stolons, pigmented rhizoids and sporangiophores, and apophysate, columellate many-spored sporangia. The differentiation into stolons and nodes with rhizoids and sporangiophores is the characteristic feature of *Rhizopus*. After spore liberation, the apophyses often collapse so that the flattened columella recalls the cap of an agaric. Sporangiospores short-ellipsoidal with almost pointed ends, brownish, striate in many species. In the study of Inui *et al.* (2661), too many species were probably distinguished on the basis of minor differences; furthermore, their species concept was based on arbitrarily chosen recent isolates and thereby often deviated from the original description; other clearly distinct species were neglected (5125). Consequently, the numerical analysis of the same physiological-morphological data of the same 34 OTUs (1255) leads to similar, sometimes unreliable, conclusions.

Chlamydomucor Bref. *1889* has often been combined with *Rhizopus*, but the type species of that genus (*Ch. racemosus* (Fres.) Bref.) in fact belongs to *Mucor*, and *Ch. oryzae* Went & Prinsen Geerligs *1896* was found to be synonymous with *Amylomyces rouxii* Calmette *1892* (≡*Mucor rouxianus* (Calmette) Wehmer *1907*). The monotypic genus *Amylomyces* Calmette *1892* is intermediate between *Rhizopus* and *Mucor*; the sporangiophores are similar to those of *Rhizopus* but arise from aerial hyphae; the sporangia are abortive and produce rather few very irregular spores (1593). Another similar genus is *Rhizopodopsis* Boedijn *1958* which has umbellate sporangiophores with short sterile spines in the centre.

Several *Rhizopus* species can attack potato, sweet potato, cotton bolls and various fruits, while others can become parasitic on animals. *Rhizopus* species are often used in food fermentations (the action of amylase leading to the production of alcohol and organic acids). Many species grow well at high temperatures. Ethanol, glycerol and adonitol are assimilated by all species, but lactose is not (5125). Thiamine is not required for growth and nitrate cannot be utilized. The fermentation patterns are of little value in species delimitation (5125).

Key to the species treated:

1	Homothallic species	2
	Heterothallic species	3

2(1)   Temperature maximum 30°C   *(Rh. sexualis)*
           Temperature maximum 46°C   *(Rh. homothallicus)*

3(1)   Temperature maximum 32°C; glucose fermentation negative; sporangiophores mostly 1·5–3 mm
           long   *Rh. stolonifer* (p. 707)
           Temperature maximum 40°C or above; glucose fermentation positive   **4**

4(3)   Sporangiospores striate; sporangiophores of variable length, to 2·5 mm long   **5**
           Sporangiospores not striate; sporangiophores not exceeding 1 mm in length   **6**

5(4)   Sporangiophores 1–2·5 mm long; sporangia mostly 160–240 μm diam   *Rh. oryzae* (p. 704)
           Sporangiophores not exceeding 0·5–0·8 mm long; sporangia mostly 80–120 μm diam
                                                             (Rh. microsporus)

6(4)   Sporangiospores globose to short-oval, 4–6(–8) × 4–5(–7) μm   *(Rh. rhizopodiformis)*
           Sporangiospores very irregular in shape, (4–)9–10(–15) × (4–)7–10(–11) μm   *(Rh. oligosporus)*

Rhizopus oryzae Went & Prinsen Geerligs *1895* — Fig. 311.

    ? = *Rhizopus arrhizus* Fischer *1892*
      = *Rhizopus japonicus* Vuill. *1902*
      = *Rhizopus nodosus* Namysl. *1906*
      = *Rhizopus delemar* Hanzawa *1912*

For numerous other synonyms see Scholer (5125).

DESCRIPTIONS: Inui *et al.* (2661), Caretta (892), Scholer (5125), Scholer and Müller (5127), and CMI Descriptions No. 525, *1977*. — Colonies very fast-growing, about 1 cm high, with some tendency to collapse, pale or dark brown-grey. Stolons hyaline, 15–25 μm wide, at their ends usually producing a rhizoid with 4–8 branches, 150–250 μm long, and 1–5 sporangiophores; other sporangiophores arising directly from stolons or aerial hyphae. Sporangiophores (0·5–)1–2·5(–3·2) mm long, 14–24 μm wide, sometimes forked, yellow-brown to dark brown. Sporangia mostly 50–200 μm diam, wall diffluent, opaque, brown-grey to black. Columella and apophysis together ± globose, soon collapsing after spore release. Sporangiospores biconical to subglobose, surface ridged, (4–)6–8(–12) × (3–)4·5–6(–8) μm, studied by SEM (149, 1591) and TEM (764, 2423). Chlamydospores always present, arising singly or in short chains, globose to oval, smooth-walled, refringent, hyaline, 10–35 μm diam. Zygospores found in *1979* (CBS). — The relative development of the *Rhizopus* architecture has little taxonomic value at the species level and is insufficient to delimit *Rh. arrhizus* (atypical) from *Rh. oryzae* (more typical stolon–rhizoid differentiation); other criteria which have been used to distinguish the latter species are a positive raffinose fermentation and rich sporulation. The original description of *Rh. arrhizus* emphasized an unusually low columella, 40–75 μm high (including the apophysis) and 60–100 μm diam, measurements never observed together subsequently; the rhizoids are said to be very poorly developed and hyaline, while in *Rh. oryzae* a more typical *Rhizopus* architecture can always be observed to some, although variable, extent. In summary, the identity of *Rh. arrhizus* is somewhat doubtful and the species is perhaps best considered as based on an atypically developed *Rh. oryzae* (5125, 5127). — Dormant and germinated sporangiospores show deep furrows and prominent ridges with a pattern different from that of *Rh. stolonifer* as seen in TEM (2423). The size of the sporangia is influenced by temperature, reaching its maximum below the optimum tempera-ture for growth (5789). The dissection of a stolon induces the formation of single sporan-giophores in the remaining part (1926). — The mycelium contains between 2·2 and 15·3% lipids (6251); these were reported to comprise traces of C_{10}- and C_{12}-saturated fatty acids, 1·2% C_{14}-, 18·4% C_{16}-, 11·0% C_{18}-, 16·2% C_{20}-saturated, and 3·7% C_{16}-, and 29·4% C_{18}-mono-unsaturated fatty acids (3232), but it must be noted that lipid content and composition change during growth (2175). — DNA analysis gave GC contents of 37·5–40·0% (5600).

Rh. oryzae is mainly distributed in the tropical and subtropical zones. In addition to very numerous reports from India (too numerous to list individually here, see list in 5058), it has been found in soils in Pakistan (4855), New Guinea (3020), Taiwan (2472), Central America (1697), Peru (2005), Argentina (1827), Namibia (1827), South Africa (1415, 1555, 4407), Iraq (92), Somalia (5048, 5049), Egypt (8, 4962), Libya (6510), Tunisia (4055), Israel (2764, 2768, 2772), Turkey (4183, 4245), Spain (3446, 3447), Italy (3450, 4183, 4538), Hungary (6551), Czechoslovakia (1700, 1703), Germany (4055), the Ukraine (5298), the British Isles (161), and the USA (1028, 1387, 2482, 2573, 2768, 3817, 4711). Its occurrence has been documented in a wide variety of soils, including uncultivated ones (3446, 4698, 4855), forest soils (2186, 2482, 2573, 2854, 4995, 5048, 5512), grassland (161, 1555, 4711), cultivated soils under lupin, corn (4538), wheat (3446), groundnuts (2768, 2772), other legumes (1317), sugar cane (3732), rice (1519), citrus plantations (2764), steppe type vegetation (4407, 5049, 5298), alkaline soils (4055), salt-marshes (8), soils treated with farm manure (968, 2764), and sewage sludge (1167, 1387); it has also been found in a uranium mine (1703), on wood in a copper mine (2664), guano in a bat cave (6551), in birds' nests (168) and feathers (2575), and in polluted still and running waters (1166, 3809). It has repeatedly been isolated from soils with a pH range 6·3–7·2 or higher (3450, 4538, 5048, 6510). — It has also been reported from senescent leaves and culms of grasses (161), decaying stems of *Heracleum sphondylium* (6462), rice roots (1282), the rhizosphere of groundnut (2768), barley (3526), *Cassia occidentalis* (6067), poplar (3452), tomato, *Trigonella foenum-graecum* and *Abelmoschus esculentus* (3733), seeds of barley, oat, wheat (1752, 3004, 4492), rice (2297), corn (57), onions (5467), cotton (2227), groundnuts (2765, 2772, 3702, 4602), stored sweet potatoes (3066), pecans (2258, 2572), Brazil nuts (2515), and frozen fruit cake (3153). — Hyphae can be parasitized by *Syncephalis californica* (2618) or invaded by *Trichoderma viride* (1729) and, to a lesser extent, *Myrothecium verrucaria* (1511); the mycelium may serve as food for mites (5380).

The optimal conditions for sporangium production are temperatures in the range 30–35°C and low water potentials (1931); sporulation is stimulated by amino acids (except L-valine) when grown in the light, while in darkness only L-tryptophan and L-methionine have a promoting effect (2171). — The germination of sporangiospores can be induced by the combined action of L-proline and phosphate ions; L-ornithine, L-arginine, D-glucose and D-mannose are also effective (6234). High percentage germination occurs on media containing D-glucose and mineral salts (1567). — *Rh. oryzae* can grow in the temperature range 15–45°C, with the optimum 34°C (or below), the minimum 5–7°C, and the maximum 44(–49)°C (517, 1827, 1931, 1999, 2661, 5125, 5789). — With the exception of L-valine, most amino acids promote growth, L-tryptophan and L-tyrosine being the most suitable. Indole-3-acetic acid is synthesized when tryptophan is supplied as a precursor (2171); together with tryptophol, this compound is also obtained by oxidation of tryptamine (1522). *Rh. oryzae* also grows well on mineral N sources (but not nitrate) and can utilize urea (3127). Proteolytic properties have been reported with the optima for this at pH 7 at 35°C; pyridoxine and thiamine favour proteinase production (3001). D-Glucose is fermented (2661, 5125) and starch utilized (1827); milk-clotting enzymes (1594) and a polygalacturonase are produced (1546); with cellobiose as C source, 1,3-β-glucanase is formed (1901, 3621, 4778); extracellular lipases (462, 2281), an alcohol dehydrogenase (1999, 2000) and phenol oxidase (934) have also been reported. *Rh. oryzae* is involved in steroid transformations (5425) and produces a number of 4-desmethyl sterols (6251). Its highest lipid content occurs when grown on fructose; the C source does influence the proportion of polar vs. neutral lipids; the highest content of unsaturated fatty acids was observed at 30°C and the lowest at 15°C (2174). *Rh. oryzae* can degrade aflatoxin A_1 to isomeric hydroxy compounds (1115) and aflatoxin G_1 to a fluorescent metabolite AF 1 (1114). The organo-phosphorus insecticide dyfonate is degraded by *Rh. oryzae* both *in vitro* and in soil (1759); phenylamide herbicides can be metabolized by oxidative transformations (5824); the systemic fungicide carboxin is converted to its sulph-oxide and sulphone derivatives (6160). Extracellular saccharides identified contain glucuronic acid, D-ribose, D-fructose, D-glucose, D-galactose, and D-mannose; the latter four are also characteristic components of the mycelium (3640). *Rh. oryzae* produces the ergot alkaloid agroclavine which is toxic to man, sheep and cattle (3231). — This fungus is listed as one of

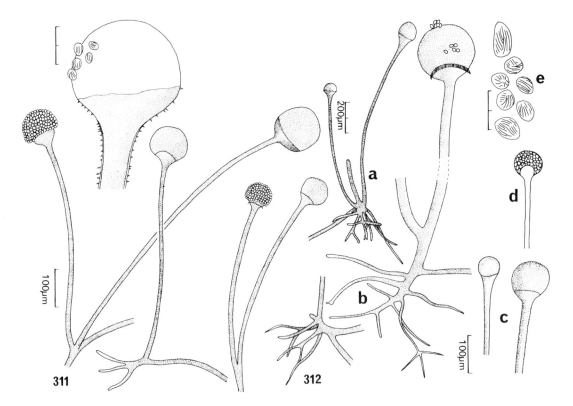

Fig. 311. *Rhizopus oryzae*, sporangiophores arising in groups from aerial hyphae and from rhizoids, dehiscent sporangia and columellae, detail of columella with sporangiospores; CBS 329.47.
Fig. 312. *Rhizopus stolonifer*. a. Group of sporangiophores arising from rhizoids; b. details of rhizoids formed on the glass surface (left) and on agar (right); c. columellae; d. an intact small sporangium; e. sporangiospores; CBS 109.76.

the causal agents of mucormycosis in both man and animals (302, 1624, 2999). *Rh. oryzae* can cause injuries to pine seeds on wounding (1975). It is said to have an antibiotic activity against some bacteria (1594, 3812, 6172). — *Rh. oryzae* can grow at a $CO_2 : O_2$ ratio of 18:1 (2096). It survived an exposure to dry heat (80–82°C) for 72 h on cured tobacco leaves (1931).

Rhizopus stolonifer (Ehrenb. ex Link) Lind *1913* — Fig. 312.

≡ *Mucor stolonifer* Ehrenb. *1818* ex Link *1824*: Fr. *1832*
≡ *Rhizopus nigricans* Ehrenb. *1820* ex Corda *1838*

DESCRIPTIONS: Inui *et al.* (2661), Caretta (892), and CMI Descriptions No. 524, *1977*. — Colonies very fast-growing and often over 2 cm high, reddish grey-brown. Stolons hyaline to brown, 13–20 μm wide, abundantly branched rhizoids (300–350 μm long) and whorls of sporangiophores produced terminally. Sporangiophores pale to dark brown, usually straight, mostly 1·5–3 mm tall, (13–)20–25(–29) μm wide. Sporangia black, mostly 100–200 μm diam; columella (including the apophysis) sublobose to oval, pale brown, mostly 70–120 μm diam. Sporangiospores subglobose, biconical to oval, ridged, (5·5–)7–12(–14) × (4·5–)6–8·5(–12) μm, mostly 4-nucleate, studied by TEM (2333) and SEM (149, 1591). Chlamydospores absent in the stolons and scarce in submerged hyphae. Zygospores are formed between compatible isolates on peptone (1%)-glucose (2%) agar at 21°C within 5–7 days (1927), on yeast extract agar or YpSs at 25°C after 1–2 weeks; zygospores black, warted, 150–200 μm diam. — The size of the sporangia and the shape of the hyphal tips were found to be influenced by temperature (5789). Immature sporangiospores are rather thin-walled, while mature ones show thick ridges surrounded by an amorphous mass of fatty material. During germination a new inner wall is laid down, while the outer ruptures (764, 2333). Germinating zygospores yielded germ sporangia with spores which reacted as either all plus, all minus, or mixtures of plus and minus mating types (1927). — Analysis of the cell wall yielded L-fucose, D-mannose, D-galactose, and D-gluconic acid, the alkali-soluble fraction contains uronic acid, glucosamine, and amino acids with small amounts of fucose, mannose and galactose (3890, 3891). — DNA analysis gave a GC content of 49% (3232).

Rh. stolonifer is one of the commonest members of the Mucorales and has a worldwide distribution, although most commonly occurring in warmer areas. It has frequently been reported from dry habitats (654, 1824, 2974, 3417, 4021, 4083). Its occurrence in India is documented by numerous reports (too many to list here individually, see list in 5058) and it has also been reported from Pakistan (4855), Sri Lanka (4021), New Guinea, the Bismarck Archipelago (3020), Indonesia (578), Taiwan (2472, 6476), China (3475), Japan (3267), Australia (3020, 3720), Central America (1697), Chile (1824), many times from the USA, Canada (234, 5363), Zaïre (4159), South Africa (1555), Somalia (5048, 5049), Iraq (92), Libya (4083, 4085, 6510), Tunisia (4055), the Sahara (2974, 3415), Egypt (8, 100, 4962), Israel (654, 2764, 2768, 2772, 4759), Italy (3445, 3538, 4538), Spain (3417), France (242, 918), Yugoslavia (2779, 5534), Austria (1823, 3418, 5678), Czechoslovakia (1703), Germany (3042, 4055, 5316), Poland (269, 272, 273, 1421, 6520), Denmark (2741), the British Isles (413, 861, 1446, 1598, 2338, 5559), and the USSR (2871, 2948, 3652, 4474, 4703). — It has been isolated equally frequently from both forest and cultivated soils, including a forest nursery soil (3908), soil in coniferous plantations (5363), many forest soils under beech (272, 273, 2719, 3538), pine (269, 1598, 2923), larch (3267), *Eugenia* (5512), teak (2854), *Eucalyptus maculata* (1555), and lowland communities with willow and cotton-wood (2008); the numerous reports from cultivated soils include ones under wheat (918), corn (4538), potato, cotton, alfalfa (2871), lupin, groundnut (2768, 2772), lentil (2164), clover (234) and other legumes (1317), sugar cane (2861, 3732, 4610), rice (1966), vineyards (242, 2719), and citrus plantations (2764, 4085). It is also known from uncultivated soils under steppe (5049) and tundra (1446) vegetation, grassland (2948, 3445, 3863, 3865, 4021, 4933), heathland (3720), peaty soils (4180, 4474, 5559, 6520), sand dunes (4371), salt-marsh (8, 413, 861, 4962) and desert soils (100, 654, 2974, 4083, 4733), and also children's sandpits (1421), a uranium mine (1703), and bat caves (6551). It has frequently been found in podzols (2923, 3414, 3416, 5534), rendzina, chernozem (3414, 3418), loess (4759), fields enriched with lime (6294) or overlying Jurassic limestone (3127), a granite outcrop (660), eulittoral and pelagic seawater (4918), estuarine sediments (655, 5316), a mangrove swamp (4700), a sewage treatment plant (1165), a river bed (3809, 3868), and a polluted pond (1166). It seems to occur mainly in the surface layers of the soil and has rarely been found to penetrate to 30–45 cm (1317, 1525, 2762, 3652, 3865, 4933, 5349). Reports of its occurrence in slightly alkaline soils predominate (273, 4055, 4152, 5048, 5049). Its occurrence was found to be affected by the application of captan, TMTD, nabam, vapam and allylalcohol (1426), and soil treatment with formaldehyde (1655). Its frequency is also reduced on plant material with high residues of the herbicide diquat (6367); its numbers are increased in manured and NPK-fertilized plots (2762). — Typical micro-habitats include fresh or decaying litter such as needles of pine and fir (681, 2344, 4984), leaves of oak, hornbeam, hazel (4817), *Eugenia heyneana* (5513), *Eucalyptus maculata* (1558), or various other deciduous trees (650), decaying shoots of *Bothriochloa pertusa* (2955), leaves of *Panicum coloratum* (1560) and cotton (5231), straw (5961), and litter of ferns (2859) and *Spartina townsendii* (5389). It has also been isolated from roots of red clover (6493), *Pinus taeda* affected by *Heterobasidion annosum* (3159), with increasing frequency from roots of *Trigonella foenum-graecum* after foliar spray with the antibiotic subamycin (2187), the rhizospheres of wheat, barley, oats (3004, 5825), broad beans (6511), peas (3870), groundnuts (2768), alfalfa (5825) and other legumes (3869), papaya (5345), manihot (4965), tobacco (3431), sugar cane (4029), coriander (1523), various sand-dune plants (4371), *Artemisia frigida* (3376), ferns (2856, 2858), senescent euphorbiaceous plants (3866), and *Aristida coerulescens* (4083). Other known substrates include peat (2719), garden compost (1424), composted municipal waste (3041), beech stumps (3568), planks in a bathhouse (5016), wood pulp (3294), tan-liquor (1260), cold-stored strawberry plants (2064) and sweet potatoes (3066), coarse fodder (3258, 4548), earthworm casts (2857), bat guano (6551), rabbit dung (2338, 3267, 5943), nests, feathers and droppings of free-living birds (2575, 5067), honeycombs of wild bees (402), seeds of *Avena fatua* (2961), cultivated grasses (3512), wheat, rice (2297, 3584, 5244), castor beans (3172), tomato (5899), groundnuts (1904, 2765, 3220, 3702, 4602), pecans (2258, 2572, 5092), Brazil nuts (2515), banana fruits (4227), cotton bolls (5227), foodstuffs (1303, 3930) and fruit juices (3428). *Rh. stolonifer* is commonly air-borne (CBS). — Spore germination is inhibited in soil by mycostatic factors (6284). In vitro it can be parasitized by *Dimargaris verticillata* (3560) and *Trichoderma viride* (1729), and inhibited by *Bacillus subtilis* (2393), *Penicillium nigricans* (5002), wortmannin from *Talaromyces wortmannii* (711), rubratoxin B (4808) and coumarin (3055). The germination of sporangiospores *in vitro* is inhibited by volatile products from leaves of *Origanum majorana* and stimulated by ones from *Ocimum basilicum* (35). *Rh. stolonifer* can serve as food for the mite *Caloglyphus redikorzevi* and *Kleemannia plumosa* (5379).

The production of sporangia was found to be independent of the wavelength of visible light (3285). Sporulation can be inhibited by aflatoxin (4806); germination requires glucose as exogenous energy source (308, 804) and O_2 (804, 6436), and the germ tubes display negative autotropism (4883, 5516); the synthesis of RNA, DNA and protein, and changes in sporangiospore dry weight, volume and respiration during germination have been studied (6042). Zygospores germinate in less than 30 days on moist filter paper (1927). — The mean doubling time for mycelium production in liquid culture at 25°C was 5·2 h; the optimal temperature for growth on agar (and peaches) was found to be 25–26°C, the minimum 4·5–5·0°C, and the maximum 32–33°C; no growth occurs at 35(–37)°C (2661, 4548, 4562, 5125, 5789); the thermal death point in apple juice was determined as 63°C for 25 min (3620). The minimum water potential for vegetative growth is −100 bars (1192); in soil it requires at least −70 bars and can grow at saturation (977, 3109); but wheat seeds were found to be colonized in soil between −210 and −70 bars (2097). Growth was found to be stimulated by light (5430), and the stolons are positively phototropic (1654). — ^{14}C from labelled KCN was incorporated into amino acids (108) and cyanide can also be utilized as sole N source (4965) but nitrate is not (5430); traces of Mn, Cu, and Zn ions are required (3709); zinc can increase the glucose utilization threefold, and in its absence growth is slowed and fumaric acid accumulates (1804). Growth on acetate is better than on glucose (3329). With decreasing O_2 (6277, 6278, 6281) and increasing CO_2 concentrations (4029, 6281), growth is progressively inhibited; in one study, however, the fungus was found to grow in 100% CO_2 (5602). — Pectin is utilized (3414, 3538), and polygalacturonase and pectinmethylesterase are produced (882, 4238, 5269, 5485, 5507); hemicellulose is degraded (3414), and cellulase can be produced (5485, 5507, 6277). *Rh. stolonifer* participates in the degradation of cotton fabrics in the soil (4186), degrades chitin (1425), and produces glycosidic (5598), proteolytic (1425, 1846, 2719, 2948) and lipolytic (1550) enzymes but, in contrast to other congeneric species, does not produce amylolytic and milk-clotting enzymes (1594); the activities of alcohol dehydrogenase (1999) and glyoxalate transaminases have been determined (1819); tryptamine can be oxidized to indole-3-acetic acid and tryptophol (1522); *Rh. stolonifer* is also involved in steroid hydroxylations (3231), can degrade usnic acid (322) and mineralize a urea-formaldehyde-based fertilizer (4010); the herbicides atrazine (2892) and simazine (2894), and the acaricide chlorphenamidine (2781) were all found to be metabolized *in vitro*; aflatoxin G1 is metabolized at a slow rate (1114). By its release of organic acids, potassium can be solubilized from orthoclase and oligoclase (4008). Known metabolites include the ergot alkaloids agroclavine, ergosine, and ergosinine (1609) and also nonanoic acid (1905). — An antagonistic action has been observed *in vitro* against some bacteria (3812), *Alternaria solani* (1283), and *Chaetomium globosum* (4427); the sclerotia of *Botrytis cinerea* and *Sclerotinia sclerotiorum* can be colonized and infected *in vitro* (1651). In experimental rabbits with acute diabetes it may cause an opportunistic mycosis (302). Toxic effects have been demonstrated in tests with brine shrimps and chicken embryos (1303). — Its sensitivity to γ-irradiation depends on the moisture content of the substrate: either 0·25 mrd at 23% moisture or 1·0 mrd at 12·5% were lethal (6231). Diurnal temperature extremes of 19·5 h at −94°C and 4·5 h at 23°C did not kill the fungus after 35 days but stopped its growth (1242); a similar effect is exerted by the exposure of pre-germinated sporangiospores to 0°C (3675); at 0°C most spores lose their ability to form colonies within four days but retain their metabolic activities, although the mitochondria gradually degenerate (3674). Germinated spores are less tolerant to the effects of critical temperatures (49 or 52°C) than dormant spores (5449).

Sagenomella W. Gams *1978*

Sagenomella diversispora (van Beyma) W. Gams *1978* — Fig. 313.

≡ *Scopulariopsis diversispora* van Beyma *1937*
≡ *Acremonium diversisporum* (van Beyma) W. Gams *1971*
= *Paecilomyces variabilis* Barron *1967*

DESCRIPTIONS: Barron (357), and Gams (1882).—The genus *Sagenomella* (type species *S. diversispora*; teleomorph *Sagenoma* Stolk & Orr *1974*) comprises the species formerly classified as *Acremonium striatisporum* series (1882) and is characterized by irregularly swollen and often sympodially proliferating phialides and catenulate phialoconidia which bear a connective at either end. This character has been used to distinguish eurotiaceous anamorphs from those of Sphaeriales (1885).

Colonies of *S. diversispora* reaching 1·4 cm diam in ten days at 20°C on MEA, olivaceous-brown, powdery. *S. diversispora* produces two kinds of conidia: (a) larger, pigmented conidia with longitudinal rows and coarse warts, 4·5–8·5 × 4·0–5·5 μm, and (b) almost hyaline and smooth-walled conidia, 3·5–5 × 1·5–2·5 μm. — The related *Sagenomella striatispora* (Onions & Barron) W. Gams *1978* (≡ *Acremonium striatisporum* (Onions & Barron) W. Gams *1971*) has longitudinal ridges along the more regularly shaped and generally pigmented conidia.

S. diversispora was originally isolated from peat soils in Ontario (357) and later from soil under red pine, white birch, Japanese larch and grassland in Japan (5844, 5846), forest soils in Georgia (2240) and Canada, particularly at greater soil depths (505).

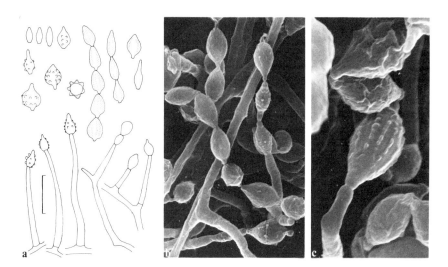

Fɪɢ. 313. *Sagenomella diversispora*. a. Phialides with smooth-walled hyaline and ornamented pigmented conidia; b, c. variously shaped conidia, SEM, × 3000 and × 6000; CBS 354.36.

Sclerotinia Fuckel *1870*

Sclerotinia sclerotiorum (Lib.) De Bary *1884* — Fig. 314.

≡ *Peziza sclerotiorum* Lib. *1837*
≡ *Sclerotinia libertiana* Fuckel *1870*
≡ *Whetzelinia sclerotiorum* (Lib.) Korf & Dumont *1972*

Anamorph (sclerotia): *Sclerotium varium* Pers. ex Gray *1821*

DESCRIPTIONS: Brefeld (693), Dennis (1347), and CMI Descriptions No. 513, *1976*; sclerotial development: Willets and Wong (6372); sclerotia in SEM: Colotelo (1125); sclerotia in TEM: Saito (4981, 4983). — Colonies broadly spreading, reaching 4·0 cm diam in five days at 20°C on OA, arachnoid, white. Sclerotia tuber-like or cushion-shaped, black, 1–8 mm or more diam, consisting of a rind of dark, palisade-like thick-walled cells, a cortex of 6–8 layers of hyaline, almost pseudoparenchymatous cells and a medulla of thick-walled hyphae not embedded in a gelatinous matrix and in which there are some intercellular spaces. Micro-conidia (spermatia) $2 \cdot 5$–$3 \cdot 2 \mu$m diam, belonging to the genus *Myrioconium* Syd. *1912*, are produced abundantly in superficial sporodochia on flask-shaped phialides when nutrients in the substrate have been exhausted; their germination *in vitro* does not proceed further than the production of a short germ tube (4723) and they contain one large nucleus and a large lipid body (853). The sclerotia mostly germinate to form apothecia (carpogenic germination). Apothecia are produced in the field in spring and can also be obtained *in vitro* on nutrient-deficient moist substrates after a period of cold storage; apothecial primordia consist of irregularly shaped cells with thin walls lacking fibrous layers; the apothecial stalks are to 3 cm long and 1–2 mm thick, the discs are concave, yellowish brown, usually 3–8 mm diam. Asci cylindrical and with an amyloid pore; ascospores ellipsoidal to almost fusiform, sometimes slightly asymmetrical, usually 9–15×4–7μm, uninucleate (2799).

   S. sclerotiorum has been removed from *Sclerotinia* because the lectotype species of this genus, *S. candolleana* (Lév. *1843*) Fuckel *1870* (≡ *Ciborinia candolleana* (Lév.) Whetzel *1945*), is a foliicolous fungus with discoid sclerotia formed within the host tissue and hence not regarded as congeneric. From the remaining *Sclerotinia* species *sensu* Whetzel *1945* the genus *Myrioclerotinia* Buchw. *1947* has been segregated for species which form their sclerotia within the culms of Cyperaceae and Juncaceae and produce immersed pycnidia for their spermatia; for *S. sclerotiorum* and related species the generic name *Whetzelinia* Korf & Dumont *1972* has been proposed (3094) but as that name is not yet universally accepted, because of both unclear species and generic delimitation, and discussions about the typification of *Sclerotinia* itself, it currently seems preferable to retain the latter, which probably will become a *nomen conservandum* (see also Kohn *in* Mycotaxon **9**: 365–444, *1979*).

   Based on a statistical analysis of ascus and ascospores dimensions, Purdy (4672) concluded that *S. sclerotiorum* is a variable species embracing also *S. trifoliorum* Erikss. *1880*, *S. minor* Jagger *1920*, *S. intermedia* Ramsey *1924* and *S. sativa* Drayton & Groves *1943*, but his view has not generally been accepted because of differences

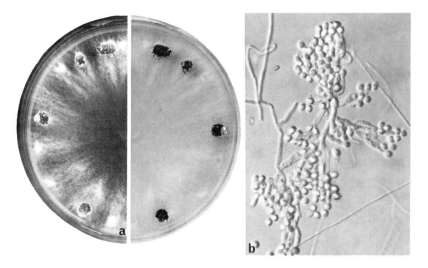

F IG. 314. *Sclerotinia sclerotiorum*. a. Three-week-old culture on OA photographed against black and white background, CBS 537.77; b. *Myrioconium*-type phialoconidia, NIC, × 400.

in biological, anatomical and biochemical properties of these taxa. Two species with large sclerotia, *S. sclerotiorum* and *S. trifoliorum* are particularly similar as their sclerotia develop terminally by the repeated branching of primary hyphae (6372); however, the former produces apothecia in the field during spring while the latter does so in the autumn (6374) and is confined to alfalfa and clover species, the former being plurivorous. The ascospores are reported to be (9–)10–14 × 4–5(–6) μm in *S. sclerotiorum* but 10–20 × (4–)6–9(–11) μm in *S. trifoliorum*. *S. minor* has smaller sclerotia, 0·3–2 mm diam, which develop laterally by the repeated branching of short aerial hyphae; initials coalesce less frequently than in the first two species (6372). The specific distinction of these taxa is further supported by the observation that hyphal contact between conspecific isolates leads to unlimited growth over the contact zone, while between *S. sclerotiorum*, *S. trifoliorum* and *S. minor* incompatibility reactions occur (6428).

The electrophoretic separation of proteins showed a clear distinction between these three species with little intraspecific variability (6427). The electrophoretic patterns of soluble proteins and the enzymes arylesterase, acid phosphatase, tetrazolium oxidase, glucose-6-phosphate dehydrogenase and NADPH phosphate dehydrogenase also differed between these three species (6429). Numerous Canadian isolates of *S. sclerotiorum* (in the broad sense of Purdy) showed a considerable morphological, pathological and physiological variability (3961), whereas isolates from different hosts in Britain showed a more continuous range of variation (4635). Anatomical studies of the sclerotia and fruit-bodies are given in the monograph by Kohn *1979* (*l.c.*). — TEM studies have been carried out on the microfibrillar structure of the different wall fractions (2797) and the organelles in hyphal tips, but no correlation between the presence of membrane-bounded vesicles, micro-bodies and multivesicular bodies and enzyme production could be established (3692). Acid phosphatase is localized in chains of vacuoles in hyphal tip cells (180). Apothecia develop from single ascospore isolates without spermatization; the stipes originate from peripheral layers of the medulla and rupture the rind during emergence (4980); their ultrastructure has been followed in SEM and TEM studies (2799). Sclerotial medullary hyphae contain numerous lipid bodies and storage vacuoles (853), multivesicular and multitubular bodies, and their walls are two-layered with a homogeneous inner layer representing the original wall and a very thick outer fibrillar layer, both of which contain 1,3-β-glucans and proteins; chitin is present only in the inner layer (2797, 4982). The 1,3-β-glucan can be degraded by the combined action of endo- and exo-1,3-β-glucanases from *Coniothyrium minitans* and a *Streptomyces* species (2797, 2803). Melanin and phenoloxidase are also present in the rind walls (2797). In wall hydrolysates of rind and/or medullar cells, amino acids, aminosugars, D-glucose, D-fructose, D-mannose, D-mannitol, D-arabitol, glycerol-*a*-*a*-D-trehalose, gentobiose and inositol have been demonstrated (1148, 1149, 2797, 3299, 5707). Sclerotia from infected plant tissues have a higher content of unsaturated lipids than those produced *in vitro*; low temperatures also favour the production of unsaturated lipids; sclerotia principally contain oleic but also smaller amounts of palmitic and linoleic acids (5641).

S. sclerotiorum is a polyphagous plant parasite with an extremely voluminous phytopathological literature. Its distribution in the temperate zones is worldwide and the most important host plants are French beans, potato, lettuce, other vegetables, sunflower and rape. Mycelial isolations from soil have rarely been reported, but the sclerotia can be recovered quantitatively from infested soils by a combination of dry- and wet-sieving and their viability tested on freshly cut carrot discs (2494). — Ascospores are the primary source of inoculum which infects above-ground parts of plants (2, 1131), for example bean petals (3), which can serve as an initial energy source, disintegrate within 72 h and facilitate the further spread of the fungus. Ascospores can survive particularly well for up to seven months (5540) at low humidities and they germinate at osmotic potentials as low as -56 bars, but not at all at -91 bars (2124). — The sclerotia afford the primary means of survival of this fungus but hardly contribute to infection of plants (2). Sclerotia reach the soil mainly with decaying and ploughed-in plant material and can subsequently form daughter sclerotia, particularly in clay soil (3135). *S. sclerotiorum* has been reported to be seed-borne in at least 27 host genera and the sclerotia can also be distributed with various seeds, particularly those of *Brassica* species (4200). The sclerotia can be formed *in vitro* at osmotic potentials between –1 and –64 bars (2124). Survival times of five or up to ten years have been observed for sclerotia in the soil (3718, 6374, 6375), particularly under dry conditions (5882). Survival is adversely affected by high soil temperatures and moisture (1131) but is determined less by temperature and humidity than by microbial activities; sclerotia buried in 5–20 cm depth of soil were still viable and produced apothecia after 3–5 years, whereas those nearer the soil surface disintegrated more rapidly (1131, 3718). However, sclerotia do not form apothecia in soil at depths greater than 5 cm and the number produced from a sclerotium decreases with depth (6374). — The most important parasite of the sclerotia is *Coniothyrium minitans* (q.v.) which can drastically reduce the percentage of viable sclerotia (2494, 2565, 2566); at the end of the growing season 59 and 76% of the sclerotia formed on root surfaces and inside the root of sunflower, respectively, were killed by this parasite (2565); *C. minitans* also penetrates into the host hyphae causing a collapse of the protoplasm (2566). A similar but less common hyperparasite is *Microsphaeropsis centaureae* Morgan-Jones (6227), and parasitism by *Rhizopus stolonifer*, *Gliocladium roseum*, *Trichothecium roseum* and *Trichoderma viride* has also been reported (1651, 2807, 3531, 4584). Numerous further fungi have been isolated from sclerotia of *S. sclerotiorum* (4699). — *Bacillus subtilis* inhibits both mycelial growth and sclerotium formation (2393). Growth is also inhibited by the phytoalexin alkaloid *a*-tomatine (182), and an *in vitro* inhibition by *Tuber melanosporum* has been reported (632). *S. sclerotiorum* can not grow more than 2 cm away from a cornmeal food base in muck soil and infects plants only in immediate contact or at a distance not greater than 2 mm from the previous host (4146);

hence, its saprophytic competitive ability is apparently very low. Ten times as many sclerotia were found in the rhizosphere soil (24 sclerotia kg^{-1}) from sunflower with root rot than in a non-infested soil (2494).

The potential for apothecium formation from sclerotia depends on the culture conditions: media either very rich or very poor in nutrients are equally unsuitable for cultivation (422). Apothecia are formed from sclerotia buried not deeper than 5 cm outdoors in soil or *in vitro* from well matured (423, 1147, 1148, 5421) sclerotia kept at low temperatures (0 to +3°C) for two to several months (5778) and subsequently incubated on a moist, nutrient-poor substrate at 10–15°C under light (422, 2408, 3718, 4140, 4980). A constant temperature of 12–21°C during 60–90 days of sclerotium maturation has been found to give the shortest delay in germination (4673), while temperatures fluctuating between 15–20°C resulted in 10% abnormal fruit-bodies (3102). Light is not only essential for disc differentiation (3102) but also strongly stimulates the development of stipe initials (3297). Apothecia are produced only in moisture-saturated or near-saturated soils (2); the minimum water potential for this in a heavy clay was −7·5 bars at 15°C (3960). The activities of tyrosinase, laccase, acid and alkaline phosphatases, esterases, malate, isocitrate, succinate and glucose-6-phosphate dehydrogenases are low in mature sclerotia but very much higher in developing apothecial initials, stipes and hypothecium (3101). The sclerotial 1,3-β-glucan is consumed during germination as 1,3-β-D-glucanase activity increases (4982). Cd^{2+} at 5×10^{-6}, Hg^{2+} at 5×10^{-5}, Co^{2+} at 10^{-4}, Ni^{2+} at 5×10^{-5}, NaCl at 5×10^{-3}, Tris buffer at $2·5 \times 10^{-3}$, EDTA at 10^{-3} molar, and glucose and other sugars at $0·25\%$ concentrations selectively inhibit carpogenic sclerotium germination as do benomyl, dicloran and some other fungicides (5541). — A method of quantitatively collecting ascospores has been developed utilizing a Millipore filter funnel attached to a vacuum pump placed above sporulating apothecia; discharge can be induced by a short air flush once in 3 h (5540). For optimal ascospore germination a high relative humidity is necessary (2124); temperatures between 5 and 10°C are suitable (6483). — In darkness, half as many sclerotia are formed as under 40–100% of normal light intensity (5883); numbers and sizes of sclerotia formed at different temperatures are inversely correlated (421, 5631) while growth rate and sclerotium formation are correlated with the diameter of the growth area; in large dishes, sclerotia form in concentric rings 0·5 cm apart (5883). Sclerotium formation can be induced *in vitro* by acid staling compounds (2609). Only actively growing hyphae from the margin of the colony give identical patterns of sclerotium production in subcultures (2610). Sclerotia form *in vitro* in the osmotic potential range −1 to −64 bars, but not at −73 bars (2124). Suitable C sources for sclerotium formation are raffinose, sucrose, maltose, lactose, D-mannose, D-glucose, D-fructose, D-galactose and L-arabinose (4702, 6177); on the other hand, some of these sugars, and also a number of inorganic ions and organic compounds, can be released from growing (1124, 1146) or dried (5421) sclerotia. During sclerotium formation a high CO$_2$ evolution and a high initial sugar (glucose) incorporation take place (1148, 1149). — The optimal temperature for mycelial growth lies in the range 15–25°C, depending on the isolate (2, 421, 881, 5739), while the optimal pH lies between 4 and 5·5 (4702). Growth is supported by araban (2843), inulin and starch (5739), in addition to the above-mentioned C sources, which also allow sclerotium formation (6176, 6177). Mycelial growth is progressively stimulated as the osmotic potential decreases from −1 to −14 bars but it is reduced below the latter value (2124); half the optimal growth rate is obtained in the range –37 to –47 bars. Growth over large distances in agar tubes is not limited by accumulating staling metabolites (3031). During its decomposition of pectin, mainly endo- but also some exo-polygalacturonase, pectin methylesterase and pectin transeliminase have been detected (1537, 3692, 3961). When grown on sterilized bean hypocotyls, an adaptive cellulase with an optimum pH of 3·0 (3433) and a phospholipase A_2 with an optimum pH of 4·0 (3434) are produced. Further enzymes investigated include an arbutine-splitting glucosidase (1340), arabanase (1840), a mycelium-bound trehalase (6180), D-mannitol-1-phosphate:NAD oxidoreductase and D-mannitol:NADP oxidoreductase (6179), pentitol oxidoreductase (detected in cell-free extracts of mycelia and sclerotia grown on D-xylose, L-arabinose or D-ribose) (6181), oxalacetate acetylhydrolase (3686), citrate synthetase, aconitate hydratase, NADP-specific isocitrate dehydrogenase, fumarate hydratase, malate dehydrogenase (1193) and acid phosphatase (180). — The following compounds have been found to be suitable N sources: aspartic and glutamic acids, alanine, serine (6178), glycine, tyrosine, peptone, sodium and potassium nitrates (1339, 5739); cysteine, cystine and methionine (6178), lysine and urea (1339, 5739), however, were not very suitable. In liquid culture, growth was depressed by deficiencies of phosphate, magnesium and trace elements and no significant growth occurred when there were nitrogen and potassium deficiencies (4674). — Oxalic (1193, 3690, 3692, 4332, 6055), fumaric, succinic and glycolic acids (1193, 6055) are abundantly produced in later growth stages. Further metabolites reported include mannitol (6177, 6179), acid phenols which can inhibit *Bacillus subtilis*, *Fusarium oxysporum* and *S. sclerotiorum* itself (4088), and the mycotoxins 8-methoxypsoralen and 4,5,8-trimethyl-psoralen, both of which may cause dermatitis in man (3231). — This fungus is said, moreover, to have antibiotic properties against *Staphylococcus aureus* (1967). Sclerin, the isocoumarin derivatives sclerotinin A and B, sclerone and isosclerone are responsible for growth-promoting effects at low concentrations and inhibition at higher concentrations in higher plants (3146, 3952, 5062, 5658). The production of sclerin is correlated with sclerotium formation and pigmentation, and this compound apparently stimulates phenoloxidase and peroxidase activities (3653). — *S. sclerotiorum* is highly tolerant to Al^{3+} ions even at a pH of 3·7; consequently, phytotoxic levels of Al^{3+} can enhance its virulence on sunflower (4308).

Scolecobasidium Abbott *1927*

= *Ochroconis* de Hoog & v. Arx *1973*

Type species: *Scolecobasidium terreum* Abbott

MONOGRAPHIC TREATMENT: Barron and Busch (365) and Ellis (1603, 1604); further taxonomic contributions: de Hoog and von Arx (2525), Mouchacca (3994), and Matsushima (3680).

Colonies mostly slow-growing (reaching less then 2·0 cm diam in 2 weeks at 20°C), brown, velvety to floccose. Conidiophores distinct but often short, unbranched, erect, ± pigmented and ± septate, forming blastoconidia in sympodial succession on narrow cylindrical denticles which seem to break across in the middle leaving part attached to the conidium. Conidia formed solitarily, simple, ellipsoidal, cylindrical, fusiform or apically bifurcate, pale to mid-brown or olivaceous-brown, smooth-walled or echinulate, with one or several transverse septa. Chlamydospores present in some species.

For many years this genus was restricted to species with 1-3-septate, small conidia, but its concept has been expended by Matsushima and Ellis to also include species with multi-septate and larger conidia, and even a species with muriform conidia. *Scolecobasidium* now contains a somewhat heterogeneous assemblage of 20 species. Moving species with darker and simple conidia and elongate conidiophores into the genus *Ochroconis* de Hoog & v. Arx does not seem to contribute to homogeneity, since these characters are not sufficiently correlated. The main distinctive feature is the conidial denticles of which a fraction remains attached to the conidium after dehiscence but without forming part of the conidial cell. In the otherwise similar genus *Dactylaria* Sacc. *1880* the conidia secede along the basal septum and the basal apiculation is a part of the conidium; *Pleurophragmium* Cost. *1888* (included in *Dactylaria* by de Hoog and v. Arx) has long, erect, thick-walled, many-septate conidiophores but is otherwise similar to *Dactylaria*.

Most species of *Scolecobasidium* are soil-borne in both cultivated and virgin soils but have infrequently been recorded. Records from plant material (1603, 1604) mostly refer to decaying leaves in contact with soil. Two species, *S. salinum* (Sutherl.) M. B. Ellis (q.v.) and *S. arenarium* (Nicot) M. B. Ellis *1976*, are adapted to saline habitats and differ from others in the genus by having larger conidia and more broadly spreading colonies.

Key to the species treated:

1	Conidia mostly 1-septate	2
	Conidia more than 1-septate	4
2(1)	Conidia pale ochraceous, Y- or T-shaped; chlamydospores brown, globose *S. terreum* (p. 722)	
	Conidia brown, ellipsoidal to cylindrical; chlamydospores usually absent	3

3(2)   Conidiophores usually exceeding $20\,\mu$m long, elongating indefinitely; conidia ellipsoidal,
        without a central constriction   *S. humicola* (p. 719)
        Conidiophores usually shorter than $20\,\mu$m and not elongating much; conidia ellipsoidal,
        sometimes slightly constricted at the septum   *S. constrictum* (p. 718)

4(1)   Conidia mostly 3-septate, coarsely verrucose, $11–20 \times 3–5\,\mu$m   *S. tshawytschae* (p. 723)
        Conidia 4–6-septate, smooth-walled, mostly $30–45 \times 4–7\,\mu$m   *S. salinum* (p. 720)

Scolecobasidium constrictum Abbott *1927* — Fig. 315.

  ≡ *Ochroconis constricta* (Abbott) de Hoog & v. Arx *1974*
  = *Heterosporium terrestre* Atkinson *1952*

DESCRIPTIONS: Barron and Busch (365), Ichinoe (2633), Nag Raj and Govindu (4069), Ellis (1603), Subramanian (5626), and Matsushima (3680). — Colonies reaching 2–3 cm diam in three weeks at 25°C, brown, aerial mycelium funiculose, reverse darker brown. Conidiophores short spherical to clavate or cylindrical, $2–22 \times 1\cdot5–2\cdot5\,\mu$m, simple, rarely septate. Conidia ellipsoidal, mostly 2-celled, usually slightly constricted at the septa, light olivaceous-brown, echinulate to verrucose, $5–12\cdot5 \times 2–4\,\mu$m, tapering to a narrow basal apiculus with part of the conidiiferous denticle attached to it. Chlamydospores absent. — The major fatty acids of the mycelium are palmitic, stearic, oleic and linoleic with a rather high proportion of saturated acids (5642).

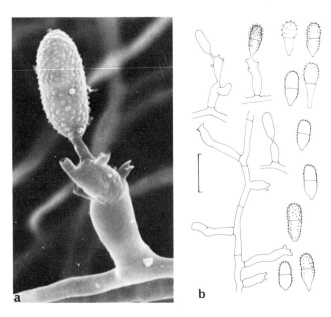

FIG. 315. *Scolecobasidium constrictum*. a. Conidiophore, SEM, × 3000, CBS 124.65; b. conidiophores and blasto-conidia produced on denticles.

This is the most common *Scolecobasidium* species in soils of Ontario and elsewhere; it has also been reported from Louisiana, Georgia and Ohio in the USA (365), the Netherlands, Germany (CBS), India (427, 4069, 4933, 5626), Japan (2633, 3680, 5626, 5846), South Africa (4407), Australia, Sri Lanka, Egypt, Malaysia, and Brazil (1603). Known habitats include cultivated soils (which predominate), savannah (4407), marsh, cedar bog and greenhouse soils (365), wheat rhizosphere (427), *Hevea* roots, *Gmelina* and *Vitis* plants (1603), cucumber (4756), decaying leaves (427), and water (4069). In Brazil it has been reported as a nematode parasite (364).

No pectolytic activity could be demonstrated in cucumbers (4756).

Scolecobasidium humicola Barron & Busch *1962* — Fig. 316.

≡ *Ochroconis humicola* (Barron & Busch) de Hoog & v. Arx *1974*

DESCRIPTIONS: Barron and Busch (365) and Ellis (1603). — Colonies reaching 4 cm diam in

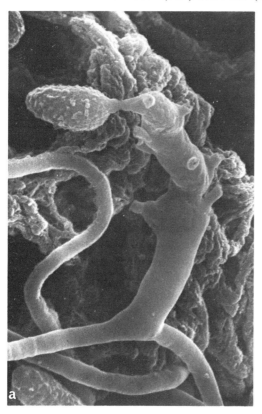
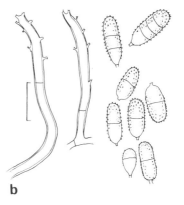

FIG. 316. *Scolecobasidium humicola*. a. Conidiophore with conidiiferous denticles, SEM, × 2750, CBS 172.74; b. sympodially elongating conidiophores and blastoconidia.

three weeks at 25°C, funiculose-floccose, olive-drab; reverse dark brown. Conidiophores cylindrical, $20–30 \times 2\cdot0–2\cdot5\,\mu$m, attaining considerable lengths (to $300\,\mu$m) by sympodial proliferation, repeatedly septate, at first light olivaceous but later becoming fuscous, irregularly undulate, bearing large numbers of conidia. Conidia mostly 1-septate, ovoid to short-cylindrical, rarely slightly constricted at the septum, pale olivaceous brown, finely echinulate, with an almost rounded base and part of the denticle attached to it, $7\cdot5–15 \times 2\cdot8–4\cdot5\,\mu$m. Chlamydospores absent. — The mycelium contains palmitic, stearic, oleic and linoleic acids as the major fatty acids with a rather high proportion of unsaturated acids (5642).

S. humicola is apparently a rare soil fungus. It was originally isolated from various soils, including marsh, peat and greenhouse soils in Ontario but at low frequencies (365). It has subsequently been obtained from soils and leaf litter of *Acacia karroo* in South Africa (CBS, 4407), *Borassus flabellifer* in India (1603), decaying needles of *Pinus densiflora* in Japan (3680), repeatedly in banana rhizospheres and on root tissue in Honduras (2035), sticks exposed to river water in the USA (5249), the toe of a mycotically infected frog (1579), and naturally infected kidneys of coho salmon (4913).

Scolecobasidium salinum (Sutherl.) M. B. Ellis *1976* — Fig. 317.

≡ *Cercospora salina* Sutherl. *1916*
≡ *Dendryphiella salina* (Sutherl.) Pugh & Nicot *1964*

DESCRIPTIONS: Pugh and Nicot (4664), Kohlmeyer and Kohlmeyer (3078), and Ellis (1604). — Colonies reaching $3\cdot0–4\cdot0(–6\cdot0)$ cm diam in 14 days at 20°C, (grey-)brown, velvety, with darker brown reverse. Conidiophores simple or branched, straight or flexuose, pale to mid-olivaceous-brown, $15–60\,\mu$m long. Conidia cylindrical to obclavate, ± straight, often somewhat constricted at the septa, pale to mid-olivaceous-brown (darker at the septa), thin- and smooth-walled, $(2–)3–5(–7)$-septate, $30–45(–70) \times 4–7(–9)\,\mu$m. — *Scolecobasidium arenarium* (Nicot) M. B. Ellis *1976* (≡ *Dendryphiella arenaria* Nicot), which occurs in similar habitats, has fusiform, mostly 3-septate, and shorter conidia, $13–20 \times 4\cdot5–6\cdot5\,\mu$m. — The genus *Dendryphiella* Bub. & Ranoj. *1914* (type species *D. interseminata* (Berk. & Rav.) Bub. & Ranoj. = *Dendryphiella vinosa* (Berk. & Curt.) Reisinger *1968*) has been distinguished from *Dendryphion* Wallr. (q.v.) by its production of conidia directly on nodes of the conidiophore axis and an indefinite sympodial elongation of the conidiophore (4798); both genera have very dark conidiophores and poroconidia. *D. salina* is not congeneric with *D. vinosa*, as the conidiophores are pale brown and little differentiated and the blastoconidia are abstricted from narrow denticles; this species has therefore been transferred to *Scolecobasidium* (1604) but its faster growing colonies and large conidia are unusual in this genus.

S. salinum is halophilic and widespread in intertidal regions along coasts of the Atlantic, Pacific and Indian Oceans and on seaweeds and test blocks immersed in seawater (1604, 3078, 4664, 5085). It has been isolated from salt-marshes in the USA (1958, 1959), a mangrove swamp in Hawaii (3265), soils and plant remains in Zaïre (2966) and, with high frequencies, on various parts of *Halimione portulacoides* in the British Isles (1372, 1378, 1379). It has also

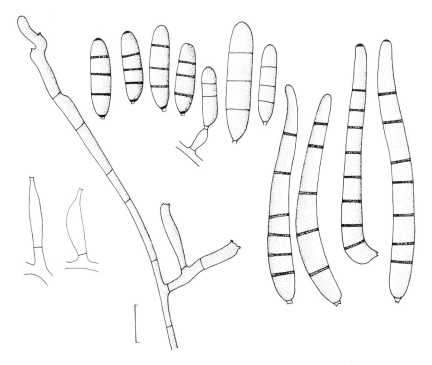

FIG. 317. *Scolecobasidium salinum*, conidiophores showing sympodial elongation and conidiiferous denticles; blastoconidia, left: formed on MEA, right: longer conidia formed on PCA with *c*. 8% salt; CBS 142.60.

been found growing on the algae *Chondrus crispus* and *Laminaria* sp. in New England (3077), on decaying *Spartina alterniflora* in the USA (1958, 1959), in the phylloplane of plants in Hawaii (296), and associated with mangrove seedling rot (2591).

The optimal temperature for the germination of the conidia is 20 or 25°C (832, 2809), while that for vegetative growth is in the range 12–20(–25)°C (5085, 5931); but 90% of the conidia also germinated at 10°C (832). Media with 100% seawater content had no adverse influence on the production of conidia (833). Optimal growth and sporulation occur at salinities of 0·6–2·7(–4)%, while good growth occurs with up to 7% NaCl, KCl or MgCl$_2$ (4664, 5931), and up to 20% NaCl is tolerated (4163). — The internal osmotic potential is regulated by the synthesis of polyhydric alcohols, inorganic ions and organic acid anions; potassium ions are more readily taken up than sodium ions (2737); in the presence of high extracellular sodium ion concentrations potassium ions are pumped actively into the cells; this process is mediated by ATP located near hyphal septa some distance from the hyphal tip (1867). — Laminarin is an excellent C source but sodium alginate is not (5931). The only soluble carbohydrate synthesized in detectable amounts after transfer from starving conditions is mannitol, accompanied, at later stages, by arabitol (2506, 2507). Arabitol synthesis from labelled C sources has been investigated (3419). The conversion of mannitol and arabitol to polysaccharides is stimulated by 3–*O*-methyl glucose (2738). Both 3-*O*-methyl glucose and L-sorbose reduce colony diameter, increase hyphal branching, but have only a slight effect on dry matter production (1868). K$^+$, Na$^+$ and Ca^{2+} ions influence the glucose uptake in various ways (105). Cellulose is decomposed (5086) but no soft rot occurs on wood (2591).

Scolecobasidium terreum Abbott *1927* — Fig. 318.

= *Humicola minima* Fassatiová *1967* (chlamydospores only)

DESCRIPTIONS: Barron and Busch (365), Ichinoe (2632), Nag Raj and Govindu (4069), Ellis (1603), Subramanian (5626), and Matsushima (3680). — Colonies similar to those of *S. constrictum*. Conidiophores small, ampulliform, mostly non-septate, sometimes proliferating with similar ampulliform structures, 2–6·5 × 1·5–2·5(–3·5) μm, hyaline to light olivaceous. Conidia 1-septate, Y- or T-shaped due to a bifurcation of the apical cell, smooth or finely roughened, pale ochraceous, with an almost pointed base and part of the conidiiferous denticle attached to it, 4–12 μm long and 1·5–2·5 μm wide at the septum. Chlamydospores sessile, arising laterally on the hyphae, globose, brown, smooth-walled, 4–6·5 μm diam. — The major fatty acids of the mycelium are palmitic, stearic, oleic and linoleic with a rather high proportion of saturated acids (5642).

S. terreum has been rarely observed in peat soils in Ontario, but more frequently in Louisiana and Ohio (365); it has also been reported from the rhizosphere of cotton in Nigeria (1701), soil and submerged decaying leaves in India (4069, 5626), cultivated soils in Japan (2632, 3680) and the rhizosphere of potato in the Netherlands (CBS).

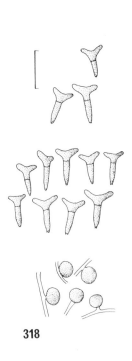

FIG. 318. *Scolecobasidium terreum*, apically divided blastoconidia and globose, *Humicola*-like conidia arising from vegetative hyphae, CBS 536.69.
FIG. 319. *Scolecobasidium tshawytschae*, conidiophore and conidia, SEM, *c.* × 3000, CBS 130.65.

Scolecobasidium tshawytschae (Doty & Slater) McGinnis & Ajello *1974* — Fig. 319.

≡ *Heterosporium tshawytschae* Doty & Slater *1946*
= *Scolecobasidium variabile* Barron & Busch *1962*
≡ *Ochroconis variabilis* (Barron & Busch) de Hoog & v. Arx *1974*
= *Scolecobasidium macrosporum* Roy, Dwivedi & Mishra *1962*

DESCRIPTIONS: Barron and Busch (365), Ichinoe (2632), Nag Raj and Govindu (4069), Mouchacca (3994), Ellis (1603), Matsushima (3680), and McGinnis and Ajello (3706). — Colonies reaching 2–3 cm diam in three weeks at 25°C, funiculose, olivaceous-black at first but becoming olivaceous-drab from the conidia; reverse dark brown. Conidiophores spherical, ovoid, clavate or filiform, hyaline at first but later becoming pale olivaceous, 4–25 × 1·5–2·5 μm. Conidia mostly 3-septate, cylindrical, sometimes slightly constricted at the septa, olivaceous-brown, minutely echinulate to verrucose, with a slightly apiculate base and part of the conidiiferous denticle remaining attached to it, 9–20 × 2·5–4·5 μm. Chlamydospores infrequently produced, lateral, verrucose or smooth-walled, multicellular, brown. — The similar *S. anellii* Graniti *1963*, originally isolated from stalactites in Italy, has sturdier conidiophores, and almost smooth (not verrucose) and slightly broader (4–7·5 μm) conidia. — The major fatty acids in the mycelium of *S. tshawytschae* are palmitic, stearic, oleic and linoleic, with a considerable proportion of unsaturated acids (5642).

This is the second most common species of *Scolecobasidium*. It has been isolated from peat of cedar bogs and tropical greenhouses in Canada (365), polder and other soils in the Netherlands (1616), peat in Ireland, composted municipal waste in Germany, the surface film of aquarium waters (CBS), desert soil (3994) and salt-marshes (8) in Egypt, a coconut grove in the Bahamas (2006), soils and decaying leaves in Japan (2632, 3680) and India (427, 4069). — *S. tshawytschae* was first observed as the causal agent of a fish mycosis in the USA (3706).

Scopulariopsis Bain. *1907*

= *Acaulium* Sopp *1912*
= *Phaeoscopulariopsis* Ota *1928* (nomen provisorium)
= *Masoniella* G. Sm. *1952*

Type species: *Scopulariopsis brevicaulis* (Sacc.) Bain.

Teleomorph: *Microascus* Zukal

MONOGRAPHIC TREATMENT: Morton and Smith (3974).
   Colonies varying from white through shades of buff, brown to black, never true green, velvety or ± funiculose or granular, not forming true synnemata; hyphae hyaline or in some species pigmented. Annellophores produced singly or in whorls on ± penicillately branched conidiophores, cylindrical, ampulliform or ellipsoidal, producing long chains of conidia in basipetal sequence, with ensuing considerable elongation of the annellated zone. Conidia globose, ovoid or mitriform, with a ± broadly truncate base, hyaline or pigmented, smooth-walled or ornamented. — Time-lapse (1106) and TEM studies (1103, 2241) of conidiogenesis have elucidated the mechanism of percurrent proliferation of the annellophores and a de novo wall apposition at the tip of the conidiogenous cell with each newly formed conidium. — Morton and Smith (3974) recognized twenty species without known teleomorphs and eleven have been described since. Most of these are soil fungi, a few (particularly *S. brevicaulis*) become human-pathogenic, and *S. brevicaulis* also occupies some particularly unusual niches (q.v.). The methods for study in this genus are the same as those applied in *Penicillium* (q.v.); but OA and CMA are also particularly suitable for cultivation.

Key to the species treated:

1   Annellophores with the base distinctly swollen, less than 15 μm long, annellated zone less than 2·0 μm wide; colonies grey   **2**
    Annellophores cylindrical or with the base slightly swollen, usually longer than 15 μm, annellated zone 2·5–4 μm wide; colonies not grey   **4**

2(1)   Ascomata usually present   see *Microascus trigonosporus* (p. 416)
       Ascomata absent   **3**

3(2)   Colonies becoming fuscous; conidia 3·0–3·5 μm broad, mostly with a rounded apex
                                                                                          S. brumptii (p. 729)
       Colonies becoming black; conidia 3·5–4·0 μm broad, often with a pointed apex
                                                                                         S. chartarum (p. 730)

4(1)   Colonies white to pale buff; conidia smooth-walled, hyaline   **5**
       Colonies avellaneous or fuscous; conidia smooth-walled or verrucose, pigmented, 5–8 × 5–7 μm
                                                                                                                6

5(4)   Conidia ovoid with a pointed apex, 8–14 × 5–6 μm   *S. acremonium* (p. 725)
       Conidia subglobose to broadly ovate, 5–8 × 4–7 μm   (*S. candida*)

6(4)   Colonies fuscous to fuscous-black; conidia smooth-walled   *S. fusca* (p. 731)
       Colonies avellaneous; conidia distinctly verrucose   *S. brevicaulis* (p. 726)

Scopulariopsis acremonium (Delacr.) Vuill. *1911* — Fig. 320.

≡ *Monilia acremonium* Delacr. *1897*
= *Scopulariopsis communis* Bain. *1907*
= *Oospora glabra* Hanzawa *1911*
= *Scopulariopsis brevicaulis* var. *glabra* (Thom *1910*) Thom *1930* [non *S. brevicaulis* var. *glabra* Raper & Thom *1949*]

DESCRIPTIONS: Morton and Smith (3974) and Matsushima (3680). — Colonies reaching 0·6–0·7 cm diam in seven days on MEA at 24°C or 2·2–2·6 cm in four weeks. Mycelium at first buff but later becoming whitish powdery as the conidia are produced; reverse pale buff. Conidiophores ± penicillate, with annellophores ± cylindrical, 10–50 μm long, basal part slightly swollen to 3·5–6 μm, annellated zones 2·5–4·0 μm wide. Conidia elongate, ovate or mitriform, hyaline to pale buff in mass, 8–14 × 5–6 μm. — The similar *S. candida* (Guég. *1899*) Vuill. *1911* has less pigmented colonies and shorter, subglobose conidia, 5–8 × 4–7 μm.

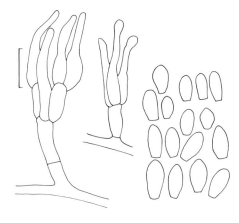

FIG. 320. *Scopulariopsis acremonium*, penicillate conidiophores and conidia, CBS 104.65.

S. acremonium has been isolated from soil many times (3974). It is known from communities of the *Oxalis-Majanthemum* and *Oxalis-Myrtillus* types in Finnish forests (5663), larch forests in Japan (3267), citrus soils in Israel (2764), wheatfield soils in Germany where it was isolated very rarely, soils in Canada (IMI), a chalk soil in the British Isles (1259), a uranium mine in Czechoslovakia (1703) and agricultural and greenhouse soils in the

Netherlands (CBS). Other known substrates include rabbit (6085) and chicken dung (3680) and the vaginal mucus of a mare (3974).

The optimal temperature for growth is about 18°C, none occurs at 5°C, and the maximum is about 37°C (3974). *S. acremonium* can utilize CMC and xylan.

Scopulariopsis brevicaulis (Sacc.) Bain. *1907* — Fig. 321.

 ≡ *Penicillium brevicaule* Sacc. *1881*
 = *Scopulariopsis rufulus* Bain. *1907*
 = *Acaulium anomalum* Sopp *1912*
 = *Acaulium insectivorum* Sopp *1912*
   ≡ *Scopulariopsis insectivora* (Sopp) Thom *1930*

For other synonyms see Morton and Smith (3974).

DESCRIPTIONS: Raper and Thom (4745), Morton and Smith (3974), Ellis (1603), Subramanian (5626), and CMI Descriptions No. 100, *1966*. — Colonies spreading broadly, reaching 4·5–5·5 cm diam in seven days on MEA at 24°C, at first whitish and somewhat funiculose but the central tuft soon becoming vinaceous buff, avellaneous or greyish sepia with a white margin; reverse honey-coloured to avellaneous. Conidiophores once or twice verticillately branched, terminating in groups of 2–4 annellophores. Conidiogenous cells 9–25 × 2·5–3·5 μm, the base sometimes swollen to 5 μm, annelated zones 2·5–3·5 μm wide. Conidia globose to ovoid, the base distinctly truncate and provided with a marginal frill, apex sometimes slightly pointed, avellaneous in mass, at maturity finely roughened or, mostly, coarsely verrucose, 5–8(–9) × 5–7 μm. Small black sclerotial bodies, 30–60 μm diam, have rarely been observed. — *Scopulariopsis stercoraria* (Link ex Pers. *1822*) Hughes *1958* would have been an earlier epithet (2602) for this species but the type specimen of that name was found to have smooth-walled conidia and has been considered as unidentifiable (3974). — The smooth-walled counterpart of *S. brevicaulis* is *S. koningii* (Oudem. *1902*) Vuill. *1911*. Another related species is *S. flava* (Sopp *1912*) Morton & G. Sm. *1963*, a common contaminant on cheese, which has white colonies. — Conidiogenesis has been studied by both time-lapse cinematography (1106) and TEM (1103, 2241). The conidial wall is continuous with a short portion of newly apposed wall material in the uppermost part of the conidiogenous cell which, after separation of the conidium, remains on the conidiogenous cell as an annellation; the conidium is delimited by a septum with a simple pore which is plugged by Woronin bodies; during maturation the conidium forms a thicker secondary, electron-transparent, wall layer; the verrucosities are an integral part of the primary, electron-dense wall layer (407), and a fine mucous pellicle has been found to surround both the conidia and the conidiogenous cells. The formation of new conidia does not always result in an elongation of the conidiogenous cell (2241), and the conidia themselves are 1–5-nucleate. — DNA analysis gave a GC content of 50% (5600).

Scopulariopsis brevicaulis is the commonest species of its genus and has a worldwide distribution. It has been reported from Finland (5663), the USSR (2871), most parts of

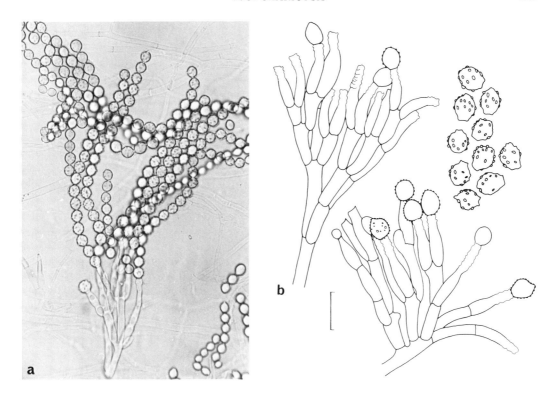

FIG. 321. *Scopulariopsis brevicaulis*. a. Conidiophore with conidial chains, × 500; b. conidiophores with annellated conidiogenous cells and conidia, CBS 273.30.

Europe, Egypt (8, 3993), Israel (2764), the Ivory Coast (4718), South Africa (1559, 3630), India (968, 2854, 4736, 5000, 5626), China (3475), Hawaii (3014), Melanesia (5983), Australia (3720, 5819), Tasmania, New Zealand (4949, 5930), Brazil (398, 4547, 6008), Canada (505) and the USA (660, 1164, 1165, 1166, 1256). It is common in various forest soils (272, 505, 1700, 1702, 1712, 2854, 3138, 5663) and also in the litter layer (2411). Soils in which it has been found include rendzina, terra fusca, pseudogley (3414, 3418), and podzol (3416), soils under steppe type (1559, 1702, 4718) and chaparral (1164) vegetation, heathland soils (3720, 5047), peat (163, 3918, 6520), grassland (159, 163, 166, 3362), dunes (745, 6414), saline (8, 3446), and arable soils (540, 2161, 3446, 4179, 4538, 4788), various plantations (2764, 3630, 3632), and banana soils (2035). Other reports include compost beds (3041), garden compost (1424), river sediments (1162, 5316), polluted running water (1154, 1166, 4429) and treated sewage (1165), a uranium mine (1703), granite outcrop (660), children's sandpits (1421), soils of tide-washed coastal areas (1256, 3014), seawater of the eulittoral zone (4918), wood pulp (3294), slime of paper mills (5060), cotton, textiles (3707), paper products (1875), and frescoes of a monastery (2666). According to numerous observations, it penetrates only the uppermost soil layers, at most to 15 or 20 cm (660, 2035, 2161, 3720, 3630). The pH of the soils investigated extends well into acid and alkaline ranges. Fertilization with NPK has increased the frequency of isolation (2764). Higher counts were obtained in soils after previous crops of clover or alfalfa than after wheat (2816). There are observations of its occurrence in the root region of *Ammophila breviligulata* (6414), groundnut (2768), banana plants (2035), various heathland plants (5819) and on dying oak roots (963). It has further been found on a wide range of organic substrates (1821, 2081, 4548, 5389), grass clippings (2338), self-heated hay (4821), dry fodder (3258), and seeds of wheat (2716, 3491, 3989, 4492, 5293), barley (1752), *Avena fatua* (2961), cultivated grasses (3512), caraway, soya bean, groundnuts and hops (3974); it has been isolated many times from feathers, nests, droppings and pellets of free-living birds (2575) and is also known to be associated with the tick *Ixodes ricinus* (5018), to occur in cultures of the collembolan *Onychiurus fimetarius*-group (1027) and on dead insects (402), though it is probably not pathogenic in all cases. However, exceptions are injuries to bee larvae and silkworms (3974). — *S. brevicaulis* can be inhibited by *Chalara elegans in vitro* (2753) and may serve as food for mites (3104, 5379, 5380, 5381).

Optimal growth occurs in the range 24–30°C, half optimal growth and poor sporulation at 18°C, the maximum temperature tolerated is 37°C, and no growth occurs at 5°C (161, 3974); optimal growth occurs *in vitro* at pH 7–8 and above (662, 3130). The minimal water potential required for vegetative growth is -140 bars (2093, 3109) or, according to another report, -210 bars (4369). Consequently, this species can be regarded as relatively xerophilic; the optimal water potential for vegetative growth is -60 bars (3109). Optimal growth occurs with 3–6(–20)% NaCl in the medium (5316), the latter concentration corresponding to an osmotic potential of -205 bars. — Good pectinase, amylase (5926) and β-fructofuranosidase (6137) activities have been demonstrated; both arabinoxylan (1757) and xylan are decomposed with little variation in this capacity between different isolates (1432). Cellulose decomposition has also repeatedly been demonstrated (3127, 3618, 4184, 4433, 5386, 5926, 6137) with a pH optimum at 6·4 (5238); *S. brevicaulis* primarily forms the enzyme C_1 and little C_x (4779). Good growth occurs on sinigrin (4774) and lignin sulphonate (1425). Its attack of the cell structures of pine wood has been investigated (3121), and losses in weight and tensile strength have been assessed in maple-wood strips (2211). It is also able to grow at high concentrations of tannic acid (3052) and decompose chitin (5926). Cyanamide (2884), glucosamine, methylamine and dimethylamine can all be utilized as N sources (3130). Proteinases are produced (3889, 5359, 5926), poly-L-lysine and poly-L-glutamic acid can be hydrolyzed (5360), progesterone 1-hydroxylated (878), and caprylate converted to 2-heptanone (1933). The relationships between N assimilation and respiration in *S. brevicaulis* have been thoroughly investigated (3506), and its reduction of nitrate (3969), uptake of ammonium ions (3507), and influences of buffers on ammonium and nitrate assimilation have also been studied (3972). After the complete absorption of inorganic nitrogen, organic N compounds are released into the medium in considerable quantities (3968) which include a peptide that cannot further serve as a substrate (3970). *S. brevicaulis* can methylate mercuric chloride (6120), arsenic oxide, arsenate, selenate or selenite and tellurite (526, 941) and is particularly resistant to As compounds, growing preferentially on As-containing paints (3974). *S. brevicaulis* can grow on keratin (3646) and attacks hairs by boring hyphae (1633, 6122). When cultured on keratin chips in a medium containing phosphate and magnesium, it forms large quantities of $NH_4MgPO_4 \cdot 6H_2O$ crystals (2880); on a corn steep liquor-glucose medium, ethylene is produced (2641). — It may cause deep-seated gummose ulcers, skin affections and commonly onychomycosis (1813, 3646, 3974, 4949, 6272). It is resistant to griseofulvin (6272) but sensitive to actidione, pimaricin and cetavlon (3182).

Scopulariopsis brumptii Salvanet-Duval *1935* — Fig. 322.

= *Masoniella grisea* (G. Sm.) G. Sm. *1952*
≡ *Masonia grisea* G. Sm. *1952*
= *Scopulariopsis melanospora* Udagawa *1959*

DESCRIPTIONS: Morton and Smith (3974), and Ellis (1603). — Colonies reaching 1·0–1·6 cm diam in seven days on MEA at 24°C, at first white, papery and with a central floccose tuft which soon becomes grey to sepia-grey or vinaceous-grey in the centre; reverse smoky grey to fuscous-black. Mycelial hyphae at first hyaline but later darkening. Annellophores mostly borne singly on simple or fasciculate aerial hyphae, or sometimes in whorls on short stalk cells. Conidiogenous cells ampulliform, 4–10 × 2·5–3·5 μm, annellated zones 0·7–1·3 μm wide. Conidia ovoid, truncate at the base, dark brown, smooth-walled or slightly roughened, 4·3–5·3 × 3·5–4·5 μm; a thin outer membrane is visible in the light microscope which often surrounds the whole chain. — Among the dark-coloured *Scopulariopsis* species with inflated annellophores, similar species are *S. croci* van Beyma *1944* and *S. chartarum* (G. Sm.) Morton & G. Sm. (q.v.) which both have darker grey colonies; the former grows faster, reaching 1·8–2·1 cm diam in seven days at 24°C, while the latter reaches only 4–6 mm in seven days and has apically ± pointed conidia.

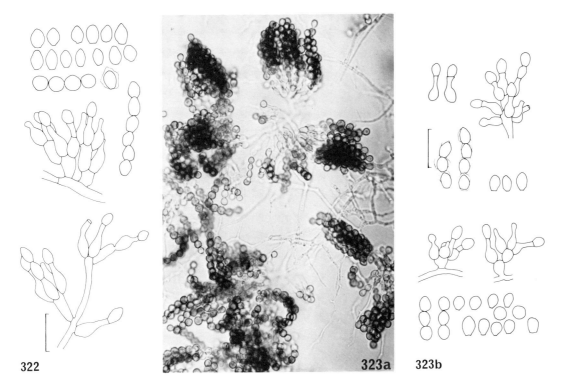

322   323a   323b

FIG. 322. *Scopulariopsis brumptii*, conidiophores and conidia; CBS 896.68.
FIG. 323. *Scopulariopsis chartarum*. a. Conidiophores with dark conidial chains, × 500; b. conidiophores and conidia; CBS 897.68.

S. brumptii was originally isolated as a contaminant in a vaccination serum but later found in milled rice (3974) and other substrates. It is a widely distributed soil fungus occurring in cultivated soils in the USSR (2871) and the southern USA (2482, 3817), teak forests in India (2854), forest soils on podzol in Hong Kong (1021), grassland in South Africa (1555, 4407), cultivated soils in India (3732, 4374), and other soils in Turkey (IMI), Egypt (8), the Bahamas (2006), Central America (1697), and Nepal (1826). It has also been found in garden compost (1424) and very rarely in wheatfield soils. Further habitats include salt-marshes (8), organic detritus in fresh water (4429) and an acid mine drainage stream (1166). It has been found in the rhizosphere of castor beans and grape, on mouldy hay, straw, ants (IMI), human hair with dandruff and in air (CBS).

The optimal temperature for growth and sporulation is 30°C; at 18°C growth is half optimal and sporulation is strongly reduced, while the minimum temperature is above 5°C and the maximum 37°C (3974). This fungus has been isolated from soil of pH 7 and showed *in vitro* equally good sugar and protein utilization at pH 7 and 5 but very little at pH 3 (1825). — Starch (1827) and cellulose (2006) are both decomposed and lignin sulphonate can also support its growth (1425). Chitin can be utilized as a N source, as can glucosamine, methylamine and dimethylamine (3130). Progesterone is 1-hydroxylated (878). Hairs are attacked by means of boring hyphae (6122). — In the lung of an addict to intravenous drug applications, granulamatous lesions by *S. brumptii* have also been found (2089).

Scopulariopsis chartarum (G. Sm.) Morton & G. Sm. *1963* — Fig. 323.

≡ *Masonia chartarum* G. Sm. *1952*
≡ *Masoniella chartarum* (G. Sm.) G. Sm. *1952*

DESCRIPTION: Morton and Smith (3974). — Colonies reaching 4–6 mm diam in seven days on MEA at 24°C, at first forming a whitish cushion but gradually turning grey and after 2–3 weeks becoming smoke-grey to mouse-grey or black, velvety or slightly floccose; reverse at first buff but later becoming fuscous; mycelial hyphae gradually darkening. Annellophores mostly borne solitarily on single or fasciculate aerial hyphae, but rarely in groups of 2–3 on short stalk cells. Conidiogenous cells 8·5–15 × 2·0–2·5 μm, annellated zones 1·0–1·5 μm wide. Conidia ovoid, with pointed or rounded apices, dark brown, 4·0–5·5 × 3·0–4·0 μm; the chains are often surrounded by a thin gelatinous membrane (as in *S. brumptii*, q.v.). — For species differentiation see *S. brumptii*.

S. chartarum was originally isolated from mouldy wallpaper (3974) but has subsequently been found in soil under *Pinus pinaster* in Yugloslavia (3798), white cedar in Canada (IMI), a teak forest in India (2854), the Sonoran desert in California (4733), wheatfields in Germany, on *Hyoscyamus albus* in Libya (CBS, IMI), a *Picea abies* branch in Czechoslovakia and isolated after heat treatment from a burnt soil in the Netherlands (CBS).

The optimal temperature for growth and sporulation is 24°C, half optimal growth and little sporulation occur at 18°C, and no growth occurs at both 5° and 30°C (3974). — Xylan is decomposed and a loss of tensile strength of maple-wood strips has been recorded. — A weak antagonistic activity towards *Gaeumannomyces graminis* has been observed.

Scopulariopsis fusca Zach *1934* — Fig. 324.

DESCRIPTION: Morton and Smith (3974). — Colonies reaching 1·9–2·5 cm diam in seven days on MEA at 24°C, rapidly becoming avellaneous to vinaceous-brown or fuscous-black. Annellophores arising singly or in groups of 2–10 on short stalk cells from mostly fasciculate aerial hyphae. Conidiogenous cells 8–27 × 2·5–4·0 μm, cylindrical or with the base swollen to 4–5 μm, annellated zones slightly narrower. Conidia globose to broadly ovate, base truncate and with a prominent frill, olive to fuscous, smooth-walled, 5–8 × 5–7 μm. — This species is the smooth-spored counterpart of *S. asperula* (Sacc. *1882*) Hughes *1958*.

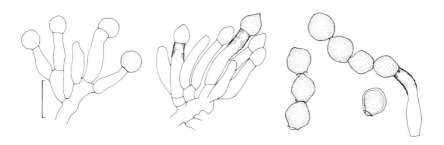

FIG. 324. *Scopulariopsis fusca*, conidiophores with annellated conidiogenous cells and conidia, CBS 872.68.

S. fusca has been isolated many times from soil, mouldy straw, paper and cheese in the British Isles, France, Sweden and Ghana (3974). It has been recovered from dune soils by selective techniques for cellulolytic fungi (4655). Further reports are available from forest soils in Hong Kong (IMI) and central Africa (3063), soils in Peru (2005), garden compost in Germany (1424), *Pteridium aquilinum* litter and air in the British Isles (IMI). — Its frequency was reduced after applications of allylalcohol, metam-sodium, captan, TMTD, and nabam to the soil (1426).

Optimal growth occurs at 24–30°C, half optimal growth with reduced sporulation at 18°C, the maximum lies above 37°C and no growth occurs at 5°C (3974). — *S. fusca* has a strong ammonification capacity and is autotrophic for growth substances (1425). Hairs are attacked by boring hyphae (6122) and progesterone is 1-hydroxylated (878). — It has been named as a causal agent of onychomycosis (CBS) and dermatomycoses, but more rarely so than *S. brevicaulis* (3974).

Sesquicillium W. Gams *1968*

Sesquicillium candelabrum (Bonord.) W. Gams *1968* — Fig. 325.

≡ *Verticillium candelabrum* Bonord. *1851*

DESCRIPTION: Gams (1880). — *Sesquicillium* (type species *S. buxi* (Schmidt) W. Gams *1968*) has been separated from *Verticillium* Nees ex Link (q.v.) because of the unusual arrangement of the terminal and subterminal phialides. Colonies are white to pink. The conidiophores are erect, hyaline, divergently verticillate over the whole length, each end bearing a terminal flask-shaped phialide usually supported by a cell with a short lateral phialidic opening. Conidia one-celled, hyaline, adhering in oblique columns.

    S. candelabrum has fairly fast-growing colonies, reaching 2·1–2·7 cm diam in ten days at 20°C on MEA, whitish, turning pale pink under light, granulose to floccose. Conidia ellipsoidal, 3·9–5·0 × 2·4–3·1 μm. — The other three known species of *Sesquicillium* are distinguished as follows: *S. buxi* (Schmidt) W. Gams *1968* (teleomorph *Nectriella coronata* Juel *1925*) has narrower, fusiform conidia 6·7–7·8 × 2·0–2·5 μm and occurs mostly on decaying *Buxus* leaves; *S. microsporum* (Jaap) Veenbaas-Rijks & W. Gams *1971* (1882) (=*S. parvulum* Veenbaas-Rijks *1970*) has subglobose, on one side flattened conidia,

FIG. 325. *Sesquicillium candelabrum*, conidiophores and phialoconidia.

$2 \cdot 0$–$2 \cdot 5 \times 1 \cdot 3$–$2 \cdot 0 \,\mu$m, and more irregularly branching conidiophores reminiscent of *Tolypocladium* W. Gams *1971* and occurs mainly on various myxomycetes; *S. setosum* Vittal *1974* has cylindrical, 2-celled conidia, 9–$16 \times 1 \cdot 5$–$2 \cdot 5 \,\mu$m, and pointed sterile setae reminiscent of *Cylindrocladium* Morgan (q.v.). The *"Sesquicillium"* state of *Gnomonia papuana* Sivan. & Shaw *1977* has discrete lateral phialides.

S. candelabrum has frequently been isolated from forest and cultivated soils in the Netherlands as well as from decaying leaves of *Buxus sempervirens, Laurus nobilis* and needles of *Pinus maritima* (1880), and other conifer litter (681). Other reports are available from soils in Nepal (1826), Japan (4269), soil under steppe vegetation in Oklahoma (1632), a sandy arable soil in Poland (3530), arable soil in the Buhara region of the USSR (2871), soils in Syria (5392) and estuarine sediments in North Carolina (655). In New Zealand its frequency increased after the long-term irrigation of grassland soils (4948). It has been found on seeds and in soil under cotton (5009) and on decaying wood, occasionally also in coarse fodder (4548).

S. candelabrum can degrade starch (1827), straw, cellulose, hemicellulose, *p*-hydroxybenzaldehyde, syringaldehyde, and ferulic acid (2165); the optimal N source for cellulose degradation was found to be $(NH_4)_2SO_4$ (5674), and the optimal C/N ratio 20:1 (5675); sodium humate is degraded in the presence of an additional C source (5673); strong decomposition of horse hair has also been observed *in vitro* (3530).

Sordaria Ces. & de Not. *1863*

Sordaria fimicola (Rob.) Ces. & de Not. *1863* — Fig. 326.

≡ *Sphaeria fimicola* Rob. *1849*
≡ *Fimetaria fimicola* (Rob.) Griff. & Seav. *1910*

For numerous other synonyms see Lundqvist (3441).

The family Sordariaceae Wint. emend. Lundq. comprises a few genera of Sphaeriales with dark brown, flask-shaped or non-ostiolate perithecia lacking paraphyses, cylindrical asci with an apical, non-amyloid ring and uniseriate ascospores with a fixed orientation in the ascus; ascospores 1- or 2-celled, brown to black, smooth or ornamented, with one or two germ pores. The genus *Sordaria* (nom. cons., type species *S. fimicola*, lectotype) is characterized by 1-celled, smooth-walled ascospores without appendages which are mostly surrounded by a gelatinous sheath. The genus has been studied by Moreau (3927), Lundqvist (3441) and Cailleux (837) and now contains about 23 generally coprophilous species.

DESCRIPTIONS: Cain (838), Moreau (3927), Lundqvist (3441), and Matsushima (3680). — *S. fimicola* is characterized by very fast-growing colonies, with a daily extension of 1·9 cm at 18°C (2279), comprising hyaline to pale brown hyphae, is homothallic, and produces perithecia within 7–10 days under light. Perithecia generally crowded, ± superficial, obpyriform, glabrous, black, 360–420×240–325μm. Asci $(155–)170$–215×14–17μm, unitunicate, non-amyloid, short-stipitate, with a truncate apex including a thickened apical ring, 8-spored. Ascospores ovoid, dark brown, binucleate (3927), with a germ pore on the apiculate basal end, mostly 19–24×11–14μm, covered with a wide gelatinous layer which swells in water (except for a basal invagination) (not visible in lactic acid or lactophenol). — The closest species is the apomictic *S. destruens* (Shear *1907*) Hawker *1951* (=*S. uvicola* Viala & Marsais *1927*) which has similar but highly granular ascospores usually lacking a gelatinous sheath. *S. humana* (Fuckel) Wint. *1872* has broader ascospores $(15\cdot5$–18μm) without, and *S. lappae* Poteb. *1907* similar ascospores with, a gelatinous sheath.

Several studies on ascoma development have been listed (3441). While hyphal septa have simple, ascomycete-type septa, the ascogenous hyphae show pores covered by an electron-dense deposit which is often connected with stacked membrane profiles (1848). A TEM study of the ascus apex showed a ring structure with lateral projections connecting it with the ascus wall; this structure is stainable with Janus green, congo red, and ink but not lugol (4783). The cleavage membrane delimiting the developing ascospores was shown to be derived from the fusion of vesicles originating from Golgi-like structures (4203). In the mature ascospores three wall layers are distinguishable in TEM (3529). The ultrastructure of chromosome pairing in meiosis has also been studied (6568). — *S. fimicola* has often been used as a model of a homothallic ascomycete suited for genetic studies because it can complete its development within eight days (3221, 4281, 4282). Tetrad analysis using spore colour mutants observed in hybrid perithecia allows the mapping of the distance of genes from the centromere and the distinction of linkage groups on the seven chromosomes (4282). Self-sterile mutants have also

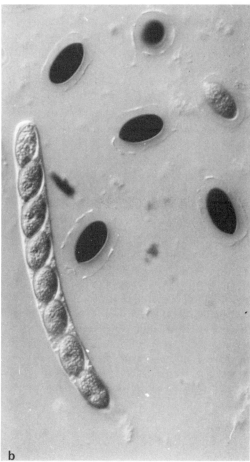

FIG. 326. *Sordaria fimicola*. a. Group of asci showing apical apparatus and ascospores of different stages of maturity; b. a single immature ascus and discharged ascospores showing slime layer; NIC, × 500, CBS 485.64.

been obtained, and the mutagen nitroso-guanine was found particularly suited to induce mutants which in combination display a balanced heterothallism (905, 2755). Crosses of mutants deviating in genes flanking the *g* (grey spores) locus gave rise to a larger number of aberrant ascus genotypes (6333). Ten mutant sites have been located in linkage group II, two of which determine self-sterility (1572).

S. *fimicola* is amongst the two or three commonest coprophilous pyrenomycetes but it has also been repeatedly isolated from soil, though never with high frequencies. Owing to its high growth rate, it can easily be isolated with "screened immersion tubes" (982) and also responds to selective techniques for cellulolytic fungi (4655). — S. *fimicola* has a worldwide distribution though predominantly in temperate regions (3441). It is mainly found on fresh droppings of horse, hare, rabbit, cow and many other mostly herbivorous animals, all over Europe, from Egypt, Canada and Argentina, with numerous additional literature records from the USA,

Iceland, Morocco, Chad, Zaïre, Tanzania, Pakistan, Indonesia, China, Japan, Tahiti, Bermuda, Puerto Rico, Mexico, Honduras, Costa Rica, Panama, Venezuela, Brazil and Paraguay (3441). — There are numerous reports from various forest soils, particularly the litter layer under pine (269, 2923, 4445), beech (270, 271, 272, 273, 3138), oak (650, 1700), *Acacia* species (2186), and other deciduous trees (2573). Other known habitats include soils from forest nurseries (6184), grassland (163, 982, 4021, 6182), heathland (3720, 5819), savannah (4407), arable soils (918, 2159, 3817, 4538), permanent wheatfields (918), banana plantations (2031), slate slopes (566), a granite outcrop (660), sand dunes (4655), desert (4733), and garden compost (1424). It generally penetrates to moderate soil depths (2923, 3720). It has often been isolated from various mouldy plant remains (163, 166, 167, 1821, 4548, 6462), achenes of sycamore (1683), seeds of various plants (842, 3512), including wheat (3491, 4492), oats, barley (1752, 3556), peanuts and pecans (2255), the rhizospheres of tea plants (41) and spruce (3573), and wood baits (beech, birch, lime) buried in the surface soil (5237). It is most commonly found on dung of wild and domesticated animals (2826, 3267, 3926, 4439, 5366, 6085, 6349). Other reports include beehives (402), and birds' nests (2577) and feathers (2575). — Desiccated spores can remain viable for over ten years (4361). *S. fimicola* is slightly inhibited (*in vitro*) by metabolites from another very common coprophilous fungus, *Stilbella erythrocephala* (5362).

The active spore discharge is an effective mechanism of dispersal (2648, 3927), lids of Petri dishes soon becoming blackened by discharged spores. The osmotic pressure in the ascus has been estimated as 1–3 MPa (2648). Discharge is enhanced by a decrease in air humidity (243) and light (2485, 2486), with the most effective wavelengths in the blue range (2648, 2651); discharge is initiated 2–3 h after a brief (1–10 min) light stimulus (2652), but then seems to continue in an endogenous rhythm when grown under either continuous light or darkness (244). — Light is also required for the induction of perithecium development, the optimum temperature for which is between 20 and 30°C (3927). Fruiting is also stimulated by a CO_2-enriched (0·5%) atmosphere (2487), C/N ratios of 5:1 or 10:1 (at KNO_3 concentrations ranging from 12–48 mmol l^{-1}); maltose was superior to glucose in supporting perithecium formation (2220). Thiamine (or pyrimidine) and biotin are both required for the development of perithecia and mycelial growth (2329, 3291, 3327). The influence of a physical growth barrier on the formation of protoperithecia has been discussed but without any clear evidence for the underlying mechanism (4588). The size of the perithecia decreases with increasing salt concentrations (833); only 2·5% of the spores germinated in a medium containing 50% seawater at 25°C, optimal spore germination being between 20°–25°C (832). Larger spores were produced at high temperatures (20°C and higher) and in the dark (6373). On rabbit excrements at 18°C the entire cycle of development and maturation is completed in nine days (2279). — The optimum temperature range for growth on agar is 30–35°C, although with some differences between isolates; 45°C for 24 h was lethal (2329). No growth occurs below pH 4 (3327). — *S. fimicola* can utilize acetate better than D-glucose (3329), and D-arabinose is apparently not utilized (3328). Numerous sugars give increasing growth with rising concentrations, although fewer perithecia are formed. Only sucrose allows fructification even at high sugar concentrations (2329). Various cellulose-containing substrates are decomposed (1425, 2736, 3618, 5386, 6085), and usnic acid can also be degraded (322). — *S. fimicola* has a low tolerance of CS_2 (3987).

Sporothrix Hektoen & Perkins *1900*

Type species: *Sporothrix schenckii* Hektoen & Perkins

Teleomorph: *Ophiostoma* H. & P. Syd.

GENERIC REVISION: De Hoog (2523). — Colonies ± slow-growing, smooth, farinose or lanose, at first whitish but later often becoming greyish or black, sometimes purplish or yellowish, consisting of narrow, hyaline (occasionally pale brown), smooth- and thin-walled hyphae. Conidiogenous cells solitary, erect or in little-branched, sometimes scarcely verticillate conidiophores which are hardly differentiated from the vegetative hyphae and completely hyaline. Conidiogenous cells bearing an apical cluster of short and narrow conidiiferous denticles (formed in sympodial succession) or a rather long, denticulate rhachis; sometimes the conidiogenous cells continue to elongate markedly and so form several clusters of denticles. Blastoconidia small, one-celled, ovoid, claviform or slightly allantoid, apiculate at the base, hyaline, smooth-walled, usually formed singly on the denticles (rarely in short acropetal chains); ± pigmented blastoconidia sometimes arise laterally on undifferentiated procumbent hyphae, but true chlamydospores are absent.

Sporothrix is easily distinguished from related genera: *Blastobotrys* Klopotek *1967* has conidiophores with a long, geniculate rhachis, where each denticle bears a vesicular ramoconidium with several small blastoconidia. *Calcarisporium* Preuss *1851* (monotypic) has long, *Verticillium*-like conidiophores with compact terminal clusters of denticles on the conidiogenous cells and mostly brown sclerotia. *Calcarisporiella* de Hoog *1974* is closest to *Sporothrix* but has more flask-shaped conidiogenous cells with long cylindrical denticles. *Raffaelea* v. Arx & Henneb. *1965* has altogether larger structures and blastoconidia which are abstricted with a broader base.

The distinction between *Sporotrichum* Link ex Fr. *1821* (basidiomycete anamorphs mostly with clamp connections) and *Sporothrix* had been a long-standing matter of debate, until Nicot and Mariat (4169) published a revised generic and specific Latin diagnosis. But in view of the rather exhaustive *descriptio generico-specifica* (Art. 42) by Hektoen and Perkins (*1900*), this is not regarded as a validation of the genus.

De Hoog (2523) recognized 24 species in *Sporothrix*, some of which are connected with *Ophiostoma* teleomorphs. The genus *Ophiostoma* H. & P. Syd. *1919* is distinguished from *Ceratocystis* Ellis & Halst. *1890* mainly by its anamorphs which in the latter belong to *Chalara* (q.v.) (2523). Most *Ophiostoma* (and *Sporothrix*) species are found on decaying wood or in association with bark beetles. Some species also form second anamorphs belonging to *Verticicladiella* Hughes *1953* or *Pesotum* Crane & Schoknecht *1973* (*Graphium* Corda s.l.) in addition to the simpler *Sporothrix* conidiophores.

Sporothrix schenckii Hektoen & Perkins *1900* — Fig. 327.

≡ *Sporotrichum schenckii* (Hektoen & Perkins) de Beurm. & Gougerot *1911*
= *Sporotrichum beurmannii* Matr. & Ramond *1905*
= *Sporotrichum jeanselmei* Brumpt & Langeron *1910*

For numerous other synonyms see de Hoog (2523).

Teleomorph: *Ophiostoma stenoceras* (Robak) Nannf. *1934*

≡ *Ceratostomella stenoceras* Robak *1932*
≡ *Ceratocystis stenoceras* (Robak) C. Moreau *1952*
? = *Ophiostoma albidum* Mathiesen-Käärik *1953*
≡ *Ceratocystis albida* (Mathiesen-Käärik) Hunt *1956*

DESCRIPTIONS: anamorph: Nicot and Mariat (4169), and de Hoog (2523); teleomorph: Griffin (2106) and Mariat and Diez (3607); conidiogenesis: Cole and Samson (1109). — Colonies reaching 0·6–2·2 cm diam on MEA in ten days at 20°C, finely floccose, velvety or lanose, at first hyaline but later often becoming greyish to dull brown; reverse often becoming grey or black. Submerged hyphae mostly 1–2 μm wide, aerial hyphae even narrower, sometimes slightly fasciculate. Conidiogenous cells usually solitary, scattered, erect, mostly 10–40 μm long and 0·7–1·5 μm wide at the base; conidiiferous denticles often in several clusters, 0·5–1·0 μm long. Conidia ovoid to fusiform with a rounded tip, straight, hyaline, mostly 2·5–5·5 × 1·5–2·5 μm, uninucleate (4169). In addition, similar conidia frequently arise laterally from undifferentiated procumbent hyphae; in some isolates pigmented globose conidia, 2·4–3·7 μm diam also occur, which are often associated with obconical or campanulate conidia of similar structure (4169). — Perithecia are often obtained in pure culture after about 11 days, particularly on beech wood blocks in a malt solution or on a synthetic medium containing 0·16–1·5% sucrose (3607). Perithecia black, with a globose base, 120–200 μm diam

FIG. 327. *Sporothrix schenckii*, erect conidiophores, fertile prostrate hyphae, and conidia of various isolates.

and with a neck 270–700(–1500) μ m long and 17–30 μ m wide at the base; with 10–25 straight, hyaline, ostiolar hairs 20–45(–60) μ m; perithecial base covered with dark, pointed hairs to 110 μ m long and 2–4 μ m wide. Asci 8-spored, evanescent. Ascospores hyaline, orange-segment shaped, without a gelatinous sheath, 2·7(–3·3) $\times$ 1·4(–1·9) μ m. — As with other fungi pathogenic to animals, this species can produce a yeast-like phase in animal tissue (5792); this also develops in liquid shake cultures on ordinary media (3607); the yeast-like cells are 2–5 $\times$ 1–3 μ m *in vivo*, somewhat larger (3·5–6·5 $\times$ 2·5–5·0 μ m) *in vitro*, and show the same polyblastic denticulate conidiogenesis as the hyphal phase, as studied by TEM (1916). The conversion to the yeast phase was also brought about rapidly by injecting the mycelium into guinea pig peritoneal macrophage tissue cultures (2373). — DNA analysis gave a GC content of 54·7–55·3% for *S. schenckii* and 52·6% for *O. stenoceras* (3753). The yeast phase showed a higher DNA content (3·7% in the total proteins) than the mycelial phase (1·9%); higher glycogen, chitin, and hemicellulose contents were found in the mycelial as compared to the yeast phase (1324).

De Hoog (2523) delimited *S. schenckii* in a very wide sense and mainly on the basis of negative characters, viz. the absence of catenulate blastoconidia, curved blastoconidia, long or inflated conidiiferous rhachides, branching conidiogenous cells, growth on daffodils or additional *Verticicladiella* or *Pesotum* states. The closest species is apparently the anamorph of the host-specific *Ophiostoma narcissi* Limber *1956* which tends to form its distinct perithecia in culture. A similar soil-borne species is *Sporothrix inflata* de Hoog *1974* (=*Humicola dimorphospora* Roxon & Jong *1974*) which has slower-growing colonies and abundantly produces globose pigmented conidia on creeping hyphae. — Amongst the synonyms listed by de Hoog, *Sporothrix pallida* (Tubaki) Matsushima *1975* ($\equiv$*Calcarisporium pallidum* Tubaki *1955*) has been subsequently separated due to its constantly white and faster-growing colonies, 50–150 μ m long, unbranched conidiophores, and the absence of conidia formed on creeping hyphae (3680). — *S. schenckii* var. *luriei* Ajello & Kaplan *1969*, the causal agent of a cranial osteolysis in South Africa, differs from var. *schenckii* in the frequent production of secondary conidia from detached conidia and dark sclerotia; var. *luriei* does not assimilate creatinine and related compounds (5519).

The purported teleomorph *Dolichoascus schenckii* Thibaut & Ansel *1970* (5794, 5795) was merely based on endoconidia released in emptied hyphal segments (4169); such endoconidia are particularly common in senescent cultures with numerous pigmented conidia and highly vacuolated hyphae, as studied by TEM (5796).

In recent years there has been a continuing discussion as to the relation of *S. schenckii* to *Ophiostoma stenoceras*. The conidial states of both *S. schenckii* and *O. stenoceras* are morphologically indistinguishable, and physiological (3605, 5764), immunological (142, 5860) and DNA hybridization tests (3753) all imply at least a very close relationship between them. Both contain palmitic, oleic and linoleic as major fatty acids in the neutral and polar lipids (1326). *S. schenckii* hyphae contain $C_{18:2}$ and $C_{18:1}$-unsaturated, and C_{16}-saturated as major fatty acids (1280); *Ophiostoma stenoceras* hyphae contain 64% $C_{18:2}$ and 20% $C_{16:0}$, while for various isolates of *S. schenckii* this proportion varies between 39–64% $C_{18:2}$: 21–28% $C_{16:0}$, but saprophytic isolates of the latter form-species showed relations of 9–28 : 36–46% (1279). It seems that non-perithecia-forming isolates can be regarded as derived from perithecial *O. stenoceras* by mutation, as they often appear after a passage of perithecial cultures through a test animal; in wild perithecial isolates the dark obconical lateral conidia are always absent, while they or dark hyphal portions can appear after an animal passage in the same isolate; repeated passages increase the pathogenicity (3607). The somatic and excreted rhamnomannans are hardly distinguishable in *S. schenckii* and *O. stenoceras* by proton and ^{13}C magnetic resonance spectroscopy (3390, 5875). An antigenically active rhamnogalactan has been extracted from cells of *S. schenckii* and purified; it cross-reacted with a peptido-rhamnomannan from culture filtrates; 1,2-*a*-linked L-rhamnose is apparently the most important constituent determining its serological specificity (4099). Antigenic properties of some *Ophiostoma* species and *S. schenckii* were also compared by immunofluorescence and only one antigen was found in *S. schenckii* that did not appear in the other fungi tested (2264). — However, it is not satisfactory to unhesitatingly refer all identical conidial isolates to *O. stenoceras* because of the essentially negative morphological delimitation of the anamorph. — Pyrolysis-gas-liquid chromatograms of perithecial and conidial isolates were almost identical, but differences between pathogenic and apathogenic isolates were found (5765), as they were in some serological tests (4196); otherwise the only critical test for the recognition of pathogenic isolates is an intraperitoneal injection into mice followed by the appearance of disseminate infection (3606, 3607).

S. schenckii originally became known as the causal agent of a chronic subcutaneous lymphatic (or rarely respiratory) mycosis in man and other mammals which has been called sporotrichosis and reviewed in detail (1624). The fungus has subsequently been found to have a widespread saprophytic phase commonly occurring on various substrates in contact with soil, such as plant litter, bark, straw and living plants (4169); it occurs particularly commonly on the wood of conifers and *Eucalyptus* (3606). Soil has been shown to be the main source of human infection (1624). — Its distribution is worldwide and apparently independent of climate or soil types (1624, 5793), although sporotrichosis is mostly confined to the tropics. Pathogenic isolates can be obtained directly from soil by injecting a soil suspension (with antibiotics) into mice; saprophytic isolates have been selectively isolated using suspensions in water and plating on Sabouraud agar, both containing 0·1% actidione and 0·1% chloramphenicol (3606, 3607). Thus, for example, sixteen (six perithecial) out of 40 soil and litter samples in Corsica, and 37 (31 perithecial) out of 422 small mammals in Alsace have yielded the fungus in areas where the disease is unknown; few of these perithecial, but most of the conidial isolates were found to be potentially pathogenic (3606). *S. schenckii* has also been reported from soils in Brazil (4897), Uruguay (3503), Israel (1732), Taiwan (6119), and tide-washed coastal areas of California (1256) and Hawaii (3014). Other known substrates include *Sphagnum* moss, beech stems, *Equisetum* (1624), timber (IMI) and cold-stored meat products (55). *S. schenckii* can be disseminated by air, water, and animals as well as with soil and plant materials (5791). The outbreak of an epidemic in the vicinity of Lake Ayarza in Guatemala was ascribed to fish vectors (3696).

Good growth occurs at pH 7 and in the temperature range 25–37°C (922, 3607); some isolates can be adapted to growth at 5°C (55). *S. schenckii* can be transformed to the yeast phase above 28°C, and *O. stenoceras* isolates produce this at 25°C in shake culture (1470, 3607). Saprophytic isolates of both perithecial and conidial types were generally found to grow well at 37°C (3606), although another report indicated a maximum temperature of 30°C for some saprophytic isolates (2551). — Suitable carbon sources are starch, D-glucose, L-fructose, sucrose, D-xylose, D-galactose, D-rhamnose, D-inositol and D-sorbitol; glucose, rhamnose and mannose are incorporated into polysaccharides (3607); in another study, *O. stenoceras* (unlike *S. schenckii*) did not utilize starch (5764). L-Alanine, L-arginine and glycine all favour the development of the yeast phase (1470). Creatine, creatinine and guanidinoacetic acid can serve as sole N sources (5520, 5522, 5523). Proteinase is produced (55), and neuraminidase (not at 37°C), fibrinolytic and lipolytic enzymes in both *S. schenckii* and *O. stenoceras* are probably essential to its pathogenicity enabling it to decompose animal tissues (4014). CO_2 is necessary for the growth of the yeast and, to a lesser extent, for the mycelial phase; it could be replaced by oxalacetic or α-ketoglutaric acids (1470, 3604); good aeration is nevertheless essential for the production of the teleomorph (3607). The pyrimidine moiety of thiamine is required by almost all parasitic and saprophytic isolates (55, 1470), while many pathogenic conidial isolates also required pyridoxin and biotin (3606).

Stachybotrys Corda *1837*

= *Synsporium* Preuss *1849*
= *Fuckelina* Sacc. *1875*
= *Gliobotrys* Höhn. *1902*
= *Hyalobotrys* Pidopl. *1947*
= *Hyalostachybotrys* Srinivasan *1958*

Type species: *Stachybotrys chartarum* (Ehrenb. ex Link) Hughes (= *Stachybotrys atra* Corda)

Teleomorph (known in one species): *Melanopsamma* Niessl

MONOGRAPHIC TREATMENT: Jong and Davis (2815), and Ellis (1603, 1604).

Vegetative hyphae usually hyaline. Conidiophores macronematous, simple or cymosely branched, with apical clusters of several ellipsoidal or subclavate phialides formed in succession. Conidia and phialides hyaline or pigmented. Conidia released in basipetal succession through an opening in the rounded phialide apex which has hardly prominent collarettes, held together in a slimy drop, one-celled, ellipsoidal, cylindrical, globose, reniform or fusiform, mostly ornamented, in some species smooth-walled, pigmented or hyaline. New conidia arise after the previous ones are mature and have been released from the phialide neck (862).

About 15 species have been recognized. — The closely related genera *Stachybotrys* and *Memnoniella* (q.v.) have a worldwide distribution but some species are restricted to (sub)tropical areas; they are commonly isolated from soil and various decaying plant substrates. Most species are able to decompose cellulose efficiently, but require biotin for growth and, particularly, sporulation (2815). The pigmented species synthesize humic-acid-like polymers (3634, 4319). A review of their importance has been given by Jong and Davis (2815). — Cellulose- or starch-containing media (OA, CMA) with some biotin are superior to media with soluble sugars in inducing sporulation.

Key to the species treated:

1	Conidia cohering in chains	cf. *Memnoniella* (p. 411)
	Conidia cohering in slime drops	**2**
2(1)	Conidia colourless, mostly 10–17 × 4–6 μm	*S. elegans* (p. 746)
	Conidia pigmented	**3**
3(2)	Conidia with longitudinal striations, ± cylindrical, 13–16 × 4–6 μm	*S. cylindrospora* (p. 746)
	Conidia coarsely roughened, 7–12 × 4–6 μm	*S. chartarum* (p. 743)

Stachybotrys chartarum (Ehrenb. ex Link) Hughes *1958* — Fig. 328.

≡ *Stilbospora chartarum* Ehrenb. *1818*
≡ *Oidium chartarum* (Ehrenb.) ex Link *1824*
= *Stachybotrys atra* Corda *1837*
= *Synsporium biguttatum* Preuss *1849*
= *Stachybotrys alternans* Bonord. *1851*

For other synonyms see Jong and Davis (2815).

DESCRIPTIONS: Ellis (1603), Matsushima (3679, 3680), Subramanian (5626), and Jong and Davis (2815). — Colonies reaching 1·4 cm diam in five days on OA at 25°C, covered with a dark powdery bloom of conidial masses; reverse uncoloured. Conidiophores simple or irregularly branched, usually about $100\,\mu$m, but to $1000\,\mu$m long and $3–6\,\mu$m wide, hyaline at the base and ± dark olivaceous and sometimes roughened towards the apex, bearing clusters of 4–10 phialides. Phialides obovate or ellipsoidal, hyaline, later becoming olivaceous, $9–14 \times 4–6\,\mu$m. Conidia aggregated in slimy masses, ellipsoidal, at first hyaline, at maturity dark olivaceous-grey, ± opaque, almost smooth-walled to coarsely roughened with warts and ridges, $7–12 \times 4–6\,\mu$m.

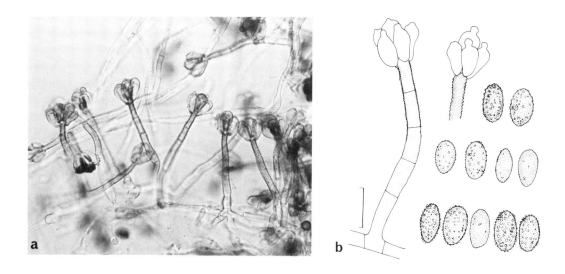

FIG. 328. *Stachybotrys chartarum*. a. Conidiophores, × 500; b. details of conidiophores and conidia.

Several rather similar species are distinguished from *S. chartarum* as follows: *S. zeae* Morgan-Jones & Karr *1976* has hyaline, always unbranched conidiophores to $85\,\mu$m long and verruculose conidia, $7–9 \times 3\cdot5–4\cdot5\,\mu$m; *S. microspora* (Mathur & Sankhla *1966*) Jong & Davis *1976* has equally short, but sometimes branched and pigmented conidiophores and coarsely roughened, short ellipsoidal conidia, $6–8 \times 4–5\,\mu$m, which later become subglobose; *S. albipes* (Berk. & Br. *1871*) Jong & Davis *1976*, the anamorph of *Melanopsamma pomiformis* (Pers. ex Fr. *1823*) Sacc. *1878*, has smooth-walled conidia, $7–9 \times 5–6\,\mu$m; *S. dichroa* Grove *1886* has colonies with an orange reverse and short-ellipsoidal, broadly apiculate, coarsely warted conidia, 7·5–10 × 5–7 μm; *S. parvispora* Hughes *1952* has small, ellipsoidal, coarsely roughened conidia, 5–6 × 3·0–3·5 μm; and *S. kampalensis* Hansf. *1943* has larger, almost cylindrical, verruculose conidia, 10–14 × 6–7 μm. — A TEM analysis of conidiogenesis in *S. chartarum* has been carried out (862). — DNA analysis gave a GC content of 49–55·5% (5600).

S. chartarum is the commonest species of its genus and has a worldwide distribution, mainly on dead plant material (1603, 3099). Due to its abundant sporulation, it is easily isolated by dilution plates, but also responds particularly well to selective media designed for cellulolytic fungi (4647, 4655, 5884, 5886). — Its occurrence has been documented in almost all of Europe and North America, New Guinea, Japan and India (2815); other reports are available from the Arctic (1750), the USSR (1257, 2871, 3850, 3856), mediterranian soils (271, 2963, 3446, 3450, 4245), Israel (2762, 2764, 2772, 3988, 4759), Kuwait (4001), Iraq (92), Egypt (8, 100, 3992, 4962), Libya (6510), Zaïre (3790), central Africa (3063), South Africa (1559), Pakistan (1963, 4855), Nepal (1826), Jamaica (4886), the Bahamas (2006), Central America (2030, 2031), Brazil (4547), Peru (2005), and Canada (505). It is common in hardwood and coniferous forests (271, 505, 1257, 2186, 2854, 2923), jungle soils (2030), soils of forest nurseries (3908, 6184), alpine grassland (6082), arable land, plantations with citrus (2764, 3632, 3988) or banana (1697, 2031, 2035), peat (3850), sand dunes (745, 3941, 4162, 4655, 6414), desert soils (100), saline soil (8, 4001), a saline lake (1292), soil under tundra (1750) or steppe type vegetation (1559), soils treated with sewage sludge (1163), compost beds (3041), garden compost (1424), pine wood exposed to seawater (831), polluted running water (1154, 1155, 1162, 1166), a uranium mine (1703) and frescoes of a monastery (2666). It has been found predominantly in the upper soil layers, the maximum recorded depth being 20 cm (2035, 2161), and is not markedly influenced by soil pH; it can occur even in chalk soils (1259). The frequency of *S. chartarum* was slightly increased by applications of mineral fertilizers in one experiment (2163) but showed no response in another (2162); however, it clearly increased after an application of farmyard manure (2161). *S. chartarum* also occurs in soils enriched with copper (4921) and has been found on wood used in a copper mine (2664). Different precrops in wheatfield soils had no influence on its frequency (2816). It is common on rotting plant remains (4548), leaf litter of *Pinus radiata* (6080) and *Gymnosporia emarginata* (6108), leaves of palms (2034, 3963), banana (3063) and *Panicum coloratum* (1560), bolls (5009), decaying leaves (5231) and roots (1965) of cotton, rotting bean pods (3063), decaying *Setaria glauca* (5233, 5234), bark and leaves of citrus (3988), decaying *Pseudoscleropodium purum* (2969), old stalks of *Heracleum sphondylium* (6462), mouldy hay (2081), fruiting-bodies of *Serpula lacrymans* (3130), sclerotia of *Sclerotinia sclerotiorum* (4699), and nests, feathers and pellets of free-living birds (2575). It is found to be accumulated in the rhizospheres of oats (3006), barley (3006, 3526) and grasses (1614), and has been isolated from the root regions of wheat (2023, 3567), corn (4553, 6563), beans (1369), *Ammophila breviligulata* (6414), banana (2035), alfalfa (3983), groundnut (2768), flax, especially in wilt-susceptible varieties (5826), *Salicornia* (4645), the rhizoplane of sugar cane (4886), and seeds of *Avena fatua* (2961), cultivated grasses (3512), wheat (3989, 4492), oats, *Tragopogon, Bromus*, cauliflower and peas (3556). Mycelium introduced into the soil is decomposed relatively slowly (3633) and the fungus has been found to contribute to the stabilization of soil aggregates (224, 225, 1451, 6229). — The germination of its conidia is inhibited by soil mycostasis (6284) and germinated conidia undergo lysis in the soil after a short period (1006). Growth *in vitro* was both inhibited by *Memnoniella echinata* (4680), and arrested by hyphal contacts with *Tuber melanosporum* (632).

With a high $NH_4Cl:K_2HPO_4$ ratio (extreme 100:1) the mycelium remains sterile, but at medium ratios conidiophores form; the concentration of glucose (in PDA, which in normal concentrations is unfavourable for sporulation) appears to mainly determine the amount of growth and sporulation (3725). CMC agar has been found to be a suitable medium for both conidium production and growth (2815). — Good growth occurs in the temperature range 10–37°C, the optimum is 23(–27)°C, the minimum (2–)7°C and the maximum 37–40°C (261, 1827, 2815, 6464); diurnal temperature extremes of −94°C (for 19·5 h) and +23°C (for 4·5 h) resulted in a growth blockage when continued for 35 days (1242). The thermal death point has been determined as between 50° and 60°C for 30 min in soil (613). *S. chartarum* can grow within the pH range 3·6–7·7 (824) and is moderately osmotolerant (4001). The optimum water potential for vegetative growth is −30 bars, and the minimum −100 bars (261); for sporulation the minimum is −60 bars (4369), and for conidium germination <−100 bars (6464). — Twenty sugars tested were all found to be utilized (4508), with maximum growth occurring on sucrose, followed by fructose and glucose (824). The glycosides scilliroside (5598) and salicin (2751) are cleft. Ammonium, nitrate, amino and imidazole groups and pyrrolidine are suitable N sources (4508) but no growth occurs on sodium nitrite (824). Biotin is required for growth and sporulation (3617). Organic acids are utilized (2747) and starch is decomposed (1750, 1827), as is cellulose in the form of filter paper (1425), textile fibres (631, 654, 3618, 6326), methyl cellulose (2878) or cellophane (5886); in the last case the fungus penetrates the substrate by elongated rooting branches. The following enzymes have been shown to be involved in cellulose degradation: β-glucosidase, cellulase, and cellobiase (2747, 2749, 2750, 6505), and the extracellular β-glucosidase has been purified (2748); the optimum temperature for cellulose decomposition is reported to be 30°C, and the degradation rate to increase with increasing N source (4433) and O_2 concentrations (6163); the optimal N source for cellulose degradation appears to be $(NH_4)_2SO_4$ (5674, 5675). During pectin degradation, polygalacturonases are produced (1089). *S. chartarum* causes a soft rot in timber (1496) and erodes birch wood (4191); phenol oxidases have been detected when grown on gentisic acid, benzidine, pyrocatechol and guaiacol (934); both *p*-hydroxybenzaldehyde (2165) and gallic acid (1750) can be utilized. Chitin (1425) and wool are decomposed (6330), hairs being attacked by boring hyphae which subsequently ramify inside them (1633). Cyanuric acid (6415) and plasticizers can be metabolized (473, 5148). Known metabolites of *S. chartarum* include a number of hydroxybenzenes and carboxylic acids (3634); the production of humic-acid-like polymers (2214, 5124) was stimulated by the presence of montmorillonite in the medium (630, 1738, 1740), but only a small proportion of the polymers produced were found to be phenolic (5123). Other known products include campesterol, β-sitosterol, and stigmasterol (3231), and, in culture, large quantities of pantothenic acid and pyridoxin can be produced (5677). — *S. chartarum* shows antagonistic activities against *Bacillus subtilis* (3599) and numerous test fungi (716, 823, 1369, 1425, 1431, 1964, 2735, 4699, 6205), and the occurrence of an unidentified antifungal metabolite has been demonstrated (824); it also suppressed the pathogenic effects of *Chalara elegans* on citrus seedlings (3635). The growth of rape, tomato and flax was significantly reduced after intense inoculations of soil with *S. chartarum* (1427). Of 73 isolates obtained from various plant substrates in Finland, 49 were found to be toxigenic in a tissue culture test with mouse fibroblasts (3099). Mycotoxins present in the mycelium can cause stachybotryotoxicosis in both horses and man (1798, 1800, 2050, 2070, 3098, 4888, 5147, 6412). Two major toxic components have been extracted from oats infested by *S. chartarum* and partially identified (3859); some isolates produce sesquiterpenoid mycotoxins classified as 12,13-epoxy-Δ^9-trichothecenes (1642). — The fungus possesses a strikingly high tolerance to UV irradiation (1512).

Stachybotrys cylindrospora C. W. Jensen *1912* — Fig. 329.

≡ *Stachybotrys atra* var. *cylindrospora* (C. W. Jensen) Rayss & Borut *1956*

DESCRIPTIONS: Barron (358), Ellis (1603), Matsushima (3679, 3680), and Jong and Davis (2815). — Colonies reaching 1·8 cm diam in five days on OA at 25°C, appearing blackish granular from the conidial masses; reverse yellowish pink. Conidiophores simple or branched, to 200 μm long and 3–5 μm wide, ± pigmented and sometimes minutely roughened in the upper part. Phialides subclavate, olivaceous near the tip, 11–16 × 4–5 μm. Conidia aggregated into slimy masses, cylindrical to ellipsoidal, at first hyaline and smooth-walled but later becoming dark olive-grey with delicate longitudinal or oblique striations, 13–16 × 4–6 μm. — The striate conidia preclude the possibility of confusion of this species with any other species of this genus.

S. cylindrospora is apparently a rather rare soil fungus but has been reported from Europe, Israel, North America (1603) including Alaska, and Japan (3680, 5846). Most reports of it come from the USA and include agricultural soil in New York State (2740), deciduous forests in Ohio (2573), forest soils in Iowa (4362), *Populus* and *Salix* communities in Oklahoma and Wisconsin (2004, 2008, 3549), and soils in Colorado (3259). It is also known from Canada (358, 505), and it has also been isolated from a *Melico-Fagetum* in Poland (273), pond mud and forest soil in India (3583), and cultivated and forest soils in central Africa (3063); it has also been reported from Central America (1697), Nepal and the British Isles (IMI). In Belgian forest soils (mor humus) it occurred in the rhizosphere of various plants (4814) and has also been discovered on *Heracleum sphondylium, Musa sapientum* (IMI), pellets of free-living birds (2575), and frescoes of a monastery (2666).

Good growth is possible in the temperature range 15–30°C. This species is apparently not cellulolytic (2815).

Stachybotrys elegans (Pidopl.) W. Gams, comb. nov. — Fig. 330.

≡ *Hyalobotrys elegans* Pidopl. *1947* (Mykrobiol. Zh. Kiev 9(2–3):55)
= *Stachybotrys bisbyi* (Srinivasan) Barron *1964*
   ≡ *Hyalostachybotrys bisbyi* Srinivasan *1958*
= *Hyalostachybotrys sacchari* Srinivasan *1958*
   ≡ *Stachybotrys sacchari* (Srinivasan) Barron *1964*
= *Stachybotrys aurantia* Barron *1962*

DESCRIPTIONS: Barron (359, 364), Matsushima (3679, 3680), and Jong and Davis (2815). — Colonies reaching to 3·3 cm diam in five days on OA at 25°C, downy, whitish, covered by salmon-coloured powdery conidial masses (but tending to lose the capacity to sporulate after several transfers); reverse pale pink to orange. Conidiophores simple or branched, hyaline, to 200 μm long and 3–4 μm wide, bearing clusters of 4–9 phialides. Phialides subclavate with a somewhat protruding apical opening, smooth-walled, 10–17 × 4–6 μm. Conidia fusiform with rounded ends, hyaline, smooth-walled, usually

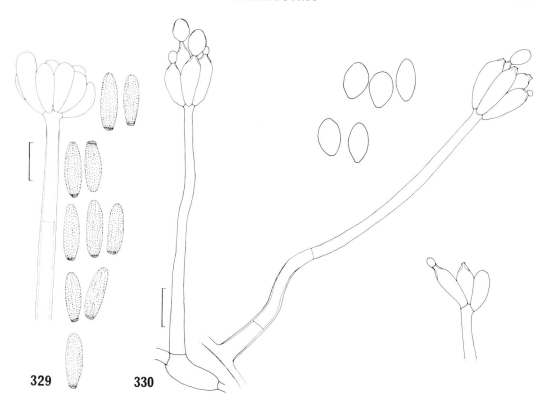

329   **330**

Fig. 329. *Stachybotrys cylindrospora*, conidiophore and striate phialoconidia, CBS 203.61.
Fig. 330. *Stachybotrys elegans*, conidiophores and conidia, CBS 868.73.

$8–14 \times 6–9\,\mu$m, but sometimes more slender and $10–16 \times 3–6\,\mu$m. — Colony habit and conidial shape and size are variable in this species. — *S. sacchari* was distinguished from *S. bisbyi* on account of its dimorphic conidia with the narrower type predominating, $7·5–11·5 \times 2·6–3·7\,\mu$m, but this difference was not recognized as sufficient for separation by Jong and Davis (2815); *S. bambusicola* Rifai *1964* and *S. palmijunci* Rifai *1974* both differ in having reddish brown conidiophores.

S. elegans has been recorded from soil and grass roots in India, Egypt, South Africa, Mozambique, Germany, Papua-New Guinea, the Society Islands (2815), cultivated soils in Japan, and tundra in Alaska (3680). It has also been isolated from soils in Canada (359), a teak forest in India (2854) and the rhizospheres of rice (IMI) and *Trifolium alexandrinum* (1963).

Growth is sparse on CzA, but good on both PDA and MEA. On a minimal synthetic medium, *S. elegans* was found to require biotin for normal growth (359).

Staphylotrichum J. Meyer & Nicot *1957*

Staphylotrichum coccosporum J. Meyer & Nicot *1957* — Fig. 331.

= *Botrydiella bicolor* Badura *1963*

DESCRIPTIONS: Nicot and Meyer (4170), Maciejowska and Williams (3496), Barron (364), Matsushima (3679, 3680), and Ellis (1603). — Colonies reaching 2·8 cm diam in five days at 25°C on OA, velvety, about 3 mm high; yellow-orange, becoming brown in reverse; hyphae hyaline to slightly pigmented. Characteristic erect conidiophores usually appear only after 2–3

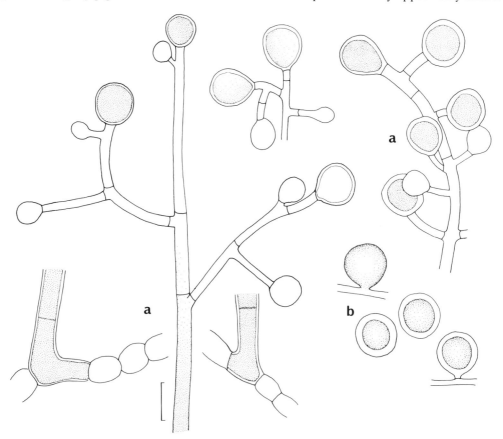

FIG. 331. *Staphylotrichum coccosporum*. a. Upper parts and foot cells of conidiophores and blastoconidia; b. *Humicola*-type chlamydospores formed on vegetative hyphae.

weeks, to 1200 μm tall, brown, lighter towards the tip, with apical racemose branching and solitary terminal, subglobose, pigmented aleurioconidia mostly 10–12 μm diam. Short-stalked or sessile aleurioconidia or chlamydospores, 10–14 μm diam, commonly occur as in *Humicola* Traaen (q.v.). — The monotypic genus *Staphylotrichum* is also similar to *Botryotrichum* Sacc. & March. (q.v.) which has white or grey colonies and unpigmented conidiophores but has sometimes pigmented setae.

S. coccosporum is a widespread soil fungus which occurs mainly in warmer climates (1603). It was originally isolated from a rain forest soil in Zaïre (4170), but has since been found in forest soils in Poland (273), Italy (271), the USA (2573), Central America (1603, 2030), arable soils in central Africa (3063), teak forests (2854), savannah with *Acacia karroo* in South Africa (4407) and grassland in India (1603), and is also known from Sri Lanka, Hong Kong, Costa Rica (IMI) and Peru (2005). This species was found to be common in greenhouses in Indiana/USA (3496) and also in cultivated organic soils in Ontario (364). It is an uncommon fungus in Netherlands soils, but has been isolated there with considerable frequency as a cellulolytic fungus from cocos fibres buried in agricultural soil (CBS). It has also been isolated from plant remains in Zaïre (2966) and sugar cane in Jamaica (IMI).

Syncephalastrum Schröt. *1886*

Syncephalastrum racemosum Cohn ex Schröt. *1886* — Fig. 332.

DESCRIPTIONS: Benjamin (450), Zycha and Siepmann (6588), Milko (3807), Raizada (4708), Watanabe (6212), and Boedijn (578); sporogenesis: Cole and Samson (1109). — Fam. Syncephalastraceae Naum. ex R. K. Benjamin, with one genus, *Syncephalastrum*. — Sporangiophores erect, ascending, or recumbent and stolon-like, often producing adventitious rhizoids, irregularly or ± racemosely branched, with septa particularly below the vesicles in mature sporangiophores; the main stalks and branches forming terminal, globose to ovoid vesicles which bear rod-like merosporangia directly over their entire surface; merosporangial wall evanescent at maturity. Sporangiospores uniseriate, globose or ovoid, formed by simultaneous cell delimitation. Zygospores globose, dark brown, rough with broad, shallow, pointed projections, formed in the aerial mycelium between nearly equal suspensors.

S. racemosum has light to dark grey, very fast-growing colonies, 0·5–1·5 cm high, primary sporangiophores 10–15 μm wide, primary vesicles 30–80 μm, secondary vesicles 10–40 μm diam; merosporangia containing (3–)5–10(–18) spores. Sporangiospores smooth-walled, globose to ovoid, 3–5 μm diam. Zygospores 50–90 μm diam. — DNA analysis gave a GC content of 48·5–50·5% (5600). — Some other species described in *Syncephalastrum* are all regarded as synonymous; isolates representative of different colony types were found to be interfertile and to represent a single variable species (450). The only other known distinct species of the genus, *S. verruculosum* Misra *1975*, has very low velvety colonies and merosporangia with 3–5 verruculose spores, 4–7 μm diam. — The merosporangia are interpreted as sporangiola (450) and a TEM study of young stages (1773) has shown that the young spores are delimited by endoplasmic membranes within a common wall, this being analogous to an ordinary sporangium.

S. racemosum is mainly distributed in dung and soil in tropical and subtropical regions. Most reports come from India (5058), but it is also known from Sri Lanka (4021), Bangladesh (2712), Indonesia (578), Hong Kong (6516), Taiwan (2472, 3377), Japan (3267), Brazil (6007), Central America (1697), mainly the southern states in the USA (450, 3817, 4918), Syria (5392), Egypt (8), Israel (2764, 2768, 2772), equatorial West Africa (1420), and central Africa (3063). Reports from temperate zones include the USA (2008, 2482), various European countries (269, 272, 1616, 6588), and the USSR (3365). Most favourable habitats are ones rich in organic matter and include forest soils (2482, 3063, 4716, 4995) under pine (269), beech (272), willow-cottonwood (2008), *Eugenia* (5512), teak (2854), and *Acacia melanoxylon* (2186), uncultivated soils (4030, 4698), grassland (3865, 4021, 4933), cultivated soils (1616, 3817, 3868, 4996, 4997), under rice (1519, 2712), sugar cane (3732, 6007), groundnut (2768, 2772), corn (4716) and citrus (2764), carst caves (3365), salt-marshes (8), a mangrove swamp (4700), seawater of the eulittoral and pelagic range (4918), and polluted running and stagnant waters (1166). It has been isolated from leaf litter of *Eugenia heyneana* (5513) and *Cyclosorus* (2859), the rhizosphere of oats, barley (3004), rice (2712), *Cassia*

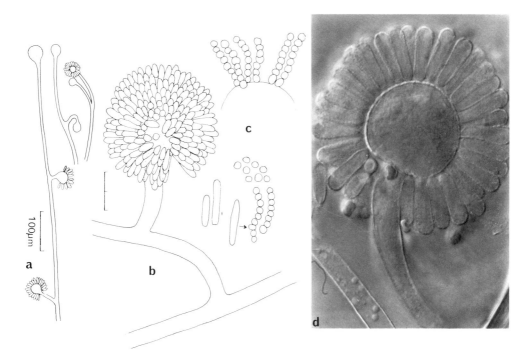

Fɪɢ. 332. *Syncephalastrum racemosum.* a. Sporangiophores at low magnification showing racemose branching; b. vesicle with young merosporangia; c. stages of maturation; d. sporangiophore vesicle with young merosporangia, NIC, × 1000; CBS 302.65.

occidentalis (6067), and *Bothriochloa pertusa* (3271), seeds of *Avena fatua* (2961), wheat (4492), soya (4975), castor beans (3172), groundnut (3220, 3702), cotton (5009), and pecans (2572), stored sweet potatoes (3066), copra and cacao (1676), fermented food (5980), rotting plant remains (4548), and also dead adult bees and honeycombs (402), droppings of free-living birds (2575), and cotton fabric (3707, 6326). — The hyphae were found to be invaded by *Trichoderma viride* (1511) and the obligate parasite *Dimargaris verticillata* (3560). *S. racemosum* can serve as food for the mites *Glycyphagus domesticus* and *Acarus siro* (5382).

Good growth occurs in the range 17–40°C (517). Zygospores form between compatible isolates after 5–7 days on YpSs agar under ordinary laboratory conditions (450). — Growth on acetate is better than on D-glucose (3329). Organic N compounds are released when it is grown on mineral media (3968), and 1,3-β-D-glucanase is produced with laminarin as the sole C source (4778). Production of β-glucosidase on a salicin-containing medium (6159) and little lipase activity (1676) have been demonstrated. The efficient conversion by *S. racemosum* of 3β-hydroxy-$5a$-androstane-7-one into its $12a$-hydroxy derivative has been described (433). — *S. racemosum* tolerates anaerobic situations but does not grow under such conditions (1244). It has been observed to parasitize the hyphae of *Aspergillus niger* (4548). Isolates resistant to phenyl-mercury fungicides have been reported (2076).

Talaromyces C. R. Benjamin *1955*

Type species: *Talaromyces flavus* (Klöcker) Stolk & Samson (=*Talaromyces vermiculatus* C. R. Benjamin)

Family Trichocomaceae Fischer (=Eurotiaceae Clem. & Shear) according to Malloch and Cain (3552). Anamorphs mostly belong to *Penicillium*, Sect. *Biverticillata-Symmetrica*.

This genus is distinguished from *Eupenicillium* Ludwig (q.v.) by soft, white to yellow ascoma walls composed of loose hyphae; the asci are formed in chains, whereas they arise singly from croziers in the related genus *Hamigera* Stolk & Samson *1971*. The genus has been monographed by Stolk and Samson (5593) who included SEM photographs of the ascospores. The genus comprises four sections with 18 species, most of which were described by Raper and Thom (4745) in the *Penicillium luteum* series (*Biverticillata-Symmetrica*). — For species identification it is necessary to observe ascoma initials in squash mounts of young developmental stages. There is still some disagreement about author's citation in the teleomorphs (6291). The five species treated here can be distinguished as follows:

Key to the species treated:

1	Conidiophores with biverticillate, asymmetrical, somewhat appressed penicilli with all elements rough-walled; conidia cylindrical (*Penicillium cylindrosporum* series), brown to creamish; more or less thermophilic: Sect. *Emersonii*
Ascomata orange-brown; initials consisting of conspicuous swollen cells, resembling chlamydospores; ascospores smooth-walled, $3 \cdot 5 – 4 \cdot 0 \times 2 \cdot 7 – 3 \cdot 5 \mu$m; strongly thermophilic
T. emersonii (p. 753)

Conidiophores with biverticillate symmetrical penicilli; conidia green; mesophilic; ascospores ellipsoidal, without ridges in the species treated here	**2**

2(1)	Ascospores smooth-walled or with few delicate spines, $2 \cdot 5 – 3 \times 1 \cdot 5 – 2 \cdot 2 \mu$m; ascomata yellow; ascogonia terminating in conspicuous coils	*T. helicus* (p. 756)
Ascospores conspicuously spinulose	**3**

3(2)	Ascomata white to creamish; initials consisting of irregularly swollen hyphae; ascospores $3 – 3 \cdot 5 \times 2 – 2 \cdot 5 \mu$m	*T. trachyspermus* (p. 756)
Ascomata yellow to orange; ascospores mostly $3 – 5 \times 2 \cdot 2 – 3 \cdot 5 \mu$m	**4**

4(3)	Colonies generally growing restrictedly; ascoma initials consisting of thick, irregularly septate hyphae, producing coiled branches	*T. wortmannii* (p. 758)
Colonies spreading broadly; ascoma initials consisting of vermiform ascogonia around which thin antheridia are coiled	*T. flavus* (p. 754)

Talaromyces emersonii Stolk *1965* — Fig. 333.

Misapplied name: *Talaromyces dupontii* (Griffon & Maubl.) *sensu* Apinis *1963* [non *Penicillium dupontii* Griffon & Maubl. *1911 sensu* Emerson *1949*: *Talaromyces thermophilus* Stolk *1965*]

Anamorph: *Penicillium emersonii* Stolk *1965*
  ≡ *Geosmithia emersonii* (Stolk) Pitt *1979*

DESCRIPTIONS: Raper and Thom (4745), Stolk (5594) and Stolk and Samson (5593). — Colonies on MEA growing rapidly, reaching 5–7 cm diam in seven days at 40–45°C. Ascomata in a dense yellow to reddish brown layer, 50–300 μm diam, often confluent, ripening within seven days; ascospores yellow, thick-walled, smooth-walled, subglobose to ovoid, 3·5–4·0 × 2·7–3·5 μm. Conidiophores arising from the funiculose aerial mycelium, particularly in the colony centre, pale brown; reverse yellow-brown or green; conidiophores usually with 1–2 branches, densely compacted, phialides cylindrical with a distinctly tapering tip; all elements conspicuously roughened. (For this type of conidiophore the *Penicillium cylindrosporum* series has been described (5593)). Conidia cylindrical to ellipsoidal, smooth-walled, hyaline to brownish, 3·5–5·0 × 1·5–2·7 μm. — The related *Talaromyces thermophilus* Stolk *1965* is distinguished by its green to pink colonies, divaricate penicilli, and the absence of ascomata on most commonly used agar media.

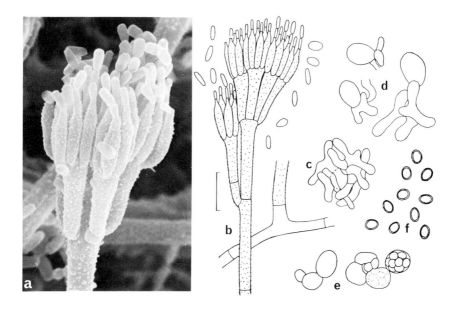

FIG. 333. *Talaromyces emersonii*. a. Penicillus, SEM, × 1500, orig. R. A. Samson; b. penicillus and conidia; c. primordium of ascoma consisting of profusely branching hyphae; d. initials; e. chains of asci; f. ascospores; orig. A. C. Stolk (from 5593).

T. *emersonii* is a ubiquitous and relatively common fungus (5593), recorded from the rhizosphere of *Cassia tora* (521) and *C. occidentalis* (6067), compost and soil in Italy, the Netherlands, the British Isles and California (5593), piles of wood chips in Sweden, Canada and the USA (5740), river banks in Japan (253, 3841), grassland, municipal waste, peat, coal spoil tips (1664, 1665, 3828) and air in Britain (1666) and India (5784), sugar cane bagasse, and palm oil kernels in Nigeria (5593), and blesbok dung in South Africa (1557).

Minimum temperature for growth is somewhat below 30°C, with the optimum at 45°C, and the maximum 55–60°C (1220, 1557, 5593). *T. emersonii* grows well and produces ascospores when incubated in sun-heated temperate soil (2699). — It degrades cellulose (5741) and wood of several pine species at a temperature optimum of 50°C (4248). Production of proteinases (3231) and formation of humic acid have been described (5240). *T. emersonii* is able to grow on a number of plasticizers which are then used as the sole C sources (3828).

Talaromyces flavus (Klöcker) Stolk & Samson *1972* var. *flavus* — Fig. 334.

≡ *Gymnoascus flavus* Klöcker *1902*
= *Talaromyces vermiculatus* (Dangeard) C. R. Benjamin *1955*
= *Arachniotus indicus* Chattopadhyay & Das Gupta *1959*

Anamorph: *Penicillium vermiculatum* Dangeard *1907* (but described inclusive of the teleomorph)

DESCRIPTIONS: Raper and Thom (4745), Udagawa and Takada (5984), Udagawa (5967), and Stolk and Samson (5593). — Colonies spreading broadly, reaching 7–8 cm diam in two weeks at 25°C on MEA. Ascomata forming a yellow or pinkish layer, reverse orange to orange-brown or purple. Ascomata 200–700 μm diam, confluent, ripening within two weeks; the initials consist of long vermiform ascogonia, 100–240 × 3·5–4·5 μm, with slender entwining antheridia; ascospores yellow, broadly ellipsoidal, spinulose, normally 3·5–5·0 × 2·5–3·2 μm, but 5–6·5 × 3·5–5·2 μm in var. *macrosporus* Stolk & Samson *1972*. Conidiophores sparsely developed, sometimes forming greyish green sectors; penicilli typically biverticillate symmetrical, all elements smooth-walled, conidia brownish green, subglobose to ellipsoidal, 2·2–3·5 × 2·0–2·5 μm. — Ascospore delimitation in young asci begins with the appearance of sheaths of endoplasmatic reticulum surrounding the nuclei together with adjacent cytoplasm; portions of the double membranes swell and form lomasomes which become the centres of new wall formation (6386). — DNA analysis gave a GC content of 49–57% (5600).

T. flavus is the commonest species of its genus and is frequently isolated not only from soil but also from other organic substrates. It appears commonly amongst the heat-resistant fungi from greenhouse and agricultural soils (614, 615). It has a worldwide distribution (5593) but is more commonly reported from warmer regions including Egypt (4745), equatorial West Africa (1420), Uganda (3839), South Africa (1555), India (2178, 2186, 3913, 4030, 5593, 5622), Pakistan (401), tropical Australia (977), New Guinea (5983), Japan (2532, 3267, 5846, 5984), Hawaii (3264), Central America (2031, 4745), Brazil (399), Argentina (489, 6445), New Zealand (4745), China (5593), Nepal (1826), the USA (2573, 3618, 3817, 4745), Canada (6352), Italy (4538), Spain (3446), France (2161), the Netherlands, Germany (5593) and Sweden (4745). It has been isolated from forest soils (2573, 3817) under *Acacia melanoxylon* (2186), *A. mollissima* (2178), *Eucalyptus maculata* (1555), and *Pinus* (6352), forest nurseries (2482), grassland (1555, 2573), a vineyard soil high in copper (4921), soil

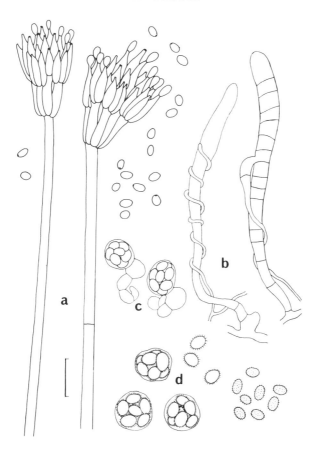

FIG. 334. *Talaromyces flavus*. a. Penicilli and conidia; b. development of ascogonia and antheridia; c. chains of asci; orig. A. C. Stolk (from 5593); d. mature asci and ascospores; a. and d. orig. J. Veerkamp.

fumigated with chloropicrin (4760), methyl bromide or ethylene dibromide (6287), paddy fields (3913, 5593), a mangrove swamp (3264) and seawater from the eulittoral zone (4918). It has also been isolated from leaf litter of *Eucalyptus maculata* (1558), wood baits (oak, lime) buried in surface soil (5237), roots of rice (1282), the rhizospheres of wheat (3567), groundnut (2532) and banana plants (2030, 2031), seeds of tomato (5899) and wheat (4492), freshly harvested pecans (2572), garden compost (1424), and cysts of the nematode *Globodera rostochiensis* (3090). — Growth *in vitro* can be inhibited by garlic extract (5744).

No growth occurs at 5°C but very good growth is obtained at 37°C, according to some (4578), but not all sources (1827); it survived a heat treatment of the soil at 70°C for 30 min (615) but ascospores were killed at 80°C for 40 min in apple juice; cleistothecia can withstand 100°C for 30 min (5502). Growth is strongly reduced at −100 bars water potential or less (3265, 4578); hair baits in soil are colonized in the range between saturation and −70 bars (977). — Degradation of starch (1827), cellulose (386, 1425, 3618, 4778) and laminarin by action of 1,3-β-D-glucanase (4778) is reported, proteinases can be formed (1425, 3231), and hydrocarbon from fuel oil utilized (4239). The mycelium contains mucilaginous polysaccharides (resembling luteic acid) (2049) and non-ascosporic isolates have been found to produce luteic acid from various carbohydrates and also malonyl-polyglucose and other laevorotatory polysaccharides, e.g. luteose (4745). — This fungus shows a marked antibiotic activity against bacteria and streptomycetes (615) and produces vermiculine (5173), an antibiotic with antibacterial and antiprotozoal properties (1850). Hyphae of *Rhizoctonia solani* can be parasitized (634), and the growth of *Rh. fragariae* (2622), *Neosartorya fischeri*, *Penicillium janthinellum*, and *Mucor plumbeus* (1526) can be inhibited by this fungus.

Talaromyces helicus (Raper & Fennell) C. R. Benjamin *1955* var. *helicus* — Fig. 335.

Anamorph: *Penicillium helicum* Raper & Fennell *1948* (but described inclusive of the teleomorph)

DESCRIPTIONS: Raper and Thom (4745), Stolk and Samson (5593), and Udagawa (5970). — Colonies spreading broadly, reaching 6 cm diam in 14 days at 25°C on MEA; ascomata forming a thin yellow layer, reverse yellow to pinkish buff; ascomata 100–300 μm diam, discrete or confluent, ripening within 1–2 weeks; initials consisting of thick, club-shaped ascogonia around which thin antheridia coil tightly, soon growing out into a large terminal coil from which the ascogenous hyphae originate. Ascospores ellipsoidal, delicately spinulose, 2·5–3·0 × 1·5–2·0 μm (ascospores smooth and 3·0–4·5 × 2·0–2·7 μm in var. *major* Stolk & Samson *1972*). — Conidiophores sparsely developed, slightly greenish in central areas of the colony; penicilli typically biverticillate symmetrical, with all elements slightly roughened; conidia pale greenish, ellipsoidal to subglobose, smooth-walled, 3·0–4·5 × 2·5–4·0 μm.

This species was originally isolated from soil in Sweden and then from compost in Switzerland and from soils in the Netherlands (1614, 1616); it has further been obtained from grassland in the USA (2573), salty soil, and from a wooden carving in France, various soils in Japan, Argentina (2012, 5593, 5970), in north-European podzols (3416), the British Isles (5429) and Sri Lanka (IMI).

No growth occurs at 5°C but good growth is obtained at 37°C; minimal growth at -100 bars water potential (4578). — Cellulolytic activity has been reported (3618, 4191).

Talaromyces trachyspermus (Shear) Stolk & Samson *1972* — Fig. 336.

≡ *Arachniotus trachyspermus* Shear *1902*
= *Talaromyces spiculisporus* (Lehman) C. R. Benjamin *1955*

Anamorph: *Penicillium spiculisporum* Lehman *1920* (but described inclusive of the teleomorph)

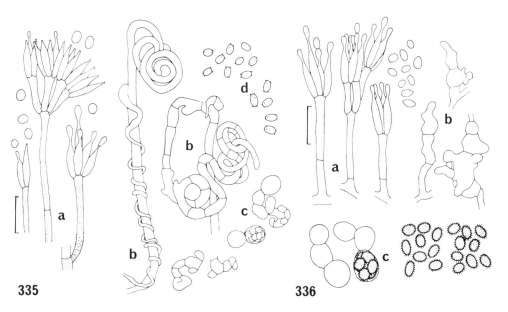

335 **336**

FIG. 335. *Talaromyces helicus.* a. Penicilli; b. ascogonium growing out with terminal coil, a later stage showing septation of the coiled part; c. chains of asci; d. ascospores; orig. A. C. Stolk (from 5593).
FIG. 336. *Talaromyces trachyspermus.* a. Penicilli and conidia; b. ascoma initials; c. chains of asci and ascospores; orig. R. A. Samson (from 5593).

DESCRIPTIONS: Raper and Thom (4745), Udagawa and Takada (5984), Udagawa (5967), and Stolk and Samson (5593). — Colonies reaching 3·5 cm diam in two weeks at 25°C on MEA, cream from the ascomata, with yellow reverse. Ascomata numerous, usually confluent, white to creamish, 50–350 μm diam, ripening within two weeks. Ascospores ellipsoidal, densely spinulose, 3–3·5 × 2·0–2·5 μm (4–4·5 × 2·8–3·2 μm in var. *macrocarpus* Wright & Loewenbaum *1973* (489)). Conidiophores usually sparsely developed, sometimes more abundant in the colony centre, greenish; conidiophores usually short, arising from aerial hyphae; penicilli biverticillate symmetrical, with all elements smooth-walled; conidia greenish, ellipsoidal, smooth-walled, 2·5–3·5 × 1·5–2·7 μm.

T. trachyspermus is a typical soil fungus with a wide distribution (4745). Soil isolations have been reported from the British Isles (IMI), Germany (5593), vineyards in France (242), flood-plain communities with maple, elm and ash in Wisconsin/USA (4226), deciduous and evergreen forests and grassland in Ohio/USA (2573), cultivated soils in Georgia/USA (3817), estuarine sediments in North Carolina (655), the rhizosphere of ferns near Darjeeling (2858), forest soil with *Acacia karroo* in South Africa (4407), *A. melanoxylon* (2186), teak (2853, 2854) and grassland (3863, 3865, 4933) in India, also known from Nepal (3840), Pakistan (401), Bangladesh, Uganda (IMI), Japan (2532), Tahiti (6084), and various soils in Argentina (489, 2012). It is also known from rape, cotton, pear, *Trichosanthes dioica, Terminalia chebula* (IMI), the rhizospheres of ferns (2858), wheat (4727), barley (3526), *Dichanthium annulatum* (3271), *Cynodon dactylon* (3864) and *Vaccinium macrocarpon* (5593), milled rice (5967), freshly harvested pecans (2572), rabbit dung (6085), bagasse and tanned hide (5429), and activated sewage sludge (1387).

No growth occurs at 5°C and very poor growth at 37°C (4578, 5967); growth is almost nil at −100 bars water potential (4578). — *T. trachyspermus* grows equally well on acetate and glucose (3329). Gallic and tannic acids can be utilized (6083). Laminarin is degraded by the action of 1,3-β-D-glucanase (4778). Citrate synthetase (3522, 3524), large amounts of ATP citrate lyase (3523), (−)decylcitric and (+)isocitric acids (680), spiculisporic, γ-ketopentadecoic (5278) and succinic acids (4745) are produced.

Talaromyces wortmannii (Klöcker) C. R. Benjamin *1955* — Fig. 337.

Anamorph: *Penicillium wortmannii* Klöcker *1903* (but described inclusive of the teleomorph)

DESCRIPTIONS: Raper and Thom (4745), Udagawa and Takada (5984), and Stolk and Samson (5593). — Colonies reaching 3–4 cm diam in 14 days at 25°C on MEA, bright yellow-orange from a basal layer of ascomata, reverse yellow to orange. Ascomata 75–300 μm diam, discrete or confluent, ripening within two weeks; ascospores ellipsoidal, spinulose, 3·7–4·7 × 2·5–3·5 μm. Conidiophores sparse, usually arising from submerged hyphae; penicilli typically biverticillate symmetrical, all elements smooth-walled; conidia hyaline to greenish, ellipsoidal or fusiform, smooth-walled to finely spinulose, 2·5–4·0 × 1·7–2·2 μm. — This species closely resembles *T. flavus* (q.v.) from which the essential distinguishing characters are the ascogonia which are thick and vermiform in *T. flavus*, and much shorter, little differentiated and of irregular shape in *T. wortmannii*; *T. wortmannii* also has a slower growth rate and more pointed conidia. — DNA analysis gave a GC content of 50% (5600).

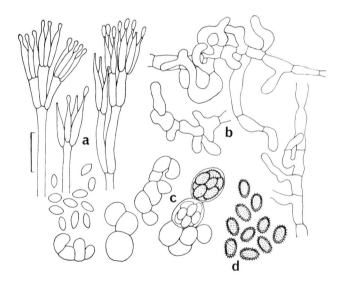

FIG. 337. *Talaromyces wortmannii*. a. Penicilli with conidia; b. ascoma initials; c. chains of asci; d. ascospores; orig. A. C. Stolk (from 5593).

T. wortmannii is the second most common soil-borne species of *Talaromyces* and has a worldwide distribution. It is easily recovered after brief soil steaming (6183). — In addition to temperate latitudes where it is recorded from the British Isles (745, 1376, 4745), Sweden (4745), Denmark (5593), Germany (1424), France (5593) and the USA (2007, 2573, 4226, 4745), it also occurs in Turkey (4245), Egypt (3993, 3997), Zaïre (3790), central Africa (3063), South Africa (4745), Sierra Leone, Nigeria, Tanzania, South Rhodesia, Trinidad (IMI), Japan (5846), New Guinea, the Solomon Islands (5983, 5984), Hong Kong (1021), Thailand (4745), Malaysia (6046), Sri Lanka (4021), India (3729, 3732, 4030, 4698, 5622), Central America (1697, 4745), Cuba, Brazil (4745), and Argentina (489, 2012). It has been isolated from uncultivated soil (4698), forest soils (432, 1021, 2007, 2008, 2573, 4225, 6046), gley, mull-like rendzina, rendzina (3413, 3414), agricultural soils (3732, 4538), bare and mulched soils of prairie stands (6347), grassland (2573, 4021), palm (6046) and rubber plantations (IMI), peat bogs (1376), desert soil (3997) and garden compost (1424). It is also known from decaying Brazil nuts (2515), decaying wood, rice roots (IMI), wheat rhizosphere (4727), seeds of celery, *Crotalaria juncea* (IMI) and wheat (4492), pecans (5092), plant parts of sisal, *Elaeis guineensis*, *Grevillea robusta*, *Trichosanthes dioica*, bagasse, cardboard (IMI), nests of free-living birds (2575), and the air in the Netherlands (5593).

No growth occurs at either 5°C or 37°C, and growth is very reduced at −100 bars water potential (4578); hair baits are colonized in the soil in the range between saturation and −40 bars (2095). — Cellulose is not attacked (3618, 6326) though 1,3-β-D-glucanase (4778) and 1,4-β-mannanase (4781) are formed. In one study it was found to be the strongest producer of β-D-xylopyranosidase of 11 fungi tested (4777). Nitrite can be used as an N source (10) and proteinases are produced (3231). The mycelium contains mucilaginous polysaccharides resembling luteic acid (2049). Metabolic products reported are the viridin-related wortmannin with antifungal but not antibacterial properties (711, 3508), wortmin (3765), flavomannin (226), the anthraquinones skyrin and rugulosin (692, 5278) and a chemically still undefined antibiotic compound (382, 383). *T. wortmannii* is, however, reported to have also some antibacterial activity (515). The herbicide metobromuron is degraded by this species (5957).

Tetracladium De Wildem. *1893*

Tetracladium setigerum (Grove) Ingold *1942* — Fig. 338.

≡ *Tridentaria setigera* Grove *1912*

DESCRIPTIONS: Ingold (2647, 2649), Petersen (4521), Nilsson (4190), Tubaki (5925), Dudka (1485), Dudka and Isachenko (1486), and Watanabe (6214).

In species of *Tetracladium* (type species *T. marchalianum* De Wildem.) several solitary tetraradiate blastoconidia are produced in sympodial succession at the ends of the conidiophores (2647). Conidia consist of 2–3 divergent tapering arms with a similarly tapering basal stipe and 2–3 central broader processes.

Colonies of *T. setigerum* are slow-growing, reaching 3·2–4·8 cm diam in 20 days at 20°C on MEA or OA, pale yellow, moist, finely granular, and often form clusters of thin-walled, swollen cells 5–8 μm diam. Moderate sporulation occurs on 2% MEA, and more abundant sporulation after transfer to water. The conidia possess three finger-like septate processes in one plane, 12–15 × 3–5 μm, one of which arises as a basal branch from another, and three setose arms, 20–40 × 3 μm, one of which stands out from the conidial plane. — The equally common *T. marchalianum* De Wildem. *1893* has only two spherical knobs instead of the three finger-like processes; *T. maxilliforme* (Rostr.) Ingold *1942* has only two finger-like processes and mostly no third setose arm. — The fatty acid content of *T. setigerum* has been determined *in vitro* (1487).

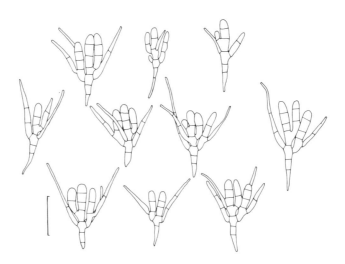

FIG. 338. *Tetracladium setigerum*, "tetraradiate" conidia.

T. setigerum was found to be the most common aquatic hyphomycete in California (4732) and the second after *T. marchalianum* in Sweden (4190) where it occurred also in forest litter. Its worldwide distribution as an aquatic fungus, particularly in temperate and cold latitudes, has been documented for many European countries (184, 185, 1483, 1484, 1614, 1889, 2338, 3655, 4189, 4190), Africa, Canada (274, 275, 321), the USA (1128, 1129, 4521, 4732, 5621, 6411), Jamaica, Australia (5930), New Guinea (1485, 4190, 4521) and Japan (3886, 5925, 5927); most observations are from foam of running water. Its terrestrial occurrence has repeatedly been reported and it has been isolated from decaying leaves which have never been in contact with open water, after submerging them in distilled water (321), or from maple leaves transferred into running water (274). Conidia of this species occurred in large quantities in a ditch filled with beech foliage close to wheatfields from which the species was very regularly isolated (1889). It has also been isolated from soil under potatoes (1616), sugar beet or grassland (1614), and wheat culms (4791). A very suitable substrate are rotting alder leaves on which it has been observed mainly in summer and autumn in the British Isles (2647), or other leaves in Sweden, particularly in autumn and winter (4188); it has also been trapped on twigs exposed to a freshwater stream (6384). In Japan it was found to be associated with roots of a *Gentiana* species and strawberries (6214). It was able to survive on dried leaves for a period of 60 days (5785).

The optimal temperature for the production of conidia is 10°C, for mycelial growth 20°C, and for germination 10–25°C (832, 1486, 3105). Production of conidia was inhibited on media with raised salinities (833) but a high percentage of germination (90%) of conidia occurred even in 100% seawater at temperatures of 20° and 25°C (832). — Most fresh isolates sporulate well on 1% MEA but not on OA, while sporulation can be strongly enhanced by transfer to water, particularly with aeration (6242). A malt-yeast extract-peptone medium has been recommended for its isolation and cultivation (321). B-vitamins and yeast extract were both found to stimulate the growth of *T. marchalianum* (4190). The utilization of D-glucose, D-fructose, sucrose, maltose, D-xylose, cellobiose and (to a lesser extent) starch by *T. setigerum* is generally very good (5805). The decomposition of pectin can be demonstrated although this ability is slight compared to other fungi (1432). *T. setigerum* does not grow well on ammonium sulphate; after growth on either L-glutamic acid or L-tyrosine the following amino acids were found in the mycelium: L-aspartic acid, L-alanine, L-serine, L-arginine, L-histidine, L-hydroxyproline, L-cysteine, L-cystine, L-methionine, L-valine, L-threonine, and γ-aminobutyric acid (5807). DDT at a concentration of $60 \mu g$ ml^{-1} led to an increased growth in liquid cultures (2488). — *Chlorella pyrenoidosa* is inhibited *in vitro* by *T. setigerum*.

Thamnidium Link ex Gray *1821*

Thamnidium elegans Link ex Gray *1821* — Fig. 339.

= *Thamnidium verticillatum* van Tiegh. *1876*

DESCRIPTIONS: Hesseltine and Anderson (2426), Zycha and Siepmann (6588), and Milko (3807). — The family Thamnidiaceae Lendner with 13 recognized genera has been revised and redefined by Benny and Benjamin (454): Sporophores erect or ascending, simple or branched, in some genera producing large terminal columellate multi-spored sporangia with a deliquescent wall; pedicellate, 1- or few-spored sporangioles with a persistent but separable wall always present; sporangia and sporangiola apophysate or non-apophysate; sporangiospores thin- and smooth-walled. Zygospores dark and roughened, borne between equal suspensors. The genus *Thamnidium* was monographed by Hesseltine and Anderson (2426) who recognized four species; in subsequent keys six (3806), four (3807), or five species have been recognized (6588). All but one of these are now excluded (454) and the genus consequently remains monotypic.

Colonies reaching 5 cm diam in 13–19 days at 20°C on synthetic *Mucor* agar and other media, 1·5–2 cm high, at first pale olive-grey, later becoming darker olive-grey with a similar reverse. Odour yeast-like. Sporangiophores erect, without rhizoids, to $32\,\mu$m wide, often reaching the lid of the Petri dish. Typical sporangiophores bearing one large yellow-olive, columellate, rapidly deliquescing sporangium, 40-$150\,\mu$m diam at the tip and several whorls of three-dimensionally dichotomous narrow branches with a 2–6-spored, greyish sporangiole on each tip, 10–$16\,\mu$m diam, with the wall breaking at maturity; both terminal sporangia and sporangioles can also occur separately on different sporangiophores. Sporangiospores of both terminal sporangia and sporangioles of similar size, oval to ellipsoidal, $6·5$–$15·5 \times 4·5$–$7·5\,\mu$m. Heterothallic; zygospores produced between compatible isolates on PDA or YpSs agar at 6–7°C (not 20°C) in the aerial mycelium, black, globose, warted, *Mucor*-like, mostly 80–$125\,\mu$m diam (2426). — The size of the sporangioles decreases with increasing temperature (5789). In a cytological study, of several substances tested only a treatment with 0·1% caffeine (for 24 h) resulted in an enlargement and dispersal of chromatids (4805); mitochondria, acid phosphatase and lipids were found in greatest amounts at sites of differentiation, i.e. sporangiophore tips, sporangia and sporangioles (4807). Protoplast cleavage and further changes occurring during sporangiospore formation were followed by TEM studies (1774). Nuclei are mostly equally distributed in the cytoplasm of the hyphae, but concentrations have been observed below the cross-walls of sporangiophores and in the tip of developing side-branches (4811). — Starch-gel zymography for ten different enzymes was used to elucidate the inter- and intraspecific relationships within five isolates of *Th. elegans* and two of *Th. anomalum* Hesselt. & P. Anderson *1956* (≡*Ellisomyces anomalus* (Hesselt. & P. Anderson) Benny & R. K. Benjamin *1975*); the enzyme patterns were almost identical within a single species but marked interspecific differences occurred (5605). — DNA analysis gave GC contents between 37 and 60·5% for different isolates (5600).

FIG. 339. *Thamnidium elegans*, sporangiophore with dichotomous branches bearing sporangioles, SEM, × 400, orig. R. A. Samson.

Thamnidium elegans is a very common psychrophilous fungus in temperate zones. It has been found in high latitudes in Finland (1712, 1713) and Norway (2206), repeatedly in the USA and Canada (2426), and various European countries (6588). Reports from more southerly countries are scarcer but include Israel (2762, 2764), the Sudan (4222) and India (3731); it has, however, been isolated relatively frequently from desert soils in the Sahara (2973, 2974). It is equally widespread in forest (269, 1713, 3184, 3416, 5663) and cultivated soils (2740, 2762, 2764, 4788), and has also been found in activated sludge (1387), moorland (1259) and terra fusca (3418). Its pH range in forest soils is given as 5–6·8 (1713). It is very often coprophilic (3358), and occurs more rarely on coarse fodder (4548), also on seeds of oat (IMI), wheat (4492) and other cereals, and flax (3556), in air, on cold-stored meat, leaf mould (509, 2426), and nests, feathers and pellets of free-living birds (2575). Isolations from the wheat rhizosphere have been reported (3567) but it occurred rarely in wheatfield soils in Germany. — The fungus can be parasitized *in vitro* by the mycoparasites *Dimargaris verticillata* (3560) and *Syncephalis californica* (2618).

The positive phototropic reaction of the sporangiophores has been investigated (3474). Sporangium formation is stimulated by light, wavelengths in the yellow and red ranges being most effective (3285). — Adaptation to low temperatures, with good growth still occurring at 1–2°C, has been observed (740); the temperature optimum is 18°C, and the maximum 27°C (5789); no growth occurs at 31°C (2426). — *Th. elegans* can utilize acetate just as well as glucose (3329); phenylamide herbicides can be metabolized by oxidative transformations (5824); β-glucosidase (6159) and alcohol dehydrogenase (1999) are produced. In cultures grown on a corn steep-glucose medium relatively large amounts of ethylene were detected (2641). No thiamine is required (4860). — Mycelial extracts were found to have a protistocidal effect on *Paramecium caudatum* (1456), and a substance antibiotic to *Mycobacterium phlei* has been recorded (3812). After soil inoculation with *Th. elegans,* the production of green matter in tomato plants was significantly reduced (1427).

Thanatephorus Donk 1956

Thanatephorus cucumeris (Frank) Donk 1956 — Fig. 340.

≡ *Hypochnus cucumeris* Frank *1883*
= *Corticium solani* (Prill. & Delacr.) Bourd. & Galzin *1911*
  ≡ *Hypochnus solani* Prill. & Delacr. *1891*
= *Pellicularia filamentosa* (Pat.) Rogers *1943*
  ≡ *Hypochnus filamentosus* Pat. *1891*
  ≡ *Ceratobasidium filamentosum* (Pat.) Olive *1957*

For further synonyms see Talbot (5709).

Anamorph (mycelium): *Rhizoctonia solani* Kühn *1858*

? = *Moniliopsis aderholdii* Ruhl. *1908*
  ≡ *Rhizoctonia aderholdii* (Ruhl.) Marchionatto *1946*

For further synonyms see Parmeter and Whitney (4455).

DESCRIPTIONS: TELEOMORPH: Warcup and Talbot (6188), Talbot (5709), Tu and Kimbrough (5921), CMI Descriptions No. 406, *1974*, Papavizas (4382), and Talbot (5708) also with key to genera and species; ANAMORPH: Parmeter and Whitney (4455), Butler and Bracker (816), Richter and Schneider (4835), and Watanabe and Shiyomi (6215). — Colonies fast-growing, usually filling a Petri dish within three days, at first colourless, submerged or with some radiating aerial hyphae, rapidly becoming brown. Hyphae subhyaline to pale brown, thick-walled, cells about 100–250 μm long, (5–)7–12(–17) μm wide, often constricted near the septa and where branching, with conspicuous dolipore septa (visible in light microscopy when stained in aniline blue or IKI), lacking clamp connections, branched at wide angles (nearly 90°), often anastomosing. Monilioid hyphae to 30 μm wide commonly present. Hyphal cells contain (2–)4–8(–18) nuclei (5040, 5242, 5922), which contrast well in aniline blue in glycerol, aceto-orcein, or under phase contrast. — Sclerotia developing from irregular agglomerations of uniform moniliform hyphae, irregular in outline, solitary and about 1 mm diam, sometimes confluent to conspicuous crusts, often in zones, particularly near the colony margin, soon turning brown. — Basidial hymenia appearing on stems or leaves just above the soil or on soil particles. They can be induced *in vitro* by covering an agar colony with partially sterilized soil (1764, 1768, 5342, 5613, 5919, 5923), adjusted to 30% moisture content and pH 8–9, and exposed to sunlight (5919), or by transfer from rich to poor nutrient media, such as SEA (4382, 5272, 6338). Fructifications resupinate, creamy, effuse, loosely attached to the substratum. Basidia arising in asymmetrical cymes or racemes from tufts of ascending hyphae (discontinuous hymenium), variable in shape, barrel-shaped to cylindrical, 10–25 × 6–12 μm, with (2–)4(–7) stout, straight sterigmata as long as or longer than the metabasidia. Basidiospores oblong to broadly ellipsoidal, unilaterally flattened, apiculate at the base, hyaline, smooth-walled, not amyloid, 6–14 × 4–8 μm; germinating by repetition with secondary spores

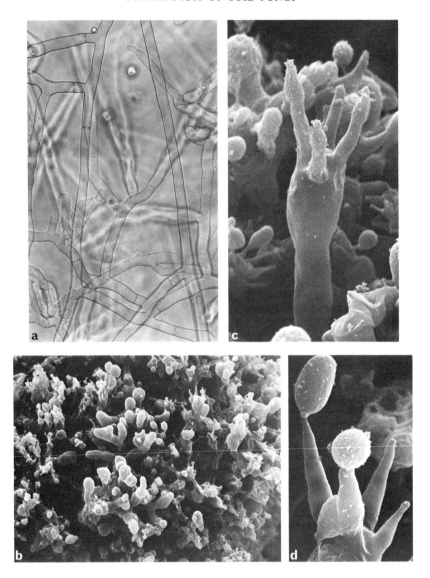

FIG. 340. *Thanatephorus cucumeris*. a. Mycelium consisting of wide, rectangularly branched hyphae, NIC, × 400; b. fluffy basidial hymenium produced by the soil-over technique, SEM, × 500; c. a single basidium, SEM, × 2000; d. sterigmata with two attached basidiospores, SEM, × 2500.

about half the size of the primary ones. The basidiospores are mostly uninucleate but sometimes to 35% may be binucleate (1767).

In the genus *Rhizoctonia* DC. ex Mérat *1821* (type species *Rh. crocorum* DC. ex Mérat) over 100 often unrelated types of sterile mycelia have been described which mostly support irregular sclerotia. The best known species, *Rh. solani* has been used to characterize the genus with wide, somewhat pigmented often monilioid hyphae and irregular sclerotia.

The genus *Thanatephorus* belongs to the Tulasnellaceae Juel emend. Talbot and is intermediate between the classical concepts of Homo- and Heterobasidiomycetes. The generic names *Corticium* Fr. *1828* and *Hypochnus* Fr. *1829* (nomina ambigua), and also *Pellicularia* Cooke *1876* (nomen confusum) cannot be taken up. *Ceratobasidium* Rogers *1935* now contains several species very similar to *Th. cucumeris* but which have binucleate hyphal cells and generally no sclerotia (4456, 5923). The closest relatives of this species are, however, *Waitea circinata* Warcup & Talbot *1962* and *Aquathanatephorus pendulus* Tu & Kimbr. *1978* which both have similar, multinucleate *Rhizoctonia*-type hyphae; in *Waitea*, isolated from wheatfield soils, the septa are more distant from the point of origin of lateral branches and the sclerotia are yellow-orange and differentiated into large central cells and 3–5 layers of elongate peripheral cells; in *Aquathanatephorus*, isolated from *Eichhornia crassipes*, the sclerotia are enveloped by a weft of thin-walled hyphae.

In common with the mycelial state named *Rhizoctonia solani* (see above), *Thanatephorus cucumeris* is regarded as a vaguely delimited species aggregate (5709). The combination of fast growth, monilioid hyphae, brown pigmentation, conspicuous dolipores, branching close to hyphal septa in young hyphae and multinucleate cells are considered as sufficient for the reliable identification of plant-pathogenic or soil isolates as *Th. cucumeris*. This name has consequently been adopted in the present compilation, but it should be remembered that some observations, particularly older ones, may refer to different taxa. *Thanatephorus* now contains about six species: *Th. orchidicola* Warcup & Talbot *1966,* isolated from *Orchis mascula*, with wider basal hyphae (frequently 15–17 μm wide), and fawn-coloured, obpyriform to obovate basidiospores, 9–12 × 7–9·5 μm; *Th. sterigmaticus* (Bourd.) Talbot *1965,* known from soil in France, with two sterigmata per basidium and elongate basidiospores, 15–18 × 5–7 μm. The similar *Th. corchori* Tu *et al. 1977*, known from jute in Taiwan, has mostly four sterigmata per basidium and basidiospores 5–12·5 × 3–6·5 μm. The following two species have recently again been removed from the synonymy of *Th. cucumeris* (5921): *Th. praticola* (Kotila) Flentje *1963* with slightly narrower hyphae, 5–12 μm wide, cells with fewer nuclei (mostly 4), more compact sclerotia, and basidia with longer sterigmata which commonly exceed 22 μm (4382, 5921), and *Th. sasakii* (Shirai) Tu & Kimbrough *1978*, known as a rice pathogen, which has slightly wider hyphae (5–14 μm) and sclerotia differentiated into a dark brown outer layer of 3–4 cells forming a weakly differentiated rind. — A further similar fungus is *Rhizoctonia cerealis* van der Hoeven apud Boerema & Verhoeven *1977*, the causal agent of sharp eyespot in cereals, which can be distinguished from *Th. cucumeris* by its slower growth (0·5–0·7 cm daily radial extension at 23°C on PDA), binucleate hyphal cells, and sclerotia which are white to yellowish at first and only become brown in old cultures.

The growth habit of *Th. cucumeris* is very variable. A limited number of groups can, however, be distinguished according to their sclerotium size and pigmentation (particularly on CzA) and their ability to anastomose with other strains (4250, 4251, 4252, 4403, 4454, 4835, 5270, 5271). The failure to form anastomoses between two isolates cannot, however, be taken as a criterion for species delimitation as some sibling single-spore isolates have been found not to anastomose (4455). When tested against each other, a high proportion of isolates showed "aversion reactions" depending on the specific genetic constitution (1516). Single-spore isolates showed either homothallism or bipolar compatibility (5612, 6339). Heterokaryons are readily obtained on PDA with 1% charcoal (815); their formation is controlled by two closely linked genes which together are referred to as the H factor (138). Field isolates (natural heterokaryons) have been paired with either each other or with homokaryons (from single basidiospores); the new heterokaryons were often morphologically and pathogenically different from their parent strains (609, 1768); detailed reviews of variations in the pathogenicity of *Th. cucumeris* have been published (298, 1768, 4403). Attempts to distinguish between formae speciales according to the spectrum of potential host plants are consequently of little value. Attention has also been drawn to the importance of true mutations in this connection (1766).

The fine structure of the hyphae (816, 4082, 5922) and sclerotia (4081) has repeatedly been investigated; the hyphal walls and septa show a multilayered structure; the dolipores are particularly thick and surrounded by a perforate pore cap consisting of two unit-membranes with electron-dense material in between them (3371, 5922); amongst the cell organelles, lamellar stacks of double membranes and whorled structures composed of endoplasmatic reticulum have been observed (816); karyogamy and meiosis have also been studied in TEM (5922). The nuclei contain six pairs of chromosomes (2341, 5922). — Hyphal walls were found to give a weak cellulose reaction ($IKI + H_2SO_4$), particularly in the sclerotia, and lipids were also concentrated in the sclerotia (5920); the cytoplasm gives a dextrinoid reaction with IKI which also facilitates the observation of the dolipores (5920). — Cell wall hydrolysates were found to contain 31·6% neutral sugars, 22·4% glucose (3232), and in addition mannose, xylose and uronic acid (2802). The significance of cytochemical reactions as additional diagnostic characters has been investigated (5920). Some isolates have been found to contain mycoviruses (674). The rate of RNA synthesis (1297), number of polyribosomes, protein synthesis (4830), soluble amino nitrogen, and ergosterol content (2054) are all markedly reduced in older mycelium.

Th. cucumeris is a polyphagous plant pathogen which attacks various parts of many plant species near the soil surface (298). Its greatest importance in temperate regions is in attacks of potatoes. Weeds can contribute to its transmission (2090). — It has a worldwide distribution and is the subject of numerous phytopathological investigations (CMI Descriptions No. 406). *Th. cucumeris* spreads in soil with a typical *Rhizoctonia* growth pattern (1266, 5886): very diffuse, fast-growing and not sporulating, but forming sclerotia. For this reason the species is rarely isolated from soil by the usual techniques. The majority of phytopathological observations are based on isolates from hypocotyls and roots. Owing to its rapid growth, both "screened immersion tubes" (982), especially if improved by the addition of radish seedling exudate (3647), and "screened immersion plates" are suitable for isolating it from soil (5810, 5812). For special purposes, bait techniques are the most favourable, among which buckwheat (4395), flax (6000), jute (6541), *Juncus* stems (1448, 1449) and filter paper discs soaked with nutrients and selective inhibitors (2414) have proved successful. Selective agar media (3056, 5762) and a wet screening through a 0·35 mm sieve, dispersing and planting the residue on water agar (6261) have also been devised. A sieving and flotation technique, using 2% aqueous solution of hydrogen peroxide, has also been developed for the extraction of sclerotia from soil (5999). Several reviews and comparisons of the available isolation techniques exist (1434, 5342, 5457). — The ecology of *Th. cucumeris* in a particular field soil has been studied in detail (4389) and an indirect immunofluorescence technique developed for identification of its mycelium in soil (3581). In general it is more frequent in grassland and arable soils than in other habitats, and more frequent in cultivated than fallow soils. It is usually not found on dead organic substrates, but was frequently reported from the shells and kernels of groundnuts (346, 1984, 2253, 3702). *Th. cucumeris* is one of the endomycorrhizal partners of orchids (87, 6189). In heavy soils it spreads horizontally more quickly than vertically (548), something which may be connected with its relatively low CO_2 tolerance (cf. below); aeration and low soil humidity favour its spread (549, 550). With an impression slide technique, the highest mycelial density was found to be at depths of 10–15 cm (5691). At 26°C and an inoculum density of only 0·76 propagules g^{-1} soil, pre-emergence damping-off of radish was most severe (457); occurrence and pathogenicity were apparently not influenced by soil pH or only in conjunction with high N manuring (389, 458, 561, 1286). Injuries are often less at high soil humidity

(388), something probably connected with competing soil saprophytes (4689). Low temperatures generally favour seedling infection. Very many experiments to control *Th. cucumeris* infestations of soil have been carried out, including crop rotation, resistance breeding, chemical treatments, green manuring and other organic amendments. In plots supplied with farmyard manure, a lower degree of infection was observed in *Brassica napus* (and a higher content of soil micro-organisms) than in plots manured with mineral fertilizer (4797, 5180). Repeated cultivation of the same crops can lead to the accumulation of specialized strains, though there are numerous exceptions (1765), and any increase in inoculum density depends very much on the soil texture (2415). Under oats, rye and rye-grass a reduction of the fungus was observed (1449). — Volatile compounds from decomposing plant residues, of which ammonia was a major component, stimulated mycelial growth and pigmentation (3311, 3314). Root exudates from radish (2942, 2943), soya bean (6458), and particularly tomatoes and potatoes (2417), were found to promote the saprophytic growth of this fungus; owing to their lack of specificity, however, these exudates cannot account for its host specificity (5325). Autoclaved baits are initially more easily colonized than living bean tissue; after four days burial, however, this proportion is reversed (5457). — For active survival, *Th. cucumeris* is largely dependent on its parasitic mode of life (4575, 5038, 5997); moreover, the duration of survival is very much dependent on the timely formation of sclerotia. Sclerotia can survive for several months (4575) or, when mature, even for some years (2418, 2419); the periods of survival also depend on the competitive saprophytic ability of the particular isolate (4078). Young hyphal tips dried at room temperature were found to survive for about one year, longevity decreasing with both age and higher temperatures used in drying (408). Surprisingly, in some cases the survival of monobasidial cultures proved to be better than that of the heterokaryotic parent strains (4285), but the competitive saprophytic ability of this species is low in comparison with many other root parasites (4734, 4961). The frequency in soil of *Th. cucumeris* was found to decrease with both increasing humidity (cf. above) and increasing available carbon (glucose, maltose, dextran, xylan, cellulose, starch), whereas it increased with increasing nitrogen (4397, 4401); $NaNO_3$ and NH_4NO_3 are particularly effective when a C/N ratio of 30 is reached (4386). The number of propagules surviving in soil for 8–80 days was found to be directly proportional to the amount of inoculum (458). A major reduction in survival of *Th. cucumeris* in buried sugar beet occurred between late April and early June (2415). Its vertical distribution has been reported to be favoured by the presence of KNO_2, which the fungus is able to utilize (2904). Low C/N ratios are favourable for the formation of sclerotia (113). — The fungus is very sensitive to antagonistic influences from other micro-organisms (1431, 3489), which may cause growth inhibition *in vitro* (numerous reports), localized lysis of hyphae (3386, 3396, 4079, 4380, 4604) by streptomycetes (4275) or bacteria (2578) or a direct parasitic action by *Talaromyces flavus*, *Trichoderma viride*, *Fusarium oxysporum* f.sp. *lini*, or *Papulaspora stoveri* (634, 813, 1511, 4080, 6204). Different soils differ greatly in their antagonistic efficiency against *Th. cucumeris* (300); the phenomenon of *Thanatephorus*-suppressive soils is not yet fully understood (299). Its means of survival have been reviewed (4387). — In reports of biological control, seed inoculations of wheat or carrots with *Streptomyces griseus* and isolates of *Bacillus subtilis*, significant yield increases occurred (3771, 3772), but when soil was inoculated with antagonists, negative or insignificant results predominate (6434), for example with *Penicillium expansum* (568), *Talaromyces flavus* and *Trichoderma viride* (109, 634, 2698), although some Russian authors have reported positive results (20, 2678, 5184, 5185). Short-term success in the control of this species was achieved by the addition of cornmeal (589, 5038). The addition of other organic substances (green manure, straw, etc.) generally causes a reduction in injury when the C/N ratio is high (cf. above) (4380, 5462); this is ascribed to a stimulation of microbial activity (1286, 1288, 4394,

4396) and an increase of antagonists (4253, 5456) in the root-free soil and the rhizosphere. Good results have also been obtained by the addition of cellulose powder, oat straw, soya bean hay with or without ammonium nitrate (4404), and chitin (2395, 5455, 5456); *n*-butanol extracts of chitin-treated soils proved highly effective in suppressing growth and pathogenicity (5453). The indirect actions of adding rice bran and oak sawdust to soil were only slightly less effective than direct control with PCNB or other fungicides (5442). — A synergistic increase in the damage to potato tubers by *Th. cucumeris* and *Fusarium solani* has been reported (1573, 1574). An accumulation of the nematode *Aphelenchus avenae* in the soil (344) reduced the injury by *Th. cucumeris* in several crop plants (344, 3037) and the fungus was found to be consumed by this nematode (344, 3579, 3580). Conversely, according to numerous reports, *Meloidogyne hapla, M. javanica* (5075, 6457), *M. incognita* (4825), *Pratylenchus* spp. (443), *Globodera rostochiensis* (1497, 1978) and *Heterodera avenae* (3728) all increased the attack of *Th. cucumeris* in various cultivated plants.

Factors determining the formation of sclerotia in *Th. cucumeris* have been thoroughly reviewed (5272). Sclerotium production *in vitro* is favoured by light (1503); even under continuous darkness an endogenous rhythm of zonation was observed (5996). — The temperature optimum for saprophytic growth is between (20–)24 and 28(–30)°C, the mean range of many measurements, while 16–25°C is given as the range for maximum injury to host plants. The temperature requirements are apparently very variable (4455, 4950); the thermal death point was determined as between 50°C and 52·5°C for 5 min (613, 5272). Optimum growth *in vitro* is in the pH range 4·5–6·5 (1574, 2834); many isolates can grow well at pH 3 or pH 8 or above (5272). A reduction of the water potential from −10 to −60 bars decreased the growth rate on agar from 7·5 to 0 mm/day (1472). Relationships between temperature requirements, growth rate and other cultural properties *in vitro* and the degree of pathogenicity have often been assumed and investigated (81, 345, 5998). — The growth of *Th. cucumeris* on various C sources (5272), enzyme activities of glycolysis, the Krebs cycle, the pentosephosphate pathway, and the electron transport systems (5855), have all been reviewed. The uptake of glucose over the entire surface of young hyphae has been demonstrated (2992). Differences in growth rate between isolates are thought to be correlated to different rates of nutrient uptake (3336). In contrast to many other fungi, *Th. cucumeris* can utilize L-arabinose (3328) and, deviating from the effects of other sugars, the utilization of arabinose was not correlated with a reduction in disease development (6262). Rhamnose was utilized mainly by highly pathogenic isolates (81); sucrose is presumably taken up only after hydrolysis (5851, 5852); fructose, maltose and starch are also used (2690). Pectin lyase, pectin methylesterase, polygalacturonase and polygalacturonate transeliminase have all been shown to be present (numerous investigations), partly with considerable differences between basidiospore isolates (4391, 5268). The sequential formation of galacturonase, cellulase (C_x) and pectin lyase has been studied (3372). Considerable portions of polygalacturonase and cellulase were found to be bound in the cell wall (3370). The formation of pectolytic enzymes is constitutive (2180). Pathogenicity was significantly correlated with the production of polygalacturonase in 22 isolates tested, but not, however, with pectin methylesterase or cellulase (345). *Th. cucumeris* can grow on D-galacturonic or D-glucuronic acid as sole C source (4749). — Its capacity for cellulose decomposition *in vitro* is, according to numerous reports, very strong, but there are considerable differences between isolates; the enzymes involved in cellulose degradation have been demonstrated (391). *Th. cucumeris* growing on buried cellophane film causes superficial etching (1266, 5886). On filter paper its cellulolytic activity is localized around terminal, short, branched hyphae (2676). *Th. cucumeris* can successfully compete in soil with other microorganisms in utilizing pure cellulose (1908), though not straw (4575). — Its capacity for cutin decomposition has been demonstrated (3361) and it can also grow on lignin sulphonate (1425). Phenol oxidases gave positive reactions with benzidine and pyrocatechol (934). The herbicides alachlor (5418) and chlorbromuron, as well as other phenylurea derivatives (6258), the insecticide aldicarb (5501) and the fungicide chloroneb (2475) can be metabolized *in vitro*. Growth on succinic acid as a C source is good, and on tartaric acid moderate (4914); a number of α-keto-acids were shown to be decarboxylated by an enzyme found in cell-free extracts but not in mitochondria (4681). Protein synthesis was increased when grown on fumaric or succinic acid in the presence of KNO_3 or NH_4Cl (937); older peripheral hyphae were shown to have an impaired mechanism of protein synthesis (4244), a lower uptake of O_2 and amino acids and lower phenylalanine synthesis (5407). Organic N compounds and nitrates are better N sources than ammonium salts, for which alterations of the pH in the nutrient solutions are probably responsible (1353, 4737). Ca (6508) and Mo (5554) ions are indispensable for growth, while Fe, Zn, Cu and Mn ions further stimulate it (1258, 4914, 5554, 6508). Transport of ^{32}P in the mycelium has been investigated (3912) and recognized as an active metabolic process (3373). Heterotrophy for thiamine, biotin and inositol (1574) is apparently a character of isolates rather than the species (4914). By the UV treatment of basidiospores, mutants have been induced which require either nicotinic acid or nitrite; both were fertile (1411). Cultivation on media containing biotin or thiamine is said to reduce pathogenicity (5341). — *Th. cucumeris* is strongly inhibited by high (10–30% v/v) CO_2 concentrations (3880); saprophytic soil isolates show a higher CO_2 tolerance than pathogenic ones obtained from plant substrates near the soil surface (1504, 2096, 4402). According to their reactions to growth with 20% CO_2 compared with air, a number of isolates were arranged in three reaction groups (4403). — After exposure to $^{35}SO_2$, the labelled sulphur became distributed through the entire fungal colony (1210). A considerable tolerance of PCNB seems to have ecological significance, since it is found to be correlated with pathogenicity; this tolerance can be generally augmented by PCNB treatment of fields (5241). Mycelial mats accumulate PCNB, DDT and dieldrin from treated soil suspensions (3057); PCNB is reduced to penta-chloroaniline (4100). *Th. cucumeris* releases amino acids into the medium (3862, 4254); qualitative difference in this exist between pathogenic and non-pathogenic isolates (4764). The release of amino acids and B vitamins also plays a role in orchid nutrition (2291, 2292). Further metabolic products are 2-furyl-hydroxymethyl ketone, rhizosolaniol (1471), 3-hydroxybenzoic acid, 3-nitro-4-hydroxyphenylacetic acid and *O*-glucosyl-2-nitrophenol (3231). — In the soil (cellophane bag technique) and *in vitro*, metabolic products can cause necrosis on the roots of various higher plants without any mycelial contact (4147, 6457, 6458); a substance with phenolic and glycosidic properties (5275), as well as various phenylacetic compounds (154) have been found in such situations. The production of phenolic compounds was found to be correlated with pathogenicity (4765, 4766), and a phytotoxin of low molecular weight has been studied in some detail (1816). A slight gibberellin activity has been reported (237). *Th. cucumeris* can also occur as an active parasite of Mucorales and Peronosporales (813).

Thelebolus Tode ex Fr. *1823* emend. Kimbr. *1967*

= *Ryparobius* Boud. *1869*
= *Ascophanus* Boud. *1869*
= *Pezizula* Karst. *1871*

Type species: *Thelebolus stercoreus* Tode ex Fr.

Contributions to a MONOGRAPHIC TREATMENT: Kimbrough (2979), Kimbrough and Korf (2980), and Otani and Kanzawa (4323).

The family Thelebolaceae (van Brummelen) Eckblad of the Pezizales (considered as a tribe or subfamily within the Ascobolaceae Sacc. or Pyronemataceae Corda by some authors), is distinguished by inamyloid asci and smooth- and thick-walled, hyaline, and non-guttulate ascospores which are often produced in large numbers within the ascus.

Thelebolus species have rather fast-growing colonies, reaching 6–7 cm diam in 14 days at 15°C on Leonian agar or other media, pink to orange, with little aerial mycelium. Ascoma initials consisting of coiled or contorted ascogonia, about 3–5 μm wide (6348). Ascomata cleistothecium-like, generally 100–250 μm diam, immersed or semi-immersed. Asci few in number, ovate to cylindrical, thick-walled, apex inamyloid, not staining in congo red above a thickened ring, rupturing the ascoma at maturity and opening with an irregular tear as the ascospores are discharged simultaneously. Ascospores less than 10 μm long, 8-2500 and more per ascus, smooth- and rather thick-walled, without de Bary bubbles.

Similar genera with rounded, pale ochraceous apothecia and a small number of many-spored asci include *Thecotheus* Boud. *1918* (Pezizaceae) with amyloid operculate asci, and cyanophilic, smooth to finely warted ascospores; *Caccobius* Kimbr. *1967* (Thelebolaceae) with 15–25 thick-walled asci enclosed in a mucilaginous sheath, staining uniformly in congo red except for a hyaline apical plug, containing 1000–1500 spores; *Trichobolus* (Sacc.) Kimbr. & Cain *1967* with cleistothecioid apothecia 300–540 μm diam, beset with septate, hyaline, thick-walled setae without basal swellings, containing mostly 1–3 ovoid asci which uniformly stain in congo red and contain 1500–7000 spores; and *Lasiobolus* Sacc. *1884* emend. Kimbr. *1967* with similar apothecia, 400–1500 μm diam, but with non-septate setae with bulbous bases, usually containing over 30 asci with 8–120 spores.

Thelebolus species are generally coprophilic and psychrophilic (6348) and widespread in the cold and temperate regions. They can be isolated from dung either by incubating it in moist chambers or by treating a dung suspension in 50% ethanol for 3 min and plating on Leonian agar with aureomycin (6348), and subsequent incubation at low temperatures.

A considerable number of species have been described, mainly in *R(h)yparobius* which are still insufficiently known because of a lack of investigations based on cultures. The number of asci in the ascomata is variable but the number of spores in the asci appears to be fairly constant in culture (6348). The species grow and sporulate readily in culture, and any serious revision of this group must therefore include extensive studies of cultures.

The following series of species might be established, bearing in mind that the name given may require revision with future critical studies:

8-spored: *Th. microsporus* (Berk. & Br.) Kimbr. *1967* (asci numerous, spores 6·8–8·2 × 3·5–4·0 µm).

32-spored: *Th. caninus* (Auersw.) Jeng & Krug *1977* (= *Th. psychrophilus* (Bergm.) Eckblad *1968*) (10–20 ovoid asci, spores 6–9 × 3·5–4·5 µm).

64-spored: *Th. crustaceus* (Fuckel) Kimbr. *1967* (q.v.) (5–12 clavate asci, spores mostly 6·5–7 × 2·5–3 µm).

128-spored: *Th. polysporus* (Karst.) Otani & Kanzawa *1970* (q.v.) (4–6 ovoid asci, spores 5–7 × 3·0–3·5 µm).

256-spored: '*Ryparobius*' *myriosporus* (Crouan) Boud. *1869* (2–4 ovoid asci, spores about 6 × 4 µm).

over 500-spored: *Th. stercoreus* Tode ex Fr. *1823* (1(–3) asci, spores 5–7 × 3–4·5 µm, to over 2500 spores per ascus).

Thelebolus crustaceus (Fuckel) Kimbr. *1967* — Fig. 341.

 ≡ *Ascobolus crustaceus* Fuckel *1866*
 ≡ *Ryparobius crustaceus* (Fuckel) Rehm *1872*
? = *Ryparobius cookei* (Crouan & Crouan *1867*) Boud. *1869*
? = *Ascophanus subfuscus* (Crouan & Crouan *1867*) Boud. *1869*
? = *Thelebolus obscurus* (Seaver) Eckblad *1968*
 ≡ *Streptotheca obscura* Seaver *1928*

DESCRIPTIONS: Kimbrough in Kobayasi *et al.* (3062), Otani and Kanzawa (4323), and Seaver (5170). — Colonies reaching 2·7 cm diam in ten days at 20°C on OA. Ascomata appearing within two weeks, scattered or gregarious, (sub)globose, later becoming discoid, glabrous, 100–130 µm diam, hyaline to pale yellowish brown. Paraphyses slender, septate, 1·5–2·5 µm wide. Asci 5–12 in number, broadly clavate, substipitate, 64-spored, mostly 40–60 × 18–25 µm, wall 2·5 µm thick. Ascospores ellipsoidal, smooth-walled, 6·5–7·0(–8·0) × 2·5–3·0(–4·5) µm.

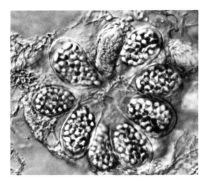

FIG. 341. *Thelebolus crustaceus,* ascoma with asci, NIC, × 400, CBS 714.69.

Th. crustaceus and *Th. polysporus* are common coprophilic fungi. *Th. crustaceus* has been reported from Alaska on dead leaves, tundra soil and the scum of a lake (3062, CBS), dog and rabbit dung in Japan (4323), the British Isles, the Netherlands and Canada (CBS), and carnivore dung in Kenya, Uganda and Tanzania (6348). It has also been isolated from a cutaneous infection in man (CBS).

In most *Thelebolus* species the optimum temperature for growth is 15–20°C (4323), but in this species the optimum is 20–24°C and the minimum below 5°C. Apothecia form on MEA, V-8 juice agar, and most readily on sterilized rabbit dung (3062). Light has no influence on fruiting (except in *Th. stercoreus*) but influences the degree of pigmentation of the apothecia (4323).

Thelebolus polysporus (Karst.) Otani & Kanzawa *1970* — Fig. 342.

≡ *Ascobolus polysporus* Karst. *1867* [non Auersw. *1868*]
≡ *Ryparobius polysporus* (Karst.) Sacc. *1880*

DESCRIPTIONS: Otani and Kanzawa (4323), and Seaver (5170). — Ascomata scattered or gregarious, superficial, sessile, glabrous, 100–200 μm diam, yellowish brown or brownish. Asci 4–6 in number, ovate to broadly ovate, sessile or substipitate, 128-spored, 50–70 × 40–52 μm,

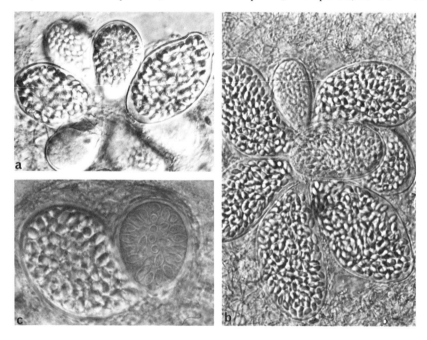

FIG. 342. *Thelebolus polysporus.* a, b. Ascomata with asci, slightly compressed, NIC, × 400; c. mature and young ascus, NIC, × 750; CBS 219.69.

wall 2·5–3·5 μm thick. Ascospores ellipsoidal, smooth-walled, 5–7 × 3·0–3·5 μm. — *Th. polysporus* is most similar to *Th. crustaceus* from which it is distinguished by the ovoid shape and lower number of the asci, more spores in the asci, and slightly broader ascospores.

Th. polysporus has been reported from Europe, Africa and North America (2979, 5170, IMI) including cow dung in Ontario, carnivore dung in Tanzania (4323), sheep and rabbit dung in the Netherlands and deer dung in Germany (CBS). It has also frequently been found in Australia (IMI), rarely in moorland soils in the British Isles (3234) and in arable soils under wheat in Germany.

As far as is known, all *Thelebolus* species are psychrophilic (or thermophobic) with an optimum temperature between 15 and 20°C (at higher temperatures ascoma size is reduced, but see *Th. crustaceus*); little growth occurs at 25–30°C, but considerable growth is possible at 0°C; mature ascomata have been obtained on V-8, Leonian-yeast, alphacel and YpSs agars (4323).

Thermoascus Miehe *1907*

Thermoascus aurantiacus Miehe *1907* — Fig. 343.

The thermophilic genus *Thermoascus* (fam. Trichocomaceae Fischer) is characterized by fast-growing colonies and the formation of conspicuous cleistothecia within five days at 40°C; the wall consists of a few layers of brown pseudoparenchymatous cells. Asci globose, evanescent; ascospores oval to ellipsoidal, yellow to pale brown, finely echinulate.

Thermoascus crustaceus (Apinis & Chesters) Stolk *1965* has a *Paecilomyces* and *Th. thermophilus* (Sopp) v. Arx *1970* a *Polypaecilum* anamorph, whilst *Th. aurantiacus* has none (5594). — The description of conidia in *Th. aurantiacus* by Cooney and Emerson (1174) is based on a confusion with *Th. crustaceus*. *Th. aurantiacus* has more definitely ellipsoidal ascospores, 5–7 × 3·5–5 μm, while in *Th. crustaceus* they are oval, 6·5–8 × 5–6·5 μm.

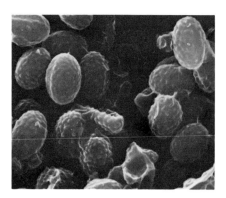

FIG. 343. *Thermoascus aurantiacus*, ascospores, SEM, × 2650, orig. R. A. Samson.

Th. aurantiacus was originally commonly isolated from self-heating hay in Germany. Further data on its occurrence are available from soils in the USA (2573, 5747, 6192, 6193), the British Isles (160), Japan (253, 3843) and the Netherlands (CBS, IMI). It is further known from composted municipal wastes in Germany (3041) and other compost in Italy (CBS), self-heated wood chips (5445, 5740) and chaff (4548), sawdust of pine (CBS, IMI), nesting material (5067) and feathers (3841) of birds, partially decomposed nesting material of an alligator (5742), tobacco (4607, 5743), tobacco leaves (1557) and air (3204, 5784).

Rapid growth and ascoma development occur on most media (253, 5742); the temperature range for growth is 22–58°C, the optimum 40–46(–51)°C, the maximum 55–62°C, and the minimum 20–35(–38)°C, no growth occurring at 18°C (160, 1174, 1220, 1557, 4548, 4903); it grows well when incubated in sun-heated soil of temperate latitudes (2699). Growth is restricted at O_2 concentrations $\leq 0.7\%$, traces of growth still occurring at 0.05%. — Palmitic, stearic, oleic, myristic and acetic acids can be utilized (4321). Amylase production has been reported (2729); optimal cellulose degradation occurs at 48°C (4903) and highest rate of wood degradation (4191) at 50°C (4247). — Mycelial extracts showed toxic effects in rats, chicken embryos and brine shrimps (1304).

Thermomyces Tsiklinsky *1899*

Thermomyces lanuginosus Tsiklinsky *1899* (sensu Miehe *1907*) — Fig. 344.

≡ *Sepedonium lanuginosum* ('Miehe') Griffon & Maubl. *1911*
≡ *Monotospora lanuginosa* (Griffon & Maubl.) Mason *1933*
≡ *Humicola lanuginosa* (Griffon & Maubl.) Bunce *1961*
= *Acremoniella thermophila* Curzi *1929*

DESCRIPTIONS: Cooney and Emerson (1174), Pugh *et al.* (4654), Barron (364), and Ellis (1603); conidiogenesis: Cole and Samson (1109). — The Hyphomycete genus *Thermomyces* is close to *Humicola* Traaen *1914* (q.v.) and has been combined with it by several authors. It can, however, be distinguished by aleurioconidia which have an ornamented surface and are generally supported by distinct stalk cells. All four species so far known are ± thermophilic.

Colonies of *Th. lanuginosus* fast-growing, reaching 2·5 to over 5·0 cm diam (1218) on various media at 45–50°C in two days, white felty, later becoming greenish grey, purplish black and eventually black, with a pink to vinaceous pigment diffusing into the agar. Aleurioconidia mostly arising on 10–15 μm long lateral stalk cells, globose, dark brown, thick-walled, with a wrinkled surface, 6–10 μm diam. — *Th. lanuginosus* was originally imperfectly described and was therefore rejected or ignored by several authors prior to its typification being settled by Pugh *et al.* (4654). — *Th. verrucosus* Pugh *et al. 1964* is distinguished from *Th. lanuginosus* in having verrucose aleurioconidia 10–17 μm diam and a less thermophilic habit; *Th. ibadanensis* Apinis & Eggins *1966* has smooth-walled aleurioconidia 4–8 μm diam which retain a basal appendage after dehiscence. — The mycelium contains 21·4% C_{16}- and 4·5% C_{18}-saturated fatty accids, and 65·2% C_{18}-mono-unsaturated fatty acids (3232). In the aleurioconidia dipicolinic acid has been found which plays an important role in heat resistance (4257). Two cytochrome components, designated as

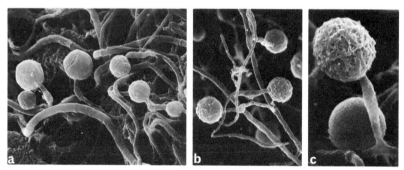

FIG. 344. *Thermomyces lanuginosus*, formation of aleurioconidia, SEM, a, b. × 1000, c. × 2500; orig. R. A. Samson.

c_A and c_B have been isolated; the major component (c_A) is a single polypeptide chain of 111 amino acid residues (3950). — When grown at high temperatures, large, electron-opaque, membrane-bounded lipid bodies are found in the hyphae (1915).

Th. lanuginosus is most commonly isolated in studies of thermophilic fungi carried out all over the world (e.g. 1174). Its occurrence appears to be worldwide (1174); it has been isolated in the British Isles (160, 161, 163, 166, 1551), Denmark (3402), Italy (2757), the USA (1174, 1218, 5745, 5747), Canada, Nigeria, Ghana (IMI), South Africa (1557), India (3518), Indonesia (2358), Brazil (398) and Japan (252, 253, 3843, 4257, 5846, 5934, 6210). It has been reported from dry and waterlogged grassland, loamy garden soil (160, 1551), and aquatic sediments (5934). This species is, however, more commonly associated with organic substrates and is known from *Ammophila* debris (163), senescent grass culms and leaves of which it is an early colonizer (161, 166), culms, roots and the rhizosphere of wheat and barley (166), *Corylus* leaf litter (IMI), rarely in the rhizosphere of wilted pineapple plants (6210), mouldy hay (IMI), decomposing nesting material of alligators (5742), self-heating grains of barley and wheat (3203, 5746), wet and/or rotting grain of corn or oat (252, 1174, 4263), the atmosphere around silos (3203), pecans (2572), rotten feedstuffs (247), tobacco products (5743), various composting materials (1174, 3402), in wheat straw compost as a colonizer during the so-called "plateau phase" at 40–50°C (953), mushroom compost (1174, 1723, 3402) during the "pasteurization phase" (1723), municipal waste (2874, 3041, 3051), self-heated and charred wood-chip piles (1756, 5445, 5740, 5746), composting bark (2757), dung of poultry, pigeon, horse, pig, and sheep (1174, 1557, 2409, 2757, 3402, 3518, 6118), and birds' nesting materials (1557, 5067). It has also been trapped from the air in Indonesia (2358) and the British Isles (2588, 3204) where it was the second most abundant thermophilic species, the commonest being *Aspergillus fumigatus* (1666).

Good growth and conidium production occur at temperatures between 40° and 55°C (1723, 2409), the minimum is about 30°C (4257), and the optimum 47·5–52·5°C (2757) or even 58°C (4257); it grows and sporulates and conidia germinate well when incubated in sun-heated soil of temperate latitudes (2699); best growth *in vitro* is obtained on PDA, glucose-peptone agar, yeast, meat broth (2409) or YpSs agar (1218). The thermal death point has been determined as about 68°C for 45 min (1727) but aleurioconidia survived 100°C for 5 min (4257); above 55°C abundant sterile mycelium is formed (2409). The pH optimum *in vitro* is (5·8–)6·8–7·3 (4257, 4911). — No growth occurs at O_2 concentrations below 0·7% on basal media and no sporulation below 1% O_2 (1350, 2874). — *Th. lanuginosus* does not grow on pure (2409) or acid-swollen cellulose (5741) or CMC (2757), but it can grow commensally with some cellulolytic thermophilic fungi (2360). Hemicellulose (952) can be utilized, in particular, arabinoxylan (1757). Amylase (348, 3402), 1,3-β-glucanase (4265), proteinase (4299) and phenoloxidase (2757) are produced. Optimal lipase production (177) occurs on 2% soluble starch and 5% corn steep liquor, and also with xylose, arabinose or glucose as C sources (3381). Suitable N sources are ammonium nitrate, urea and asparagine (1217). It also utilizes myristic, oleic, palmitic, stearic acids and glycerol (4321). Humic acid-like substances are produced when it is grown on wood substrates (5240). An acid phosphatase with a remarkable thermostability but similar to that of some mesophilic organisms has been found (1219). Several plasticizers can be attacked by the esterases of *Th. lanuginosus* (3829). — It has been implicated in bovine abortion and "farmer's lung" and found to be toxic to brine shrimp, chicken embryos and rats (1304). Some antibiotic activity towards *Staphylococcus aureus* and no toxicity to mice has been observed in another report (2757).

Thielavia Zopf *1876*

= *Thielaviella* v. Arx & Tariq Mahmood *1968*

Type species: *Thielavia basicola* Zopf *1876*

Anamorphs: *Myceliophthora* and *Acremonium*-like

Contributions to a MONOGRAPHIC TREATMENT: Booth (638), Mouchacca (3995), Malloch and Cain (3553), Pidoplichko *et al.* (4552), Belyakova (440), and v. Arx (207); ascoma development: Doguet (1413).

Thielavia is regarded as allied to the Sordariaceae (207). Mycelium hyaline, rather fast-growing. Ascomata developing from spiral ascogonia, non-ostiolate, globose, glabrous, setose or tomentose, with a brown or pale wall of textura epidermoidea. Asci formed on croziers, fasciculate, clavate or ellipsoidal, unitunicate. Ascospores ± ellipsoidal or fusiform, 1-celled, brown, smooth-walled, with one distinct apical or subapical germ pore. Blastic or phialidic conidia are produced in a few species. — About 20 species are currently recognized (207).

Similar cleistothecial genera include: *Corynascus* v. Arx (q.v.), with ascospores with two germ pores and a *Sepedonium*-like anamorph; *Corynascella* v. Arx *1975*, with pale ascomata covered with dark brown hyphae and ascospores with two prominent germ pores; *Apodus* Malloch & Cain *1971*, with thick-walled ascomata, thickened ascus tips and sometims 2-celled ascospores; and *Chaetomidium* (Zopf) Sacc. (q.v.), which is a non-ostiolate counterpart of *Chaetomium*, and has ascomata with a thicker dark pseudoparenchymatous wall covered with variously shaped hairs and ascospores with a rather inconspicuous germ pore. *Boothiella* Lodhi & Mirza *1962* was originally separated from *Thielavia* because of hyaline, several-layered ascoma walls, but subsequently combined with *Thielavia* because of the existence of intermediate species (207) and separated again because of verruculose ascospores (5976).

Two species are treated here, both of which have brown ascomata. *Th. terrestris* (Apinis) Malloch & Cain *1972* is treated in a similar way by Samson *et al.* (5025).

Key to the species treated:

1	Ascospores with two germ pores; *Sepedonium*-like conidia present	
		cf. *Corynascus sepedonium* (p. 232)
	Ascospores with one germ pore; *Sepedonium*-like conidia absent	**2**
2(1)	Thermophilic; ascospores oval, ellipsoidal or pyriform; conidia present	**3**
	Mesophilic; ascospores fusiform; conidia absent	**4**

3(2)   Ascospores 7·5–11 × 4·5–7 µm; *Myceliophthora* conidia abundantly produced, obovoid to
        pyriform, 4·5–11 × 3·0–4·5 µm   *Th. heterothallica* (p. 781)
        Ascospores 5·5–7 × 4–5·5 µm; conidia produced on scattered *Acremonium*-like phialides,
        ovoid, with truncate base, 3–6 × 2–3 µm   (*Th. terrestris*)

4(2)   Ascoma wall ± hyaline; ascospores 10–12 × 5·5–6·5 µm   (*Th. basicola*)
        Ascoma wall brown; ascospores 12–16 × 7–9 µm   *Th. terricola* (p. 782)

Thielavia heterothallica Klopotek *1976* — Fig. 345.

Anamorph: *Myceliophthora thermophila* (Apinis) van Oorschot *1977*

≡ *Sporotrichum thermophilum* Apinis *1962* (as "*thermophile*")
≡ *Chrysosporium thermophilum* (Apinis) Klopotek *1974*

DESCRIPTIONS: Apinis (160), Semeniuk and Carmichael (5201), Hedger and Hudson (2359),
Klopotek (3043), and van Oorschot (4303); teleomorph: Klopotek (3044). — Colonies
fast-growing, reaching 5–9 cm diam in three days at 45°C, floccose to powdery, cinnamon-
fulvous to fawn-hazel. Ascomata forming at 30–35°C on MEA and cellulosic media in some
isolates, solitary, submerged or partly so, globose, non-ostiolate, 70–180 µm diam; ascospores
ellipsoidal, brown to black, smooth-walled, with one terminal germ pore, 1-guttulate,
7·5–11 × 4·5–7·0 µm. Conidia abundantly produced, solitary or formed in a random succession terminally and laterally on slightly inflated conidiogenous cells, sometimes producing
secondary conidia and forming irregular clusters, ovate to pyriform, with a truncate base,
mostly coarsely warted but rarely smooth-walled, pale to dark brown, mostly
5–7 × 3–4 µm. — *Th. heterothallica* is described as being heterothallic (3044), but on mating
of certain isolates ascomata appear all over one of the partners and not in the contact zone. —
The anamorph is very similar to *Myceliophthora lutea* Cost. *1892* (≡ *Chrysosporium luteum*
(Cost.) Carm.) but has considerably higher cardinal temperatures. The similar *Myceliophthora
fergusii* (Klopotek) van Oorschot *1977*, anamorph of *Corynascus thermophilus* (Fergus &
Sinden) Klopotek *1974*, has rosy-buff colonies, more frequently solitary and also larger
(mostly 9·5–11 × 5·5–6·0 µm), smooth-walled conidia. The type culture of *Sporotrichum
thermophilum* has smooth-walled conidia and does not form ascomata with *Th. heterothallica*;
isolates with roughened conidia are, however, more common. — Neither *Sporotrichum* Link
ex Fr. *1821*, the anamorph of some Basidiomycetes, with broad hyphae and clamp connections, nor *Chrysosporium* Corda (q.v.), the anamorph of some Gymnoascaceae, with broadly
truncate arthroconidia, are suitable genera to accommodate these blastoconidial
fungi. — Hyphae of *Th. heterothallica* contain C_{18}-mono- and double-unsaturated fatty acids in
various proportions and C_{16}-saturated as major fatty acids (1280); the composition of fatty
acids was found to be significantly influenced by the growth temperature, the degree of
unsaturation rising at lower temperatures (1278).

Th. heterothallica is commonly encountered in studies of thermophilic fungi and occurs
mainly on a variety of composted substrates (3043). — It has been isolated from soil in Japan
(3843), Ghana (IMI), Senegal (CBS), the USA (2573, 5747), Germany (CBS), and the
British Isles (160, 163, 166, 1551). It has been found in dry or waterlogged grassland soil
(160, 166, 1551), on *Ammophila* debris (163), plant debris in swamps (160), dead *Phragmites*,

barley straw and wheat straw compost (2359, CBS), as a late colonizer of self-heated wheat straw but not before the temperature decreases after the "plateau phase" to about 30°C (953). Other isolates have been obtained from silage (5201), coal tips (1665), self-heated wood-chips (5201, 5445, 5740) and wood pulp (CBS), and nesting material of birds (5067); it appears rather sporadically in the air in Britain (1666, 2588). — Cultures on agar slants survived in the dry state for four years (5201).

Th. heterothallica has a minimum temperature for growth of 23–25°C, a maximum of up to 55°C and an optimum in the range 35–48°C (160, 517, 952, 958, 1174, 1220, 3043, 4903); a treatment of 59°C for 24 min was tolerated (1727) and it grows and sporulates well when incubated in sun-heated soil in temperate latitudes (2699). — *Th. heterothallica* degrades cellulose particularly intensely (952, 1725, 4903, 5201, 5741), but not starch (3043) except for one isolate; the maximum production of C_1 and C_x cellulases occurred at 45°C in 2–4 days with nitrate or urea as N sources (1201, 4903). Following cultivation on cellulosic media, this fungus is a relatively good degrader of wood from pine, spruce (4248), aspen, birch and beech (4191). *Th. heterothallica* produces a humus-like substance *in vitro* when growing on wood (5240). — Mycelial extracts exhibited toxicity against chicken embroys (1304).

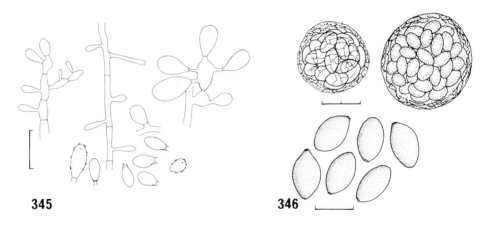

FIG. 345. *Thielavia heterothallica*, anamorph *Myceliophthora thermophila*, CBS 357.69.
FIG. 346. *Thielavia terricola*, two mature ascomata and ascospores, CBS 358.53.

Thielavia terricola (Gilman & Abbott) Emmons *1930* — Fig. 346.

≡ *Coniothyrium terricola* Gilman & Abbott *1927*
= *Anixiopsis japonica* Saito & Minoura *1948* (nomen inval., Art. 36)

DESCRIPTIONS: Booth (638), Mouchacca (3995), Malloch and Cain (3553), v. Arx (207), Minoura *et al.* (3840), Doguet (1413), Udagawa (5970), and Matsushima (3679, 3680). — Colonies reaching 7·0–8·0 cm diam in 14 days on MEA or OA at 28°C, forming a white mat which darkens with the development of the dark brown to black ascomata; reverse initially white to very pale orange, later also darkening. Ascomata mostly submerged, globose, dark brown, (60–)80–200 µm diam, with a semi-transparent wall of textura epidermoidea, surface cells 6–16 × 4–12 µm, smooth-walled, not setose. Asci pyriform to ellipsoidal, 8-spored, 24–40 × 14–20 µm. Ascospores broadly fusiform with slightly rounded ends, brown when mature, with a distinct germ pore at one end, 12–16 × 7–9 µm, 2-nucleate (1413). Conidia lacking. — Young hyphal cells contain up to seven nuclei, older ones mostly one (1413). — *Th. minor* (Rayss & Borut) Malloch & Cain *1973* is not only distinguished by its smaller ascospores (10–12 × 5·5–6·5 µm) but also by the paler ascomata and has been considered identical to the type species of the genus, *Th. basicola* Zopf *1876* (207).

Th. terricola is a cosmopolitan species (3553) found on soil, dung, seeds, decaying plant material, etc. (3553). It is mainly distributed in the warmer regions of the world and has been documented from various parts of Africa (1413, IMI, CBS) including the Sudan, Sierra Leone, Nigeria (638), central (3063) and South Africa (1555, 1556, 1559), Israel (654, 4759), Kuwait (4001), Pakistan (638, 3861), Nepal (3840), India (638, 1519, 2164, 2186, 2854, 3863, 3865, 4030, 4651, 4698, 4995), New Guinea, New Britain (3679), Japan (2532, 3267, 3680, 5970), Central America (1697), Turkey (4245) and Italy (3913), but there are also reports from the Netherlands (3526), the British Isles (5221) and the USA (655, 2007, 2482, 3817, 4733). It has been found in uncultivated soils (4030, 4698), forest soils (1555, 2007, 2186, 2482, 2854, 3267, 3817, 3965, 4995), grassland (3863), grass plots with *Dichanthium annulatum* (3865), heathland (5220, 5221), cultivated soil (3817) with rice (1519, 3913), lentils, *Cicer arietinum* (2164) and barley (3526), soils with steppe type vegetation (1559), saline soil (4001), desert soil (654, 4733, 4759), and estuarine sediments (655). It has also been isolated from decaying strawberry plants (4133), rotting fruits of *Cordia dichotoma* (5735), leaves of *Panicum coloratum* (1560), shoots of *Bothriochloa pertusa* (2955), the rhizoplane of barley, the rhizospheres of groundnuts (2532, 2768), *Trifolium alexandrinum* (1963) and *Dichanthium annulatum* (3271), seeds of wheat, groundnuts (1984, 2253, 3220), pecans (2258), peas and cotton (CBS), dung of monkey, sheep (3401, 3861), cow, deer and elephant (CBS), and chitin buried in tropical soil (4264).

The optimal temperature for growth is 37°C, the minimum 15°C, and the maximum 46°C; good growth occurs between pH 3·9 and 6·0, but none at pH 2·9 or above pH 7·9 (2203). — *Th. terricola* decomposes both chitin (4264) and cellulose (4651), and can utilize several other poly-, tri-, di- and monosaccharides as well as alcohols; nitrate, ammonium and organic N compounds support growth equally well; a metabolite with antibacterial activity is known (2203).

Torula Pers. ex Fr. *1821*

Torula herbarum Pers. ex Gray *1821* f. *herbarum* — Fig. 347.

≡ *Monilia herbarum* Pers. *1801*

For numerous older synonyms see Joly (2793) and Subramanian (5626).

DESCRIPTIONS: Mason (3661), Ellis (1603), Joly (2793), Tubaki (5926), Subramanian (5626), Malone and Muskett (3556), and Matsushima (3680); conidiogenesis: Hashmi *et al*. (2309), Ellis and Griffiths (1588), and Cole and Samson (1109).

Species of the genus *Torula* (type species *T. herbarum*) have been keyed out recently by Ellis (1603, 1604) and Vasant Rao and De Hoog (4741). *Torula* is a genus of dematiaceous Hyphomycetes with phragmoconidia arising from ± globose, monoblastic or polyblastic conidiogenous cells. Blastoconidia forming branched chains, cylindrical or fusiform, constricted at the septa, ± warted, brown, the end-cells usually paler. Seven species are now recognized, while the very many others described in *Torula* are not considered congeneric (4741). Conidiogenesis is blastic (1588), contrary to previous reports of poroconidia (2599), and the conidia have an acrosporous differentiation (2309). — A similar genus but with non-catenulate conidia borne on vesiculose and often polyblastic conidiogenous cells is *Polyschema* Upadhyay *1966* (cf. 1604).

Colonies of *T. herbarum* reaching to 2·7 cm diam in ten days at 20°C on MEA, dark brown, powdery due to conidium formation. Aerial hyphae 2–6 μm wide, conidiogenous cells 7–9 μm diam. Conidia cylindrical, constricted at the septa, (3–)4–5(–10)-septate, olive-brown, verruculose, 20–70 × 5–9 μm. — *T. herbarum* f. *quaternella* Sacc. *1913*, common in the tropics, has smoother, mostly (1–)2(–3)-septate conidia 10–17 × 5–7 μm. The similar *T. graminis* Desm. *1834*, common on grasses in Europe, has conidiogenous cells to 5 μm diam and conidia of very irregular length, 1- to many-celled, minutely verrucose, cells 4–6 μm diam, often broader than long.

T. herbarum is a cosmopolitan fungus but appears to be most frequent in temperate regions (1603). It is very common on dead herbaceous stems, and occasionally also on leaves, wood, etc., and has also been isolated from air and soil (1603, 4548). — Records include ones from the British Isles (1603, 2338), the Netherlands (1614, 1616), Poland (273, 1421, 3571), Italy (2726), Canada (505), the USA (2004, 4540, 6351), Panama (3963), Brazil (6008), Venezuela, Nigeria (IMI), central Africa (3063), South Africa (2816), India (2852, 2854, 3868, 5626), Nepal (1826) and Japan (2532). Isolates from soil are not very numerous and mainly come from the uppermost layers; they include forest (273, 505, 2004, 2852, 2854, 3571, 6351) and arable soils (1614, 1616, 2726, 2816, 3868), uncultivated soil (3868), deserts (2974), sand blows (4540), and children's sandpits (1421). It has been isolated from sugar beet roots (4559), the geocarposphere of groundnuts (2532), and seeds of oats (3556) and cultivated grasses (3512). Other known plant substrates include dead leaves of palms (3963), and bananas, bean pods, *Lobelia* stems (3063), leaves of *Dactylis glomerata* (6243), litter of

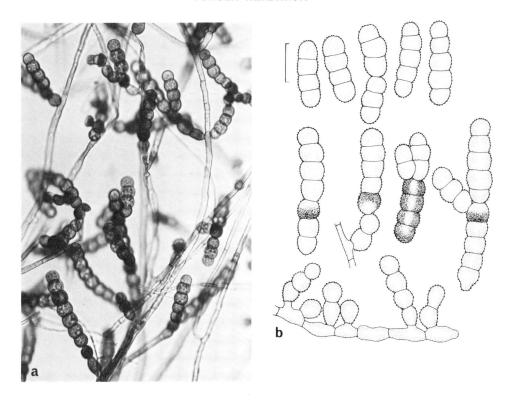

FIG. 347. *Torula herbarum*. a. Conidiogenous cells and conidia, × 500; b. conidiogenous cells and conidia, CBS 220.59.

Atlantia monophylla and *Gymnosporia emarginata* (6108), stems of *Heracleum sphondylium* (6462, 6464), *Phragmites australis* (5713), *Urtica dioica* (6463, 6464), *Panicum coloratum* (1560), *Setaria glauca* (5233), *Andropogon sorghum* and wheat (3219). Other known substrates are pecans (2258, 2572), bamboo garden stakes (IMI), sticks exposed to river water (5249), running water (1157), seawater (831), and birds' nesting material (2575, 2577).

For conidium germination vapour saturation or at least −100 bars water potential are required (6243, 6464). Optimal growth was reported at 35°C (1827). — Cellulose degradation increases with increasing concentration of the N source (4433), and maple-wood strips are decomposed (2211). Peroxidase, tyrosinase (2516) and the naphthoquinones herbarin (2832) and dehydroherbarin (3231) are produced. Herbarin shows weak antibacterial and antifungal activities (2832). — *T. herbarum* was found to damage pea roots *in vitro* (1430).

Torulomyces Delitsch *1943*

Torulomyces lagena Delitsch *1943* — Fig. 348.

= *Monocillium humicola* Barron *1961*

DESCRIPTIONS: Barron (357, 364), Christensen and Backus (1035), and Hashmi *et al.* (2306). — The genus *Torulomyces* is distinguished from *Penicillium* by its solitary phialides. The subglobose conidia are formed in dry chains, attached to each other with connectives as in *Penicillium*. The synonymy of *T. lagena* with *Monocillium humicola* was discovered by Barron (363). Hashmi *et al.* (2306) also placed *Monocillium indicum* S. B. Saksena *1955* in *Torulomyces* as both fungi form solitary phialides with conidial chains (as seen in time-lapse photomicrography). *Monocillium indicum*, however, has conidial chains in which the conidia adhere to each other by a truncate end as in some *Acremonium* species, while other clearly related species form conidial heads (1882). — In *Monocillium humicola* two varieties were described from different habitats (1035): var. *brunneum* Christensen & Backus *1964*, isolated from upland oak and pine communities in Wisconsin, producing a red-brown pigment in reverse on honey-peptone agar and MEA, and var. *humicola*, common in conifer bog soils, becoming grey to black.

T. lagena has slightly swollen, thin-walled phialides subtended by a short stalk arising from submerged hyphae, and more or less roughened, globose conidia, 2–3 μm diam. Colonies reaching <1 cm diam in ten days at 20°C on MEA.

T. lagena is a widely distributed but evidently not very common soil fungus. Isolations have been reported from conifer swamps (1039) and forest soils in Wisconsin (1032) and Canada (6243, 6352), from soils in Japan (5846), Peru (2005), equatorial West Africa (1420), savannah with *Acacia karroo* in South Africa (4407), sand dunes in Poland (1423), peat in Canada (IMI) and Ireland (1376, 1438), and forest litter in Belgium (CBS). It has also been found on the mycorrhiza of *Lactarius* on *Castanea sativa* (1794), in the rhizospheres of wheat (IMI) and some peat bog plants (1376), on bark of *Cyathea* in New Zealand, and in cornmeal in Yugoslavia (CBS).

The optimum temperature for growth is 20°C. Suitable C sources include galactose or sucrose, and glycine is a good N source (1375). Conidium germination is reported to be inhibited on peat (1436). — In a chronically irradiated soil it was still found at intensities of 120 R per day (2009).

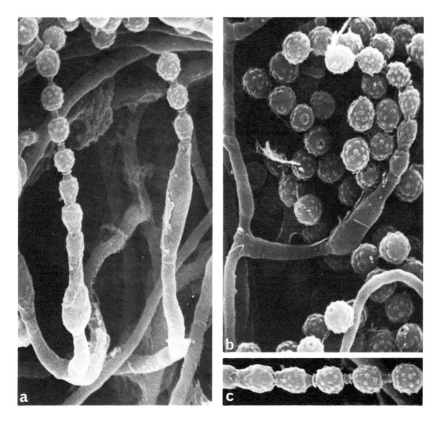

FIG. 348. *Torulomyces lagena.* a, b. Phialides with conidial chains, SEM, × 4500; c. conidial chain, SEM, × 5800; CBS 185.65.

Tricellula van Beverw. *1954*

Tricellula aquatica Webster *1959* — Fig. 349.

DESCRIPTIONS: Webster (6239), Nilsson (4190), Petersen (4526), Dudka (1485), and Dudka and Isachenko (1486).

The genus name *Tricellula* (type species *T. inaequalis* van Beverw.; conidia with one lateral branch) has priority over *Volucrispora* Haskins *1958* (type species *V. aurantiaca* Haskins; conidia mostly with two lateral branches); the two genera cannot, however, be separated on this morphological feature. — Colonies pink, slimy, and slow-growing (<2 cm diam in ten days at 20°C on MEA). Conidiogenesis is polyblastic with several conidia arising in succession at the widening tips of clustered conidiogenous cells. Conidia consisting of a main axis with two elongate cells, constricted at the septum, and 1–2 lateral branches arising from the basal cell below the septum; the conidia do not usually exceed 20μm in length.

T. aquatica sporulates abundantly on commonly used media when freshly isolated, and transfers to water do not enhance sporulation further. Conidia are Y-shaped, with one lateral branch attached rather closely to the primary apical cell which is somewhat longer than the branch cell; total length 9–16 μm, basal cell $4·5–6(–8) \times 2·2–2·5(–3·5) \mu$m. — *T. aquatica* can be distinguished from the other species known by the forward-directed lateral branch, whilst in *T. inaequalis* van Beverw. *1954* and *V. aurantiaca* Haskins *1958* the branches have a forward-directed hump but the major part is directed backwards; in *T. curvatis* Haskins *1958* the basal and apical cell do not form a straight line but the terminal cell is bent away from the lateral branch. — The fatty acid content of *T. aquatica* has been determined (1487).

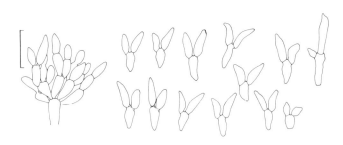

FIG. 349. *Tricellula aquatica*, conidiophore and conidia.

T. aquatica has been collected in some scum and foam samples in Sweden but is perhaps not truly aquatic (4190). It is known from the British Isles (4432, 6239), the Netherlands (CBS), Germany (184), Sweden (4190), Finland (3513), the USSR (1485), Canada (275) and the USA (4526). In the original description, *T. aquatica* was reported on leaves of alder and horse chestnut in a river-bed (6239), but it has since been found in moist soil (4190), wheatfield soils and other agricultural polder soils (CBS), and isolated from filter paper exposed to river water (4432), on tulips (IMI) and living grasses (3513).

The temperature range for the production of conidia, their germination, and mycelial growth is 10–25°C (3105), with the optimum at 18–20°C (1486). — Very good decomposition of xylan has been reported (1432).

Trichocladium Harz *1871*

Type species: *Trichocladium asperum* Harz (lectotype)

Contributions to a MONOGRAPHIC TREATMENT: Hughes (2597), Dixon (1409), Ellis (1603), and Pidoplichko and Kirilenko (4551).

Trichocladium is a genus of the Hyphomycetes with hyaline (or in some species pigmented) vegetative hyphae and short-stalked or sessile, 2- or more-celled, ellipsoidal to clavate, pigmented aleurioconidia. A phialidic state is known in two species, including the type of the genus.

Nine species have been described; these were isolated from soil, decaying wood and other plant material. *Zalerion* R. Moore & Meyers *1962* is similar but has helicoid or irregularly coiled, more-celled aleurioconidia.

Key to the species treated:

1	Hyphae pigmented; aleurioconidia mostly 2-4-celled, smooth-walled, $20–40 \times 10–16\,\mu$m	
		T. opacum (p. 792)
	Hyphae hyaline	**2**
2(1)	Aleurioconidia mostly two-celled, coarsely warted, mostly $16–25 \times 10–15\,\mu$m	
		T. asperum (p. 790)
	Aleurioconidia mostly 2-3-celled, smooth-walled, mostly $12–20 \times 6–8\,\mu$m	(*T. canadense*)

Trichocladium asperum Harz *1871* — Fig. 350.

≡ *Dicoccum asperum* sensu Lindau *1907* [non *Dicoccum asperum* (Corda) Sacc. *1886* ≡ *Sporidesmium asperum* Corda *1838* ≡ *Monodictys aspera* (Corda) Hughes *1958*]

DESCRIPTIONS: Hughes (2597), Mason (3660), Groves and Skolko (2144), Malone and Muskett (3556), Barron (364), Hennebert (2399), and Matsushima (3680). — Colonies effuse, cottony, at first white but later becoming grey, reaching 5·5–6·0 cm diam in ten days on MEA at 20°C; reverse olive-green. Aleurioconidia borne on scarcely differentiated lateral branches, which are hyaline, 1–4-celled, to $25\,\mu$m long and $1·5–3·5\,\mu$m wide; conidia (1–)2(–3)-celled, somewhat clavate or ellipsoidal with a tapered basal cell, slightly constricted at the septum, brown, coarsely warted, with an apical germ pore, (12–)16–25(–28) × 10–15 μm. A phialidic state similar to that of *Humicola* Traaen has been observed in old cultures (2399). — A smooth-walled counterpart, also with apical germ pores in the aleurioconidia and a phialidic state, is *Trichocladium canadense* Hughes *1959*.

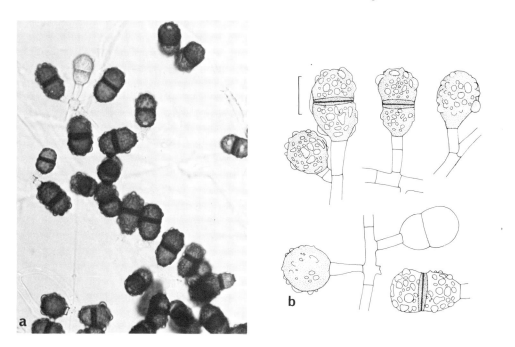

FIG. 350. *Trichocladium asperum.* a. Aleurioconidia, × 500; b. formation of aleurioconidia.

T. asperum has a worldwide distribution but predominates in the temperate zones. It has been reported from the British Isles (861, 1376, 6182), Norway (5874), the Netherlands (1614, 1616), Poland (269, 1421, 1422), Germany (1736, 3042), Canada (505), the USA (1163), Israel (2764), Zaïre (2966), central Africa (3063), Mauritius (1603), Australia (1603, 3720), New Zealand (1603), India (1519), Japan (3267, 3680, 5846) and Central America (1697). It can be isolated with selective techniques for cellulolytic (4655) and keratinolytic fungi (1422) and screened immersion tubes (5221). *T. asperum* is distributed in both forest and cultivated soils and there are particularly numerous reports of isolations from soils in beech forests (272, 273, 3138), and also coniferous forests (269, 505, 989, 1736, 2068, 2923, 3267). Other habitats reported include grassland (6182), heathland (3720, 5221, 5819), peat (1376, 6520), podzols (3414, 3416), sand dunes (745, 4655), arable soils (861, 1614, 3042, 4703, 6138), rice fields (1519), citrus plantations (2764), garden compost (1424), children's sandpits (1421), and soil treated with sewage sludge (1163). It commonly occurs in deciduous (2411) and coniferous litter (2923). It has been found on plant debris in soil (2705, 3113), on roots of broad beans (6134), in the rhizospheres of oat (4443), wheat (2023), *Stipa sareptana* (3376), alfalfa (3982) and clover (5815), on hyacinth and lily bulbs (2597), rarely on oat (3556) and other seeds (2144, 3512), wood baits in soil (5237), buried beech twigs (850), pellets and feathers of free-living birds (2575, 6223), and mouse (4917) and rabbit dung (2279). — Conidium germination was found to be highly sensitive to soil mycostasis (1407).

The optimal temperature for growth is 20–25°C; reduced growth occurs below pH 4·0. D-Glucose, D-fructose, D-mannose, D-galactose and L-arabinose are well utilized, but D-arabinose and D-ribose poorly so; most disaccharides tested and inulin and starch also support good growth (6191); pectin (3416, 6191) and cellulose (1425, 2021, 2736, 4433, 4434, 5238) can also be attacked. There are contradictory reports as to its capacity for chitin decomposition (1425, 2068). No growth substances are required (6191). When accumulated to high inoculum densities in soil, it clearly encouraged root development in wheat (1427).

Trichocladium opacum (Corda) Hughes *1952* — Fig. 351.

≡ *Sporidesmium opacum* Corda *1837*
≡ *Clasterosporium opacum* (Corda) Sacc. *1886*
= *Clasterosporium claviforme* (Preuss) Sacc. var. *leptopus* Sacc. *1886*

DESCRIPTIONS: Hughes (2597), Kendrick and Bhatt (2924), and Matsushima (3680). — Colonies effuse, brown-black, reaching 2·5–3·5 cm diam in ten days at 20°C on MEA, mycelium immersed and superficial, composed of hyaline to pale brown hyphae 2–4 μm wide. Aleurioconidia borne either on little differentiated (to 12 μm long) lateral branches or sessile on prostrate hyphae; aleurioconidia (sometimes regarded as chlamydospores) pyriform to clavate, straight or somewhat bent, (1–)2–4(–5)-celled, hardly constricted at the septa, olive-brown, with the basal cell slightly paler, smooth- and thick-walled, 20–40 × 10–16 μm; the fine structure of conidia and "chlamydospores" has been demonstrated and discussed (1585). — The brown hyphae and the absence of phialoconidia in this species indicate that it is not closely related to *T. asperum*. Its variability, particularly the occurrence of many-celled chlamydospores submerged in the agar, has been described by Kendrick and Bhatt (2924). — *T. pyriforme* M. Dixon *1968* is a comparable fungus with 3-celled, straight, sessile aleurioconidia and a striking heat resistance.

T. opacum probably has a worldwide distribution but it is locally usually infrequent and is only seldom determined. Records of its isolation from soil are available from the Arctic (1750), Alaskan tundra soil (frequently) (3680), Northern Ireland (4429), Canada (505, 3959), the USA (1603), Peru (2005), France (2963), Italy (3976), central Africa (3063), India (4106), Pakistan (1603), Nepal (1826), Japan (3680, 5846), Australia (2091), and New Zealand (5812, 5814). It has been reported from alpine habitats (1876, 3445, 3976, 6082), forest soils (505, 850, 989, 2411, 3063, 3680, 3959, 3962), grassland (5812, 5814, 6082), garden (4451) and arable soils (1889, 3063, 3450, 6563, 6564), podzols and leached brown soil (2963), and organic detritus in fresh water (4429). Its frequency in arable soils under wheat was found to be significantly increased after a previous wheat crop but not so after peas and rape (1433). It has been isolated from the root region of dwarf bean (4452), cabbage (4451), wheat (3567), oat, barley (3006, 4450, 4451) and corn (6563), but no appreciable rhizosphere effects have been observed. Other substrates known include many kinds of deciduous wood and herbaceous stems (1603, 2597), beech twigs buried in forest soil (850), preservative-treated wood in contact with ground (4193), oak and birch stumps (4757), pieces of wood exposed in fresh (4955) or seawater (3171), decaying stems of *Heracleum sphondylium* (6462) and *Urtica dioica* (6463), and rotting and healthy strawberry plants (4133).

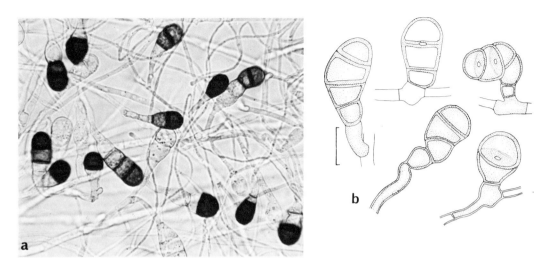

FIG. 351. *Trichocladium opacum*. a. Mycelium with aleurioconidia, × 500; b. details of aleurioconidia, CBS 355.62.

Starch (1750, 1827) and xylan are decomposed, the latter with relatively marked variability between isolates; its pectolytic activity is low compared with other fungi (1432). Its ability to decompose cellulose increases with the concentration of the N source (4433). Birch and pine wood are attacked and eroded (4191). Weak growth occurs on gallic acid (1750). — *T. opacum* was found to reduce the growth of pea roots *in vitro* while those of wheat remained unaffected (1430).

Trichoderma Pers. ex Fr. *1821*

= *Aleurisma* Link ex Fr. *1821*
= *Pachybasium* Sacc. *1885*

Type species: *Trichoderma viride* Pers. ex Gray

Teleomorph: *Hypocrea* Fr.

MONOGRAPHIC TREATMENT: Rifai (4842); electron-micrographs: Hashioka *et al.* (2304); review of the ecology: Danielson (1267).

The Hyphomycete genus *Trichoderma* is characterized by fast-growing hyaline colonies bearing repeatedly branched conidiophores in tufts with divergent, often irregularly bent, flask-shaped phialides. Conidiophores may end in sterile appendages with the phialides only borne on lateral branches in some species. Conidia are hyaline or, more usually, green, smooth-walled or roughened. Hyaline chlamydospores are usually present in the mycelium of older cultures. If the phialides are strongly convergent the conidial states of otherwise similar cultures are placed in *Gliocladium* (q.v.), while if they are straight and moderately divergent, in *Verticillium*. — Ascomata of *Hypocrea* are not produced in culture in most species.

Rifai (4842) distinguished nine species "aggregates" based on microscopic characters; the macroscopic features are very similar in most of them. The teleomorphs of *Hypocrea* and related genera were thoroughly studied by Doi (1414, and other publications), who also carefully described the associated anamorphs. From his work, it appears that it is generally not possible to correlate the anamorphs with single teleomorphs, as each of Rifai's "aggregates" can be connected with several teleomorphic species, the imperfect states of which can hardly be distinguished in culture. Moreover, some of the characters used by Rifai for distinguishing his "aggregates" have proved to be hardly reproducible or strongly dependent on the medium used. Therefore, the separation of further taxa in *Trichoderma* does not seem warranted at present. It is also not justifiable in most cases to use the name of a teleomorphic species for these conidial isolates. The teleomorph names listed under each conidial name here are indicated only as possibilities and suggestions for further study. — The numerous reports compiled here have to be regarded with particular reservation with respect to species identification; in particular, the identity of almost all taxa reported prior to Rifai's monograph require restudy.

Trichoderma cultures are best grown on OA (or 2% MEA) in daylight or under near-UV; in the dark they quickly loose the capacity to sporulate. For the examination of phialide structures, very young cultures (usually five days old) must be employed, while conidial roughening is best judged in ones two weeks old. — A selective medium has been developed for quantitative recovery from soil; besides peptone-glucose agar with 17 ppm rose bengal and 30 ppm streptomycin sulphate, it contains 200 ppm formaldehyde (3705).

Key to the species:

1   Conidiophores long and thick, often with sterile hyphal elongations; side branches mostly short and thick, bearing crowded, short and plump phialides; colonies white or whitish green to green, generally with compact tufts of conidiophores   **2**
     Conidiophores and their side branches long and slender, without sterile hyphal elongations; phialides not crowded, rather slender; colonies yellowish, bright, dull to dark green, floccose or with compact conidiophore tufts   **5**

2(1)   Sterile hyphal elongations absent; conidia globose, hyaline   *(T. piluliferum)*
     Sterile hyphal elongations present or modified or rarely absent; conidia not globose   **3**

3(2)   Conidia green, short-ellipsoidal, surrounded by a wide irregular veil   *(T. saturnisporum)*
     Conidia smooth-walled or finely punctuated   **4**

4(3)   Conidia hyaline, small, $2 \cdot 4$–$3 \cdot 8 \times 1 \cdot 8$–$2 \cdot 2\,\mu$m   *T. polysporum* (p. 801)
     Conidia green, small to large, $3 \cdot 8$–$6 \cdot 0 \times 2 \cdot 2$–$2 \cdot 8\,\mu$m   *T. hamatum* (p. 795)

5(1)   Conidia roughened, $3 \cdot 6$–$4 \cdot 8 \times 3 \cdot 5$–$4 \cdot 5\,\mu$m   *T. viride* (p. 803)
     Conidia smooth-walled   **6**

6(5)   Conidiophores with a complicated dendroid branching system; phialides ± regularly disposed in numbers of 3 or more   **7**
     Conidiophores with a simpler branching system; phialides irregularly laterally disposed, often arising singly   **9**

7(6)   Conidia ellipsoidal or oblong, often appearing angular, $3 \cdot 0$–$4 \cdot 8 \times 1 \cdot 9$–$2 \cdot 8\,\mu$m
         T. koningii (p. 799)
     Conidia shorter with a length:width ratio of less than $1 \cdot 5$ **8**

8(7)   Conidia obovoid, with truncate base, $3 \cdot 0$–$4 \cdot 8 \times 2 \cdot 0$–$3 \cdot 0\,\mu$m; reverse of colonies often discoloured; colonies reaching 3 cm diam in 5 days at 20°C on OA   *(T. aureoviride)*
     Conidia globose, subglobose or short-obovoid, with length:width ratio of less than $1 \cdot 25$, $2 \cdot 8$–$3 \cdot 2 \times 2 \cdot 5$–$2 \cdot 8\,\mu$m; colony reverse uncoloured; colonies reaching >9 cm diam in 5 days at 20°C on OA   *T. harzianum* (p. 797)

9(6)   Conidia subglobose to ovoid, $3 \cdot 8$–$4 \cdot 5 \times 2 \cdot 5$–$3 \cdot 0\,\mu$m   *(T. reesei)*
     Conidia ellipsoidal   **10**

10(9) Phialides usually only slightly attenuate at the base; conidia large, partly dark green, to $7\,\mu$m long, mostly ellipsoidal   *(T. longibrachiatum)*
     Phialides usually distinctly attenuate at the base; conidia smaller, pale green $2 \cdot 8$–$4 \cdot 8\,\mu$m long, mostly oblong ellipsoidal   *(T. pseudokoningii)*

Trichoderma hamatum (Bonord.) Bain. *1906* — Fig. 352.

≡ *Verticillium hamatum Bonord. 1851*
≡ *Pachybasium hamatum* (Bonord.) Sacc. *1885*

Teleomorph: ?*Hypocrea semiorbis* (Hook. *1840*) Berk. *1860*

DESCRIPTIONS: Rifai (4842), and Komatsu (3086). — Colonies reaching over 7 cm diam in five days at 20°C on OA. *T. hamatum* is often recognizable macroscopically by the formation of greyish green pruinose pustules from a cover of mostly curled sterile conidiophore ends. Branches and phialides are particularly broad (the latter 3–4 μm wide) and the conidia short-cylindrical, green, smooth-walled by light microscopy and of variable size in different isolates. — CBS 244·63 was received as the anamorph of *H. semiorbis* and is a small-spored *T. hamatum* but it does not agree with Dingley's (1403) description. — *T. saturnisporum* Hammill *1970* has subglobose conidia ornamented with a wide undulating sheath. — The conidia of *T. hamatum* are slightly verrucose in SEM (2302).

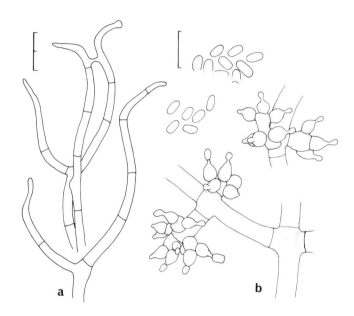

Fɪɢ. 352. *Trichoderma hamatum*. a. Sterile conidiophore ends; b. clusters of phialides and conidia; CBS 244.63.

Whereas a vast amount of data exists for *T. koningii*, *T. hamatum* has been rarely reported. Nevertheless, the few reliable determinations suggest a worldwide distribution. It is known from soils in the USA (1269, 2573, 4784), the British Isles (2344, 2345, 4028), the Netherlands (1617, 4658), Germany (1827, 1889), Poland (269, 272, 273), the USSR (5252), Italy (270, 271), Turkey (4245), Zaïre (2966, 3790), equatorial West Africa (1420), Namibia (1827), Nepal (1826), Australia (3720), Tasmania, New Zealand (4946, 4947, 5930), and Japan (3086, 5846). It has been isolated from forest soils (1269, 2573) under beech and pine (269, 270, 271, 272, 273, 2344, 2345), cultivated (1617, 1827, 1889, 4658) and heathland soils (3720). Most *Trichoderma* species accumulate in the soil after disinfection treatments (4028). In one study in a clay loam in England, *T. harzianum* was generally predominant, but it was found to be replaced by *T. hamatum* after liming (4028). *T. hamatum* was also found to be the commonest *Trichoderma* on pine roots from alkaline sites in Britain (1973). It has also been isolated from the rhizosphere of wheat (3567), rotting wood (2338), sticks exposed to soil (1956, 2833), earthworm casts (1429), and sewage sludge (1165, 1167). — Forty-seven percent of chloroform-fumigated mycelium was mineralized within ten days at 22°C (133). Growth *in vitro* was inhibited by *Memnoniella echinata* (4). *T. hamatum* can serve as food for the mites *Pygmephorus mesembrinae* and *P. quadratus* (3104).

Germination of conidia between -70 bars and vapour saturation (3086) in a nutrient-depleted system was best at pH $4 \cdot 3$ (1272). The optimum pH at normal CO_2 partial pressure is in the range $3 \cdot 7$–$4 \cdot 7$ and the optimal growth temperature is 24°C and the maximum 34°C or more, depending on the geographic origin of the isolate (1270, 1825, 1827, 3086); very slow growth occurs at 0°C (1267). — The best growth (at pH 4–4·5) occurs with amino-N, followed by ammonium, nitrate and urea; suitable C sources include starch (1827), D-fructose, D-galactose, D-xylose, D-cellobiose, D-trehalose, glycerol, *meso*-erythritol, D-mannitol, and dextrin (1271). The decomposition of CMC by *T. hamatum* is good, but there are differences in degree between different isolates (1432). High weight losses of loblolly pine needles and dogwood leaves occur after incubation with *T. hamatum* (1271). This fungus was also found to be involved in the decomposition of diesel-oil components (5630). It is markedly inhibited on media with either HCO_3^--ions in the alkaline range, or 5% NaCl (1270, 1991), corresponding to *c.* -40 bars, but maximal growth was obtained on media containing 1% NaCl (1428). — Antifungal activity of volatile and non-volatile metabolites from this species has been described (1343, 1344). Hyphal contact between *T. hamatum* and several saprophytic or parasitic fungi is accomplished by "curling" of hyphae (1345). The growth *in vitro* of *Candida albicans* (3599), *Heterobasidion annosum* (1973) and *Lentinus edodes* (3086) is inhibited and some isolates are antibiotically active against *Rhizoctonia solani*; in soil, however, the growth of *Rh. solani* is reported to be stimulated in the presence of *T. hamatum*. This fungus, in contrast to other *Trichoderma* species, is totally ineffective against *Armillaria mellea* (4028). Root growth of peas and wheat has been found to be inhibited *in vitro* (1430); *T. hamatum* can also cause lesions in corn seedlings (3705).

Trichoderma harzianum Rifai *1969* — Fig. 353.

Teleomorphs: *Hypocrea albofulva* Berk. & Br. *1873, Hypocrea* sp.I. J. Webster *1964, Hypocrea microrufa* Doi *1972, Hypocrea gelatinoperidia* Doi *1972, Hypocrea pseudogelatinosa* Komatsu & Doi *1973, Hypocrea subalbocornea* Doi *1973*

DESCRIPTIONS: Rifai (4842), and Komatsu (3086). — *T. harzianum* has been described as the smooth-spored counterpart of *T. viride*. Colonies reaching over 9 cm diam in five days at 20°C on OA. Conidia subglobose to short-oval, measuring $2 \cdot 8$–$3 \cdot 2 \times 2 \cdot 5$–$2 \cdot 8 \, \mu$m. To be sure of the smoothness, conidia of two-week-old colonies must be examined with an immersion lens. If there is only a slight roughening, an isolate should be referred to *T. viride* according to Rifai. — The distinction of *T. harzianum* from the anamorph of *Hypocrea vinosa* Cooke *1879* and *Trichoderma aureoviride* Rifai *1969*, which should have a more truncate conidial base, may also cause considerable difficulties, but the latter grows much more slowly *in vitro*.

Although *T. harzianum* has been rarely reported in the literature, it is very frequently received at CBS and CMI for identification and is perhaps the commonest species of the genus. Worldwide occurrence on the most varied substrates can be assumed. Soil isolations are recorded with high frequencies in the Netherlands (1614, 1616), the British Isles (4028), Canada and India (IMI), where it has been found in clay loam, a sandy ridge and a wheat field, respectively. In forest soils in the USA (Colorado, Ohio, North Carolina, Virginia, Washington) it was most frequently found in relatively warm regions (1269, 2573) but also up to

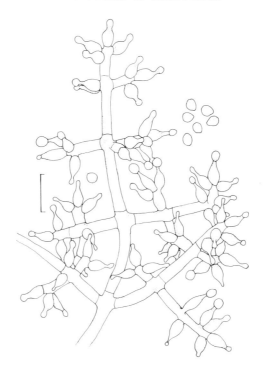

FIG. 353. *Trichoderma harzianum*, conidiophore and conidia, CBS 819.68.

3450 m elevation (4784); it has also been isolated from paddy fields in Italy (3913), light sandy soils in Libya (6510), forest soils in central Africa (3063), tobacco-field soil in Nigeria (1682), and the rim of a crater on the volcanic island of Surtsey (2406). It was isolated with increased frequency from soil after treatment with allyl alcohol (4028). Forty-nine percent of chloroform-fumigated mycelium was mineralized in soil within ten days at 22°C (133). *T. harzianum* was also commonly isolated from washed mineral soil particles (1614, 1615). Examples of its further substrates are roots of poplar (3452) and *Pinus contorta*, the rhizospheres of tobacco (1682), potato, sugar beet, wheat and grasses (1614), straw, wood, timber (IMI), often dominating on bed-logs of *Lentinus edodes* where it causes severe injury (3086), also on *Agaricus bisporus* (IMI), sclerotia of *Athelia rolfsii* (6276), rotting tubers of cassava (1568), moss, pea seeds (IMI), grains of *Sorghum vulgare* (57), pecans (2258), sewage sludge (1165, 1167), paper, textiles and jet fuel (IMI). — It can serve as food for the mites *Pygmephorus mesembrinae* and *P. quadratus* (3104).

Conidia germinate between (–100 to)–70 bars and vapour saturation or optimally in saw-dust with 30% moisture (3086); germination requires an external source of nutrients, and CO_2 under poor nutrient conditions; the percentage germination was much higher under acid conditions than under neutral ones (1272). — The optimum temperature for growth is in the wide range 15–35°C (1827) but 30°C represent a good average for most isolates; the maximum is 30–36°C, and the optimum pH is between 3·7 and 4·7 at normal CO_2 partial pressure (1270). An osmotic potential of *c.* −40 bars (5% NaCl) in the medium is tolerated;

the addition of HCO_3^- ions to an alkaline medium results in strong inhibition (1270, 1991). — At pH 4–4·5 the best growth occurs with amino-N, followed by ammonium, urea and nitrate; good growth also occurs with D-mannose, D-galactose, D-xylose, sucrose, cellobiose and D-mannitol; high weight losses of loblolly pine needles and dogwood leaves occur when incubated with *T. harzianum* (1271). *T. harzianum* is able to degrade starch (1827) and cellulose (4434); the degradation of the herbicide diallate in soil was slower than in a liquid medium (132). — In studies on the antagonistic properties of *T. harzianum*, it was observed that hyphae coil around or invade the hyphae of several tested fungi (1345, 1682). Some isolates have been found to be antagonistic to *Candida albicans* (3599), *Rhizoctonia solani*, *Armillaria mellea* (4028), and *Lentinus edodes* (3086). Carbon dioxide and ethanol production by *T. harzianum* were considered responsible for the growth and sporulation inhibition found in *Aspergillus niger* and *Pestalotia rhododendri* (2626). In soils infested with *Athelia rolfsii* (6276) or *Pythium aphanidermatum* (1682), successful biological control has been accomplished. *T. harzianum* can cause lesions in corn seedlings (3705).

Trichoderma koningii Oudem. *1902* — Fig. 354a.

Teleomorphs: *Hypocrea ceramica* Ellis & Everh. *1892, Hypocrea brunneo-lutea* Doi *1975*

DESCRIPTIONS: Rifai (4842), and Komatsu (3086). — Colonies reaching 3–5 cm diam in five days at 20°C on OA. Conidiophore branching as in *T. viride*, but conidia smooth-walled, slightly roughened under TEM (6161), short-cylindrical with a truncate base, $3–4·8 \times 1·9–2·8\,\mu$m. — *T. koningii* may be confused with *T. pseudokoningii* Rifai *1969* which has similar phialides and conidia, but in that species the phialides mostly arise singly and laterally and the whole conidiophore system is elongate rather than pyramidal (Fig. 354b).

While formerly almost all *Trichoderma* isolates with elongate conidia were generally identified as *T. koningii*, it has become evident since Rifai's monograph (4842) that *T. koningii* in the stricter sense is also a very common soil fungus. The following reports should be treated with caution, particularly if published before Rifai's monograph. — The recorded distribution includes Norway (5874), the British Isles (165, 413, 1446, 1598, 4182), the Netherlands (1614, 1616), Poland (269, 273, 1978, 3530, 4137, 6520), the USSR (1637, 2871, 2948, 3652, 3850, 4474), France (242, 1828, 2963, 4178), Germany (1736, 3042), Austria (1823, 3413, 3418, 4179, 5678), Czechoslovakia (1702, 1703), Yugoslavia (5534), Canada (540), the USA (13, 655, 1269, 2573, 2740, 3550, 4784, 5671), Italy (270, 271, 3445, 3450, 4537, 4538, 5047, 6082), Spain (3417, 3446, 3447), Turkey (4183, 4245), Syria (5392), Israel (2764, 2768, 2772, 4758), the Sahara (2974), Chad (3415), South Africa (1559, 4407), Pakistan (4855), India (968, 1317, 1519, 1525, 2186, 4700), Nepal (1826), Singapore (3331), China (3475), Japan (3086), Tahiti (6084), Central America (1697), Chile (1824), Peru (2005), and New Zealand (4947, 4948). *T. koningii* is most frequently isolated from forest soils: for example, under pine (269, 1598, 1637, 3447) or other conifers (4179), beech (270, 271, 273, 2719, 4537), *Nothofagus* (4947), *Acacia* (2186), or other broad-leaved trees (1702, 1736, 2573), in a forest nursery (1978), fertilized forest plots (1828), under pioneer vegetation in successional forest communities (3550); also very frequently isolated from cultivated soils, (too numerous to list here individually), but also grassland (165, 1614,

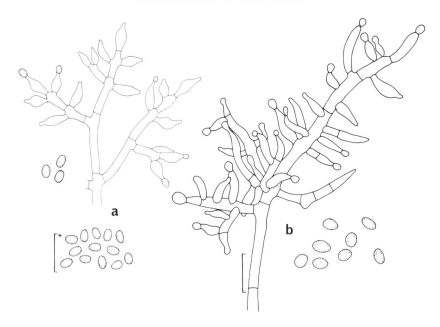

Fig. 354 a. *Trichoderma koningii*, conidiophore and conidia, CBS 817.68; — b. *Trichoderma pseudokoningii*, conidiophore and conidia, CBS 489.78, orig. J. Veerkamp.

2948, 3417, 3445, 4184, 4948, 6082), soils with steppe type vegetation (1559, 3415, 4407, 6347), a citrus plantation (2764), garden soil (968), vineyards (242, 4178), heathland (4178, 5047), peat bogs and peat fields (3530, 4180, 4474, 6520), swamps (2719, 4700), salt-marsh (413), sandy soils (2974, 4178, 4184), soils on sandstone and mergel (3127), leached brown soil (2963, 3414), podzols (3413, 3414, 3416, 5534), rendzina, chernozem, loess, terra fusca (3414, 3418), terra rossa (3417), on coastal cliffs (4178), in estuarine sediments (655), a uranium mine (1703) and caves (4137). — This fungus occurs relatively frequently in the surface (litter) layer of soils (272, 273, 540, 1736, 3445, 3572, 4947), but also at a soil depth of 120 cm (5671), and generally in slightly acid habitats (269, 271, 2963, 3331, 3445, 3450, 4538, 5047). — It was present in high frequencies in soils treated with allyl alcohol (4028) or various fungicides (6135, 6136). — Forty-one percent of chloroform-fumigated mycelium was mineralized in soil within ten days at 22°C (133). — It is common on leaf litter, e.g. of *Abies firma* (4984), *A. grandis, Picea sitchensis, Pinus sylvestris* (2343), beech, *Ginkyo biloba* (3448), and *Castanea sativa* (3449). It has also been isolated from *Spartina townsendii* (5389), decaying stems of *Heracleum sphondylium* (6462), peat (3391), stumps of ash, alder, birch, willow (3568), wood exposed to soil (1956) or seawater (4546), bed-logs of *Lentinus edodes* under shady humic conditions during the hot season (3086), chitin buried in soil (4264), roots of wheat (5330), broad beans (6134), sugar beet (4559) and pine (3572), the rhizosphere of clover (3429), groundnut (2768), alfalfa (5825), coriander (1523), coffee (5313), potato, sugar beet, wheat (1614), oats, barley (3006) and grasses (1614, 3376, 3864), *Suillus luteus* mycorrhiza on *Pinus strobus* (1794), seeds of cotton (5009), groundnuts (2765) and *Avena fatua* (2961), achenes of sycamore (1683), pecans (2572), corn silages (1092), self-heated wood chips (1756), birds' nests (168), dead adult bees (402), cotton tissue of conveyor belting (6163), and wood pulp (3294, 4857). — Conidium germination is relatively insensitive to soil mycostasis (1407, 5144). Growth *in vitro* has been seen to be stopped by hyphal contacts with *Tuber melanosporum* (632).

Germination of conidia under poor nutrient conditions requires an external source of nutrients and CO_2 and is better under acid than neutral conditions; the acidity effect can be eliminated by the use of malt extract (1272). — The optimum temperature for growth is 26°C or higher, depending on the origin of the isolate (3086, 6513), the maximum 32–40°C, and the optimum pH 3·7–6·0 under normal CO_2 partial pressure. Growth is strongly inhibited by HCO_3^- ions (1270); it grows well at –30 bars water potential but not at all below –110 bars (1991, 6513). — Optimal growth occurs on amino-N, followed by ammonium, urea and nitrate; known suitable C sources include D-fructose, D-galactose, D-ribose, cellobiose, melibiose, D-trehalose, glycerol, *meso*-erythritol, dextrin (1271), D-mannose, D-glucose, D-xylose, maltose fructosan and fumaric acid (6513). The degradation of cellulose (825, 1271, 2224, 2718, 3127, 4184, 5872, 5873) with production of C_1, C_x, β-glucosidase and (purified) 1,4-β-glucan cellobiohydrolase has been described (1688, 2225, 2226, 5873, 6435); amylase, β-fructofuranosidase, a-glucosidase (5873), and endo-1,3-β-xylanase (3231) have also been identified. *T. koningii* will also grow on sodium polypectate and pectin (825, 1688, 3127, 4183, 4184), and two endopolygalacturonases have been identified (1687); chitin (825, 4264) and tannin (3127) are utilized, and ribonuclease (3231), phenol oxidase (934) and proteolytic enzymes are produced (2741, 3429, 5873). Dieldrin is said to be degraded by *T. koningii in vitro* (541). — The formation of a staling substance with mycostatic properties has been described (4425) and antifungal substances were noticed but not identified by other workers (716, 1343, 1344, 2753, 2754, 3570). In soils of a citrus grove, a negative correlation between the occurrence of *T. koningii* and *Fusarium* spp. was observed (2763). *T. koningii* is able to coil around the hyphae of *Lentinus edodes* (3086) and several other saprophytic or parasitic fungi tested (1345). Some isolates were found to stimulate the production of oospores in *Phytophthora cinnamomi* (4623). In soils containing 10^4–10^5 propagules g^{-1}, *T. koningii* produced lesions in 70–100% of corn seedlings (3705). — The fungus shows a high tolerance to CO_2 (2096).

Trichoderma polysporum (Link ex Pers.) Rifai *1969* — Fig. 355.

$\equiv$ *Sporotrichum polysporum* Link ex Pers. *1822*; Fr. *1832*
$=$ *Trichoderma sporulosum* (Link ex Pers.) Hughes *1958*
$\equiv$ *Sporotrichum sporulosum* Link ex Pers. *1822*
? $=$ *Trichoderma album* Preuss *1851*
$=$ *Pachybasium candidum* (Sacc.) Peyronel *1913*
$\equiv$ *Pachybasium hamatum* (Bonord.) Sacc. subsp. *candidum* Sacc. *1855*

Teleomorph: *Hypocrea pachybasioides* Doi *1972* (others not yet described (6247))

DESCRIPTIONS: Rifai (4842), and Komatsu (3086). — Colonies relatively slow-growing, reaching 7·0 cm diam in ten days at 20°C on OA, formimg some discrete white pustules which are covered with sterile conidiophore ends. Conidiophore branches and phialides relatively broad (the latter 3–3·5 μm wide); conidia short-ellipsoidal, in *Hypocrea pachybasioides* apparently roughened, but in SEM (2302) smooth-walled, 2·4–3·8 × 1·8–2·2 μm, compacted into slimy heads. — The other known white species, *T. piluliferum* J. Webster & Rifai *1969*, has a conidiophore branching pattern similar to *T. viride* and globose conidia.

FIG. 355. *Trichoderma polysporum*, conidiophores, one with sterile end, and conidia, CBS 147.69.

T. polysporum is a widely distributed species, though apparently not frequent anywhere. Records are available from Europe and North-America, and also from Nepal (1826), Australia, New Zealand (IMI), equatorial West Africa (1420), central Africa (3063) and Japan (1414, 3086, 5846). It ranges in the north up to Sweden (5465) and Finland (5663), and also the alpine zone in Europe (1876, 3445). In soils in the USA it is restricted to those with low temperature (1267, 1269, 1446). It has been isolated from soils under beech (272, 273, 3138), mixed hardwood (2573, 3448, 3549), pine (269, 1598, 3798), spruce (3975, 5465, 5663), mixed conifer-hardwood (6350) and other (swampy) coniferous forests (1039, 5363, 6138). It is also known in moorland soils (3234), *Sphagnum* peat (1267), arable soils (540, 2161, 2949, 3450, 3982), estuarine sediments of varying salinity (655), polluted running water (1154, 1157), sewage sludge (1165), owl pellets (IMI), earthworm casts (1429), bed-logs of *Lentinus edodes* where it causes severe damage under shady conditions in the cool season (3086), rotten wood (IMI), wood pulp, slime of paper mills (699), *Pteridium aquilinum* litter (1821), and leaf litter of oak, maple and pine (IMI). In some investigations it was found only in the superficial soil layers (3975) while in others it occurred down to 20 cm (2161) or more (540); apparently it is more prevalent in the finer textured soils (6350). Its occurrence is favoured by mineral fertilizers in the order PK>NP>NPK>NK (2163). It has been isolated from the rhizospheres of wheat (2949, 3567), alfalfa (3982), poplar (5898) and spruce (3573), and from seeds of corn (57) and groundnuts (1016). — It can serve as food for the mites *Pygmephorus mesembrinae* and *P. quadratus* (3104).

The germination of the conidia requires vapour saturation (3086), an external source of nutrients and CO_2, and is strongly inhibited under neutral or alkaline conditions (1272). Isolates from arctic areas are reported to still grow at 0°C (1267); otherwise the temperature range for growth is 10–30°C with an optimum at 20–25°C and the maximum at 28–31°C; the pH range for growth is pH 3·5–5·6 and the optimal pH (at normal CO_2 partial pressure) 3·4–4·7; growth is strongly inhibited by HCO_3^- ions and 5% NaCl in the medium (1270, 3086), corresponding to *c.* −45 bars. — *T. polysporum* grows best with amino-N, followed by ammonium, urea and nitrate; good C sources include D-mannose, D-galactose, D-xylose, cellobiose, glycerol and D-mannitol (1271). Decomposition of CMC is very good and little differences in performance occurs between isolates (1432); cellulose decomposition increases with the concentration of the N source (4433); birch wood is eroded (4191) and chitin moderately decomposed (2706). — Hyperparasitism (1017), hyphal coiling around (1345) and antagonistic activity towards *Lentinus edodes* (3086) and several other test fungi have been shown (1431). A non-volatile antibiotic has been reported (1343), and the metabolites reported include trichodermin, trichodermol (=roridin C) (28), pachybasin and chrysophanol (5278).

Trichoderma viride Pers. ex Gray *1821* — Fig. 356.

= *Trichoderma lignorum* Tode ex Harz *1871*
? = *Trichoderma glaucum* Abbott *1926*

Teleomorphs: *Hypocrea rufa* (Pers. ex Fr.) Fr. *1849, Hypocrea aurantiaca* P. Henn. *1900, Hypocrea coprosma* Dingley *1952, Hypocrea atrogelatinosa* Dingley *1956, Hypocrea muroiana* Hino & Katsumoto *1958, Hypocrea albo-medullosa* Doi *1972*

DESCRIPTIONS: Rifai (4842), and Komatsu (3086). — Colonies reaching 4·5–7·5 cm diam in five days at 20°C on OA. Conidiophores typically pyramidally branched, i.e. short branches occurring near the tip and longer ones with repeated branching in the lower part. Phialides arranged in divergent groups of 2–4, slender and irregularly bent. Conidia almost globose, 3·6–4·5 μm diam (or to 4·8 μm long), mostly distinctly roughened. In the anamorphs of *H. rufa* and *H. albo-medullosa* the roughening is coarse and recognizable under low power, whereas other isolates require oil immersion optics to discern the sculpturing. — *H. atrogelatinosa* was described as having roughened conidia, but the representative isolate, CBS 237.63, has practically smooth-walled conidia, while in *H. aurantiaca* the roughened conidia are elongate. — TEM studies of *T. viride* with intact (6161) and freeze-etched conidia (2798) show the coarse surface structure, while Hashioka (2302) apparently used *T. harzianum* instead of *T. viride* for his SEM work. — SEM of intact colonies (1870) allows the distinction of several hyphal types. Conidial cell walls contain 1,3-β-glucans as do mycelial walls, with no chitin but melanin occurring in the outermost layer (445). — DNA analysis of *T. viride* gave a GC content of 49·5–51% (5600).

The data on ecology and physiology compiled here have to be judged with considerable reserve in view of the frequent confusion of this species with other species of the genus. — The *H. rufa* teleomorph is frequently encountered in the temperate zone of Europe (e.g. 6240) on

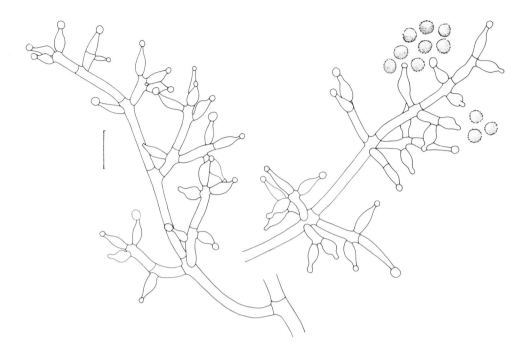

FIG. 356. *Trichoderma viride*, conidiophores and conidia, CBS 189.79.

decaying wood (1538). *T. viride* in a broad sense can, however, be ranked as one of the most widely distributed of all soil fungi. On account of its rapid growth, the fungus is easily isolated by all techniques, particularly using immersion tubes and plates. — It appears in all investigations of habitats in the extreme north, in alpine areas as well as in numerous analyses from tropical regions. There are numerous references for a wide range of forest, grassland and cultivated soils which cannot be cited here individually. The fungus has been shown to predominate in the soil of very humid mixed hardwood forests (2007, 2008, 3549, 4225, 4716), but occurs also very generally in salt-marshes (413, 4645, 4646, 4987), polder soils (4658), saline soils (3446), mangrove swamps (3264, 3265, 4700), dunes and deserts (745, 1423, 1512, 2973, 2974, 4162, 4540, 6414), soils with tundra (1446) or savannah type vegetation (1559, 4407, 4718, 4719, 5049), soils under burnt and unburnt chaparral (1164), heathland (989, 3720, 5047, 5220, 5221, 5222, 5811, 5813), and very frequently in peat and moorland soils. Its occurrence has also been reported in polluted streams (1154, 1155, 1157, 1166, 1482), a sewage treatment plant, a waste stabilization pond (1166), composted household refuse (1132), activated sludge (1172) and fields treated with sludge (1163), children's sandpits (1421), fresh water (56, 1166, 3809, 4229, 4850) as well as in estuarine water and sediments (56, 4850), seawater in eulittoral and oceanic zones (4850, 4918), coastal soils (4477), on driftwood and in soil of a crater rim on the volcanic island of Surtsey (2406), and in clay deposits of a carst cave (3364). It is in particular an early colonizer of the litter layer in the most varied forest soils (269, 983, 989, 2344, 2411, 3449, 3801, 6351). Its greatest density in alpine podzol was attained in the F_2 horizon (2923) and, under conditions of excessive moisture, in the F and H horizons of a forest soil (1269); generally it decreases with increasing soil depth; however, it has been found by numerous workers to occur at least to a depth of 60 cm or into the C horizon. *T. viride* occurs in soils at all pH values between 3·1 and 8 (6182), though it prefers the more acid part of this range (745, 1204, 1713, 1973, 3417, 3530, 5758, 6182). It has also been shown to be present in soil in which copper had accumulated (2922, 4921).—The growth in soil is apparently little influenced by the incubation temperature (160, 5758), but in comparison with other *Trichoderma* species it is said to occur most frequently in cool temperate regions (1269). Stimulation by manuring with mineral fertilizers and farmyard manure (2161, 2163) has been reported and disinfected soils are rapidly recolonized (6184). The numbers of propagules increased in the rhizosphere of *Picea glauca* after soil treatment with the fungicide captan but decreased after methylbromide fumigation (6026). It is apparently insensitive to sulphite liquor (1161). — *T. viride* occurs as a colonizer of numerous plant materials, for example, timber used in a copper mine (2664), wood of *Betula alleghaniensis, Acer saccharum, Fagus grandifolia* (1295) and *F. sylvatica* (3568), bed-logs of *Lentinus edodes* where it causes severe injury under shady humid conditions during the hot season (3086), decaying twigs of *Alnus sinuata* (1152), litter of *Eugenia heyneana* (5513), *Pinus sylvestris* (2345, 2925), *P. radiata* (6080), other conifers (2343), *Castanea sativa* (3449), beech (273), *Nothofagus truncata* (4946), other broad-leaved trees (650), wood baits buried in soil (5237), litter of *Spartina alterniflora* (1958), leaf litter of cotton (5231), *Pseudoscleropodium purum* (2969), leaves of *Adiantum, Polypodium* (2859) and *Brachypodium pinnatum* (2970); roots of numerous plants, the rhizoplane of sugar cane (2861), *Trichodesma amplexicaule, Abutilon indicum* (3864) and *Anthoceros* (2860). It has also been isolated from wood pulp (3294), accumulated slime of paper mills (5060), marine algae (1256), and rabbit dung (6085). Negative rhizosphere effects have been reported only for *Salicornia* (4645) and *Lolium perenne* (2611); otherwise, marked accumulations have been observed, particularly with beans (4451, 4452, 5758, 5759), groundnuts (2768, 4735), clover (3429, 5815), wheat (1614, 2816, 2949, 3567, 4727, 5330, 5825), corn (1330, 3823), rice (2712), sugar cane (2861) and sugar beet (1614). It has also been isolated from the rhizosphere of wilted pineapple plants (6210), coriander (1523), carrot, *Brassica oleracea* (3733), tomato (264), papaya (5345), coffee (5313), *Nicotiana tabacum* (3431), *Polypodium, Cyclosorus, Adiantum* (2856), peat bog plants (1376), several euphorbiaceous plants (3866), various grasses (1614, 3376), *Picea glauca* (6026), *Quercus, Carpinus, Corylus* (4817) and *Halimione portulacoides* (1379), as well as on the mycorrhiza of certain forest trees (1794). In barley, dead roots were colonized more strongly than live roots (4450). From washed mineral soil particles it was isolated more frequently than from organic ones (1614). Its occurrence in stored grain of wheat and oats (1752, 2716, 3491, 3556, 4492, 4591), barley (1753), seeds of *Avena fatua* (2961), cultivated grasses (3512), Brazil nuts (2515), pecans (5092), groundnuts (2765, 2768), stored sweet potatoes (3066) and tomatoes (5899) is also reported. *T. viride* has also been isolated from birds (4652), birds' nests (682, 2577, 5067), feathers, droppings and pellets (2575, 6223), and mummified bees (402, 2117). — Relatively rapid spread was observed in sterile soil (4220). Conidia introduced into the soil germinate only to a limited extent (633, 849), and mycostasis against *T. viride* is correlated with soil alkalinity (5144) and the soil water content (5181); ten isolates examined differed strongly in their sensitivity to soil mycostasis (3879). Germination of conidia can also be inhibited by a sporostatic factor produced by *Fusarium oxysporum* (4882), the effect of which is reversed by lowering the pH, or adding DL-a-alanine and other amino acids. *T. viride* is relatively insensitive to antagonistic effects from other micro-organisms (2260, 3213). In chitin-enriched soil an extractable inhibitor active against *T. viride* is produced (5453). This fungus is also inhibited *in vitro* by oak leaf litter tannins (2282), the gas phase of terpinolene, *p*-cymene and *a*-pinene (1333) and some volatile bacterial metabolites (3920); it is further slightly inhibited by cochliodinol, an antibiotic metabolite from various *Chaetomium* spp. (704) and *in vitro* in the presence of

Stachybotrys chartarum (1964). It can serve as food for the mites *Suidasia medanensis* (5380), *Acotyledon redikorzevi* (5379), *Pygmephorus mesembrinae* and *P. quadratus* (3104). — It can survive in the deeper layers of burnt forest soil (2711) and recolonize the upper layer quickly. Only about 25% of the conidia survived 12 days incubation in non-sterile soil (4421). *T. viride* can survive in the chlamydospore state in soil for at least 20 months. Straw, once colonized by this fungus, can be occupied by other fungi only with difficulty (753, 3213), but the competitive saprophytic ability apparently decreases in the presence of the herbicide diquat (6367). Since *Fusarium solani* is particularly sensitive to *T. viride*, a negative correlation has been shown between the occurrence of these two species (2763).

The optimum temperature for linear growth on agar and for mycelium production is in the range 20–28°C; good growth has, however, still been recorded at 6°C and 32°C (191, 1204, 1827, 2193, 2698, 4517, 5874, 6191), but the maximum for most isolates is 30°C (191, 1270); some exceptional isolates are able to grow at 37°C (160); no growth occurs at 0°C (1267). The thermal death point in soil has been determined as between 49° and 55°C for 30 min (613, 6453, 6454). Growth is possible in the pH range 1·5–9·0, the optimum being in the range pH 4·5–5·5 (744, 2193, 3086, 6191). A high salt content in the medium inhibits conidium germination but not mycelial growth (655, 1270, 1991); no growth occurs in media with 10% NaCl (4987). The minimum water potential *in vitro* for vegetative growth is −110 bars and for sporulation between −70 and −100 bars (2093, 2193, 4369), but *T. viride* was also reported to grow at −240 bars when incubated at 10°C (6344). — An atmosphere with 10% CO_2 hardly affects growth (784, 4029); the response to CO_2-enriched atmospheres was, however, shown to be pH-dependent (1270), and good growth was even obtained under anaerobic conditions (1425). — Two phases of conidium germination have been described which are characterized (a) by water uptake and (b) increasing protein synthesis and heat sensitivity (5494). Conidium germination occurs in the range −100 to −70 bars and vapour saturation (3086); it requires CO_2 and exogenous C and N sources (1272): without nitrogen, swelling of conidia occurs, but there is no germination, while without carbon source, neither process takes place (3625); the optimum temperature for conidium swelling is 30°C (3726). With little available nutrients, germination was much better under acid (pH 4·3) than neutral conditions (1272); germ tubes of conidia germinating in groups have been found to show negative autotropism (4883). The behaviour of respiratory enzymes during conidium germination has been followed (4296); the shape and numbers of mitochondria change during conidium maturation and the conidial wall becomes thicker, while on germination it becomes more permeable again (1872, 4909). — *In vitro*, a low N content promotes the production of conidia, while a decrease due to high nitrate concentrations can be compensated for by the presence of magnesium sulphate; Cu ions promote the development of conidium pigmentation, while those of iron are important for good sporulation (717). Under submerged conditions, conidia are produced in media supplemented with trace elements (6088). Light favours conidium production (5539) in logarithmic proportion to its intensity (1654, 1872); according to different reports, wavelengths from 430 to 490 nm (543, 3818) and/or 320 and 380 nm (3170) are the most effective; alternating and narrow-beam illumination shows that induction of conidiogenesis is restricted to areas of growing mycelia (1869, 1872). — Production of perithecial stromata of *H. rufa* has been accomplished once *in vitro* by five weeks incubation on cornmeal agar at approx. 20°C under daylight (6244). — The incorporation of (^{3}H-)*N*-acetylglucosamine in growing hyphal tips has been used to follow its growth dynamics *in vitro* (1871). Most sugars are good C sources (1271) but D-arabinose (6191), and, in contrast to most fungi, urea (3414) are less suitable substrates. When grown on glucose under submerged conditions at 28°C, the yield coefficient was 0·4 g cells g^{-1} substrate (744). The decomposition of starch and amylase production by *T. viride* are both very good (655, 1425, 1750, 1827, 3471, 3472); Ca^{2+} ions reduce the amylase activity (3563). The production of an inducible β-D-xylopyranosidase (4777) has been described and 1,3-β-D-glucanase production found to be weak (4778); sclerotan is broken down by an enzyme of this latter type (267, 2803), an exo-glucanase (2298) is produced, and endo-1,3-α-D-glucanase, which can hydrolyse cell walls of *Fusicoccum amygdali* (2299) and numerous basidiomycetes (6126), was isolated. Good pectinase production is documented (655, 3416, 3471, 4779), and little differences seem to occur between isolates (1432). Xylan is also utilized well (3416, 3471, 3472, 5427) and arabinoxylan degraded by both endo-1,3-β-xylanase and cellulase (1753). According to numerous reports, cellulose is attacked in the most varied forms and this fungus has thus often been employed for biochemical investigations on cellulases in biodeteriogenic testing (though several studies citing *T. viride* deal in fact with *T. reesei* Simmons (strain QM 6a)): conditions for their formation, induction (3403), depression by sugars, optimum N sources (2177), C/N ratio (3303), stability, range of temperature and pH, substrate specificity, course of decomposition in cotton fibres (2224), ultrastructural changes in cells during growth on cellobiose and cellulose fibres (465), cofactors of enzyme activity (3563) and enzyme components (470, 471, 472, 3322, 4779, 5194, 5195, 5288, 5857) have all been investigated. The cellulolytic activity has been used to process liquid feedlot waste into a stable, odourless substrate (2107). Chitin is decomposed in the soil by *T. viride*, which has also been isolated from buried chitin strips (2068, 2069) and been successfully tested *in vitro* for chitinase activity (1425, 2902, 4264). The existence of a proteolytic enzyme system decomposing β-fructofuranosidase of fungal origin has been assumed (4892). There are reports on the breakdown of synthetic dipeptides (3699) and fats (2719) and the ability to attack various plasticizers (473), dihexyladipate and dioctylsebacate (5859). *In vitro* it apparently can attack the insecticides carbaryl (610), dieldrin (3677), bromophos (5558), dichlorvos, DDT, and some others (444, 3676, 3678, 4464); cell-free preparations show a degrading activity on ^{14}C-DDT (134), and PCNB is metabolized *in vitro* within six days to less than 10% (935). *T. viride* can grow on simazine and atrazine as sole C or N sources (3799, 5616) with the chain-C in simazine being preferentially utilized (2894); atrazine is degraded by N-dealkylation of the alkylamino-group (2892); in soil in the presence of atrazine an increased uptake of nitrate by *T. viride* was observed (1239). Chlorphenamidine, dicryl, propanil, propachlor, propham and solan are metabolized to the aniline moiety and/or chloride ions (2893). Mono- and dichloroacetates are dechlorinated (2742, 2743). Allylalcohol is utilized optimally with 0·1% glucose in the medium and metabolized to acrylate and acetate via an α-oxidative pathway (2702); cinerone can be hydroxylated (5681). — This fungus tolerates only 0·25% tannin in the nutrient solution (3052) and displays no particular tolerance to phenol and gallic acid (1153), but good growth on phenol lignin, calcium lignin sulphonate (1744) and wood (3769, 3956) occurs; wood is eroded (4191) and soft rots may be induced (3323, 5072). Wood decomposition has been confirmed by observations of the cell wall structure (3768) and measurement of losses in weight and tensile strength (3323, 3767, 3770). Hair keratin is attacked (3530) by boring hyphae (1633). Mineral N sources are well utilized (6191) but particularly good growth is achieved with asparagine (1153). Chemotropic growth towards phosphate granules *in vitro* occurs without their dissolution (3074). A maximum uptake of [134]Cs was found to be correlated with a potassium deficiency and a maximum uptake of ^{60}Co occurred at pH 4·0 (6410). There appears to be no requirement for growth substances (1153, 6191), except, perhaps in special circumstances where there is a high nitrate content in the culture solution (565). The presence of clay minerals (2%) in nutrient solutions was found to cause a marked inhibition in respiration (5604). With glucose or starch as C sources, the fungus produced 60–80% of the theoretically possible amount of citric acid (5775). On starch- and glucose-containing media the production of humic acid-like substances has been observed (5240). In ageing cultures an uncharacterized acidic substance with a morphogenic effect on fungal hyphae has been demonstrated (4435). The coconut-like aroma of *T. viride* has been traced to 6-pentyl-*a*-pyrone (1123), which may have a stimulatory influence on the oospore production in *Phytophthora cinnamomi* (3984, 4786). Other metabolic products known include pachybasin, chrysophanol and emodin (5410). Neither gliotoxin (436, 715) nor viridin (710, 721, 6100) are produced by *T. viride*, but by *Gliocladium virens* (q.v.) (6245). — Nevertheless, of a series of *Trichoderma* species investigated, isolates of *T. viride* proved to have the strongest antibiotic activity (4028). Antibiotic properties and a structure related to trichothecin are found in the sesquiterpene trichodermin (2016, 2017). Other substances isolated include a cyclic polypeptide (3783, 4822), the cyclic peptide trichotoxin A with mammalian toxicity (2547), the antibacterially active peptide suzukacillin (4302), and the unsaturated monobasic acid dermadin active against gram positive and gram negative bacteria (3782, 4679, 6593), a volatile antibiotic active against *Rhizoctonia solani, Fusarium oxysporum, Mucor hiemalis* and *Pythium ultimum* (1344), and non-volatile substances antagonistic against *Lentinus edodes* (2303, 3086), *Rhizoctonia solani, Heterobasidion annosum* and other test fungi (1343). *T. viride* is regarded a lethal parasite on *Pholiota nameko* (3085). The antagonistic activity of this fungus has led to numerous experiments in culture; these will not be detailed here; in experiments with partially sterilized soils, the following pathogens were suppressed by *T. viride: Pythium* sp. (2078, 5370); *Macrophomina phaseoli* (4220), *Verticillium albo-atrum* (232), *V. dahliae* (920), *Gaeumannomyces graminis* (5409), *Fusarium oxysporum* f.sp. *lini, Colletotrichum lini, Helminthosporium* sp., *Alternaria* sp. and *Rhizoctonia solani* (1711), provided that the soil pH was kept low (1354, 6260); *Athelia rolfsii* (2260, 4890) and *Armillaria mellea* are also suppressed, but this is closely dependent on the density of the *Trichoderma* inoculum (262, 1907). The success is especially evident after soil disinfection owing to its rapid recolonization (556, 1655, 6184). On inoculating non-sterilized soil, the results of this biological control were, as to be expected, less certain. — There are, however, positive reports on stump treatment for control of *Heterobasidion annosum* (4121), control of *Rhizoctonia solani* attack in potato (5184) and cabbage (2698) in the field, and a reduction of *Pythium* and *Rhizoctonia* attack in cucumber seedlings (109). These antagonistic and parasitic activities have been confirmed in a critical study with several isolates of *T. viride* (4028), but they also occur in several other *Trichoderma* species. Scattered pellets (grown on barley grain, mixed with ground bark etc. and containing *T. viride* and *T. polysporum*) have been used to introduce the fungus as an antagonist against pathogens and decay fungi in wood, trees and vegetable crops, where protection is required (4829); the fungus involved in this work, however, has now been reidentified as *T. harzianum* (J. W. Veenbaas-Rijks, pers. comm.). Seed treatment with *T. viride* (or with a conidial preparation termed "trichodermin 3") is said to reduce injuries by *Cochliobolus sativus* on barley (5962), *Leptosphaeria maculans* in cabbage (6011), *Pythium ultimum* in beets (3378), and *Fusarium oxysporum* f.sp. *lini* and *Collectotrichum lini* in flax (1710). — Hyperparasitism has been observed on numerous fungal hosts including sclerotia of *Sclerotinia sclerotiorum* (420, 1729, 2807, 3531, 4584, 4699). *T. viride* has also been isolated from sclerotia of *Sclerotium cepivorum* (3990), *Sclerotinia trifoliorum* (2886), *S. borealis,* and *Claviceps purpurea* (3531); it also parasitizes the mycelia of *Fusarium solani, F. oxysporum, Cochliobolus sativus, Rhizoctonia solani* (868, 1511, 5186, 6259), *Athelia rolfsii, Pythium* sp., *Phytophthora parasitica* (6259), *Lentinus edodes* (3087), *Armillaria mellea, Phaeolus schweinitzii* (262), *Gaeumannomyces graminis* (5409), *Rhizopus oryzae, Mucor plumbeus, M. hiemalis, Actinomucor elegans, Phycomyces nitens, Circinella muscae, Zygorrhynchus moelleri* and *Syncephalastrum racemosum* (1511). Hyphal contact with a host fungus is frequently accomplished by means of entwining narrow hyphae

(1345). — Negative influences on various cultivated plants have often been detected in pure-culture tests, in sterilized soils, after accumulation in the soil and after seed inoculation (233, 293, 2611, 2645, 3272, 3273, 5961, 6028, 6116, 6450). — This fungus has been described also as a wound parasite (1975) and as the causal organism of storage rots in sweet potatoes (1133) and citrus fruits (2192, 2193). The attack of alfalfa roots by *Pratylenchus penetrans* was increased in the presence of *T. viride* due to a stimulation of root respiration and exudation (1543). — Conidia are relatively resistant to γ- irradiation (5748); in a chronically irradiated soil, *T. viride* appeared with increasing numbers at 230 R per day, but no viable cells occurred at 1400 R per day (2009).

Trichophyton Malmsten *1845*

Type species: *Trichophyton tonsurans* Malmsten

Teleomorph: *Arthroderma* Berk.

DEFINITION OF THE GENUS: Ajello (73). — Descriptions of teleomorphic species: Padhye and Carmichael (4343); including other species: Dvořák and Otčenášek (1521). — Colonies spreading, whitish, producing both smooth-walled macro- and micro-conidia. Macro-conidia mostly borne laterally directly on the hyphae or on short pedicels, thin- or thick-walled, clavate to fusiform; in some species rarely produced. Micro-conidia spherical, pyriform or cylindrical or of irregular shape. — The solitary thallic conidium formation has been analysed by Galgóczy (1862). The presence of micro-conidia distinguishes this genus from *Epidermophyton* Sab. *1910*, and the smooth-walled, mostly sessile macro-conidia separate it from *Microsporum* (q.v.). Twenty species are recognized (73) and those treated here are found under the names of their *Arthroderma* teleomorphs. Isolates of the *T. terrestre* complex can hardly be distinguished by the anamorphs alone.

Key to the species treated:

1 Macro-conidia predominating, well differentiated, fusiform with rounded apices, wall mostly over 2 μm thick, 18–60 × 8–12 μm   *Arthroderma uncinatum* (*T. ajelloi*) (p. 73)
  Macro-conidia less predominant, cylindrical-clavate, wall less than 1·2 μm thick (*T. terrestre* complex)   **2**

2(1) Macro-conidia predominantly 2-celled, 6–20(–30) × 2–5 μm; colony reverse yellow-brown, or pink to vinaceous; peridial cells asymmetrically constricted and without protuberances
                          A. insingulare (p. 69)
  Macro-conidia predominantly 3–6-celled, 15–20 × 4–6 μm   **3**

3(2) Colony reverse mostly yellowish brown; peridial cells asymmetrically constricted with 2 protuberances at each end   *A. quadrifidum* (p. 70)
  Colony reverse mostly yellow, peridial cells symmetrically constricted *A. lenticulare* (p. 69)

Trichosporiella Kamyschko ex W. Gams & Domsch *1970*

Trichosporiella cerebriformis (de Vries & Kleine-Natrop) W. Gams *1971* —
Fig. 357.

≡ *Sporotrichum cerebriforme* de Vries & Kleine-Natrop *1957*
= *Trichosporiella hyalina* Kamyschko ex W. Gams & Domsch *1970*

DESCRIPTIONS: De Vries and Kleine-Natrop (6125), and Gams and Domsch (1887); nomenclature: v. Arx (198). — Colonies slow-growing, reaching 2·0 cm diam in ten days at 20°C on MEA; whitish, the surface moist and slightly granular. Vegetative hyphae prostrate, somewhat fasciculate, 1–3 μm wide. Blastoconidia sessile, borne singly through narrow lateral openings on undifferentiated hyphae either on the agar surface or submerged; conidia globose, hyaline, simple, smooth-walled, guttulate, 3·5–6 μm diam.

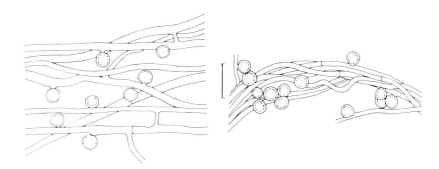

FIG. 357. *Trichosporiella cerebriformis,* pleurogenously formed blastoconidia, CBS 244.68 (from 1887, reproduced with permission J. Cramer).

T. cerebriformis was originally isolated from a seborrhoic-pityriasis-like lesion of a human scalp (6125). Although it has very rarely been reported, we know it as a frequent soil fungus in both Germany (1888) and the Netherlands (1614, 1616), and in the Netherlands it has also been isolated from the rhizospheres of potato, wheat and grasses (1614). Other known habitats are arctic tundra soils (1750) and organic detritus in fresh water (4429).

Pectin (1432), starch and cellulose (1750, 4433) are degraded and birch wood can be eroded by *T. cerebriformis* (4191).

Trichosporon Behrend *1890*

Trichosporon beigelii (Küchenm. & Rabenh.) Vuill. *1902* — Fig. 358.

≡ *Pleurococcus beigelii* Küchenm. & Rabenh. *1867*
= *Trichosporon ovoides* Behrend *1890*
= *Trichosporon cutaneum* (De Beurm. *et al.*) Ota *1926*
= *Endomyces laibachii* Windisch *1965*

For numerous other synonyms see Do Carmo-Sousa (904), King and Jong (2991), and von Arx *et al.* (213).

DESCRIPTIONS: Do Carmo-Sousa (904) and King and Jong (2990, 2991); SEM: Cole (1101); TEM: Kreger-van Rij and Veenhuis (3125).

The genus *Trichosporon* (type species *T. beigelii* = *T. ovoides* Behrend) was generally characterized as an asporogenous yeast-like group of hyphomycetes with true mycelium and both arthro- and blastoconidia. (Arthroconidia must be looked for on solid media). The repent hyphae fragment into arthroconidia at an early state, usually before any blastoconidia are formed; blastoconidia form in clusters near the ends of the arthroconidia or by the apical budding of lateral branches (1101). After the basidiomycetous nature of *Trichosporon* was understood, i.e. dolipore septa and multilayered cell walls (3125), a high GC content (1500), and presence of xylose-containing polymers in the cell walls (CBS), many of the described species had to be excluded from it and the number now accepted has been reduced to about six (2991).

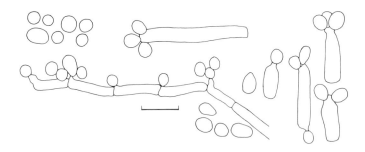

FIG. 358. *Trichosporon beigelii*, blastoconidia formed from fragmenting hyphae.

T. beigelii has rather slow-growing colonies, reaching 2·0 cm diam in ten days at 20°C on MEA, white to cream, butyrous, smooth or wrinkled. Arthroconidia cylindrical to ellipsoidal, mostly 5–20 × 3–4 µm; blastoconidia subglobose, attached by a narrow but distinct scar, 3–4·5 µm diam. Large, globose, ovoid or ellipsoidal, hyaline, thick-walled cells, 9·5–27 × 8–21 µm, and resembling chlamydospores are occasionally found (6165). — The fatty acids have been examined by gas-liquid chromatography (4792). DNA studies gave a GC content of 55 (1500) or 59% (4101). — *T. beigelii* is a ubiquitous and heterogeneous species defined mainly by physiological characters, such as absence of both nitrate utilization and glucose fermentation. The closely related *T. pullulans* (Linder) Diddens & Lodder *1942* can assimilate nitrate. A physiologically similar fungus which regularly forms endospores but no exogenous blastoconidia is *T. dulcitum* (Berkh.) Weijman 1979 (≡ *Protendomycopsis dulcita* (Berkh.) W. Gams & Domsch *1970* = *P. domschii* Windisch *1965*), which is an equally common soil fungus (6401); comparable endospores have, however, also been reported from fresh isolates of *T. beigelii* after three months (904). Neither the endospores nor clusters of blastoconidia (cf. *Endomyces laibachii* Windisch *1965*) can be regarded as asci.

T. beigelii is by far the commonest species of the genus, and has a worldwide distribution. It is well-known as the causal agent of white piedra hair infections, skin infections (2155, 3398), and more rarely gummatous lesions, pulmonary and systemic infections in man and dogs, and in bovine mastitis, for over a century (1624). — It is nevertheless much more frequently found as a saprophyte in soil, fresh and seawater, and plant detritus (2991). Diseases caused by it mainly occur in the temperate to tropical regions of Europe, Asia, South America, and, rarely, the southern USA (2991, 4854). *T. beigelii* has also been isolated from garden compost (1425), sewage sludge (1165), intertidal sand (3014) and mud (IMI), eutrophicated river sediments, running water (1162), and cultures enriched with phenol (4143), vanillin or *p*-hydroxybenzaldehyde (2375). It has also been found in fermented corn dough (2416). — *In vitro, T. beigelii* can be inhibited by *Pseudomonas chlororaphis, P. fluorescens, Streptomyces* spp., *Trichoderma viride* (1397) and *Pseudocercosporella herpotrichoides* (1431).

The temperature maximum for growth varies between 29 and 41°C for different isolates (904, 1812). *T. beigelii* was found to utilize all disaccharides tested, soluble starch, D-xylose, D-ribose, arbutin, acetic and succinic acids, glycine, and L-proline as C sources (1812, 2991), while fumaric, gluconic, *a*-ketoglutaric, oxalacetic and tartaric acids are less suitable (1396); glucose is not fermented; arthroconidium formation is best observed on a glucose-salts medium (2991). Lactic and citric acids are reportedly not utilized (2991). *T. beigelii* produces urease (2991), protease (3231), xylanase (5566), and considerable amounts of exo- and endo-cellulases (1341). The inability to assimilate nitrate has been demonstrated by many authors and used as a taxonomic criterion, and the enzymes involved in ammonia assimilation have been thoroughly studied (2780). Phenols and phenol derivatives can be oxidized by adapted isolates, and the presence of several hydroxylating enzymes is indicated (4142, 4143). *T. beigelii* grows well on cheese whey (227) and is heterotrophic for thiamine and other vitamins (904, 4363). Pentosylmannan is produced in shake cultures (2048). — *T. beigelii* was found to promote the growth of *Rhizoctonia solani in vitro* (1427).

Trichothecium Link ex Gray *1821*

Trichothecium roseum (Pers.) Link ex Gray *1821* — Fig. 359.

≡ *Trichoderma roseum* Pers. *1801*
= *Cephalothecium roseum* Corda *1838*

DESCRIPTIONS: Barron (364), Meyer (3788), Tubaki (5926), Subramanian (5626), and Matsushima (3679, 3680); conidiogenesis: Kendrick and Cole (2927), and Cole and Samson (1109).

The genus *Trichothecium* is characterized by pinkish, dusty colonies and erect, unbranched conidiophores, which at the tip bear basipetal, imbricate zig-zag chains (connected by some mucilage) of ellipsoidal, 2-celled, thick-walled conidia with truncate bases. As was shown by time-lapse cinematography (2927), the conidiogenous cells are progressively shortened with each conidium formed; this structure is sometimes termed "meristem arthroconidia", or, preferably, a "retrogressive succession of blastoconidia".

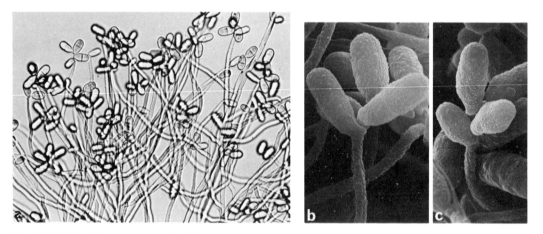

FIG. 359. *Trichothecium roseum*. a. Conidiophores with imbricate conidial chains, × 350; b, c. details of basipetal chain formation, SEM, × 2000, CBS 250.57, orig. R. A. Samson.

T. roseum forms rather fast-growing colonies, reaching to 9 cm diam in ten days at 20°C on MEA, pinkish, zonate in diurnal rhythm (1654, 4973), powdery from conidia. Conidiophores erect, to 2 mm long (base difficult to trace), 4–5 μm wide, often with three septa in the lower part. Conidia ellipsoidal to pyriform with an obliquely prominent truncate basal scar, 2-celled, the upper cell slightly larger, hyaline, smooth- and thick-walled (*c.* 1 μm), 12–23(–35) × 8–10(–13) μm.

Although some additional species have been described in *Trichothecium*, their status is questionable. Comparable 2-celled conidia are found in *Arthrobotrys* (q.v.) but they are thin-walled and formed in a sympodial succession on broad denticles; a similar conidiogenesis occurs in the mycogenous species of *Cladobotryum* Nees ex Steud. *1824*, but there the conidiophores are verticillately branched and the conidia thin-walled. — Cytological investigations showed the conidia of *T. roseum* to be multinucleate, each cell containing 4–12 or more nuclei (3534). The details of nuclear division have been studied (2307). Good germinability of the conidia is found to depend on the equal distribution of the numerous nuclei in the cell (4367). — The lipid fraction of young hyphae was found to consist of mono-, di- and triglycerides, lecithines, phosphatidyl-ethanolamines and cardiolipids; the most important fatty acids are palmitic and linoleic, followed by oleic and linolenic acids (5034). — DNA analysis gave 25% adenine, 24·2% guanine, 25·5% cytosine, 25·3% thymine (3232), or a GC content of 49·7%.

T. roseum has a worldwide distribution (4843) and is commonly found on decaying plant substrates and macromycete sporocarps (5924) which it can cover with a pale pink conidial dust. The conidia are frequently air-borne. It is known from lichenized fungi (D. L. Hawksworth, pers. comm.) and as a destructive mycoparasite (635) but is not characteristically a soil fungus (4421). — It has been reported from soils in Poland (272, 273), Denmark (2745), France (3941), the USSR (2871, 2948, 3850), Turkey (4245), Israel (2764, 4758), Egypt (8), the Sahara (2974), Chad (3415), Zaïre (2966, 2967), central Africa (3063), Australia (3720), Polynesia (6084), India (1317, 2854, 4030, 4698, 5512), China (3475), and Panama (3963). Known habitats include uncultivated soils (4030, 4698), forest nurseries (3908), forest soils under beech (272, 273, 2745), teak (2854) and *Eugenia* (5512), cultivated soils (2948) with legumes (1317), a citrus plantation (2764), soil under steppe vegetation (3415), heathland (3720), podzols (3416), dunes (3941), salt-marshes (8), and garden compost (1424). It has been isolated very often from the leaf litter of various deciduous trees (650), birch (3449), *Eugenia heyneana* (5513), pine (2344), fir (681), cotton (5231), bananas (3063), and palms (402), but also from the rhizospheres of wheat (3567), tobacco (3431) and sand-dune plants (4371). Other habitats include broad bean roots (6134), seeds of corn (4831), coffee (3063), groundnuts (1984), barley (1752), oats (3097), *Avena fatua* (2961), *Poa pratensis* (1110) and some cultivated grasses (3512), pecans (2572), termite nests (3963), bark of *Acacia* (4758) and sycamore (2338), maple and willow twigs (2338), slime of a paper mill (5060), sticks exposed to flowing water (5249), wood pulp (3294), fresh water (3809), sewage sludge (1166), jute bags (4106), coarse fodder (4548), stored apples (5504), fruit juices (5205), flour-type foodstuffs (5980), prepupae of wild bees (402), nests and feathers of free-living birds (2575), and frescoes of a monastery (2666). — Its growth was found to be inhibited *in vitro* by *Memnoniella echinata* (4) and tannins from freshly fallen oak leaves (2282). The mycelium may serve as food for mites (5379, 5380). Conidium germination has been reported to be both affected by (5181) and completely insensitive to (1407), soil mycostasis. — Aerial dispersal is favoured by movements of dry air (2648).

The optimal temperature for growth is 25°C, the minimum 15°C, the maximum 35°C, and no growth occurs at 37°C; a wide pH range is tolerated, the optimum for growth being at pH 6, but at pH 4·0–6·5 excellent sporulation occurs (2845). Low temperatures (15°C) and high glucose concentrations (10%) can cause an increase in the size of conidia (2846, 6373). Treatment with colchicin increased the number of nuclei in the conidia, growth rate, and biosynthetic activities (3533). Optimal growth occurs between saturation and −70 bars water potential (4493); *T. roseum* sporulates in submerged culture with 2% NaCl and corn steep liquor in the basal medium (6088). — *T. roseum* can utilize the following sugars: D-fructose, sucrose, maltose, lactose, raffinose, D-galactose, D-glucose, arabinose, and D-mannitol (4945). Good growth occurs on L-methionine, L-isoleucine, L-tryptophan, L-alanine, L-norvaline, and L-norleucine (4366). Starch, pectin, chitin (5926), and cellulose are decomposed (1425, 2718, 2948, 5926); arabinoxylan is more effectively degraded by the cellulase than by the xylanase extracted from *T. roseum* (1753). Some hydrolytic enzymes produced have a pH optimum of 4·6 and 5·6 (5014), and proteinases are also formed (3536, 5926). Cinerone (5681) and progesterone (1607, 5012) can be hydroxylated. Identified metabolites include the mycotoxic trichothecin (1831, 3493, 3535, 5482) which can even inhibit the conidium germination of *T. roseum* itself (3532), the sesquiterpenes cyclonerodiol (872, 4232), trichodiol-A, trichodiene (1667, 4233, 4234), and crotocin (3231), the diterpenes rosenonolactone (= rosein I), rosenololactone, rosololactone (= rosein II), 7-deoxyrosenonolactone (19, 3231), 11β-hydroxyrosenonolactone (= roscin III) (2194, 3007), the toxic 6β-hydroxyrosenonolactone (2520), isorosenolic acid (3231), the mycotoxic cyclodepsipeptide roseotoxin B (1636), the siderochromes fusigen and fusigen B (3231), and carotenoid pigments (3535). — *T. roseum* can colonize sclerotia of *Sclerotinia sclerotiorum* (4584, 4699), *S. borealis* and *Claviceps purpurea* (3531), has antifungal properties (716, 1369, 3599), and is toxic to young ducklings fed with contaminated cereals (5158). *T. roseum* severely inhibited the growth of rape, tomato, beets, wheat and flax in greenhouse experiments (1427), and has been considered a primary pathogen of stored apples (647) and tomatoes in a greenhouse (6271).

Trichurus Clem. & Shear *1896*

Trichurus spiralis Hasselbr. *1900* — Fig. 360.

= *Trichurus gorgonifer* Bain. *1907*

DESCRIPTIONS: Lodha (3399), Ellis (1603), Wright and Marchand (6446), Matsushima (5626), Fahmy and Yusef (1679), and Fungi Canadenses No. 100, *1977*; synnema development: Swart (5666). — Colonies and synnema development and structure are very similar to those of *Doratomyces* (q.v.) (1679, 5666). Colonies reaching 1·0 cm diam in five days at 20°C on OA. Synnemata to 3(–5) mm tall, grey-brown, with a well developed stipe and an ellipsoidal or cylindrical sporulating portion; the ultimate synnematal branches bear the annellophores which are often less densely aggregated than in *Doratomyces* and may give rise to curled sterile setae at the same level as the annellophores or at one or two branchings below them. Conidiogenous cells slender barrel-shaped, $5–12 \times 3–4\,\mu$m. Conidia ovoid to mitriform with a truncate base and rounded apex, smooth-walled, $4–6 \times 3·5–4·2\,\mu$m, uninucleate (1679). Sterile setae subhyaline or brown, unbranched, coiled once or twice, even in diameter, with blunt ends, very variable in length and abundance, to $150 \times 2–4\,\mu$m. The formation of setae on the synnemata is very variable and dependent on both the individual isolate and the culture conditions; OA has proved particularly favourable to their development. — A TEM study of conidiogenesis (1679) showed more distinct annellations than those found in *Doratomyces*, and an elongation of $4\,\mu$m showed up to 65 annellations. — The type species of the genus, *T. cylindricus* Clem. & Shear *1896*, which has apparently not been found since its original description, is said to have conidia $8–9 \times 3\,\mu$m; the apparently rare *T. terrophilus* Swift & Povah *1929* has straight, gradually tapering, and sometimes dichotomously branching setae and smaller conidia $3–6 \times 2·0–3·5\,\mu$m (3399). — If only a few sterile setae are formed, *T. spiralis* can easily be confused with *Doratomyces purpureofuscus* (q.v.).

T. spiralis has a worldwide distribution, though it is not a very frequent species. It appears to be commonest in temperate zones but has also been found in Israel (2764, 2772), Egypt (1679), Iraq (IMI), South Africa (4407), Pakistan (4855), India (3399, 3863, 4030, 4997), Nepal (1826), Polynesia (6084), Argentina (6446), Japan (5626), Jamaica, Nigeria, Tonga (1603, IMI), and central Africa (3063). Known habitats include soils of an alpine pasture (3445), under *Pinus nigra* in Spain (3447), of a mixed forest in Canada (IMI), and savannah with *Acacia karroo* (4407), but it has been most frequently isolated from cultivated soils (2161, 4538, 4997); for example, soils under *Citrus* (2764) or groundnut (2772). *T. spiralis* has also been found in Spanish saline soils (3446). It has been found at soil depths down to 20 or 40 cm (2161, 3447). In India, its greatest frequency is reported to be during the summer (3863). Its numbers may also increase after manuring (2161); organic substrates recorded include rotten potatoes (IMI), groundnuts (1016), rabbit (6085), cow (3399) and deer dung (5626), free-living birds (2575), compost beds (3041), wood of cooling towers (1199), and sticks exposed to river water (5249). It has been isolated from the rhizosphere of broad beans, cotton (6511), tomato, sugar cane, and *Dioscorea alata* (IMI).

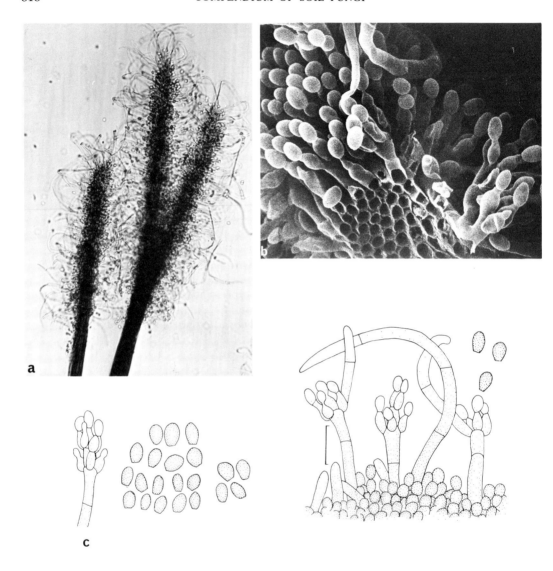

FIG. 360. *Trichurus spiralis*. a. Three synnemata, × 200, CBS 877.68; b. a fractured synnema, showing conidiogenous cells and conidia, SEM, × 1500; c. details of conidiophores with origin of setae and conidia.

Long synnemata are formed at 20° and 25°C, much shorter ones at 30 and 35°C, none at all at 40°C, and ones of medium length at 15°C. The longest synnemata appeared at pH 5 and in the pH range 8·6–9; short periods of light (1–3 h per day) gave the relatively longest synnemata, whereas none developed under continuous light (1680). Maltose, galactose or starch as sole C sources induced long synnemata, while with mannitol, glucose, lactose, sucrose, fumaric, or succinic acids, shorter synnemata were produced. The optimal maltose concentration was found to be 2–5% at 20°C. No sporulation occurred on acetate or citrate. Nitrate or glycine proved to be the most suitable N sources for synnema formation; ammonium sulphate or tartrate gave fewer and shorter synnemata, NaCl concentrations above 0.1 mol l^{-1} reduced the synnema length and 1 mol l^{-1} NaCl suppressed it (1680). — *T. spiralis* degrades cellulose (6085), can play a part in wood decay (1199, 2211, 2940), and has been named as one of the causal agents of soft rot in timber (1494, 1496, 4191, 5072). — The growth of the alga *Chlorella pyrenoidosa* was promoted by *T. spiralis* (1430).

Typhula Pers. ex Fr. *1821*

= *Pistillaria* Fr. *1821*
= *Pistillina* Quél. *1880*

Type species: *Typhula phacorrhiza* Reichard ex Fr. (lectotype)

MONOGRAPHIC TREATMENT: Berthier (483). — The concept of the clavariaceous basidiomycete genus *Typhula* has been broadened by Berthier (483) to include the genera *Pistillaria* and *Pistillina*, previously treated as distinct, since the presence or absence of sclerotia is not acceptable as a generic criterion. The genus is characterized by generally small, claviform, unbranched fruit-bodies, which are differentiated into a fertile clavula and a slender stipe; basidia cylindrical to clavate, 2–4(–6)-spored; cystidia absent in the hymenium but present on the stipe; basidiospores slightly allantoid, amyloid or inamyloid. Sclerotia, when present, consisting of gelatinized and non-gelatinized hyphae, covered with a pigmented epidermoidal layer and, in some species, by a hyaline cuticle. In the recent monograph by Berthier (483), seven subgenera are defined mainly on the base of sclerotial and carpophore structures; 63 species are fully described and about 80 species can be recognized.

Most species are epiphytes on decaying plant remains on the soil in temperate and cold regions; fruit-bodies usually appear in the autumn in moist atmospheres, whilst the sclerotia appear in the spring and remain inactive during the summer. Distribution occurs by basidiospores which may remain viable for several weeks at low temperature (1233), while the sclerotia and vegetative hyphae account for local spread.

Pure cultures can be obtained from basidiospores discharged by fruit-bodies attached to the lid of a Petri dish or from surface-sterilized sclerotia. They are cultivated at low temperatures (below 18°C), on media such as dilute MEA, and fruit-bodies can be induced by near-UV (3280, 4819) or by exposure to non-sterile outside conditions in autumn (754). — Transverse sections through the fruit-bodies and sclerotia provide the essential characters needed for species identification. The most reliable species concept in this genus is obtained by mating unknown isolates with defined "tester" strains. The species so far studied have tetrapolar incompatibility with multiple alleles (483, 754, 1232, 3018, 4893); the most efficient technique is to mate an unknown dikaryon with a known monokaryon and transfer some mycelium of the monokaryon 1 cm from the contact zone 4 days after they make contact; in conspecific isolates, the monokaryon becomes dikaryotized through nuclear migration and all hyphae growing out will have clamp connections (754).

The two plant-parasitic species considered here are placed in the subgenus *Microtyphula* Berthier *1974* which is characterized by a distinct clavula supported by a long stipe; the sclerotia in this subgenus are not or hardly gelatinized, and have a typical superficial epidermoidal layer; the fruit-bodies are hardly gelatinized, and the spores are amyloid.

Key to the species treated:

1   Fruit-bodies differentiated into a pink clavula and greyish white stipe; sclerotia reddish brown, slightly gelatinized   *T. incarnata* (p. 821)
      Fruit-bodies greyish white to light brown, clavula lighter than the stipe; sclerotia dark brown, not gelatinized   *T. ishikariensis* (p. 822)

Typhula incarnata Lasch ex Fr. *1838*

    = *Typhula elegantula* Karst. *1870*
    = *Typhula graminum* auct. [non Karst. *1876*]
    = *Typhula itoana* Imai *1929*

DESCRIPTIONS: Remsberg (4819), Corner (1190), Berthier (483), Lehmann (3279), and Årsvoll (191); synonyms: Røed (4893), and McDonald (3703). — Colonies reaching 3·0 cm diam in ten days at 15°C on MEA. Fruit-bodies appearing in dense clusters on grass rhizomes and leaves, 3–30 mm tall, clavula pointed, pink; stipe greyish white; basidiospores 8·5–10·5 × 3·7–4·5 μm, weakly amyloid; sclerotia pinkish orange, later becoming reddish brown, subglobose to elongate, 0·5–1·5(–3) mm diam, covered by a normal epidermoidal layer and a 3–4 μm thick cuticle. — *T. graminum* Karst. *1876* is distinguished from *T. incarnata* by having white fruit-bodies with a very short clavula (<1 mm tall) and sclerotia to 1·2 mm long, and is apparently restricted to *Molinia*.

    T. incarnata is one of the causal agents of grey snow-mould in snow-covered grasses and cereals. Dikaryons are more virulent than monokaryons (3278). The distribution of this fungus extends further south than that of *T. ishikariensis* and its varieties and it occurs mainly in the southern parts of Finland (2713) and in parts of southern and central Norway with at least 90 days of snow cover and relatively mild winters (191); unlike *T. ishikariensis*, its distribution extends into central Europe where it has been reported from light soils with wheat and barley in Germany (3280); there are also reports of it from Japan (4819) and the USA (4893). A mutual antagonism between *T. incarnata* and *T. ishikariensis* in infecting host plants has been observed (192). *T. incarnata* causes less damage to cereals than *T. ishikariensis* but is not restricted to graminaceous hosts and can occur on various other crops and ornamentals (191). No host-specific "special forms" have been found (3278). On clover it occurs less frequently than *T. trifolii* Rostr. *1890* and *T. ishikariensis* (6494). Sclerotia can be found down to 1–3 cm in soil where they remain dormant during the summer (2266, 3278). At greater soil depth they are unable to infect host plants; this effect is more pronounced in loamy than in sandy soils. Sclerotial germination is independent of external C and N sources, better on sandy than on heavy loam or peaty soils and not influenced by light (3280). — *Trichoderma koningii, T. viride* and *T. harzianum* reduce the germinability of sclerotia drastically both *in vitro* and in soil (2266), but such antagonism is apparently less noticeable in light soils (3280). — Basidiospores do not germinate at all in direct contact with soil (3280).

    Basidiospores germinate *in vitro* without any special nutritional requirements and optimally at 12–17°C (3280). The minimum temperature for growth is −6 to −5°C, the optimum 8–15° or 9–12°C, and the maximum 20–21°C (191, 1334, 2713); host plants are not attacked below

0°C (191). — Laccase and cellulase activities have been demonstrated (483). *T. incarnata* grows in the pH range 3–9, with an optimum of pH 6–7·6 (1334), and has a relatively high osmotolerance (752).

Typhula ishikariensis Imai *1930* — Fig. 361.

Typhula ishikariensis Imai *1930* var. *ishikariensis*
Typhula ishikariensis var. *idahoensis* (Remsberg) Årsvoll & J. D. Sm. *1978*
 ≡ *Typhula idahoensis* Remsberg *1940*
Typhula ishikariensis var. *canadensis* J. D. Sm. & Årsvoll *1978*

 = *Typhula borealis* Ekstrand *1937*
? = *Typhula hyperborea* Ekstrand *1955*

DESCRIPTIONS: Remsberg (4819), Corner (1190), Årsvoll (191), Berthier (483), and Årsvoll and Smith (193); discussion of synonyms: McDonald (3703), Bruehl *et al.* (754), and Årsvoll and Smith (193).

 After *T. ishikariensis* and *T. idahoensis* had been synonymized by numerous authors (e.g. 2713), detailed microscopical studies of the sclerotia (483) and mating experiments (754) suggested that they were closely related but specifically distinct. The finding of isolates which are interfertile with both types (193) proves, however, that all belong to one species in which three varieties can now be distinguished.

 Fruit-bodies mostly 3–14 mm tall, greyish white to light brown, clavula mostly 0·5–1 mm wide, stipe slightly darker. Basidiospores ovoid to ellipsoidal, hyaline, smooth-walled, flattened at one side, mostly 6–9 × 3–5 μm. Sclerotia globose to flattened, light brown to almost black, smooth-walled or roughened, not gelatinized, mostly 0·3–1·5 μm diam. — The var. *ishikariensis* has smooth or roughened sclerotia with rind cells of regular outline, moderately lobate, and basidiospores 6–8 × 3–4 μm; var. *idahoensis* has roughened sclerotia with rind cells of irregular outline, lobate and frequently digitate, and basidiospores mostly 7–9 × 3–5 μm; var. *canadensis* has the smallest fruit-bodies and sclerotia, the latter mostly 0·3–0·8 mm diam, smooth, often with attached hyaline hyphae, rind cells regular or irregular in outline, and basidiospores as in var. *ishikariensis*. Var. *ishikariensis* is now known from northern Europe, the Alps, northern Asia, and northern N-America, var. *idahoensis* and var. *canadensis* both only from northern N-America (193). At least 99% of the basidiospores were uninucleate when ejected from the basidium (1232), as in most species of the genus (483). Dikaryotic mycelia grow faster than monokaryons (1232), and virulence was found to be determined by several genes in numerous dikaryons tested, but there was only a weak correlation with growth rate and none with incompatibility genes (3018).

 T. ishikariensis with its varieties grows mainly on Gramineae but can also be found on many other field crops (6494) and ornamentals (191, 2713). Var. *ishikariensis* is the most important pathogen of all fungi found in winter cereals in Finland (2713), where its distribution ranges further north than that of *T. incarnata* (2713); similarly, in Norway it extends throughout the country but is generally absent near the coast south of 69°N and most often occurs in places with drier climates and very cold winters (191). Var. *idahoensis* is also most virulent at low temperatures, particularly to very young wheat (752); fruiting starts in early November and reaches a maximum by mid-November (1233). — Distribution is both by means of basidiospores, which remain viable for 52–67 days at −1 to +10°C and germinate at temperatures up

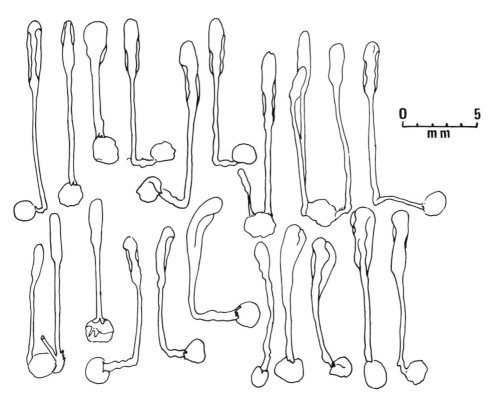

FIG. 361. *Typhula ishikariensis*, sclerotia with basidiomata, (from 193, reproduced with permission Canad. Res. Council).

to 15°C (1233), and, more importantly, through sclerotia. — In var. *idahoensis* little competitive saprophytic ability (1233) was observed on weathered wheat straw, and the survival of its sclerotia was drastically reduced by *Cladosporium herbarum* (1296). Sclerotial survival is not normally much affected in soil at about 2°C except in the rhizosphere of clover and alfalfa; at higher temperatures (24°C) pea plants also depress it (2580). Sclerotia cannot infect plants when these are buried at more than 2 cm down. Infection is proportional to sclerotium density up to 300 sclerotia per kg soil (1296). The germination of sclerotia can be drastically increased by the addition of nutrients (1296).

This particularly psychrophilic fungus has a temperature minimum for growth of −6 to −5°C, an optimum of 4–10°C, and a maximum of 15–18°C (191, 483, 1334). The temperature optimum for O_2 uptake is higher than that for mycelial growth (1334), and greatest virulence is observed between −1·5 and +1·5°C (2713). It can grow in the pH range 4–9 with an optimum at pH 6–7. — A number of hexoses and pentoses are assimilated to different extents (1334). The enzymes observed include laccase and cellulase (483), pectinase (4032), lipase, amylase, esterase, polyphenoloxidase and proteinases (4033). Exogenous respiration was stimulated by acetate at 5°C, and was greater at 5°C than at 20°C with glucose in non-starved cultures (1335). — Var. *idahoensis* was found to be fairly osmotolerant but to withstand KCl concentrations less than ones of sucrose with similar osmotic potentials (752).

Ulocladium Preuss *1851*

Ulocladium consortiale (Thüm.) Simmons *1967* — Fig. 362.

≡ *Macrosporium consortiale* Thüm. *1876*
≡ *Stemphylium consortiale* (Thüm.) Groves & Skolko *1944*
≡ *Alternaria consortialis* (Thüm.) Groves & Hughes *1953*
≡ *Pseudostemphylium consortiale* (Thüm.) Subram. *1961*
= *Stemphylium ilicis* Tengw. *1924*

DESCRIPTIONS: Neergaard (4118), Joly (2792), Ellis (1603), Subramanian (5626), Matsushima (3680), and Simmons (5332), with a key to the species of the genus. — The genus *Ulocladium* (= *Pseudostemphylium* (Wiltsh.) Subram. *1961*; type species *U. botrytis* Preuss) is characterized by usually strongly geniculate conidiophores arising by repeated sympodial elongations and the production of numerous, usually solitary, poroconidia. Poroconidia typically obovoid, with a conical base and more broadly rounded tip; two species form conidial chains through apical, paler "false beaks"; conidia muriform with several transverse and in each portion 1–2 longitudinal septa, the wall commonly dark brown and coarsely warted. Seven species have so far been described. Conidiogenesis has been studied by TEM (908).

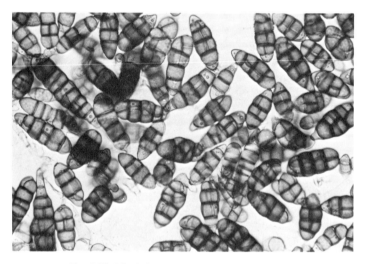

FIG. 362. *Ulocladium consortiale*, conidia, × 500.

Besides *U. botrytis* Preuss *1851* and *U. atrum* Preuss *1852* with solitary, very dark brown conidia and with mostly three and one transverse septa, respectively, *U. consortiale* is rather common and characterized by paler brown, almost smooth-walled, obovoid or ellipsoidal conidia, 16–34 × 10–15 μm, with 1–5 transverse and some longitudinal or oblique septa and a tendency to form short chains. Colonies reaching 5·0–6·8 cm diam in six days at 25–28°C on MEA. — Longer conidial chains are found in *U. chartarum* (Preuss) Simmons *1967* which therefore resembles *Alternaria alternata* (q.v.). *Alternaria radicina* Meier *et al. 1922* is also somewhat similar but has less densely geniculate conidiophores, and somewhat darker and larger (more *Alternaria*-like) conidia, 27–57 × 9–27 μm. — The fungus described under the name *Stemphylium consortiale* by Groves and Skolko (2142) seems to be identical with *U. atrum* Preuss (5332). Consequently numerous subsequent determinations are probably erroneous.

U. consortiale is known from temperate, subtropical, and tropical zones (1604), from several plants, leaf litter, seeds and soil (1604). Because of often uncertain identifications, data compiled below are to be accepted with reservation. Reports include ones from the British Isles (6184), the Netherlands (1614, 1616), Germany (3041), Czechoslovakia (1703), the USSR (2871), France (2159, 2161), Turkey (4245), Israel (2764, 2772, 4759), Iraq, Libya (1604), South Africa (4408), Kuwait (4000, 4001), Afghanistan (5332), India (1604, 5626), the USA (1166, 1512, 3631), and Panama (3963). It has been isolated from forest nurseries (6184), cultivated soils (1614, 2159, 2161, 2772, 4759), citrus plantations (2764, 3631), desert soils (1512), saline soil (4000, 4001), eutrophicated river sediments and running water (1162, 1166), a uranium mine (1703) and the air (2792). Its abundance in soil was found to be reduced by the addition of farmyard manure (2764). *U. consortiale* can grow saprophytically on the most varied of substrates; it has been isolated from wheatfield debris (4408), pine (2344) and oak litter (IMI), palm leaves (3963), composted municipal waste (3041), sugar beet roots (4559), and the rhizospheres of wheat (2492), alfalfa (3982), groundnuts (2768), pear (5499), and *Hyoscyamus albus* (IMI). It is found very often on a very wide range of seeds (1110, 1752, 2792, 3512, 3556, 5940) and has further been noted on wood pulp (6168), paper, textiles and excrements (2792, 6074).

U. consortiale survives and grows well even after exposure to extreme ($-94/+23$°C) diurnal temperature changes (1242). Cellulose is decomposed (6326); arabinoxylan is degraded by cellulases and, to a lesser extent, by xylanases (1753). Birch and pine woods are eroded (4191) and plasticizers can be utilized (473). — Spore germination in dwarf bunt and smut of rye was found to be promoted *in vitro* by *U. consortiale* (1922). In infection tests, it proved pathogenic to seedlings of *Avena fatua* (2961, 2962). — It can grow in media containing 20–40% sucrose (4001) and has a very good resistance to UV irradiation (1512).

Varicosporium Kegel *1906*

Varicosporium elodeae Kegel *1906* — Fig. 363.

DESCRIPTIONS: Kegel (2916), Bessey (493), Ingold (2647, 2649), Tubaki (5925), Petersen (4522), Nilsson (4190), and Miura (3886).

This is the type species of *Varicosporium*, of which six species are now known. The conidia in this genus have a main axis with 2–3 straight lateral branches arising at right angles and which may branch again; the branches are constricted at their points of origin, but do not taper towards the ends. Conidiogenesis is blastic and acropleurogenous and occurs in the upper parts of the conidiophores (2647). — *Varicosporium* resembles *Tricladium* Ingold *1942*, which differs in that the conidia have only first order branches which taper towards their ends.

V. elodeae forms rather fast-growing colonies, reaching 2·5 cm diam in ten days at 20°C on OA, whitish but later becoming greenish or ochraceous to light brown in the centre, particularly in reverse, and with felty mycelial strands. Sporulation occurs on OA, particularly after transfer to water. Conidiophores little differentiated from the vegetative hyphae, 100–200 μm long. Main axis of the conidia slightly curved, 60–120 × 3 μm, consisting of 4–8 cells; 1–3 lateral branches of the same width develop along the convex side of the main axis progressing from the centre towards the apex; each lateral branch may branch again in the same one-sided manner. — Other species of *Varicosporium* are: *V. delicatum* Iqbal *1971* which has a narrow conidial axis and branches 1·5 μm wide; *V. giganteum* Crane *1968* with an angular main axis 260–470 × 2–3 μm and clavate end cells to the branches; *V. macrosporum* Nawawi *1974* with a curved main axis 190–320 × 2·0–2·5 μm; *V. splendens* Nawawi *1973*, a basidiomycete, with secondary branches wider than the main axis and binucleate cells; and *V. helicosporum* Nawawi *1974* with a strongly curved main axis. The anamorph of *Hymenoscyphus varicosporioides* Tubaki *1966* has larger conidia with the main axis 300–500 × 7–7·5 μm.

V. elodeae was originally described as a soil fungus (2916) but has commonly been found in studies on aquatic Hyphomycetes mainly in foam and on submerged leaves. Its geographical distribution extends from Poland (917), Czechoslovakia (3655), Germany (185, 917, 2916), Sweden (4188, 4190), and the British Isles (99, 2647, 6131, IMI), to some northern states of the USA (1213, 6411), Canada (321, 493), and Japan (3886, 5659, 5925, 5927). In the USSR it has been reported from the Leningrad area, the Ukraine and Armenia (1485). It has been isolated from both soil (493) and dead leaves of various plants away from any open water (321), glass slides buried in soil (4431), the rhizosphere of beans (4453) and rye (CBS), root surfaces of beech (6131), frost-damaged grass leaves (191), moist leaves on the ground (4190), preservative-treated pine and beech poles in contact with the ground (4193) and snow at 1000 m in Sweden (4190). Observations in aquatic habitats range from stagnant waters (4188, 4190), ponds, lakes and ditches (185, 6411), to well aerated slowly running waters (3655), springs, and streams (917, 6411); it is also often found on scum and foam (4190, 6384). In aquatic habitats it is commonly associated with decaying and/or submerged vascular

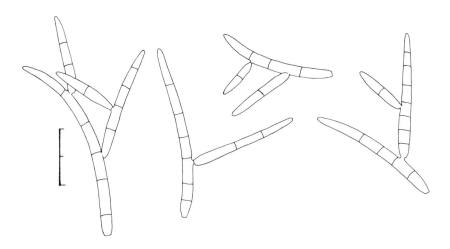

Fig. 363. *Varicosporium elodeae*, "tetraradiate" conidia, CBS 952.72.

plant debris (1213), such as leaves of maple, birch, willow, oak (185), alder (185, 917, 2647, 5925), beech (185, 917), elder, several Gramineae, Cyperaceae and Umbelliferae (917), *Nuphar ozeense, Drosera rotundifolia* (5925) and *Elodea canadensis* (2916).

The optimal temperature for growth is 21°C, the maximum 28°C, and the minimum −6°C (191). The fungus seems to sporulate, germinate and grow well in a temperature range from 10–20°C (–25°) (321, 3105, 5925), while no growth occurs at 30°C (5925). Strong aeration of the water increases sporulation within 18 h but decreases the average number of conidiogenous areas (6242, 6248). Mycelium on agar blocks placed in distilled water develops conidia within 6–10 days. When the fungus grows in shallow water, conidia are formed just above the water surface (4190), but those produced under submerged conditions are indistinguishable (2647). — Good growth is obtained *in vitro* with decoctions of *Elodea* or prune, malt, peptone or meat extract (2916); an increased growth response was also observed in the presence of DDT (2488). Good C sources are glucose, fructose, sucrose, maltose, cellobiose, and starch, but not xylose (5805). Cellulose is degraded (4433, 5926) and strong amylase and pectinase activities have been observed (5925, 5926). On media without glucose, aspartic acid, glutamic acid, tyrosine or arginine can be utilized as C sources (5806). NH_4^+ ions are more efficiently utilized as N sources than NO_3^- ions; the fungus is partially auxotrophic for biotin (5805). It can also slightly liquefy gelatine (5925).

Verticillium Nees ex Link *1824*

= *Acrostalagmus* Corda *1838*
= *Acrocylindrium* Bonord. *1851*
= *Diheterospora* Kamyschko *1962*
= *Pochonia* Batista & Fonseca *1965*

Type species: *Verticillium tenerum* (Nees ex Pers.) Link (teleomorph *Nectria inventa* Pethybr.)

Teleomorphic genera: *Nectria* Fr., *Cordyceps* (Fr.) Link, *Torrubiella* Boud.

Contributions to a MONOGRAPHIC TREATMENT: van Beyma (494); type species: Hughes (2596); reviews: Isaac (2667, 2673); section *Prostrata*: Gams (1882); general review of the pathogenic species: Pegg (4491).

Verticillium is a genus of Hyphomycetes with predominantly hyaline vegetative hyphae and colonies with moderate growth rates. Conidiophores usually well differentiated and erect, verticillately branched over most of their length, bearing whorls of slender flask-shaped or aculeate divergent phialides with inconspicuous collarettes. Conidia hyaline or brightly coloured, mostly one-celled, borne in slimy heads, exceptionally in chains. — Hyphal cells, phialides and conidia of the species so far examined are uninucleate (2496, 2673, 3488), but hyphal tips may contain up to 15 nuclei (3488). Conidial nuclei of four blackening species contain six chromatin granules (3488). — The taxonomy of the genus as a whole is not yet settled. The five most important plant-pathogenic species with blackening cultures have, however, repeatedly been studied in detail (2673, 4491). Some white, often entomogenous species have been monographed by Gams (1882) and placed in a separate section, *Prostrata*, on account of their little differentiated and often prostrate conidiophores. But there are many other, often soil-borne species the delimitation and nomenclature of which has not yet been elucidated.

Verticillium is distinguished from *Gliocladium* Corda (q.v.) and *Clonostachys* Corda *1839* by the absence of appressed penicillate conidiophores and the usually straight conidia with rotational symmetry; *Sarocladium* W. Gams & D. Hawksworth *1975* is intermediate between *Verticillium* and *Gliocladium* with very irregularly arranged but often appressed phialides; *Sesquicillium* W. Gams (q.v.) commonly has, in addition to the terminal phialides, subterminal phialides with a narrow lateral opening; *Harziella* Cost. & Matr. *1899* has broadly spreading colonies and very dense terminal clusters of phialides; *Sibirina* G. Arnold *1970* has similar clusters in more regularly verticillate conidiophores but consistently 2-celled conidia; *Phaeostalagmus* W. Gams *1976* has very slow-growing dematiaceous colonies, pigmented conidiophores and phialides with *Phialophora*-like collarettes, and *Stachylidium* Link ex Gray *1821* has faster growing dematiaceous colonies and broad phialides with a constricted tip.

Verticillium comprises numerous saprophytic species and several plant-pathogenic species which mostly cause vascular wilts; of these *V. albo-atrum* and *V. dahliae* are amongst the best-known root-inhabiting fungi, whilst other darkening species are primarily soil inhabitants. The section *Prostrata* W. Gams (1882) comprises several species which attack insects or grow as hyperparasites on fungal substrates (particularly rusts). — Microscopic observation of undisturbed cultures in Petri dishes is essential for the description of conidiophore branching. This is commonly carried out on various media such as 2% MEA, OA or PDA; a prune extract medium with yeast extract and lactose (5710) or dilute soil extract agar with 0·2% polygalacturonic acid (2075), are particularly suitable for inducing microsclerotium formation in *V. dahliae*; the development of resting structures may require a prolonged incubation of several weeks; SEA or PCA are suitable for the study of branching and dictyochlamydospores in species of section *Prostrata*.

Key to the species treated:

1	Colonies intensely orange to orange-brown due to the pigmented conidiophores and conidia; conidia ellipsoidal, 3·5–5 × 2·0–2·5 μm	*Nectria inventa* (p. 501)
	Colonies white, yellowish or darkening	**2**
2(1)	Colonies darkening in reverse	**3**
	Colonies remaining entirely white or yellowish	**7**
3(2)	Colonies producing discrete microsclerotia but no chlamydospores; conidiophores hyaline; conidia 2·5–6(–8) × 1·4–3·2 μm	*V. dahliae* (p. 836)
	Colonies usually not producing discrete microsclerotia, but in some species dark resting mycelium and/or chlamydospores occur	**4**
4(3)	Only chlamydospores present	**5**
	Dark torulose resting mycelium present	**6**
5(4)	Chlamydospores 5·5–8(–10) μm diam, usually produced singly; conidia mostly 1-celled, 4–8·5 × 1·5–2·5 μm	*V. nigrescens* (p. 841)
	Chlamydospores 8·5–17 μm diam, commonly in chains; conidia often 2-celled, 4–10 (–12·5) × 2·5–3·5 μm	*V. nubilum* (p. 843)
6(4)	Chlamydospores present, 7·5–11 μm diam, in addition resting mycelium and some microsclerotia; reverse golden yellow in young cultures; conidia 3·5–10 × 1·5–3·5 μm	(*V. tricorpus*)
	Chlamydospores and microsclerotia absent; reverse white to cream in young cultures; conidia 3·5–10·5(–12·5) × 2–5 μm	*V. albo-atrum* (p. 830)
7(2)	Dictyochlamydospores present at least on SEA (cf. Fig. 366)	**8**
	Dictyochlamydospores always absent	**9**
8(7)	Dictyochlamydospores predominant; conidia short-cylindrical, 3–4 × 1·5–2·0 μm, cohering in slimy heads	*V. chlamydosporium* (p. 834)
	Dictyochlamydospores scanty; conidia subglobose, 2·9–3·3 × 1·9–2·2 μm, cohering in chains	*V. catenulatum* (p. 833)
9(7)	Well-differentiated, thick-walled conidiophores present with globose conidial heads; conidia slightly fusiform, straight or curved, 3·8–7·2 × 1·2–2·4 μm	(*V. fungicola*)
	Phialides on little-differentiated, often prostrate conidiophores; conidia commonly transversely attached to the phialide tip	**10**
10(9)	Conidia cylindrical with rounded tips	*V. lecanii* (p. 840)
	Conidia with pointed tips	**11**
11(10)	Conidia straight, 3–10(–12) × 0·8–1·2 μm	(*V. lamellicola*)
	Conidia ± curved, 5·8–8·0 × 1·5–1·8 μm	*V. psalliotae* (p. 844)

Verticillium albo-atrum Reinke & Berthold *1879* — Fig. 364.

= *Verticillium albo-atrum* var. *caespitosum* Wollenw. *1929*
= *Verticillium albo-atrum* var. *tuberosum* Rudolph *1931*

DESCRIPTIONS: van Beyma (494), Isaac (2667), Smith (5436), Skadow (5396), Campbell and Griffiths (869), Devaux and Sackston (1362), and CMI Descriptions No. 255, *1970*; reviews : Isaac (2673), and Pegg (4491). — Colonies growing moderately fast on PDA or MEA, reaching 2·5 cm diam in 10 days at 20°C, or 5·8–6·0 cm diam in 15 days at 20–22·5°C (2667), at first white to greyish, flocculose, and with a whitish to cream reverse, but after 2–3 weeks becoming brownish cream to black centrally due to the formation of dark resting mycelium. Conidiophores ± erect, hyaline but sometimes with a darkened base (particularly when grown on plant tissue), bearing several whorls of 2–4 phialides; phialides usually 20–30(–50) × 1·4–3·2 μm. Conidia ellipsoidal to short-cylindrical, hyaline, mainly 1-celled, occasionally 1-septate, 3·5–10·5(–12·5) × 2–5 μm. Resting mycelium starts to appear after 10–15 days, particularly on some natural media (869), and comprises swollen darkened and often torulose hyphal cells, 3–7 μm wide; no budding or lateral septation occurs (the production of resting mycelium tends to cease after prolonged subculturing). Microsclerotia and chlamydospores absent. White sectors lacking resting mycelium commonly appear in the colonies.

FIG. 364. *Verticillium albo-atrum*, verticillate conidiophores with detail of apical whorl and conidia, CBS 454.51.

Since *1913* the relationship between *V. albo-atrum* and *V. dahliae* has been the subject of much controversy; several investigators, e.g. van den Ende (1627), regarded them as representatives of one variable species (called *V. albo-atrum*), but most recent workers, e.g. Isaac (2667, 2673), Skadow (5396), and Smith (5436), consider them as distinct, mainly because of the generally dark conidiophore stipes and lack of discrete microsclerotia, lower temperature maximum, and slightly larger conidia (5436) in *V. albo-atrum* though the latter character is rather variable (4495). The most convincing evidence for their specific distinction is perhaps the discovery that heterozygous diploids obtained between both species have little tendency to segregate into haploids (2318), contrary to the situation found in intraspecific diploids which easily segregate (2317). — In this compilation all informations on microsclerotial isolates have been treated under *V. dahliae*.

Suitable media to distinguish the two species are a prune extract agar (5710) or dilute SEA with 0·2% polygalacturonic acid (2075). — Another species with dark, but taller, much-branched conidiophores and narrow cylindrical conidia ($4·5–8 \times 1·5–2·5 \mu$m) is *V. theobromae* (Turconi) Mason & Hughes *1951*, which is commonly found on bananas in tropical countries. — The fine structure of conidia and conidiogenesis is almost identical in *V. albo-atrum* and *V. nigrescens* (765). In cultures grown in the dark on PDA, intrahyphal hyphae and abundant endoconidia of various shapes and sizes were observed (235). Nuclear division in the hyphae is perpendicular to the axis, and diploid vegetative nuclei have been seen (758). The dark resting mycelium has been thoroughly investigated on a variety of media and in different isolates (869). — In genetic studies anastomoses and parasexuality (2316) have been demonstrated, also between isolates of *V. albo-atrum* and *V. dahliae*, and prototrophic colonies were obtained from the combination of two deficient mutants (1797); other genetic observations (see above), however, point to a more distant relationship between the two species. From black parent strains three types of progeny could be obtained: (a) with reduced black pigmentation, (b) a pale brown type, and (c) a white type; in dual cultures (black plus white) the white type did not change, while 86% of the progenies from the black type had changed to white; the operative mechanism is not yet fully understood (4920). On media containing 10^{-4} mol l^{-1} acriflavine, hyaline variants can be induced (5962). — Electrophoretic studies of protein patterns appear contradictory: in one study (4496) the two species appeared to be closely related, but in others appeared more distinct (2219, 6343); phosphatases and esterases are apparently not specific, but some different protein bands were found in both species by acrylamide-gel-electrophoresis and immunoserology (5200). Electrophoretic enzyme patterns of virulent and less virulent isolates of *V. albo-atrum* did not differ (6232). — The formation of dark pigments is controlled by cytoplasmic factors and does not seem to be inherited by a single gene (2316, 2349). Conidia contain a higher proportion of lipids than does the mycelium, the main components being palmitic, stearic and oleic acids (6155). The dominating L-amino acids in the cells are threonine (14·0%), proline (13·2%), glycine (10·0%), serine (8·6%), glutamine (8·5%), valine (8·4%), asparagine (8·2%), and alanine (7·9 mol % of the total amino acids) (3232). — The GC content (59·5%) did not differ significantly from that of *V. dahliae* (5299).

V. albo-atrum was originally recognized as a causal agent of potato wilt but is now known to cause also wilt in hop, tomato, alfalfa and, less commonly, strawberry, sainfoin, runner bean, broad bean, pea, clover, and cucumber (2352, 2673, 5436). In these hosts, *V. albo-atrum* is usually more virulent than *V. dahliae*; vascular infection leads to wilt with or without obvious flaccidity and the infected xylem vessels are commonly browned. — *V. albo-atrum* is generally not host-specific, notwithstanding the fact that some alfalfa isolates can attack this host as well as other plants, including some weeds, while others cannot (2352). — *V. albo-atrum* is far less common than *V. dahliae* and has a rather northern distribution in Europe and the USA (2673), although it has been found rarely in Greece (2673), India, Honduras (IMI), and Australia (869). It is a strictly root-inhabiting fungus (2669, 2673) and its hyphae do not usually spread far beyond roots into the soil, although densely planted seedlings can easily be infected through root contact (2352), but infection of cut stems with aerial conidia and pieces of diseased material is considered as the most important source of infections. Dissemination through the air is indicated by the observation that infected alfalfa fields showed a higher conidium content in the air than non-infected fields (1300). *V. albo-atrum* has been found as part of the air spora in Pakistan (IMI). Selective isolation is only possible by removing conidia from host roots incubated in a moist chamber (3715). — The following reports of soil isolates have to be considered with reservation: northern Scotland (163), Germany (5316, 6546), France (918), Pennsylvania (1163), Japan (3267), equatorial West Africa (1420), and Peru (2005), while ones from British Honduras and India have been verified at IMI. — In addition to occurrences in grassland and cultivated soils, it has been found in soil treated with sewage sludge (1163), and in fallow soils (918). Soils rich in humus or clay (with a low rate of heat exchange) are more suitable for the survival of *V. albo-atrum* than ones which either have low humus contents or are sandy (6546). It is reported to occur in soils of pH 5·6–6·5 (163), and grow in soil in the temperature range 7–27°C (6546). High soil moisture seems favourable to both the presence and pathogenicity of the fungus (3318). It survived storage in sterile water for ≥ 7 years *in vitro* (605). Weeds such as *Matricaria chamomilla, Lithospermum arvense* (5397), *Senecio vulgaris, Capsella bursa-pastoris, Plantago lanceolata,* and *Polygonum persicaria* (2284) have all been observed to serve as hosts. The *V. albo-atrum* population in soil decreases rapidly in the absence of suitable host plants (2673, 3318, 5224). Resting mycelium was found to be viable in soil after nine months when buried at 30 cm, seven months at 15 cm and five months at the surface (2352). In Ontario it was found to overwinter only in infected potato tubers (4491). — *V. albo-atrum* has a weak competitive ability in soil (649) and is susceptible to a number of antagonists (6397), such as *Gliocladium roseum* and *Trichoderma viride* (232, 236); *in vitro* it was also found to be inhibited by *Tuber melanosporum* (632), *Streptomyces rimosus* (998) and *Erwinia carotovora* (3036). It can serve as food for the nematode *Aphelenchus avenae* (3579, 3580). An increased pathogenic activity towards *Solanum melongena, Fragaria vesca* and *Impatiens balsamina* was found in combination with the nematode *Pratylenchus penetrans* (469, 4020). — Soil volatile factors were found to stimulate conidium germination on an agar layer (2105).

The optimal temperature for growth is 21°C (5396) or 20–24°C (966, 1362, 2667), and the minimum 10°C; no growth occurs at 30°C (2673, 6546), and no pathogenicity above 25°C (2673). The pH optimum for growth is 8·0–8·6 (2667, 2673). The production of resting structures and melanin is inhibited under near-UV (300–320 nm) (1856). In shake cultures, after initial mycelium production the fungus occurs mainly as conidia, particularly if there was a high initial inoculum density ($1·6 \times 10^8$ conidia ml^{-1}) (2915). Conidia germinate when placed in contact with soil; anastomoses between germ tubes, conidia and hyphae start to occur immediately (5137). At −6 bars water potential the germination of conidia was most rapid at 25°C and inhibited at 30°C, while at −30 bars it was greater at 30°C (2157). The soaking of resting mycelium in water may activate the germination process to some extent (2675). — *V. albo-atrum* grows well on a variety of media (966); a special medium for obtaining high yields of heterozygous diploids has been described (2646). Trehalose and glycerol are the most suitable C sources for both *V. albo-atrum* and *V. dahliae*, while the former grows better than the latter on D-mannitol, D-glucitol and D-ribitol, and can also become adapted to growth on D-erythritol. D-Galactitol, D-xylitol, D- and L-arabitol and *meso*-inositol are poor C sources (6056), while starch is another very suitable substrate (1678). When cultivated on glucose or glycerol as C source, *V. albo-atrum* requires CO_2, which was not essential when it was grown on succinate or acetate (2289). Pyruvate carboxylase and phosphoenolpyruvate carboxykinase activities have been demonstrated in cell fractions (2286). D-Glucose and D-xylose were found to be especially effective in stimulating conidial respiration, but metabolic intermediates were less suitable in this respect (417). Zinc ions are required for optimal dry weight production (5818). Glucosidase production (1340) was found to be twice as high in haploids as in diploids of *V. albo-atrum* (2553). The fungus produces pyruvate carboxylase (2287) and phosphoenolpyruvate carboxykinase, with high yields when grown on malate (2288). *V. albo-atrum* can grow on D-galacturonic or D-glucuronic acids (4749), cellobiose (2350) and CMC (6342) as sole C sources. The hydrolytic enzymes endo-polygalacturonase, exoarabinase, endo-pectintranseliminase, endo-xylanase and cellulase appear in this order after 2–9 days, and β-D-galactosidase, galactanase, and β-D-glucosidase are also formed (1179, 2912). In studies on the nature of the wilt toxin (5589), phytotoxic proteins (4056), a fructosan, pectinase and cellulase C_x (2351) were implicated, but endo-polygalacturonase together with some polysaccharides are probably the most important (4491). The production of endo-polygalacturonase is found to be correlated with the pathotype (4057), though this may not be the case in all isolates (2909); its production can be stimulated by a pectin substrate and it can be inhibited by oxidized phenols (4465, 4466). Pectintranseliminase and pectinmethylesterase are also constitutive and their production can be increased by an appropriate substrate (2351). The kinetics of chitinase production has been investigated (6086). Proteolytic enzymes are produced (4058) and polyphenoloxidases were found in infected potato tubers (4491). The C/N ratio of the substrate had little influence on the development *in vitro* of *V. albo-atrum* (5198). In a medium with glucose as the only C source, a,a-D-trehalose, D-mannitol, D-arabitol and sometimes also *meso*-erythritol and *meso*-inositol were found in both this species and *V. dahliae* (3300); on the other hand, glucose, fructose, traces of raffinose and tri-, tetra- and pentasaccharides were found when grown on sucrose (3298). The amino acids L-threonine, L-valine and L-arginine were more suitable for the growth of *V. albo-atrum* than for *V. dahliae*, contrary to the situation with L-lysine (5199). A growth substance with gibberellin-like properties (237), thiourea (5278) and a protein-lipopolysaccharide complex (2910) are produced. — *V. albo-atrum* can become hyperparasitic on *Rhopalomyces elegans* (367). — *V. albo-atrum* is much more sensitive to UV irradiation than *V. dahliae* (4666). It can grow at 20% CO_2 *in vitro* (1504).

Verticillium catenulatum (Kamyschko ex Barron & Onions) W. Gams *1971* — Fig. 365.

≡ *Diheterospora catenulata* Kamyschko ex Barron & Onions *1966*

DESCRIPTIONS: Barron and Onions (368), Hammill (2240), and Gams (1882). — Colonies reaching 1·3–2·0 cm diam in ten days at 20°C on MEA. *V. catenulatum* is very similar to *V. chlamydosporium* (q.v.), but the phialoconidia are predominant; these are catenulate, subglobose, with slightly apiculate ends and 2·9–3·3 × 1·9–2·2 μm. Dictyochlamydospores (as in *V. chlamydosporium*) scanty, often only produced on SEA.

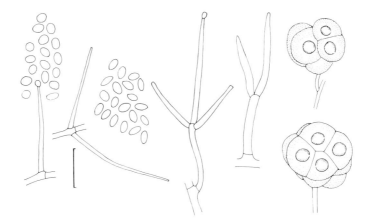

FIG. 365. *Verticillium catenulatum*, solitary and verticillate phialides with conidia and dictyochlamydospores.

This species has a worldwide distribution in various types of soil and is reported from the USSR, Czechoslovakia (1014), the Netherlands, Denmark, Germany (1882), the British Isles, Canada (368, 505), Chile (4569), Trinidad, Australia (368), Georgia/USA (2240) and Italy (488). Its known habitats include forest (505, 2240), grassland (1014) and cultivated soils (1882), *Picea* roots (1882), healthy and rotted strawberry roots (4133), remains of potato plants (3113), slug eggs and millipede droppings (368).

V. catenulatum has been observed to decompose paper (6072).

Verticillium chlamydosporium Goddard *1913* — Fig. 366.

≡ *Diheterospora chlamydosporia* (Goddard) Barron & Onions *1966*
= *Stemphyliopsis ovorum* Petch *1939*
= *Diheterospora heterospora* Kamyschko *1962*
= *Pochonia humicola* Batista & Fonseca *1965*
= *Dictyoarthriniopsis kelleyi* Dominik & Majchrowicz *1966*

DESCRIPTIONS: Barron (364), Barron and Onions (368), Gams (1882), and Campbell and Griffiths (871). — Colonies reaching 1·3–2·0 cm diam in ten days at 20°C on MEA, white or ochre-yellow; reverse similarly coloured. Phialides produced singly or in one or two whorls on aerial hyphae, slender, 22–26 μm long, tapering from 1·2–1·5 μm to 0·4–0·5 μm towards the tip. Conidia forming slimy heads, ellipsoidal, usually with a slightly apiculate base, 3–4 × 1·5–2·0 μm. Dictyochlamydospores abundantly produced in the aerial mycelium on 9–12 μm long stalks, consisting of a cluster of 6–9 thick-walled cells, overall 20–25 μm diam.

The two closely related species, *V. chlamydosporium* and *V. catenulatum* (q.v.), provide a remarkable example of the transition between conidial chain and head formation. Since such a

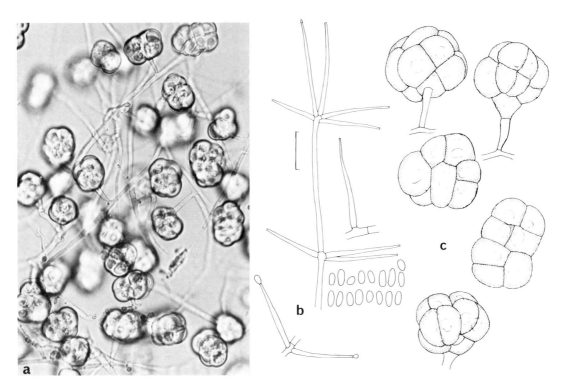

FIG. 366. *Verticillium chlamydosporum*. a. Dictyochlamydospores, × 500; b. verticillate conidiophores and conidia; c. dictyochlamydospores; CBS 103.65.

transition, however, is known in a considerably greater number of species which are all placed in the same section *Prostrata* of *Verticillium* (1882), the presence of dictyochlamydospores alone does not justify segregation as a separate genus. — The rather similar *V. catenulatum* (q.v.) produces comparable dictyochlamydospores, although less abundantly (and mainly on SEA) but has catenulate conidia; a third dictyochlamydosporous species, *V. bulbillosum* W. Gams & Malla *1971*, has more irregular dictyochlamydospores and slightly curved conidia; these latter are less pointed than those of *V. psalliotae* (q.v.). — The development of the dictyochlamydospores involves a secondary wall thickening and has been studied by TEM and SEM; the cells were found to be uninucleate (871).

V. chlamydosporium has been isolated from the most varied soils. A worldwide distribution is evidenced by isolations from soils in France (4178), the Netherlands, Germany, Guinea, Brazil (1882), Chile (4569), the USA (2573), the British Isles, the USSR, Canada, Sri Lanka (368), Easter Island (75), Japan (3267, 5846), India (427, 4698, 5366), Australia (977), New Zealand (368, 2705, 5815), Brazil (379), the Bahamas (2006), Zaïre (2966), and South Africa (4407). *V. chlamydosporium* has been isolated from forest (977, 2006, 2573, 3267, 4179), grassland (2573, 4179), savannah (4407), arable (2816, 3414) and garden soils (2010), and the rhizospheres of *Lolium perenne*, clover (5815) and flax (5826). Other known substrates include the cysts of *Heterodera schachtii* (798, 6368), snail eggs (368), and *Hyaena striata* dung (5366). It can colonize hair baits in the soil only at vapour saturation (977).

V. chlamydosporium can degrade both cellulose (2006, 3414, 4184) and chitin (2706), and also attack wool (2463). — It showed an antibiotic activity against several bacteria *in vitro* (3600) and it has been found to parasitize oospores of *Phytophthora cactorum* (5454).

Verticillium dahliae Kleb. *1913* — Fig. 367.

= *Verticillium albo-atrum* sensu Wollenw. *1929* et auct. [non Reinke & Berth.]
? = *Verticillium tracheiphilum* Curzi *1925*
= *Verticillium ovatum* G. H. Berk. & A. B. Jackson *1926*
= *Verticillium albo-atrum* var. *medium* Wollenw. *1929*
= *Verticillium dahliae* var. *longisporum* C. Stark *1961*

DESCRIPTIONS: Heale and Isaac (2353), Mamluk (3558), and CMI Descriptions No. 256, *1970*; see also under *V. albo-atrum*. — Colonies growing moderately fast on PDA or MEA, reaching 1·8–2·7 cm diam in ten days at 20°C, or 5·5–5·8 cm diam in 15 days at 20–25°C (2667), white, flocculose; reverse whitish to cream but later becoming black with the formation of microsclerotia. Conidiophores ± erect, hyaline throughout, with several whorls of 3–4 phialides. Phialides subulate, mostly 16–35 × 1·0–2·5 μm. Conidia ellipsoidal to short-cylindrical, hyaline, mainly 1-celled, occasionally 1-septate, 2·5–6(–8) × 1·4–3·2 μm (5–12·5 × 1·6–3·5 μm in var. *longisporum*). The only resting structures produced are microsclerotia which are dark brown to black, torulose, and consist of almost globose cells; they arise from single or adjacent hyphae by repeated budding and multilateral septation, are ± elongate, and 50–200 × 15–50(–100) μm. White sectors without microsclerotia commonly appear.

For a discussion of the delimitation of this species see under *V. albo-atrum*, where its separation from that species and comparative physiological studies conducted are summarized. Reports of "*V. albo-atrum*" isolates known to have had microsclerotia are treated as for *V. dahliae* here.

The effects of heterokaryosis are limited to the production of binucleate anastomosing cells which permit the development of complementary auxotrophs at 21°C, while at 30°C only the development of heterozygous diploids is possible (4667). *V. dahliae* var. *longisporum* is a stable diploid (2654). — Ultrastructural studies have been carried out on hyphal structures (6443) but most have been primarily concerned with the microsclerotia (2108, 6318, 6443). Pigment granules are located in the walls and in a fibrillar network encapsulating the microsclerotia (6318). Ultrastructural details of sclerotial germination have proved somewhat controversial (748, 4062). Paramural bodies arising as invaginations of the hyphal plasmalemma were observed in freeze-etched preparations (2108). The sclerotial pigment itself is an indole melanin formed with scytalone (6318), the naphthalenon derivative (−)-vermelone (5581) and some other naphthole derivatives (5582) as precursors. — Cell walls, as prepared from shake cultures, contained a complex of polysaccharides, proteins and lipids: D-glucuronic acid (1·2–1·3%), D-glucose (41–56%), galactose (8–8·5%), mannose (6–6·4%), glucosamine (2–3·3%), chitin (7·6–10%), protein (11–14%), extractable lipids (0·4–0·6%), and bound lipids (2·3–2·8%) (6174). — The pyrimidine sequences of DNA were studied in three isolates and showed both some similarities and differences (2188). DNA analysis gave a GC content of 58·7%, a figure almost identical to that obtained for *V. albo-atrum* (5299).

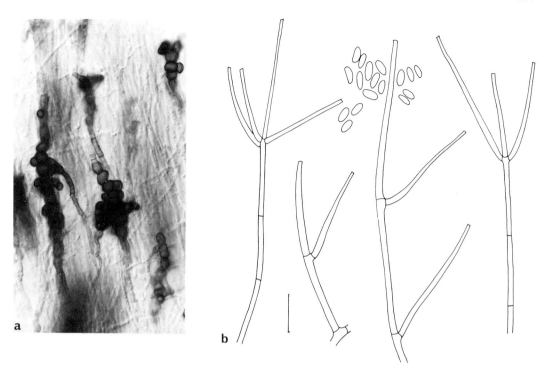

FIG. 367. *Verticillium dahliae*. a. Microsclerotia, NIC, × 400; b. verticillate conidiophores and conidia; CBS 205.26.

V. dahliae is a very common causal agent of wilt diseases in many crop plants, and the symptoms usually involve flaccidity or chlorosis of leaves followed by permanent wilting, but are often less severe than with *V. albo-atrum*. Known host plants include cotton, dahlia, mint species, vine, tomato, potato, eggplant, olive, pistachio, stone fruits, brussel sprouts, ground-nut, horse-radish, red pepper, strawberry, and others (1659, 4491). On mint, brussel sprouts, and pepper, some host specificity has been observed (2673). — Its distribution ranges from northern climates down to the subtropics in all continents (CMI Descriptions No. 256; 3558). In common with *V. albo-atrum*, *V. dahliae* is strictly a root-inhabiting fungus and no or very little growth occurs through soil (1659, 2673, 5694), in which it is a poor competitor (2673, 3650, 4734). Its occurrence in soil is especially low at temperatures of 6–8°C, above that its frequency increases with peaks in the summer months (4068). — Microsclerotia are the principal form of survival (2074, 3756, 5136, 6363). The density of propagules in soils with different textures ranged from 0·03 to 250 microsclerotia per gram of soil (219, 222, 418). They can also survive in the roots or rhizomes of weeds (2284, 5398), but whilst conidia or mycelial fragments are short-lived in soil (1858, 6363), these can also remain in soil without the host plants for many years (2669, 3649): microsclerotia can in fact survive almost indefinitely in soil and reports range from 4–14 years or longer (2607, 2673), although the population density diminishes rapidly. The survival of microsclerotia is prolonged at temperatures of 5–15°C (4061, 4064). However, microsclerotia are reduced or killed by flooding the soil for six weeks, incubating it under N_2 for three weeks, or by the addition of alfalfa meal or sucrose under N_2 even after five days (2663, 3755). Crop rotation (with moderate success)

(2607) and the breeding of resistant plant varieties were consequently the most important control measures used prior to the advent of systemic fungicides. Inoculum densities of 6–12 microsclerotia per gram of soil were found to be sufficient to infect eggplants (1660) but a population of 100 microsclerotia or 50 000 conidia per gram of soil proved necessary for a 100% infection of tomato plants (2074). A straight log-log relationship was observed between the inoculum density in soil and the infection of cotton plants (221). *V. dahliae* has been found in soil depths of 0·5–15(–75) cm (2073); after cropping, a higher number of microsclerotia can be found in the deeper layers (10–20 cm) (220); no propagules were found below 75 cm (6362). Seasonal changes of the sclerotial density have been repeatedly observed with a gradual decline over the vegetation period followed by an increase during the harvest due to the release of sclerotia from damaged roots (1660, 1662). Transmission can also take place with the seeds of cotton, sunflower (CMI Descriptions No. 256) and spinach (5491). — A number of selective isolation techniques have been described (1363, 1434, 3715) which involve the incorporation of ethanol into water agar together with streptomycin (4063), washing and decanting (1662), selective sieving of the soil for microsclerotia with 37 and 125 μm mesh sizes (2604, 2605, 2606), plating on PDA with PCNB (1662), a flotation technique employing a cesium chloride solution (461), a dilute soil-extract agar with antibiotics (3756) and 0·2% polygalacturonic acid (2075), or a sodium polygalacturonate-guanidine-biotin medium (2605). The developing colonies can be gently washed to remove the aerial mycelium and facilitate the observation of microsclerotia. To discover very low inoculum densities, plant roots themselves are still the most sensitive bait; the number of sclerotia formed within 20 days on root lengths of *Datura stramonium* was found to be correlated with the inoculum density in the soil (1661). — In the rhizosphere of cotton rather low numbers of microbial antagonists were observed, while the proportion of fungi amongst them was increased under diseased plants (223). Several antagonistic fungi were found to stimulate microsclerotium formation *in vitro* (265). — The germination of microsclerotia which had been stored in the dry state for about six months (2863) at $-10°C$ (2675) was incited by remoistening alone, but in the soil the addition of exogenous carbon and nitrogen sources was required for germination to overcome antagonistic effects induced by adding oat or alfalfa residues (4609). Microsclerotia were found to germinate and sporulate each time after up to nine consecutive cycles of air-drying and re-wetting (1694). Amendments of glucose, sucrose and $NaNO_3$ caused an initial increase in sclerotium populations followed by a decrease, while ribose, alfalfa and oat residues caused a marked decrease after 12 weeks (2075). Substrates such as chitin, laminarin or wheat and clover residues can inhibit their germination (451, 2819), but a narrow C/N ratio was found to stimulate it (2075). For further data on germination from *in vitro* studies see below. — Faeces of the bulb mite *Rhizoglyphus echinopus* fed on *V. dahliae* contained both viable conidia and microsclerotia (4634). *Verticillium* infection was influenced synergistically by the nematodes *Pratylenchus penetrans* on tomato (3999), *P. minyus* on peppermint (1708), and *Globodera rostochiensis* on potato (1183, 2283).

The optimal temperature for growth is 23·5°C (5396) or 23–25°C (1362). Growth is still possible at 30°C (2667). The thermal death point was found to be at 57·5°C for 30 min in soil (613) but this species is also reported to survive extreme temperatures in the range $-30°C$ to 80°C in soil (5469). In water the thermal death point was lower, at 47°C for 5 min for conidia and hyphae, and 10 min for microsclerotia (4127). Mycelial growth is reduced to 50% at water potentials of -49 bars and completely suppressed at -119 bars (1126, 1991). Conidial production is increased under white, blue (478 nm) and fluorescent light (2838, 3697). — The formation of resting structures is inhibited under blue (478 nm) (2838) and near-UV

(300–320 nm) (1856) but is optimal under red or far-red light or in the dark (2836, 3697). The optimal pH range for growth is 5·3–7·2 (2667). Microsclerotia form in 2–5 days at 18–30°C; none are produced at 32°C (724, 6363). Good production of microsclerotia was obtained on a medium containing sucrose and Mn^{2+} ions (683) or fructose and $NaNO_3$ at pH 7–7·5 and 22·5°C (2863). Increased carbon concentrations and constant nitrogen content enhance microsclerotium formation up to a C/N ratio of 50:1 (2221), while sporulation decreases (5198). — Germinability was found to decrease in aged microsclerotia (2673). A near to 90% germination of microsclerotia was obtained on CzA after 18 h (4678) but mycostatic influences seem to inhibit germination in soil to a considerable extent (1619). Microsclerotia germinated within a pH range of 4·2–9·2 in a phosphate buffer, whereas other buffers were inhibitory below pH 5 (2037); germination of microsclerotia is optimal at 25–30°C and decreases at water potentials lower than −10 bars; a complete inhibition usually occurs at −100 bars (1126). A self-inhibitory, diffusible substance is also produced *in vitro* by microsclerotia which suppresses germination, conidium production and hyphal growth (684). The growth of new mycelium appears to originate from old hyaline hyphae associated with microsclerotia; germination of the dark, thick-walled microsclerotial cells themselves has not been observed (5115). — Sucrose, D-glucose, and maltose (2667) were found to be suitable C sources; the best growth occurred on trehalose and glycerol, while D-mannitol, ribitol (6056) and starch (6092) were less suitable. Ammonium is a suitable N source but an organic N source was better than an inorganic one (2667). Some isolates of *V. dahliae* were found to be auxotrophic for biotin (2837, 3833, 4920). In studies on the nature of the wilt toxin, endo-polygalacturonase was repeatedly found to be the agent in the presence of Mg^{2+} ions (5197), but extracellular (lipo-)polysaccharides were also implicated (2911, 2914, 6553). The endo-polygalacturonase has been purified and characterized (6175). Cotton root exudates containing L-alanine and choline reduced growth and suppressed endo-polygalacturonase production (642). *V. dahliae* produces three fractions of endo- and one exo-1,4-β-glucanase (4953). The production of catalase, cellulases, and β-fructofuranosidase was reported to be positively correlated with the degree of pathogenicity (4970), and there seems to be no difference in this property between white mycelial and black microsclerotial isolates (2189). At 10^{-6} g ml^{-1}, both indole-3-acetic and gibberellic acids can stimulate cellulase production (4952). On a mineral medium with D-glucose as the only C source, trehalose, mannitol and arabitol were detected, together with some traces of sugar alcohols (3300). Several oligosaccharides were analysed in cultures grown on sucrose (1026). *V. dahliae* can degrade catechol (which also stimulates microsclerotium formation) and caffeic acid by means of a peroxidase, but concentrations exceeding 5×10^{-4} mol l^{-1} inhibit its growth (3301). *V. dahliae* produces kojic acid (512). It detoxifies the antifungal compound sanguinarine by reducing it to its dihydroxy derivative (2554), produces an antibiotic active against *Micromonospora vulgaris* (3122), and a growth substance with gibberellin-like properties (237). — It can become a mycoparasite of *Rhopalomyces elegans* (367) and some other fungi (5668).

Verticillium lecanii (Zimm.) Viégas *1939* — Fig. 368.

≡ *Cephalosporium lecanii* Zimm. *1898*
= *Acrostalagmus aphidum* Oudem. *1902*
= *Acrostalagmus coccidicola* Guéguen *1904*
= *Cephalosporium lefroyi* Horne *1915*
= *Cephalosporium coccorum* Petch *1925*
= *Cephalosporium aphidicola* Petch *1931*
= *Verticillium hemileiae* Bouriquet *1939*

For many other synonyms see Gams (1882).

DESCRIPTIONS: Gams (1882), and Skou (5404). — Colonies reaching 1·8–2·2 cm diam in ten days on MEA at 20°C, white to pale yellow, cottony, rarely with fasciculate hyphae; reverse uncoloured, yellow or ochraceous. Phialides solitary or in scant whorls arising from erect conidiophores or little differentiated prostrate aerial hyphae, aculeate, extremely variable in size, usually 12–40 × 0·8–3·0 μm. Conidia in heads or parallel bundles, cylindrical with rounded tips or ellipsoidal, in the range 2·3–10 × 1·0–2·6 μm. Chlamydospores absent. —*V. lecanii* is interpreted here as a heterogeneous species aggregate. The distinction of species according to host insects or intensity of ramification as proposed by Petch in various publications cannot, however, be retained on morphological grounds (1882). — The vegetative mycelium and conidia are uninucleate and hyphal anastomoses have been observed (5377).

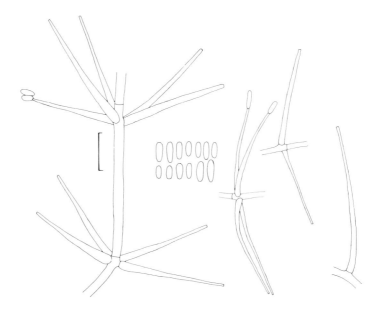

FIG. 368. *Verticillium lecanii*, verticillate and solitary phialides and conidia.

V. lecanii is one of the most important and common entomogenous hyphomycetes, occurring in all climatic regions on coccids, aphids, thrips, Diptera, Homoptera, Hymenoptera (5404), Lepidoptera and mites (1882, 3628, IMI). Other important substrates are rusts and other fungi; it is a consequence of this habit that the species is not infrequently isolated from soil (1882). There appears to be no specific difference between entomogenous and fungicolous isolates, as equal aphid pathogenicity in four isolates originating from rusts has been observed (R. A. Hall, unpublished). There are reports from uncultivated soils (4698) and grassland (IMI) in India and the Netherlands (CBS), garden composts in Germany (1424), and tundra soils in Alaska (1750); it has also been isolated from leaf litter of oak, ash and birch (2411), tea leaves, barley seed (IMI), baker's yeast, beet seed and bursting corn kernels (1882).

V. lecanii can decompose cellulose and chitin and has a pronounced proteolytic activity (1425); pectin and starch are also utilized (1750). Uredo- and other spores of rusts are degraded with high pectolytic activity (3255). This is the only *Verticillium* capable of attacking uredospore germ tubes (18); the enzyme system involved in this is thermolabile. A standardized bioassay for aphid pathogenicity under high humidities has been developed (2223). Known metabolites include aphidicolin which is active against a virus (766), and the insecticidal cyclo-depsipeptide bassianolide (5653). *V. lecanii* has an unusually high tolerance for fluorides in solution, and at concentrations up to 0·211 mol l^{-1} NaF growth was found to be possible, though reduced and with usually larger conidia (3296).

Verticillium nigrescens Pethybr. *1919* — Fig. 369.

= *Verticillium dahliae* var. *zonatum* van Beyma *1939*

DESCRIPTIONS: Pethybridge (4527), Isaac (2667, 2668), and CMI Descriptions No. 257, *1970*; chlamydospores: Hoes (2493). — Colonies slow-growing, reaching 2·4–2·8 cm diam on PDA at 20°C in nine days (3·0–3·6 cm at 25°C), at first white, velvety, soon becoming dark grey and almost black in reverse with the formation of chlamydospores, sometimes also centrally orange-brown; no hyaline sectors occurring. Vegetative hyphae and conidiophores hyaline; conidiophores erect, little differentiated from the vegetative hyphae, bearing 1–2 nodes with 1–2(–3) phialides, but also sporulating within the agar on reduced phialides. Phialides subulate, mostly 20–35(–50) × 1·5–3 μm. Conidia ellipsoidal to short-cylindrical, mostly simple, only rarely 1-septate, 4–8·5 × 1·5–2·5 μm. Chlamydospores abundantly produced after ten days, terminal or intercalary, singly or in chains, with or without intermittent hyaline cells, olivaceous brown, ± globose or pyriform, 5–8(–10) μm diam; sometimes longer chains of more cylindrical pigmented cells present in old cultures. Microsclerotia and torulose resting mycelium absent. — Similar species are *V. nubilum* Pethybr. (q.v.), which has larger chlamydospores, and *V. tricorpus* Isaac *1953*, which also forms some large microsclerotia and has a characteristic yellow pigmentation in young colonies. — The frequently sparse branching of the conidiophores of *V. nigrescens* can lead to it being confused with *Acremonium* species. — *V. nigrescens* has been investigated both cytologically (3488) and genetically (2349). TEM and light-microscopical studies of the conidia did not reveal any significant differences between those of *V. nigrescens* and *V. albo-atrum* (765), but gel electrophoresis of the proteins and serological tests showed characteristic differences between this and other species of *Verticillium* (3834, 4496, 6459).

V. nigrescens is an extraordinarily common soil fungus and a typical soil inhabitant, mainly in Europe and Canada (2669, 2673). It was originally isolated from potato tubers (4527), but is also commonly isolated from subterranean parts of potato and tomato, and, less commonly, other plants (CMI Descriptions No. 257), including eggplant and red pepper (3750). It has been found to attack spearmint and peppermint to which it provides immunity against attack by *V. dahliae* (3751). Its pathogenicity has been rated as intermediate between that of *V. dahliae* and *V. nubilum* (2673). — *V. nigrescens* forms phialoconidia and chlamydospores freely, which ensures its isolation from dilution plates (1425, 2161), soil plates (1379), and washed soil particles (307, 1888, 1889). A moist-chamber method has been described for its isolation from roots (3715). *V. nigrescens* has so far not been reported from soils in warmer climates, but is found in high frequencies in soil analyses from the British Isles (745, 1379, IMI), Germany (1425, 1888, 5316), the Netherlands (1614, 1616, 4658), France (2161), and Canada (234). It has been most commonly isolated from agricultural soils (234, 1616, 1888, 1889, 2161), but is also known from garden compost (1425), salt-marsh (1379), estuarine sediments (5316), sand dunes (745), and a sandy podzol under *Pinus nigra* (4445). In agricultural soils it reaches its maximum density at a depth of 10 cm, but can penetrate to over 40 cm (2161). Its distribution in soil is little influenced by pH (745), is not influenced by manuring (2161), but was found to be stimulated by growing peas (1433) or grass (1617); in one analysis of washed soil samples, *V. nigrescens* was exclusively isolated from washed organic particles (307) while in another study it appeared on mineral particles with higher frequencies than on organic particles (1888). It has frequently been reported from solanaceous and less frequently from other plants, and also from millipede pellets (CMI Descriptions No. 257). Other reports include the rhizospheres of potato, sugar beet, wheat and grasses (1614), although without an accumulation on roots (2669), and on seeds and seedlings of flax (2739). — It is a typical soil inhabitant with a slight, though detectable tendency to spread in the soil (2670) and with a good capacity for indefinite survival (2669); conidia did not, however, survive one week's storage at 50°C (2675). *V. nigrescens* has a relatively high sensitivity to soil fungicides (1426). Forty-two per cent of chloroform-fumigated mycelium was mineralized in soil within ten days at 22°C (133).

The optimal temperature for growth and chlamydospore production is 22·5–25·5°C, and the minimum is 5°C; fair growth occurs at 32–34°C but none at 37°C (2493, 2667, 3750, 5749). The optimal pH is in the range 5·3–7·2 (2667). — Weak growth occurs on glycerol (2667). Starch (6092) and pectin (1432) are efficiently utilized and endopolygalacturonase, pectin-transeliminase and pectinmethylesterase are produced (5711). Utilization of xylan and CMC is also very good, but sometimes marked differences in this capacity occur between isolates (1432). Erosion of birch wood has been observed (4191). Ammonium ions were the most suitable of several N sources tested (2667). Mn^{2+} ions are oxidized (5151). Some gibberellin-like activity has also been reported (237).

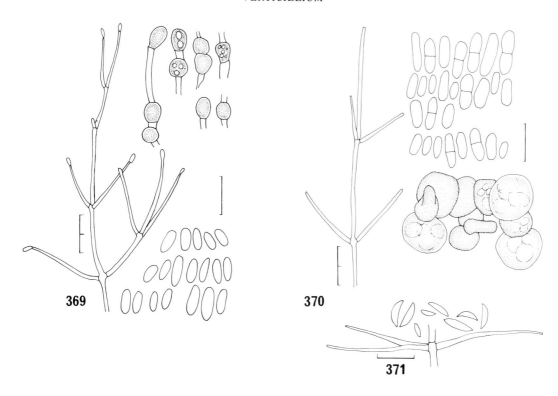

FIG. 369. *Verticillium nigrescens*, verticillate conidiophore, chlamydospores and conidia.
FIG. 370. *Verticillium nubilum*, verticillate conidiophore, conidia and chlamydospores.
FIG. 371. *Verticillium psalliotae*, whorl of phialides and conidia, CBS 154.70.

Verticillium nubilum Pethybr. *1919* — Fig. 370.

DESCRIPTIONS: Pethybridge (4527), Isaac (2668) and CMI Descriptions No. 258, *1970*; chlamydospores: Hoes (2493). — Colonies rather slow-growing, reaching 6·0 cm diam in 14 days on PDA at 20°C, at first white, flocculose, soon becoming brownish or black in reverse with the formation of chlamydospores. Hyaline sectors sometimes occurring. Vegetative hyphae and conidiophores hyaline; conidiophores erect, little differentiated from the vegetative hyphae, bearing 1–3 nodes with 1–3 phialides on each. Phialides subulate, mostly 25–35(–45) × 1–2·5 μm. Conidia cylindrical to ellipsoidal, hyaline, simple or 1-septate, 4–10(–12·5) × 2·5–3·5 μm. Chlamydospores abundantly produced, singly or in (sometimes branched) chains of 2–7 dark brown, ± globose cells, 8·5–17 μm diam, terminal or frequently intercalary due to terminally continued growth, often subtended by a slender, ± pigmented pedicel. Microsclerotia and dark resting mycelium absent. — For a discussion of the circumscription of this species see under *V. nigrescens*. — One per cent of the conidia were found to be binucleate (3488). Separation of the proteins by gel electrophoresis showed characteristic differences from other *Verticillium* species (3834).

V. nubilum was originally isolated from potato tubers but has subsequently been recognized as a true soil inhabitant (2669, 2673) although it is rarely listed in published analyses. It becomes incidentally weakly pathogenic on potato (twisted sprout), tomato and *Antirrhinum* (2672, 2673, CMI Descriptions No. 258). — It is so far known only from Europe and Canada (5330) and has been reported from garden soil under beans (4451, 4452, 5801), potato fields (418, 1614, 1616), and the root surface of wheat (5330); but it also occurred, although very rarely, in wheatfield soils in Germany. In a forest nursery soil it reappeared 22 months after a steam treatment (3908). Its chlamydospores permit an almost indefinite survival in the soil (2673).

The optimal temperature for growth is 19·5–22·5°C; almost no growth occurs at 29°C and higher temperatures are lethal (2493, 2668); the optimal pH range is 7·2–8·6 (2668). — *V. nubilum* produces endopolygalacturonase, pectin transeliminase and pectin methylesterase (5711).

Verticillium psalliotae Treschow *1941* — Fig. 371.

= *Cephalosporium curtipes* var. *uredinicola* Sukapure & Thirum. *1966*

DESCRIPTIONS: Treschow (5879), Gams (1882), Dayal and Barron (1314), and CMI Descriptions No. 497, *1976*. — Colonies reaching 2·3–2·5 cm diam in ten days at 20°C on MEA, white cottony, reverse mostly pale or dark red to purple or brown, the pigment often coagulating in dark globules. Phialides formed on prostrate aerial hyphae, 17–35 μm long, tapering from 1·0–1·7 to 0·3–0·6 μm. Conidia few in number, often transversely positioned on the phialide tips, sickle-shaped, with pointed ends, occasionally 2-celled, typically 5·8–8·0 × 1·5–1·8 μm, but secondary conidia often much shorter. Chlamydospores absent. — *V. psalliotae* as interpreted by Gams (1882) is probably a somewhat heterogeneous taxon and comprises isolates with widely differing conidial measurements. — The somewhat similar *V. fungicola* (Preuss) Hassebr. *1936* (=*V. malthousei* Ware *1933*), causal agent of dry bubble in mushrooms and many wild agarics, has often been confused with *V. psalliotae*; it can be distinguished by its more clearly differentiated, erect conidiophores and less pointed conidia formed in globose heads. *V. fusisporum* W. Gams *1971* and *V. lamellicola* (F. E. V. Sm.) W. Gams *1971* are rather similar to *V. psalliotae* but have straight fusiform conidia of 4·4–5·5 × 1·7–2·5 and 3·0–10 × 0·8–1·2 μm, respectively.

V. psalliotae was originally described as the causal agent of a brown-spot disease in cultivated mushrooms (5879), but has subsequently become known from rust sori (IMI), *Endogone* sporocarps, *Tuber* species, textiles buried in soil, cow dung, collembolan exuviae (1882), soil mites (3576), the tick *Ixodes ricinus* (5018), and as a mycoparasite on *Rhopalomyces elegans* (1314). Reports extend from Europe and North America (1882, CMI Descriptions No. 497) to tropical Africa, Malaysia, India, Australia (1882), and Israel (CBS). *V. psalliotae* has rarely been encountered in soil investigations; it has been found in an evergreen forest in Ohio/USA (2573), a *Casuarina* forest and a coconut grove in the Bahamas (2006), a forest nursery and lead mine spoil heaps in the British Isles (IMI), on calcareous stone fragments in France (4166), rock on the volcanic island of Surtsey (2406), and soils in

Australia (IMI) and Japan (5846). It is also known from cicads and lepidopteran insects in Ghana (CBS). — The mushroom disease was found to be most severe in mild foggy weather, and a synergistic effect with *Pseudomonas tolaasii* has been observed (CMI Descriptions No. 497).

The optimal temperature for growth is 23–24°C (5879), the minimum 8–10°C, and the maximum 35°C (4166). No infection of mushrooms occurs below 22°C (5879). The optimal pH is 6·5–7·0, but good growth is possible between pH 3·9 and 7·4, the latter being conducive to acid production (5879). *V. psalliotae* produces the yellow pigment oosporein (=3,3',6,6'-tetrahydroxy-4,4'-dimethyl-2,2'-bibenzoquinone) (3929). The production of a red pigment is enhanced by adding L-glutamic acid to the medium but not by ammonium chloride (5879). — In a chronically irradiated soil its frequency declined significantly between intensities of 53 and (230–)820 R per day (2009).

Volutella Fr. *1832*

Volutella ciliata Alb. & Schw. ex Fr. *1832* — Fig. 372.

≡ *Tubercularia ciliata* Alb. & Schw. *1805*
≡ *Atractium ciliatum* (Alb. & Schw.) Link *1816*
≡ *Fusarium ciliatum* (Alb. & Schw.) Link *1825*
= *Psilonia rosea* Berk. *1837*

DESCRIPTIONS: Chilton (1005), and Matsushima (3679, 3680). — The genus *Volutella* (nom. conserv. *1975*; lectotype species *V. ciliata*) is characterized by conspicuous sessile or short-stalked sporodochia with simple, hyaline, slimy phialoconidia and straight sterile setae. — A comparable genus with curled and warted setae is *Kutilakesa* Subram. *1956*.

V. ciliata forms moderately fast-growing colonies, reaching 3·0 cm diam in ten days at 20°C on OA; colonies almost white, cream to pale pink, pustulate from the sporodochia. Conidia produced on solitary, ± compacted *Verticillium*-like conidiophores, but mainly in sessile or somewhat stipitate sporodochia consisting of an almost stromatic base and densely fasciculate cylindrical phialides, 10–15 × 2–3 μm. Sterile setae arising in variable numbers from a level a few cells below the phialides, hyaline, 5–10 μm wide, thick-walled (to 2·5 μm), with pointed or blunt tips. Conidia ellipsoidal to cylindrical, hyaline, non-septate, smooth-walled, 4–6(–9) × (1·5–)2·0–2·5 μm.

Over 110 species have been described in *Volutella*, but most are still only vaguely delimited. It is doubtful whether species with darkening conidia, such as *V. piriformis* Gilman & Abbott *1927* and *V. piracicabana* Verona & S. Joly *1956*, can be accommodated here and might be best included in *Myrothecium* Tode ex Fr. (q.v.). — Samuels (5033) regarded *Nectria consors* (Ellis & Everh.) Seaver *1909* (= *N. ignea* Höhn. *1909*) as the teleomorph of *V. ciliata*. His isolates from the USA differ, however, from typical *V. ciliata* in having rod-shaped conidia, 5–6 × 1·0–1·5 μm (fitting *V. minima* Höhn. *1909*), and red sclerotia, and others from New Zealand in pyriform conidia, and almost no sterile setae; these are consequently not regarded as conspecific with *V. ciliata*.

Some comparable species are described in Table 3, but it must be emphasized that the specific epithets are liable to revision when herbarium specimens of older species have been examined.

While *Nectria consors* has been reported mainly from New Zealand and further from North Carolina, Java, Colombia and Panama (5033), *V. ciliata* is mainly known from the temperate zones and has only been rarely reported from warmer regions, for example, Italy (270, 271), Greece (4647), Israel (4758), central Africa (3063), Zaïre (3790), South Africa (1559, 4407), and India (IMI). It is as easily isolated by dilution and Warcup's soil plates (4647) as from washed soil particles (1889). It has been reported from forest soils under beech (270, 271, 272, 273, 3138, 3448), aspen, poplar (6101), mixed deciduous stands (2004, 2008, 2573, 3549, 3790), white cedar (505), and forest nurseries (6184), grassland (1614, 1700, 4313, 6182), steppe (1559), arable soil (540, 918, 1614, 1889, 2159, 2816, 3790), a fen (5559),

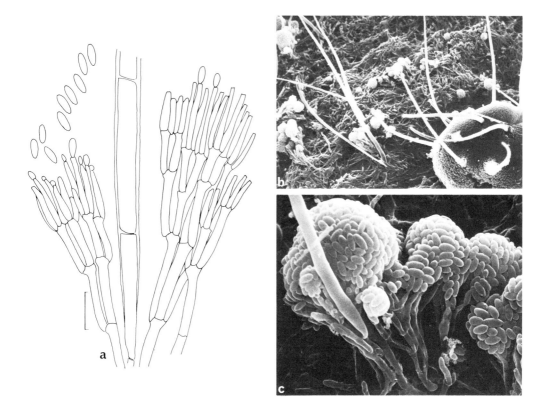

Fig. 372. *Volutella ciliata*. a. Part of young sporodochium with basal part of a seta; b. fully developed and young sporodochia, SEM, × 200; c. penicillate conidiophores with seta of a young sporodochium, SEM, × 1000; CBS 426.54.

dunes (4655) and garden compost (1424). It is widely distributed, especially in damp soils (4313). According to the available data, it seems to occur preferentially in alkaline soils. It is commonly found on plant remains (4548, 4758), for example, leaf litter of *Panicum coloratum* (1560) and birch (3449), decaying stems of *Urtica dioica* (6463) and *Pteridium aquilinum* (1821), decaying branches of willow (IMI), submerged *Hedera* leaves (2338), rotten wood (3063), and sticks exposed to river water (5249). *V. ciliata* has been isolated from tomato roots (1531, IMI), occurs at markedly increased frequencies on decaying bean roots (6134), and colonizes the bean rhizoplane in a late stage (1406); it is also found on beet roots, and *Gerbera* (IMI), dead rhizomes of *Iris* (983), and in the rhizospheres of flax (5826), potato (1614), onion (2159), grasses (1614, 5815), and wheat (2816, 3567). Under wheat, however, it was more frequent in the surrounding soil than in the rhizosphere (2816). Other known substrates include germinating seeds of cucumber and legumes (4120), and the dung of rabbits (2279), opossum and horses (2338, IMI). — In the soil the germination of conidia is generally inhibited by mycostasis (1407); *in vitro* it can be inhibited by metabolites from *Stilbella erythrocephala* (5362).

CMC is rapidly decomposed, with little difference in this ability between isolates (1432), starch is utilized (1827), and hairs are attacked (1633). No particular growth substances are required; a low antagonistic activity to *Phytophthora cactorum* has been observed (1425). — Significant reductions in shoot and root weights in wheat and pea seedlings have been observed *in vitro* (1430).

TABLE 3. Species comparable with *Volutella ciliata*

	Conidia		Peculiarities
	Shape	Size	
V. buxi (DC. ex Link) Berk. *1850* (teleomorph *Pseudonectria rousseliana* (Mont.) Wollenw. *1930*)	fusiform	6·5–9 × 2–3·5 μm	
V. gilva (Pers. ex Fr.) Sacc. *1881*	cylindrical-ellipsoidal	10–13 × 1·5–2·5 μm	setae flexuous
V. minima Höhn. *1909* (= *V. ramkumarii* Sarbhoy *1967*; teleomorph *Nectria consors* (Ellis & Everh.) Seaver *1909*)	rod-shaped	3–5 × 1·0–1·5 μm	
V. colletotrichoides Chilton *1954*	navicular	5·2–10 × 2·0–3·5 μm	setae dark
V. lini Mukerji *et al. 1968*	fusiform	8–14 × 1·2–1·6 μm	red sclerotia
V. keratinolytica Dominik & Majchr. *1964*	ellipsoidal-lemon-shaped	4·5-7 × 2·0–2·7 μm	setae pale grey

Wallemia Johan-Olsen *1887*

Wallemia sebi (Fr.) v. Arx *1970* — Fig. 373.

≡ *Sporendonema sebi* Fr. *1832*
= *Torula epizoa* Corda *1829*
    ≡ *Sporendonema epizoum* (Corda) Cif. & Red. *1934*
= *Torula sacchari* Corda *1840*
= *Torula rufescens* Fres. *1863*
= *Wallemia ichthyophaga* Johan-Olsen *1887*
= *Bargellinia monospora* Borzi *1888*
= *Hemispora stellata* Vuill. *1906*
= *Oospora d'agatae* Sacc. *1918*
= *Torula pulchra* Tengw. *1924*

For further synonyms see Ciferri (1060).

DESCRIPTIONS: Redaelli and Ciferri (4761), Wang (6168) and Ellis (1603); detailed studies on conidiogenesis: Cole and Samson (1109), Hashmi and Morgan-Jones (2308), Hill (2451), and Madelin and Dorabjee (3509).

The monotypic genus *Wallemia* (= *Hemispora* Vuill. *1906* = *Mauginiella* Cav. *1925* = *Bargellinia* Borzi *1888*) has a unique method of conidiogenesis: short and slender, closely packed, erect conidiophores develop a basipetal succession of cylindrical conidiogenous cells from an apical meristem (no trace of a phialidic collarette is visible); each conidiogenous cell subsequently falls apart into a regular quartet of initially cuboid, but later almost globose, conidia which resemble arthroconidia (sometimes termed meristem arthroconidia). Sympodial proliferation of the conidiophores sometimes occurs (2308, 3509).

W. sebi grows best on media with high osmotic values, but even on these growth is very slow, reaching only about $0 \cdot 5$ cm diam in ten days at 23–25°C. Colonies fan-like or stellate, powdery, orange-brown to dark blackish-brown. Conidiophores cylindrical, smooth-walled, pale brown, $1 \cdot 5$–$1 \cdot 8 \mu$m wide below but tapering below the meristem to *c.* 1μm and then widening again into the cylindrical meristem, about $15 \times 1 \cdot 7 \mu$m; conidiogenous cells and subtending meristem finely warted; conidia pale brown, finely warted, $2 \cdot 5$–$3 \cdot 5 \mu$m diam, uninucleate (2308). — The hyphal septa have a basidiomycete-like dolipore surrounded by an electron-dense material but without any endoplasmic pore cap (5777).

The conidia contain an isoprenoid quinone (5004) and more (13%) glycolipids than the mycelium ($8 \cdot 1$%) and some phospholipids; the former consist of sterol glycoside and three acyl glycoses (3409); in the soluble lipids of conidia and mycelia the unsaturated ones represent about 60% of the total fatty acids which also contain considerable amounts of short-chain fatty acids (3410); free lipids contain mainly squalene, the main steroid component is ergosterol followed by zymosterol and two other compounds (1847). Cytochrome c from *W. sebi* has been purified and investigated (5005).

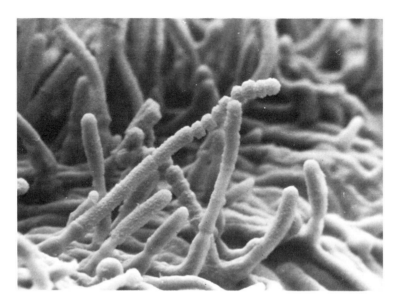

FIG. 373. *Wallemia sebi*, formation of meristem arthroconidia, SEM, × *c.* 2000, orig. R. A. Samson.

W. sebi is a cosmopolitan osmophilic, commonly air-borne fungus, often observed con-
taminating drying agar cultures (6396) and documented for the British Isles, Germany,
Denmark, Sweden, Norway, Italy, Israel, Uganda, Rhodesia, Ghana, S. Africa, Malaysia,
Brazil, and Australia (IMI). Its main substrates include rock salt, salted fish, salted beans,
bacon, jam and fruit jellies, bread, malt extract, marzipan, dates, fruits, textiles, man and
animals (1603, 3509, 4761). It has been isolated from a forest soil under white cedar in
Canada (IMI); other substrates include rotting paper (6168), pecan nuts (2572), dry fodder
(3258), stored hay, and a dead spider (3063). — *W. sebi* is known as causal agent of ulcerating
abscesses in man and has been isolated from the lungs, bones, lacrymal canals, skin scales and
other parts of the human body (4761).

W. sebi is highly osmophilic (4311, 6031). Optimal growth is obtained on 3% MEA with
10% NaCl followed by 3% MEA with 30% sucrose (3509). The optimal concentration is
$1·8–2·4\,mol\,l^{-1}$ NaCl, while no growth occurs below $0·8\,mol\,l^{-1}$ or above $4·5\,mol\,l^{-1}$ (4311).
Even in 50% sucrose some growth has been observed (4761). — The osmotic effect of NaCl
can be replaced by D-glucose, sucrose, glycerol or sodium itaconate; several sugars and
carboxylic acids can be utilized but fatty acids are not (4311). Ammonium and urea are
suitable N sources, whilst nitrate is not utilized and has toxic effects (4311). Growth is possible
in the pH range 5·5–7·5 with an optimum pH of 6–7 (4311). The optimal temperature is
23–25°C, while at 18°C growth is somewhat retarded; some isolates have their maximum at
36°C, others somewhat below (4761). Potassium and phosphate are required for growth, and
high concentrations of both magnesium and calcium are tolerated; growth is stimulated by the
addition of yeast extract, tryptone or L-proline (4311). A medium for the induction of large
conidial masses has been developed (5004). *W. sebi* produces several pyrrol-2-yl polyenes
(268) and shows some antifungal activity *in vitro* (6396).

Wardomyces Brooks & Hansf. *1923*

= *Hennebertia* Morelet *1969*
= *Gamsia* Morelet *1969*

Type species: *Wardomyces anomalus* Brooks & Hansf.

Second anamorphs: *Scopulariopsis* Bain., ?*Gymnodochium* Massee & Salm.

Teleomorph (known in one species): *Microascus* Zukal

MONOGRAPHIC TREATMENT: Hennebert (2396, 2398), and Ellis (1603). — Vegetative hyphae thin, hyaline. Conidiophores arising singly on the sides of the vegetative hyphae, short, hyaline, ± branched; the primary and secondary branches arising successively from the main stalk or its branches. Conidiogenous cells terminal, inflated, ± constricted below the conidium. Aleurioconidia (blastoconidia) formed singly or in succession in small numbers on the conidiogenous cells, the first apically, the others laterally. Conidia ellipsoidal or ovoid or mitriform, brown, smooth-walled, provided with a longitudinal germ slit. — Similar aleurioconidia are found in *Mammaria* Ces. (q.v.) which, however, has pigmented hyphae with darker septa and occasionally forms a *Phialophora*-like state. *Wardomycopsis* Udagawa & Furuya *1978* forms similar pigmented conidia with germ slits in chains on annellophores.

The eight species so far known in *Wardomyces* are relatively uncommon soil fungi. Two groups can be distinguished: one (which includes the type species) with almost penicillate conidiophores, and the other with simple conidiogenous cells or conidiophores, and additional annellophores with catenate conidia. If these latter conidia are one-celled they agree with *Scopulariopsis* Bain., if two-celled *Gymnodochium* Massee & Salm. *1902* may be the adequate genus; *Gymnodochium fimicola* Massee & Salm., which because of the absence of type material is of doubtful identity, may be the same as the annellophoric state of *Wardomyces aggregatus* Malloch *1970* which has sporodochially aggregated annellophores. The genera *Hennebertia* and *Gamsia* were defined by the presence of both aleurioconidia and annellophores and distinguished in having one-celled and two-celled conidia, respectively, borne on annellophores. Since neither the conidial septation nor the aggregation of sporodochia seem to warrant generic separations we prefer to retain all these species in *Wardomyces* as long as no other, more satisfactory generic arrangement can be made. Moreover, the anamorph of *Microascus giganteus* Malloch *1970* is intermediate between the two groups in having only aleurioconidia borne on sometimes branched conidiophores.

Key to the known anamorphic species:

1	Conidiophores repeatedly branched in an almost penicillate manner; annellophores with catenate conidia absent	**2**
	Conidiophores usually unbranched; annellophores present	**5**

2(1)   Elements of the conidiophores strongly constricted at the septa and ramifications; conidia
        ellipsoidal, with broadly rounded tips, 4·2–6·0 × 3·0–3·6 μm   (*W. inflatus*)
        Elements of the conidiophores moderately constricted at the septa and ramifications; conidia
        with ± pointed tips   **3**

3(2)   Conidia ovoid, 4·8–8·0 × 3·6–6·0 μm (length/width ratio 1·5–2·0)   (*W. anomalus*)
        Conidia more elongate   **4**

4(3)   Conidia seceding as 2-celled, brown units, the top cell 5–8 × 2·5–4·0 μm and the basal cell about
        half as long   *W. humicola* (p. 853)
        Conidia seceding as single cells but often retaining part of a torn subtending cell after dehiscence,
        apex strongly apiculate and pointed, 5·5–10·0 × 3·0–4·5 μm   (*W. pulvinatus*)

5(1)   Catenate conidia 1-celled   **6**
        Catenate conidia 2-celled   **7**

6(5)   Catenate conidia hyaline, 6·0–7·5 × 2·5–4·0 μm; aleurioconidia pyriform with ± pointed tips,
        6·0–9·5 × 4·5–5·5 μm   *W. columbinus* (p. 852)
        Catenate conidia subhyaline to pale brown, 5·5–8·0 × 3·5–5·0 μm; aleurioconidia subglobose to
        pyriform, 8–11 × 4–5 μm   (*W. ovalis*)

7(5)   Annellophores aggregated in sporodochia and forming compact columns of greyish brown
        conidia; conidia 8·0–10·5 × 3·5–5·0 μm; aleurioconidia broadly ellipsoidal, 4·0–6·5
        × 3·5–5·0 μm   (*W. aggregatus*)
        Annellophores formed in groups on solitary conidiophores, very similar to those of *W.
        columbinus* but conidia 2-celled, 6·5–8·5 × 3·0–5·5 μm; aleurioconidia with a somewhat less
        pointed apex than *W. columbinus*, 6·5–11 × 4–5 μm   *W. simplex* (p. 854)

Wardomyces columbinus (Demel.) Henneb. *1968* — Fig. 374.

≡ *Trichosporium columbinum* Demel. *1923*
≡ *Hennebertia columbina* (Demel.) Morelet *1969*

DESCRIPTIONS: Hennebert (2398), Gams (1879), and Ellis (1603). — Colonies reaching
3–5 cm diam in 30 days at 20°C on MEA, mostly prostrate, sometimes with small aerial tufts,
at first white, but later becoming grey to black due to conidium formation. Aleurioconidia
arising laterally, sessile on creeping aerial hyphae or on short, erect lateral branches,
1–2-celled and 3–15 μm long, to 8–10 in the apical clusters. Aleurioconidia broadly ovoid,
with a subacute apex and flattened base, dark brown, smooth-walled, with a bilateral
longitudinal germ slit, 6·0–9·5 × 4·5–5·5 μm. Conidiophores of the *Scopulariopsis* type
formed solitarily and bearing annellophores in whorls of 2–3; conidiogenous cells slender, ±
cylindrical, 20–45 × 2–5 μm, with the annellated zone 1–2 μm wide and increasing to 10 μm
in length. Conidia ovoid to mitriform, hyaline, smooth-walled, 6·0–7·5 × 2·5–4·0 μm. — This
species is very close to *W. simplex* (q.v.) which can be distinguished reliably only by the
occurrence of two-celled catenate conidia.

W. columbinus was originally described from plum jelly in Austria. It is also now known
from loamy and sandy soils in Germany (2398), arable soils in the Netherlands (CBS),
salt-marsh in the British Isles and decaying wood in Germany (2398).

It can cause losses in weight and tensile strength in maple-wood strips (2211).

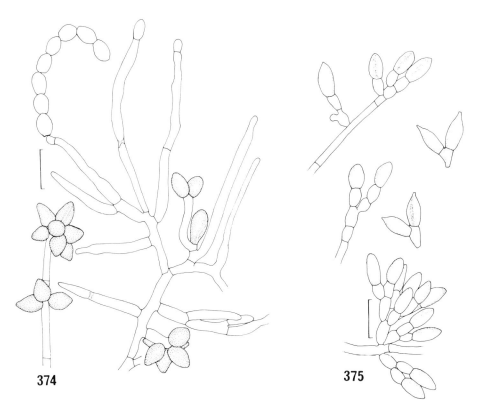

374   375

FIG. 374. *Wardomyces columbinus*, formation of aleurioconidia and *Scopulariopsis*-type conidia.
FIG. 375. *Wardomyces humicola*, conidiophores and liberated conidia, CBS 176.78.

Wardomyces humicola Henneb. & Barron *1962* — Fig. 375.

DESCRIPTIONS: Hennebert (2396), Ellis (1603), and Matsushima (3680). — Colonies reaching 3 cm diam on MEA and 4·5 cm diam on PDA in 30 days at 20°C, at first white but later becoming grey to black due to conidium formation. Conidiophores usually much branched, vaguely penicillate, 1–4 cells long, bearing up to 100 conidia in a cluster about 50–60 μm wide. Conidiogenous cells globose to barrel-shaped, 2–5 × 2–4 μm, each producing 1–2(–3) conidia. Conidia separating as 2-celled bodies, smooth-walled, constricted at the septum, the apical cell about twice as long as and darker than the basal one, with a longitudinal germ slit, apex subacute, together 9–12 × 2·5–4(–5·5) μm.

W. humicola was originally isolated from a greenhouse soil in Ontario (2396) but has subsequently been found in forest soils in central Africa (3063) and under *Thuja occidentalis* in Canada (IMI), and single isolates were obtained from an arable soil under wheat in Germany. It has also been reported from soil in Italy (6081), forest soil in Japan, tundra in Alaska (3680), and organic detritus in fresh water (4429).

W. humicola produces humus-like substances *in vitro* (5555).

Wardomyces simplex Sugiyama *et al. 1968* — Fig. 376.

= *Wardomyces dimerus* W. Gams *1968*
≡ *Gamsia dimera* (W. Gams) Morelet *1969*

DESCRIPTIONS: Gams (1879), Sugiyama (5635, 5636), and Ellis (1604); conidiogenesis of aleurioconidia: Cole and Samson (1109). — Colonies very similar to those of *W. columbinus*, reaching 2·5 cm diam in two weeks on MEA at 20°C. Aleurioconidia irregularly pyriform, dark brown, mostly provided with a longitudinal germ slit, 6·5–11 × 4–5(–6·5) μm. Annellophores 10–15 μm long in their cylindrical part, annellated zone to 30 μm long. Conidia regularly 2-celled, ellipsoidal, with a broadly truncate base, hyaline, smooth-walled, 6·5–8·5 × 3·0–5·5 μm. — The name *W. simplex* was published a few months before that of *W. dimerus* on the basis of the aleurioconidial form only, but a study of the type isolate (CBS 546.69) revealed that 2-celled conidia also occurred in it (5636).

W. simplex was originally isolated from milled rice in Japan (5635). Three isolates from arable soil under wheat in Germany, and one as an atmospheric contaminant in Belgium have since been obtained. It has also been isolated from porcupine dung in Canada (CBS).

Xylan decomposition by this species is, in comparison with other fungi, very strong (1432).

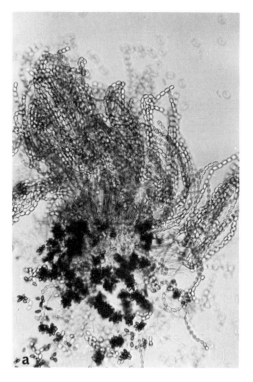
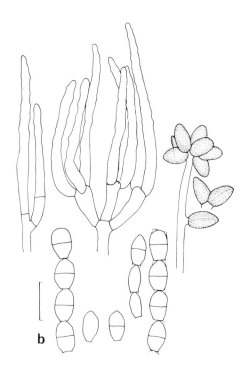

FIG. 376. *Wardomyces simplex*. a. Clusters of aleurioconidia and chains of *Scopulariopsis*-type conidia, × 200; b. details of both kinds of conidium formation; CBS 235.66.

Zygorrhynchus Vuill. *1903*

Type species: *Zygorrhynchus heterogamus* (Vuill.) Vuill.

The spelling *Zygorrhynchus* is regarded as an orthographically necessary correction of the original and commonly used spelling *Zygorhynchus*. — In the MONOGRAPHIC TREATMENT of Hesseltine (2428) seven species were recognized; one, *Z. psychrophilus* Schipper & Hintikka *1969*, has been described subsequently.

Fam. Mucoraceae Dumort. — Sporangiophores irregularly branched, bearing both zygospores and sporangia. Sporangia without apophyses, many-spored, columellate. Sporangiospores small, smooth or finely roughened, hyaline or greyish. Homothallic. Progametangia (and subsequently suspensors) very unequal, the larger arising as a lateral swollen branch, most commonly from a sporangiophore and less commonly from another hypha, the smaller cut off from the same hypha by a septum. Zygospores dark and ornamented as in many *Mucor* species (5102).

Zygorrhynchus species are distinguished from homothallic *Mucor* species (particularly *M. genevensis* Lendner *1908*) by the unequal suspensors commonly arising from the sporangiophores. The two species treated here have sporangial walls which do not split open if they are undisturbed (as occurs in *Z. exponens* Burgeff *1924*).

Key to the species treated:

1 Larger suspensor enlarging abruptly; zygospores black, 30–70 μm diam, projections blunt; sporangia 15–35 (–55) μm diam; columellae applanate to almost globose; sporangiospores oval, 4–6·5 × 2–4 μm   *Z. heterogamus* (p. 855)

   Larger suspensor enlarging gradually; zygospores ± dark brown, mostly 30–35 μm diam, projections pointed; sporangia 12–40 μm diam; columellae applanate; sporangiospores ellipsoidal, 3–6·5 × 2–3·3 μm   *Z. moelleri* (p. 857)

Zygorrhynchus heterogamus (Vuill.) Vuill. *1903* — Fig. 377 a.

≡ *Mucor heterogamus* Vuill. *1886*

For some further synonyms see Hesseltine *et al.* (2428).

DESCRIPTIONS: Hesseltine *et al.* (2428), Zycha *et al.* (6588), and Milko (3807). — Colonies covering the surface of a 9 cm Petri dish in less than ten days on MEA or synthetic *Mucor* agar at 20–25°C, sometimes very low but sometimes to 1 cm high, at first white but later becoming grey to olive-grey; reverse yellowish in some isolates. Sporangiophores arising from the substrate mycelium, erect, to 18 μm wide, ± sympodially branched. Sporangia almost hyaline, globose, 15–35(–55) μm diam, on erect or nodding sporangiophore tips, the wall breaking or deliquescing at maturity; columella applanate to subglobose, 6·5–26 μm diam. Sporangio-spores oval, hyaline, smooth-walled, 4–6·5 × 2–4 μm. Chlamydospores mostly produced on mycelium within the substrate, hyaline, thick-walled, globose to oval, 6·5–21 μm diam. Zygospores most abundantly produced just above the surface of the substrate, dark brown to black, 30–70 μm diam, covered with blunt projections to 4·5 μm high. — The blunt and black zygospore projections are the most outstanding feature of this species. — The allied *Z. psychrophilus* Schipper & Hintikka *1969* is most easily distinguished by the rod-shaped to fusiform sporangiospores, 5·4–8·8 × 1–2 μm.

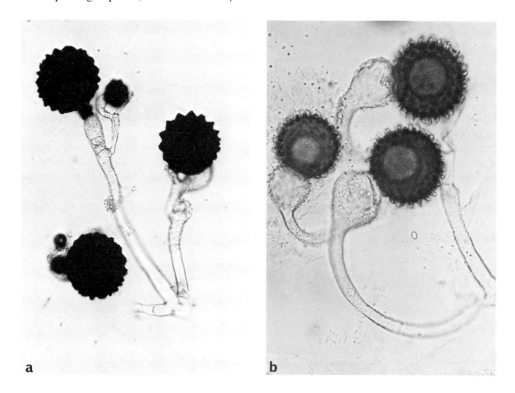

a   b

FIG. 377 a. *Zygorrhynchus heterogamus*, zygospores, × 200. — b. *Zygorrhynchus moelleri*, zygospores, × 500.

As in all species of its genus, *Z. heterogamus* is typically a soil fungus (2428) which, although rarely recorded outside western Europe and Canada (540, 6588), has a wide range which includes Czechoslovakia (1702), the USSR (3184, 3652, 3807, 4703), Chile (1824), Nepal (1826), India (4030, 4997, 5058), and Japan (3267). It is known from soils in oak (1700, 1702) and larch forests (3267), a forest nursery (6184), dry grassland (164), salt-marsh (IMI), periodically flooded soils (2428), cultivated soils (540, 4030, 4997) and houshold refuse (1132). It has been isolated down to depths of 20–25 cm in soil (164, 3267).

Z. heterogamus grows well in the range 0–15°C but not at 20°C. It can utilize L-xylose, D-glucose and glycerol, but not lactose, starch or cellulose (2428). The herbicide paraquat can be accumulated in the mycelium *in vitro* (5446).

Zygorrhynchus moelleri Vuill. *1903* —Fig. 377 b, 378.

≡ *Mucor moelleri* (Vuill.) Lendner *1908*
= *Zygorrhynchus vuilleminii* Namysl. *1910*
= *Mucor saximontensis* Rall *1965* (agamous strain)

For further synonyms see Hesseltine *et al.* (2428).

DESCRIPTIONS: Hesseltine *et al.* (2428), Zycha *et al.* (6588), Watanabe (6212), and Milko (3807). — Colonies covering the surface of a 9 cm Petri dish in less than 11 days on MEA or synthetic *Mucor* agar at 20–25°C, very low in some isolates to over 1 cm high in others, at first white but later becoming pale to deep olive-grey; reverse similarly coloured, often showing irregular zonation. Sporangiophores arising from the substrate mycelium, erect, to $15\,\mu$m wide, ± sympodially branched. Sporangia white to grey, globose to slightly flattened, $12–40(–60)\,\mu$m diam, mostly on nodding sporangiophore tips; wall breaking at maturity; columellae generally applanate, $8–25(–45)\,\mu$m diam. Sporangiospores ellipsoidal to ovoid, hyaline, smooth-walled, $3\cdot0–6\cdot5 \times 2\cdot0–3\cdot3\,\mu$m. Hyaline chlamydospores sometimes present in the aerial mycelium, intercalary, oval, to $13\,\mu$m diam. Zygospores most abundant just above the surface of the substrate, brown to dark brown, mostly $30–35\,\mu$m diam, covered with stellate projections to $5\,\mu$m high (5102). —This is a very variable species. *Z. vuilleminii* was considered to be a separate species for a long time on account of its higher aerial mycelium. *Mucor saximontensis* Rall and other agamous isolates were induced to produce zygospores by combinations with some *Mucor* species at 25°C (5096). None of the varieties formerly distinguished are now regarded as sufficiently distinct to merit formal recognition. — A similar but much rarer species is *Z. japonicus* Kominami *1914* characterized by pyriform to globose columellae and sporangiospores $3–10 \times 1\cdot5–6\,\mu$m. — DNA analysis gave a GC content of 39% (5600). Hyphal hydrolysates were found to contain 31·5% glucosamine, 16·0% glucuronic acid, 6·8% fucose, 5·1% galactose, and traces of mannose (3232).

Z. moelleri is one of the commonest members of the Mucoraceae and has a worldwide distribution. It is almost exclusively found in soil (2428, 6588) and responds very well to all usual isolation techniques, particularly "screened immersion tubes" and their modifications (5812) and soil washing (307, 989, 4449). It has been reported from northern Norway (2206), the British Isles (165, 861, 4182, 5222), the USSR (432, 3807, 4474), Poland (1978), Denmark (2745), Czechoslovakia (1702), Austria (1823, 3418), Romania (4379), Italy (3445, 4537, 5047), central Africa (3063), South Africa (1559), the USA (1163, 1166, 2573, 2822), Canada (6352), Hong Kong (6516), Taiwan (2472), Japan (3267), New Zealand (5814, 5930), India (5058) and Nepal (1826). It occurs in alpine habitats (1876, 3445), was isolated frequently in the Rocky Mountains up to 3600 m (4711), and is also known from Himalayan forest soils (4995). It is recorded from a wide range of soil (3414) and humus types (2745, 4815). There are reports from forest, grassland and arable soils but these are too numerous to cite here individually. Other known habitats include heathland (2736, 5047, 5222,

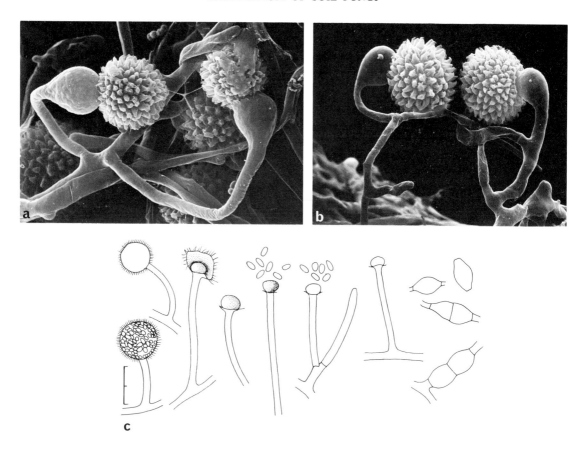

Fig. 378. *Zygorrhynchus moelleri.* a, b. Zygospore formation, SEM, × 600; c. sporangiophores with sporangia, columellae, sporangiospores and chlamydospores; CBS 501.66.

5813), peat (432, 2779, 3119, 4474), dunes (745, 861), salt-marsh (413), open savannah (1559), managed prairie sites (6347), forest soils after prescribed burning (2822), fields treated with sewage sludge (1163) and sewage sludge itself (1165, 1167, 1172), river (1162) and estuarine sediments (413, 5316) even with high salinities (655), and polluted running water (1154, 1155, 1157, 1166, 4429). In contrast to other Mucoraceae, it has been isolated from considerable soil depths, down to 100 cm (159, 2779, 4995, 6182, 6516). The pH range is very great, though with a clear preference for the acid side (745, 1823, 1825, 6182), the range being given as 4·8–6·6 (861) or 3·9–6·5 (3871); growth *in vitro* has been obtained between pH 3·1 and 8·8 (2741). In the field, areas with medium humidities have proved optimal for its development (4313), while in laboratory experiments it was found to be hydrophilic (3109) with a minimum water potential of −45 bars and optimal growth occurring at −5 bars (3109). An increased occurrence in the summer has been observed in New Zealand (5812). It has been isolated from the rhizospheres of numerous plants, particularly in early growth phases in barley and oats (3004, 3429, 4443), partly also on the culm base (4136), but also on wheat (3567), *Lolium perenne* and *Trifolium repens* (5815), poplar (5898) and fir

(3635). It is also known on the roots of *Pinus taeda* (3159), peat (2719), brown coal (4474), wood exposed to soil (1956, 5237), birds' roosts (682), stored wheat grain (4492), and straw compost (1528). — Spore germination can be inhibited by morphogenically active metabolic products from ageing cultures of *Fusarium oxysporum* (4435, 4882); its growth can also be inhibited by volatile metabolites from bacterial cultures (3920, 3922). Zygospore formation can be inhibited by a streptomycete metabolite (3011). — The sporangiospores were relatively resistant to the lytic action of a streptomycete enzyme preparation (1049). Hyphae can be invaded by *Trichoderma viride* (1511) and the obligate parasites *Dimargaris verticillata* (3560) and *Syncephalis californica* (2618).

Optimal growth occurs at 25°C, and the maximum is 32°C (1827, 6588); the thermal death point was determined as 50–55°C for 30 min in soil (613). Optimal growth was obtained in a liquid medium of the following composition: 4% D-glucose, $0 \cdot 1$ mol l^{-1} Tris buffer, $6 \cdot 25$ mmol l^{-1} $(NH_4)_2SO_4$, 3–12 mmol l^{-1} $MgSO_4 \cdot 7H_2O$ (different for various isolates), 2 mmol l^{-1} KH_2PO_4, 2 ppm Fe, Zn and Mn cations (1807). Suitable C sources include glycerol (2428), acetate (3329), D-glucose, D-fructose, D-mannose, D-xylose, and arabinose (1807) but not lactose, starch or cellulose (2428). Good N sources are L-glutamine, L-asparagine, glycine, and $(NH_4)_2SO_4$ (1807). The influence of ammonium ions (3977) and the $K^+:Ca^{2+}$ ratio (3978) on glucose metabolism (369, 3980) have been investigated. The endogenous respiration was found to be hardly affected by glucose, acetate and succinate, raised by $NaNO_3$ (3981), but not by NH_4Cl (3977). Pectin (655, 825, 1432, 3414) and hemicellulose (825, 3414) are decomposed. No thiamine is required (4860). An homothallic isolate with a minus-tendency showed the ability to complete trisporic acid synthesis from precursors produced by *Mucor mucedo* $(+)$ and also produced small amounts of a precursor material very similar to that of *M. mucedo* $(-)$ (6295). The production of glycolic and tartaric acids has been reported (4009) and the metabolism of citric acid has been investigated (3979). β-Glucosidase (6159), alcohol dehydrogenase (1999) and phenoloxidase (934) are produced. Humic-acid-like substances have been obtained *in vitro* (5555). — Seedling roots of peas were damaged *in vitro* by pure cultures of *Z. moelleri* (1430), and barley and oat seedlings were also somewhat inhibited (3005). — *Z. moelleri* can grow at very high CO_2 concentrations (784, 1244, 2096). Spore germination is slightly inhibited by $3 \cdot 5\%$ NaCl, but mycelial growth is not (655). *Z. moelleri* is very sensitive to soil steaming (6183) and soil treatment with formaldehyde (1655), CS_2 (5001), captan, dicloran and triarimol (6136).